International Symposium on Resource Exploration and Environmental Science (REES 2017)

IOP Conference Series: Earth and Environmental Science
Volume 64

Ordos, China
14 – 16 April 2017

ISBN: 978-1-5108-4071-3
ISSN: 1755-1307

Printed from e-media with permission by:

Curran Associates, Inc.
57 Morehouse Lane
Red Hook, NY 12571

Some format issues inherent in the e-media version may also appear in this print version.

Printed by Curran Associates, Inc. (2017)

For permission requests, please contact the Institute of Physics at the address below.

Institute of Physics
Dirac House, Temple Back
Bristol BS1 6BE UK

Phone: 44 1 17 929 7481
Fax: 44 1 17 920 0979

techtracking@iop.org

Additional copies of this publication are available from:

Curran Associates, Inc.
57 Morehouse Lane
Red Hook, NY 12571 USA
Phone: 845-758-0400
Fax: 845-758-2633
Email: curran@proceedings.com
Web: www.proceedings.com

Table of contents

Volume 64

International Symposium on Resource Exploration and Environmental Science

14–16 April 2017, Ordos, China

Accepted papers received: 2 May 2017
Published online: 19 May 2017

Preface

Resource Exploration and Related Technologies

Environmental Engineering and Sustainable Development

Energy and Power Engineering

Biochemical Engineering and Science

IOP Conf. Series: Earth and Environmental Science **64** (2017) 011001 doi:10.1088/1755-1315/64/1/011001

Preface

The organizing committee of REES 2017 is proud to present the proceedings of the 2017 International Symposium on Resource Exploration and Environmental Science (REES 2017: http://www.isrees.org/), held on April 14-16, 2017 in Ordos, China.

REES 2017 was a platform for presenting excellent results and new challenges facing the field of earth and environmental sciences, resources and environmental engineering, energy and electricity engineering and related topics. It brought together experts from industry, governments and academia, experienced in engineering, design and research.

REES 2017 received 378 manuscripts, and 123 submissions have been accepted by our reviewers. By submitting a paper to REES 2017, the authors agreed to the review process and understood that papers would undergo a peer-review process. Manuscripts were reviewed by appropriately qualified experts in the field selected by the conference committee, who took detailed comments and-if the submission was accepted-the authors would submit a revised version that took into account this feedback. All papers were reviewed using a double-blind review process: authors declared their names and affiliations in the manuscript for the reviewers to see, but reviewers did not know each other's identities, nor did the authors receive information about who had reviewed their manuscript. The committees of REES 2017 invested great efforts in reviewing the papers submitted to the conference and organizing the sessions to enable the participants to gain maximum benefit.

Hopefully, all participants and other interested readers will benefit scientifically from the proceedings and also find it stimulating in the process.

With our warmest regards,

Lin Liu

Conference Organizing Chair
Wuhan, China

IOP Conf. Series: Earth and Environmental Science **64** (2017) 011002 doi:10.1088/1755-1315/64/1/011002

Committees

Chairmen:
Prof. Y.T. Xiong, Wuhan University, China;
Prof. G.X. You, Wuhan University, China;
Prof. Z.B. You, Wuhan University of Technology, China;

Members:
Prof. H. Davis, Boya Century Publishing Ltd., Hong Kong;
Prof. K. Georgiou, University of Waterloo, Canada;
Prof. W.L. Xiao, Wuhan University of Science and Technology, China;
Prof. H.W. Hua, Wuhan University of Technology, China;
Prof. J.X. Zhang, Wuhan University of Technology, China;
Prof. J.Y. Zhong, Huazhong University of Science and Technology, China;
Prof. J.D. Wang, Wuhan University of Science and Technology, China;
Prof. T. Cui, Guizhou Institute of Technology, China;
Prof. W.L. Su, China University of Geosciences, China;
Prof. H.F. Liu, Jiangsu University, China;
Prof. H.Y. Zhu, Chongqing University, China;
Prof. X.J. Gao, Hefei University of Technology, China;
Prof. X.P. Han, Jiangsu University of Science and Technology, China;
Prof. H. Chen, Shanghai University of Engineering Science, China;
Dr. J. Hu, Ohio State University, USA;
Dr. W. Zhong, New York State University at Stony Brook, USA;
Dr. C.W. Yang, Wuhan University of Science and Technology, China;
Dr. W.M. Luo, Wuhan University of Science and Technology, China;
Dr. G.W. Yang, Shanghai University, China;
Dr. H.T. Zhang, South China University of Technology, China;
Mr. Q. Yun, Chi Shun Chemical Co., Ltd., China.

Organizing Committee:
Prof. Z.B. You, Wuhan University of Technology, China;
Prof. S. Hu, Wuhan University of Technology, China;
Prof. D.G. Yang, Wuhan University, China;
Prof. T. Liu, Wuhan University of Science and Technology, China;
Dr. C.W. Yang, Wuhan University of Science and Technology, China;
Dr. S.F. Zhao, Wuhan University of Science and Technology, China;
Dr. W.M. Luo, Wuhan University of Science and Technology, China;
Dr. G.L. Liu, Wuhan University of Technology, China;
Dr. X. Xiao, Wuhan University of Technology, China;
Ms. Y.F. Ma, Hubei New Wen Sheng Conference Co., Ltd., China;
Mr. Z.P. Ma, Hubei New Wen Sheng Conference Co., Ltd., China;

IOP Conf. Series: Earth and Environmental Science **64** (2017) 011002 doi:10.1088/1755-1315/64/1/011002

Mr. B.T. Zhou, Hubei New Wen Sheng Conference Co., Ltd., China;
Mr. H.T. Ke, Hubei New Wen Sheng Conference Co., Ltd., China;
Mr. X.L. Jiang, Hubei New Wen Sheng Conference Co., Ltd., China;
Mr. Y.J. Shi, Hubei New Wen Sheng Conference Co., Ltd., China;
Ms. W.H. Qin, Hubei New Wen Sheng Conference Co., Ltd., China;
Ms. D.W. Liu, Hubei New Wen Sheng Conference Co., Ltd., China.

IOP Conf. Series: Earth and Environmental Science **64** (2017) 011003 doi:10.1088/1755-1315/64/1/011003

Peer review statement

All papers published in this volume of *IOP Conference Series: Earth and Environmental Science* have been peer reviewed through processes administered by the proceedings Editors. Reviews were conducted by expert referees to the professional and scientific standards expected of a proceedings journal published by IOP Publishing.

IOP Conf. Series: Earth and Environmental Science **64** (2017) 012001 doi:10.1088/1755-1315/64/1/012001

Study on the influence of the electrode model on discharge characteristics in High-voltage Pulsed Deplugging Technology

Bingnan Yan[1, a], Zhou Jing[1, b] and Zhao Liang[2, c]

[1]School of Electronic Engineering, Xi'an Shiyou University, Xi'an, China
[2]Xi'An Thermal Power Research Institute Co.Ltd, Xi'an, China

[a]bnyan@xsyu.edu.cn, [b]814227078@qq.com, [c]zhaoliang@tpri.com.cn

Abstract. In the oil-field development, blocking caused by impurities leads to a decline in oil production. The high-voltage pulsed deplugging technology can be applied successfully in oil deplugging. One of the key problems in this technology is the influence of the electrode model on discharge characteristics. In this paper, the electrode structure was studied. Firstly, the influence of the electrode gap on electric-field intensity was studied by using ANSYS simulation. Secondly, a high-voltage pulsed discharge experiment system was built and the discharge characteristics were studied under different static pressure when the electrode gap varied. The results show that the larger the electrode gap, the lower the electric strength and the longer the time delay. Short breakdown time delay would make greater energy of impact waves and better blockage relieving effect.

1. Introduction

In the oil-field development, oil layer is polluted by kinds of impurities caused by drilling, fracturing, and so on. The blocking leads to a decline in oil production. Conventional approaches are releasing the chemical agent to relieve clogged wells [1], but the normal chemical technologies bring secondary pollution. How to relieve the blockings is the problem which must be solved in oil extraction. The pulsed high voltage deplugging technology has the advantages, such as environmental protection, high efficiency, and reusability .It is the deplugging technology with broad prospects.

The new pulsed high-voltage discharge technology uses electrohydraulic effect [2, 3] to generate mechanical action. Electrohydraulic effect essentially rapidly transforms electric energy into mechanical energy in impact waves, which will then act on an oil layer. These waves which have extremely strong penetrating power bombard plugging substances and pollutants near the wellbore zone, resulting in the oscillation and crushing of these blockages. Then, cavitation effect and pressure of the oil layer itself are used to suck and push these crushed blockages. Finally, the oil layer plug is removed, and permeability characteristics near the wellbore zone at the oil layer are improved.

In order to apply this technology more effectively on oil exploration, many scholars studied the pulsed high voltage system and discharge characteristics to improve power efficiency and enhance shock wave energy [4-7]. Ref. [4] studied the changing of the properties of the liquid surface. Ref. [5] studied the resistance characteristics of the plasma channel. Ref. [6] investigated the spark discharge parameters for material processing with high power ultrasound. Ref. [7] developed a high-energy arc fracturing device to carry out a series of fracture experiments.

IOP Conf. Series: Earth and Environmental Science **64** (2017) 012001 doi:10.1088/1755-1315/64/1/012001

However, the electrode structure, the electrode gap affect the discharge characteristics of the high-voltage pulse, thus they affect whether the technology can be effectively used in the oil block. In this paper, the influence of the electrode model on electric-field intensity was studied by using ANSYS simulation firstly. Because of the complex downhole operating environments, the high static pressure and the crude oil medium have a significant impact on the characteristics of high-voltage pulsed discharge. Thus a high-voltage pulsed discharge experimental system was designed to simulate the downhole working environment with high pressure to study the influence of the electrode model. The experiment will lay a theoretical foundation for the development of the downhole equipment.

2. The influence of the electrode model on electric-field intensity based on ANSYS simulation

The breakdown of matter is related to the strength of the electric filed, and the electrode structure and the gap between the positive and the negative affect the strength. In this paper, column-column model is chosen to study.

Theoretically, the higher the working voltage, the higher the voltage between the discharge electrodes, the easier the breakdown of the medium, and the better the Deplugging effect will be. But the volume of the supply voltage will increase when the voltage increases and insulation requirements for equipment will increase significantly. Considering most well operating environment in China, the dimensions of the underground device was limited. So, the working voltage was chosen in the range of 10-25kV in this paper.

The column-column model shown in Fig.1 was built by using ANSYS simulation, and the model parameters were set as follows: the radius was 2mm and the height was 5mm. Three electrode gaps were chosen as d=1.5mm, d=2mm, d=3mm. A voltage of 20kV was applied between the electrodes of the column-column model.

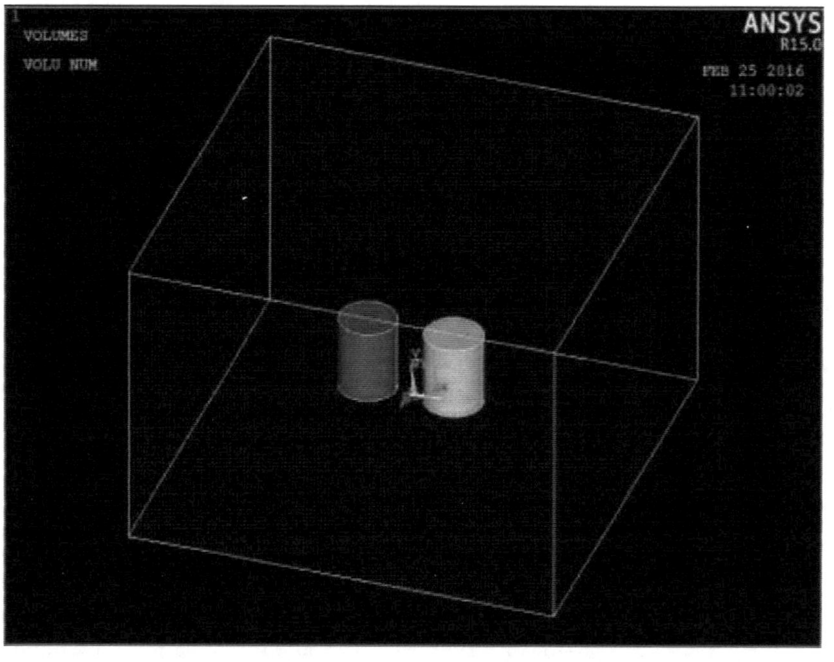

Figure 1. The column-column model

When d=1.5mm, the destribution of the electric-field intensity between the two electrodes was shown in Fig. 2. From the cloud picture, the strongest electric field was concentrated near the electrodes, and the maximum value was 143.52kV/cm.

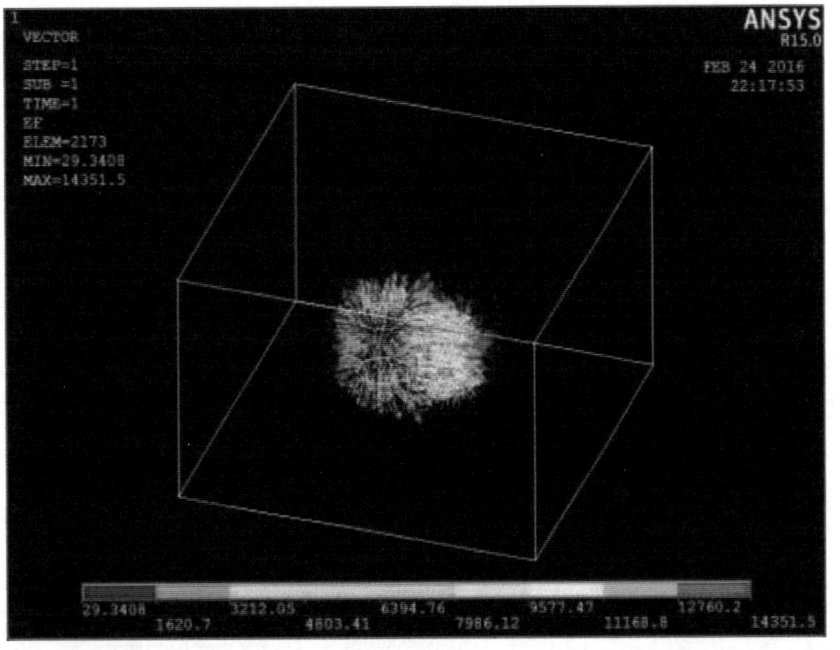

Figure 2. The distribution of the electric-field intensity between the column-column electrodes when d=1.5mm

Adjust the electrode gap and simulate several times. According to the results, the relationship of the electrode gap and the maximum electric-field intensity was got in Fig. 3. From Fig.3, we can get that the maximum electric-field intensity between the electrodes decreases when the electrode gap increases. When the electrode gap is not sufficient to make discharge breakdown phenomenon occur, the high voltage pulse discharge fails. If the phenomenon occurs, the larger electric-field intensity is needed, thus the larger working voltage is needed. Hence, parameter requirement in the high-voltage pulsed discharge system will also increase. This phenomenon amplifies the requirements for insulation and size of a pulsed high-voltage discharge device.

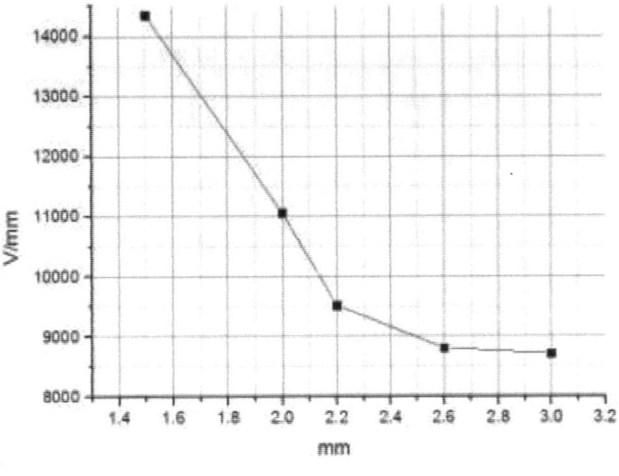

Figure 3. The relationship of the electrode gap and the maximum electric-field intensity

3. Design of Experimental System and Results Analysis

3.1. Establishment of Experimental Platform and Parameter Setting

In order to make the technology applied better in the well deplugging, the influence of the electrode model on discharge characteristics was studied when the static pressures were different and the mediums were water or oil-water mixture.

The experiment system was shown in Fig.4. It was mainly composed of voltage, boost-up circuit, rectifier, capacitor with high specific energy storage, vacuum switch and control circuit, electrode device, a high pressure sealed drum(HPSD) and some measuring equipments. In the figure, the boost-up circuit was composed of voltage regulator and high-tension transformer, and the range of the output voltage was 0-50kV adjustable. A HPSD of low-carbon steel material was designed in order to simulate the downhole working environment with high pressure. The device can be able to withstand the maximum pressure of 40MPa and its tightness is very good. The measuring equipments were composed of high voltage probes P6015A measuring the voltage between the anode and the cathode and PEM Probes measuring the current of the discharge circuit. The voltage and the current were captured by oscilloscope with Model TDS2022. The parameter settings in the experiment were shown in Table.1, where the working voltage was the charging voltage of capacitor.

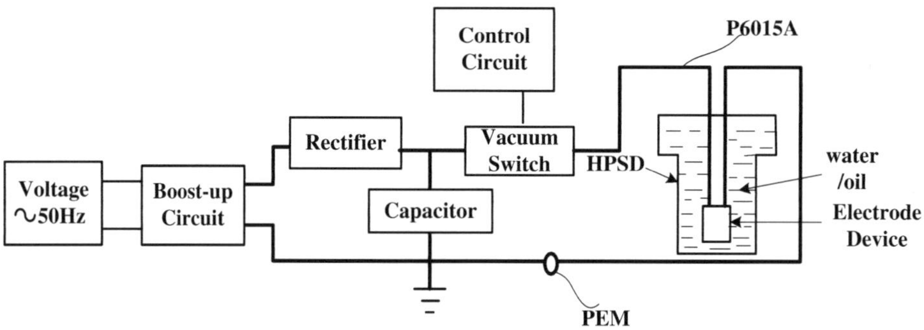

Figure 4. The experiment system of pulsed high voltagedischarges

Figure 5. The electrode structure

In this experiment, the electrode structure was column-column shown in Fig. 5 in order to withstand the high voltage and make the discharge easier to produce. The electrode device was 300mm high, which center electrode was an anode and the cathode was connected to the housing.

Electrode gap can be adjusted according to the discharge channel .The electrode device was placed in the tank by using plastic rods, and the distance was 280mm from the bottom. The high-voltage line and the ground line of the electrode were introduced by seal pins connected in the tank.

Table 1. The key patameters of the experiment

Parameters	Data Range
Supply Voltage (V)	220
Working Voltage (kV)	10-25
Capacitor (μF)	12
Electrode Type	Column-Column
Distance of Electrode Gap (mm)	1.6-1.8
Pressure (MPa)	0-30

3.2. Experimental verification and analysis

3.2.1. Experimental verification. Underground operating environment is more complex, in which high static pressure and discharge medium on the high-voltage pulsed discharge characteristics have a key impact. These downhole parameters must be considered for the successful application of high-pressure pulsed deplugging techniques in the downhole. In the experimental verification of the electrode gap, two types of experiments were carried out considering the parameters. Based on the simulation results of ANSYS above, the electrode gap was chosen between 1.6mm-2.0mm and the working voltage was 20kV.

Experiment 1 Discharge medium was water, electrode gap was 1.6mm and the high static pressures were set to 5, 10 and20 MPa. Under these conditions the experiments were successful and the discharge medium was breakdown. The experimental data were shown in Table 2 and the discharging voltage and discharging current under 20MPa pressure were shown in Fig. 6.

Table 2. Discharge data under different static pressures

No.	Electrode gap (mm)	Static pressure (MPa)	Working voltage (kV)	Whether breakdown	Discharge voltage (kV)	Discharge current (kA)	Time delay (us)
1	1.6	5	20.0	Y	18.0	16.8	112.0
2	1.6	10	20.0	Y	20.2	16.8	52
3	1.6	20	20.0	Y	19.8	18.7	52.0

Experiment 2 Discharge medium was water, high static pressure was 20 MPa and electrode gaps were set as 1.6mm and 1.8mm. Under these conditions the experiments were successful and the discharge medium was breakdown. The experimental data were shown in Table 3.

Table 3. Discharge data under different electrode gaps

No.	Electrode gap (mm)	Static pressure (MPa)	Working voltage (kV)	Whether breakdown	Discharge voltage (kV)	Discharge current (kA)	Time delay (us)
1	1.6	20	20.0	Y	19.8	18.7	52.0
2	1.8	20	20	Y	19	18.8	73.0

From Table 3, we can get: when the discharge medium, the static pressure and the working voltage are unchanged, the larger the electrode gap, the longer the delay. Because the slow capacitor discharge process occurs. The time delay when electrode gap was 1.8mm is longer by 21us compared with that at gap of 1.6mm. And the capacitor experienced a slow discharging process. So the breakdown is more random and the phenomenon is more difficult.

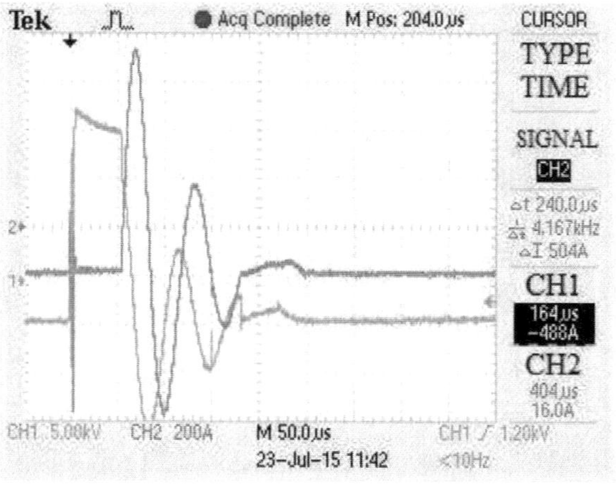

Figure 6. The discharging voltage and discharging current under 20MPa

3.2.2. Field devices pilot experiment. In order to study the deplugging effect, the cement tube was used for pilot experiment. The cement tube was put into the HPSD and the experiment environment was set as follows: the medium was water and the pressure was 30MPa. The results were shown in Fig. 7. From the figures, the cement tubes were damaged by the shock waves and the damage was obvious.

(a) (b)

Figure 7. The damage of cement tube

4. Conclusion

In order to study whether high-voltage pulsed discharge technology could effectively remove plugs under downhole operating environment, the ANSYS simulation and experimental verification were respectively adopted on the study of the influence of electrode models on discharge characteristics. The following conclusions could be drawn:

1) The column-column model was chosen to study. The results show that the maximum electric-field intensity between the electrodes decreases when the electrode gap increases. The high-voltage pulsed discharge maybe fails. If the phenomenon occurs, the larger electric-field intensity is needed, parameter requirement in the high-voltage pulsed discharge system will increase. This phenomenon amplifies the requirements for insulation and size of a pulsed high-voltage discharge device.

2) An experimental system of high-voltage pulsed discharge was designed. The discharge characteristic was studied with different static pressures and different electrode gaps. These results indicate that as static pressure, discharge medium and working voltage are unchanged, when electrode gap is larger, the discharge voltage reduced, and the time delay increases. The breakdown is more difficult. Parameter requirement of the whole high-voltage pulsed discharge system is amplified.

3) A cement tube impact experiment was conducted. The tube was broken when the impact waves generated in the discharging medium acting on the tube. Increasing working voltage or decreasing electrode gap are beneficial for improving the strength of impact waves.

In this paper, the results provide a foundation for subsequent development of high-voltage pulsed discharging device. In the next study, other factors will be combined with the developed experimental system.

Acknowledgments
This research is supported by National Key Science & Technology Special Projects of China (2011ZX05021-005) and Youth Science and Technology Innovation Fund of Xi'an Shiyou University (2013QN002).

References
[1] Zhenjun Wang, Yuanming Xu and Suman Bajracharya, The comparison of removing plug by ultrasonic wave, chemical deplugging agent and ultrasound-chemical combination deplugging for near-well ultrasonic processing technology, Ultrasonics Sonochemistry, 2015, pp. 339-344.
[2] A.I. Zhakin, Electrohydrodynamics, Physics-Uspekhi, 2012, pp. 465-488.
[3] F. Golovashchenko, J. Gillard, and Tang Zejun, Electrohydraulic trimming of advanced and ultra high strength steels, Journal of Materials Processing Technology, 2014, pp. 1027-1043,.
[4] A. F. Gutsol and W. R. Pyle, Electro-Hydrodynamic and Plasma Phenomena in Corona Liquid Interaction, in Proc. 2013 IEEE International Pulsed Power Conf.
[5] M. J. Given, I. V. Timoshkin, and J.M. Lehr, Analysis of The Current Waveforms Observed in Underwater Spark Discharges, in Proc. 2007 IEEE 34th International Conference on Plasma Science.
[6] B. M. Kovalchuk, A.V. Kharlov, and V. N. Kiselev, High-voltage pulsed generator for dynamic fragmentation of rocks, Review of Scientific Instruments, 2010, pp. 308-311.
[7] M.P. Wilson, L.Balmer, and I.V, Timoshkin, An investigation of spark discharge parameters for material processing with high power ultrasound, Minerals Engineering, 2007, pp. 1159-1169.
[8] Fu Rongyao, Zhou Jian, and Yan Ping, "Application of high voltage pulse discharge in rock fracturing without confining pressure," High Voltage Engineering, 2015, pp. 4055-4059.
[9] Li Zhiyuan, Liu Yi, and Lin Fuchang, Erosion mechanism research of metal electrode under pulse current in water, High Power Laser and Particle Beams, 2016, pp. 045007-045007-7.

IOP Conf. Series: Earth and Environmental Science **64** (2017) 012002 doi:10.1088/1755-1315/64/1/012002

The study on the results of hydrogen pipeline leakage accident of different factors

Haonan Chen and Zhanli Mao*

The Chinese People's Armed Police Forces Academy, Langfang, China

*Corresponding author e-mail: maozhl@mail.ustc.edu.cn

Abstract. Pipeline is frequently used in long-distance hydrogen transport, while pipes can be disturbed in some conditions, prone to leakage accident, causing severe combustion or explosion. This paper analyzed the influences of different factors on the consequences of hydrogen pipeline leakage through the simulation software ALOHA. The results show that wind speed, ground roughness, tube pressure and gap area have a great impact on the consequences of the accident; wind speed and ground roughness have negative correlation to danger distance; tube pressure and gap area have positive correlation to danger distance. These results are due to the release amount and release rate influenced by the four factors.

1. Introduction

Hydrogen, as one of the most important chemical materials, has been widely used in industrial production. But the explosion limit of hydrogen is wide, and the minimum ignition energy is only 0.019 mJ with the volume concentration of 30%. When leakage occurs, collision, tiny static electricity or weak spark will be able to ignite it, causing flash fire, detonation or explosion accident [1-3].

The studies on hydrogen leakage have been conducted and many methods were proved effective. Numerical simulation is usually established to analyze the consequences of gas pipeline leakage accident. ALOHA software is used to simulate accident consequences. Besides, a lot of experiments have been carried out. Hydrogen and air premixed flame in a container was studied by Huahua Xiao, Zhanli Mao [4]. Dynamic thickening flame was used to study propane and hydrogen premixed flames by Huahua Xiao, Zhanli Mao [5]. A hydrogen leakage accident in simple closed space was analyzed by Michael R. Swain, Patrick Filose [6]; Numerical simulation of hydrogen was carried out by Hector A. Olvera and Ahsan R. Choudhury [7] in stably stratified atmosphere.

ALOHA software was used in this paper and variables such as wind speed, the inner tube pressure, and the gap size were selected as the factors to simulate the consequences of hydrogen pipeline leakage accidents [8]. Through the diffusion model, the vapor cloud explosion model and the sustained combustion model, the impacts of each variable on accident consequences were analyzed for obtaining the reasonable and effective reference in actual accidents and conducting fire risk assessment [9, 10].

2. Simulation Process

2.1. Types of Hydrogen Pipeline Leakage Accident

Hydrogen at room temperature is colorless, odorless gas. When the pipeline ruptures, hydrogen gas will rapidly diffuse outside, flow with air, and gradually rise. If hydrogen does not encounter an ignition source, but form a cloud of steam, and then a fire source ignites the vapor cloud in the explosion limit, it may cause a flash fire or explosion. If the hydrogen gas encounters an ignition source during early leakage, it is possible to form a stable combustion region.

2.1.1. Hydrogen Diffusion Only. When hydrogen pipeline is damaged, hydrogen continues to spread into atmosphere. The hydrogen diffusion region is formed within a period of time, which is a hazardous area where explosion might occur. The hydrogen concentration is in its explosive limit 4.1% ~ 74.2%, in the event of sufficient ignition energy, explosion will occur.

2.1.2. Sustained Combustion. Hydrogen is ignited by the ignition source in the early diffusion. Although the hydrogen still spreads out, surrounding environment only forms a stable combustion zone because of the continuing consumption of hydrogen and oxygen in the air. If pipeline is not damaged further, combustion will be maintained at a relatively stable level. It is unlikely to explode.

2.1.3. The VCE (Vapor Cloud Explosion). After hydrogen pipeline is damaged, hydrogen in the atmosphere diffuses to a certain extent, and stays in the range of its explosive concentration limit. In the event of an ignition source the whole piece vapor cloud will be ignited, and then vapor cloud explosion occurs. Vapor cloud explosion products a variety of effects, such as shock waves, thermal radiation, which could cause serious damage in a large area.

2.2. Initial Conditions of Simulation

Suppose leakage occurred at a certain section of "Baling-Changling" hydrogen transfer line with a total length of 42 km. The accident site was 300 m high, and was 113°07′E, 29°23′N. It was at 12:13 on April 1, 2016. The accident occurred in the flat region, and there was stable wind speed as well. Other basic data is shown in Table 1.

Table 1. Basic data of simulation condition.

Length of Pipeline/m	1000
Wind Speed/(m/s)	2
Ambient Temperature/℃	25
Humidity/%	50
Ground Roughness/cm	0.1
Pipe Diameter/cm	50
Tube Temperature/℃	25
Tube Pressure/atm	20
Gap Area/cm^2	10

3. Simulation Results and Discussion

As described above, three kinds of parlous situations may occur in the hydrogen leakage accident. What causes serious casualties and losses is the overpressure in vapor cloud explosion. And whether a man or a building is safe in a VCE depends on overpressure distance. Besides, the diffusion distance and combustion distance are also the standards to measure the degree of safety in a VCE. A set of scenarios were designed respectively to observe how the hazardous distances were changed with the increasing of wind speed, ground roughness, gap area and tube pressure[11]. In the last comparison among these results was conducted.

3.1. The Influences on Diffusion Distance

3.1.1. Effect of Wind Speed. The maximum distances were respectively recorded from the source of the leakage to corresponding 40000 ppm (Lower Explosive Limit, LEL), 24000 ppm (60% LEL) and 4000 ppm (10% LEL) concentration area within 40 minutes. The simulation results show that wind speed has a significant impact on the diffusion distance. The higher wind speed leads to a shorter diffusion distance at three concentrations of hydrogen. And more details can be indicated in Fig. 1.

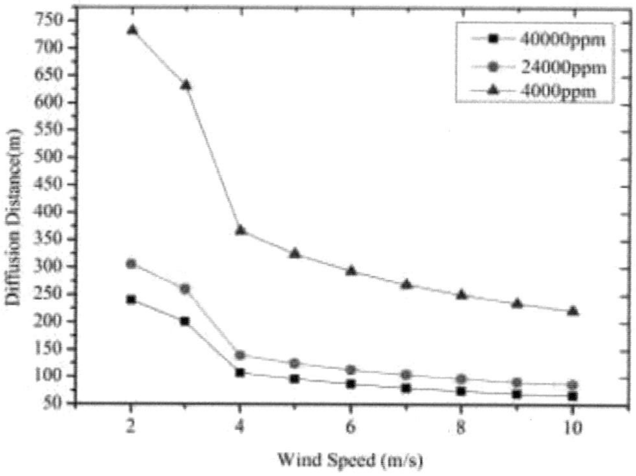

Figure 1. Relationship between wind speed and diffusion distance.

As can be seen, the relation between two parameters is not linear. At the low wind speed (<4 m/s), as wind speed increases, the diffusion distance decreases rapidly; at the high wind speed (>4 m/s), as wind speed increases, the diffusion distance decreases slowly. On the one hand, wind quickens the diffusion of hydrogen; on the other hand, it reduces the hydrogen concentration windward. When the wind speed is 2 m/s, hydrogen diffusion range within 40 min is described by a sector with the release source point as a central point. Because of the effect of wind for long time, the maximum distance of LEL is downwind 200 m from release point. Within the range of LEL, an ignition source will lead to VCE. When the wind speed changes, the shape of leakage area is the same.

3.1.2. Effect of Ground Roughness. Ground roughness refers to the height where the average wind speed is 0 m/s, which is used to indicate the rough degree of ground. Table 2 lists a group of representative values of the ground roughness.

Table 2. Representative values of the ground roughness.

Ground Cover	Roughness Z_0(cm)
Water Surface	0.1
Sparsely Populated Grasslands and Fields	10
Pasture and Field with Dwelling House	20
Fields with Many Houses and Shrubs	40
House Area in Suburb	60
Urban and Forest	100~500

Ground roughness value was taken from Table 2 respectively. Z_0 for urban and forest was 200 cm. The maximum distances from the source of the leakage to the regions with the concentration of 40000 ppm, 24000 ppm and 4000 ppm within 40 minutes were recorded, and results are indicated in Fig. 2.

IOP Conf. Series: Earth and Environmental Science **64** (2017) 012002 doi:10.1088/1755-1315/64/1/012002

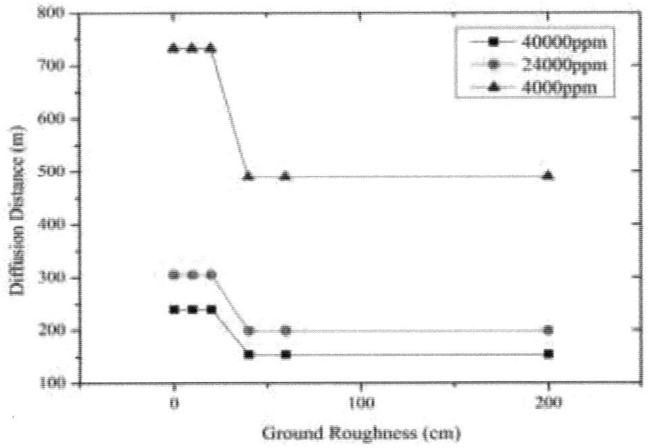

Figure 2. Relationship between ground roughness and diffusion distance.

It can be observed there is a steep decrease when the roughness increases approximately from 20 cm to 40 cm. However, it seems that there is no effect on the diffusion distance when the roughness is below 20 cm or above 40 cm. To a certain extent, the influence of ground roughness is limited.

3.1.3. Effect of Tube Pressure. Hydrogen pipeline pressure is usually 1 ~ 3 MPa. Other basic data unchanged, hydrogen pressure was taken from 10 atm to 30 atm, and maximum distances from the leakage source with the concentration of 40000 ppm, 24000 ppm and 4000 ppm within 40 minutes were recorded. The relation between tube pressure and diffusion distance is indicated as follows.

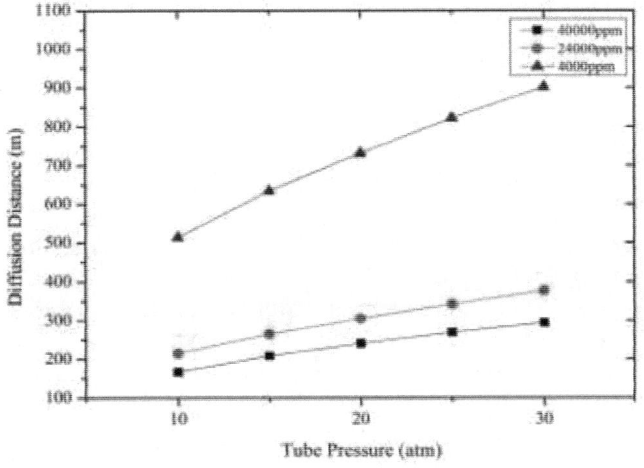

Figure 3. Relationship between tube pressure and diffusion distance.

Fig. 3 shows that diffusion distance is almost linear with tube pressure. By calculating the slopes of the three curves are 6.9, 8.85 and 21.2. K value is a representative under the accident condition indicated by the basic data, but its versatility needs to be discussed. And for a certain value of the pressure, the greater the K is, the greater the risk distance increases.

3.1.4. Effect of Gap Area. In the basic data, the tube diameter is 50 cm, and the corresponding cross-sectional area is 1963 cm^2. Other basic data unchanged, cross-sectional area was taken from 10 cm^2 to

1963 cm^2 respectively, and maximum distances were recorded from the source of the leakage to the regions with the concentration of 40000 ppm, 24000 ppm and 4000 ppm within 40 minutes. During the simulation, when the gap area was greater than 350 cm^2, the diffusion distance was no longer changed. So gap area in Fig. 4 is only shown to 350 cm^2.

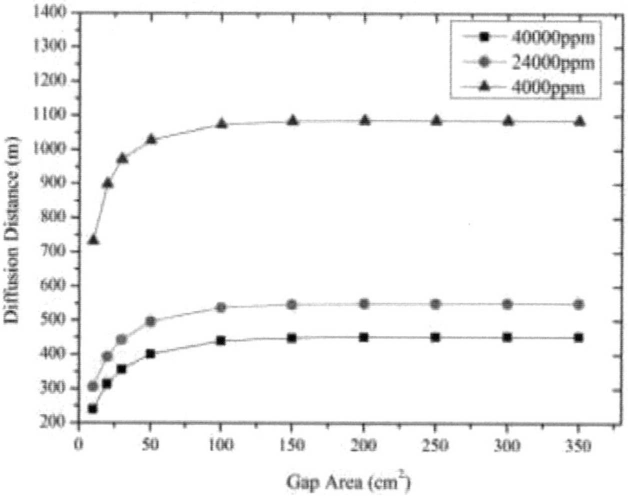

Figure 4. Relationship between gap area and diffusion distance.

As can be seen from Fig. 4, when the leakage gap area is less than 150 cm2 (less than 7.6% pipe cross-sectional area), with the gap area increasing, the diffusion distance increases rapidly and then levels off. When the area is greater than 150 cm2, the diffusion distance does not increase any longer.

3.2. The Influences on Combustion Distance

3.2.1. Effect of Wind Speed. The model was changed into a combustion model, and the wind speed was taken from 2 m/s to 10 m/s. The maximum distances from leakage source to the corresponding combustion region were recorded. Simulation results indicate that the wind speed does not affect the maximum distance from the source of the leakage to the combustion area. For the leakage accident where hydrogen burns at the beginning, to some extent the degree of danger is limited.

3.2.2. Effect of Ground Roughness. Ground roughness value was taken from Table 2 respectively. The main influence of rough ground is to make released gas move slowly in the flow due to concentration gradient. But there is usually not so much effect, especially in some complex conditions.
Simulation results show that the ground roughness does not affect the corresponding maximum distance in combustion area. We can deduce from 3.2.1 and 3.2.2 that the environmental factors don't have any effect on the combustion distance. That is because the gas consumption has a greater impact on distance than the diffusion. The release source factors significantly affect the combustion distance.

3.2.3. Effect of Tube Pressure. Tube pressure value was taken from 10 atm to 30 atm respectively in the simulation. Then the maximum distances from leakage source to the corresponding combustion region of 10.0 kW/m^2, 5.0 kW/m^2 and 2.0 kW/m^2 were recorded.

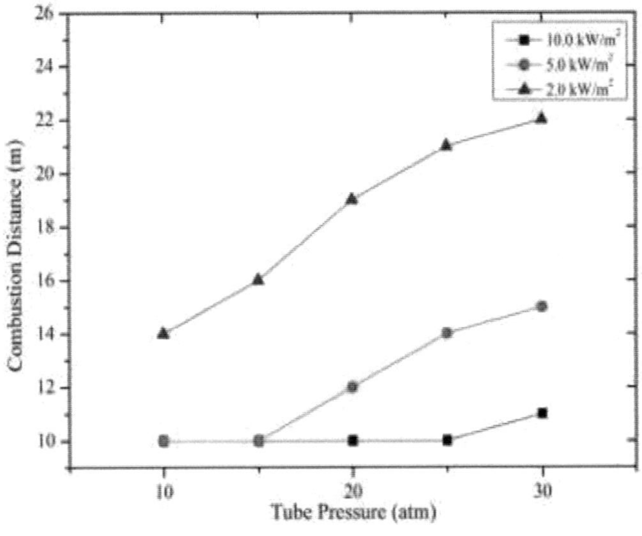

Figure 5. Relationship between tube pressure and combustion distance.

Fig. 5 shows that with the increase of pressure, the combustion distance also goes up, but later its growth trend levels off. The effect of the tube pressure on the combustion distance is limited. Finally, even if the inner tube pressure increases a lot, the combustion distance will tend to an upper limit.

3.2.4. Effect of Gap Area. Gap area value was taken from 10 cm^2 to 1960 cm^2 respectively. Similar to the result above, the growth of gap area has a negative influence on the accident result. And no matter how large it is, the influence is effective.

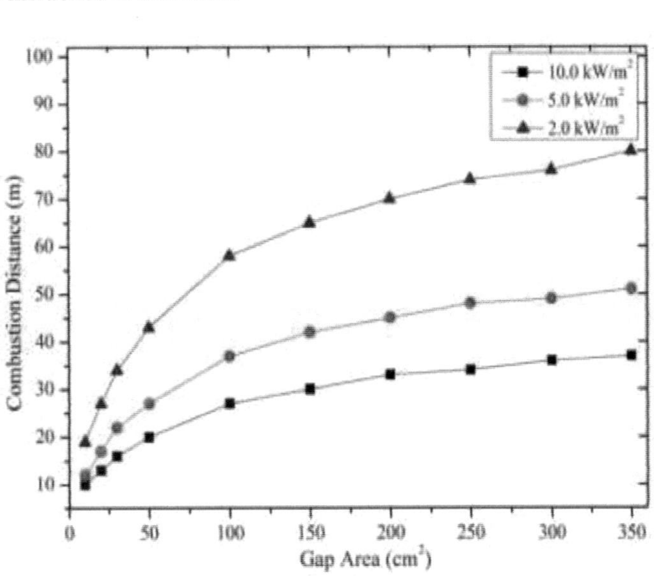

Figure 6. Relationship between gap area and combustion distance.

Fig. 6 shows that with the increase of gap area the combustion distance continuously increases. The growth trend is smooth and predictable according to the existing data. And so, it is very necessary to cut off the gas source with all methods when leakage occurs for fear that the scope of risk is expanded.

3.3. The Influences on Overpressure Distance

3.3.1. Effect of Wind Speed. Select the VCE overpressure hazard threshold value, the simulation results show that when the wind speed is 2 m/s, the maximum distance downwind area is 337 m and the maximum distance upwind is 103 m. The maximum distance crosswind is 211 m. Other basic data unchanged, when wind speed ranged from 2 m/s to 10 m/s, record maximum distance downwind in these areas. The results are indicated in Fig. 7.

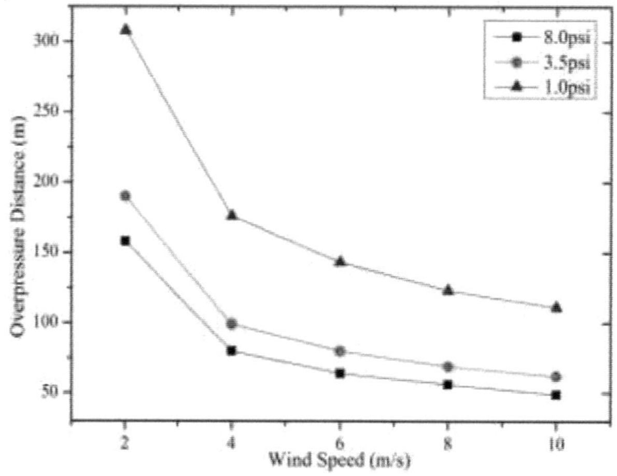

Figure 7. Relationship between wind speed and overpressure distance.

Fig. 7 shows that as the wind speed increases, overpressure distance of the VCE decreases rapidly. And its value is much larger than the maximum distance of combustion area.

3.3.2. Effect of Ground Roughness. Ground roughness value unchanged, the simulation model was set to vapor cloud explosion model. Record the maximum distances from source of the leakage to the overpressure region of 8.0 psi, 3.5 psi and 1.0 psi, and the results are indicated as follows.

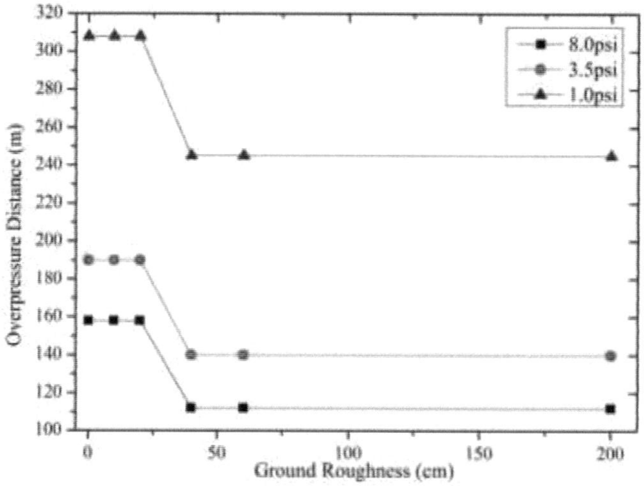

Figure 8. Relationship between ground roughness and overpressure distance.

As can be seen from Fig. 8, when ground roughness Z_0 is more than 20 cm, the overpressure distance for VCE product a mutation that is about a quarter to one-third of the original value. Thus, in the flat and open area, warning distance should be increased a third of the original alert distance.

3.3.3. Effect of Tube Pressure. When the tube pressure changes, release rate of gas will significantly be affected. Besides, the initial momentum of flammable gas depends on the pressure inside. Once the tube pressure increase, the gas will flow faster and further. The detailed results are shown in Fig. 9.

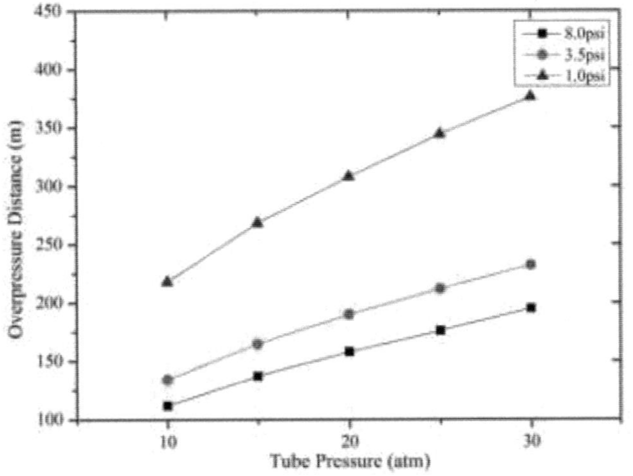

Figure 9. Relationship between tube pressure and overpressure distance.

Compare Fig. 3 with Fig. 9. It can be seen that the curve of diffusion distance and tube pressure is consistent in trend with the curve of overpressure distance and tube pressure. With the increase of pipe length, the overpressure distance increases linearly. And the approximate slope values are 8.65, 4.55 and 5.4, little difference between the three, which means the impact of tube pressure on the three overpressure distance curves is close, but is much less than the *K* values of Fig. 3. It means the impact of tube pressure on overpressure distance is much smaller than the impact on diffusion distance.

3.3.4. Effect of Gap Area. The same initial simulation environment using explosion model was chosen and then the maximum distances from leakage source to the corresponding overpressure range of 8.0 psi, 3.5 psi and 1.0 psi were recorded. The results are as follows.

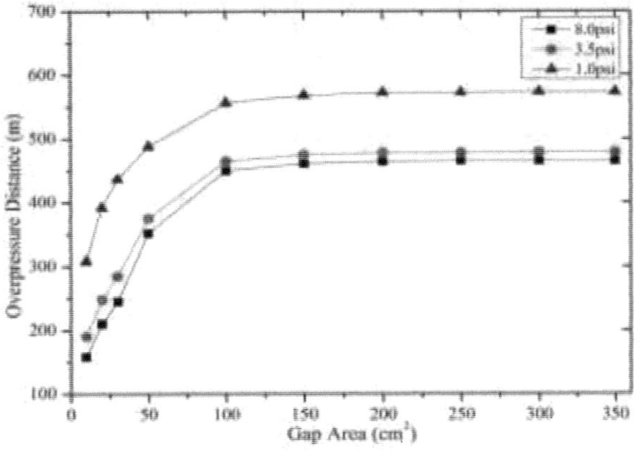

Figure 10. Relationship between gap area and overpressure distance.

As can be seen from Fig. 10, when the leakage gap area is less than 150 cm^2 (less than 7.6% pipe cross-sectional area), with the gap area increasing, the overpressure distance increases rapidly and then levels off. When the area is greater than 150 cm^2, the overpressure distance does not increase. So there may be two kinds of condition about release source gap area. Whichever condition we stay in, the most important thing is to plug the leak and cut off gas source as soon as possible.

A comparison was conducted after the change trend of diffusion distance, combustion distance and overpressure distance had been calculated. The environment factors have a beneficial influence on the leakage accident consequences, and the release source factors can increase the risk of accident.

4. Conclusion

In this article, the affecting degree about danger distance of different variables in hydrogen leakage accident can be obtained. Conclusions are as follows: (1) Wind speed, ground roughness, and tube pressure and leakage gap area have a great influence on the diffusion distance. And of them the most influential factor is the wind speed; (2) Wind speed, tube pressure and leakage gap area have a great influence on the overpressure distance. And the most influential factor is the gap area; (3) Gap area has a significant impact on the combustion distance. But the impact of other variables on the combustion distance is very little or no; (4) the diffusion distance and overpressure distance reduce as the wind speed and ground roughness increase. Wind speed and ground roughness do not affect combustion distance.

Acknowledgments

This paper is supported by National Natural Science Foundation of China (51406191).

References

[1] Verhelst S, Wallner T. Hydrogen-fueled internal combustion engines [J]. Progress Energy Combustion, 2009, 35(6): 490-527.

[2] Bychkov VV, Liberman MA. Dynamics and stability of premixed flames [J]. Physics Reports, 2000, 325(4-5): 115-237.

[3] Chunna Liu. Hydrogen energy: future of green energy [J]. Power Technology, 2010, 134(6): 536-538.

[4] Huahua Xiao, Zhanli Mao, Weiguang An, et al. Numerical simulation of premixed propane/air flame propagation using a dynamically thickened flame approach [J]. Applied Mechanics and Materials, 2013, 444-445: 1574-1578.

[5] Huahua Xiao, Zhanli Mao, Weiguang An, Jinhua Sun. Experimental and LES investigation of flame propagation in a hydrogen/air mixture in a combustion vessel [J]. Chinese Science Bulletin, 2014, 59(20): 2496-2504.

[6] Michael R. Swain, Patrick Filoso, Eric S. Grilliot, et al. Hydrogen leakage into simple geometrics enclosures [J]. International Journal of Hydrogen Energy, 2002, 28(28): 229-248.

[7] Hector A. Olvera, Ahsan R. Choudhuri. Numerical simulation and Hydrogen dispersion in the vicinity of a cubical building in stable stratified atmospheres [J]. International Journal of Hydrogen Energy, 2006, 31(15): 2356-2369.

[8] Zhu H, Mao Z, Wang Q, et al. The Influences of Key Factors on the Consequences Following the Natural Gas Leakage from Pipeline [J]. Procedia Engineering, 2013, 62:592-601.

[9] Utgikar V P, Thiesen T. Safety of compressed hydrogen fuel tanks: Leakage from stationary vehicles [J]. Technology in Society, 2005, 27(3):315-320.

[10] Rosi-Marshall E J, Kincaid D W, Bechtold H A, et al. Simulation Experiment for Gas Pipeline Leakage Detection[J]. Advanced Materials Research, 2013, 805-806(3):1502-1506.

[11] Choi J, Hur N, Kang S, et al. A CFD simulation of hydrogen dispersion for the hydrogen leakage from a fuel cell vehicle in an underground parking garage[J]. International Journal of Hydrogen Energy, 2013, 38(19):8084-8091.

Temperature-controlled 3-phase oil-water-oil continuous separation by PNIPAM-modified stainless steel mesh

Baiyi Chen, Jianhui Qiu *, Kazushi Ito, Eiichi Sakai and Nobuhiro Kanazawa

Faculty of Systems Science and Technology, Akita Prefectural University, Akita 015-0055, Japan

*Corresponding author e-mail: qiu@akita-pu.ac.jp

Abstract. In recent years, the frequent occurrence of oil spills has been a serious threat to marine and aquatic ecosystems. The environmental and economic demands also emphasize the need for materials which can effectively separate oil and water. In order to handle these environmental issues caused by oil spills, all kinds of oil-water separation methods have been developed. In this work, we have fabricated a thermosensitive device with a simple and versatile method: we sprayed the spray adhesive onto a stainless steel mesh box, then amino-P25 nanoparticles were sprayed onto the device directly. Finally, PNIPAM was synthesized on the surface of device. The wettability of PNIPAM modified separation device can be controlled by temperature and finally a 3-phase oil-water-oil continuous separation was realized.

1. Introduction

The smart material with conversion of wettability on external stimuli has attracted widely attention because of their promising potential in fabricating intelligent controllable materials. [1-5] These smart responsive materials can display controlled changes in response to physical or chemical changes of the external environment, such as temperature, pH, light, electricity, and magnetism. [6-12] The smart responsive materials have numerous applications in many fields, for example, in biomedical field, light-, pH-, and temperature-responsive dendrimer systems with different chemical compositions and architecture can be used to encapsulate and release drugs. In the meantime, they also have potential application in many other fields such as sensors, enzyme immobilization, and changing wettability. For example, some typical smart responsive materials are utilized to fabricate responsive surfaces with reversibly switchable wettability to realize oil-water separation.

In recent years, the frequent occurrence of oil spills has been a serious threat to marine and aquatic ecosystems. The environmental and economic demands also emphasize the need for materials which can effectively separate oil and water. In order to handle these environmental issues caused by oil spills, all kinds of oil-water separation methods have been developed to recover spilt oil from marine environments, such as physical diffusion, bioremediation13, 14, oil skimmers[15] and so on. Owing to the difference of interfacial effects for oil and water, utilizing the wetting behavior of solid surfaces to design an oil-water separating material has attracted a great deal of attention in fundamental research and for potential application in the field of oil-water separation.

In this work, we utilized stainless steel mesh as a substrate, spray adhesive was sprayed on the substrate directly. Following, amino-P25 nanoparticles were dispersed in absolute ethanol forming a paint-like solution and then sprayed onto the adhesive coated substrate directly. Finally, poly (N-isopropyl acrylamide) (PNIPAM) was generated on the surface of amino-P25 coated substrate. Consequently, thermosensitive surface that can change from super hydrophilic to super hydrophobic by heating was obtained and used in 3-phase oil-water-oil separation.

2. Experimental

2.1. Materials and Instruments

Oil Black and Scarlet 4GE (Askul Co., Ltd., Japan), Stainless steel mesh (Anping Huirui Wire Mesh Manufacture Co., Ltd., China), Spray adhesive (Sumitomo3M Co., Ltd., Japan). Toluene, ethanol, dichloromethane, methyl, APTES, N-isopropyl acrylamide and trimethylamine were purchased from Nacalai tesque. 2-Bromo-2-methyl-propionyl bromide, CuBr and PMDETA were purchased from Sigma Aldrich. All chemicals were analytical grade reagents and were used as received.

Scanning electron microscopy (SEM) measurements were carried out on an EVO MA 25 instrument at 20.0 kV. Photographic images were taken with a Nikon camera (D5000).

2.2. Fabrication of the Amino-P25 Coated Stainless Steel Mesh

P25 nanoparticles and 10%wt APTES was added into absolute ethanol and refluxing for 12h, then the amino-P25 nanoparticles were vacuum dried for 12h under 60 °C. Stainless steel mesh was folded into an open and rectangular box with an edge length of 3 cm×3 cm×1 cm. The as-prepared box was ultra-sonicated in ethanol and deionized water for three times, and dried in oven. Spray adhesive was sprayed onto the box directly, following, amino-P25 nanoparticles were dispersed in absolute ethanol forming a paint-like solution and then sprayed onto the adhesive coated substrate directly.

2.3. Synthesis of the PNIPAM Coating

400 μL trimethylamine was added into 20 ml dichloromethane and stirred for 30 min, and then the amino-P25 coated box was put into the solution, following, 1 ml of 2-Bromo-2-methyl-propionyl bromide was added and stirred for 1h. Finally, the treated as-prepared stainless steel box was take out and washed for 3 times with deionised water.

For the synthesis of PNIPAM coating on the as-prepared box, as-prepared box was added into a mixture of 20 ml methyl and 20 ml deionized water. Then, 1.6 g N-isopropyl acrylamide and 25 mg CuBr was added into the solution and stirred. Finally, 560 μL PMDETA was added and stayed for 1 h under 60 °C. After wash and drying, a thermosensitive oil-water separation device was obtained.

2.4. Oil-Water-Oil 3-Phase Separation

In order to distinguish oil phase and water phase from oil-water-oil mixture, we dyed the dichloromethane layer with Solvent Black and toluene layer with Solvent Red to indicate the directional delivery. The labeled dichloromethane (5 mL) and toluene (5 mL) were added into a pipette with 5 mL of deionised water forming a stratified oil-water-oil mixture solution. After the as-prepared box put on the top of the supporting beaker, the mixture solution was discharged slowly. In the meantime, environment temperature was controlled to match up the separation process. Consequently, oil phase and water phase were separated.

3. Results and Discussion

3.1. Surface Modification and Characterization

Figure 1. SEM images of the amino-P25 nanoparticles.

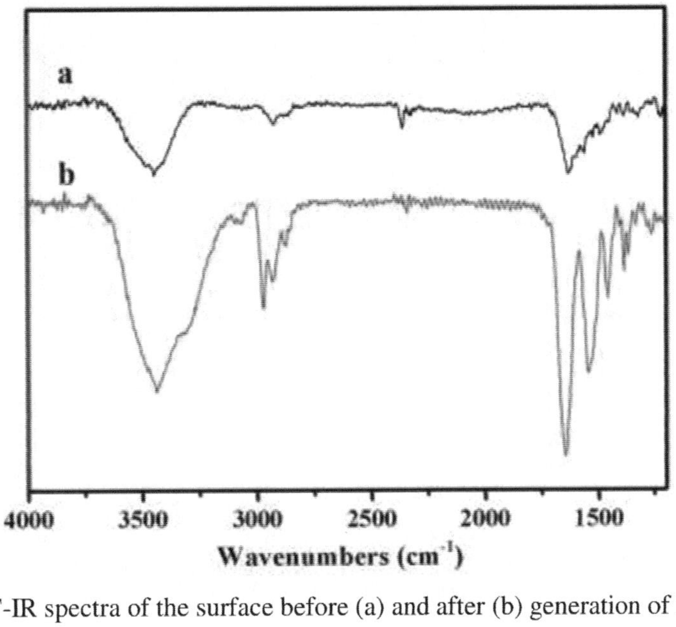

Figure 2. FT-IR spectra of the surface before (a) and after (b) generation of PNIPAM.

During modification, evident changes in wettability of the surface occurred in the process of preparing the thermosensitive oil-water separation device. After being coated by amino-P25 nanoparticle, the surface showed superhydrophilicity. As can be seen from figure 1, the amino-P25 nanoparticle showed an average particle size of 20 nm. In the meantime, as we can see from the figure 2, the FT-IR spectra of the device surface after generation of PNIPAM showed four obvious strong peaks from about 1380 cm^{-1} to about 1650 cm^{-1}. These four peaks belonged to the amido bond between PNIPAM and substrate which indicated PNIPAM has generated on the substrate surface successfully.

When the PNIPAM was generated on the surface of amino-P25 nanoparticles, the surface wettability emerged a temperature-responsive property. When temperature was higher than 37 °C, the device surface showed superhydrophobicity, otherwise, the surface showed superhydrophilicity. Take advantages of this phenomenon, we designed a simple separation process to separate a 3-phase oil-water-oil mixture continuously.

3.2. 3-Phase Oil-Water-Oil Separation

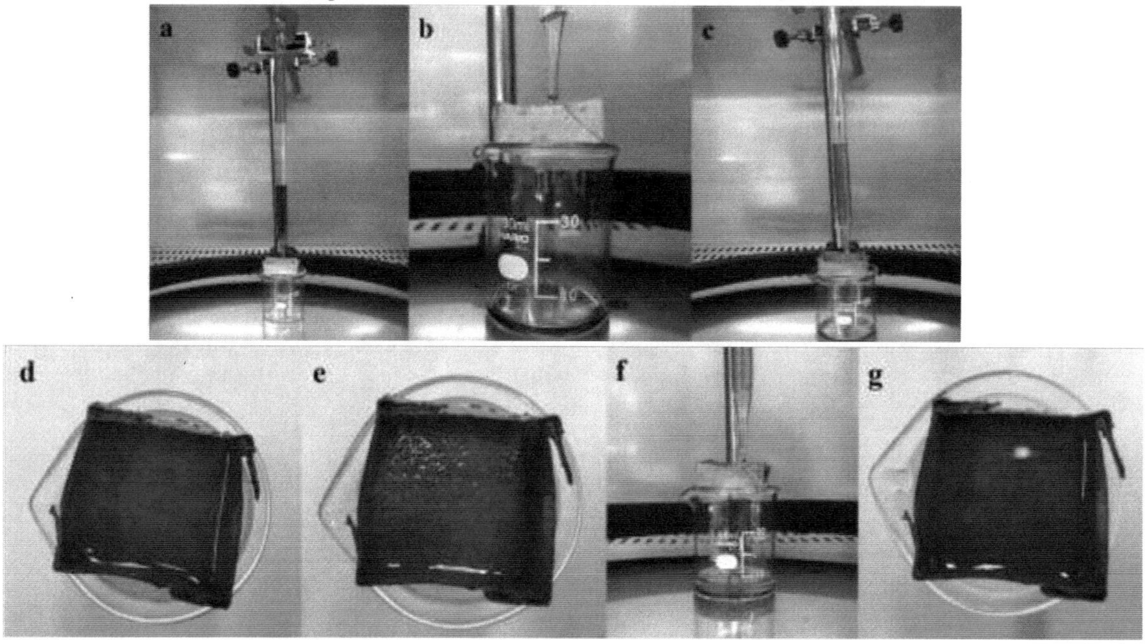

Figure 3. A 3-phase oil-water-oil separation process using the thermosensitive device.

We dyed the dichloromethane layer with Solvent Black and toluene layer with Solvent Red to indicate the directional delivery and distinguish oil phase and water phase from oil-water-oil mixture. In order to separate the oil phase and water phase continuously and respectively, the body temperature of temperature-responsive separation device was regulated through controlling environment temperature. As can be seen from figure 3a, we began the separation process above 37 °C, thus the device surface was superhydrophobic. When dichloromethane dropped into the device, it was immediately taken up by the device, penetrated through the modified mesh, and finally dropped into the flask (figure 3b). Following, the water was dropped into the device (figure 3c), because of the superhydrophobicity, water was retained in the device (figure 3d). Next, we regulated the body temperature of device below 37 °C, thus the wettability of device surface transformed into superhydrophilicity. As a result, the remained water penetrated through the device rapidly (figure 3e) and dropped into the flask. Finally, the red dyed toluene was dropped into the device (figure 3f). Because the device surface has transformed into superhydrophilic, the toluene was retained in device (figure 3g). Thereby, the 3-phase oil-water-oil continuous separation has been realized.

4. Conclusion

In this work, we have fabricated a thermosensitive device with a simple and versatile method: we sprayed the spray adhesive onto a stainless steel mesh box, then amino-P25 nanoparticles were sprayed onto the device directly. Finally, PNIPAM was synthesized on the surface of device. The wettability of PNIPAM modified separation device can be controlled by temperature and finally a 3-phase oil-water-oil continuous separation was realized.

References

[1] P. Brochu, Q. Pei, Advances in dielectric elastomers for actuators and artificial muscles, Macromol. Rapid Commun. 31 (2010) 10–36.
[2] C. Ohm, M. Brehmer, R. Zentel, Liquid crystalline elastomers as actuators and sensors, Adv. Mater. 22 (2010) 3366–3387.
[3] T. Mirfakhrai, J. D. W. Madden, R. H. Baughman, Polymer artificial muscles, Mater. Today 10

(2007) 30-38.

[4] N. Bassik, B. T. Abebe, K. E. Laflin, D. H. Gracias, Polymer artificial muscles, Polymer 51(2010) 6093–6098.

[5] T. G. Leo nga, C. L. Randallb, B. R. Bensona, N. Bassika, G. M. Sterna, D. H. Gracias,Tetherless thermobiochemically actuated microgrippers, Proc. Natl. Acad. Sci. U.S.A. 106 (2009), 703–708.

[6] S. Das, P. Ranjan, P. S. Maiti, G. Singh, G. Leitus, R. Klajn, Dual - Responsive Nanoparticles and their Self-Assembly, Adv. Mater. 25 (2013) 422–426.

[7] D. Wang, P. Jiao, J. Wang, Q. Zhang, L. Feng, Z. Yang, Fast photo - switched wettability and color of surfaces coated with polymer brushes containing spiropyran, J. Appl. Polym. Sci. 125 (2012) 870–875.

[8] S. Liu, S. P. Armes, Polymeric Surfactants for the New Millennium: A pH-Responsive, Zwitterionic, Schizophrenic Diblock Copolymer, Angew. Chem., Int. Ed. 41 (2002) 1413–1416.

[9] G. Shi, S. Jin, G. Xue, C. Li, A conducting polymer film stronger than aluminum, Science 267 (1995) 994–996.

[10] Y. Deng, W. Yang, C. C. Wang, S. K. Fu, A novel approach for preparation of thermoresponsive polymer magnetic microspheres with core-shell structure, Adv. Mater. 15 (2003) 1729–1732.

[11] W. Sun, S. Zhou, B. You, L. Wu, A facile method for the fabrication of superhydrophobic films with multiresponsive and reversibly tunable wettability, J. Mater. Chem. A 1 (2013) 3146–3154.

[12] W. Sun, S. Zhou, B. You, L. Wu, Polymer brush-functionalized surfaces with reversible, precisely controllable two-way responsive wettability, Macromolecules 2013, 46 (2013) 7018–7026.

[13] R. Z. Hoff, Bioremediation: An Overview of its Development and Uses for Oil Spill Cleanup. Mar. Pollut. Bull. 26 (1993) 476-481.

[14] R. M. Atlas, Microbial Hydrocarbon degradation-Bioremediation of Oil Spills. J. Chem. Technol. Biotechnol. 52 (1991) 149-156.

[15] M. Fingas, Oil spills and their cleanup. Chem. Ind. 24 (1995) 1005-1008.

Visible study of water shutoff techniques for reservoirs with high permeability channels

Yanyong Xin[1, 2, *], Xuan Zhao[2]

[1]Energy Resources College of China University of Geosciences, Beijing, 100083, China
[2]Sinopec International Petroleum E&P Company, SINOPEC, Beijing, 100012, China

*Corresponding Author e-mail: xinliu102019@sina.com

Abstract. This paper describes a compound-linking gel system consisting of Polymer, Chromium (+3), Phenolic resin and rubber particles, which has good performance for water blocking. Indoor experiments demonstrate that the breaking through pressure of compound gel system is 4.92 MPa/m, which is much higher than 2.5-3 MPa/m of conventional, and its flooding resistance is better. The pressure of sand-pack with compound gel system reaches stability at 2.5MPa after 7PV water flooding while others stay at 1.1-1.6 MPa after 3-4PV water displacement. Visible flooding models are developed to optimize the injection volume and investigate the best location to deploy plugging agents. The results show that there is a limit on injection volume, and when it equals to 0.1-0.2 PV, the application may have economic success. For models with worm-shaped channels, the best location for plugging agents is just after "inflection point" near injecting well, which can make full use of the fluid diverting to achieve highest oil recovery.

1. Introduction

In China, many oilfields have stepped into the late development stage, where "fingering" of injected water becomes a common problem due to the reservoir heterogeneity [1, 2]. A recent survey indicated that the amount of water-intake in high permeability zone would account for 90% of total injected if the permeability ratio reached 5, which seriously impacts the economics and recovery [3].

Cross-linked polymer gels are currently used in profiling and water plugging application [4~6]. One choice is chromium (+3) cross-linked polymer gels, cross-linking mechanism of which is the complexation of Cr (+3) ions with carboxylate groups on the polymer chains [4-5]. Another commonly used gel system is phenol/formaldehyde cross-linker system [6-7].The cross-linking mechanism involves hydroxymethylation of the amide nitrogen, with the subsequent propagation of cross-linking by multiple alkylation on the phenolic ring. Reddy et al [8] described an environmentally-preferable cross-linker to produce gel. Dai [9] et al had a lab experiments and developed gel foam water-plugging system aiming at the severe water production in the East China Sea Gas Field. The effectiveness of polymer gel is obvious; however, its strength, stability and washing resistance are still the challenges. In addition, few scholars had systematically research on the best position to deploy plugging agents.

In this paper, a compound cross-linked gel system is introduced, which consists of Polymer, Chromium, and Phenolic resin and rubber particles. The compound gel system has better strength, stability and washing resistance. Also, visible plate models are designed to study the relationship between blocking effects and position to deploy plugging agents.

2. Experimental section

2.1. *Chemical and fluids*

The following chemicals were used in this study: Polymer (Dongying Guangzheng Products, molecular weight 15-18 million); Water-soluble phenolic resin (Wuhan Dahua Products, Content 99%); Organic Chromium cross-linker (Panjin Dongmeng Products, Content 40%); Rubber particles (diameter 0.25mm --0.3mm)

The oil this study used comes from Gudong reservoir (Shengli oilfield). The API of oil is 24 and its viscosity is 245 mPa·s at 60 °C (the experimental temperature). All solution is prepared by simulated formation water, whose salinity is 6778 mg/L.

2.2. *Gel system formulation optimization*

Strength is an important index of gel system properties. Figure 1 shows the "vacuum pressure breaking through" method to evaluate the gel strength [10]. The measuring procedure was shown as follows: 1) Link the colorimetric tube and measuring device with U-tube and rubber horse; 2) Locate U-tube end at 1-2 CM below the gel surface; 3) Start vacuum pump and boost slowly; 4) Record the maximal number on the gauge as breaking through vacuum (BV) pressure when the air breaks through the gel; 5) Each gel sample is measured three times and average the numbers as the final BV.

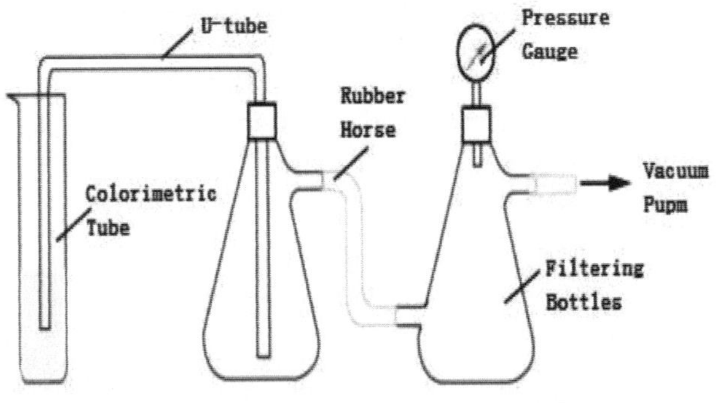

Figure 1. Measuring equipment of vacuum pressure breakthrough.

In this study, the influence factors of strength of compound gel system were discussed in detail, including concentration of Polymer, Chromium, and Phenolic resin and rubber particles. The optimal formulation was determined by orthogonal experiments. All of these tests were conducted at 60℃. The parameters and results of different tests are shown in Table 1.

Table 1. Orthogonal experiment of compound cross-linking gel system optimization.

No.	1	2	3	4	5	6	7	8
Polymer, ppm	2000	2000	2000	2000	3000	3000	3000	3000
Chromium, %	0.01	0.02	0.04	0.06	0.01	0.02	0.04	0.06
Phenolic resin,%	0.225	0.45	0.675	0.9	0.45	0.225	0.9	0.675
Rubber particle,%	0.02	0.04	0.06	0.08	0.06	0.08	0.02	0.04
BV pressure, MPa	0.034	0.056	0.061	0.065	0.076	0.075	0.052	0.067
No.	9	10	11	12	13	14	15	16
Polymer, ppm	4000	4000	4000	4000	5000	5000	5000	5000
Chromium, %	0.01	0.02	0.04	0.06	0.01	0.02	0.04	0.06
Phenolic resin,%	0.675	0.9	0.225	0.45	0.9	0.675	0.45	0.225
Rubber particle,%	0.08	0.06	0.04	0.02	0.04	0.02	0.08	0.06
BV pressure, MPa	0.09	0.088	0.072	0.061	0.078	0.074	0.09	0.084

2.3. *Evaluation of plugging performance*

In order to investigate the plugging performance of gel system, a series of sand-pack flooding tests were done. A core holder of 1 in. (2.54 cm) in diameter and 1.8 ft (45 cm) in length was used for the sand-pack. Fresh quartz sand of different mesh was packed to get different permeability cores [11]. The cores were packed as follows [12]: the core holder, which was filled with formation brine, was positioned vertically and the quartz sand was added to fill it in 10 steps. In each step after sand was packed, the core holder was vibrated for 5 minutes. The evaluation experiments were conducted as the following procedures: Step 1: Evacuate the core and saturate with formation water. Step 2: Sand-pack core is water-flooded at the rate of 0.3 mL/min until the pressure reached stability, and the permeability (k_b) of the sand-pack core is calculated. Step 3: 0.15PV chemical slugs were injected and gelling 24 hours at 60℃ conditions. Step 4: Restart water-flooding at 0.3 mL/min rate and get the results of displacement pressure versus PV. The maximal value during flooding is the breaking through pressure. Step 5: Measure the permeability (k_a) of core after plugging when the pressure reached stability and get the plugging efficiency: $E = \frac{k_b - k_a}{k_b} \times 100\%$.

The results of experiments are shown in Table 2.

Table 2. Sand-pack flooding tests for plugging performance.

No.	Initial Permeability /$10^{-3}\mu m^2$	Permeability after plugging /$10^{-3}\mu m^2$	Plugging efficiency,%	Breakthrough Pressure (MPa/m)	Agents
1	632	38.2	94.0	2.53	0.4% P+0.01%Cr
2	601	35.1	94.2	2.70	0.4% P+0.67%Pr
3	625	9.6	98.5	3.03	0.4% P+0.01%Cr+0.67%Pr
4	613	7.9	98.7	4.92	0.4% P+0.01%Cr+0.67%Pr+0.08%R

(For the agents, P= Polyer, Cr=Chromium, Pr= Phenolic resin, and R=Rubber particles)

2.4. *Micro flooding models*

Two ideal porous micro-models are prepared by cementing quartz on glass plate with epoxy resins [13]. This paper used the conventional channel model to optimize the injection volume of gels and the position to deploy plugging agents, while the worm-shaped channel is used to simulate reservoir conditions. The general procedure is as follows: (1) Clean micro-models with laboratory cleaner. (2) The models are saturated with oil and then dry in an oven at 60℃ 24 hours. (3)0.05-0.4PV Plugging agents are injected at certain position of channel. (4) The subsequent water displacement was conducted until oil production became negligible (water cut >95%). (5)The sweep efficiency was calculated through "Grid-Box-Counting" method. The flooding tests were carried out at 60℃.

3. Results and discussion

3.1. Compound cross-linked gel system performance

The optimized compound gel formulation consisted of 4000 ppm polymer, 0.01% Chromium, 0.675% Phenolic resin and 0.08% Rubber particles. Figure2 shows that, all the agents have good plugging efficiency. The breaking through pressure of sand-pack No.4 with the optimized gel formulation is 4.92 MPa/m, much higher than others. The pressure of sand-pack No.4 reaches stability after 7PV water flooding, while the others only 3-4PV.

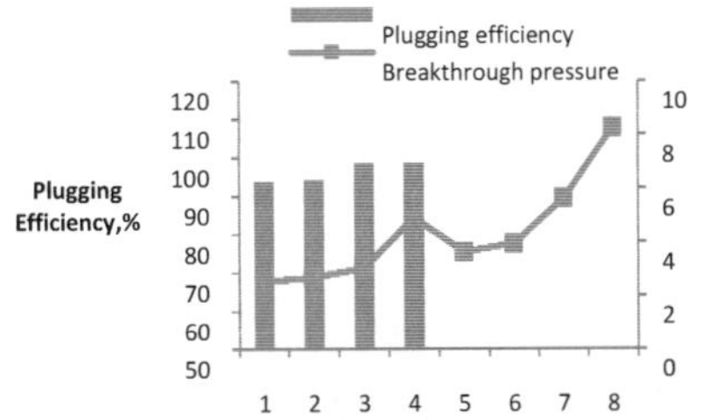

(a) Plugging efficiency and breaking through pressure

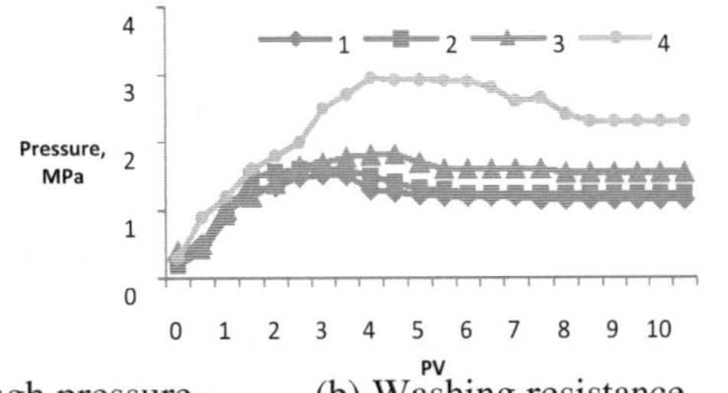

(b) Washing resistance

Figure 2. Plugging performance of different agents.

3.2. Optimization of the amount of plugging agents

Generally, the larger amount of blocking agents applied, the more effective plugging performance can be gained, and diversion are more likely to occur for long-term water-flooding. Experiments performed in micro-models investigated the relation between sweep efficiency and the amount of plugging agent, as Figure3 shows.

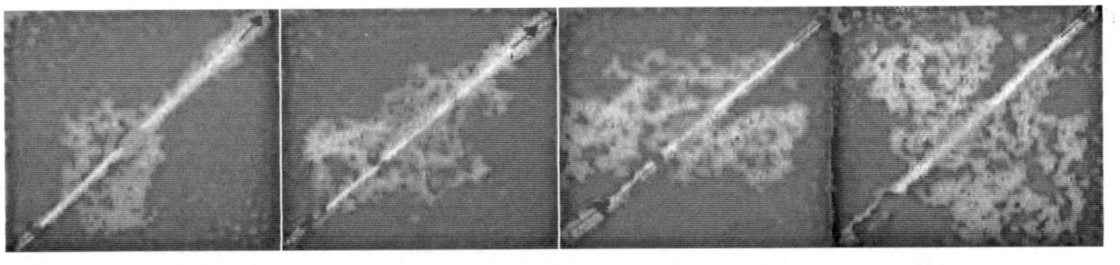

0.05PV 0.10PV 0.15PV 0.20PV

Figure 3. Flooding models of different volumes of plugging agents.

As Figure4 shows, the sweep efficiency increases quickly with the injected PV at early stage and becomes gentle afterwards and the optimal scope is 0.1-0.2 PV.

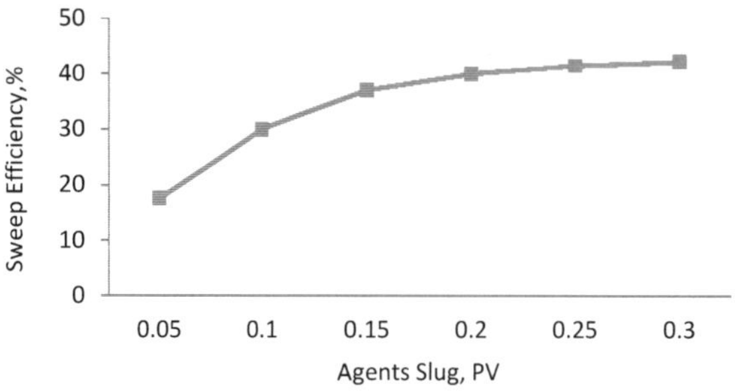

Figure 4. Relationship of sweep efficiency and PV of agents slug with fixed deploying position (mid of interval).

3.3. Optimization of position to deploy plugging agents

3.3.1. Conventional channel models. Observation of Figure5 and Figure6 shows that diversion appears when the water front reaches blocking agents, however, water returns to the channels quickly after bypassing. The "wide and long" swept zone formed at the deploying position of 1/2 channel length distance to the injection well, nearly 50% sweep efficiency.

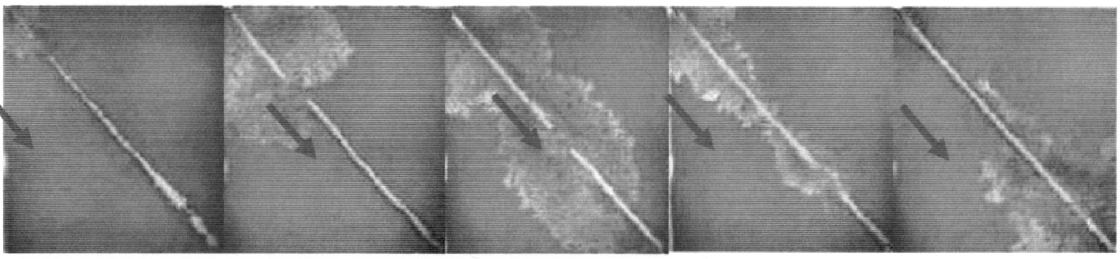

1/6 1/3 1/2 2/3 5/6

Figure 5. Micro flooding models of different deploying position (distance of 1/6、1/3、1/2、2/3 and 5/6 channel length to the injection well) with 0.15 PV agents slug.

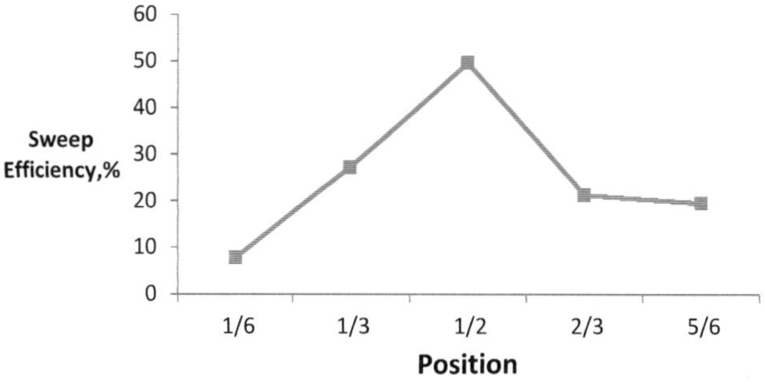

Figure 6. Relationship of sweep efficiency and different deploying position with 0.15PV aggents slug for conventional channel models.

3.3.2 Worm-shaped channels models. Worm-shaped channels models are developed to simulate reservoir characteristics with 2 "inflection points" are located near the injection and production wells respectively. Five deploying positions are designed (front of the inflection point near injection well, at inflection point near injection well, behind of the inflection point near injection well, middle place of channel and front of the inflection point near production well). These experiments demonstrate that there is significant disparity in terms of oil recovery for different deploying positions. The case of location after the "inflection point" near injection well where agents are deployed has highest oil recovery, as Figure7 shows.

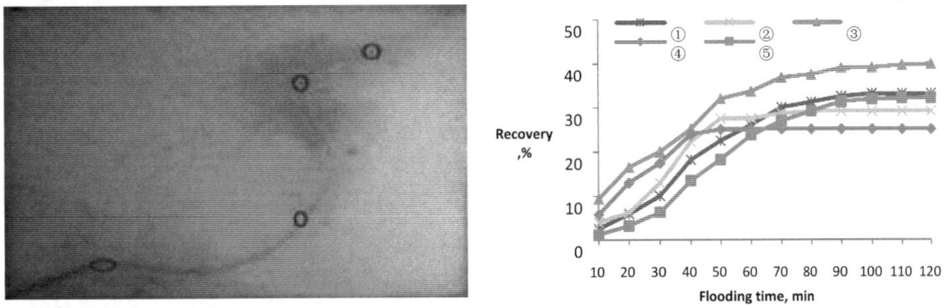

Figure 7. Relationship of recovery and different deploying positions with 0.15PV agents slug for worm-shapeded models.

4. Conclusion

The optimized compound-linked gel formulation has good plugging performance with higher breaking through pressure and flooding-resistance than conventional gel system. 0.1-0.2 PV is the best scope with respect of the amount of agents injected.

For conventional channel models, the deploying position of 1/2 channel length distance to the injection well has largest swept efficiency. For worm-shaped channels, blocking agents should be positioned at the location just after the "inflection point" near injection well in order to achieve highest recovery.

5. Notes

The authors declare no competing financial interest.

Acknowledgments

China University of Petroleum is gratefully acknowledged because of the generous support during this study.

References

[1] Mao, Z.; Gu, J.; Dong, P. Study on occurrence mechanism and prediction of big channels in loose cemented sandstone reservoir, Petroleum Geology and Recovery Efficiency, 2011,18 (1) :77-79.

[2] Li, B.; Zhao G. Remaining oil distribution rules for multi dimension sandstone oil reservoir after plymer flooding, Petroleum Geology and Oil Field Development in Daqing, 2006, 25(2):45-48.

[3] Wang, X. High-capacity channel features of the high water-cut stage in the Gudong oilfield. West China Petroleum Geosciences, 2006, 2(1):101-103.

[4] Liu, X. Development and Application of the Water Control and Profile Modification Technology in China Oil Fields. SPE 29907, 1995.

[5] Southwell, G.P.; Posey, S.M. Applications and Results of Acrylamide Polymer/Chromium (III) Carboxylate Gels, SPE 27779, 1994.

[6] Lerma, M.K., Giuliani, M.A. Cost-Effective Methods of Profile Control in Mature Water flood Injectors, SPE 27849,1994.

[7] Albonico, P. Studies on Phenol-Formaldehyde Crosslinked Polymer Gels in Bulk and Porous Media, SPE 28983, 1995.

[8] BR, Reddy. Larry, Eoff. A Natural Polymer-Based Crosslinker System for Conformance Gel Systcms, SPE 75163, 2002.

[9] Dai, C.; Feng, H. A selective water-plugging system with heat-resistance gel foam: A case study from the East China Sea Gas Field, Natural Gas Industry, 2015, 35(3): 60-66.
[10] Zhou, H.; Jia, H. Quick measurement of gel strength by breakthrough vacuum value method through viscosity quantification, Oil Drilling and Production Technology, 2015, 37(5): 116-119.
[11] Vasquez J.; Civan F. Laboratory evaluation of high temperature conformance polymer systems, SPE 80904, 2003.
[12] Bai, Y.; Xiong, C. Experimental study on Ethanolamine/Surfactant flooding for enhanced oil recovery, energy & fuels, 2014,28,1829-1837.
[13] Liu, J. Simulation experiments of the formation process of high capacity channel in a sandpack with glass plate, Petroleum Geology and Recovery Efficiency, 2008,15 (5) :95-97.

IOP Conf. Series: Earth and Environmental Science **64** (2017) 012005 doi:10.1088/1755-1315/64/1/012005

Simulation analysis of jiujiang river basin runoff based on swat model

Peipei Tian, Cheng Zhang, Yizhong Chen, Honghai Zhao, Jia Liu, Hongwei Lu*
State Key Laboratory of Alternate Electrical Power System with Renewable Energy Sources, North China Electric Power University, Beijing, 102206, China.

*Corresponding author: luhw@ncepu.edu.cn

Abstract. This paper discussed the feasibility of application the Soil and Water Assessment Tools (SWAT) model in the Yangtze River Basin of Jiujiang. Two statistical indicators were used for model evaluation; i) coefficient of determination (R2) and ii) Nash-Sutcliffe efficiency index (Ens). The software of SWAT-CUP was used to calibrate and validate the model based on the measured data from 2000 to 2015 in JiuJiang hydrological station, which could assist in determining the applicability of the model to the basin. Results indicated R2 being 0.82 and 0.76; Ens being 0.78 and 0.72 in terms of the calibration and verification periods, both of which meeting the requirements, It suggested that the SWAT model could be able to well simulate the Yangtze River Basin of Jiujiang watershed.

1. Introduction

The hydrological model is widely used in hydrological field. Calibration and validation are recognized as vital processes of modeling accomplishments before they are applied into a special case. Previously, several methods have been proposed for calibration and validation during the decision-making process [1, 2].

The main purpose of this study is to calibrate and validate the SWAT hydrological model for Jiujiang river basin, and measure its performance by using two statistical metrics, namely, coefficient of determination (R2) and the Nash-Sutcliffe efficiency index (Ens).

2. Methodology

Distributed hydrological model has a strong physical mechanism, which can effectively use the space data information to simulate the river basin in the complicated hydrological physical process [3]. The SWAT hydrological model is a kind of distributed hydrological model, which is based on ArcGIS. It is has been proved as a reliable model in simulating hydrological characteristics at basins scale and is widely used in Stream flow simulation, simulation of pollutants migration, watershed underlying surface conditions change and climate change response simulation, etc [4, 5].

The establishment and operation the SWAT model in Jiujiang basin were based on the processes including loading of digital elevation model (DEM), sub basin delineation, loading of soil and land use data, meteorological data input, hydrological response unit (HRU) identification, input file generation and editing, the SWAT model running, sensitivity analysis, parameter calibration, as well as model validation. Runoff dataset in JiuJiang hydrological station of 16 years were used for running the SWAT model in this study. Data of the year 2000 were applied for warming up the SWAT model. After it, date from 2001 to 2010 were chosen for modelling calibration and data from 2011 to 2015 for modelling validation.

Moreover, R2 and Ens are selected as statistical indicators to evaluate the applicability of the proposed model. The expressions of R2 and Ens can be presented as follows:

$$R^2 = \{\frac{\sum_{i=1}^{n}(Q_{0i} - Q_{avg})(Q_{pi} - Q_{pavg})}{\sqrt{\sum_{i=1}^{n}(Q_{0i} - Q_{avg})^2} * \sqrt{\sum_{i=1}^{n}(Q_{p} - Q_{pavg})^2}}\} \quad (1)$$

$$Ens = \{1 - \frac{\sum_{i=1}^{n}(Q_{0i} - Q_{pi})^2}{\sum_{i=1}^{n}(Q_{0i} - Q_{avg})^2}\} \quad (2)$$

where O_{0i} is the observed runoff discharge at time i; Q_{avg} is the average observed runoff discharge; Q_{pavg} is the average simulated runoff discharge; Q_{pi} is the simulated runoff discharge at time i and n is the number of registered runoff discharge data. Generally, the value of Ens nearly 1 corresponds to the higher the credibility of the model. The average annual error between the simulated and measured values should be less than 15% of the measured value, which is the monthly average R2> 0.6 and Ens> 0.5 [6].

3. Results and discussion

This study combines the use of automatic rate of the SWAT-CUP software with manual parameter determination to achieve a rapid and efficient rate verification work. Fig.1 illustrates the sensitivity analysis of parameters based on the SWAT-CUP. In this simulation, the most sensitive parameter is V__CH_N2 that is defined as the runoff curve coefficient, and the least sensitive parameter is R__SOL_AWC that represents the soil available water content. Results show that runoff simulation outputs are the first-rate compared with the observed runoff discharge (Figs.2 and 3). In this study, the validation results are acceptable compared to the calibration results. It suggests an outstanding performance when the values of R2 are 0.82 and 0.76; the values of Ens are 0.78 and 0.72 in terms of the calibration and verification periods.

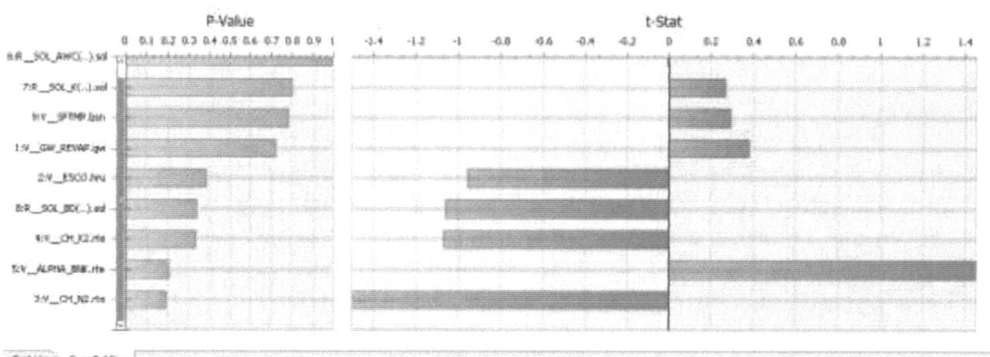

Figure 1. Sensitivity analysis of parameters based on the SWAT-CUP.

During the calibration period (Fig.3), it is obvious that the results of wet seasons are better than those of dry seasons. The major reason arises from the effect of lake in basin. But the simulation results in the validation period are significantly preferred, In the case of the same trend, the main error may be caused by the accuracy of the input data.

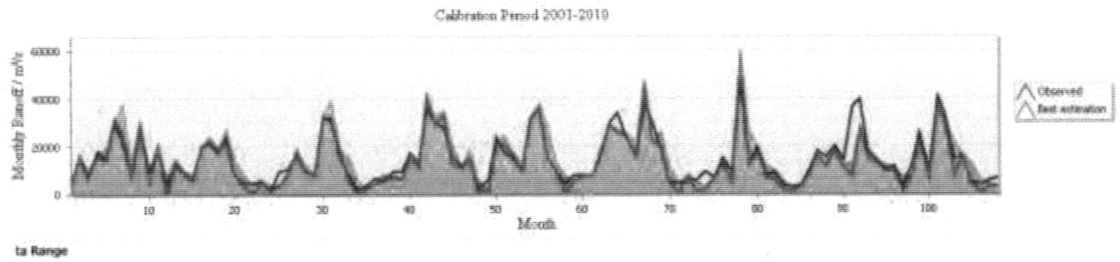

Figure 2. Solutions of Calibration Period from to2001-2010.

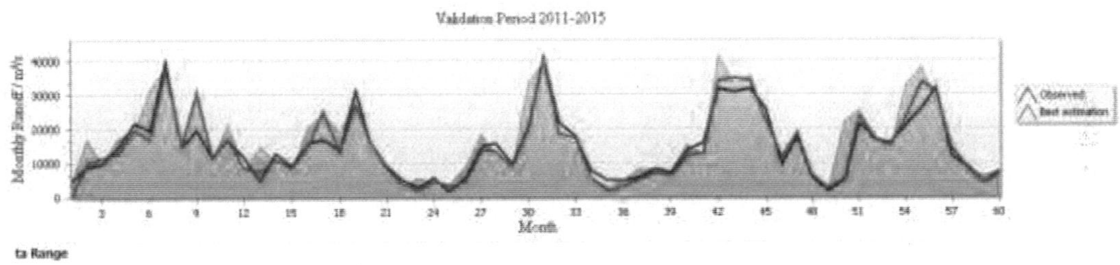

Figure 3. Solutions of Validation Period 2011-2015.

4. Conclusion

In this study, the SWAT model is applied to a sub basin of the Yangtze River Basin, The model was calibrated and validated by using the measured runoff data obtained from the JiuJiang hydrological station from 2000 to 2015. Results indicate that the value of Ens and the value of R2 are higher than the criteria values, which suggests that the SWAT model is applicable to the Jiujiang basin of the Yangtze River Basin.

A further direction for this study is to improve the accuracy of modeling inputs. A series of soil parameters such as soil moisture, soil saturated water conductivity should be achieved through field experiments instead of calculated by running the SPAW software. The improvement of original data will make the results more reliable.

Acknowledgements

This research was supported by the China National Funds for Excellent Young Scientists (51422903), National Natural Science Foundation of China (41271540), and Fundamental Research Funds for the Central Universities.

References

[1] M. W. Van Liew, J. G. Arnold and D. D. Bosch, American Society of Agricultural Engineers 48 (3), 1025–1040 (2005).
[2] E. Rouholahnejad, K. C. Abbaspour, M. Vejdani, R. Srinivasan, R. Schulin and A. Lehmann, Environmental Modelling & Software 31, 28-36, (2012).
[3] Mohd M S F, Juneng L, Tangang F, et al. Simulation of upper Kuantan River basin streamflow using SWAT model. 020021, (2016).
[4] Duan Z, Song X, Liu J. Application of SWAT for sediment yield estimation in a mountainous agricultural basin International Conference on Geoinformatics. IEEE Xplore, 1-5, (2009).

IOP Conf. Series: Earth and Environmental Science **64** (2017) 012005 doi:10.1088/1755-1315/64/1/012005

[5] Nor Faiza Abd Rahman, Mohd Fozi Ali, Khairi Khalid, Junaidah Ariffin, Muhamad Radzali Mispan, J. Appl.Sci. & Agric. 9 (18) 156-162, (2014).
[6] D. N. Moriasi, J. G. Arnold, M. W. Van Liew, R. L. Binger, R. D. Harmel, T. L. Veith. American Society of Agricultural and Biological Engineers. 50 (3), 885-900. (2007).

IOP Conf. Series: Earth and Environmental Science **64** (2017) 012006 doi:10.1088/1755-1315/64/1/012006

Research on Coupling Method of Watershed Initial Water Rights Allocation in Daling River

Jinhua Liu[1, 2,*] and Fengping Wu[1, 2]

[1]Business School, Hohai University, Nanjing, China
[2]Planning and Decision Institute, Hohai University, Nanjing, China

*Corresponding author e-mail: liujinhua89@hotmail.com

Abstract. As a typical abnormal, nonlinear and multidimensional system decision-making problem, watershed initial water rights allocation involves various resource distribution, economic, social and environment objectives. Because of the large subjectivity of weight determination, different from the traditional methods, we adopt a new coupling method which is a dimension reduction method in this study to solve the watershed initial water rights allocation problem. The data allocated in Daling watershed can be calculated in optimum projection direction to gain the watershed initial water rights allocation scheme in low-dimensional space.

1. Introduction

Water is one of the basic resources for agricultural development and plays a strategic position in the agricultural production. As one of the increasingly scarce strategic resources, water resources have a global and long-term impact on food security and agricultural economic development. Nowadays, water resources scarcity is becoming much severe due to the increasing population, climate change and water pollution problems. Groundwater resources in many regions have been exploited to irrigate crops in dried years which result the drop of the water tables. Many issues lead to the severe damage to the environmental and social efficiency decreased. The contradiction between water supply and demand has become increasingly prominent, especially the excessive consumption and indiscriminate abuse of the severe problem of water supply and demand. The situation of the water shortage and water pollution is very grim for the sustainable use of water resources and water environment. How to achieve the rational use of water resources, improve the efficiency and equity of water resources utilization has become the focus of economic theorists and policy makers concerned.

Although the excessive use and waste of water resources are the main reasons leading the water shortage, structural defects of the existing system of water rights can be called the deep-seated reason of current water shortage. Due to the unclear water property, economic agents regard the water resources as the free "public resources" which lead to excess use and waste of water. If the current water rights system in China continues without fundamental change, only by increasing people's awareness of water conservation and expansion of the limited water supply is impossible to solve the current low water use efficiency fundamentally.

Some researchers have studied the hydrology, ecology and other subject models. Most of them focused on one factor but ignored the multi-factor analysis, it is still a challenge to directly solve the water rights allocation which is a multi-factor cluster problem. How to get a single-factor question

IOP Conf. Series: Earth and Environmental Science **64** (2017) 012006 doi:10.1088/1755-1315/64/1/012006

from the multi-factor one is the key issue. As a consequence, we adopt Projection Pursuit (PP) introduced by Friedman and Tukey (1974) to solve the above problem. Projection pursuit (PP) is an effective method based on different functions of projection indexes has been presented in this paper when standard deviation and partial density are used to structure the function of projection indexes, it deduces and acquires the empirical formula of calculation of exclusive parameter the density window breadth of projection pursuit cluster model. Fu (2003) also studied this model to solve the nonlinear questions. By transferring high-dimensional data into a low dimensional space (e.g., three-dimension, two-dimension or one-dimension), PP model can be directly driven by the low-dimensional data. Except the PP model, some researchers studied some derived models to analyze the high-dimensional data. Such as projection pursuit regression (PPR), projection pursuit density estimation (PPDE), projection pursuit cluster (PPC), projection pursuit learning network (PPLN), projection pursuit wavelet learning network (PPWLN), etc. (Friedman and Stuetzle. 1981, Friedman et al. 1984, Hall 1989, Hwang et al. 1994, Lin et al. 2003). The method of PP model, possessing advantages of good stability, strong anti-interference and high accuracy, has been widely used in many areas.

Coupling method has been used successfully in different areas to evaluate the multi-factor problems. But this method just provides the projected characteristic value remaining the major characteristics of data according to the projection index (Wang et al. 2002). In this paper, we propose a mathematical model by projection pursuit technique to solve the above mentioned problems, and find the optimal projection direction according to the Coupling optimization algorithm. Set up a new projection index to overcome the difficulties of weights determination. The method owns the advantages of complying with watershed initial water rights allocation mechanism and meeting the control requirements of water quantity, water quality and water utilization efficiency, which help to achieve effective allocation of water resource.

2. Study Area

As the largest single flow in the western of Liaoning province in China, Daling watershed stretches across the provinces of Liaoning, Inner Mongolia and Hebei. This region belongs to the typical continental monsoon climate characterized by hot and rainy summers, cold and dry winters, which results in uneven amount of precipitation of year and rainfall is concentrated in July and August. Meanwhile the average rainfall of years increases by degrees from north to south. The annual mean precipitation of this watershed is between 400mm and 600mm. The per capita possession of water resources of this watershed is merely 392m^3, which accounts for 18% of the national level. In spite of the shortage and conflict of water resources, the frame work, such as the comprehensive scheme of social and economic development of this watershed, and the comprehensive scheme of water resources, are relatively complete. Hence, this article selects the initial water rights allocation scheme of Daling watershed as a case to study. Choose the fairness, efficiency and sustainability as three basic principles, show as table 1.

IOP Conf. Series: Earth and Environmental Science **64** (2017) 012006 doi:10.1088/1755-1315/64/1/012006

Table 1. The Index System of Water Rights Allocation

Principle layer	Index layer
Fairness(B_1)	Rate of water shortage(C_1)
	Per capita water resources(C_2)
	Per capita water consumption(C_3)
	Per capita cultivated land area(C_4)
	Per capita effective irrigation area(C_5)
	Population per unit area(C_6)
	Livestock numbers per unit area(C_7)
	Urbanization rate(C_8)
Efficiency(B_2)	Per Capital GDP(C_9)
	Ratio of GDP in the whole watershed(C_{10})
	Ratio of industrial production in the whole watershed(C_{11})
	Ratio of agriculture production in the whole watershed(C_{12})
	Economic growth rate(C_{13})
	GDP per cubic water(C_{14})
	Agriculture production per cubic water(C_{15})
	Industrial production per cubic water(C_{16})
	Utilization rate of industrial water duplication(C_{17})
	Utilization rate of Irrigation water(C_{18})
	Rate of water resources development(C_{19})
Sustainability(B_3)	Standard rate of water quality(C_{20})
	Per capita emissions of pollutants(C_{21})
	Unit GDP sewage discharges(C_{22})
	Rate of vegetation coverage(C_{23})
	Rate of population growth(C_{24})
	Per capita net income(C_{25})

3. Research methods

3.1. Coupling optimization algorithm

Coupling is a random phenomenon with seemingly rule in deterministic system, because of its randomness and regularity properties, chaotic motion could go through on its own rule within a certain range. Thus, by using chaotic variables, we can obtain globally optimal solution with high efficiency. The basic idea of coupling optimization algorithm is to linearly map chaotic variables to the value interval of optimized variables, and then search iteratively to get the optimal solution. Generally, the mathematical model of nonlinear programming can be expressed as:

$$\begin{cases} \min f(X) \\ g_i(X) \geq 0 \\ h_j(X) = 0 \end{cases} \tag{1}$$

Where $X \in E^n$, $(i = 1, 2\cdots, m)$ and $(j = 1, 2\cdots, n)$, $f(X)$ is the objective function. $g_i(X)$ And $h_j(X)$ are constraint functions of inequality and equality respectively, m and n are the number of inequality and equality constraint functions. The constraint condition can be expressed as:

$$S=\{X\geq|g_i(X)\geq 0, i=1,2\cdots,m; h_j(X)=0,1,2,\cdots,n\} \quad (2)$$

S is called feasible set and the elements included are feasible points.

Logistic model is one of the most classical model of fuzzy research, chaotic variables generated by the model can be used to search the optimal solution, and the equation is:

$$X_{k+1,j}=\lambda\cdot X_{k,j}\cdot(1-X_{k,j}) \quad (3)$$

Where λ is the control parameter, and it is selected in the range of 0 and 4, when λ =4, the system is in the fuzzy state.

3.2. Coupling Model

Coupling model sets high dimensional data into a low dimensional data space, finds out structure or characteristics projection can reflect the original high dimensional data and studies the purpose of high dimensional data. The problem of structuring projection index function and its optimization is the key of applying method of projection pursuit successfully. This problem is very complex, traditional projection pursuit has considerable amount of calculation, it limits the method of thorough research and widely use to some extent. The concrete steps of the model are as follows:

Step 1: Structuring coupling model index function

In order to eliminate the effect of different ranges of values of cluster factors, the initial data are standardized before it is used in the PPDC model. Samples experience indexes set as $\{x^*(i,j)|i=1,2,\cdots n; j=1,2,\cdots,p\}$, where $x^*(i,j)$ is the j th value of the i th sample, n , p are sample number (sample size) and index number respectively. To eliminate the dimensions of index value and range of unity each index, following formula can be adopted for extreme normalized.

In case the index the bigger the better:

$$x(i,j)=\frac{x^*(i,j)-x_{\min}(j)}{x_{\max}(j)-x_{\min}(j)} \quad (4)$$

In case the index the smaller the better:

$$x(i,j)=\frac{x_{\max}(j)-x^*(i,j)}{x_{\max}(j)-x_{\min}(j)} \quad (5)$$

Where, $x_{\max}(j)$ and $x_{\min}(j)$ is the max and min value of the j th indicator respectively.

Step 2: Construct coupling model index function $Q(a)$.

Essentially, projection is used to observe data characteristic from all angles. The main purpose is to find the hidden structure from high-dimensional data sets by searching through all their low-dimensional projections (Cui 1997). Using projection pursuit technique, $\{x^*(i,j)|j=1,2,\cdots,p\}$ can be projected into the direction $a=\{a(1),a(2),a(3)\cdots a(p)\}$, that is

$$Z(i)=\sum_{j=1}^{p}a(j)x(i,j) \qquad (i=1,2,\cdots,n) \quad (6)$$

Projection index function is:

$$Q(a) = S_Z D_Z \tag{7}$$

$$S_Z = \sqrt{\frac{\sum_{i=1}^{n}[Z(i) - E(Z)]^2}{n-1}} \tag{8}$$

$$D_Z = \sum_{i=1}^{n}\sum_{j=1}^{n}(R - r(i,j)) \cdot u(R - r(i,j)) \tag{9}$$

Where S_Z is the standard deviation of projection value $Z(i)$; D_Z is the local density of $Z(i)$; $E(Z)$ is the mean of $\{Z(i)|i = 1,2,\cdots,n\}$; R is the window radius of local density, normally is set to $0.1\,S_Z$; $r(i,j)$ is the distance between samples, $r(i,j) = |Z(i) - Z(j)|$; $u(t)$ is the unit step function, it equals to 1 when $t \geq 0$, otherwise 0.

Step 3: Optimize the coupling index function.

When the sample set of indicators is given, $Q(a)$ only depends on the projection direction a, the best projection direction is the one that exposures the characteristics of high-dimensional data structure of maximum possible, which can be obtained by solving the max value of projection index function:

$$\max Q(a) = S_Z D_Z \tag{10}$$

Constraint is:

$$\sum_{j=1}^{p} a^2(j) = 1 \tag{11}$$

The model is a complex nonlinear problem with variables $\{a(j)|j = 1,2,\cdots,p\}$, and can be solved by Coupling optimization algorithm.

Step 4: Initialize parameters.

Assume X in (1) has p dimensions, $X = (x_1, x_2, \cdots, x_p)$, $x_i = [a_i, b_i]$. Set control variable of iterations $k = 0$, generate p initial values $x_{i,k}$, $i = 1,2,\cdots,p$, $X_k = (x_{1,k}, x_{2,k}, \cdots, x_{p,k})$, let

$$x'_{i,k} = a_i + (b_i - a_i)x_{i,k} \tag{12}$$

Where $X'_k = (x'_{1,k}, x'_{2,k}, \cdots x'_{p,k})$, when $X'_k \in S$, let $X^* = X'_k$ and $f^* = f(X^*)$. Otherwise, go to step (4) until X_k is found, which has the according X'_k that satisfies constraints of (1).

Step 5: Coarse search.

Set the max iteration step N, generate fuzzy sequence X_k, $k = 1,2,\cdots,N$, and then test its feasibility, if successful and $f(X'_k) < f(X^*)$, let $X^* = X'_k$, $f(X^*) = f(X'_k)$; otherwise test X'_{k+1}, repeat step (5),until $f(X'_k)$ satisfies N or precision requirement.

Step 6: Fine search.

IOP Conf. Series: Earth and Environmental Science **64** (2017) 012006 doi:10.1088/1755-1315/64/1/012006

Set the max iteration steps N', let $Z_0 = X^*$, and $Z_k = Z_{k-1} + \alpha Z'_{k-1}$, $k = 1, 2, \cdots, N'$. Where Z'_{k-1} is the chaotic sequence generated by Logistic mapping, α is the given small real number. Test the feasibility of Z_k, $k = 1, 2, \cdots, N'$, if successful and $f(Z_k) < f(X^*)$, let $X^* = Z_k$, $f(X^*) = f(Z_k)$, otherwise test Z_{k+1}, repeat step (6) until $f(X^*)$ satisfies precision requirement. Finally, X^* is the optimal projection direction a^* of (10).

Step 7: Determine the proportion of allocated water.

Substitute the optimal projection direction a^* in step (6) into (6), and projection value $Z^*(i)$ of every sample can be obtained. Normalizing $Z^*(i)$ can get the proportion of allocated water, multiplied by the total amount of water is the final allocated water.

4. Results and analysis

In addition, the result obtained through coupling optimization algorithm in this study is an optimal solution. This is because coupling optimization algorithm due to its ergodicity characteristic as well as enlarging the fuzzy sequences to the value interval of optimized variables to carry on iterative optimization. As long as the fuzzy sequences are long enough, we can obtain the globally optimal solution through the coarse search and the fine search. However, the solution of genetic algorithm is uncertain because many random numbers are used during calculation process, and it is easy to be trapped in the local optimal solution even divergent. Table 2 and table 3 are the curves between optimization iterations and projection index values of Coupling optimization algorithm and genetic algorithm.

As shown in the following tables, the optimization iterations of Coupling optimization algorithm and genetic algorithm are given. With this comparison, coupling optimization algorithm can obtain the optimal projection value and calculation speed is faster and the result is more stable, which proves the effectiveness of the model in this study fully.

Table 2. Normalized Index Values of Water Rights Allocation

Index	Region			Index	Region		
	Inner Mongolia	Liaoning	Hebei		Inner Mongolia	Liaoning	Hebei
C1	0.289	0.390	0.321	C14	0.185	0.192	0.623
C2	0.448	0.415	0.137	C15	0.248	0.224	0.528
C3	0.359	0.391	0.250	C16	0.191	0.198	0.611
C4	0.464	0.298	0.238	C17	0.311	0.315	0.374
C5	0.463	0.393	0.144	C18	0.325	0.307	0.362
C6	0.365	0.416	0.219	C19	0.310	0.362	0.328
C7	0.394	0.504	0.102	C20	0.410	0.312	0.278
C8	0.231	0.215	0.554	C21	0.356	0.365	0.279
C9	0.198	0.165	0.637	C22	0.405	0.392	0.203
C10	0.188	0.190	0.622	C23	0.385	0.330	0.285
C11	0.221	0.201	0.578	C24	0.198	0.195	0.607
C12	0.357	0.382	0.261	C25	0.199	0.143	0.658
C13	0.305	0.312	0.383				

Table 3. Water Rights Allocation Scheme of Daling Watershed

Region	Projection Value	Proportion of water allocation
Inner Mongolia	1.6356	35.2%
Liaoning	1.8965	39.3%
Hebei	1.3575	25.5%

Table 4. Comparative Analysis of the Results of Various Models

Region	Proportion of water use in history	Results of PP	Results of AHP
Inner Mongolia	34.9%	35.2%	34.1%
Liaoning	41.8%	39.3%	38.9%
Hebei	23.3%	25.5%	27.0%

5. Conclusion

With the rapid development of the economy of our country, the demand for water resources is increasing every day. The shortage of water resources has become an important restrictive factor for the sustainable development of the economy of our country, which causes water use conflicts between the upstream and downstream in the watershed or between different areas, so it's really important to allocate water rights fairly. This study proposed a new coupled fuzzy optimization projection pursuit model, taking advantage of projection pursuit to map high-dimensional data to one-dimensional space, using coupling optimization algorithm to find the optimal projection direction. This model is objective and fair in weights determination, which overcomes the difficulties of traditional methods that the objective functions and constraints must be continuous and differential and avoids being trapped in the local optimal solution during the process of optimization. Through an example analysis, the result of initial water rights allocation obtained through the model is reasonable and effective, simultaneously enriching and developing initial water rights allocation theory and method. However, the window radius of local density of projection pursuit in the model is determined by empirical formula, lack of theoretical basis and the optimal solution obtained through coupling optimization algorithm has great dependence on the length of chaotic sequences and the number of iterations, which need to be improved in the future.

Acknowledgments

We would like to thank the reviewers for their valuable comments and suggestions. The first author acknowledges the support provided by China Scholarship Council. This study was supported by the National Natural Science Foundation (Grant No. 41271537) and Research Innovation Program for College Graduates of Jiangsu Province [Grant No. KYZZ15_0159], P.R. China.

References

[1] B. Srdjevic, Linking analytic hierarchy process and social choice methods to support group decision-making in water management. Decision Support Systems, 42(2007) 2263-2272.
[2] Friedman JH, Stuetzle W, Projection pursuit regression. J Am Stat Assoc 76(1981) 817–823.
[3] Friedman JH, Stuetzle W, Schroeder A, Projection pursuit density estimation. J Am Stat Assoc 79(1984) 599–608.
[4] F. Pan, C. Liang, Z. L. Wang, Q. Fu, Fuzzy matter-element model for evaluating sustainable utilization of regional water resources. Advances in Water Science, 14(2003) 271-275.
[5] H. Friedman, J. W. Turkey. A projection pursuit algorithm for exploratory data analysis. IEEE Trans On Computer, 23(1974) 881-890.
[6] Hwang JN, Lay SR, Maechler M, Martin RD, Schimert J, Regression modeling in back-propagation and projection pursuit learning. IEEE Trans Neural Netw 5(1994) 342–353.
[7] Lin W, Tian Z, He F, Study On improving unsupervised restoration of image with PPWLN.

Journal of Northwestern Poly technical University, 21(2003) 344–347.
[8] Hall P. On polynomial-based projection indices for exploratory projection pursuit. Ann Stat, 17(1989) 589–605.
[9] Q. Fu, H. Fu, Applying PPE model based on RAGA in the investment decision-making of water saving irrigation project. Nature and Science, 11(2003) 72-77.
[10] Wang SJ, Zhang XL, Ding J, Projection pursuit cluster model and its application. Journal of Yangtze River Scientific Research Institute, 19(2002) 53–55.
[11] Zhang XL, Ding J, Li ZY, Jin JL, Application of new projection pursuit algorithm in assessing water quality. China Environ Science, 20(2000) 187–189.

IOP Conf. Series: Earth and Environmental Science **64** (2017) 012007 doi:10.1088/1755-1315/64/1/012007

Optical fiber pressure sensor based on F-P cavity in the oil and gas well

Qingchao Zhao [1], Xiaohui Liu[1], Long Ma[1], Wenan Zhao[1] and Hongzhong Wang[2]

[1]Laser Institute, Shandong Academy of Sciences, Jinan, China
[2]Xinsheng petroleum exploration technical service co., Ltd, Shengli oilfield, Dongying, China

*Corresponding author e-mail: zhaoqc1988@163.com

Abstract. To meet the need of the measurement in high temperature and high pressure in oil and gas well, an optical fiber pressure sensor based on extrinsic Fabry Perot (F-P) cavity is developed. The sensor is fabricated by hydrogen and oxygen flame thermal bonding in high temperature technique, and the change of the senor's cavity length is compensated in temperature changing with cascading fiber bragg grating, and the sensor has characteristics of large dynamic range, high resolution, high repeatability, long term operation stability and high temperature resistance. The pressure measuring range is 0~69MPa, the repeatability is 0.01% F·S,the hysteresis is less than 0.01% F·S, the long term stability is less than 0.02MPa/y, and the sensor can satisfy the requirement of the measurement in the oil and gas well. Practical application proved that the relative deviation of the pressure measurement value is less than 0.01% compared with the traditional electronic pressure sensor.

1. Introduction

In the process of oil and gas exploitation, it is one of the important measures to understand the physical condition of oil and gas wells, optimize the oil recovery technology scheme and improve the oil and gas recovery and production [1, 2]. But the oil and gas environment is extremely poor underground, the most is high temperature, high pressure and strong corrosion, etc. The traditional electronic sensors cannot run in this harsh environment stably in long-term. Fiber optic sensor with its small size, high precision, high temperature resistance, corrosion resistance and ther advantages [3], has a wide range of applications in the mining process of the oil and gas field.

In recent years, the use of optical fiber sensors to monitor the pressure parameters in the oil and gas well are mainly based on fiber bragg gratings and optical extrinsic Fabry Perot (F-P) cavity [4-6], because the temperature and pressure cross-sensitivity of fiber bragg grating is very large, it cannot operate reliably in oil and gas well; And for the fiber F-P cavity, in current the fiber and quartz capillary is adhesived by epoxy glue, because the expansion coefficient of different material in the bonding point is different, which can lead to thermal mismatch phenomenon, resulting in cavity length the occurrence of temperature crosstalk of the cavity length, and because the epoxy glue can creep at high temperatures[7], the accuracy, repeatability and long-term stability of the sensors are affected, and the sensors cannot operate reliablly in the oil and gas wells.

This paper has developed a kind of fiber-optic F-P cavity pressure sensor which can be applied to the harsh environment in the oil and gas wells. The sensor has characteristics of large dynamic range, high resolution, high repeatability, long term operation stability and high temperature resistance, and it has been successfully applied and achieved good results.

2. Design of fiber optic F-P cavity pressure sensor

2.1 Working principle of the sensor

Fiber-optic F-P cavity pressure sensor is based on Fabry-Perot (F-P) interferometer, and its structural diagram is shown in Figure 1, mainly consist of the incident fiber, the reflective fiber and the quartz capillary in total of three parts, two parallel end faces of the incident fiber and the reflective fiber form an optical fiber FP interferometer (also referred to as a F-P cavity) inside the capillary. When a bunch of monochromatic light (wavelength λ, intensity I_0) transmit the incident fiber to the fiber F-P cavity, the incident beam will be reflected several times in the two parallel ends of the fiber, and it will occur interference effect when the beam was reflected several times[8], the spectral information was shown in Figure 2.

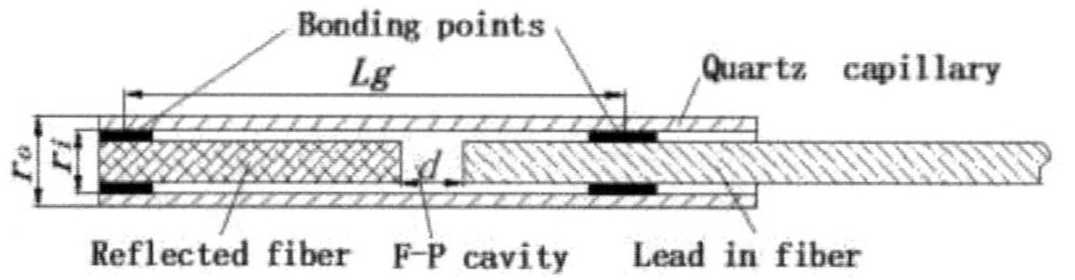

Figure 1. The structure diagram of optical fiber F-P cavity.

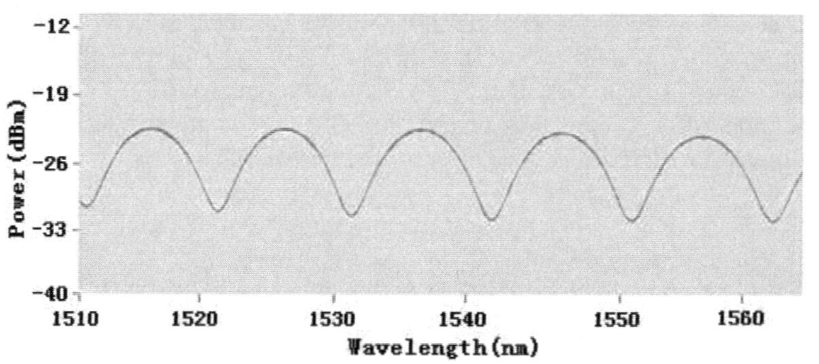

Figure 2. The light spectrum of the optical fiber F-P cavity.

When the static pressure is applied to the F-P cavity structure, the quartz capillary is crushed to produce deformation, which causes the change of the length of the F-P cavity. The change of the d will cause the reflection spectrum to change, and the spectral signal is measured by the signal demodulation system, which calculated the length *d* changes of F-P cavity, and we can get the pressure change information.

2.2 The design of fibre optic F-P cavity pressure sensor

In this paper, the sensor is fabricated by hydrogen and oxygen flame thermal bonding in high temperature, the thermal expansion coefficient of each material is basically the same at the connection point, which avoids the occurrence of thermal mismatch at the connection point, completely solves the problem of temperature crosstalk at the connection point. At the same time, the connection point material is quartz, which solves the high temperature creep problem which exists in the gelatinized package, and greatly improves the precision, repeatability and long term stability of the sensor.

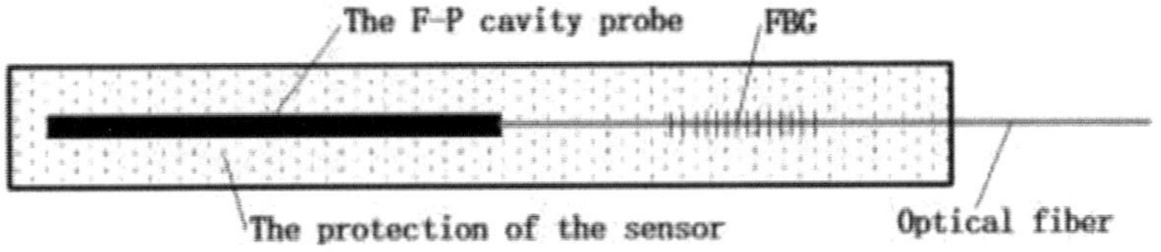

Figure 3. The structure diagram of temperature compensation

In this paper, a fiber grating is connected with fiber F-P cavity in series, which is shown in Figure 3, is used to monitor the temperature and compensate the influence of temperature change on the cavity length of the fiber optic F-P cavity, and greatly reduces the temperature and pressure cross sensitivity of the fiber optic F-P cavity pressure sensor.

As shown in Figure 1, the inner diameter of the quartz capillary which constitute the F-P cavity is r_i, the outer diameter is r_o, and the sensor cavity length change Δd caused by the external application of the pressure P is [9]:

$$\Delta d = \left[\frac{L_g r_0^2}{E\left(r_0^2 - r_i^2\right)}(1 - 2\mu)\right] \cdot P \qquad (1)$$

In the formula, μ is the Poisson's ratio of quartz glass, E is the Young's modulus, and the change of the external pressure can be measured by measuring the cavity length. In addition, it can be seen from the above formula we can adjust the pressure sensor measurement sensitivity and dynamic range within a certain range by adjusting the capillary diameter r_i, outer diameter r_o and gauge L_g of the three geometric parameters.

3. The performance test of the fibre optic F-P cavity pressure sensor

The performance test system of fiber optic F-P cavity pressure sensor which is designed in this paper is shown in Figure 4, mainly including pressure gauges, thermostats, pressure pipes, pressure devices, demodulators, and computers and so on. We use the GE 3100 piston pressure gauge of United States which can provide high precision and stable pressure, whose range is 0 ~ 110MPa and the accuracy is one ten thousandth. We use the FM841-3 oven of Wu jiang fei ma company whose temperature range is room temperature ~ 250 ℃ and the accuracy is ± 2 ℃. We use the high-pressure pipe of Shanghai Dayton Electrical and Mechanical Technology Co., Ltd. The pressure device is designed by ourselves whose maximum pressure is 150MPa and which is use to place the sensors.

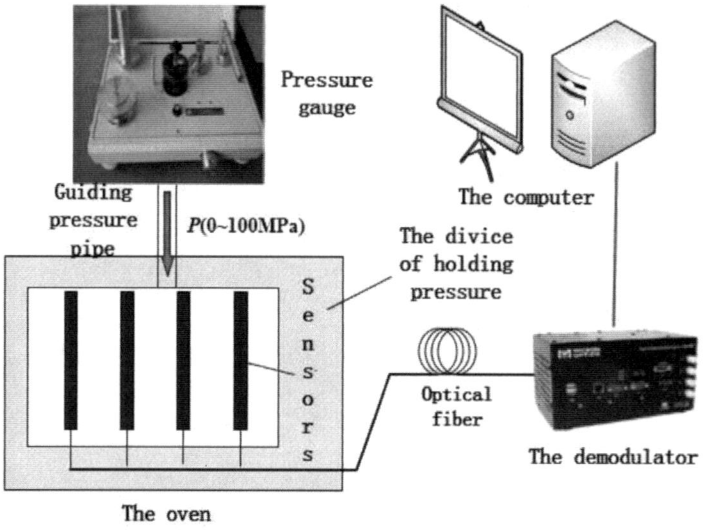

Figure 4. The Schematic diagram of the sensor testing system

3.1 Test of the linearity

For the quartz material: E = 74GPa, μ = 0.17, and L_g, r_i *and* r_o whose change is very small in the monitoring process which can be considered constant, so from the formula (1) we can see that the change of the cavity length Δd is proportional to the pressure P applied, and we can obtain the relationship between the cavity length d and the pressure P by calibrating the cavity length of the sensor.

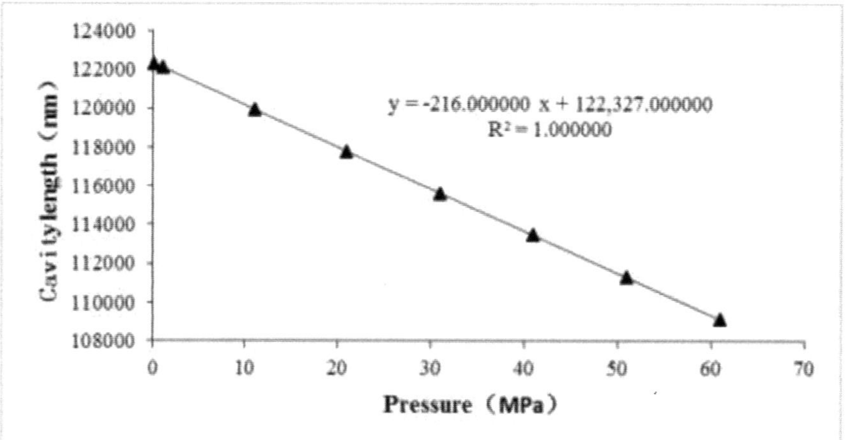

Figure 5. The linearity of the optical fiber sensor based on F-P cavity

During the test, the sensor was pressurized by positive stroke (from 0 MPa to 62 MPa) or reverse stroke (from 62 MPa to 0 MPa), and the F-P cavity length was recorded and then was linear fitted. The figure show that the linearity of the test of the fiber optic F-P cavity pressure sensor, for linear fit degree R^2> 0.999999, the relationship between cavity length and pressure is y = -216.000000x +122327.000000, the sensitivity is 216nm/MPa, and the resolution is greater than 0.005MPa.

3.2 Test of the repeatability

Repeatability is defined as the inconsistency (volatility) of the resulting characteristic curve which is got by the continuously repeated measurements under the same operating conditions within the full range of input within the forward or reverse stroke.

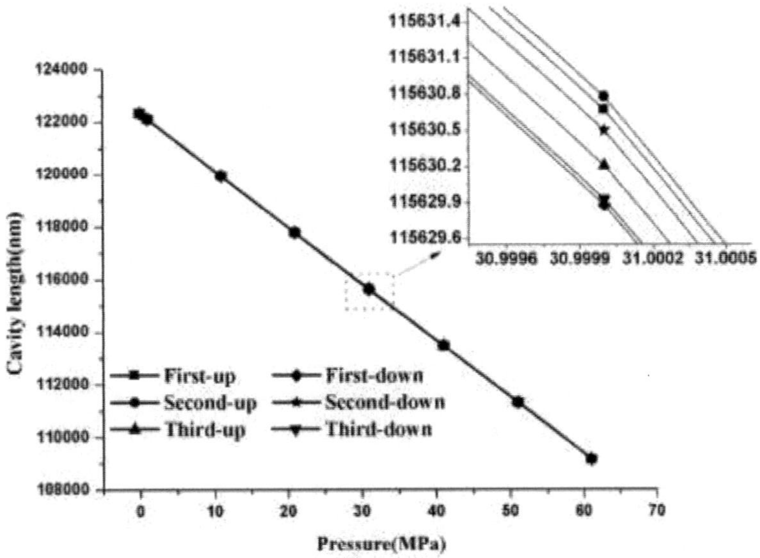

Figure 6. The repeatability of the optical fiber sensor based on F-P cavity

Figure 6 shows the repeatability test results for the fiber optic F-P cavity sensor. Keep the temperature constant in the process of testing and calibrate it three times from 0 MPa to 62 MPa and in reverse stroke (from 62 MPa to 0 MPa) to measure its repeatability. The results show that the maximum deviation of F-P cavity length is 1nm and the repeatability is less than 0.01% F.S (Full scale) in three measurements, and the repeatability is very good.

3.3 Test of the hysteresis

The hysteresis is defined as the degree of inconsistency between the forward and reverse travel outputs when the sensor is tested over the full range under the same measurement conditions.

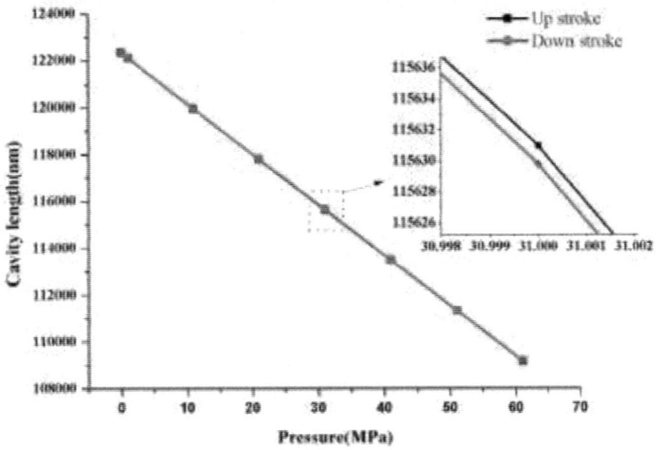

Figure 7. The hysteresis of the optical fiber sensor based on F-P cavity

Figure 7 shows the hysteresis test results for the fiber optic F-P cavity sensor. Keep the temperature constant of 100℃ in the process of testing. Do the step-up and step-down pressure test whose forward stroke is from 0MPa to 62MPa and reverse stroke is from 62MPa to 0MPa. The results show that the hysteresis is very small and the maximum deviation of F-P cavity length is 1.2nm and the hysteresis is less than 0.01% F.S (Full scale).

3.4 Test of the long-term stability

The long-term stability is defined as the ability of the sensor to maintain its performance parameters over a long period of time. In this paper, the length of the fiber optic F-P cavity is 115um ~ 125um. Place the fiber optic F-P cavity in the test system, keep the pressure constant of 21MPa and the temperature constant of 100 ℃, and the record the length of the fiber F-P cavity which change with time, and then observe the long-term stability of the sensor, and the test results is shown in Figure 8.

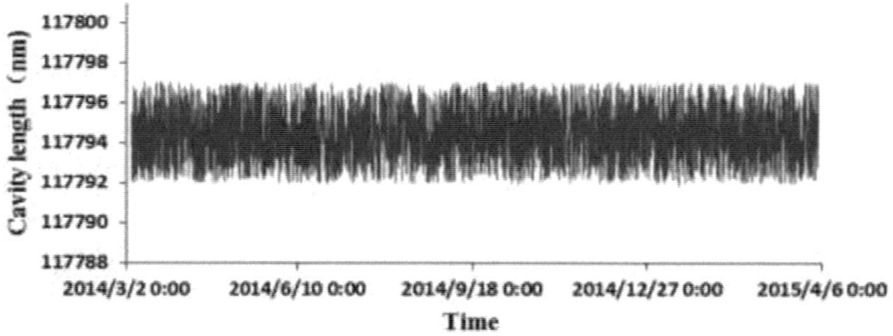

Figure 8. The long term stability of the optical fiber sensor based on F-P cavity

Figure 8 show that the change of the cavity length of the fiber optic F-P cavity pressure sensor in this paper is only 4nm and about 0.02MPa in 12 months, and the long-term stability is very good under high temperature and high pressure.

4. Application test

In order to test the performance of fiber-optic F-P cavity pressure sensor under the condition of actual field conditions, a test experiment was carried out in a test well of CNOOC Shanghai Branch, and the monitoring date was compared with the synchronous data obtained with the more mature electronic pressure sensor. The depth of the test well is 2000 meters and the bottomhole temperature is about 75 ° C. The fiber F-P chamber pressure sensor and the electronic pressure sensor are simultaneously placed into the bottom of the well and the well is continuously monitored for about 42 days.

Figure 9. Optical fiber pressure sensor based on F-P cavity prepared to installed in the well

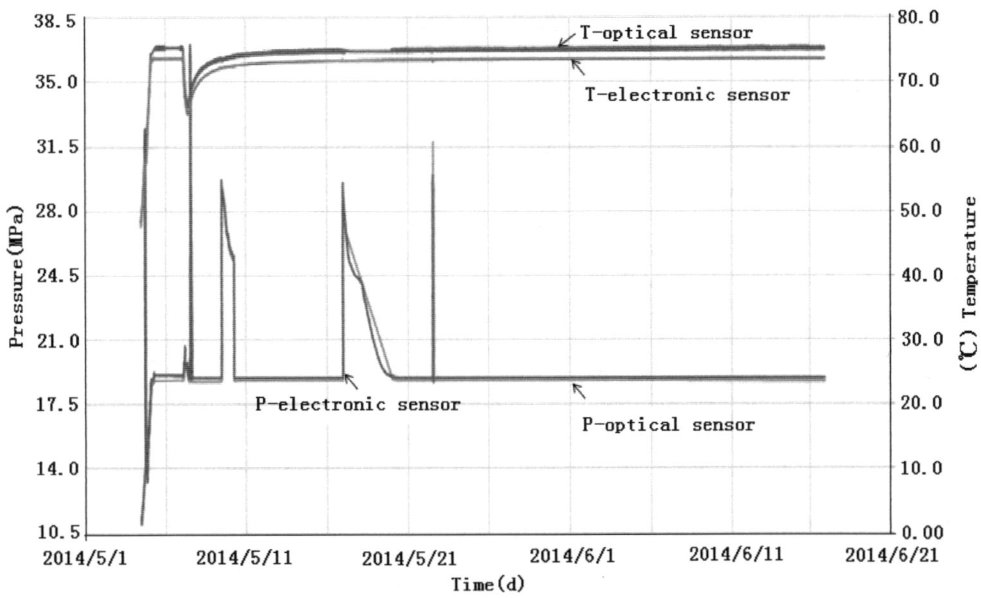

Figure 10. Test results of the optical fiber pressure sensor and electronic pressure sensor

Figure 10 shows the temperature and pressure monitoring results of the electronic pressure sensor and the fiber pressure sensor on the bottom of the well. The results show that the bottomhole pressure value monitored by the fiber F-P cavity pressure sensor is 18.45MPa during the 42-day monitoring process, the pressure value monitored by electronic pressure sensor is 18.67MPa, the deviation is about 0.01%, the temperature deviation does not exceed 1.0 ℃, the measurement accuracy is high; and the wellhead operation causes the pressure to change three times, which can be measured by the fiber optic F-P cavity pressure sensor.

5. Conclusion

A kind of pressure sensor based on fiber FP cavity was developed for the harsh environment of high temperature, high pressure and corrosive of oil and gas wells. The sensor is fabricated by hydrogen and oxygen flame thermal bonding in high temperature technique, and the change of the senor's cavity length is compensated in temperature changing with cascading fiber bragg grating. The designed sensor's pressure range is 0 ~ 69MPa, resolution is greater than 0.005MPa, linear fit is greater than 0.999999, repeatability is 0.01% FS, hysteresis is less than 0.01% FS, the fluctuation of the cavity length is within 4nm (pressure is less than 0.02MPa); and it showed a good stability in 100 ℃ and 21MPa environment. In the field application, it can accurately measure the pressure in the 2000m deep well. Compared with the mature electronic pressure sensor, the monitoring pressure deviation is about 0.01% and the temperature deviation is less than 1℃.

Acknowledgments

This work was financially supported by the National Natural Science Foundation of China (Grant NO. 61307101).

References

[1] Normann R,Weiss J.Krumhansl development of fibers optic cables for permanent geothermal wellbore deployment[A].Proceedings,Twenty-Sixth Workshop on Geothermal Reservoir Engineering Stanford University[C].2001,29-31.
[2] FU Jian wei,Xiao Li zhi,ZHANG Yuan zhong. Progress of permanent fiber optic sensor

applications to oil and gas well[J].Progress in Geophysics,200,19(3):515-523.

[3] GRATTANKTV, MEGGITT B T.Optical fiber sensor technology; fundamentals[M]. Boston:Kluwer Academic, 2000.

[4] YU Da kuan, QIAO Xue guang, JIA Zhen an,et al. Experiment study of novel fiber bragg grating pressure sensors[J]. Optoelectrionics Laser,2006,17(5):513-516.

[5] LI li tong, Zhang Dong sheng, Wen Xiao yan an.et al. Micro fiber F-P sensor used in the measurement of thermal expansion coefficient[J]. Optoelectrionics Laser,2014,17(5):513-516.

[6] YU Qing xu,WANG Xiao na,SONG Shi de et al. Fiber optic pressure sensor system based on extrinsic fabry perot interferometer for high temperature oil well measurement[J] Optoelectrionics Laser,2014,25(11):2131-2135.

[7] WANG Xiao na. Study on the optical fiber EFPI sensor system and its applications in gas and oil well[D].Da Lian: Dalian University of Technology, 2007,78-83.

[8] ZHAO Qing chao,GUO Shi sheng, LI Shun shui et al. Optical fiber sensor testing the temperature and pressure under the oil well[J].ShanDong Science,2014,27(4):57-61.

[9] WANG YY, WANG C, LIUXH, et al, Research on optical fiber grating sensor based on fabry-perot cavity applied in oilfield[J].Applied Mechanics and Materials, 2013,336/338:153-157 .

The secondary explosion phenomenon of gasoline-air mixture in a confined tunnel

Peili Zhang[1,*], Jianxiang Li[2] Yanbo Guo[2] and Yang Du[1]

[1]Department of Petroleum Supply Engineering, Logistical Engineering University, Chongqing, China

[2]Unit 72617 of Shenyang joint logistics support center, Wuhan logistics support base, Shandong, China

*Corresponding author e-mail: zpl612323@163.com

Abstract. In this article, experimental evidences of the special phenomenon of the secondary explosion of gasoline-air mixture were presented in the first place, and then, the causes for the secondary explosion were discussed. At last, effects of concentration on the secondary explosion were studied in detail through multiple experiments carried out in a cylinder tunnel with a solid heating device. Based on the experimental results analysis, the critical concentration for the secondary explosion of the gasoline-air mixture in the confined tunnel were discussed. Results indicated that main causes for the secondary explosion could be summarized as 4 points. Whether the secondary explosion occurred or not was determined by critical gasoline vapour and oxygen concentration even if the temperature was maintained in a reasonable scope. When the concentration of the gasoline vapour was as low as 0.45% and the oxygen as low as 10.4%, the secondary explosion still could be triggered.

1. Introduction

Fires and explosions are extremely dangerous in oil depots and coal mines. In fact, hundreds of these serious accidents have happened in the past ten year [1]. It is found that the special phenomenon of secondary or even multiple deflagrations induced by heat are very common in these accidents. For example, a serious explosion occurred in a southwestern oil depot in 2006 in China and, in that accident, a deflagration (explosion) still emerged many times after the flame of thermal spontaneous combustion was being extinguished. As a result, this occurrence made a common fire accident turn into a serious fire and explosion accident, and the oil depot was destroyed thoroughly by explosions and the resulting fast-spreading fire. In order to prevent and reduce the hazards of fire and explosion, the research on producing condition and influencing factors of the gasoline-air mixture's secondary thermal ignition in a confined tunnel has great importance in science and practice.

For the (first) gas explosion, many experimental and theoretical studies on its ignition laws, propagation characteristics, chemical reaction process and other mechanisms have already been undertaken [2-3], but the studies on secondary or multiple explosions induced by a heat source, however, are seldom reported. Harrison et al. [4], Ponizy et al. [5] Catlin [6] and Forcier et al. [7] have studied the secondary explosions caused by vented explosions in a variety of situations. In addition, experimental and numerical simulation research on the inducing process, influencing factors and mathematical model of the multiple explosions of combustible gas and dust have also been carried out by Jiang et al.[8], Fan et al. [9], Ou et al. [10] and Du et al. [11] etc. However, due to the differences among flammable materials, the mechanism and influencing factors in multiple ignition processes will vary from one situation to another. From surveying the data we collected, explosive mixtures from flammable gases of simple composition, such as hydrogen, methane, propane and other single industry gases mixed with air or oxygen [12-14], have been studied in detail. However, research on mixtures involving multi-component explosive gases, such as gasoline, diesel oil and other materials, which can trigger spontaneous thermal combustion in the air easily are very rare.

In this study, through experiments carried out in a cylinder tunnel with a solid heating device, the basic experimental data and occurrence laws of the secondary gasoline-air mixture explosion will be obtained, and then, the concentration effects on the secondary explosion of gasoline-air mixture will be explored by means of results analysis.

2. Experimental apparatus and methods

Visualization research on the secondary occurrence of gasoline-air explosion in confined tunnel was carried out during the experimental process. The experimental apparatus used in this study mainly consisted of experimental tunnel (containing two visualization windows), high speed camera, gasoline evaporation apparatus, vacuum circulating pump, data acquisition system and computer, as shown schematically in Figure 1.

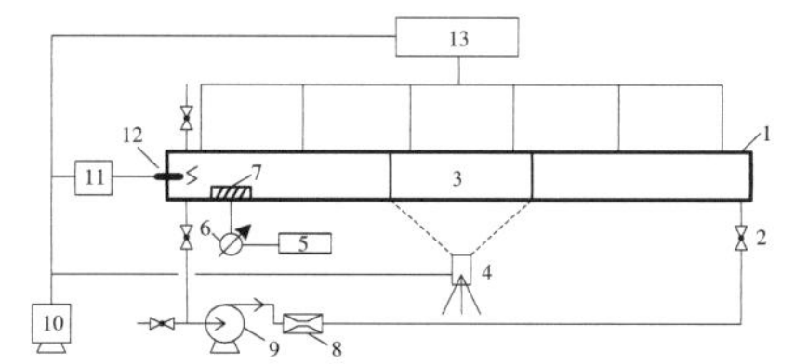

1. Test tunnel; 2.Valve; 3. Visualization section; 4. High speed camera; 5. Power; 6. Transformer; 7. Heating device; 8. Gasoline evaporation apparatus; 9.Vacuum circulating pump; 10.Computer; 11. Ignition system; 12. Spark plug; 13. Data acquisition system.

Figure 1. Schematic of experimental equipments

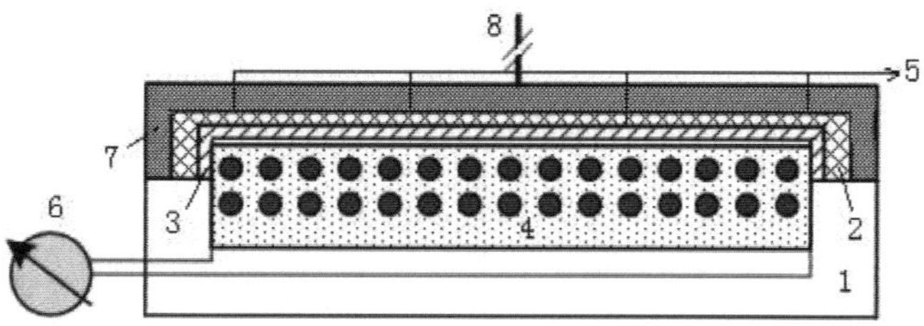

1. base support 2. copper sheet 3. heating-insulating layer 4. heat source 5. thermocouples 6. transformer 7. heat-insulating cover 8. rope

Figure 2. composition of the heating device

The heating device was set horizontally in the tunnel and could be seen from the right observation window. The heating device mainly consists of a transformer, a heat source, a copper sheet, a base support, a heat-insulating layer and five thermocouples, as shown in Figure 2. The copper sheet is square (170×170 mm) with thickness of 2 mm. In the experiments, the copper sheet was heated by the heat source (its heating rate is 0-3000 W and can be adjusted by the transformer) and can get a temperature region of 300-900K. In order to prevent influences of local high temperature on heating device surface on the concentrations of the gas composition during the heating process, the heating rate was set at a lower value of 500 W and the gas mixture in the tunnel was also circulated by the circulation pump, when the gas mixture was heated by the heating device. When the heating device was used as an igniter in the experiments, the heating rate was set at a higher value of 2000 W, which can increase the temperature of copper sheet quickly and ignite the gas mixture.

In order to better prevent influences of heating device on the concentrations of the gas composition during the heating process, a heat-insulating cover was set on the surface of the heat source to separate the heat source and the gasoline-air mixture. The heat-insulating cover also can be lifted easily by a rope when the contact of gasoline-air mixture with the heat source is needed.

Experimental data needed to be collected mainly includes gaseous concentration, temperature and humidity etc. The GXH-1050 infrared analyzer was used to obtain gasoline vapor concentration, and other gaseous concentrations were respectively acquired by NHA-502 automotive emission analyzer.

Temperature in the experiments was collected by a Monitor and Control Generated System. Humidity in this present paper was gathered by a TRH-AZ temperature-humidity meter.

The temperature of gas mixture was measured by eight chrome-nickel-silicon thermocouples, which were arranged at intervals of 200mm along the experimental tunnel, while the heat device temperature is measured by five thermocouples located on the copper sheet (one is arranged at the center of copper sheet and the other four at the four corners of the sheet). Temperature data from these thermocouples were averaged as the final temperatures of gasoline-air mixture and heating device, respectively.

3. Results and Analysis

3.1. Phenomenon of the secondary explosion

Multiple experiments indicated that the flame behavior of the gasoline-air explosion in the closed tunnel was affected by various factors, such as the initial concentration, temperature, humidity, etc. The flame in the test tunnel roughly had undergone ignition, development, attenuation and extinguishment process, respectively, and the lasting time was always less than 500 ms. However, both explosion overpressure curve and high speed photos were showed that the secondary explosion was occurred when the initial gasoline vapor concentration was maintained at a higher level (vol. 2.55%). Figure 3 and Figure 4 show the explosion overpressure curve and high speed photos of the secondary gasoline-air explosion process. In this experiment, the sampling frequency of high speed camera was 250 fps and the shutter speed was 1/250 s, so the time interval between the neighbor photos was 4 ms.

Figure 3 shows that the overpressure rose rapidly after the ignition of the mixture at about 1.81 s, and the overpressure peak was appeared at about 2.22 s and then the overpressure curve dropped sharply. However, the overpressure curve increased again at about 2.97 s and climbed sharply to another peak at about 3.09 s, and then attenuated down once again until the end of the explosion. The time interval between these two overpressure peaks was 0.846 s, which was longer than the lasting time of the normal experiment we carried out. This special two-peak phenomenon of the overpressure curve indicated that a secondary explosion might happen in the test tunnel. This conclusion was proved by the followed high speed photos.

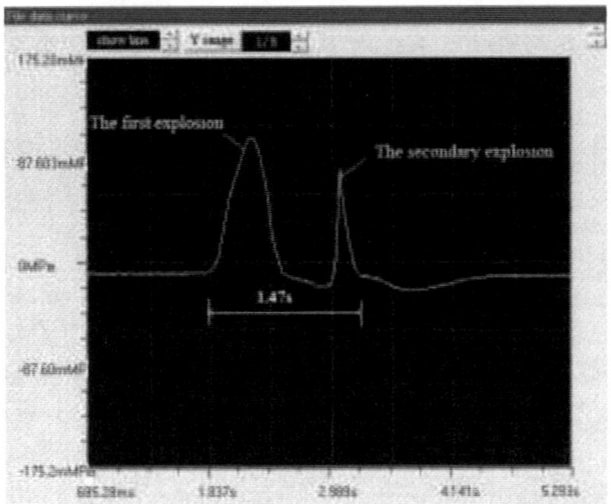

Figure 3. Overpressure curve of the secondary explosion

It can be seen from Figure 4 that the flame propagated forward after the ignition of the gasoline/air mixture (1.82 s, ignition delay). The flame area and the brightness were also increased as the flame propagated forward. The color of the flame was orange when the flame entered the observation window and then became bright yellow-white color. The flame brightness reached maximum at about 2.285 s and the visualization window was fully covered by bright yellow flame at this time. Then, the flame began to decay and quenched at about 2.451 s. After 0.5 s (i. e. at about 2.973 s), however, combustion flame of a secondary explosion was appeared. From Figure4, it can been easily seen that the flame of the secondary explosion had in turn experienced resurgence (2973-2989 ms), development (2989-3053 ms), attenuation (3053-3085 ms) and extinguishing (after the time of 3149 ms) process.

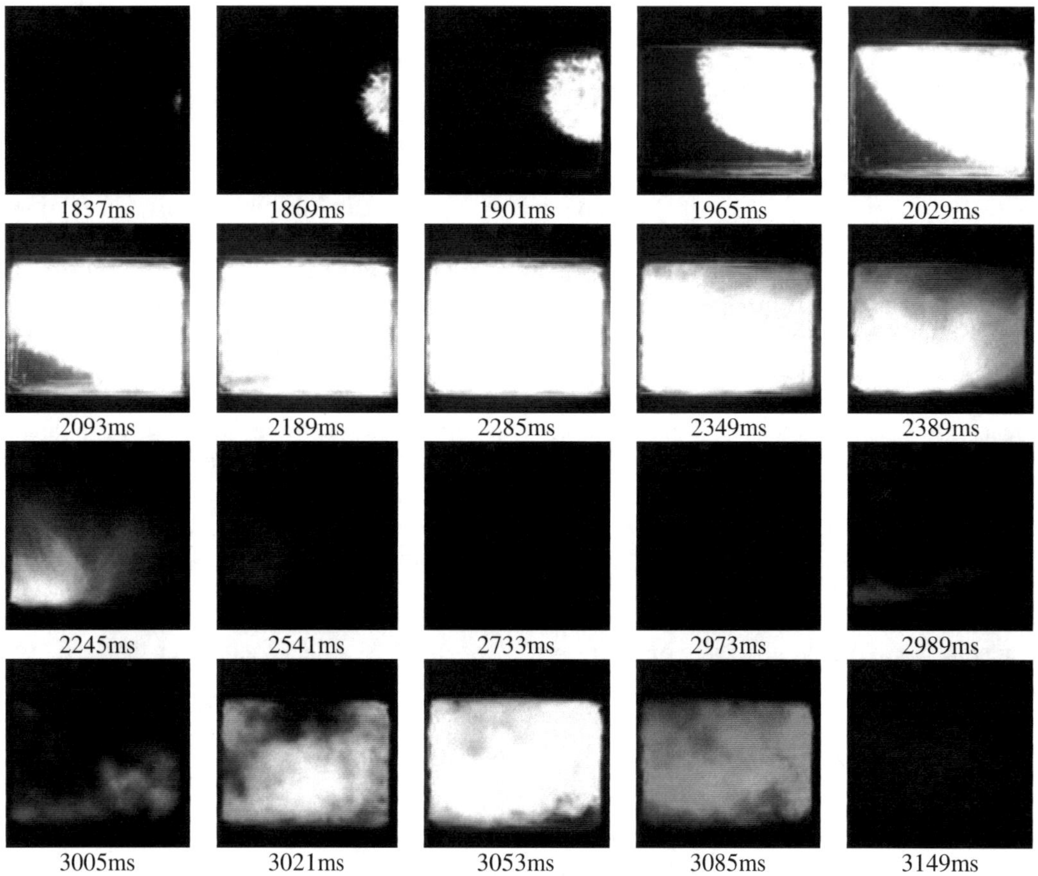

Figure 4. High speed photos of the secondary explosion process

Based on the comprehensive analysis of the experimental data before and after the secondary explosions, we realized that the main causes for the secondary explosion could be summarized as follows: (1) the high temperature environment and the existence of a large number of active group formed after the first explosion were the necessary conditions. (2) Due to the high initial gasoline vapor concentration, gasoline-air mixture was still remaining after the first explosion. (3) The experimental tunnel could not be absolutely airtight, part of the fresh air outside the tunnel might enter into the test tunnel so that the oxygen in the tunnel had been added. (4) In addition, it was worth noting that the occurrence of the secondary explosion had undergone a longer induction period (about 1.47 s) after the first explosion. Although no flame was appeared in this induction period, some important pre-ignition chemical reactions [3, 15] might possibly happened in the test tunnel, these pre-ignition chemical reactions produced a lot of intermediate products or active groups, which provided favorable conditions for the secondary explosion of the gasoline-air mixture. Causes (2) and (3) had ensured that the concentration of combustible gas mixture after the first explosion in the test tunnel was maintained in the range of explosion limit, cause (1) had provided ignition conditions for the secondary explosion, while, cause (4) had provided subsidiary condition for the occurrence of secondary explosion.

3.2. Effects of concentration on the secondary explosion of gasoline-air mixture

Although the influence factors for the secondary explosion maybe included humidity, temperature, concentration, etc. [16, 17]. The most important influence factor is concentration. So multiple experiments were carried out to investigate the critical concentration for the secondary explosion.

Multiple experiments indicated that whether the secondary explosion occurred or not was determined by critical gasoline vapor and oxygen concentration even if the temperature was maintained in a reasonable scope. A comparison of components' concentration before and after ignition is shown in Table 1. From Table 1, it can be seen that the secondary explosion in the confined tunnel would still be triggered even if the gas vapor concentration was as low as 0.45% and the oxygen concentration was as low as 10.4%. This value was much lower than the critical value (1%) of the first explosion reported by some references [18]. The reason behind this observation was that the first explosion of the gasoline-air mixture usually produced incomplete combustion and the chemical reaction was not complete either. As a result, there were a large number of intermediate products and active free radicals in the confined tunnel. It was shown that the highest concentration of CO had reached 6.06% before the secondary explosion, however, after the secondary explosion, this concentration was nearly 0 at some time and the highest CO_2 concentration had increased to 6.32%. This phenomenon indicated that intermediate products continued to react and released heat after encountering a heat source at high temperature, leading to the occurrence of the secondary explosion. In addition, due to the increase of the gas mixture temperature and pressure, the decrease of humidity in the confined tunnel after the first explosion, all components became so active that the chemical reaction would be triggered easily, and then the flame appeared. Based on the analysis mentioned above, the components of gasoline-air mixture after the first explosion became more complex, and the criterion on minimum oxygen and gas vapor concentration used for the first explosion was no longer appropriate to judge whether a secondary explosion would be triggered.

Table 1. Comparison of components concentration before and after the secondary explosion in the confined tunnel

Components concentration after the first explosion (%)				**Components concentration after the secondary explosion (%)**			
HC	O_2	CO	CO_2	HC	CO	CO_2	O_2
0.60	12.9	4.21	2.51	0.19	0.01	4.01	7.80
0.47	10.8	2.35	3.22	0.25	0.00	5.32	8.20
0.45	10.4	2.50	3.78	0.10	0.00	4.01	9.00
0.57	12.8	1.35	2.78	0.18	0.00	4.03	7.10
0.76	11.5	5.30	2.19	0.09	0.00	4.53	6.90
0.82	10.4	5.52	2.46	0.11	0.01	4.83	7.00
0.99	12.3	5.43	4.55	0.05	0.02	5.80	7.09
0.67	14.9	3.78	5.14	0.02	0.00	6.90	7.50
0.77	15.3	6.06	4.53	0.11	0.01	6.32	6.40
0.86	14.4	4.78	3.10	0.14	0.01	5.81	13.00
0.72	15.1	2.51	4.80	0.07	0.01	6.20	11.00
0.63	11.6	2.64	4.23	0.11	0.00	6.22	9.00
0.57	11.9	4.12	3.93	0.16	0.01	5.46	8.70

4. Conclusions

The main causes for the secondary explosion could be summarized as follows: (1) the high temperature environment and the existence of a large number of active group formed after the first explosion were the necessary conditions. (2) The oxygen in the tunnel had been added. (4) Some important pre-ignition chemical reactions might possibly happened in the test tunnel, these pre-ignition chemical reactions produced a lot of intermediate products or active groups, which provided favorable conditions for the secondary explosion of the gasoline-air mixture. Causes (2) and (3) had ensured that the concentration of combustible gas mixture after the first explosion in the test tunnel was maintained in the range of explosion limit, cause (1) had provided ignition conditions for the secondary explosion, while, cause (4) had provided subsidiary condition for the occurrence of secondary explosion.

The secondary explosion could still be triggered even if the gas vapor concentration was as low as 0.45% and the oxygen concentration was as low as 10.04%. After the first explosion, due to the increase in the gas mixture temperature and pressure, all components became more active and they could participate in the chemical reactions more easily, which was the main reason for the secondary occurrence of the gasoline-air explosion.

Acknowledgments
This work was financially supported by LEU Young Investigator fund (YQ16-420802).

References

[1] J. I. Chang and C. C. Lin, A study of storage tank accidents, J. Loss Prevent. Process Indust. 19 (2006) 51-59.
[2] A. A. Pekalski, E. Terli, J. F. Zevenbergen, S. M. Lemkowitz, H. J. Pasman, Influence of the ignition delay time on the explosion parameters of hydrocarbon-air-oxygen mixtures at elevated pressure and temperature, Proc. Combust. Inst. 30 (2005) 1933-1939.
[3] D. Razus, C. Movileanu, V. Brinzea, D. Oancea, Explosion pressures of hydrocarbon–air mixtures in closed vessels, J. Hazard. Mater. 135 (2006) 58-65.
[4] A. J. Harrison and A. J. Eyre, External explosions as a result of explosion venting, Combust Sci Tec. 52 (1987) 91-106.
[5] B. Ponizy and J. C. Leyer, Flame dynamics in a vented vessel connected to a duct: 1. mechanism of vessel-duct interaction., Combust. Flame 116 (1999) 259-271.
[6] C. A. Catlin, Scale effects on the external combustion caused by venting of a confined explosion, Combust. Flame 83 (1991) 399-411.
[7] T. Forcier and R. Zalosh, External pressures generated by vented gas and dust explosions, J. Loss Prevent. Process Indust. 13 (2000) 411-417.
[8] X. H. Jiang and B. C. Fan, Experimental study of secondary explosions in methane-air flow field, Exp. Fluid Mech. 2 (2005) 12-19.
[9] B. C. Fan, X. H. Jiang, Secondary explosion induced by vented explosion, Exp. Shock Waves 1 (2005)11-16.
[10] Y. H. Ou, Y. Du, X. S. Jiang, D. Wang, J. J. Liang, Study on the thermal ignition of gasoline-air mixture in underground oil depots based on experiment and numerical simulation, J. Thermal Sci. 19 (2010) 173-181.
[11] Y. Du and Y. H.Ou, Thermal ignition phenomenon of gasoline-air mixture induced by hot wall, Exp. Shock Waves 3 (2009) 268-274.
[12] A. E. Dahoe, Laminar burning velocities of hydrogen–air mixtures from closed vessel gas explosions, J. Loss Prevent. Process Indust. 18 (2005) 152-166.
[13] D. Razus, V. Brinzea, M. Mitu, C. Movileanu, D. Oance, Inerting effect of the combustion products on the confined deflagration of liquefied petroleum gas–air mixtures, J. Loss Prevent. Process Indust. 22 (2009) 463-468.
[14] D. Razus, V. Brinzea, M. Mitu, D. Oance, Temperature and pressure influence on explosion pressures of closed vessel propane–air deflagrations, J. Hazard. Mater. 174 (2010) 548–555.
[15] R. Edenhofer, K. Lucka, H. Köhne, Proc. Low temperature oxidation of diesel-air mixtures at atmospheric pressure, Combust. Inst. 31 (2007) 2947-2954.
[16] D. Yang, Z. P. Li, O. Y. Hong, Effects of humidity, temperature and slow oxidation reactions on the occurrence of gasoline-air explosion, J. Fire Pro. Eng. 23 (2013) 226–238.

[17] P. Zhang, Y. Du, S. Wu, Y. Zhou, J. Zhou, J. Xu, Experiments of the secondary ignition of gasoline－air mixture in a confined tunnel, J. Therm. Anal. Calorim. 118 (2014) 1773-1780.
[18] H. Y. Zhao, Explosion theory of gas and dust. Beijing Institute of Technology Press, Beijing, 1996.

IOP Conf. Series: Earth and Environmental Science 64 (2017) 012009 doi:10.1088/1755-1315/64/1/012009

Study on driving forces of wetland change in the Western Liaohe River basin based on random forest model

Menghong Wu[1,*], Changbao Yang[1], Yanhong Zhang[1] and Nan Lin[2]
[1]School of College of Geo-exploratin Science and Technology, Jilin University, Changchun, China
[2]School of College of Surveying and Prospecting Engineering, Jilin Jianzhu University, Changchun, China

*Corresponding author e-mail: wxn920911@126.com

Abstract. Based on the platform of RS and GIS, random forest progression model is used for study driving force of wetland change in western Liaohe river basin, five influencing factors which include elevation, slope, temperature, precipitation and population density are chosen to establish random forest progression model about the wetland change and the driving factors. Using the the mean value of the prediction accuracy outside the bag calculated by the model to evaluate the importance of the variables. The result indicates that the coefficient of partial correlation between precipitation and wetland density is the largest among the five influencing factors, followed by temperature, population density, elevation and slope is smallest. The influence of natural factors on the change of wetland density is mainly reflected in precipitation and temperature factors, and the precipitation is obviously higher than that of temperature, under the influence of human factors, the influence of population density factor on wetland density is higher than that of elevation and slope factor. The result shows that in the past 40 years, the human activities in the study area have increased the density of wetland to some extent, but it is not the main factor.

1. Introduction

The study area is in the Western Liaohe River basin, which is located in agriculture and animal husbandry in Western Northeast Plain staggered along the eastern edge, the climate belongs to the transition band from warm temperate and semi humid to moderate temperate and semi-arid climate. The area is one of the largest wetland area in Northeast China, is also an important commodity grain producing areas in our country [1]. Therefore, the analysis of the wetland ecological condition health has a great significance in explore main influence factors of wetland change. The natural factors and human factors on the study area is chosen to establish random forest regression model about the wetland change and the driving factors, the natural factors contains topography and meteorology, the human factors contains density of population. Mean value of out of pocket prediction accuracy is used to measure the degree of influence of each driving factor on wetland change, the grid overlay method is used to integrate the wetland data into a specific spatial unit to calculate, stack and evaluate the values, the quantitative analysis of driving forces of wetland change can provide reference basis for protect wetland of western Liaohe river basin effectively [2].

IOP Conf. Series: Earth and Environmental Science **64** (2017) 012009 doi:10.1088/1755-1315/64/1/012009

2. Study Area

The Western Liaohe River is seated in the southern part of China's northeast, flowing through 22 county-level administrative units in Inner Mongolia, Liaoning, Jilin and Hebei. The study area is located at 40°8'18.21"-45°1'20.36"N, 115°9'11.79"-123°00'13.93"E, the total area is 16.73 * 10^4km^2. The administration includes Tuquan, Horqin Right Wing Middle Banner, Jarud Banner, Horqin Left Middle Banner, kailu, tongliao, Horqin Left Back Banner, Ar Horqin Banner, Baarin Left Banner, Baarin Right Banner, Naiman Banner, Hure Banner, linxi, Hexigten Banner, Ongniud Banner, Aohan Banner, chifeng, jianping and Harqin Banner. The transportation in this area is convenient, railway and highway net is spread by radial pattern, Tongliao and Siping is the center of net.

3. The extraction of wetland change information and influencing factors

(1)Wetland change information

Choose MSS remote sensing image data in 1975, ETM remote sensing image data in 2000 and OLI remote sensing image data in 2015 as the remote sensing data source, artificial visual interpretation is used to obtain the distribution information of wetland in the past 40 years based on GIS software platform, using superposition analysis and transfer matrix to analysis spatiotemporal dynamic change rule of wetland change transformation in past 40 years in Western Liaohe River basin from temporal dynamic change, spatial dynamic change and wetland type transformation respectively[3]. The dynamic change of wetland can be divided into three types: the increase of wetland, the stability of wetland and the decrease of wetland, the spatial distribution of wetland spatial dynamic change is as shown in Figure 1:

Fig.1 The wetland dynamic change figure in western Liaohe river basin

(2)Topographic factor

Topographic information is mainly extracted by DEM data, extracting terrain factors by DEM in an automatic way to spatial analysis has characteristics like fast, scientific, objective, which means that can replace the complex field observation in traditional operation. Elevation and slope factors which influence wetland distribution larger are selected in this study, as shown in figure 2, 3:

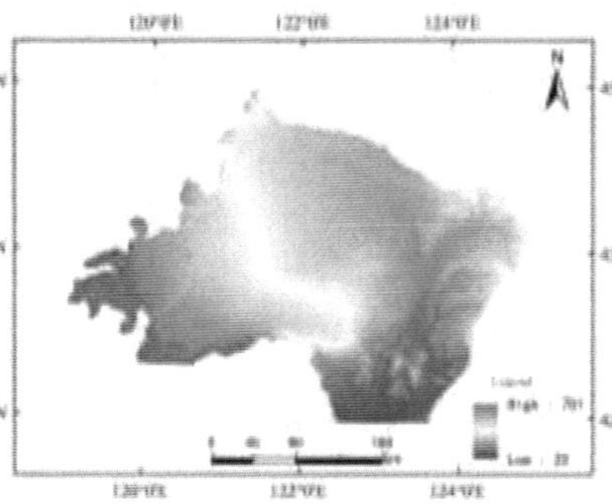

Fig.2 Elevation interpolation figure In western Liaohe river basin

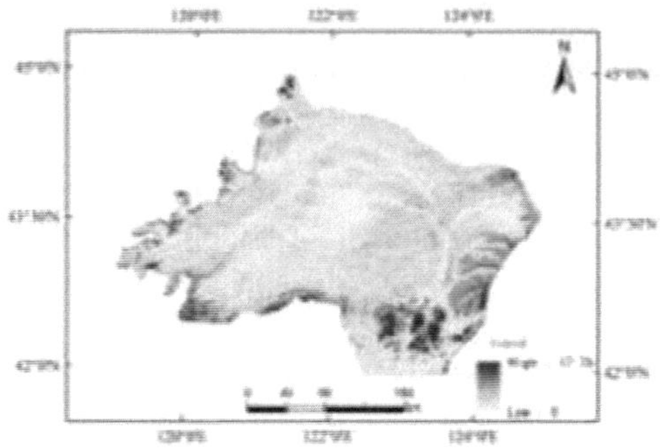

Fig.3 Slope interpolation figure in western Liaohe river basin

(3)Meteorological factor

Temperature and precipitation which have a strong relationship with wetland growth are selected as study object, using kriging interpolation method to process interpolation of temperature and precipitation in study area[4-6], the temperature interpolation figure and precipitation interpolation figure are shown in figure 4,5:

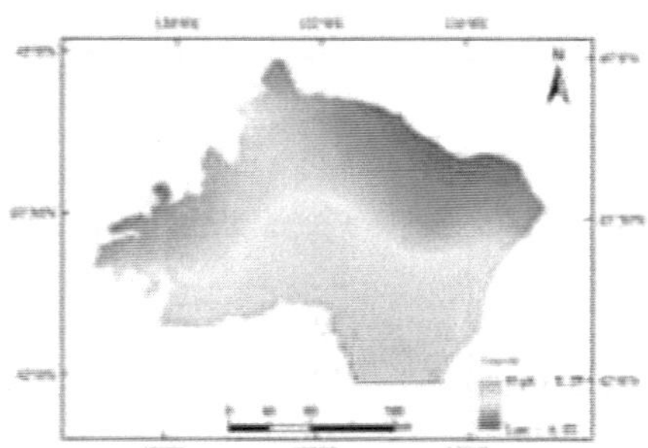

Fig.4 Temperature interpolation figure In western Liaohe river basin

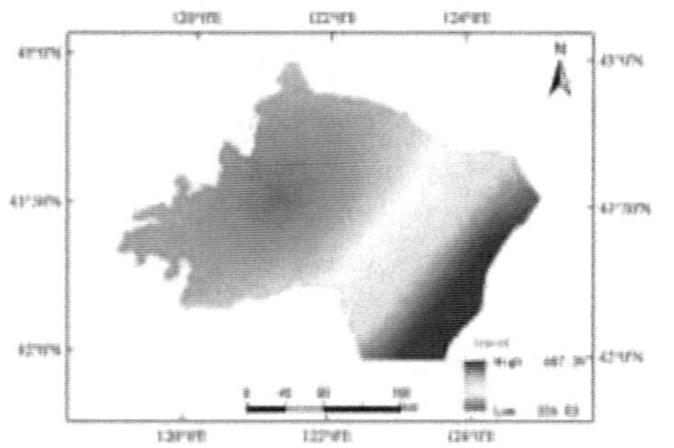

Fig.5 Precipitation interpolation figure in western Liaohe river basin

(4)Population density factor

In a short term, human activity is the main driving factor. According to the actual situation of the study area, the paper selects a representative population density factor to explore the impact of human factors on wetland change. As shown in figure 6:

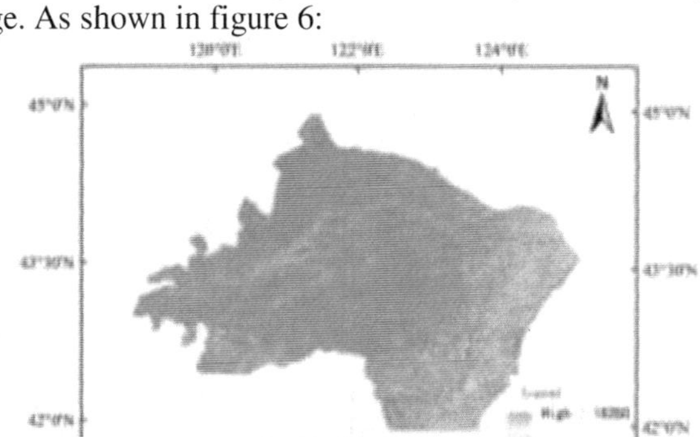

Fig.6 Population density interpolation figure in western Liaohe river basin

4. Random forest progression model

Random forest is ensemble classifier contains multiple non pruning classification and regression tree (CART), Breiman proposes the concept in 2001, Breiman combines the classification tree into a random forest, and each decision tree has two random processes, the generation process is as follows: assuming that the dependent variable Y has N samples, there are P independent variables associated with the dependent variable Y, In the process of constructing the regression tree, random forest will use Bootstrap sampling method for extracting k sample training set (T_1, $T_2 ... T_k$) randomly in the dependent variable Y, the size of each training set is about 2/3 of the original data set, the data which is not extracted is used for test sample set as out-bag data; Establishing classification and regression tree to generate K th trees $\{T(x_n,\theta_1),(x_n,\theta_2)\cdots(x_n,\theta_k)\}$ for each data set, since each regression tree is likely to be random in nature, random forests are typically able to generate hundreds or even thousands of classification trees randomly [7,8]. The second stochastic process is to extract m features from the P features as the feature set of the current node, each classification regression tree grow recursively from top to bottom, setting the NTree value as the termination condition of the growth of the regression tree, all decision trees are combined into a random forest.

The importance evaluation of variables is an outstanding feature of random forest algorithm, among them, the evaluation of the importance of variables based on the prediction accuracy of out-bag data is widely used, the idea is to evaluate the importance of the variables by changing the value of the independent variables outside the bag randomly and to calculate the change of the accuracy of the model, the mean of the accuracy is proportional to the importance of the variable, and the larger the error caused by the change of the characteristic variable, the more important the variable is. The specific process is as follows:

(1)The prediction accuracy MSE_{OBB} of the corresponding bag data (OOB) is calculated for each training sample set, assuming that the number of samples outside the bag is q, the formula is:

$$MSE = \frac{1}{q}\sum_{i=1}^{q}(y_i - \hat{y}_i)^2 \tag{1}$$

In the formula, y_i represents the true value of the dependent variable in the outside bag data, $\hat{y}_i$ represents the predict value of the regression model;

(2)The value of the j characteristic variable X_J is changed randomly, and the accuracy of the new outside bag error MSE_j is recalculated;

(3)The calculate formula of the importance of variable X_J is:

$$V(X^j) = \frac{1}{q}\sum_{j=1}^{q}(MSE - MSE_j) \tag{2}$$

5. Results

In order to quantitatively analyze the driving forces of wetland change, the grid superposition method is used to integrate the wetland data into a specific spatial unit to calculate, stack and evaluate. The size of the mesh affects the boundary range and the accuracy of the spatial unit directly. The study area is divided by orthogonal kilometer network, according to the actual situation of the study area; the grid size is 10km * 10km. After the grid is divided, the dynamic change trend of the wetland can be refined to each mesh, which is represented by discrete points. The following formula is to calculate the wetland change intensity:

$$P = \frac{\left[\sum U_b - \sum U_a\right]}{M_0} \tag{3}$$

Using the TIN model in DEM analysis in MAPGIS for contouring tracing analysis, P in the formula is the intensity of wetland change per unit area during the study period. $\sum U_a$ Represents the total area of wetland in each grid at the beginning of the study, $\sum U_b$ represents the final phase of the study, M_0 represents unit area. 1000 samples are selected randomly in the study area, While wetland change density is the dependent variable , elevation, slope, temperature, precipitation, population density are independent variables, the random forest progression model of wetland dynamic change is established to calculate the mean value of prediction accuracy out of pocket, the results are shown in Figure 7:

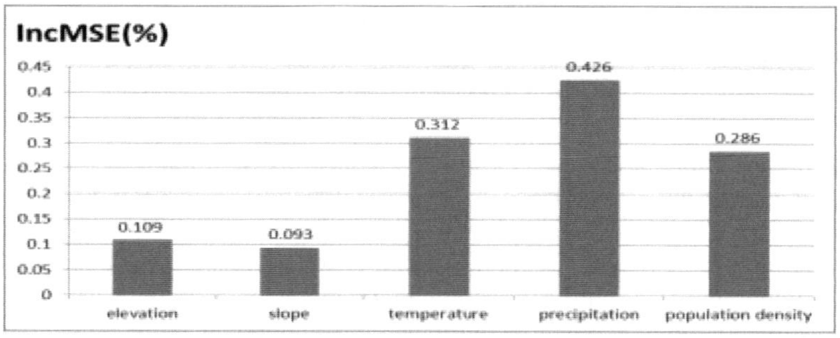

Fig.7 Analysis figure of driving factors

6. Conclusions

(1)According to the statistical results of the partial correlation coefficient of each factor and the wetland density, it can be seen that the temperature and precipitation factor under the meteorological factors have the greatest impact on the wetland change. Followed by the population density factor, the smallest impacts are the terrain factor elevation and slope factor. The high correlation coefficient of temperature and precipitation shows that the meteorological factors play an important role in controlling wetland change. Because the area is far away from the sea, the climate of the study area is more complex, which belongs to the transition zone of warm temperate semi humid climate to moderate temperate semi-arid climate. The continental climate is significant, which makes the temperature and precipitation play an important role in the development and utilization of wetland, and thus becomes the limiting condition of wetland change.

(2)Population density, elevation and slope also have a certain impact on wetland change, the impact of population density on vegetation is mainly reflected in the impact of human activities on Wetlands. The increase of population, the development of regional economy and the intensification of urbanization are the concrete manifestation of the influence of human activities on Wetlands. However, the partial correlation coefficient between elevation, slope and the density of wetland is low, because the terrain factors (elevation, slope) will not make a big change in a short period of time. Therefore, the change of wetland in the study area is relatively stable for a long time. It also shows that the elevation and slope of the whole study area are appropriate, and the adjustment ability of elevation and slope is strong.

Acknowledgements
This work was financially supported by Graduate Innovation Fund of Jilin University (No.2016199).

References

[1] LI Yanqing. Spatial-temporal change of vegetation in the western Liaohe River Basin,Northeast China:master's degree thesis[D]. Changchun: Northeast Normal University,2008.
[2] Detenbeck N E, Taylor D L, Lima A, et al. Temporal and spatial variability in water quality of wetlands in the Minneapolis/ST. Paul, MN metropolitan area: implications for monitoring

strategies and designs [J]. Environmental Monitoring and Assessment, 1996, 40 (1): 11-40.

[3] Moorhead K K, Brinson M M. Response of wetlands to rising sea level in the lower coastal plain of North Carolina [J]. Ecological Applications, 1995, 5 (1): 261-271.

[4] Acreman M, Blake J, Booker D, et al. A simple framework for evaluating regional wetland ecohydrological response to climate change with case studies from Great Britain [J]. Ecohydrology, 2009, 2(1): 1-17.

[5] Burkett V, Kusler J. Climate Change: Potential impacts and interactions in wetlands of the United States, Virginia [J].Journal of the American Water Resources Association, 2000, 36(2): 313-320.

[6] LIN Nan. Study on the change and driving force of vegetation cover in eastern Jilin Province based on RS and GIS: master's degree thesis [D]. Changchun: Jilin University, 2010.

[7] Wang Dong, Yue Cairong, Tian Chuanzhao, et al. Classification of TM remote sensing image based on random forests of Dayao country[J]. Forest Inventory and Planning, 2014,39(2):1-5.

[8] Lei Zhen. Random forest and is application in remote sensing. [D]. Shanghai: Shanghai Jiao Tong University, 2012.

IOP Conf. Series: Earth and Environmental Science **64** (2017) 012010 doi:10.1088/1755-1315/64/1/012010

Potassium dichromate method of coal gasification the study of the typical organic compounds in water

Jiankang Quan[1, a], **Guangfei Qu**[b,*], **Zhanneng Dong**[c], **Pei Lu**[d], **Yingying Cai**[e], **Shibo Wang**[f]

School of Kunming uniwersity of science and technology, Kunming, China

[a]1391039409@qq.com, [*b]Corresponding authore-mail: qgflab@sina.com
[c]zhndong@126.com, [d]854930287@qq.com ,[e]caiyymail@qq.com
[f]772325862@qq.com

Abstract. The national standard method is adopted in this paper the water - digestion spectrophotometry for determination of the chemical oxygen demand (COD), after ultrasonic processing of coal gasification water for COD_{Cr} measurement. Using the control variable method, measured in different solution pH, ultrasonic frequency, ultrasonic power, reaction conditions of different initial solution concentration, the change of coal gasification water COD_{Cr} value under the action of ultrasonic, the experimental results shows that appear when measurement is allowed to fluctuate, data, in order to explain the phenomenon we adopt the combination of the high performance liquid chromatography and mass spectrometry before and after ultrasonic coal gasification qualitative analysis on composition of organic matter in water. To raw water sample chromatography - mass spectrometry (GC/MS) analysis, combined with the spectra analysis of each peak stands for material, select coal gasification typical organic substances in water, with the method of single digestion, the equivalent COD_{Cr} values measured after digestion. Order to produce, coal gasification water contained high concentration organic wastewater, such as the national standard method is adopted to eliminate the organic material, therefore to measure the COD_{Cr} value is lower than actual COD_{Cr} value of the emergence of the phenomenon, the experiment of the effect of ultrasound [9-13] is promote the complex organic chain rupture, also explains the actual measurement data fluctuation phenomenon in the experiment.

1. Introduction

With the development of economy and the improvement of people's life, and the rapid industrial development, population growth and persistent drought problems, led to a nationwide shortage of clean water, forcing more and more people begin to pay close attention to environmental protection, especially the problem of water environment pollution control and protection [1]. Coal as main energy in our country, the use of coal and its related technology brought about by the related environmental problems have to be reckoned with, especially in high temperature dry distillation process of coal and gas purification, chemical products, such as refining process of waste water[2], containing phenol tar, cyanide and ammonia, pollutants, phyletic and various, and the pollutant concentration is high, high chromaticity, high toxicity, is a typical refractory organic wastewater[3-5].

At present all sorts of treatment technology is not entirely mineralization of organic matter, coking wastewater after treatment still contains a lot of low concentration of pollutants in wastewater with outer drainage into the environment, most of these pollutants has carcinogenic, teratogenic and mutagenic effects and toxicity, likely the ecological security and human health risk, and on the qualitative and quantitative research of coking wastewater organic matter less [6-8]. As a complex matrix of organic matter enrichment, separation and pretreatment technology development and the popularization of high performance analysis test instrument, make know the components of coking wastewater.

By national standard method the water quality, rapid digestion spectrophotometry for determination of the chemical oxygen demand (COD), as measured in high concentration organic wastewater measurement is allowed to fluctuate, aiming at to explain the phenomenon.

2. Experimental materials and methods

4.5 ~ 5.0 L coal gasification of water in the reactor, the transducer above the highest point 12 ~ 15 mm shall prevail, considering the experimental process to sampling accordingly index measurement, so the initial liquid level as transducer above the highest point of 15 mm. With dry beaker every 30 min take a watery, mild mixing solution before sampling, measuring the COD_{Cr}. Coal gasification water from anning city coking plant, the COD_{Cr} initial concentration of 3532 mg/L.

With rapid spectrophotometric method of COD_{Cr} in ultrasound before and after the change of the concentration of quantitative analysis; Using the combination of the high performance liquid chromatography and mass spectrometry before and after ultrasonic coal gasification qualitative analysis on composition of organic matter in water. To raw water sample chromatography-mass spectrometry (GC/MS) analysis [14-17], combined with the spectrum for each peak stands for material analysis, the results are shown in table 1.

Tab.1 Coal gasification suwon wateryGC-MS spectrum diagram

number	matter	number	matter
1	methylbenzene	13	quinoline
2	4roxy-4-metilo-2-pentanone	14	2-amino-benzonitrile
3	cyclohexanone	15	indole
4	Amino-benzene	16	Cinnamyl aldehyde
5	phenol	17	1H-indenol
6	o-cresol	18	2,6-bis-4-methylphenol
7	p-cresol	19	2-isonaphthol
8	2-metilo-indoline	20	2,5-dimetil-benzimidazole
9	2,4-methylphenol	21	5-hydroxy-isoquinoline
10	3,5-methylphenol	22	Pyridine
11	3,4-methylphenol	23	5H-indeno[1,2-b]piridina
12	1-hydroxy-benzocyclobutene		

3. Results and discussion

3.1. The influence on the effect of initial pH on CODCr degradation

Raw water initial COD_{Cr} 3529.25 mg/L, the pH value of 8.7, the water temperature of 26 ℃, respectively in the raw water with 1 mol/L HCl solution (HCl) and 1 mol/L sodium hydroxide solution (NaOH) adjusting the reaction liquid pH 2.5 and 2.5, in 40 KHz, 300 w, the sound intensity under the condition of 57.09 w/cm^2 ultrasound alone 180 min, once every 30 min take sample, COD_{Cr} after diluted 10 times its absorbance measurement, the COD_{Cr} concentration change under different initial pH value were studied. Due to the raw water pH value neutral alkaline, subsequent experiments not adjusted pH value. Different initial pH value on the effects of the COD_{Cr} degradation experiment result is shown in figure 1.

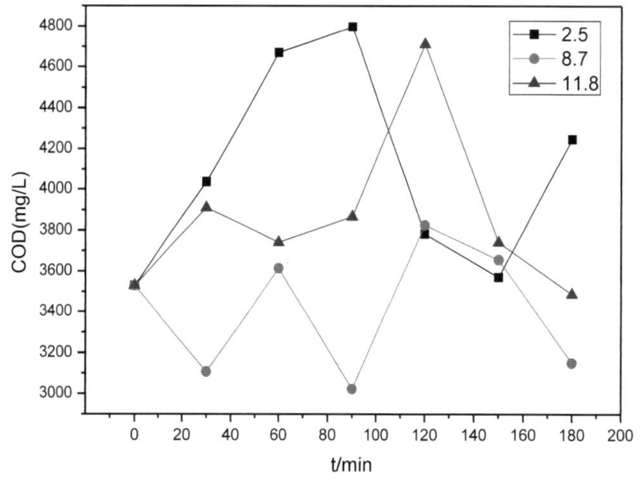

Fig.1 The influence on the effect of initial pH on CODCr degradation

3.2. The change of CODCr concentration under different ultrasonic power

Raw water initial COD_{Cr} 3529.25 mg/L, the pH value of 8.7, the water temperature of 26 ℃, under the condition of 40 KHz ultrasonic frequency ultrasound alone 180 min, experiments, respectively, set the size of the power ultrasonic generator to 180 w, 240 w, 300 w, once every 30 min take sample, COD_{Cr} after diluted 10 times its absorbance measurement, the COD_{Cr} concentration change under different initial pH value were studied. Due to the raw water pH value neutral alkaline, subsequent experiments not adjusted pH value. Under the condition of different ultrasonic power on the effects of the COD_{Cr} degradation experiment result is shown in figure 2.

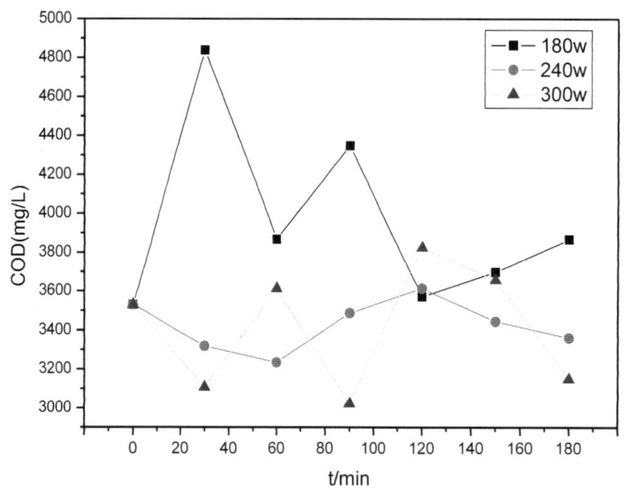

Fig.2 The change of CODCr concentration under different ultrasonic power

IOP Conf. Series: Earth and Environmental Science **64** (2017) 012010 doi:10.1088/1755-1315/64/1/012010

3.3. The change of CODCr concentration under different ultrasonic frequency

Raw water initial COD_{Cr} 3529.25mg/L, the pH value of 8.7, the water temperature of 26 ℃, under the condition of ultrasonic power is 300 w, ultrasound alone 180 min, the experiment set the frequency of ultrasonic generator size respectively for 20 KHz, 28 KHz, 40 KHz, once every 30 min take sample, COD_{Cr} after diluted 10 times its absorbance measurement, the COD_{Cr} concentration change under different initial pH value were studied. Due to the raw water pH value neutral alkaline, subsequent experiments not adjusted pH value. Under the condition of different ultrasonic frequency influence on COD_{Cr} degradation effect, the experimental results are shown in figure 3.3.

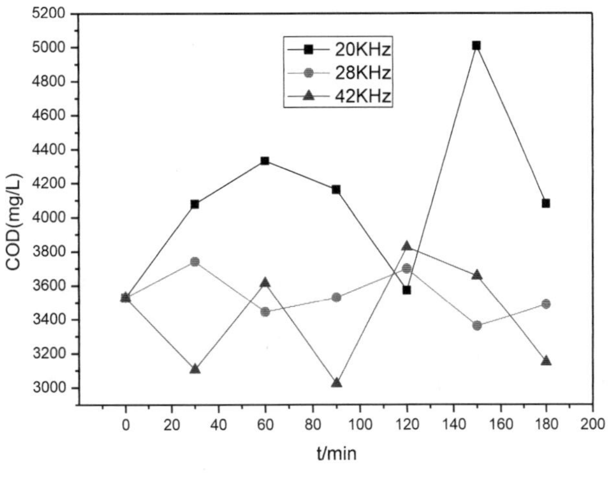

Fig.3 The change of CODCr concentration under different ultrasonic frequency

3.4. The change of CODCr concentration under different initial concentrations

In order to ensure the water samples were used as raw water samples, sample pretreatment and dilution. In ultrasonic power is 300w, ultrasonic frequency of 40 KHz, under the condition of single ultrasound, 180 min, once every 30 min take sample, COD_{Cr} after diluted 10 times its absorbance measurement, the COD_{Cr} concentration change under different initial pH value were studied. Due to the raw water pH value neutral alkaline, subsequent experiments not adjusted pH value. A research on the ultrasonic function under different initial concentration of COD_{Cr} degradation effect, the experimental results are shown in figure 4.

IOP Conf. Series: Earth and Environmental Science **64** (2017) 012010 doi:10.1088/1755-1315/64/1/012010

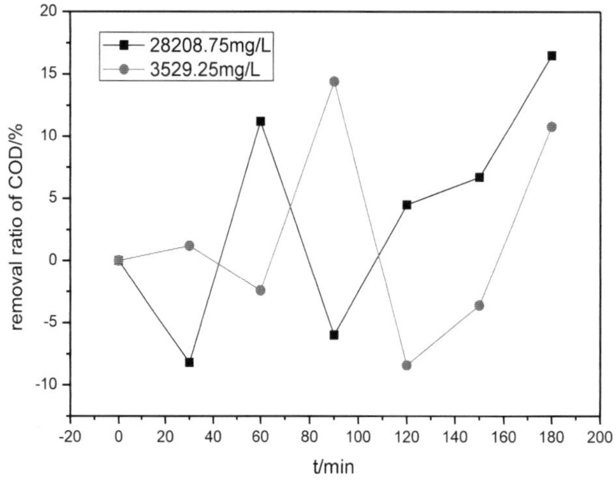

Fig.4 The change of CODCr concentration under different initial concentrations

In the above four figure show that the different initial concentration and different ultrasonic power, ultrasonic frequency, ultrasonic degradation process of COD_{Cr} under the conditions of different initial pH, the COD_{Cr} degradation regularity is not obvious, with no obvious trend of rise and fall. Experiments with rapid spectrophotometry (HJ/T399-2007) the principle is: the reagent adding known quantity of potassium dichromate solution, in strong acid medium, with silver sulfate as the catalyst, after high temperature resolution, by spectrophotometer method determination of COD_{Cr} value. But there are also some disadvantages: under the condition of acid potassium dichromate, some aromatic hydrocarbon organic matters, such as pyridine compounds [14-17] difficult to oxidation, the oxidation efficiency is low, and even cannot be oxidized. Therefore, will exist in the COD_{Cr} measurement to measure the COD_{Cr} value does not response the real COD_{Cr} value

3.5. Equivalent to phenol was dispelled, measured CODCr value

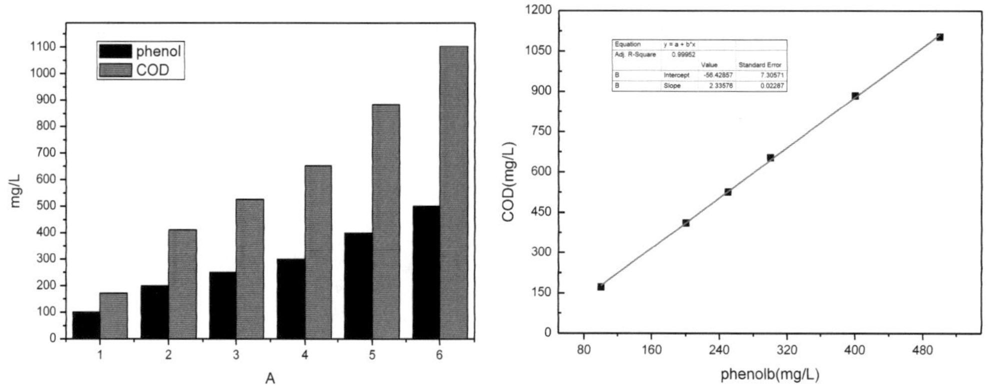

Fig.5 Equivalent to phenol was dispelled, measured COD_{Cr} value

IOP Conf. Series: Earth and Environmental Science **64** (2017) 012010 doi:10.1088/1755-1315/64/1/012010

3.6. Equivalent to quinoline was dispelled, measured CODCr value

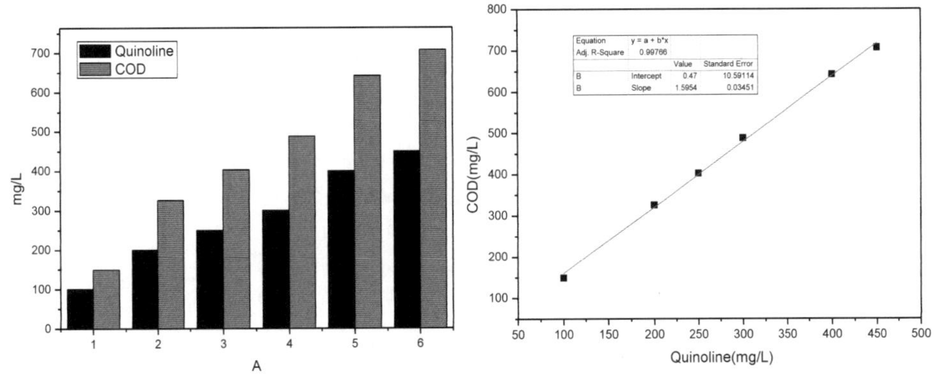

Fig.6 Equivalent to quinoline was dispelled, measured COD_{Cr} value

In potassium dichromate digestions of pyridine with different concentration gradient, eventually couldn't measure the equivalent COD_{Cr} value. To use potassium dichromate digestion of 50 mg/L pyridine solution product analysis, because the product concentration is too high, so adopt the method of deionized water diluted 104 times were measured, using spectrophotometer wavelength scanning method in 190-600 nm in the scan after digestion solution, the determination results as shown:

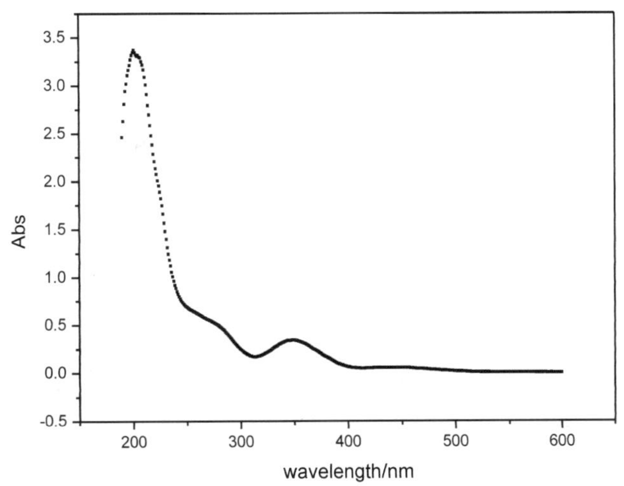

Fig.7 50 mg/L pyridine by potassium dichromate solution after digestion with spectrophotometer spectrum scan

By as shown in figure: the 50 mg/L pyridine solution after digestion solution using the wavelength scanning spectrophotometer, measuring the product. Because the concentration of certain substances after digestion is exorbitant, absorbance measurement range of the instrument is limited, need to dilute the solution after digestion. Appeared a peak at 200 nm, shows the peak as the peak of total nitrogen, depending on the absorbance to reflect the content of how much; Appeared a peak at 350 nm, the peak peak value of the display of flavonoids, according to difference of absorbance to reflect how much content.

IOP Conf. Series: Earth and Environmental Science **64** (2017) 012010 doi:10.1088/1755-1315/64/1/012010

3.7. Uses the method of ultrasonic degradation of 50 mg/L, pyridine solution measure equivalent CODCrvalue of the solution

Uses the method of ultrasonic degradation of 50 mg/L pyridine solution, experimental condition is: the ultrasonic power is 300 w, ultrasonic frequency for 28 KHz, in the process of the reaction is not adjusting test of pH, temperature and other conditions, over time, and reaction results shown:

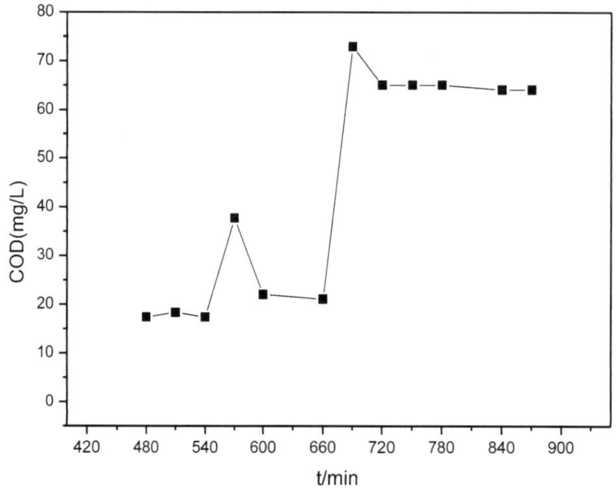

Fig.8 The method of ultrasonic degradation of 50 mg/L pyridine solution

Equivalent COD_{Cr} value

The results show that under the action of ultrasonic, over time, pyridine in the reaction of ring fracture, the generated intermediate can be potassium dichromate digestion. Therefore, in the actual containing pyridine class refractory organic wastewater, such as using potassium dichromate method is difficult to determine the inside of the organic matter to times by potassium dichromate, also it will be difficult to ensure the accuracy of the data of the measurement of COD_{Cr}, there will be a COD_{Cr} data measured values than the phenomenon of low COD_{Cr} value in actual production.

4. Condusions

Ultrasonic degradation of coal gasification by above knowable, reuse water raw water measuring COD_{Cr} value, the volatility, considering the reasons as follows:

(1)The determination of COD_{Cr} by potassium dichromate digestion spectrophotometry to measure, because of the complexity of coal gasification of water inside the type of organic matter, and the organic matter degradation characteristics, some organic matter cannot be digested with potassium dichromate, ultrasonic wave, hydroxyl radicals can make some complex organic compounds produced by the open loop or bounce the oxidation and then transformed into other oxidizable substances digestion.

(2)COD_{Cr} measurement is still behind some fluctuation, because the material is difficult to degradation degree under the effect of ultrasonic, some material is easy to ultrasonic degradation, some are difficult to ultrasonic degradation, so there will be a variety of fluctuation.

(3)Phenol, quinoline, the two substances to digest by potassium dichromate.

(4)The heterocyclic pyridine organic matter cannot be eliminated by potassium dichromate.

(5)Also explains why along with the advancement of ultrasonic processing time of coking wastewater, the determination of COD_{Cr} value volatility, started to use potassium dichromate in determination of COD_{Cr} value, the inside of the pyridine (and other organisms) equivalent of the

COD_{Cr} value has not been measured, with time, the effect of ultrasonic wave can be converted into other substances that pyridine, can be eliminated by potassium dichromate, and can measure equivalent COD_{Cr} value.

(6) in an actual containing pyridine class refractory organic wastewater, such as using potassium dichromate method is difficult to determine the inside of the organic matter to times by potassium dichromate, also it will be difficult to ensure the accuracy of the data of the measurement of COD_{Cr}, there will be a COD_{Cr} data measured values than the phenomenon of low COD_{Cr} value in actual production.

Acknowledgements

The authors acknowledge the research grant provided by Major Special Science and Technology Project of Yunnan Province for new energy (Project No. 2012ZB005).

References

[1] Meng Nan Chong, Bo Jin, Christopher W.K. Chow, Chris Saint. Recent developments in photocatalytic water treatment technology: A review [J]. Water Research, 2010, 44, 2997-3027.

[2] Hefa Cheng, Weipu Xu, Junliang Liu, Huanjun Wang, Yanqing He, Gang Chen. Preatment of wastewater from triazine manufacturing by coagulation, electrolysis, and internal microelectrolysis [J]. Journal of Hazardous Materials, 2007, 146(1-2), 385-392.

[3] Liu Lang, Cao Yan, Liu Qingcai. Kinetics studies and structure characteristics of coal char under pressurized CO2 gasification conditions [J]. Fuel, 2015, 146, 103-110.

[4] Pandey R A, Parhad N M, Kumaran P. Biological treatment of low-temperature carbonization waste-water by activated-sludge process acase-study [J]. Water Res, 1991, 25(12) : 1555-1564.

[5] Li Y M, Gu G W, Zhao I. Treatment of coke-plant wastewater by biofilm systems for removal of organic compounds and nitrogen [J]. Chemosphere, 2003, 52(6) : 997-1005.

[6] REN Yuan, WEI Chaohai, WU Chaofei. Organic compounds analysis by GC/MS during the biological fluidized bed A/O~2 process of coking wastewater [J]. College of Environmental Science and Engineering, 2006, 26(11) :1785-1791.

[7] Chen Zhengfu, Xu Sihua. Speciation analysis of polycyclic aromatic hydrocarbons in coking wastewater [J].Journal of Tongji University(Natural Science). 1995, 23(4) : 473-476.

[8] He Miao, Zhang Xiaojian, Qu Fuping. Removal characteristics of aromatic and hetercyclic organic compounds of coke plant wastewater treated by activated sludge process[J]. China Water & Wastewater, 1997, 13(1) : 14-17.

[9] Kotronarou A, Mills G, Hoffmann M R. Decomposition of parathion in aqueous solution by ultrasonic irradiation[J]. Envion.Sci.Technol,1992,26(7):1460-1462.

[10] Liang S,Can W, Min J,et a1.Achieving biodegradability enhancement and acute biotoxicity removal through the treatment of pharmaceutical wastewater using a combined internal electrolysis and ultrasonic irradiation technology[J],Frontiers of Environmental Science&Engineering in China.2011,5(3):481-487.

[11] Wang Zunbo, Sun Yuchuan, Liang Zuobing, et a1. Contamination and migration of polycyclic aromatic hydrocarbons in water of Qingmuguan Karst underground river catchment in Chongqing,China[J], Acta Scientiae Circumstantiae, 2016(3): 812-819.

[12] Mahvi A H. Application of ultrasonic technology for water and wastewater treatment.[J]. Iranian Journal of Public Health, 2009, 38(2):1-17.

[13] And I H, Hoffmann M R. Optimization of Ultrasonic Irradiation as an Advanced Oxidation Technology[J]. Environmental Science & Technology, 1997, 31(8):2237-2243.

[14] Jiewei Deng,Yunyun Yang,Xiaowei Wang,Tiangang Luan. Strategies for coupling solid-phase microextraction with mass spectrometry[J]. Trends in Analytical Chemistry . 2014,55:55-67.

[15] XiaoweiWang,RuohongChen,TiangangLuan,LiLin,ShichunZou,QingshuYang. Full automatic

determination of chlorophenols in water using solid-phase microextraction/on-fiber derivatization and gas chromatography-mass spectrometry[J]. J. Sep. Science . 2012, 35:1017-1026.

[16] Li N,Lu X,Zhang S. A novel reuse method for waste printed circuit boards as catalyst for wastewater bearing pyridine degradation[J]. Chemical Engineering Journal, 2014, 257(8):253-261.

[17] Elsayed M A. Ultrasonic removal of pyridine from wastewater: optimization of the operating conditions[J]. Applied Water Science, 2015, 5(3):221-227.

IOP Conf. Series: Earth and Environmental Science **64** (2017) 012011 doi:10.1088/1755-1315/64/1/012011

Rheological properties of soil: a review

Guangli Zhu*, Long Zhu and Chao Yu

Transportation Design Co. Ltd. Wuhai Shi-da Inner Mongolia, Wuhai, China

*Corresponding author e-mail: 262427532@qq.com

Abstract. Recently rheological methods have been applied to investigate the mechanical properties of soil micro-structure. Rheological techniques have a number of quantitative physically based measurements and offer a better understanding of how soil micro-structure behaves when subject to stress. Rheological material is refers to deformation properties similar to the solid and flow properties similar to the liquid of bound water and colloidal substances under stress. Soil rheology is divided into fluid rheology and plasticity rheology. Fluid rheology is produced by rheological material. Plasticity rheology mainly refers to the sliding and peristaltic between soil solid particles under shear stress. It is generally believed that the soft soil rheology mainly belongs to fluid rheology, while the rheology of sand and other coarse grained soil mainly belongs to plasticity rheology. Thus, rheology mechanisms of soft soil and sand are different. This paper introduces the methods of the research progress on the rheology of soil, in the soil rheological mechanism, rheological model and rheological numerical aspects of the research at home and abroad were summarized and analysed, discussed the problems existed in related research, and puts forward some suggestions for the future study on the rheology of soil.

1. Introduction

Rheology is the deformation characteristic of the time related to the deformation of the object. The soil has rheological properties, and its common rheological phenomena mainly include creep, relaxation, flow, strain rate effect and long-term strength effect. It is usually used to study the rheological phenomena, to reveal the rheological law and to establish the corresponding theory and calculation method. In recent years, due to the need of engineering construction, the calculation of deformation of soil structure or foundation and strength analysis of the increasingly high demand, must consider the rheology of soil, the soil rheological research becomes one of the hot issues in recent years in the field of geotechnical engineering. The rheological properties of the soil are studied first from the beginning of the experiment, the related research is a lot, its summary or review is also more. On the contrary, the research methods and the constitutive models of the rheological properties of the soil are relatively few, and the analysis of the rheological mechanism is helpful to the understanding and in-depth study of the rheological model. Therefore, this paper mainly analyzes the rheological mechanism of soil, and summarizes the research methods of soil rheology, the rheological model of soil, the solving method and so on.

Because the soil rheology exists, which must be given adequate attention in the design and construction of the geotechnical engineering projects, many projects because of the security risks caused by the rheology of soil also exist in the world. As Japan's first runway of Kansai International

IOP Conf. Series: Earth and Environmental Science **64** (2017) 012011 doi:10.1088/1755-1315/64/1/012011

Airport was built in saturated soft soil, started in 1986, completed in 1990, the design of settlement prediction 5.7-7.5m, completion settlement of 8.1m, 1991 to predict the final settlement of 11m, by December 2006, the settlement has reached 12.5m, still continue to produce serious cracks leads to surface subsidence, ground settlement, and maintenance. The cost greatly exceeds the budget, the timeliness of the deformation and foundation of; Shanghai industrial exhibition hall in Shanghai on the soft soil foundation, started in May 1954, completed in June 1957, and on the surface of the central hall of the average settlement has exceeded 160cm, indoor have serious cracks, deformation and rheology of the soil surface. Therefore, the aging deformation characteristics of soil is one of the key research topics in geotechnical engineering.

As early as 1940, Taylor and Merchant took the lead in the consolidation analysis to consider the rheological properties of the soil, the first to have a systematic study of the country is Holland, and the researchers are scholars Geuze. At the time of Holland, some large engineering projects such as the Vlaggeman Bridge, Zuiderzee for seawall rheodestruction, since scientists began to pay close attention to and studied. During the great development of soil rheology can be traced back to 1953, was held in the Third International Conference on soil mechanics and Foundation Engineering Conference (ICSMFE), many scientists of soil rheology problems were reported, since then, the rheology of soil has become a new development direction. Since then, more and more reports about the rheology of soil have become more and more. With the development of science and technology and the continuous increase of the scale of practice, the rheology of soil has been listed as an important scientific research project in many countries such as Holland, Portugal, Japan and so on.

2. Research methods of rheology

The rheological properties of soil can be studied from the view of micro, meso and macro performance. On the micro level, mainly through the intrinsic link between the microscopic characteristics and spatial distribution of contact electron microscope and X ray diffraction instrument of soil particles and aggregates, stacked dimer and micropore micro structure unit connection characteristics and rheological properties of soil. The domestic and foreign scholars have conducted a lot of research work, Chen Zongji put forward the frame structure theory, Shi Bin established a micro mechanical model for simulating viscous soil creep, Mitchell J K, based on the theory of rate processes of the clay stress strain time relation are studied, such as Bazant are used in theory to establish normal transversely isotropic consolidated soil and clay visco elasto plastic constitutive equation. However, due to the complexity and quantitative testing technology of microstructure of clayey soil the limit, plus some physical and chemical properties are difficult to determine, the only micro structure from theory research to make qualitative analysis, it is very difficult to be used in engineering practice.

At the mesoscopic level, the research of soil pore water, types, mineral particles and water adsorption combined with rheological properties of meso structure parameters and soil correspondence and correlation analysis of the quantitative relationship between material factors, rheology of soil physical mechanism, mechanical behavior and rheological properties of the control factors and the parameters, the present study mainly concentrated in the pore water and the bound water, clay mineral types and effects of colloidal substances in the pores on the rheological properties of soil.

In addition, with the rapid development of our country urbanization, the state of infrastructure construction and development of rich marine resources seriously, the coastal towns and port construction land resources are becoming increasingly scarce and precious, at the same time, from a practical sense, the full development of resources and comprehensive utilization, is particularly important to solve and to alleviate Chinese contradiction is much less. In the construction and maintenance of these projects, the cost and environmental pressure mainly include a large amount of silt and dredging soil. Therefore, in order to protect the environment and resources, and maintain the sustainable development of the economy, a lot of silt can be used to engineering construction and maintenance in the process of land reclamation, the benefits of which can alleviate the lack of fill material in coastal areas and can solve a lot of dredging mud way, so it will have a huge environmental and social benefit. Due to the reclamation silt soil and coastal soft soil with soil fine particles, high water content, high void ratio, low strength, low permeability, consolidation slow, big deformation of engineering properties, so the key techniques of dredger reclamation and coastal soft soil is to overcome the technical problem of the following: how to make the soft soil foundation deformation can be in the water content, consolidation and stability, they can meet the requirement of the superstructure on the bearing capacity of soft soil foundation and long-term settlement.

In the macroscopic aspect, the soil is regarded as a homogeneous continuum, which does not consider the physical foundation of the soil rheology and the factors of the fine structure of the soil. In the rheological model of the structure, the irreversible thermodynamics and continuum mechanics are used, and the rheological test results are used to estimate the rheological deformation of the soil. The characteristics of this kind of research are as follows: in the simulation of the macroscopic phenomena of soil mass, the research on this level is more emphasis on the former.

The applicability of rheological measurements as a tool for studying the development of soil structure on the micro-scale is well described in the literature. The fundamental investigations of Mezger, Markgraf et al. and Markgraf and Horn demonstrated the sensitivity of the rheological approach to differ-entiate between the rigidity and the strengthening effects of various substrates. They showed that micro-structural stability is influenced by clay mineralogy and texture, which lead to different shear behavior. Rheological parameters allow a detailed semi-quantitative classification of stiffness degradation, when micromechanical effects on the particle-to-particle scale need to be described. Barré and Hallett investigated the rheological characteristics of wet soils and in particular the effects of soil root, fungal exudates, and clay mineralogy on the stabilization of soil micro-structure. They found that analogs of biological exudates increased viscosity by up to 10-fold in some soils dominated by 2:1 clay and at saturated water contents, but the relative effect diminished as soils dried. The stabilizing effect of biological exudates also decreased with mechan-ical damage.

3. Rheology Mechanism of soil

Is that the rheological mechanism of soil in skeleton stress (effective stress) under the action of soil particles adsorbed on the surface of the water (gas) with viscosity, so that the rearrangement of particles and the skeleton of the dislocation with delay time effect, soil deformation, deformation and time; on the other hand the soil deformation by boundary this constraint, constraint from the creep deformation trend of development, therefore, the soil stress will gradually adjust, namely the stress change with time. In early 1925, people began to realize that we must establish a true and can reflect the stress-strain relationship of the formula or model clay or other materials fully, according to the micro mechanical deformation on Microstructure on the level of the physical process are described. The concept of physical concepts of clay or other material creep should be belong to the "activation energy".

Bazant and other application rate process theory, a series of constitutive model of soil is established. Shi Bin studied the theory of rate processes in clayey soil creep simulation, introduces the derivation of rate process theory and the corresponding creep model of cohesive soil, the selection of Huaiyin three different types of clay creep test were carried out on, were prepared by the sample structure anisotropic and isotropic, obtained the corresponding results of creep then, the theoretical model of rate process to fit the test results. The results show that the creep process and creep model to trend rate process theory based can effectively simulate different clay and deformation characteristics, reflect the essence of the cohesive soil, soil and soil mechanics closely together. The effects of water, mineral types and organic matter on the rheological properties of soft clay were studied by Gu Renguo and the real light. The direct shear creep apparatus carried out a series of tests on a group of modified clay samples, the test results show that the organic matter and mineral composition and its content is an important factor affecting the rheological properties of soft clay, the viscous coefficient of soil decreases with increasing the content of clay minerals, which decreases with the increase of deformation resistance of organic matter and clay the mineral content, and the influence of organic matter on the rheological properties of soft soil is more obvious. It is concluded that the clay mineral water by combining the rheological properties of soil particles, the strong bound water is the main factor affecting creep of soft soil, and the weakly bound water is a relatively minor factor. The experimental results can help to further understand the cause of soil rheology, and it is helpful to improve and perfect the existing calculation theory of rheological deformation.

4. Rheological model of soil

Soil rheological model is the soil in the process of mechanical stress and strain - the mathematical description of the time.Just as the soil constitutive relationship plays important role in soil mechanics, rheological constitutive relation of soil is the key to study the rheological properties of soil.In long-term research, people have put forward many kinds of rheological constitutive model of geotechnical material is suitable for analysis of existing models, the rheological model of induction for component model, the yield surface model, damage the rheological model and empirical model.

In engineering practice, the rheological phenomenon of rock soil including creep, stress relaxation and long-term strength reduction, in geotechnical engineering, time must be taken into consideration, in order to ensure the long-term safety of engineering facilities, rheology of rock and soil has been paid more and more attention. On the whole, the study of rheological phenomena is divided into two aspects: microscopic and macroscopic. The former is from the micro structure of rock and soil of rock and soil has rheological properties mainly study the causes and effects of the rheological properties of rock and soil, can only make qualitative analysis on the rheology of rock and soil; the latter assumption of rock and soil heterogeneity, using physical and mechanical means more intuitive to simulate the structure of rock and soil, through the establishment of mathematical model the related formulas are derived, on the rheological properties of rock and soil and its influence on the practical engineering of qualitative research. After decades of development, the geotechnical engineering community has accumulated a large number of data of rheological model, through the comparative analysis found that the rheological model of many can be roughly divided into the following four categories: component model, yield surface model, endochronic model and empirical model.

Element model by Newton N, Hooke dashpot spring H and S nansu SWAI and other basic components reflect the properties of the material, to describe the rheological properties of rock and soil through these basic elements of "series" and "parallel", the establishment of geotechnical constitutive model should reflect the relationship of stress strain time. In component model of two elements model famous Bing ham model, Maxwell model and Kelvin model; multi element model is the mother of the mountain village correction model, tension model, Huang Xiyuan model, Burges model, generalized Maxwell model and generalized Kelvin model, in the village of Liu Bao model, model, and model and Ma Ming Chen Sofield model. In order to reflect the characteristics of three-dimensional geotechnical materials displayed, based on linear rheological model, according to the principle proposed by Perzyna, using the method of analogy, the one-dimensional soil rheology analysis is extended to 3D, and the model belongs to linear rheological model.

The yield surface rheological model is a study of the variation law of the three elements of the yield surface, the association rule and the hardening law. Olszak and Perzyna first proposed the flow model, which is based on the theory of elastic-plastic theory, the concept of the yield surface, the position and the flow stress, the viscoplastic strain represents the strain history and time and other factors. The flow surface model is usually a hat type model.

Based on the concept of equivalent time and the flow surface theory, Yin, Graham, Zhu Jun advanced a three-dimensional elastic visco plastic (EVP 3D) model. The model assumes that the soil meets the small strain and strain rate can be divided into elastic strain and plastic strain rate of the two part, the elastic strain rate obtained from the generalized Hooke's law, the plastic by the associated flow rule to determine the rate of strain at the same time, it is assumed that the plastic potential function is equal to the current Q function F, to introduce the concept of equivalent time the equivalent time to control the rate of development of visco plastic deformation. The form of the model is simple, the model parameters are small, and the parameters of each model can be determined by the conventional test.

The damage rheological model is according to the basic theory of damage mechanics, the damage considering the mechanical parameters of soil with time equivalent strain, the damage evolution equation is established, and the damage variable is introduced into the constitutive model to establish the rheological constitutive model. For different material properties, stress state, as well as different research methods, research purposes, scholars put forward a lot of damage to the rheological model of rock and soil materials. According to the different research methods of material damage variable, these models can be divided into macro damage model, micro damage model and statistical damage model.

For the empirical constitutive model of rock and soil rheology, it is generally based on the actual engineering test or the laboratory test data to give the function expression of the rheological equation. The experience constitutive model can be divided into empirical function type function relationship between stress and strain or stress strain relationship experience function rate (i.e. the rate type constitutive relation of the letter). The latter is divided into two models, which appear in the constitutive relation and time implicit (including the strain rate) in the constitutive relation. The empirical constitutive relation of soil mass reflects only the external performance of the rheology, which cannot reflect the internal characteristics and mechanism of the rheology, and the difference of the general nature. But it is more intuitive and can be used directly, so loved by engineering designers.

5. Solution of rheological problem of soil mass

In the study of rheology of soil, considering time factors, problems should be solved between the rheological stress strain times than the general elastoplastic calculation of stress strain relationship is more complex, the rheological method needs to be further improved. The solution of the rheological problem of soil mass includes analytical solution and numerical solution.

The analytical solution is the use of the correspondence principle, using the integral transform technique, the specific approach is to first obtain the elastic solution, then Laplasse transform, which must use the elastic constants of the viscoelastic constitutive relation of viscoelastic constants after Laplasse transform are replaced from obtained viscoelastic Laplasse transformation solution by Laplasse inverse transform of viscoelastic final result. The analytic solution is an exact solution, which is based on the application of algebra, calculus and other mathematical principles. The analytical solution of inverse Laplasse transform only some simple and special function can be found, many problems can get a viscoelastic Laplasse transformation solution, but cannot get another analytic inverse solution, rheological problems are relatively complex, it can really get little analytical solution.

With the rapid development of the electronic computing technology, the numerical solution method of the rheological problem is developed. The application and development of numerical solution of soil rheological problem has brought a vital force to the practical application of Rheological Study of soil mass. The basic methods for numerical solutions of the main sometimes step viscous initial strain method, the viscous strain as the initial strain, calculated in each time step viscous initial strain viscous additional load caused by the balance equation is added to the time step in revising and solving, step by step, finally get the solution to the problem of soil rheology. As for the soil nonlinear rheological problem, is the only numerical solution, is generally used when the step incremental nonlinear iterative method, for each load increment, through continuous iteration, with a series of linear rheology to approximate the nonlinear rheological, the nonlinear rheological problem is simplified to linear rheological problems. The numerical solution of the soil rheological problem is mainly used in the finite element method, finite difference method, boundary element method, discrete element method, infinite element method, manifold element method, and so on. The rapid development of numerical calculation method for geotechnical engineering provides a more adequate design basis for complex geotechnical engineering problems. Numerical simulation technology not only substitute and complement the prototype test, model test play, but also to bridge the gap between laboratory test and practical engineering, showing great advantages in the geotechnical engineering nonlinear. However, rock discontinuity, heterogeneity, anisotropy, initial stress and natural boundary conditions such as used in the calculation of rheological constitutive relationship is difficult to accurately grasp the physical and mechanical parameters of rock mass determination has become a bottleneck problem solving.

6. Conclusions

The soil itself is a heterogeneous system, its rheological property is a more complex problem, research some characteristics only data fitting intuitive, understanding of its internal characteristics and the mechanism is not clear, need to be further in-depth research, to promote and improve the rheological analysis theory and method of soil. With the innovation and development of the related observation and testing technology, it is possible to study the rheological properties of the soil from the perspective of fine and even micro level. At present, the rheology of the soil is mainly focused on the soft soil, the recent development of the study of the rheological properties of sand, the results are less, the depth is not enough. Reflect the rheological constitutive model is not mature, the retaining wall displacement, the settlement of tunnel construction, the excavation process of horizontal and vertical displacement with time development and change, the slope will failure due to long-term strength decreasing with time, and this kind of engineering problems need to be solved urgently study on the rheological model and the calculation method of reasonable. Therefore, the soil rheological still need to test reveal deep rheological properties, and a comparatively perfect model and establish reasonable and reliable method of calculating and analyzing the three aspects at the same time, research, analysis results can provide accurate calculation of Engineering required.

References

[1] Anonymous, 2006. Rothamsted Research: A guide to the classical and other long-term experiments, data sets and sample archives. Lawes Agricultural Trust, Harpenden,UK. 51 pp.
[2] Avery, B.W., Catt, J.A., 1995. The soil at Rothamsted. Lawes Agricultural Trust. IACRRothamsted,Harpenden, UK.
[3] Barré, P., Hallett, P., 2009. Rheological stabilization of wet soils by model root and fungalexudates on clay mineralogy. Eur. J. Soil Sci. 60, 525–538.
[4] Blake, L., Goulding, K.W.T., Mott, C.J.B., Johnston, A.E., 1999. Changes in soil chemistry accompanying acidification over more than 100 years under woodland and grass at Rothamsted Experimental Station, UK. Eur. J. Soil Sci. 50, 401–412.
[5] O. Jenck, D. Dias, R. Kastner, Two-dimensional physical and numerical modeling of a pile-supported earth platform over soft soil, J. Geotech. Environ.Eng. (ASCE) 133 (3) (2007) 295–305.
[6] Geuze E. C. W. A.&Tan Tjong Kie., The Mechanical behavior of clays[A]. Proc. Of the 2nd international Congress on Rheology[C].Oxford. Geotechnique (1):1953.
[7] Taylor D.W.,Merchant W.. A theory of clay consolidation accounting for secondary compression[J].J.of Mathematics and Physics,1940,19(3):167–185.
[8] Abelev M. Y.&Kubtski V. L., et al.. Investigation of soils creep[A]. Proc. of the 10th ICSMFE[C].Stockholm,1981,1:523–526.
[9] [9] Bishop A. W.&Lovebury H. T., Creep characteristics of two undisturbed clays[A], Proc. of the 7th ICSMFE[C]. Mexico City,1969,1:29–37.
[10] Christie I. F.&Tonks P. M..Developments in the time lines theory of consolidation[A]. Proc. of the 11th ICSMFE[C]. San Francisco, 1985,1A8: 423–426.
[11] Christie I. F., Secondary compression effectsduring one-dimensional consolidationtests[A]. Proc. of the 6th ICSMFE[C], Montreal, 1965,1:198–202.
[12] Esu F.. Creep Characteristics of overconsolidatedjointed clay[A]. Proc. of the 9th ICSMFE[C]. Tokyo, 1977,1: 93–100.
[13] Kavazanjian E.. Time-dependent deformations in clay soils[A]. Proc. of the 11th ICSMFE[C]. San Francisco, 1985, 1A30:535–538.
[14] Murayama S., Shibàta T. ,Rheological properties of clays[J]. Proc., 5th Int. Conf., S.M.F.E., 1961,1 :269–274.

[15] Mitchell J K. Shearing Resistance of Soils as a Rate Process[J]. Journal of the Soil Mechanics and Foundations Division,ASCE, 1964,90(SM1) : 29 – 61
[16] Mitchell J K,Campanella R G,Singh A. Soil Creep as a Rate Process[J]. Journal of the Soil Mechanics and Engineering Foundations,1968: 231–253
[17] Mitchell J K,Singh A,Campanella R G. Bonding Effective Stresses and Strength of Soils[J]. Journal of the Soil Mechanics and Foundations Division,ASCE,1969,95 (SM5) : l219 – 1246
[18] Bazant Z P,Oh B H. Microplane Model for Progressive Fracture of Concrete and Rock [J]. Journal of Engineering Mechanics,ASCE,1985,3(4) : 559 – 582
[19] Bazant Z P,Ozaydin K,Krizek B J.Microplane Model for Creep of Anisotropic Clay [J]. Journal of Engineering Mechanics Division,ASCE,1975, l0l(EMl) : 57 – 78

Experimental research on coalbed gas drainage effect and economy of long directional borehole in roof

Huiming Yang[1, 2], Liangping Hu[1, 2,*]

[1]State Key Laboratory of Gas Disaster Detecting, Preventing and Emergency Controlling, Chongqing, China
[2]Chongqing Research Institute Co.,Ltd. of China Coal Technology & Engineering Group, Chongqing, China

*Corresponding author e-mail: hlp_1120@hotmail.com

Abstract. In order to study the coalbed gas drainage effect and economy of long directional roof borehole, 2 boreholes were laid out in Xinji No. 2 mine to analyze its gas drainage and investment costs comparing with high position roof borehole and high position roof roadway. The result indicates that the long directional roof borehole save investment by 44.8% and shorten the construction period by 30% , comparing with high position roof roadway for controlling gas in the working face. Investment slightly less and shorten the construction period by 47.5%, comparing with the roof high position borehole. Therefore, the method of the long directional roof borehole to drain coalbed gas in working face is the most cost-effective.

1. Introduction

The mining-induced strata fracture field is excellent flowing channel of coalbed gas drainage. It is an effective way to control the gas limit by utilizing mining-induced fracture field to carry out gas drainage and control in working face and goaf.

At present, the common gas drainage methods used in coal mine underground roof crack field include the followings: roof high position drill, roof high position roadway and long directional roof borehole. The roof high position borehole method [1] has the characteristics of well-targeted, small construction quantities and operation relatively simple. But as working face of coal mining goes on and the roof falls, high borehole will be cut off, plugged or leaked, so there exists some defects including short extraction time, reduced utilization and so on. The high position roof roadway [2] has the characteristics of large extraction capacity and high extraction concentration. But the high drainage roadway is needed constructing before working face roadway for guaranteeing drainage of working face roadway and safety mining, and the roadway construction time is relatively long, construction and safety mining operation can't be carried out at the same time, which makes the shortage of mining face and increasing the mining cycle. Directional long roof borehole [3-5] has the characteristics of large amount of extraction, short construction period, low cost of extraction and so on. In addition, through the "borehole replace roadway", it reduces the amount of roadway excavation, borehole construction, mining operation which can be carried out at the same time and ease the mining face replacement tensions. Meanwhile, because the borehole is longer than conventional borehole, the concentration of gas extraction gets bigger and time of drainage gets longer.

Recently, along with the development of drainage equipment and directional drilling technology, the new technology of gas drainage with long borehole directional drainage gradually had been popularized [6]. Zhao [3] had been successful proposed long-bore directional gas drainage technology in the test, which provided a new way for gas control of high yield and high efficiency. Yao [7] proposed the development characteristics of strata pressure relief fractures in mining roof overburden coal seam, which made the difficulty of borehole orientation and construction gradually solved. However, it still needs to conduct in-depth analysis on the effect advantages of long borehole in roof, the difficulty of the project, the cost of extraction which compared with the traditional roof borehole and high roadway. In this paper, 210100 working face of Xinji No.2 Mine was tested with long directional roof borehole, and compared with other roof gas drainage methods from the aspects of extraction effect and economic factors, so as to provide guidance for popularization and application of the technology.

2. Gas drainage technology principle of long directional roof borehole

With the advance of working face, mining goaf around balance of original rock stress is destructed, the stress distributed, roof rock caving lead to overlying strata shear slip, separated or other forms of violent rock movement. According to the "masonry beam" mechanics model proposed by academician Qian[8], the mining fracture field is divided into caving zone, roof fracture zone and bending zone in vertical direction. An annular mining fracture developing area is formed around the roof of the goaf, which is called "O" shape circle [9-10]. Under the action of mining induced stress, a large number of roof cracks develop, so that the permeability of coal seam is greatly improved, and the effect of increasing permeability and releasing pressure is obtained. The gas moving to goaf mainly come from mining coal seams, adjacent coal seams and goaf left coal, pressure released gas aggregate and move into mining fracture area, and finally gathered in the goaf roof mining fracture circle. Based on the above principle, through arranging long drainage borehole in the roof mining fracture "O" shape circle, and make the borehole always be in roof fracture zone, so orifice is always in the negative state, so we can make full use of the mining fracture channel to drainage.

3. The long directional roof borehole field test

3.1. General situation of the engineering

The test position is 210100 working face of the first mining area in Xinji No. 2 mine, coal seam Num. The coal seam 1 is majority. Working face coal seam thickness is on average 3.67m, coal seam solidity coefficient - is about 0.78 ~ 1.43, coal seam angle is between 1 ~ 10°, average of 6°, original gas content is 4.0m3/t, up to 12.46m3/t, and it is located in no outburst dangerous area. The roof is mainly sandstone, in which the old roof is fine sandstone, thickness is 15.3 ~ 24.7m, average of 21.7m, pseudo roof generally does not develop, direct roof is mainly medium sandstone, thickness is 6.7 ~8.0m, average of 7.3m.

3.2. Scheme of borehole layout

With conditions of coal seam occurrence and production technology of Xinji No. 2 mine, according to the coal seam overburden "vertical three zones" distribution theory [11-12], Xinji No. 2 mine Num.1 coal seam overburden "three vertical zones" distribution height respectively is: caving zone is 10.11m, fracture zone is 47.98m (including caving zone). Therefore, according to the height of caving zone and drainage practical research, we combined with the conditions of coal roadway layout work and then layout 2 roof long boreholes; the borehole depth was not less than 400m and 500m respectively. The specific location was shown in Fig.1~3.

Azimuth angle of opening borehole 1# was 122.45°, opening dip angle of 7°. Horizontal distance from 210100 working face return-air roadway was 12m, height from roof was 15~30m. Borehole depth was not less than 400m and terminal point through the open-off cut.

IOP Conf. Series: Earth and Environmental Science **64** (2017) 012012 doi:10.1088/1755-1315/64/1/012012

Azimuth angle of opening borehole 2# was 111.45°, opening dip angle of 8°. Horizontal distance from 210100 working face return-air roadway was 30m, height from roof was 30~40m, the main borehole terminal position arrived on the upper cut and constructed branch borehole behind 200m, the branch borehole moved towards machine roadway, the terminal borehole position distance to extracellular return-air roadway was 90m, the maximum depth of the borehole was not less than 500m, the branch borehole terminal position height from roof was 5-10m.

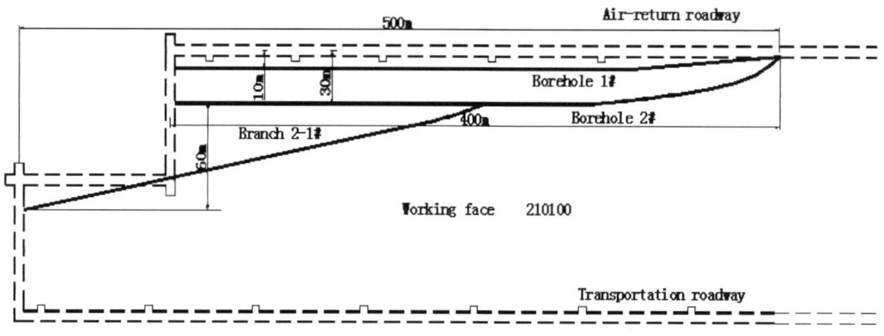

Fig.1 borehole Sketch of horizontal track projection

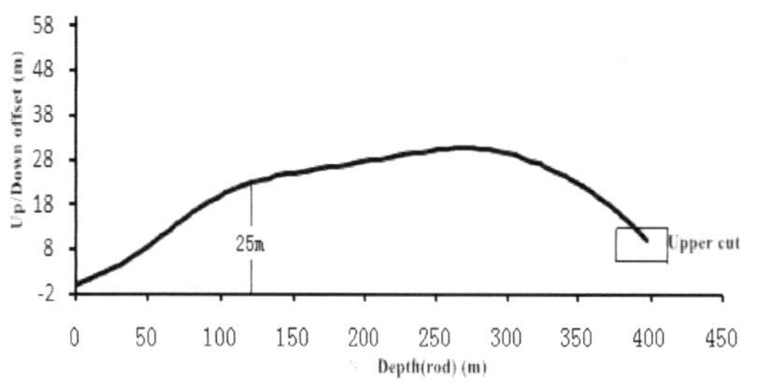

Fig.2 borehole 1# Sketch of track vertical profile

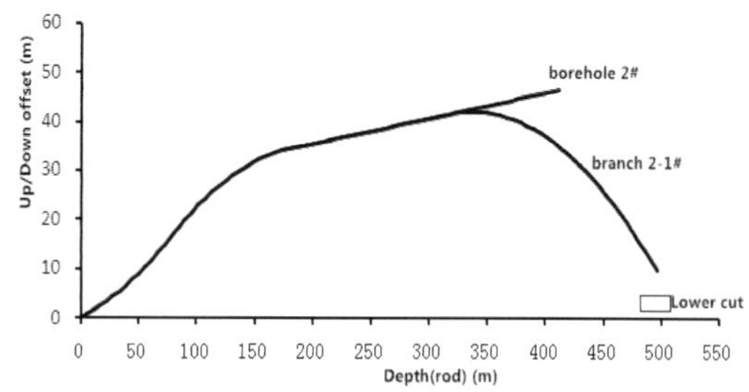

Fig.3 borehole 2#Sketch of track vertical profile

3.3. Effect of long directional roof borehole 1# and 2# gas drainage

The test time of borehole 1# was from May 7th, 2015 to June 16th, gas drainage monitoring was carried out in 210100 working face of 2 directional boreholes about 30d respectively, the borehole gas drainage effect as shown in Figure 4: during the test, from the beginning of May 21th, mixed gas drainage quantities of borehole 1# were between 9.09 ~ 10.23m3/min ,which were average of 9.6m3/min, pure mixed quantities ware between 0.17 ~ 0.69m3/min, and basically stable. The average pure quantities were 0.48m3/min, and the average extraction concentration was 5%.

Borehole 2# gas extraction data began to measure from May 10th, mixed gas drainage quantities were between 4.4 ~ 8.5m3/min, which were average of 7.84m3/min. Before May 18th, gas concentration of 2# borehole were less than 3%, pure quantities of drainage was less than 0.15m3/min. Since the beginning of May 19th, the concentration jumped to 11.7%, pure quantities jumped to 0.75 m3/min, which were increased triple, before May 19th, pure quantity was between 0.75 ~ 1.82m3/min, from the beginning of May 19th, pure quantity rose significantly. Gas concentration rose to 21.39% on May 22th, pure quantity rose to 1.82m3/min. During the test, Average pure quantity was 1.18m3/min and average drainage concentration was 15%.

Comparative analysis of drainage data indicated that the mixed quantities of borehole 1# were more than 2#borehole, however, the average gas drainage pure quantities of borehole 2# are about 2.4 times as much as borehole 1#, the average drainage concentration were 3 times as much as borehole 1#. The reason was that borehole 1# which was nearby air-return roadway leaked.

IOP Conf. Series: Earth and Environmental Science **64** (2017) 012012 doi:10.1088/1755-1315/64/1/012012

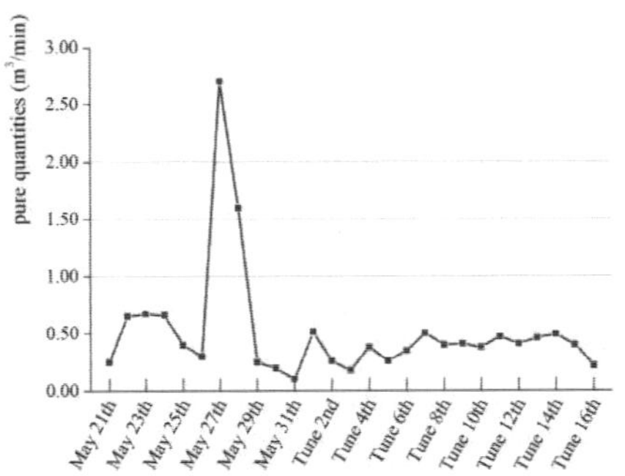

(a) Graph of borehole 1# mixed quantities

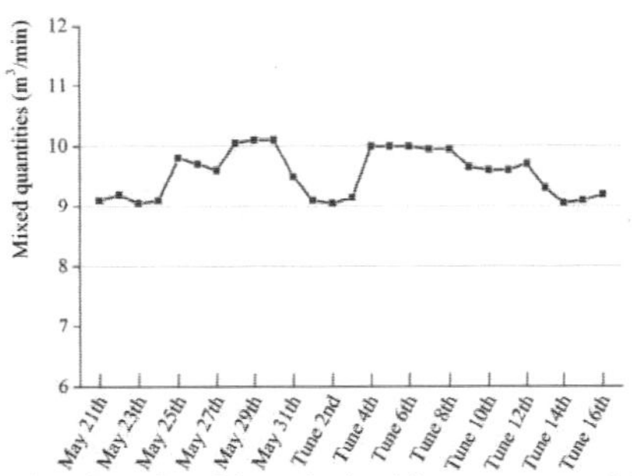

(b) Graph of borehole 1# pure quantities

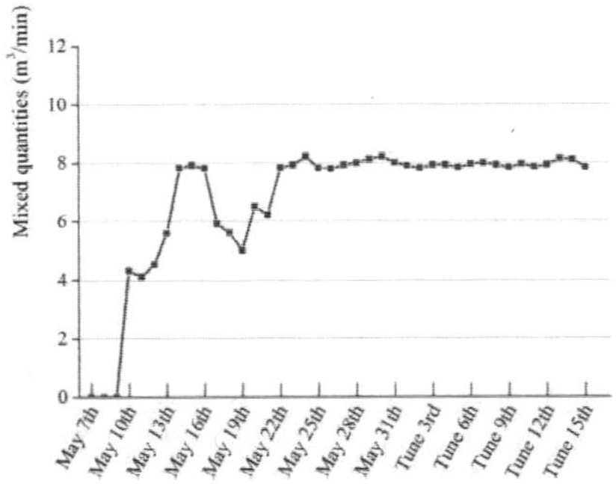

(c) Graph of borehole 2# mixed quantities

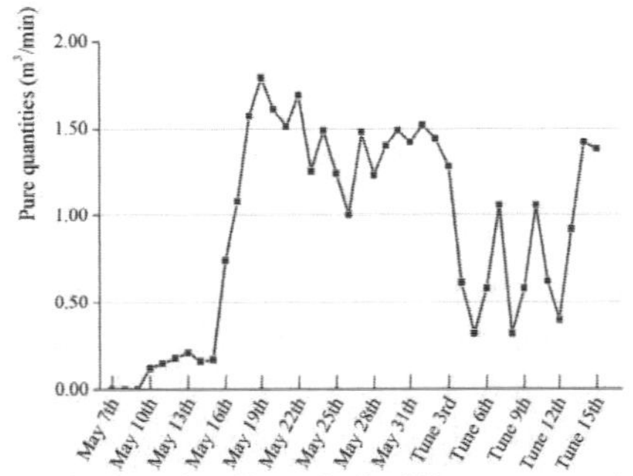

(d) Graph of borehole 2# pure quantities

Fig.4 Borehole gas drainage effect

4. Drainage comparison of long directional roof borehole, high position roof roadway and high position roof borehole

4.1. Basic situation

Compared between the same mining area roof gas drainage measures, there were mainly roof gas drainage following categories in mining area 2100: long directional roof borehole, high position roof roadway, high position roof borehole. Working face 210106, 210108 were arranged with high position roof roadway, and depth was 500m. Working face 210100, 210102, 210106 laid out high position roof borehole, a drilling field was arranged at intervals of 80m, arranged a total of 6 drilling field, each drilling field laid out 12 boreholes, and average depth of borehole were 120m. Mining face 210100 was arranged with long directional roof borehole 1#, 2#.

4.2. Comparative analysis of gas drainage effect

Roof gas drainage effect statistics of mining area 2101 are shown in figure 5. Compared with the other working face of mining area 2101, the proportion of directional borehole drainage quantities and absolute gas emission rate were all lower than working face in high position roof roadway and high position roof borehole, the main reason was the load flows of directional borehole were limited, thus if we increase the number of directional borehole and further optimize the borehole parameters to improve the drainage ability, it will be possible that the long directional roof boreholes replace high position roof roadway or high position roof borehole. The actual mixed quantities of two directional long boreholes in the test were above 7m3/min, however, considering the economic quantity, the average quantities of single borehole were 5.6m3/min. The mixed quantities of traditional high position roof borehole were 20m3/min. Therefore, to achieve the same effect, 4 directional boreholes were needed. The mixed quantities of high position roof roadway were above 30m3/min, 6 directional boreholes can meet the requirements.

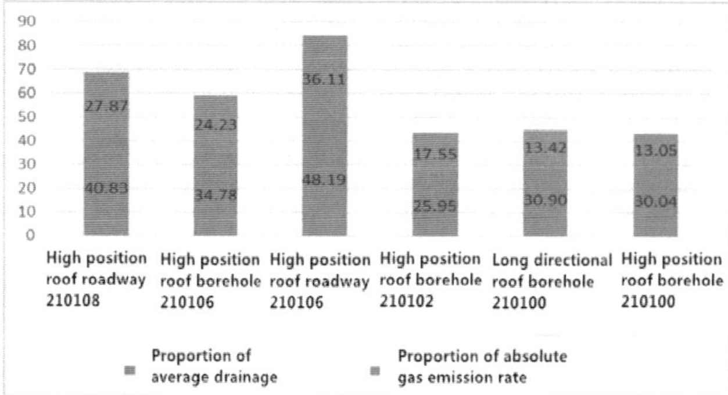

Fig.5 roof gas drainage effect chart in mining area 2101

4.3. Economic analysis

From the upper section, based on reaching the same gas drainage effect, 6 directional boreholes equal to high position roof roadway, 4 directional equal to high position roof borehole, and then analyze three kinds of drainage way of investment cost, time and human resources. The economy of these three methods were compared and analyzed. The statistical analysis of each gas drainage method is shown in Table 1. It can be seen from Table 1:

(1) In the same gas drainage effect conditions, laying out the long directional roof borehole of controlling gas which were compared to the high position roof roadway saved investment of 44.8%, shortened the construction period of 30%, and saved human resources of 28 people.

(2)In the same gas drainage effect conditions, laying out the long directional roof boreholes of controlling gas which were compared to the high position roof borehole nearly had not saved investment, but shortened the construction period of 47.5%.

In summary, laying out long directional roof borehole of controlling gas is cost-effective.

Table 1. Statistical analysis of each drainage method

Drainage measure	Unit price (*Yuan/m*)	Number of borehole or roadway	Footage (*m*)	Engineering cost (*10K Yuan*)	Engineering cost of drilling Field (*10K Yuan*)	Total cost (*10K Yuan*)	Human resources (*People*)	Time (*Month*)
high position roof roadway	8000	1	500×1	400	/	400	45	4.5
Long directional roof borehole	736	6	500×6	220.8	6.5	227.3	17	3.15
long directional roof borehole	736	4	500×4	153.7	6.5	153.7	17	2.1
high position roof borehole	163	12	120×12	140.832	15	155.832	16	4

5. Conclusions

(1) The development of roof fractures under mining-induced stress provides channels for gas flowing of coal seam, increased roof permeability and decreased pressure. Meanwhile, destressing gas concentrated in the "O" shape circle, the formation of a series of favorable conditions will promote gas drainage effect in roof.

(2) Field test result indicates that in the same gas drainage effect conditions, the long directional roof borehole saved investment by 44.8% and shortened the construction period by 30% which compared with high position roof roadway for controlling gas in the working face, and investment slightly decreased and shortened the construction period by 47.5% which compared with the high position roof borehole. The arrangement of the long directional roof borehole to control gas in working face was cost-effective; it had achieved safety and efficient production in Xinji No. 2 mine working face 210100.

(3) As a new gas drainage method, roof directional long borehole has high drainage efficiency, low investment funds, and short construction period, which compared with the traditional coal seam roof gas drainage method. It will have greater application prospects.

Acknowledgments

This work was financially supported by the National Natural Science Foundation of China (51604298), the State Key Research Development Program of China (2016YFC0801402), the Science and Technology Innovation Foundation Program of China Coal Research Institute (2016ZYMS016), and the Key Laboratory Open Foundation of Anhui Province (KLDCMERDPC14101).

References

[1] Zhai Cheng, Lin Boquan, Wu Haijin. Gas drainage with high level borehole in roof applied to gas control, Coal Engineering, 2005, (9): 4-6.

[2] Sun Rongjun, Wei Gang, Wu Zhang, Liu Jinjun, Wu Di. Experimental research on using approximate horizontal long-hole instead of closed high extracted roadway to drain gas ,Coal Geology & Exploration, 2005, 33(5):19-22.

[3] Zhao Xusheng, Sun Dongling. Comprehensive gas drainage method with long directional borehole,Coal Science and Technology, 2001, 29(3): 13-15.

[4] Yao Ginguo, Liang Minfu, Yuan Baoning, Yan Banghua. Gas drainage technology with super-long directional roof drilling, Coal Engineering, 2015, 47(3): 61-64.

[5] Zhao Xusheng, Zhou Yinhui. Experimental research on gas drainage from adjacent layers by rock horizontal long boreholes, Mining Safety and Environmental Protection, 2000,27:13-15.

[6] Yao Ningping. Development trend of nearly horizontal directional drilling technology in coal mines of China, Coal Geology & Exploration, 2008, 36(4): 78-80.

[7] Yao Shangwen. Improving method and increasing effect of gas drainage, Journal of China Coal Society, 2006, 31(6): 721-726.

[8] Qian Minggao, Miu Xiexing, He Fulian. Analysis of key block in the structure of voussoir beam in longwall mining, Journal of China Coal Society, 1994, 19(6): 557-563.

[9] [9]Qian Minggao, Xu Jialin. study on the “O-shape” circle distribution characteristics of mining-induced fractures in the overlaying strata [J]. Journal of China Coal Society, 1998, 23(5): 466-469.

[10] Xu Jialin, Qian Minggao. Study of features of distribution of overlaying stratum fractures caused by mining, Mine pressure and roof management, 1997, (4): 210-213.

[11] Qian Minggao, Liu Tingcheng. Mining pressure and strata control, China University of Mining and Technology press, Xuzhou, 2003.

[12] Liang Yunpei, Yu Bufan. Study on horizontal long borehole distribution in roof strata, Mining Safety and Environmental Protection, 2000, 27: 3-5.

IOP Conf. Series: Earth and Environmental Science **64** (2017) 012013 doi:10.1088/1755-1315/64/1/012013

Comparison between steady state method and pulse transient method for coal permeability measurement

Yongkai Qiu[1, 2,*], Zhiping Li[2] and Shiyan Hao[3]

[1] Beijing E-Hualu Information Technology Co., Ltd., Beijing, China
[2] School of Energy Resources, China University of Geosciences, Beijing, China
[3] Yanchang Oil Field Co., Ltd., Yan'an, China

*Corresponding author e-mail: qiu-yk@qq.com

Abstract. Coal permeability measurement is significant for research on reservoir characteristics and development of coalbed methane. In order to find a favourable method to obtain exact data of coal permeability, two different methods are operated to investigate coal permeability evolution. Steady state method is applied with constant upstream gas pressure and downstream pressure to keep constant pressure difference. Permeability is measured when velocity of gas flow is constant. Pulse transient method is applied with a pulse upstream pressure increment to discharge gas through coal sample to downstream gas reservoir. Permeability is calculated when the upstream reservoir and downstream reservoir reach a new equilibrium pressure. Experiments by different methods are both conducted on 2.5 cm diameter, 2.5–5 cm long cylindrical samples at same constant confining pressure and axial pressure, using both non-adsorbing (helium (He)) and strongly adsorbing (carbon dioxide (CO_2)) gases. Every single coal sample is measured by both methods, while all samples are collected from Southeast Ordos Basin in China. The results are compared to investigate the advantages and disadvantages of the two methods. Permeability measured by steady state method is slightly larger than that measured by pulse transient method, which indicates that pulse transient method is preferable for measuring low permeability measurement. Factors that may influence the results of measurement are analyzed. Steady state method may destroy the structure of coal and slightly enlarge the pore fracture, with constant pore pressure, which results in higher permeability. On the contrary, pulse transient method keeps samples close to in situ condition, which obtains more exact data. However, several improvements are needed to minimize the error of the pulse transient methods. Temperature has important influence on permeability measurement.

1. Introduction

Coal permeability measurement is significant for research on reservoir characteristics and development of coalbed methane (CBM). Gas permeability varies during the process of CBM recovery, since the stress condition would change and coal fracture-pore system would change accordingly. Gas transport in coal seams is significantly different from that of other rock types due to

IOP Conf. Series: Earth and Environmental Science **64** (2017) 012013 doi:10.1088/1755-1315/64/1/012013

gas sorption and coal swelling. The stress level, gas pressure and fracture geometry of coal are related to the processes of gas sorption, diffusion, transport, and coal swelling [1].

Significant experiments have been applied to investigate gas permeability and its evolution in coal. It is known that laboratory measured permeability of coal to sorting gases such as CH_4 and CO_2 are lower than permeability to nonsporting gases such as Helium (He) [2]. Sorbing gas permeability decreases with increasing pore pressure due to coal swelling [3-4], and increases with decreasing pore pressure due to matrix shrinkage [5-6], under constant total stress. Experiments have shown that CO_2 is adsorbed preferentially relative to CH_4 in most instances [7-8].

Several methods have been applied for measuring coal permeability, such as steady state method, pulse transient method and pore pressure oscillation method. However, results from different measurement methods of gas permeability have significantly differences. It is because structures of coals are usually compact with low permeability, while coals with large fractures have higher permeability. These differences lead to different results of the same coal, due to different applied ranges of permeability measuring methods [9]. In order to find a favorable method to obtain exact data of coal permeability, two different apparatus are operated with steady state method and pulse transient method to investigate coal permeability evolution.

2. Experimental methods

2.1. Sample Preparation

Coal samples were collected as bituminous coal from Southeast Ordos Basin in China, with depth of 400-800 meter. Samples were drilled parallel to the bedding plane. Experiments by different methods are both conducted on 2.5 cm diameter, 2.5-5 cm long cylindrical samples at same constant confining pressure and axial pressure, using both inert (helium (He)) and strongly adsorbing (carbon dioxide (CO_2)) gases.

2.2. Experimental setup

The experiments in this study are operated in two standard triaxial apparatuses for gas flow-through and permeability testing, which can also load separate confining pressure and axial pressure on samples by water or oil (Figure 1). Coal samples are sandwiched within triaxial core holders between two cylindrical stainless steel loading platens with through-going flow connections and flow distributors. The sample and axial platens are isolated from the confining fluid by a polyvinyl chloride (PVC) rubber jacket.

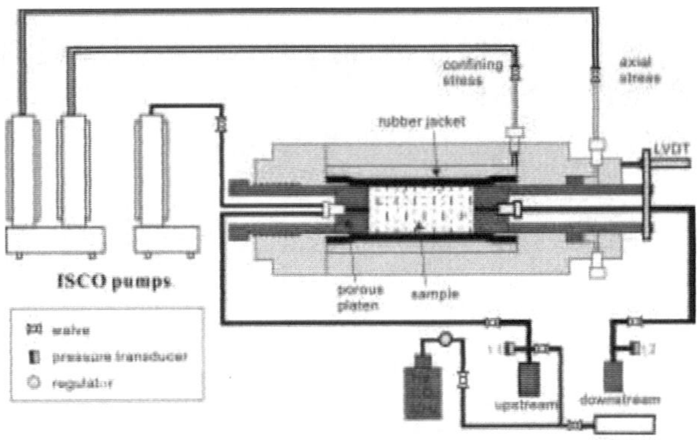

Figure 1. Permeability testing tri-axial setup.

Steady state method injects gas from a gas tank directly through the core holder, applying constant upstream pressure to keep a constant gas flow rate, with the downstream reservoir open to the atmosphere to measure gas permeability. Pulse transient method uses two low-volume stainless steel

IOP Conf. Series: Earth and Environmental Science **64** (2017) 012013 doi:10.1088/1755-1315/64/1/012013

tubes with certain volumes as upstream and downstream reservoirs.While gas reservoirs are kept in water baths to get temperature constant. The gas-pressurized upstream reservoir is discharged through the sample to the downstream reservoir with equilibration time defining permeability of the sample [10-11]. Pressure transducers measure upstream and downstream fluid pressures.

2.3. *Measurement procedure*

2.3.1. Permeability measured by steady state method. Steady state method is applied with constant upstream gas pressure and downstream pressure to keep constant pressure difference. Permeability is measured when velocity of gas flow is constant. The permeability to gas is commonly calculated with the following equation.

$$k_g = \frac{2p_o q_g \quad {}_g L}{A(p_1^2 \quad p_2^2)} \tag{1}$$

where k_g is the coal permeability to gas; p_o is the standard atmospheric pressure; q_g is the gas flow rate at standard atmospheric pressure; p_1 and p_2 are the upstream and downstream gas pressures, respectively; μ_g is the coefficient of kinetic viscosity for the gas at the mean pressure $((p_1+p_2)/2)$ at the experimental temperature; L is length of the sample; and A is the cross-sectional area of the core.

2.3.2. Permeability measured by pulse transient method. Pulse transient method is applied with a pulse upstream pressure increment to discharge gas through coal sample to downstream gas reservoir. Permeability is calculated when the upstream reservoir and downstream reservoir reach a new equilibrium pressure. The governing equation for the pressure pulse through the coal sample is written as follows.

$$P_{up}(t) \quad P_{dn}(t) = \left(P_{up}(t_o) \quad P_{dn}(t_o)\right) e^{\quad t} \tag{2}$$

$$= (kA / \quad L)\left(1 / V_{up} + 1 / V_{dn}\right) \tag{3}$$

Where $P_{up}(t)-P_{dn}(t)$ is the pressure difference between the upstream and downstream reservoirs at time t; and$(P_{up}(t_0)-P_{dn}(t_0))$ is the initial pressure difference between the upstream and downstream reservoirs at time t_0. α is the slope of the line when plotting the pressure decay $P_{up}(t) - P_{dn}(t)$ on semi-log paper against time (Figure 2). A and L are the cross sectional area and length of the sample, respectively, which define the dimensions of the sample. μ, β, V_{up}, and V_{dn} are the dynamic viscosity and compressibility of the gas, and the volume of the upstream reservoir and downstream reservoir, respectively. Permeability k is calculated from Eq. (3) where it is the only unknown.

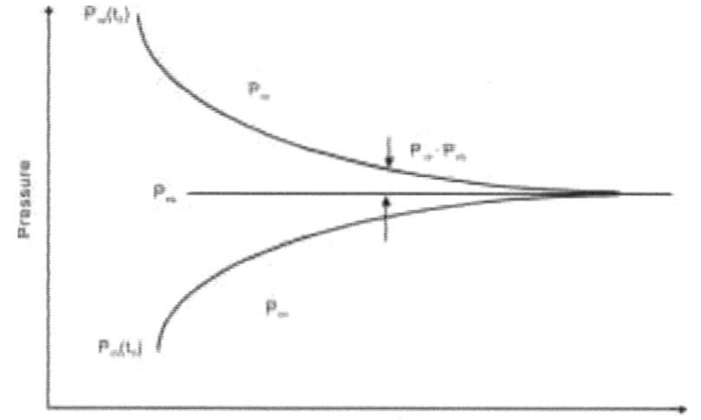

Figure 2. Typical pressure decay plot of upstream and downstream pressure vs time

3. Results

3.1. Results of permeability with two methods

Samples were applied with constant confining and axial stresses at same value by steady state method and pulse transient method. Permeability was measured at different pore pressures while using He and CO_2. Figure 3 shows He permeability evolution of Sample 1 as a function of applied pore pressure for constant confining and axial stresses at 4 MPa. Permeability increases as pore pressure increases by both of the two methods. However, it indicates that permeability measured by steady method is higher than that measured by pulse transient method for almost 3-4 times. As applied with constant upstream pressure, gas may enlarge the pore fracture, which leads gas flow through the sample faster than pulse transient pressure. Sample 1 is embedded with small fractures, due to which gas at lower pressure is hard to flow through. Thus permeability increases faster at higher pressure when it was measured by pulse transient method.

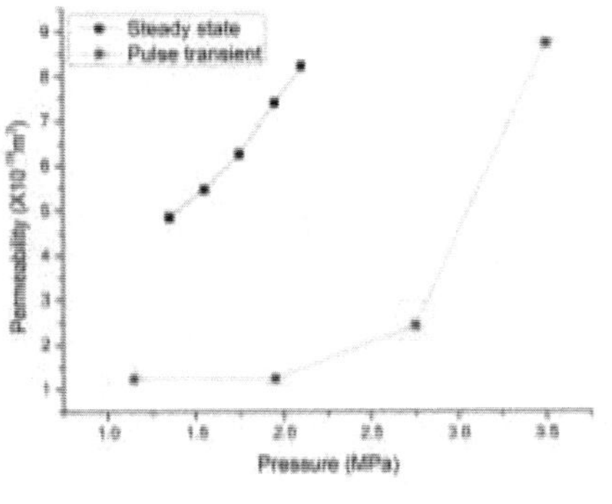

Figure 3. He permeability evolution of Sample 1 with two methods

3.2. Results of permeability with two gases

Permeability was measured with He as non-adsorbing gas and CO_2 as adsorbing gas. Fig. 4 shows permeability evolution of Sample 2 with pulse transient method. Permeability was measured as pore pressure increase with constant confining and axial stresses at 6 MPa. He permeability is higher than CO_2 permeability, as a result of CO_2 adsorption. As CO_2 will be adsorbed into pores of coal at first, the permeability decreases first, and increases when the pore pressure increases above the threshold Langmuir pressure. The trend of CO_2 permeability evolution is affected by gas sorption and coal swelling [1, 3-4].

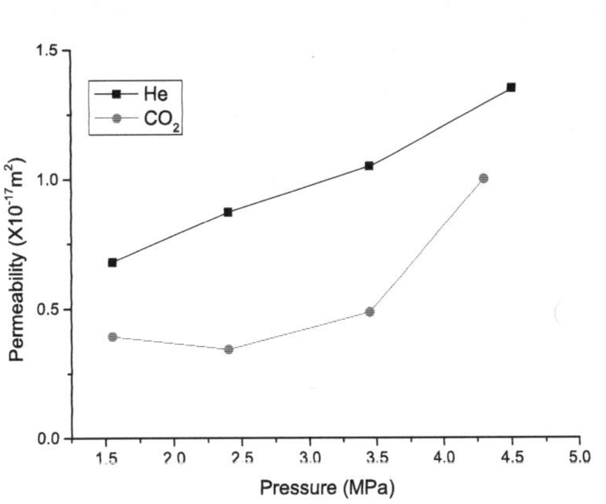

Figure 4. Permeability evolution of Sample 2 with two gases.

4. Results

Steady state method is based on constant pressure difference applied by constant upstream pressure and downstream pressure equal to atmosphere pressure. This large difference between upstream and downstream pressure may destroy the structure of coal and slightly enlarge the pore fracture, with constant pore pressure, which results in higher permeability. On the contrary, upstream and downstream pressure of pulse transient method have smaller difference range in the process of coming to equilibrium pore pressure. This method keeps samples under nearly in-situ condition, which obtains more exact data. Furthermore, temperature has an important influence on permeability measurement. Gas volume may easily change due to the high compressibility of gas, which makes steady method hard to keep gas at constant temperature. However, with pulse transient method, gas reservoirs are kept in water baths to get temperature constant, thus volume change effects in the reservoirs are minimized. Permeability measured by steady state method is slightly larger than that measured by pulse transient method, which indicates that pulse transient method is preferable for measuring low permeability measurement.

5. Conclusion

Coal permeability to gases was measured by steady state method and pulse transient method in this study. It indicates that results by steady state method are higher than that by pulse transient method. Several reasons have been discussed. Conclusions can be drawn as following.

1) Pulse transient method is preferable for measuring low permeability measurement, as it can be applied under nearly in-situ condition, while steady state method is easier to destroy coal micro fractures.

2) Pulse transient method can minimize the errors resulted from temperature, by using water bath to keep reservoirs at constant temperature.

3) Permeability to He is higher than CO_2 due to adsorption capacity of CO_2. CO_2 permeability may decreases first due to its adsorption and coal swelling, and increases when the pore pressure increases above the threshold pressure.

Acknowledgments

This work was financially supported by China Postdoctoral Science Foundation.

References

[1] S. Wang, D. Elsworth, J. Liu, Permeability evolution in fractured coal: The roles of fracture geometry and water-content, International J. Coal Geology 84 (2011) 39–48.

[2] H. Siriwardane, I. Haljasmaa, R. McLendon, G. Irdi, Y. Soong, G. Bromhal, Influence of carbon dioxide on coal permeability determined by pressure transient methods, International J. Coal Geology 77 (2009) 109–118.

[3] Z. Pan, L.D. Connell, M. Camilleri, Laboratory characterisation of coal reservoir permeability for primary and enhanced coalbed methane recovery, Int. J. Coal Geology 82(2010) 252–261.

[4] S. Wang, D. Elsworth, , J. Liu, Evolution of permeability in coal to sorbing gases, A Preliminary Study, 44th U.S. Rock Mechanics Symposium Salt Lake City, Utah, 2010, pp. 209

[5] X.J. Cui, R.M. Bustin, Volumetric strain associated with methane desorption and its impact on coalbed gas production from deep coal seams, AAPG Bulletin 89 (9)(2005) 1181–1202.

[6] S. Harpalani, G. Chen, Influence of gas production induced volumetric strain on permeability of coal, Geotechnical Geological Eng 15(1997) 303–325.

[7] S. Harpalani, B.K. Prusty, P. Dutta, Methane/CO2 sorption modeling for coalbed methane production and CO2 sequestration, Energy Fuels 20 (2006) 1591–1599.

[8] M. Mastalerz, H. Gluskoter, J. Rupp, Carbon dioxide and methane sorption in high volatile bituminous coals from Indiana, USA, Int J. Coal Geology 60 (2004) 43–55.

[9] M. Wu, X. Yang, J. Chen, The calibration of ultralow permeability measurement apparatus and

preliminary experimental results, Seismology and geology 33 (2011) 719–735.
[10] W.F. Brace, J.B. Walsh, W.T. Frangos, Permeability of granite under high pressure, J. Geophys. Res. 73 (1968) 225–2,236.
[11] P.A. Hsieh, J.V. Tracy, C.E. Neuzil, J.D. Bredehoeft, S.E. Silliman, A transient laboratory method for determining the hydraulic properties of 'tight' rocks-I. Theory, Int. J. Rock Mech. Min. Sci. Geomech. Abstr. 18 (1980) 245–252.

IOP Conf. Series: Earth and Environmental Science **64** (2017) 012014 doi:10.1088/1755-1315/64/1/012014

Stability analysis and optimum reinforcement design for an intense weathered rock slope

Kuan Qi[1], **Zhuoying Tan**[1, 2,*] **and Wen Li**

[1]School of Civil & Environmental Engineering, University of Science and Technology Beijing, Beijing, China
[2]Key Laboratory of Ministry of Education of China for Efficient Mining and Safety of Metal Mines, University of Science and Technology Beijing, Beijing, China

*Corresponding author e-mail: markzhy_tan@163.com

Abstract. In view of the complex structural characters of Chengmenshan copper mine slope, the slope stability should be analyzed and additional reinforcement measures need to be considered to ensure mining safety. In this paper, the slope model was built and its stability was analyzed by numerical simulation method under nature and dynamic loading state. After that the design of orthogonal experiment was discussed for the key factors which influence the reinforcement effect of anchors with SPSS software, and the primary and secondary relation of factors and the optimal combination were obtained using the range analysis method. Finally, the slope stability with optimal reinforcement measure was tested. The results show that the safety factor of slope under nature state is low and it is in the critical instability condition. Under dynamic loading state, the failure probability of slope increases from 0 to 18% as the seismic magnitude varies from 6 to 8. Primary and secondary sequence of factors that influence the anchor reinforcement effect is the bonding length, anchor installing angle, anchor length at 3rd bench, anchor length at 2nd bench and anchor length at 1st bench .The safety factor of slope reinforced with anchors is larger than 1.1, which could ensure the safety and stability of the slope.

1. Introduction

The main task of slope stability analysis is to calculate the safety factor, to evaluate the condition of slope at present and the probable change and development trends in the future, which can be used as the technical foundation of slope renovation engineering [1]. Slope is one of the main dangerous structures in open-pit mine, geological disasters such as landslide and collapse are often occurred, which seriously threaten the safety of workers and engineering equipments and affect the normal production of mine [2]. Therefore, effectively analyzing of the slope stability in open-pit mines and proposing scientific reinforcement techniques are the technical problems that should be urgently solved at present. At present, numerous scholars have done investigations on slope stability, and kinematics, limit equilibrium and numerical modelling techniques are commonly used methods for rock slope stability analysis [3]. Meanwhile, numerical modelling techniques are much more suitable for estimating the failure dimensions of complex rock slope [4].

Chengmenshan is a multi-metal open-pit mine. The dominated ore elements of ores are copper and sulphur. The rock physics mechanics property of the slope in the mining boundary is weak, weathering and the soil erosion are mightiness. The stability of the slope is directly affected by the fault distribution. The engineering geology group of the slope is granodiorite-porphyry [5].Because of the strong weathering, the top of the slope shows granular structure. The potential or progressive failure pattern of the slope is circular failure. Based on detail data of the engineering geological exploration, a numerical simulation model of slopes was established and the slope stability was analyzed with the numerical method under nature and dynamic loading state. According to orthogonal design, anchor reinforcement scheme was determined and the priority of each factor for the stability of open mining slope was obtained through range analysis method. Finally, the slope safety stability with optimal reinforcement measure was tested.

2. Slope model and rock mass parameter definition

The open-pit mine slope model was built based on the previous geological survey, and shown in Fig. 1. The stratums were partioned and assigned physical-mechanics parameters [5] according to Table 1.

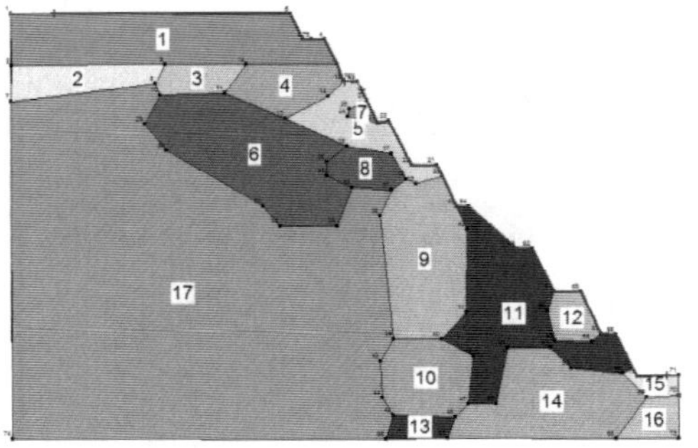

Figure 1. Slope model of open-pit mine.

Table 1. Rockmass physical-mechanics parameters of slope.

Rock region and description		Rock mass physical-mechanics parameter				
Region	Rock description	Density γ/KN/m^3	Cohesion c/KPa	Internal friction angle Φ/°	Elasticity modulus E/GPa	Poisson ratio μ
1	surface soil	27.4	80	25	0.0002	0.33
2	Sludgy loam	26.46	27	15.3	0.0002	0.33
3	dacite	24.5	239.3	15.7	18.7	0.29
4	granodiorite-porphyry (intense weathered)	24.6	316	13.1	20.7	0.27
5	limonite	37.34	241.4	13.6	26.8	0.24
6	limestone (intense weathered)	26.1	212.2	7.7	10	0.28
7	breccia	28.8	78	26.2	23.7	0.25
8	quartz-porphyry	24.5	976.6	18.4	18.7	0.29
9	skarn	33.12	836.3	14.8	32.9	0.25
10	magnetite	33.7	1002.8	19.1	31.5	0.23
11	pyrite	33.7	1002.8	19.1	31.5	0.23
12	limestone (weathered)	26.1	704.6	11.6	14.9	0.28
13	granodiorite-porphyry	24.6	969	18.2	28.7	0.27
14	quartz-porphyry (intense weathered)	24.5	325	13.2	10	0.29

3. Stability analysis of mine slope

The key factors that influence the stability of slope mainly include slope shape, geo-stress, underwater, engineering geology, earthquake and blasting vibration [6]. As drainage works have been done around the mining boundary before mining, the influence of water was not considered in the analysis of slope stability [7]. The safety conditions of slope under static and dynamic loading were carried out in this article.

3.1. Rock sample preparation

The stability analysis results of slope under static condition are shown in Table 2. And the critical sliding surface of slope is shown in Fig. 2.

Table 2. Safety factors of slope under natural condition.

	Stability analysis method			
	Ordinary	Bishop	Janbu	M-Price
Safety factor	1.001	1.031	0.97	1.038

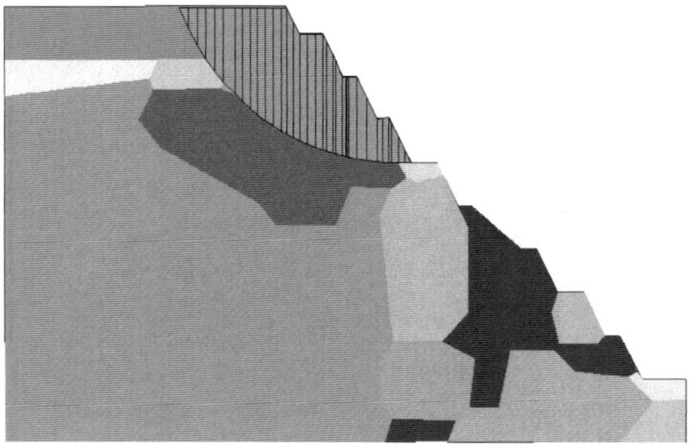

Figure 2. Critical sliding surface of slope.

The data in Table 2 show that safety factors of slope under static loading are all lower than 1.10 which is the required minimum value according to the national regulation. The critical sliding surface occurs at the upper four benches from the Fig. 2. So, reinforcement measurements should be taken to the unstable benches in order to improve its stability and ensure the safety of slope.

3.2. Stability analysis of slope under dynamic loading

The influence of dynamic loading is very important to structures which may cause serious damages or destructions [8]. So the stability of slope under dynamic loading should be analyzed and the possible damage should be evaluated.

3.2.1 Quasi-static method. By using quasi-static method, a seismic inertia force is imported into the analysis of slope stability. Due to its simplicity and applicability, quasi-static method is widely used in the stability analysis of slope, a larger amount of engineering experiences have accumulated and the method has been written into some related regulations about slope stability analysis[9]. Safety factors of slope under different seismic intensities are listed in Table 3.

IOP Conf. Series: Earth and Environmental Science **64** (2017) 012014 doi:10.1088/1755-1315/64/1/012014

Table 3. Safety factors of slope under different seismic intensities.

Safety factor		Stability analysis method			
		Ordinary	Bishop	Janbu	M-Price
Seismic intensity	6	0.920	0.947	0.877	0.958
	7	0.849	0.875	0.797	0.888
	8	0.75	0.756	0.721	0.769

According to the results of slope stability analysis using quasi-static method, safety factors of slope under different seismic intensities are all lower than 1.As the deficiencies of quasi-static method's assumptions, H. B. Seed once put it: even though the safety factor of slope is less than 1 for the moment, global instability will not always happen to the slope, and just a certain permanent deformation will occur [9].

3.2.2 Time-history method. During the period of earthquake, the minimum safety factor of slope under seismic loading only occurs at a certain moment, which is not suitable to evaluate the seismic stability of slope. The probability of occurrence that safety factor is lower than 1 during the seismic loading is considered to be a good index to evaluate the stability of slope under seismic loading [10].

The acceleration time-history curves of different seismic intensities are shown in Fig. 3, and the time-history curves of safety factors were obtained after loading seismic acceleration curves onto the slope, and shown in Fig. 4.

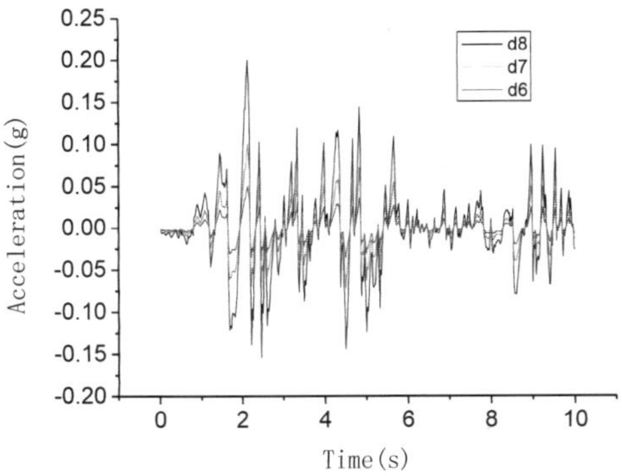

Figure 3. Acceleration time-history curves of different seismic waves.

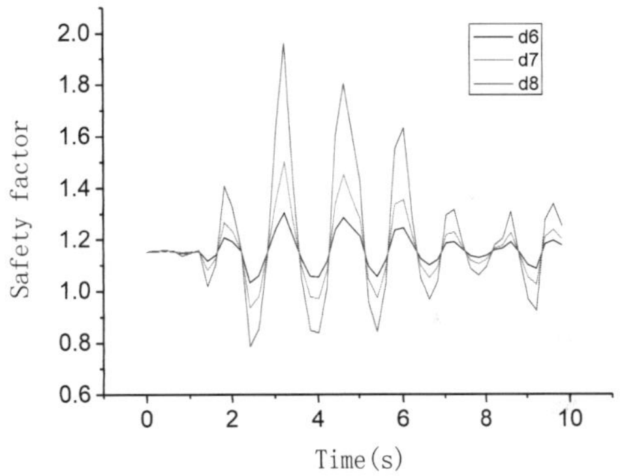

Figure 4. Safety factor time-history curves of slope under different seismic intensities.

From Fig.4 it appears that slope safety factor changes along with the changes of loading stress waveforms. The initial, minimum and maximum value of the safety factors and failure probabilities [11] are shown in Table 4.It can be seen from these data that under different seismic intensities, safety factors of slope fluctuate near the initial value and the range increases as the seismic intensity increases, meanwhile, the failure probability also increases. There is an 18% probability of slope instability when the seismic intensity is 8.

Table 4. Safety factors and failure probabilities of slope under different seismic intensities.

Seismic intensity	Safety factor			Probability of instability
	initial	minimum	maximum	
6	1.152	1.032	1.304	0
7	1.152	0.934	1.502	10%
8	1.152	0.785	2.156	18%

4. Acoustoelastic theory of rock based on continuous medium hypothesis

Based on the analysis above, some measures should be taken to reinforce the slope to improve its stability.

4.1. Optimum scheme of anchor reinforcement

According to the position of the critical sliding surface of slope (as shown in Fig.2), anchors are arranged in three benches (Fig.5), three anchors in 1stand 2nd bench and two in 3rd bench, the spacing is 5m, 3m and 5m. The parameters of anchor are as follows: elastic modulus E=201Gpa, Poisson's ratio μ=0.26, diameter d=0.1m, cohesion (tangential and normal) C=0.18GN/m, angle of internal friction φ=29°, stiffness (tangential and normal) K=1GN/m^2.

The reinforcement effect mainly depends on anchor length, length of anchorage segment and anchorage direction. Based on these factors, SPSS was used to design orthogonal experiments to calculate and analyze the slope stability [12].The results are shown in Table 5.

Table 5. Orthogonal experiment schemes and results of anchors reinforcement.

Test number	Influencing factor					Safety factor			
	Anchor length of first bench A/m	Anchor length of second bench B/m	Anchor length of third bench C/m	Length of anchorage segment D/m	Anchorage direction E/°	Ordinary	Bishop	Janbu	M-Price
1	40	35	20	8	40	1.198	1.242	1.142	1.247
2	40	40	25	10	45	1.189	1.231	1.138	1.236
3	40	45	30	12	50	1.198	1.242	1.142	1.247
4	40	50	35	14	55	1.2	1.244	1.149	1.233
5	45	35	25	12	55	1.165	1.208	1.112	1.213
6	45	40	20	14	50	1.229	1.277	1.088	1.272
7	45	45	35	8	45	1.148	1.187	1.102	1.193
8	45	50	30	10	40	1.221	1.263	1.167	1.268
9	50	35	30	14	45	1.232	1.281	1.096	1.275
10	50	40	35	12	40	1.243	1.292	1.106	1.287
11	50	45	20	10	55	1.136	1.176	1.087	1.182
12	50	50	25	8	50	1.128	1.167	1.081	1.173
13	55	35	35	10	50	1.162	1.204	1.111	1.209
14	55	40	30	8	55	1.108	1.146	1.063	1.152
15	55	45	25	14	40	1.242	1.291	1.104	1.285
16	55	50	20	12	45	1.235	1.284	1.098	1.278

Results of orthogonal experiment were analyzed by using range analysis method. Firstly, the algebraic sum of the indexes K(i), average value k(i) and range R of all factors and levels were calculated. According to the range(R), the influence order of the parameters was determined. Finally the optimal combination was determined by the principle of maximum safety factor. The results of different analysis methods are shown in Table 6-9.

Table 6. Test results of Ordinary method.

Index	Result				
	A	B	C	D	E
*K*1	4.785	4.757	4.798	4.582	4.904
*K*2	4.763	4.769	4.724	4.708	4.804
*K*3	4.739	4.724	4.759	4.841	4.717
*K*4	4.747	4.784	4.753	4.903	4.609
*k*1	1.19625	1.18925	1.1995	1.1455	1.226
*k*2	1.19075	1.19225	1.181	1.177	1.201
*k*3	1.18475	1.181	1.18975	1.21025	1.17925
*k*4	1.18675	1.196	1.18825	1.22575	1.15225
Range *R*	0.0115	0.015	0.0185	0.08025	0.07375
Primary and secondary order	D>E>C>B>A				
Optimal combination	A1B4C1D4E1				

Table 7. Test results of Bishop Method.

Index	Result				
	A	B	C	D	E
*K*1	4.959	4.935	4.979	4.742	5.088
*K*2	4.935	4.946	4.897	4.874	4.983
*K*3	4.916	4.896	4.932	5.026	4.89
*K*4	4.925	4.958	4.927	5.093	4.774
*k*1	1.23975	1.23375	1.24475	1.1855	1.272
*k*2	1.23375	1.2365	1.22425	1.2185	1.24575
*k*3	1.229	1.224	1.233	1.2565	1.2225
*k*4	1.23125	1.2395	1.23175	1.27325	1.1935
Range *R*	0.01075	0.0155	0.0205	0.08775	0.0785
Primary and secondary order	D>E>C>B>A				
Optimal combination	A1B4C1D4E1				

Table 8. Test results of Janbu method.

Index	Result				
	A	B	C	D	E
*K*1	4.571	4.461	4.415	4.388	4.519
*K*2	4.469	4.395	4.435	4.503	4.434
*K*3	4.37	4.435	4.468	4.458	4.422
*K*4	4.376	4.495	4.468	4.437	4.411
*k*1	1.14275	1.11525	1.10375	1.097	1.12975
*k*2	1.11725	1.09875	1.10875	1.12575	1.1085
*k*3	1.0925	1.10875	1.117	1.1145	1.1055
*k*4	1.094	1.12375	1.117	1.10925	1.10275
Range *R*	0.05025	0.025	0.01325	0.02875	0.027
Primary and secondary order	A>D>E>B>C				
Optimal combination	A1B4C3(4)D2E1				

Table 9. Test results of M-Price method.

Index	Result				
	A	B	C	D	E
*K*1	4.963	4.944	4.979	4.765	5.087
*K*2	4.946	4.947	4.907	4.895	4.982
*K*3	4.917	4.907	4.942	5.025	4.901
*K*4	4.924	4.952	4.922	5.065	4.78
*k*1	1.24075	1.236	1.24475	1.19125	1.27175
*k*2	1.2365	1.23675	1.22675	1.22375	1.2455
*k*3	1.22925	1.22675	1.2355	1.25625	1.22525
*k*4	1.231	1.238	1.2305	1.26625	1.195
Range *R*	0.0115	0.01125	0.018	0.075	0.07675
Primary and secondary order	E>D>C>A>B				
Optimal combination	A1B4C1D4E1				

The primary and secondary order of factors that affect the reinforcement effect obtained from the above analysis method is as follows: Ordinary method, D>E>C>B>A; Bishop Method, D>E>C>B>A; Janbu method, A>D>E>B>C; M-Price method, E>D>C>A>B. From the above results we can see that the important order differs across these methods. The results from these four analyses suggested that the most reasonable order is the bonding length, anchor installing angle, anchor length at 3rd bench, anchor length at 2nd bench, anchor length at 1st bench. Finally, A1B4C1D4E1 was taken as the optimized combination. The result is as follows: anchor length at 1st bench is 40m, anchor length at 2nd bench is 50m, anchor length at 3rd bench is 20m, bonding length is 14m and anchor installing angle is 40°.

4.2. Checking calculation of slope stability after reinforcement

The slope was reinforced by the arrangement which determined by the orthogonal experiment (Fig.5) and the stability was calculated and checked. The slope safety factors are shown in Table10 and the critical sliding surface is shown in Fig. 6.

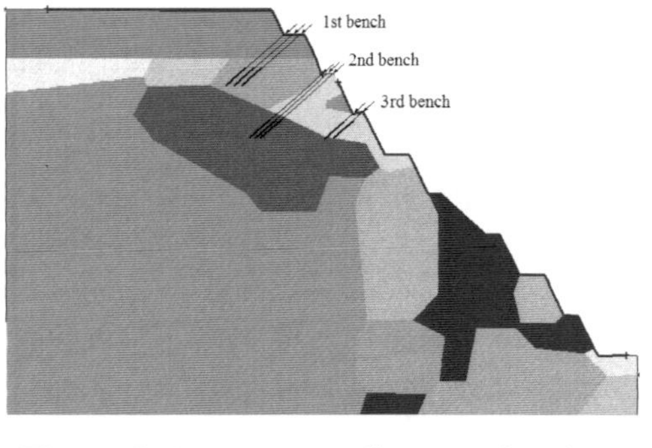

Figure 5. Arrangement diagram of anchors reinforcement.

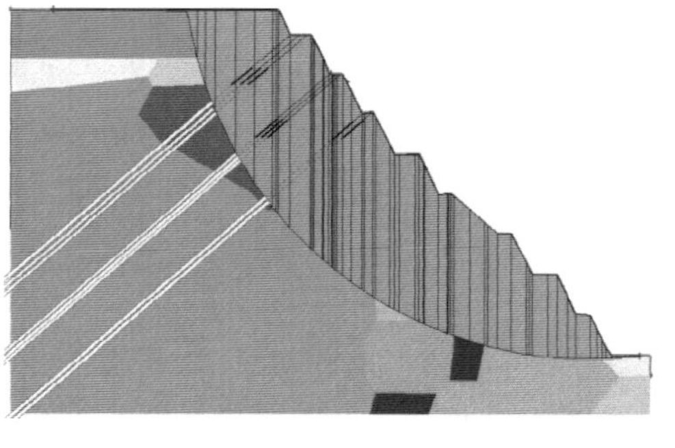

Figure 6. Critical sliding surface of slope with anchors reinforcement.

Table 10. Safety factor of slope with anchor reinforcement.

Stability analysis method	Safety factor		
	Before reinforcement	After reinforcement	Increment
Ordinary	1.001	1.242	24.1%
Bishop	1.031	1.291	25.2%
Janbu	0.97	1.104	13.8%
M-Price	1.038	1.285	23.8%

As seen from the table, the safety factors are all above 1.1 after reinforcement, which could meet the demands of security and stability. By comparison with natural state, the safety factor increases 13.8%~25.2%.Anchor reinforcement plays a decisive role in the stability of the slope.

5. Conclusions

The slope is in a state of critical instability and the safety factor is low in nature condition. Under dynamic loads, the probability of slope instability increases from 0 to 18% as the seismic intensity varies from 6 to 8.

The primary and secondary order of factors that affect the reinforcement effect is the bonding length, anchor installing angle, anchor length at 3rd bench, anchor length at 2nd bench, anchor length at 1st bench. The optimum anchor reinforcement scheme: anchor length at 1st bench is 40m, anchor length at 2nd bench is 50m, anchor length at 3rd bench is 20m, bonding length is 14m and anchor installing angle is 40°.

The safety factors are all above 1.1 after reinforcement, which could meet the demands of security and stability.

Acknowledgments

This work was financially supported by National Natural Science Foundation of China (51174013, 51574015).

References

[1] Ma Jianxun, Lai Zhisheng, CaiQing'e, et al, 3D FEM Analysis of Slope Stability Based on Strength Reduction Method, J. Chinese Journal of Rock Mechanics and Engineering. 2004, 23(16): 2690-2693.

[2] VermaDhananjai, Thareja Rahul, KaintholaAshutosh, et al, Evaluation of open pit mine slope stability analysis, J. International Journal of Earth Sciences and Engineering. 2011, 4(4): 590-600.

[3] ZulfuGurocak, SelcukAlemdag, Musharraf M. Zaman, Rock slope stability and excavatability assessment of rocks at the Kapikaya dam site, Turkey, J. Engineering Geology. 2008, 96(1): 17-27.

[4] Changqing Qi, Jimin Wu, Jin Liu, et al, Assessment of complex rock slope stability at Xiari, southwestern China, J. Bulletin of Engineering Geology and the Environment. 2016, 75(2): 537-550.

[5] Tan Zhuoying, Zhong Wen, Hu Tianshou, et al, Engineering Case Study of Weak Interfaces Detection by EH-4 for Highly Weathered Rock Slope, J. Metal Mine. 2013, (10): 84-87.

[6] ChenPeng, XuBohou, Reliability Analysis of Slope Stability Based on Factors Sensitivity, J. China Journal of Highway and Transport, 2012, 25(4): 42-48.

[7] GeWeixiang, Stability Analysis of the Lake Mud Slope of Chengmenshan Copper Mine, J. Metal Mine. 2002, (8): 24-25.

[8] TolonMert, Ural Derin N, Slope stability during earthquakes: A neural network application, J. International Review on Computers and Software. 2010, 5(4): 454-459.

[9] LiHaibo, Xiao Keqiang, Liu Yaqun, Factor of Safety Analysis of Bedding Rock Slope Under Seismic Load, J. Chinese Journal of Rock Mechanics and Engineering. 2007, 26(12): 2385-

IOP Conf. Series: Earth and Environmental Science **64** (2017) 012014 doi:10.1088/1755-1315/64/1/012014

2394.

[10] Liang Li, Yu Wang, Zijun Cao, Probabilistic slope stability analysis by risk aggregation, J. Engineering Geology. 2014, 176: 57-65.

[11] Griffiths D. V., Fenton Gordon A, Probabilistic slope stability analysis by finite elements, J. Journal of Geotechnical and Geoenvironmental Engineering. 2004, 130(5): 507-518.

[12] JiangZhaorong, Wang Lehua. The affecting factors of slope stability based on orthogonal design[C]. 4th International Conference on Manufacturing Science and Engineering, ICMSE 2013, Dalian China, 756-759.

Research Progress in Metallogenic Rules of Pb-Zn Ores in Western Guizhou, China

Tao Cui

College of Resources and Environmental Engineering, Guizhou Institute of Technology, Guiyang, 550003, China

cuitao1203@163.com

Abstract. Current situation of research on metallogenic rules of Pb-Zn ores in western Guizhou Province is summarized in combination with previous research materials. The results suggest that many scholars have explored metallogenic rules of Pb-Zin ores and achieved abundant outcomes, generally concluding that Pb-Zn deposits of Western Guizhou Province are similar to MVT and ore-forming material sources are diverse, details about metallogenic epoch, material sources and metallogenic mechanisms of Pb-Zn ores are still quite controversial and remain to be confirmed with new methods or materials through further research.

1. Introduction

Western Guizhou Province is rich in Pb-Zn ores, which have been investigated by many experts and scholars (Jin, 2006; Nie et al., 2007,2014; Zhang, 2005; Li et al.,2012), who have gained abundant outcomes in their research. Although plenty of experts and scholars have explored Pb-Zn ores of northwestern Guizhou Province, so far categories of Pb-Zn deposits are still rather controversial. In this paper, the author summarizes previous research materials to know about research progress in metallogenic rules of Pb-Zn ores in western Guizhou Province and discuss metallogenic rules of these ores in that area.

2. Geological Background of Metallogenesis

Sichuan, Yunnan and Guizhou provinces are the most potential areas of super large Pb-Zn ores and gold deposits in China (Tu et al., 2000). Located in the southwestern margin of Yangtze Para platform and in the east of the Xikang Yunnan axis, western Guizhou Pb-Zn ores are upper Yangtze met allogenic sub provinces of the met allogenic provinces on Yangtze paraplatform as integral parts of areas with polymetallic ores in Sichuang, Yunnan and Guizhou provinces (Jin, 2006). Arranged in order from the old to the new, exposed strata include Simian Dengying Formation, cambrian, silurian, devonian, carbonic, Permian, Triassic, Jurassic, tertiary and quaternary, among which Permian Emeishan Basalt is distributed most widely within the province.

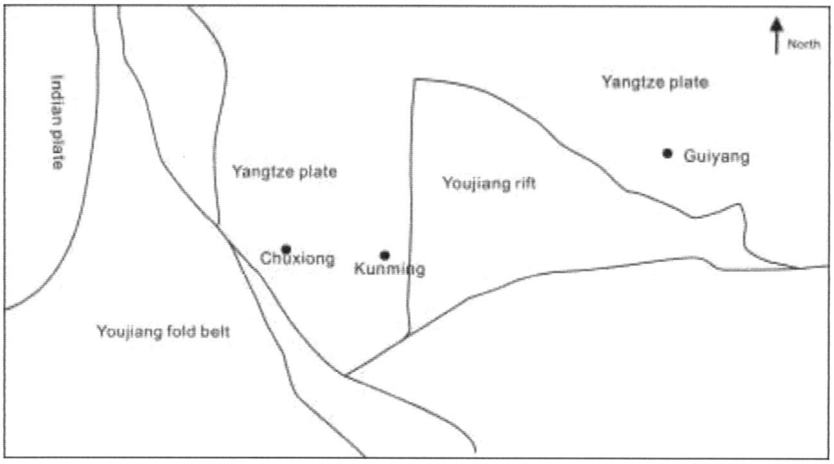

Fig.1 Geotectonic map of western Guizhou (modified from Gao, 2002)

3. Features of Ores in Pb-Zn Ores

In these ores, ores are mostly massive, vein-shaped and impregnated mainly with idiomorphic-hypidiomorphic textures, allotriomorphic granular textures, metasomatic intersertal textures, pressure textures and framboidal textures (Jin et al., 2006; Zhang, 2016). Sphalerite, galenite and pyrite are major minerals, followed by chalcopyrite, limonite, hydrozincite and malachite and so on, while the gangue minerals mainly include dolomite, calcite, barite, quartz and fluoride and so on (Zhang, 2005).

4. Metallogenic Epoch

Mineragenetic epoch of Pb-Zn ores is still controversy, concerning which there are mainly several views as follows: 1) Period of hercynian orogeny - Yanshan orogeny (Liu, 1995); 2) late permian period (Huang et al, 2001); 3) period of Yanshan orogeny (Ou, 1996; Qin, 2016; Zhang, 2016); 4) period of Himalayan oroge (Zhou, 1998); there is a lack of effective symbols for dating Pb-Zn ores, so related problems still remain to be solved although metallogenic epoch of Pb-Zn ores of northwestern Guizhou Province have been explored by numerous experts and scholars. In the future, more efforts can be made to perform related research and look for suitable authigenic minerals for dating Pb-Zn ores, or else it is impossible to identify metallogenic epoch of these ores.

5. Geochemical Characteristics

Being Mn, In, Sn and Cd deficient, trace elements are rich in Ce, Ga (content changes within a narrow scope), Pb, Ag and as (content changes within a wide range), which indicates that deposits are metallogenic at low temperature (Li et al., 2016). Pb, Zn and Ag have significant positive correlations to As, Sb and Cd in ores of all deposits, where the content of rare earth elements is much lower than ore-bearing wall rocks, which indicates that source beds are not major sources of ore-forming materials (Jin, 2006). The geochemical characteristics of Pb-Zn ores suggest that apart from lead and zinc, no other elements are found to be apparently rich in these ores. For instance, the trace elements haven't been discovered in relatively common strata rich in rare earth elements in bauxites of WZD area- Zunyi of northern Guizhou Province (Fig 2, a), areas like Kaili - Huangping in middle Guizhou Province and iron ores in northwestern Guizhou Province (Fig 2, b), which suggests that neither rare earth elements nor other elements have heavily precipitated and enriched during metallogenesis of Pb-Zn ores.

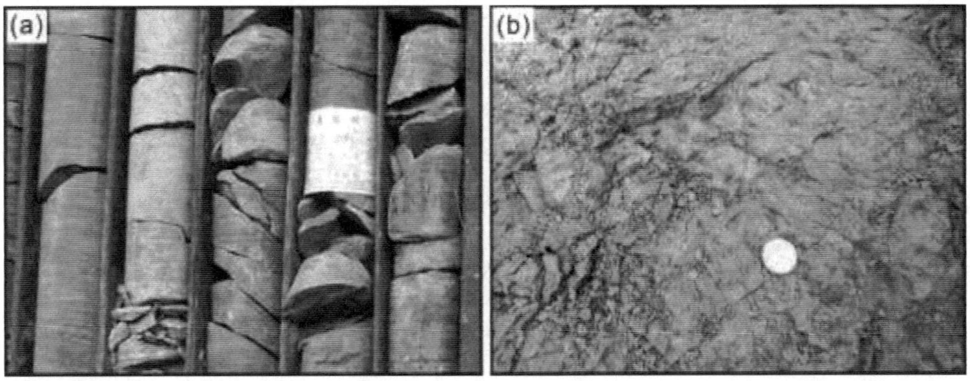

Fig.2 Bauxite layer with high content REE in northern Guizhou (a), Iron layer with high content REE in north-western Guizhou (b)

6. Types and Met allogenic Causes of Deposits

Pb-Zn deposits are mainly divided into 3 categories, including volcanogenic massive sulfide (VMS) deposits, MVT Pb-Zn deposits and SEDEX (sedimentary exhalative deposits) Pb-Zn deposits. Except three main types, there are some transitional types. Zhang (2005) considers that Pb-Zn ores of Western Guizhou Province are MVT deposits, while these ores are thought to be similar to MTV deposits by Jin (2006) who deems that they differ from each other to certain extent. Qin et al. (2016) thinks that all above three types of deposits are seen in adjacent areas of Sichuan, Yunnan and Guizhou provinces, but they are mostly MVT deposits in western Guizhou.

The metallogenic mechanism of Pb-Zn ores in this area has been studied by lots of experts and scholars, among whom Liu (1995) regards that the ores are epigenetic after sedimentation and regeneration. In other words, they are sedientary deposits, but regenerated in the later stage; Chen (1986) deems that the deposits have formed after sedimentation and in-situ regeneration, while Zhang (2016) and Peng et al.(2016) thinks that the metallogenesis of the deposits shall be attributable to hydrothermal regeneration and superimposition during diagenesis. Zhao (1995) and Chen (2012) have reported that Pb-Zn ores form through sedimentation and regeneration. Scholars such as Zhang (2005), Qin et al. (2016) and Li (2016) consider that like MVT Pb-Zn ores, these ores also belong to MVT Pb-Zn deposits. Liu (2016) deems that the Pb-Zn deposits are volcanic hotwater sedimentary deposits. Although theory of mantle plume has aroused great concern and led to intensive research on mantle plume, many scholars think that the genesis of Pb-Zn ores in northwestern Guizhou Province is related to Emei mantle plume activity (Huang, 2001; Han, 2001; Nie 2014). Based on existing research, it may be discovered that there are multiple views about metallogenic causes of deposits of Pb-Zn ores in northwestern Guizhou Province. Although their ore body features and metallogenic rules suggest that the Pb-Zn ores in western Guizhou are similar to MVT Pn-Zn deposits, the final conclusions can't be reached until intensive research, on the grounds that deposits with similar ore body features and metallogenic rules would also appear among different categories of deposits. The author considers that the metallogenesis of Pb-Zn ores in western Guizhou is attributed to multiple factors, repeated metallogenesis or alternation. The geochemical characteristics of Pb-Zn ores suggest that sedimentation and regeneration are possibly not the major metallogenic causes of these ores.

7. Conclusions

After a comprehensive summary and analysis of previous research materials, following conclusions are reached: 1) Ore features, ore body characteristics and distribution laws of Pb-Zn ores have been explored relatively clearly. 2) Met allogenic epoch and deposit type of Pb-Zn ores are still quite controversial that they remain to be investigated more deeply. 3) The metallogenesis of Pb-Zn ores is attributable to superimposed metallogenesis, whereas sedimentation wouldn't be the major metallogenetic cause of Pb-Zn ores. 4) The enrichment of associated elements like rare earth elements is not evident in Pb-Zn ores, so there is no great potential for comprehensive or independent development of these ores in view of existing research.

Acknowledgements
This work was supported by the initializing fund projects of Guizhou institute of technology (No.XJGC20131204; No.XJGC20140702), Guiyang, 550003, China, the provincial project of characteristics of REE and its transferring disciplinarian in mineralizing process of the bauxite in northern Guizhou, China (No.LH[2014]7358), and the Key support disciplines of Mineral prospecting and Exploration from Guizhou Province(ZDXK[2014]20).

References

[1] Jin Zhongguo. Research on the ore-controlling factors, metallogenic regularity and prediction of lead-zinc ore district in Northwest, Guizhuo, Central south university, 2006, 79-111.
[2] Nie Aiguo, Qin Dexian, Guan Daiyun, Huang Zhiyong, Zhang Zhuru. A research on regional metallogenic contribution to gushing Emeishan basalt magma in western of Guizhou province . Geology and prospecting, 2007, 43(2):50-54.
[3] Nie Aiguo, Kang geng. Research on the Metallogenic difference of Emeishan basalt in Guizhou. Guizhou technology press, 2014, 42-45.
[4] Zhang Changqing. Distribution, characteristics and genesis of mississippi Valley-type lead-zinc deposits in the triangle area of Sichuan-Yunnan-Guizhou Provinces. China university of Geosciences, 2005, 73-77.
[5] Li bo, Gu Xiaochun, Wen Shuming et al. Effect of Emeishan bsalt in northesat Yunan on lead and zinc mineralization, mineral resources and geology, 2012, 26(2):95-100.
[6] Tu Guangzhi. Super large deposits, Science press, 2000,350-390.
[7] Gao Zhengmin. The formation and exploration of major gold in Yunnan and Guizhou province, Geology Press, 2002:78-96.
[8] Zhang jian. Geological features and ore genesis for the Qingganglin Pb-Zn deposit in Hezhang, Guizhou, Acta Sichuan geology sinica, 2016, 36(2): 253-257.
[9] Liu Hechang. The relationship between Emeishan basalt and Pb-Zn metallogenesis, Geology and exploration, 1995, 4:1-6.
[10] Huang Zhilong, Chen Jin, Liu Chongqiang, Hu Runsheng, Li Wenbo. A preliminary discussion on the genetic relationshio between Emeishan basalt and Pb-Zn deposits as exemplified by the Huize Pb-Zn deposit, Yunnan province, Acta mineralogica sinica. 2001, 21(4):681-688.
[11] Ou Jinxiu. Geological characteristics and ore-control geological conditions of Qingashan lead-zinc deposit, Guizhou, Journal of Guilin institute of technology, 1996, 16(3): 277-282.
[12] Qin Jianhua, Liao Zhenwen, Zhu Sibao, Lai Yang. Mineralization of the carbonate-hosted Pb-Zn deposits in the Sichuan-Yunnan-Guizhou area, southwestern China, Sedimentary geology and tethyan geology, 2016,36 (1):1-13.
[13] Zhou Chaoxian. The source of mineralizing metals, geochemical characterization of ore-forming solution, and metallogenetic mechanism of Qilingchang Zn-Pb, northeastern Yunnan Province, China, Bulletin of mineralogy, Petrology and geochemistry,1998,17 (1):34-36.
[14] Li Zhenli, Ye Lin, Huang Zhilong, Nian Hongliang, Zhou Jiaxi. Primary research on trace elements in sphalerite from Tianqiao Pb-Zn deposit, northwestern Guizhou province, China, Acta mineralogical sinica, 2016, 36 (2): 183-188.
[15] Chen Shijie. Research on the genesis of lead-zinc ore-deposits in western Guizhou and northeastern Yunnan, Guizhou geology, 1986, 3:211-222.
[16] Peng song, Jin Zhonguo, Lin Guisheng et al..Analysis of ore-controlling factores and metallogenic model of Wuzhishan lead-zinc deposit, Guizhou: a case study of Nayougzhi deposit, Mineral exploration, 2016, 7 (3):463-470.
[17] Zhao Zhun. Metallogenic model of Pb-Zn deposit in eastern and northeastern Yunnan, Yunnan

geology, 1995, 14(4):350-354.
[18] Chen Suihai, Cheng Heming, Wen Deixiao. Enrichment mechanisms and implication of Ti-Cd-Ga in sulfides of Shaojiwan Pb-Zn deposit, Guizhou province, China, Acta mineralogical sinica, 2012, 32(3):425-431.
[19] Liu Shuwen, Li Rongxi, Liu Yunhua, Zeng rong. Geochemical characteristics and metallogenic mechanism of the Mayuan Pb-Zn deposit on the northern margin of Yangtze plate, Acta geoscientica sinica, 2016, 37(1):101-110.
[20] Han Runsheng, Liu Chongqiang, Huang Zhilong, Chen Jin. Genesis model of Huizhe lead-zinc ore deposit in Yunnan, Acta mineralogica sinica. 2001, 21(4): 674-680.

IOP Conf. Series: Earth and Environmental Science **64** (2017) 012016 doi:10.1088/1755-1315/64/1/012016

Supply-demand 3D dynamic model in water resources evaluation: taking Lebanon as an example

Hong Fang and Zhimin Hou*

College of science, Huazhong Agricultural University, Wuhan, China

*Corresponding author e-mail: 2000fanghong@sina. com

Abstract. In this paper, supply-demand 3D dynamic model is adopted to create a measurement of a region's capacity to provide available water to meet the needs of its population. First of all, we draw a diagram between supply and demand. Then taking the main dynamic factors into account, we establish an index to evaluate the balance of supply and demand. The three dimension vector reflects the scarcity of industrial, agricultural and residential water. Lebanon is chosen as the object of case study, and we do quantitative analysis of its current situation. After data collecting and processing, we calculate the 3D vector in 2012, which reveals that agriculture is susceptible to water scarcity. Water resources of Lebanon are "physical rich" but "economic scarcity" according to the correlation chart and other statistical analysis.

1. Introduction

Water covers 71% of the Earth's surface and it is vital for all known forms of life. Although there is a tremendous amount of water, many observers have estimated that by 2025 more than half of the world population will be facing water-based vulnerability [1]. The reason is that a reliable supply of water for future growth must be available to local populations in sufficient quantity and quality, while not compromising local ecosystems. Because of increasing population and climate change, many countries are exposing to the risk of severe water scarcity and its subsequent challenges. According to Human Development Report, more than one out of every six people in the world is water stressed, meaning that they do not have access to potable water. Thus, it is of great significance to find a way to alleviate this water scarcity problem and reestablish a better world for today's generation.

Climate change, urbanization and water pollution cause adverse effects on water resources and rehabilitation costs may exceed the carrying capacity of countries or regions. A standardized indicator framework is necessary to assess water resources. Water Stress Index is an advanced concept (ranging from 0 to 1), serving as a characterization factor for a suggested midpoint category "water deprivation" in LCIA [2]. Afterwards, some other researchers gave a critical review of the City Blueprint Approach regarding the assessment of the sustainability of water resources [3].

2. Supply-demand 3D Dynamic Model

To simplify the problem, we make three basic assumptions, each of which is properly justified. At first, during the study periods, the population growth and water consumption per capita are independent from each other. The population growth follows its historical trend: individual's water consumption does not change because of population change. Water consumption per capita is affected by other dynamic factors which are explained in the following part. The next one is that the dispersal of pollution is not influenced by wind speed or water velocity. The data about these natural factors are unavailable, so we don't consider them in our model to simplify the complex problem. At last, a country's development goal is try to make balance between economies, ecology, and security and ease the water stress. A country needs to project a sustainable development plan and reestablish a better world for today's generation and next generations.

2.1. A Diagram of Water

Before analysis, we create a diagram of the main factors that influence the total demand of water and total supply of water.

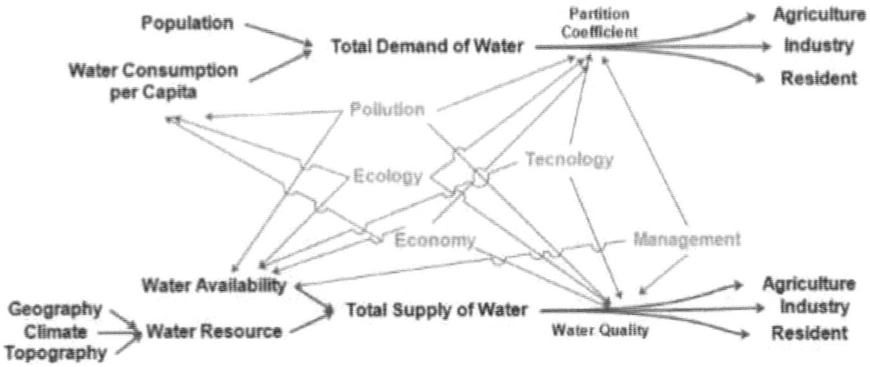

Figure 1. Supply and demand diagram of water.

In the diagram, the main factors are pollution, ecology, economy, technology and management. And all these macro factors are decomposed to micro factors. Lines link two factors which have relationship.

For the sake of measuring a region's capacity to provide clean water to meet the needs of its population, we consider the demand and supply factors respectively. Then we define and establish a Supply-Demand Index (SDI) which reflects the balance level between water supply and water demand. Through this index, we will judge whether a region is suffering from water scarcity or not.

2.2. Water Supply

To measure the sustainable water supply, we need to take both physical factors and socio-economic factors into account. Our water supply model is shown as below:

$$TS(t) = WR(w_t^1, w_t^2, \cdots, w_t^n) \cdot \eta(t),$$

where TS(t) is the total supply of water, WR is the quantity of water resources, $w_t^1, w_t^2, \cdots, w_t^n$ are different types of water resources: surface water, ground water, seawater desalination, etc. $\eta(t)$ is the conversion coefficient from water resources to available water.

Then, we allocate the total supply to three applied aspects: industrial water, agricultural water and residential water:

$$[S_t^i \; S_t^\alpha \; S_t^\gamma] = TS(t) \cdot [\sigma_t^i \; \sigma_t^\alpha \; \sigma_t^\gamma],$$

where $[S_t^i \; S_t^\alpha \; S_t^\gamma]$ is a matrix revealing the water supply for industrial, agricultural and residential purposes, $[\sigma_t^i \; \sigma_t^\alpha \; \sigma_t^\gamma]$ is the transfer matrix from total supply to three sectors.

2.3. Water Demand

As for water demand, we consider two factors: population and water consumption per capita.

About population, logistic model is a common model to simulate population growth [4]. This model is based on the assumption that there is a maximum of population named Environmental Capacity, constraint by environmental conditions. When the population reaches Environmental Capacity, it won't increase any more. Given the hypothesis that environmental conditions have negative impacts on the population, the impact is scale up as the population density. The logistic model is shown as following:

$$\begin{cases} \frac{dN(t)}{dt} = [\alpha - \beta N(t)] \cdot N(t) \\ N(t_0) = N_0 \end{cases}, \quad (1)$$

where $P(t) = \alpha - \beta N(t)$ is the net growth rate of population, α is the natural growth rate of population, β is the coefficient of life, $N(t)$ is the population at time t, and $N(t_0)$ is the population at initial time t_0.

Equation (1) is a Bernoulli Equation. However, parameter α and β are difficult to determine as they may influence the accuracy of predicted results. For this reason, we introduce a net growth rate model:

$$P(t) = at + b \quad (ab < 0).$$

In this case, solving the equation, we get:

$$N(t) = N_0 \cdot \exp[-(bt_0 + at_0^2 / 2)] \cdot \exp(bt + at^2 / 2).$$

Then, we begin to consider the water consumption per capita. A person's demand of water depends on many dynamic factors. According to previous studies [5, 6], water price and pricing manner can affect people's water consumption. Water saving technology and apparatus also greatly influence the amount of water demand per person. In addition, personal life style, social culture and some other elements do have relationships with people's demand of water.

Considering the features of these factors, we divide them into three kinds: trend-cycle factors, seasonal factors and random factors. Referring to Park's method [7], water consumption per capita can be decomposed as follows:

$$Y(t) = TC(t) \cdot [S(t) + R(t) - 1],$$

Where $Y(t)$ the water consumption is per capita, $TC(t)$ is trend-cycle factors, $S(t)$ is seasonal factors, and $R(t)$ is random factors.

Accordingly, we can calculate the total demand of water:

$$TD(t) = N(t) \cdot Y(t).$$

Also, we allocate the total demand to three applied aspects: industrial water, agricultural water and residential water:

$$[D_t^i \ D_t^\alpha \ D_t^\gamma] = TD(t) \cdot [\rho_t^i \ \rho_t^\alpha \ \rho_t^\gamma],$$

IOP Conf. Series: Earth and Environmental Science **64** (2017) 012016 doi:10.1088/1755-1315/64/1/012016

Where $[D_t^i\ D_t^\alpha\ D_t^\gamma]$ is a matrix revealing the water demand for industrial, agricultural and residential purposes? And $[\rho_t^i\ \rho_t^\alpha\ \rho_t^\gamma]$ is the transfer matrix from total demand to three sectors, it relied on the proportion of a region's demand for three different fields.

2.4. Supply-Demand 3D Dynamic Model

In this part, we measure a region's ability of providing available water to satisfy its citizens' demand. The Supply-Demand Index (SDI) can be calculated:

$$(SDI_t^i\ SDI_t^\alpha\ SDI_t^\gamma) = (\frac{S_t^i}{D_t^i}\frac{S_t^\alpha}{D_t^\alpha}\frac{S_t^\gamma}{D_t^\gamma}),$$

Where SDI_t^i, SDI_t^α and SDI_t^γ are Supply-Demand Index for industrial water agricultural water and residential water? The range of each index is (0, 1).

Therefore, the three dimension vector $(SDI_t^i\ SDI_t^\alpha\ SDI_t^\gamma)$ reflects the balance between supply and demand. We select some countries which have different levels of water scarcity to test our 3D model, and then get reasonable ranges for each level.

Table 1. Range of index.

range	0~0. 5	0. 5~0. 8	0. 8~1. 0
type	safe	warning	severe

3. A Case Study on Lebanon

Lebanon is a country in Western Asia. It is bordered by Syria to the north and east and Israel to the south. Lebanon's location at the crossroads of the Mediterranean Basin and the Arabian hinterland facilitated its rich history and shaped a cultural identity of religious and ethnic diversity.

Lebanon is referred to as the source of water in the Middle East. Its water resources are far more than its neighbors Israel and Syria. As estimated, it can obtain 4. 3 billion cubic meters of water per year,

Among which 0. 51 billion cubic meters flows to the north and 0. 44 billion cubic meters flows underground to Syria and Israel, while 0. 75 billion cubic meters inflows the sea.

3.1. Data Sources and Processing

We collect data from World Bank, NOAA Central Library, WHO and Lebanon Statistical Yearbook. The data are about Lebanon's climate, hydrology and social economy. Then we delete the indicators which are seriously lack (over 20%) and do data interpolation with the rest of indicators, in order to make data available for our study and ensure the accuracy.

3.2. Water Scarcity Analysis

Adopting supply-demand dynamic 3D model, we can calculate the 3D vector of Lebanon in 2012:

$$(SDI_t^i\ SDI_t^\alpha\ SDI_t^\gamma) = (0.51, 0.89, 0.65).$$

The result is intuitively shown in Figure 2.

IOP Conf. Series: Earth and Environmental Science **64** (2017) 012016 doi:10.1088/1755-1315/64/1/012016

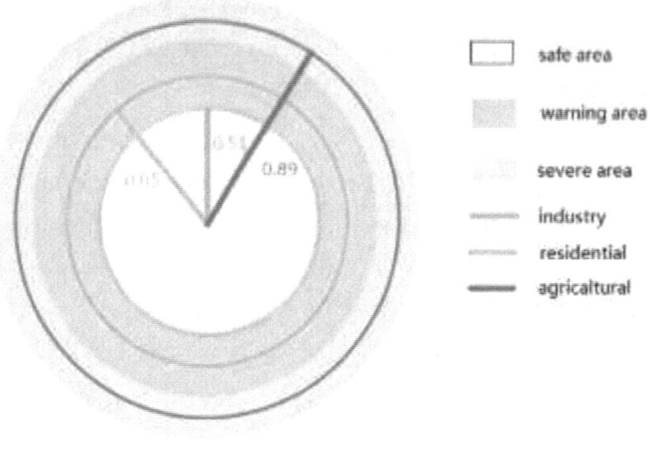

Figure 2. Result of 3D model in Lebanon.

As is shown in this figure, agricultural water is in the sever area (0. 8~1. 0); residential water is in the warning area (0. 5~0. 8); industrial water is in the safe area (0~0. 5).

Thus, we know that agricultural water in Lebanon is susceptible to water scarcity. However, Lebanon's agricultural economy occupies an important position in the domestic economy, which has a great influence on Levant area economy. Agriculture's contribution to GDP is 7%, and about 15% percent of the population is engaging in farm production. As for residential water, it's also significant for people's every life. So it's high time to change the current situation of water scarcity, especially for the need of agricultural and residential purposes.

We choose both environmental and socio-economic factors to analysis their correlation with the level of water scarcity. The specific indicators are shown in Table 2.

Table 2. Indicators table.

water scarcity level	environmental factors	socio-econmic factors
industrial water(SDI1)	annual rainfall(V1)	technology investment(V4)
agricultural water(SDI2)	CO_2 emissions(V2)	population density(V5)
residential water(SDI3)	temperature(V3)	CPI(V6)
		GDP per capita(V7)
		education investment(V8)

Calculate the Pearson correlation coefficient between two factors, and the results is in Figure 3. Obviously, both the environmental factors and socio-economic factors have correlation with SDI. Technology and education investment have negative correlations with SDI, which means that if Lebanon invests more on technology or education, the water scarcity can be mitigated to some extent. Others have positive correlations with water scarcity.

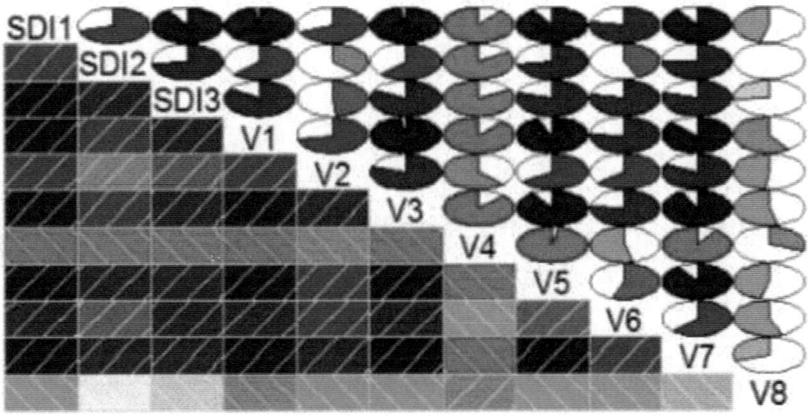

Figure 3. The correlation chart.

To interpret this graph, start with the lower triangle of cells (the cells below the principal diagonal). By default, a blue color and hashing that goes from lower left to upper right represents a positive correlation between the two variables that meet at that cell. Conversely, a red color and hashing that goes from the upper left to the lower right represents a negative correlation. The darker and more saturated the color, the greater the magnitude of the correlation. Weak correlations, near zero, will appear washed out.

4. Conclusion

In our research, we consider all the macro factors which can influence the water supply or water demand, and then divide them into micro factors. So the system is comprehensive and can describe the relationships of these factors clearly. The three dimension vector reflects the scarcity of industrial, agricultural, and residential water. Thus, we can know about the situation of each sector.

Usually, there are two primary causes for water scarcity: physical scarcity and economic scarcity. Physical scarcity is where there is inadequate water in a region to meet demand. Economic scarcity is where water exists but poor management and lack of infrastructure limits the availability of clean water. In this case, knowing about a country's main causes for water scarcity, we can find more effective ways to deal with the problem. Through our analysis, water resources of Lebanon are "Physical Rich" but "Economic Scarcity".

References

[1] Kulshreshtha, A global outlook for water resources to the year 2025, Water Resources Management. 12 (1998) 167-184.

[2] S. Pfister, A. Koehler, S. Hellweg, Assessing the environmental impacts of freshwater consumption in LCA, Environmental Science&Technology. 14(2009)4098-4104.

[3] S. H. A. Koop, C. J. Van Leeuwen, Assessment of the sustainability of water resources management:a critical review of the city Blueprint Approach, Water Resources Management. 29(2015)5649-5670.

[4] G. F. Mulligan, Logistic population growth in the world's largest cities, Geographical Analysis. 38(2006)344-370.

[5] OECD, Household energy&water consumption and waste generation, Paris, 2001.

[6] OECD, Household water pricing in OECD countries, Paris, 1999.

[7] S. O. Park, Seasonal adjustment method of employment statistics:X-12_ARIMA, Journal of Korean Official Statistics. 14(2009)40-68.

Design and mechanical analysis of a new automatic vertical drilling tool used in a slim borehole

Lixin Li[1, 2], Qilong Xue[1, 2,*], Baolin Liu[1, 2], Liudong Zhao[1, 2] and Xinyu Li[1, 2]

[1]China University of Geosciences, Beijing, China
[2]Key laboratory on Deep Geo-Drilling Technology of the Ministry of Land and Resources, Beijing, China

*Corresponding author e-mail: xqlfly@gmail.com

Abstract. A new kind of electronic-controlled automatic vertical drilling tool is developed to satisfy the requirements of slim-hole, which has a unique push-actuator to push-the-bit. This paper introduced the mechanical structure and working process of the automatic tool, and analyzed in detail one of the most critical component - the push-actuator. The analysis reveals that, with the pitch-row of studdle increasing form 20mm to 24mm, the ratio of push-out force to forward force gradually decreases from 2.04 to 0.78, and the completion time of push-out process reduces from 75.8s to 42.8. This result indicates that, using studdles with shorter pitch-rows, larger straightening force could be obtained. On the contrary, studdles with longer pitch-rows could realize more quickly sensitive response of deviation control. These unique characters make the vertical drilling being able to be adapted to different geological conditions and technological requirements, the applicability and feasibility of the drilling tool remarkably improved. Thus this vertical drilling tool owns a great promotional value for oil & gas exploration.

1. Introduction

Facing to the reality that the exploitation of traditional oil & gas is going to the end and the development of unconventional oil & gas is forwarding to more complex geological conditions, the challenge of drilling engineering is becoming more and more arduous, advanced technology and equipment is pressing needed by the requirements of modern drilling engineering. Under such a background, owning to majority of new oil & gas wells are drilled in the blocks with high-dip formations, conventional packed hole assembly and pendulum drill assembly could not meet the technological requirements of modern vertical drilling engineering, the primary problem nowadays is how to increasing penetration rate while preventing deviation. Deviation control has always been one of the hardest technical difficulties in oil & gas drilling, until the end of 21 century, the barrier of passive borehole deviation mechanism and control method had broken through by automatic vertical drilling technic [1, 2], which is originated from Germany's superdeep drilling program (KTB) [3, 4] and can obtain the maximum WOB (weight on bit) & ROP (rate of penetration) [5, 6]. Nowadays, automatic vertical drilling technic has been widely used in drilling deep & super-deep wells and complex-structure wells, making a great contribute to solve the problem of preventing deviation in high dipping blocks [7, 8].

For the past few decades, owning to low cost and high efficiency, slim-hole drilling has been rapidly developed and shown great economic benefit [9], especially for oil prices depressed today. Though more than 10 kinds of automatic vertical drilling tools, such as Power V by Schlumberger [10], Verti Trak by Baker Hughes [11] and SL-AVDS by Sinopec [12], have been developed and obtained successful commercial application at present, these vertical drilling tools always owning a relative larger diameter, which could not be satisfied for slim hole drilling. Therefore, a new automatic vertical drilling tool specifically used in slim borehole needs to be urgently designed.

For this purpose, a new kind of automatic vertical drilling tool with electronic closed loop control system is designed by China University of Geosciences. Considering the drilling requirements of maximizing weight on bit (WOB) and rate of penetrate (ROP), push-the-bit model is selected for the vertical drilling tool because it is unconstrained by rotating speed. To service in a slim borehole, a unique push-actuator machinery using linkage mechanism is designed to minimize the radial dimension. In this paper, the mechanical structure and the deviation control progress of the new vertical drilling tool are introduced, furthermore, as the most critical component directly influences the deviation correction performance, the mechanical characteristic of the unique push-actuator are analyzed in detail.

2. Overall design

2.1. Mechanical structure

Figure 1 is the structural schematic of automatic vertical drilling tool. The device has two joints, respectively connect to drill bit and drill pipe, and there is an inner drill pipe mounting between the joints to deliver pressure and torque. These three parts are rotating together with upper drill pipes, to ensure the work part of the drill tool can maintain relative static. There are two bearing units connect push-pad base and bearing connector with inner drill pipe. On push-pad base, there are three push-actuators circumferential distribute uniformly with 120°, mainly include wearing plate, push pad, two studdles, tackle and connecting rod, which constitute link mechanism by pin rolls, and each of push-actuator connects with its independent electronic control section. In one push-actuator, prevented by restrict screws, the tackle can only move along the groove of push-pad base with its eight wheels from beginning to end, and the spring is compacted by the tackle along spring axle. Between push-pad base and bearing connector, an annular space is separated by inner sleeve and outer sleeve, which is a sealing workspace of the drill tool and fulfilled with insulating oil. In consideration of oil leakage, a compensating mechanism is introduced in with a piston to keep oil pressure, the piston can be pressed

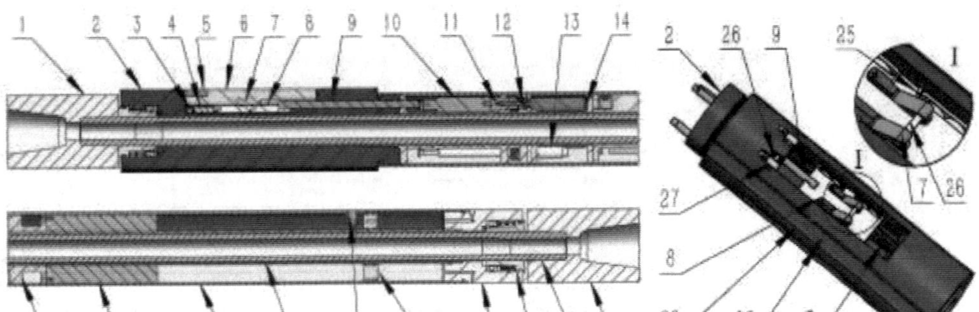

1-drill bit joint 2-push-pad base 3-spring 4-spring axle 5-wearing plate 6-push-pad 7-studdle 8-tackle 9-connecting rod 10-gas spring 11-transmission shaft 12-flange 13-displacement sensor 14-electromotor 15-chip section 16-battery 17-outer sleeve 18-inner sleeve 19-non-return oil valve 20-piston 21-bearing connector 22-bearing unit 23-internal drill pipe 24-drill pipe joint 25- restrict screw 26-pin roll 27-lock block of pin roll 28-flow-conveying groove

Figure 1. Structural schematic of automatic vertical drilling tool and its push-actuator.

IOP Conf. Series: Earth and Environmental Science **64** (2017) 012017 doi:10.1088/1755-1315/64/1/012017

By drilling fluid from the through-hole to annulus in bearing connector. In the annular space, there are transmission shaft, flange, displacement sensor, electromotor, chip section and battery. In chip section, there are gyroscopic sensors, gravity accelerometers and single chip microcomputers (SCM), these parts constitute the electric control section of the drilling tool. The transmission shaft is connected with the electromotor and flange respectively by lead screw and plat key, holding the transmission shaft can only move along axial but not rotation. To avoid the damage by unexpected sudden reactive force from borehole wall, a gas spring is mounted between each connecting rod and transmission shaft as a buffer part. Through these optimal structural design, traditional push-pistons are deprecated, thus the outside diameter is minimized to 145mm to satisfy the requirements of oil & gas slim-hole vertical drilling.

2.2. Working process

The deviation control of this vertical drilling tool is shown in figure 2. When well deviation occurs, the gyroscopic sensors and gravity accelerometers in electronic control section can measure the inclination and give the signal to SCM, by analysis inclination signals, the SCM actuates electromotor run clockwise, driving the transmission shaft move forward to push the gas spring and connecting rod, while displacement sensor is concomitantly moving. The connecting rod push the tackle sliding on the groove platform of push-pad base and compress the spring, while the studdles move upward and pushing the push-pad out, until the wearing pad touches the upper side of borehole wall. Under the reactive force form the borehole wall, the rock cutting of drill bit on the low side of hole-bottom is intensified, the well path gradually return to vertical. After the inclination restoring in expectation range, the electric control section command the electromotor rotates reversely, together with the back force from compressed spring, the push-pad goes back into the base groove to initial non-operating state, then execution of push-the-pit is completed once and the vertical drilling tool turn to prepare for next deviation decreasing time. During this process, the displacement sensor can continuously supervise the motion of transmission shaft, delivering the signal to SCM to control the electromotor run or stop. The spring axle not only can relieve the spring radial run-out, but also work as a mechanical position limitation part to avoid the tackle sliding out of its stroke.

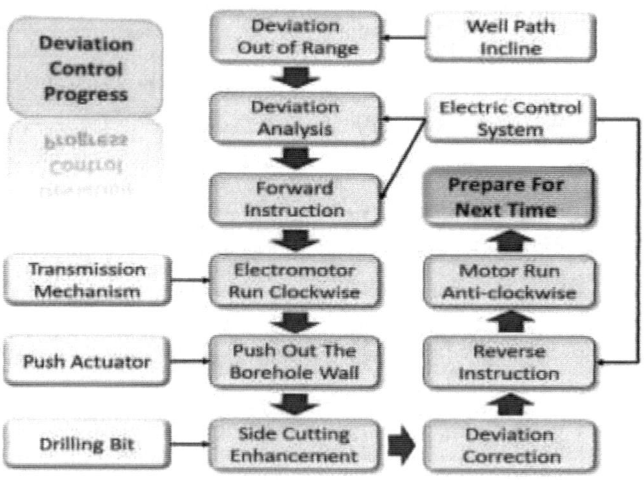

Figure 2. Deviation control flow chart of automatic vertical drilling tool.

3. Mechanical analysis of push-actuator

3.1. Simplified two-dimensional calculation model

As the most critical part under stress, the reliability and performance of the push-actuator directly determine the deviation control behavior of the vertical drilling tool. In order to analyze the approximate characteristics, the push-actuator are simplified as a two-dimensional calculation model, which is shown in figure 3 and 4 respectively represents the initial non-operating state and the maximum pushing state during its working process. On the initial state, the push-pad is closed in the groove of the push-pad base. When the deviation occurs, the electric motor rotates under the instruction of SCM and give a forward force F_0 to connecting rod and tackle, pushing the tackle sliding forward on the groove plat of push-pad base. In the meantime, because of the structure of link mechanism, the studdle moves forward while rotating θ degree and pushing the push-pad rising α degree. At this point, the push-pad gives a push-out force Fp to the high side of borehole wall and the drilling tool has moved to the low side of the borehole, the maximum stroke of the push-pad is Δd (the difference between the diameter of the borehole and drilling tool outside diameter).

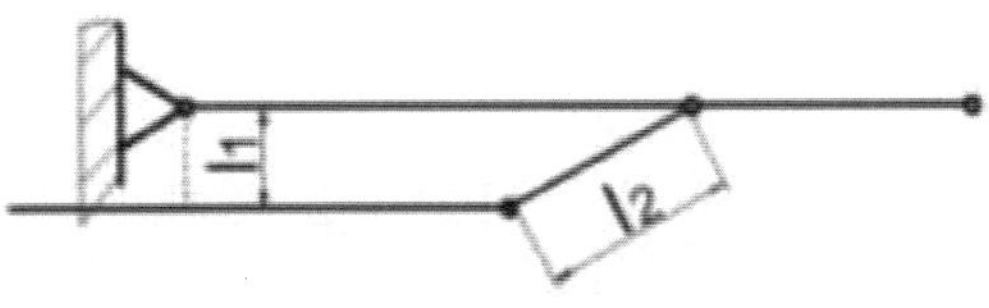

Figure 3. Initial state of push-actuator.

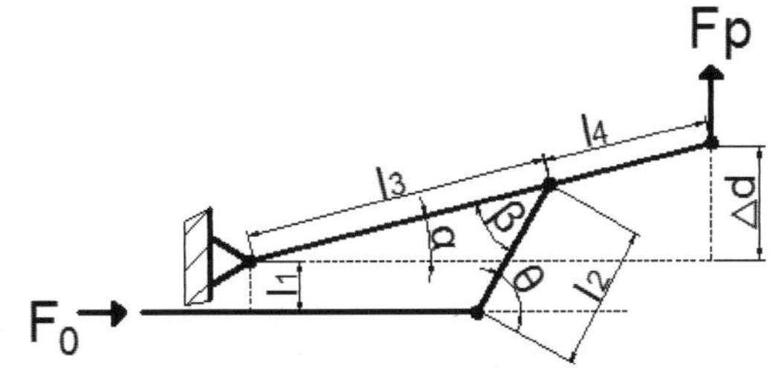

Figure 4. Maximum pushing state of push-actuator.

According to lever principle and moment equilibrium, the geometrical and mechanical relationship of the push-actuator are as follow:

$$\sin\alpha = \frac{\Delta d}{l_3+l_4} \tag{1}$$

$$\sin\theta = \frac{l_3\Delta d+(l_3+l_4)l_1}{(l_3+l_4)l_2} \tag{2}$$

$$K = \frac{F_P}{F_0} = \frac{l_3\sin(\theta-\alpha)}{(l_3+l_4)\cos\alpha\cos\theta} \tag{3}$$

Where l_1 is the distance between the connecting rod and push-pad on initial state, l_2 is the pitch-row of studdle, l_3 and l_4 respectively represent the distance from the pin hole to the front-end and rear-end. The angles α and θ are shown in figure 4, and K is the ratio of push-out force to forward force. For this drilling tool, the parameters of l_1, l_3 and l_4 are respectively 15mm, 150mm and 100mm, the external diameter is 145mm to match a 152mm PDC drill bit, thus the maximum Δd is 7mm. By using different studdles with various pitch-row, different pushing force Fp can be attained, the value of ratio K can be calculated according to formula (1)-(3).

3.2. Mechanical analysis of push-out process

Due to the linkage design of this unique push-actuator, the push force can be adjusted by selecting different studdles with various pitch-row for different geological conditions and technological requirements, which remarkably improves the applicability of this automatic vertical drilling tool. In order to fabricate conveniently and avoid self-locking phenomenon, the pitch-row of studdle l_2 are selected from 20mm to 24mm on integer numbers. Figure 5 shows the relation curves between the values of ratio K and push-out distances using different studdles. As can be seen from the relation curve, the value of K increases effectively with the increasing of push-out distance. And the larger the push-out displacement, the faster the growth rate of K increasees. As the increasing of the studdle pitch-row l_2, the value of K reduces obviously. At the maximum push-out state (Δd=7mm), with the pitch-row increases from 24mm to 20mm, the value of K drops from 2.04 to 0.78, which decreases 61.8%. When the push-actuator using studdles with 22mm pitch-row, the K value is 1.06 and the push-out force Fp is nearly equal with the forward force F_0. So with the studdle pitch-row longer than 22mm, the push force can be reduced by the linkage mechanism, on the contrary, the push force can be enlarged if the pitch-row shorter than 22mm. Thus the feasibility of vertical drilling tool was effectively improved by this unique character of the push-actuator, the studdle can be easily exchange by actual working conditions. If the drilling formation is soft, studdles with relative longer pitch-row could be selected to reduce sticking, on the contrary, if the formation is hard, a shorter pitch-row could be chosen to gain larger straightening force.

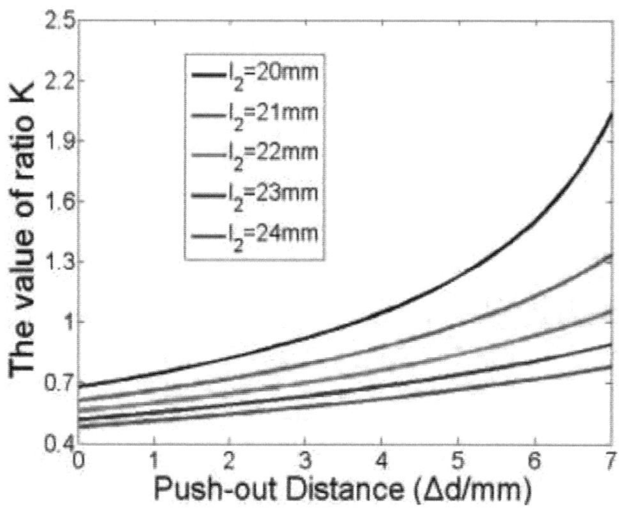

Figure 5. Relation curves between the values of ratio K and push-out distances.

The phenomenon mentioned above is due to the linkage mechanism design of push-actuator, during the push-out process, the angle θ of studdle gradually growth and the push-out force changed. Table 1 shows the angle variations during push-out process using different studdle pitch-row, with the studdle pitch-row increasing from 20mm to 24mm, the start angle θ_S (Δd=0mm) and end angle θ_E (Δd=7mm) both decreasing fast, respectively from 48.59° and 73.74° drop to 38.68° and 53.13°, and the angle stroke of studdle decreasing from 25.15° to 14.45°, reducing more than 42.5%. According to trigonometric function system of the linkage mechanical analysis, the larger the end angle, the larger the K value, thus a larger push-out force can be obtained. When the end angle θ_E increases to 60°, the push-out force F_P is equal with the forward force F_0, therefore studdles with a pitch-row shorter than 22mm are selected, the push-out force F_P could be enlarged.

Table 1. Angle variations of push-out process using different studdle pitch-row.

Pitch-row of studdle l_2 (mm)	Start angle θ_S (°)	End angle θ_E (°)	Angle stroke θ_S-θ_E (°)
20	48.59	73.74	25.15
21	45.58	66.10	20.52
22	42.99	60.78	17.79
23	40.71	56.59	15.88
24	38.68	53.13	14.45

In order to analysis the process of push-out behavior, as the time goes on, the changing of push-out distance and push-out force are shown in figure 6. Under a constant rotate speed of electromotor, the connecting rod and the tackle can move along the axial direction with an invariable velocity, which is set as 0.1mm/s. With the studdle pitch-row increasing from 20mm to 24mm, the push-out velocity decreasing obviously, and the completion times using different studdles are respectively 75.8s, 61.4s, 53s, 47.2s and 42.8s, the completion time reducing 43.5% between the pitch-row of 20mm and 24mm. Because of the exist of angle θ, the push-out velocity gradually drops along with the push-out process carry through, thus the longer the studdle pitch-row, the faster the push-out velocity. This means if a quick response of deviation control is needed, a relative longer studdle pitch-row should be chosen.

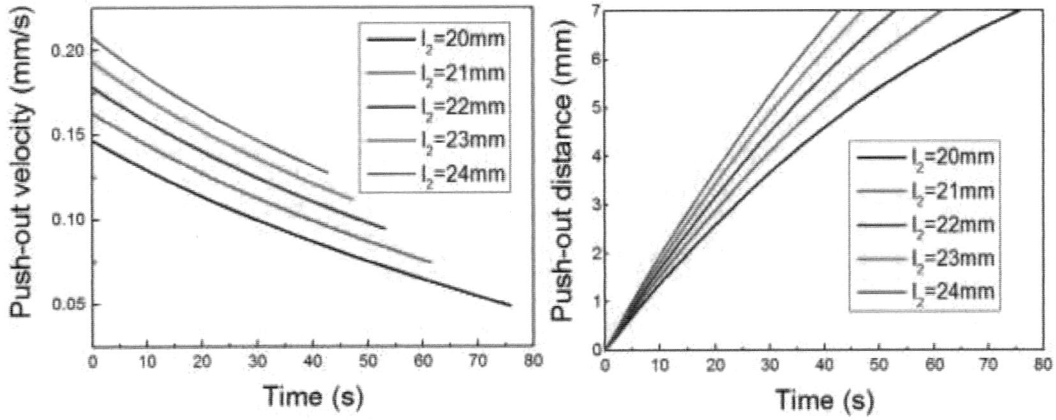

Figure 6. Relation curves between velocity coefficients and push-out distances.

4. Conclusion

As a new kind of electronic-controlled automatic vertical drilling tool, the outside diameter is shrunken effectively by optimal structural design for slim borehole drilling. In order to obtaining better performance and reliability, each push-actuator are controlled by its independent electronic control system, and hard links of the drive system are ameliorate by buffer spring. Besides, the applicability and feasibility of this drilling tool are remarkably improved by its optimal designed push-actuator with link mechanism, which can be able to adjust to different geological conditions and technological requirements. Based on the mechanical analysis using different studdles with five pitch-rows from 20mm to 24mm, due to the exist of θ, with the pitch-row increases, the ratio of push-out force to forward force gradually decreases from 2.04 to 0.78 and the completion time of push-out process reduces from 75.8s to 42.8s. These unique characters makes the vertical drilling can be feasibly adapted to actual working conditions, studdles with relatively shorter pitch-rows can be chosen to gain larger straightening force, by which a larger WOB and ROP can be exerted in hard formations, on the contrary, studdles with relatively longer pitch-rows are suitable for soft formations and obtained more quickly sensitive response of deviation control. As a results, this electronic-controlled automatic vertical drilling tool owning a great promotional value for oil & gas exploitation.

Acknowledgments

This work is financially supported by special scientific research fund for public welfare industry of ministry of land and resources of the People's Republic of China (No.201411054).

References

[1] Information on http:// www.bakerhughes.com.
[2] Barr J D, Clegg J M, Russell M K. Steerable Rotary Drilling With Experimental System. SPE/IADC Drilling Conference, Amsterdam, Netherlands, 28 February-2 March, 1995. SPE

IOP Conf. Series: Earth and Environmental Science **64** (2017) 012017 doi:10.1088/1755-1315/64/1/012017

29382.
[3] Chur C & Oppelt J. Vertical Drilling Technology: A Milestone in Directional Drilling. SPE/IADC Drilling Conference, Amsterdam, Netherlands, 22-25 February, 1993. SPE 25759.
[4] Hans de Pater C J, Zoback M D, Wright C A, et al. Complications with Stress Tests Insights from a Fracture Experiment in the Ultra Deep KTB Borehole. SPE Annual Technical Conference and Exhibition, Denver, Colorado, 6-9 October, 1996. SPE 36437.
[5] Kurt B & Johann D. 1995. The KTB Borehole-Germany's Superdeep Telescope into the Earth's Crust. Oilfield Review, 17(1), 4-22.
[6] Ligrone A, Oppelt J, Calderoni A, et al. The Fastest Way to the Bottom: Straighthole Drilling Device-Drilling Concept, Design Considerations, and Field Experience. European Petroleum Conference, Milan, Italy, 22-24 October, 1996. SPE 36826
[7] Liu Z, Drilling technology HTHP ultradeep well and the crucial technique application. International oil & gas conference and exhibition, Beijing, China, 8-10 June, 2010. SPE 131799.
[8] Oppelt J, Chur C, Feld D, et.al. New concept for vertical drilling of boreholes. SPE/IADC Drilling Conference, Amsterdam, Netherlands, 11-14 March, 1991. SPE 21905.
[9] Zhang, K., Li, M., and Liu, X.: Advances of slim-hole cementingtechnologies at home and abroad, Drill. Product. Technol., 38, 23–26, 2015.
[10] Reich M, Oesterberg M, Montes H, et al. "Straight Down to Success: Performance Review of a Vertical Drilling System." SPE Annual Technical Conference and Exhibition, Denver, Colorado. 5-8 October, 2003. SPE 84451.
[11] Shinmoto Y, Wada K, Sawada I, et al. Case study of push-the-bit vertical drilling in shallow sediments of the Nankai-Trough seismogenic zone. Offshore Technology Conference, Houston, Texas, USA, 3-6 May, 2010. OTC 20716
[12] Xue Q, Wang R, H L, et al. 2012. The Strap-down Automatic VerticalDrilling System Design and Field Applications. The Electronic Journal of Geotechnical Engineering. 17: 3009-3018.

Multi-objective optimization design of rocker arm on crown-mounted compensator

Zhiqiang Huang[1], Ziyang Xu[1,*], Chunping Liang[2, 3], Xinming Mu[2, 3] and Lingfeng Huang[1]

[1]MOE Key lab of Oil and Natural Gas Equipment, Southwest Petroleum University, Chengdu, 610500, China
[2]National Oil and Gas Drilling Equipment Research Center, Baoji, 721002, China
[3]China Baoji Oilfield Machinery CO., LTD, Baoji, 721002, China

*Corresponding author e-mail: swpuxzy@163.com

Abstract. Aiming to solve the severe wear of wire of rocker arm on crown-mounted compensator, the working principle of rocker mechanism is analyzed and mechanical properties of wire is calculated by a model based on Simulink. To reduce the contact stress amplitude between the wire rope and three pulleys on the rocker mechanism, the multi-objective optimization design model of rocker mechanism is constructed. In addition, the length of rocker and linkage arm are considered as design parameters by using MATLAB software. The results showed that the contact stress of wire rope decreased by 21.6%, when the length of rocker and linkage arm is respectively 3450mm and 5900mm, which significantly reduced the possibility of fatigue failure and the wear loss of wire rope.

1. Introduction

While floating drilling platform explore oil and gas in the deep sea, heave compensators must be equipped to ensure the safety and efficiency of drilling or other operations [1, 2]. Rocker mechanism is the key mechanisms of crown-mounted compensator (CMC) to ensure the tension of wire rope [3]. The movement of crown block caused the fluently changes of the contact angle between wire rope and three pulleys, which results in the severe wear of wire rope [4, 5], threatening the safety of rocker mechanism and even the normal operation of CMC.

With high technical reliability, CMC is early and widely used as compensation system on drilling platforms or ships [6]. Burns [7] first proposed the use of rocker mechanism on the CMC to reduce the number of turns between wire rope and the winch; Rocker mechanism on passive heave compensator are designed by Yanting Zhang [8] but the optimization was not involved. S Takagawa [9] applied the rocker mechanism to ROV heave compensator, founding that the winding method of the rope would affect the wrap angle between wire and pulleys, which causes the contact stress changes but not taken into consideration. Through the combination of experiment and finite element analysis, the failure form of drilling wire rope is studied by S Moradi [10], who founded that the changes of alternating stress caused by wire rope overload is the main reason for its wear failure.

Aiming to solve those problems, the working principle of rocker mechanism is analyzed and mechanical properties of wire is calculated through a model established by Simulink. To reduce the contact stress amplitude between the wire rope and three pulleys on the rocker mechanism and to improve the stability and service life, the multi-objective optimization design model of rocker mechanism is constructed, in which the structural parameters of rocker mechanism is optimize.

2. Working principle of rocker mechanism

CMC is illustrated as Fig.1, which consists of floating crown block, rocker arm, supporting structure, compensator cylinder, accumulator and high-pressure vessels. Rocker mechanism is composed of two sets of rods and guide wheels, whose role is to tighten the wire rope and reduce the number of turns between wire rope and winch, thereby enhancing the service life of wire rope.

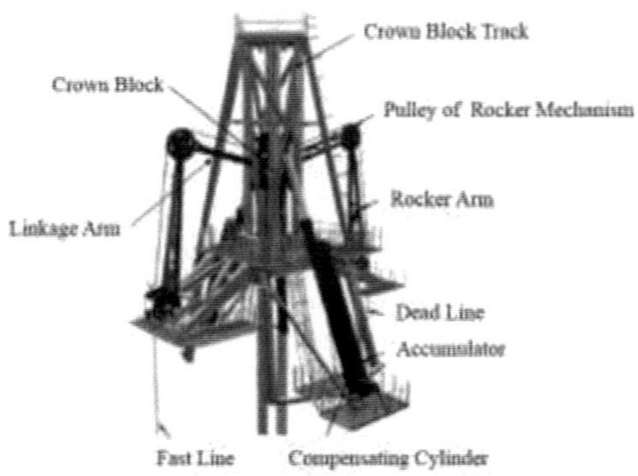

Figure 1. Crown-mounted Compensator

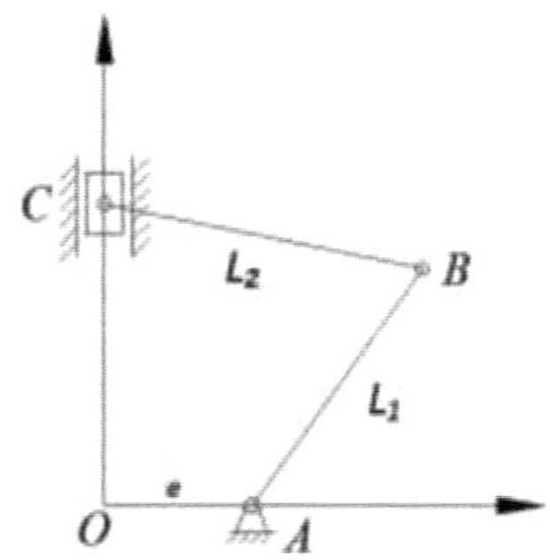

Figure 2. Simplified schematic diagram of rocker mechanism

When the platform (ship) rises, the crown block moves downward relative to the track, as the piston of compensator cylinder retracts and compress the gas of the main accumulator, compensating for the displacement of the platform (ship) and storing energy. When the platform (ship) sank, it moves in the opposite state.

The linkage arm connects with the crown block while the rocker arm with the water table. During the movement of the crown block along the rail, the linkage arm and crown block move like an offset rocker mechanism as illustrated in Fig.2.

3. Mechanical properties Analysis of rocker mechanism

3.1. Contact analysis of arms with pulleys

The winding method of wire rope is shown in Fig.3. Wire rope pass through the winch and guide pulley at the fast line then wrap on the crown block and travelling block, leading to the fixed point at another side.

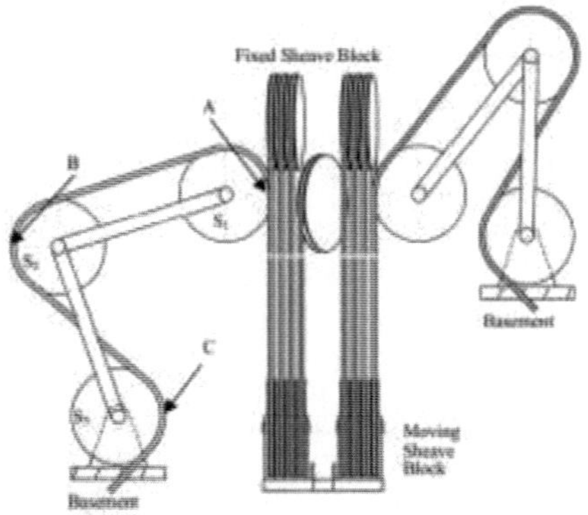

Figure 3. The winding method of wire rope on rocker mechanism

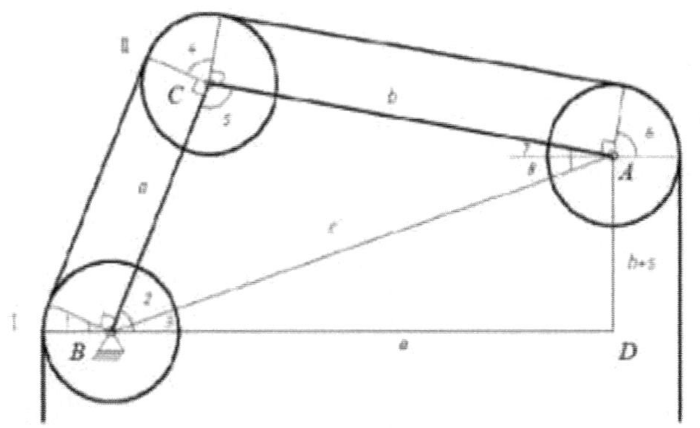

Figure 4. The wrap angle between wire rope and pulley of rocker mechanism

As illustrated in Fig 4, there are three wrap angles between wire rope and three pulleys. Pulley 1 is installed at the hinge of rocker arm and the water table, while pulley 2 at the hinge of linkage arm and rocker arm, pulley3 at the hinge of linkage arm and crown block.

According to the Fig. 4, angle equations between contact angle and internal angle are established.

$$\angle 1 = \pi/2 - (\angle 2 + \angle 3) \quad (1)$$

$$\angle 4 = \pi - \angle 5 \quad (2)$$

$$\angle 6 = \pi/2 - \angle 7 \quad (3)$$

$$\angle 3 = \angle 8 \quad (4)$$

Where :

$\angle 1$ is the wrap angle between pulley 1 and the wire rope;

$\angle 4$ is the wrap angle between pulley 2 and the wire rope;

$\angle 6$ is the wrap angle between pulley 3 and the wire rope.

According the law of cosines, equation between internal angle and length of rocker are established in ΔABC and ΔABD.

$$c = \sqrt{(h+s)^2 + e^2} \quad (5)$$

$$\angle 3 = \angle 8 = \arctan\left(\frac{h+s}{e}\right) \quad (6)$$

$$\angle 2 = \arccos\left(\frac{a^2 + c^2 - b^2}{2ac}\right) \quad (7)$$

$$\angle 5 = \arccos\left(\frac{a^2 + b^2 - c^2}{2ab}\right) \quad (8)$$

$$\angle 7 + \angle 8 = \arccos\left(\frac{c^2 + b^2 - a^2}{2cb}\right) \tag{9}$$

Where:
a is the length of the rocker arm;
b is the length of the linkage arm;
e is the horizontal distance between the pulley I and the pulley III hinge;
c is the distance between the pulley I and the pulley III hinge;
h is the initial position of the crown block;
s is displacement of crown block, which is similar to sinusoidal motion s=$A\sin(2\pi ft + \varphi)$.
Contact angles between the wire rope and three pulleys on the rocker mechanism are as follow

$$\begin{cases} \theta_1 = \frac{\pi}{2} - \arccos(\frac{a^2 + (h+s)^2 + e^2 - b^2}{2a\sqrt{(h+s)^2 + e^2}}) - \arctan\left(\frac{h+s}{e}\right) \\ \theta_2 = \pi - \arccos(\frac{a^2 + b^2 - (h+s)^2 - e^2}{2ab}) \\ \theta_3 = \frac{\pi}{2} - \arccos(\frac{b^2 + (h+s)^2 + e^2 - a^2}{2b\sqrt{(h+s)^2 + e^2}}) + \arctan\left(\frac{h+s}{e}\right) \end{cases} \tag{10}$$

Where: θ_1. θ_2. θ_3 are the wrap angles between three pulleys and wire rope, respectively.
To determine the wrap angle between three pulleys and wire rope, a Simulink model is built as seen in Fig.5 to calculate those angles, in which a=3050mm, b =4900mm, h = 700mm, e=3250mm, and the simulation time is set to 2 cycles (24 seconds). Output simulation data to the workspace by the oscilloscope and formed a curve described the wrap angle changes over time shown in Fig.6. As seen, while the crown block moves up and down, the maximum of wrap angel ups to 106.8°. Due to the changes of wrap angle and the heavy load the wire rope carries, contact stress between the pulley and wire has a significant change over time.

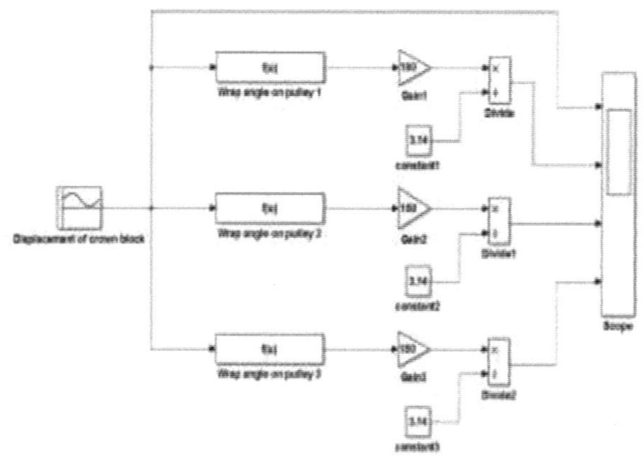

Figure 5. Simulink calculation model of contact angle

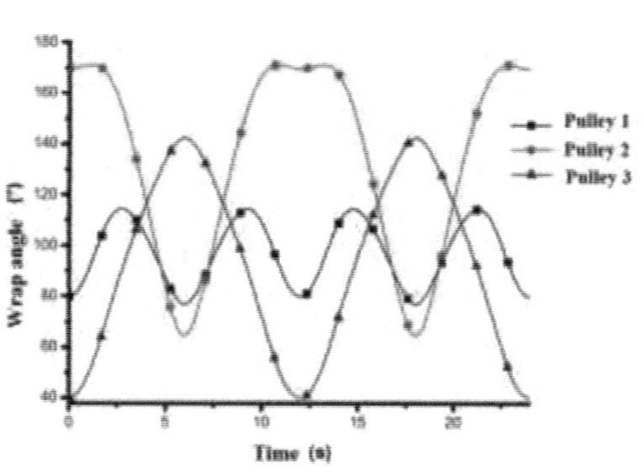

Figure 6. The curve of contact angle

3.2. Contact stress analysis of wire rope and pulleys

Due to the low rotate speed of rocker mechanism at the process of heave compensation, the bearing efficiency and friction efficiency is not taken into consideration. That is, the wire rope tension forces equal on both sides of the pulley. Take any pulley on the rocker mechanism, studying its relationship with the role of wire rope.

Take the pulley and the wire rope in contact as a separate body as illustrated in Fig 7 (a) (not consider the gravity) and build XOY Cartesian Coordinate System.

$$F_n=2T\sin(\theta/2) \tag{11}$$

Where:
θ is the wrap angle between wire rope and pulley;
T is the tension force of wire rope;
F_n is the resistance of the hinge to the pulley.
The contact stress of wire rope and pulley can be calculated by Hertz formula [11] (Line contact situation as shown in Fig7b) , according to the contact state between wire rope and pulley.

$$\sigma_H=\sqrt{\frac{P}{\pi L}\frac{1/R_1-1/R_2}{(1-\mu_1^2)/E_1+(1-\mu_2^2)/E_2}} \tag{12}$$

Where:
P is the load, $P=F_n$;
R_1 is the radius of wire rope;
μ_1 is the Poisson ratio of wire rope ;
E_1 is the elasticity modulus of wire rope;
R_2 is the radius of wheel groove;
D is the diameter of pulley;
μ_2 is the Poisson ratio of pulley ;
E_2 is the elasticity modulus of pulley;
L is the length of contact, $L=D\times\theta$;
σ_H is the maximum contact stress.

(a) **(b)**

Figure 7. Interaction model of wire rope and pulley
(a) global mechanic analysis; **(b)** contact model of wire rope and pulley

Simulink model is built to calculate the contact stress between wire rope and pulley as shown in Fig8. Take the drill string load 450t, Wire rope radius R_1 = 254mm, Wheel groove radius R_2 = 28mm, and the Pulley diameter D = 1600mm. Set two cycles to carry out the simulation and the curves of contact stress is shown in Fig 9.

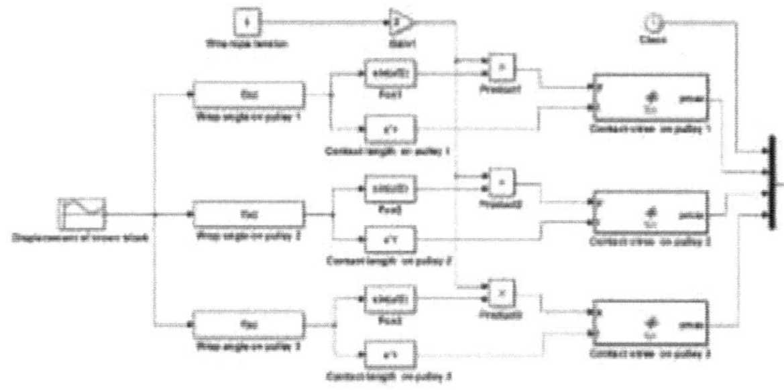

Figure 8. Simulink calculation model of contact stress between wire rope and pulley of rocker mechanism

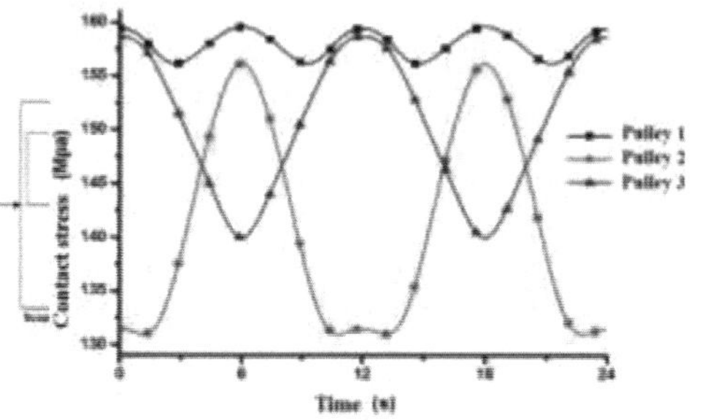

Figure 9. The curve of contact stress between wire rope and pulley

As illustrated in Fig 9, contact stress change with time periodically. During one heave cycle, the pulley 1, which has the smallest contact stress amplitude in three pulleys of 7.24MPa. While the pulley 2 has the highest amplitude of 25.15MPa, which indicates that the pulley 2 is where the most serious parts of wear happens. Under the alternating contact stress, the place with biggest stress on the wire rope would form a crack first, then gradually expanded, leading to fatigue broken eventually [12].

4. Multi - objective Optimization Design Model of Rocker Mechanism

4.1. Design variable analysis
According to Eq. (10), we can see that the value of the intermediate variable θ is related to a, b, s, h, e, and substituting θ into Eq. (12) shows that the maximum contact stress is related to the variables a, b, s, h, e, T, R_1, R_2, D. T depends on the drill string load. h , e depends on the arrangement of the rocker mechanism. R_1, R_2, D are selected by configuration principle of wire rope and pulley. s depends on the heave displacement of the floating drilling platform, $s = A\sin(2\pi ft + \varphi)$, where, t is process control variables. Then determine the design variables as $X = (x(1), x(2)) = (a, b)$.

4.2. Objective function construction
Reducing the contact stress amplitude of the wire rope can reduce the fatigue and wear on the surface of the wire rope. As can be seen from Fig. 9, the contact stress between the wire rope and the pulley.1 reaches the peak at t=0s and t=6s and reaches the trough at 2.72s. Set the contact stress difference between t=0s, t=2.72s and t=6s, t=2.72s as the objective function $f(1)$, $f(2)$. The contact stress of the wire rope and pulley.2 reaches the maximum at t = 6s and achieves the minimum at t=1.18s. The contact stress amplitude is the difference between the two extreme values, which is set as the objective function $f(3)$. The contact stress of the wire rope and the pulley.3 reaches the maximum at t=0s and achieves the minimum at t=6s. The contact stress amplitude is the difference between the two extreme values, which is set as the objective function $f(4)$.

4.3. Constraint analysis and established
The rocker mechanism should satisfy the movement condition of offset rocker slider and the condition of the compensative travel, and should not exist the crank. That means when the rocker mechanism is working, there should not only can't find the dead point position ($e < a+b$) and the ΔABC always exist, $a+b>\mathrm{sqrt}(e^2+(s+h)^2)$, but also satisfy the condition of $a+e>b$.
Using MATLAB language to describe the mathematical model is

$$\begin{cases} g(1) = x(2) - x(1) - e \\ g(2) = e - x(1) - x(2) \\ g(3) = c - x(1) - x(2) \end{cases} \quad (13)$$

4.4. Multi-objective optimization model solving.

This optimization design of crown heave compensation rocker mechanism is a nonlinear multi-objective programming problem. Using the multi-objective programming problem function of fgoalattain of the MATLAB to solve the problem. Call the format for [x,fval,attainfactor,exitflag]=fgoalattain(@myfun,x0,goal,w,A,b,Aeq,beq,Lb,Ub,@mycon). Which the x represent the optimal solution of multi-objective optimization model, the favl represent the value of the multi-objective function in the optimal solution x, the attainfactor represent the planning factor in the optimal solution x, and attainfactor > 0 represent the target without overflow. The exitflag is the exit flag of the multi-objective programming solution, the @myfun and @mycon is the name of the invoking objective function and the function of constraint conditions.

x0 is the initialization vector of the design variable. In this model x0= [3050; 4900].

Goal is the value vector of the designed objective function. The length of the vector is equal to the number of the objective function. In this model goal= [152; 150; 148; 146].

Weight is the weight value vector of the objective function. The length of the vector is equal to the number of the objective function. In this model weight=goal.

A is the coefficient of the linear inequality constraints, while b is the corresponding item in the right of the inequality. And Aeq is the coefficient of the linear equation constraints, while beq is the corresponding item in the right of the equation. Because there don't have the linear inequality constraints and the linear equation constraints. The A, b, Aeq and beq all is empty matrix.

The lb is the lower limit of the design variable, while the ub is the upper limit. In this model lb= [2900; 4800], while ub= [5400; 6100].

Through iterative calculation, the optimal solution X= (3450, 5900). It means that through optimized the length of the rocker arm is 3450mm, and the length of the linkage arm is 5900mm. The final rocker mechanism optimization results is shown in Table 1.

Table 1. The contrast of the rocker mechanism after optimization

Parameter name	Original parameter	Optimize parameter
The length of the rocker arm [mm]	3050	3450
The length of the linkage arm [mm]	4900	5900
The amplitude of contact stress on Pulley.1[Mpa]	7.24	7.02
The amplitude of contact stress on Pulley.2[Mpa]	25.15	22.21
The amplitude of contact stress on Pulley.3[Mpa]	18.75	14.51

It can be seen from the table 1 that, the wire rope wear situation has been improved after the optimization. The amplitude of the contact stress that wire rope applying on the pulley .1 was reduced by 3.04%. The amplitude of the contact stress of the pulley.2 which is the most serious wear and tear was reduced by 11.69%. And the amplitude of the contact stress of the pulley.3 which is the second wear and tear was reduced by 22.61%.

5. Conclusion

The contact angle between wire rope and pulleys on the rocker mechanism is analysed, and contact stress between wire rope and pulleys is calculated through a model established by Simulink. The results showed that contact stress change with time periodically. During one heave cycle, the pulley 1, which has the smallest contact stress amplitude in three pulleys of 7.24MPa. While the pulley 2 has the highest amplitude of 25.15MPa, which indicates that the pulley 2 is where the most serious parts of wear happens.

IOP Conf. Series: Earth and Environmental Science **64** (2017) 012018 doi:10.1088/1755-1315/64/1/012018

The optimization goal is that reducing the contact stress amplitude between the wire rope and three pulleys on the rocker mechanism, and the design variable, length of rocker and linkage arm, is ascertained by analysing the effect of the structural parameters of rocker mechanism to the contact stress. The multi-objective optimization design model of rocker mechanism is constructed through MATLAB software and calculated by fgoalattain function which is one of the multi-objective nonlinear programming function.

The contact stress of wire rope decreased by 21.6% with optimization of structural parameters(rocker arm of 3450mm and linkage arm of 5900mm), which significantly reduced the possibility of fatigue failure and the wear loss of wire rope.

Acknowledgments

This work was financially supported by Ministry of Industry and Information Technology of the People's Republic of China.

References

[1] Qingyou Liu, Tao Xu. Domestic Status and Thoughts on the Development Ideas of Deepwater Drilling Heave Compensation System [J]. Journal of Southwest Petroleum University (Science & Technology Edition), 2014, 36(3):1-8.

[2] Weixu Wang, Tianming Zhou, Xingjun Yu, et al. Analysis of Heave Motion of Floating Drilling Platform [J]. Oil Field Equipment, 2011, 40(9):36-38.

[3] Keren Ren, Dachun Shen, Dingya Wang, et al. Technology Analysis of Drilling Heave Compensation System [J]. China Petroleum Machinery, 2009, 37(9):125-128.

[4] Jiquan Hu, Zhengquan Hu. Wire Rope Mechanical Characteristics Influence on Fatigue Life [J].Port Operation, 2005, 1:10-12.

[5] Yidong Hou. The Influence Factors on the Service Life of Wire Rope for Crane [J]. Construction Machinery & Maintenance, 2014, 20(9):149.

[6] Tao Xu, Qingyou Liu. Design and Analysis of Passive Compensation Wire Rope Rocker Mechanisms on Deep Water Drilling Platforms [J]. China Offshore Oil and Gas, 2015, 27(6):92-97.

[7] BURNS J, GREEN W. Motion compensated crown block system [P]. US, 3791628.1974-02-12.

[8] Yanting Zhang, Yingfeng Qu, Zhendong Liu, et al. Design of Swing Device for Crown-Block Heave Compensation System [J]. Journal of Zhejiang University (Engineering Science), 2012, 46(12): 2267-2269.

[9] S Takagawa. A New Concept Design of Heave Compensation System for Longer Life of Cables [J]. Oceans IEEE-SYDNEY, 2010, 159(1):1-5.

[10] S Moradi. Failure Analysis of a Drilling Wire Rope [J]. Journal of Failure Analysis & Prevention, 2012, 12(5):558-566.

[11] Guoqiang Zhou, Changxiang Qi. Effect of the Sheave Material on the Contact Stress between the Sheave and the Wire Rope [J]. China Petroleum Machinery, 1992, 20(9):20-23.

[12] Keyong Yu, Wenjin Xu. Study on fatigue life of elevator hoist wire rope [J]. Metal Products, 2014, 40(4):59-67.

A study of well test data interpretation model for water-bearing reservoirs with phase redistribution

Junxiang Zhang[1], Rui Deng[2], Haipeng Liang[2,*] and Jing Yang[2]

[1]Gas field Exploitation Department of PetroChina Changqing Oilfield Company, Xi'an, Shanxi, China
[2]School of Energy Resource, Chengdu University of Technology, Chengdu, Sichuan, China

*Corresponding author e-mail: 303453310@qq.com

Abstract. In China, plentiful marine reservoirs exist. Net pay thickness in individual gas reservoirs where partial penetration was performed can be hundreds of meters. Due to the influence of condensate water and formation, water phase separation phenomenon, where gas rose up and liquid moved down, and a morsel of water production emerged in some gas wells, which makes the build-up curves distorted and thus unable to be interpreted. On the basis of seepage theory and Laplace transformation, a seepage mathematical model and a well test interpretation model for gas wells with phase separation considered are developed to analyze the impact of such various elements as phase separation and partial penetration on the pressure and pressure derivative log-log plot. With practical data of well test in Xihu Sag, reliability analysis of the mathematical model mentioned above was demonstrated. Theoretical research results proposed in our study substantially improved the accuracy of well test interpretation for thick water-bearing gas reservoirs and laid a technical foundation of development of the similar oil & gas reservoirs.

1. Introduction

Such transient pressure tests as pressure buildup tests have been prevalently used as a reliable tool to estimate reservoir characteristics and other properties for years. Admittedly, pressure data acquired during the test are impacted by wellbore-related effects as well as reservoir properties. Phase redistribution is one of the wellbore effects, which exerts its main influence on wellbore storage. While wellbore storage is almost omnipresent, wellbore phase redistribution occurs in wells where two or more phases exist. It has been accepted that phase redistribution can cause anomalous pressure data in some oil and gas wells. A more sophisticated method is needed to avoid mistaking these effects as reservoir properties and achieve higher accuracy of well test interpretation for wells of this type.

As the gas reservoir exploitation are absorbing lasting attention in China, more and more importance is being attached to the research on production and exploitation of gas reservoirs while little has been made if any significant progress towards an agreeable investigation into multiphase flow in gas reservoirs. Hu Yong (2000) et al. have investigated the influence of multiphase flow of gas and water on the seepage in reservoirs. Li Xiaoping (2001) have addressed the methodology of research on productivity of reservoirs of gas-water multiphase. An analysis on steady and transient well test has been performed by Cheng Suimin *et al.* (2006). Li Chengyong et al. (2010) have addressed a mathematical model for low-permeability water-bearing gas reservoirs with the phase redistribution taken into account.

A considerable number of gas wells in Xihu Sag where an abundant resource of gas reservoirs with thick layers exist present distorted well test curves due to phase redistribution, which encumbers the interpretation of well test data and thus the exploitation of the gas field. With the discussion about the effects of phase redistribution on well test data and a validation based on factual data from Xihu Sag, a comprehensive investigation on the well test features of wells in Xihu Sag.

Fair[1] has presented a model for the transient pressure test data by considering phase redistribution as a parameters of wellbore storage. We have fused the model proposed by Fair with Duhamel's superposition principle.

2. The mathematical seepage model for water-bearing gas reservoirs

Fair et al. (1981) argued that the problem of phase redistribution was equivalent to the problem of variable bottom-hole storage. Therefore, it's a workable solution to revise the pressure data by superposition of pressure deviation function during phase redistribution. And the pressure deviation function for transient well test interpretation is given by

$$p_{\varphi D} = C_{\varphi D}(1 - e^{-t_D/\alpha_D}) \tag{1}$$

Where,

$$p_{\varphi D} = \frac{khp_{\varphi}}{1.842\times10^{-3}qB\mu}, \quad C_{\varphi D} = \frac{khC_{\varphi}}{1.842\times10^{-3}qB\mu}, \tag{2}$$

And

$$t_D = \frac{3.6kt}{\varphi\mu C_t r_w^2}, \quad \alpha_D = \frac{3.6k\alpha}{\varphi\mu C_t r_w^2} \tag{3}$$

Most of various theoretical researches referring to phase distribution are dependent on the exponential function proposed by Fair or the error function suggested by Hegeman since it has been proved that these formations are adequately consistent with laboratory data and factual data. Consequently, based on Fair Model and Hegeman Model the mathematical model in this paper for well test interpretation of wells with liquid loading is given by

$$\begin{cases} \dfrac{\partial^2 p_D}{\partial r_D^2} + \dfrac{1}{r_D}\dfrac{\partial p_D}{\partial r_D} = \dfrac{\partial p_D}{\partial t_D} \\ p_D(r_D,0) = 0 \\ p_{wD} = \left[p_D - S\left(\dfrac{\partial p_D}{\partial r_D}\right)\right]_{r_D=1} \\ C_D\left(\dfrac{dp_{wD}}{dt_D} - \dfrac{dp_{\varphi D}}{dt_D}\right) - \left[r_D \dfrac{\partial p_D}{\partial r_D}\right]_{r_D=1} = 1 \\ \lim\limits_{r_D-\infty}[p_D(r_D,t_D)] = 0 \end{cases} \quad (4)$$

Where,

$$p_{\varphi D} = C_{\varphi D}(1 - e^{-t_D/\alpha_D}) \qquad \textit{Fair Model} \quad (5)$$

And, if we let erf(x) is the error function

$$p_{\varphi D} = C_{\varphi D} erf(t_D/\alpha_D) \qquad \textit{Hegeman Model} \quad (6)$$

3. Analysis of the pressure behaviour

With Stehfest numerical inversion, the semi-analytical solution of the model can be obtained by solving simultaneously and thus analyses on effects of $C_{\varphi D}$, α_D the perforated thickness, the location of perforation *et al.* on the phase redistribution. Five flow periods, listed as following, can be clearly recognized on a pressure and pressure derivative log-log type plot (Fig.1).

1) I is named as well the storage period where the pressure curve and pressure derivative overlap each other and share a unit slope since it is influenced by the well storage effect.
2) II is the phase redistribution period where gas rose up and liquid moved down. The derivative curve meander as a hump appending to the unit slope of the previous period, which is a distinctive feature of phase distribution.
3) III is the local radical flow period. The curve presents a linear portion characteristic of a slope of negative one, which is the result of the radical flow through the perforation.
4) IV is the period of spherical flow where a linear segment with slope -1/2. This stage indicates layers, in addition to perforated layers, begin to give a considerable contribution to the overall flow, turning plane radical flow to spherical flow.
5) V is the pseudo-radial flow, distinguished by its horizontal linear portion with value 0.5, which is a remarkable sign that the whole system has obtained a state of equilibrium.

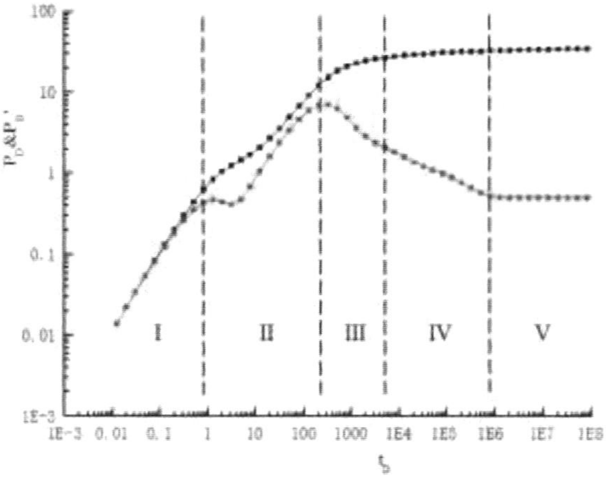

Fig. 1 A typical pressure and pressure derivative log-log plot of vertical wells with phase redistribution in thick water-bearing reservoirs.

4. Field Application in Xihu Sag

Production well HY1-1-3, spud-in on 6th Dec. 2013, was completed on 5th Nov. 2014 with 4468m total depth. Details of various parameters are listed as Table 1.

Table 1. Parameters of HY1-1-3 according to logging data interpretation

Parameters	Values	Parameters	Values
Apparent Net Pay Thickness(m)	69.1	Liquid Density(g/cm^3)	1.154
Perforation Thickness(m)	61.5	Mud Density(g/cm^3)	1.2
Porosity	7.1%	Mud Viscosity(S)	45
Water Saturation	54.9%	Formation Temperature(℃)	157
Permeability(mD)	0.38	Formation Pressure(MPa)	45.8
Concentration of Cl^-(mg/L)	38500	Formation Pressure Coefficient	1.15

The pressure and pressure derivative separated from each other at the early time, which conforms to the characteristic of phase redistribution in well bores (Fig2). According to this and the declination of the curve at the late stage, we, taking the geological settings of the reservoir into account, adopt the well test interpretation mathematical model for homogeneous gas reservoirs with closed outer boundary and phase redistribution considered to analyze the practical well test data, and the result was as the following Table 2.

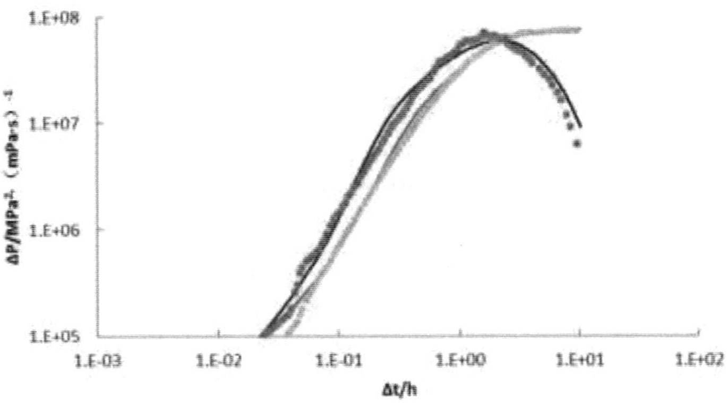

Fig. 2 Diagnostic log-log plot of HY1-1-3 well test data.

Table 2. Well test interpretation result of HY1-1-3

Parameters	Values
Extrapolated Formation Pressure (MPa)	52.9596
Formation Coefficient (mD•m)	2.56
Skin Coefficient	10.3
Well Storage Coefficient	0.0963
α_D	0.48

5. Conclusion

(1) An well test interpretation model for wells affected by phase redistribution was developed in this investigation, and the result shows that the hump and the negative pressure derivative on the log-log plot are the critical features to identify the occurrence of the phase redistribution.

(2) For thick water-bearing gas reservoirs, the phase redistribution is influenced, mainly, by $C_{\varphi D}$, α_D and the perforation thickness.

References

[1] Li Chengyong, Yi Xiangyi, Deng Yuanzhou, et al. Well test interpretation method for low permeability gas reservoirs considering phase separation. Drilling & Production Technology, 2012; 06(10):10.
[2] Yan xingyu, Zhang Luoling, Zhang Xinyuan. Using point source function to establish low permeability reservoir we1l testing interpretation model of straight. Petrochemical application, 2012; 6(10):10.
[3] Sun Laixi, Li Chengyong, Li Cheng, et al. Application of boundary element method in the flow of partially perforated well in complex boundary reservoir. Geophy-sical and geochemical exploration technology. 2008; 05(381):385+348-349.
[4] Lin Jiaen, Sun Hedong. Understanding the response characteristics of pressure curve in Fair cylinder storage model. Well Testing, 2006; 04(1):4+75.
[5] Li Xiaoping, Zhang Linen, Qu Yunfang. Mathematical model and interpretation method of gas well test for variable well. Oil Drilling & Production Technology, 1993; 03:71-78.

Analysis of connectedness between oil and water wells by numerical simulation technique

Wenbo Zhang*

No.2 Oil Production Plant, Daqing Oilfield Limited Company, Daqing 163511, China

*Corresponding author e-mail: 2760913691@qq.com

Abstract. For water flooding reservoirs, accurately identify the connectivity of oil and water wells, this is essential for designing a reasonable reservoir development program, it is also the basis for the success of oilfield development and management. Previous studies have focused on the connectivity between oil and water wells, and there is no concrete to the oil and water wells at the simultaneously shot layers. In this paper, numerical simulation technique is used to analyze the connectivity between oil and water wells, and the connectivity relationship of simultaneously shot layers between oil and water wells is also determined. Based on the fine modeling of reservoirs, the connectivity analysis was carried out by using the tracer flow simulation technique in ECLIPSE software and the results of stratified output concentration of tracer in PRT file. The results show, the method can be used to accurately determine the connectivity of oil and water wells. The oil well A in Daqing Oilfield have high water cut layer, using this method, on the 27th floor of well C with the best connectivity of target layer were screened out, the target layer is blocked for wells C, thereby the injecting water is optimized, reduce the amount of inefficient injection of water.

1. Introduction

The reservoir is a dynamic equilibrium system. The change of water injection in the well caused the oil production fluctuation is the characteristic reflection of the connection between the oil and water well, and the fluctuation range of oil production is related to the degree of connectivity of oil and water wells. At present, this field is based on this idea, qualitative judgment of oil and water wells connectivity, but the results is subjective. In addition, in the production of oilfields, each well is not isolated, and its output changes are associated with the combined action of all the water inflow wells that are connected to it, which may also enhance or weakened the signal obtained by the well.

As a general numerical simulation software, the software of Eclipse can accurately simulate the actual situation of the reservoir, and the software also provides a tracer flow simulation technology, which provides a better choice for the judgment of the connectivity between the oil and water wells.

2. Fine modeling of reservoirs

The PETREL software was used to model the blocks and import the model into ECLIPSE software for tracer flow simulation.

IOP Conf. Series: Earth and Environmental Science **64** (2017) 012020 doi:10.1088/1755-1315/64/1/012020

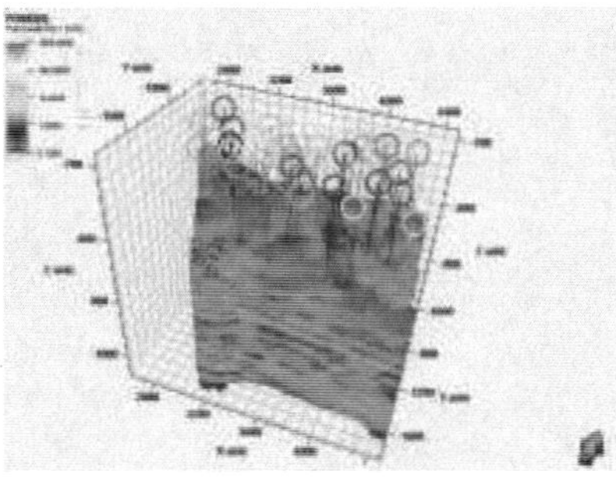

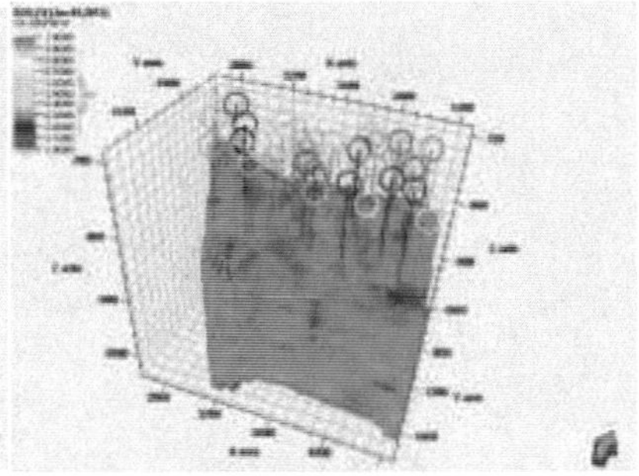

Fig. 1 The model of permeability field **Fig. 2** The model of residual oil saturation field

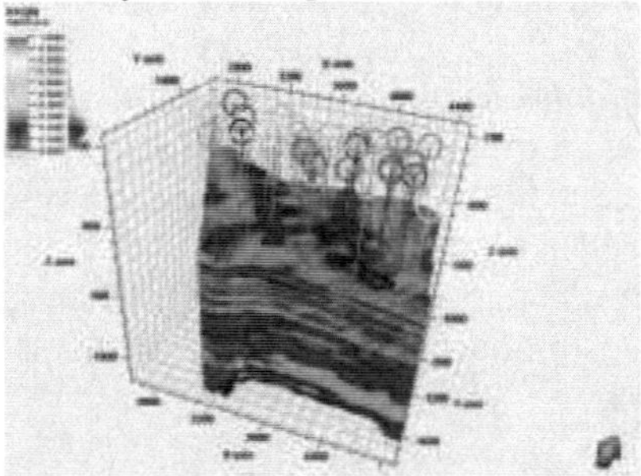

Fig. 3 The model of static ratio field

3. Judgment of connectivity of oil and water wells

3.1. Judgment of connectivity between oil and water wells

In the actual production, the method of using tracer in oil field is: by detecting the concentration curve of the tracer concentration in the production wells, reflecting the information about the characteristics of the reservoir, observing the producing characteristic of tracer in the oil wells, such as the parameters of breakthrough time, the size and number of peaks and corresponding injection of total fluid, which can further study and understand the movement law of injection fluid and reservoir heterogeneity.

For numerical simulation technology, the tracer can be injected from the early stage of the oil field development without the limitation of the time of injection tracer. The tracer output curve cannot be explained without the interpretation error. Real - time analysis of water. The incoming water of oil well can be detected in real time.

Taking the in Daqing Oilfield as an example, there is a high water level in, a total of six water wells are located around the oil well A. Analysis of the tracer output curve of oil well a shows that the main water direction of a well is different at different times. At present, the main water supply direction of oil well A is water well C and E.

Tab.1 The incoming water list of well A

Oil well	water well	The percentage of incoming water (%)
A	B	4.16
	C	42.29
	D	1.24
	E	18.9
	F	3.59
	G	0.51

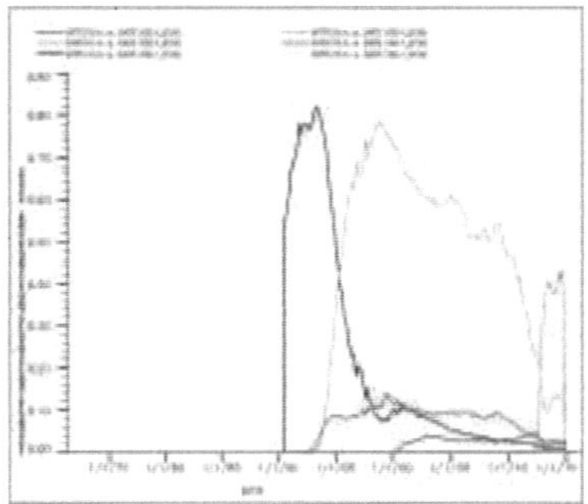

Fig. 4 Variation curve of tracer concentration of well A

3.2. Connectivity judgment of simultaneously shot horizon of oil and water wells

According to the tracer flow simulation results, combined with the results of stratified tracer output concentrations in the Eclipse results file, it is possible to judge the simultaneously shot horizon connectivity of the oil and water wells.

According to the A well tracer output curve, combined with the use of the Eclipse results file in the A well stratified tracer output concentration, the 27 layers in simultaneously shot horizon of well A and well C is the best connectivity layer.

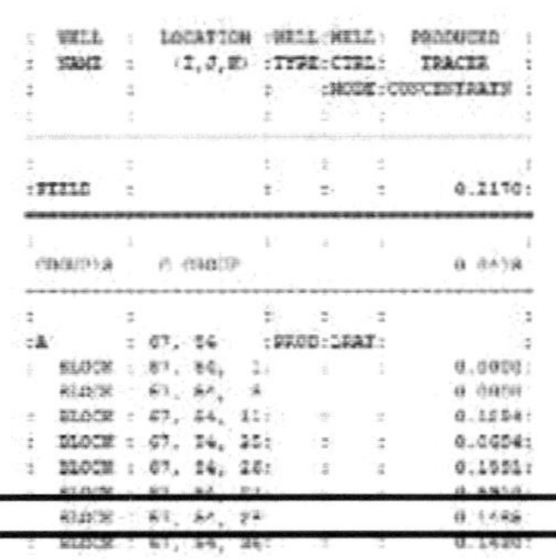

Fig. 5 The output stratified tracer Concentration of well A

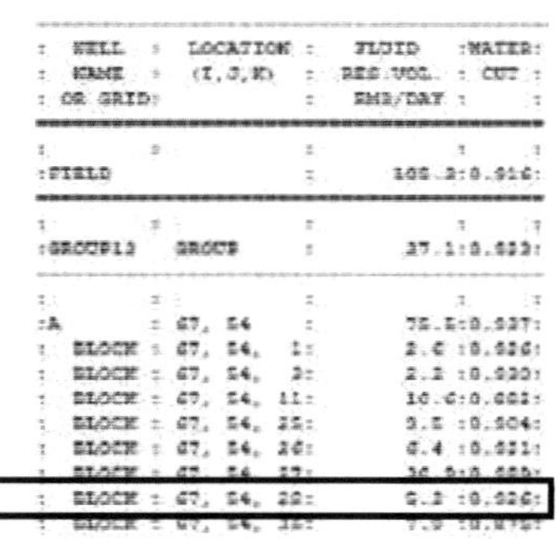

Fig. 6 Stratified liquid and water cut of well A

4. Verification of results

According to the judgment results, the 27th layer is a reservoir with water cut of up to 98.8%., which is the perforation of the oil well, the main water direction of the layer is C wells. Profile control work for the 27 layer in the C well, oil production is basically no change (from 4.74 t / d to 4.50 t / d), water cut decreased by 3.7%. Reduce the amount of inefficient injection of water, and achieved a better development effect

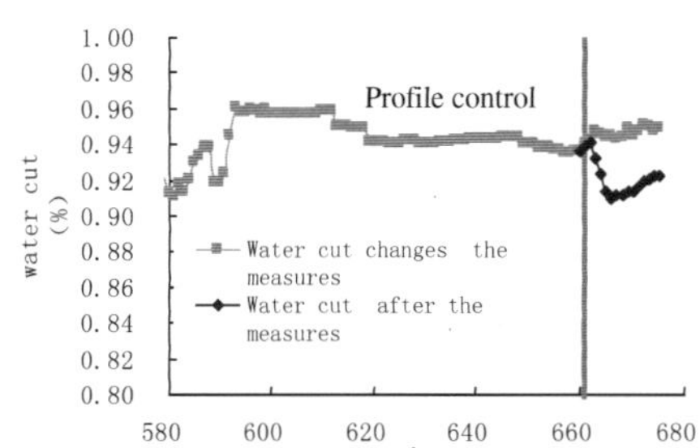

Fig. 7 The correlation curve of water cut before and after measures

5. Conclusion

(1) For water drive reservoir, accurate identification of oil and water well connectivity is not only for the design of a reasonable reservoir development program is essential, but also the oil field development and management of the success of the foundation;

(2) The tracer flow simulation technique can be used to analyze the connectivity between oil and water wells;

(3) Based on the fine modeling of reservoirs, the connectivity analysis of oil and water wells was carried out by using the tracer flow simulation technique in ECLIPSE software and the results of stratified output concentration of tracer in PRT file;

(4) Based on the method adopted in this paper, the C wells will be blocked at the target level, the water cut can be reduced by 3.7%, reduce the low injection rate, and obtain good development effect.

References:

[1] Jiang Ping, Zhao Changlei, Li Shunming, et al. Dynamic inversion and the geological controlling factors of interwell connectivity in reservoirs of lower Karamay Formation in Liuzhongdong block, Junggar basin [J]. Petroleum Geology and Recovery Efficiency. 2015(01).

[2] Wang Shuoliang, Jiang Hanqiao, Zhang Yuan, et al. Application of ridge regression analysis in the research of reservoir injection-production connectivity [J]. Fault-Block Oil and Gas Field. 2010(02).

[3] LI Xiaobo, PENG Xiaolong, SHI Ying, et al. Application of Interwell Tracer Testing in Fracture-cavity Reservoirs [J]. Journal of Oil and Gas Technology. 2008(06).

[4] JAMALI Ali, ETTEHADTAVAKKOL Amin. Application of capacitance resistance models to interwell connectivity of large-scale mature oil fields [J]. Petroleum Exploration and Development. 2017(01).

[5] Tan Yiran. Study on Well - bed Connectivity of AD4 Well in Tahe Cave - type Reservoir [D]. Southwest Petroleum University. 2016.

[6] Cui Mingyue, Jiang Weidong, Wang Qinghua, et al. Detecting method of the tracers for the oil reservoirs injected by the sea water[J]. Petroleum Geology & Oilfield Development in Daqing. 2016(05).

[7] Yang Jiao, Tang Fang, Zhaoweichao, et al. Study on reservoir connectivity of K reservoir in eastern M oil field of South China Sea[J]. Petroleum Geology and Engineering. 2016(03).

[8] Zhang Haifeng, Liu Shijie, Feng Yi, et al. Research and application of interwell dynamic connectivity[J]. Journal of Yangtze University (NATURAL SCIENCE EDITION). 2016(16).

[9] Krisanne L. Edwards, Sean Weissert , Joshua Brandon Jackson. Marcellus Shale Hydraulic Fracturing and Optimal Well Spacing To Maximize Recovery And Control Costs. SPE-140463-MS, SPE Hydraulic Fracturing Technology Conference, 24-26 January, The Woodlands, Texas, USA,2011.

[10] Hassan Bahrami , M. Sadegh Asadi. Stress Anisotropy, Long-Term Reservoir Flow Regimes and Production Performance in Tight Gas Reservoirs. SPE-136532-MS, SPE Eastern Regional Meeting, 13-15 October, Morgantown, West Virginia, USA, 2010.

[11] A. Kumar , D. Camilleri, M. Brewer. Comparative Analysis of Dual Continuum and Discrete Fracture Simulation Approaches to Model Fluid Flow in Naturally Fractured, Low-Permeability Reservoirs. SPE-180221-MS, SPE Low Perm Symposium, 5-6 May, Denver, Colorado, USA, 2016.

[12] zhao hui, kang zhijiang, zhangyun, et al. An interwell connectivity numerical method geological parameter characterization and oil-water two phase dynamic prediction [J]. Acta Petrolei Sinica. 2014 (05).

IOP Conf. Series: Earth and Environmental Science **64** (2017) 012021 doi:10.1088/1755-1315/64/1/012021

Study on immobilization and migration of nuclide u in superficial soil of uranium tailings pond

Zhe Chang and Shukui Zhou *

School of Civil Engineering, University of South China, Hengyang, China

*Corresponding author: zhoushukui@usc.edu.cn

Abstract. The uranium tailings in southern China was used as the object of study to study the fixation and migration characteristics of nuclide U in shallow tailings. The results showed that the precipitation of tailings in the tailings soil was not linearly related to the depth during the acid rain leaching process. Tailings soil in the role of fixatives, when the lime as a fixative, the tailings of different soil uranium in 20 days after the re-precipitation. However, when lime and ammonium phosphate were used as fixing agents, the cumulative precipitation of U had a significant effect, and the migration of uranium was inhibited.

1. Introduction

Since the 1940s, nuclear energy has been developed and exploited, uranium demand is increasing, resulting in a large number of uranium tailings. It is expected that by 2020, Chinese nuclear uranium demand will reach 3180 to 6540 t [1]. The diffusion of radionuclides in the uranium mining tailings has become a serious problem that needs to be solved in contemporary society because of the lack of understanding of the pollution problem after the retreat of uranium and metallurgical tailings.

Uranium tailings during the long-term open dumping process, the residual radioactive elements and toxic heavy metals susceptible to rain, surface and groundwater and other environmental factors to a certain extent precipitated, resulting in uranium tailings into a Radioactive-heavy metal composite pollution sources[2,3]. The acid leaching process is becoming more and more widely used in the hard rock uranium mine, Waste residue produced during the mining process is used before the entry into the tailings reservoir, and the lime is neutralized and then transported to the tailings reservoir [4-6]. With the passage of time, the nuclides that have been deposited are re-precipitated due to the acid reflux phenomenon. The radionuclides can disperse into the air with the weathering effect and reach the surface water from the surface soil around the polluted reservoir area by lateral migration, But also through the vertical migration of infiltration to reach the groundwater and with the flow of land around the migration, thus the reservoir area of the ecological environment poses a serious threat [7, 8].

Domestic research on the migration of radionuclides mainly focuses on the geological work of low and medium-sized radioactive waste repository, and few studies have been reported on the migration and control of radionuclide migration in waste disposal sites. In this study, the leaching test of uranium tailings was carried out to study the uranium removal of uranium tailings in acid rain and the fixation effect under the action of fixing agent. The reference was made to the safe operation and decommissioning of uranium tailings as well as the basis for studying the chemical and kinetic behavior of radionuclides in uranium tailings.

IOP Conf. Series: Earth and Environmental Science **64** (2017) 012021 doi:10.1088/1755-1315/64/1/012021

2. Experimental Part

2.1. Collection and Treatment of the Test Samples
The soil used for the test was taken from uranium tailings in southern China. Sampling point takes a point every 20 cm in vertical sampling, just a total of 4 samples. Samples is collected on the sample bag, and edit the corresponding number from 1 to 4, which is naturally air-dried. And then the collection of each sample is broken, mixed, shrinked, and other processes to deal with.

2.2. Test Reagents
Test simulate acid rain in acidic liquid with c ($S0_4^{2-}$): c (NO_3^-) =5:1 and adjust pH at 4.8 with deionized water [9].

2.3. Test Methods

2.3.1. pH Determination of Uranium Tailings. According to the International Standard for the Determination of Soil Quality-pH (ISO10390: 2005) [10]. The sample was air-dried at room temperature, a sample of 1: 5 by volume and a suspension of water, weighing the dried sample of 2 mm diameter sieve 20 g in 100 ml high beaker, adding 100 ml of deionized water to remove CO_2, stirring the stirrer for 1 min. After standing for 30 min, the pH was measured with a PH-3C acidity meter.

2.3.2. Leaching experiment design. Three parallel leaching columns were made of 10mm Plexiglas with an inner diameter of 7 cm and a height of 130 cm. Sampling ports were set at 20 cm, 40 cm, 60 cm, 80 cm. Loading 30 cm high gravel firstly, and then getting the 4 to 1 soil samples of the uranium tailings into it, height 80 cm, after compaction, the bulk density of the packed soil was consistent with the measured soil bulk density as 1.41 $g{\cdot}cm^{-3}$.Leaching device was controlled by peristaltic pump, the flow rate of the peristaltic pump was set to 17 $ml{\cdot}min^{-1}$ and the time was set to 60 min. The total amount of leaching was 1020 ml. A part of the supernatant was taken out for sampling and filtering, the solution was measured with PH-3C acidity meter, and the 5-Br-PADAP spectrophotometric method for the determination of uranium, the schematic diagram of the test device is shown in Figure 1.

No. 1 column for direct leaching uranium tailings sampling soil, No.2 column surface covered with 50 grams lime and leaching uranium tailings sampling soil, No.3 column surface covered with 25 grams of lime and 25 grams of diammonium phosphate and leaching uranium tailings sampling soil, Three leaching columns were leached for 50 days at simulated acid rain (pH =4.8). Eight sampling ports on both sides of tailing soil at the depth of 20 cm, 40 cm, 60 cm, 80 cm, each sample port collects 50 ml of leachate. Take two symmetrical sampling ports to determine the average uranium concentration and pH value as the final uranium concentration and pH for each depth, collecting totally 400 ml per day, the bottom sampling port collects 600 ml of leachate per day.

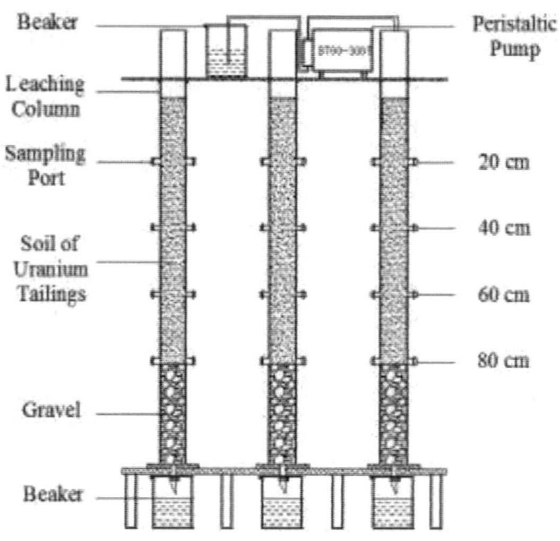

Figure 1 Diagram of experimental device.

3. Experimental results and analysis

3.1. Chemical composition

Table 1 for uranium tailings samples using inductively coupled plasma emission spectrometry (ICP-MS) measured uranium and X-ray diffraction analyzer measured part of the heavy metal content analysis results. The total amount of tailings per kilogram of tailings U is 142 mg·kg^{-1}, and the initial pH values of the samples 1 to 4 of the uranium tailings were 4.38, 4.36, 4.37, 4.22 respectively.

Table 1. The contents of uranium and heavy metals in the uranium tailings. (Mg·kg^{-1})

Element	U	SiO_2 (%)	Al_2O_3 (%)	Fe_2O_3 (%)	CaO (%)	K_2O (%)	Na_2O (%)	MgO (%)
Average Value	142	84.15	8.31	1.54	0.25	1.58	0.12	0.53

3.2. Uranium Tailings in the Nuclide Uranium Content and the Dynamic Changes

After 50 days of continuous leaching of simulated acid rain on uranium tailings in different depths, three column precipitation out of uranium cumulative precipitation shown in Figure 2.

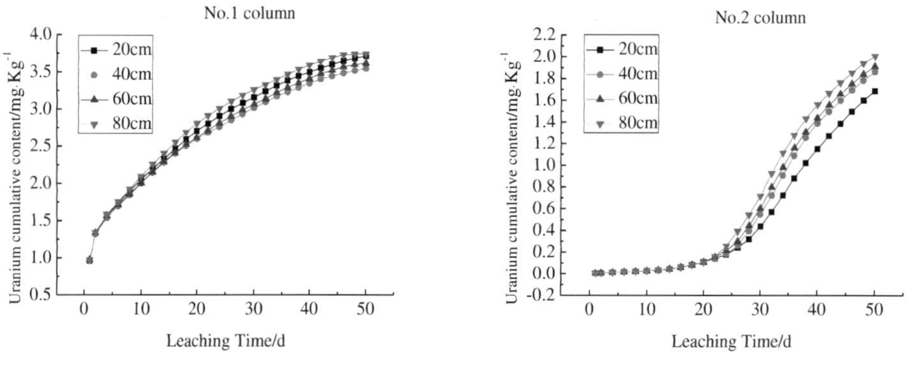

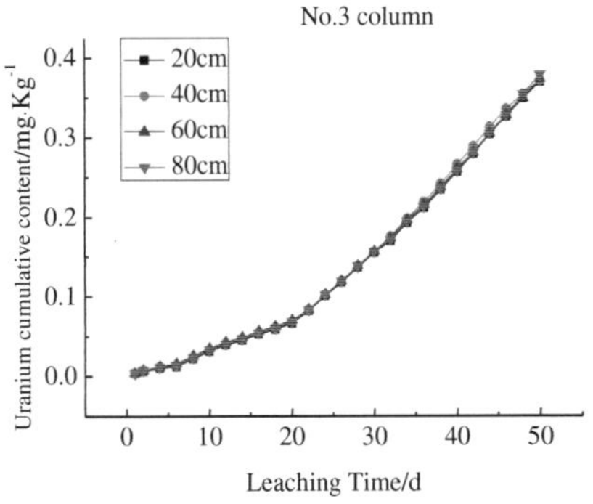

Figure 2. The changing curve of cumulative release of uranium with time

From Figure 2, the cumulative precipitation of uranium over time shows that: different leaching column in the 50 days acid rain leaching, the precipitation law of uranium at different depths is similar, but only the total amount of precipitation is different. No.1 column direct leaching in the uranium tailings soil. Uranium precipitation increased rapidly in the initial period of the leaching cycle, and precipitation tends to be stable and gradually reduced in late period. Due to simulated acid rain leaching tailings, Precipitated the surface of the surface covered by the aged oxide film to produce a new surface.

It improves the adsorption capacity of the tailings, and the precipitation is controlled by internal diffusion, and the precipitation rate is slowed down. No.2 column cover 50 grams lime in the uranium tailings soil, 20 days before the cumulative precipitation of uranium close to zero. Adding lime makes the soil in the alkaline environment, uranium through complex precipitation, uranium element activity is inhibited. So that the nuclide in the alkaline environment deposition and not spread, 20 days after the cumulative precipitation of uranium gradually increased. The precipitation of uranium in the surface of the No.3 uranium tailings is 25 grams of lime and 25 grams of diammonium phosphate, and the precipitation concentration of uranium is increasing. However the amount of uranium precipitates is very small. It may be due to the fact that the soil is in the alkaline environment. And at that time, the addition of diamines causes the strong adsorption of UO_2^{2+} by phosphate ions to form a precipitate of uranium phosphate, which prevents the migration of uranium.

3.3. Vertical distribution of nuclides in uranium tailings

The distribution of uranium in uranium tailings can reflect the degree of uranium pollution and the potential environmental risk of uranium in uranium tailings soil. In this experiment, uranium tailings soils were treated with simulated acid rain leaching under different fixatives uranium content precipitated after 50 days and shown in Figure 3.

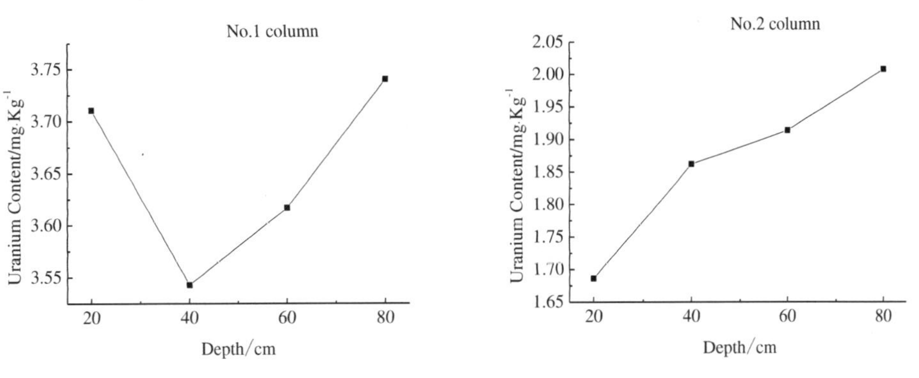

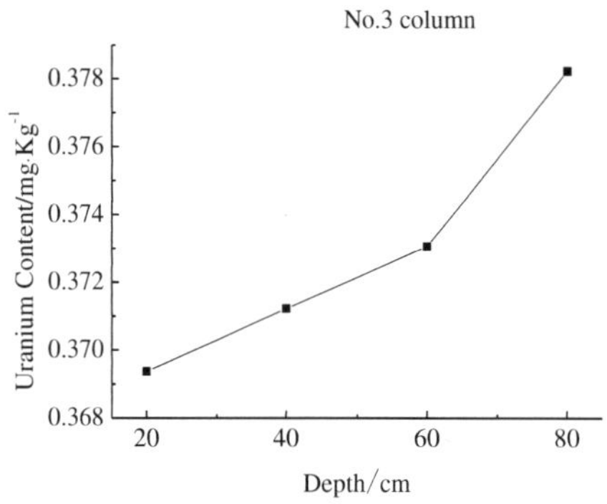

Figure 3. Uranium content at different depths

It can be seen from Figure 3 that the distribution of uranium in No.1 column soil is not uniform and does not have a linear relationship with depth. The content of uranium in 80 cm tailings is the highest, and Uranium content in 20cm tailings soil followed, The precipitated concentrations of uranium in the different depths were 3.71 mg·Kg^{-1},3.54 mg·Kg^{-1},3.61 mg·Kg^{-1},3.74 mg·Kg^{-1} respectively. The soluble uranium in the simulated acid rain will dissolve and migrate downwards, Wu X Y et al.[11] have shown that rainfall has seriously affected the migration and diffusion of radionuclide uranium in groundwater, Cations in acid rain (such as H^+) also accelerate uranium replacement and precipitation, So that the uranium in the tailings soil at 80 cm the highest content. 20 cm tailings soil in the soil surface, with high oxygen content, uranium is prone to migration in the oxidation conditions. The contents of No.2 column and No.3 column uranium increased with the depth, the fixative was placed in the tailings soil surface, after 50 days of acid rain leaching, the fixative dissolved with the infiltration, and uranium content in the different depths is in the range of 0.68 to 2.02 mg·Kg^{-1}, and the uranium content in No.3 column is in the range of 0.36 to 0.38 mg·Kg^{-1} at different depths. Through the intuitive analysis of the data, when lime and diammonium phosphate were used as fixing agents, more conducive to control the precipitation of nuclides uranium.

4. Conclusion

The vertical distribution of nuclide U in tailings soil is not regular, and the maximum amount of radioactive precipitates is reached at 60cm to 80 cm. The amount of uranium precipitate in the soil surface to 20 cm deep may be in the oxidized environment or in the soil, resulting in uranium precipitation is higher than 20 cm to 60 cm precipitation. In the leaching experiment, the cumulative release and the release rate of uranium are larger and the precipitation increases with time, and it is easy to leach out in the process of simulated acid rain. When the lime is used as the fixing agent, the initial fixation effect is good. Uranium has been deposited over time has been re-precipitated due to the acid reflux phenomenon. When the lime and diammonium phosphate are used as the fixing agent, the migration of uranium is inhibited, suitable for long-term treatment of uranium tailings.

Acknowledgments

This work was financially supported by National Natural Science Foundation of China (51174117), Hunan Provincial Department of Education Research Project (14K106).

References

[1] Xu L, Qian J P, Tang Z W. Characteristics and Treatment of Uranium Waste Residue in China [J]. China Mining, 2013, 22 (1): 61-64.
[2] Liu J, Li H C, Wang J. Environmental quality of surface waters in a uranium industrial site in South China [J]. Environmental Chemistry, 2012,31 (7): 981-987.
[3] Zhang B, Zhang X W, Li M, et al. Study on characteristics of uranium tailings pollution and

comprehensive treatment technology [J]. China Mining Industry, 2015 (4): 58-62.
[4] Zhang X W, Zhou Y H, Kui L C, et al. Study on filling treatment of heap leaching tailings of uranium ore [J]. Environmental Engineering, 2004, 22 (1): 51-53.
[5] Qin J F, Li X Y, Zhou Y S, Li X J, Hu P H. The effect of radiation in tailings filling process in uranium mine [J]. Uranium mining and metallurgy, 2015,34 (1), 44-47.
[6] Cai L P, Tan K X, Shi W G, et al. Experimental study on ion exchange and neutralization of acid leaching uranium wastewater [J]. Modern Mining, 2010, 26 (11): 21-22.
[7] Yang M T. Study on the Status of Radionuclide Migration Research [J]. Nuclear Electronics and Detection Technology, 2006, 25 (1): 878-880.
[8] Jurjovec J, Ptacek C J, Blowes D W. Acid neutralization mechanisms and metal release in mine tailings: a laboratory column experiment[J]. Geochimica Et Cosmochimica Acta, 2002, 66(9):1511-1523.
[9] Xu W. Characteristics and Mechanisms of Ion Releases of Several Soils Collected from Northern Guangdong under Simulated Acid Rain [D]. Journal of Hunan Agricultural University, 2005.
[10] Li Q, Zhao X L, Hu C R. ISO10390: 2005 Soil Quality-Determination of pH [J]. Pollution Control Technology, 2006 (1): 53-55.
[11] Wu X Y, Xiong Z W, Peng X Y, et al. Multi-simulation study on the effect of rainfall on nuclide migration in groundwater of uranium tailings pond [J]. Journal of Safety and Environment, 2013 (1): 92-95.

A brief analysis on contribution of Emeishan basalt to the formation of deposits in Guizhou

Tao Cui

College of Resources and Environmental Engineering, Guizhou Institute of Technology, Guiyang, 550003, China

cuitao1203@163.com

Abstract. All Based on previous data and field research, we consider that Emeishan basalt is important for genesis of numerous mineral resources in the southwest and northwest of Guizhou Province, and distribution of different types of mineral resources may be used as reference for geological surveys. With the advancement of technologies, aluminium and rare earth elements will have great potential for development.

1. Introduction

At present, Emeishan basalt is the only one that is recognized by the international academic circle as large igneous province in China (Courtillot et al., 1999). It has been studied by Mei Houjun and Huang Kainian et al since the 1980s(Huang,1988; Nie, 2014), investigated by many scholars from the perspective of its distribution features, formation time, petrology, geochemistry and isotopes in their studies, where numerous outcomes have been gained (Zhang et al.,2003; Song et al., 2005; Xu et al., 2002; Hu et al., 2005; Ni, 2007, 2014; Qian et al., 2007; Zhu et al., 2011; Meng et al., 2015; Li, 2014). The content of many elements in Emeishan basalt is above the background value of the crust. Emeishan basalt has gone through multistage denudation during its eruption and genesis. Owing to its relatively high content of elements, weathering, denudation and enrichment, the formation of multiple minerals has become possible. In spite of numerous previous research outcomes, there has been very little research about metallogenic regularity of Emeishan basalt across the Southwest China. According to previous research outcomes, multiple mineral resources of Emeishan basalt are statistically analyzed in a comprehensive manner, and their contributions are discussed to analyze causes of metallogenic differences in Emeishan basalt among different areas.

2. Distribution of Emeishan Basalt

Emeishan basalt originally meant the basalt on the Maokou Formation in mountainous areas of Emeishan in southwestern Sichuan Province, whereas it was generally known as trap extensively distributed in three provinces of southwestern China, especially represented by late Permian basalt (Fig 1), and widely utilized by a formal lithostratigraphic unit of the Upper Permian (Bureau of Geology and Mineral Resources of Sichuan Province, 1991). With a volume that appropriately equaled to $0.3x10^6KM^3$, Emeishan basalt is differs in thickness in different areas, distributed along hundreds and even fifty thousand meters (Xu et al, 2001; Ali et al, 2005). In Southwestern China, Emeishan basalt is irregularly distributed approximately from the northwest to the east. There is

Longmenshan fault zone in northwestern China, Honghe fault zone in southwestern China and large fault zones in central China (Fig 1, Fig.2, a). Limestone of the Maokou Formation underlies under Emei basalt, which is covered by late Permian strata and tends to be increasingly thinner from the west to the east (He et al., 2003).

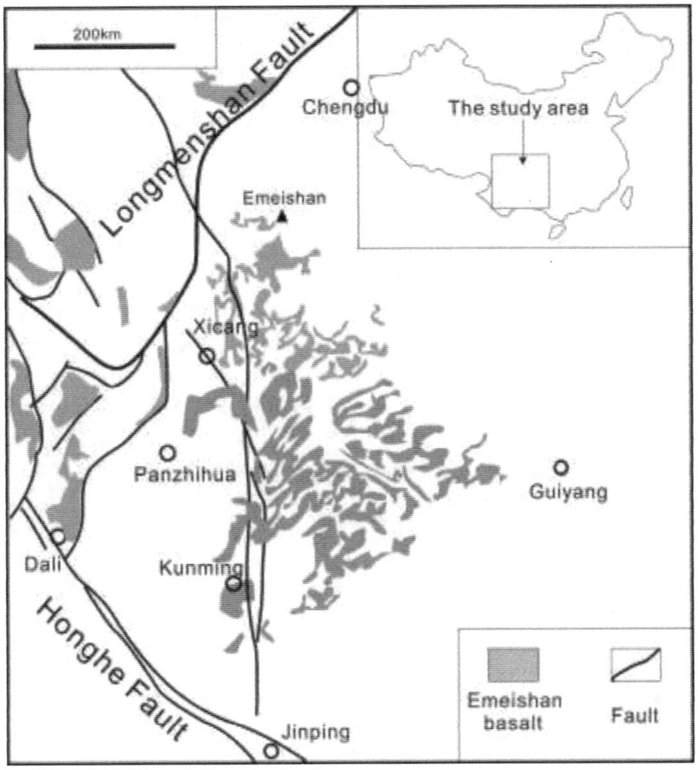

Fig.1 Distribution of Emeishan Basalt [Modified from xu (2008)]

3. Deposits Related to Emeishan Basalt

Emeishan basalt has created conditions for the formation of many mineral resources, and deposits related to Emeishan basalt mainly include: 1) Cu-Ni-PGE deposits with direct connections with mantle plume activities of Emeishan (Nie , 2007; Tang, 2013);2) V-Ti-Mt deposits associated with magmatic differentiation of basic-ultrabasic rocks (Li et al., 2002; Zhang et al., 2004); 3) Cu deposits connected with hydrothermal action (Li et al., 2005; Zhang et al., 2008), Cu-Fe deposits formed for weathering/deposition (Wang et al., 2006; Wang et al., 2011; Li, 2014), anatase associated with weathering/deposition (Nie, 2014); 4) mesothermal and epithermal hydrothermal gold/antimony deposits (Jiang et al., 2009; Nie et al., 2009; Liao et al., 2010); 5)Ore deposits of northwestern Guizhou Province include bauxitic strata, where there is a high content of REE (Fig 2, b). In some areas, both aluminum and REE are likely to be enriched and form ores. 6) Basalt is widely used in engineering construction of railways and roads as stone materials (Fig 2, a).

Since surveying of mineral resources is so complex and uncertain that mineral resources related to basalt of Emeishan basalt would not just include the metal minerals mentioned above. With the advancement of surveying technologies and progress of surveying, new mineral resources related to the basalt will be inevitably discovered. In addition, industrial grades of some minerals or elements will decline with the advancement of extraction technologies. These resources will also become integral parts of mineral resources of Emeishan basalt.

IOP Conf. Series: Earth and Environmental Science **64** (2017) 012022 doi:10.1088/1755-1315/64/1/012022

Fig.2 Emeishan basalt (a) and REE layer (b)

4. Contribution of Emeishan Basalt to Metallogenesis and Metallogenic Differences

The genesis of numerous minerals is related to Emeishan basalt, which provides provenance for plenty of mineral resources. This is just the tremendous contribution of Emeishan basalt to the Metallogenesis in three provinces of southwestern China. It is generally uncontroversial that Emeishan basalt offers provenance to multiple types of deposits, whereas the roles of the basalt in lead-zinc ores are highly controversial. Some scholars consider that Emeishan basalt provides provenance for lead-zinc ores (Han et al., 2001; Huang et al., 2001) and some others think both of them just have overlaid spatial relationships (Zhou, 2002). In addition, some other scholars believe that Emeishan basalt provides thermodynamically activated and extracted materials during metallogenesis of lead-zinc ores, acting as “shielding strata” and “protective strata” (Zhang et al., 2003; Li et al., 2012). The author reports that even if Emeishan basalt is not the provenance of lead-zinc ores, it has created certain conditions for the genesis of these ores, because such large-scale spatial overlay would be not accidental, they are critical for metallogenesis even if they are just employed for “blocking” or “protection”.

Metallogenic zonation is detected in areas where Emeishan basalt is distributed, typically including “northwestern copper-lead-zinc and southwestern Au-As-Hg-Sb” in western Guizhou Province (Nie, 2014), and there is abundant iron resource in northwestern Guizhou. Except for lead and zinc, multiple types of mineral resources such as copper, iron and antimony need to be migrated for metallogenesis. The mineral resources with weak migration capabilities deposit easily, while those with strong migration capabilities may move for a relatively long distance until they deposit in appropriate paleogeographic environment. The differences in sea-land layouts contribute to two different kinds of metallogenic environment, so the author considers that paleogeography is an important factor for zoning of mineral resources. From the distribution of mineral resources on different belts, it may be discovered that future surveying work can be carried out according to metallogenic distribution laws: copper, lead and zinc are approximately negatively correlated to gold, arsenic, antimony and mercury. Gold, arsenic and antimony would be contained in copper, lead and zinc-free areas, while copper, lead and zinc would exist in gold, arsenic and antimony-free areas. The rare earth elements were possibly negatively correlated to gold.

5. Conclusion

After a comprehensive analysis, following conclusions are reached:1) The research of numerous experts and scholars has suggested that Emeishan basalt is important for genesis of many metallic mineral resources, it does not only provide metallogenic materials for several categories of metallic minerals, but also possibly creates certain conditions for lead-zinc ores. 2) With the advancement of surveying technologies and intensification of geological work, aluminum and rare earth elements will have potential value for development and utilization.

Acknowledgements

This work was supported by the initializing fund projects of Guizhou institute of technology (No.XJGC20131204; No.XJGC20140702), Guiyang, 550003, China, the provincial project of characteristics of REE and its transferring disciplinarian in mineralizing process of the bauxite in northern Guizhou, China (No.LH[2014]7358), and the Key support disciplines of Mineral prospecting and Exploration from Guizhou Province(ZDXK[2014]20).

References

[1] Courtillot V. Evolution catastrophes: The science of mass extinctions. Cambridge: Cambridge university press,1-237.

[2] Huang Kainian, Yang Ruiying, Wang Xiaochun et al.. A preliminary study on trace element geochemistry of Emeishan basalts from SW China. Acta petrologica sinica, 1988,21(4):49-60.

[3] Nie Aiguo, Kang geng. Research on the Metallogenic difference of Emeishan basalt in Guizhou. Guizhou technology press, 2014, 86-87.

[4] Zhang Zhaochong, Wang Fusheng. Sr, Nd and Pb isotopic characteristics of Emeishan basalt province and discussion on their source region. Earth science- Journal of China university of geosciences, 2003,28(4):431-438.

[5] Song Xieyan, Hou Zengqian, Wang Yunliang et al.. The mantle plume features of Emeihshan basalts. Journal of mineralogy and Petrology, 2002,22(4):27-32.

[6] Xu Yigang. Mantles plumes, large igneous provinces and their geologic consequences. Earth science frontiers,2002,9(4):341-353.

[7] Hu Ruizhong, Tao yan, Zhong hong, Huang Zhilong, Zhang Zhengwei. Mineralizaiton system of a mantle plume: A case study from the Emeishan igneous province, southwest China. Earth science frontiers, 2005,12(1):42-54.

[8] Nie Aiguo, Qin Dexian, Guan Daiyun, Huang Zhiyong, Zhang Zhuru. A research on regional metallogenic contribution to gushing Emeishan basalt magma in western of Guizhou province . Geology and prospecting, 2007,43(2):50-54.

[9] Qian Zhuangzhi, Xu Cuiling, Zhang Zhengjun, Jiang Changyi, Hou Shuguang, Tang Dongmei. Ore-formation source of copper deposits in Emeishan basatlt area in northeast Yunnan province. Journal of mineralogy and Petrology, 2007, 27(1):78-82.

[10] Zhu Jiang, Zhang Zhaochong, Hou Tong, Kang Jianli. La-ICP-MS zircon U-Pb geochronology of the tuffs on the uppermost of the Emeishan basalt succession in Panxian County, Guizhou Province: Constrains on genetic link between Emeishan large igneous province and the mass extinction. Acta Petrologica Sinica, 27(9):2743-2751.

[11] Meng Changzhong, Chen Yang, Zhang Yinghua, Wu Hui, Lin Wenli. Unconfig of Emeishan LIP:-factor of polymetallic deposit: isotope constraint of U-Pb of zircon. Science China, Vol. 45(2015) No.10, p.1469-1480.

[12] Li Songtao. Geological and Geochemical characteristics of iron-copper deposits in Lushan area, Weining, Northwest Guizhou(Doctor, Chendu university of technology,(2014), p.57-80.

[13] Sichuan Geology bureau. Sichuan Regional geological records.Geology press, Beijing, 1991,1310-1312.

[14] Xu Y G, Chung S L, Jahn B M, et al. Petrologic and geochemical constraints on the petrogenesis of Permian-Triassic Emeishan flood basalts in southwestern China. Lithos, 2001,58: 145-168

[15] Ali J R, Thompson G M, Zhou M F, et al. Emeishan large igneous province, SW China. Lithos, 2005,79: 475-489.

[16] He Bin, Xu Yigang, Xiao long et al.. Generation and spatial distribution of the Emeishan large igneous province:new evidence from stratigraphic records, Acta geologica sinica,2003,77(2):194-202.

[17] Xu Yigang, Luo Zhengyun, Huan Xiaolong, He Bin et al.. Zircon U-Pb and Hf isotope

constraints on crustal melting associated with the Emeishan mantle plume. Geochim cosmochim Acta,2008,72:3084-3104.

[18] Tang Qinyan. Comparison of ore-forming systems of Cu-Ni-PGE Sulfide deposits associated with the Emeishan and Siberian mantle plumes, and Jinchuan Cu-Ni-PGE sulfide deposit. Lanzhou university, 2013,56-63.

[19] Li Hongyang, Niu Shuyin, Wang Lifeng et al.. Mantle plume structure. Beijing, Earthquake press, 2002.

[20] Li Houming, Mao Jingwen, Wang Denghong et al.. PGE and trace element characteristics of copper mineralization in Emeishan basalts of Yunnan-Guizhou border area. Mineral deposits,2005,24(3):285-291.

[21] Zhang Qian, Zhu Xiaoqing, Zhang Zhengwei, Wang Dapeng. Discussion on the ore-formation prospecting of the Emeishan basalt-type native copper- chalcoctte deposits in the Weining district of Guizhou Province, China. 2007,27(3);379-383.

[22] Wang Juli, Guo Jian, Liu Zhongkui, Zhang Yunfeng, Zhang Rong et al.. Sedimentray copper deposit in Emeishan basalt, northeastern Yunnna Province.Mineral deposits,2006,25(6):663-671.

[23] Wang Fudong, Zhu Xiaochun, Han Tao, Wang Zhonggang. An experimental study on genesis of Emeishan basalt-related native copper deposit in the Siuchuan-Yunnan-Guizhou area. Acta mineralogica sinica, 2011,31(1):322-327.

[24] Jiang Jie, Li Weifeng. Geological characteristics and prognosis in the periphery of dachang antimony orefield in Guizhou Province. Geology and resources, 2009,18(4):288-291.

[25] Liao Zhenwen. Comparision of geological characteristics of the copper and gold deposits related to Emeishan basalt in the borders area between Yunnna and Guizhou. Journal of Jilin university(Earth Science Edition).2010,40(4):821-827.

[26] Han Runsheng, Liu Chongqiang, Huang Zhilong, Chen Jin. Genesis model of Huizhe LEAD-ZINC ore deposit in Yunnan, Acta mineralogica sinica. 2001,21(4):674-680.

[27] Huang Zhilong, Chen Jin, Liu Chongqiang, Hu Runsheng, Li Wenbo. A preliminary discussion on the genetic relationshio between Emeishan basalt and Pb-Zn deposits as exemplified by the Huize Pb-Zn deposit, Yunnan province, Acta mineralogica sinica. 2001,21(4):681-688.

[28] Zhang Changqing. Distribution, characteristics and genesis of mississippi Valley-type lead-zinc deposits in the triangle area of Sichuan-Yunnan-Guizhou Provinces. China university of Geosciences,2005,73-77.

[29] Zhang ming. Geochemistry characteristics and enrichment mechanism of Ga, Ge in Lead-zinc ore at Huize of Yunnan province, Chengdu university of technology, 2003,47-58.

[30] Li bo, Gu Xiaochun, Wen Shuming et al.. Effect of Emeishan bsalt in northesat Yunan on lead and zinc mineralization, mineral resources and geology, 2012,26(2):95-100.

Research progress in metallogenic chronology of Mississippi valley-type (MVT) Pb-Zn deposits

Huiguo Liu

Guizhou Vocational Technology Institute, Guiyang, Guizhou 550023, China.

huiguo_liu@outlook.com

Abstract. MVT Pb-Zn deposits are one of world's important Pb-Zn deposits. Their lead and zinc account for approximately 20% of global lead and zinc resources. After constant explorations, researchers have made substantial progress in geological features, geochemical characteristics, ore-forming material sources, and transport and deposition mechanisms for ore-forming fluids, chronology restriction, and geochemical kinetic background for ore genesis of MVT Pb-Zn deposits over the past decades. This paper discusses problems with chronological research by summing up previous outcomes.

1. Introduction

As a type of Pb-Zn deposits with striking epigenetic characteristics produced in carbonatite, MVT deposits are well-known for their typical development in the Mississippi River of the United States. This kind of deposits has been studied since one century ago. Earlier studies mainly recorded features of hydrothermal genesis in investigating these deposits and defined the deposits as post-magmatic hydrothermal deposits or tele-hydrothermal deposits. By 1970s, many scholars further explored these deposits according to theories of strata-bound ore deposits. They deeply analyzed and summarized their metallogenic background, their associations with magma rocks, spatial distribution, ore-controlling factors and features (Garven, 1985; Ge et a1., 1992; Appold et a1., 1999). In particular, many new data, including lead isotopes, sulfur isotopes, fluid inclusions and metallogenic fluids, were obtained from the perspective geochemistry and mineralogy of deposits by performing hydrothermal experiments. Lots of geological and geochemical data of deposits have fully demonstrated that the genesis of these deposits is not associated with magma or hydrothermal liquids of magma. Nowadays, it is acknowledged that these deposits are epigenetic under the control of stratigraphic horizons, arising from filling and metasomatism of underground brine between space of carbonatite, including voids, cracks, karst caves, surface of unconformity and intercalated crushed zones. With the increase of market demands for lead and zinc in the world, people have paid more attention to explore MVT Pb-Zn deposits over the past years. In this paper, MVT deposits are studied from the perspective of their metallogenic chronology, in hope of better understanding these deposits.

2. Structural Settings and Metallogenic Environment

MTV Pb-Zn deposits were considered to be irrelevant to plate tectonics and it was considered that these deposits would be metallogenic as long as there was carbonatite on meseta. However, since 1980s, more and more studies have suggested that most MVT Pb-Zn deposits of the world have

IOP Conf. Series: Earth and Environmental Science **64** (2017) 012023 doi:10.1088/1755-1315/64/1/012023

formed when extensive ore-forming fluids flow through foreland basins with the drive of gravity of adjacent orogenic belts and their metal sulfide deposits (Garven, 1985; Ge et al, 1992, Appold et al, 1999). Although surrounding rocks, basin brine, driving forces of fluids and areas where ore body deposits are critical for genesis of deposits, all of these conditions are finally caused by activities of tectonic structures inside these structures, so it is important for studying genetic causes of deposits and guiding prospecting by exploring tectonic backgrounds. These years, some research has suggested that MVT Pb-Zn deposits mostly form in foreland basins of orogenic belts, a few of them form in thrust-nappe belts and rarely in extensional tectonic structures of continents. This view was questioned at first when it was just proposed, but has been widely acknowledged at last.

3. Research on Metallogenic Chronology

With the development of high-precision high-resolution mass spectrometers and advanced super clean laboratories, people have studied dating techniques of deposits increasingly more intensively and several radioisotope dating methods have been made available for more widespread uses. Concerning many problems about metallogenic epoch of MVT Pb-Zn deposits, relatively satisfactory results have been obtained. The dating methods are not only applicable to certain extent, but also have their respective limitations. Hereunder, some major dating techniques will be discussed.

(1) Paleomagnetic method

The application of paleomagnetic techniques is a major progress in dating of MTV Pb-Zn deposits. Further magnetization occurs in chemical reactions related to fluids, so paleomagnetic fields may be recreated in surrounding rocks with the transport of ore-forming fluids of MVT Pb-Zn deposits. More favorably, these deposits generally develop in unperturbed and structurally stable on carbonatite meseta, which is ideal for palingenestic dating. However, precision of magnetometer was always just measured based on theories for certain period. The magnetometers haven't been put into practices and widely utilized until the emergence of low-temperature magnetometers. Symons et al (1996) summarized paleomagnetic dating methods and several practical instances. They pointed out that these methods would be applicable if only there were magnetic minerals for recording directions of geomagnetic field such as magnetic iron ores, red iron ores and pyrrhotite ores whatever the content.

(2)^{40}Ar-^{39}Ar(^{40}K-39Ar) Method

Over the past decades, it has been the most widespread for dating hydrothermal genesis of K-included hydrothermal minerals by ^{40}Ar-^{39}Ar. concerning this dating method, the measured objects mainly include alkali feldspar, mica,hornblende, illite and quartz. ^{40}Ar-^{39}Armethod performs many other methods in following aspects: Evolution histories of minerals can be usually reflected from plateau ages on age spectra obtained from multi-stage heating. ^{40}Ar-^{39}Ar method is incomparable to other methods in this aspect. ^{40}Ar-^{39}Ar is more precise than ^{40}K-^{39}Ar in dating. Only a small amount of samples is needed by this method that very few problems are caused by non-homogeneity of samples. For ^{40}Ar-^{39}Armethod, the experiment performed by Kontak et al. (1994) is the most typical case of dating MVT Pb-Zn deposits. In their experiments, they finally exactly confirmed the metallogenic age of the MVT deposits to be±27Ma by measuring plateau ages of sedimentary clastic rocks related to the deposits in Gays River, which was consistent with 300-320Ma determined by paleomagnetic measurements. This indicates that the deposits formed in the paleozoic period of Pennsylvania and their genesis was associated with Alleghanian orogeny.

(3) U-Pb Method

In the metallogenic epoch, uranium-bearing minerals such as zircon, rutile, edwardsite, allanite, sphene and apatite involved in hydrothermal genesis in altered belts or ore bodies were mostly measured by single-grain U-Pb method. However, very few uranium-bearing minerals suitable for U-Pb dating form during metallogenic genesis, so this method isn't used so widely as ^{40}Ar-^{39}Ar, Rb-Sr and Sm-Nd for studying metallogenic epoch. Many scholars measure epoch of MVT Pb-Zn deposits by U-Pb method. Brannon et al (1992) measured that the age of calcite (a gangue mineral) in MVT Pb-Zn deposits of east Kansas and Canning basins by^{232}Th-^{208}Pband^{238}U-^{206}Pb, confirmed to be351±15Ma.

(4) Rb-Sr Method

Rb-Sr method has been proven to be effective for directly measuring the mineralization age of MVT Pb-Zn deposits. Nakai et al. (1990) firstly reported Rb-Sr isochron age of sphalerite in Coy Deposit of East Tennesee Mine to be377±29Ma. This result was in disagreement with previous view that the genesis of MVT Pb-Zn deposits in middle and easten parts of the United States was associated with late paleozoic alleghanian orogeny before330-250Ma. Besides, Nakai et al(1990) proposed a new metallogenic model for basin compaction. Furthermore, they measured the Rb-Sr isochron age of sphalerite in Immel Deposit of east Tennessee Mineto be347±20Ma, which was in line with the metallogenic epoch of the Coy Deposit, so this demonstrated that the deposit of that mine formed between 350-380Ma and its genesis was connected with mid paleozoic acadian orogeny. Brannon et al (1992) measured the Rb-Sr isochron ages of sphalerite in Upper Mississippi Valley by Rb-Sr method and concluded that they were 269±6Ma and 270±4Ma, which formed a sharp contrast with Nakai's 377±29Ma. Nevertheless, Brannon et al (1992) considered that the completely different age of two metallogenic provinces in North America indicated that in those areas, MVT Pb-Zn deposits formed through multiple stages, including at least two metallogenic epochs, and they were mostly associated with major tectonic movements of the same stage. In China, Li et al. (2004) clarified the mineragenetic epoch of No.6 Ore Body of Qilin Plant in Huize Pb-Zn Deposit of Yunnan Province to be 225.11±2.9Ma and 225.9±3.1Ma respectively. Thus, they concluded that Pb-Zn metallogenesis inside metallogenic provinces of Sichuan, Yunan and Guizhou provinces was associated with movement of basaltic magma in Emei Mountain

However, isochron age of most MVT Pb-Zn deposits hasn't been reasonably determined because sphalerite samples are not closed after their metallogesis, or inclusions of clay minerals exist in sphalerite, or fluid inclusions still exist in residual facies of sphalerite. So far, the causes of this phenomenon are still unclear. At present, Institute of Geology and Geophysics, Chinese Academy of Sciences has successively measured single-grain mica and single-grain pyrite from the perspective of Rb-Sr isotopes and according to chemical procedures of ultra-low backgrounds using advanced and high-precision IsoProbe-T thermal ionization mass spectrometers. To certain extent, this measurement method not only satisfies basic closure requirement for Rb-Sr dating, but also increases precision of Rb-Sr measurement and reduces lower limit of tests. With the promotion of this research outcome, worthy lessons may be learnt from it for Rb-Sr dating of low-temperature hydrothermal Pb-Zn deposits.

(5) Sm-Nd Method

Although this is a relatively new and effective dating method, it is not commonly used for dating MVT Pb-Zn deposits. Halliday et al (1990) reported the application of Sm-Nd method in exploring metallogenic causes of MVT Pb-Zn deposits in Pennines in the north of England. Although they failed to clarify the exact metallogenic age, they concluded after comparing ^{40}Ar-^{39}Ar and Rb-Sr ages that the metallogenic age was about 200Ma, and the metallogenesis of these deposits was associated with regional tectonic movements between the carboniferous period and carbonic period. By analyzing Sm-Nd isotopes of fluorite in MVT Pb-Zn deposits of Illinois-Kentucky mine, Chesley et al.(1994) concluded that the metallogenic age of this mine was 277±16Ma and thus inferred that the metallogenesis of the deposit was attributable to large-scale fluid movement arising from the movement of the Allegheny River. Li et al. (2004) measured metallogenic epoch of No.6 and No.1 ore bodies of Qilin Plant with Rb-Sr isochrone of calcite (a gangue mineral). Their results suggested that the metallogenic epoch of both mine concessions was the same. The metallogenic epochs were 226±15Ma and 225±38Ma respectively for the deposit of the Qilin Plant and the deposit of the mine. The measured age was close to the metallogenic epoch of extensive basalt in the southwest of Emei Mountain within the error scope. They provided chronology evidences for proving associations between metallogenesis of Pb-Zn deposits and movement of basaltic magma on Emei Mountain.

(6) Fission Track Dating

Fission track dating is simple, feasible, sensitive and applicable to widespread objects, particularly suitable for measuring young samples. This dating method shall be adopted on the premise that uranium and fission tracks are closed and measured minerals are generally hypothesized to be associated with mineralization in terms of their annealing.

4. Conclusion

In spite of some doubts about dating of MVT Pb-Zn deposits, Leach et al (2001) discovered in summarizing ages of major MVT Pb-Zn deposits in the world that among 10 deposits measured by paleomagnetism and isotope to identify their ages, 8 results were consistent. This suggests that the paleomagnetic and isotopic dating results are generally the same among most deposits. Only evidences haven't been found for demonstrating that the research data couldn't reflect metallogenic ages in 3 deposits measured by isotopic dating and 4 deposits measured by paleomagnetic dating. Thus, it is clear that after decades of explorations, some progress has been made in exploring metallogenic ages of MVT Pb-Zn deposits. The methods introduced above are effective for exactly measuring ages of the deposits under certain conditions.

5. Conferences

[1] Garven G. The role of regional fluid inclusions: a new tool in constraining source regimes and its implications for the genesis of Mississippi Valley-type deposits. Economic Geology, 1985, 80: 307-324.

[2] Ge S and Garven G. Hydromechanical modeling of tectonically-driven groundwater flow with application to the Arkoma basin. J. Geophys. Res., 1992, 97: 9119-9144.

[3] Appold M S and Garven G. The hydrology of ore formation in the Southeast Missouri district: numerical models of topography-driven fluid flow during the Ouachita orogen [J]. Economic Geology, 1999, 94: 913-936.

[4] Symons D T A, Sangster D F, Leach D L. Paleomagnetic dating of Mississippi Valley-type Pb-Zn-Ba deposits. In: Sangster D F, ed. Carbonate-hosted lead-zinc deposits. Society of Economic Geologists Special Publication, 1996, 4: 515-526.

[5] Kontak D J, Mcbride S, Farrer E. 40Ar-39Ar dating of fluid migration in a Mississippi Valley-type deposit: Gays River Zn-Pb deposit, Nova Scotia. Economic Geology, 1994, 89: 1501-1517.

[6] Brannon J C. Allcghenian age of the Upper Mississippi Valley zinc-lead deposits determined by Rb-Sr dating of sphalerite. Nature, 1992, 356: 509-511.

[7] Nakai S, Halliday A N, Kesler S. Rb-Sr dating of sphalerites from Tennessee and the genesis fo Mississippi Valley-type ore deposits. Nature, 1990, 346: 354-357.

[8] Halliday A N, Shepherd T J, Dickin A P. Sm-Nd evidence for the age and origin of a Mississippi Valley-type ore deposit. Nature, 1990, 344: 54-56.

[9] Chesley J T, Halliday A N, Kyser T K. Direct dating of Mississippi Valley-type mineralization: use of Sm-Nd in fluorite. Economic Geology, 1994, 89: 1192-1199.

[10] Li W B, Huang Z L, Chen J, Han R S.Rb-Sr dating of mineral assemblage from the Huize giant Zn-Pb deposit, Yunnan province. Acta mineralogical sinica, 2004,24 (2):112-116.

[11] Leach D L, Bradley D C, Lewchuk M T. Mississippi Valley-type lead-zinc deposits through geological time: implications from recent age-dating research . Mineralium Deposita, 2001, 36: 711-740.

IOP Conf. Series: Earth and Environmental Science **64** (2017) 012024 doi:10.1088/1755-1315/64/1/012024

Landsat-based long–term LUCC mapping in Xinlicheng Reservoir Basin using object-based classification

Wei Su[1,*], Dongmei Liang[1], Gula Tang[2], Zundong Xiao[1], Jingxin Li[1], Zhengyu Wan[1] and Ping Li[1]

[1]Jilin Provincial Academy of Environmental Sciences, Changchun 130012, China
[2]Key Laboratory of Agricultural Remote Sensing, Ministry of Agriculture/Institute of Agricultural Resources and Regional Planning, Chinese Academy of Agricultural Sciences, Beijing 100081, China

*Corresponding author e-mail: suwei.e@foxmail.com

Abstract. Rapid urbanization dramatically changes the local environment around Xinlicheng Reservoir Basin. Landsat images are suitable for the land use change caused by human impact. In order to obtain consistent land cover products, a hybrid classification method combining object-based classification and pre-classification alteration detection method was developed and applied to long-term multi-temporal Landsat images to obtain land cover change information. Object-based classification method was combined with Random forest (RF) classifier to classify the Landsat image in 2008. Then the changed areas in 2000, 2004, 2012, and 2016 were identified by comparing with the images in 2008 via the re–weighted multivariate alteration detection transformation method. The images in 2000, 2004, 2012 and 2016 were classified by RF classifier. Land cover maps for 2000, 2004, 2012, and 2016 were produced by combining the unchanged area in 2008 with the new classes of the changed areas in 2000, 2004, 2012 and 2016. According to the accuracy assessment, the overall accuracy of the land covers of the four periods are all greater than 93%. The accuracy assessment indicates that this hybrid method can produce consistent land cover datasets for a long time period.

1. Introduction

Land use / land cover change (LUCC) is an important part of global climate change and global environmental change research [1]. LUCC is the most direct signal that expresses the influence of human activities on the natural ecological system of land surface. LUCC is the link between human social and economic behavior and natural ecological process. LUCC have an important impact on the regional ecological environment, and affect global environmental change in a cumulative manner [2-4]. It is generally believed that natural conditions and anthropogenic activities are important drivers of LUCC, which together affect the manner and intensity of LUCC. LUCC has become a hot topic at science research.

Traditionally, most LUCC identifications involve time-consuming and laborious field surveys. In recent years, with the development of remote sensing technology, a growing number of studies have focused on utilizing satellite data to detect the phenology of different land use types because the frequent remotely sensed images have significant potential for monitoring LUCC dynamics [5-7].

Currently, the methods based on pixels ignore the useful context information, thus leading to "salt and pepper" effects on the classification result [8]. On the contrary, object-based classification methods are currently gaining more attentions [9, 10]. Compared with a pixel, an object is a group of homogeneous pixels that represent a meaningful object in the real world, and contains richer information regarding the shape, context and spatial relationship of the object. To retrieve information from these images, geographic object-based image analysis has been widely used and represents an evolving paradigm [11, 12]. The object-based methods are less sensitive to registration errors and can effectively solve the "salt and pepper" effect. The spatial features can significantly improve the classification performance. More advanced methods for object identification, referred to as Object Based Image Analyses (OBIA), have been developed and applied to e.g. land-use classification where they have provided substantial improvement [13, 14]. This class of techniques builds on segmentation, edge-detection, feature extraction and classification concepts that are combined to improve the performance.

Xinlicheng Reservoir Basin is an important ecological barrier and an important water source in Changchun City. The water quality of Xinlicheng Reservoi has become a problem related to resident health and national economic development. The monitoring and evaluation of reservoir water quality has always been a relatively important research areas. With the continuous expansion of Changchun city scale, the population has increased dramatically, and the problem of urban drinking water supply has been paid more and more attention by all sectors of society. With the development of urbanization, the structure and function of land use in Xinlicheng Reservoir Basin changed drastically. The land use change will inevitably lead to the change of structure, type and regional configuration of vegetation cover, and then the stability and equilibrium of the basin ecosystem have a far-reaching impact. In addition, the existence of salinization in the region seriously restricts the development of local and its surrounding areas. At present, the study of this basin is focused on water quality and pollutants, and there is little research on long-term high-resolution land use.

In the study, the combination of the object-based classification method and IR-MAD change detection method were used to classify the Landsat satellite data from 2000 to 2016 to provide the high-resolution and long-term LUCC dataset [15, 16]. Based on the analysis of the results, the LUCC characteristics of the Xinlicheng Reservoir Basin are presented, and the LUCC can provide the basis to support the environment protection and rational utilization of land.

2. Study area and data pre-process

2.1. Study area

Xinlicheng Reservoir Basin lies in the middle of Jilin province, south of Changchun City. The outlet of the Xinlicheng Reservoir Basin is 16km away from Changchun City. The area covered by Xinlicheng Reservoir Basin is between northern latitude 43°3′~43°44′ and eastern longitude 125°0′~125°37′. The area of Xinlicheng Reservoir Basin is 1970 km^2. Xinlicheng Reservoir is located at the end of the Xinlicheng River Basin. Xinlicheng Reservoir is a large reservoir that is equipped with a comprehensive river channel for water supply, flood control, irrigation and tourism in Changchun City. Xinlicheng Reservoir east-west width of 15-30 m, north-south extension of 16 km, is a shallow lake. The study area is located in the temperate continental monsoon climate zone. In Xinlicheng reservoir Basin, the main rivers include Yitong River, Yidan River, Nadanbo River, Jiaguan River, Xinxing River, which are all seasonal rivers. Yidan River and Nadanbo River are the main upstream rivers of the Xinlicheng Reservoir.

IOP Conf. Series: Earth and Environmental Science **64** (2017) 012024 doi:10.1088/1755-1315/64/1/012024

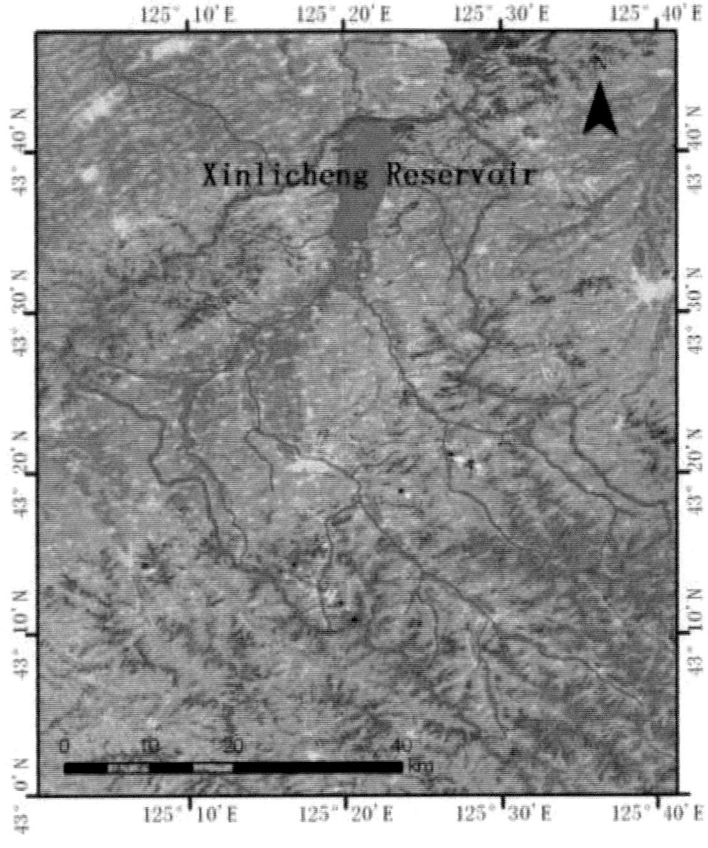

Figure 1. The study area.

2.2. Data preprocess

2.2.1. Classification system and sampling. Taking into account the spatial resolution of Landsat images and application requirements, the classification system includes 7 categories of existing classification systems: forest, grass, dry farmland, paddy field, bare land, building, and water body. The training samples were collected primarily through a visual interpretation Landsat images in 2008 referring to Google Earth. The sample points are randomly distributed in the study area.

2.2.2. Landsat images reprocessing. The land use change caused by human impact is limited, which is far lower than that caused by climate change and natural disasters. In China, this limitation is more obvious due to the small-scale individual-owned farmland (usually smaller than 1 ha). Therefore Landsat images with 30m resolution are particularly important [5]. The Landsat satellite series has continually observed Earth since 1972 and has accumulated an enormous number of time series images [17]. Landsat images have been widely used in land-cover classification because of their stable imaging quality [5, 18-21]. In this study, Level 1 terrain corrected (L1T) Landsat images were selected as the primary data source for the land-cover classifications. The Landsat images were atmospherically corrected to produce the surface reflectance by using the Landsat Ecosystem Disturbance Adaptive Processing System.

Table 1. The acquired Landsat images.

Date	path/row	Sensor
2000/09/26	118/30	Landsat 5
2004/09/21	118/30	Landsat 5
2008/09/16	118/30	Landsat 5
2011/06/05	118/30	Landsat 5
2016/09/22	118/30	Landsat 8

The acquired Landsat images were shown in table 1. However, in study area, there was not acquired Landsat image in 2012. In order to accomplish the land over every 4 years, the STARFM was used to predict the Landsat 5 image acquired on September 16, 2012 [22]. Therefore the MODIS daily surface reflectance (MOD09GA) data in June 5, 2011 and September 16, 2012 were re-projected and resampled to Landsat resolution and extend. Finally, the STARFM produced the predicted Landsat5 images in September 16, 2012.

3. Methods

The Object-oriented classification method was used in this study. First, random forest classifier was used to classify the Landsat 5 images in 2008 and then the land cover of 2008 was produced. In order to obtain consistent land cover products, the pre-classification method was used to detect the changed area. In the specific operation, the Landsat 5 image in 2008 was used as the reference image, and the IR-MAD method was used to detect the change area of the Landsat images in 2000, 2004, 2012 and 2016 compared with the Landsat 5 image in 2008. After detecting the changed areas, the random forest classifier was used to classify the changed areas. The land covers of the changed area were then integrated with the land cover in 2008, and the land cover products of 2000, 2004, 2012 and 2016 were obtained. The process was shown in Figure 2.

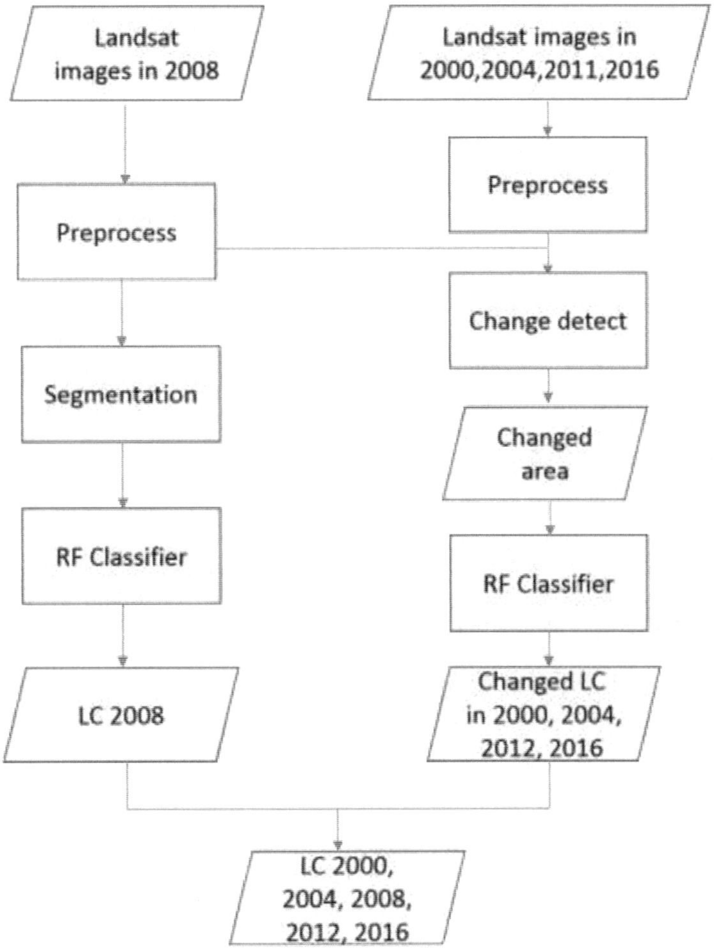

Figure 2. The flowchart.

3.1. Image segmentation

3.1.1 Mean shift segmentation method. Image segmentation is the foundation of the object-based classification [23]. The objective of image segmentation is to partition an image into spatially contiguous and homogeneous regions [24]. The segmented regions are viewed as image objects, which represent the core of object-based analysis. Generally, some parameters must be set for image segmentation, such as the spatial scale, bandwidth and shapes [25]. The mean shift algorithm has been widely used in image segmentation and identification tracking [26,27]. As a kernel density estimate method, the mean shift does not require prior knowledge of the number of segmentations and sets no limits on the segmented shape [28]. In this study, the mean shift method was used to segment the Landsat 5 image in 2008. The reflective bands with 30m resolution of the Landsat 5 images were segmented by mean shift. The Gaussian function was taken as the kernel function, and the bandwidths of the geometry space and color space were set to 10 and 8, respectively.

3.1.2 Object properties. The classification of objects requires more properties. In order to present the characteristics of the object, mean, variance, median, maximum, and minimum values of the objects on each band were counted, and the circumference and area of the object were also calculated. In addition, Normalized Difference Vegetation Index (NDVI) is a typical index of vegetation detection and is also applied to this classification. So each object had a total of 8 characteristics to participate in the classification.

3.2. Random Forest classification

Random Forest is a supervised learning method that can be used to solve the problem of classification or regression [29]. It is made up of a combination of tree predictors so that each tree independently obtains the value of the random vector and has the same layout for each generated vector. Random forest is a collection of classifiers, which are structured as trees, each tree produces a vote according to the input. The random vector represents a set of random numbers that determine the structure of each tree. The RF classifier has been of interest in geo-mapping because of the small input parameters, the less sensitivity to noise and over fitting, and the ability to determine variable importance [30]. In FR, a random subset (image channels) of the variable are selected for each tree, and a random sample (with substitution) of the whole training samples are used to generate a single tree classification, and the remaining samples are used to verify the process. This process is referred to as bagging and the validation training areas provide an out-of-bag accuracy assessment for each tree. The RF classifier uses a voting system that votes the given class from all trees and forms the final classification map [31].

3.3. Change detect

The IR-MAD algorithm was used to detect the changed area, and then, the Spot5 images of the changed area were classified by the RF classifier. IR-MAD judged the unchanged pixels by iteratively calculating the chi-square distribution of the difference in the bi-temporal images [15, 32]. The IR-MAD linearly transformed the original images R and T to new images U and V based on the canonical correlation analysis (CCA) in Equation (1):

$$U = a^T R,\ V = b^T T, \quad (1)$$

Where U and V are the canonical correlation images, a^T and b^T are the transform matrixes and R and T are the bi-temporal images.

The bands of the canonical correlation images were referred to as the canonical variates. The canonical variates were arranged according to the correlations. If the correlation was relatively low, this pair of canonical variates had more change information. The differences in each pair of canonical variables were mutually uncorrelated.

According to the central limit theorem, the differences in the canonical variates approximately corresponded to a Gaussian distribution, and the sum of squares of the differences in the canonical variates corresponded to a chi-square distribution [15, 32]. The differences in the canonical variates and the weights of unchanged pixels were calculated by using Equations (2)–(4):

$$M_i = U_{N-i+1} - V_{N-i+1},\ i = 1,2,\ldots,N, \quad (2)$$

$$Z = \sum_{i=1}^{N} \left(\frac{M_i}{\sigma_{M_i}} \right)^2, \quad (3)$$

$$\Pr(no\ change) = 1 - P_{\chi^2;N}(Z), \quad (4)$$

where i is the band of the image, M_i is the difference in the variates, σ_{M_i} is the standard deviation of M_i, N is the total number of bands, $P_{\chi^2;N}(Z)$ is the χ^2 distributed with N degrees of freedom and $\Pr(no\ \ change)$ is the weight.

$\Pr(no\ \ change)$ is used to weigh each pixel. Then, Equations (1)–(4) are iterated until no significant changes observed in the canonical correlations. These pixels could then be determined as unchanged pixels if their chi-square distribution probability was lower than a threshold [15].

In this study, we calculated the chi-square images for 2004, 2008, 2016 based on the images from 2012. The threshold was slightly lower than 0.95, which was set by the developer, so some unchanged pixels were selected in the identified area. Although some extra unchanged pixels were selected, mostly changed pixels were chosen, so the omission error was reduced.

4. Result

4.1. Accuracy assessment

At present, the most common precision evaluation method is the confuse matrix method, that is, through the confusion matrix to calculate the various statistics and build the accuracy evaluation index, and finally give the accuracy of classification. The samples were selected by stratified random sampling. By referring to the Google Earth image for visual interpretation, the evaluation samples were collected to assess the classification accuracy. The confuse matrixes were constructed to calculate the overall accuracy, Kappa coefficient, user's accuracy, and producer's accuracy for2000, 2004, 2008, 2012, and 2016. The overall classification accuracy is expressed the consistent in probability between the classification results and test samples types for each random sample. The user's accuracy is the proportion of correctly-classified pixels with regard to all of the pixels that are classified as this class in the classified image. The producer's accuracy is the proportion of correctly-classified pixels with regard to all of the pixels of that ground truth class.

Table 2. The error matrix.

	2000		2004		2008		2012		2016	
	PA	UA	PA	PA	UA	UA	PA	UA	UA	UA
Forest	91.23	92.33	91.23	93.34	94.52	95.48	91.22	92.31	90.32	91.67
Grass	87.56	86.54	85.67	89.92	88.16	86.54	89.43	88.21	88.65	87.17
Dry Farmland	96.15	94.45	96.10	95.32	96.15	95.40	94.41	94.56	95.16	94.24
Paddy Field	94.27	90.12	94.48	93.67	94.43	91.34	94.27	89.28	94.34	87.54
Bare Land	95.87	94.56	96.34	95.66	95.36	95.65	95.32	95.29	95.43	95.16
Building	91.45	87.52	91.56	92.13	91.12	88.34	90.84	87.56	90.44	87.43
Water Body	95.13	95.89	90.43	95.66	96.23	95.91	95.43	95.47	95.32	95.18
Overall	93.02		93.11		94.72		94.12%		93.83	
Kappa	0.90		0.91		0.92		0.92		0.91	

The land cover in 2008 had the highest overall accuracy and Kappa coefficient, which were 94.72% and 0.92 respectively. The overall accuracies and Kappa coefficients were 93.02%, 93.11%, 94.12% and 93.83% and 0.90, 0.91, 0.92 and 0.91 for 2000, 2004, 2012, and 2016 respectively. Overall, the accuracy for 2008 was higher than other years. However, the difference was minor.

4.2. . Land cover and change

The land cover in the Xinlicheng Reservoir Basin has dramatically changed over the past 16 years. The sum of the areas of dry farmland, forests and impervious surfaces always exceeded 80% of the total study area during the 16 years. The most remarkable changes were the reduction in arable land. The arable land changed from 1117.54 km^2 to 1071.48 km^2 from 2000 to 2016, shrinking by 46 km^2.

IOP Conf. Series: Earth and Environmental Science **64** (2017) 012024 doi:10.1088/1755-1315/64/1/012024

The decrement of the arable land was mainly caused by Paddy Fields. The Paddy Fields decreased from 126.44 km^2 to 97.13 km^2 about 29.31km^2. The changes of the rest classes were far less than the arable land. The land cover areas in 2004 were different between other years because the decrement of water storage in Xinlicheng Reservoir in 2004. So the area of water body increased remarkably from 2000 to 2004 and increased from 2004 to 2008. The area of building was increased continually from 2000 to 2016, but the change was slow. From 2000 to 2016 only 4.70 km^2 about 0.24% of the total area. The areas and proportions of the classes are shown in Table 3.

Table 3. Areas of the classes for the five periods. (Unit: km^2)

	2000	**2004**	**2008**	**2012**	**2016**
Paddy Fields	126.44	151.31	115.71	113.54	97.13
Dry Farmland	991.10	961.99	987.56	959.91	974.35
Forests	573.18	576.29	574.29	580.38	581.87
Grass	48.24	79.25	54.23	80.75	79.78
Water Bodies	72.93	42.50	77.37	72.63	73.73
Building	129.50	129.98	132.25	133.85	134.21
Bare Land	0.30	0.38	0.29	0.64	0.64
Sum	1941.71	1941.71	1941.71	1941.71	1941.71

IOP Conf. Series: Earth and Environmental Science **64** (2017) 012024 doi:10.1088/1755-1315/64/1/012024

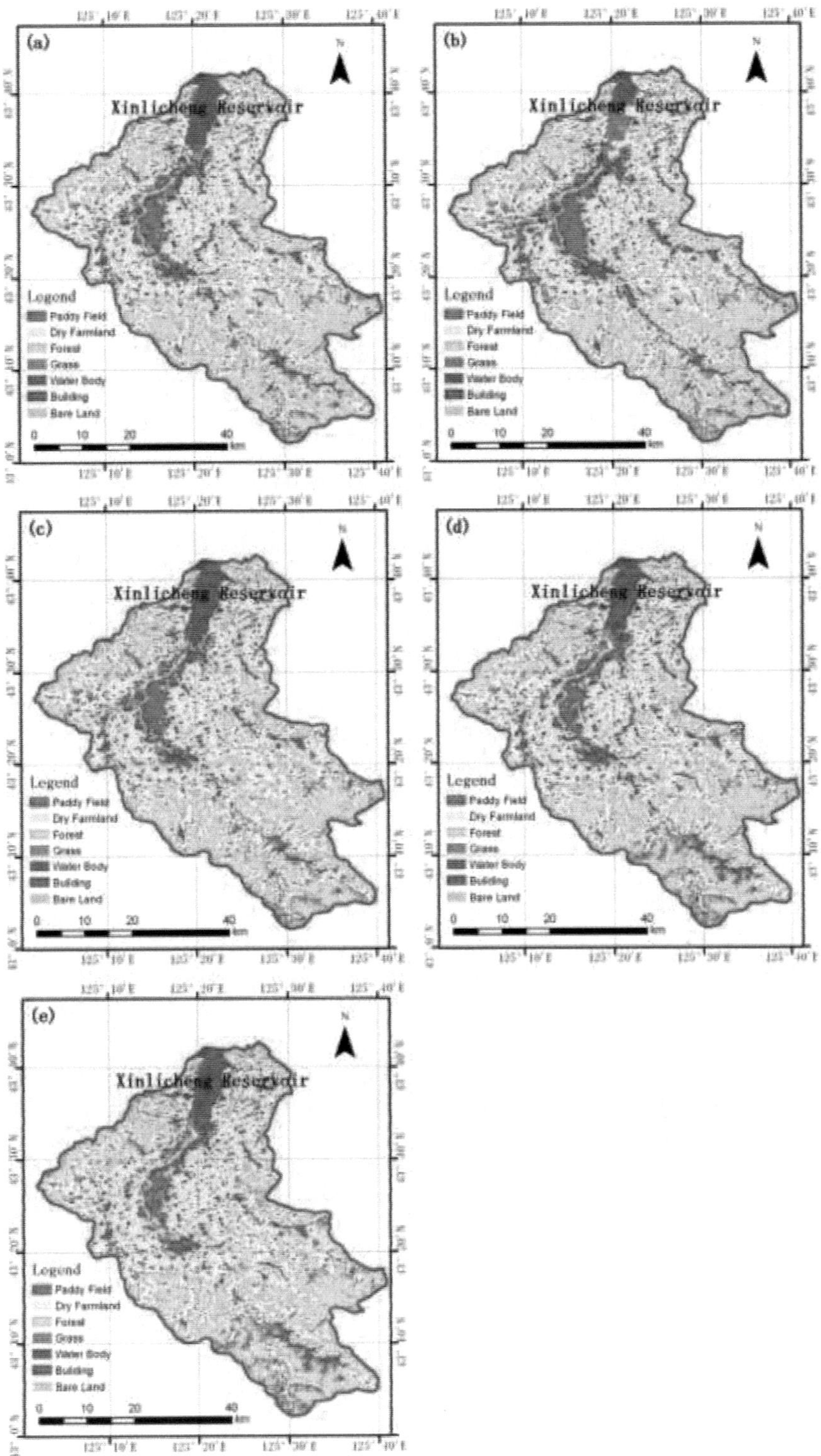

Figure 3. Land covers every 4 years from 2000 to 2016. From (a) to (e) corresponding the year from 2000 to 2016.

5. Discussion

The use of a consistent classification method is essential to ensure that the classification results at different times are comparable. The spectral characteristics of the different terrestrial order usually have some confusion or overlap, and are not always separated by a clear boundary. The transition between some classes, such as between bare and sparse grasslands and between wetlands and water bodies, is continuous. Classification errors usually occur near the differences between these classes, for their spectral similarity. The image has a unique Sun well-known geometric and weather characteristics. The same classifier is used to classify the images observed on each date. Happily, the boundary of unaltered land cover in the transitional area is difficult to identify stains. This type of discrepancy in boundary-adjacent areas can result in errors when comparing land covers between different periods. Generally, only a portion of the land cover in a region changes over a given period. If the identified change area is substantially larger than the actual change area, the updated effect is the same as the whole image of the classification. On the contrary, if the identified change area is a unique area, the error increase is omitted, and the unique area of the new area coverage product is the same as the basic product.

In this study, the software developer sets the threshold for changing the pixel to the inferred threshold, so the determined change area is slightly larger than actually changed region. This choice is to reduce omission errors. In general, the method of updating the change area reduces the computations required for classification and improves the consistency of land coverage between different periods.

6. Conclusion

Xinlicheng Reservoir Basin is an important ecological barrier and drinking water source for the residents in Changchun City. Rapid urbanization dramatically changes the local environment around Xinlicheng Reservoir Basin. Landsat images are suitable for the land use change caused by human impact. In order to obtain consistent land cover products, a hybrid classification method combining object-based classification and pre-classification alteration detection method was developed and applied to long-term multi-temporal Landsat images to obtain land cover change information. Object-based classification method can avoid the "salt and pepper" effect, and segment is the critical process in 0bject-based classification. As a kernel density estimate method, the mean shift method was used, which requires no prior knowledge of the number of segmentations and sets no limits on the object shape. Random forest (RF) classifier is less sensitivity to noise and over fitting and can determine variable importance. Pre-processed Landsat image in 2008 was segmented using the mean shift method, and then random forest (RF) classifier was used to classify the segmented objects. The changed areas in 2000, 2004, 2012, and 2016 were identified by comparing with the images in 2008 via the re–weighted multivariate alteration detection transformation method. The images in 2000, 2004, 2012 and 2016 were masked as changed areas and then were put into the RF classifier. Land cover maps for 2000, 2004, 2012, and 2016 were produced by combining the unchanged area in 2008 with the new classes of the changed areas in 2000, 2004, 2012 and 2016. In the study area, arable land remarkably decreased from 1117.54 km^2 to 1113.30 km^2, 1103.27 $km^{2,}$ 1073.45 km^2 and 1071.48 km^2 in 2000, 2004, 2008, 2012 and 2016. Meanwhile, the building area continually increased from 129.50 km^2 to 129.98 km^2, 132.25 km^2, 133.85 km^2 and 134.20 km^2. Forest and grass all decreased in 2008 and then increased. The decrements were mainly caused by increasing of water storage in Xinlicheng Reservoir. Correspondingly, the water body area increased in 2008. Overall, the building increment area mainly from dry farmland and the sprawl speed is declining. According to the accuracy assessment, the overall accuracy of the land covers of the four periods are all greater than 93%. The accuracy assessment indicates that this hybrid method can produce consistent land cover datasets for a long time period.

Acknowledgments
This work was funded by the Research on Comprehensive Management Technology and Engineering Demonstration of Water Pollution in Yitong River Basin (2012ZX07201-001) fund, and Technical Research and Comprehensive Demonstration of Water Pollution Management and Water Quality Improvement in Yinma River Basin (2014ZX07201-011) fund.

References

[1] Du H, Wang D, Wang Y, Zhao X, Qin F, Jiang H and Cai Y 2016 Influences of land cover types, meteorological conditions, anthropogenic heat and urban area on surface urban heat island in the Yangtze River Delta Urban Agglomeration Science of The Total Environment 571 461-70
[2] Liu Q and Gong F 2013 Monitoring land use and land cover change: a combining approach of change detection to analyze urbanization in Shijiazhuang, China
[3] McCallum I, Obersteiner M, Nilsson S and Shvidenko A 2006 A spatial comparison of four satellite derived 1 km global land cover datasets International Journal of Applied Earth Observation and Geoinformation 8 246-55
[4] Saadat H, Adamowski J, Bonnell R, Sharifi F, Namdar M and Ale-Ebrahim S 2011 Land use and land cover classification over a large area in Iran based on single date analysis of satellite imagery ISPRS Journal of Photogrammetry and Remote Sensing 66 608-19
[5] Chen J, Chen J, Liao A, Cao X, Chen L, Chen X, He C, Han G, Peng S, Lu M, Zhang W, Tong X and Mills J 2015 Global land cover mapping at 30 m resolution: A POK-based operational approach Isprs Journal of Photogrammetry and Remote Sensing 103 7-27
[6] Friedl M A, Sulla-Menashe D, Tan B, Schneider A, Ramankutty N, Sibley A and Huang X 2010 MODIS Collection 5 global land cover: Algorithm refinements and characterization of new datasets Remote Sensing of Environment 114 168-82
[7] Hansen M C, Defries R S, Townshend J R G and Sohlberg R 2000 Global land cover classification at 1 km spatial resolution using a classification tree approach International Journal of Remote Sensing 21 1331-64
[8] Blaschke T 2010 Object based image analysis for remote sensing ISPRS Journal of Photogrammetry and Remote Sensing 65 2-16
[9] Blaschke T, Hay G J, Kelly M, Lang S, Hofmann P, Addink E, Queiroz Feitosa R, van der Meer F, van der Werff H, van Coillie F and Tiede D 2014 Geographic Object-Based Image Analysis – Towards a new paradigm ISPRS Journal of Photogrammetry and Remote Sensing 87 180-91
[10] Geneletti D and Gorte B G H 2003 A method for object-oriented land cover classification combining Landsat TM data and aerial photographs International Journal of Remote Sensing 24 1273-86
[11] Benz U C, Hofmann P, Willhauck G, Lingenfelder I and Heynen M 2004 Multi-resolution, object-oriented fuzzy analysis of remote sensing data for GIS-ready information ISPRS Journal of Photogrammetry and Remote Sensing 58 239-58
[12] Dorren L K A, Maier B and Seijmonsbergen A C 2003 Improved Landsat-based forest mapping in steep mountainous terrain using object-based classification Forest Ecology and Management 183 31-46
[13] Wang L, Sousa W P and Gong P 2004 Integration of object-based and pixel-based classification for mapping mangroves with IKONOS imagery International Journal of Remote Sensing 25 5655-68
[14] Vieira M A, Formaggio A R, Rennó C D, Atzberger C, Aguiar D A and Mello M P 2012 Object Based Image Analysis and Data Mining applied to a remotely sensed Landsat time-series to map sugarcane over large areas Remote Sensing of Environment 123 553-62
[15] Canty M J and Nielsen A A 2012 Linear and kernel methods for multivariate change detection Computers & Geosciences 38 107-14

[16] Nielsen A A and Canty M J 2010 Image and Signal Processing for Remote Sensing Xvi, ed L Bruzzone

[17] Sexton J O, Urban D L, Donohue M J and Song C 2013 Long-term land cover dynamics by multi-temporal classification across the Landsat-5 record Remote Sensing of Environment 128 246-58

[18] Hansen M C, Egorov A, Potapov P V, Stehman S V, Tyukavina A, Turubanova S A, Roy D P, Goetz S J, Loveland T R, Ju J, Kommareddy A, Kovalskyy V, Forsyth C and Bents T 2014 Monitoring conterminous United States (CONUS) land cover change with Web-Enabled Landsat Data (WELD) Remote Sensing of Environment 140 466-84

[19] Gong P, Wang J, Yu L, Zhao Y, Zhao Y, Liang L, Niu Z, Huang X, Fu H, Liu S, Li C, Li X, Fu W, Liu C, Xu Y, Wang X, Cheng Q, Hu L, Yao W, Zhang H, Zhu P, Zhao Z, Zhang H, Zheng Y, Ji L, Zhang Y, Chen H, Yan A, Guo J, Yu L, Wang L, Liu X, Shi T, Zhu M, Chen Y, Yang G, Tang P, Xu B, Giri C, Clinton N, Zhu Z, Chen J and Chen J 2013 Finer resolution observation and monitoring of global land cover: first mapping results with Landsat TM and ETM+ data International Journal of Remote Sensing 34 2607-54

[20] Yang L, Huang C, Homer C G, Wylie B K and Coan M J 2003 An approach for mapping large-area impervious surfaces: synergistic use of Landsat-7 ETM+ and high spatial resolution imagery Canadian Journal of Remote Sensing 29 230-40

[21] Liu J, Kuang W, Zhang Z, Xu X, Qin Y, Ning J, Zhou W, Zhang S, Li R, Yan C, Wu S, Shi X, Jiang N, Yu D, Pan X and Chi W 2014 Spatiotemporal characteristics, patterns, and causes of land-use changes in China since the late 1980s Journal of Geographical Sciences 24 195-210

[22] Gao F, Masek J, Schwaller M and Hall F 2006 On the blending of the Landsat and MODIS surface reflectance: predicting daily Landsat surface reflectance IEEE Transactions on Geoscience & Remote Sensing 44 2207-18

[23] Johnson B and Xie Z 2011 Unsupervised image segmentation evaluation and refinement using a multi-scale approach ISPRS Journal of Photogrammetry and Remote Sensing 66 473-83

[24] Myint S W, Gober P, Brazel A, Grossman-Clarke S and Weng Q 2011 Per-pixel vs. object-based classification of urban land cover extraction using high spatial resolution imagery

[25] Yu W, Zhou W, Qian Y and Yan J 2016 A new approach for land cover classification and change analysis: Integrating backdating and an object-based method Remote Sensing of Environment 177 37-47

[26] Fukunaga K and Hostetler L 1975 The estimation of the gradient of a density function, with applications in pattern recognition Information Theory, IEEE Transactions on 21 32-40

[27] Cheng Y 1995 Mean shift, mode seeking, and clustering Pattern Analysis and Machine Intelligence, IEEE Transactions on 17 790-9

[28] Comaniciu D and Meer P 2002 Mean shift: a robust approach toward feature space analysis Pattern Analysis and Machine Intelligence, IEEE Transactions on 24 603-19

[29] Shiraishi T, Motohka T, Thapa R B, Watanabe M and Shimada M 2014 Comparative Assessment of Supervised Classifiers for Land Use–Land Cover Classification in a Tropical Region Using Time-Series PALSAR Mosaic Data IEEE Journal of Selected Topics in Applied Earth Observations and Remote Sensing 7 1186-99

[30] Yu L, Wang J and Gong P 2013 Improving 30 m global land-cover map FROM-GLC with time series MODIS and auxiliary data sets: a segmentation-based approach International Journal of Remote Sensing 34 5851-67

[31] Pelletier C, Valero S, Inglada J, Champion N and Dedieu G 2016 Assessing the robustness of Random Forests to map land cover with high resolution satellite image time series over large areas Remote Sensing of Environment 187 156-68

[32] Canty M J and Nielsen A A 2008 Automatic radiometric normalization of multitemporal satellite imagery with the iteratively re-weighted MAD transformation Remote Sensing of Environment 112 1025-36

IOP Conf. Series: Earth and Environmental Science **64** (2017) 012025 doi:10.1088/1755-1315/64/1/012025

Synthesis and analysis of foam drainage agent for gas well in Jilin Oilfield

Sanyuan Qiao, Qingwang Liu, Zhenzhong Fan, Jigang Wang and Jianjun Xu
Northeast Petroleum University, DaQing, Heilongjiang, China 163318

Abstract. The gas well in Jilin oil field has the characteristics of large temperature variation range and high condensate oil content. So the foam drainage agent of the gas well in Jilin oil field needs to have the performance of oil resistance and less effected by temperature. In this paper, a main foaming agent named lauramidopropyl betaine (LAB) and two kinds of auxiliary foaming agent named sodium alcohol ether sulphate (AES) and lauramidopropylamine oxide (LAO). Through the evaluation of the static foaming capacity and dynamic liquid carrying capacity, the AES is more suitable for LAB. The foaming agent with 70% LAB and 30% AES has 138mm foam height with ROSS-Miles equipment; stirring foam volume can reach 480mL, the half-life of foam is 520s. When the ventilation volume is 8L/min the liquid carrying capacity of 10% of the condensate oil content reached 82g. When the foaming agent concentration is 2%, the liquid carrying capacity of 10% of the condensate oil content reached 75g. When the aeration rate reaches 8-10L/min, the liquid carrying capacity of foam drainage agent can reach the best. The foam drainage agent can retain the performance after 120℃ aging for 12h, these performances above can satisfy the requirements for gas well foam drainage in Jilin Oil Field.

1. Experimental part

1.1. Experimental materials
Experimental Materials: Lauric acid, N,N'-Dimethyl-1,3-propanediamine, Propylene diamine, Hydrogen Peroxide(30%), Sodium chloroacetate, Lauryl alcohol, Epoxyethane, Sulfamic acid, Sodium hydroxide, Potassium hydroxide, Concentrated sulfuric acid,both are analytical pure.
The experimental water is the water sample of gas wells in Jilin oil field (4000mg/L); Condensate oil samples from Jilin Oilfield.

1.2. Experimental instruments
(1)Liquid carrying test equipment, (2) Shanghai Lin Lin Electronic Technology Co.Ltd. GHX series oil bath pot, (3) Shanghai Bank Equipment Co. Ltd. 2151 standard Roche foam tester.

1.3. Synthesis of blowing agent
The dodecanol was chosen as the starting agent, and the starting agent was introduced into the autoclave, 0.5% KOH was added as catalyst, and after evacuation under nitrogen protection conditions, ethylene oxide was added dropwise, heat preservation for 30 minutes after the addition until the pressure is

negative. The product of small molecule was removed by vacuum cooling, then the fatty alcohol polyoxyethylene ether was obtained. Weigh the prepared fatty alcohol polyoxyethylene ether into the three-necked flask, using sulfamic acid as sulfation reagent, and then adding 0.5% concentrated sulfuric acid as a catalyst, after 2.5h, the reaction was terminated, after cooling, sodium hydroxide was added at room temperature to neutralize the excess sulfamic acid and the ammonium salt was converted into sodium salt. Finally, the effective content of AES solution was 30%.

2. Evaluation of foaming performance

2.1. Evaluation of Roche Foaming Performance

Roche foaming performance is using Ross-Miles foam tester on the impact of surfactants to evaluate the ability to foam, in the experimental method, 297.0 g of distilled water was weighed, and then added in 3.0 g of the foaming agent, followed by gentle stirring to disperse the solution. The 50 mL of the formulated foaming agent solution was slowly poured into the bottom of the Roche foam tester with glass rod, another 200mL solution was placed in the funnel device of the foam tester, the funnel placed in the bubble above the instrument to open the valve, so that the solution evenly outflow, until the solution flowing out then reading the foam height. And at 65 ° C for 15 min, the foam height was read again. The experimental data is shown in Fig.1.

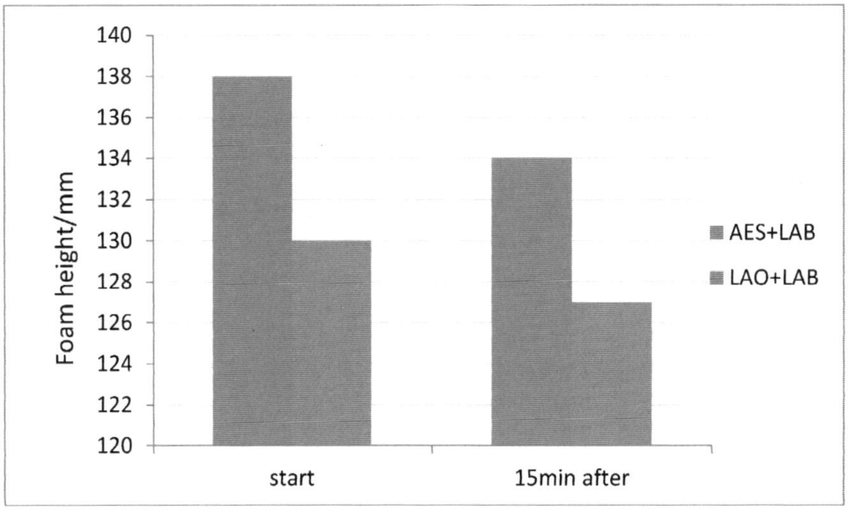

Figure 1 Roche foam height for different foaming agents

As can be seen from Figure 1, after joining AES, LAB has higher foam height and foam stability compared with LAB added LAO, It can be concluded that after joining the LAB, AES is better than LAO and LAB in the foaming capacity and stability.

2.2. Foaming ability and half-life test

The foaming agent can quickly generate a large amount of foam under the condition of high speed stirring, the foaming and stabilizing ability of the foaming agent can be judged by the amount of foam produced and the rate of foaming under the same conditions. The experimental procedure is to take 100.0 g of distilled water and add 1.0 g of foaming agent, then put the foam into the 500mL cylinder, at this time the foam volume is the foaming volume. We record the time at which the solution at the bottom of the graduated cylinder reaches 50 mL.,this time is half-life, the experimental data are shown in table 1.

Table 1 Foaming volume and half-life of different foaming agents

Foaming agent	Foaming volume /mL	Half-life /s
AES+LAB	480	520
LAO+LAB	450	550

As can be seen from Table 1,the foam formed by high-speed stirring compared to the large foam formed by the Roche foam will form more subtle micro-foam, So the water loss of this bubble is slower than the big bubble, and the liquid film formed between the foams has a larger relative surface area, therefore, the stabilizing effect of LAO on LAB is more obvious than that of AES on LAB, this is due to the high degree of similarity between the lipophilic groups of LAO and the lipophilic groups of LAB, the formation of double-layer film on the surface of the liquid film will form a symmetrical structure, more balance21d. So foam has a longer half-life which formed by LAO and LAB.

2.3. Foaming agent dynamic carrying capacity test

The dynamic liquid carrying capacity test of foam drainage agent is a method to simulate the formation process of foam drainage agent bring formation water into the ground, it is an important index to evaluate the performance of foam drainage agent. The oil samples and condensed oil samples from the gas wells of Jilin Oilfield were selected as experimental liquids, According to the conditions of the bottom of oil well in Jilin oilfield as the experimental condition of dynamic liquid carrying experiment: The experimental temperature is 60, The condensate oil content is 10% mass fraction. Take 100.0g prepared oil water samples, add 1.0g foaming agent after mixing slowly, ventilation volume is 8L/min, and record the quality of liquid within 15min, the experimental data shown in Table 2.

Table 2 The amount of liquid in different foaming agents

Foaming agent type	Liquid carrying capacity /g
AES+LAB	82
LAO+LAB	68

3. Formula optimization of foam drainage agent

3.1. Determination of the concentration of foaming agent

The oil samples from Jilin Oilfield were selected as water-based liquids, the dynamic carrying capacity of the foam drainage agent was tested when the condensate oil content was 10%, 20%, 30% and 40%, respectively. The experimental temperature was 60 ° C, the concentration of foaming agent was 0.5%, 1%, 1.5%, 2%, 2.5%, and 3% respectively, ventilation is 8L / min. The experimental data are shown in figure 2. Indicating that the concentration of 2.0% is the appropriate concentration.

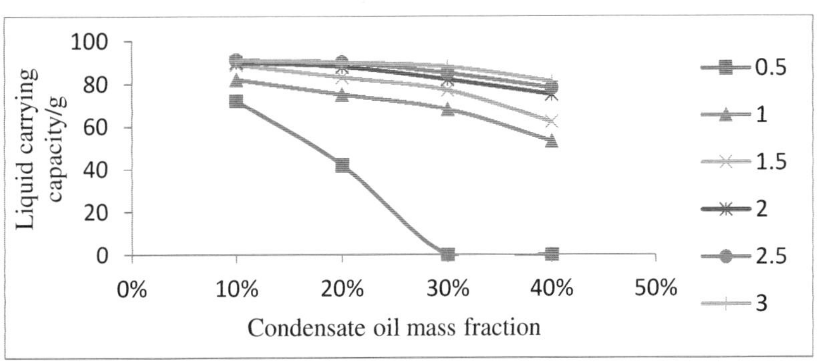

Figure 2 The effect of different oil content and the amount of foaming agent

As can be seen from Figure 2, with the increase of condensate oil content, foaming agent carrying capacity will be affected, if foaming agent is added less, when the condensate oil content is high, the liquid cannot be taken out of the container, at this time, liquid carrying capacity of 0. With the increase in the amount of foaming agent used, the influence of condensate oil on liquid carrying capacity is smaller, when the foaming agent reaches 2.0g, in 40% of the condensate oil conditions can still reach 75g carrying liquid. Indicating that the concentration of 2.0% is the appropriate concentration.

3.2. *Effect of ventilation on carrying capacity*

In the process of foam drainage, the vapor liquid ratio has a great influence on the production of foam, if the vapor liquid ratio is too small, will lead to the production of less foam, and the foam generation rate is less than the rate of death of the foam ,which cannot bring liquid; if the vapor liquid ratio is too large, It will cause the gas slippage and reduce the gas lift efficiency, So the appropriate vapor liquid ratio has a great influence on the carrying capacity of the foaming agent. The change of ventilation volume from 4L/min to 10L/min was used to study the effect of ventilation volume on the liquid carrying capacity of foaming agent, the experimental temperature was 60 ° C, the concentration of foaming agent was 1.0 %, and the condensate mass fraction is 10%.

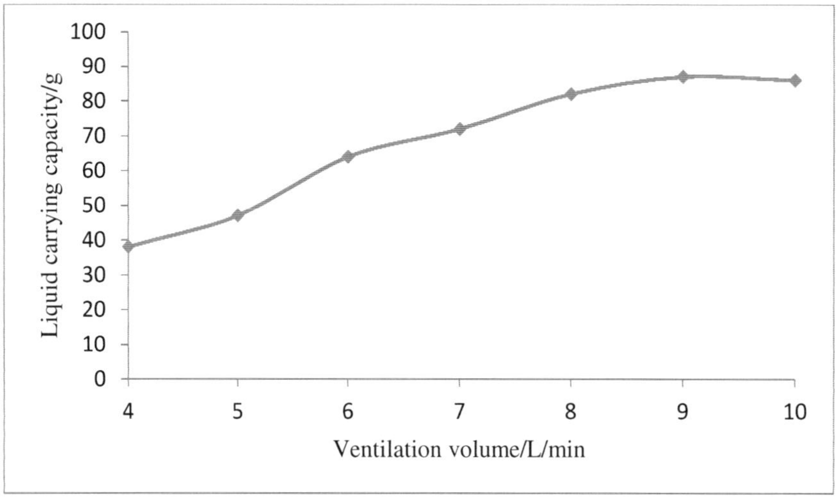

Fig. 3 The effect of different ventilation on liquid carrying capacity

As can be seen from Figure 3 that with the increase in ventilation, the liquid carrying capacity of the foam drainage agent increases, but when the air volume exceeds 9L/min, the liquid carrying capacity will decrease slightly, so the most suitable ventilation should be 8~10L/min.

3.3. *Foaming agent temperature resistance*

As the temperature of the bottom of the gas field in Jilin oil field changes greatly, some gas wells bottom temperature can reach 80 ~ 90 ℃, so the foam drainage agent needs to have the ability to resist aging of 120 ℃, the prepared foaming agent was placed in a high temperature aging tank and heated at 120 ° C for 12 h, and then determine the performance change between the foaming agent has been aging and the foaming agent before aging. The mass fraction of the foaming agent used in the experiment was 1%, the condensate oil content was 10%, and the experimental temperature was 60 ° C. The experimental data are shown in Table 3.

Table 3 Performance changes of foaming agents before and after aging

Experimental conditions	Liquid carrying capacity/g	Foaming volume/ml	Half-life/s	Roche bubble height/mm
Before aging	82	480	520	138
After aging	80	480	510	130

4. Conclusion

1, in this paper, through the synthesis of three kinds of surfactants and complex, obtained a low price, green, temperature and oil-resistant foam drainage agent, and has a good foaming ability and foam stability.

2, The effect of different condensate content on the liquid carrying capacity of foam solution with different mass fraction was studied experimentally, and the relationship between the amount of ventilation and the amount of dynamic foaming liquid carrying capacity. When the foaming agent reaches 2.0g, in 40% of the condensate oil conditions can still reach 75g carrying liquid. Indicating that the appropriate concentration is 2.0%. And with the increase in ventilation, the liquid carrying capacity of the foam drainage agent increases, but when the air volume exceeds 9L/min, the liquid carrying capacity will decrease slightly, so the most suitable ventilation should be 8~10L/min.

Acknowledgements

This research is supported by China National Natural Funds project "Selection of Oil Soluble Resin Double Crosslinking Gel Plugging Agent" (No. 51374071).

This work is supported by PetroChina Innovation Foundation (2016D- 5007- 0201). Corresponding Author is XU Jianjun.

References

[1] Xu Jianjun, Xu Yan-chao, Yan, Li-me,et.al. Research on the method of optimal PMU placement. International Journal of Online Engineering, v9, S7, p24-29, 2013

[2] Xu Jian-Jun, Y. Y. Zi., Numerical Modeling for Enhancement of Oil Recovery via Direct Current. International Journal of Applied Mathematics and Statistics, 2013, 43 (13):318-326

[3] Longchao, Zhu Jianjun, Xu; Limei, Yan. Research on congestion elimination method of circuit overload and transmission congestion in the internet of things. Multimedia Tools and Applications, p 1-20, June 27, 2016

[4] Yan Limei, Zhu Yusong, Xu Jianjun,et.al. Transmission Lines Modeling Method Based on Fractional Order Calculus Theory. TRANSACTIONS OF CHINA ELECTROTECHNICAL SOCIETY, 2014 ,Vol.29,No. 9:260-268 (In Chinese)

[5] YAN Li-mei, CUI Jia, XU Jian-jun,et.al. Power system state estimation of quadrature Kalman filter based on PMU/SCADA measurements. Electric Machines and Control. 2014, Vol.18 No.6, 78-84. (In Chinese)

[6] YAN Limei,XIE Yibing, XU Jianjun, et.al. Improved Forward and Backward Substitution in Calculation of Power Distribution Network with Distributed Generation. JOURNAL OF XI'AN JIAOTONG UNIVERSITY, 2013, Vol.47, No.6, p117-123. (In Chinese)

[7] Xu J.J., Gai D., Yan L.M. A NEW FAULT IDENTIFICATION AND DIAGNOSIS ON PUMP VALVES OF MEDICAL RECIPROCATING PUMPS. Basic & Clinical Pharmacology & Toxicology, 2016,118 (Suppl. 1), 38-38.

Stabilization of SiO_2 nanoparticle foam system and evaluation of its performance

Chong Sun, Zhenzhong Fan, Qingwang Liu, Jigang Wang, Jianjun Xu*

Northeast Petroleum University, Daqing, Heilongjiang, China 163318

*Corresponding Author Email: 123939274@qq.com

Abstract. As tertiary recovery is applied in the oil field, foam flooding technology plays an important role in the oil field. Steam flooding is easy to generate a series of problems such as excessive pressure, gas channelling, heat loss ect. The foam flooding can be better used in the formation of plugging and profile control. However, the foam is not stabilizing in thermodynamics and breaks easily while it encounters oil. So the emphasis of the research is how to make the foam stable. The Warning Blender method is used to evaluate the foam In the course of experiment, which verifies that the modified Nano SiO_2 solid not only works very well in coordination with SDS solution but also contributes to the generation of stable foam in solution. The optimum concentration of SDS is determined by 0.5%, and the best concentration is 1.4% of H20 type SiO_2 particles that the concentration is 79.26°. Finally, the 0.5%SDS+1.4%H_2O type SiO_2 is chosen as the complete foam flooding system, and the performance of salt tolerance and oil displacement of composite foam system is evaluated. It is concluded that the stability of foam is the key to improve the oil recovery.

1. Experiment Condition

1.1. Experiment reagent

SDS(Sodium dodecyl sulfate, anionic surface active agent); H15,H18,H20,N15,FM150 five types SiO_2 grain (particle size is 20nm, white powder with negative charge on surface); air; distilled water; NaCl(AR); $CaCL_2$(AR); $MgCL_2$(AR).

1.2. Laboratory apparatus

PL203 electronic balance ; ultrasonic dispersing instrument ; C-J type high speed agitator ; powder compressing machine ; GT10 type high speed centrifuge.

2. Experiment method

2.1. Contact angle measurement

The experiment method of measuring the contact angle between SiO_2 nanoparticles and the interface of water is sessile drop method [1-3]. The experiment sets the pressure parameter of tablet press is 20MP and takes the right amount of the five-type SiO_2 nanoparticles to press them into a pie shape by using powder compressing machine. The experiment uses the contact angle measurement function of

IOP Conf. Series: Earth and Environmental Science **64** (2017) 012026 doi:10.1088/1755-1315/64/1/012026

the full automatic interfacial rheometer to measure the contact angle between the five-type SiO_2 nanoparticles and the interface of water. The temperature of the experiment is indoor temperature. The expetiment takes 3 different measuring points of the same sample during the experimentation. The average value of the result is taken as the final result. The results are shown in the following table.

Table 1. The measurement of SiO_2 particle contact angle

Sample type of SiO_2	measuring point			average value of contact angle
	First	Second	Third	
H15	67.56	67.63	67.58	67.59
H18	121.44	121.47	121.77	121.56
H20	79.12	79.45	79.21	79.26
N20	39.32	39.63	39.31	39.42
FM150	51.16	51.32	51.63	51.37

2.2. Foam evaluation experiment

The concentration of five group SDS solution is 0.25g/L should be prepared in the 22℃ temperature condition. The expetiment uses the electronic balance to weigh the fraction of the five-type SiO_2 nanoparticle which is 0.15g. And five-type SiO_2 nanoparticles are added to the SDS solution. The experiment disperses the mixed solution 3minutes by ultrasonic dispersing [11] instrument. The 100g dispersion is weighed in a high stirring cup and then the sealing is covered. The air is put into the high stirring cup about 3minutes. The speed of the high speed mixer is adjusted with 8000r/min and the mixture is stirred 3minutes. Then we pour the generated foam into the cylinder rapidly and recor the dinitial foam volume and the time that the foam precipitates 50ml liquid. The time is seen as the half-life $T_{1/2}$. During the experiment, each sample is measured 3 times and the average value is taken as the final result. The results are shown in the following table.

Table 2. The volume and half-life of SiO_2 solid particles with different contact angles

Sample type of SiO_2	Foaming volume (v/ml)	Half-life (min)
H15	330	16.5
H18	325	15.6
H20	435	27.6
N20	314	6.7
FM150	318	9.6

3. The quality optimization of SDS and SiO_2

3.1. The quality optimization of SDS

The foam evaluation experiment is carried out by Warning Blender method. The experiment uses distilled water to prepare the anionic surfactant SDS solution that the mass fraction are 0.1%, 0.3%, 0.5%, 0.7%, 0.9% .The solution is stirred for 3 minutes by high speed mixer that its speed is 8000r/min. The foam is poured into the cylinder in time. The volume and half-life $T_{1/2}$ are recorded at the same time. The result of each group shoule be measured three times and the average value is taken as the final result.The results are shown in the following table.

Table 3. Foaming volume and half-life of SDS solution

Content of SDS (g)	Foaming volume (v/ml)	Half-life (min)
0.1	545	6.8
0.3	586	7.0
0.5	635	8.1
0.7	628	8.8
0.9	640	7.6

The experiment result shows when the content of SDS solution is 0.5% the foaming volume and half-life are the best.

3.2. Content optimization of H_20 type SiO_2

The experiment of foam evaluation is carried by Warning Blender method. Firstly, 10 groups SDS solution that the mass fraction is 0.5% are prepared and 0.3g, 0.5g, 0.8g, 1.0g, 1.2g, 1.4g, 1.6g, 1.8g, 2.0g H20 type SiO_2 particles are obtained by electronic balance. These SiO_2 particles are added into the 0.5% SDS solution one by one. The mixing solution is mixed into the high stirring cup 3minutes and the speed of high-speed mixer is 8000r/min. The foam is produced into a measuring cylinder firstly and the foaming volume and half-life $t_{1/2}$ are recorded. The production of foaming volume V and half-life $t_{1/2}$ are regarded as the comprehensive value of foam.The result of each group shoule be measured three times and the average value is taken as the final result. The results are shown in the following table.

Table 4. The height and half-life of foam system

Content of SiO_2 (g)	Foaming volume (v/ml)	Half-life (min)	comprehensive value of foamv$t_{1/2}$(ml min)
0.0	635	8.1	5143.5
0.3	345	10.5	3622.5
0.5	353	14.8	5224.4
0.8	380	20.4	7752
1.0	410	25.7	10537
1.2	395	34.5	13627.5
1.4	390	45.6	17784
1.6	375	48.2	18075
1.8	365	49.8	18177
2.0	355	50.2	17821

4. Evaluation of optimization formula performance

It is necessary to evaluate the effect of inorganic salts on the SiO_2+SDS foam system because the formation water contains a certain degree of mineralization. The laboratory prepared 0.5%SDS and 1.4%SiO_2 composite foam solution. Warning Blender method is used to evaluate the foam performance of composite foam solution by adding NaCl, $MgCl_2$, and CaCl2. The foam is poured into the cylinder after we fully stir the solution. The changings of foam volume and half-life are observed under the different condition of concentration and salinity. The effects of these three ions on the foam system are mainly evaluated because the content of Na^+, Mg^{2+} and Ca^{2+} in the formation water is more. The experiment results are shown in the following figure.

IOP Conf. Series: Earth and Environmental Science **64** (2017) 012026 doi:10.1088/1755-1315/64/1/012026

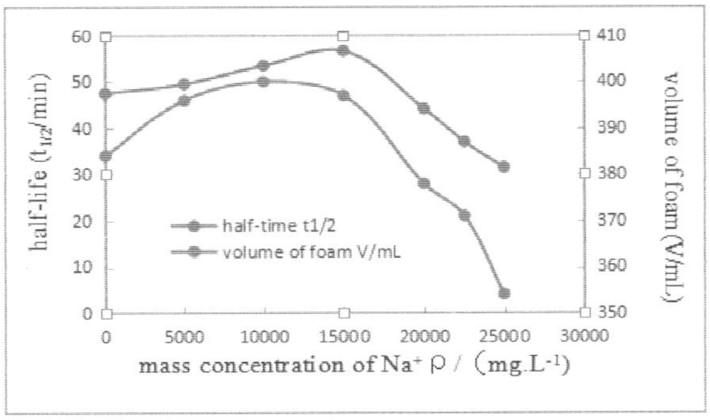

Figure 1. The influence of Na^+ effects on SiO_2+SDS system

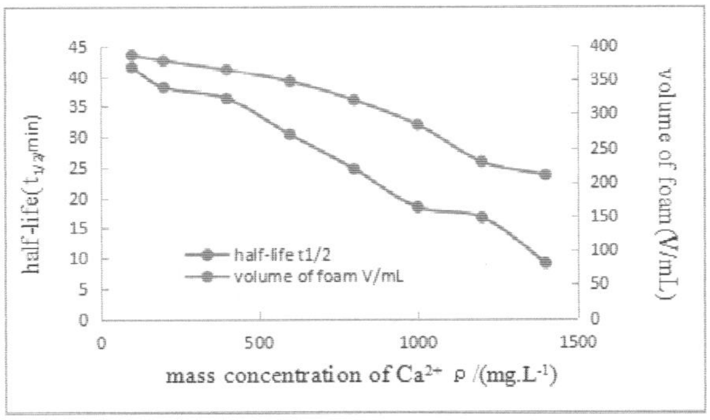

Figure 2. The influence of Ca^{2+}effects on SiO_2+SDS system

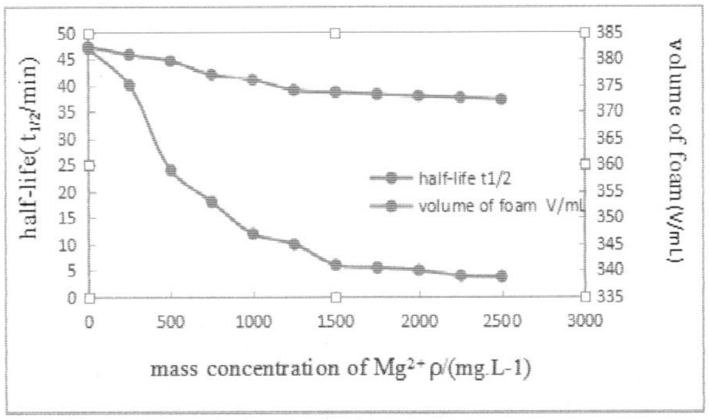

Figure 3. The influence of Mg^{2+}effects on SiO_2+SDS system

The figure 1 shows that with the increase of Na^+ content, the height and half-life of foam increase at first but then decrease. It can be seen that the appropriate amount of Na^+ plays an important role in the foaming volume and half-life. There are three reasons can analysis above phenomenon. The first one is the appropriate amount of Na^+ can reduce the interfacial tension of SDS and enhance its surface activity. The second one is Na^+ which can reduce the repulsion between particles, which dues to the negative charge of SiO_2 itself. The particles will be better adsorbed on the interface in this way. The third one is the negative electricity can exact the surface of SiO_2 nanoparticles and SDS hydrophobic group at the same time. The Na^+ can make the combination between hydrophobic chain of SDS and SiO_2 hydrophobic group, so that the SiO_2 particles will be pulled to the foam surface in this way [14].

The figure 2 and 3 show that Mg^{2+} and Ca^{2+} are not like Na^+ can promote the foam volume and half-life of the foam in a certain amount. With the increase of the content of Mg^{2+} and Ca^{2+}, the height and half-life of foam decreases gradually. The foaming volume and half-life of Mg^{2+} foam solution decrease slowly, it shows that the foam system of SiO_2+SDS is better at confrontation of Mg^{2+}. The Ca^{2+} greatly destroys the hydration properties of SDS and reduces the content of SDS solution, which causes more SiO_2 particles break away from the interface of the liquid film and reduces the interaction between SDS and SiO_2[15].

5. Conclusion

1.Finally, the 0.5%SDS+1.4%H20 type SiO2 foam flooding system is selected as the complete system.

2. The SDS+SiO_2 composite foam system is better than single SDS foaming system in terms of improving oil recovery and plays an active role in resisting inorganic ions.

3.The SiO_2 solid particles can increase the mechanical strength of the foam liquid film in order to stabilize the foam and the stable foam is the key to improve the recovery rate of foam flooding.

Acknowledgement

This research is supported by The National Natural funds project of China “Selection of Oil Soluble Resin Double Crosslinking Gel Plugging Agent” (No. 51374071)

This work is supported by Petro China Innovation Foundation (2016D- 5007- 0201). Corresponding Author is XU JianJun.

Reference

[1] Xu Jianjun, Xu Yan-chao, Yan, Li-me, et.al. Research on the method of optimal PMU placement. International Journal of Online Engineering, v9, S7, p24-29, 2013

[2] Xu Jian-Jun, Y. Y. Zi., Numerical Modeling for Enhancement of Oil Recovery via Direct Current. International Journal of Applied Mathematics and Statistics, 2013, 43 (13) :318-326

[3] Longchao, Zhu Jianjun, Xu; Limei, Yan. Research on congestion elimination method of circuit overload and transmission congestion in the internet of things. Multimedia Tools and Applications, p 1-20, June 27, 2016

[4] Yan Limei, Zhu Yusong, Xu Jianjun,et.al. Transmission Lines Modeling Method Based on Fractional Order Calculus Theory. TRANSACTIONS OF CHINA ELECTROTECHNICAL SOCIETY, 2014, Vol.29, No. 9:260-268 (In Chinese)

[5] YAN Li-mei, CUI Jia, XU Jian-jun,et.al. Power system state estimation of quadrature Kalman filter based on PMU/SCADA measurements. Electric Machines and Control. 2014, Vol.18 No.6: 78-84. (In Chinese)

[6] YAN Limei,XIE Yibing, XU Jianjun, et.al. Improved Forward and Backward Substitution in Calculation of Power Distribution Network with Distributed Generation. JOURNAL OF XI’AN JIAOTONG UNIVERSITY, 2013, Vol.47, No.6, p117-123. (In Chinese)

[7] Xu J.J., Gai D., Yan L.M. A NEW FAULT IDENTIFICATION AND DIAGNOSIS ON PUMP VALVES OF MEDICAL RECIPROCATING PUMPS. Basic & Clinical Pharmacology & Toxicology, 2016,118 (Suppl. 1), 38-38

IOP Conf. Series: Earth and Environmental Science **64** (2017) 012027 doi:10.1088/1755-1315/64/1/012027

Relationship between sandstone-type uranium deposits and hydrocarbon in the northern ordos basin

Feifei Wang*, Jianqiang Wang[1], Pei Guo[1], Xianghu Cheng[1], Bei Li[1], Zisheng Song[2], Anjun Wei[3]

[1]State Key Laboratory of Continental Dynamics, Northwest University, Shaanxi 710069, China
[2]No. 272 Geological Team of Guangxi, Nanning, Guangxi 530028, China
[3]Sino Shaanxi Nuclear Industry Group Geological Survey Co. Ltd, Shaanxi 710100, China

*Corresponding author E-mail: doublefly2006@163.com

Abstract. The Ordos Basin is one of the largest proliferous basins and also one of the most important uranium-bearing basins in China. It is characterized by the coexistence of petroleum and uranium in the northern part of the basin. To understand the coexistence mechanism, more studies are called for on the genesis of sandstone-type uranium deposits in the northern part of the basin, especially on the role of hydrocarbon in uranium mineralization. In this study, we investigated the relationship between uranium and hydrocarbon in the northern Ordos Basin using the methods of petrology, mineralogy and geochemistry. Our results show that the hydrocarbon seepage plays an important role in the mineralization of sandstone-type uranium deposits. It greatly affects the reduction in mineralization and the environmental rehabilitation to protect the ore body from being destroyed by the groundwater after mineralization.

1. Introduction

The organic-inorganic interaction has been a hot and challenging research topic in earth science [1-3]. In the current uranium mineralization theory, one important issue concerning the interaction is the relationship between hydrocarbon (organic) and sandstone-type uranium deposit (SUD) (inorganic). In fact, that uranium deposits coexisting with hydrocarbon or bitumen has been found worldwide, such as in some basins of Texas, Colorado Plateau, Europe, and Tarim Basin of China [4-5]. The study on the role of hydrocarbon in uranium mineralization has important scientific significance and application value for in-depth understanding of the relationship of organic-inorganic symbiotic accumulation and mineralization.

The Ordos Basin is one of the largest oil-gas basins in China. Oil, gas, coal and uranium coexist in the basin. In recent years, many world-class SUDs have been discovered. However, there is some controversy about the origin of ore deposits [6-7]. It is not yet known whether the oil and gas take part in the uranium mineralization process or not. This not only has become a difficult scientific problem, but also has hindered the expansion of deposit scale. The present study was therefore conducted to investigate the relationship between hydrocarbon and SUDs using the methods of petrology, mineralogy and geochemistry. We aimed to better understand the role of hydrocarbon in uranium mineralization and to understand the uranium mineralization mechanism. The potential exploration direction was also proposed.

2. Geological setting and deposit profile

The Ordos Basin is a large depression-type basin located in the west of the North China craton. It is in the first quadrant of "Cross Tectonic Belt", spanning over 25x10^4 km. The Ordos Basin is surrounded by mountains, with the Cenozoic graben basins in between separating them (Fig. 1). The Paleozoic and Meso-Cenozoic sedimentary strata have been formed in the basin. The Ordos Basin produces a variety of energy minerals such as oil and gas, coal and SUDs, and therefore being a large energy base of China. The SUDs in the northern Ordos Basin were mainly produced in Zhiluo Formation sandstone of Jurassic, with a (super) large scale (≥10000 t), and even world-class scale in some parts. The Zhiluo Formation can be divided into upper and lower units from the bottom upward. The uranium mineral was produced in the second section of the lower unit ($J_2Z_1^2$) (Fig. 1), and the ore-bearing rocks are gray medium-sized sandstones.

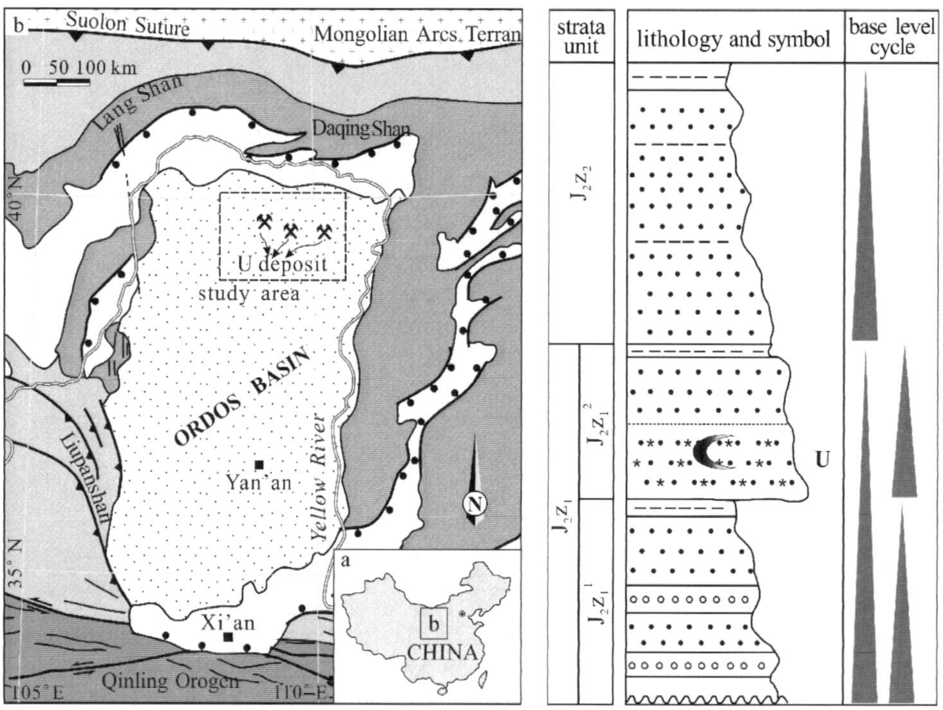

Figure 1. Geological setting and Stratigraphic column of Middle Jurassic Zhiluo Formation in the Ordos Basin (left: modified from [8])

3. Petrologic characteristics

3.1. Characteristics of field outcrop and core

The geological characteristics of ore-bearing rocks can be seen from Shenshangou field outcrop in the northern Ordos basin. Zhiluo Formation is mainly composed of grey and green sandstones, and the red sandstone is visible in the outcrop and well core. The gray sandstone contains a large number of plant charcoal chips and coal debris, and some green sandstones have red patches (Fig. 2). It can be inferred from the structural morphology that the red patches might be residues of the green altered sandstone, which suggests that the early red sandstone has been altered and changed into green sandstone by certain fluid in a later stage, but there remained a portion of red sandstones that has not been altered completely.

Zhiluo Formation has also developed lots of calcareous sandstones, which were mainly cemented by calcite. These calcareous sandstone nodules were arrayed in a discontinuously line along the strata (Fig. 2). Their density is significantly higher than that of the surrounding rocks. This is because that the calcareous sandstones foamed when they encountered hydrochloric acid.

Figure 2. Outcrop profiles and core characteristics of the northern Ordos Basin

A: outcrop of bleached sandstone; B: altered sandstone; C: linear arrayed calcareous sandstone concretion; D: green sandstone including remaining red sandstone; E: grey sandstone; F: grey sandstone with coal debris; G: red sandstone with green spot

3.2. Mineralogy characteristics

The thin slices of various altered sandstones and uranium ores were observed. The results show that the color of the red sandstone is mainly caused by the minerals such as limonite, and the color of the green sandstone is mainly caused by the chlorite film. Moreover, the calcareous sandstone is cemented. Its particles, floating in the cements, are composed mostly of quartz and feldspar. Under the optical microscope, the calcareous sandstone is shown to have the inclusions, which contain hydrocarbon substances, as evidenced by the fluorescence.

The results of the Electron Probe Microanalysis (EPMA) show that uranium ores and pyrites are closely symbiotic. Uranium ores are surrounded by pyrites (Fig. 3). This indicates a possible link of causes between the two. The uranium minerals are mainly coffinite (U $(SiO_4)_{1-\omega}$ $(OH)_{4\omega}$) and uraninite $(mUO_2 \cdot nUO_3)$.

Figure 3. Calcite cement and uranium mineral in sandstone

4. Geochemical characteristics

The essence of alteration is the interaction between fluids and rocks, which causes the change of the composition of rocks and minerals. In fact, the change of the composition can be characterized by geochemical characteristics. They can indicate the migration and emigration of rocks, and thus can reflect the fluid alteration of rocks. In this study, we selected four different types of rocks (10 pieces per type) to perform the major element analysis (Fig. 4). The results for the rocks of the same type were averaged.

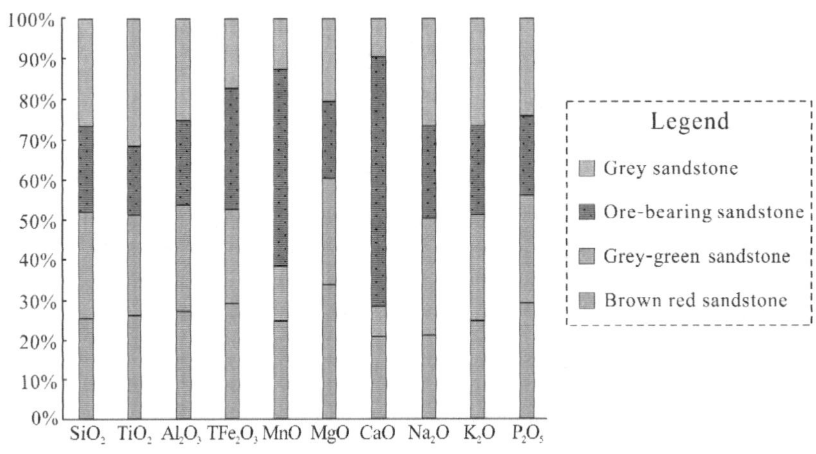

Figure 4. Distribution characteristics of major element content of four types of altered sandstone

1) The amount of TFe_2O_3 in the three altered sandstones increases slightly compared to that in the grey sandstone, indicating that there is the migration of Fe in the process of fluid alteration. Moreover, the amount of TFe_2O_3 in the gray-green sandstone is lower than that in the brown red sandstone, which indicates the changes of the valence state of Fe and the migration of Fe in the later stage of the secondary reduction of oil and gas.

2) The amount of Al_2O_3 in the four rocks (more than 11%) is almost the same, which is higher than that of the normal lithic sandstone (8.1%) and the feldspar sandstone (8.7%). This indicates that the four

rocks contain a high proportion of clay minerals, and the hydrolysis of feldspars and transformation of clay minerals did not migrate or emigrate Al in the sandstones considerably.

3) The amount of CaO in the ore-bearing sandstone is the highest among the four rocks, with an average proportion of 7.97%. This is primarily because of the presence of the carbonate cements in the ore-bearing sandstone. The corresponding hydrocarbon fluid, along with the alkaline fluid, caused an important mixed fluid alteration in the process of epigenetic alteration. This alteration process is closely related to the uranium mineralization. The resulting brown red altered sandstone contains the most MgO, followed by the green altered sandstone.

5. Discussion on the relationship between uranium mineralization and hydrocarbon

5.1. Spatial distribution relationships between uranium deposits and hydrocarbon

Hydrocarbon and uranium deposits coexist in the Ordos basin. They are distributed closely in the spatial dimension. Uranium deposits are mainly distributed on the edge of the basin, concentrating in Dongsheng-Hangjingqi of the north basin, Diantou-Huangling of the south basin and Huianbao-Ciyaobao of the west basin. Oil reservoirs are mainly distributed in the southern basin, mostly concentrating in the Chang 7 period deep lake area of the basin and nearby. Gas reservoirs are mainly distributed in the north region, the Yulin-Inner Mongolia area, and also the central and southern Yan'an region, according to the recent discoveries. Some shale gas reservoirs with exploration potential have also been found in the area bearing Chang 7 hydrocarbon source rocks. Oil is primarily produced in Yanchang Formation (T_3y) and Yanan Formation (J_2y), while natural gas in the Upper Paleozoic and part of Lower Paleozoic and the SUDs in the Jurassic Zhiluo Formation. That is, the minerals rank in the order of natural gas reservoir, oil reservoir and uranium deposit from the bottom up.

SUDs in the Ordos Basin mainly concentrate in Yimeng Uplift area (the north of the basin), which is also the direction and seepage area of Paleozoic gas. There formed faults in the northern Ordos Basin, and part of which cut through the Upper Cenozoic rock and gas reservoir. The BoerJianghaizi Fault remained active during the Cenozoic. It could provide the natural gas a channel for the upward migration and northeast-ward seepage.

As seen above, the distribution of hydrocarbon and uranium deposits in the Ordos Basin have a certain spatial pattern. This pattern is not only determined by their own characteristics of accumulation or metallogenic, but also by the interaction between various minerals.

5.2. Age comparison between hydrocarbon accumulation and uranium mineralization

The maturing of the Upper Paleozoic source rocks and the mass generation of hydrocarbon in the Ordos Basin can be traced back to the end of the early Cretaceous, about 120~100 Ma. According to the thermal history of the basin, there were significant thermal anomalies in the basin during this period. Consequently, the hydrocarbon source rocks of different ages became mature simultaneously, making this period the peak of hydrocarbon expulsion and the main period of hydrocarbon accumulation in the basin. We tested the ore-forming age of Dongsheng-Hangjingqi SUD and found that the uranium deposits were formed mainly in three periods, i.e., 120~90 Ma, 80~60 Ma and 30 Ma~10 Ma [9]. The age before 30 Ma should be the pre-enrichment age of the sedimentary period, whereas the age after 30 Ma is the late and ultra-late mineral age and the age of reformation. Note that the 120~90 Ma is about the same period as or slightly later than the formation and large-scale loss of natural gas reservoirs in the basin. It therefore seems reasonable to assume that the formation of SUD is in sync with or slightly later than the formation and dissipation of natural gas in the Paleozoic. They are very likely to have a link of causes.

5.3. Uranium metallogenic mechanism

In the preceding context, we analyzed the petrological and geochemical characteristics of the hydrocarbon and uranium. We also explored the spatial distribution of SUDs and oil and gas reservoirs. Furthermore, we compared their reservoir-forming and ore-forming age. The results show that in the process of mineralization of SUDs in the northern Ordos Basin, the hydrocarbon seepage occurred and the oil and gas concentrated along the shallow strata. They interacted with the uranium-bearing fluid, causing the reduction of the uranium element. After being precipitated in the Zhiluo Formation, the uranium element concentrated to form the uranium deposit. Since the formation of Hetao Fault Depression during Late Cenozoic, the oil and gas that have seeped to the northeastern part of the basin have created a reducing environment for the uranium deposits formed earlier. They protected the ore body from being destroyed by the groundwater after mineralization. And Yimeng Uplift, the uplift of the northern part of the basin, has become a favorable prospecting area for SUD.

6. Conclusion

(1) The green alteration in Zhiluo Formation results from the secondary reduction of oil and gas. Its predecessor is the red sandstone formed in the oxidizing environment. The oil and gas seeping to the northeastern part of the basin could form green alteration, white alteration and pyrite as a result of the fluid-rock interaction.

(2) The oil and gas plays an important role in uranium mineralization. The seepage of oil and gas interacts with the uranium-containing and oxygen-containing fluid, reducing U^{6+} to U^{4+}. After being precipitated in Zhiluo Formation, the uranium element would concentrate to form the uranium deposit.

(3) The reduction environment due to the hydrocarbon seepage can protect the ore body from being destroyed by the groundwater after mineralization.

(4) In the northern part of the basin, the Yimeng Uplift area serves as a good indicator of the hydrocarbon seepage, where the sandstone-type uranium deposits are likely to form. Its exploration potential is tremendous.

Acknowledgments

This work was financially supported by National Natural Science Foundation of China (41330315), MOST Special Fund from the State Key Laboratory of Continental Dynamics, Northwest University and Postgraduate Independent Innovation Program of Northwest University (YZZ14019).

References

[1] Seewald J S. Organic-inorganic interaction in petroleum-producing sedimentary basins. Nature, 426 (20) (2003): 327~333
[2] Zhou Shi Xin, et al. Organic-Inorganic interactions during the formation of oils in sedimentary basin. Natural Gas Geoscience 17.1(2006):42-47.
[3] LIU C Y, Mao G Z, Qiu X W, et al. Organic-inorganic energy minerals interactions and the accumulation and mineralization in the same sedimentary basins. Chinese Journal of Nature, 35(1) (2013):47-055.
[4] QIN Mingkuan and ZHAO Ruiquan. New understanding in genesis of uranium deposit bashblak in Tarim Basin. Uranium geology, 16(1) (2000):26-30.
[5] Landais P. Evidence for bacterial degradation of hydrocarbons in uranium deposits. Terra Nova, 1(2) (2010):163-171.
[6] WANG Feifei. Forming Environment and Main Controlling Factors of the Sangstone-type Uranium Deposit in Ordos Basin Hangjinqi Area. Northwest University. (2013)
[7] LI Ziying, Fang Xiheng, Chen Anping, et al. Origin of gray-green sandstone in ore bed of sandstone type uranium deposit in north Ordos Basin. Science China Earth Sciences, 50(2) (2007):165-173.
[8] Darby, B. J, and B. D. Ritts. Mesozoic contractional deformation in the middle of the Asian tectonic collage: the intraplate Western Ordos fold–thrust belt, China. Earth & Planetary Science Letters 205.1(2002):13-24.
[9] LIU H B, Xia Y L, Tian S F. Study on geochronology and uranium source of sandstone-type uranium deposit in Dongsheng area. Uranium Geology, 23(1) (2007):23-29.

IOP Conf. Series: Earth and Environmental Science **64** (2017) 012028 doi:10.1088/1755-1315/64/1/012028

Research on the information security system in electrical gis system in mobile application

Chao Zhou[1], Renjun Feng[2], Haitao Jiang[1], Wei Huang[1], Daohua Zhu[1]

[1]State Grid Jiangsu Electric Power Research Institute, Nanjing, 211100, China
[2]State Grid Suzhou Power Supply Company, Suzhou, 215004, Chian

Abstract. With the rapid development of social informatization process, the demands of government, enterprise, and individuals for spatial information becomes larger. In addition, the combination of wireless network technology and spatial information technology promotes the generation and development of mobile technologies. In today's rapidly developed information technology field, network technology and mobile communication have become the two pillar industries by leaps and bounds. They almost absorbed and adopted all the latest information, communication, computer, electronics and so on new technologies. Concomitantly, the network coverage is more and more big, the transmission rate is faster and faster, the volume of user's terminal is smaller and smaller. What's more, from LAN to WAN, from wired network to wireless network, from wired access to mobile wireless access, people's demand for communication technology is increasingly higher. As a result, mobile communication technology is facing unprecedented challenges as well as unprecedented opportunities. When combined with the existing mobile communication network, it led to the development of leaps and bounds. However, due to the inherent dependence of the system on the existing computer communication network, information security problems cannot be ignored. Today's information security has penetrated into all aspects of life. Information system is a complex computer system, and it's physical, operational and management vulnerabilities constitute the security vulnerability of the system. Firstly, this paper analyzes the composition of mobile enterprise network and information security threat. Secondly, this paper puts forward the security planning and measures, and constructs the information security structure.

1. Introduction

The development of information technology is affecting the development and progress of the whole society, and the communication and network technology are the subjects of information technology. The popularization and application of Internet, rise of mobile communication, and the increasing integration of efforts of satellite communication and various kinds of communication technologies and Internet technologies make the modern communication technology and network technology has made the rapid development. With the development of optical fibre transmission technology and the continuous update of access network technology, the network level of the whole world is increasing day by day [1]. It also promotes the progress and development of various fields of the national economy. At present, the technology development of the world telecommunication industry has entered a new stage of development, appearing the new trend of integration, adjustment, and reform. In particular, the development and application of broadband technology has become a hot spot for global development in the coming period.

The rise of the information society has further brought opportunities for the rapid development of global information technology. The application of information technology has caused great changes in people's production, life style and concept, which greatly promotes the development of human society and the progress of human civilization, and brings the human into a new era. However, while people are enjoying the enormous benefits brought by information network, they also face the severe test of information security. The existing information security has been a threat to national security, economic security, military security and social security [2]. Because the mobile is the technology dependent on computer network and mobile communication network, the information security has attracted more and more attention. An important geographical location, the leaked military targets and so on are the major threat to national security. From this point of view, this paper discusses the composition of the information network and its application in the process of application, and the relevant security threats and solutions in all links of computer network and communication network.

2. Composition of mobile enterprise network and information security threat

2.1. Importance of information network security

With the continuous development of Internet, it also brings some problems of enterprise network security. With the further development of information technology, both the telecommunications network and other network information infrastructure have become a key technology of information society. They not only service information transmission, but also become the important national infrastructure. Important information systems have been related to the people's livelihood. Information security is not only the thing related to network community and information industry, but has been related to the overall national security. Whether it is economic security, political security, social stability, national security and information security, they are closely related [3]. As a result, to understand the informatization from this global aspect, the guarantee of network and information security work is more important.

The development of information security and information technology is a complementary relationship. The national information security strategy and national information security related deployment clearly pointed out the need for security in the development, to promote the development with security, which is a highly summary for the relationship between the development of information security and information technology. Information security is closely related to the development of information technology, and the development of information security technology is closely related to the overall development of information technology [4]. In addition, information security equipment is also closely related to general information, technology equipment, and information industry equipment. Form network equipment, communication equipment, computing devices to software, many security features and technical equipment itself are closely related, let alone the close relationship between the depth and existence of information security and information technology application. With the deepening of application, it is deeper and extending, so there is close relationship between information guarantee and information security. The relationship suggests that, to do well the management of information security, it is necessary to closely connect the information security with the development stage of information technology. Moreover, it needs certain strategies and practice for information security guarantee at a certain stage of development. Therefore, information security must be understood and practiced from the overall situation of information technology.

Information security technology and concept are continuously developing, and it achieves the security through the network isolation and information encryption in the early stage. After a stage of development, it is found that it needs for coordination for how to balance the application and development by isolation and encryption. In consequence, it entered a new stage. The so called information security stage is to use the security protection detection response, such life cycle, to

IOP Conf. Series: Earth and Environmental Science **64** (2017) 012028 doi:10.1088/1755-1315/64/1/012028

achieve the security work. However, with the development of information technology, the popularization of the application is not enough. Because when the network infrastructure and critical information systems need to continue running when they are attacked by human factors or making blockage and obstacles because of natural factors, it converts into the so-called turning survivability stage. Regardless of virus, or attack or natural disaster, the important system needs to continue running, which the process of understanding and development of information security. With the development of information technology and the further development of information security, our theory and technology, concepts and models are deepening and developing.

2.2. Security hidden trouble in Mobile Enterprise Intranet

1) Compared with the external network, the internal network is faster, which can prevent omissions and has simple security measures. Enterprise internal network has the following security risks:

The speed of the internal LAN is fast, the information is easy to be stolen quickly, and the monitoring time is lost quickly;

Internal LAN access is easy. In general, as long as the general information equipment is used, it can be connected to the network;

Internal LAN access authentication measures are simple;

The vast majority of local area network information is not encrypted, making use of plaintext transmission;

Users in the internal network often have the right to operate directly on the database and the server. There is an opportunity to misuse the key data, and to intend to steal or destroy;

Improper use of external devices in computer systems can cause leaks.

Many internal network users do not pay attention to the password, but using a weak password, so the cracking procedure of hackers in the internal network is easier to work;

The internal network is based on the Client / Server way, the client directly operates the server, and the transmission of information in the internal network is very insecure.

2) The internal network data management itself is usually not very strict. Core confidential data generally uses simple protection authority password. In the product research and development process, the various core technology information and job information even have no protective measures. At least for the entire development group, all of the development results and data are transparent and shared. Almost no measures are taken for protection mis-operation, anti-theft and anti-destruction. This makes it very easy to obtain authorization and access to information in the network. The information system with no internal network security management makes anyone may intentionally or unintentionally cause security risks and lead to disasters.

3) Lack of effective management mechanism. Enterprises often lack a more effective management mechanism to effectively manage the internal security [5]:

The internal security management mechanism is imperfect. There are a number of effective management mechanisms within the enterprise, but there are some problems and deficiencies in the management mechanism, which also led to no "institutional" protection of the internal security management.

Internal security management mechanism cannot be effectively implemented. The enterprise lacks effective implementation mechanism of the management system. As a result, the management system is not implemented, and many management systems are still only on the surface.

4) Lack of Intranet security solutions. The lack of internal network security systems and programs has brought some obstacles on the intranet security management:

Most security systems are only from external intrusion. According to the intrusion from the outside, it has a large number of mature security systems to prevent, but the internal security threats and risks have rarely been noticed, or have noticed. Whereas, there is no perfect security system and security scheme to solve the internal security problems of enterprises.

Internal security system is not mature enough. In a small number of internal security systems, most of them just focus on a specific access, such as dial-up connection control and file protection. It is simply a product in the technical sense, which cannot put forward the comprehensive scheme for internal security management.

2.3. *Mobile GIS information security*

As the mobile network is a combination of mobile communication network and technology, the information security includes hardware and software problems [6]. First of all, we start from the form of internal LAN of mobile communication to analyze the problem of network security. Secondly, from the point of view of security of the data itself, we analyze data security issues.

2.3.1. Composition of mobile GIS network The mobile is mainly composed of wireless communication network, mobile terminal equipment, geographic application server, and spatial database, as shown in Figure 1.

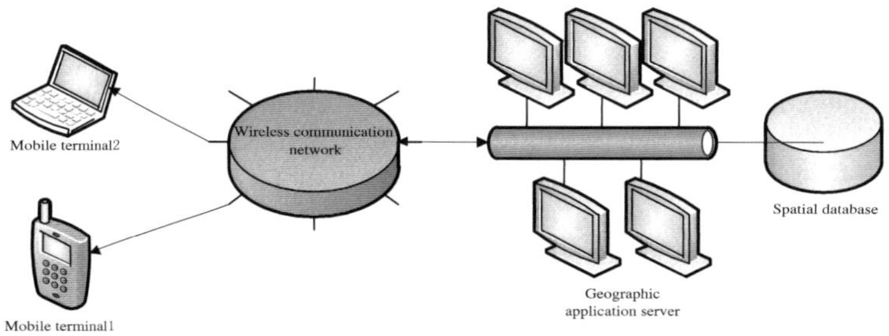

Figure 1. Composition of mobile GIS

Wireless communication networks: GSM, GPRS, and CDMA based on cellular communication systems.

Mobile terminal equipment: portable computers, PDA, WAP mobile phones, handheld GPS machines, including the display, RAM, high-speed processor and other components.

Geographic application server is the key part of the whole mobile GIS, and it is the GIS engine of the system. It is located in a fixed place, offering a wide range of geographic services and potential spatial analysis and query service for mobile GIS users. It has the following characteristics: to provide high quality maps, data and spatial query and analysis services; to handle a large number of service requests and uninterrupted access requests; to deal with the application request of huge data sets and a large amount of data; to increase the processing capacity without interrupting the operation; to be extensible, so as to adapt to the increase of data amount and access of new equipment; and uninterrupted feature of geographic service.

The spatial database is used to organize and store the spatial data related to geographic spatial data and corresponding attribute description information. It is the data storage center of mobile GIS, and it is able to manage the data, to provide a variety of spatial location data for the mobile application. It the data source for the geographic application server to realize geographic information service [7]. Mobile spatial database, in mobile GIS, plays the role of a data pump. It can make mobile devices interact with a variety of data sources, shield the difference of fixed network environment, and optimize the query conditions, so the entire mobile GIS has good flexibility and adaptability.

2.3.2. Information security problem of mobile GIS Because the mobile GIS independently exists not separates from the computer and network system, it also has a series of security problems, such as database security, data secrecy and so on. It mainly includes the following aspects:
Security and confidentiality of data transmission. The main contents include the determination of the security target, the encryption methods and algorithms in the network transmission, and the security level.

Security and confidentiality of data access, including for the different levels of users, through the different operating authorities, to achieve data access restrictions; for the different categories of data, to set different access rights; to establish operation log files for tracking the system operation; to encrypt the data and ensure the data security through the data dump, data backup and recovery.
The physical security design of the system is mainly to meet the technical safety requirements and safety requirements of equipment [8].

Personnel security. It is mainly to regulate the security and confidentiality of institutions and personnel, to determine the appropriate authorities and management practices, to develop contingency plans and a variety of safety rules and regulations, while to strengthen the safety awareness education. In addition, it is necessary to have the ability to defend the virus, and deploy anti-attack device firewall and real-time monitoring system.

3. Mobile GIS enterprise information security architecture

From a technical point of view, a complete network security system should be three-dimensional. Different levels reflect the different security issues. According to the structure of the network and the application of network, it can be divided into the physical layer security, system layer security, network layer security, application layer security and security management. It is an effective way to realize the network and information security system through the good coordination between the technique aspect and management aspect [9]. Among them, the information security technology is achieved through the adoption of host computer system and safe network system containing the construction security, and equipped with appropriate security products. In the management level, it is realized through the framework of information security management system, as shown in Figure 2.

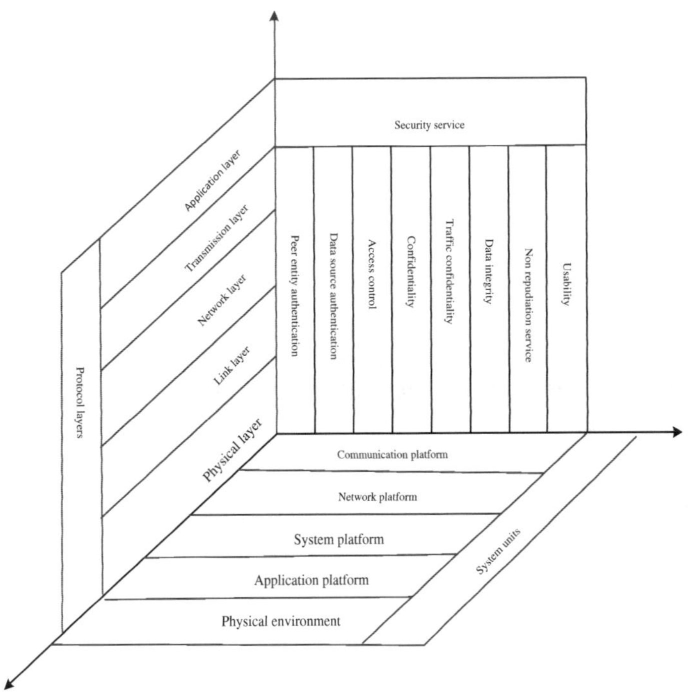

Figure 2. Three dimensional security technology framework

3.1. System layer security

3.1.1. Reasons for operating system vulnerabilities Programmers' human factors, in the process of programming, in order to achieve ulterior purpose, preserve the back door in the hidden place of the program code.

Restricted by the programmer's ability, experience and security technology at that time, in the process, there will inevitably be inadequate, and it will affect the efficiency of the program, but also lead to the authority enhancement of unauthorized users.

Due to the hardware, the programmers cannot make up for the vulnerability of the hardware, so the problem of hardware is reflected through the software.

Of course, everything is not perfect. As the desktop operating system - Windows is also in this way. And because of its monopoly in the desktop operating system, the problems will soon be exposed. In addition, compared with the operating systems of Linux and other open source code, Windows is a black box operation that ordinary users cannot get the source code [10]. As a result, the security issues are resolved by Microsoft itself.

3.1.2. Corresponding security measurements Install the latest version and patch program of the operating system in time. There will always be a number of system vulnerabilities being found. Usually, the software vendors will release a new version or patch program to patch security vulnerabilities, so as to maintain the used latest version can make the security threat minimal. Make the necessary security configuration. In the system configuration, close the services with security risks and unnecessary services, such as FTP, Telnet, finger, login, shell, BOOTP, TFTP and so on [11]. All of these protocols have security risks. Prohibit certain r orders, such as rlogin, rsh and so on. Strengthen the authentication in login process, set up a complex and not easy to guess password, carefully protect the account password and often change, prevent unauthorized users to easily guess the password, ensure the legitimacy of users, and restrict unauthorized access to the host. Strictly restrict the use permission of key documents in the system as follows, /.rhost, etc/host, shadow, group and so on, and strictly control the logged in visitor's operating authority, and limit the completion operation to the minimum range. Make full use of the log function of the system itself, record all the accesses of users, and regularly check the system security log and system state for early detection of possible illegal intrusion behaviors in the system, so as to provide the basis for decision-making for administrators and provide the basis for post review.

3.2. Security management

3.2.1. Security risks of management Management is an important part of network security. The imperfection of safety management system is one of the important sources of network risk. Security vulnerabilities caused by the network administrator, improper allocation or not in time upgrading of operating system and network application patch program, weak user passwords, freely using software downloaded by ordinary web site, setting up dial-up server inside the firewall but no authentication of the account, not strong user security awareness and so on, these management problems cannot be solved no matter how sophisticated the security strategy and security system are, and the network will be in danger.

3.2.2. Security measurements Information security technology is an important attribute of information security. The purpose of secrecy is to prevent the confidential information in the information system intercepted by the enemy. Encryption is an important means to realize the confidentiality of information. The so-called encryption is the use of mathematical methods to transform the message so that it is difficult for anyone else, in addition to the legitimate receiver, wants to restore the original message [12]. The process of converting cipher text into plaintext is called decryption. It can be seen that encryption technology can make some important data stored on an insecure computer or an insecure channel without leakage.

Information authentication technology: information authentication is another important attribute of information security. The purpose of certification contains three aspects. The first one is to verify whether the information sender is real. The second is to verify the integrity of information, that is to say, to verify that the authentication information has not been tampered with, replayed or delayed in the transmission or storage process. The last one is to verify the non-repudiation of information, that is, to prevent the communication party denying having participated in an activity afterwards. Authentication is one of the most important techniques to prevent the adversary from making an attack on the system and falsification of information. The commonly used identification methods have access to the word and the way of holding, digital signatures and so on.

4. Conclusion

With the development of computer hardware and software, continuous improvement of network technology, and mobile devices with faster information transmission, the mobile terminal can browse a large number of multimedia contents. This paper analyzes the mobile compositions. In the paper, the security planning and solutions of the whole network are analyzed. In addition, for security threats that may occur, the corresponding solutions and security measures are deployed. More importantly, the framework for information security is constructed, which provides basis for the follow-up researches on the mobile GIS information security.

References

[1] Mai, B., Parsons, T., Prybutok, V., & Namuduri, K. (2017). Neuroscience Foundations for Human Decision Making in Information Security: A General Framework and Experiment Design. In Information Systems and Neuroscience (pp. 91-98). Springer International Publishing.

[2] Vithanwattana, N., Mapp, G., & George, C. (2016, September). mHealth-Investigating an Information Security Framework for mHealth Data: Challenges and Possible Solutions. In Intelligent Environments (IE), 2016 12th International Conference on (pp. 258-261). IEEE.

[3] Tupia, M., Bruzza, M., & Rodriguez, F. (2016, September). An information security framework for ubiquitous services in e-government structures: a peruvian local government experience. In Computer Science and Information Systems (FedCSIS), 2016 Federated Conference on (pp. 1309-1316). IEEE.

[4] Lee, J., Yoo, B., Lee, H., Cha, G. D., Lee, H. S., Cho, Y., ... & Kang, M. (2017). Ultra - Wideband Multi - Dye - Sensitized Upconverting Nanoparticles for Information Security Application. Advanced Materials, 29(1).

[5] Bansal, G., Hodorff, K., & Marshall, K. (2016). Moral Beliefs and Organizational Information Security Policy Compliance: The Role of Gender. Proceedings of the Eleventh Midwest United States Association for Information Systems, 1-6.

[6] Kearney, W. D., Kearney, W. D., Kruger, H. A., & Kruger, H. A. (2016). Theorising on risk homeostasis in the context of information security behaviour. Information & Computer Security, 24(5), 496-513.

[7] Tsai, N., & Xiong, Y. (2016). An investigation of the information system security issues in Taiwan. International Journal of Business Information Systems, 21(3), 309-320.

[8] Zhou, C., Guo, Y., Huang, W., Jiang, H., Li, B., & Chen, J. (2016). Information security defense method of electric power control system based on digital watermark. system, 2, 4.

[9] Hameed, M. A., & Arachchilage, N. A. G. (2016). A Model for the Adoption Process of Information System Security Innovations in Organisations: A Theoretical Perspective. arXiv preprint arXiv:1609.07911.

[10] Lakhno, V. A., Petrov, O. S., Hrabariev, A. V., Ivanchenko, Y. V., & Beketova, G. S. (2016). Improving of information transport security under the conditions of destructive influence on

the information-communication system. Journal of theoretical and applied information technology, 89(2), 352.

[11] Fenz, S. (2016). researcher at Vienna University of Technology and SBA Research, an Austrian research center for information security, and founder of Xylem Technologies GmbH, a company supporting SMEs in identifying optimal information security strategies. From 2012 to 2015, Stefan was an appointed member of the European. Understanding Complex Urban Systems: Integrating Multidisciplinary Data in Urban Models, 129.

[12] Joo, J., & Hovav, A. (2016). The influence of information security on the adoption of web-based integrated information systems: an e-government study in Peru. Information Technology for Development, 22(1), 94-116.

Numerical studies on surrounding rock deformation controlled by pressure relief groove in deep roadway

Chaoke Liu*, Jianxi Ren, Kun Zhang, Shaojie Chen

School of Civil Engineering and Architecture, Xi'an University of Science and Technology, Xi'an 71005, China

*Corresponding author e-mail: liuchaoke8@163.com

Abstract. After entering deep mining, the roadway is in a high stress state, the deformation of surrounding rock becomes larger, and the roadway floor is particularly significant under unsupported state, which brings great difficulty to the safe production and support of the coal mine. Pressure relief method can change the stress field of surrounding rocks so that the surrounding rock can be in stress-reducing area. The present paper studied the deformation law of the roadway and the changes in the stress state and plastic zone of the surrounding rocks around the roadway before and after the excavation of pressure relief groove on the bottom floor of the high-stress roadway by using FLAC under the engineering background of one mine in Binchang, analyzed the influence of different groove depths and widths on the floor heave, convergence on both sides and roof subsidence. The simulation results show that: after the roadway floor was grooved in the high stress roadway, a larger stress-relaxed area will be formed near the roadway floor, the stress will be transferred to the deep roadway floor, and the pressure relief groove plays a better control effect on the deformation of the high-stress roadway. With the increase of the width and depth of the pressure relief groove, the convergence of the top and bottom of the roadway will be decreased accordingly, but the effect is not significant, while its influence on the convergence on both sides is relatively significant. After applying the simulation results to the engineering practice, the practice shows that: the combined support of anchor rod, anchor rope plus pressure relief groove can control the deformation of the roadway well and the conclusion obtained can provide some reference values for the study and design of the grooving pressure relief control technology.

1. Introduction

With the rapid economic development, the demand for energy is increasing, which results in that the shallow coal becomes depleted gradually and the burial depth of the roadway depth becomes increased year by year, according to statistics, the mining depth of coal is progressively increased at an annual rate of 10-20 meters and the average mining depth of coal has reached 600 meters and those of partial coal mines have reached more than 1000 meters [1].After entering the deep part, the deformation of the surrounding rock around roadway becomes remarkable, and the roof fall, rib spalling, floor heave and other phenomena often occur on the roadway support so that it needs frequent maintenance and reinforcement, the maintenance workload is heavy and the support is expensive, which brings serious influence on the normal safe mining of the mine, thus the stability control of the surrounding rocks in deep roadway is the most important problem to be solved at present [2-5].

After long-term study by experts and scholars, a number of effective support methods have been put forward currently, in general, according to the different mechanisms, it can be divided into two categories: reinforcement method and pressure relief method [6-9]. Pressure relief method is to make the original continuous rock body in a discontinuous state with some artificial measures, change the stress state of the surrounding rock around the roadway and transfer the concentrated stress around the roadway to the deep part of the surrounding rock around the roadway so as to make the roadway floor rock formation in the stress reducing area, hereby ensuring the reliable state of the roadway floor, it is especially suitable for high-stress roadway. At present, in the underground engineering, the applied pressure relief method has joint-cutting, drilling, blasting, roadway digging pressure relief and other forms. With the application and promotion of the grooving machine in the mine engineering, the pressure relief groove construction problem has been solved, which lays the foundation for the promotion of the pressure relief method. The present paper studied the grooving pressure relief and deformation mechanism on the roadway floor, the influence of the changes in the pressure relief groove width and depth on the roadway deformation and plastic zone by using FLAC under the engineering background of one mine in Binchang, which provides the basis for the scheme engineering design and a certain reference value for the study and design of the grooving pressure relief.

2. Engineering Overview

There are three main haulage roadway between the auxiliary shaft and the panel, they are no.1 auxiliary transport roadway, no.2 auxiliary transport roadway and the belt transport roadway. The roadway lies in the east-west direction, the average length is 2300 meters, the average burial depth is 760 meters and the slope is 3.5 degrees. The actual measurement distance from the nearby coal worked-out section to the belt transport roadway is 130 meters and the working face width is 100 meters. According to the geological survey data, the transport roadway passes through the complex rock, mainly including mudstone, siltstone, fine sandstone, coal seam and argillaceous sandstone.

The cross section of the belt transport roadway is a straight wall semicircle arch with a roadway width of 5640mm and a roadway height of 4720mm (straight wall:1900mm,arch height:2820mm),the thickness of concrete floor is 300mm and the spaying layer thickness is 120mm.The roadway support method is bolting and concreting with wire mesh, roof and side anchor bolt diameter is 20mm,it's length is 2700mm,inter-row spacing is 800mm.Anchor cable diameter is 17.8mm,it's length is 7300mm,spacing is 1600mm,row spacing is 2400mm;the weld reinforcing mesh is manufactured using 6mm diameter Q235 steel wire rod, grid size is 150mm.

The floor heave occurred in the belt transport roadway two months after the completion of the roadway support, the maximum floor heave amount reached 300mm, the phenomena of rib spalling and partial roof fall occur, which severely impact on the coal mine safety and roadway stability.

3. Numerical Simulations

3.1. Computational modelling

In accordance with the Saint-Venant's principle, influence range is generally from 3 to 5 times of the hole scope after excavation.in order to ensure the calculation accuracy, the model selection range is larger than 5 times of the hole size. The working surface width of the nearby coal mine is 200m and the overlying rock formation within a range of 100m only on one side near the coal pillar in the nearby coal mined-out area influences the deformation of the roadway, thus, only 100 meters is taken from the nearby coal mine working surface. The calculating principle of unit division is to take the unit number as small as possible without affecting the accuracy of the calculation result, which is convenient to improve the efficiency of numerical simulation calculation. In the modeling process, in order to simplify the calculation, the secondary rock formation in minor thickness is ignored and the

IOP Conf. Series: Earth and Environmental Science **64** (2017) 012029 doi:10.1088/1755-1315/64/1/012029

fine sandstone, siltstone, mudstone, coal and argillaceous sandstone in the model are in different colors respectively, as shown in Figure 1.

3.2. Boundary condition

Figure 1. Established model for the FLAC

Fix the left and right displacement and the front and back displacement of the model in horizontal direct and fix the model bottom horizontal and vertical displacement, the upper boundary is the free boundary, in order to reduce the number of units and save the calculation time, the external load of 17MPa is applied on the upper part of the model to simulate the self-weight of the overlying rock formation.

3.3. Constitutive model selection

The Mohr-Coulomb yield criterion is used to determine the failure mode of the rock mass in the numerical simulation of rock mechanics elastic-plastic problem, as shown in formula1, after the load reaches the ultimate strength limit, the rock mass is destroyed, and the residual strength of the rock mass is decreased gradually with the development of deformation in the process of plastic flow after peak [10-14].

$$f_s = \sigma_1 - \sigma_3 \frac{1+\sin\varphi}{1-\sin\varphi} - 2c\sqrt{\frac{1+\sin\varphi}{1-\sin\varphi}} \tag{1}$$

Where σ_1, σ_3 is the maximum and minimum principal stress, c, φ is the adhesive force and friction angle respectively, when f_s is greater than zero, the material will be sheared failure, under normal stress state, the tensile strength of the rock mass is very low, thus, whether the tension failure is produced can be judged according to the tensile strength criterion.

3.4. Determination of calculation parameters

The values are taken for the physical and mechanical parameters of each rock formation in accordance with indoor rock test results, and the specific values are shown in Table 1.

Table 1. Physical and mechanical properties of stratum

Rock type	Shear modulus (GPa)	Cohesion (MPa)	Tensile strength (MPa)	Internal friction angle (degree)	Density (kg/m^3)
Fine grain Sandstone	5.08	10.1	9.6	40.7	2700
siltstone	5.82	7.9	2.4	39.7	2500
Mud stone	1.3	0.5	0.6	28	1900
# 4 coal	0.76	3.64	0.11	31.55	1700
mudstone	5.91	10.7	2.2	40.6	2480

3.5. Working condition simulation and monitoring scheme

In order to study the grooving pressure relief mechanism and the influence of the pressure relief groove shape on the roadway deformation, a total of nine working conditions are designed, the first group simulates two kinds of working conditions respectively: not excavating pressure relief groove and excavating pressure relief groove; the second group designs four kinds of working conditions to study the influence of the depth of the pressure relief groove on the deformation of the roadway, pressure relief groove size is 600×800mm,600×1200mm,600×1600mm and 600×2000mm;The third group designs three kinds of working conditions to study the influence of the width of the pressure relief groove on the deformation of the roadway, pressure relief groove size is 400×1200mm,600×1200mm and 800×200mm.

In order to monitor the deformation of the roadway, thirty-five monitoring points are set on the roadway surface, among which seven monitoring points are set on the roof to monitor the roof convergence. Eight monitoring points are set on the floor to monitor the floor heave quantity. Seven monitoring points are set on sides to monitor rib sides convergence. Six monitoring points are set along the depth of the pressure relief groove on the floor at the depth of 0.2, 0.4, 0.6, 0.9, 1.4 and 2.0 meters to monitor the stress and displacement of the rock formation on the floor.

3.6. Numerical simulation result analysis

Table 2. The deformation of roadway

Working condition	Bottom heave quantity (mm)	Subsidence of arch roof (mm)	Convergence on both sides (mm)
Before grooving	306.9	333.7	564.6
After grooving	162.1	85.02	214.3
Deformation reduction quantity after grooving	144.8	248.68	350.3

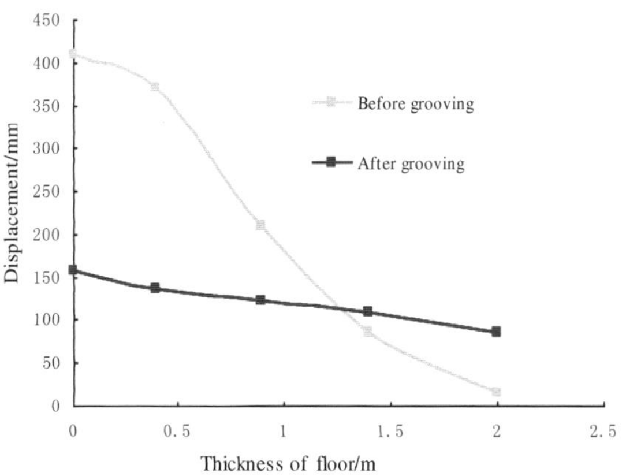

Figure 2. Displacement of the floor at different depths

Table 2 shows that the roadway vault subsidence quantity is 333.7mm, the floor heave quantity is 260.9mm, and the rib sides convergence is 564.6mm, and the deformation of the surrounding rock around the roadway is larger before grooving. The numerical simulation results show that the maximum vertical stress of the surrounding rock around the roadway is more than 40MPa, which makes the roadway in a high stress state. After the roadway floor is grooved, the floor heave quantity of the roadway is 162.1mm, reduced by 47.18%.The vault subsidence is 85.02mm, reduced by

74.52%.The convergence on two sides is 214.3mm, reduced by 62.04%.The above data show that the pressure relief groove has a good effect on the control of roadway deformation, and its effect on the vault subsidence and the rib sides convergence is especially remarkable [15].

According to Figure2,when the floor is not grooved, the floor heave quantity is drastically decreased with the increase of the depth of the surrounding rock of the floor, and the deformation of the floor heave is mainly concentrated in the range of 1 meter above the roadway floor, the roadway floor is basically in a stable state at 2 meter, and the displacement is close to zero; The floor heave quantity of the roadway after grooving is obviously reduced, and the floor heave quantity is decreased accordingly with the increase of the depth of the floor, but the disease is more gentle and nearly close to the linear decrease, which indicates that the deformation of the rock formation along the floor depth tends to be uniform after grooving.

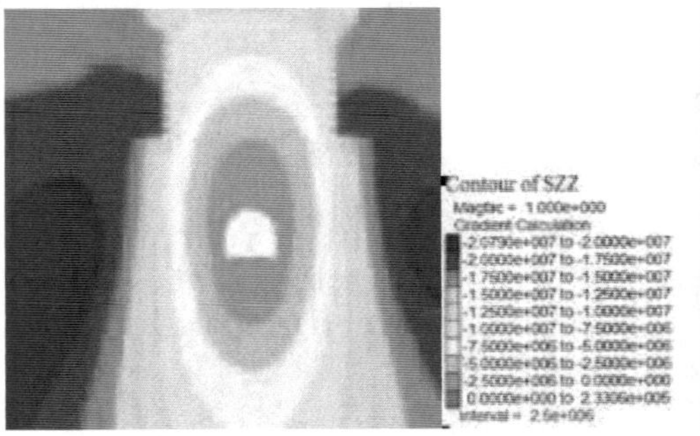

Figure 3. Stress before grooving

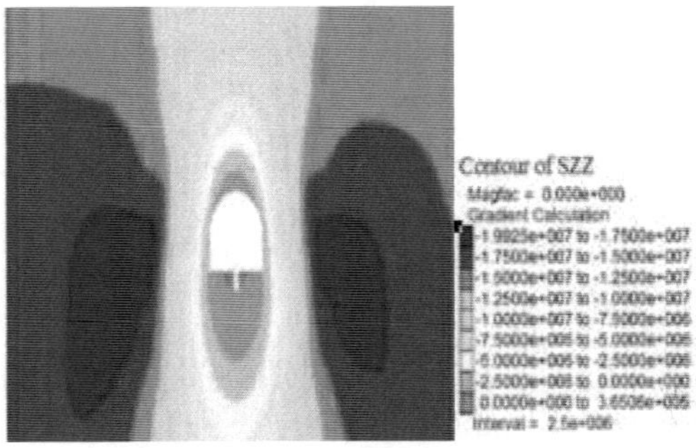

Figure 4. Stress after grooving

From Figure 3 and Figure 4, after the roadway floor is grooved for pressure relief, the monitoring point on the left side of the floor is reduced from 1.66MPa (when not grooved) to 0.98Mpa, reduced by 40.96%.The monitoring point on the right side of the roadway floor is reduced from 1.69MPa originally to 0.92MPa, reduced by 45.56%.The above analysis shows that the roadway floor stress is obviously reduced after the roadway floor is grooved for pressure relief. It can be seen from the stress diagram that the range of the stress reducing area near the roadway floor after grooving is obviously increased and the stress is transferred to the deep part.

In summary, the pressure relief groove plays a significant role in controlling the deformation of the roadway, improving the stress state of the surrounding rock and treating the bottom heave.

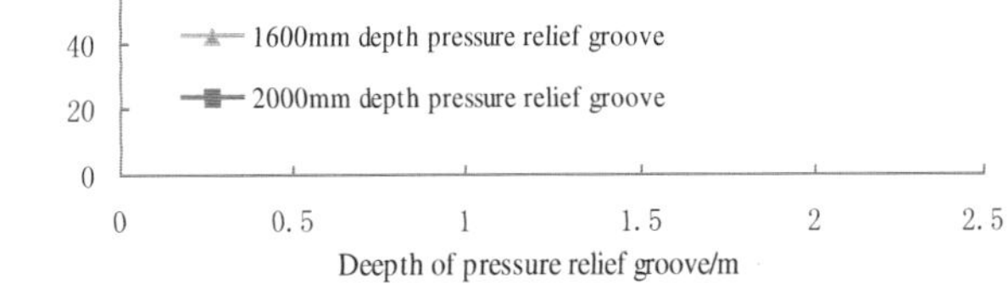

Figure 5. Displacement of different depths of the floor

IOP Conf. Series: Earth and Environmental Science **64** (2017) 012029 doi:10.1088/1755-1315/64/1/012029

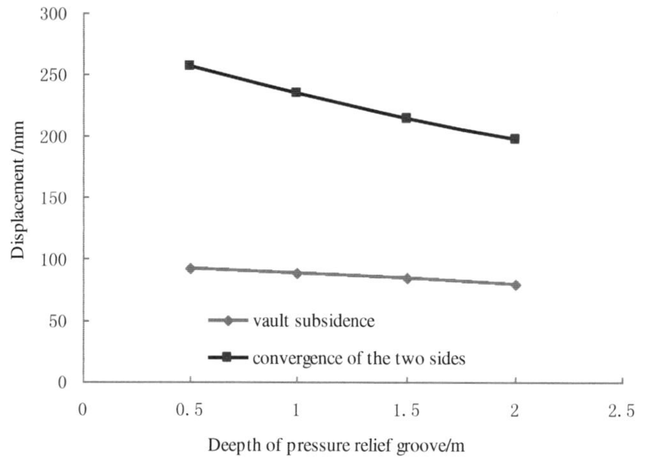

Figure 6. Displacement of vault subsidence and the convergence

It can be seen from Figure 5 and Figure 6 that floor heave and vault subsidence quantity are decreased with the increase of the depth of the pressure relief groove, but the change of the floor heave quantity and the vault subsidence quantity are not significant. With the increase of the depth of the pressure relief groove, the convergence of the two sides is decreased relatively. It shows that the increase of the depth of the pressure relief groove has a relatively good effect on the convergence on two sides of the roadways and does not influence and is not insensitive to the floor heave and vault subsidence.

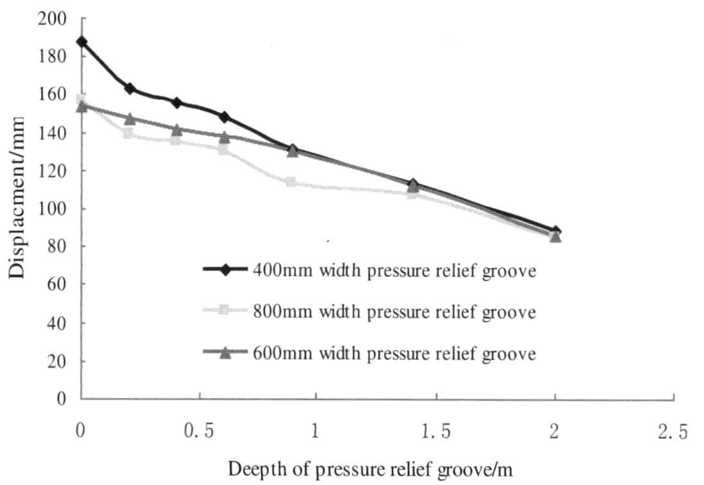

Figure 7. Displacement of different width pressure relief

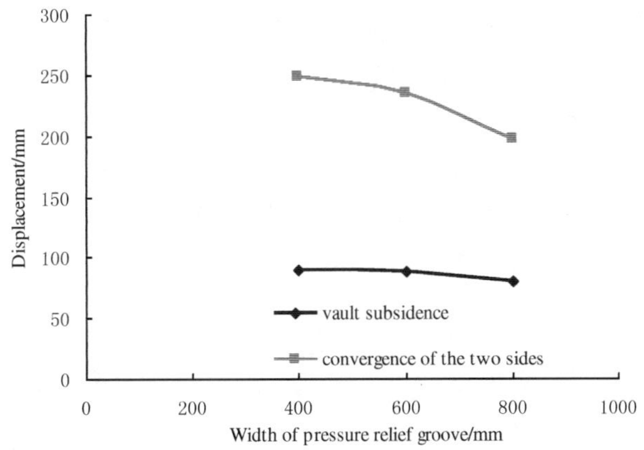

Figure 8. Vault subsidence and the convergence of the two sides

It can be seen from Figure 7 and Figure 8,with the increase of the pressure relief groove width, the roadway floor heave quantity is decreased accordingly, but not significant, when the pressure relief groove width is greater than 600mm,with the increase of the width, the floor heave quantity basically remains unchanged, thus, the width of the pressure relief groove has little effect on the floor heave quantity. With the increase of the roadway width, the convergence on two sides is decreased significantly, and the convergence slope on two sides is increased accordingly, indicating that the width of the roadway has a significant effect on the rib sides convergence. The arch crown subsidence curve is gentle and close to the horizontal line, indicating that the width of the pressure relief groove has little effect on the arch crown subsidence.

4. Engineering Practices

The average burial depth of the transport roadway for one coal mine in Binchang is 760m,belonging to the deeply buried, the deformation of the surrounding rock around the roadway is larger under high stress, in order to control the deformation of the roadway,60 meters test section is chosen in the auxiliary transport roadway and the test section uses anchor rod, anchor rope and pressure relief groove combined support scheme. The numerical study results show that the pressure relief groove width has a significant and the pressure relief groove width has little effect on the deformation of rib sides, roof and floor. Therefore, 1000mm width is taken, 1200mm depth is taken considering the construction factor.

The test section monitoring data show that the maximal convergence value of the roof and floor is 120mm, and the maximum convergence value of rib sides is 113mm during the observation period, and convergence value of the roadway is basically stable after two months observation. The maximum convergence speed rate of the roof and floor is 2.8mm/d, and the average deformation speed rate is 0-0.5mm/d after deformation becomes stabilized. The maximum deformation speed rate of the rib sides is 2.5mm/d, and the average deformation speed is 0-0.5mm/d after deformation becomes stabilized. The deformation of the roadway is effectively controlled after adopting this scheme. Figure 9 is the deformation diagram for the pressure relief groove after 60 days.

Figure 9. Pressure relief groove

5. Conclusion

Based on the previous papers of this study the following conclusions were pointed out.

(1)After the floor is grooved in the high stress roadway, a larger stress reducing area will be formed near the floor, the stress will be transferred to the deeper floor, and the deformation scope will increase towards the deep floor, the deform becomes even and pressure relief groove plays a better control effect on the deformation of the high-stress roadway.

(2)With increasing width and depth of the pressure relief groove, the convergence of the roof and floor will be decreased accordingly, but the effect is not significant, while its influence on the rib sides convergence is relatively significant.

(3)The engineering practice shows that the combined support of anchor rod, anchor rope and pressure relief groove can effectively control the stability of the surrounding rock in deep roadway.

Acknowledgments

This work was financially supported by the National Nature Science foundation of China (No.11402195), the collaborative innovation in shaanxi province (No.2015XT-15), and Scientific research project in Shaanxi province department of education (No.16JK1512).

References

[1] Zhou Hongwei, Xie Heping, Zuo Jianpin. Developments in researches on mechanical behaviors of rocks under the condition of high ground pressure in the depths, Advances in Mechanics, 2005, 35(1), 91-99.

[2] Cao Rihong,Cao Ping,Lin Hang.Support technology of deep roadway under high stress and its application, International Journal of Mining Science and Technology, 2016, 26(5), 787-793.

[3] Bai Qingsheng, Tu Shihao, Zhang Cun, Zhu Defu. Discrete element modeling of progressive failure in a wide coal roadway from water-rich roofs,International Journal of Coal Geology, 2016, 167(10) 215-229.

[4] He Manchao, Xie Heping, Peng Suping, et al. Study on rock mechanics in deep mining engineering, Chinese Journal of Rock Mechanics and Engineering, 2005, 24(16), 2803-2813.

[5] Xiao Tongqiang,Wang Xiangyu,Zhang Zhigao.Stability control of surrounding rocks for a coal roadway in a deep tectonic region, International Journal of Mining Science and Technology, 2014, 24(2), 171-176.

[6] He Yongnian, Han Lijun, Shao Peng, et al.Some problems of rock mechanics for roadways stability in depth, Journal of China University of Mining &Technology, 2006, 35(3), 288-295.

[7] Yuan Liang, Xue Junhua, Liu Quansheng, et al. Surrounding rock stability control theory and support technique in deep rock roadway for coal mine, Journal of China Coal Society, 2011, 36(4), 535-543.

[8] Liu Quansheng, Lu Xingli. Research on nonlinear large deformation and support measures for broken surround-ing rocks of deep coal mine roadway, Rock and Soil Mechanics, 2010, 31(10), 3273-3279.
[9] Cai Chengong.Simulation and testing study on outbust prevention measure of pressure-relief slots, Chinese Journal of Rock Mechanics and Engineering, 2004, 23(22), 3790-3793.
[10] Liu Jun, Kong Xianjing.Numerical simulation of behavior of jointed rock masses during tunneling and lining of tunnels, Rock and Soil Mechanics, 2007, 28(2), 321-325.
[11] Han Xu, Wang Ning-wei,Su Jing-yu, and Wang Wei.Numerical Analysis on Deep Foundation with Spatial Effect, Electronic Journal of Geotechnical Engineering, 2015, 20(18), 11017-11030.
[12] Qijun Hu, Yahui Xu, Zhiwei Long, and Pengcuo Zaxi.Impact Analysis of Deep Foundation Pit Excavation on Adjacent Pipeline in the Sand Gravel Area, Electronic Journal of Geotechnical Engineering, 2015, 20(2), 681-694.
[13] Zhou Chuanbo, Guo Liaowu, Yao Yingkang. Numerical simulation of wall rock deformation mechanism of mining tunnel, Rock and Soil Mechanics, 2009, 30(3), 654-658.
[14] Kang hongpu, Study applition of strata destressing method in coal roadway, Coal science and technology, 1994, 22(5), 14-16.
[15] Sun Jin, Wang Lianguo.Numerical simulation of grooving method for floor heave control in soft rock roadway, Mining Science and Technology (China), 2011, 21(1)49-56.

Studies on water resources carrying capacity in Tuhai river basin based on ecological footprint

Chengshuai Wang, Lirong Xu* **and Xin Fu**

School of Resources and Environment, University of Jinan, Jinan 250022, China

*Corresponding author e-mail: stu_xulr@ujn.edu.cn

Abstract. In this paper, the method of the water ecological footprint (WEF) was used to evaluate water resources carrying capacity and water resources sustainability of Tuhai River Basin in Shandong Province. The results show that: (1) The WEF had a downward trend in overall volatility in Tuhai River Basin from 2003 to 2011. Agricultural water occupies high proportion, which was a major contributor to the WEF, and about 86.9% of agricultural WEF was used for farmland irrigation; (2) The water resources carrying capacity had a downward trend in general, which was mostly affected by some natural factors in this basin such as hydrology and meteorology in Tuhai River Basin; (3) Based on analysis of water resources ecological deficit, it can be concluded that the water resources utilization mode was in an unhealthy pattern and it was necessary to improve the utilization efficiency of water resources in Tuhai River Basin; (4) In view of water resources utilization problems in the studied area, well irrigation should be greatly developed at the head of Yellow River Irrigation Area(YRIA), however, water from Yellow River should be utilized for irrigation as much as possible, combined with agricultural water-saving measures and controlled exploiting groundwater at the tail of YRIA. Therefore, the combined usage of surface water and ground water of YRIA is an important way to realize agricultural water saving and sustainable utilization of water resources in Tuhai River Basin.

1. Introduction

It is necessary to study water resources carrying capacity in order to resolve the contradiction between water supply and human needs, and the research on water resources carrying capacity has attracted great attention and become a key and hot research topic in the field of water resources science. The water resources carrying capacity, the organic combination of carrying capacity, water environment and water ecology, can reflect the social and natural attributes and environmental value of water [1]. Studies on water resources carrying capacity are popular in China, which is, however, likely taken into the theory of sustainable development in other countries. The concept of water resources carrying capacity was firstly put forward by Shi et al in the studies on water resources carrying capacity of Urumqi River Basin in Sinkiang based on conventional tendency method [2]. Since then the connotation of water resources carrying capacity has been enriched and improved and many methods, such as fuzzy synthetic evaluation, system engineering theory, multi-objective analysis, have been widely used in the study of regional water resources carrying capacity [3].

Compared with other evaluation methods of sustainable development, the method of ecological footprint is more closely linked with the sustainable development theory, with more simple and operable calculation [4]. Accordingly the water ecological footprint (WEF) method is quite useful in assessment on water resource utilization. The constructed water resources account can be integrated into the total ecological footprint and used to evaluate not only the sustainable development of water resources, but also regional sustainable development [5, 6]. Since its introduction, the WEF method has been used in assessing the water resources carrying capacity in some administrative regions [7-9]. It is clear that the integrity of the basin was ignored in the case of evaluating the regional water resources carrying capacity from administrative regionalization standpoint. However, there is little research about the application of WEF for water resources carrying capacity in the basin scale, besides in typical watershed of Liaohe Basin, Shulehe Basin, Shiyanghe Basin, Hunhe Basin and other typical basins [10-13].

Tuhai River Basin in Shandong Province is a typical basin with severe water shortage, large water consumption, high degree exploitation and utilization, and rapid adjustment of industrial structure. It is necessary and practical to assess and analyze water resources carrying capacity under the new context of the full implementation of the most stringent water management system and the optimal allocation of water resources in Shandong Province. Therefore, the method of the WEF was used to analyze and calculate water resources carrying capacity and ecological deficit in this paper to provide scientific basis for sustainable development of Tuhai River Basin in Shandong Province.

2. Data sources and Research Methods

2.1. General situation of Study Area

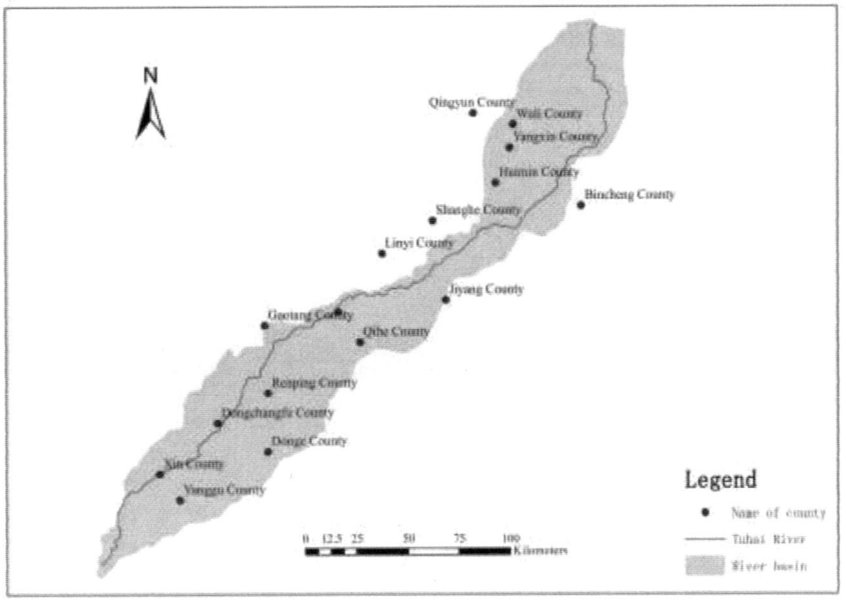

Figure 1. The Geographical Location Map of Tuhai River Basin.

Tuhai River Basin in Shandong Province was used as the research zone (Figure 1). Tuhai River, located to the south of the Haihe River Basin, flows through four cities, such as Liaocheng, Dezhou, Jinan and Binzhou, from upstream to downstream, and the catchments area of Tuhai River basin, with the main channel 406 km long, is 13296 km^2 in Shandong Province, including Shenxian, Yanggu, Dongchangfu, Chiping and so on 13 counties (or cities). Our research zone has a transitional zone of warm temperate semi-humid and semi-arid temperate monsoon climate, where the precipitation is mostly distributed in the flood season from June to September and the average annual precipitation is 564mm. The conflict of water supply and demand is more obvious in the spring due to less precipitation, more evaporation and water demand for irrigation.

2.2. Data Sources
Data of water consumption sub-accounts such as agricultural water consumption, industrial water consumption, and so on, were collected from "Water Resources Bulletin in Shandong Province" from 2003 to 2011, social and economic data from "Shandong Year Book ", the cultivated area and annual rainfall data of each county from "Statistical Yearbook of Jinan", "Statistical Yearbook of Liaocheng", "Statistical Yearbook of Binzhou" and "Statistical Yearbook of Dezhou" from 2003 to 2011, and Table 1 was obtained through the analysis and calculation of collected data.

Table 1. Socioeconomic Status and Water Consumption in Tuhai River Basin from 2003 to 2011.

		2003	2004	2005	2006	2007	2008	2009	2010	2011
Population / 10^4		971.8	977.1	971.9	979.1	987.1	991.3	992.3	985.2	1070.2
GDP / 10^9 RMB Yuan		82.9	99.3	123.1	145.1	175.1	217.5	235.4	274.0	314.9
Annual Precipitation / mm		638.4	614.5	584.3	413.8	516.0	441.1	631.0	566.8	549.1
Cultivated Land Area/10^3 km^2		10.08	10.21	10.18	10.11	10.10	10.78	10.51	10.44	10.15
Water Availability / 10^8 m^3		27.3	18.7	21.9	9.3	12.1	8.9	23.0	32.8	15.4
Water Consumption / 10^8 m^3	Agricultural	36.2	30.7	34.2	39.1	36.3	35.1	34.3	33.9	29.9
	Industrial	1.75	1.80	1.92	2.03	2.29	2.36	2.69	2.55	2.86
	Domestic	1.95	2.29	2.30	2.21	2.45	2.52	2.38	2.36	2.14
	Urban Public	0.26	0.23	0.20	0.42	0.35	0.31	0.39	0.38	0.34
	Ecological	0.10	0.14	0.18	0.19	0.29	0.34	0.28	0.38	0.37

2.3. Research Methods
The WEF method is widely used to measure the degree of sustainable utilization of water resources by comparing ecological surplus with ecological deficit of water resources. Water ecological surplus will appear when ecological carrying capacity is more than regional WEF, which indicates that the utility condition of water resources is in sustainable state. On the other hand, water ecological deficit will appear when ecological carrying capacity is less than regional WEF, which indicates that the utility condition of water resources is in unsustainable state.

2.3.1. Calculation Method of Water Resources Carrying Capacity. The calculation model of water resources carrying capacity [5] is:

$$EC_{\omega} = 0.4\gamma_{\omega}\varphi_{\omega}W / p_{\omega} \quad (1)$$

where EC_{ω} is the regional water resources carrying capacity, φ_{ω} is the water resources yield factor in the study area, γ_{ω} is a global water balance factor, W is the total amount of water resources in the study area, and p_{ω} is the world's average production capacity of water resources.

(1) Water resources yield factor (φ_{ω})

The water resources yield factor in various regions can be obtained by the ratio of the average production capacity of water resources in the study area to that in the world, under the assumption that the water yield factor is one in the world. In this paper, the calculation formula of regional water resources yield factor is:

$$\varphi_{\omega} = P / p_{\omega} \tag{2}$$

where φ_{ω} and p_{ω} are the same as in Formula (1), P is the yield per unit area. The average production capacity of water resources is $29.46\times10^4 m^3/km^2$ in the world and 12.01×10^4 m^3/km^2 in Tuhai River Basin, so φ_{ω} of Tuhai River Basin is 0.41.

(2) Global water resources balance factor (γ_{ω})

Global water resources balance factor is the ratio of the average ecological productivity of water resources in regional biological production to the average ecological productivity in the world, and this paper followed 5.19 as its value for comparative purpose [5].

2.3.2. Calculation Method of WEF. In the current paper, the WEF was divided into five sub-accounts based on the characteristics of water utilization: domestic WEF, agricultural WEF, industrial WEF, urban public WEF and ecological WEF, in which domestic WEF resulted from water for urban and rural domestic, agricultural WEF mainly from water for farmland irrigation and forestry, animal husbandry and fishery, urban public WEF mainly from water for commercial and service industry, ecological WEF mainly from water for green area and dilution pollution. The WEFs of different sub-accounts are calculated as follows:

$$\begin{aligned} EF_1 &= \gamma_{\omega} W_1 / p_{\omega} \\ EF_a &= \gamma_{\omega} W_a / p_{\omega} \\ EF_i &= \gamma_{\omega} W_i / p_{\omega} \\ EF_c &= \gamma_{\omega} W_c / p_{\omega} \\ EF_e &= \gamma_{\omega} W_e / p_{\omega} \end{aligned} \tag{3}$$

where EF_1 is the domestic WEF(km^2), W_1 is domestic water consumption, EF_a is the agricultural WEF, W_a is agricultural water consumption, EF_i is the industrial WEF, W_i is industrial water consumption, EF_c is the urban public WEF, W_c is urban public water consumption, EF_e is the ecological WEF, W_e is ecological water consumption. φ_{ω} and p_{ω} are the same as in Formula (1).

3. Results and Discussion

3.1. The Dynamic Characteristics of WEF in Tuhai River Basin

The WEF in Tuhai River Basin can be obtained by formula (3) as shown in Table 2.

Table 2. Water Resources Carrying Capacity and WEF in Tuhai River Basin. km^2

		2003	2004	2005	2006	2007	2008	2009	2010	2011
WEF	Agricultural	44087.8	36734.3	41172.4	48909.9	44277.2	42738.5	41508.9	40575.6	35662.5
	Irrigation	37673.5	30633.4	35600.8	43466.0	38203.4	37503.3	36339.1	35614.7	31530.7
	Industrial	2174.6	2223.1	2453.1	2579.8	3172.0	2967.1	3441.5	3106.6	3523.8
	Domestic	2312.6	2752.1	2814.0	2657.0	3041.6	3100.3	2874.5	2879.1	2687.6
	Urban Public	330.7	298.3	241.4	546.8	474.5	399.2	525.1	477.3	429.3
	Ecological	124.0	183.9	176.3	190.3	327.7	424.6	345.3	400.5	372.0
Total WEF		49029.7	42191.7	46857.2	54883.8	51293.0	49629.7	48695.3	47439.1	42675.2
WCC		7875.1	5416.5	6323.0	2686.1	3507.7	2566.1	6651.6	9489.2	4456.1
WED		41154.6	36775.2	40534.2	52197.7	47785.3	47063.6	42043.7	37949.9	38219.1

Note: WRCC is water resources carrying capacity; WRED is water resources ecological deficit.

The total WEF in Tuhai River basin decreased from 49029.7 km^2 in 2003 to 42675.2 km^2 in 2011, with a downward trend in overall volatility, in which the upward trend from 2003 to 2006 is more significant, reaching the highest value of 54883.8 km^2 in 2006, followed by a downward trend. Agriculture is the primary water consumer in Tuhai River basin, accounting for more than 83% of the total water consumption. Though the proportion of agricultural water consumption has been decreasing from 89.9% in 2003 to 83.6% in 2011, it is still too high due to its slow declining, consequently, the contradiction of WEF lies in water utilization structure in view of the fact that agricultural water is the main contributor to WEF in Tuhai River Basin.

The average agricultural WEF was 41740.79km^2, accounting for 86.8% of the total WEF, in which 86.9% is used for irrigation. The analysis showed that WEF was positively correlated with irrigation WEF while negatively correlated with annual precipitation. The precipitation of Tuhai River Basin is 18.2% less than the average annual precipitation in 2006, when different degrees of droughts were popular throughout China, so there was a sudden increase in irrigation water, leading to the highest value of WEF during the study period.

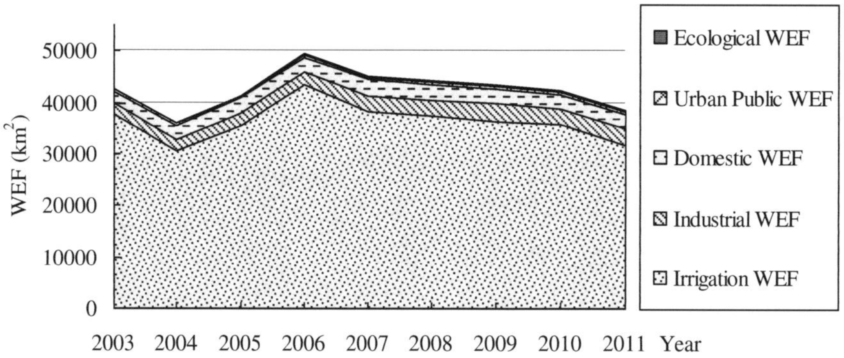

Figure 2. The Area Curve Map of WEF in Tuhai River Basin.

The average industrial WEF was 2849.1km^2, accounting for 5.93% of the total WEF. The industrial WEF showed a significant growth trend due to rapid increase in economic development and industrial water demand from 2003 to 2011. The average domestic WEF was 2791.0 km^2, accounting for about 5.81% of the total WEF. Domestic WEF showed a slow growth trend due to improvement of people's living standard, especially for rural residents. The average of urban public WEF was 562.8km^2, accounting for 0.86% of the total WEF, and the urban public WEF showed a fluctuating ascending trend from 2003 to 2011. Compared with other water consumptions, ecological WEF had the lowest proportion and the fastest growth, from 0.3% in 2003 up to 0.9% in 2011. The changes of WEF area can demonstrate the changes of WEF of each sub-account, as illustrated in Figure 2.

3.2. Dynamic Characteristics of Water Resources Carrying Capacity in Tuhai River Basin

It can be seen from Table 2 that the water resources carrying capacity in the basin changed remarkable in downward trend. The water resources carrying capacity was 7875.1 km^2 in 2003 and continued to decline to 2 566.1 km^2 in 2008, however, in high level in 2009 and 2010 mainly due to the increase of precipitation in the basin. The analysis shows that water resources carrying capacity of Tuhai River Basin have a significant positive correlation with precipitation (Figure 3). To some extent, water resources carrying capacity is mostly affected by natural factors such as hydrology and meteorology.

IOP Conf. Series: Earth and Environmental Science **64** (2017) 012030 doi:10.1088/1755-1315/64/1/012030

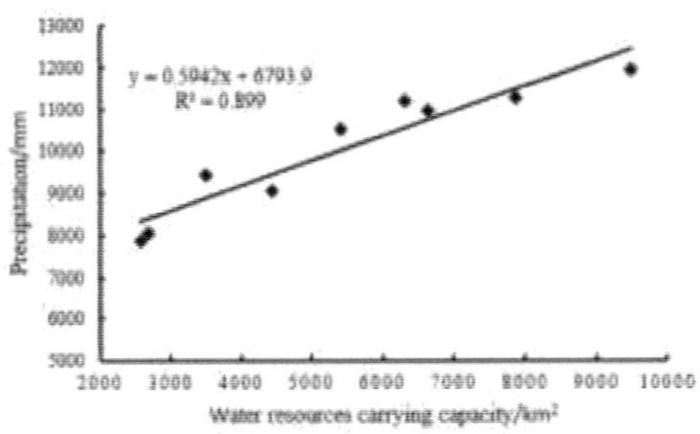

Figure 3. The Relationship between Water Resources Carrying Capacity and Precipitation.

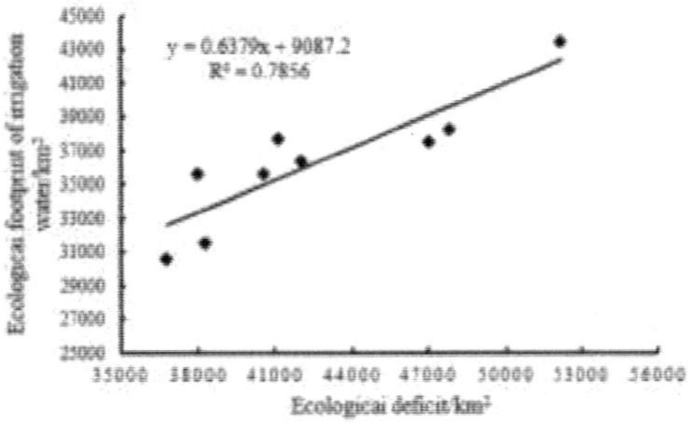

Figure 4. The Relationship between Ecological Deficit and Irrigation WEF.

3.3. Water Ecological Deficit in Tuhai River Basin

The fact that WEF was more than water resources carrying capacity showed that water supply was inadequate and the utility condition of water resources was in unsustainable state in Tuhai River Basin from 2003 to 2011.The water ecological deficit fluctuated upward trend from 2003 to 2006, with a peak of 53540.6 km^2 in 2006, followed by a decrease to 38219.2km^2 in 2011, which was consistent with the trend of WEF in the basin. The analysis showed that WEF had a positive correlation to irrigation WEF of Tuhai River Basin (Figure 4), which indicated that the water ecological deficit was mainly caused by agricultural water consumption. It can be forecasted that there would be a further increase in the water ecological deficit with the increase of water consumption judging by the present economic improvement. Accordingly, it is urgent to find effective ways of rational utilization of limited water resources for the sustainable development in Tuhai River Basin.

4. Conclusions and Discussions

The WEF showed an overall downward trend, in which agricultural WEF accounted for more than 83% in Tuhai River Basin from 2003 to 2011. Agricultural water consumption is the main contributor to WEF, which proportion has decreased but not much, and about 86.9% of agricultural WEF was used for irrigation water. The water resources carrying capacity had a downward trend in general, a drop of 43.42% in 2011 compared with 2003, which was mostly affected by the hydrology, meteorology and other natural factors. The analysis of water resources ecological deficit showed that the degree of water resources exploitation in this basin is very high with little potentiality. There was a big contradiction between water demand and water supply, and the water resources utilization mode of Tuhai River Basin was in an unhealthy pattern, so it was necessary to improve the utilization efficiency of water resources.

As a major agricultural district, agricultural water is the main contribution to the water ecological deficit in Tuhai River Basin, and local agricultural irrigation mostly depends on water form Yellow River, supplemented by well-canal combined irrigation. The main problems of water utilization in the Yellow River Irrigation Area (YRIA) are: irrigation relies on water form Yellow River, with lower groundwater utilization at the head of irrigation area; On the other hand, the ecological environment has been deteriorated due to the lack of water from Yellow River and over exploitation of groundwater at the tail of irrigation area, such as Shenxian, Huimin and other counties. Therefore, the combined usage of surface water and ground water of YRIA should be an important way of water saving in the basin. Specific measures are as follows: well irrigation should be greatly developed and shallow groundwater resources should be exploited as much as possible at the head of YRIA, and more water from Yellow River should be transferred to the tail area for irrigation and supplement groundwater with lining the main channel; however, water from Yellow River should be utilized for irrigation, combined with agricultural water-saving measures and controlled exploiting groundwater at the tail of YRIA. In conclusion, combined usage of surface water and ground water of Yellow River Irrigation

District is an important way to realize agricultural water saving and sustainable utilization of water resources in Tuhai River Basin.

Acknowledgments
This work was financially supported by Major Projects for Social Livelihood of Jinan City "Research and demonstration on ecological control technologies of water environment in Jinan"(SHMS2015-301).

References

[1] C. Zeng, Y.F. Liu, W.S. Zhang and D.W. Tang, Origination and development of the concept of aquatic ecological carrying capacity in the basin, Resources and Environment in the Yangtze Basin. 20 (2011) 203-210. [In Chinese]
[2] Y.F. Shi, Y.G. Qu, The carrying capacity of water resources and its reasonable use of Urumqi river, Science Press, Beijing, 1992. [In Chinese]
[3] Z.J. Yao, J.H. Wang, D. Jiang and C.F. Chen, Advances in study on regional water resources carrying capacity and research on its theory, Advances in Water Science. 13 (2002) 111-115. [In Chinese]
[4] C. Hao, H.Y. Li, W.Q. Meng, Comparison and analysis on different methods of sustainability evaluation at home and abroad. China Population, Resources and Environment. 20 (2010) 161-166. [In Chinese]
[5] L.N. Huang, W.X. Zhang, C.L. Jiang, and X.Q. Fang, Ecological footprint method in water resources assessment, Acta Ecologica Sinica. 28 (2008) 1279-1286. [In Chinese]
[6] G. Stoeglehner, P. Edwards, P. Daniels and M. Narodoslawsky, The water supply footprint (WSF): a strategic planning tool for sustainable regional and local water supplies, Journal of Cleaner Production. 19(2011) 1677-1686.
[7] C.H. Sun, L.Y. Xue, The analysis of water resources' ecological footprint in Jiangsu province, Journal of Yunnan Normal University (Natural Science Edition), 30(2010) 56-61. [In Chinese]
[8] J. Wang, C.X. Zhang, Y.T. Yu, F.Y. Li and F. Ma, Calculation model of urban water resources ecological footprint and its application: a case study in Shenyang city of northeast China, Chinese Journal of Applied Ecology, 23 (2012) 2257-2262. [In Chinese]
[9] L.H. Meng, Z.P. Ye, S.F. Yuan and G.X. Liu, Dynamic characteristics of ecological footprint and ecological carrying capacity of water resources in Jiangxi province during 2007-2011, Bulletin of Soil and Water Conservation. 35(2015) 256-261. [In Chinese]
[10] S. Wang, F.L. Yang, L. Xu and J. Du, Multi-scale analysis of the water resources carrying capacity of the Liaohe Basin based on ecological footprints, Journal of Cleaner Production.53 (2013):158-166.
[11] J. Zhang, R.Z. Zhang, D.M. Zhou, A study on water resource carrying capacity in the Shule river basin based on ecological footprints, Acta Prataculturae Sinica. 21(2012) 267-274. [In Chinese]
[12] Y. Jia, J. Zhang, R.Z. Zhang, Ecological footprint of water resources in the Shiyang river basin from 2001-2011, Acta Prataculturae Sinica. 25 (2016) 10-17. [In Chinese]
[13] Y. Liu, Study on water resources carrying capacity of Hunhe river basin, Shenyang Agricultural Uinversity, Shenyang, 2010. [In Chinese]

IOP Conf. Series: Earth and Environmental Science **64** (2017) 012031 doi:10.1088/1755-1315/64/1/012031

Investigation on velocity distribution of TFM and DEM phase in hybrid model of CBFB in mining

Tianyu Zhang[1,*], Ying Feng[2] and Zhening Zhao[1]

[1]North China Electric Power Research Institute, Beijing 100032, China
[2]Institute of Economics and Technology of State Grid Chengdu power supply company, Chengdu 610000, China

[*]Corresponding author e-mail: zzztttyyy456@126.com

Abstract. As a novel model for gas solid flow simulation, the investigation of TFM-DEM hybrid model is far from completely, including mutual interaction of TFM and DEM phase, selection of DEM portion and coherence of the predicted results from both phases. Therefore, in present study, the consistency of velocity distribution between TFM and DEM phase is investigated. The correlation of instantaneous and time-averaged velocity distribution of TFM and DEM phase in specific area in CBFB for mining is studied. And the differences of the axial and radial velocity between the particles of different sizes are discussed. The influence of particle diameter and the ratio of DEM and TFM phase on the correlation of velocity, both instantaneous and time-averaged, are taken into consideration.

1. Introduction

Gas-solid flow is widely used in various industrial applications, such as pharmacy, chemical industry, food processing, energy and environment [1]. Therefore, gas-solid flow has drawn the attention of great number of researchers on its interaction between gas and particle, rate of reaction and the dynamic characters of flow field, which can provide the basic theory for prediction and control of gas-solid flow process [2].

There are two methods of simulation in gas solid flow: Euler-Euler and Euler-Lagrange method. Due to the assumption of taking particles as continuous media, the Euler-Euler method can provide relatively reliable prediction with less computational consumption. However, Euler-Euler method traces particles from the meshing level, failing to capture the detailed information of individual particle. On the contrary, Euler-Lagrange method (namely Discrete Element Method) can trace particles by particle level. But the computational cost increases sharply as the number of particles increases, limiting its applications in industrial process [3].

Under the premise mentioned above, the concept of TFM-DEM hybrid model is brought up. As shown in Fig. 1, the solid is divided into two phases for calculating and post-processing, namely TFM and DEM phase. The solids solved by TFM model is called TFM phase, so is DEM phase. The TFM model can significantly reduce the consumption in computation and DEM model can capture the information (location, velocity and force) from particle level. However, hybrid model is still in its infancy whose parameters and applications have not been fully studied [4].

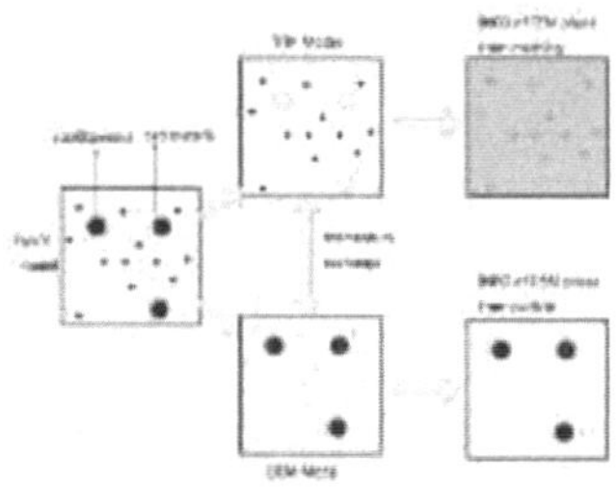

Figure. 1 Diagram of classification of solid in CBFB

2. Governing equations

Based on statement above, the main focus lies on interaction between gas, TFM and DEM phase, which should be defined explicitly. Since separate TFM model and DEM model have already been elaborated fully, only TFM and DEM momentum equations are listed.

$$\frac{D}{Dt}(\varepsilon_{cs}\rho_{cs}v_{cs}) = \nabla \overline{\overline{S}}_{cs} + \varepsilon_{cs}\rho_{cs}g + I_{gc} - I_{cd} \tag{1}$$

Where S_{cs}—— stress tensor in continuous phase (Pa);

I_{gc} —— momentum exchange term between gas and continuous phase;

I_{cd} ——momentum exchange term between continuous phase and discrete phase;

$$\frac{dx^{(i)}(t)}{dt} = v^{(i)}(t) \tag{2}$$

$$m^{(i)}\frac{dv^{(i)}(t)}{dt} = F_T^{(i)} = m^{(i)}g + F_{gd}^{(i)}(t) + F_{cd}^{(i)}(t) + F_c^{(i)}(t) \tag{3}$$

$$I^{(i)}\frac{dw^{(i)}(t)}{dt} = T^{(i)} \tag{4}$$

Where $m^{(i)}$, $I^{(i)}$ —— particle's mass (kg)and moment of inertia (kg·m^2);

$T^{(i)}$, $F_T^{(i)}$ —— total moment of force (N·m) and external force on the particle (N);

$F_{gd}^{(i)}$ —— drag force exerted on particle i in its meshing, including pressure item and viscious item (N);

$F_{cd}^{(i)}$ —— force exterted on particle i by continuous phase (N);

$F_c^{(i)}$ —— interaction between the DEM particles

The drag force model is used to calculate the interaction in momentum between the gas and TFM phase. Take Gidaspow model for instance, actually it is a combination of Wen-Yu drag model and Ergun equation with the formula below.

$$\beta = 150\frac{\varepsilon_s^2(1-\varepsilon_g)\mu_g}{\varepsilon_g^2 d_s^2} + 1.75\frac{\rho_g \varepsilon_s |v_g - v_s|}{\varepsilon_g d_s} \quad \varepsilon_g \le 0.8$$
$$= \frac{3C_d \varepsilon_g \varepsilon_s \rho_g |v_s - v_g|}{4d}\varepsilon_g^{-2.65} \quad \varepsilon_g > 0.8 \tag{5}$$

The drag force between the gas and DEM phase (particle *i*) is expressed below:

Where $P_g(x^{(i)})$—— average gas pressure in particle *i* meshing (Pa);

$v_g(x^{(i)})$—— gas velocity in particle *i* meshing (m/s);

$V^{(i)}$—— particle volumn (m^3);

β_{gd}—— momentum exchange coefficient between gas and DEM particles

The momentum interaction terms between the TFM and DEM phases share similar form. However, the gas velocity is taken by the center of the mesh, instead of actual particle location. Meanwhile, the DEM phase particles' velocity is calculated approximately as the mesh average, not the individual. Under the premise above, the coefficient of momentum interaction between TFM and DEM phase can be referred in according literature.

In present study, e_{cd} and C_f are set as 0.9 and 1. Among the terms in the right part of Equ.9, the first term means taking consideration of momentum exchange due to the collision and sliding, and the second term indicates the influence of the existence of cluster in dense gas-solid flow.

3. Configuration and simulation

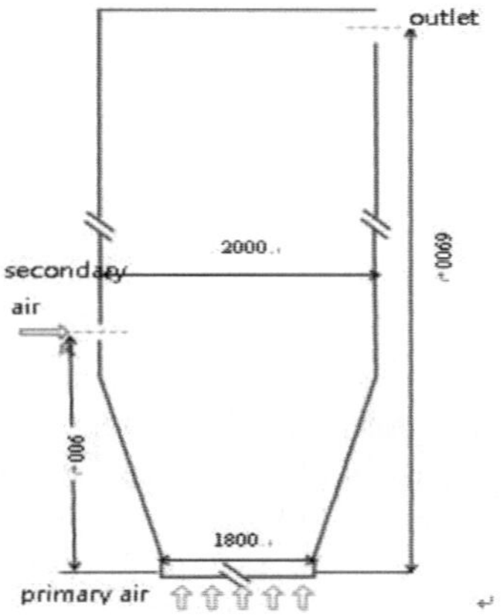

Figure. 2 Simplified configuration of furnace

The simplified configuration of 10t/h furnace is shown in Fig. 2. Primary air is pumped into the furnace from bottom evenly with velocity of 3.0m/s. In order to enhance the mixing in furnace, secondary air is injected from the single side at the height 0.9m with velocity of 35m/s. The air flow ratio of primary and secondary air is 3:1. The meshing in the configuration is uniform and the size is 0.05m×0.05m.

In present simulation, the particle consists of bio-component with same density 1200kg/m3 but different diameter 1.5mm and 4.0mm, respectively. The particles are well mixed and equal in number. The initial bed height is 0.6m with void fraction 0.4. The DEM phase is homogeneously distributed in TFM phase, and other parameters are listed in Table 1.

Table 1 Configuration and parameters

	Unit	Value
height	m	7.3
width	m	2
primary air velocity	m/s	3.0
secondary air velocity	m/s	35
inlet diameter of secondary air	m	0.05
furnace outlet diameter	m	0.9
the number ratio of particles 1&2 in TFM phase	—	1:1
the number ratio of particles 1&2 in DEM phase	—	1:1
Volume ratio of DEM and TFM phase	—	0.078 (1:3.2)

In order to demonstrate the relation of TFM and DEM phase velocity, the time of 10s, 12s and 14s and height of 0.5m, 1.0m and 1.5m are picked to do the analysis and comparison, in which the TFM phase velocity is extracted from meshing level Then the program is MATLAB is used to trace DEM particles to get the instantaneous and time-averaged velocity. Moreover, these figures below are also drawn in the environment of MATLAB.

3.1. Comparison of instantaneous velocity distribution

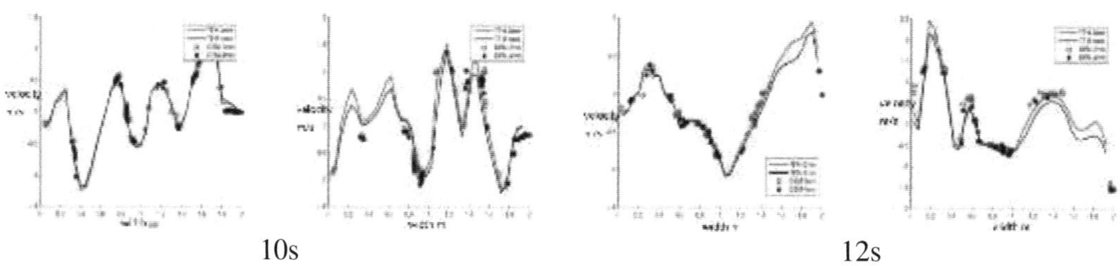

10s 12s

Figure. 3 The instantaneous radial and axial velocity distribution at different time (y=0.5m)

The figure above shows the velocity comparison of two phases (TFM phase in meshing, and DEM particle in its location). The DEM particles' velocity is presented as spots in figure. The time point of 14s shows the same inclination, thus it is not presented in present study.

As for the instantaneous radial distribution, the TFM phase velocity varies significantly. It can also be seen that the velocity distribution curves of 1.5mm and 4.0mm are very close to each other, indicating they have similar moving state. This phenomenon also applies to DEM phase. Except for specific DEM particles, most TFM and DEM particles' radial velocity agree well with each other, with average error of 5.6% and 1.7%, respectively.

As for the instantaneous axial velocity distribution, the small particles are more prone to be influenced by gas flow, leading to a large velocity value in both TFM and DEM phase. As is in radial velocity distribution, the error in axial is small as well: 5.4% and 3.5%. In addition, in the scope investigated, the diameter does not make great difference on the velocity coherence. The comparison results show though the solids are divided into two phases, hybrid model can give similar instantaneous result.

IOP Conf. Series: Earth and Environmental Science **64** (2017) 012031 doi:10.1088/1755-1315/64/1/012031

3.2. The influence of height on the velocity comparison

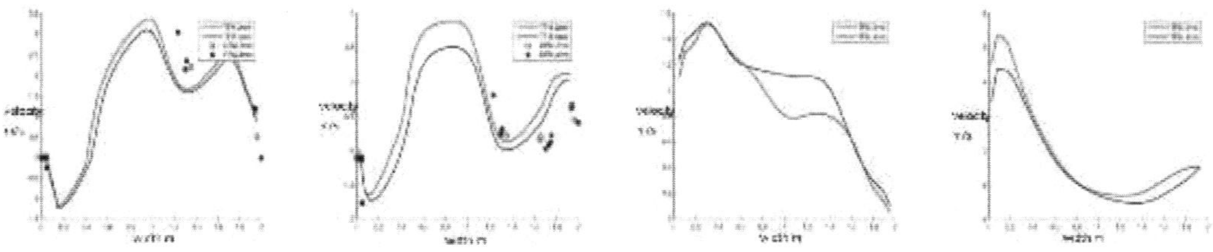

y=1m y=1.5m

Figure. 4 The instantaneous radial and axial velocity distribution at different heights

As shown above, the studied region lies at the height of y=1m, above the secondary air inlet, thus the flow condition is much influenced by secondary air, leading to higher maximum velocity compared with y=0.5m in general. Because the secondary is pumped into the furnace from singe side, the velocity distribution appears the inclination of 'one side high, the other low'. As for the difference in radial and axial velocity, no matter TFM and DEM phase, the velocity of 1.5mm and 4mm particles are close to each other at same velocity, with average error of 18% and 16.1% at 10s. Compared with y=1m, the error in y=0.5 is less, indicating the lower the studied region is, the more coherent TFM and DEM phase is. Probable reasons are:

(1) In the meshing of y=1m, the DEM particles are less than in y=0.5m. Take 10s for instance, the numbers are 23 and 102, respectively. The decrease of particles results in the declination in regulation.

(2) The error in simulation calculation also makes contribution to some extent.

However, in the scope investigated, the velocity distribution of TFM and DEM phase shows good agreement. Furthermore, the phenomenon of no DEM particle in y=1.5m meshing validates the conclusion that in such apparatus, the solid concentration is close to 0 above the height of 1.5m. Therefore, no discuss and analysis is made in such condition.

3.3. Comparison of time-averaged velocity distribution

The time-averaged velocity can significantly make the conclusion more applicable. Therefore, the program in the environment of MATLAB is coded to do the time-averaging of velocity of both TFM and DEM phase with formula.The axial time-averaged velocity shares the same.

After time-averaged processing, the velocity distribution of TFM and DEM phase is shown:

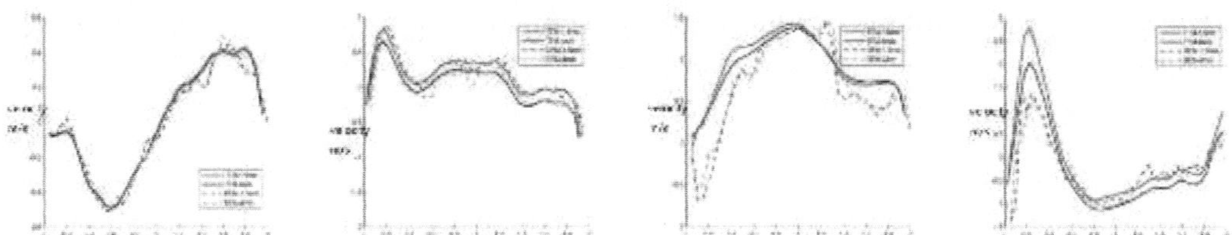

y=0.5m y=1m

Figure. 5 The time-averaged radial and axial velocity distribution

We can clearly conclude that the 1.5mm and 4.0mm particles have same moving inclination in radial direction, as it is in instantaneous case. As for axial direction, 1.5mm particle velocity is more than 4.0mm, which leads to stratification fluidization in bed. Compare y=0.5m with y=1.0m, the coherence in velocity distribution is better, and the reason is stated above, namely more particles cluster at the lower part.

The identical point of instantaneous and time-averaged velocity distribution lies on (1) radial velocity of different particles is nearly the same; as for axial velocity, small particles are more prone to move upwards than big ones. (2) With the increase of height, the coherence in velocity distribution gets worse both in instantaneous and time-averaged. The diverse point: the fluctuation in instantaneous velocity distribution is violent while in time-averaged it is stable.

However, we can also observe there exists some error between the TFM and DEM phase velocity distribution. It is mainly because:

(1) The hybrid model only takes consideration of the drag force between TFM and DEM phase, neglecting the equivalent collision.

(2) The particles number of DEM is relatively small, failing to occupy every meshing at every time point. These data in 'empty' meshing will lead to curve interruption in time, thus causes fluctuation in curve in the figure.

3.4. The influence of ratio of DEM phase on velocity distribution
Keep other parameters constant, only change the ratio of DEM and TFM phase to 1:12.8, then redo the simulation and results are shown below:

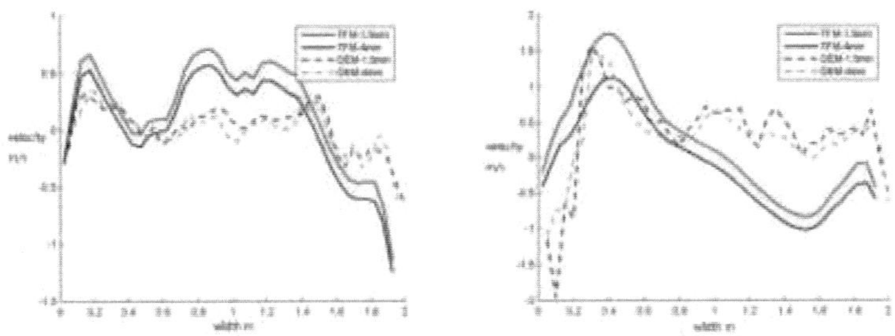

y=0.5m y=1m

Figure. 6 The time-averaged axial velocity distribution after decrease in DEM phase

As shown above, with the decrease of DEM phase particles, more error between the two curves can be obviously observed, which means in the scope investigated, the increase in DEM phase ratio can help improve the velocity distribution coherence to some degree. It can also indicate that the trace of DEM particles and their other information can represent all particles in the flow filed, aiming to figure out the flow, mixing and mass transfer phenomenon with less computation cost.

4. Conclusion
This present research investigates the gas solid flow in furnace with the help of TFM-DEM hybrid model. The velocity distribution coherence of TFM and DEM phase is studied.

1. In instantaneous velocity distribution, the velocity in radial and axial direction shows good agreement with small error. In the scope investigated, diameters do not have conspicuous effect on the distribution relation. With increase in height, the coherence condition gets worse due to particle concentration begins to fall.

2. In time-averaged velocity distribution, good agreement in velocity distribution can also be evidently observed, but with little fluctuation.

3. Under the premise of constant simulation condition, the increase in ratio of DEM can help enhance the velocity coherence. Meanwhile, the simulation result obtained in such condition can be more representative.

References
[1] Tang Y, Kriebitzsch S H L, Peters E, et al. A methodology for highly accurate results of direct numerical simulations: Drag force in dense gas–solid flows at intermediate Reynolds number [J]. International journal of multiphase flow, 2014, 62: 73-86.

[2] B Sarker M R H, Chowdhury A S M R, Love N. Prediction of gas–solid bed hydrodynamics

using an improved drag correlation for nonspherical particles [J]. Proceedings of the Institution of Mechanical Engineers, Part C: Journal of Mechanical Engineering Science, 2015: 0954406215622652.
[3] Wang S, Yang X, Lu H, et al. CFD studies on mass transfer of gas-to-particle cluster in a circulating fluidized bed [J]. Computers & Chemical Engineering, 2009, 33(2): 393-401.
[4] Tsuji Y, Kawaguchi T, Tanaka T. Discrete particle simulation of two-dimensional fluidized bed [J]. Powder technology, 1993, 77(1): 79-87.

IOP Conf. Series: Earth and Environmental Science **64** (2017) 012032 doi:10.1088/1755-1315/64/1/012032

Support design and practice for floor heave of deeply buried roadway

Chaoke Liu*, Jianxi Ren, Bingli Gao, Yongjun Song

School of Civil Engineering and Architecture, Xi'an University of Science and Technology, Xi'an 71005, China

*Corresponding author e-mail: liuchaoke8@163.com

Abstract. Aiming at the severe floor heave of auxiliary haulage roadway in Jianzhuang Coal Mine, the paper analysed mechanical environment and failure characteristics of auxiliary haulage roadway surrounding rock with the combination of mechanical test, theoretical analysis, industrial test, etc. The mechanical mechanism for deformation and failure of weak rock roadway in Jianzhuang Coal Mine was disclosed by establishing a roadway mechanical model under the effect of even-distributed load, which provided a basis for the design of inverted concrete arch. Based on complex failure mechanism of the roadway, a support method with combined inverted concrete arch and anchor in floor was proposed. The result shows that the ground stress environment has extremely adverse influence on the roadway, and the practice indicates that the floor heave countermeasures can effectively control the floor heave. The obtained conclusion provides a reference for the research and design on control technology of roadway floor heave in the future.

1. Introduction

The mushroom growth of the economy causes an increasing energy demand, which leads to deep mining of coal mine successively and increases the buried depth of roadway year by year [1-3]. Generally, deep mining pertains to the coal mine with a depth of 600 m. The mining depth increasing not only increases the mining cost, but also faces many challenges such as high gas, high ground stress, deformed surrounding rock and difficult roadway support, which has a direct effect on coal mine safety production [4-5], wherein roadway floor heave is the most common challenge faced by deep coal mining.

For the deformation mechanism and support technology of roadway floor heave, experts and scholars have carried out numerous beneficial researches and practices, and the research on support theory is gradually deepened. Starting from the initial pressure arch hypothesis only considering the rock weight in caving arch and the Terzaghi K. formula considering limit equilibrium arch, the theory research now has developed into the Kastner formula considering plastic zone support outside the surrounding rock of fracture zone at ultimate equilibrium [6]. Multiple control methods including anchoring and grouting, floor beam with inverted arch, concrete pouring, etc. have been formed, and a certain effect has been achieved under the specific engineering conditions [7-9]. These theories and practices have accelerated the roadway support technology but requiring further researches on how to determine the reasonable support parameters.

IOP Conf. Series: Earth and Environmental Science **64** (2017) 012032 doi:10.1088/1755-1315/64/1/012032

2. Engineering Overview

The auxiliary haulage roadway, as a horizontal development roadway in this mine, will be continuously used for ventilation and haulage in more than ten years with such a noticeable importance. The roadway surrounding rock is featured with low strength, poor integrity and high degree of polarization, and easy to weathering and deliquescence, which are all the features of weak rocks.

The average buried depth of auxiliary haulage roadway is about 700 m, the cross section shape is semicircular arch with a vertical wall height of 1.8 m, an arch height of 2.4 m and the net width of 4.8 m. At the beginning, rock bolt, anchor cable and metal mesh were used to construct a permanent support. After three months from the roadway construction, severe floor heave occurred, and even in some sections, the floor heave reached more than 1000 m. The floor was then renovated and reinforced by adding anchors in floor with a size of Φ20 mm×L2500 mm and an inter-row distance of 2400 mm×2400 mm based on the original support. Meanwhile, a 500 mm thickness of concrete was poured on the floor to increase the stiffness. Nevertheless, the floor heave in roadway was not mitigated, and the repeated renovation and reinforcement made the roadway trapped in a viscous cycle of endless digging and repairing. Thus, it is urgent to make deep research and realize optimization on its failure mechanism and support design.

The exposed rocks during tunneling of auxiliary haulage roadway were mainly composed of mudstone, sandy mudstone, muddy siltstone and fine sandstone. Rock samples on site were classified into siltstone, fine sandstone, medium sandstone and mudstone by the stone natured, and the regular physical mechanics test was carried out on rock samples by rock servo pressure tester. The test result is shown in Table 1.

Table 1.physical and mechanical properties of stratum

Rock type	Shear modulus (GPa)	Cohesion (MPa)	Uniaxial compressive strength (MPa)	Internal friction angle (degree)	Density (kg/m^3)
Fine grain Sandstone	5.08	10.1	38.2	40.7	2700
Siltstone	5.82	7.9	31.5	39.7	2500
Mudstone	1.3	3.5	18.6	32.1	1900
# 4coal	0.76	3.64	12.0	31.55	1500
Mudstone	5.91	8.2	22.4	34.6	2480

3. Roadway Floor Support Design

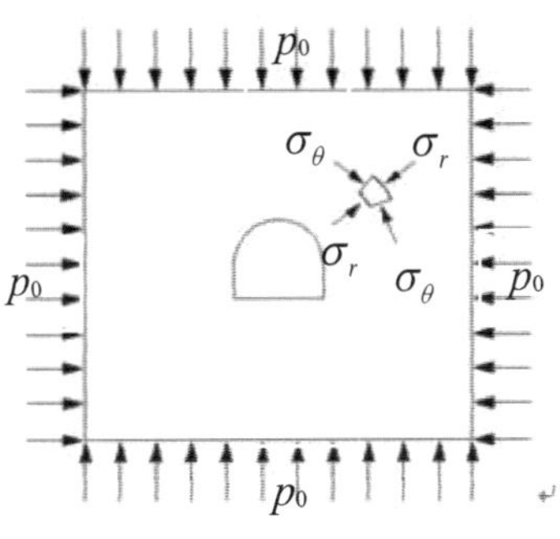

Figure 1. Mechanics model

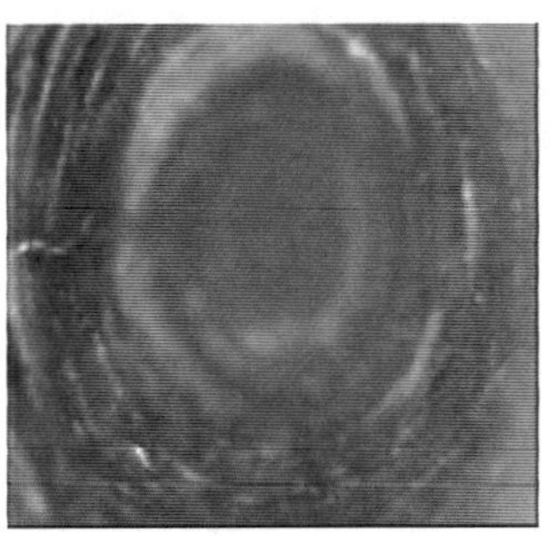

Figure 2. Loose circle

3.1. Mechanical model establishment

According to the theories of rock mechanics and elastic mechanics, a roadway mechanical model under the even-distributed loads from two directions was built as shown in Figure 1. The buried depth of roadway was so great that the self-weight of rock stratum on the roadway roof and floor was much less than q which could be neglected. For convenience of obtaining the analytical solution, the following assumption was made:

1. The cross section of auxiliary haulage roadway was calculated by the shape of equivalent circumcircle;
2. It was assumed that lateral pressure coefficientλ=1 because the measured horizontal stress is closed to the vertical stress;
3. The roadway wall had the support action, so the support resistance to surrounding rock was assumed as p_i.

According to the equilibrium equation, geometric equation and physical equation, the elastic stress equation is obtained [10]:

$$\sigma_r = p_0\left(1-\frac{r_a^2}{r^2}\right) \tag{1}$$

$$\sigma_\theta = p_0\left(1+\frac{r_a^2}{r^2}\right) \tag{2}$$

The elastic stress equation is obtained:

$$\sigma_{rp} = \left(p_i+\frac{\sigma_c}{\xi-1}\right)\left(\frac{r}{r_a}\right)^{\xi-1}-\frac{\sigma_c}{\xi-1} \tag{3}$$

$$\sigma_{\theta p} = \xi\left(p_i+\frac{\sigma_c}{\xi-1}\right)\left(\frac{r}{r_a}\right)^{\xi-1}-\frac{\sigma_c}{\xi-1} \tag{4}$$

$$\xi = \frac{1+\sin\theta}{1-\sin\theta} \tag{5}$$

$$\sigma_{\theta p}+\sigma_{rp} = 2P_0 \tag{6}$$

From the formula (3), (4) and (6), the supporting force of plastic region are expressed in the following form when the supporting force exists.

$$P_i = \frac{1}{\xi^2-1}[2p_0(\xi-1)+2\sigma_c]\left(\frac{r_a}{R_p}\right)^{\xi-1}-\frac{\sigma_c}{\xi-1} \tag{7}$$

Where r_a is roadway radius, σ_r is radial elastic stress, σ_θ is tangential elastic stress, σ_{rp} is radial plastic stress, $\sigma_{\theta p}$ is tangential plastic stress, p_0 is surrounding rock pressure, r is radius, P_i is supporting resistance, c is cohesion, R_p is radius of plastic zone. θ is internal friction angle.

3.2. Mechanism analysis of roadway floor heave

The average buried depth of auxiliary haulage roadway was 700 m, then the average unit weight of upper overlaying stratum was considered as 25 kN/m^3, and the self-weight stress should be 17.5MPa. Based on the mechanical model, the cross section of roadway chambers was equivalent to its circumcircle with a radius of 2.83 m. It could be calculated based on the elastic theory that the shear stress of floor was 25.74 MPa when the floor was not supported, which was greater than the shear strength of floor mudstone. Shear slip occurred to the floor, and the floor mudstone crushed, resulting in the volume expansion of crashed mudstone on the floor; and the squeezing action of two sides aggravated the floor heave. It can deferred from the above theory that if the floor is not supported when the deeply buried roadway surrounding rock is subject to the high stress, and the floor shear stress is bound to be larger than its shear strength, so that floor heave occurs. Therefore, the floor must be supported to control roadway floor heave.

3.3. Inverted arch design

Currently, method for dealing with floor heave mainly includes reinforcement method, relieving method and combination method [11-15]. The reinforcement method is mainly suitable for the floor heave roadway with poor floor rock nature, this method is further divided into floor grouting, floor anchoring and inverted concrete arch. Floor anchoring and inverted concrete arch are widely used in the rock roadway, while floor grouting is difficult to achieve the desired results. Considering the importance and service life of auxiliary haulage roadway, the support scheme with combined concrete inverted arch and anchor in floor is adopted.

With support force obtained by the formula 7 as the load on inverted arch, radius of plastic zone was gotten by actual measurement as shown in Figure2. The maximum axial force N_{max} on inverted arch can be obtained by the structural mechanics.

$$N_{max} = \alpha_1 f_c b h_0 + f_y A_s \tag{8}$$

Rebar arrangement can be adopted for the concrete inverted arch. The inverted arch span was the roadway width 4.8 m and the arch height to ground was 1.5 m; the concrete grade of inverted arch was C30, rebar grade was RRB400, inverted arch thickness was 400 mm, inverted arch length of each truss was 4 m; and the span of temperature joint was 150 mm. The area of main rebar in the inverted arch could be figured out by formula 8; the diameter of Φ20 mm was adopted by referencing to the table, and the vertical layout space was 250 mm. Fill the inverted arch with coal slag and then compacted it, and finally pave concrete with a thickness of 200 mm on the coal slag to harden the surface.

Two left-hand thread steel bolts with a diameter of Φ20 mm and a length of 2500 mm were used as the anchor in floor on each side with an inter-row space of 800×800 mm and cement paste was grouted. The anchor in arch forms an angle of 45° with the floor, imposing a priestess of 50 kN. The Φ18 mm vertical rebar was laid out longitudinally along the roadway with an interval space of 200 mm. The concrete strength was graded as C30, and 425# ordinary Portland cement mixed with waterproof agent blended.

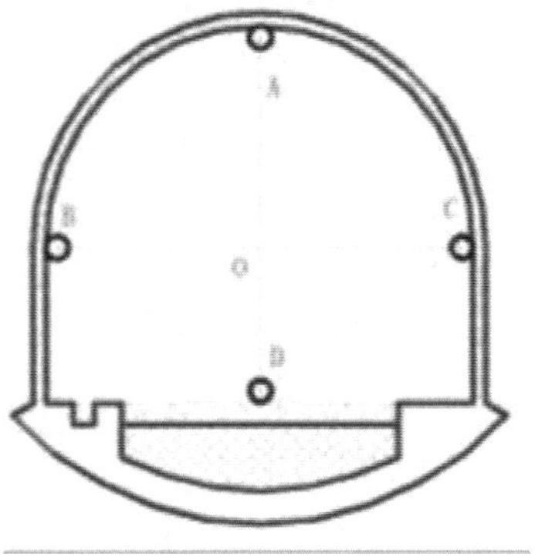

Figure 3. Monitoring profile

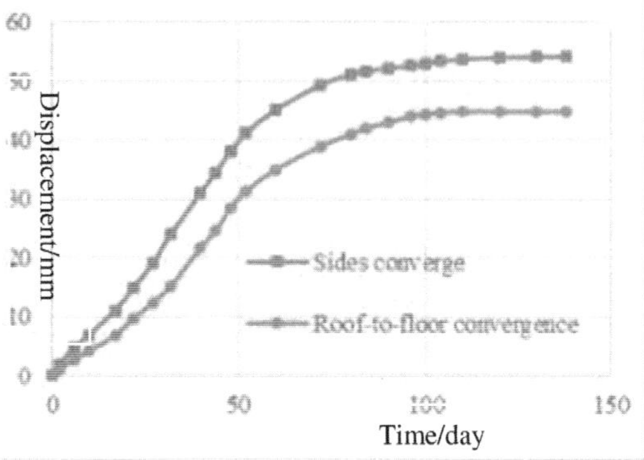

Figure 4. Roadway displacement monitoring curve

4. Engineering application effect

The above research results were applied to the incomplete surrounding rock support for haulage roadway to check the effectiveness and reasonability of the support. To accurately know the convergence and deformation of roadway surrounding rock, 3 observation stations were set up with an interval of 30 m with each other, monitoring profile is shown in Figure 3. The deformation observation on the roadway was carried out for 5 months, and the obtained typical curves of displacement and time are shown in Figure 4. It can be observed from Figure 4 that the overall roadway support is good, and the deformation of surrounding rock is effectively controlled, according to the two-side contraction of 56 mm and the maximum roof-floor convergence of 48 mm. The deformation of surrounding rock at the initial stage is not obvious, but during the period of 22～72d, the displacement change is relatively large and the two-side contraction is larger than the roof-floor convergence with the slow release of surrounding rock and the influence of construction disturbance. When the roadway tends to be stable, the deformation of two sides and the roof and floor are controlled within the allowable range and meet the engineering requirements, indicating that the support technology is successful in controlling the stability of deeply buried roadway floor.

5. Conclusion

Aiming at the floor heave of auxiliary haulage roadway in Jianzhuang Coal Mine, the paper proposed specific control measures by considering the theoretical analysis, value calculation, indoor test and site test data and comprehensively analyzed the deformation failure mechanism of roadway floor heave. The following conclusion is obtained by this study:

(1) Based on the stress environment of auxiliary roadway in Jianzhuang Coal Mine, a force model for the roadway under the even-distributed load from two directions is created, and the research discloses the mechanical mechanism of roadway deformation and failure, drawing a conclusion that the shear failure of floor surrounding rock is the key factor of roadway floor heave, because the floor shear stress is certain to be larger than the roadway shear strength if the floor is not supported in the roadway subject to high stress;

(2) The paper also proposes a support scheme combined inverted concrete arch and anchor in floor. The application of site construction and deformation observation results indicate that the surrounding rock deformation is mitigated and good support effect is obtained after adopting the above mentioned support scheme;

(3) The results mean that the technology is successfully applied in the engineering practice of floor heave control for the deeply buried roadway, which provides a beneficial reference for the stability control of surrounding rock in other works under the similar conditions and worthies wide promotion.

Acknowledgments
This work was financially supported by the National Nature Science foundation of China (No.11402195), the collaborative innovation in shaanxi province (No.2015XT-15), and Scientific research project in Shaanxi province department of education (No.16JK1512).

References
[1] Zhou Hongwei, Xie Heping, Zuo Jianpin. Developments in researches on mechanical behaviors of rocks under the condition of high ground pressure in the depths, Advances in Mechanics, 2005, 35(1): 91-99.
[2] Cao Rihong, Cao Ping, Lin Hang.Support technology of deep roadway under high stress and its application, International Journal of Mining Science and Technology, 2016, 26(5):787-793.
[3] Bai Qingsheng, Tu Shihao, Zhang Cun, Zhu Defu.Discrete element modeling of progressive failure in a wide coal roadway from water-rich roofs, International Journal of Coal Geology, 2016, 167(10): 215-229.
[4] He Manchao, Xie Heping, Peng Suping, et al.Study on rock mechanics in deep mining engineering, Chinese Journal of Rock Mechanics and Engineering, 2005, 24(16):2803-2813.
[5] Xiao Tongqiang, Wang Xiangyu, Zhang Zhigao.Stability control of surrounding rocks for a coal roadway in a deep tectonic region, International Journal of Mining Science and Technology, 2014, 24(2): 171-176.
[6] He Yongnian, Han Lijun, Shao Peng, et al.Some problems of rock mechanics for roadways stability in depth, Journal of China University of Mining & Technology, 2006, 35(3): 288-295.
[7] Yuan Liang, Xue Junhua, Liu Quansheng, et al.Surrounding rock stability control theory and support technique in deep rock roadway for coal mine, Journal of China Coal Society, 2011, 36(4): 535-543.
[8] Liu Quansheng, Lu Xingli.Research on nonlinear large deformation and support measures for broken surround-ing rocks of deep coal mine roadway, Rock and Soil Mechanics, 2010, 31(10): 3273-3279.
[9] Cai Chengong.Simulation and testing study on outbust prevention measure of pressure-relief slots, Chinese Journal of Rock Mechanics and Engineering, 2004, 23(22): 3790-3793.
[10] Shen Mingrong. Rock Mechanics [M], Bei Jing, Tong Ji university press, 2006:153-156.
[11] Li Dawei, Hou Chaojiong, Bai Jianbiao, Control mechanism and application of doubly supported roadway swith large rigidity and high strength, Chinese Journal of Geotechnical Engineering, 2008(7): 1072-1078.
[12] Wang Xiangyu, Bai Jianbiao. Research on the mechanism of pressure relief based on the analysis of compressive stress deformation, Journal of China University of Mining& Technology, 2010(3): 313-317.
[13] Hou Chaojiong, Zhang Shudong, Zou Xizheng. Discussion on the steel main types of yieldable entrysupport. Journal of China University of Mining &Technology, 1987(2): 11-18.
[14] Guo Zhibiao, Hu Yongguang, Ren Aiwu, et al. Roadway repair technology in swell soft rock with highdepth and high stress. Journal of Mining & Safety Engineering, 2006, 23(3): 316-319.
[15] Jiang Yaodong, Zhao Yixin, Liu Wengang, et al. Research on floor heave of roadway in deep mining.Chinese Journal of Rock Mechanics and Engineering, 2004, 23(14): 2396-2401.

IOP Conf. Series: Earth and Environmental Science **64** (2017) 012033 doi:10.1088/1755-1315/64/1/012033

The comprehensive application of test data in the test area of lamadian oilfield

Yanbo Quan

Interpretation and Evaluation Center on Monitored Information of Daqing Logging & Testing Services Company. China

2296795605@qq.com

Abstract. In order to improve the oil production of the buffer area in Lamadian oilfield, slow down the decline rate of oil production and explore effective methods for a gas cap oilfield development, a test of injecting polymer in buffer area was carried out in the gas cap oilfield of south Sa II 2 + 3 for the technical reserves of gas cap development and buffer area development in Lamadian oilfield. As this block is the key test area in an oil production plant, the comprehensive application of the test data provided a strong technical support for adjusting the injection production measures of the wells and improving the development effect of the test area.

1. Introduction

Lamadian Oilfield is a layered sandstone gas cap oilfield controlled by its structure. The gas cap reservoir is mainly distributed in 17 layers of Sa I ~ Sa III4-7, and its geological reserves is 99.59 × 10^8m3. Each layer has its own independent oil and gas boundary due to its structure. In the early stage of development, a 450m ~ 600m buffer strip for preventing hydrocarbon migration was reserved outside the hydrocarbon boundary. Figure.1 is a structure chart of the test area with polymer injection barrier. In order to improve the oil production of the buffer area in Lamadian oilfield, explore the effective methods for a gas cap oilfield development, a test of injecting polymer in buffer area was carried out in the gas cap oilfield of south Sa II 2 + 3 for the technical reserves of gas cap development and buffer area development in Lamadian oilfield. The idea of the test was to drill a row of wells nearby the hydrocarbon boundary of gas cap oilfield and the oil and gas areas were separated by a barrier which was formed by injecting polymer into wells. The buffer area was directly developed with polymer flooding, and the gas area was not mined. The block is located in the structural axis of the south oilfield and is composed of three parts including the gas area, the buffer area and the three rows of old oil area with perforations. The target layer of the test was Sa II 2 + 3 oil layer and the geological reserves of the buffer area was 179.8 × 10^4t. The total number of wells in the test area was 168, including 81 injection wells and 87 producing wells. After the polymer barrier wells injecting the polymer in the test area, the oil and gas layers could be separated to make the oil in the buffer area flow smoothly [1]. Wang Hua [2] studied the geological characteristics and development techniques for the oil and gas buffer area of the gas cap in south Lamadian oilfield. Fan Wen ke[3]analyzed the practice and understanding of the development of buffer area with polymer injection for the gas cap in Lamadian oilfield. These studies provide a strong technical support for the development of the buffer area with polymer barrier wells. But in the process of development, it is also needed to study the

IOP Conf. Series: Earth and Environmental Science **64** (2017) 012033 doi:10.1088/1755-1315/64/1/012033

comprehensive maintenance technology of polymer barrier in the test area, ensure the sealing and stability of polymer barrier, analyze the pressure distribution and injection-production status, adjust the production parameters of the production wells and control the balance of pressure system in the test area.

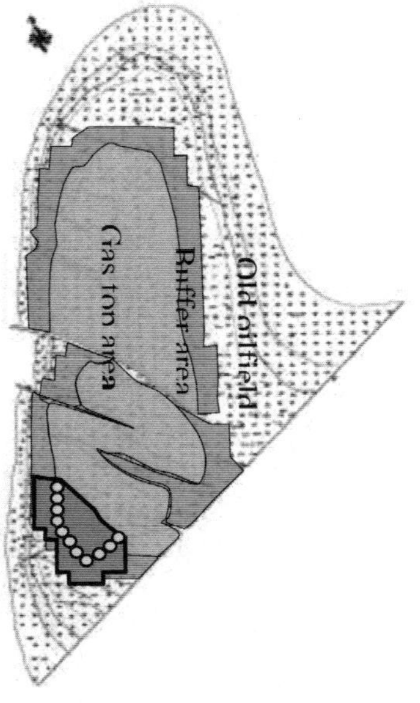

Figure 1. Structure of polymer injection barrier in test area

2. Sealing evaluation of polymer barrier strip

After the polymer barrier was formed, according to the monitored polymer barrier morphology and the distribution of its leading edge, a maintaining method of polymer barrier that injecting the barrier with two batches was established. The strength of the barrier at the equilibrium point between adjacent barrier wells was enhanced, and the extension of the barrier to the gas cap area and buffer area was reduced. In the process of forming the barrier, part of the gas cap gas in the oil and gas transition area was separated from the barrier strip because the barrier wells were located in the gas cap.

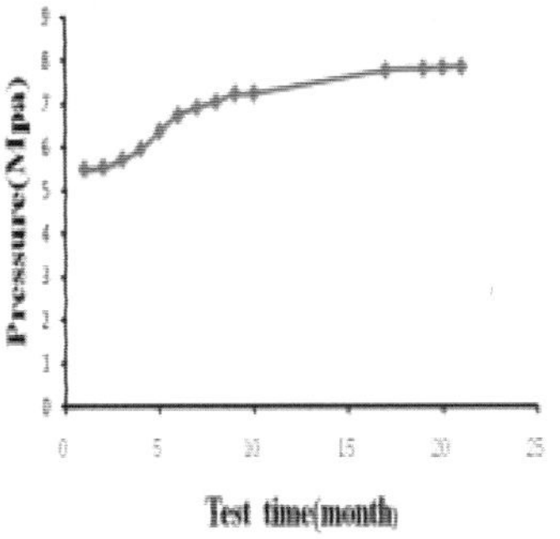

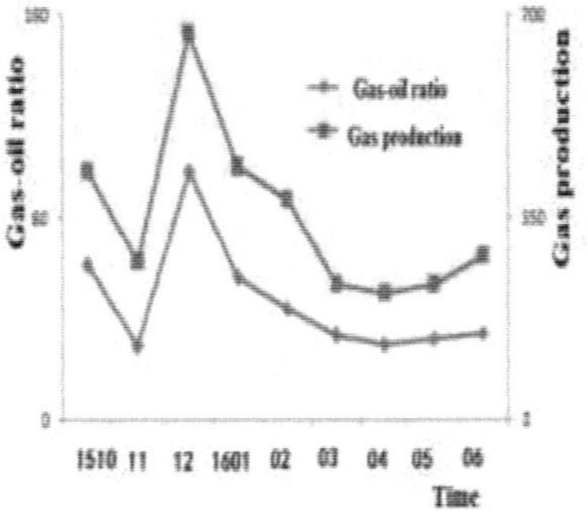

Figure 2. Test report on pressure wells **Figure 3.** Curve of gas production and gas to oil ratio of the first row of oil wells in the buffer area

Monitoring well A in gas cap area. By monitoring the pressure change of well A and analyzing the stability of gas cap area, Figure 2 shows the continuous pressure change in well A. As can be seen from the chart, the pressure of the A well in the gas cap area is continuously monitored for up to 21 months. With the increase of the time for the polymer barrier, the formation pressure rose. As the gas cap was compressed when the polymer barrier was initially formed, the pressure of the gas cap area increased obviously, and the pressure increased 0.8MPa in 2 months. After entering the maintenance phase in May 2015, the pressure increased 0.5MPa in the first 3 months, with the reduction of the maintenance for the barrier well, the pressure in the gas cap area remained stable and increased 0.5MPa in 8 months. At present, the pressure of the gas cap area is stable at about 7.8MPa, which is in the stable stage, which fully showed the sealing property of the polymer barrier strip. The formation pressure in buffer area was 9.2MPa and the formation pressure in old oil area was 10.8MPa. The pressure in the three zones was relatively stable, indicating that the gas cap area and the buffer area have been separated into two independent pressure systems. The pressure change during normal production of the buffer area did not affect the pressure of the gas cap area. According to the data above, the oil and gas can be separated by injecting polymer barrier nearby the oil and gas inter face, in the case of gas temporarily not being mined, the buffer area can be injected and developed normally without oil and gas migration.

In order to further verify the accuracy of the test pressure data, the gas production and gas-oil ratio out of the gas cap area were counted, as shown in Figure.3. In the early stage of production, the average single well gas production rate reached 429m3, and the production gas was mainly the gas separated by the polymer barrier.

With the extension of production time, the daily gas production reduced gradually. After 3 months of production, the average daily production of single well reduced down to 381m3. At present, the average single well daily gas production is 283m3.The gas production tends to be stable. By observing the oil and gas ratio change of the producing wells at the first row, the oil and gas ratio was not high but was stable, the current average single well oil and gas ratio was 34m3 / t.

The analysis above shows that after the oil wells in buffer area putting into production, the amount of gas separated by polymer barrier outside the barrier strip decreases, and the gas in the strip does not pass through the polymer barrier to the buffer area outside the strip and the polymer barrier strip is closed. Through a comprehensive analysis of the test data, the pressure of the test data can be used to accurately determine the sealing of the barrier strips.

3. The main contradiction in the development experimental zones

Using the test data, comprehensively analyze the change of the water absorption capacity in the buffer area, the influence of the interlayer difference on the production and the development status of the wells.

3.1. Analysis of water absorption capacity at the producing position

After the injection of the polymer in the test area, the water absorption capacity of each layer was determined by isotope measurement. Statistics of isotope water injection profile of 45 wells, there were 12 wells of the Sa II2 + 3 layer unabsorbed water in the upper half, and the ratio for the wells without water absorbed capacity was nearly 30%. In order to confirm the reason why these parts do not absorb water, the geological department analyzed the connectivity of the oil layer.

The results show that the connectivity of oil wells and water wells were good, and there was no possibility for not absorbing water due to the poor connectivity of oil layer. Furthermore, the permeability of these parts which could not absorb water was quiet high which was caused by slush entering these parts during the drilling process [4-5].

Further statistics on the well test data for the 45 wells in the buffer area indicated that there were 10 wells whose skin coefficients were greater than zero, accounting for 22.2% of the wells, the value of the average skin factor S was up to 9.24, which was mainly the injection wells with unusual water

IOP Conf. Series: Earth and Environmental Science **64** (2017) 012033 doi:10.1088/1755-1315/64/1/012033

absorption capacity shown on the water absorption profile. And this further indicated that the formation contamination of some wells in the test area was quiet serious.

3.2. Some injection wells have high permeability and inefficient circulation strip in the buffer area

Analysis of the well test data showed that the double logarithmic well pattern of 45 wells in the block reflected two reservoir characteristics [6-7], one was the typical double-seepage reservoir characteristics, there were 12 wells; the other was a clear homogeneous reservoir characteristic, there were 33 wells.

The double logarithmic well pattern indicated that there was a high permeability zone in this part of the wells, and the permeability difference was large and the contradiction in the layers was prominent. The homogeneous reservoir characteristic curve shows that the water absorption capacity was quiet uniform in this part of the well, there was no strong water absorption site, or there was a strong single layer of the phenomenon, the injected water only entered the high permeability zone, low permeability deposition unit did not absorb water.

The PI (90) data of the 45 wells above were recorded with an average value of 7.6 MPa.(PI (90) value reflects the pressure drop velocity after injection wells were closed, the better the formation seepage characteristics is, the stronger the water absorption capacity is, the smaller the PI value is; or PI value is large) The PI index of 15 wells in 45 wells was lower than the average value of 7.6MPa, accounting for 33.3% of the total number of wells, indicating that this part of the injection wells had a strong water absorption zone with high permeability, the contradiction within the layer was prominent.

3.3. Abnormal of well pressure curve, reflect low formation pressures in some well groups

During the interpretation of oil well pressure data in the test area, some liquid recovery wells appeared to rise at the beginning and then fall, and even the flow pressure was higher than the static pressure. For this anomaly, in order to ensure the accuracy of the data, we selected five typical wells which had significant fluid level fluctuation to retest, and proved that the test equipment was reliable.

Table 1 shows the depth changes of the five wells. It can be seen from Table 1 that these five well all had an phenomenon that, in the short time after the closure of the wells, the liquid surface rose and then fell until the end, the fluid level of well 5 which had the largest fluid level drawdown declined 540m after the closure of the wells, and the pressure declined from 0.923 to 5.886MPa. Reflecting it on the pressure curve, the pressure rose and dropt suddenly at the beginning, after the well was shut down for 11.3 hours, the pressure rose to 10.1MPa, and latterly there was a downward tendency, when the well was shut down for 71.3 hours, pressure declined down to 7.53Mpa. We analyzed that the underground deficit was serious, and had no energy to supply and resulted in serious degassing in the well.

Table 1. Change of well liquid level

Well	Turn off time (h)	Initial level (m)	Final level (m)	Liquid level drop (m)	Flow pressure (MPa)	Terminal pressure (MPa)
J1	68.7	600	993	393	6.86	7.58
J2	68.7	309	662	353	7.2	8.26
J3	68.7	326	584	258	5.29	7.66
J4	68.7	319	702	383	8.04	9.43
J5	71.3	313	853	540	7.82	7.53

4. Adjustment of injection and production wells in buffer area

In the first half year of 2016, according the formation contaminant shown by the test data, the existence of high permeability and inefficient circulation strip and the contradiction of low formation pressure in some well groups, the geological department had an adjustment in time

IOP Conf. Series: Earth and Environmental Science **64** (2017) 012033 doi:10.1088/1755-1315/64/1/012033

4.1. Injection well acidification to solve plugging

The 12 injection wells with abnormal water absorption were selected for acidizing treatment. After the treatment, the thickness of the formation which did not absorb water decreased from 10.5m to 1.1m, the skin factor decreased from 5.65 to -0.37; deep profile control measures were carried out for 15 wells, and the relative water absorption of the main water absorbing unit of high permeability decreased from 55% to 23% after the treatment, At the same time, in order to restore the formation pressure of the well group, the 40 injection wells have been carried out with super water injection scheme. Figure 4 shows the double logarithmic fit curve of the E-well before and after acidification. The test data PI (90) was 8.01 and the skin coefficient was 5.211, indicating that there was contaminant near by the well bore. Figure 4 shows that, before the acidification, the pressure derivative curve hump was high which meant that the formation contaminant was serious; the pressure derivative curve hump position after acidification slowed, and the skin coefficient was -4.72, which reflected that the acidizing effect was good, the degree of pollution had been improved.

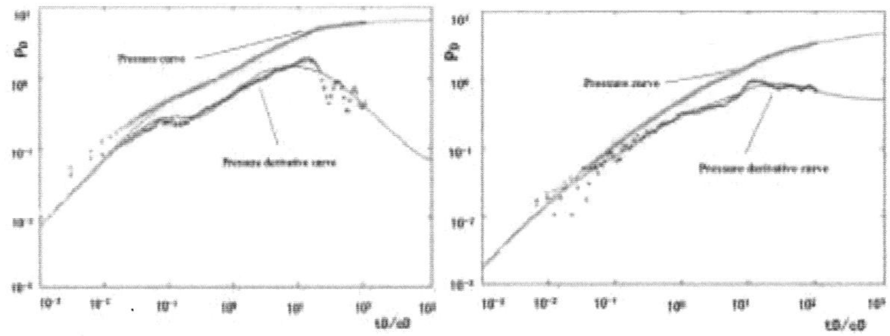

Before the acidizing treatment After the acidizing treatment

Figure 4. Double logarithmic fit analysis chart before and after measures

4.2. Improvement of oil well fracturing

For the production wells with low liquid production, low water content and poor reservoir property, choosing the mid-low permeability area which have better connectivity with surrounding injection wells to fracture to improve oil production, there were 10 wells designed, and there were 8 wells had been operated. F was a fractured well, this well was fractured in July 2016. Before fracturing, the production was13.8t/d, the water content was 54.8%, the formation pressure was 8.43MPa; after fracturing, the production was 52.0t/d, the water content was 73.9%and the formation pressure was 9.25MPa. It can be analyzed from the double logarithmic curve of the well that after fracturing, there were cracks at the bottom of the wells, the pressure curve and the derivative curve did not coincide in the early stage, it can be seen that the fracturing received good results.

5. Conclusions and recommendations

(1) By monitoring the pressure in the three areas including the gas-cap area, the buffer area and the old oil area, it is helpful for the geological department to analyze the pressure distribution and injection and production conditions in the test area, adjust the production parameters of the production wells in the test area and control the balance of pressure system.

(2) By comprehensively analyzing the test data, understand the development contradiction of the test area, adjusting the measures with targeted purpose.

(3) Aiming at the production of the first row of well outside the gas cap, we can reasonably arrange the test, reduce the impact on the stability of the polymer barrier, determine the reasonable production parameters of the first row of oil wells under the stable condition of in the barrier strip, ensure the closure and stability of the polymer barrier in the buffer development.

(4) According to the monitoring data of pressure and gas production, we analyzed the sealing and stability of polymer barrier strips, determined the reasonable timing and injection quantity for the maintenance of the polymer barrier wells in the middle and later stage of the buffer development.

IOP Conf. Series: Earth and Environmental Science **64** (2017) 012033 doi:10.1088/1755-1315/64/1/012033

Acknowledgments
The project fund for this article is National Natural Science Foundation of China "High Concentration Polymer Flooding Mechanism and Optimization Design in Second-class Reservoir" (project number 51374076).

References
[1] Fu Xing 2012 Separated from oil and gas layer to buffer smooth flow of oil N. Daqi daily. 2012- 02- 09012.
[2] Hua Wang 2012 Research on the development and geological characteristics of oil and gas buffer on the gas cap of the South block in Lamadian oil field D. Northeast Petroleum University.
[3] Wenke Fan 2014 Practice of the barrier-injection development in the buffer area of Lamadian oilfields gas cap. J. Petroleum Geology and Oilfield Development in Daqing.2014-04:121-125.
[4] Daoxiang Mu 2014 The research of the drilling completion technology of gas barrier area in Lamadian oilfield D.Northeast Petroleum University.
[5] Yunbo Li 2013 The research of the pressure maintaining drilling fluid technology of gas barrier area in Lamadian oil field J. China Petroleum and Chemical Standards and Quality. 2013.12:65-66.
[6] YunfangZhai 2003 Seepage mechanics (second edition) Beijing: Petroleum Industry Press.
[7] Nengqiang Liu 2008 Practical modern well test interpretation method (fifth edition) Beijing: Petroleum Industry Press.

IOP Conf. Series: Earth and Environmental Science **64** (2017) 012034 doi:10.1088/1755-1315/64/1/012034

Geochemical Characteristics of Longxing Bauxite in Northern Guizhou, China

Changqian Du

Geological team 106, Guizhou bureau of geology and mineral exploration and development, Zunyi, Guizhou 563000, China

Duchangqian2017@163.com

Abstract. Revealed the chemical characteristics of Longxing bauxite through field geological research and elements analysis: 1) Longxing bauxite displays an overall U-shape distribution; 2) Main minerals contained in the ores are diaspores, clay minerals, iron minerals and titanium minerals; 3) The formation of Longxing bauxite is the process of enrichment of Al, Si and loss of Si, Fe; 4) Elements with aluminum affinity from parts of the area such as Ga and Li may concentrate to form ores.

1. Introduction

Longxing in the northern area of Guizhou province is rich in bauxite, which is part of the Wuchuan-Zheng'an-Daozhen bauxite. A large number of researches and reports focus on the bauxite in the northern region of Guizhou (Wu et al.,2008;Du et al., 2013, 2014; Liu et al., 2008, 2010; Wang et al., 2013; Zhang et al., 2013), while independent studies of Longxing areas are relatively less. The ores from Longxing bauxite are of good quality, and Longxing bauxite is located in the central part of Wuzhengdao bauxite metallogenetic region. Researches based on its deposit basic characteristics and geochemical characteristics reveal the bauxite ore-forming mechanism, and provide new information for studies on bauxite in northern part of Guizhou, which is of great instructive meanings for the next exploration work.

2. Geological background

Longxing bauxite is located in northern Guizhou (Fig.1). The study area is exposed from the old to the new: Cambrian, Ordovician, Silurian, Carboniferous, Permian and Jurassic. The tectonic lines in the area are NNE, and the Cambrian strata are mainly distributed in the core of anticlines. The Triassic and Jurassic are mainly distributed in the core of synclines. Overlying strata of Longxing bauxite is Permian Liangshan Formation (P_2l) or Qixia Formation (P_2q), and the Lower strata is Carboniferous Huanglong Formation (C_2h) or the Silurian shale of Hanjiadian Formation, all of them are parallel unconformity. The spatial distribution of bauxite in Wuchuan-Zhengan-Daozhen area is controlled by Daizheng, Datang, Luchi - Liyuan, Taoyuan, Anchang, Huanxi, Xinmo and Zhangjiayuan synclines.

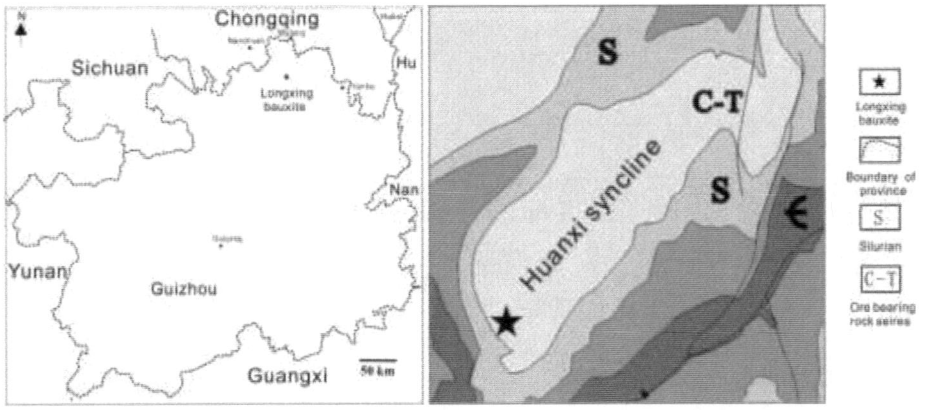

Fig.1 The location of Longxing bauxite[modified form Wang et al.,2013]

3. Ore body characteristics

Longxing bauxite is a sedimentary type of diaspore bauxite deposit. The ore body is generated from HuanXi syncline both wings and its southern twist, which displays an overall "U"-shape distribution. The output layer is located in the middle-upper parts of Dazhuyuan formation of Lower Permian (P_1d). Ore body internal structure is of medium complexity, with stable thickness variation, which belongs to medium sized bauxite deposits.

The ore block is 0.86-1.66m thick, with an average thickness of 1.30m, which contains Al_2O_3 of 56.01-65.05% (average 58.84%), SiO_2 of 8.25-14.82% (average 12.24%), A/S of 3.89-7.89 (average 4.81), Fe_2O_3 of 4.11-10.83% (average 7.26%), and TS of 0.04-4.08% (average 3.32%).

4. Ore structures and tectonics

The ore structures and tectonic features of Longxing bauxite are studied according to the core and outcrop samples (Fig. 2).

4.1. Ore structures

Ore structures can be classified into clastic texture, mud-crystal structure and pink-crystal structure based on the mineral collection shape, as described below:

Clastic texture: is the main ore structure, composed of debris and cement, whose boundaries are clear. The debris is of angular and sub-angular shape, some of which is round or ellipsoidal shaped. The particle size is of large differences, which is generally 0.005-0.008mm, and the bigger one can be 5-8mm. It can be divided into sand-like structure (<0.1mm) and gravel-like structure (>2mm) according to the particle size. The primary composition of the debris is granular semi-automorphic crystal diaspore, flake and scale like kaolinite, hydromica and other clay minerals, and pyrite, hematite, trace amounts of chlorite and anatase. Cement composition is of <0.005mm clay minerals or bauxite, whose debris content accounts for more than 70%.

Pink-crystal structure: is composed of 0.005-0.008mm granular diaspore, which contains debris of irregular shapes. The debris composition is semi-automorphic crystal diaspore combined with lamellar and flake like kaolinite and hydromica mixture. It is the secondary structural types of census area.

Mud-crystal structure: is composed of <0.005mm granular diaspore, whose particle size is uniform, forming automorphic and semi-automorphic crystal combined with little hydromica and limonite, which is the secondary structural types of census area.

4.2. Ore tectonics

Block tectonics: is the main tectonics type. The ore shape is even block, whose particle size is uniform, which has no clear orientated and laminated striation.

IOP Conf. Series: Earth and Environmental Science **64** (2017) 012034 doi:10.1088/1755-1315/64/1/012034

Semi-earthy tectonics (Fig.2, b): The mineral is grey, dark grey and is of mud-crystal structure, with debris content of generally <5%. The ore section is smooth and dense, with good hardness and weak water absorption.

Massive tectonics: is the rare tectonics type and distributes locally. The mineral is grey, light grey and yellow-gray, most of which is debris structure. The debris is angular and sub-angular shape with poor sorting, a few of which is mud-crystal or pink-crystal structure.

Oolitic tectonics: is another important structure types after the debris structure, consisting of beans, oolitic and cement. Oolitic has core and the concentric layers, and generally the concentric layers is 1-3 layers. Core of oolitic and beans is composed of diaspore, clay minerals, pyrite or siliceous. Concentric layers consist of crossed bauxite and clay minerals. Cement consists of <0.005mm granular semi-automorphic crystal bauxite or flake like hydromica. Beans content is 15-30%, while oolitic content is 30-50%.

Fig.2 a. core of Longxing bauxite b. high quality ore

5. Chemical constituents of the ores

5.1. Mineral component of the ores

By test analysis, mining bauxite is composed of 23 kinds of minerals, of which the diaspore is the main type, followed by clay minerals, then the iron and titanium minerals. The four kinds of above mineral accounted for more than 98% of the total ore minerals (table.1). 1) diaspore: the main ore mineral, which is needle like or plate-shaped, cylindrical or granular, and vertical lines are common cylinder. Diaspore is automorphic and semi-automorphic crystal aggregation, with particle size of general 0.005-0.008mm and up to 0.6mm. 2) hydromica: A scale and flake-like output. The mineral is colorless and transparent, and the interference color is bright. Particle size is less than 0.005mm, mainly in the clay minerals. The highest mineral content is 10%. 3) Kaolinite: small particle size, with aphanitic, a grainy and film output. Kaolinite is mainly silicon containing minerals in bauxite, with the mineral content of 0-10%. 4) Chlorite: including iron chlorite, leptochlorite, oolitic chlorite and so on. It has flaky and grainy output, mainly concentrated in the bottom of the ore-bearing rock series, making up the chlorite rock and chlorite-clay rock. 5) Pyrite: rhombohedral crystal, a nodule or granular aggregate output, is the main gangue minerals in bauxite.

Table 1 Ores mineral component of Longxing bauxite

bauxite	diaspores
clay minerals	Hydromica, kaolinite, chlorite, white mica
iron minerals	limonite, hematite and magnetite
titanium minerals	anatase, rutile, sphene, leucosphenite, jurinite
zirconium minerals	Zircon and baddeleyite
sulfide mineral	Pyrite, pyrrhotite
Other minerals	Barite, taltalite, hussakite, kietyoite, akanthikon, edwardite

5.2. Characteristics of main elements

Main chemical constituents of bauxite are Al_2O_3, SiO_2, Fe_2O_3, TiO_2, TS. Minor chemical constituents are MgO, CaO, K_2O, Na_2O, V_2O_5, P_2O_5, RE_2O_3, and CO_2 and so on. Trace amount elements are Li, Ga, Ge, Ba, Sr, Nb, Ta, Zr, Cr, Mn, Pb, Cu, Zr, V, Sn, Be and Au.

According to statistics of 149 marginal grade samples, the main chemical constituents in the bauxite region demonstrate significant or obvious correlation. Among them, Al_2O_3 and SiO_2 are significantly negatively related with coefficient of correlation of -0.777 (Fig. 3), and Al_2O_3 and Fe_2O_3 are significantly negatively related with coefficient of correlation of -0.508 (Fig. 4), and Al_2O_3 and TiO_2 are significantly positively related with coefficient of correlation of 0.462 (Fig. 5).

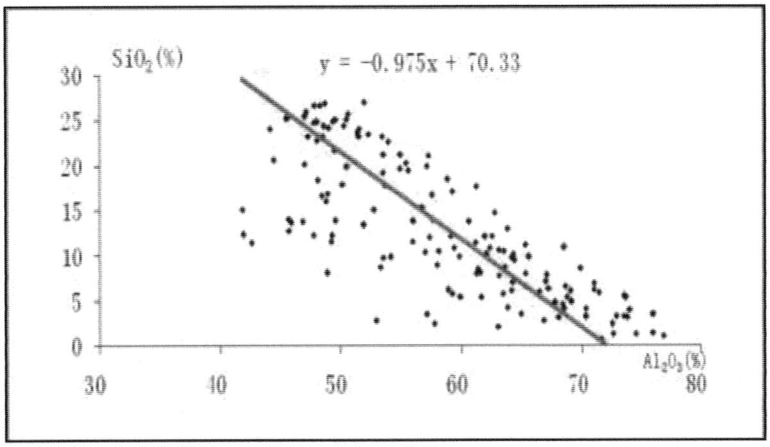

Fig.3. Regression curves and scatters plots of Al_2O_3-SiO_2 in Longxing bauxite (According to 149 samples)

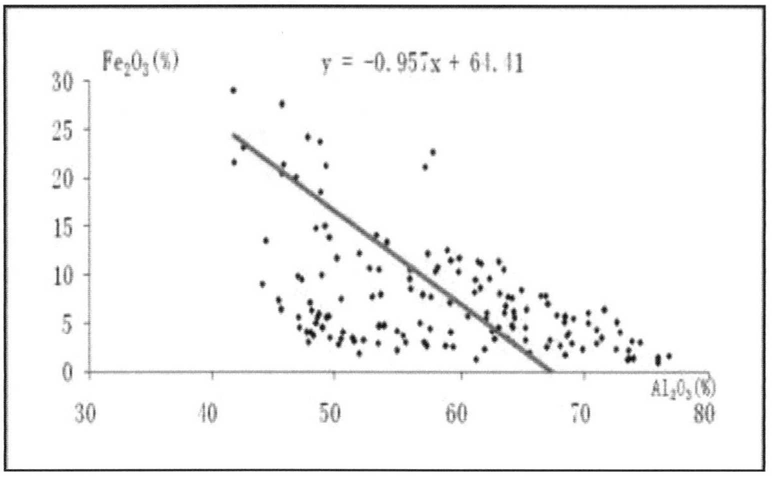

Fig.4. Regression curves and scatters plots of Al_2O_3- Fe_2O_3 in Longxing bauxite (According to 149 samples)

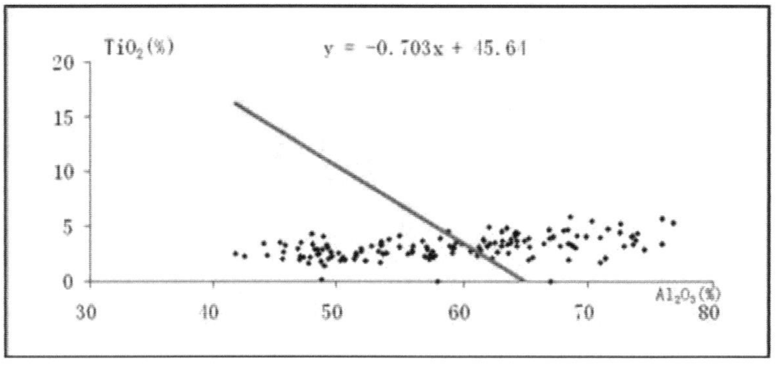

Fig.5 Regression curves and scatters plots of Al_2O_3- TiO_2 in Longxing bauxite (According to 149 samples)

6. Discussion

The characteristics and geochemical characteristics of Longxing bauxite ore show that the formation is a process of continuous leaching and leaching. In this process, the loss of elements such as Al and Ti is small, and the loss of Si and Fe is large. The different structures and tectonic characteristics of the ore show that the metallogenic progress of the bauxite in the Longxing area is inconsistent. Due to the difference of the metallogenic progress, the different structure and tectonic of the ores occur simultaneously. The structure and tectonic characteristics of Bauxite show that it leaching process plays an important role in the formation of bauxite similar to that of bauxite in the rest of Wuchuan-Zheng'an-Daozhen area(Wang et al., 2013; Yu et al., 2013; Huang et al., 2014). The formation of bauxite is Al, Ti, Si, Fe and other elements of the migration process, pro-aluminum elements such as Ga, Li, etc. in some areas may be integrated into an independent deposit, and mineralization area has a greater comprehensive development and utilization value.

7. Conclusion

According to the comprehensive analysis, the following conclusions are drawn: 1) Ore body of longxing bauxite is produced in the inclined section of the Huanxi syncline and its southern turning end, and the whole is "U" spreading. 2) Longxing bauxite mineral is mainly diaspore, clay minerals, iron minerals, titanium minerals, four minerals accounted for 98% of the total mineral minerals. 3)

IOP Conf. Series: Earth and Environmental Science **64** (2017) 012034 doi:10.1088/1755-1315/64/1/012034

Al_2O_3 is positively correlated with TiO_2, Al_2O_3 is negatively correlated with SiO_2 and Fe_2O_3, indicating that the formation process of Longxing bauxite is the relative enrichment of Al and Ti elements and the relative loss of Si and Fe. 4) Pro-Aluminum elements may be enriched in mineralization.

8. Conferences

[1] Wu Guohui, Jin Zhongguo, Bao Miao, Mao ZL.Bauxite metallogenic regularly in the Wuchuan-Zheng, an -Daozhen Area, Northern Guizhou. Geology and Prospecting, 2008, 44(9):31-35.

[2] Du Yuansheng, Zhou Qi, Jin Zhongguo, Lin Wenlin, Zhang Xionghua. Adcances in basic geology and metallogenic regularity study of bauxite in Wuchuan-Zheng,an-Daozhen area, northern Guizhou, China. Geological science and technonlogy information, 2013, 32(1):1-6.

[3] Du Yuansheng, Zhou Qi, Jin Zhongguo, Lin Wenlin, Wang Xiaomei, Yu Wenchao. Mineralization model for the early permian bauxite deposits in Wuchuan-Zheng,an-Daozhen area, northern Guizhou Province. Juornal of palaeogeography, 2014, 16(1):1-8.

[4] Liu Xuefei, Wang Qingfei, Zhang Qizuan. Thermal analysis of core of the ore body VII in the Xinxu bauxite deposit, Jingxi county, Guangxi province. Mineral Petrol, 2008, 28:53-58.

[5] Liu Xuefei, Wang Qingfei, Deng Jun, Zhang Qizuan, Sun Silei, Meng Jianyin. Mineralogical and geochemical investigations of the Dajia Salento-type bauxite deposits, western Guangxi, China. Journal of Geochemical exploration, 2010, 105(9):137-152.

[6] Wang Xiaomei, Jiao Yangquan, Du Yuansheng, Ling Wenlin, Wu Liqun. REE mobility and Ce anomaly in bauxite deposit of WZD area, northern Guizhou, China. Journal of Geochemical Exploration, 2013, 133: 103-117.

[7] Zhang Yinghua, Lin Wenli, Wu Hui, Zhang Yanan, Ding Xiaoying. Geochemistry of varied type ores of northern Guizhou bauxite and its implication for mineralization. Geological science and technology information, 2013, 32(1):71-79.

[8] Wang Denghong, Li Peigang, Qu WJ, Yin LJ, Zhao Z, Lei ZY, Weng SF. Discovery and preliminary study of the high tungsten and lithium contens in the Dazhuyuan bauxite deposit, Guihou, China. Science China: Earth Science,2013, 56(1):145-152.

[9] Yu Wenchao, Du Yuansheng, Wang Xiaomei, Huang Xing. Combustion-derived polycyclic aromatic hydrocarbons in bauxite deposit of Wuchuan-Zheng,an-Daozhen area, northern Guizhou province and significance of paleoclimate. Geology science and technology information, 2013, 32(1):57-61.

[10] Huang Zhilong. Metallogenic theory and metallogenic prognosis of bauxite in WZD area, northern Guizhou, China. Beijing, Science press. 2014:252-255.

IOP Conf. Series: Earth and Environmental Science **64** (2017) 012035 doi:10.1088/1755-1315/64/1/012035

Hydrocarbon potential of the jurassic coal-bearing strata in the Liupanshan area, china

Pei Guo[1], Changzhi Li[2], Feifei Wang[1]

1Department of Geology, Northwest University, Xi'an, China
2Shengli Oilfield Dongsheng Jinggong Petroleum Development Group, Dongying, China

*Corresponding author e-mail: nwupeig@126.com

Abstract. The Liupanshan area is situated to the southwest of the Ordos Basin. Because of larger stratigraphic thickness and thicker overlying sedimentary coverage, the coal-bearing Yan'an Formation in the Liupanshan area is supposed to have higher hydrocarbon potential than that in the Ordos Basin. 25 samples collected from 10 sedimentary sections in the Liupanshan area have been tested to evaluate the quality of the coals and dark mudstones of the Yan'an Formation. The Yan'an Formation in the Liupanshan area generally bears thick coals and dark mudstones, though the stratigraphic thickness and coal-bearing coefficient vary greatly. Most of the source rocks have high total organic carbon and chloroform bitumen "A", which can be classified to the category of "good to excellent source rocks". The main organic matter type of the source rocks is II_2 and III kerogen, having larger potential to generate gas. The maturity is the key factor because the coals in the Ordos Basin are mostly immature to low-mature. Results show that the maturity of the source rocks in the Liupanshan area is higher, most reaching mature to high-mature stage. Therefore, the hydrocarbon potential of the Yan'an Formation in the Liupanshan area is very high and this formation deserves further exploration.

1. Introduction

The Early-Middle Jurassic Yan'an Formation is the most significant coal-bearing succession in the Northwest China [1], especially in the Ordos Basin [2]. Numerous high-quality Jurassic coal seams have exposed or been uplifted in very shallow subsurface in the northern and western margin of the Ordos Basin [2], acting as the major economic source for local people. While in the interior of the basin, this coal-bearing succession was buried in relatively deep subsurface and has potential to generate coal-derived gas. However, previous study has shown that the maturity of these coals in the interior were not high enough to generate large-scale gas [3].

The Liupanshan area, situated to the west of the Ordos Basin (Fig. 1), was in the same sedimentary depression as the Ordos Basin during the Early-Middle Mesozoic [4]. It also contains the Jurassic coal-bearing succession [5]. Different from the Ordos Basin, the thickness of the Yan'an formation in the Liupanshan area is supposed to be larger and currently this formation is covered by thick Cretaceous and Cenozoic strata, indicative of possibly higher maturity for the Jurassic coals and mudstones. Therefore, the Yan'an Formation in the Liupanshan area may have great potential to generate hydrocarbon. Previous work mostly focused on the Cretaceous source rocks in the Liupanshan area, most of which are only in immature to low-maturity stage [6, 7]. However, the hydrocarbon potential of the Jurassic coal-bearing strata has remained poorly studied.

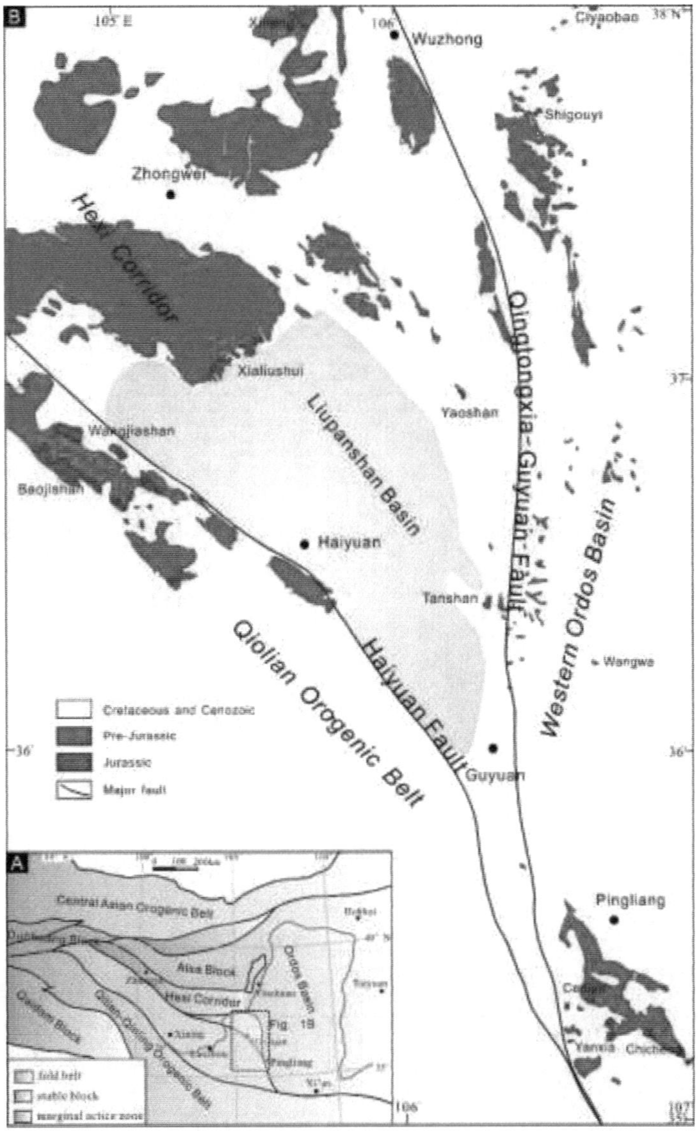

Figure 1. (A) The tectonic setting of the Liupanshan area in China. (B) The sample locations of Jurassic source rocks, which can belong to three tectonic units: the Western Ordos Basin, Hexi Corridor and Qilian Orogenic Belt. Note that the Liupanshan area is mainly situated in the Hexi Corridor.

At present, there is only one well that has drilled the Early-Middle Mesozoic strata inside the Liupanshan area and the stratigraphic classification of this well has great controversy. In this paper, sampling of source rocks are mainly carried out in outcrops. Almost all the Jurassic outcrops in the Liupanshan area have been sampled to test the source-rock quality of the Jurassic coal-bearing strata.

2. The distribution of source rocks

As shown in Table 1, the thickness of the Yan'an Formation surrounding the Liupanshan area shows great variations, ranging from 100 m to 700 m, much distinct from that in the interior of the Ordos Basin and indicative of the subsidence instability during the Jurassic. In the Western Ordos Basin, from the north to south, the thickness of the Yan'an Formation and the number of coal seams decrease but the coal-bearing coefficient increases. In the Hexi Corridor, the thickness of the Yan'an Formation is mostly varied and the coal-bearing coefficient is smallest. Especially in the Xinjing area, the Yan'an Formation is only ca. 100 m and its coal-bearing coefficient varies from 0~38.5%. Three sections, the Tanshan, Yaoshan and Xialiushui, situated on the margins of the Liupanshan area, contain ca. 400~700 m coal-bearing strata. This means that the thickness of the Yan'an Formation inside the Liupanshan area may also reach 400 m. However, the coal-bearing coefficient is very small. In the Qilian Orogenic Belt, the Yan'an Formation is less than 200 m and the number of coal seams is 1~3. But the thickness of coal seam can reach to 37m, much higher than the other parts.

Another potential source rocks are the dark mudstones. According to the wells of coal fields, the thickness of dark mudstones in the Tanshan section can reach 327 m and that of the Yaoshan section can reach 50% of the whole thickness. This suggests that even the sections in the margins of the Liupanshan area is very low, its high proportion of dark mudstones can make up this defect. In summary, the Yan'an Formation surrounding the Liupanshan area generally bear coal and dark mudstones and their proportion of the strata can be very high.

Table 1. Thickness of the Yan'an Formation and its bearing coals. Location of the sections is referred to the Fig. 1B.

Region	Section		Strata thick(m)	Number of coal seam	Number of minable coal seam	Average thickness of coal(m)	Coal-bearing coefficient (%)
Western Ordos Basin	North	Ciyaobao	370	17-22	7	17	4.6
		Suishijing	300	17-26	13	21.2	7.1
		Yuanyanghu	294	16-23	2	4.47	1.5
		Majiatan	281	19-22	12	23.38	8.3
		Shigouyi	694		9	15.05	2.2
	Wangwa	North	160	4	1	19.25	12.0
		Middle	282	4	1	61.85	21.9
		South	114	4	1	17.85	15.7
	South	Cicheng	263	6~12	6	25.52	9.7
		Yanxia	146		6	22.77	15.6
		Cedipo	148.22		5	22.84	15.4
Hexi Corridor	Tanshan		395.62	10~25	5~10	26.68	6.7
	Yaoshan		>562.18	16	9	48.7	8.7
	Xialiushui		>729	11~17	1~5	25.92	3.6
	Xinjing		96	3	1	0-37	0~38.5
Qilian Orogenic Belt	Baojishan		110	5	3	26.64	24.2
	Wangjiawashan		175	6	3	37.73	21.6
	Honghui		70	2	1	15.68	22.4

3. Organic geochemical analysis

Generally, the quality evaluation of source rocks include three aspects: the organic matter abundance, type and maturity. In total, 25 samples of the Yan'an Formation have been tested, including 13 coal samples and 10 dark mudstones samples (Table 2).

3.1. Organic matter abundance

As shown in Table 2, the total organic carbon of the mudstone samples ranges from 1.06~26.7%, averaging at 9.347% and all higher than the lower limit value of source rocks [8]. Most of the dark mudstones are even in the category of "good or excellent source rocks", except the samples in the Mahuanggou section. The chloroform bitumen "A" ranges from 0.77% to 5.19%, averaging at 1.66%. The hydrocarbon potential (S_1+ S_2) ranges from 0.116 to 142.344‰, averaging at 33.705‰ and many falling in the category of "excellent source rocks [9].

Table 2. The organic geochemistry characteristics of the source rocks of the Yan'an Formation in the Liupanshan area. S_1-soluble hydrocarbon, S_2-pyrolysed hydrocarbon, T_{max}- peak temperature of pyrolysis, Pg- hydrocarbon potential, PI- production index

Region	Sample	Location	Lithology	TOC (%)	SO_2 (%)	S_1 (‰)	S_2 (‰)	T_{max} (℃)	Pg=S_1+S_2(‰)	PI=S_1/(S_1+S_2)	(S_2×100) /TOC	Ro (%)
Western Ordos Basin	NX-218	Shiganggou	mudstone	26.7	0.386	0.900	32.780	432	33.680	0.027	195	0.518
	NX-222		mudstone	11.9	0.167	0.090	15.550	437	15.640	0.006	225	0.622
	NX-133	Shigouyi	mudstone	21.3	0.14	0.130	0.190	449	0.320	0.406	1	0.463
	NX-207		coal	-	-	0.510	9.290	437	9.800	0.052	53	0.446
	NX-137		coal	-	-	0.550	62.760	399	63.310	0.009	213	0.546
	NX-10	Wangwa	coal	-	-	0.460	45.070	507	45.530	0.010	185	0.509
	NX-48		coal	-	-	0.870	64.570	435	65.440	0.013	164	0.591
Hexi Corridor	NX-54	Tanshan	coal	-	-	0.340	52.560	443	52.900	0.006	258	0.569
	NX-128		mudstone	1.9	0.0271	0.078	1.190	438	1.280	0.061	63	0.570
	NX-32		coal	-	-	1.408	140.364	422	142.344	0.010	-	0.560
	NX-87		coal	-	-	0.020	6.400	446	6.420	0.003	17	0.551
	NX-87		coal	-	-	0.100	7.040	449	7.140	0.014	38	0.551
	NX-92	Yaoshan	coal	-	-	0.190	37.740	347	37.930	0.005	137	0.453
	NX-92		coal	-	-	0.420	25.950	441	26.370	0.016	92	0.453
	NX-239		coal	-	-	0.340	42.440	444	42.780	0.008	122	0.626
	NX-04		coal	-	-	0.564	26.545	435	27.179	0.021	-	0.480
	NX-118	Xinjing	coal	-	-	0.500	81.280	459	81.780	0.006	154	0.859
	NX-280	Mahuanggou	mudstone	3.79	0.116	0.190	3.340	347	3.530	0.054	133	0.512
	NX-12		mudstone	1.06	0.00672	0.035	0.087	500	0.139	0.250	8	0.980
	NX-13		mudstone	6.18	0.0131	0.066	0.363	414	0.444	0.148	6	0.970
	NX-39	Xialiushui	mudstone	1.67	0.0148	0.027	0.072	505	0.116	0.234	4	1.690
	NX-49		mudstone	6.87	0.156	0.061	0.916	442	0.994	0.061	13	0.640
	NX-50		mudstone	12.1	0.343	0.113	4.261	434	4.394	0.026	35	1.140
Qilian Orogenic Belt	NX-150	Honghui	coal	-	-	0.882	78.010	437	78.920	0.011	-	1.110
	NX-25	Baojishan	coal	-	-	1.835	92.354	443	94.239	0.019	-	1.220

3.2. Organic matter type

According to the microscopic observation of kerogen maceral, rock pyrolysis, steroid terpene biomarkers and elementary composition of kerogen, the organic matter of the Jurassic mudstones is mainly type II_2 and III kerogen, accounting for 83.3%. Organic macerals are mainly vitrinite and inertinite, which can generally make up 60~90%, while the proportions of the sapropelinite and exinite are generally under 40%, indicative of low potential to generate oil. The hydrogen index ranges from 14~928 mg/g, averaging at 108 mg/g. All of these suggest that the organic matter sources of the Jurassic mudstones mainly come from terrigenous higher plants and minor plankton and have high potential to generate gas.

3.3. Organic matter maturity

According to the results of the Ro and rock pyrolysis (Table 2), the kerogens of Jurassic mudstone and coals are immature to high-mature, showing great variations (Fig. 2). In the western Ordos Basin, the Ro mostly falls in the range of 0.45~0.60%, suggestive of the immature property, just like the Jurassic source rocks in the interior of the Ordos Basin. The two sections (Tanshan and Yaoshan), on the boundary of the Ordos Basin and Hexi Corridor, also bear immature source rocks. As in the Xialiushui and Mahuanggou sections, on the northern margin of the Liupanshan area, most mudstones and coals reach the oil threshold with Ro ranging from 0.51 to 1.69%. The coals in the Xinjing section also reach the oil threshold, with Ro at 0.859%. While in the two sections in the northern margin of the Qilian Orogenic Belt, the Ro of the Jurassic coals exceeds 1.0%, showing higher maturity. The results of the peak temperature of pyrolysis (T_{max}) also suggest most of the source rocks in the SW Ordos Basin are in low-mature stage, with several exceptions.

It is noteworthy that bounded by the Qingtongxia-Guyuan Fault (Fig. 1), source rocks in the east part are mostly immature or low-mature while those in the west part are mostly mature to high-mature.

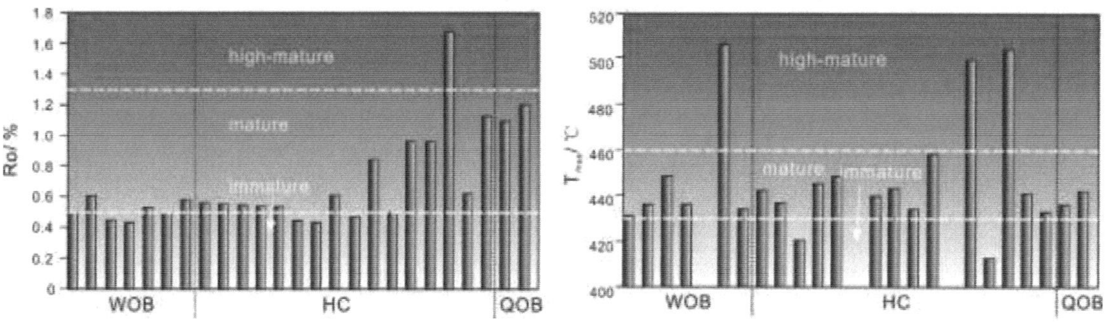

Figure 2. The maturity of the Jurassic source rocks. WOB- Western Ordos Basin; HC- Hexi Corridor; QOB- Qilian Orogenic Belt. The classification is based on Wu, 1986[10].

3.4. Hydrocarbon potential

As discussed above, the Yan'an Formation in the Liupanshan area generally bear coals in different degrees. Although the three sections on the margins of Liupanshan Basin have the lowest coal-bearing coefficient, the Yan'an Formation there have higher proportion of dark mudstones. Based on the thickness of the Yan'an Formation in above three sections, this formation in the Liupanshan Basin are supposed to reach 400~700 m. And the source rocks mainly contain type II_2 and III kerogen. It is suggestive that the coal-bearing strata in the Liupanshan have great potential to generate gas if the source rocks reach maturity. Although the maturity of the source rocks in the Tanshan and Yaoshan sections is not very high, just as that in the Western Ordos Basin, the other sections in the Hexi Corridor contain mature to high-mature source rocks. Moreover, the Cretaceous source rocks inside the Liupanshan area has partly reached the oil threshold (Ro=0.5%), implying that the maturity of the underlying Jurassic source rocks may be higher. Therefore, the Yan'an Formation in the Liupanshan area has great hydrocarbon potential.

4. Conclusion

25 samples collected from 10 sedimentary sections in the Liupanshan area have been tested to evaluate the quality of the coals and dark mudstones of the Yan'an Formation. The Yan'an Formation in the Liupanshan area generally bears thick coals and dark mudstones, though the stratigraphic thickness and coal-bearing coefficient vary greatly. Most of the source rocks have high total organic carbon and chloroform bitumen "A", which can be classified to the category of "good to excellent source rocks". The main organic matter type of the source rocks is II_2 and III kerogen, having larger potential to generate gas. The maturity is the key factor because the coals in the Ordos Basin are mostly immature to low-mature. Results show that the maturity of the source rocks in the Liupanshan area is higher, most

reaching mature to high-mature stage. Therefore, the hydrocarbon potential of the Yan'an Formation in the Liupanshan area is very high and this formation deserves further exploration.

5. Acknowledgments

This work was financially supported by the China Geological Survey (No. 12120113039900).

6. References

[1] H. Zhang, The coal-bearing strata and coal accumulation of the Jurassic in Northwest China, Geological Publishing House, Beijing, 1988.
[2] S.M. Wang. Coal accumulation and resource evaluation of the Ordos Basin, Coal Industry Publishing House, 1996.
[3] S.P. Yao, J.R. Zhang, K.R. Wang, Q. Luan, Organic petrology study on the Yan.an Formation coal measures in Ordos Basin, Acta Sedimentolo. Sin. 17(1999) 291-3000.
[4] S.F. Liu, The coupling mechanism of basin and orogen in the western Ordos Basin and adjacent regions of China, J. Asian Earth Sci. 16(1998) 369-383.
[5] P. Guo. Reconstruction of South Ningxia-Liupanshan original basin during Yan'an Period, Northwest University, Xi'an, 2015.
[6] C.H. Li. Structure and hydrocarbon potential of the southwestern margin in the Liupanshan area of China, Journal of Chengdu University of Technology (Science & Technology Edition) 36(2009) 243-248.
[7] D.F. Li, Characteristics of Petroleum Geology and Prospecting Targets in Liuoanshan Basin. Xinjing J. Petrol. Geol. 22(2001) 27-31.
[8] J.P. Chen, D.F. Huang, J.J. Chen, The hydrocarbon generation and migration in the Jiudong Basin, Petroleum Industry Press, 1996.
[9] G.D. Liu, Petroleum geology, Petroleum Industry Press, 2010.
[10] L.Y. Wu, Quantitative evaluation of source rocks by rock pyrolysis, Science Press, Beijing, 1986.

IOP Conf. Series: Earth and Environmental Science **64** (2017) 012036 doi:10.1088/1755-1315/64/1/012036

New advances in oil and gas survey of the Qianjin depression in Sanjiang basin, north china

Wenhao Zhang[1], Shizhen Li[1], Xingui Zhou[1], Dandan Wang[1], Yanhua Lin[1], Weibin Liu[1], Yuanlin Meng[2], Ximing Zhou[3]

1Oil & Gas Survey, China Geological Survey, Beijing 100029, China
2 Northeast Petroleum University, Daqing 163328, Heilongjiang, China
3 No. 814 Geological Party, East China Geological &Mining Bureau for Nonferrous Metals in Jiangsu Province, Zhenjiang 212005, Jiangsu, China

Abstract. With more difficult of the development of oil and gas exploration in the large oil and gas bearing basins in Northeast China, the development level of old oil field has entered the late stage, and new oil and gas area is an urgent need to find. As one of the seven primary prospective basins in the periphery of the Songliao Basin, Sanjiang Basin has a great potential for oil and gas exploration. Through a series of oil and gas geological survey work carried out in Sanjiang Basin recently, we made some new progress in biogas resources, Late Paleozoic strata dark mudstone and the hydrocarbon potential of silica rock. These advances are significant to evaluate oil and gas potential resource in Northeast China.

1. Introduction

Sanjiang Basin is located in the Sanjiang plain in northeast of Heilongjiang Province, west from Jiamusi, east to the Wusuli River and Wanda Mountain, north of Heilong River and south of Shuangyashan watershed. It is the same basin with Middle Amur Basin in Russia, extends in the north-east direction with an area of 33730 km^2 in China (Fig. 1).

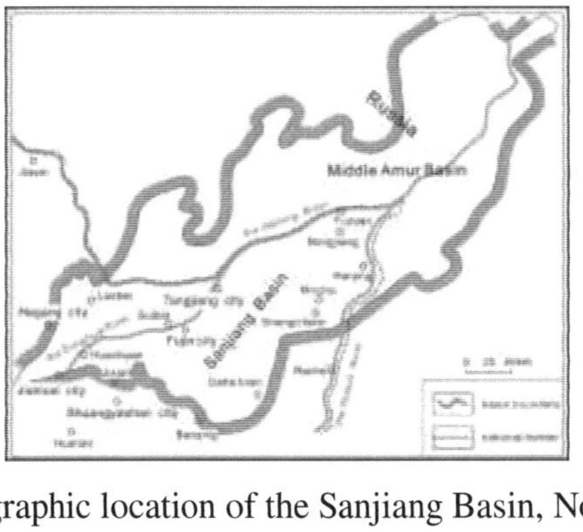

Figure 1 Geographic location of the Sanjiang Basin, Northeast China

The oil and gas exploration work in the Sanjiang Basin began in 1959, after years of unremitting exploration (Table 1), several oil and gas shows have been found in the basin and its surroundings, indicating the oil and gas generation and migration occurred in Sanjiang Basin(Hu et al., 2006; Liu et al., 2006; Wu et al., 2004, 2009).

Under the urgent need of strategy replace area for Chinese exploration area of gas and oil today, small and medium sized sedimentary basins in the periphery of Songliao Basin have become the important fields of oil and gas exploration. In recent years, a series of basic research and exploration work in the small and medium-sized basins being carried out, made some important oil and gas discoveries, such as new breakthroughs in oil and gas exploration in Zhangwu Basin, oil and gas discoveries of Tucan1 well in Tuquan Basin, Fang 4 and Fang 6 well in Fangzheng Rift, Yan 4 well and Yan 10 well in Yanji Basin, and these show that the small and medium-sized basins around the Songliao Basin have good prospect for exploration. Based on the application of the basic geological framework, basin simulation and oil and gas geological conditions in the basin, we sorted the oil and gas prospect of the outer basin of Songliao area, and Sanjiang Basin is one of the seven first-class prospect basins with the characteristics of large area, deep burial, well developed hydrocarbon source rocks and large amount of resources. The oil resources of Sanjiang Basin is 2.81×108t, showing the great oil and gas exploration potential.

Table 1 Statistical table of oil and gas exploration in Sanjiang Basin

Structural belt	Seismicexploration		Well drilling		Aeromagnetic survey (km)	MT (point)	Gravity prospecting (point)	Magnetic prospecting (point)
	Simulate (km)	Digitization (km)	Well	Total footage (m)				
Suibin depression		4202	4	7902.15		233	7566	
Fujin uplift							4111	3149
Qianjin depression	320.8	1529.3	7	15044.08	14359	89	27110	11418
Total	320.8	5731	8	21428.07	14359	322	34676	

2. Geological background

The Sanjiang Basin is located in the east of Heilongjiang Province in Northeast China. The boundary faults are the Yitong-Shulan and the Dunhua-Mishan fault zones (Fig. 2). The present structures belong to the Pacific Ocean tectonic domain. The multi-stage tectonic evolution determines the particularity and complexity of the evolution of the basic tectonic framework in Sanjiang Basin.

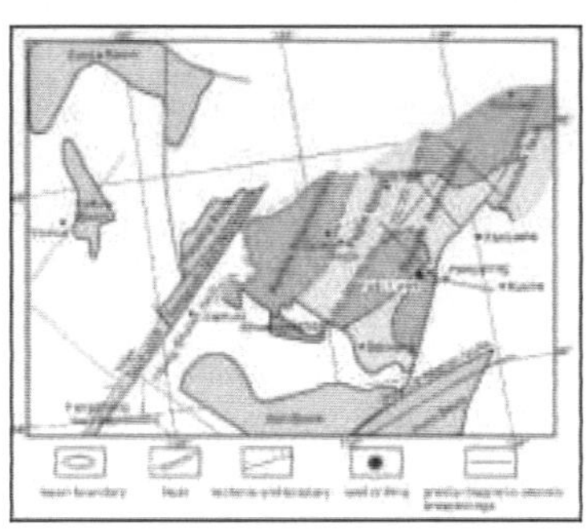

Figure 2 Division of tectonic units of Sanjiang Basin, Northeast China

Sanjiang Basin is a Meso-Cenozoic superimposed residual basin. The structural unit from west to east is Suibin depression, Fujin uplift, and Qianjin depression (Zhang and Ma, 2010; Liu et al., 2006; Fig. 2). The uplift and depressions are arranged in sequence with the characteristics of east-west zoning and the north-south blocking. Suibin depression in the west of the basin, with the area of 6440km^2, and mainly composed of Lower Proterozoic metamorphic rocks (Mashan Group) and intrusive granites, with the development of late Jurassic and early Cretaceous strata. The basement of Fujin uplift is the Paleozoic variscan folded rock series that are dominated by the Upper Paleozoic shallow metamorphic rocks and the Variscan granitic intrusion. The Qianjin depression (ca.22000km^2), between the Fujin uplift and the Dahezhen fault, is mainly composed of the Cenozoic sediments and Mesozoic strata.

3. Progress of Petroleum geology survey

3.1. Progress in the study ofMeso-Cenozoic mud shale

Strata developed in Qianjin depression of Sanjiang Basin from bottom to top is Lower Cretaceous Dongshan Formation, Upper Cretaceous Hailang Formation, Qixinghe Formation, Yanwo Formation, Tertiary Baoquanling Formation, Fujin Formation and Quaternary(Bi, 2011). The depth of "Well Fudi 1" in the Qianjin depression is 1518.16m, and its ending formation is Qixinghe Formation, to reveal the thick dark mudstone formed in Baoquanling Formation, Yanwo Formation and Qixinghe Formation(Fig. 2, Fig. 3).The mudstone thickness of Yanwo Formation is 115.34m, accounted for 34% of the formation. Qixinghe Formation developed carbonaceous mudstone, whose thickness up to 45m.

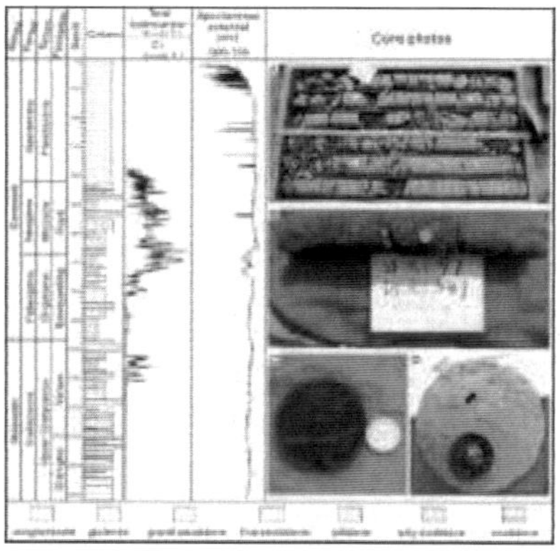

Figure 3 Integrated histogram of Fudi 1 well

During Fudi1 well drilling, we found continuousgas logging abnormal in Neogene Fujin Formation and Paleogene Baoquanling Formation, with characteristics that total hydrocarbon content more than 1%, dry gas, $C_1/(C_1\sim C_5)>0.9$, and small content of heavy hydrocarbon and non-hydrocarbon. Statistics can be seen, gas logging abnormal in total 220 m/15 layers, the highest total hydrocarbon content anomalies is 2.78% measured at 658m(Fig. 3). Through observing the corresponding cores of Fujin Formation and Baoquanling Formation, they are mainly composed of grayish black mudstone and carbonaceous mudstone which display gas logging abnormal, some of the core visible carbon chip particles(Fig. 3A-C). The organic geochemical test results show that gas source rock is the dark mudstone, carbonaceous mudstone and coal seam mainly(Table. 2). The TOC content of the dark mudstone ranging from 0.36% to 2.64%, good organic matter type(II 1- II 2), Ro value between 0.32%-0.33%, these reflect the source rock in the immature stage (Table. 2).

According to the specific characteristics of the gas logging, we know it has obvious biological characteristics, and biogas is an important unconventional gas reservoir now, generation and accumulation on a large scale(Lin et al., 1997, 2007). Research and exploration results indicate that the formation of biogas is mainly affected or controlled by temperature, reduction-strong reduction environment, salinity, pH value of water medium, type and abundance of organic matter, rapid deposition and burial, good cap rock, favorable traps and many other chemical and geological conditions, and the favorable combination of these factors can contribute to the formation of rich biogas resources (Shurr and Ridgley, 2002).

Table 2 Fudi 1 well core geochemical indicators

Sample Number	Stratum	Lithology	TOC/%	S_1+S_2/ mg·g-1	Tmax/℃	Ro/%	Abundance of organic matter	kerogen type
FD1-449	Fujin Formation	mudstone	0.47	0.43	428	0.32	poor	II_1
FD1-457	Fujin Formation	mudstone	2.64	3.03	437	0.34	high	II_1
FD1-547	Fujin Formation	coal	49.21	146.80	371	0.33	high	II_2
FD1-915	Baoquanling Formation	carbonaceous mudstone	31.59	144.88	419	0.33	high	II_1
FD1-1187	Yanwo Formation	mudstone	1.11	0.84	434	0.35	high	II_1
FD1-1309	Qixinghe Formation	mudstone	2.07	4.63	426	0.35	high	II_2
FD1-1317	Qixinghe Formation	carbonaceous mudstone	19.02	48.17	416	0.37	high	II_2
FD1-1326	Qixinghe Formation	coal	49.73	173.65	418	0.35	high	II_2
FD1-1345	QixingheFormatin	coal	50.59	169.04	417	0.36	high	II_2
FD1-1368	Qixinghe Formation	carbonaceous mudstone	34.61	114.52	419	0.37	high	II_2
FD1-1387	Qixinghe Formation	carbonaceous mudstone	38.47	115.08	418	0.37	high	II_2
FD1-1429	Qixinghe Formation	coal	43.81	207.84	411	0.39	high	II_2
FD1-1475	Qixinghe Formation	mudstone	0.36	0.59	449	0.36	not	II_2

The Sanjiang Basin has favorable conditions for the formation of biogenic gas reservoirs, such as cold weather in favour of the preservation of organic matter, if the surface temperature is high, large number of organic matter may be oxidized by bacteria degradation in oxidation zone to no formation of biogas; Quaternary lacustrine clay layer as cap rock is widely distributed, continuous and thick, conducive to the preservation of biological gas; black, dark gray mudstone layer and coal in Neogene Fujin Formation and Paleogene Baoquanling Formation, which can be used as good gas producing strata; well-developed reservoir stratum in Quaternary and Tertiary are favorable for biogas accumulation.

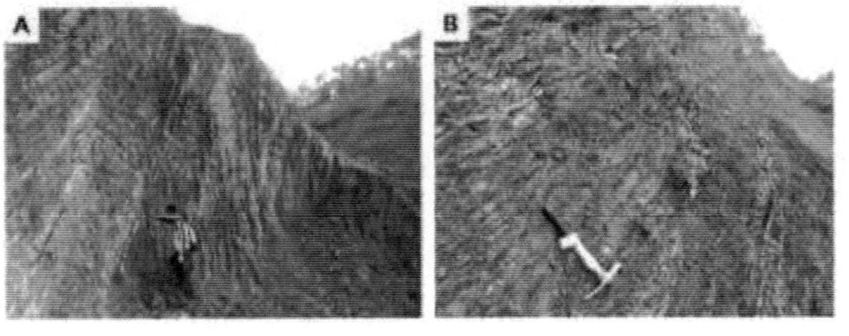

Figure 4 Black shale with sandstone of Zhenzishan Formation

A, Carbonaceous mudstone and coal seam of Zhenzishan Formation, Baolong coal mine; B, Black mudstone with thin siltstone of Zhenzishan Formation on the east side of highway 308, Baoqing country

3.2. Progress in the study of Paleozoic mud shale

With the continuous discovery of large and medium-sized oil and gas fields in the Paleozoic strata, Paleozoic oil and gas resources have become one of the important successor areas of oil and gas exploration in China (Kang, 2006). The Late Paleozoic stratum in northeast China has long been regarded as the Hercynian fold belt or the collisional orogenic belt, a metamorphic base of the Mesozoic and Cenozoic basins, so it was a restricted area for oil and gas exploration before. But recent studies have indicate that Late Paleozoic stratumin the northeastern region is the first caprock with regional distribution on the composite block basement, to a new stratigraphy for oil and gas exploration in northeast China (Zhou et al., 2009). Through oil-gas exploration and evaluation of Paleozoic strata in Sanjiang Basin recently, we know Sanjiang basin develops Paleozoic, Mesozoic, Cenozoic and other hydrocarbon source layers. Paleozoic hydrocarbon source beds are mainly developed in the Devonian and Carboniferous-Permian. Mesozoic hydrocarbon source beds are developed in the Middle-Upper Jurassic system, the upper and lower Cretaceous system. Cenozoic source rocks are mainly developed in the Paleogene.

The upper Paleozoic stratum in Sanjiang Basin are exposed in Yanshan town east of Fujin city, Baoqing country in southern margin of the basin and other places in Heilongjiang province in different degrees, or was revealed by a few shallow wells. Among them, Heitai Formation (D1-2h) is a set of coastal-shallow sea terrestrial clastic-carbonate rock construction, and the source rocks are marine limestone, bio-limestone and mudstone. Zhenzishan Formation (C2P1z) which is mainly distributed in Fujin, Mishan and Baoqing areas in Heilongjiang province, is a set of normal continental clastic sedimentary rock with coal seams and a small amount of tuffaceous slate assemblage (Fig. 4). The cumulative thickness of dark mudstone of Zhenzishan Formation in Baoqing area is 56.34m, with 21.5m thick coal seam, and the TOC of dark mudstone is 2.88-3.87%; vitrinitere flectance (Ro) is 2.19-4.15%; its organic matter type is III, small amounts of II 2. The TOC of dark mudstone of Zhenzishan Formation in Mishan is 0.57-1.83%, vitrinitere flectance (Ro) is 1.12-1.28%.

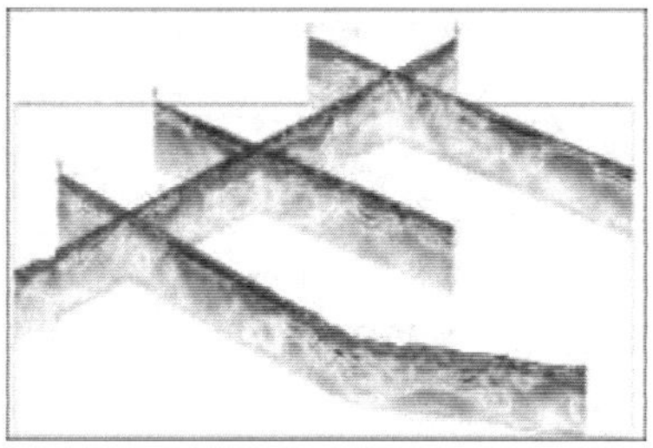

Figure 5 Three-dimensional view of electricity residual section

In order to initially determine the depth and fluctuation characteristics of Paleozoic strata in Sanjiang Basin, to further explore the distribution of Paleozoic mudstone, Oil & Gas Survey of China Geological Survey deployed gravity-magnetic-electric prospectings engineering which total 415km in 2015 (Fig. 2). Through the systematic research of petrophysical properties, include the stratum and different kinds of rock mass, we analyse the difference of physicalproperties among the responses of gravity, magnetic and the electrical, then, to establish the internal contact between these anomaly and physical properties. In order to separation of gravity and magnetic source, we use various methods include the upward continuation, tendency analysis and wavelet transform to extract the exception information of deep-seated geological mass, tectonic framework and distribution, speculated that strata and rock mass distribution; Meanwhile using the residual to extract the electrical abnormal of mudstone; Based on electrical layers, build the geological and geophysical model, then carries on the forward and inverse modeling fitting calculation and integrated geological explanation. At last, we determine the distribution and buried depth of the Paleozoic pluton and mudstone in section line.

We are taking four gravity-magnetic-electric profiles for the skeleton (Fig. 2), while using the results of the previous electrical profile and combining with gravity anomaly, can initially determine that the Paleozoic mudstone in this area is mainly developed in the middle-upper part of the Zhenzishan

Formation of Carboniferous-Permian system. Figure 5 is the three-dimensional perspective of residual method sections of this area, and the low value electrical anomaly in the middle-lower part of the picture can roughly reflect the distribution of mudstone in the Carboniferous-Permian strata in this area. From the plane point of view, Paleozoic mudstone mainly distributed in the northwest, northeast and eastern regions of the study area (Fig. 6). The distribution of Paleozoic mudstone has a tendency to further expand in the northwest, northeast and east direction. Due to the multi-period magmatic intrusion in the middle and south of the study area, the distribution of mudstone in the Paleozoic is relatively limited. The distribution area of Paleozoic mudstone in the study area is about 4900km2. The thickness of Paleozoic mudstone is generally 200-800m, the maximum of which can reach about 1100m in the northern part of the Qianjin farm, the thickness in the southwestern part of the area is about 1000m. Therefore, according to the results of gravity-magnetic-electric prospectings engineering, the strata of Zhenzishan Formation is widely distributed in the middle and eastern part of Sanjiang Basin, and the dark mudstone is developed well, which is the new stratum of deep oil and gas exploration in this area.

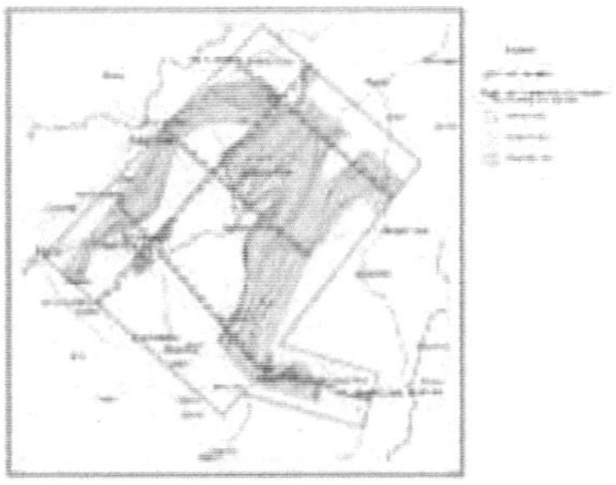

Figure 6 Distribution and thickness of Paleozoic mudstone in the Qianjin depression

3.3. Progress in the study of siliceous rocks

The distribution of oil and gas in siliceous rocks is very extensive in the world, but the hydrocarbon generating potential of siliceous rocks needs to be further studied. First, this stratigraphy rich in micro-fossil fossils with siliceous shells (such as Radiolaria, diatoms and sponges) is often excellent hydrocarbon source rocks in petroliferous basins; Secondly, because the siliceous rocks are dense, hard and fragile, it is often in the development of cracks, which can be used as a good oil and gas reservoir. Therefore, there is a significant relationship between some siliceous rocks and oil/gas fields in spatial and temporal distribution (Lu et al., 2011). At present, oil and gas field from siliceous mudstone have been found in the world, such as the Caspian Sea Basin, the Volga-Ural Basin and the Timan–PechoraBasin in Russia, which are top oil fields in the world currently under development. However, there is no report on this kind of oil and gas fields in China, so it is necessary to explore the hydrocarbonpotential of siliceous rocks in Sanjiang Basin.

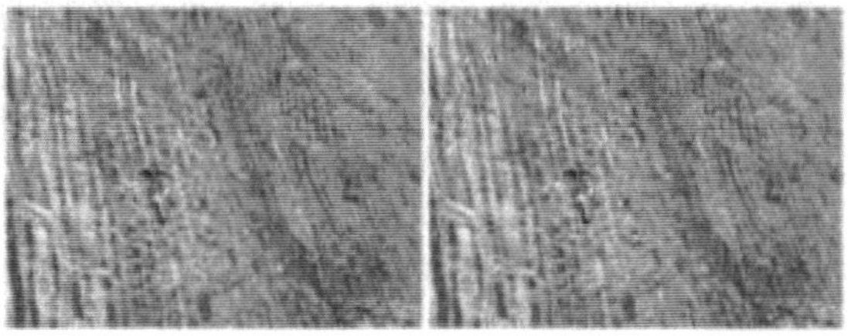

Figure 7 Intercalation of siliceous rocks and mudstone in Wandashan complex

The siliceous rocks of late Triassic-early Jurassic are extensively developed in the Fuyuan and Wandashan area (Fig.7). In order to further evaluate the hydrocarbon generation potential of siliceous rocks in Sanjiang area, we collected samples from Fuyuan, Raohe, Hongqiling three regions in Sanjiang

Basin and did some geochemical tests(Fig.2). The results of the identification of the whole rock XRD (X-raydiffraction) show that in the Fuyuan area of Sanjiang basin the siliceousrock(quartz content> 50%) and siliceous shale(clay content> 50%) are developed in the Dajiashan Formation (T3-J1d). The brittle mineral content in siliceous rocks is mainly distributed in 59-95%, with an average value of 75%; siliceous shale has a brittle mineral content of 42%, both satisfying the conditions as a shale gas reservoir. In Raohe and Hongqiling areas, the siliceous rocks (quartz content> 50%) are developed in the Dajiashan Formation (T_3-J_{1d}), and the brittle mineral content ranged from 75.6% to 82.5%, with an average value of 79.4%.

Test results of organic geochemical indicat, the organic carbon content of siliceous rocks is between 0.09-1.89% with the average value of 0.45%, and the organic carbon content of siliceous rocks in Fuyuan area is high with the average value is 0.64%, which belongs to medium source rock; the hydrocarbon generation potential (S_1+S_2) is between 0.01mg/g and 1.55 mg/g with the average content of 0.16 mg/g; organic type of silicalite is mainly type III, a small part of type II 1. The maturity of organic matter (Ro) is 2.06%-6.26%withthe average content of 4.67%; the Tmax value ranges from 451°C to 566 °C with an average value of 480 °C, and the average crystallinity of illite is 0.39, so siliceous rocks are clearly at high and overmature stages(Table 3). Research Institute of Exploration and Development of Daqing Oilfield Company Ltd. has also carried out the organic carbon content test of the radiolariansilicalite from Wandashan area, the organic carbon content is generally between 0.5%-0.6%.

Table 3 Organic geochemical characteristics of siliceous source rock inRaohe, Fuyuan, Hongqiling area

Region	Stratum	Lithology	TOC/%	S1+S2 /mg·g-1	Chloroform asphalt "A"	Tmax/℃	Ro/%	Abundance of organic matter
Raohe	T3-J1d	Siliceous rock	0.09~0.81 0.43(14)	0.01~0.07 0.04(14)	0.0036-0.0127 0.0083(4)	450~556 484(14)	4.74-6.3 5.7(3)	poor
Fuyuan	J1d	Siliceous rock	0.07~1.89 0.64(5)	0.04~1.55 0.38(5)	0.0051~0.0067 0.0059(2)	446~492 463(5)	6.26	medium
Hongqiling	T3-J1d	Siliceous mudstone	0.06~0.55 0.28(5)	0.02~0.09 0.05(5)	0.0070(1)	460~525 493(5)	2.06	not

On the whole, the organic geochemical indicators of siliceous rocks in Sanjiang area are somewhat low, and of course, long-term weathering may lead to a decrease in some organic geochemical indicators. At present, it can't be used as an effective exploration layer in this area. But in view of the discovery of siliceous oil and gas fields in the world, the hydrocarbon generating potential of marine siliceous rocks in Sanjiang area is worth further exploration and research.

4. Prognosis Prospective Province

According to the gravity-magnetic-electric prospecting results, Zhenzishan Formation in the Paleozoic are widely distributed in the central and eastern Sanjiang Basin, especially in the north and east of the Qianjin depression, where the dark mudstone is developed and with good conditions for source rocks; the local tectonic form is faulted anticline, which is favorable for oil and gas reservoir; the upper Middle Cenozoic mudstone can be used as a good caprock, therefore, the Upper Paleozoic strata can be used as a new stratum for deep oil and gas exploration in this area. The northern and eastern parts of Qianjin despression in Sanjiang Basin are prospective area for deep Paleozoic oil and gas exploration. There are some cases of large oil and gas fields that have been discovered and mined in the siliceous rocks so far. The research and exploration of siliceous rocks in late Triassic-Early Jurassic, which extensively developed in Fuyuan and Wandashan area, can better evaluate its hydrocarbon generation potential.

5. Conclusions

(1) Thick dark mudstone is developed in Fujin and Baoquanling Formation in Qianjin depression of Sanjiang Basin, which has high organic matter content, good organic matter type and abundant gas

logging abnormal(220 meters/15 layers), and these are similar to the characteristics of biogas. As an important unconventional gas reservoir, biogas reservoir has the characteristics of wide distribution and shallow burial, and has been paid more and more attention. Sanjiang Basin has favorable conditions for the formation of biogenic gas reservoirs, which is worthy of further exploration and exploration.

(2) Late Paleozoic strata as a new layer of oil and gas exploration in Northeast China, is developed extensively in the eastern part of Sanjiang Basin. Dark mudstone of Paleozoic Zhenzishan Formation in Qianjin depression of Sanjiang Basin has a large thickness, high organic content, with the value of TOC is 2.88%-3.87%, the value of Ro is 2.19%-4.15%, and the organic matter is mainly ofIIItype. The results of latest high-precision gravity-magnetic-electric prospectings exploration show that the mudstone in Paleozoic Zhenzishan Formation in Qianjin depression of Sanjiang Basin has a wide distribution area, up to about 4900km2, thickness of 200-800m.

(3) As a special set of source rocks, a large suite of siliceous rocks of Late Triassic-early Jurassic in Fuyuan area of Sanjiang Basin and Wandashan area, which need to be systematically and comprehensively studied and explored to evaluate its hydrocarbon generation potential. Organic geochemical indicators show that the organic carbon content of siliceous rocks is between 0.09%, and 1.89%, with an average value of 0.45%; the hydrocarbon generation potential (S_1+S_2) is between 0.01mg/g -1.55 mg/g with an average value of 0.16 mg/g; the type of organic matter in is mainly III type, and a small part is II 1 type; the maturity of organic matter Ro is 2.06mg/g -6.26%,with an average value of 4.67%. In a word, although the organic geochemical indicators of siliceous rocks are low, in view of a number of large oil and gas fields have been found in this type of strata in world and the research is still very limited in China, it is necessary to intensify the research and exploration on siliceous rocks, and further determine its hydrocarbon potential.

Acknowledgements
The study was supported by the National Natural Science Foundation of China (No. 41572135), the China Geological Survey (No. 121201021000150005), the Science Foundation of Heilongjiang (QC2016049) and University Nursing Program for Young Scholars with Creative Talents in Heilongjiang Province. Thanks are given to Mr. Jia Rui and Mr. Wang Peiye for their field assistance.

Reference

[1] Bi Dianhui. Research progress of petroleum geology in the Qianjin depression of Sanjiang Basin. Inner Mongolia Petrochemical Industry, 2011, 11: 134~135.
[2] Kang Yuzhu. Prospects of petroleum exploration in the Paleozoic of China. Journal of Geomechanics, 2006, 12(1): 1~5.
[3] Lin Chunming, Li Yanli, Qi Binwen. Research status and exploration potential of biogenic gas. Journal of Palaeogeography, 2006, 8(3): 317~330.
[4] Lin Chunming, Qian Yizhong. Gas source rock features and controlling factors for biogas formation in Holocene Series in Zhejiang coastal plain. Acta Sedimentologica Sinica, 1997, 15(Suppl.): 70~75.
[5] Lu Fang, Liu Jiaqi, Li Yahui, et al. Types and characteristics of chert reservoirs in oversea petroliferous basins. Petroleum Exploration and Development, 2011, 38(5):628~636.
[6] Shurr G W, Ridgley J L. Unconventional shallow biogenic gas systems. AAPG Bulletin, 2002, 86(11): 1939~1969.
[7] Wu Heyong, Liu Wenlong. Evaluation and optimization of peripheral basins. Petroleum Geology & Oilfield Development in Daqing, 2004, 23(5): 20~22.
[8] Wu Heyong, Yang Jianguo, Wang Shihui, et al. Development of oil and gas exploration in the greater Sanjiang area. Petroleum Geology & Oilfield Development in Daqing, 2009, 28(5): 49~53.
[9] Zhang Xingzhou, Ma Zhihong. Evolution of Mesozoin-Cenozoicin the eastern Heilongjiang Province, Northeast China. Geology and Resources, 2010, 19(3): 191~196.

IOP Conf. Series: Earth and Environmental Science **64** (2017) 012036 doi:10.1088/1755-1315/64/1/012036

[10] Zhou Jianbo, Zhang Xingzhou, Ma Zhihong, et al. Tectonic framework and basin evolution of Northease China. Oil & Gas Geology, 2009, 30(5): 530~538.
[11] Qu G, Meng Y, Shen A, et al. Experimental study of the feasibility of air flooding in an ultra-low permeability reservoir. Energies, 2016, 9(10):1-13.

Genetic mechanism and stability assess of the Boai landslide in northeastern Yunnan province

Hao Wang*, **Kaijun Ni and Yanguo Fang**

99/Optics Valley Chuangye Street, East Lake High-tech Zone, Wuhan, Hubei, P. R. China

*Corresponding author e-mail: wanghao198241@163.com

Abstract. Field and drilling data shows that, the boai landslide, located in Funing country, northeast Yunnan province, has length of 800m and width of 250-570m, covering square of $40\times10^4 m^2$ and volume of $1600\times10^4 m^3$. On basis of investigation and experimental data, we illustrate the basic characteristics and formation progress of the Boai landslide, analyze its safety assess, and put forward some feasible treatment measures.

1. Introduction

The Boai landslide was located in Funing county, northeast Yunnan province. Affected by a long term heavy rain, ground surface began to deform since 2008, residential buildings and roads began to crack, then the Boai landslide breakout. According to the monitoring data, there were no signs to show that the deformation was slowing down. The Boai landslide was threatening the safety of 1400 lives and property of the local residents.

The Boai landslide goes to the narrow width of the funnel-shaped in the plane, which has length with 800m and width with 250-570m, covering square of $40\times10^4 m^2$ and volume of $1600\times10^4 m^3$. In this paper, we will systematically analyze geological characteristics and the basic features of the Boai landslide, discuss its causes and stability. Finally, based on the actual investigation, we propose the appropriate engineering control measures.

2. Geological background conditions

2.1. Topography

Regional geomorphology is a hilly region, with some basins. The symbiotic relationships between the slope and slow stage, slope and gully are the overall performance.

2.2. Lithology and structure

Exposed strata in the study area are the Quaternary and the Triassic rocks. Drilling data results showed that the Quaternary silty clay scattered in the surface of the landslide, with thickness of 2.0-10.0m. While the Triassic siltstone and clay distributed widely in the range of the landslide, with thickness of 4.0-58m.

Study area is located in Yunnan western edge of plate fold belt, the crustal movement rise slowly. The main territory faults are the Wenshan-Malipo and Napo faults, and some north-trending secondary activity fractures.

2.3. Hydrogeology
There are three surface gullies in study area, and its cutting depth varies (4-20m). The trend of gullies is 100-130°, and perennial water is found in these gullies.

Rainfall infiltration, town water supply and drainage and mountain side recharge are the main source of supply of the landslide groundwater [1]. Groundwater is mainly bedrock fissure water and pore water. Most of the former is diving water, the latter overflow from low lying areas as springs due to water effect of slip zone.

3. Characteristics of the Boai landslide

3.1. Spatial distribution
The Boai landslide is the funnel-shaped in planar (Fig.1), the length is 800m and the width is 250-570m, with distribution square of $40\times10^4 m^2$ and volume of $1600\times10^4 m^3$. The main slip direction is 115°, average thickness of landslide is 40m, and leading edge was thicker (64m).

Figure 1. The panorama image of the Boai landslide

3.2. Material component
Drilling sonic test results indicate that the slider longitudinal wave velocity is in the range of 1.61-2.98 km/s, with large amplitude, suggesting landslide composition is extremely uneven and significant overhead localized. From bottom to up, the material component is gravel soil with shale stone, siltstone fragments of stone, mudstone with gravels and silty clay with gravels.

3.3. Slip zone and slip bed
Slip zone soil consists of clay minerals and debris, which have high active minerals content (illite), indicating that slip zone soil has a certain expansion (Fig. 2), it will be a water-swellable, and intensity decrease significantly.

IOP Conf. Series: Earth and Environmental Science **64** (2017) 012037 doi:10.1088/1755-1315/64/1/012037

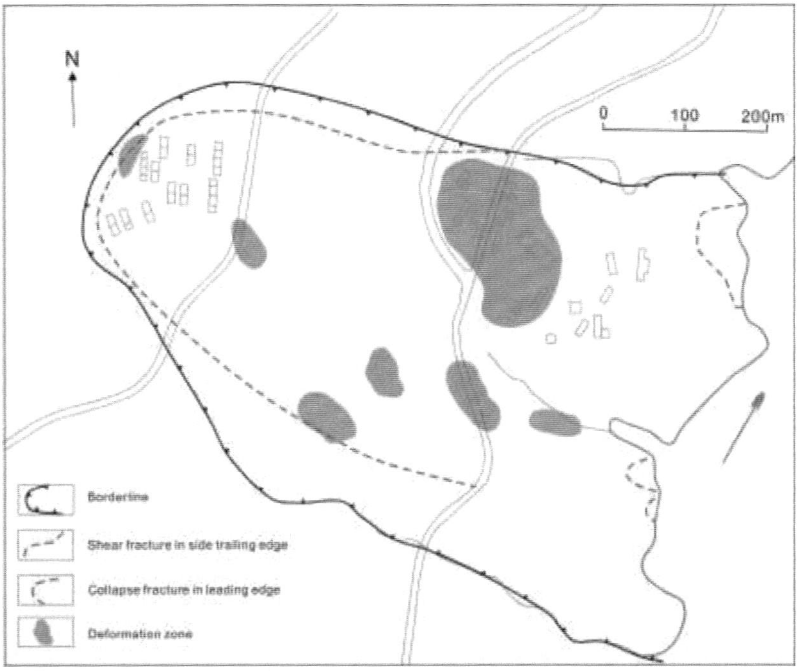

Figure 2. Schematic diagram of displacement and deformation of the Boai landslide

The bedrocks in slip bed are the Triassic mudstone and siltstone, dark gray, medium-thick layer. The lithology is weak, develops the interlinear shear zone. Longitudinally, trailing edge is steeper (20°), front area is gradual (5-7°), while leading edge is plane and reverse rock-layer (-3- -5°). Laterally, it presents an asymmetric groove.

4. Discussions

4.1. Genetic mechanism

Firstly, slippery rocks provide the material basis for the formation of the landslide [2]. The combination of the exposed strata in the landslide area is hard in upper and soft in bottom, the upper part is mainly fine sandstone and siltstone, while the lower part is weak mudstone. And the interlayer shear zone develops due to effect of structure, all the causes constitute the typical slippery layer. Secondly, monoclinic, gently inclined and forward slope geological structure forms unfavourable geological structure. Thirdly, occurrence of rock-layers is 150°-170°∠17°-25°, while the general trend of slope is 20°-25°. Moreover, fractures and terrain cutting provide favourable condition to form airport and landslide boundary.

Therefore, we indicate that the formation of the Boai landslide is as following: (1) the landslide was located in monoclinic structure, apparent slope position. As consequent erosion and cutting of the Jiacun river, the upper hard rocks had been cut, the soft layer rocks exposed (Fig. 3a). (2) Under the long-term weight action of the slope body, the through slip surface in the landslide formed progressively (Fig. 3b). (3) With the erosion of the Jiacun river, the airport surface of the accumulation front was increasing (Fig. 3c). (4) Many factors including heavy rainfall infiltration, urban water supply and drainage system leakage and reservoir pressure, made the landslide deformation and displacement, and outbreak in 2008 (Fig. 3d).

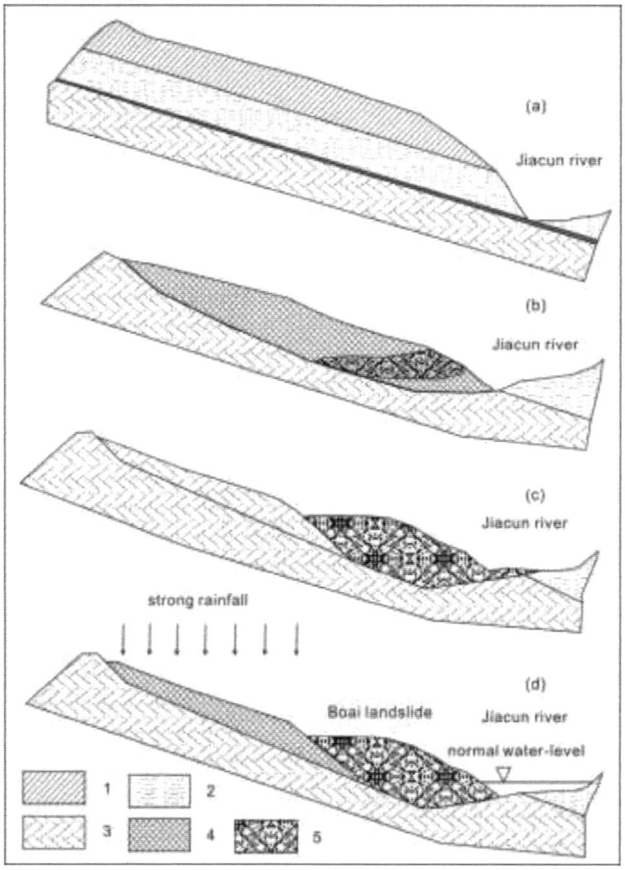

Figure 3. Cartoons of the Boai landslide formation mechanism (1-sandstone, 2-mudstone, 3-sility clay, 4-alluvial materials, 5-slope accumulation)

4.2. Stability calculation

The rigid body limit equilibrium method has been frequently used for stability calculation of landslide [3, 4]. In this paper, we will adopt the unbalance method or transfer coefficient method [5] to evaluate the stability of the Boai landslide, with the concrete calculation formula as follows:

$$F_S = \frac{\sum_{i=1}^{n-1}(Ri\prod_{j=i}^{n-1}\psi_j) + Rn}{\sum_{i=1}^{n-1}(Ti\prod_{j=i}^{n-1}\psi_j) + Tn}$$

Based on the results mentioned above, we inverse the gravity calculation of landslide, the results are listed in Table 1.

Table 1. Collected gravity values of inversion calculation from the Boai landslide

sequence	lithology	Native / saturated wet
V	silty clay	19.5/21.0 11.0
III	mudstone	20.5/22.0 12.0
II	siltstone fragments of stone	22.0/23.0 13.0
I	gravel soil	22.0/23.5 13.5

We select a section (Z2-2′) to calculate the stability coefficient (F_s) under different working conditions (weight, weight and rain, and weight and earthquake). The result shows that, the F_s values under different working conditions are 1.00, 1.05 and 1.09, respectively. So, we consider the Boai landslide is in unstable to under-stable, and it will breakout once heavy rain encountered.

5. Engineering treatment

According to some special factors, including the sliding body thickness of leading edge, steep slope and, long and gentle sliding bed and reverse rock-layer, we believe that the leading edge of the landslide is the key to prevent the overall deformation and instability. Therefore, it is necessary to carry out the backfill and surface drainage engineering, meanwhile, anti-slide retaining projects should be implemented in middle parts.

Acknowledgments

This work was financially supported by project of the Jinsha River geological disasters survey (K139E101) fund.

References

[1] Rahardjo, H., Li, X. W., Toll, D. G. The effect of antecedent rainfall on slope stability. Geotechnicaland Geological Engineering, 19 (2001) 371-399.

[2] Peng, Y. L, Guo, D. L., Hu, X. W., Gu, C. Z., Zhou, J. Characteristics and stability analysis of a landslide in Kuancheng country, Hebei province. Journal of Mountain Science, 29 (2011) 591-597. (In Chinese)

[3] Specification of geological investigation for landslide stabilization. DZ/T0218 -2006. Chinese Standard Press, Beijing, 2006.

[4] Zhang, X. Z., He, J. D. The application and compare of the rigid limit balance method and finite element approach in the slope stability analysis. Jilin Water Research, 8 (2007) 32-35. (In Chinese)

[5] Code for investigation of geotechnical engineering. GB50021-2001, Beijing, 2001.

IOP Conf. Series: Earth and Environmental Science **64** (2017) 012038 doi:10.1088/1755-1315/64/1/012038

Study on microwave induced pyrolysis of low metamorphic coal and liquefaction residue

Zhao Li, Kunyao Wu, Jing Cao, Yongfeng Wang

School of Materials Engineering, Xi'an Aeronautical University, Xian 710077 China

Abstract. This paper mainly studies pyrolysis characteristic in the different heating of microwave of low metamorphic coal and Liquefaction Residue, which was focused on the effects of particle sizes, pyrolysis reaction time, and microwave powers. The product are analyzed by Gas chromatography- mass spectrometry (GC-MS), etc. The results showed that when heating time is 40 min and 800 W, yield of tar is about 15.51%.

1. Introduction

China's energy is featured with large quantity of coal but small quantity of oil and gas. With rapid development of China's economy, there is an enormous demand for energy especially for petroleum. Import of crude oil will not only increase our reliance on foreign energy resources, but also will directly affect safety of our economic development and energy supply. Therefore, the coal-to-oil project is of vital importance to development of our national economy. Low metamorphic coal accounts for an extremely large proportion in China's coal resources. The technology of retorting low metamorphic coal in Northern Shaanxi, which prioritizes comprehensive utility rate of resources, is of great significance to quality improvement and comprehensive process & utility of low metamorphic coal in Northern Shaanxi. At the same time, it can reduce environmental pollution. So in recent years, it has received wide attention as a special technology in the field of clean coal technology.

Production of tar and combustible gases by coal pyrolysis has evoked interest in science researchers. Microwave heating, as a volumetric heating, can turn microwave energy absorbed by materials in the electromagnetic field into thermal energy. As compared with conventional heating, microwave heating has the following advantages: instant and pollution-free. As an all-new heating method, microwave heating has been widely used in many fields. Huang Xin, etc examined effect of pyrolysis temperature on occurrence mode of nitrogen in pyrolysis products by rapid pyrolysis experiment towards sewage sludge of Xuzhou in the drop-bed quartz tube reactor. Liu Hongzhen [12] studied pyrolysis characteristics of wheat straws in a microwave field. Based on the preliminary studies, this paper analyzed and characterized low metamorphic coal and tar, gases and semi coke produced by microwave pyrolysis of liquefaction residuals by adopting the chromatograph-mas spectrometer (GC-MS) and gas analyzer in order to explore the affecting rules of microwaves on coal pyrolysis & destructive distillation.

2. Experiment

2.1. Experiment Materials & Equipments

The materials in the experiment are residues directly liquefied from coal supplied by National Institute of Clean-and-low-carbon Energy and Shendong coal. Industrial analysis and element analysis of coal and liquefaction residuals followed GB/T 2001-1991 *Analysis and Determination Method in Coking Industry* and GB/T 214-2007 *Method of Determining Total Sulfur in Coal*. Contents of C, H, N, S and M_{ad}, A_{ad}, V_{ad} are taken as averages of two or three parallel samples. The content of FC_{ad} is calculated by minusing. The detailed formula is as follows:

Table 1 shows industrial and element analyses on coal and liquefaction residuals.

raw material	Industrial analyses				element analyses			
	M_{ad}	A_{ad}	FC_{ad}	V_{ad}	C_{ad}	H_{ad}	N_{ad}	$S_{t,ad}$
SDM	4.64	7.26	54.92	33.18	70.71	4.32	0.84	0.22
Liquefaction Residue	0.14	17.74	48.37	33.75	75.00	4.22	0.79	2.16

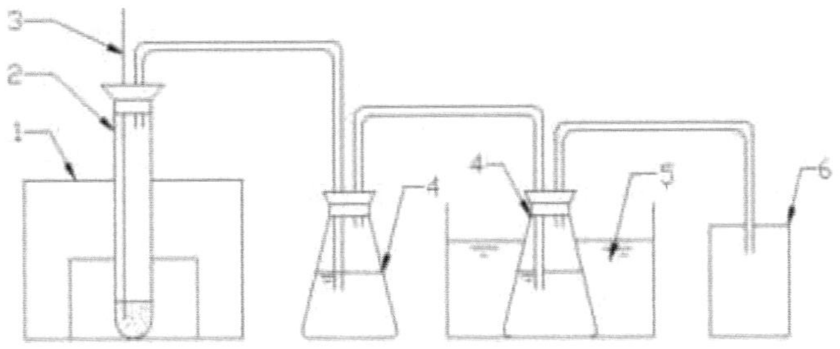

Figure 1 shows the experiment equipments.
1 – Microwave Oven; 2 – Quartz Tube; 3 – Thermocouple; 4 – Oil Washing Bottle; 5 – Cooling Liquid for Circulating Pump; 6 – Gas Collecting Bag.

2.2. Experiment Method

Under different microwave powers (320 w, 480 w, 640 w, 800 w & 960 w), condensation temperature 0 ℃ and pyrolysis durations (10 min, 20 min, 30 min, 40 min & 50 min), microwave pyrolysis experiments are performed to 50 g of mixture of coal and liquefaction residuals in different proportions (9:1, 8:2, 7:3, 6:4, 5:5) and yields, compositions and properties of the products are detected and analyzed. Industrial and element analyses of coal samples were performed by Shaanxi Energy Institute. Contents of different gases in coal gas were measured with the Gasboard-3100P portable six-component gas analyzer developed by Wuhan Sifang Scientific & Technological Co., Ltd. Constituent analysis of coal tar was done by GCMS-QP2010 Plus gas chromatograph-mass spectrometer of Daojin Company.

3. Results & Discussion

3.1. XRD Analysis of Liquefaction Residuals

Industrial and element analyses on coal and liquefaction residuals show that the residuals are featured with high ash and low moisture. Reason for high ash: Minerals in the coal, including liquefaction catalysts, will concentrate in the residues after organic matters in the coal are liquefied. Reason for high volatile: There are large quantities of heavy oil and asphalting in the residues. Reason for low moisture: The distillation temperature in the solid-liquid separation process of the coal direct liquefaction process is about 425℃, so the liquefaction residuals contain less water than the raw coal. Due to use of sulphurous catalysts and promotors in the liquefaction process, the sulphur content of the residues is far higher than that of the raw coal. Through analysis on Figure 2 "XRD Diagram of Liquefaction Residuals", we can see that inorganic components in the liquefaction residuals are mainly SiO_2, $CaCO_3$ and $Fe_{1-x}S$ (pyrrhotite), with contents 32.4%, 42.6% and 24.9% respectively.

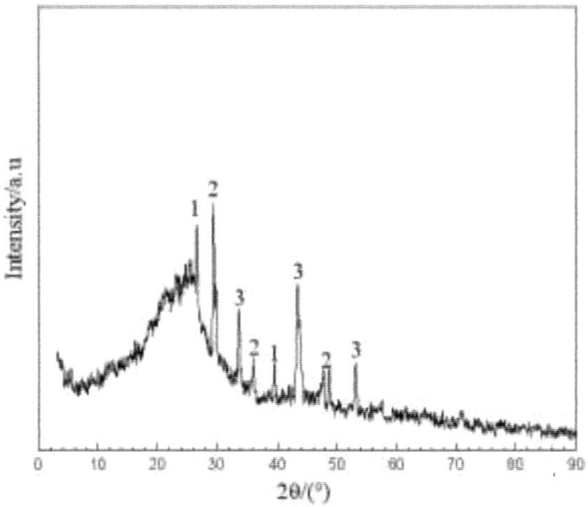

Figure 2 XRD pattern of Liquefaction Residuals

3.2. Analysis on Coal and Liquefaction Residuals Pyrolysis Experiment

Under microwave power 800 W, condensation temperature 0℃ and pyrolysis duration 40 min, effects of different proportions of Shendong coal and liquefaction residuals on products yields are investigated. Raw material granularity <1mm. The mixture of coal and liquefaction residuals weighs 50 g. Proportions of coal and liquefaction residuals are 9:1, 8:2, 7:3, 6:4 and 1:1 respectively. Table 2 shows yields of pyrolysis products. From it, we can see that when the proportion of Shendong coal and liquefaction residuals is 6:4, the yield of the pyrolysis oil reaches the maximum value 15.51%. The possible reason for this result is as follows: When the proportion of liquefaction residuals is excessively high, the liquefaction residuals will turn from a solid state into a liquid state in the low temperature section. Large quantities of liquefaction residuals will cover surface gaps of the low metamorphic coal, preventing separation of gases and pyrolysis oils from the coal.

Table 2 Product Yields (%) in Pyrolysis of Shendong Coal and Liquefaction Residuals in Different Proportions

Product	Coal and Liquefaction Residuals in Different Proportions				
	9:1	8:2	7:3	6:4	1:1
coal	83.84	69.47	68.44	69.63	69.60
oil	7.21	12.61	13.21	15.51	12.21
gas	8.95	17.92	13.35	14.86	18.19

Table 3 shows product yields (%) in pyrolysis of Shendong coal and liquefaction residuals in the proportion of 6:4 under different microwave powers. From it, we can see that when the microwave power for pyrolysis of Shendong coal and liquefaction residuals is 800w, the yield of the pyrolysis oil reaches as high as 15.51%. When the microwave power continues to rise, the amplification of the tar yield slightly decreases. So the optimal microwave power is 800w. Microwave power mainly affects temperature rising rate of the pyrolysis process. The higher the power, the higher the temperature rising rate. Volatiles can be separated out instantly. So the gaseous products have a short retention time in the reactor, leading to low incidences of secondary reactions.

Table 3 Product Yields (%) in Pyrolysis of Shendong Coal and Liquefaction Residuals in the Proportion of 6:4 under Different Microwave Powers.

Product	Powers(W)				
	320	480	640	800	960
coal	92.51	90.62	68.61	69.63	67.90
oil	1.22	5.31	12.24	15.51	15.72
gas	6.27	4.07	19.15	14.86	16.38

Under proportion of Shendong coal and liquefaction residuals 6:4, microwave power 800 W, pyrolysis durations 10min, 20min, 30min, 40min & 50min, effects of different microwave pyrolysis durations on products yields are investigated. The experiment result is shown in Table 4.

Table 4 Product Yields (%) of Pyrolysis of Shendong Coal and Liquefaction Residuals in the Proportion of 6:4 under Different Heating Durations (%)

Product	Time (min)				
	10	20	30	40	50
coal	91.6	72.01	68.40	69.63	67.90
oil	1.01	10.61	12.01	15.51	15.32
gas	7.39	17.38	19.59	14.86	16.78

From Table 4, we can see that as heating duration increases, the yield of pyrolysis oil gradually increases and the yield of solid coke gradually decreases. When the heating duration is 40min, the yield of pyrolysis oil reaches as high as 15.51%. However, if the microwave pyrolysis duration continues to be prolonged, the increase of the yield of pyrolysis oil is not obvious. So the optimal heating duration is 40min.

4. Conclusion

From the experiment on the mixture of Shendong coal and liquefaction residuals in microwave pyrolysis and in conventional pyrolysis, we can conclude the following: The pyrolysis effect is optimal when the microwave power is 800W and the microwave duration is 40min. When liquefaction residuals account for 40%, the pyrolysis oil yield in microwave pyrolysis of Shendong coal and liquefaction residuals can reach the maximum value 15.51%, which is 2.28% higher than that in conventional pyrolysis. In microwave pyrolysis as compared in conventional pyrolysis, there are more holes with greater diameters, showing that microwave pyrolysis is more adequate.

References

[1] Jue Huang, Xilan Lu, Dexiang Zhang et al.A Study on the Hydrotreating of Coal Hydroliquefaction Residue and its Kinetics [J]. Energies, 2010, 3(9): 1576-1585.
[2] Jun Li, Jianli Yang, Zhenyu Liu.Hydro-treatment of a direct coal liquefaction residue and its components [J]. Catalysis Today, 2008, 130(2-4): 389–394.
[3] Tao-Xia Wang, Zhi-Min Zong, Jia-Wei Zhang et al.Microwave-assisted hydroconversions of demineralized coal liquefaction residues from Shenfu and Shengli coals [J]. Fuel, 2008, 87(4-5): 498-507.
[4] Robin A.M, Yang H.L. Gasification of Residual Materials from Coal Liquefaction [R]. Fossil Energy: Energy and Development Administration, 1977.
[5] Nie Yi, Bai Lu, Li Yi et al.Study on Extraction Asphaltenes from Direct Coal Liquefaction Residue with Ionic Liquids[J].industrial and engineering Chemistry Research, 2011, 50(17): 10278-10282.
[6] Zhang Xiangping, Lai Shiyao et al.Ionic liquids to extract valuable components from direct coal

liquefaction residues [J]. Fuel, 2012, 94: 617-619.
[7] Li Yi, Xiangping Zhang, Heifeng et al. Efficient extraction of direct coal liquefaction residue with the [bmim] Cl/NMP mixed solvent [J]. RSC Advances, 2011, 1(8):1579-1584.

Occupational hazard evaluation model underground coal mine based on unascertained measurement theory

Quanlong Deng, Zhongan Jiang*, Yaru Sun and Ya Peng

School of Civil and Resource Engineering, University of Science and Technology Beijing, Beijing 100083, China

*Corresponding author e-mail: jza1963@263.net

Abstract. In order to study how to comprehensively evaluate the influence of several occupational hazard on miners' physical and mental health, based on unascertained measurement theory, occupational hazard evaluation indicator system was established to make quantitative and qualitative analysis. Determining every indicator weight by information entropy and estimating the occupational hazard level by credible degree recognition criteria, the evaluation model was programmed by Visual Basic, applying the evaluation model to occupational hazard comprehensive evaluation of six posts under a coal mine, and the occupational hazard degree was graded, the evaluation results are consistent with actual situation. The results show that dust and noise is most obvious among the coal mine occupational hazard factors. Excavation face support workers are most affected, secondly, heading machine drivers, coal cutter drivers, coalface move support workers, the occupational hazard degree of these four types workers is Ⅱ mild level. The occupational hazard degree of ventilation workers and safety inspection workers is Ⅰ level. The evaluation model could evaluate underground coal mine objectively and accurately, and can be employed to the actual engineering.

1. Introduction

Coal mine occupational hazards have always been the focus of occupational disease prevention and control[1], underground coal mining with limited space, narrow vision, long air replacement cycle characteristics, there are a variety of occupational hazards[2,3] that will bring serious harm to miners' health. The study of how to evaluate the occupational hazards is the primary work of occupational hazards control. Many scholars have done some research on mine occupational hazards assessment. For example, Guo J.P. et al [4] combined the binary semantics and analytic hierarchy process (AHP), and applied it to the mine occupational hazards risk assessment. Based on Fuzzy Mathematics Method, Wang X.N. et al [5] constructed multi-level evaluation model for underground mine occupational hazards. Li H. et al [6] combined the analytic hierarchy process with the set pair analysis to conduct a comprehensive evaluation of coal mine occupational hazards. Based on fuzzy hierarchy method combining with the TOPSIS method, Xiong L.X. et al [7] proposed an identification method of occupational hazard degree in deep well. How to scientifically qualitatively and quantitatively analyze the uncertainties of occupational hazard in coal mine is the focus of evaluation, unascertained measurement theory provides the appropriate approach. Unascertained measurement theory has been widely developed in the last decade, especially in the field of risk assessment, such as coal mine safety [8], dumping landslides [9], gob collapse [10], sulfide spontaneous combustion [11] and so on. Based on the unascertained measurement theory and the coal mine actual situation, this paper establishes the evaluation model of occupational hazards, and implements the proceduralization through programming technology, and finally applies it to the occupational hazards evaluation of six posts in the actual coal mine.

2. Unascertained Measurement Theory

Set the number of evaluation object X to m and use $X_1, X_2, \ldots, X_m$ to represent them, so the evaluation object space $X=\{X_1, X_2, \ldots, X_m\}$. Each Evaluation object $X_i(i=1, 2, 3, \ldots, m)$ all has n single evaluation index and $I_1, I_2, \ldots, I_n$ are used to represent them, so the evaluation index space $I_j=\{I_1, I_2, \ldots, I_n\}$. Suppose x_{ij} represent the measured value of the ith evaluation object $X_i(i=1, 2, \ldots, m)$ about the jth evaluation index I_j $(j=1, 2, \ldots, n)$, so $X_i=\{x_{i1}, x_{i2}, x_{i3}, \ldots, x_{in}\}$. If x_{ij} has p evaluation levels $C_1, C_2, \ldots, C_p$, use U to represent evaluation space, so $U=\{C_1, C_2, \ldots, C_p\}$. Set the harm degree of k lower than $k+1$ which is recorded as $C_k<C_{k+1}$, so $C_1<C_2<\ldots<C_p$. $\{C_1, C_2, \ldots, C_p\}$ is an ordered partition on the evaluation space U[12,13].

2.1. *Single parameter unascertained measure and its matrix*

If $u_{ijk}=u$ is the measured value which represents the kth evaluation level of x_{ij} and u conforms to formula (1)~(3), then u is called single unascertained measure, namely, single index measure.

$$0 \le u(x_{ij} \in C_k) \le 1 \tag{1}$$

$$u(x_{ij} \in U) = 1 \tag{2}$$

$$u\left| x_{ij} \in \bigcup_{l=1}^{k} c_l \right| = \sum_{l=1}^{k} u\left(x_{ij} \in c_l \right) \tag{3}$$

Among them: the formula (1) indicates that index measure is non-negative bounded, the formula (2) indicates that the index measure has the normalization in the rank space and the formula (3) is the accumulation of the level space of u and in the formula $i=1, 2,\ldots, m;\ j=1, 2, \ldots, n;\ k=1, 2, \ldots, p$. A certain index measure value of X_i is u_{ijk} and the matrix $(u_{ijk})_{n\times p}$ is composed of all measure values u_{ijk}, which is called single index measure matrix. The matrix is shown as formula (4).

$$(u_{ijk})_{n\times p} = \begin{bmatrix} u_{i11} & u_{i12} & \cdots & u_{i1p} \\ u_{i21} & u_{i22} & \cdots & u_{i2p} \\ \vdots & \vdots & \ddots & \vdots \\ u_{in1} & u_{in2} & \cdots & u_{inp} \end{bmatrix} \tag{4}$$

2.2. *Unascertained measure function*

In order to get unascertained measure accurately and rapidly, the concrete expression of the measure function needs to be constructed and the measure value can be calculated by this function. The constructed function must follow the principle of "non-boundedness", "normalization", "accumulation", otherwise the result is wrong. The classical methods of constructing the measure function include the straight line method, the quadratic curve method and the exponential curve method etc. the straight line method is widely used, practical and acceptable and the expressions are shown in formula (5). In the formula, x is the index measure value, $u_k(x)$ and $u_{k+1}(x)$ are the measured values of x in the C_k, C_{k+1} states. a_k and a_{k+1} respectively represents the intermediate value of the measured value x in the C_k and C_{k+1} state.

$$\begin{cases} u_k(x)=\begin{cases} \dfrac{-x}{a_{k+1}-a_k}+\dfrac{a_{k+1}}{a_{k+1}-a_k}, a_k < x \le a_{k+1} \\ 0, x > a_{k+1} \end{cases} \\ u_{k+1}(x)=\begin{cases} 0, x \le a_k \\ \dfrac{x}{a_{k+1}-a_k}-\dfrac{a_k}{a_{k+1}-a_k}, a_k < x \le a_{k+1} \end{cases} \end{cases} \tag{5}$$

2.3. *Information entropy index weight*

If w_{ij} represents the importance of evaluation index I_j of object X_i compared with other index, namely, w_{ij} is the weights of evaluation index I_j of X_i and it should be consistent with $0 \le w_{ij} \le 1$, $\sum_{j=1}^{n} w_{ij}=1$. The index weight is determined by the information entropy theory. According to the formula (6) and (7) into the known index measure value is u_{ijk}, we can calculate the index weight w_{ij}. For an evaluation object, the index weight vector of evaluation object X_i is recorded as $W_i=\{w_{i1}, w_{i2}, \ldots, w_{in}\}$.

$$v_{ij}=1+\frac{1}{lg\,p}\sum_{j=1}^{n}(u_{ijk}\times lg\,u_{ijk}) \tag{6}$$

$$w_{ij}=v_{ij}/\sum_{j=1}^{n}v_{ij} \tag{7}$$

2.4. *Multi index unascertained measure*

If $u_{ik}=u(u_i \in C_k)$ is the measured value which represents the kth evaluation level of X_i, w_{ij} is the weights of evaluation index I_j of object X_i and it conforms to non-boundedness: $0 \le u_{ij} \le 1$, normalization: $\sum_{k=1}^{p} u_{ik}=1$, accumulation: $u(u_i \in U)=\sum_{k=1}^{p} u_{ik}$, then u_{ik} is the multi index unascertained measure of evaluation object X_i and the calculation formula is shown as formula (8). $u_i=\{u_{i1}, u_{i2}, \ldots, u_{ip}\}$ is used to represent the multi index unascertained measure of a certain object.

$$u_{ik}=\sum_{j=1}^{n} w_{ij}u_{ijk} \tag{8}$$

2.5. *Confidence criterion*

If $\{C_1, C_2, \ldots, C_p\}$ is an ordered partition of the evaluation space and satisfies $C_1 < C_2 < \ldots < C_p$, then λ is set as the confidence according to confidence criterion, the range of λ is $\lambda \ge 0.5$. The evaluation object X_i belongs to the k_0th rating, namely C_{k0}.

$$k_0=min\left| k:\sum_{i=1}^{k} u_i > \lambda, k=1,2,\ldots,p \right| \tag{9}$$

2.6. *Sorting*

If the evaluation results of multiple objects are sorted by the degree of occupational hazard, then the q value of the relative degree of occupational hazard is defined as the comparison parameter. If

$C_1 < C_2 < \ldots < C_p$, let the score of C_i be I_i, there are $I_i < I_{i+1}$ and $q = \sum_{i=1}^{p} I_i u_i$ .In the formula u_i is the unascertained measure belongs to the ith level of evaluation object, q represents the parameter of occupational hazard degree and different objects can be sorted according to the value of q.

3. Occupational Hazards Evaluation Model of Coal Mine

3.1. Evaluation index system of occupational hazards in coal mine

Through the coal mine underground workplace occupational hazards on-site research, based on the relevant research standards[14,15] and the relevant literature research[4,5], eight indicators were selected to establish the evaluation index system of occupational hazards in coal mine. Quantitative and quantitative indicators were quantified, as shown in Table 1 and Table 2.

Table 1. The quantitative indicators' grading standards of occupational hazard evaluation

Indicator	Ⅰ level, relatively harmless	Ⅱ level, slight harm	Ⅲ level, moderate harm	Ⅳ level, Serious harm
Respirable dust, I_1(mg/m^3)	<3.5	3.5-7	7-16	>16
Hazardous gas index, I_2	<1.0	1.0-1.5	1.5-3.0	>3.0
Noise, I_3(dB)	<85	85-90	90-95	>95
Temperature, I_4 (℃)	15-25	25-28 or 10-15	28-34 or 5-10	>34 or<5
Humidity, I_5 (%)	40-60	30-40 or 60-75	20-30 or 75-85	<20 or>85
Wind speed, I_6(m/s)	0.5-1.5	0.35-0.5 or 1.5-3.0	0.25-0.35 or 3-4	<0.25 or>4

Table 2. The qualitative indicators' classification and assignment about occupational hazard evaluation

Indicators	Ⅰ level, relatively harmless	Ⅱ level , slight harm	Ⅲ level, moderate harm	Ⅳ level, serious harm
Value	1	2	3	4
Work space, I_7	Spacious	Limited	Narrow	Extremely narrow
Labour intensity, I_8	Light	Medium	Heavy	Extremely heavy

3.2. The solution process of coal mine occupational hazard evaluation model

(1) Construct the occupational hazard evaluation index system. Eight occupational hazards were selected as the discriminant indexes, and the occupational hazard evaluation index system was constructed and classified.

(2) Quantify the evaluation index of each object. Through the measurement of the evaluation indexes of the coal mine, the evaluation index of the occupational hazards in the mine is quantified.

(3) Establish an unascertained measure function. This paper use the straightness measure function, according to the index classification criteria of Table 1 and Table 2, the indicators measurement function was created, as shown from Figure 1 to Figure 8.

(4) Solve the index unascertained measurement matrix. According to the results of step (2) and step (3), the unascertained measurement of each index could be solved and the unascertained measure matrix was composed.

(5) According to the information entropy theory to calculate the weight coefficient of each index.

(6) Calculate the multi-index measure vector. The formula is shown in (8).

(7) According to the confidence degree identification criteria, the evaluation results were obtained.

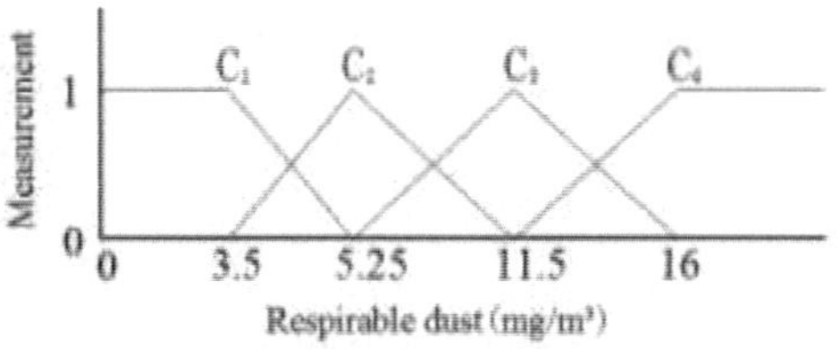

Figure 1. Uncertainty measurement function of respirable dust

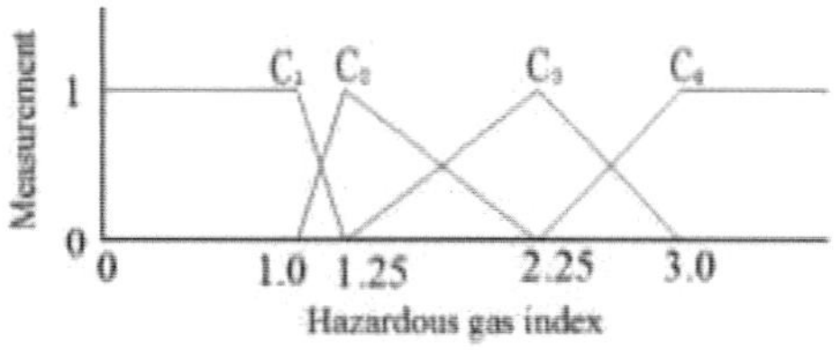

Figure 2. Uncertainty measurement function of Hazardous gas index

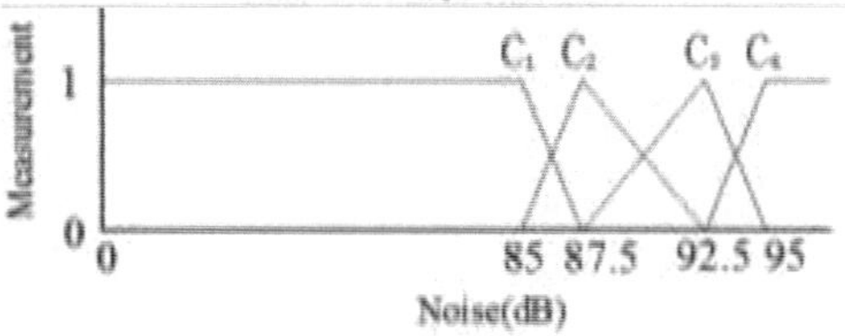

Figure 3. Uncertainty measurement function of noise

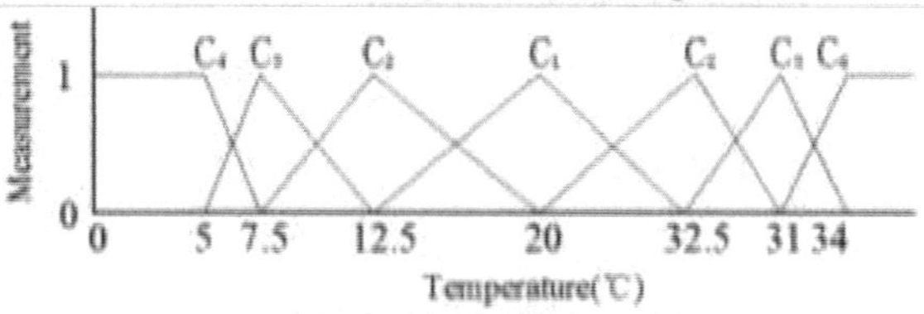

Figure 4. Uncertainty measurement function of temperature

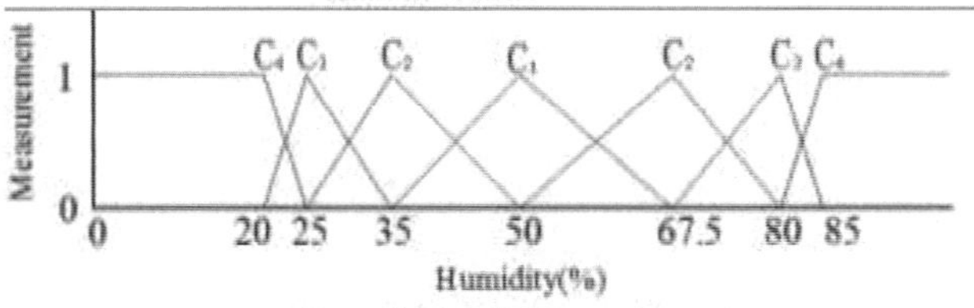

Figure 5. Uncertainty measurement function of humidity

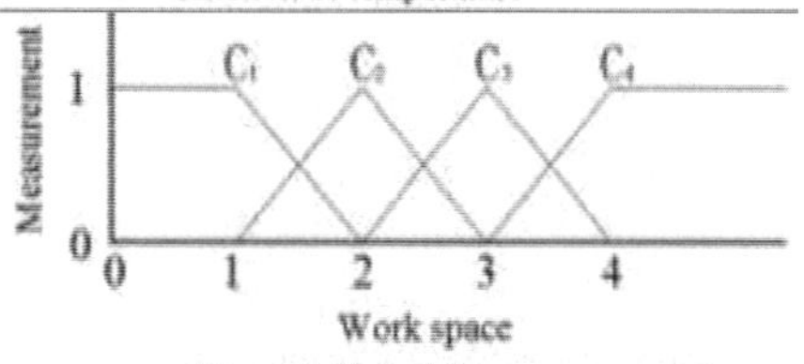

Figure 7. Uncertainty measurement function of labour work space

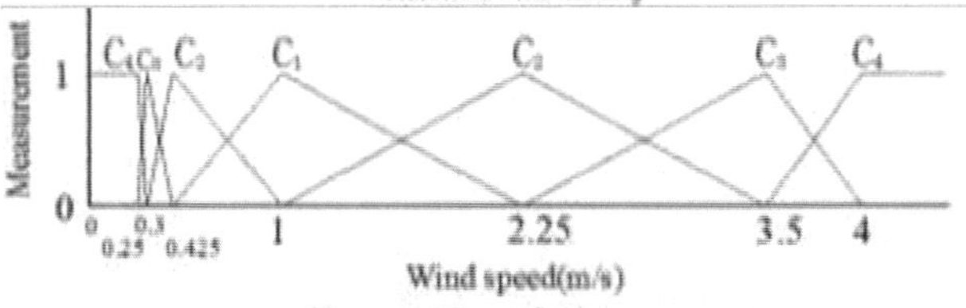

Figure 6. Uncertainty measurement function of wind speed

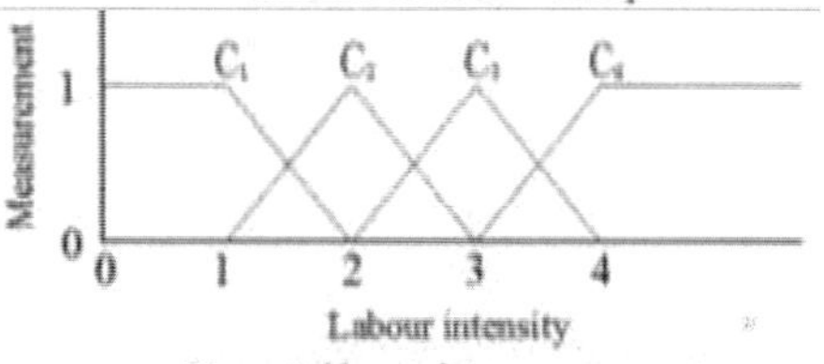

Figure 8. Uncertainty measurement function of labour intensity

3.3. Programming of occupational hazard evaluation model for coal mine

In order to realize the programming of the evaluation model, the Visual Basic language is used to program, and the VB good visual interface is used to link the Access database, and the application software of coal mine occupational hazard evaluation was compiled, as shown in Figure 9.

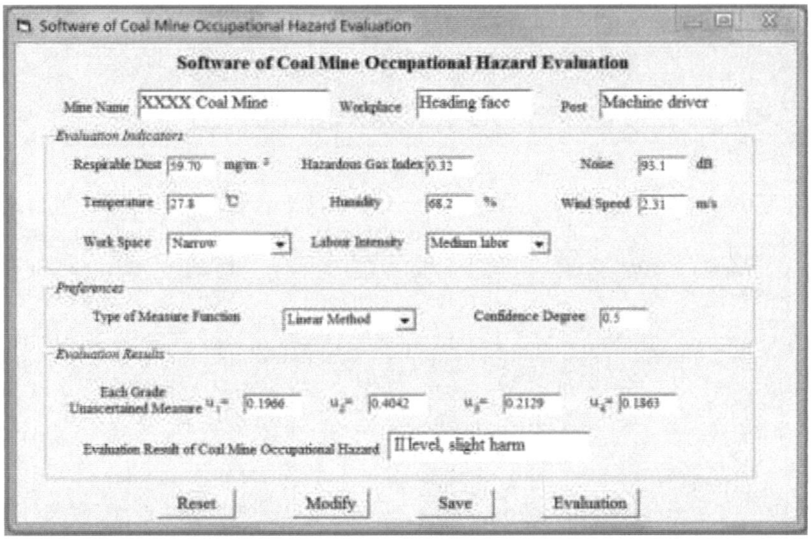

Figure 9. The software of coal mine occupational hazard evaluation

4. Example Application

By investigating the occupational hazards of coal mine operators, combined with the relevant drawings and materials of the mine, a total of six posts, including heading machine drivers, coal cutter drivers, coalface move support workers, ventilation workers, excavation face supporting workers and safety inspection workers. The eight occupational hazards indicators were measured and analysed during these six posts work period respectively, the occupational hazard indicators of the measured results shown in Table 3.

Table 3. The occupational hazard evaluation indicators' value of operation posts underground coal mine

Posts	Respirable dust, I_1 (mg/m^3)	Hazardous gas index, I_2	Noise, I_3(dB)	Temperature, I_4 (℃)	Humidity, I_5 (%)	Wind speed, I_6(m/s)	Work space, I_7	Labour intensity, I_8
Heading machine drivers	59.70	0.32	93.1	27.8	68.2	2.31	3	2
Coal cutter drivers	49.50	0.34	92.0	28.5	63.0	1.54	3	2
Coalface move support workers	37.40	0.20	88.4	28.3	64.5	1.30	3	3
Ventilation workers	10.80	0.07	90.1	23.1	62.4	2.25	2	2
Excavation face supporting workers	27.10	0.16	94.3	27.4	65.2	2.14	3	3
Safety inspection workers	11.50	0.03	83.5	24.5	61.1	0.86	2	1

In Table 3, the measured data of the occupational hazards indicators of the six posts are input into the unascertained measure evaluation software for occupational hazards in coal mine. The selection method is linear method and the confidence degree is 0.5. The evaluation result can be obtained, as shown in Table 4. The results of the evaluation model are consistent with the present situation of the coal mine.

Table 4. The evaluation results of occupational hazard underground coal mine

Results	Heading machine drivers	Coal cutter drivers	Coalface move support workers	Ventilation workers	Excavation face supporting workers	Safety inspection workers
Evaluation results	II level, slight harm	II level, slight harm	II level, slight harm	I level , relatively harmless	II level, slight harm	I level , relatively harmless

Ranking the occupational hazards degree of the six posts. Due to C_1<C_2< C_3<C_4, the scores of C_1, C_2, C_3 and C_4 are 1, 2, 3 and 4 respectively. According to the formula (10), The relative degree of occupational hazards of each post calculated, $\{q_1, q_2, q_3, q_4, q_5, q_6\}$ = {2.531, 2.447, 2.324, 2.302, 2.587, 1.479}. According to the post by the severity of occupational hazards, the results are: excavation face supporting workers>heading machine drivers> coal cutter drivers > coalface move support workers > ventilation workers > safety inspection workers.

5. Conclusion

(1) Based on the comprehensive consideration of coal mine occupational hazards, eight occupational hazard evaluation indexes were selected and the evaluation index system was established. The unascertained measure theory was introduced, the information entropy was applied to determine the index weight, and the coal mine occupational hazard evaluation model was established, which provides a new method for coal mine occupational hazard evaluation.

(2) Based on the theoretical basis of the unascertained measure evaluation model of occupational hazards in coal mine, Visual Basic language is used to design and visualize the interface design, link the Access database, compile the mine occupational hazard evaluation software, and facilitate the mine occupational hazard evaluation model Promotion and application.

(3) Taking the actual situation of a coal mine as an example, through the investigation of six posts underground coal mine, the self-compiled evaluation software was applied to measure the evaluation index, the results of the evaluation model are consistent with the present situation of the coal mine.

(4) In order to further improve the universality of coal mine underground occupational hazard evaluation model based on unascertained theory, it is necessary to carry out research on different coal mine and establish a more widely applicable measure function, so that the model could be maturely applied in practice.

Acknowledgments
This work was supported by the National Natural Science Foundation of China (No. 51574016).
*Corresponding author. Tel.: +86 010- 62313972; E-mail addresses: jza1963@263.net.

References

[1] Wang Z.X., Qin R.L., Li Y.Z., et al, A study of occupational hazards in 23 coal mine enterprises in China and their protection status. Journal of Environmental & Occupational Medicine, 2009, 26(6): 565–568.
[2] Liao H.J., Sun Q.Y., Zhang Z.B., et al, Analysis for cross-sectional survey data of occupational hazards in coal mine. Journal of Safety Science and Technology, 2009, 5(5): 152–156.

IOP Conf. Series: Earth and Environmental Science **64** (2017) 012039 doi:10.1088/1755-1315/64/1/012039

[3] Shen B.H., Lei Y., Guo Y.H, Progress of coal science and technology in China. Journal of China Coal Society, 2011, 36(11): 1779–1783.
[4] Guo J.P, Shang J.Y., Wang X.N., Comprehensive safety evalution method for mine occupational hazards based on two-tuple linguistic information and APH. China Safety Science Journal, 2011, 21(8): 116–122.
[5] Wang X.N., Guo J.P., Study on the occupational hazards evaluation system of underground mine. Metal Mine, 2011 (4): 130–133.
[6] Li H., Tian Y.Q., Zhong X.R., Comprehensive evaluation of the occupational hazards in coal enterprises by using identical discrepancy contrast model. Journal of Safety and Environment, 2012, 12(3): 216–219.
[7] Xiong L.X., Luo Z.Q., Wu C., et al, Evaluation method for occupational hazards of confined space of deep mines. Metal Mine, 2014(11): 132–137.
[8] Cao Q.K., Yang Y.L., Yu R.L., The coal mines safetyappraisal based on unascertained set. Journal of China Coal Society, 2007, 32(2): 181–185.
[9] Luan T.T., Xie Z.H., Wu Z.Z., et al, Risk evaluation model of waste dump landslide based on uncertainty measurement theory. Journal of Central South University (Science and Technology), 2014, 45(5): 1612–1617.
[10] Gong F.Q., Li X.B., Dong R.J., et al, Underground goaf risk evaluation based on uncertainty measurement theory. Chinese Journal of Rock Mechanice and Engingeering, 2008, 27(2): 323–330.
[11] Yang F.Q., Wu C., Risk assessment on spontaneous combustion of sulfide ore dump in stope based on uncertainty measurement theory. Journal of Central South University (Science and Technology), 2010, 41(6): 2373–2380.
[12] Wang G.Y., Uncertainty information and its mathematical treatment. Journal of Harbin Architecture and Engineering Institute, 1990, 23(4):52–58.
[13] Liu K.D., Wu H.Q., Wang N.P., et al, Uncertainty mathematics. Huazhong University of Science and Technology Press, 1997: 199–206.
[14] GBZ 2.1—2007 Occupational exposure limits for hazardous agents in the workplace Part 1: Chemical hazardous agents [S]. People’s Medical Publishing House, 2008.
[15] GBZ 2.2—2007 Occupational exposure limits for hazardous agents in the workplace Part 2: Physical hazardous agents [S]. People’s Medical Publishing House, 2008.

Research on evaluation method for water saturation of tight sandstone in Suxi region

Hong Lv[1], Fuqiang Lai[1,*], Liang Chen[2], Chao Li[1], Jie Li[1], Heping Yi[1]

[1]Key Laboratory of Complex Oil and Gas Field Exploration and Development in Chongqing University of Science and Technology, Chongqing, China
[2]PetroChina International Middle East West Qurna1 Project, Beijing, China

[*]Corresponding author e-mail: laifq1982@163.com

Abstract. The evaluation of irreducible water saturation is important for qualitative and quantitative prediction of residual oil distribution. However, it is to be improved for the accuracy of experimental measuring the irreducible water saturation and logging evaluation. In this paper, firstly the multi-functional core flooding experiment and the nuclear magnetic resonance centrifugation experiment are carried out in the west of Sulige gas field. Then, the influence was discussed about particle size, porosity and permeability on the water saturation. Finally, the evaluation model was established about irreducible water saturation and the evaluation of irreducible water saturation was carried out. The results show that the results of two experiments are both reliable. It is inversely proportional to the median particle size, porosity and permeability, and is most affected by the median particle size. The water saturation of the dry layer is higher than that of the general reservoir. The worse the reservoir property, the greater the water saturation. The test results show that the irreducible water saturation model can be used to evaluate the water floor.

1. Introduction

The Sulige gas field is located in the Yisha slope of the Ordos basin, and its tectonic form is the westward slope of the slope. The main oil and gas layers are the $Shan_1$ segment and He_8 segment. The lithology is mainly composed of quartz sandstone and clastic quartz sandstone. Because of braided delta deposit, reservoir characteristics are more complex [1]. The irreducible water saturation is one of the important reservoir properties, which plays a key role in reservoir oil and gas evaluation, reserve calculation and capacity prediction [2-3]. The irreducible water saturation is the critical parameter of gas phase and gas-water two-phase seepage for low porosity and low permeability reservoirs, and it is also an important parameter to calculate its gas saturation [4-5]. Therefore, the tight sandstone water saturation Degree evaluation has important practical significance.

At present, the methods of measuring the irreducibility of water are semi-permeable partition method, adsorption method, desorption method [6], humidification gas displacement method [7], centrifugal capillary pressure method [8], mercury intrusion and infiltration experiments [9], unsteady constant pressure gas flooding method [10], low magnetic field NMR T_2 spectrum technique and so on [3,12]. Existing studies have shown that nuclear magnetic resonance and core flooding can be more accurately. The experiment of unsteady displacement experiment simulates the process of gas migration and accumulation, and the displacement pressure is easily determined, but the time of displacement is different under different experimental conditions [6-7]. The measured values measured by NMR techniques avoid the experimental errors and are closer to true values caused by unfavorable factors such as hard rock density, low porosity, and low permeability [13].

Experiments about multi-functional core flooding and nuclear magnetic resonance centrifugal were carried out firstly. Then analyze the effect of median particle size, porosity and permeability on the saturation water saturation [4]. Finally, combined with logging data the evaluation model of irreducible water saturation is established, and the evaluation of irreducible water saturation is carried out. The relationship between formation water saturation and aquifer formation is analyzed and summarized.

2. Regional Geological Survey

2.1. Reservoir lithology characteristics

The Upper Paleozoic developed the Delta Plateau in the Sulige Gas Field. Top-down developed the Carboniferous Benxi Formation, the Taiyuan Formation, the Permian Shanxi Formation, the Lower Stone Box Group, the Upper Stone Box Group and the Shiqianfeng Formation. The main oil and gas production are from the $Shan_1$ and He_8 segment.

The lithology of the Shan is mainly fine-medium granulite, lithic quartz sandstone and mudstone. The mudstone is dark gray, grayish black and gray. The sandstone mostly is gray, light gray (Fig. 1).

The lithology of the 8th section is mainly composed of light gray gravel coarse sandstone, gray and coarse sandstone, gray green lithic quartz sandstone and gray green mudstone. The color of mudstone is variegated, as gray-green, gray, brown-based, dark gray and gray black mudstone. For sandstone is mostly gray. Gray-green mudstone is often mixed with variegated or brown mudstone (Fig. 2).

Fig. 1 Core photos of the $Shan_1$ segment

Fig. 2 Core photos of the and He_8 segment

According to the data of more than 1000 pieces of block, it is indicated that the sandstone of the $Shan_1$ and He_8 are mainly quartz sandstone, lithic quartz sandstone and lithic sandstone. It has the characteristics of low feldspar, high quartz and high detritus. The particle size is generally between 0.1mm and 2.55mm, with medium coarse grain structure. The separation is mainly medium deviation, and some samples are well sorted. The degree of roundness is mainly sub - angular, small - to - round - sub - circular and sub - round, showing a high - low maturity. The structure type of the interstitial is mainly cement or miscellaneous matrix or mixed with cement and cement.

2.2. *Reservoir physical characteristics*

The core porosity of Su 54-20-86 well is between 7% and 8%, the average permeability is between $0.4\times10^{-3}\mu m^2$ and $0.6\times10^{-3}\mu m^2$. He_8 segment's porosity is concentrated in 8% to 10%, it's permeability is mainly distributed in $0.4\times10^{-3}\mu m^2$ to $0.6\times10^{-3}\mu m^2$, the porosity of $Shan_1$ segment and $Shan_2$ segment is distributed in 6% to 8%, the permeability is distributed in $0.2\times10^{-3}\mu m^2$ to $0.4\times10^{-3}\mu m^2$, and the reservoir is better in the He_8 segment (Fig. 3).

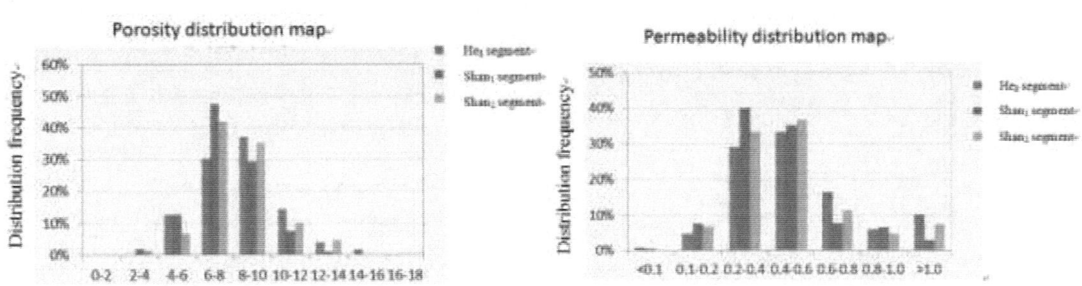

Fig. 3 Histogram of pore permeability distribution about $Shan_1$, $Shan_2$ and He_8 segment

The porosity and permeability is correlated based on the analysis of the He_8 segment (236 core analysis samples) and the Shan1 segment (58 lithological analysis sample points) showed good correlation (as shown in Fig. 4).

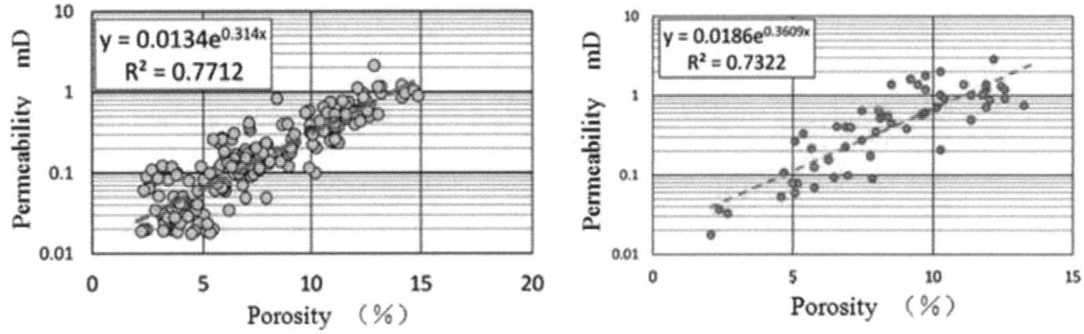

Fig. 4 Analysis of porosity correlation about $Shan_1$ and He_8 segment

2.3. Storage space characteristics

According to the analysis of cast thin section data, the main pore types of He_8 segment is matrix pore, intergranular pore, and intergranular pore, intergranular dissolved pore, occasionally pores, micro cracks; the main pore types of $Shan_1$ segment is intergranular pore and intragranular dissolved pore, secondly intergranular dissolution pores. According to the statistical analysis of mercury injection parameters in the west of Sulige gas field (Table 1), the pore throat structure of the upper Paleozoic Sandstone has the following characteristics:

The mercury pressure parameters of sandstone in $Shan_1$ and He_8 segment are close. The displacement pressure range is from 0.01MPa to 7.76MPa, and the average values are 1.19MPa and 1.20MPa respectively. The median pressure range respectively is between 0.24MPa and 43.76MPa, with an average of 6.76MPa and 12.06MPa. Sandstone median distribution of throat radius in the range of 0.01μm to 3.19μm, the average was 0.25μm and 0.16μm; He_8 segment's is from -3.51 to 1.89. Pore skewness' average value is -0.01, $Shan_1$ section changes from -2.02 to 1.98, the average value is -0.2.

Table 1 Table of mercury injection parameters about Shan1 and He8 segment's reservoir

Horizon	Mercury pressure (MPa)	Displacement pressure (MPa)	Median pressure	Separation pressure	Variation coefficient	Mean coefficient	Skewness coefficient	Median radius (μm)	Maximum mercury saturation (%)	Exit efficiency (%)
$Shan_1$ segment	Average value	1.19	6.76	1.57	0.67	4.81	-0.01	0.25	80.54	41.34
	Maximum	5.49	36.38	3.27	2.11	12.48	1.89	0.47	93.9	60.24
He_8 segment	Average value	1.2	12.06	1.53	0.82	3.68	-0.2	0.16	84.13	50.3
	Maximum	7.76	43.76	2.6	1.88	12.55	1.98	3.19	94.73	61.3

3. Irreducible water saturation test

3.1. Preparation of rock samples

Collect 9 block core diameters of Su 54-20-86 well in the sample preparation of laboratory equipment. Make 27 rock samples for the following experiments done to the core preparation with the diameter of 1in (2.54cm), the length of 2.5-5cm.

3.2. Measure rock bound water by steady state constant pressure gas drive method

3.2.1. Experimental principle. The micro pore structure of rock has a strong heterogeneity, and some water exists in the form of film or bubble, which is difficult to be replaced. The gas with different velocity has different carrying capacity, and the irreducible water of the gas reservoir is a physical quantity relative to the velocity of the gas flow. The core standard for water saturated, for device for pressure drive using different multifunctional core flooding, the nitrogen displacement standard saline rock, until the displacement speed is stable, then weigh the core, further calculation of the irreducible water saturation under the corresponding condition.

3.2.2. Experimental process and results. The treated cores were vacuum saturated with standard brine, and the displacement experiments were carried out in the core of saturated brine. The displacement pressure was 1.0MPa, 1.5MPa, 2.0MPa, 2.5MPa, 3.0MPa. Under each displacement pressure, the volume of gas is 60 times than that of pore. The quality of the samples is measured before and after the displacement. The mass difference before and after the displacement is divided by the density of the standard brine and then divided by the pore volume.

Table 2 Experimental data of irreducible water saturation

Sample number	Core length (cm)	Displacement pressure (MPa)	Pressure gradient (MPa/cm)	Irreducible water saturation (%)
9-2	3.408	1.06	0.31	82.60
	3.408	1.58	0.46	80.47
	3.408	2.00	0.59	79.69
	3.408	2.50	0.73	79.03
117-1	3.861	2.08	0.54	86.54
	3.861	2.49	0.64	83.32
	3.861	3.03	0.78	81.26
	3.861	3.51	0.91	80.19
117-2	3.768	2.05	0.54	84.71
	3.768	2.52	0.67	81.54
	3.768	3.01	0.80	79.12
	3.768	3.49	0.93	78.26
124-1	3.411	2.05	0.60	74.02
	3.411	2.52	0.74	71.03
	3.411	3.02	0.89	69.34
	3.411	3.50	1.03	68.23
185-1	3.904	0.55	0.14	72.85
	3.904	0.99	0.25	71.46
	3.904	1.52	0.39	69.78
	3.904	2.01	0.51	69.10

At the same time, the irreducible water saturation of the core is shown in Table 2, which shows that the irreducible water saturation of the core is mainly distributed between 68.23% and 86.54%. We can see that the irreducible water saturation shows a decreasing trend with the increase of displacement pressure when the displacement water saturation of each core is different under the different displacement pressure (Fig. 5).

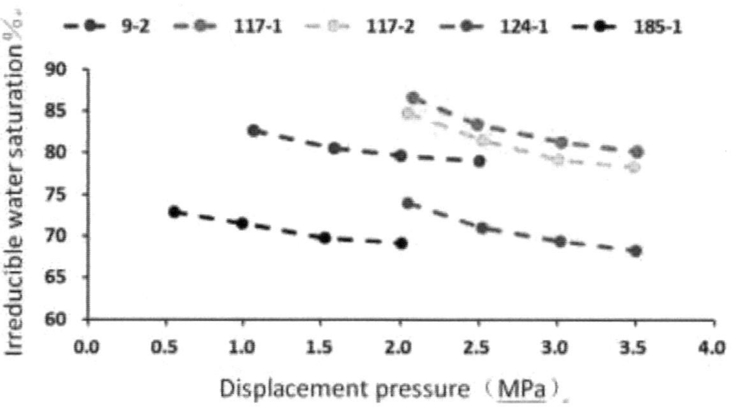

Fig. 5 Relationship between irreducible water saturation and core displacement pressure

3.3. Measure the bound water of rocks by low magnetic resonance T_2 spectroscopy

3.3.1. Experimental principle. Relaxation is the process of recovering from equilibrium state when magnetic resonance is deviated from the excitation of the RF field. The relaxation velocity is determined by the physical properties and fluid characteristics of rocks. The relaxation velocity is only determined by the petro physical properties of the same fluid, and the T_2 relaxation time is usually measured by the present NMR. The bound water of the solid surface force is strong, the relaxation time is short, the movable water is weak and the relaxation time is long. According to the apparent difference between the relaxation time of the constrained water and movable water analysis and image system measurement relaxation time by the large aperture NMR. The T_2 cut-off value of the movable fluid is used to distinguish the movable water and the bound water, calibrated by centrifugal experiment thus obtain the irreducible water saturation of the core.

3.3.2. Experimental process and results. Corresponding to the rock specimen NMR T_2 spectra curves, nuclear magnetic bound water saturation is equal to the T_2 spectra less than T_2 cutoff value T_2cutoff of the non-dynamic peak under the packet area and the whole T_2 spectrum of the area ratio. The volume of irreducible water is equal to the accumulation of irreducible water saturation and porosity, and the volume of movable water equals the difference of pore volume and bound water volume.
The standard T_2 test results of the 2-32/206 samples in the well Su 54-20-86 with a depth of 3571.11m, as an example, are shown in the following illustration:

The results of all samples are shown in table 3, which can be seen that the maximum bound water saturation is 92.57%, the minimum value is 62.38%, the median is 77.48%, the average is 77.29, and the main distribution is between 69% and 85%.

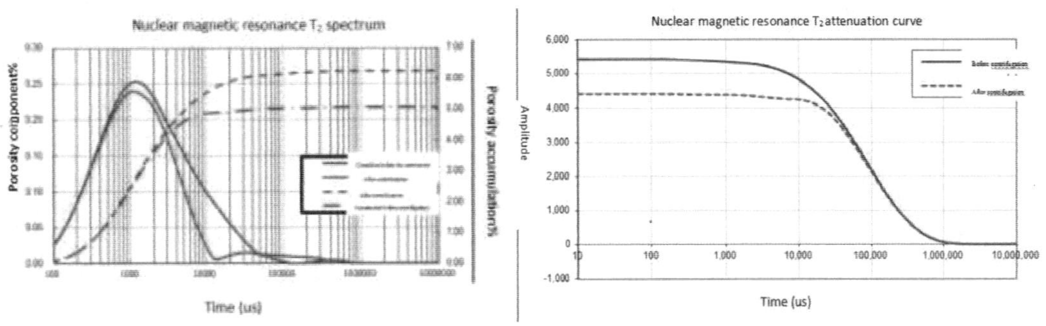

Fig. 6 The standard T2 test results of the 2-32/206 samples in the well Su 54-20-86

Table 3 The result table of nuclear magnetic resonance measurement of irreducible water saturation

Bore specimen	sample	Bound fluid saturation (%)	Cutoff values Taken
185-2	10	62.38	5
71-5	34	73.01	7.5
195-1	37	92.57	33
71-4	40	73.53	7.5
32-4	50	80.50	5.5
117-3	52	72.34	4.5
71-2	54	75.80	10
40-1	55	83.83	8.5
117-4	57	73.71	5
40-3	62	84.96	8.5
32-3	64	85.52	8.5
9-1	66	69.28	1.2

3.4. Experimental Comprehensive analysis

The maximum and minimum value of the irreducible water saturation measured by the two methods are close can see in table 4, the median and the average and the main distribution range are also close. It can be considered that both methods are credible and can be used as the basis parameters of reservoir development.

Table 4 Statistics results of two methods for binding water saturation

Test	Max	Min	Average	Main distribution range
Gas measurement method	80.19	68.23	74.96	68%~80%
nuclear magnetic resonance	92.57	62.38	77.29	69%~85%

4. Evaluation of irreducible water saturation

4.1. Analysis of influencing factors of irreducible water saturation

According to statistical data of irreducible water saturation with granularity, porosity and permeability, do the value scatter plot as shown:

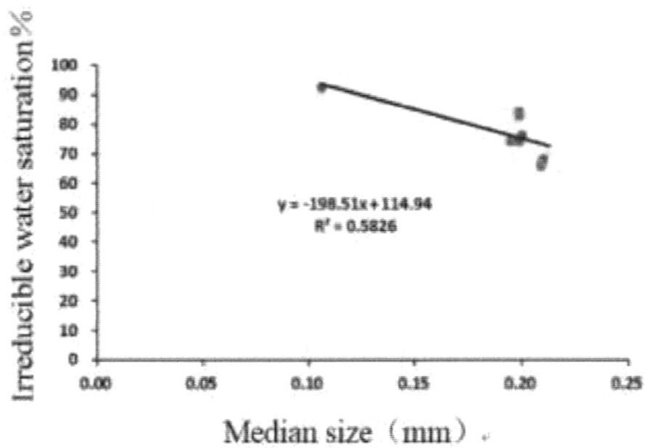

Fig. 7 Relationship curves of irreducible water saturation and Median size

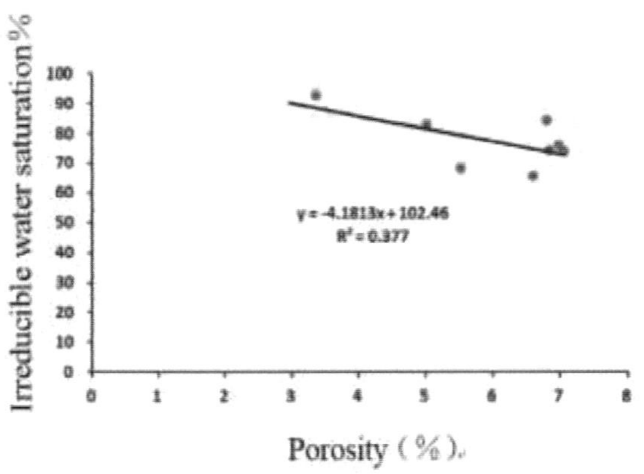

Fig. 8 Relationship curves of irreducible water saturation and porosity

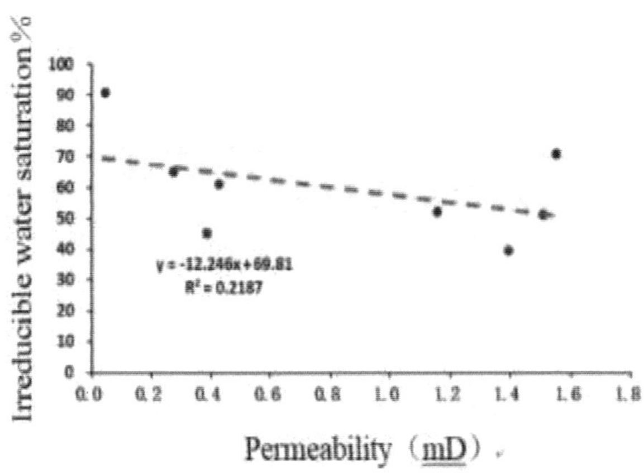

Fig. 9 Relationship curves of irreducible water saturation and permeability

4.2. Model of irreducible water saturation

Through the analysis of influencing factors, the saturation of irreducible water is closely related with permeability, porosity and other physical parameters. The model of irreducible water saturation is established by using the synthetic physical parameter when evaluating the irreducible water saturation in the actual single well (fig. 10). Based on the relationship between the irreducible water saturation and the pore structure coefficient in Suxie's area, the model of the irreducible water saturation interpretation was established jointly with the core sample points of Su 54-20-86 well. The model for calculating the irreducible water saturation is:

$$S_{WII}=74.136\times(\sqrt{K/\phi})^{-1.782},\ R^2=0.7693$$

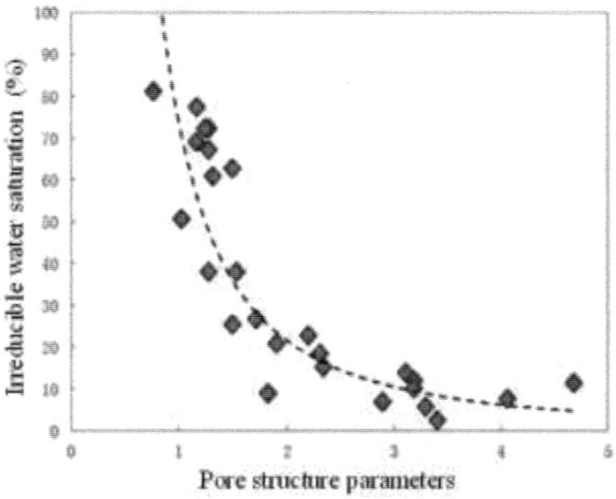

Fig. 10 Relationship between irreducible water saturation and pore structure parameters

4.3. Logging evaluation and model testing

The depth of the data point of the analysis of the core of Su 54-20-86 well is plotted, and the well logging interpretation comprehensive map of Su 54-20-86 well in depth 3555m to 3600m is shown in Fig. 11. It can be seen from the graph that the curve of irreducible water saturation (Swii-Calc curve) has a good consistency with the results of core analysis (Swii-core line segment), which shows that the constrained water saturation model meets the requirements of precision.

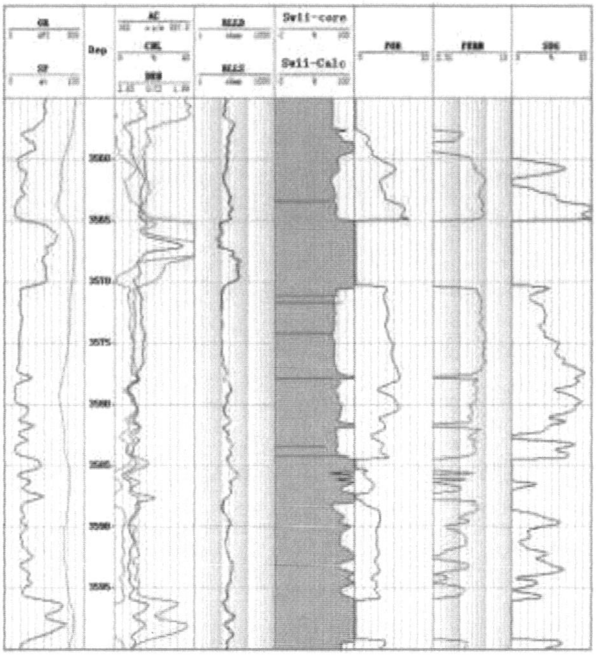

Fig. 11 The well-bound water saturation logging interpretation results map of SU 54-20-86 well

IOP Conf. Series: Earth and Environmental Science **64** (2017) 012040 doi:10.1088/1755-1315/64/1/012040

5. Application of irreducible water saturation model

Taking the study of the district SU 373 well as an example (Fig. 12,13), in the 3426.5~3430m Wells and 3476~3479 wells, the natural gamma value is lower, the natural potential curve obviously negative abnormal .The neutron value is small, the excavation effect is obvious, the resistivity is higher, the depth resistivity shows the obvious difference. The saturation of irreducible water saturation is from 60% to 80%, the movable water is less and the synthetic interpretation is the gas reservoir. For the test gas separately, the water saturation of the perforating Wells is 60.9%, the test gas results are: Nissan gas 7.2785 and 43,967 Square, not producing water. At present, Nissan 1.6 fang, daily water 0.7 square, moisture ratio 0.43, after production and troubleshooting, for the normal production wells. The result accords with the production result, so it is better to establish the model of irreducible water saturation with porosity structure index for low porosity and low permeability reservoirs.

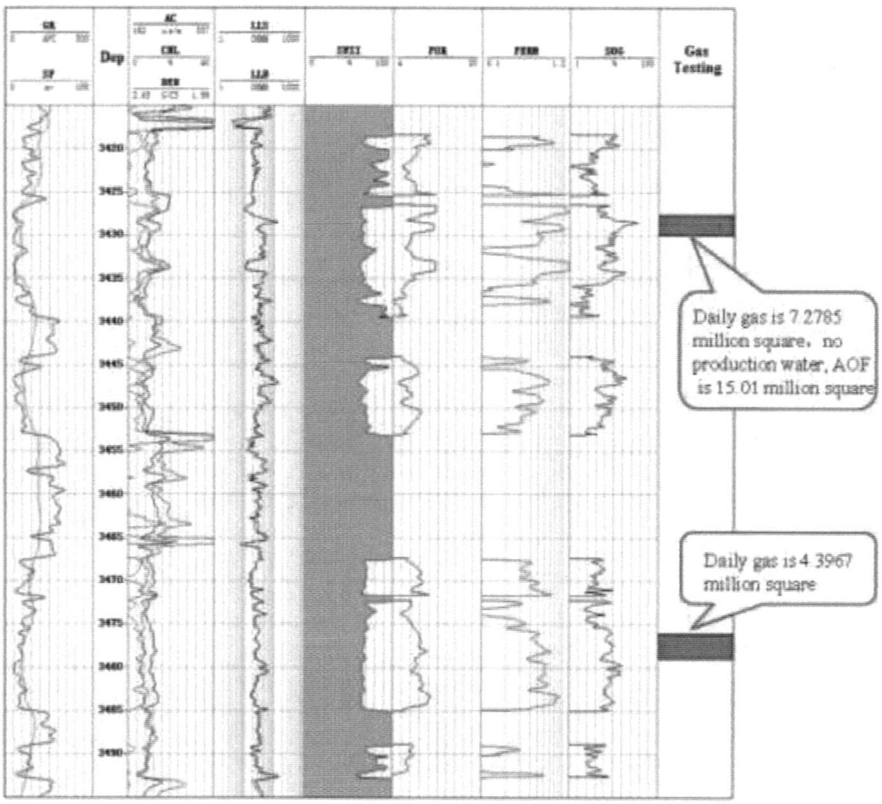

Fig. 12 The well-bound water saturation logging results of SU 373 well

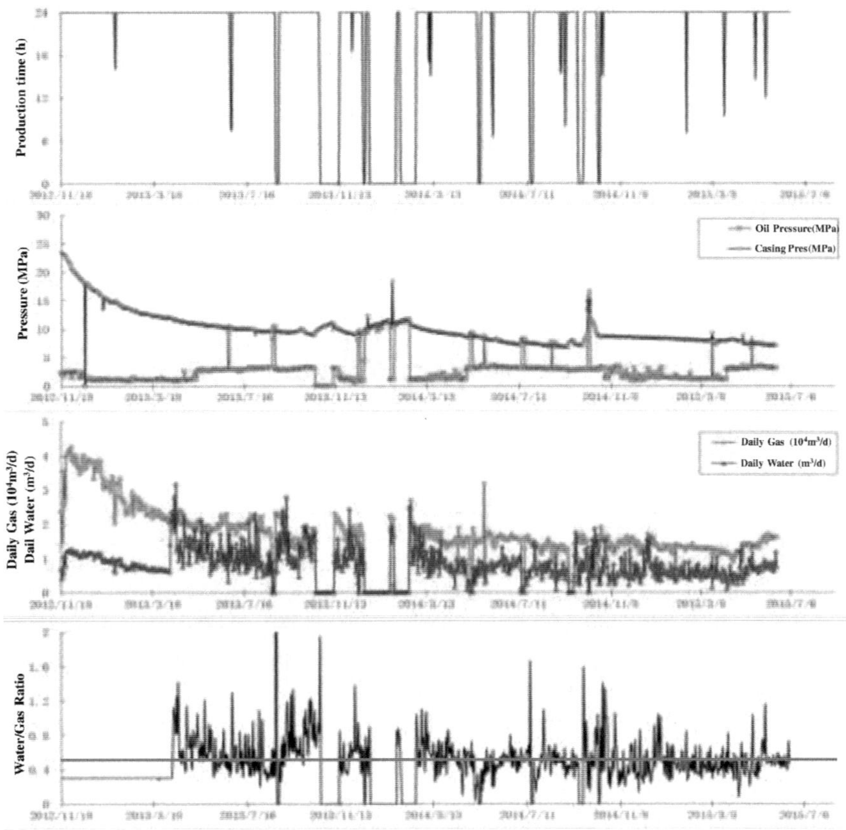

Fig. 13 Production curve of Su 373 well

6. Conclusion

(1) The experimental results of the core-bound water saturation measured by NMR and core displacement methods are close, and the results of the two methods are credible, and they can be used as the basis parameters of reservoir development, but are relatively large.

(2) The irreducible water saturation and the size of the rock, porosity, and permeability are negative correlation, with the value of the particle size change the fastest, with the permeability and the slowest change. The irreducible water saturation of tight sandstone reservoirs is generally higher, and the irreducible water saturation of dry layer is higher than that of the general reservoir; the worse the reservoir property, the greater the saturation of the irreducible water.

(3) Compared with the results of core analysis, we can find that the irreducible water saturation model has been established to meet the precision requirement. The model was used to predict and evaluate, and the model was found to be reliable by gas test.

Acknowledgements

The authors would like to thank Sulige Research Center for providing the data and technical assistance. The Project was supported by National Natural Science Foundation of China (Grant No. 41402118), and the innovation talents training program of Chongqing University of Science and Technology (201508).

References

[1] Wang Xiaomei, Zhao Jingzhou, Liu Xin, Society Characteristics of stratigraphic water distribution in tight sandstone reservoirs in the west of Sulige Gasfield, J, Petroleum experimental Geology, 2012,33 (5): 802-808

IOP Conf. Series: Earth and Environmental Science **64** (2017) 012040 doi:10.1088/1755-1315/64/1/012040

[2] Huang Zhilong, Pu Xiugang, Liang Chunxiu etc, Characteristics and formation mechanism of low resistivity reservoirs in southern Songliao basin, J, Natural gas industry, 2006, 26 (8): 27~29

[3] Yang Zhengming, Jiang Hanqiao, Shuei, etc, The saturation degree of irreducible water in low permeability and moisture reservoirs is measured by NMR technique , J, Oil drilling and mining technology, 2008,30 (3): 56-59

[4] Sun Yi. Influence of accumulation power on irreducible water saturation, J, Oil and gas geology and recovery rate, 2007,14 (2): 64-66.

[5] Yang Shenglai and so on. Oil industry publishing house physics, Beijing: Petroleum Industry Press, 2004

[6] Li Ning, Zhou Keming, Zhang Qingxiu, etc. Experimental study on the saturation of irreducible water , J, Natural gas Industry, 2002,22 (S1): 109-112

[7] Cui Yingchun, Zhang Yan. Indoor realization of the irreducible water saturation of rock specimens in low permeability gas reservoirs , J, Oil drilling and mining technology, 2000,22 (4): 11-14

[8] Guo Ping, Huang Wei Gang, Jiang Yi Wei, etc. Research on the restraint and movable water of tight gas reservoirs , J, Natural gas industry, 2006,26 (10): 99-101

[9] Shangju. Determination of irreducible water saturation based on mercury pressure and phase infiltration experiment , J, Domestic and foreign logging technology, 2009, 169: 51-52

[10] Liu Jiankun. Study on the damage mechanism of fracturing fluids in low permeable sandstone gas reservoirs , D. Langfang: Graduate School of the Chinese Academy of Sciences (Institute of Seepage Fluid Mechanics) Ma dissertation, 2011

[11] Guo Hekun, Liu Jiankun, Li Taiwei, etc. Study on the irreducible water saturation of low permeability sandstone gas reservoirs , J, Daqing Petroleum Geology and Development, 2012,31 (2): 91-97

[12] Zhu Qi. Study of water sensitivity damage of tight sandstone reservoirs by NMR, M. Petrochemical Applications, 2014, 33 (5).

IOP Conf. Series: Earth and Environmental Science **64** (2017) 012041 doi:10.1088/1755-1315/64/1/012041

Study on fracture identification of shale reservoir based on electrical imaging logging

Zhou Yu[1], Fuqiang Lai[1, *], Lei Xu[2], Lin Liu[1], Tong Yu[1], Junyu Chen[1], Yuantong Zhu[1]

[1]Key Laboratory of Complex Oil and Gas Field Exploration and Development in Chongqing University of Science and Technology, Chongqing, China
[2]PetroChina Halfaya FZ-LLC, Beijing, China

[*]Corresponding author e-mail: laifq1982@163.com

Abstract. In recent years, shale gas exploration has made important development, access to a major breakthrough, in which the study of mud shale fractures is extremely important. The development of fractures has an important role in the development of gas reservoirs. Based on the core observation and the analysis of laboratory flakes and laboratory materials, this paper divides the lithology of the shale reservoirs of the XX well in Zhanhua Depression. Based on the response of the mudstone fractures in the logging curve, the fracture development and logging Response to the relationship between the conventional logging and electrical imaging logging to identify the fractures in the work, the final completion of the type of fractures in the area to determine and quantify the calculation of fractures. It is concluded that the fracture type of the study area is high and the microstructures are developed from the analysis of the XX wells in Zhanhua Depression. The shape of the fractures can be clearly seen by imaging logging technology to determine its type.

1. Introduction

In recent years, the research on fractured reservoirs has been paid more and more attention. With the huge demand for energy and the complexity of oil and gas exploration, fractured reservoirs have been continuously displayed in oil and gas exploration and development. Out of importance. Especially for the shale reservoir, the development of the shale fractures is an important factor in determining the quality of the shale oil and gas reservoirs. Generally speaking, the higher the fractures, the higher the enrichment degree of the reservoirs and the fractures more developed gas reservoir, its quality is relatively good.

In the exploration and development of shale oil and gas reservoirs, it is very important to study the shale fractures. At present, scholars at home and abroad have done a lot of research work, there are many ways to identify mud shale fractures, in theory and methodologies and other aspects have made important achievements, awareness and progress. It mainly includes the types and causes of fractures, the qualitative and quantitative identification methods of fractures, the characteristics of fracture development and the performance parameters of fractures and the distribution prediction of fractured reservoirs [1]. At present, foreign scholars have made important achievements in mud shale identification and characterization, genetic type and distribution law, basic parameters and physical

IOP Conf. Series: Earth and Environmental Science **64** (2017) 012041 doi:10.1088/1755-1315/64/1/012041

parameters research, reservoir performance evaluation and so on [2]. The fractures can be identified by geological methods, logging methods, seismic methods, drilling logging methods, and intertrochanical interference pressure tests. The use of logging data to identify and evaluate the fractures is an effective method, in the oil and gas exploration work has been widely used. However, due to the uneven development of fractures and the characteristics of reservoir anisotropy, the evaluation of such oil and gas reservoirs is difficult. Imaging logging provides reliable and intuitive image data for accurate identification and evaluation of fractures, but is constrained by economic benefits. Imaging logging is not as extensive as conventional logging. In this paper, the core types of Zhanhua Depression in Shengli Oilfield are used to analyze the types of fractures and identify the shale fractures through the technical means of imaging logging technology.

2. Overview of the study area

Shengli Oilfield Zhanhua Depression is a tectonic unit in the Jinan depression in the Bohai Bay Basin. Depression in the north near the north east of Yihe Zhuang raised its south near the east and west to Chenjiazhuang raised. The study area as a whole constitutes a north-east direction of the open mountain basin-like basin [3]. The well XX is located in the western wing of the Luojia Nose-shaped structural belt in Zhanhua Depression, Jiyang Depression. The Luo Nian-shaped tectonic belt is a large underwater uplift near the north-south, north of Bonan sag, south of Chenjiazhuang raised. The Luo-tang-shaped tectonic belt deposited several hundred meters of dark mud shale during the third trimester, and a thin layer of limestone and ointment was deposited in the upper part of the sand. The strata and the lower South super complex on the Paleozoic unconformity, north into the Bonansag [3]. At the same time, the Neoproterozoic Paleogene strata in this area are cut by several northeast or near-eastward extension of the basin.

3. Electrical imaging logging to identify mud shale fractures

In the process of oil and gas exploration and interpretation, it is necessary to identify the type of fractures, which can be divided into the fracture development section, so as to determine a better reservoir section, and then combined with laboratory core analysis data and field test data to determine Oil and gas layers. FMI imaging logging is an effective method for identifying fractures and is a new logging technique developed to accommodate complex heterogeneous reservoirs. The core analysis results are compared with the imaging log images, and the fractures can be accurately identified and evaluated. In the process of imaging comprehension, the data need to be processed to obtain the static and dynamic imaging of the borehole wall after the single well imaging logging data is processed, and the fractures around the wellbore are extracted. The emphasis is on the qualitative and quantitative identification of the electrical imaging logging Method of Fractures in Shale Reservoir.

3.1. Qualitative identification of fractures

The fractures are the rupture of the rock formation, and the rock is not obvious on both sides of the rupture surface. It is divided into natural fractures and induced joints by type. The natural fractures are usually formed by multi-stage tectonic movement, which is affected by precipitation, The fracture is usually less regular and the slit width changes greatly. The induced slit is due to the influence of the drilling and the formation is drilled to cause the formation stress to be released in the wellbore induced by the fractures. Therefore, the induced joints are neat and regular Strong, no change in the width of the slot, radial extension of small.

(1) Natural fracture electrical imaging response characteristics

According to the conductivity of the division, the natural fractures are divided into high-profile joints and high-resistance joints, high-profile joints are generally not filled with other substances to open the fractures in the drilling process by the mud intrusion, resistivity smaller than the surrounding rock Dark or black; and high-resistance joints are generally due to tectonic stress after filling the cementing of other minerals, the image shown as light or bright yellow sine lines.

According to the shape, the natural fractures are divided into five types: high angle seam, oblique seam, low angle seam, horizontal seam and irregular mesh seam. On the imaging chart, there are different side lines with different angle of inclination angle. And light-colored, mesh ically controlled generally appear in carbonate rocks, caused by a strong dissolution, different shapes of high-profile profiles shown in Figure 1 a-d, and high-resistance see Figure 1-e.

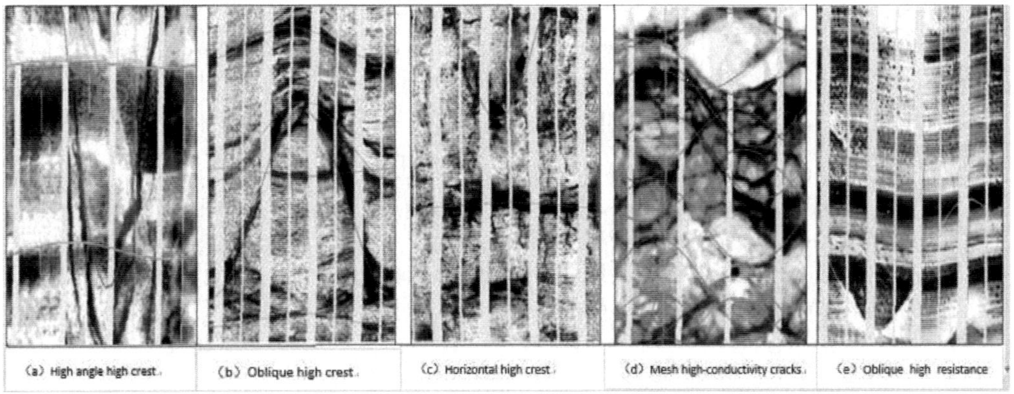

Figure 1. Electronographic image interpretation template for natural fractures

(2) The characteristics of induced seam electrical imaging response

The induced joints are generally divided into drilling tools, vibrating geese, fractures, stress relief joints, respectively, described below in the imaging image on the characteristics of:

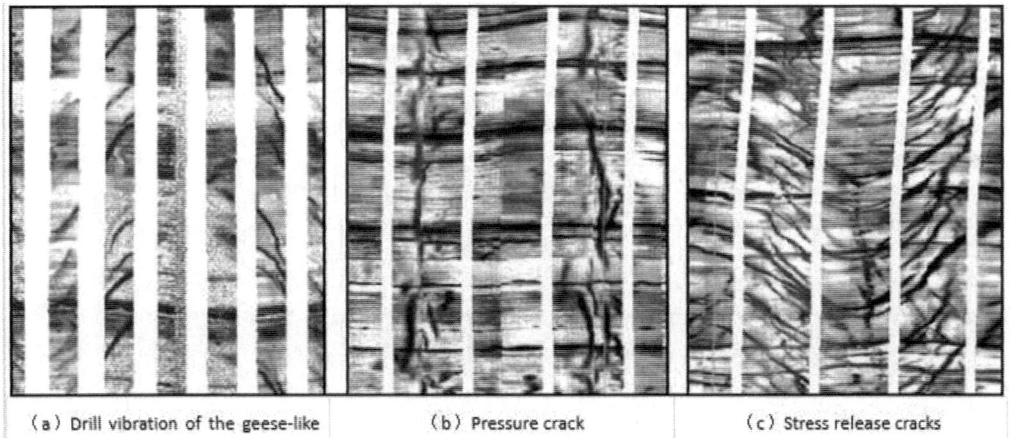

Figure 2. Electronographic image interpretation template for induced fractures

Ge-like seam: fractures caused by drilling tools, the seam is very small and the extension is shorter, showing a geese or feathers, see Figure 2-a.

Fracture: Fractures caused by heavy mud or unbalanced geostresses, with shorter radial extensions, but with larger openness, longer longitudinal extension, and two dark vertical bars with 180 degree azimuths on the imaging image. Figure 2-b.

Stress release joints: generally appear in dense carbonate rocks, are produced by the release of paleo-tectonic stress and modern tectonic stress release. The stress relief slit is clearly visible on the borehole wall image, which is a group of rules close to the parallel seam High angle fractures, see Figure 2-c.

3.2. Quantitative evaluation of fractures

The basic parameters of the fracture are the important parameters of the reservoir, and the quantitative evaluation of the fracture mainly includes the orientation, inclination, fracture width, fracture length, fracture density, average hydrodynamic width and fracture porosity (fracture porosity).

(1) Fracture width (FVA)

Fracture width refers to the fracture as the opening degree. The fracture width in each direction of the fracture trajectory and the direction of the fracture is the single point fracture width, and the average of all the single point fracture widths on the fracture track is the fracture width. The accuracy of the calculation results of the fracture width quantitatively affects the calculation of other fracture parameters. In this paper, the calculation formula of the fracture width of Schlumberger's finite element method is adopted.

According to the finite element method, the anomaly of the conductivity at the fracture is related to the width of the fracture. The abnormal value of the conductivity can be expressed by the curve. The integral area A of the curve is determined by the opening degree of the fracture and the resistivity of the intrusion zone near the borehole wall. The following calculation of the fracture width is given:

$$W = C \cdot A \cdot R_m^b \cdot R_{xo}^{(1-b)} \tag{1}$$

In formula (1):

W-- Fracture width, in units (mm);

R_{xo}-- Flush with resistivity (LLS after calibration plate data instead), unit ($\Omega \cdot m$);

A-- The area of the electrical conductivity caused by the fracture, the unit (mm/$\Omega \cdot m$);

R_m-- mud resistivity, unit ($\Omega \cdot m$);

C, b -- Instrument-related constants, = 0.004801, b = 0.863, constants associated with instrument structure

(2) Fracture length (FVTL)

The length of the fracture is the sum of the length of the fracture seen per square meter of the borehole wall. The unit is m/m^2 or 1/m, the path of the fracture is picked up by means of human-computer interaction, and then the length of the fracture track is calculated by means of the method of elemental integration. Long range (0.6096m) statistics obtained fracture length.

(3) Fracture density (FVDC)

The fracture density is the number of fractures in the unit section.

(4) Average hydrodynamic width (FVAH)

The average hydrodynamic width is a measure of the width of each fracture in the unit section, and then the value obtained by opening the cubic is the fitting of the average hydrodynamic effect of the fracture in units of mm.

(5) Fractures Porosity (FVPA)

The apparent porosity of the fracture is the area of the fracture that can be identified on the 1m borehole wall image divided by the whole image of the 1m borehole wall. Since the imaging image mainly reflects the borehole wall feature, the fracture porosity can be expressed by the porosity,

$$\phi_f = \frac{S_{\phi_f}}{S} \tag{2}$$

$$S_{\phi_f} = \sum (W_i \cdot L_i) \tag{3}$$

In the formula (2): ϕ_f - the fracture surface rate, fraction, W_i - the width of a single fracture, L_i - the length of a single fracture, S_{ϕ_f} - the pore area of the fracture, m^2 and S - the total surface area of the rock,m^2.

3.3. Application of fracture identification

Based on the qualitative identification of the fractures in the third sub-section of the XX wells, the microstructures of the microstructures with high angle are also developed, and the microstructures are also developed (Figure 3). The relative development of the total fractures is 2970-2971m, 2981-2983m, 3005-3013m, 3028-3032m, 3038-3058m, 3075-3077m, 3081-3082m, 3109-3111m, 3115-3122m, 3128-3132m, total 51m, 3005-3013 m, 3038-3058 m, 3075-3077 m, 3115-3122 m, 3128-3132 Thickness, high fracture parameters, is the most favourable fractured reservoir section, lithology is mainly gray matter mudstone, gray matter mudstone and Oil shale (Figure 4). Through the quantitative identification of fractures, it is concluded that the depth of the well is 2996-3060 m in the middle and low angle seam, mainly in the depth of 3060-3125 meters, the fractures in the high angle of the main, the overall direction of the fractures better. The poor oil layer appears in layers 1, 3, 4, 5 and 7, the oil layer appears in the layer 2, 6, the maximum number of fractures is 16, the minimum is 2; the maximum fracture density is 1.63, the lowest is 0.40; The maximum fracture width is 275.32μm, the lowest is 40.08μm, the porosity is between 0.010-0.078 (Table 1).

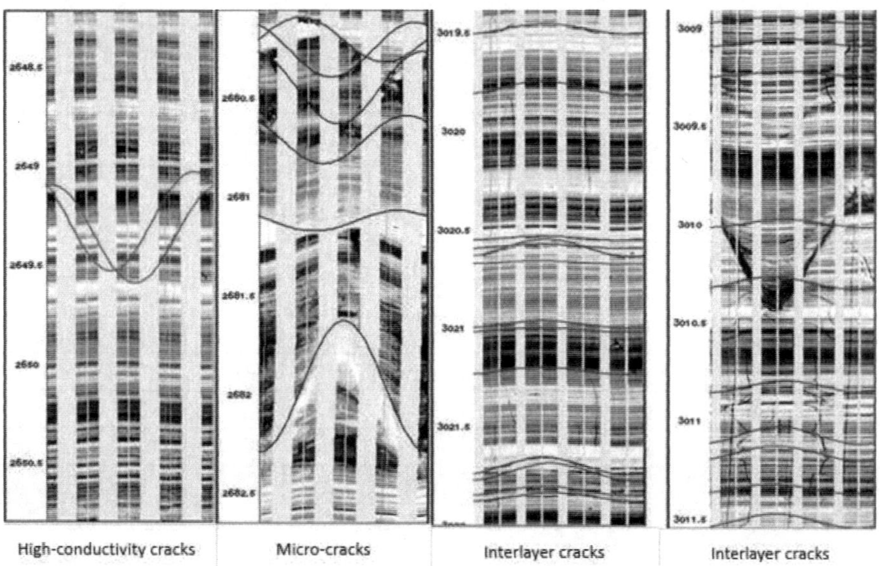

Figure 3. Electronographic image of the fracture of the well

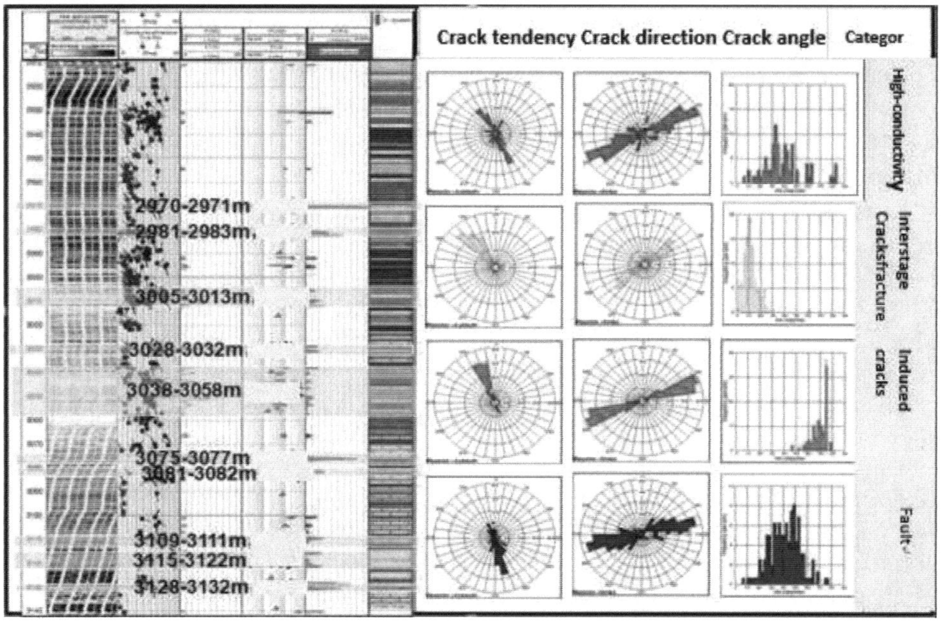

Figure 4. The quantitative identification results based on electrical imaging

Table 1. Calculate the results of quantitative calculation of fractures

Layer No.	Top deep (m)	Bottom deep (m)	Thick-ness (m)	FMI process result					Overview
				No. of fracture	Fracture angle	FVDC ,1/m	FVAH, 1/m	FVPA, %	
1	2996	3019.7	23.7	13	Low-medium	0.78	82.15	0.021	Poor oil layer
2	3019.7	3042	22.3	8	Low-medium	0.58	105.9	0.022	Oil dry layer
3	3042	3057	15	16	Low-medium	1.63	40.08	0.01	Poor oil layer
4	3057	3067.2	10.2	2	Medium - high	0.4	66.98	0.012	Poor oil layer
5	3067.2	3081.1	13.9	3	high	1.17	92.36	0.057	Poor oil layer
6	3081.1	3123.8	42.7	14	Medium - high	0.65	118.86	0.024	Oil dry layer
7	3123.8	3135.7	11.9	5	Medium	0.58	275.32	0.078	Poor oil layer

4. Conclusion

(1) In general, the fractures of the wells in the XX wells are sporadic, and the microstructures are mainly developed. The fractures of the whole well are not developed, and the development of the fractures is 7.2 m. The direction of the high crevices and micro faults is mainly northeast and southwest.

(2) From the fractures identified in the imaging logging, the fractures of the wells in the XX well are sporadic, and the microfacies are mainly developed. The velocity of the high trough and microfractures is mainly northeast-southwest, the fracture line density is 0.04 ~ 1.63 / m and the width is 0.40 ~ 2.75 mm.

(3) Thickness between 10m-45m, more than 15m thick thickness of the shale accounted for more than 80%. The average porosity is between 1.0% and 6%, and the average porosity is 5.7% in the 3067.2m-2081.1m shale section. The overall average effective porosity is 4.6%. The effective porosity shows the characteristics of "middle high, top and bottom low".

IOP Conf. Series: Earth and Environmental Science **64** (2017) 012041 doi:10.1088/1755-1315/64/1/012041

Acknowledgements
The Project was supported by National Natural Science Foundation of China (Grant No. 41402118), and the innovative training program for college students in Chongqing (201611551310).

References

[1] Ding Wenlong, Xu Changchun, Jiukai, et al. Research progress of shale fractures [J]. Advances in Earth Science, 2011, 26 (2): 135-144.
[2] Long Pengyu, Zhang Jinchuan, Tang Xuan, et al. Characteristics of slime fracture development and its influence on shale gas exploration and development [J]. Natural Gas Geoscience, 2011, 22 (3): 525-532.
[3] Wang Yong Shi, Li Zheng, Gong Jianqiang, and so on.Evaluation method of shale oil and gas in Jinan depression - A case study of Luojia area in Zhanhua depression [J]. Journal of Petroleum Science, 2013, 34 (1): 83-90.
[4] Wang Yong Shi, Wang Weiqing, Hao Yun light.Shale reservoir characteristics of the Paleogene Shahejie Formation in Luojia area, Zhanhua Depression, Jinan depression [J]. Journal of Palaeogeography, 2013, 15 (5): 657-662.
[5] Liu Huimin, Zhang Shoupeng, Wang Pu, and so on.Petrographic characteristics of the lower member of Shahejie Formation in Luojia area, Zhanhua Depression [J]. Oil and Gas Geology and Oil Recovery, 2012, 19 (6): 11-15.
[6] Zhang Lin Ye, Zhu room, Li Zheng, and so on. The possibility of shale gas resources in the Paleogene of Jinan depression [J]. Natural Gas Industry, 2008, 28 (12): 26-29.

IOP Conf. Series: Earth and Environmental Science **64** (2017) 012042 doi:10.1088/1755-1315/64/1/012042

Sequence stratigraphy and sedimentary study on Mishrif formation of Fauqi Oilfield of Missan in south east Iraq

Hua Sang[1, *], Changsong Lin[2] and Yiming Jiang[3]

[1]CNOOC International Limited, Beijing, China
[2]School of Energy Resource, China University of Geosciences, Beijing, China
[3]School of Geoscience, South west petroleum University, Sichuan, China

[*]Corresponding author e-mail: sanghua@cnoociraq.com

Abstract. The reservoir of Mishrif formation has a large scale distribution of marine facies carbonate sediments in great thickness in central and south east Iraq. Rudist reef and shoal facies limestones of the Mishrif Formation (Late Cenomanian - Middle Turonian) form a great potential reservoir rocks at oilfields and structures of Iraq. Facies modelling was applied to predict the relationship between facies distribution and reservoir characteristics to construct a predictive geologic model which will assist future exploration and development in south east Iraq. Microfacies analysis and electrofacies identification and correlations indicate that the limestone of the Mishrif Formation were mainly deposited in open platform setting. Sequence stratigraphic analyses of the Mishrif Formation indicate 3 third order depositional sequences.

1. Introduction

The Missan Oilfields are located to the south east of Iraq, adjacent to the border of Iraq and Iran, the transitional zone between Zagros Mountain and Arabian Plate, the southern Mesopotamian basin, which is about 350km of the southeast of Baghdad and approximately 175km to the north-north east of Bashra. It includes 3 on production oilfields: Buzurgan, Fauqi and Abu Ghirab. The target Fauqi oilfield across the border of Iraq and Iran. (Figure 1.1) There are two oil-bearing sequences: the upper is Tertiary Asmari formation and the lower is Cretaceous Mishrif formation. The Mishrif formation is a typical reef and shoal facies carbonate reservoir with a huge thickness. Geological and geophysical studies indicated that the sedimentary landform here is almost flat, the carbonate deposition is sustained and steady. Vertically it was devided into 3 oilgroups which is named as MA, MB, MC (the Abb. of oilgroup A, B and C of Mishrif formation). Sedimentary studies shows a facies characteristics of intra-platform of shoal and open sub-tidal deposits vertically, Especially the intra-platform of shoal and open sub-tidal deposits were mainly deposited in MC2 (the Abb. of oilgroup C; layer 2 of Mishrif) and MB21 (the abb. of oilgroup B; layer 2 sublayer 1of Mishrif), which are the two enriched and much thicker reservoirs compare with MA and other sub-oilgroups. Especially, The bioclastic shoal reservoirs of upper to middle interval of MB21 have good vertical connection and lateral extension, with such a stable and huge thickness of about 80m across the area, and is easily to do lateral tracing and comparision. It is the main target reservoir and oil-produced pay in Mishirf formation of Fauqi oilfield.

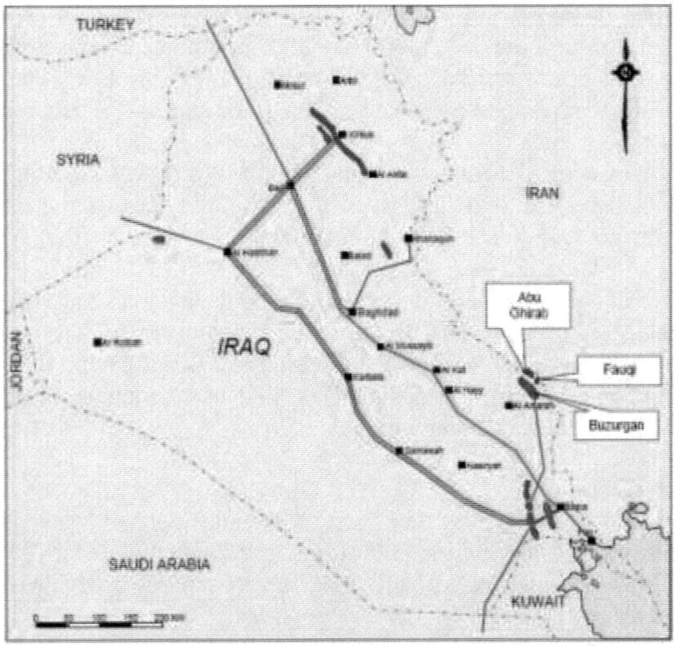

Figure 1.1 Location map of Missan Oilfield

2. Regional Geological Setting

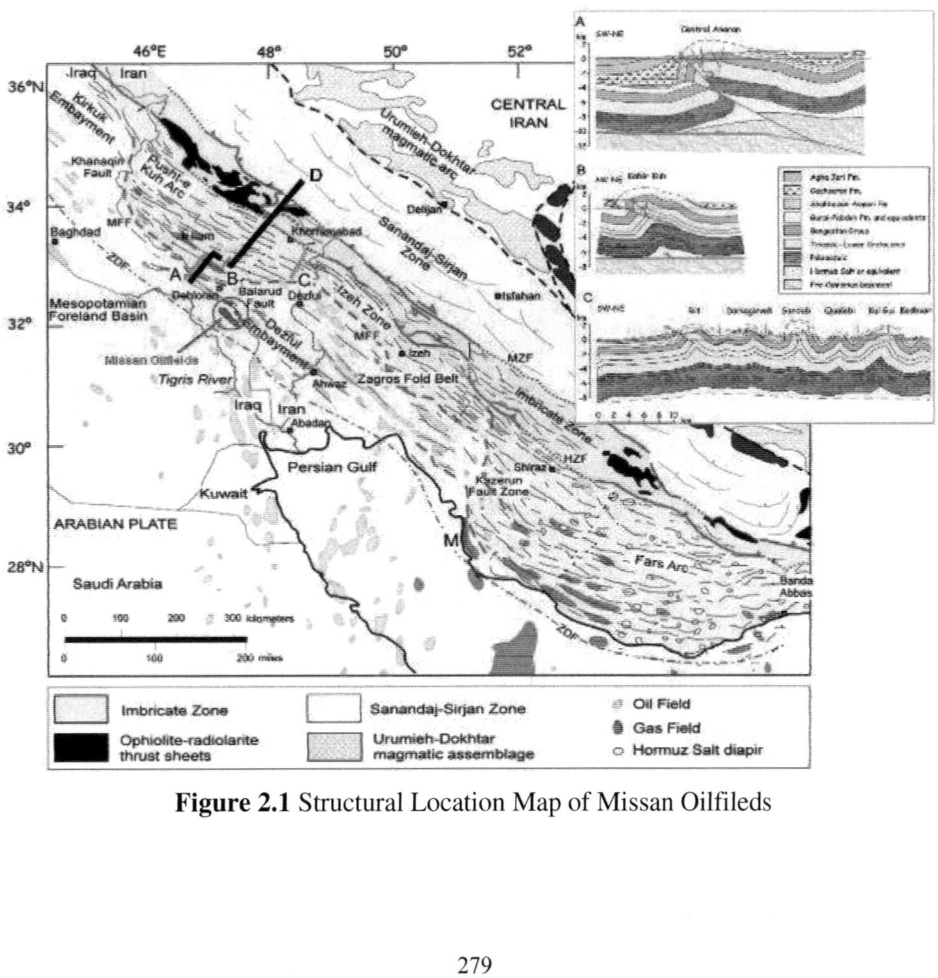

Figure 2.1 Structural Location Map of Missan Oilfileds

IOP Conf. Series: Earth and Environmental Science **64** (2017) 012042 doi:10.1088/1755-1315/64/1/012042

Missan oilfield is located in the transitional zone between Zagros Mountain and Arabian Plate, the southern Mesopotamian basin (Figure 2.1). More than 3500m of Cretaceous sediments were deposited in this basin under the influence of tectonic, eustatic and climatic controls. During Cretaceous period, the basin formed part of a larger-scale carbonate platform located on the NE passive margin of the Arabian Plate.

There are two anticlines with NW-SE orientation developed in the Missan Oilfields, which namely eastern anticlinal belts and western anticlinal belts. Fauqi oilfields rest on the eastern anticlinal belt which dipped southward gradually. The Formation develops inheritantly from Asmari Formation to Mishrif Formation.

A thick carbonate succession was deposited varying from pelagic basinal facies to grain-rich reef and foraminifera-rich lagoonal facies. The Mishrif group was deposited during generally high sea levels of the late Cenomanian to middle Turonian. Sedimentary background of Mishrif formation is mainly open marine carbonate platform deposits. This kind of epicontinental deposit featured as a steady reservoir distribution with good continuity.

3. Reservoir Characteristics

The Mishrif Formation was dominated by limestone. The main reservoir lithology of interval MB21 consists mainly of bioclastic (mainly rudists; grainstone) limestone, micrite, chalky limestone and etc. It was divided into 3 members of MA, MB and MC, which can be subdivided into 6 zones and 6 subzones as well as 7 layers.

The three main pore types of Mishrif formation are dissolved vug, matrix pore and micro fracture-pore. Base on the fracture study, Due to low density of the fractures, the pore type remains as the major storage space for oil while the fractures act as an important conduction that is the key to improve permeability. Fractures, especially the high angle fractures, are less developed and mainly made up of filled fractures in Mishrif reservoir. Microfractures zone only serves as a communication channel between the upper and lower layers.

The reservoir of main oil intervals MB21 and MC in Buzurgan oilfield is characterized by middle to high porosity and low to middle permeability, whereas that in Fauqi oilfield is middle porosity and extreme low to low permeability.

The OWC (Oil & Water Contact) of MB21 sub-zone have not been drilled, but it was confirmed in southern block of MB22 sub zone and both southern and northern block of MC sub zone, which is 4125mSS. Since the net pay thickness of Mishrif is over 100m, the faults could not be the barriers of the oil flow. According to the analyses above, the reservoir trap type of Fauqi area is defined as anticline –structure trap.

The reservoir quality was mainly controlled by facies and diagenesis. The best reef shoal reservoir was deposited at the high energy area where the relative sea level decrease (Drain1, Drain2 & Drain3, the three high perm zones of MB21). The influence of diagenesis to reservoir growth and heterogeneity mainly is differential dissolution and differential cementation, which has close correspondence to the relative sea level changes. [1, 2]

4. Sequence Stratigraphy

The Cretaceous succession in the Mesopotamian basin can be divided into six super sequences or so-called 2nd order sequences, each of these super sequence is separated by regional-scale unconformities and maximum flooding surfaces which are related to regional-scale tectonic deformation and eustatic sea-level changes. According to this division, Mishrif group is considered to be part of the fourth super sequence. The Mishrif group represents the uppermost part of this super sequence and overlain by the Khasib formation with a sharp contact corresponding to an early-middle Turonian unconformity. By contrast at its lower boundary, the Mishrif group passes gradually into the underlying Rumaila formation, and the two formations cannot be easily distinguished in many wells in Missan oilfield and even across the southern Mesopotamian basin [3, 4].

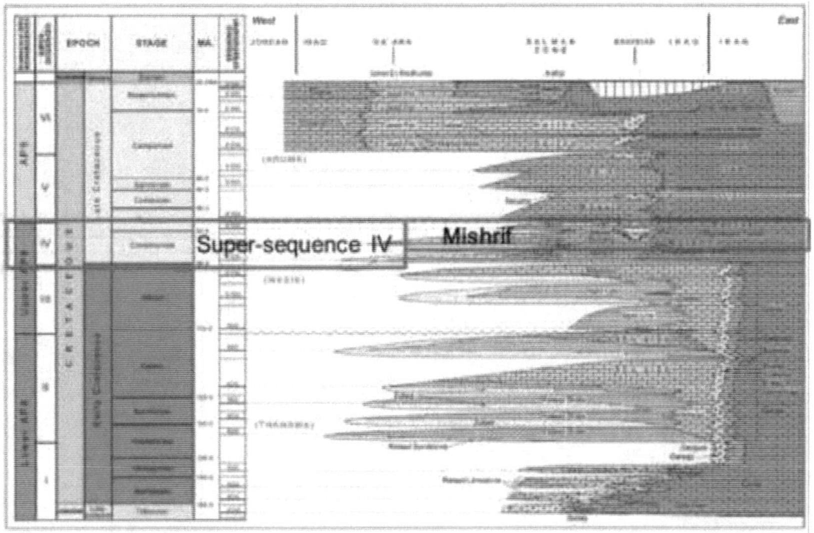

Figure 4.1 Super sequence of Cretaceous in the southern Mesopotamian basin

The Mishrif group has an age gap of about 6 Ma, and the carbonates are heterogenous and include rudistids, bioclastics, alge and foraminifera-rich facies deposited in setting ranging from shallow open marine to lagoonal facies. Reulet (1982) and Aqrawi et al. (1998) suggested that the formation should be divided into two long-term sequences or so-called 3rd order sequences. This division is based on facies evolution and their identification of a regional-scale intra-formational disconformity surface separating the two sequences (Figure 4.1).

However, according to change trend of the fine grains, here it is preferred here that the Mishrif group be divided into three 3rd order sequences. They are named SQ1, SQ2 and SQ3 from the bottom up (Figure 4.2). It should be pointed out that the part of the lowest sequence SQ1 is included in the Rumaila group, precisely, the maximum flooding surface MFS1 is located in Rumaila group. Stratigraphically, the Mishrif and Rumaila group together consist of three 3rd order sequences. Each of them is composed of a lower transgressive hemi-cycle and an upper regressive hemi-cycle. Each of the 3rd order sequences is separated by a disconformity and/or a conformity surface, and all the maximum flooding surfaces are characterized by relatively finer grain, which is mudstone and wackestone. The boundary between SQ1 and SQ2 is the conformity surface on top of MC11, and sequence boundary SB1 is not recognizable on some wells in Fauqi and Buzurgan oilfield. The sequence boundary between SQ2 and SQ3, which is SB2, is the disconformity on top of MB21. It can be evidently identified on cores and well logs, and is marked by the leaching vugs observed on the cores. At present sequence boundary SB2 is thought to be a shortly exposed surface after deposition. Sequence boundary SB3 is Mishrif-Khasib regional unconformity and has an obvious log response. SB3 is underlied by the Mishrif carbonate and overlain by Khasib greenish shale. On the seismic profiles, SB2 and SB3 are identified as MB21 top and MA top respectively, but SB1 cannot be recognized.

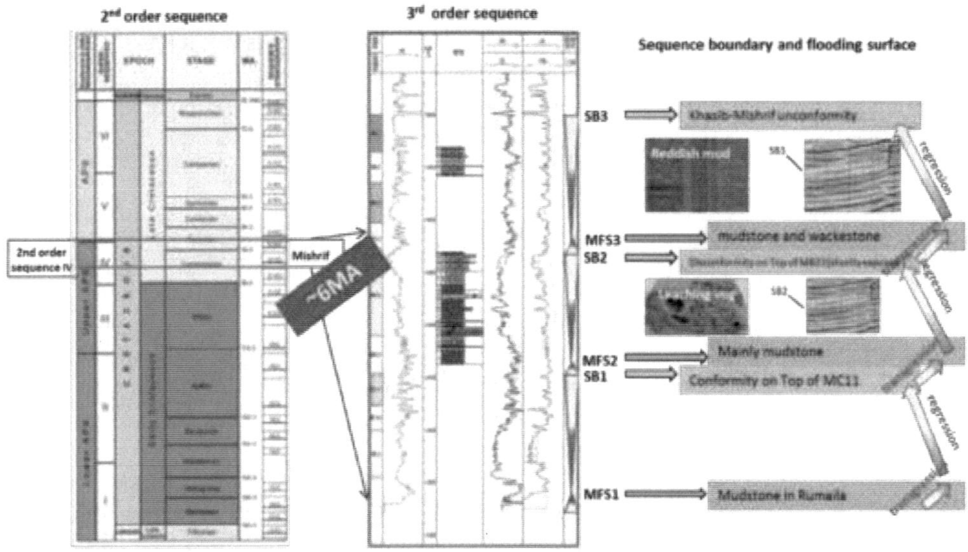

Figure 4.2 The 3rd order sequence division of the Mishrif formation in Fauqi oilfield

The division above could be correlated regionally (Figure 4.3, Figure 4.4). According to the tripartite division, the Mishrif group of Amara oilfield in Iraq and Dehluran oilfield can be correlated to that of Missan. Some authors, including Ali Asghar Taghavi (2006) and T. A. Mahdi(2014), have adapted this tripartite division to Mishrif group in Mesopotamian basin before.

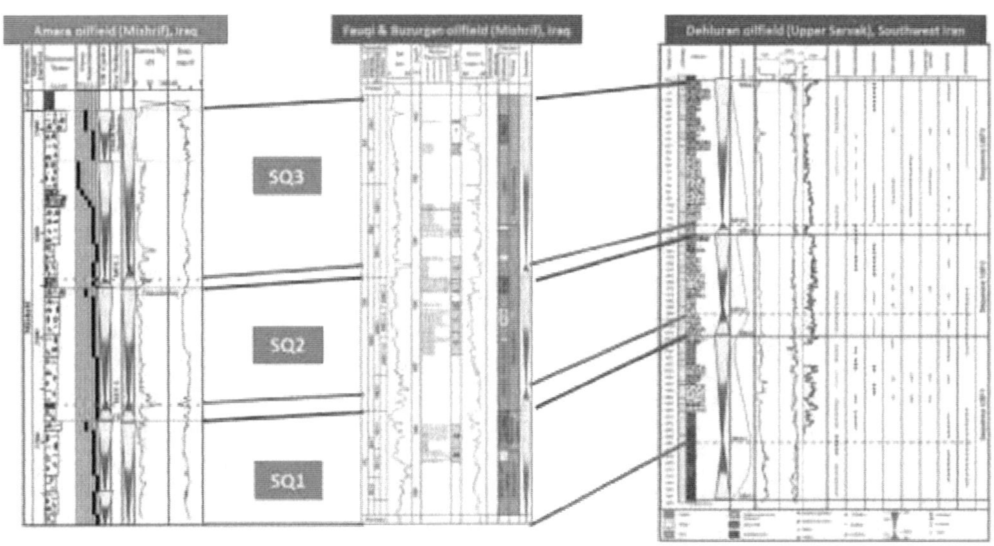

Figure 4.3 Third order sequence correlation profile of Mishrif group across Amara, Fauqi, Buzurgan and Dehluran oilfield

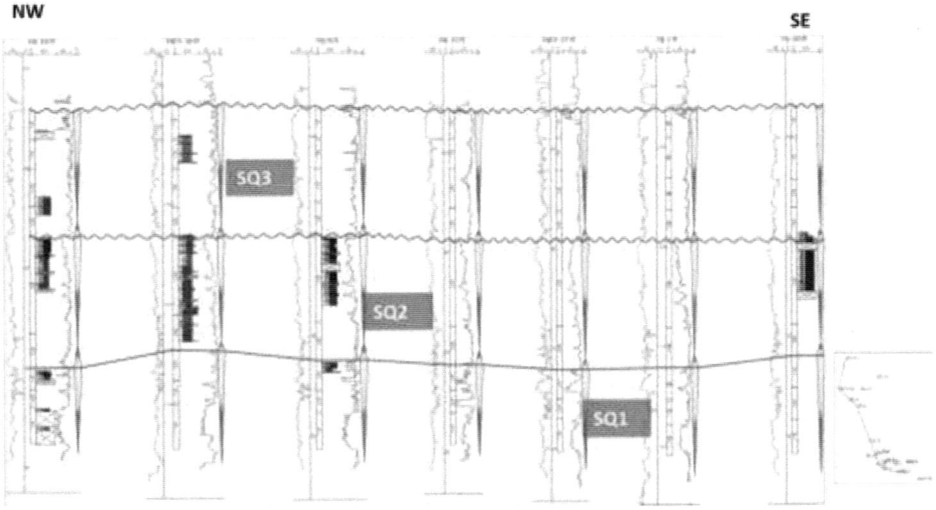

Figure 4.4 The third order sequence correlation profile of Mishrif group in Fauqi oilfield

The division and correlation of the 3rd order sequence would provide general guide and be useful to analyze the sedimentary facies evolution within the whole Mishrif formation. But it will not be enough when the main production zone (sub-member MB21) is studied in details. Therefore, the 4th and 5th order sequence division and correlation should be carried out.

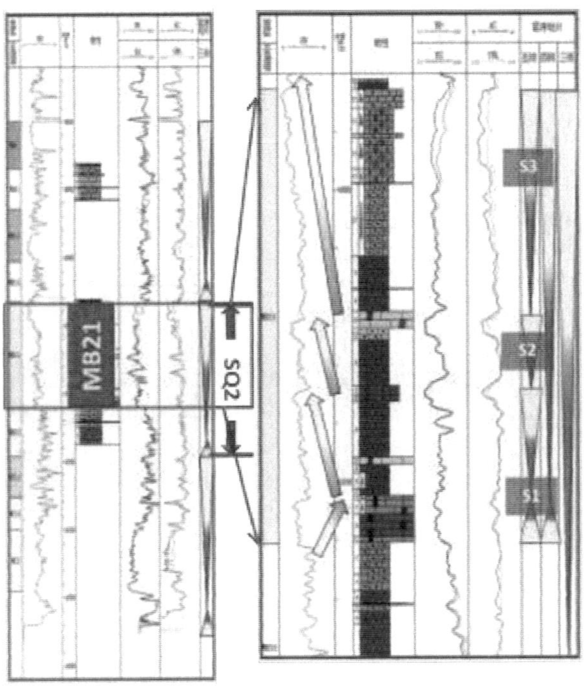

Figure 4.5 The 5th order sequence division of the Mishrif group in Fauqi oilfield

Based on the 3rd order sequence division, the Mishrif group is divided into five 4th order sequences. Sub-member MB21 just takes up one whole 4th order sequence. Consequently, three 5th order sequences S1, S2 and S3 are divided within sub-member MB21 (Figure 4.5). S1 consists of a relative sea-level rising hemi-cycle and a falling one. The relative sea-level rising or water-deepening hemi-cycle is common on all the wells of Fauqi and Buzurgan oilfield. However, the falling or water-shoaling one is not obvious on some wells. This indicates that the 5th order sequence is likely to be autogenetic cycle. On the contrary, the 3rd sequence is allogeneic because of its regional correlation across Missan oilfield and even the southern Mesopotamian basin [5, 6]. Different from sequence S1, the 5th order sequence S2 and S3 merely includes a relative sea-level falling hemi-cycle respectively. Both the core and log data shows evident upwards water-shoaling and grain-coarsening trend. In particular, the lithology changes gradationally from wackestone to peloidal pack stone or even grain stone within each cycle. As a result, it can be concluded that MB21 is dominantly made up of three upwards water-shoaling and grain-coarsening cycle's stratigraphic ally (Figure 4.6).

The 5th sequence division of Mishrif group has been carried out in Fauqi and Buzurgan oilfield, even though the relative sea-level rising hemi-cycle does not always show an obvious upwards grain-fining trend.

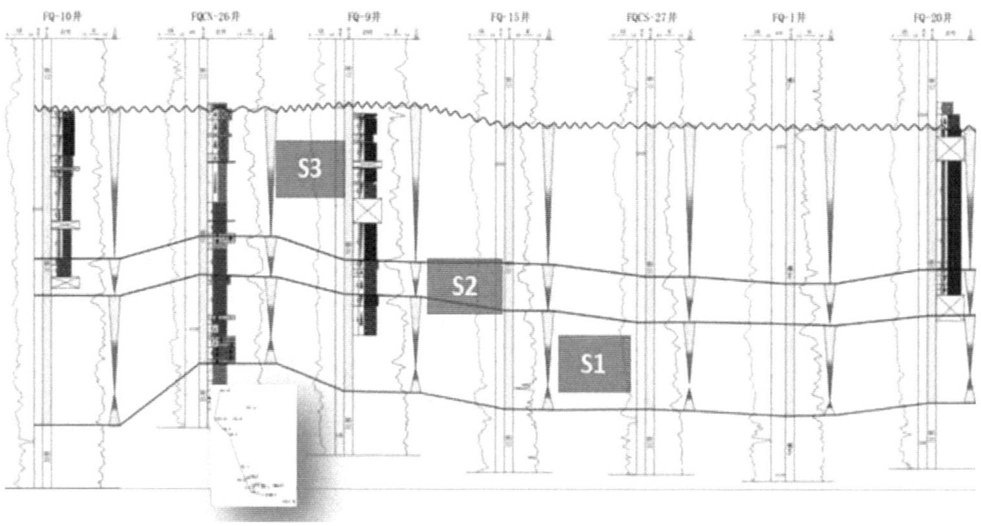

Figure 4.6 The 5th order sequence correlation profile of the Mishrif group in Fauqi oilfield

5. Sedimentary Facies Identification

Lots of literatures has discussed the sedimentary model in the southern Mesopotamian basin, and two different types could be summarized: the reef-rimmed carbonate platform and the carbonate ramp. Some authors like James and Macintyre (1985) and T. A. Mahdi and and A.A.M. Aqrawi (2014) recommend the reef-rimmed carbonate platform. In this depositional model, rudist congregations occupied the platform margin without forming a continuous rim, and may pass laterally into or be replaced by shoal facies, depending on the precise palaeo geographical location within the southern Mesopotamian Basin. The ancient slope break is the important evident characteristic for the reef-rimmed carbonate platform. However, after precisely interpreting the 2D and 3D seismic data which collecting grid covered almost the whole Missan oilfield, no such break have been observed on the seismic profile. The carbonate ramp depositional model for the Mishrif group in Iraq or the upper Sarvak formation in Iran, which was comprehensively discussed by Trevor P. Burchette(1993), A.A.M. Aqrawi(1998) and Ali Asghar Taghavi(2006), is adapted here (Figure 5.1).

The paleo-environment varied from open platform to restricted platform during the Cenomanian–early Turonian Mishrif formation period. Four principal facies are recognized corresponding to palaeo-depositional settings ranging from shallow open marine to lagoon deposit. These facies are summarized as below (Figure 5.2).

IOP Conf. Series: Earth and Environmental Science **64** (2017) 012042 doi:10.1088/1755-1315/64/1/012042

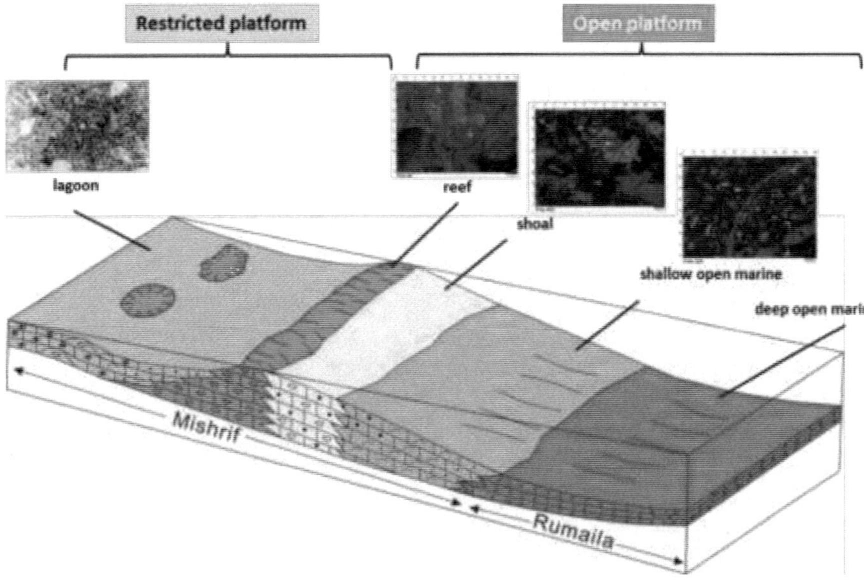

Figure 5.1 Proposed ramp carbonate platform model for the Mishrif formation

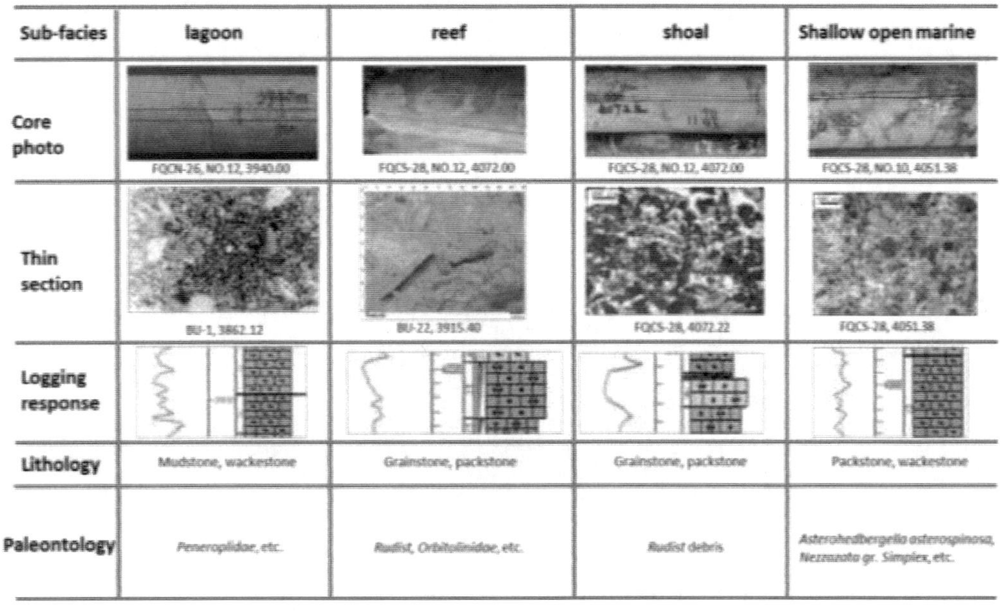

Figure 5.2 Characteristics of sedimentary facies of Mishrif group in Fauqi oilfield

5.1. Shallow Open Marine Facies

Shallow open marine facies comprising massive, thick limestone units consist of bioclastic packstones and wackestones. Bioclasts include mixed debris of rudists, green algae, echinoids and bivalves, together with benthonic and plantonic foraminifera (Figure 5.2). Colonial corals and algae also occur.

Shallow open marine facies become more grain-rich and rudist-rich upwards within the studied well successions. Grain-rich bioclastic limestones in this facies have high porosities and typical blocky motifs on the porosity logs, although stylolites and cementation zones may cause local irregularities in log pattern.

5.2. Shoal Facies

Shoal facies composed predominantly of grain-dominated grainstones and packstones (Figure 5.2). Bioclasts include rudists, echinoderms and algae together with pellets. Allochthonousmolluscs and small benthonic foraminifera also occur locally, although these latter components are poorly preserved indicating long transport.

The shoal facies deposits are interpreted to have been deposited in high energy settings above fair-weather wave base, and as a result contain only little clay. Therefore gamma-ray signatures show only minor deflections, although porosity logs indicate the occurrence of high porosity intervals.

5.3. Rudist Biostromes or Reef Facies

Rudist biostromes are dominated by autochthonous rudists within grainstones and packstones (Figure 5.2). Rudist bioclasts are poorly sorted, frequently broken but hardly rounded, and are often highly compacted.

Within the Mishrif group, rudists do not show features characteristic of reefs. According to Gili et al. (1995), rudist-rich rock units here are referred to as lithosomes and the dense faunal associations as congregations. Intact rudist congregations in the Mishrif group are volumetrically less extensive than the rudist-dominated debris found in slope and shoal locations.

5.4. Lagoon Facies

Lagoon facies were deposited in restricted areas within the Mishrif platform, and are characterized by a diverse benthonic foraminifera fauna including miliolids, alveolinids and textulariids (Figure 5.2). Gastropods, ostrapods, rudist fragments and sponge spicules occur locally, showing little breakage indicating short transport distances. Lagoon facies deposits may be more than 30m thick in Fauqi oilfield.

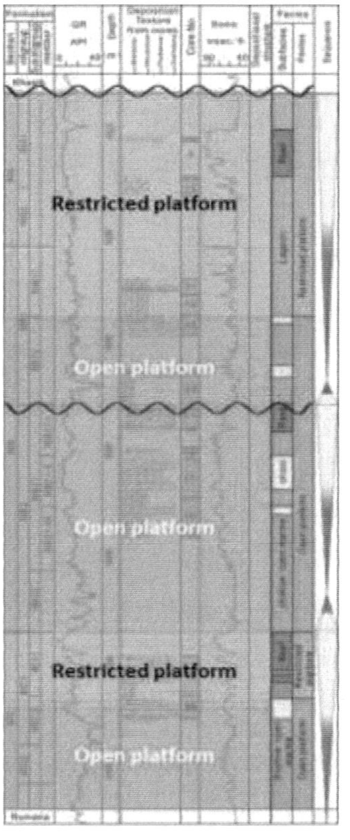

Figure 5.3 The 3rd order sequence division and sedimentary facies of the Mishrif formation

Within the lagoonal successions, thin beds of peloidal and rudistidpackstones and grainstones may occur, interpreted as small-scale shoals and biostromes which developed within the lagoonal environment. Log responses for lagoon are characterized by high gamma-ray reflections and relatively low porosity readings compared with shoal facies.

6. Conclusion

The rock type of Mishrif formation especially MB21 are mainly bioclastic rudist limestone, bioclastic grainstone, bioclastic peloidal packstone. The reservoir space type is primary intergranular pore, secondary intergranular dissolved pore, moldic pore, matrix micropore, a few small dissolved pore and micro-fracture. The sedimentary environment is carbonate open platform facies, which include intra-platform of reef, intra-platform of shoal and open sub-tidal deposits. The reservoir quality was mainly controlled by facies and diagenesis. The Mishrif formation was divided into two long-term sequences or so-called 3rd order sequences. This division is based on facies evolution and their identification of a regional-scale intra-formational disconformity surface separating the two sequences. Based on the 3rd order sequence division, the Mishrif group is divided into five 4th order sequences. Sub-member MB21 just takes up one whole 4th order sequence. Consequently, three 5th order sequences S1, S2 and S3 are divided within sub-member MB21.

Acknowledgments

This work was financially supported by CNOOC Iraq Limited. The main achievement of this work is part of ***Missan Oilfield Comprehensive Geological Research on Fauqi Oilfield***, which is an Appendix of ***Enhanced Redevelopment Plan on Missan Oilfield***. I would like to appreciate Mr. Xinglong Wang, who has given a great help on this paper.

References

[1] Sadooni F N, Alsharhan A S. Stratigraphy, microfacies, and pe-troleum potential of Mauddud Formation (Albian-Cenomanian) in the Arabian Gulf Basin [J]. AAPG Bulletin, 2003, 87(10): 1653-1680.

[2] Al-Qayim B. Sequence stratigraphy and reservoir characteristics of the Turonian-Coniacian Khasib formation in Central Iraq [J]. Journal of Petroleum Geology, 2010, 33 (4): 387-403.

[3] Wilgus. C.K., Posamentier, H.W., Sarg, J.F. etc. Editors, Translaters: Huaida Xu, Kuisheng Wei and weidong Hong: *The Comprehensive analysis of sea level change* of *The Sequence Stratigraphy Theory*. Petroleum Industry Press, [M] Beijing, 1995, pp. 124~125

[4] Alsharhan, A S. and Nairn, A. E. M. A Review of the Cretaceous Formations in the Arabian Peninsula and Gulf, Part II. Mid Cretaceous (Wasia Group) Stratigraphy and Palaeo-geography [J]. Journal of Petroleum Geology, 1988, (11): 89~112.

[5] Sadooni. F.N, Aqrawi A. A .M. Cretaceous sequence stratigraphy and petroleum potential of the Mesopotamian Basin, Iraq [J] Special Publications, 2000: 315-334.

[6] MA Wenxin.Carbonate reservoir characteristics and formation mechanism of Cretaceous Cenomanian-Turonian in MA oilfield, Iraq [D]. Chengdu: Chengdu University of Technology, 2014.

IOP Conf. Series: Earth and Environmental Science **64** (2017) 012043 doi:10.1088/1755-1315/64/1/012043

Study on influencing factors of carbon emissions for industrial energy consumption in Dalian based on LMDI model

Bo Lai, Hongbo Zheng*

School of Environmental Science & Technology, Dalian University of Technology, Dalian, Liaoning, China

*Corresponding author e-mail: dlhongbo@126.com

Abstract. After calculating and analysing the carbon emissions for industrial energy consumption in Dalian during 2007 to 2014, the LMDI model is built to decompose the influencing factors on carbon emissions from the perspectives of energy structure, energy intensity, industrial structure and output scale. As many as 17 types of energy are involved in the calculation of emissions, and based on the specific industrial structure of Dalian, the analysis for the influence on carbon emissions from industrial structure is made from the perspective of light and heavy industry. The result shows, output scale effect is a major contributor for the growth of the carbon emissions for industrial energy consumption in Dalian, while energy intensity effect is the principal inhibitor. Energy structure and industrial structure influence relatively weakly on the energy-consumption emissions, which indicates that the emission reduction potentials from these two perspectives are still not achieved.

1. Introduction

The global warming caused by carbon emissions is one of the world top ten environmental issues. As the world's fastest growing developing country, China has made rapid development in economy, but its energy consumption becomes high simultaneously. Among all the economic sectors of China, the energy consumption of industrial sector accounts for 75 percent of the total energy consumption, and the carbon emissions caused by industrial production makes up 83% of total emissions. This means that industry should be strongly concerned if China wants to reduce energy consumption and carbon emissions.

Dalian, which is located in northeast old industrial base, is one of the most important industrial cities in Liaoning province. While Dalian's industry is showing strong growth, the concern about the industrial carbon emissions is increasingly raised. Therefore, it is necessary to calculate and analyze the industrial carbon emissions in Dalian, so as to find a way to control and manage the emissions.

Emission factor method, mass balance method and actual measurement method are the most popular methods for estimating carbon emissions. As the first method proposed by IPCC for calculating carbon emissions, emission factor method is the dominating calculation method. In this method, estimation results are obtained by multiplying emission factors by activity data [1]. Seeing that mass balance method takes too many intermediate processes into consideration and actual measurement method is too detailed [2], emission factor method is the most suitable one for calculating industrial carbon emissions in Dalian.

Decomposition analysis methods could be split into two categories, structure decomposition analysis (SDA) and index decomposition analysis (IDA). While the method of SDA is based on input-output table, and its calculating process is complicated, IDA only needs a smaller amount of data to conduct a multi-level analysis. IDA could be divided into Laspeyres index and Divisia index. Since Divisia index is relatively flexible and it could decompose influencing factors more completely, it is widely used by researchers. Logarithmic Mean Divisia Index (LMDI) is one of the Divisia index methods. With a firmer theoretical foundation, stronger computability and more interpretable result [3], LMDI is considered as an effective decomposition method.

LMDI is widely used to study carbon emissions by researchers in China. Guo[4] uses LMDI method to decompose China's carbon emissions in 1995-2007 at industrial and regional levels. The results show that expansion of economic scale is the most important factor for the continuous carbon emissions growth and the improvement of energy efficiency is the most important to inhibit carbon emissions' growth. Liu et al. [5] decompose carbon intensity by LMDI method into ten driving factors, which covering energy and non-energy related emissions from industrial and household sectors. Wu et al. [6] use the LMDI decomposition method to decompose the factors of energy carbon emissions in Beijing from the year 1995 to 2010. There are also some researchers focusing on the industrial carbon emissions. Shi et al.[7] first discuss characteristics of carbon emissions from carbon emissions and intensity and some other indicators of the industrial economy in Nanjing for the period 2000-2009, and then build LMDI model to unravel the carbon emission mechanism of the energy sector in Nanjing from the perspectives of industrial scale, energy intensity, energy structure, and energy emission intensity.

In this paper, emission factor method is adopted to calculate the carbon emissions for industrial energy consumption in Dalian. And seeing that Dalian relies more on heavy industry than light industry, the carbon emissions generated from these two kinds of industry are calculated separately. Then, LMDI model is built to decompose the influencing factors on carbon emissions from the perspectives of energy structure, energy intensity, industrial structure and output scale.

2. Carbon emissions for industrial energy consumption in dalian

2.1. Method

The calculation of carbon emissions for industrial energy consumption in Dalian is based on the calculation methods provided by IPCC.

The formula for calculating carbon emissions is shown as the following:

$$\text{Emissions} = \text{AD} \times \text{EF} \qquad (1)$$

Where Emissions is the total carbon emissions come from burning a certain type of fossil fuels in industrial process, EF is its emission factor, and AD is its activity data.

2.2. Result

During 2007 to 2014, the carbon emissions for industrial energy consumption in Dalian increase steadily in early period, and turn to fluctuate later. From 2007 to 2014, the annual emission is 2701.28, 3006.21, 3228.16, 3271.11, 2990.28, 3065.34, 2930.43, 2873.86 $\times 10^4$ t respectively. Carbon emissions rise an average 1.1% annually during 2007 to 2014, and emissions grow fastest in 2007 by 11.3%. As shown in Figure 1, the industrial carbon emissions show a meagre growth. In 2011, 2013 and 2014, the growth rates are even negative.

During 2007 to 2014, the carbon emissions generated from the energy consumption of light industry averagely make up 1.6% of total emissions. However, the output value made by light industry averagely makes up 23.3% of the gross output. This suggests that, to achieve an equal output, the production heavy industry creates more carbon dioxide than that of light industry. Therefore, adjusting the proportion of light and heavy industry in Dalian and optimizing the industrial structure are beneficial for reducing the carbon emissions for industrial energy consumption.

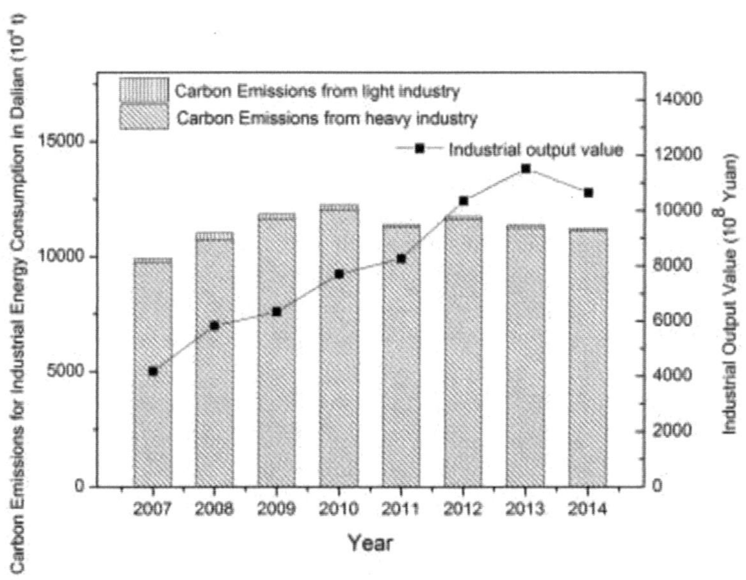

Figure 1. Carbon Emissions for Industrial Energy Consumption and Gross Industrial Output Value in Dalian

3. Decomposition of carbon emissions for industrial energy consumption by LMDI

3.1. Method

Carbon emissions could be influenced by many factors, and each factor has a different influence. In this paper, LMDI model is built to decompose the influencing factors on carbon emissions from the perspectives of energy structure, energy intensity, industrial structure and output scale.

The decomposition formulas are shown as follow:

$$\Delta C_{es} = \sum_i^{17}\sum_j^{2}\frac{c_{ij}^t - c_{ij}^0}{lnc_{ij}^t - lnc_{ij}^0}\cdot\left[ln\left(\frac{E_{ij}}{E_j}\right)^t - ln\left(\frac{E_{ij}}{E_j}\right)^0\right] \quad (2)$$

$$\Delta C_{ei} = \sum_i^{17}\sum_j^{2}\frac{c_{ij}^t - c_{ij}^0}{lnc_{ij}^t - lnc_{ij}^0}\cdot\left[ln\left(\frac{E_j}{GVIO_j}\right)^t - ln\left(\frac{E_j}{GVIO_j}\right)^0\right] \quad (3)$$

$$\Delta C_{is} = \sum_i^{17}\sum_j^{2}\frac{c_{ij}^t - c_{ij}^0}{lnc_{ij}^t - lnc_{ij}^0}\cdot\left[ln\left(\frac{GVIO_j}{GVIO}\right)^t - ln\left(\frac{GVIO_j}{GVIO}\right)^0\right] \quad (4)$$

$$\Delta C_{os} = \sum_i^{17}\sum_j^{2}\frac{c_{ij}^t - c_{ij}^0}{lnc_{ij}^t - lnc_{ij}^0}\cdot\left[ln(GVIO)^t - ln(GVIO)^0\right] \quad (5)$$

Where ΔC_{es} represents the change of carbon emissions caused by energy structure effect; ΔC_{ei} represents the change of carbon emissions caused by energy intensity effect; ΔC_{is} represents the change of carbon emissions caused by industrial structure effect; ΔC_{os} represents the change of carbon emissions caused by output scale effect.

3.2. Result

The result is shown in Table 1.

Table1. Result of industry energy-related emission decomposition for Dalian, 2007-2014

Unit: 10^4 t C

Year	Energy Structure	Energy Intensity	Industrial Structure	Output Scale	Total
2007-2008	11.40	-626.51	-25.56	945.61	304.94
2008-2009	-12.34	10.46	-32.47	256.48	222.13
2009-2010	-5.71	-538.22	12.49	640.52	109.08
2010-2011	30.26	-492.02	7.79	226.88	-227.09
2011-2012	2.95	-555.93	-63.92	708.65	91.77
2012-2013	6.86	-438.23	-7.04	336.50	-101.91
2013-2014	10.80	109.10	81.31	-241.60	-40.40
Total	44.22	-2531.35	-27.40	2873.04	358.51

As shown in Table 1, during 2007 to 2014, carbon emissions for industrial energy consumption in Dalian increase by 358.51×10^4 t, to which energy structure effect, energy intensity effect, industrial structure effect and output scale effect contribute 44.22, -2531.35, -27.40, 2873.04×10^4 t, respectively. It suggests that energy intensity and output scale are major influencing factors to the carbon emissions for industrial energy consumption in Dalian.

As shown in Figure 2, during 2007 to 2014, output scale effect is a major contributor for the growth of the carbon emissions for industrial energy consumption in Dalian, while energy intensity effect is the principal inhibitor. The decline of carbon emissions caused by energy intensity effect could buffer the increase caused by output scale effect.

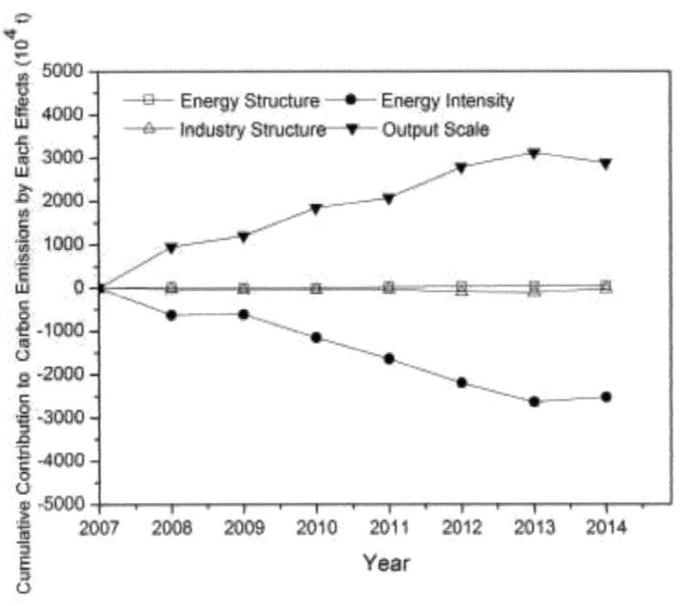

Figure 2. Decomposition of Cumulative Growth in Carbon Emission for Industrial Energy Consumption

Energy structure and industrial structure influence relatively weakly on the energy-consumption emissions, which indicates that the emission reduction potentials from these two perspectives are still not achieved.

4. Conclusion

In this paper, the carbon emissions for industrial energy consumption in Dalian during 2007 to 2014 is calculated, and the result shows that the industrial carbon emissions do not monotonically increase any more, instead they turn to fluctuate. It suggests the energy conservation policies have achieved an initial success.

The result of LMDI shows that, while output scale effect is a major contributor for the growth of the carbon emissions for industrial energy consumption in Dalian, energy intensity effect could buffer the increase caused by output scale effect. Energy structure and industrial structure show relatively meagre influences on the energy-consumption emissions, which indicates that the emission reduction potentials from these two perspectives are still not achieved.

References

[1] IPCC/UNEP/OECD/IEA. Revised 1996 IPCC guidelines for National Greenhouse Gas Inventories, Intergovernmental Panel on Climate Change, Paris, 1997.
[2] M. Liu, J. Meng, B. Liu.: Tropical Geography, 34(2014) 248-258.
[3] B. W. Ang: Energy Policy, 32(2004) 1131-1139.
[4] C. Guo: China Population, Resources and Environment, 20(2010) 4-9.
[5] Y. Liu, X. Li, J. Lin, et al.: Acta Ecologica Sinica, 34(2014) 2378-2387.
[6] Z. Wu, J. Shi, S. Wang: Forum on Science and Technology in China, 2(2014) 85-91.
[7] A. Shi, M. Li: Resources Science, 33(2011) 1890-1896.

The evaluation of the coordination degree between resource-environment and economic development in the yellow river delta-a case study on Dongying city

Kuifeng Wang[1, 2, 3 *], Meng Xu[1]

[1]Shandong Institute and Laboratory of Geological Sciences, Shandong Key Laboratory of Geological Process and Resource Utilization in Metallic Minerals, Key Laboratory of Gold Mineralization Process and Resource Utilization Subordinate to the Ministry of Land and Resource, Jinan China

[2]Key Laboratory of Carrying Capacity Assessment for Resource and Environment, Ministry of Land and Resources (Chinese Academy of Land and Resource Economics, China University of Geosciences Beijing), Beijing, China

[3]School of Civil Engineering, Shandong University, Jinan, China

*Corresponding author e-mail: maplewkf@126.com

Abstract. Dongying city is located in the mouth of the Yellow River and it is the most important central city in the Yellow River Delta. This area's ecological environment is very weak. When we maintain the economic development, we must coordinate the relationship between the environment and economy. This paper used principal component analysis and coordination degree evaluation index system and the calculation model to assess the coordination degree of economic and environmental development in Dongying city from 2011 to 2015.The results showed that the coordinate degree of Dongying's environment and economic development belonged to the type of coordinated development. However, the economic level was significantly higher than its level of the environment. The coordination of the economy and environment in Dongying was not stable, with a certain degree of volatility. This paper analyzed the causes of the coordination degree situation in Dongying, and put forward some suggestions of coordinated development of the ecological environment and economic in Dongying city.

1. Introduction

With the increasingly serious global environmental problems, the sustainable development becomes the consensus of the world. As we all know, the environment is not only the resources but also the restriction conditions of development. Today, the essence of environmental problems is that the speed of human asking to the environment for resources in the economic activities is faster than the regeneration speed of resource itself and its substitutes significantly. And the quantity of waste discharged into environment more than the self-purification ability of environment. Therefore, coordinated development has become the best choice to deal with the relationship between environment resources and economic development. It is the only way to realize the sustainable development strategic goal of the human society (Zhou Detian, et al, 2010; LU Ning, et al, 2015; Li N, Wang Kuifeng,2016, DU Zhongchao, B Wenting, 2015.).

Yellow River delta is located in the northeast of Shandong province. It is the only river delta which has not been fully developed and it is a piece of young land. Because of the Yellow River delta became land time is short, its ecological environment is fragile and the environmental bearing capacity is low. The environmental quality must been improved to realize the sustainable development of the Yellow River delta regional economic. Promote coordinated development of environment and economy. Therefore, this article selects environmental and economic data of Dongying City in 2011-2015, using principal component analysis (PCA) method and the data model to calculate the coordination degree between them, to fully recognize the coordination degree of Dongying City resources environment and economic development situation. To make corresponding policy adjustment according to the changes of the coordination degree in a timely manner, in order to promote the efficient ecological economic zone of Dongying City and even the sustainable development of the Yellow River delta. The coordination level between the two states were calculated with coordination degree C model, Countermeasures and suggestions were put forward based on the evaluation results, the index and flow chart of research were shown in the following table 1, table 2.

2. Construction of Index System and Model

2.1. Index Selection Principle

The sustainable development evaluation index system of oil and gas resource-based city reflected the development present situation and trends of economy, society, resource, environment four systems in complex system, as well as the coordination of the four systems. In addition to consider the basic ideological and the general principles of sustainable development, it should reflect the characteristics of oil and gas resources city. The system of urban economic development and resources environment is complex, so it should be evaluated objectively and right. We must establish a set of quantitative factor index system that could reflects regional economy and environmental quality from different sides and possess statistical significance under the guidance of scientific, comprehensive, operability, dynamics and stability. According to the specific condition of Dongying City and the difficulty to obtain index, we selected Xi (i = 1, 2, 3,..., 11) as the indexes of economic characteristics to set up the economic evaluation index system (table 1), and selected Yi (I = 1, 2, 3,..., 8) as the indexes of resources environment characteristics to set up the resources and environment index system (table 2). The index system of resources environment and economy coordinated development of Dongying City has been constructed.

The resources and environment indexes in this system included qualitative and quantitative indexes, and the social and economic indexes in this system were all quantitative indexes. Quantitative indexes were calculated with the data which mainly referred to Shandong Statistical Yearbook, Dongying Statistical Yearbook, Dongying Yearbook, Yearbook of Counties of Dongying, National Economy and Society Developed Statistical Bulletin of Dongying, China Environmental Statistics Yearbook, Dongying Water Gazette and Dongying Environment Gazette in 2011-2016. Qualitative indexes were mainly graded through various industries (e.g., land, environmental protection, land, forestry, Marine and other) investigation and study reports in recent years, their credibility and reliability were high.

Table.1 Resource and Environmental indicators, Dongying City, 2011-2015

year indicators	2011	2012	2013	2014	2015
Resources abundance of oil and gas	2.8	3.2	3.4	3.4	3.8
Per capita area of available cultivated land (Ha/per)	0.152	0.15	0.142	0.14	0.136
Land development intensity (The proportion of construction land area) (%)	17.702	18.024	18.536	19.166	19.646
Total water consumption (10K m^3)	100212	93300	90274	92354	97478
Repeat utilization rate of industrial water (%)	50.792	45.32	54.27711	72.594	68.04
Irrigative water effective utilization coefficient	0.627	0.621	0.63904	0.6438	0.65
Qualified rate of industrial waste water discharge (%)	98.2	97.8	98.8	99.2	99.6
Decontamination rate of urban refuse (%)	100	100	100	99.4	98.4
Qualified rate of centralized drinking water sources quality (%)	100	100	100	99.6	98.8
Fineness rate of air quality (%)	90.8	89.6	86.6	83.4	80.4
Mean annual concentration of inhalable particle (mg/m^3)	0.1196	0.1232	0.1446	0.1434	0.1572
Mean annual concentration of SO_2 (mg/m^3)	0.4244	0.4524	0.7414	0.7924	0.8466
Mean annual concentration of NO_2 (mg/m^3)	0.0448	0.053	0.0546	0.0602	0.0658
Solid waste comprehensive utilization rate (%)	94.72	96.114	97.096	97.158	98.124
Per capita area of urban greenbelt (m^2)	21.232	24.68	26.76	27.318	27.848
Forest coverage rate (%)	22.698	21.71	23.466	24.694	24.616
Wetland area relative ratio	2	2.2	2.4	2.6	3
Shoreline development intensity	2.2	2	1.6	1.6	1.6
Regional crustal stability	2.8	2.4	2.2	2.2	1.8
Geological disasters risk	2.8	2.4	2.4	1.8	1.8
Relative area of all kinds of ecological reserve	1.8	2	2.4	2.8	2.8
Pollution level of water and soil caused by oil	3	3	2.8	2.4	2

Remarks: These seven indexes including resources abundance of oil and gas, wetland area relative ratio, shoreline development intensity, regional crustal stability, geological disasters risk, relative area of all kinds of ecological reserve, pollution level of water and soil caused by oil were qualitative indexes.

Table.2 Economic indicators, Dongying City, 2011-2015

year / indicators	2011	2012	2013	2014	2015
Environmental governance investment as a share of GDP (%)	1.468	1.642	1.648	1.662	1.78
Per capital GDP (million yuan)	8.0794	9.364	10.026	11.048	11.424
Fiscal revenue as a share of GDP (%)	9.512	9.952	7.556	6.98	6.672
Growth rate of investment in fixed assets (%)	27.06	23.54	20.648	16	10.904
Disbursement of foreign capital (million dollar)	11806.3	12005.4	15448	18735	11725
Total export-import volume (hundred million dollar)	67.8186	75.64	82.44	86.02	80.35
Urban per capita disposable income (yuan)	24961.64	28119.16	30870.6	33556.4	33614.4
Per capita living space	36.354	37.252	41.03	41.03	39.76
industrial output (hundred million yuan)	5936.02	7629.1	9015.23	10444.76	10661.74
Engel coefficient (%)	28.97	28.664	28.888	30.03	28.746

2.2. Standardization of Data

The raw data differ in dimension and quantity size, therefore, it need to be standardized by extremum method. The purpose is to eliminate dimension and to make all the data comparability, so that they can reflect the relative of the situation of Dongying City each year. It is important to note that the resource and environmental indexes including mean annual concentration of inhalable particle, mean annual concentration of SO2, mean annual concentration of NO2, shoreline development intensity, regional crustal stability, geological disasters risk and pollution level of water and soil caused by oil were negative indexes, others were positive indexes. In certain value range, positive indexes were the larger, the better, and negative indexes were as small as possible. The formula as:

$$x_{ij} = \begin{cases} x_{ij}^{/} = (x_{ij\max} - x_{ij}) / (x_{ij\max} - x_{ij\min}) & x_{ij} \quad \text{Negative index} \\ x_{ij}^{/} = (x_{ij} - x_{ij\min}) / (x_{ij\max} - x_{ij\min}) & x_{ij} \quad \text{Positive index} \end{cases}$$

Where Xi is initial data, $X_i'$is the data which has been standardized, $X_{i\ min}$ is the minimum value in initial data, $X_{i\ max}$ is the maximum value in initial data.

2.3. Calculation of Comprehensive Evaluation Indexes

In order to minimize the artificial influence, principal component analysis (PCA) was used to calculate integrated development indexes in economic system and resource environment system, respectively. The standardized data was analysed by PCA using the SPSS analysis software (IBM SPSS Statistics 21), to determine the characteristic root correlation matrix, the contribution rate of each principal component and cumulative contribution rate. The cumulative contribution rate of the former k composition which is greater than 85% reflects the basic original indexes of main information.

2.4. Coordination Degree Model

Coordination degree is the degree of harmony in the development and evolution process between systems or components in system. It could reflect the trend of the system from disorderly to orderly. Coordination degree is the strength of synergy between the parameters in the process of system development, is the quantitative index to confirm the coordination between the metric systems or the internal system. From mathematical point of view, the coordination degree between two systems can be measured by the deviation between the representative function, the smaller the deviation is, the higher the coordination degree is; on the other hand, the lager the deviation is, the lower the coordination degree is. Coordination degree C between two systems was defined as:

$$C=\left[\frac{f(x)\cdot g(y)}{\left[\frac{f(x)+g(y)}{2}\right]^2}\right]^k$$

where C is coordinate degree; f(x) represents comprehensive evaluation index of the resources environment system; g(y) represents comprehensive evaluation index of economic system; k is adjustment coefficient, it means when the level of economic development and resources environment under certain conditions, in order to maximize the comprehensive coordination degree between economy and environment , k≥2, in this study k=3.The range of C values is between 0 and 1, obviously, 0≤C≤1, C=1 reflect the best coordination states, and C=0 reflect the worst coordination states.

On the basis of the coordination degree results of resources environment system and economic system, to establish coordination degree level evaluation system. According to the meaning of the coordination degree, it can be divided into several grades, and graded into the subordinate level according to the score (table 3).

Table.3 Coordination degree classification

Coordination degree (C)	Grade	Meaning	Coordination degree (C)	Grade	Meaning
0 ~ 0.1	1	Extreme imbalance	0.5 ~ 0.6	6	Reluctant coordination
0.1 ~ 0.2	2	Serious imbalance	0.6 ~ 0.7	7	Primary coordination
0.2 ~ 0.3	3	Moderate imbalance	0.7 ~ 0.8	8	Medium coordination
0.3 ~ 0.4	4	Mild imbalance	0.8 ~ 0.9	9	Benign coordination
0.4 ~ 0.5	5	Endangered imbalance	0.9 ~ 1.0	10	Superior coordination

Table.4 Economic, Environmental Systems comprehensive evaluation index and the coordination degree of Dongying City, 2011—2015

Year	f(x)	g(y)	C
2011	-5.450663461	-2.633999245	0.678272
2012	-3.114121556	-1.553201278	0.700588
2013	0.514995932	0.634175293	0.968079
2014	2.750274686	2.366748164	0.983241
2015	5.299514398	1.186296175	0.21364

3. Resources Environment and Economic Development Coordination Degree Evaluation System

According to economic system comprehensive evaluation index f(x) and environmental system comprehensive evaluation index g(y) of Dongying City in 2011-2015, using coordination degree model and adjustment coefficient k=3, the coordination degree C was calculated and shown in table 4. Table 4 showed that economic system comprehensive evaluation index and resources and environment system comprehensive evaluation index of Dongying City were increasing in 2011-2015. The economic and resources environmental coordination degree develops stably. The minimum coordination degree was 0.21364; the maximum coordination degree was 0.98324. According to the grade division of coordination degree in the table 3, the economic and resources environmental coordination degree level of Dongying City is the primary coordination in 2011, medium coordination in 2012, superior coordination in 2013 and 2014, and moderate imbalance in 2015, respectively.

Overall, economic and environmental development coordination degree of Dongying City belongs to the coordinated development type. The ecological environment improved gradually and the social economy developed steady. The coordination degree level of economic and social is higher than its level of resources and environment in the initial stage significantly, resources and environment level is above the economic and social level obviously late. Especially in 2015, economic and social development level lags behind the level of resources and environment development obviously, this lead to imbalances between them, which showed the moderate imbalance state. The main reason is that some of economic and social indexes present the negative growth trend in 2015. This is not in conformity with the trend of social development to a certain extent. The reasons for this phenomenon may be related department report the fuzzy statistics data, but it should belong to the short-term phenomenon. Overall, the resources environment and economic coordination degree development of Dongying City showed good momentum, it mainly in line with the coordinated development and gradually optimized trend.

4. Conclusion

Regional development focuses on coordination. In order to obtain the maximum economic benefits and social benefits, the relationship between environment and economy must be coordinated and seek a reasonable mode of development to maximize the efficiency of the resource environment.

The relationship between the resource environment and economic development is an interactive one. Once a threshold is exceeded however, it becomes the limiting factor for regional development. Economic development is not always passively subject to environmental factors. When economic activity is very high, there is commensurate level of investment in the protection of resources and the environment. In turn, this will assist in the protection and development of the resource environment. A low level of economic activity is usually paired with more eagerness to develop and use resources to further develop, but such an economy is also unable to protect environmental resources.

Acknowledgments

This work was financially supported by National Natural Science Foundation of China (41602356), Open project of Key Laboratory of resources and environment carrying capacity evaluation of Ministry of land and resources (CCA2016.08), Shandong Provincial Geological Prospecting Fund Project (2013(55), 2016(07)).

References

[1] Zhou Detian, Huang Bingjie,Dong Ruizhi. Thestudy of sustainable development of the Yellow River delta. Beijing: China industrial and commercial press, 2010,pp.1-7.

[2] LU Ning,Z Xu,Z Shingqing,et al. Evaluation of coordinated development between construction and resource-environment in China'30 provinces during 2008~2012.Journal of Arid Land Resources and Environment ,2015, pp. 13-18

[3] Li N, Wang K. Evaluation of coordinated development of regional resources and economy around Shandong Peninsula urban agglomerations.Journal of Groundwater Science and Engineering. 2016, pp. 220-230.

[4] DU Zhongchao,B Wenting. Analysis of Water Recourse, Environment and Social Economy as well as Coordinated Development in Guanzhong-Tianshui Economic Zone.Journal of Xianyang Normal University ,2015, pp. 70-75.

[5] Z Ju-hua. Integrative Evaluation of Coordinated Development in Population, Resource, Envi ronment and Economy of Guangxi. Journal of Guangxi Teachers Education University (Natural Science Edition) ,2014,pp.78-83.

IOP Conf. Series: Earth and Environmental Science **64** (2017) 012045 doi:10.1088/1755-1315/64/1/012045

Spatial and temporal variations of nutrients composition and structure in the main estuaries of Jiaozhou Bay

Pingli Zhong, Ge Yu*, Yang Zheng, Xue Sun and Youxiao Wang

Key Laboratory of Marine Environmental Science and Ecology, Ministry of Education, Ocean University of China, Qingdao, China

*Corresponding author e-mail: yuge@ouc.edu.cn

Abstract. Spatiotemporal distribution of nutrients were investigated in the main estuaries of Jiaozhou Bay in January, April, June, August, October and December, 2016, and the relationships between nutrients and temperature, salinity, pH and DO were analyzed as well. Results showed that, the main estuaries of Jiaozhou Bay total dissolved nitrogen (TN) showed a time trend of decreasing after the first increasing, small changes of TN content in each estuarine water body, organic nitrogen (DON) was the main component of TN, its relative content was about 58.6-98.5%. NH_4^+-N was the main component of inorganic nitrogen (DIN) in most estuaries, followed by NO_3^- - N, NO_2^{2-} - N lowest relative content. The content of total dissolved phosphorus (TP) was low (averaged 0.15 mg/L), mainly showed a higher spring and summer, autumn and winter low time trends, the change of TP content in the water body of the estuary was also small, the relative content of inorganic phosphorus (DIP) in most estuaries was more than organic phosphorus (DOP). Temperature, salinity, pH and DO with different forms of nutrient analysis showed that there are strong primary productivity and microbial regeneration activities in the study area, environmental factors affected the microbial production activities in estuarine water affecting the spatial and temporal changes in nutrient composition and structure.

1. Introduction

Jiaozhou Bay is a typical semi enclosed bay in the southern part of the Shandong Peninsula, China. In recent years, a large number of land-based pollutants such as domestic sewage, agricultural sewage and industrial wastewater have been washed into the ocean through surface runoff, the content of nutrients in Jiaozhou Bay has been seriously exceed the standard [1]. The river input is an important source of nutrients in Jiaozhou Bay, and Jiaozhou Bay surrounded by dozens of rivers into the sea, these rivers carried large amounts of pollutants into Jiaozhou Bay every year [2]. Into the sea estuary as a link between the river and Jiaozhou Bay, bear a huge terrestrial nutrient load [3]. At present, the studies on Jiaozhou Bay were mainly concentrated in the sea area [4]-[7], and the study of the nutrients in the estuary was relatively few, which leads to the lack of understanding of the temporal and spatial characteristics of nutrients in the estuary. Therefore, the temporal and spatial characteristics of nutrients in the main estuaries of Jiaozhou Bay, not only reveal the biogeochemical processes of nutrient elements in estuaries region, but also provide basic data for the accurate prediction of the transport fluxes of nutrients to the coastal environment.

2. Survey areas and research methods

2.1. Station settings and sample collection
Based on the investigation in the Jiaozhou Bay coastal zone, taking into account the representation and integrity of the monitoring sites, in January (winter), April (spring), June (summer) and October (autumn), 2016, in the main estuaries of Jiaozhou Bay set 10 sites were sampled, such as the Yang River, Dagu River, LouShan River, Banqiaofang River, Licun River, Haibo River and Baisha River etc. Site settings shown in Figure 1. Water samples were pretreated for different forms of nutrient analysis, synchronized monitoring of temperature, salinity, pH and DO (HACH HQ40d).

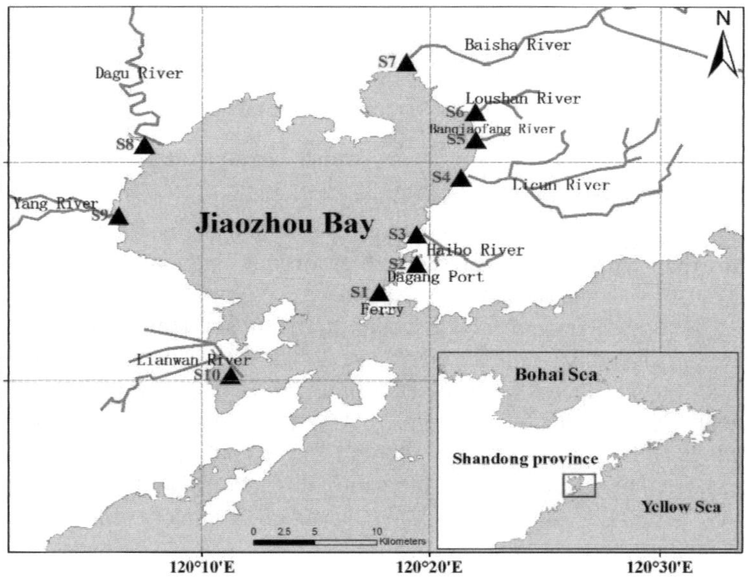

Figure 1. Study area and sampling site in the main estuaries of Jiaozhou Bay

2.2. Analytical methods
The water samples were filtered through a 0.45 μm acetate filter (precoated) and stored with saturated $HgCl_2$ for the dissolution of inorganic nutrients and dissolved total nitrogen (TN) and total phosphorus (TP) analysis.

After the sample was thawed at room temperature, the use of automatic nutrient analyzer (QuAAtro) analysis of nitrate(NO_3^--N), nitrite (NO_2^{2-}-N), ammonia nitrogen (NH_4^+-N) and phosphate (PO_4^{3-}-P), followed by the method of Cd-Cu reduction, diazo - azo, indophenol blue and phosphomolybdenum blue[8]. The sum of NO_3^--N, NO_2^{2-}-N and NH_4^+-N is dissolved inorganic nitrogen (DIN). After the samples were digested with potassium disulfide, the dissolved organic nitrogen and phosphorus were changed into inorganic nitrogen and active phosphate. Then, TN and TP were obtained by nutrient analyzer. Total dissolved nitrogen minus dissolved inorganic nitrogen is dissolved organic nitrogen (TON), total dissolved phosphorus minus dissolved inorganic phosphorus is dissolved organic phosphorus (TOP).

3. Results and discussion

3.1. Spatiotemporal variation of nutrient composition and structure in the main estuaries of Jiaozhou Bay

3.1.1. Total dissolved nitrogen and total dissolved phosphorus

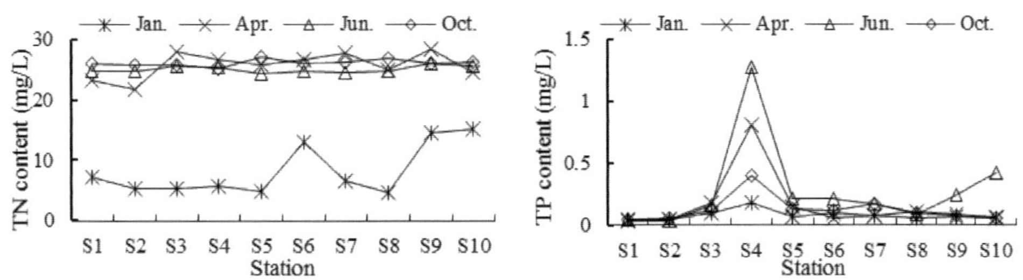

a. TN b. TP

Figure 2. Spatiotemporal distribution of TN and TP in the main estuaries of Jiaozhou Bay

The seasonal variations of TN and TP contents in the main estuaries of Jiaozhou Bay were shown in Figure 2. As could be seen from Figure a, from January to October, the content of TN in the main estuaries in Jiaozhou Bay showed a trend of increased first and then decreased, and the TN content in the water body was higher in the summer. From the space, the content of TN in the water body of each estuary (site) changes little, but in a few estuaries (sites) appear larger value. From Figure b, the TP content was generally low, the average content was 0.15 mg/L, and the time variation is not large, mainly presented the change trend of spring and summer was higher, lower in autumn and winter. From the space, the content of TP in the water body of each estuary (site) changed little, only the TP content in the water of the Licun estuary (S4) was higher in different months.

3.1.2. The composition of dissolved nitrogen

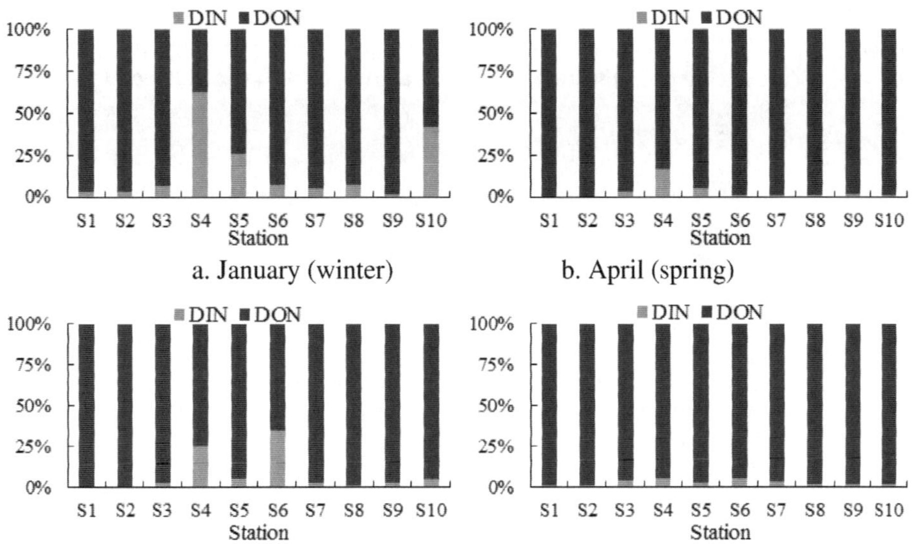

a. January (winter) b. April (spring)

c. June (summer) d. October (autumn)

Figure 3. The composition of TN in the main estuaries of Jiaozhou Bay at different months (seasons)

It can be seen from the Figure 3 that although the structure of TN in the main estuaries water bodies was constantly changing, DON is always the main component of TN, the relative content is about 58.6-98.5%, the relative content of DIN is always low. From January to October, the relative content of DIN in the estuaries decreased first and then increased, the main reason may be the impact of temperature and precipitation.

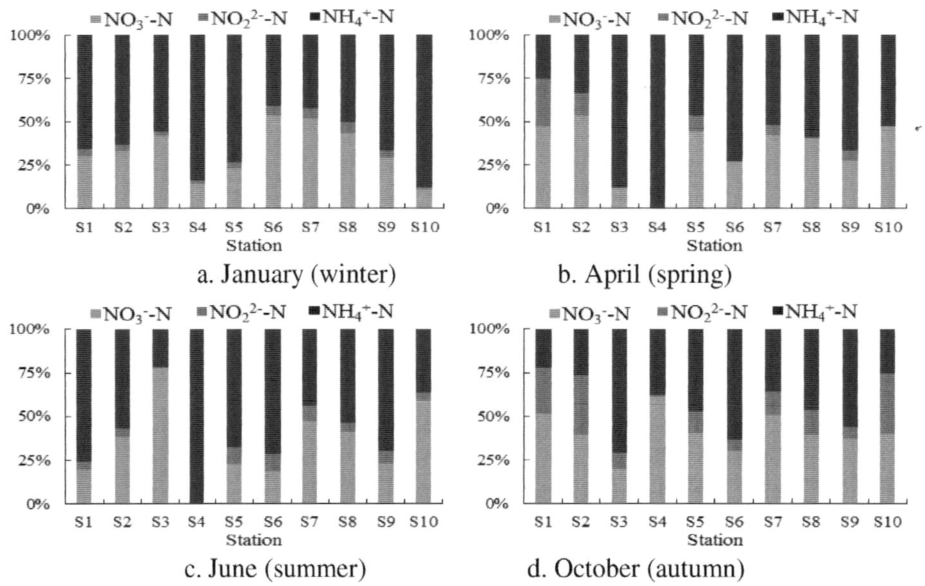

Figure 4. The composition of DIN in the main estuaries of Jiaozhou Bay at different months (seasons)

It can be seen from the Figure 4 that, the structure of DIN in the main estuaries water bodies was changed significantly, but the size of DIN components in most estuaries (site) water bodies was NH_4^+-N> NO_3^--N >NO_2^{2-}-N, only a few of the estuary (site) DIN component size: NO_3^--N >NH_4^+-N> NO_2^{2-}-N. From January to October, the proportion of NO_3^--N increased firstly and then decreased, the proportion of NH_4^+-N was first decreased and then increased, the proportion of NO_2^{2-}-N changed little, the basic stability of about 7%. The main reason is the change of water temperature, made the intensity of nitrification in the water changes, resulting in ammonia nitrogen to nitric acid conversion process.

3.1.3. The composition of dissolved phosphorus

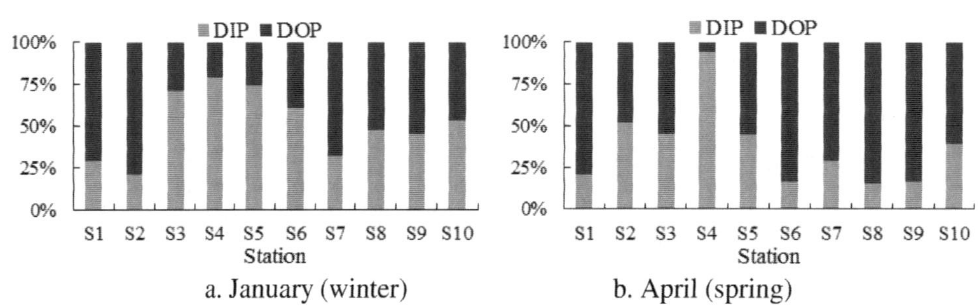

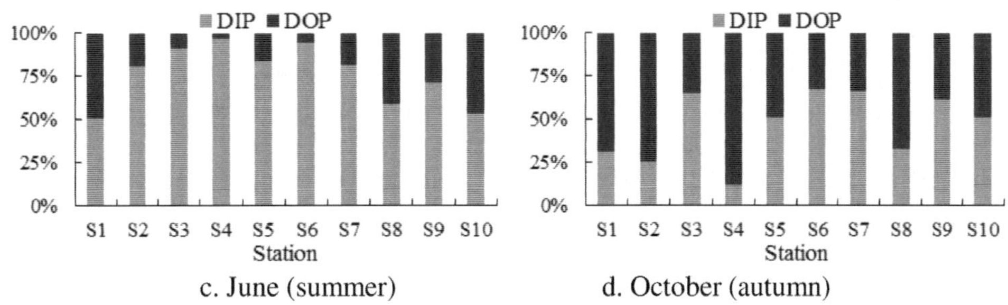

Figure 5. The composition of TP in the main estuaries of Jiaozhou Bay at different months (seasons)

It can be seen from the Figure 5 that, the structure of TP in the main estuaries water bodies was changed significantly, the relative content of DIP in the most estuaries (sites) is much higher than that of DOP, only a few of estuary (site) DOP relative content more than DIP. From January to October, the proportion of DIP in the estuarine water body increased first and then decreased, DOP on the contrary, the main reason was the impact of rainfall, a large number of inorganic phosphorus pollutants washed into the river.

3.2. The relationship between the main physical and chemical factors and the composition and structure of nutrients

In this paper, the Pearson correlation analysis of different forms of nitrogen and phosphorus nutrients and main environmental factors in the main estuaries of Jiaozhou Bay in the four seasons was shown in Table 1.

Table 1. Pearson correlation coefficient of nutrients and environmental parameters

	NO_3^--N	NO_2^{2-}-N	NH_4^+-N	DIN	DON	PO_4^{3-}-P	DOP
temperature	-0.030	0.076	0.014	0.205**	0.223*	-0.416	0.015
salinity	-0.428	-0.159	-0.567	0.551**	-0.504**	0.174	-0.554
pH	-0.111	0.118	-0.142	0.168**	-0.154**	0.121	-0.149
DO	-0.093	-0.073	-0.447**	0.350	-0.589**	0.531	-0.485**

* showed significant correlation (P<0.05, bilateral test); ** showed very significant correlation (P<0.01, bilateral test)

The analysis results show that, the temperature was positively correlated with DIN and DON at very significant levels, showed that the chemical reaction increased with the increase of water temperature in the study area, the DIN and DON nutrients in the water body gradually increased. Salinity was positively correlated with DIN at extremely significant level, and negatively correlated with DON at extremely significant level, but the change of salinity in the study area had little effect on the composition and structure of nutrients. pH and DIN appear in the very significant level of weak positive correlation, and DON at very significant level showed a weak negative correlation, because pH was closely related with temperature, the change of DIN in water is related to the temperature, DON the changes may have a great relationship with biological decomposition behavior. DO was negatively correlated with NH_4^+-N, DON and DOP at extremely significant level, while NH_4^+-N and DOP were higher in water usually showed that the degradation of organic matter was stronger, showed that with the increase of biological activity, the degradation of organic matter is enhanced, more NH_4^+-N and DOP were released, and a large amount of DO would be consumed.

4. Conclusion

The content of TN in the main estuaries of Jiaozhou Bay showed a trend of increasing first and then decreasing, the content of TN in each estuary was smaller. DON was the main component of TN, its relative content was about 58.6-98.5%. NH_4^+-N was the main component of DIN in most estuaries, followed by NO_3^- - N, NO_2^{2-} - N lowest relative content. The content of TP was low, the average content was 0.15 mg/L, mainly presented time trend was higher in spring and summer, lower in autumn and winter, the content of TP in each estuary was small. The change of TP content in the water body of the estuary was also small, the relative content of DIP in most estuaries was more than DOP.

The correlation analysis between physicochemical factors and different forms of nutrients in the main estuaries of Jiaozhou Bay showed that there are strong primary productivity and microbial regeneration activities in the study area. Environmental factors affected the composition and structure of nutrients by affecting the production of microorganisms in the estuary water body.

Acknowledgments

This work was financially supported by the National Natural Science Foundation of China (41201569), the Shandong Provincial Natural Science Foundation (ZR2011DQ003), and the Qingdao Municipal Science and Technology Program (12-1-4-1-(15)-jch).

References

[1] Y. Yao, S.Q. Zheng, Z.L. Shen, Study on the Mechanism of Eutrophication in the Jiaozhou Bay , Marine Sci. Bulletin, 26(2007):91-98.
[2] S.M. Liu, J. Zhang, H.T. Chen, G.S. Zhang, Factors influencing nutrient dynamics in the eutrophic Jiaozhou Bay, North China, Progress in Oceanography, 66 (2005):66-85.
[3] S.J. Painting, M.J. Devlin, S.J. Malcolm, et al., Assessing the impact of nutrient enrichment in estuaries: Susceptibility to eutrophication, Marine Pollution Bulletin, 55 (2007): 74-90.
[4] Shen Z L. Long-term changes in nutrient structure and its influences on ecology and environment in Jiaozhou Bay, Oceanologia Et Limnologia Sinica, 03 (2002):322-331.
[5] P.X. Sun, Z.L. Wang, R. Zhan, et al., Study on Dissolved Inorganic Nitrogen Distributions and Eutrophication in the Jiaozhou Bay, Advances in Marine Sci. 23(2005): 466-471.
[6] X.X. Sun, S. Sun, Z.X. Zhao, Z.L. Shen, Long-term Changes in Nutrient Concentration and Structure in the Jiaozhou Bay, Oceanologia Et Limnologia Sinica, 05 (2011):662-669.
[7] X.J. Li, M.H. Kang, P.F. Hou, X. Gao, Spatial and temporal distribution of nutrient in Jiaozhou bay in 2013, Marine Environmental Sci. 03(2016):334-342.
[8] K. Grasshoff, K. Klaus, E. Manfred, Methods of seawater analysis, Wiley online library, (2007)

IOP Conf. Series: Earth and Environmental Science **64** (2017) 012046 doi:10.1088/1755-1315/64/1/012046

Recent enlightening strategies for co₂ capture: a review

Peng Yuan, Ziyang Qiu and Jia Liu*

School of North-eastern University, Shenyang, China

*Corresponding author e-mail: 417850081@qq.com

Abstract. The global climate change has seriously affected the survival and prosperity of mankind, where greenhouse effect owing to atmospheric carbon dioxide (CO_2) enrichment is a great cause. Accordingly, a series of down-to-earth measures need to be implemented urgently to control the output of CO_2. As CO_2 capture appears as a core issue in developing low-carbon economy, this review provides a comprehensive introduction of recent CO_2 capture technologies used in power plants or other industries. Strategies for CO_2 capture, e.g. pre-combustion, post-combustion and oxyfuel combustion, are covered in this article. Another enlightening technology for CO_2 capture based on fluidized beds is intensively discussed.

1. Introduction

Energy is a major foundation for a country's economic development. Now 90% of global energy consumption comes from fossil fuels [1]. However, the excessive utilization of fossil fuels has led to accumulation of greenhouse gases, which contribute to global warming. Global warming will lead to the rise of sea levels and the ablation of glaciers. It will not only break ecological balance, but also influence the development of human. According to the Intergovernmental Panel on Climate Change (IPCC) Fifth Assessment Report [2], the globally averaged temperature increased by 0.89[0.69 to 1.08] ℃ between 1901 and 2012. It has been predicted that the global average temperature will rise by 0.3-0.7℃ between 2016 and 2035.

The warming rate of earth's surface is proportional to the atmospheric CO_2 concentration. Scientific studies [3] show the concentration of CO_2 in the atmosphere has increased by more than 20% since 1958. By 2013, global CO_2 emissions reached 32.2 $GtCO_2$, up 2.2% from the previous year. Figure 1 shows the top ten emitting countries in 2013 [4]. China produced the most CO_2 (about 9 $GtCO_2$), occupying 28% of global emissions. The accumulation of CO_2 makes the largest contribution to global warming, so a series of down-to-earth measures need to be urgently implemented to control the output of CO_2. Riahi et al. had a modeling study of CO_2 capture incorporating factors of energy demand, demographic, economic and alternative policy [5]. They concluded CO_2 capture was one of the most reliable policies for CO_2 reduction.

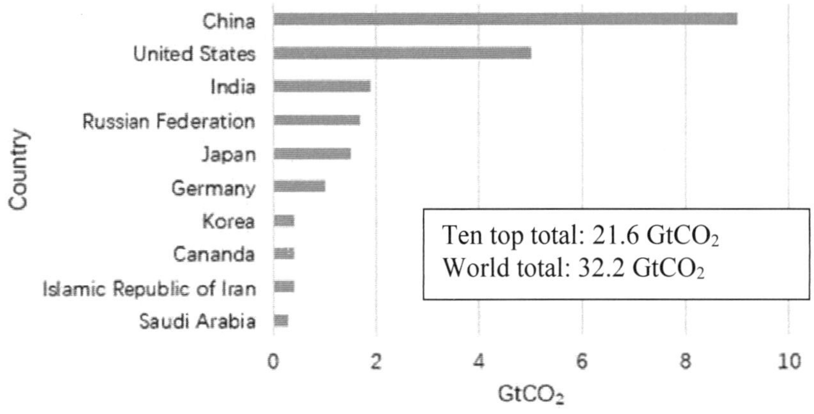

Figure 1. Top ten emitting countries in 2013

2. CO_2 Capture

CO_2 capture is a process of capturing and collecting CO_2 from flue gas of power plants or other factories, such as chemical fertilizer plants, cement plants and so on. The collected CO_2 will be transported to a suitable location by several means of transportation such as motor carriers, railway, ship and pipelines [6]. Then, the CO_2 can be used in caustic soda industry, sugar industry, food industry and so on.

Three different approaches of capturing CO_2 are widely used: pre-combustion, post-combustion and oxyfuel combustion. Each method has its own advantages and disadvantages (Table 1).

Table 1. Comparison of three approaches

Method	Advantages	Disadvantages
Pre-combustion	High CO_2 levels, easy separation	Complicated process, high cost
Post-combustion	Simple process, more branches	Low CO_2 levels
Oxyfuel combustion	High CO_2 levels, low cost of separation	High cost of oxygen supply

2.1. Pre-combustion.

Pre-combustion capture technology can be found in typical integrated gasification combined cycle (IGCC) power plants [7]· The fossil fuel is first gasified to produce syngas, a mixture of H_2 and CO. In the process of gasification, a mixture of CO_2 and H_2 is produced before the combustion reaction [8]. If CO_2 is completely separated herein, only water will be produced after combustion to reducing CO_2 emissions significantly. Figure 2 shows the illustrative diagram of pre-combustion [9].

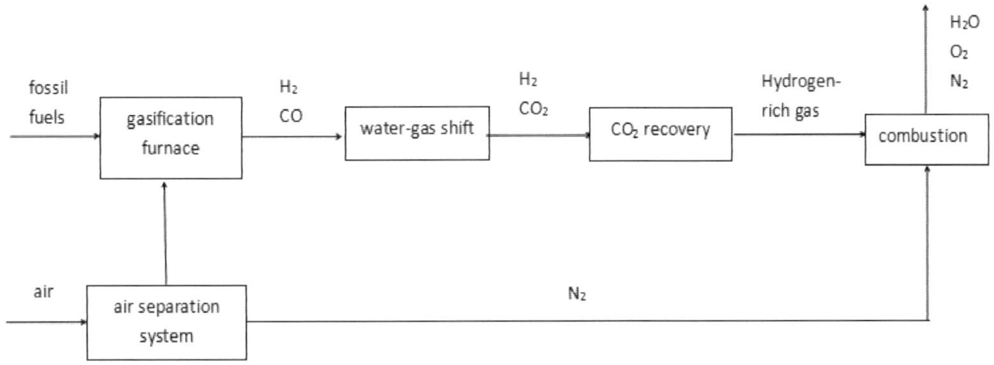

Figure 2. CO_2 pre-combustion capture

At present, pre-combustion capture has been adopted by a raft of proposed IGCC and poly-generation power plants, such as Future Gen in America and Green Gen in China [10]. In general, there are three commercialized methods: Rectisol method, Selexol method and MDEA method.

Allied Chemical Company (now owned by Norton) developed Selexol technology successfully in 1960s [11]. This process uses Union Carbide Selexol solvent, a physical solvent made of dimethyl ether polyethylene glycol [$CH_3(CH_2CH_2O)_nCH_3$]; where n is between 3 and 9 [12]. Selexol process can be used to remove CO_2 and H_2S under a low temperature condition and to regenerate absorbent by depressurization [13]. Absorption is accomplished under low temperature (0-5 ℃) operation. Desorption of the rich Selexol solvent takes place at low pressure by vacuuming or stripping. The solvent can be applied to remove SO_x, CO_2, H_2O as well as aromatic compounds (BTEX) as required. Desiccation of the flue gas is necessary before entering the Selexol unit [13]. According to Southern Company, they planned to convert partial CO into CO_2 and H_2 in a 582 MW IGCC project in Kemper County, using Selexol technology to capture 65% of total CO_2 emissions. About 1.1-1.5 million CO_2 would be sold for oil displacement every year [14]. The advantages of this method are low steam tension, less toxicity and low corrosiveness while the disadvantage is the hydrocarbon losses.

MDEA is a weakly basic solvent which will desorb after absorbing acid gases. It can be regenerated by reducing the pressure and flashing. MDEA method was first used in desulfurization until BASF company in Germany developed a novel process to capture CO_2 using MDEA in 1980s [15]. Then MDEA method was introduced into China and developed rapidly. Between late 1980s and early 1990s, Nanjing Chemical Industry Research Institute pushed out a characteristic decarburization system by MDEA method [16]. The first industrial installation went into production in 1991. So far, more than 100 devices have been put into production [17]. Many researchers pay close attention to the composition of MDEA. Zhang Xuemo chose primary amine and secondary amine as activators to increase the absorption speed of solution [18]. Double activators can also reduce CO_2 partial pressure on liquid phase surface, which is conducive to CO_2 absorption. Another innovation point is based on double physical and chemical performance of MDEA. Making full use of its physical property can reduce energy consumption. Semi lean solution is used to absorb most of the CO_2 while tiny amounts of lean solution is used to guarantee cleanliness. Under normal circumstances, amine decarburization process can reach the highest proportion of up to 8:1 when the ratio of semi lean solution and lean solution is 3-4:1 [18]. Application results [19] show that MDEA method can not only reduce energy consumption, but also achieve an excellent absorption effect. In addition, MDEA has many intrinsic advantages, such as good stability, small volatility, and low hydrocarbon solubility and so on.

Rectisol process uses chilled methanol as a solvent, the process is usually operated at -34 to -73 ℃ [20] due to high pressure of methanol. It is best suited when quantities of ethane and heavier components are limited. Rectisol process can be tracked back to the research work of Lurgi Company and Linde Company in 1950s. In 1954, Lurgi Company built up the world's first rectisol industry project at Sasol Company [21]. Rectisol process gets great attention in China. Although it is a relatively mature technology, it still has a bigger research space in energy conservation. In 2013, Shenhua Group invented a coupling method of rectisol and CO_2 capture [22]. The method includes the following steps: desorption will be operated after methanol absorbs acid gases. Then a steam of CO_2 will be separated. After purification and concentrated capture, CO_2 concentration in the exhaust gas can reach up to 95%. This process combines and optimizes rectisol process and CO_2 membrane separation and capture process, resulting to investment reduction. In the same year, Cui Sijing came up with a coupling method of rectisol and CO_2 liquefaction separation [23]. Firstly, partial CO_2 in the exhaust gas will be liquefied. Then methanol will be used to absorb the rest of CO_2. By this way, methanol circulating volume can be reduced, which can reduce energy consumption of liquid delivery in the production process. In 2014, a kind of shunting type low temperature methanol washing technology was developed independently by Shanghai International Engineering Consulting Company [24]. This system can deal with impurities such as BTX and has low energy consumption. Rectisol process is most applicable when dealing with the combination process with desulfurization and decarbonization. The merits include more stable absorbent, less corrosive as well as high thermal and chemical stability [25].

2.2. Post-combustion

Post-combustion capture removes CO_2 from waste gas, which mainly consists of CO_2 and N_2, using chemical or physical methods. This type of technology can be directly used in traditional power plants, so it saves a large amount of construction cost. Figure 3 shows the illustrative diagram of post-combustion [9].

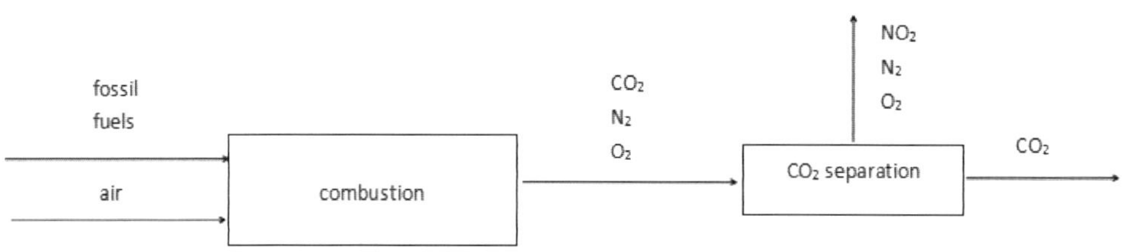

Figure 3. CO_2 post-combustion capture

Post-combustion contains many kinds of adsorption methods: physical absorption, chemical absorption, adsorption and membrane [26]. Among the many options used for CO_2 post-combustion capture from waste gases, there is an increasing interest in using adsorption process as a long-term alternative for CO_2 capture. It has many potential advantages [27], such as greater capacity, low energy regeneration, selectivity and so on. Therefore, adsorption progress will be intensively discussed in the following. Adsorption processes can be broadly classified into two types according to the principles of adsorption and desorption: by regulating the pressure or the temperature [28].

Pressure swing adsorption (PSA) is a novel method of CO_2 adsorption and separation technology. It was processed by karstrom and Guerrin Dumine simultaneously in 1960s [29]. In China, the technology was first developed by Southwest Chemical Research and Design Institute [30]. The basic principle of PSA is the selective adsorption of adsorbents, which is represented in different capacity, velocity and adsorption force.CO_2 will be adsorbed at high pressure while it will be released at low pressure in order to achieve the purpose of CO_2 capture and adsorbents regeneration. In the primary stage of PSA, it was generally believed that this method with higher cost and energy consumption was difficult to be promoted and applied. With the improvement of technology, structure of adsorption tower, circulation design as well as adsorbents [31] have been promoted. Current multi-tower circulation devices are based on Skarstrom circulation [31]. Because of high concentration of CO_2, low energy consumption and investment, PSA technology has been widely applied. However, there are still some problems should be solved urgently.

①Improve circulation design or develop novel adsorbents with stronger adsorption capacity of CO_2 [32].

Normally, CO_2 adsorbents include molecular sieve, activated carbon, silica gel, activated alumina, etc [33]. Regardless of types of adsorbents, the adsorption capacity of CO_2 is stronger than other components in the mixed gas, which is caused by the intrinsic molecular structure and molecular polarity of CO_2. Actually, there are still 5% water vapor cannot be removed in the treated flue gas [32]. And adsorbents tend to adsorb water vapor ahead of CO_2, so that the adsorption of CO_2 is reduced. In conclusion, it is necessary to improve circulation design or develop novel adsorbents with stronger adsorption capacity of CO_2. Based on four bed molecular sieves, American researchers substituted a novel type of hydrophobic carbon molecular for original 5A zeolite molecular sieve [34]. The whole process was improved by omitting air predrying sector, in which only two adsorption beds were needed. This breakthrough not only improved the adsorption capacity of CO_2, but also reduced the operation complexity greatly.

②The negative effect caused by pressure drop [32].

The existence of pressure drop has a negative effect on pressing process, pressure reducing progress and flushing progress. The reduction of system pressure will lead to high energy consumption and low recovery of CO_2.

Temperature swing adsorption (TSA) is the earliest physical adsorption method used in industry. In general, solid adsorbents have a strong adsorption capacity of CO_2 at high temperature while it becomes weak at low temperature. And the adsorption capacity of other gases (such as N_2) is small in the two temperature conditions. Based on this characteristic, adsorption progress is carried out at a lower temperature and then adsorbents can be regenerated by heating. At last, adsorbents will be cooled for next circulation. The adsorption progress can be continuously carried out by the use of a plurality of adsorption beds [35]. When some strong adsorption components cannot be separated completely by PSA, TSA shows obvious advantages. J. Merel et al. [36] used zeolite as adsorbents when they studied the separation performance and energy consumption of TSA technology. Results show that the energy consumption of TSA is a little higher than that of amine absorption method. This disadvantage can be made up by making full use of waste heat from power plants. Another advantage of TSA is slow heating rate and cooling rate, limiting the circulation time [37]. In order to avoid the disadvantages of TSA when dealing with hot gas, the concept of electrical thermal swing adsorption (ETSA) has been proposed by Fabuss and Dubois [38]. The adsorbent needs to be an electrical conductor. The process uses heat generated by Hall Effect to release CO_2. Grande C.A. and Rodrigues A.E. [39] carried out a systematic study on ETSA. Rodrigues research group [40, 41] studied to capture CO_2 from flue gas by ETSA progress, using honeycomb integrated activated carbon as adsorbent. Under this operation, the recovery of CO_2 was over 89% and the purity was up to 16%.

A coupling method of PSA and TSA has also been developed. Pugsley et al. [42] systematically studied the performance of a pressure–temperature swing adsorption (PTSA) process in a circulating fluidized bed. PTSA progress combines vacuuming and heating together, which is suitable for strong adsorption components. Compared with VPSA, PTSA enhances the driving force effectively and reduces the mass transfer resistance, so that it can adopt a higher adsorption pressure and reduce the power consumption of the vacuum pump. Compared with TSA, PTSA can lower regeneration temperature to reduce the heat consumption [35]. Tlilil et al. [28] tested different effects of CO_2 capture by 5A zeolite molecular sieve. The recovery and purity were nearly 99% by PTSA progress, far higher than PSA and TSA. M. G. Plaza et al. [43] adopted different progress to test CO_2 production capacity and recovery rate of activated carbon. Results showed that PTSA had the best effect in primary circulation with the maximum production capacity of 1.9mol/ (kg·h).

2.3. Oxyfuel combustion

Oxyfuel combustion chooses pure oxygen or oxygen enriched air instead of ordinary air as combustion medium. Principal combustion products are CO_2, water vapour and redundant oxygen. After condensation, concentration of CO_2 in flue gas can reach to 80%-98%, so the separation cost is low. Figure 4 shows the illustrative diagram of oxyfuel combustion [9].

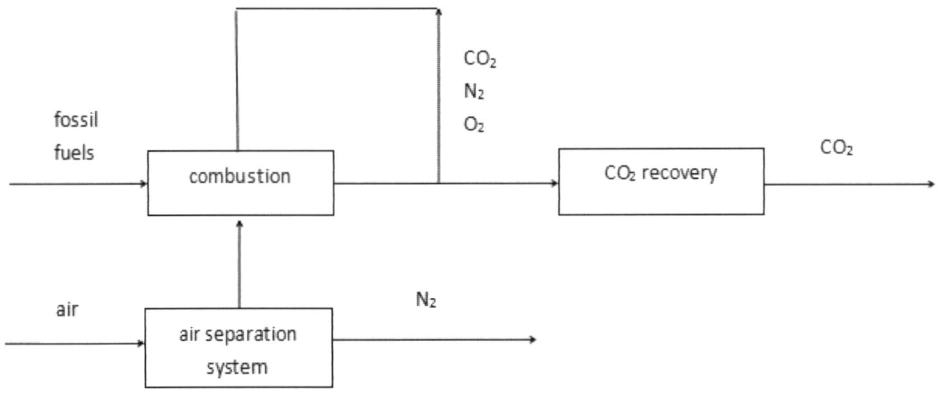

Figure 4. CO_2 oxyfuel combustion capture

The system of an oxyfuel combustion power plant mainly includes: air separation unit (ASU), boiler or gas turbine, flue gas processing unit and CO_2 processing unit (CPU) [44]. A nearly pure steam of CO_2 and water vapor will be produced after oxyfuel combustion; and then CO_2 can be separated from the steam by desiccation and low temperature purification processes. The application of oxycombustion for coal fired power plants are mainly two types: oxy-PC and oxy-CFB [44]. However, the application of oxyfuel combustion faces a series of challenges [45]. The first one is design of next generation oxyfuel boiler. The combustion process needs to be optimized for reduced formation of NOx. The second one is the energy consumption of O_2 production. Cryogenic air separation is only applicable to large scale oxygen production. In conclusion, oxyfuel combustion is still in a stage of demonstration.

3. Summary

Reduction of atmospheric CO_2 is becoming an urgent issue as global temperature is increasing overwhelmingly. In order to control CO_2 concentration, CO_2 capture has been widely adopted. Through this view, the following conclusions can be drawn:

①Pre-combustion capture is usually applied in IGCC power plants. CO_2 separation was finished before the combustion, so the size of separator can be reduced, which helps to improve system efficiency. However, the regeneration section will cause energy penalties. So it is necessary to optimize the conversion process and invent high temperature adsorbents.

②Post-combustion capture separates CO_2 from waste gas after combustion. This method can be applied to traditional power plants directly, so it saves a large amount of construction cost. The engineering design challenge is the partial pressure of CO_2 is relatively low while the flue gas temperature is relatively high.

③Oxyfuel combustion capture is a promising method with a better effect. It applies to different kinds of power plants, including large coal-fired and gas-fired units. But this method still has certain problems. The first one is to lower furnace temperature to avoid corrosion. The second one is to improve related technology to reduce the cost.

References

[1] Jiang Peng. Study on decarbonization by amine process simulation and device running[D]. Dalian University of Technology, 2009.
[2] IPCC. Working Group Ⅰ Contribution to the IPCC Fifth Assessment Report (AR5). Climate Change 2013: The Physical Science Basis. Final Draft Underlying Scientific-Technical Assessment[R/OL]. [2013-10-30].
[3] Alexander L, Allen S, Bindoff N L. Climate Change 2013: The Physical Science Basis[J]. Intergovernmental Panel on Climate Change, 2013.
[4] Nejat P, Jomehzadeh F, Taheri M M, et al. A global review of energy consumption, CO2, emissions and policy in the residential sector (with an overview of the top ten CO 2, emitting countries)[J]. Renewable & Sustainable Energy Reviews, 2015, 43:843-862.
[5] Riahi K, Rubin E S, Taylor M R, et al. Technological learning for carbon capture and sequestration technologies[J]. Energy Economics, 2004, 26(4): 539-564.
[6] Gao L, Fang M, Li H, et al. Cost analysis of CO2, transportation: Case study in China[J]. Energy Procedia, 2011, 4(22): 5974-5981.
[7] Babu P, Linga P, Kumar R, et al. A review of the hydrate based gas separation (HBGS) process forcarbon dioxide pre-combustion capture[J]. Energy, 2015, 85: 261-279.
[8] Benson S M, Orr F M. Carbon dioxide capture and storage[J]. Environmental Policy Collection, 2010, 33(04): 303-305.
[9] Tian Heyong, Wang Wanfu, Wang Renfang, et al. Study on carbon dioxide capture technology[J]. Energy Environmental Protection, 2012, 26(6): 39-41.
[10] Bu Xuepeng. Carbon dioxide capture technology and application analysis[J]. Clean Coal Technology, 2014(5): 9-13.
[11] Gao Yunyi. Study on treatment technology of high carbon dioxide natural gas in Changling gas field[D]. Northeast Petroleum University, 2012.
[12] Johnson JE, Homme Jr AC. Selexol solvent process reduces lean, high-CO2 natural gas treating costs[J]. Energy Prog.; (United States), 1984, 4: 4.
[13] Olajire A A. CO2 capture and separation technologies for end-of-pipe applications - a review.[J]. Energy, 2010, 35(6): 2610-2628.
[14] Kuuskraa V, DiPietro P. Carbon dioxide flooding: the main driving technology of CCS[J]. Cornerstone: Chinese version, 2013, 1(4): 40 -45.
[15] Gao M. Application and prospect of domestic MDEA solution[J]. Journal of Chemical Industry & Engineering, 2007.
[16] Zhang Xuemo, Wu Jilin, Lu Feng. Polyamine (modified MDEA) decarbonization process[C]// National Gas Purification Information Technology Exchange in 2005. 2005: 11-15.
[17] Zhang Xuemo. Application of polyamine method (modified MDEA) decarbonization process[G]. China Nitrogen Fertilizer Industry Association.Proceedings of the Symposium on Nitrogenous Fertilizers in China (two), 2004: 18—21.
[18] Zhang Xuemo. Removal of CO2 and sulfide from synthetic ammonia gas by polyamine method[J]. Nanjing Chemical Technology, 1994(2): 1--5.
[19] Jia Xiaoyu. Application analysis of MDEA decarbonization technology[J]. Chemical Industry Management, 2014(3): 64-65.
[20] Weiss H. Rectisol wash for purification of partial oxidation gases[J]. Gas Separation & Purification, 1988, 2(4): 171-176.
[21] Stiegel G J, Maxwell R C. Gasification technologies: the path to clean, affordable energy in the 21st century[J]. Fuel Processing Technology, 2001, 71(1–3): 79-97.

[22] Bu Xuepeng, Zhang Qi, Suo Ya,et al. Low temperature methanol washing technology and CO2 capture coupling method and system, CN103157346A[P]. 2013.
[23] Cui Jingsi. Discussion on energy saving improvement of low temperature methanol washing process[C]// National Gas Purification Technology Exchange and Seminar on New Coal Chemical Gas Purification Technology in 2013, 2013.
[24] Yang Zhendong, Zhang Huayong, Jin Liqiang,et al. A kind of shunting type low temperature methanol washing technology[J]. Nitrogen Technology, 2015, 36(2).
[25] Yu C H, Huang C H, Tan C S. A Review of CO2 Capture by Absorption and Adsorption[J]. Aerosol & Air Quality Research, 2012, 12(5): 745-769.
[26] Yu C H, Huang C H, Tan C S. A review of CO2 capture by absorption and adsorption[J]. Aerosol Air Qual. Res, 2012, 12(5): 745-769.
[27] Samanta A, Zhao A, Shimizu G K H, et al. Post-Combustion CO2 Capture Using Solid Sorbents: A Review[J]. Industrial & Engineering Chemistry Research, 2011, 51(4): 1438-1463.
[28] Tlili N, Grévillot G, Vallières C. Carbon dioxide capture and recovery by means of TSA and/or VSA[J]. International Journal of Greenhouse Gas Control, 2009, 3(5): 519-527.
[29] Tang Li, Wang Baolin. Separation and recovery of CO2 by pressure swing adsorption[C]// Mechanical Gas Separation Equipment Science and Technology Information Network, China Refrigeration Institute Second Specialized Committee. 1998: 47-51.
[30] Lv Lu, Li Ning, Mi Tie,et al. Research progress of carbon dioxide adsorption technology for flue gas[J]. Industrial Safety and Environmental Protection, 2012, 38(12): 16-18.
[31] Guerin M P, Domine D. Process for separating a binary gaseous mixture by adsorption: US, 3155468[P]. 1964.
[32] Zhang Shiling. Research progress of PSA separation of CO2 from industrial waste gas[J]. Acetaldehyde Acetic Acid Chemical, 2015(8): 8-9.
[33] Juan C. Abanades,,†, Edward S. Rubin,‡ and, Edward J. Anthony§. Sorbent Cost and Performance in CO2 Capture Systems[J]. Industrial & Engineering Chemistry Research, 2004, 43(13): 3462-3466.
[34] Kimble M C, Nacheffbenedict M S, Dallbauman L A, et al. Molecular Sieve CO2 Removal Systems for Future Missions: Test Results and Alternative Designs[J].
[35] Wang Lu. Study and application of adsorption and separation of CO2-N2 mixture by zeolite 13XAPG[D]. East China University of Science and Technology, 2013.
[36] Mérel J, Clausse M, Meunier F. Carbon dioxide capture by indirect thermal swing adsorption using 13X zeolite[J]. Environmental Progress, 2006, 25(4): 327-333.
[37] Shen Chunzhi. Study on CO2 post-combustion capture by carbon materials[D]. East China University of Science and Technology, 2011.
[38] Fabuss B M, Du B W C. APPARATUS AND PROCESS FOR DESORPTION OF FILTER BEDS BY ELECTRIC CURRENT: US, US 3608273 A[P]. 1971.
[39] Grande C A, Rui P L R, Oliveira E L G, et al. Electric swing adsorption as emerging CO2, capture technique[J]. Energy Procedia, 2009, 1(1): 1219-1225.
[40] Grande C A, Ribeiro R P, Rodrigues A E. Challenges of electric swing adsorption for CO2 capture[J]. Chemsuschem, 2010, 3(8): 892.
[41] Grande C A, Rodrigues A E. Electric Swing Adsorption for CO2, removal from flue gases[J]. International Journal of Greenhouse Gas Control, 2008, 2(2): 194-202.
[42] Pugsley T S, Berruti F, Chakma A. Computer simulation of a novel circulating fluidized bed pressure-temperature swing adsorber for recovering carbon dioxide from flue gases[J]. Chemical Engineering Science, 1994, 49(24): 4465-4481.
[43] G, García S, Rubiera F, et al. Post-combustion CO2, capture with a commercial activated carbon: Comparison of different regeneration strategies[J]. Chemical Engineering Journal, 2010, 163(1–2): 41-47.
[44] Hu Y, Yan J. Oxyfuel Combustion for CO2 Capture[J]. International Journal of Greenhouse Gas Control, 2015, 40: 55-125.

[45] Jordal K, Anheden M, Yan J, et al. Oxyfuel combustion for coal-fired power generation with CO2, capture—Opportunities and challenges[J]. Greenhouse Gas Control Technologies, 2005: 201-209.

The ecological damage compensation for hydropower development based on trade-offs in river ecosystem services

Bing Yu*

School of Business, Ningbo University, Ningbo, China

*Corresponding author e-mail: yubing1@nbu.edu.cn

Abstract. Hydropower development is a kind of trade-offs in river ecosystem services, while increasing provision services but reduce some regulation or supporting services. However, some ecosystem services we lost are difficult to recover and thus affect the regulated river and riparian ecosystems inevitably. In this context, an ecological compensation framework for damaged river in hydropower development was proposed based on the trade-off in river ecosystem services. And the accounting system for affected rivers was established based on river eco-service alteration, for the hydropower services cannot replace the lost services. Using the ecosystem services valuation methods, the trade-offs of river ecosystem services were quantified and illustrated by some hydropower projects in Lancang River, Yalung River and Min Chiang in China. According to the survey data, the annual compensation standard of per unit installed capacity in above cases may range from 27 to 206 RMB Yuan.

1. Introduction

Hydropower development in China has shown its huge development potential in last decades, for its low carbon emission and the increasing energy demand. However, hydropower projects construction would bring a series of negative impacts to nature ecosystems, and the affected river could be the most vulnerable and complex one.

In the view of watershed ecosystem services, Wang et al [1] proposed an evaluation models for hydropower development, and estimated the impacts using ecosystem services valuation methods. Some studies also pointed that the non-market value of environment was under evaluated during hydropower projects cost-benefit assessment [2]. Moreover, considering the cumulative impacts due to hydropower dam construction, the researchers have extended their focus into the riparian ecosystem and the whole regional ecosystem, the studies included the riparian vegetation effects analysis, the prediction of river ecological integrity, the impact of the earthquake frequency in the reservoir area, as well as the changing of the regional ecological carrying capacity were all contributed to assess a hydropower [3, 4]. Moreover, some developed countries had launched dam removing because of environmental considerations and the economic cost [5]. The MA acknowledges us that people have to make difficult trade-offs to maximize human well-beings when utilizing the nature environment and resources [6]. In this context, hydropower is a finite resource. When we exploit this service from the regulated river, some other ecological service would be decreased or damaged. That creates an increasing call for an ecological compensation for the habitats that damaged from the hydropower development [7, 8].

The ecological compensation was an effective environmental management method, which can promote ecological protection by internalized environmental externalities [9]. Although the ecological compensation policies are deficient in hydropower development, the researches at home and abroad have always continued. An ecological compensation standard for river basin water environment was calculated based on total pollutants control [10]; an ecological compensation for inundated habitats was made due to carbon stock balance [11]; moreover, the payment for ecosystem (PES) scheme was designed in Cambodia between hydropower operators and forest managers [12]. From the respective of the trade-offs in river ecosystem services during hydropower development, this research make efforts to propose an ecological damage compensation framework for the regulated river and riparian ecosystem. It aims to provide effective suggestions for some office and policy-makers.

2. Framework and methods

2.1. Framework

In this paper, we proposed an ecological damage compensation framework for the regulated rivers during hydropower development (Figure.1). In this framework, the ecological impacts due to hydro projects were analyzed based on the related river ecosystem services' trade-off. According to the four categories of ecosystem services changing, the benefited and impaired which caused by the environmental externality in hydropower development can be recognized.

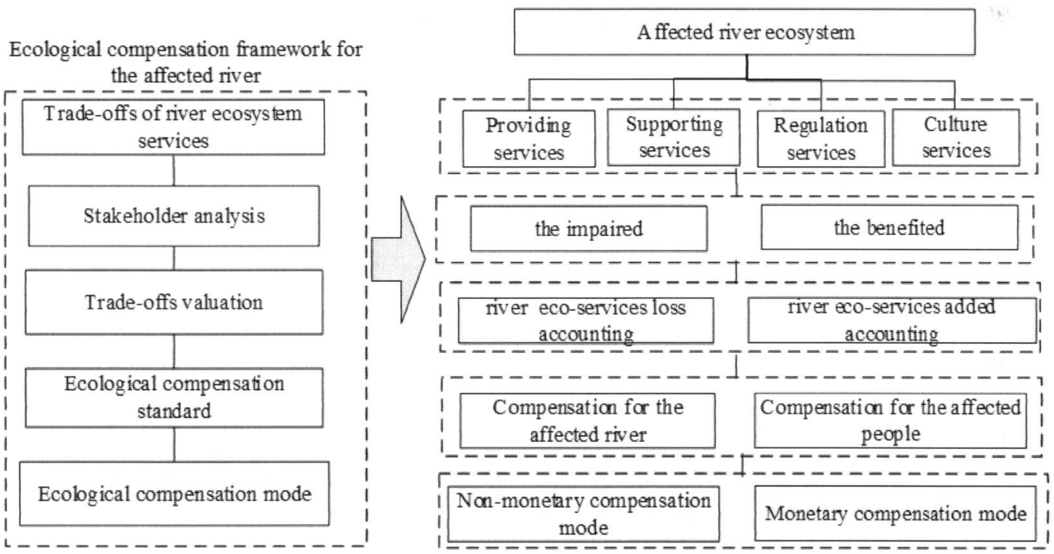

Figure 1 A framework of ecological compensation for the affected river in hydropower development

2.2. Methods

The variation of affected river ecosystem services by hydropower would be calculated through ecosystem service valuation methods (Table 1). Thus, the trade-off services can be quantified which can be as a basis for the standard of the ecological compensation.

Table 1 The river ecosystem services valuation methods

Ecosystem service type	Service item	Method
Providing service	Hydroelectricity	Market value method
	Water supply	Market value method
Regulation service	Flood control	Shadow price method
	Fluvial transportation	Project restoration method
	Soil conservation	Project restoration method
	Environment purify	Project restoration method
Supporting service	Habitat supporting	Project restoration method
		Project restoration method
Culture service	Cultural	Contingent valuation method

3. Results and analysis

3.1. Stakeholder analysis

According to the river ecosystem services' alteration, the possible related stakeholders in hydropower development would be recognized, as shown in Table 2.

Table 2 The possible stakeholders in hydropower project

	Ecosystem service type	Stakeholders	
The added	Hydroelectric generation	Hydropower developers	Human beneficiaries
	Water supply	Water users	
	Flood control	Downstream residents	
	Other provision services	Other beneficiaries	
The lost	Biodiversity	River and Riparian ecosystem	Environment benefits losers
	Fluvial transportation		
	Soil conservation		
	Environmental decontamination		
	Water regulation		

3.2. Trade-off analysis

We selected three representative rivers in China with hydropower development and calculated the annual variation of each river ecosystem service value caused by per unit installed capacity. The results were shown in Table 3 and Figure 2.

Table 3 The annual variation of river ecosystem service value caused by per unit installed capacity (RMB Yuan/Kw)*

Regulated river Ecosystem service type		Lancang River	Yalung River	Min Chiang
The added	Hydroelectric generation	714	723	828
	Water supply	0.57	/	37.78
	Flood control	0.32	4.88	23.14
The lost	Biodiversity	13.90	2.47	36.26
	Fluvial transportation	188	7.85	49.12
	Soil conservation	3.01	12.39	61.35
	Environmental decontamination	0.67	4.50	11.34

Note: the parameters for calculation came from Ref. [13-15].

IOP Conf. Series: Earth and Environmental Science **64** (2017) 012047 doi:10.1088/1755-1315/64/1/012047

It seems that the trade-offs in river ecosystem services during hydropower development are advantageous and profitable, for the added ecosystem service value is higher than the lost. However, the concern cannot just stay at the increased 'value'. On the one hand, the river ecosystem service loss would be a conservative estimate; on the other hand, the added services cannot substitute for the lost from the perspective of ecological functions. In this concern, the ecological compensation for the damaged river ecosystem should be established and the compensation standard could be estimated based on the river ecosystem services value loss. Take the cases in this study as an example, the annual compensation standard could be ranged from 27 to 206 RMB Yuan/KW due to the degree of river ecosystem services loss.

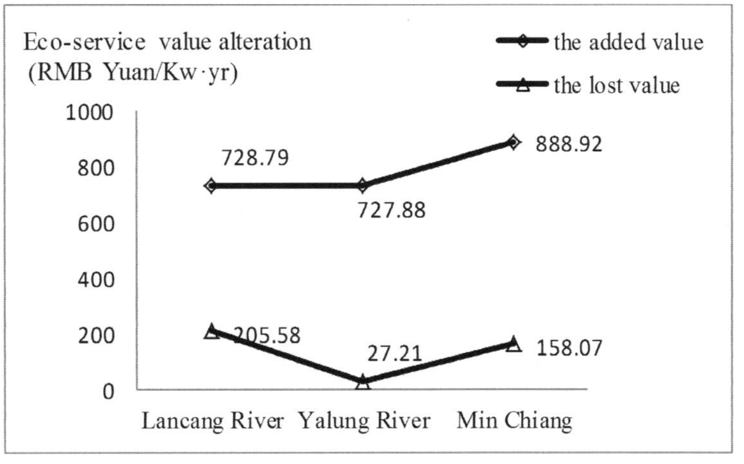

Figure 2. The river ecosystem services value alteration generated by hydropower development

3.3. Eco-compensation mode analysis
The different compensation objects should be compensated in different modes. For the affected human, the provided ecological compensation should meet their actual requirements. Thus, various compensation ways could be combined, such as policy compensation, fund compensation, and material compensation may be selected for their benefits loss. For the habitat loss, project compensation method, such as afforestation projects could be built in watershed, which offset the lost ecosystem services. The damaged aquatic river could be also compensated by off-site compensation mode, which means that the beneficiaries could take some measures to improve other pollutant water bodies in local watershed, to realize regional water system ecological balance.

4. Conclusion
In this paper, we proposed a hydropower development ecological compensation framework to promote water resource sustainable exploitation. According to the trade-off analysis, the compensation objects were identified, and standard was calculated based on the loss of affected river ecosystem services by hydropower development. With the case study of Lancang River, Yalung River and Min Chiang hydropower development, the annual compensation standard could be ranged from 27 to 206 RMB Yuan/Kw due to the degree of river ecosystem services loss.

This study was a preliminary attempt for hydropower development trade-off analysis in the view of affected river ecosystem services. Due to the data acquisition restrictions, the specific standard setting and how to carry out the ecological compensation need further studies.

Acknowledgments
This work was financially supported by K.C.Wong Magna Fund in Ningbo University.

IOP Conf. Series: Earth and Environmental Science **64** (2017) 012047 doi:10.1088/1755-1315/64/1/012047

References

[1] G. H. Wang, Q. H. Fang, L. P. Zhang, W. Q. Chen, Z. M. Chen, H. S. Hong, Valuing the effects of hydropower development on watershed ecosystem services: Case studies in the Jiulong River Watershed, Fujian Province, China. Estuar. Coast. Shelf S. 86(2010) 363-368.

[2] C. Gowan, K. Stephenson, L.Shabman, The role of ecosystem valuation in environmental decision making: Hydropower relicensing and dam removal on the Elwha River. Ecol. Econ. 56 (2006)508-523.

[3] H. J. Zhai, B. S. Cui, B. Hu, K. J. Zhang, Prediction of river ecological integrity after cascade hydropower dam construction on the mainstream of rivers in Longitudinal Range-Gorge Region (LRGR), China, Ecol. Eng. 36(2010)361-372.

[4] J. P. Li, S. K. Dong, Z. F. Yang, M. C. Peng, S. L. Liu, X. Y. Li, Effects of cascade hydropower dams on the structure and distribution of riparian and upland vegetation along the middle-lower Lancang-Mekong River, Forest Ecol. Manag. 284(2012) 251-259.

[5] E. P. Anderson, M. C. Freeman, C.M. Pringle, Ecological consequences of hydropower development in Central America: impacts of small dams and water diversion on neotropical stream fish assemblages, River Res. Appl. 22(2006)397-411.

[6] H. Caroline, S. Helen, V. Bhaskar, Creating win-wins from trade-offs? Ecosystem services for human well-being: A meta-analysis of ecosystem service trade-offs and synergies in the real world, Global Environ. Chang. 28 (2014) 263–275.

[7] R. Home, O. Balmer, I. Jahrl, Motivations for implementation of ecological compensation areas on Swiss lowland farms, J. Rural Stud. 34(2014) 26-36.

[8] B. Yu, L. Y. Xu, Review of ecological compensation in hydropower development, Renew. Sustain. Energy Rev. 55(2016)729-738.

[9] B. Yu, L. Y. Xu, Study of Eco-compensation inHydropower Development in China, Procedia Environ.Sci,13(2012) 891-898.

[10] X. Guan, W. Liu, M. Chen, Study on the ecological compensation standard for river basin water environment based on total pollutants control, Ecol. Indic. 69(2016)446-452.

[11] B. Yu, L. Y. Xu, Z. F. Yang, Ecological compensation for inundated habitats in hydropower developments based on carbon stock balance, J. Clean. Prod. 114(2016)334-342.

[12] M. E. Arias, T. A. Cochrane, K. S. Lawrence, T. J. Killeen, T. A. Farrell, Paying the forest for electricity a modelling framework to market forest conservation as payment for ecosystem services benefiting hydropower generation, Environ. Conserv. 4 (2011) 473-484.

[13] G. L. Wei, B. S. Cui, S. K. Dong, Z. F.Yang, Impact of hydropower development on river ecosystem service :a case study from the Manwan Hydropower Project. Acta Sci. Circumstantiae,28(2008)235- 242.

[14] J. H. Xiao, S. J. Chen, Q. D. Yu, D. J. Chen, H. P. Liu, Ecological compensation standard forZhaoshi water conservancy project based on river ecosystem services, Res. Envi. Yangtze Basin, 5(2012)611-617

[15] M. Chen, S. C. Li, H. L. Sun, T. Tian, Impact of cascade hydropower development in the lower-reach of Yalung river on river ecosystem services, J. Hydroelectric Eng.30(2011)89-107.

Enhanced photocatalytic activity of zeolitic imidazolate framework-8 by modification with phosphor tungstic acid

Jiahui Guo[1], Lili Yang[2,*], Ruhuai Guo[2], Shuo Yang[2], Jianbin Chen[2]

[1]Centre for Training, Guangdong Polytechnic of Environmental Protection Engineering, Nanhai District, Foshan 528216, China
[2]School of Materials and Environment, Beijing Institute of Technology, Zhuhai 519088, China

*Corresponding author e-mail: yllgdut@126.com

Abstract. Phosphor tungstic acid (PTA)-modified zeolitic imidazolate framework-8 (ZIF-8) catalysts were synthesized for photocatalytic degradation of methylene blue under UV irradiation. The as-prepared catalysts were characterized by the powder X-ray diffraction, N_2 adsorption-desorption, Fourier transform infrared spectroscopy, Raman spectroscopy and UV-vis diffuse reflectance spectroscopy techniques. The optimum reaction conditions for photocatalytic degradation of methylene blue on the obtained materials were investigated. The results show that PTA evidently improved the activity of ZIF-8 catalysts because of its fast reversible multi-electron redox transformations. In addition, the reaction mechanism underlying the process was proposed.

1. Introduction

Organic dyes are major sources of environmental contamination, and the removal of these organic pollutants is very important for both academic and industrial societies. Since the first artificial photo catalyst TiO_2 was discovered for pollutants degradation,[1] many metal oxides including ZrO_2 and WO_3 have been identified as photo catalysts for solving environmental problem.[2, 3] The degradation capacity towards organic pollutants using these developed metal oxides as catalysts, however, is still moderate at present. Thus, it is of great interest to search for the novel higher effective photo-catalysts.

Metal organic frameworks (MOFs), composed of organic and inorganic building blocks, have emerged as a novel kind of porous crystalline materials. Their excellent properties, such as uniform microspores, accessible pore volumes, and large surface areas,[4] have opened the door for MOFs in many potential applications.[5-11] In particular, MOFs have the ability to behave as semiconductors when exposed to light, making them unique platforms for light harvesting and photoinduced catalysis.[12-14] Some MOFs can act as a charge-carrier transporting system because of the existence of the open metal sites and the catalytically active organic linkers.[15-17] Majima investigated the photoinduced charge-transfer process between excited MOF-5 nanoparticles and organic compounds by using time resolved emission and diffuse reflectance spectroscopies.[18, 19] In addition, as the host porous materials for photocatalytic reactions, MOFs can also provide extra pathways for the migration of photoinduced electrons and thus facilitate charge-carrier separation, increasing the photocatalytic efficiency.[23] Besides, the presence of coordinatively unsaturated metal sites in MOFs allows their post-functionalization by grafting of active species. Recently, an increasing number of papers have appeared to indicate that MOFs, such as MOF-5, provide a unique opportunity for integrating different molecular functional components to achieve light harvesting and to drive photocatalytic hydrogen evolution and CO_2 reduction.[14] However, the slight exposure of MOF-5 to the ambient atmosphere or residual water in the solvent remarkably lowers their photocatalytic activity because of the destruction of the pore structures.

The zeolitic imidazolate frameworks (ZIF-8, $(Zn(MIM)_2$, MIM=2-methylimidazole), a subclass of MOF, was selected in this study as the photo catalyst because of its high stability.[22] The secondary building units of ZIF-8 are $Zn_6(MIM)_6$ and $Zn_4(MIM)_4$ rings, and each Zn(Ⅱ) ion is tetrahedrally coordinated by four nitrogen atoms from bridging MIM to furnish an infinite 3D frameworks.[20-21] To the best of our knowledge, photocatalytic degradation of methylene blue (MB) by ZIF-8 has rarely been studied. Unfortunately, the obtained results show that the degradation capacity towards MB using ZIF-8 as catalyst in this paper is still moderate. Phosphotungstic acid (PTA), which consists of heteropolyanions $PW_{12}O_{40}^{3-}$ with W-O octahedra as the basic structure unit, has received largely attentions in the past decades.[24] The properties such as extremely high proton mobility, fast reversible multi-electron redox transformations and well stability make PTA an excellent catalytic candidate. Several compounds that PTA was introduced into the MOF segments have been reported.[25] However, highly efficient, truly heterogeneous catalysis over well-defined PTA /MOF materials had not been presented until now. Maksimchuk et al.[26] recently demonstrated that PTA /MIL-101 could be utilized as oxidation catalysts, but they did not determine their crystal structures. Herein, we try to anchor highly dispersed PTA on the ZIF-8 frameworks via a facile room-temperature synthetic strategy and expect that the as-synthesized PTA/ZIF-8 can act as a solid-state photo catalyst with high activity. To illustrate the activities of PTA/ZIF-8 catalysts, photo degradation of methylene blue has been carried out. Interestingly, the results show that the PTA/ZIF-8 catalysts exhibit an excellent reusability and higher catalytic performance for methylene blue degradation than that of bare ZIF-8 under UV light illumination. It is expected that our current research could provide a new way to design ZIF semiconductor photo catalysts for organic pollutants degradation.

2. Experimental Section

2.1. Preparation of ZIF-8 and PTA/ZIF-8

ZIF-8 was prepared according to the procedure described by Moises A. Carreon et al.[27]: a solution of 2-methylimidazole ($C_4H_6N_2$, $6.4x10^{-3}$ mol) in methanol (1.4 mol) was added to a solution of zinc nitrate hexahydrate ($[Zn (NO_3)_2]\cdot 6H_2O$, $8x10^{-3}$ mol) in methanol (1.4 mol), then vigorously stirred for 10 min. White polyhedral crystals appeared immediately. Finally, this solution was centrifuged at 2000 rpm and washed 3 times with methanol, then dried overnight at 60 °C.

For the preparation of PTA/ZIF-8 (impregnation of PTA in ZIF-8), PTA (0.1 g) was dispersed in acetone (10 mL), while ZIF-8 (0.1 g) was added and stirred for 12 h at room temperature. The resulting materials were recovered by filtration and washed twice with acetone and dried overnight at 60 °C.

2.2. Page Numbers Photo degradation of MB

The photo catalytic activities of ZIF-8 and PTA/ZIF-8 were evaluated by the photo degradation of MB dye under a 30W Xe lamp irradiation at room temperature. The distance between the light and the beaker containing reaction mixture was about 10 cm. The original solution was prepared by MB solution (20 mg/L, 100 mL) and the catalyst (4 mg) in a 250 mL beaker. Prior to irradiation, the mixture was stirred in dark for 1 h to ensure the establishment of an adsorption-desorption equilibrium. Then the mixture reacted at room temperature under stirring. A small quantity of reaction mixture (3 mL) was withdrawn at definite intervals until the end of the reaction and immediately centrifuged to remove the catalyst before analysis.

3. Results and Discussions

3.1. Catalyst and Characterizations

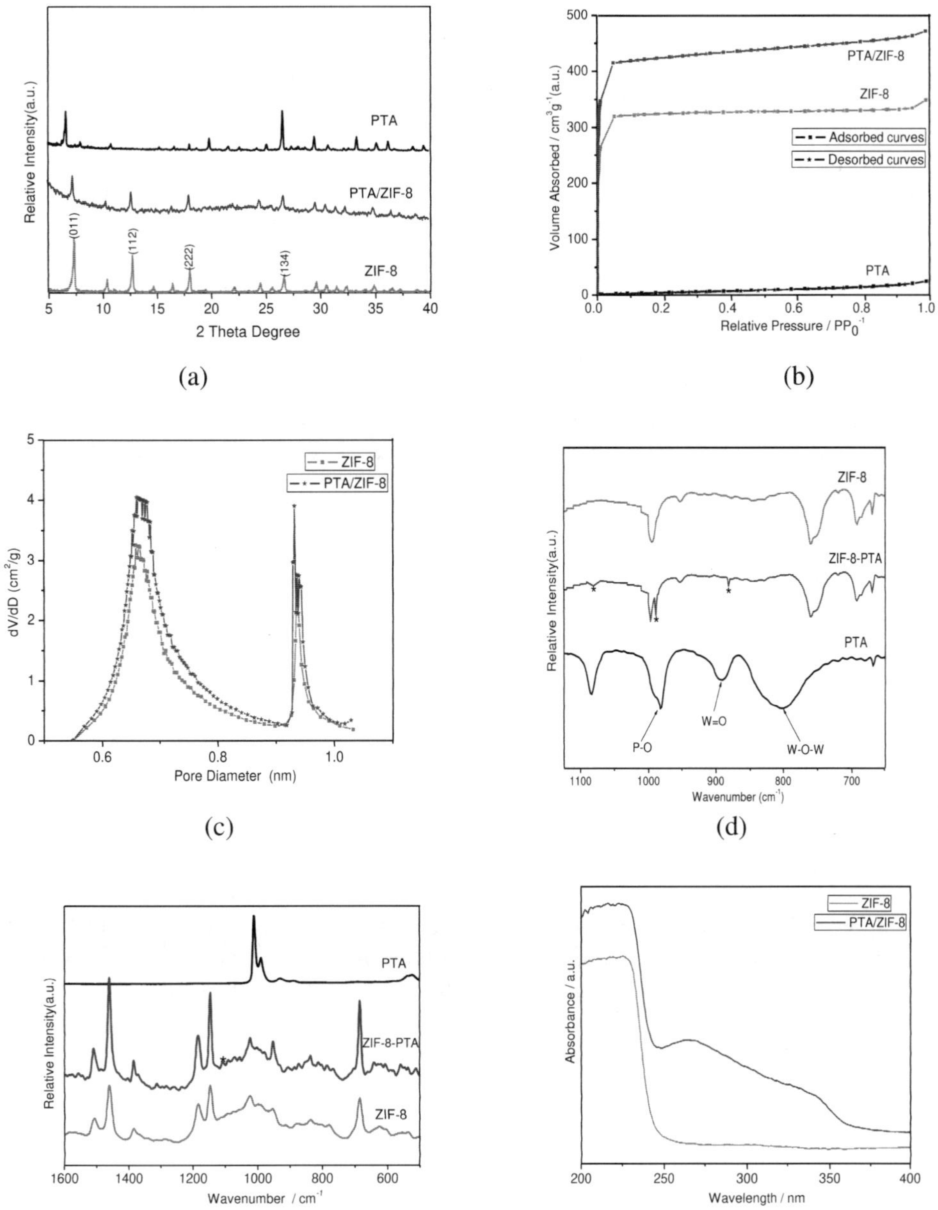

Fig 1. Characterizations of ZIF-8, PTA and PTA/ZIF-8 materials: **(a)** Powder X-ray diffraction patterns of ZIF-8, PTA and PTA/ZIF-8. **(b)** N_2 adsorption-desorption isotherms of ZIF-8, PTA and PTA/ZIF-8 at 77 K. **(c)** Pore size distribution curves of ZIF-8 and PTA/ZIF-8. **(d)** FT-IR spectra of ZIF-8, PTA and PTA/ZIF-8. **(e)** Raman spectra of ZIF-8, PTA and PTA/ZIF-8. **(f)** UV-Vis diffuse spectra of ZIF-8 and PTA/ZIF-8 at ambient temperature.

IOP Conf. Series: Earth and Environmental Science **64** (2017) 012048 doi:10.1088/1755-1315/64/1/012048

The XRD patterns of ZIF-8, PTA, and PTA/ZIF-8 materials are illustrated in Fig 1-a. The sharp peaks of ZIF-8 indicate its excellent crystallinity, and all the diffraction peaks are well matched with the previous results.[28] The as-prepared PTA/ZIF-8 materials possess similar XRD patterns as that of ZIF-8. After the impregnation of PTA, no obvious loss of crystallinity of ZIF-8 is found in the XRD patterns, suggesting that the integrity of the ZIF-8 framework is maintained. Furthermore, the characteristic diffractions of PTA are not found from the XRD patterns of PTA/ZIF-8. This is probably due to the fact that the content of PTA in the synthesized PTA/ZIF-8 materials is relatively low, which is further proved by the ICP-AES. ICP-AES is used to quantify the amounts of W produced as a function of PTA loading in the PTA/ZIF-8 (2.690 *wt%*).

The surface area and porosity of the obtained materials are analyzed by N_2 sorption experiments at 77 K. As shown in Fig1-b, the sorption isotherms of ZIF-8 and PTA/ZIF-8 indicate that micropores are retained. The langmuir surface areas of PTA/ZIF-8, ZIF-8 and PTA were calculated to be 1799 cm^2/g, 1485 cm^2/g and 18 cm^2/g, respectively. The surface area of ZIF-8 is also in good agreement with the recently reported values.[28] The higher surface area of PTA/ZIF-8 is probably attributed to the fact that PTA with small particle size distributed on the ZIF-8 surface uniformly, instead of blocking of the open pores in ZIF-8. The pore size distribution curve of ZIF-8 (Fig.1-c) calculated by Horvath-Kawazoe (HK) method gives two types of microspores located at 6.7 Å and 9.4 Å, respectively, which consistent with the reported literature. Furthermore, the addition of PTA increased the volume of these two types of microspores, corresponding to the larger surface area of PTA/ZIF-8.

To confirm the presence of PTA in the PTA/ZIF-8 catalysts, FT-IR spectroscopy and Raman spectroscopy are performed. Fig.1-d shows the FT-IR spectra curves of ZIF-8, PTA and PTA/ZIF-8, suggesting that three characteristic peaks of PTA are observed in the PTA/ZIF-8. A small peak at around 880 cm^{-1} is attributed to a P-O bond, and a strong peak at approximately 990 cm^{-1} is corresponded to the W=O bond. Furthermore, an additional peak at around 1080 cm^{-1} is observed, which might have resulted from the stretching vibration of W-O-W.[29] The Raman spectra of ZIF-8, PTA and PTA/ZIF-8 in Fig.1-e further demonstrated the above results. And for the crystalline PTA,[30] stretching vibration of P-O lied in 1007 cm^{-1}, stretching vibration of W=O lied in 986 cm^{-1}, stretching vibration of W-O_C-W lied in 900 cm^{-1}, and stretching vibration of O-P-O lied in 550 cm^{-1}. Besides the characteristic bands of ZIF-8, the spectra of PTA/ZIF-8 also exhibited a new small stretching vibration locating at 1100 cm^{-1}, which was assigned to W=O.

The photoabsorption property plays a key role in determining the photocatalytic activity, the UV-vis diffuse reflectance absorption spectroscopy (DRS) of the samples are measured. Fig.1-f displays the DRS of the as-prepared ZIF-8 and PTA/ZIF-8 materials. As can be observed, the main optical absorption band of them lied at approximately 212 nm, which was assigned to ligand-to-metal charge transfer.[13] In other words, under the interference of the ligands and UV irradiation, electron transfer occurred in the d orbital of transition metal Zn ions. In addition, the introduction of PTA enhanced the light absorption intensity of ZIF-8 ranging from 200 nm to 350 nm, which is expected to improve the photocatalytic performance of ZIF-8. This inference is well identified by the photocatalytic testing of ZIF-8 and PTA/ZIF-8 toward degradation of MB under UV light irradiation.

3.2. Photo degradation of MB

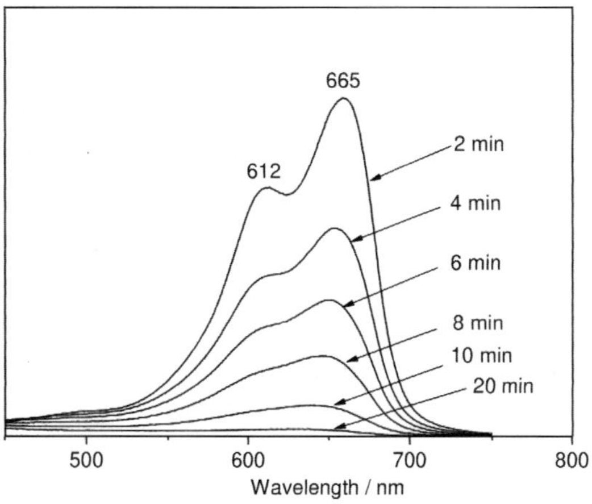

Fig 2. UV-Vis absorption spectra of MB with PTA/ZIF-8 (10 mg) and H_2O_2 (0.5 mL).

The photo catalytic performance of PTA/ZIF-8 is evaluated by monitoring the decolorization of the UV-vis absorption spectra of MB solutions (λ_{max}=665 nm). To determine the optimum reaction conditions, a number of experiments have also been carried out to demonstrate the photo catalytic nature of the reactions. Specially, H_2O_2 plays double roles in the reaction: on one hand, when the concentration of H_2O_2 is relatively low, catalytic efficiency become higher with the increasing of concentration; On the other hand, when the concentration of H_2O_2 is much more, H_2O_2 will become the scavenger of ·OH. In addition, ·OH, ·O_2H even between ·OH and ·O_2H are able to occur the coupling reaction, [31] resulting in the reduction of radicals. Results also showed that the reaction went well when the H_2O_2 volume is equal to 0.5 mL. All the reactions were carried out with the optimum reaction conditions in the following studies. Fig.2 presents the typical temporal evolution of the spectral changes of MB solutions at various intervals in the presence of PTA/ZIF-8. With the increase of irradiation time, the absorption peak at 665 nm ascribed to MB decreases gradually, and it almost disappears after 20 min of irradiation.

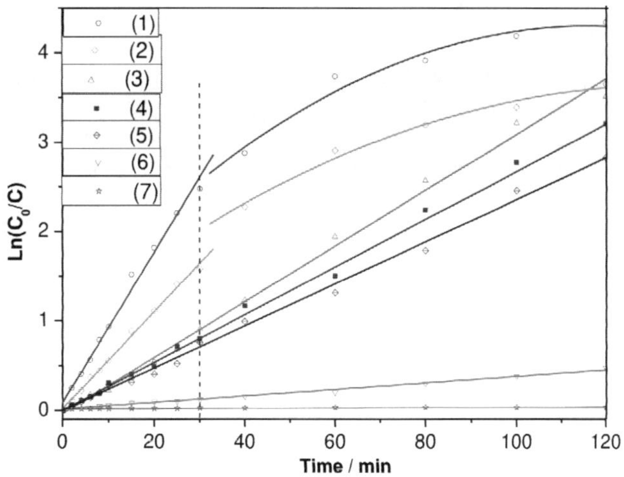

Fig 3. Reaction kinetic plots: (1)with the presence of PTA/ZIF-8 and H_2O_2, (2)with the presence of ZIF-8 and H_2O_2, (3)with the presence of PTA and H_2O_2, (4)with the presence of H_2O_2 only, (5) with the presence ZIF-8 only, (6)with the presence of PTA only, (7) without any catalyst or H_2O_2.

IOP Conf. Series: Earth and Environmental Science **64** (2017) 012048 doi:10.1088/1755-1315/64/1/012048

As shown in Fig 3, only a small margin of MB was degradated after 120 min light irradition in the control experiments without any catalysts. Instead, the degradation of MB proceeds smoothly in the presence of photo catalyst. In addition, PTA/ZIF-8 exhibits higher activity than that of ZIF-8 (Table 1), the degradation ratio is rapidly increased to 99% after UV light illumination for 20 min. The degradation rate of catalysts followed the order: PTA/ZIF-8 > ZIF-8 > PTA. It's significant that impregnation of PTA improved evidently the activity of catalysts, and all of these results confirmed that the PTA/ZIF-8 showed extremely high catalytic activity. The enhanced photo catalytic activity of PTA/ZIF-8 may be ascribed to the intimate interfacial contact between PTA and ZIF-8, which provides short distance for photo-excited charge carrier transfer and reduce the electron-hole pair recombination efficiently. We also noted that the degradation process could be divided into two stages (Fig3): the demethylation of MB chromophoric group and the degradation of aromatic miscellaneous ring of MB.[31] The first stage in the first 30 minutes conformed to first-order kinetics model (1); the second stage, occurred from 40 minutes to 120 minutes, were consistent with second-order kinetics mode (2).

$$Ln(C/C_0) = Kt \tag{1}$$

$$1/C - 1/C_0 = Kt \tag{2}$$

Wherein, C_0 is the initial concentration of MB, C is the final concentration of MB, t is the reaction time, and K is the kinetic rate constant which was obtained from the slope of the linear plot.

Table 1. Effect of different catalysts on the MB degradation rate constant under UV irradiation.[a]

Entry	Catalyst	K^b/min^{-1}
1		-
2	ZIF-8	0.0749
3	PTA/ZIF-8	0.2216
4	PTA/ZIF-8 (1st)	0.2213
5	PTA/ZIF-8 (2nd)	0.2212
6	PTA/ZIF-8 (3rd)	0.2206

[a] Reaction conditions: m (catalyst) = 10 mg, V (H_2O_2) =0.5 mL;
[b] The apparent rate constant (K).

The stability and reusability of the photo catalyst are crucial. To evaluate the reusability of PTA/ZIF-8, three successive cycles for the photo-degradation of MB were carried out over the recycled catalysts. Photo catalyst PTA/ZIF-8 can be simply recovered by filtration, washing and drying. The XRD patterns of the samples before and after the reaction show no significant difference, suggesting the superior stability of PTA/ZIF-8 catalyst. As shown in Table 1, the PTA/ZIF-8 could retain its photo catalytic activity fairly well even after three cycles of application, indicating the high reusability of PTA/ZIF-8.

3.3. Photo degradation of MB Mechanism of Degradation

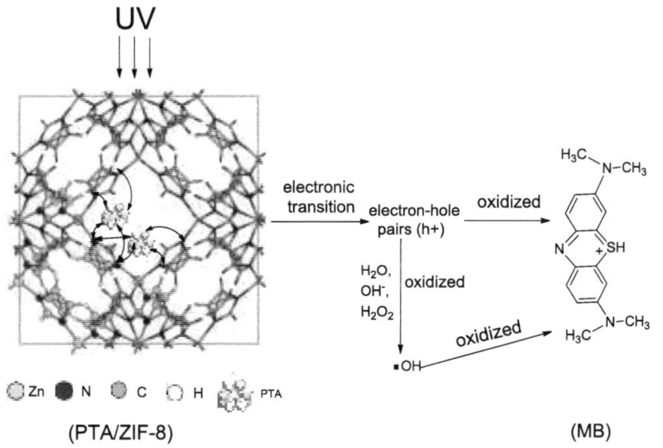

Scheme 1. Mechanism for methylene blue (MB) degradationin when irradiated by UV light in presence of PTA/ZIF-8.

Considering that the conduction band of TiO_2 was constructed by empty Ti 3d orbital1, ZIF-8 containing transition metals Zn as structural nodes is also expected to be an active photo catalyst because the empty Zn d orbital mixed with the organic linkers would form the conduction band. Based on the reported mechanism for MB degradation [31] and the present observations, the photo catalytic reaction mechanism was proposed as shown in Scheme 1. Upon UV light irradiation, it took place that the electronic transition from valence band to conduction band when ZIF-8 received photons with energy, generating electron-hole pairs (h^+). Secondly, electron-hole pairs (h^+) with strong oxidant capacity could directly oxidized organic molecules or reacted with H_2O or OH^- to generate hydroxyl radical(·OH) with stronger oxidant capacity. Consequently, the introduction of external electron acceptor H_2O_2 could enhance the activity of photo-catalyst by generating more ·OH. Moreover, the impregnation of PTA improved clearly the activity of catalysts due to the fast reversible multi-electron redox transformations. The present study indicated that ZIF-8 and PTA/ZIF-8 can be used as the novel catalyst for photo degradation of MB dye pollutants.

4. Conclusion

A novel PTA/ZIF-8 photo-catalyst was prepared by impregnation of phosphortungstic acid at room temperature and applied in the photo degradation of methylene blue with the assistance of electron acceptor hydrogen peroxide. PTA evidently improved the activity of ZIF-8 photo-catalyst due to its fast reversible multi-electron redox transformations. The performance of PTA/ZIF-8 catalysts in the photo degradation of methylene blue was outstanding, indicating the suitable interaction between PTA and ZIF-8. We hope that our current work can widen the application range of the ZIF and provide a feasible way for tuning their photocatalysis properties. Further studies are in progress to design robust and efficient ZIF photo catalysts for visible light-driven degradation of methylene blue.

Acknowledgments
This work was financially supported by the Foundation of Guangdong Province (NO.10251009001000003), Fund of Higher Education of Guangdong Province (NO.ZX-2016-012), and Scientific Program of Beijing institute of technology Zhuhai (NO.XK-2015-08).

References

[1] Daghrir, R., Drogui, P., and El Khakani, M.A, Photo electrocatalytic oxidation of chlortetracycline using Ti/TiO2 photo-anode with simultaneous H2O2 production,

Electrochem Acta, 87 (2013) 18-31.

[2] S.S.K. Ma, K. Maeda, K. Domen, Modification of TaON with ZrO2 to improve photocatalytic hydrogen evolution activity under visible light: influence of preparation conditions on activity, Catal Sci Technol. 2(2012):818-823.

[3] Nithya Vaidyanathan, David M. Hercules, Marwan Houalla, Surface characterization of WO3/ZrO2 catalysts, Anal Bioanaly Tech. 373(2002):547-554.

[4] M.E. Hailian Li, M.O'Keeffe,O.M.Yaghi, Design and synthesis of an exceptionally stable and highly porous metal-organic framework, Nature, 402 (1999) 276-279.

[5] H.T. Kwon, H.K. Jeong, In situ synthesis of thin zeolitic-imidazolate framework ZIF-8 membranes exhibiting exceptionally high propylene/propane separation, J Am Chem Soc. 135(2013): 10763-10768.

[6] Y. Hu, Z. Liu, J. Xu, Y. Huang, Y. Song, Evidence of pressure enhanced CO2 storage in ZIF-8 probed by FTIR spectroscopy, J Am Chem Soc. 135(2013): 9287-9290.

[7] Y. Zhou, J. Song, S. Liang, S. Hu, H. Liu, T. Jiang, B. Han, Metal-organic frameworks as an acid catalyst for the synthesis of ethyl methyl carbonate via transesterification, J Mol Catal A: Chem. 308(2009):68-72.

[8] M. Zahmakiran, Iridium nanoparticles stabilized by metal organic frameworks (IrNPs@ZIF-8): synthesis, structural properties and catalytic performance, Dalton T.41(2012):12690-12696.

[9] Xueling Yi, and Xiaodan Zhang, MIL-53(Fe) MOF-mediated catalytic chemiluminescence for sensitive detection of glucose, Anal Bioanal Tech.408(2016):8805-8812.

[10] T.T. Dang, Y. Zhu, J.S.Y. Ngiam, S.C. Ghosh, A. Chen, A.M. Seayad, Palladium Nanoparticles Supported on ZIF-8 As an Efficient Heterogeneous Catalyst for Aminocarbonylation, ACS Catal. 3(2013):1406-1410.

[11] O. Karagiaridi, M.B. Lalonde, W. Bury, A.A. Sarjeant, O.K. Farha, J.T. Hupp, Opening ZIF-8: a catalytically active zeolitic imidazolate framework of sodalite topology with unsubstituted linkers, J Am Chem Soc. 134(2012):18790-18796.

[12] M. Alvaro, E. Carbonell, B. Ferrer, F.X. Llabres i Xamena, H. Garcia, Semiconductor behavior of a metal-organic framework (MOF), Chemistry.2007;13:5106-5112.

[13] C.G. Silva, A. Corma, H. García, Metal-organic frameworks as semiconductors, J Mater Chem. 20(2010):3141-3156.

[14] J.-L. Wang, C. Wang, W. Lin, Metal-Organic Frameworks for Light Harvesting and Photocatalysis, ACS Catal.2(2012):2630-2640.

[15] D. Jiang, T. Mallat, F. Krumeich, A. Baiker, Copper-based metal-organic framework for the facile ring-opening of epoxides, J Catal.257(2008):390-395.

[16] F. Llabresixamena, O. Casanova, R. Galiassotailleur, H. Garcia, A. Corma, Metal organic frameworks (MOFs) as catalysts: A combination of Cu2+ and Co2+ MOFs as an efficient catalyst for tetralin oxidation, J Catal. 255(2008):220-227.

[17] S.H. Shinpei Hasegawa, Ryotaro Matsuda, Shuhei Furukawa,, Y.K. Katsunori Mochizuki, and Susumu Kitagawa, Three-Dimensional Porous Coordination Polymer Functionalized with Amide Groups Based on Tridentate Ligand: Selective Sorption and Catalysis, J. Am. Chem. Soc.129(2007):2607-2614.

[18] J.R.C. Takashi Tachikawa, Mamoru Fujitsuka, and Tetsuro Majima, Photoinduced Charge-Transfer Processes on MOF-5 Nanoparticles: Elucidating Differences between Metal-Organic Frameworks and Semiconductor Metal Oxides, J Phys Chem C. 112(2008):14090-14101.

[19] T.T. Jun Rye Choi, Mamoru Fujitsuka, and Tetsuro Majima, Evaluating Host-Guest Interactions in a Metal-Organic Framework Using a Polarity-Sensitive Probe, J Phys Chem Lett.1(2010):1101-1106.

[20] H.F. Rahul Banerjee, David Britt, Carolyn Knobler, Michael O'Keeffe, and, O.M. Yaghi, Control of Pore Size and Functionality in Isoreticular Zeolitic Imidazolate Frameworks and their Carbon Dioxide Selective Capture Properties, J. Am. Chem. Soc.131(2009):3875-3877.

[21] X.-C. Huang, Y.-Y. Lin, J.-P. Zhang, X.-M. Chen, Ligand-Directed Strategy for Zeolite-Type Metal-Organic Frameworks: Zinc(II) Imidazolates with Unusual Zeolitic Topologies, Angewandte Chemie. 118(2006):1587-1589.

[22] Alba González, Jessica Avivar, and Fernando Maya, In-syringe dispersive μ-SPE of estrogens using magnetic carbon microparticles obtained from zeolitic imidazolate frameworks, Anal Bioanal Tech. 409(2016):225-234.

[23] L. Shen, S. Liang, W. Wu, R. Liang, L. Wu, CdS-decorated UiO-66(NH2) nanocomposites fabricated by a facile photodeposition process: an efficient and stable visible-light-driven photocatalyst for selective oxidation of alcohols, J Mater Chem A.1(2013):11473-11482.

[24] I.V. Kozhevnikov, Catalysis by Heteropoly Acids and Multicomponent Polyoxometalates in

[25] Liquid-Phase Reactions, Chem Rev.98(1998):171-198.

[26] N. Maksimchuk, M. Timofeeva, M. Melgunov, A. Shmakov, Y. Chesalov, D. Dybtsev, V. Fedin, O. Kholdeeva, Heterogeneous selective oxidation catalysts based on coordination polymer MIL-101 and transition metal-substituted polyoxometalates, J Catal. 257(2008):315-323.

[27] L. Yang, H. Naruke, T. Yamase, A novel organic/inorganic hybrid nanoporous material incorporating Keggin-type polyoxometalates, Inorg Chem Commun.6(2003):1020-1024.

[28] M. Zhu, D. Srinivas, S. Bhogeswararao, P. Ratnasamy, M.A. Carreon, Catalytic activity of ZIF-8 in the synthesis of styrene carbonate from CO2 and styrene oxide, Catal Commun.32(2013): 36-40.

[29] K.S. Park, Z. Ni, A.P. Cote, J.Y. Choi, R. Huang, F.J. Uribe-Romo, H.K. Chae, M. O'Keeffe, O.M. Yaghi, Exceptional chemical and thermal stability of zeolitic imidazolate frameworks, PNAS. 103(2006):10186-10191.

[30] C.M. Granadeiro, A.D.S. Barbosa, P. Silva, F.A.A. Paz, V.K. Saini, J. Pires, B. de Castro, S.S. Balula, L. Cunha-Silva, Monovacant polyoxometalates incorporated into MIL-101(Cr): novel heterogeneous catalysts for liquid phase oxidation, App Catal A: Gen. 453(2013):316-326.

[31] E. Caliman, J.A. Dias, S.C.L. Dias, A.G.S. Prado, Solvent effect on the preparation of H3PW12O40 supported on alumina, Catal Today. 107-108(2005):816-825.

[32] J.J. Du, Y.P. Yuan, J.X. Sun, F.M. Peng, X. Jiang, L.G. Qiu, A.J. Xie, Y.H. Shen, J.F. Zhu, New photocatalysts based on MIL-53 metal-organic frameworks for the decolorization of methylene blue dye, J Hazard Mater. 190(2011):945-951.

The variation of atmospheric particulate matter concentration in the suburb areas over Fuzhou city

Weicong Fu[1], Ziru Chen[1], Zhipeng Zhu[1], Shuping Huang[1], Jianwen Dong[1, 2*], Guochang Ding [1, 2]

[1]College of Arts & Landscape Architecture, Fujian Agriculture and Forestry University, Fuzhou, China
[2]State Forestry Administration National Forest Park Engineering Research Center, Fuzhou, China

*Corresponding author e-mail: fjdjw@fafu.edu.cn

Abstract. The atmospheric particulate matter concentration (APMC) and meteorological factor were supervised and analyze the variation and characteristics in Fuzhou National Forest Park. The result showed that: 1) the atmospheric particulate matter concentration is generally low in summer and high in winter, while in spring and autumn, the APMC with different diameters changed between summer and winter. 2) The daily variation of APMC in different seasons has different characteristics; 3) maximum wind speed, average wind speed, air pressure, wet bulb index, thermal index, humidity and temperature were the meteorological factors that affect APMC. Improving the site condition play a positive role on reducing the APMC and enhancing environmental functions.

1. Introduction

With the economic developed, life quality of residents has been generally improved. Health promotion has become an important part of tourism and residents when they are tripping. Due to convenient transportation and beautiful environment, suburban recreation areas have gradually become the most popular tourist sites [1-3]. Air quality as one of the criteria to evaluate the quality of recreation, people paid more and more attention on it. Due to atmospheric particulate matters (PMs) are the primary pollutant in city areas over China, the concentration of PMs are the certain standard to determine the level of recreational environment quality.

At present, the studies of PMs in various recreational areas were mainly concentrated in the different subsurface effects on PMs [4], the effect of plant configuration on dust removal effect [5-9] and the distance effect from different sources of pollution [10-11]. Suburban recreation should be plan by what, which season, which period, which type of recreation is the most suitable for visitors is a question worth discussion. Fuzhou is located in the southeast coast, the rapid economic development, industrialization and urbanization caused the increase of the concentration of PMs. Coupled with the basin near the formation of inversion and other special weather conditions, urban air pollution was increased [12]. This study focused on the concentration of PMs in Fuzhou, the diurnal variation of PMs and the meteorological factors were analyzed. The contents of PMs, including total suspended particulate (TSP, d≤100 μm), PM_{10} (d≤10), $PM_{2.5}$ (d≤2.5μm) and $PM_{1.0}$ (d≤1.0μm) were investigated.

Meteorological factors e.g. wind speed, air pressure, wet bulb index, heat index, temperature and humidity were also calculated and to explore the relationship between different types of PMs and meteorological factors to provide basic data for suburban recreational land planning and design, and enrich the relevant background of Fuzhou air quality research For Fuzhou.

2. Materials and methods

2.1. Overview of the study site

The annual temperature is 16 ~ 20 ℃, the average annual sunshine is 1700 ~ 1980h, the annual precipitation is 900 ~ 2100mm, the winter is short, cold and dry while summer is long, warm and humid. The coldest month are January-February with average temperature of 6-10 ℃, the extreme minimum temperature -1.2 ℃, the hottest months are July-August with average temperature of 24-29 ℃, the extreme maximum temperature of 42.3℃, relative humidity is around 77%.

2.2. Research methods

Days with typical weather were selected to monitored the data in spring, summer, autumn and winter, during the study period. Observation time are 7: 00, 9:00, 11:00, 13:00, 15:00, 17:00, 19: 00. The monitors were set with the height of 1.2 ~ 1.5 m, and 3 times were repeated at each observation point.

3. Result and analysis

3.1. Variation of seasonal mean value of atmospheric particulate matter concentration

The concentration of atmospheric particulate matter in the recreational area of Fuzhou National Forest Park has obvious seasonal variation, and the atmospheric particulate matter concentration with different particle size shows different variation characteristics. In general, the concentration of TSP in the total suspended particulates in autumn forest park was significantly lower than that in the other three seasons. The concentration of TSP was in the rake of spring, winter, summer, autumn (Figure 4-1). $PM_{2.5}$ and $PM_{1.0}$ showed different seasonal trends, and the order of the large particle size (TSP, PM_{10}) showed the same seasonal variation, and the order of the small particle size ($PM_{2.5}$, $PM_{1.0}$) , Spring, autumn, summer (Figure 1).

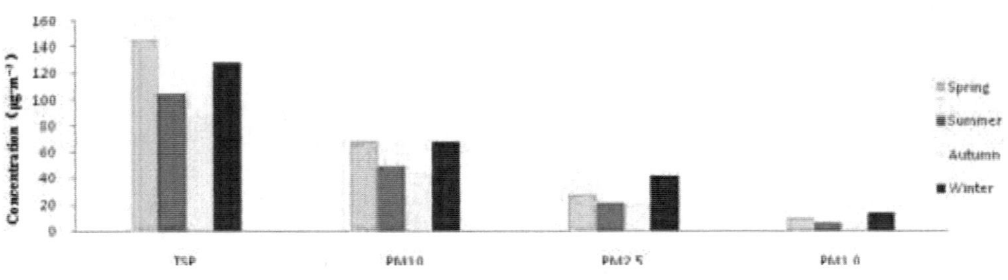

Fig. 1 The seasonal variation of concentration of PMs

The seasonal variation of atmospheric particulate matter at different particle sizes was also significantly different. Inhaled particulate matter and other two fine particulate matter concentrations were highest in winter, followed by spring, autumn and summer, PM_{10} concentration was the lowest in autumn, and the lowest concentration of two fine particles ($PM_{2.5}$, $PM_{1.0}$) was spring (Fig. 2), the proportion of particles with smaller particle size is the lowest in spring, the highest in winter, and the difference between summer and autumn (Fig. 2). The main reason for this is that the contribution of spring dust to coarse particles is larger than that of fine particles. The contribution of winter coal to coarse particles is coarse.

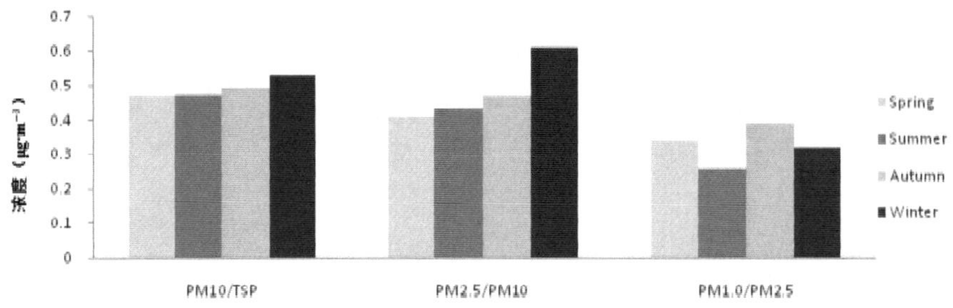

Fig. 2 The seasonal variations of the rate of smaller size PMs in the larger size PMs

3.2. Diurnal variation of atmospheric particulate matter at different seasons

3.2.1 Spring In spring, TSP and PM_{10} diurnal curves were "V" shape, $PM_{2.5}$ and $PM_{1.0}$ diurnal curve was "U" type, the lowest value of the four particle size atmospheric particulate concentration were appears at noon (Fig. 3). From 7:00 am to 13:00 pm, PMs of each particle size were decreasing, and it starts to rise from 15:00 to 19:00 in the afternoon. At 7:00 and 9:00, the concentrations of PMs reached the peak. The concentration of atmospheric particles increases rapidly during 7:00 to 13:00, and the lowest is at 13:00, and then the upward trend is also larger, the whole variations showed "V" type change; and the fine particle is relatively large at 7:00 to 11:00, although it reaches the lowest value at 13:00, but the downward trend from 11:00 to 13:00 was slight, and the rising trend after 13:00 is also less obvious, the overall was oblique "U word" type.

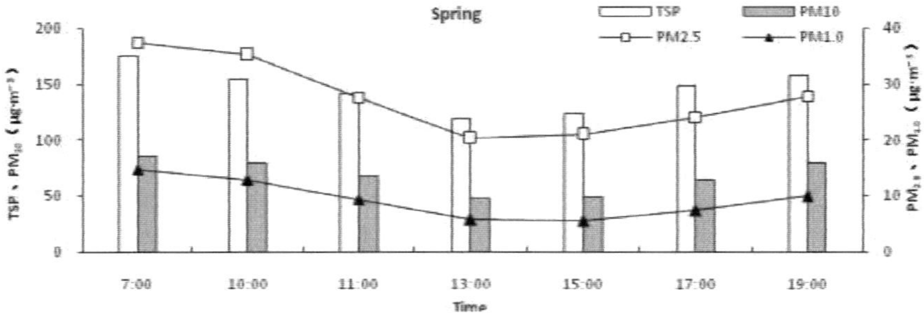

Fig. 3 The variations of diurnal concentration of PMs in spring

3.2.2 Summer The concentration of atmospheric particulate matter in summer was as high as that of summer and early afternoon (Fig. 4). The specific trends are: from 7:00 am to 13:00 pm, the atmospheric particulate matter concentration decreases from 15:00 to 19:00 in the afternoon, and the increase rate is small, and the monitoring time 7:00 in the segment is the peak. 7:00 to 13:00 to reach the lowest value, and a larger decline, followed by a slow upward trend.

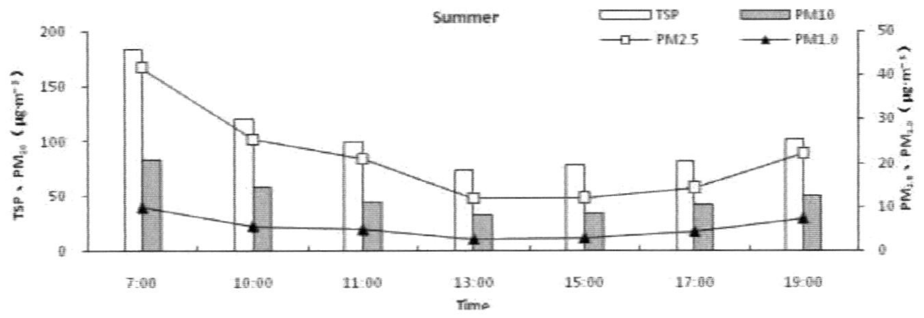

Fig. 4 The variations of diurnal concentration of PMs in summer

3.2.3 Autumn In the autumn, the diurnal variation and characteristics of TSP, PM_{10}, $PM_{2.5}$ and $PM_{1.0}$ were similar to summer, and the whole curve showed "U" type, and the concentration of atmospheric particles was high at 7:00, and the lowest appeared at 13:00, in the afternoon the concentration of PMs increased slightly (Fig.5). Specific trends are similar to spring.

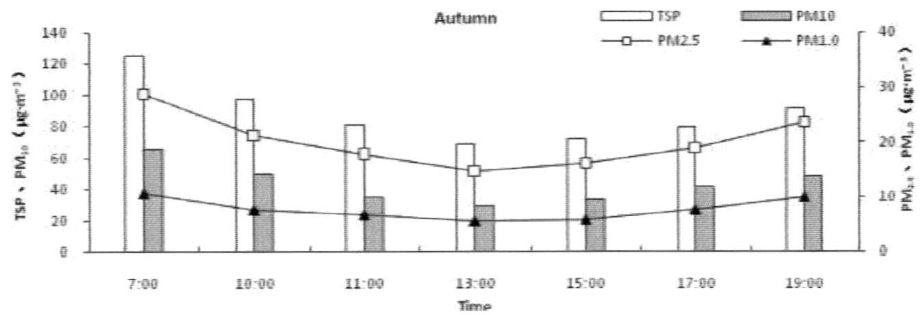

Fig. 5 The variations of diurnal concentration of PMs in autumn

3.2.4 Winter In winter, the diurnal concentrations of TSP and PM_{10} curves were "V" shape, $PM_{2.5}$ and $PM_{1.0}$ diurnal curve showed "oblique L word" type, the lowest of the four particle size of the atmospheric particulate concentration were appeared at 13:00 and increased in the afternoon (Fig. 6). All the four size of the particular meter showed two peaks, which were between 7: 00 and 9:00 in the morning and 19:00 pm, respectively. At 7: 00- 13:00, the $PM_{2.5}$ and $PM_{1.0}$ were showed a downward trend, while the increase trend in the 13: 00-19: 00 was less obvious.

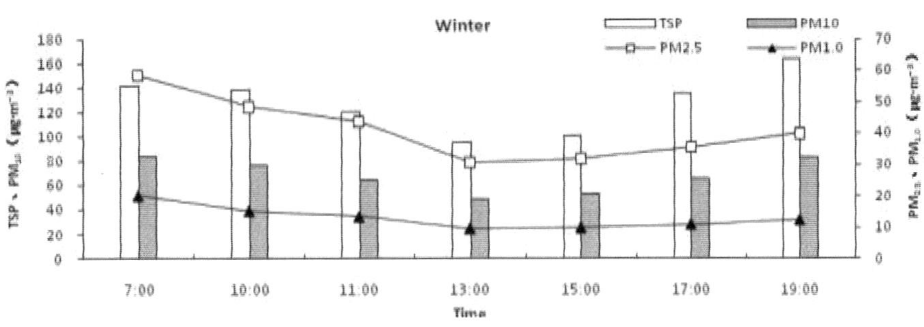

Fig. 6 The variations of diurnal concentration of PMs in winter

3.3. Correlation between meteorological factors and concentration of PMs

The concentrations of the atmospheric particulate matter are affected by the environmental factors such as meteorological factors. We found that the atmospheric particle concentration and meteorological factors (including wind speed, temperature, cold, relative humidity, Dew point temperature, wet bulb temperature) have a correlation. The results show that the environmental factors which affecting the atmospheric particles were wind speed, air pressure, wet bulb index, heat index, humidity and temperature. Among them, the relative humidity, heat index and air pressure have a significant positive correlation with the particles (TSP, PM_{10}, $PM_{2.5}$ and $PM_{1.0}$) with different particle sizes in the air, and the influence of the thermodynamic index on the TSP was obviously. The effect of atmospheric pressure on the particle size of each particle size was slight, and the atmospheric particulate matter concentration increases with the increase of air pressure and humidity. Wind speed, wet bulb index, temperature and atmospheric particulate matter were significantly negatively correlated. The wind speed was only related to the concentration of large particle size, while the wet bulb index and temperature were negatively correlated with the atmospheric particle concentration.

Table 1. Correlation coefficients of PMs and meteorological factors

PMs	correlation	Wind Speed	Air pressure	Wet bulb	Thermal Index	Humility	Temperature
TSP	Pearson	-0.19*	0. 36**	-.054**	0.40**	0.91**	-0.53**
	Significance	0. 32	0.00	0.00	0.00	0.00	0.00
PM_{10}	Pearson	-0.017	0.45**	-0.53**	0.19*	0.72**	-0.87**
	Significance	0.27	0.00	0.00	0.09	0.00	0.00
$PM_{2.5}$	Pearson	0. 20	0.122	0.59**	-0.81**	-0.13	0.58**
	Significance	0.83	0.26	0.00	0.00	0.14	0.00
$PM_{1.0}$	Pearson	0.10	0.57**	-0.81**	-0.16*	0.67**	-0.75**
	Significance	0.11	0.00	0.00	0.03	0.00	0.00

4. Conclusion

The results show that the atmospheric particulate matter concentration was met the national II emission standards by CAAQS, and the air quality was relatively good, by the consistent of the results of previous studies [9-11]. It also showed that, the seasonal variation of air quality was obvious with winter and spring was high while in summer and autumn air quality was better in Fuzhou National Forest Park. Combined with the experimental results with the major cities of atmospheric particulate comparison [6-10], indicating that Fuzhou ecological environment was good, long-term investment in environmental protection has made a significant effect. The construction of the recreational area has improved the living environment of the public. Based on the study of the seasonal variation of atmospheric particulate matter in this study, we will further study the variation and characteristic of atmospheric particulate matter concentration in a wide range of recreational land in Fuzhou, and provide some guidance for improving the landscape planning and construction of Fuzhou air ecology. For relative humidity, heat index and air pressure have significant positive correlations with different particle size in air, the meteorological factors are worth to concern to improve the air quality.

References

[1] Che, H., Zhang, X., Li, Y., Zhou, Z., & Chen, Z. (2006). Relationship between horizontal extinction coefficient and PM10 concentration in Xian, China, during 1980–2002. *China Particuology, 4*, 327–329.

[2] Wan, J. M., Lin, M., Chan, C. Y., Zhang, Z. S., Engling, G., Wang, X. M., Chan, I. N., & Li, S. Y. (2011). Change of air quality and its impact on atmospheric visibility in central-western Pearl River Delta. *Environmental Monitoring and Assessment, 172*, 339–351.

[3] Wei Huang, Jianguo Tan, Haidong Kan, et al. 2009. Visibility, air quality and daily mortality

in Shanghai, China. Science of the Total Environment 407(2009):3295-3330.

[4] Du K., Mu C., Deng J., & Yuan, F. (2013). Study on atmospheric visibility variations and the impacts of meteorological parameters using high temporal resolution data: an application of environmental internet of things in China. International Journal of Sustainable Development & World Ecology, 20, 238–247.

[5] Tsai, Y. I. (2005). Atmospheric visibility trends in an urban area in Taiwan 1961–2003.Atmospheric Environment, 39, 5555–5567.

[6] Husar, R.B., Holloway, J.M., 1984. The properties and climate of atmospheric haze. In: Ruhnke, L.H., Deepak, A. (Eds.), Hygroscopic Aerosols. Deepak Publ., Hampton, Va, pp. 129e170.

[7] W.C. Malm, J.F. Sisler, D. Huffman, et al. 1994 Spatial and seasonal trends in particle concentration and optical extinction in the United StatesJ. Geophys. Res., 99 (D1) (1994), pp. 1347–1370

[8] Nishanth T, Praseed KM, Rathnakaran K, Kumar MKS, Krishna RR, Valsaraj KT. Atmospheric pollution in a semi-urban, coastal region in India following festival seasons. Atmospheric Environment. 2012; 47(1):295–306.

[9] Molnar, A., Meszaros, E., Imre, K., et al., 2008. Trends in visibility over Hungary between 1996-2002. Atmos. Environ. 42, 2621-2629.

[10] Song Y, Wan XM, Bai SX, Guo D, Ren C, et al. (2017) The Characteristics of Air Pollutants during Two Distinct Episodes of Fireworks Burning in a Valley City of North China. PLoS One

[11] Liu XG, Li J, Qu Y, Han T, Hou L, et al. (2013) Formation and evolution mechanism of regional haze: a case study in the megacity Beijing, China. Atmos Chem Phys 13: 4501-4514.

[12] Xu LL, Chen XQ, Chen JS, Zhang FW, He C, et al. (2012) Seasonal variations and chemical compositions of PM2.5 aerosol in the urban area of Fuzhou, China. Atmos Res 104: 264-272.

IOP Conf. Series: Earth and Environmental Science **64** (2017) 012050 doi:10.1088/1755-1315/64/1/012050

Migration of DEHP from plastic to food simulants under microwave heating

X Zhu[1, 2], F Li[1, 2,*], Z Z Qiu[1, 2] and J W Huang[1, 2]

[1]Research Center for Environment and Health, Zhongnan University of Economics and Law, Wuhan 430073, China
[2]Department of Environmental Science and Engineering, Zhongnan University of Economics and Law, Wuhan 430073, China

*Corresponding author e-mail: lifei@zuel.edu.cn

Abstract. The migration of plasticizer DEHP from the plastic products (4 kinds of commonly used plastic food containers under microwave heating: plastic wrap, food bags, ordinary plastic boxes, microwave special plastic boxes) through food contact materials to food simulants (isooctane, 10% ethanol-water solution (*v/v*), 3% acetic acid-water solution (*w/w*) and distilled water) was studied under microwave heating (power levels of 400 W) .The results shows that the DEHP mobility increases with the increase of microwave heating time, DEHP mobility in isooctane and 3% acetic acid-water solution (*w/w*) is significantly greater than in 10% ethanol-water solution (*v/v*) and distilled water; the order of DEHP mobility in isooctane is plastic wrap>food bag>common plastic box>microwave-safe plastic box, while in 3% acetic acid (*w/w*), the order is food bag>common plastic box>microwave-safe plastic box>plastic wrap.

1. Introduction

DEHP is a typical one of Phthalate Esters (PAEs) which spread widely in the environment. It is often used as plasticizer of polyvinyl chloride plastics in industrial production [1]. DEHP has toxic effect on the reproductive system, respiratory system, heart and liver of animals, furthermore, clinical studies have found that DEHP also has physiological toxicities to human [2,3]. DEHP connects with plastic molecules by hydrogen bond or Van Edward force instead of firm covalent bonds, therefore, it's easy to enter the room, atmosphere, food, drinking water and other substances through ways of leaching, migration or evaporation [4].

Nowadays, microwave heating has become an important way of food cooking. Additionally, a certain group of people are forced to use plastic containers to keep and heat food by microwave for a long time because of fast pace of life. So food safety issues under the conditions of microwave heating have been paid more and more attentions [5-10]. At present, the study of migration under microwave heating is mainly focused on the migration from food packaging materials to food, while there is few researches about migration from food container plastic to food in such a short and extreme condition such as microwave heating [11-17].

In present study, 4 kinds of commonly used plastic food containers were selected to do a series of experiments under microwave heating, and isooctane, 10% ethanol-water solution (*v/v*), 3% acetic acid-water solution (*w/w*), distilled water were selected respectively as the fat simulants, wine simulants, acid simulants, water simulants. Major objectives of the present study were to explore the influences of microwave heating time, plastic type, and food properties on migration rates of DEFP from plastic. Results obtained can provide reference for the safe use of food contact materials under microwave conditions.

2. Materials and methods

2.1. Material and reagents
4 kinds of commonly used plastic food containers were selected: New microwave plastic boxes made of polypropylene (PP); new ordinary plastic boxes made of polypropylene (PP); food special bags made of high density polyethylene (HDPE); food wrap films made of polyethylene (PE).
According to GB 23296.1-2009, SN/T 2037-2007 choose food simulants, as shown in Table 1.

Table 1. Food simulants.

type of simulants	food simulants
Aquatic food (pH>4.5)	distilled water
Acidic food(pH≤4.5)	3% acetic acid (*w/w*)
Alcohol food	10% ethanol (*v/v*)
Oily food	isooctane

2.2. Instruments and equipment
10% ethanol-water solution (*v/v*); 3% acetic acid-water solution(*w/w*); distilled water; isooctane(AR); methanol (GR); ethanol (AR); acetic acid (AR); hexane (AR); diethylhexyl phthalate (AR);

GC-4000A gas chromatograph; P70D20TJ-D3Galanz microwave oven; HH-2 thermostat water bath; DHG-9240AS electrothermal constant-temperature drying box; SB-300DTY ultrasonic cleaning machine; TG16-II centrifugal machines; EL204 electronic balance; ultrapure water filter (R > 18.2 MΩ).

2.3. Sample preparation
The plastic samples, which had been washed with aether and DI water in turn, were heated in microwave oven by submerging in food simulants. After microwave heating (power levels of 400 W), samples tested with isooctane were directly heated in water-bath, while samples tested with the three other food simulants need to be extracted by hexane and centrifugal separated firstly. When samples were heated in water bath to nearly dry, they were transferred into volumetric flask and constant-volumed.

2.4. Detection conditions
The GC-FID was chosen as the sample detection method for DEHP, and the chromatographic conditions were shown in Table 2.

Table 2. Working conditions of gas chromatography.

Detector type	Column model	Column temperature	Vaporization temperature	Hydrogen flame temperature
flame ionization detector (FID)	SE-54 capillary column	250℃	270℃	260℃
Protection temperature	Nitrogen pressure	Air pressure	Partial pressure of hydrogen	Sample volume
300℃	0.3MPa	0.2MPa	0.05MPa	1μl

2.5. Migration rate
The migration rate was calculated as follows:

$$R = \frac{S \times V}{a \times W}$$

Where, R is the migration rate of DEHP; S is peak area of sample (μV×s); V is the volume of sample after constant volume, here is 1 ml for all samples; a is the correlation coefficient of concentration and peak area, in this study, its 57791.35; W is the weight of plastic sample.

3. Results and discussion

3.1. Migration rates of DEHP to fat simulants
The change trend of DEHP mobility from different plastic food containers to fat simulants with microwave heating time was presented in Figure 1. For each kind of plastic container, migration rate of DEHP increased with the increase of microwave heating time, especially for food wrap films and food special bags. The migration rates of food wrap films and food special bags were obviously higher than that of microwave plastic boxes and ordinary plastic boxes. The migration rate of DEHP from food wrap films increased rapidly in the first 7 minutes, and then decreased.

The food wrap films and food special bags selected in this study are made of PE, while ordinary plastic boxes and microwave plastic boxes are made of PP. Usually plastic food containers made of PP or PE contain various levels of phthalic acid esters (PAEs) which can enhance the flexibility and ductility of plastics. DEHP is one of the most frequently used PAEs, which is added into most of the plastic products. The requirement of DEHP in PP is relatively lower than that in PE, which can explain the results aforementioned [18].

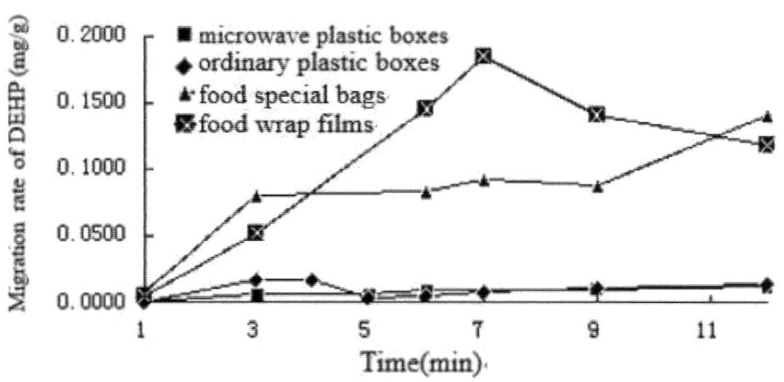

Figure 1. Migration rates of DEHP from different plastic food containers to fat simulants.

3.2. Migration rates of DEHP to acid simulants
The change trend of DEHP mobility from different plastic food containers to acid simulants with microwave heating time was presented in Figure 2. For each kind of plastic container, migration rate of DEHP increased with the increase of microwave heating time, especially for food special bags. The migration rates of food special bags were obviously higher than that of food wrap films, microwave plastic boxes and ordinary plastic boxes. Different with the migration regularity of DEHP to fat simulants, the migration rate from food wrap films to acid simulants remained stable until 6 minutes, and then increased in slow growth from 6 to 9 minutes while increased rapidly after 9 minutes.

Previous researches have shown that when ester groups of DEHP break down under the external conditions, they will combined with H+ or OH-. And the stronger the acidity or alkalinity of external conditions, the more favorable for the hydrolysis of DEHP, in other words, acidic conditions can promote the hydrolysis of DEHP to a certain extent [19]. At the same time, acidic conditions can accelerate the delamination and aging of the plastic products, resulting in decline of mechanical strength, to a certain extent, it can also promote the precipitation of DEHP [20]. Therefore, for the three kinds of non-oily food simulants (wine simulants, acid simulants and water simulants), the migration rate of DEHP from acid simulants is higher than the other two.

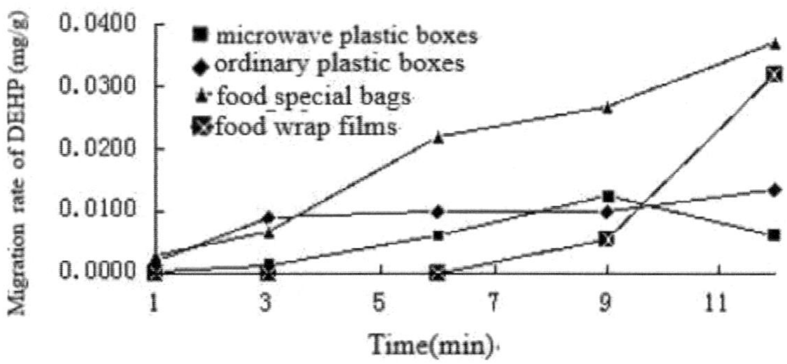

Figure 2. Migration rates of DEHP from different plastic food containers to acid simulants.

3.3. Migration rates of DEHP to wine simulants

The change trend of DEHP mobility from different plastic food containers to wine simulants with microwave heating time was presented in Table 3. Similarly, for each kind of plastic container, migration rate of DEHP increased with the increase of microwave heating time. However, due to the low migration rate of DEHP to wine simulants, the increasing trend was not obvious. In general, the migration rate of food special bags was slightly larger than that of the other three plastics.

Table 3. Migration rates of different kinds of plastic food containers to wine simulants (mg/g).

Heating time	microwave plastic boxes	ordinary plastic boxes	food special bags	food wrap films
1min	0.0016	---	0.0032	---
3min	---	0.0003	0.0051	0.0014
6min	0.0006	0.0023	0.0024	0.0024

3.4. Migration rates of DEHP to water simulants

The change trend of DEHP mobility from different plastic food containers to water simulants with microwave heating time was presented in Table 4. DEHP is the lipophilic organic compounds, so it's molecular structure is quite different from that of H_2O. According to the theory of similar dissolve mutually, the solubility of DEHP in water is very poor. As shown in Table 3, overall the migration rate of DEHP from the plastic to water simulants is fairly low, therefore, there is no significant change trend of migration rates.

Table 4. Migration rates of different kinds of plastic food containers to water simulants (mg/g).

Heating time	microwave plastic boxes	ordinary plastic boxes	food special bags	food wrap films
3min	---	---	---	0.0032
6min	0.0024	0.0016	0.0022	---
9min	---	0.0009	---	0.0046

3.5. Migration rates of DEHP to different food simulants under the same heating time

Heating time of 3 and 6 minutes were selected to investigate the variation laws of migration rate with time, as shown in Figure 3 and Figure 4. The results indicated that basically the migration rate of DEHP with 6 minutes heating time was higher than that of 3 minutes; the migration rate of DEHP to fat simulants was significantly higher than the other three food simulants. DEHP is a moderately low polar compound and easily soluble in solvents with similar polarity such as isooctane, so DEHP in plastic is more likely to migrate to fat simulants (isooctane). Under the condition of non-microwave heating, the regularity also exists. For example, when the food-grade plastic inner materials was soaked in the four kinds of food simulants for 6 days under 60℃, respectively, the migration rate of DEHP to fat simulants was the biggest, which can reach 22% [21].

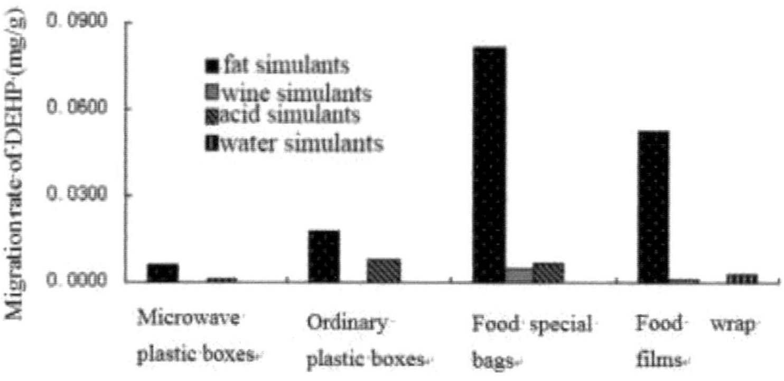

Figure 3. Migration rates of DEHP from different plastic food containers to simulants when heating time is 3 minutes.

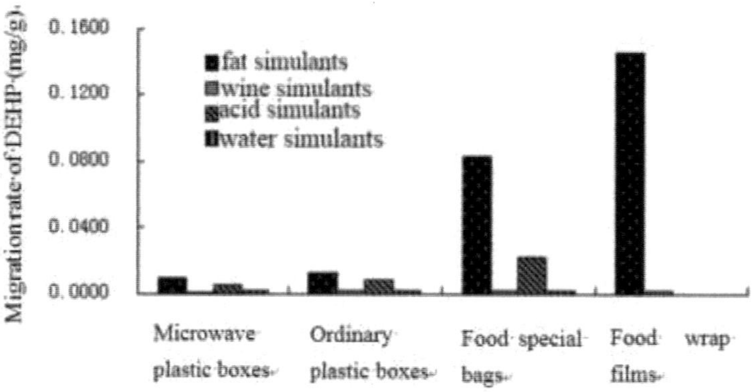

Figure 4. Migration rates of DEHP from different plastic food containers to simulants when heating time is 6 minutes.

4. Conclusions

With the increase of microwave heating time, the migration rate of DEHP in the four kinds of plastic food containers increased in varying degrees, that is, the migration rate of DEHP was positively correlated with the microwave heating time. The migration rate of DEHP to fat simulants was the largest, and was the smallest in water simulants. In addition, DEHP migration rate of all kinds of plastic food containers except food wrap films, was second in acid simulants behind fat simulants.

Moreover, DEHP migration rates of different plastic food containers had the similar regular which decreased in the order of following order: food special bags>ordinary plastic boxes>microwave plastic boxes; Only the migration rate in food wrap films was unique, as the results aforementioned indicated, for food wrap films, the migration rate to fat simulants was the biggest in the four kinds of plastic food containers while that of acid simulants was the lowest.

Acknowledgements
This study was supported by the National Natural Science Foundation of China (51308076; 71503268), the Youth Foundation of Zhongnan University of Economics and Law (31541311313) and the Open Fund from Institute of Wuhan Studies (IWHS2016106).

References

[1] Wang J, Dong S J 2012 Asian J. Ecotox. 7 25-34
[2] Rusyn I, Peters J, Cunningham M 2006 Crit. Rev. Toxicol. 36 459-479
[3] Deutschle T, Reiter R, Butte W, Heinzow B, Keck T, Riechelmann H 2008 Environ. Health Persp. 116 1487-1493
[4] Autian J 1973 Environ. Health Persp. 4 3-26
[5] Ehlert K, Beumer C, Groot M 2008 Food Addit. Contam. 25 904-910
[6] Liu Z, Lu L, Wang Z, Liu Q, Xu W 2009 Food Sci. 30 64-66
[7] Wang Q, Ying L, Dai S, Liu Z 2009 Food Sci. 30 246-250
[8] Alin J, Hakkarainen M 2010 J. Appl. Polym. Sci. 118 1084-1093
[9] Alin J, Hakkarainen M 2012 Polym Degrad Stabil. 97 1387-1395
[10] Alin J, Hakkarainen M 2011 J. Agric. Food Chem. 59 5418-5427
[11] Li D, Hu C, Cheng J, Wu Y, Wang Z 2013 Food Sci. 34 41-45
[12] Jiang X, Zeng M, Hao Y, Xu Z, Wang J 2013 Chin. J. Inorg. Anal. Chem. 3 69-72
[13] Wu J, Wang Z, Wang N, Cheng J, Wu Y 2013 Food Sci. 34 25-28
[14] Gao S, Wang Z, Hu C 2014 Food Sci. 35 6-9
[15] Cao X 2010 Compr. Rev. Food S. F. 9 21-45
[16] Aurela B, Kulumala H, Soderhjelm L 1999 Food Addit. Contam. 16 571-577
[17] Zulch A, Piringer O. 2010 Food Addit. Contam. 27 1306-1324
[18] Zhang L, Wu Q, Liang J 2012 Food Sci. 33 184-188.
[19] Gang X, Li F, Wang H 2006 Water Wastewater Eng. S1 178-181
[20] Gang X, Li F, Wang H 2006 J. Agro-Environ. Sci. 25 1625-1630
[21] Xie L, Li X, Zhang G, Guo Z, Wang S, Yu J, Ren P 2012 J. Xi'an Univ. Technol. 27 83-87

Saving energy and protecting environment of electric vehicles

Lina Yuan*, Huajun Chen and Jing Gong

College of Data Science, Tongren University, Tongren, China.

*Corresponding author e-mail: 893422817@qq.com

Abstract. With the concept of low carbon economy, saving energy, and protecting environment spread, the development of the electric promotes the research pace of wireless charging electronic vehicles, which will become the best choice of energy supply in the future. To generalize and exploit the corresponding alternative fuels and the research and development, and promotion of electric vehicles, becomes the effective means to directly reduce the consumption of fuel, effectively relieves the problem of nervous energy and environmental pollution, and really conforms to the requirements of the national strategy of sustainable development in China. This paper introduces the status of electronic cars and wireless charging, expounds the principle of wireless charging, and concludes the full text.

1. Introduction

In recent years, the development of new energy industry, especially the pure electric vehicles are increasingly growing rapidly [1-4]. The deputy director of pollution prevention and control department in Ministry of Environmental Protection Jian Wang said, motor vehicle's amount of China continues to be sustainable growth. In 2014, the production and sales of nationwide cars reached 3.723 million and 23.492 million, respectively. Compared with 1980, the national motor vehicles owned increased by 33 times, and reached 245.772 million, among which, 144.522 million are cars, 9.72 million cars at low speed, and 91.53 million motorcycles. Classified by emissions standards, the achieved Ⅳ and above standard cars accounted for 22.7% of total car ownership, the Ⅲ standard cars 52.5%, the Ⅱ standard cars 10.4%, the Ⅰ standard cars 10.6%, and the remaining 3.8% of the cars cannot reach the Ⅰ standard. According to the classification of environmental mark, the "green-label cars" accounted for 93.2%, and high emissions of "heavy-polluting vehicles" still 6.8% [3]. In 2016, the energy department website in the United States (U.S.) announced a novel wireless charging technology, which can be up to 20 kW charging powers, is about three times of the current charging pile, is expected to replace the traditional charging pile, and becomes the charging infrastructure of the electric vehicles. The energy department of Oak Ridge National Laboratory (ORNL) recently cooperates with Toyota, cisco systems, Evatran and Clemson university international center for automotive research in Tennessee to show the new wireless vehicles' charging technology, tested vehicle model is Toyota RAV4 (10 kWh battery) that is filled with 80% electric quantity only to need 18 minutes, while 100% 30 minutes. Now the research and development of 50 kW system has begun, if successful, the fast charging operation of general electric cars only need 7 minutes [4].

Therefore, it is for the charging modes' diversification and convenience of electric vehicles to put forward higher requirements. There are three energy supply ways of electric vehicles: battery replacement, slow charging alternating current (AC), and fast charging direct current (DC). The form of battery replacement faces two key challenges: the realization of standardizing the battery box and the practical utilization of fast battery replacement; The shortcoming of slow charging AC is the lengthy of filling charging time; The form of fast charging DC will greatly shorten the service life of batteries. The changes of large current will impact on power grid. While using wireless charging can make charging equipment stealth, the wear rate of equipment is lower, the applied range is wide, public charging area is relatively small, which can realize non-contact charging. Even to realize that when electric cars are driving, they can be non-contact charging. Wireless charging form not only has the convenience, but also can adapt to the rapid development of the electric car industry. Meanwhile national policies can be also the support and planning of electric cars, which has a great impetus role on wireless charging. The wireless charging technology of electric vehicles belongs to a kind of brand-new energy supply technology, which fully conforms to carry out the demand of saving energy, protecting environment and developing low carbon economy initiated by the electric vehicles, which will become a main direction of electric cars for power supply technology [5].

2. The principle of wireless charging technology

Wireless charging technology refers to the devices with batteries do not rely on the electric wires, using the principle of the electromagnetic induction (EMI), or other relevant AC induction technology, in the senders and receivers with the corresponding equipment to send and receive the induced AC signals, is a technology for charging, deriving from the technology of wireless electronic power transmission. Mainly analyze from the perspective of near field charging and far field charging as follows [1, 6, 7].

2.1. Near field charging

2.1.1. Near field inductive charging Nikola Tesla proved energy delivered by oscillating magnetic field in the early 1880s, which can charge the battery by near field or magnetic resonance (MR) [8]. The near field induction charging system is shown in Fig. 1 [9]. The current and voltage from the transmitter to the receiver must be AC. By converting AC network voltage to step-down and DC, provide the drive and controller circuit of transmitters for bias. Drive and controller generates switch signals and adjusts switching frequency, which converts DC into AC, and then inputs to the primary coil. At the receiver, to rectify AC signals, and then adjust by synchronized transformation, for the battery charging. According to the needed watt level of the receiver, the coil of frequency changes. Through signals superimposed on the power signals, so both know equipment is placed on the charging mat. Inductive charging efficiency is higher, but whether coil alignment is sensitive. Need to adjust coupling coil to slightly deviate from the resonant frequency, in order to optimize power transmission.

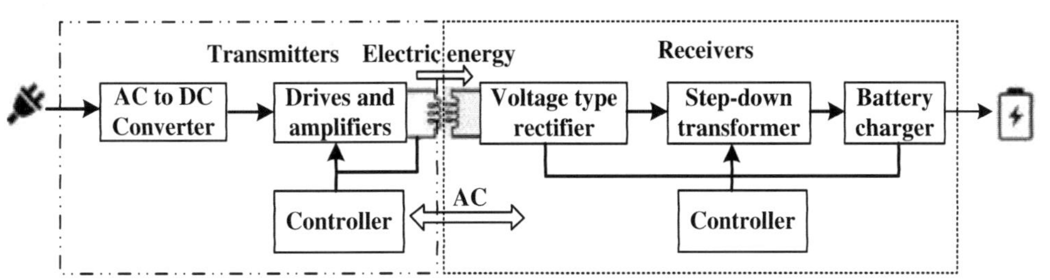

Figure 1. The near field inductive charging system.

2.1.2. Near field resonance charging Resonance charging is another kind of near field charging form, whose working principle is the same as that of the electromagnetic field, but the former needs the front end of the resonator. The standard dominated by the AirFuel Resonant, allows the shorter distance between the transmitter and receiver. A single 6.78 MHz transmitter can support multiple receivers, without physical alignment. However, the strict frequency match is required between the receiver and transmitter, and to farthest extend the power transmission distance in the specific coil size. With the increasing number and extending distance of connected equipment, transmission power will go down. At the same time, the standard requires has independent two-way communication channels (Bluetooth) between the transmitter and each receiver [10].

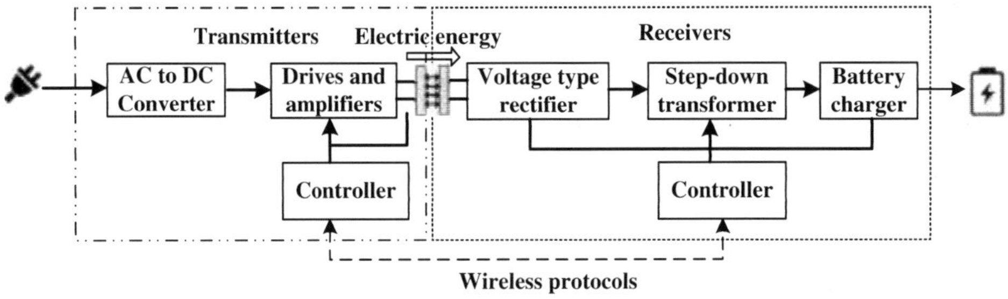

Figure 2. Qi-compliant wireless power transfer model.

AirFuel Inductive is another standard, and Powermat is a good example of bridging technology that provides a common ring, can be used with the charging mat, to recharge portable devices. Because there are two standards, general motors (GM) company announced that its cars will simultaneously support AirFuel Inductive and Qi standard [11, 12]. Samsung Company also decided its mobile phones will support two standards. The related technical indicators of near field charging standards is as shown in Table 1 [9].

Table 1. The comparison of three near-field charging standards: Qi, AireFuel Inductive, and AirFuel Resonant

Technical indicators	Qi (WPC)	AireFuel Inductive	AirFuel Resonant
Power frequency	100~205 kHz	277~357 kHz	6.785 MHz
Data communication	Superposition	Superposition	Bluetooth: 2.4 G
Free receiver position	×	×	√
Multiple receivers	×	×	√
Modes	Induction	Induction	Induction
Power transmission	Good (~5 W)	Weak	Weak

2.2. *Far field charging*

Both inductive charging and resonance charging have requirement for the distance between the transmitter and receiver. In the far field charging, energy needs to be transferred from power hub to a specific device. Bluetooth, WiFi, ultrasonic and infrared ray, etc,, were ever explored to be used. Based on radio frequency (RF) system (for example, WattUpTM and CotaTM) uses one or more antenna to broadcast energy and proceed to communication. In October 2015, Energous company announced it can provide the first radio frequency power receiver IC, which will make radio frequency rectify DC signals, this integration technology-CotaTM, in fact, use the existing WiFi and Bluetooth antenna to realize data communication and wireless power reception, and then these micro-signals are added to the battery's charging current. The receiving device of response and specific battery charging characteristic data transfer back to the router of power source in the way of wireless [12-14].

After continuous link is established, the router of power source emits energy beam to the receiver's position. In the ultrasonic system represented by uBeam, the signal generator in the router generates electrical signals, which are sent to the amplifier, and then the amplified signals are connected to the transducer, producing ultrasonic wave, and through focusing sent to the receiver. Ultrasonic wave imposes on piezoelectric sensor, resulting in a charging current. Sensors used by the two sides of the system are needed to support high efficiency and high power. Another entrepreneurial firm-Wi-Charge, concentrates on converting the line of sight light to energy. The transmitter uses laser diode to accurately transmit infrared beam to the receiver, and then photo-voltaic cells of the receiver convert light back to electric energy. This company is for system to function demo in February 2015 in San Francisco [9]. Infrared technology has an obvious advantage of no EMI radiation.

3. The latest achievements achieved by wireless charging technology

As early as in 2011, Perry, at the University of Pennsylvania in the U.S., invented her first wireless charging system, which is mainly composed of boxes with the size of two electric ovens, and can implement electricity transmission within the scope of 2 inches. Thereafter, Perry showed its working performance and principle of wireless charging system at the summit of science and technology, called her equipment as uBeam, and subsequently, UBeam Company continues to the research and development of wireless charging technology and shows it will launch its research and development of prototypes in the near future. For some existing wireless charging equipment on the market, Perry thinks they cannot reach the technology level of uBeam. Nowadays, there are quite a number of electronics manufacturers that have applied wireless charging technology in the electronic products, but it is mainly adopted by one-to-one charging mode, which do not have the charging ability of many sets of electronic equipment, while American manufacturers roll "Wait Up" wireless charging equipment that can be realized to charge the electronic equipment within the scope of the emitter as the center for radius of 4.6 meters (m), and can drive the 12 sets of electronic equipment at the maximum and its transmission is not higher than 0.25 W power supply, or four sets no more than 4 W, and its charging speed is inversely proportional to the charging distance, this device depends on its internal Bluetooth to realize for the positioning of the required charging equipment and electricity detection, after detecting that the battery is low, emitting wireless signals about 5.8 Hz implements the power transmission. With the constant depth of wireless charging technology, WaitUp Company is committed to research and development of the corresponding hardware equipment to improve its industrial chain, and strengthen the competitiveness of its products [9-11].

Meanwhile, in early 2015, the U.S. Patent Office has received a patent application submitted by Samsung company, its research content is mainly about the research and development of the electromagnetic induction coil of wireless charging technology, Samsung's wireless charging technology has the basic function of charging, at the same time, the most important characteristics is the wireless charger embedded in the function of point of ultraviolet radiation, to make the surface of the charging devices disinfect, the U.S. Patent Office exhibited the appearance, structure, and working principle, etc., of the wireless charging equipment submitted by Samsung. At the same time, Microsoft is also unwilling to feel lonely, formulating automatic positioning smart phones, called AutoCharge, of research and development planing for wireless charging technology, through charging its researched prototype installed on the ceiling indoors to cover a wide range for indoor mobile phones, so as to realize wireless charging for mobile phones. The prototypes of wireless devices are mainly built-in a monitoring module and a charging module, in collecting monitoring module adopting technology is based on image processing to monitor and track indoor phones, through continually testing the indoor equipment such as similar mobile phones and after detected, to use the beam of light produced by the solar power with power technology provides remote wireless charging service for smart phones, at the same time, AutoCharge detects the electronic quantity of electronic equipment through its Bluetooth, and after completion of charging, stops battery charging [1,6,12,13].

In brief, wireless charging technology is an important direction of the electronic technological development nowadays and even for a period of time in future, through the application of wireless charging technology that can make people liberated from multifarious wires, which to plays an important role in better implementing the popularization and utilization of the electronic equipment and new energy vehicles [14, 15].

4. Conclusion

In two sessions of 2016, topics related automobiles once again become a significant issue for discussion. Keqiang Li, the prime minister, not only mentioned the "active used-car market", "speeding up the construction of the urban parking lot", and "constructing the charging infrastructure of new energy vehicles", etc., in the government report, when talking about the problem of environmental pollution, more put forward "put forth effort to reduce emissions of fire coal and vehicles", "comprehensively generalizing vehicles to use the V standard of fuel oil", and "weeding out 3.8 million yellow-label cars and old cars".

In a word, how to implement green development, i.e., saving energy, protecting environment, and low carbon, is the problem for automobile industry to face. In the future energy conservation and environmental protection will be the backbone of automobile industry in China.

Acknowledgments

This work was supported by the Collaborative Fund Project of Science and Technology Agency in Guizhou Province Marked by the word LH on 7487[2014], and the reform project of teaching contents and curriculum system in colleges and universities of Guizhou Province on 2014SJJGXM003, and the project of education and cooperation for talent team word in Guizhou in 2015(NO: [2015]67), and partly supported by the National Natural Science Foundation of China (NO.61562703).

References

[1] Xiao Lu, Ping Wang, Dusit Niyato, Dong In Kim, and Zhu Han, Wireless charging technologies: fundamentals, standards, and network applications, IEEE Commun. Surveys & Tutorial, 2 (2016) 1413-1450.
[2] W. Khan-ngern, H. Zenkner, Wireless power charging on electric vehicles, in Proc. Int. Elect. Eng. Congr. (iEECON), Chonburi,Thailand, Mar. 2014.
[3] Qiyi Wen, Energy conservation and environmental protection is the main line of Chinese vehicles' makers, Energy research & utilization, 1 (2016) 20-20 (Chinese).
[4] http://news.bjx.com.cn/html/20160408/723360.shtml.
[5] Liwen Lin, Li Zhai, Chao Song, The wireless charging technology of automobile cars, New energy source of cars, 6 (2016) 1-5 (Chinese).
[6] J. Garnica, R. A. Chinga, J. Lin, Wireless power transmission: From far field to near field, Proc. IEEE, 6 (2013) 1321-1331.
[7] Meng He, The development of wireless charging technology, Electronic Engineering & Product World, www.eepw.com.cn, 2 (2016) 28-30 (Chinese).
[8] Tesla coil [Online], http://www.teslauniverse.com/nikolatesla-article-inventor-of-dreams.
[9] Hong Qiu, The application status and development prospect of wireless charging technology, Science and technology innovation and application, 15 (2016) 75-75 (Chinese).
[10] S. Hached, A. Trigui, I. El Khalloufi, M. Sawan, O. Loutochin, J. Corcos, A Bluetooth-based low-energy Qi-compliant battery charger for implantable medical devices, in Proc. IEEE Int. Symp. Bioelectron. Bioinf. (ISBB), Chung Li, Taiwan, Apr. 2014.
[11] S. Y. Hui, Planar wireless charging technology for portable electronic products and Qi, Proc. IEEE, 6 (2013) 1290-1301.
[12] Y. Xu, Heating system based on wireless charging QI standards, CN 203554077 U, Apr. 2014.
[13] Jian Sun, Study efficient wireless charging technology of the electric vehicles, Autotime, www.cnautotime.com, 1 (2016) 41-43 (Chinese).
[14] H. H.Wu, A. Gilchrist, K. Sealy, P. Israelsen, J. Muhs, A review on inductive charging for electric vehicles, in Proc. IEEE Int. Elect. Mach. Drives Conf. (IEMDC), Niagara Falls, ON, USA, May 2011.
[15] S. Bi, C. K. Ho, R. Zhang, Wireless powered communication: Opportunities and challenges, IEEE Commun. Mag., 4 (2015) 117-125.

IOP Conf. Series: Earth and Environmental Science **64** (2017) 012052 doi:10.1088/1755-1315/64/1/012052

Study on tourist carrying capacity of sustainable tourism---by taking Qingliang Mountain in the south-eastern china as an example

Shenglang Jin*, Ting Liu

College of Tourism, Huangshan University, Huangshan 245021, China

*Corresponding author e-mail: jinshenglang@163.com

Abstract. This paper analyzed measurable problems of tourist carrying capacity and then studied the corresponding measures by adopting theories and methods of environment sciences. Tourist carrying capacity of Qingliang Mountain was studied by analyzing the system of tourism environment capacity. The results showed that the level of service was 9726, which was smaller than the ecological carrying capacity of 12894. The facilities and spatial capacity were identified as key factors to limit the tourist carrying capacity in Qingliang Mountain. Some advices and measures including newly built and rebuilt basic facilities, construction of management methods for slack and peak tourist season respectively, construction of new sight spot and touring path were discussed to improve the tourist carrying capacity of Qingliang Mountain and solve the problems of overloading of tourists in the peak season.

1. Introduction

Sustainable tourism means to visit a place as a tourist and try to make a positive impact on the environment, society and economy. Nowadays, the environmental problem is very important for all the countries in the world. The resource shortage, environmental pollution and ecological imbalance have become global hot issues [1]. The contradiction among economic development, resource use and environmental protection has confined economic progress and endangered human's life, which has been widely concerned by the world. Tourism, as an industry with fast growth, high output values as well as a large capacity of employment, has attracted great attention due to its notable economic benefits. Further, it is known as a "Rising Sun Industry" because of its powerful momentum in growth. In the past, people thought that there was no contradiction and conflict between tourism and the environment. Tourism was regarded as an "anthracitic industry" without any pollution [2]. Nevertheless, as tourism develops, recently, environmental problems arise frequently which have influenced sustainable development of tourism. People have to reexamine the relationship between tourism development and environmental protection [3]. Researches concerning sustainable development and environmental protection are launched; influences of tourism development on the environment is studied according to ecological principles; sustainable development of tourism is widely concerned.

Tourism, as an economic activity, has contributed extremely to development of many geographically isolated regions as well as other countries [4]. In the last years, the lacking of organization tourism development of many regions in combination with massiveness of tourism causes new problems in the organization of regional space. The main reasons can be summarized as follows: (a) regional inequalities because of excessive concentration of tourism in specific areas; (b) progressive abandonment of other productive activities and specific tourism as the only or main activity in a number of regions; (c) degradation of allocated tourism resources and reduction of tourist attractiveness due to overexploitation of natural and structured environments [5].

Tourist carrying capacity can be regarded as the maximum allowable number of tourists in a scenic area under restrictions of ecological environment and basic facilities such as transport or accommodation [6]. At present, many researchers study this problem from the view of ecological environment protection. Undoubtedly, ecological protection is important to tourism areas. In fact, it is the premise to develop tourism industries [7]. However, the above researches ignored one issue that tourism activities are interaction of tourism areas and tourists. The carrying capacity is related to ecological factors and service quality. Enough facilities such as transport vehicles, restaurants and hotels must be guaranteed to provide good service for tourists. Otherwise, the number of tourists will be decreased and the tourism activities will be vanished.

Tourism is a global socioeconomic phenomenon based on human needs for recreation by getting in contact with nature and acquainting with other places and cultures, which is regarded as an activity of important economic, social, cultural and environmental consequences [8]. The economic or social dimension is stressed, but the environmental dimensions is underestimated or ignored among many studies and developmental programs. The balance between tourism and environment imposes configuration of suitable policies for tourism development of some area and imposition of essential limits. Thus, the tourism will not lead to environmental degradation or tourist disdain of some area in a long term by focusing on its protection or its conservation. In other words, the environment will have an influence on tourism and vice versa [9]. The environment of some region is negatively influenced by the increase of tourism in most of times, whereas the tourism development depends on quality and characteristics of environments. Therefore, a new approach is established to calculate tourist carrying capacity by considering both level of service of tourist and ecological environments. First of all, a level of service model is developed; then, the tourist carrying capacity is derived corresponding to critical values of level of service. Later, compared with ecological carrying capacity, the final tourist carrying capacity is determined. Finally, this paper calculates the tourist carrying capacity and evaluates the tourism management methods by taking Qingliang Mountain Resort as an example.

2. Calculating Model of Level of Service

2.1. Influencing Factors

Level of Service means service quality of provided by tourism areas, which is directly reflected by perception of tourists to their activities in tourism areas. For example, if the tourism area can provide enough and convenient transport vehicles, restaurants, hotels and clean environments, tourists will enjoy their travels. Then level of service is high.

Except for the above mentioned, other factors such as sanitations, route planning, weather condition will also affect level of service. This section will study level of service of established facilities in tourism areas. Nevertheless, only important factors including transport, hotels, and restaurants will be studied in this paper.

2.2. Level of Service Model of Single Factors

2.2.1. Level of Service of Transport The spatial accessibility of transport network is important to traveler. If the transport network can cover every spot, the tourists can reach their destinations conveniently. Besides, the crowd degree inside the vehicle carriage is another important factor that will affect tourist's perception. A crowed carriage will reduce service quality. Therefore, the level of service model of transport is established by considering the above two mentioned factors.

2.2.2. Level of Service of Accommodation Tourists will spend more than one day in many famous tourism areas such as HuangShan. Thus, tourists have to stay one night or more at hotels in some areas. Two factors including the number of beds and price of accommodation will affect LOS of accommodation. Tourists will choose hotels based on redundant beds and reasonable price.

Tourists who come from different regions with different economic levels will have discrepant expectations on the price of accommodation. If the accommodation cost in scenic areas exceeds their expectation or affordability, tourists will cancel trips. The accommodation cost can be divided into K (k=6) levels with each level corresponding to one economic level of tourists. The datas are shown in Figure 1.

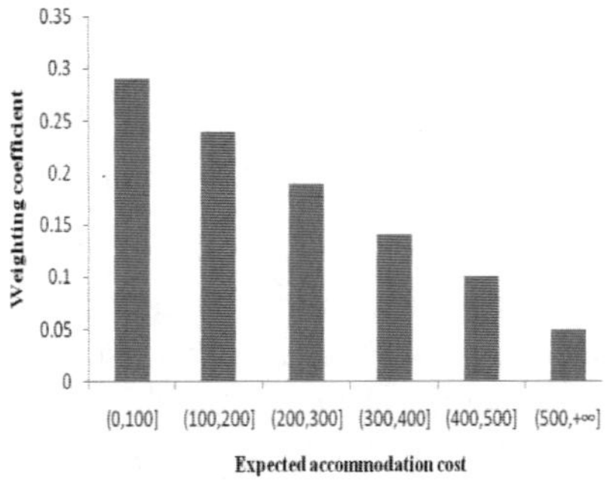

Figure 1. Six levels of the cost of accommodations

A_i Denotes the number of tourists that need accommodation corresponding to level i in tourism areas and E_i denotes the number of beds in economic level i. Generally, the lower of accommodation cost, the more demands of the hotel. Therefore, the lower cost of beds has more important effect on level of service of accommodation.

2.2.3. Level of Service of Restaurants Dining is another important thing for tourists. Like accommodation, the number of seats for dining and the price are two critical factors that affect level of service of restaurants. Therefore, the dining cost is divided into Q (Q=5) levels according to economic levels (as shown in Figure 2).

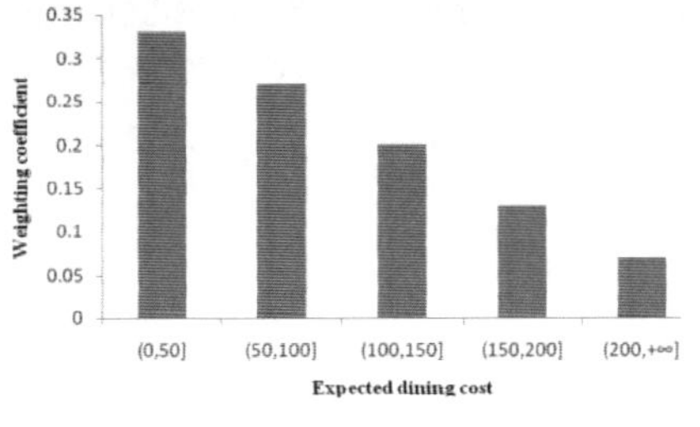

Figure 2. Five levels of dining cost

2.3. Integrated Level of Service Model

Three level of service models of single factors are established in section B, but they cannot reflect the integrated level of service of tourism areas. Thus, an integrated level of service model is established based on three single models.

IOP Conf. Series: Earth and Environmental Science **64** (2017) 012052 doi:10.1088/1755-1315/64/1/012052

2.3.1. Determination of weighting factors of single factors as discussed above, three factors have an influence on level of service of tourism areas but with different levels. The degree of influence of each factor shall be determined by using weighting coefficients.

Generally, the transport is more important than accommodations and restaurants, because a large number of tourists have to take vehicles to spots in many tourism areas. The transport factors are closely related to tourist activities, while hotels and restaurants are extra factors of tourism. This paper determines weighting coefficients of transport, accommodation and restaurant factors as 0.4, 0.3 and 0.3 respectively.

2.3.2. Development of Integrated Level of Service Model From equation (1), (2) and (3) , it can be found that with the increasing number of tourists, PI is increased, which indicates deteriorated level of service. The values of PI_t, PI_z and PI_c are larger than 0 but without upper bound. In most cases, we need a value between 0 and 1 that can directly show the degree of LOS. Thus, we conduct a normalization process to three single level of service models.

2.3.3. Tourist Carrying Capacity Based on Level of Service Model The value of PI is between 0 and 1 now. The more tourists come into the tourism area, the larger of the value will be, which indicates worse level of service. When PI increases to a certain value, the level of service will deteriorate badly and tourists feel severely comfortless. In this case, the number of tourists of the tourism area is equal to its tourist carrying capacity. In this paper, we define the critical value of PI as 0.90 with its corresponding number of tourist's equivalent to tourist carrying capacity T_{LOS}.

3. Final Calculating Model of Tourist Carrying Capacity

3.1. Calculating Model Based on Ecological Environments

Many researchers studied this problem from the view of ecological environment. This paper uses the model established in references 8 directly.

3.2. Final Model of Tourist Carrying Capacity

The T_{CC} is used as the tourist carrying capacity, whose determination should consider both level of service of service facilities and ecological environments.

When $T_{EE} > T_{LOS}$, if the number of tourist is larger than T_{LOS} and smaller than T_{EE}, though the ecological environment can be protected, tourists who come into the tourism area will have a bad level of service and cannot get a glad trip. Otherwise, when T_{EE} is smaller than T_{LOS}, if the number of tourists is larger than T_{EE} but smaller than T_{LOS}, then the ecology will be damaged. Thus, it can be concluded from the analysis that the two factors must be considered when determining tourist carrying capacity.

4. Example Analysis

In this section, we take Qingliang Mountain scenic resort as an example to explain the model established in this paper. Qingliang Mountain, located in Anhui Province, is a national forest and geology park.

4.1. Data collection

In August 2015, when the resort was in peak season, the investigated data were shown in Figure 3 and 4.

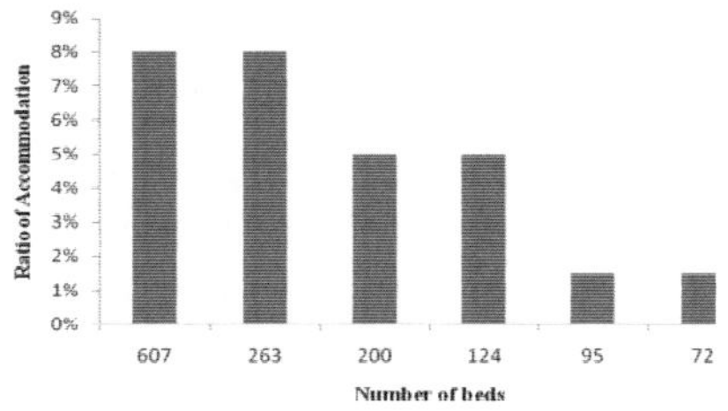

Figure 3. Statistical data about accommodation

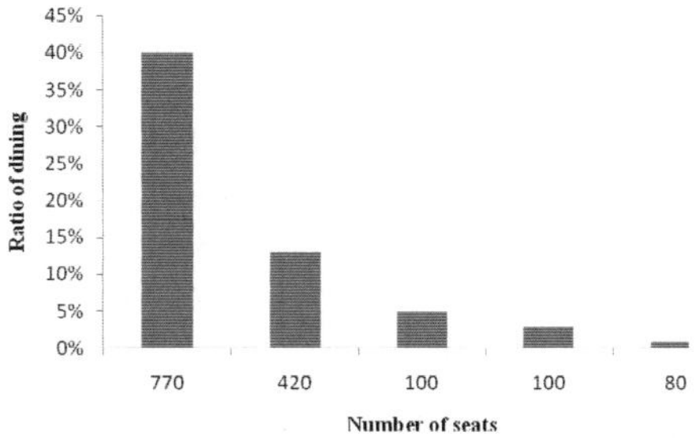

Figure 4. Statistical data about dining

Besides, 20 YuTong buses are available in the tourism area to pick up tourists in each spot for free. Each bus can carry 40 tourists.

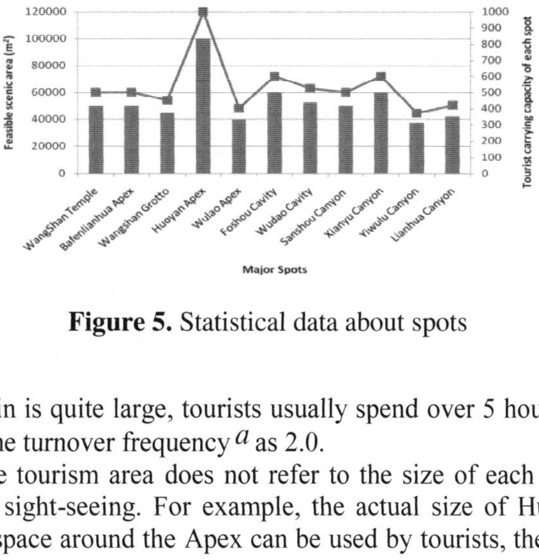

Figure 5. Statistical data about spots

The Qingliang Mountain is quite large, tourists usually spend over 5 hours on traveling every day. Therefore, we determine the turnover frequency a as 2.0.

In Figure 5 the feasible tourism area does not refer to the size of each spot but the size that can accommodate tourists for sight-seeing. For example, the actual size of Huoyan Apex is more than 1000000m2, but only the space around the Apex can be used by tourists, the size is 100000m2, which is 10% of the actual size.

4.2. Calculation of Tourist Carrying Capacity

The dates in Figure 3, 4, 5 are used and then substituted into equation 11 and 12, we can get $T_{LOS} = 9726$, $T_{EE} = 6447 \times 2 = 12894$. The final tourist carrying capacity equals to:

$$T_{CC} = \min(9726, 12894) = 9726$$

Therefore, the tourist carrying capacity of Qingliang Mountain is 9726 tourists every day.

It can be found from the above mentioned results that the ecological carrying capacity T_{EE} is larger than level of service carrying capacity T_{LOS}, which indicates the development of basic facilities (transport, accommodation and restaurant) cannot afford enough tourists. If the tourism area can allow 12894 tourists, though the ecological environment will not be damaged but tourists will get a low level of service. Thus, a further investment on these facilities shall be continued to promote the level of service.

5. Conclusions

There have been conflicts between tourism exploitation and environmental protection. The conflicts lead to obviously worsening environmental quality. The worsened water quality has especially restricted sustainable development of tourism there. However, the tourism development in the new century poses higher demands on Qingliang Mountain. On one hand, the number of tourists needs to be settled. On the other hand, while reducing polluted drainage and solving pollution, the self-purification ability of environment shall be particularly stressed. This paper establishes a new tourist carrying capacity model by considering both ecological environments and basic facilities such as transport, accommodation and restaurant. Then the Qingliang Mountain Resort is taken as an example for further validation. The results show the ecological carrying capacity is 12894 and the carrying capacity of basic facilities is 9726. Thus, the tourist carrying capacity is determined as 9726. The resort can increase its capacity of basic facilities aiming to provide better service to tourists.

We should advocate environment-friendly travels. According to environment conditions and pollution characteristics in Qingliang Mountain, the environment-friendly products should be advocated and detergents should be reduced. The use of detergents with phosphorus is prohibited. The environment behaviors of tourists lead to sustainable development of environments and tourism. The tourism industry in Qingliang Mountain shall achieve its sustainable development by combining pollution prevention with strict pollution administration so as to have a glorious future. The area possesses necessary resources (natural and cultural) for a powerful, balanced and long-term tourism development. Undoubtedly, the sustainable development model shall be reinforced; the local factors are required along with modernization of the existing tourism products; the additional services and products as well as modernization of the existing infrastructures and services shall also be improved.

Acknowledgment

This work is supported by AnHui Province Social Sciences Research Program (NO. AHSKQ2014D59), Humanities and Social Sciences Research Project of Department of Education of Anhui Provincial (NO. SKHS2015B06), the Project of Cultivating Tourism Youth Professionals of Anhui Province (NO.AHLYZJ201613).

References

[1] Yang Meixia, Zhou Guohai. Study on Effective Measures of Environmental Protection for Sustainable Development of Tourism Industry in the World Natural Heritage Wulingyuan. Ecological Economy. Vol.1, No.2, p.83-88. (2005)

[2] Zhuang Dachuan. Environment Quality Evaluation in Wulingyuan scenic spot[J]. Geography Research. Vol.23, No.2, p.192-200. (2004)

[3] Shenglang Jin , Jing Sai , Hongxing Xu. Calculating Model of Tourist Carrying Capacity for Scenic Area. Energy Procedia. Vol.11, p.2886-289. (2011)

[4] O'Reilly A M. Tourism carrying capacity concept and issues. Tourism Management. Vol.7, No.4, p.254-258. (1986)

[5] Ryan C. Recreational Tourism. London: Routledge,(1991).

[6] Alexis Saveriades. Establishing the social tourism carrying capacity for the tourist resorts of the east coast of the Republic of Cyprus. Tourism Management. Vol.21, No.2, p.147-156. (2000)

[7] Tony Prato. Modeling carrying capacity for national parks. Ecological Economics. Vol.39, No.3, p.321-331. (2001)

[8] Steven R Lawson,Robert E,Manning W A et al. Proactive monitoring and adaptive management of social carrying capacity in Arches National Park: an application of computer simulation modeling. Journal of Environmental Management. Vol.68, No.3, p.305-313. (2003)
[9] Simon F J G,Narangajavana Y,Marques D P. Carrying capacity in the tourism industry: a case study of Hengistbury Head. Tourism Management. Vol.25, No.2, p.275-283. (2004)

IOP Conf. Series: Earth and Environmental Science **64** (2017) 012053 doi:10.1088/1755-1315/64/1/012053

Low carbon transition and sustainable development path of tourism industry

Hongbing Zhu[1,*], Jing Zhang[2], Lei Zhao[1] and Shenglang Jin[1]

[1]School of Tourism, Huangshan University, Anhui, China
[2]School of Tourism, Huangshan Vocational and Technical College, Anhui, China

*Corresponding author e-mail: thomas1985@hsu.edu.cn

Abstract. The low carbon transition is as much a transformative technology shift as it represents a response to global environment challenges. The low carbon paradigm presents a new direction of change for tourism industry. However, the lack of theoretical frameworks on low carbon transformation in tourism industry context provides a significant knowledge gap. This paper firstly investigates the relationships between low carbon and sustainable development, followed by exploring the existing challenges of tourism sustainable development. At last, this paper presents a sustainable development path framework for low carbon transition of tourism industry, which include accelerating deployment of renewable energy, energy-saving green building construction, improving green growth investment, and adopting a sustainable consumption and production system, in order to promote energy and water efficiency, waste management, GHG emissions mitigation and eventually enhance its sustainability.

1. Introduction

The current fossil fuel-based energy system is at the root of global climate change. Currently, 82% of energy uses are still coming from fossil-fuel based sources while only 13% from renewable energy. Although there has some important improvements in resource efficiency, the overall use of resources continues to increase. For example, in 2010 the global per capita Ecological Footprint (EF) was 2.6 gha, which is higher than the per capita bio capacity of 1.7 gha. However, the sustainable development of humankind depends on a stable natural and ecological environment. It seeks to improve human development while maintain natural resources and ecosystem services for use by future generations.

Tourism sector is a significant component of the global economy, which ranks as the fourth largest economic sector after fuels, chemicals, and food. Currently, it accounts for 10% of GDP (direct, indirect and induced), 6% of the global exports, and one in 11 jobs. However, most energy use in tourism industry is still based on the traditional fossil fuels and this sector continues to emit increasing levels of greenhouse gas (GHG) emissions [1]. The tourism sector's current contribution to global GHG emissions accounts for 5% of global anthropogenic emissions (1302 Mt CO_2), which is expected to grow substantially under a business-as-usual (BAU) scenario [2]. The sustainability of tourism industry depends in part on new development patterns, like energy efficiency, renewable energy using, and energy-saving green building construction in tourism areas in reducing the GHG emissions. Thus, this research aims to investigate how to promote the sustainable development of tourism industry through using the low carbon related development theory and technology. This paper will discuss the new idea of low carbon and green economy and their potential application in tourism industry, and further suggesting ways to promote sustainable development of tourism industry.

2. Low carbon transition and sustainable development challenges

The need to better coordinate sustainable development of the economy, society and environment has become a hot topic area in the world. For example, the construction of an ecological civilization is now a national strategy and top priority for China. However, the main challenge for China in constructing an ecological civilization lies in its efforts to cope with the tightening resource constraints, environmental pollution and ecosystem degradation. China needs to transform to a new model of consumption which is much more resource-efficient. Currently, the global economy is still based on resource-intensive consumption and production patterns. However, in a low carbon economy scenario, economic growth should be driven by public and private investments that improve resource and energy efficiency, reduce carbon emissions and environmental pollution, and prevent the loss of biodiversity and ecosystem services.

Tourism is one of the important economic sectors in the world. In 2015, international tourists have reached 1186 million while the international tourism receipts earned by destinations worldwide have increased to US$ 1260 billion. According to the UNWTO, international arrivals are expected to reach 1.8 billion in 2030. However, like many other sectors, tourism sector still faces a multitude of significant sustainability-related challenges, like energy and GHG emissions, water consumption, waste management and loss of biological diversity. Table 1 presents an estimated global average number for direct energy use and emissions by different accommodation categories [3]. For example, tourism has become a main contributor of GHG emissions at the global scale because of the growth of energy consumption in tourism transport, accommodation and related activities and its dependency on fossil fuels. It is estimated that CO2 emissions from tourism sector (excluding aviation) will grow at 2.5% per year until 2035. It contributes to climate change through emissions of GHG, including nitrous oxides (NOx), hydrofluorocarbons (HFCs), methane (CH4), sulfur hexafluoride (SF6) and perfluorocarbons (PFCs). Tourism-related energy use and associated emissions of GHG emissions can be categorized into three sections: tourism transport, accommodation and tourism-related activities. Transport and accommodation are the most energy-intensive components, which accounts for the 75% and 21% of GHG emissions [4].

The tourism sector can play a significant role in meeting the UN's Millennium Development Goals (MDGs). In fact, many international organizations and governments have set up many policies to help reduce GHG emissions in the tourism context. For instance, the 10YFP Sustainable Tourism Programmed catalyzes changes in tourism operations. It supports transformation for sustainability through resource efficient and low-carbon planning, conserving ecosystems, and adapting to the reality of a changing climate.

Table 1. Estimated energy use and CO_2 emissions by type of accommodation

Type of accommodation	Energy use per bed night (MJ)	CO_2 emissions per guest-night (Mt)
Hotels	130	55.7
Campsites	50	7.9
Pensions	25	2.7
Self-catering	120	11.6
Holiday villages	90	1.8
Vacation homes	100	0.8
Estimated average	98	13.4

Source: Gossling (2002)

3. Sustainable development path based on low carbon theory and practice

There are many synergies between the low carbon transition and the pursuit of environmental goals, such as reducing water pollution and conserving biodiversity. Table 2 presents the examples of policy misalignments that undermine low-carbon investment [2]. This section presents some pragmatic low carbon development measures in order to build a sustainable roadmap of tourism sector.

Tabel 2. Facilitators and implications of investment in sustainable tourism strategic areas

Strategic area	Sustainability facilitator factors	Potential implications
Energy	- Increasing energy costs - Potential carbon surchages - Progress of low-carbon technology - Low cost of enewable energy - Legislation on energy efficiencey and low-carbon buildings	- Low costs for tourism enterprises through energy efficiency - Investments in energy retrofist - New energy-effficient investment stock - More investments in energy efficient tourism products and services
Climate Change	- Costs of GHG emissions - Consumers' concern on footprint - Host government policies and priorities - Climate change impact on tourism attractions	- Increased substitution of fuels toward electricity, particularly increased investment in solar collectors and alternative fuels for vehicles - Demand for carbon offsets and policies to compensate for residual emissions
Water	- Water sarcity - Price for water and conflicts - Expectations from the major tour operators	- Reducing water costs from internal water efficiency - Investments in water saving technology in rooms, facilities and attractions - Increased use of water treatment systems at different levels
Waste	- Public opinion - Presser from major tour operators - Consumer demand for clean destination	- Lower pollution and natural resource - Improved solid waste management - Lower sewage and clean-up fees - Investment in sanitary landfills and solid waster recycling capaicity

Source: UNWTO and UNEP (2012)

3.1. Accelerating deployment of renewable energy to improve energy-efficiency

There is an emerging need to deployment of low carbon sustainable fuels in tourism sector in order to help promote energy and resource efficiency. In hotels there is considerable scope for additional investment in energy-efficient features and services, such as green building construction, efficient refrigeration, air conditioning and heating in order to reach lower energy profiles. Hotel owners and managers have recognized the need for change and have already made certain improvements in using renewable energies alongside the adoption of low carbon building technology. Meanwhile, tourism sector can cooperate with transport department in order to accelerate the decarbonisation of tourist transport through using cleaner fuels (e.g. biofuels, fuel cells) and more efficient vehicles, like electric cars.

3.2. Encouraging government support on low carbon technology

Government policy has s critical role in enhancing incentives for investigating in renewable energy use. In order to increase the adoption of renewable energies, the tourism sector policy makers should set appropriate market incentives to foster the adoption of next-generation biofuels. For example, time-bound subsidies can be given on the purchase of green tourism equipment that reduces waste and improves energy and water efficiency [5]. In addition, governments can also use tax incentives and credits to promote private investment in green tourism.

3.3. Improving green growth investment to enhance environmental sustainability

A green economy substitutes clean energy and low-carbon technologies for fossil fuels. UNEP defines green growth as one that results in improved human well-being and social equity, while significantly reducing environmental risks and ecological scarcities. In its simplest expression, a green economy can be described as one which is low carbon, resource efficient and socially inclusive. Compared with the BAU development pattern, under a green economy scenario, tourism can make a more large contribution to economic growth, while significant environmental benefits include reductions in CO2 emissions (52%), energy use (44%), and water consumption (18%) [5]. For example, tourism sector can drive the improvement of resource utilization efficiency through the introduction of new green technologies, developing more green products, and green taxes and subsidies [6].

3.4. Adopting a sustainable consumption and production system

Sustainable consumption and production (SCP) is a holistic approach to minimizing the negative environmental impacts from consumption and production systems. SCP advocates and encourages low-carbon and environment-friendly consumption that use fewer materials and less hazardous substances, which include choosing environment-friendly tourism products, encouraging moderate consumption, and adopting economic incentives for certified green tourism products. In addition, tourism sector can make full use of social media, NGOs, and academic institutions to promote the ideas and practices of SCP. Further, tourism sector should advocate the wide application of SCP indicators by local governments and tourism companies to promote transformation towards sustainable development.

4. Conclusion

In conclusion, the low carbon paradigm presents a new direction of change for tourism industry. By adopting and encouraging the low carbon technologies and the green investment, the tourism sector can make a great progress in energy and water efficiency, waste management, emissions mitigation and eventually enhance its sustainability.

Acknowledgments

This work is financially supported by the following Anhui Social Sciences Key Projects funding (SK2015A526; SK2016A0893; AHSKZ2015D15; AHSKQ2014D59).

References

[1] L. Alexandra, D.L. Terry, L. Geoffrey, M. Jiang, Transitioning to a green economy: the case of tourism in Bali,Indonesia. J. Clean. Prod. 111(2016)295-305.
[2] UNEP, UNWTO, Tourism in the Green Economy – Background Report, UNWTO, Madrid,2012.
[3] S. Gossling, Global environmental consequences of tourism. Glob Environ Change.12(2002) 283-302
[4] D. Scott, C.M. Hall, S. Gossling, Tourism and Climate Change: Impacts, Mitigation and Adaption, Routledge, London, 2012.
[5] World Economic Forum (WEF), Towards a Low Carbon Travel & Tourism Sector, Davos, World Economic Forum, 2009.
[6] Information on http://www.oecd.org/greengrowth/greeninnovationtourism.htm

Human-environment sustainable development of rural areas in China

Lei Zhao*, **Hongbing Zhu, Shanfeng Hu**

School of Tourism, Huangshan University, Anhui, China

*Corresponding author e-mail: zhaolei@hsu.edu.cn

Abstract. Human-environment sustainable development has become the important issue of rural transformation development in China. This paper analyses the development status of rural sustainability in China, and also presents the challenges facing the sustainability from the economic, social and environmental levels, including land and energy efficiency, solid waste, water and other types of environmental pollution. At last, the paper proposes the measures to establish the sustainable and liveable rural areas in China, like raising rural community awareness of sustainable development thinking; improving resource efficiency and new energy; and creating rural green industries and green products.

1. Introduction

Rural development is the process of improving the well-being and life quality of rural people. In additional to the traditional economic development, it also involves social and environmental development. In recent years, China has made great progress in terms of the economic development of rural areas [1]. The living standards of the rural community have improved greatly. However, China also faces many economic, social and especially environmental challenges to making rural areas more sustainable and liveable, such as high cost of developing new energy, the presser to increase energy conservation, an increasingly ageing population, environmental burdens, natural resource constraints and other related issues. For example, the industrialization and modernization of the rural areas still largely depended on the consumption of the traditional fossil fuels, which contributes to the climate change. Furthermore, as industrialization, urbanization and agricultural modernization continue, the rural places are often faced by biodiversity loss, air and water pollution issues. These issues pose adverse risks to natural environment and local community development, eroding the natural and human capital upon which rural future development depends.

Currently, China aims to build a "Beautiful China" and enhance the ecological civilization development with emphasize on the quality and efficiency of rural economic development. This requires the rural areas to explore a new sustainable growth pattern, which combine human development and environmental sustainable development objectives [2]. Resource consumption, energy security and critical environmental issues are increasingly integrated in rural development and planning. In fact, the rural areas should pay more attention on green development and low carbon development to achieve poverty eradication, changing unsustainable and promoting sustainable patterns of consumption and production and protecting and managing the natural resource base of economic and social development.

2. Status and challenges of sustainable development of rural areas in China

The theory of harmony between nature and man is one significant component of the traditional values of Chinese civilization. During the past several years, China has proposed some comprehensive and sustainable development concepts such as resource-saving and environment-friendly society, ecological civilization and green development, innovation–oriented country, and continuously put these concepts into practice. These substantive actions have transformed the sustainable development of rural areas in terms of resource conservation and environmental protection [3]. For example, over the past few decades, China has put in great efforts to improve energy efficient, and also intensified air and water pollution control and comprehensive waste management in rural areas. As a result, the total emissions of Sulphur dioxide and chemical oxygen demand have dropped by 14.29% and 12.45% in 2010. In 2009, China agrees to reduce CO2 emissions per unit of GDP in 2020 by 40% to 45% compared with 2005 levels. In fact, China has become the world's largest producer in terms of renewable energy using, like hydro power, wind power installed capacities and solar water heater collector area (See Table 1) [4].

Table 1. Installed Renewable Energy and power-Generating Capacities in China.

	Installed capacity(MW)			Power-Generating Capacity(100 billion kWh)		
	2006	2007	2008	2006	2007	2008
Nuclear	6,850	-	8,850	543	626	684
Hydro	128,570	145,260	171,520	4,167	4,867	5,633
Wind	1,870	3,304	8,940	27	56	128
Biomass	84	268	596	1.04	7.00	18.04

Source: P.J. Li (2009)

As rural areas developed in China, both challenges and opportunities will arise. Thus, the benefits of rapid growth in rural areas have also caused some serious costs, including damages to rural environment and natural resources that could undercut sustainable future growth and development. Rural people face many interconnected environmental challenges. First, In terms of economic development, most of the rural places are still depending on the solely agricultural development, which are is not sustainable in the long term. Also, this can lead to more poverty in some areas. Second, rural people often lack of the knowledge of environment sustainability, which forms another challenge for the rural sustainable development. Residents and local rural leaders have limited awareness of the importance of saving energy and reducing emissions. Third, environmental damage has particularly important implications for increases in poverty and lack of inclusion. In some rural areas, land and energy, solid waste, water and other types of environmental pollution threaten people's lives, health and productivity and also bring frequent occurrences of extreme climate events. For example, in some rural areas, people used marginal lands, destroyed forests for fuel or production, over-extracted and polluted of water resources, which brought long-term consequences. Other examples of environmental threats from rural development include dietary shifts towards meat and dairy, increased demand from urban populations for environmental services, and increased waste associated with changing consumption patterns. However, People in rural areas are particularly reliant upon natural resources for their livelihoods, with most engaged in farming.

In conclusion, rapid rural development causes the depletion of natural resources, threatens the live ability of rural areas, and impedes sustainable human development in China. Environmental degradation and the current unsustainable growth model add pressure on human health and ecosystems in rural areas. China needs a new transformative grow path to shift away from its GDP growth-driven development goals to sustainable development, green economy, and human development in its socio-economic blueprint in rural areas in order to improve rural people's living standards and environmental quality [5].

3. Towards more sustainable and liveable rural areas in China

3.1. Raising rural community awareness of sustainable development thinking

China needs to establish a robust public awareness-building mechanism and education programs in energy saving and emission reduction, especially in rural areas, to raise public awareness about sustainable concept and promote public participation in environmental protection. Nationwide campaigns need to be launched to educate the public about sustainable concepts and to promote environmentally friendly consumption and lifestyles. [6] Meanwhile, public awareness needs to be fostered of low carbon and energy-efficient products, technologies, methods, and practices, and people should be encouraged to adopt sustainable consumption and production practices. The government also needs to ensure the fullest public involvement in scrutinizing the environmental management process.

Furthermore, the currently performance appraisal and review process and system for officials needs to be reformed to shift the focus away from only economic growth and infrastructure construction to include measuring their performance based on environmental protection in addition to improvement in living standards and social economic progress.

3.2. Creating rural green products and improving resource efficiency

Rural planning should aim at ensuring that rural areas are people-centered, which means that rural development should improve the quality of life and economic well-being of farmers. It involves rising agricultural productivity, commercialization and diversification of production patterns and livelihoods within the agricultural sector and the rural non-farm sector [7]. In addressing the sustainable development of rural areas, the rural should effectively adopt green production system and produce more green products.

China should transition from its heavy dependence on energy and resource consumption to improving energy and resource efficiency. The rural areas also should transfer from its heavy dependence on fossil fuel use to use new energy, such as solar energy, wind energy, and water energy to improve energy efficiency (see table 2). Such a transformation is not only for the economic development of rural areas, but also for the long-term sustainable growth.

Table 2. Future of China's renewable energy development

	2010	2015	2020
Annual consumption of renewable energy(100 million tons coal equivalent)	2.55	4.78	-
Proportion of renewable energy in total energy consumption (%)	7.8	9.5	15
Annual bio-energy consumption(10,000 tons coal equivalent)	550	5,000	-
Solar power capacity(100 million kilowatts)	80	2,100	5,000
Accumulated area of solar heating(100 million square meters)	0.17	4	8
Grid wind power capacity(100 million kilowatts)	0.31	1	2
Hydropower capacity(100 million kilowatts)	2.16	2.9	4.2

Source: National Energy Administration, 12th Five-Year Plan of Renewable energy Development

3.3 Promoting ecological civilization of rural areas

Ecological civilization is a new concept, which shifts from the past emphasis on economic construction as the core of social development, to one based on development that is more balanced between man and nature. The ecological civilization process of rural areas encompasses all material, spiritual and institutional achievements in easing the increasingly intense conflicts between humans and the environment. It includes mitigating ecological crises and disasters, relieving pressures on natural resources, and improving the balance between the environment and economy, and between individuals and society.

4. Conclusion

In conclusion, China's rural areas have achieved great development during the past few decades. Nevertheless, this paper demonstrates that there is still a need to move to a green development mode which is not only environmentally sustainable, but also including creating the conditions for greater job creation, greater resource efficiency and energy security.

Acknowledgments

This work is financially supported by the following Anhui Social Sciences Key Projects funding (SK2015A526; SK2016A0893; AHSKZ2015D15).

References

[1] World Bank, World Development Report 2008, Agriculture for development, s.l.: World Bank Publications, 2007.
[2] H.L. Long, Y.S. Liu, X.B. Li, and Y.F. Chen, Building new countryside in China: A geographical perspective. Land Use Policy 27 (2010)457-470.
[3] Z. Xu, N. Li, Impacts of trade liberalization on China's agriculture and poverty in China.International Journal of Bio-Resource & Stress 3(2005)464-467.
[4] P.J. Li, The Resesrch of the China Renewable Energy Policy, Beijing, 2009.
[5] J. Byrne, B.Shen, W. Wallace, The economics of sustainable energy for rural development: A study of renewable energy in rural China. Energy Policy 26(1998) 45-54.
[6] UNEP, UNWTO, Tourism in the Green Economy – Background Report, UNWTO, Madrid,2012.
[7] J.G. Wang, X.P. Wang, New urbanization: A new vision of China's urban–rural development and planning, Frontiers of Architectural Research 4 (2015) 166–168.

IOP Conf. Series: Earth and Environmental Science **64** (2017) 012055 doi:10.1088/1755-1315/64/1/012055

Visibility characteristics over Fuzhou during 1956-2016

Weicong Fu[1,2], Zhipeng Zhu [1,2], Shuping Huang [1,2], Ziru Chen [1,2], Guochang Ding[1, 2] and Jianwen Dong[1, 2 *]

[1]College of Landscape and Architecture, Fujian Agriculture and Forestry University, Fuzhou, China
[2]State Forestry Administration National Forest Park Engineering Research Center, Fuzhou, China

*Corresponding author e-mail: fjdjw@fafu.edu.cn

Abstract. This research analyzes atmosphere visibility characteristics in Fuzhou. For residents to choose the best travel time and improve the quality of their life. The results show that the average atmospheric visibility in 1957-2016 is 17.98 km and the trend is -2.86 km / decade. The highest atmospheric visibility occur in the summer, the lowest value appears in the winter, further observation shows that atmospheric visibility in July showed the maximum value, and in March shows the lowest value. The "bad" level of atmospheric visibility showed a modest upward trend between 1956 and 2006, and showed a high growth after 2006, indicating that effect of governance to improve the air quality is not obvious, and the trend of "good" level of is decline from nearly 90% in 1956 to nearly 20% in 2006. In 1956-1987 Ridit values are greater than 0.5, the highest value of 0.77, in the upward trend in the fluctuation, in 1987-2005 Ridit value of 0.5 up and down, in a steady period of time the trend is not obvious, and in 2006-2016 Ridit value is less than 0.5, the lowest value of 0.28, indicating that the trend of atmosphere is decline. The value of atmospheric visibility about worst 20%, 50% and best 20% is in a downward trend.

1. Introduction

Atmospheric visibility is defined as the maximum horizontal distance between person's eye and the visible target[1], which not only reflects the regional atmospheric quality[2-3], but also closely related to the life quality of the inhabitants. For the significance of life, in recent years, atmospheric visibility has become the most popular issue of public [4]. During the haze days in London [5], Brimblecombe do the first research of the impact of air pollutants on atmospheric visibility. Followed by the research reports of atmospheric visibility continue to emerge. According to previous studies, the most influencing factors of atmospheric visibility are the aerosol scattering effect, especially particular meter (PM) [2]. Some scholars have reported that atmospheric pollutants such as nitrogen dioxide (NO_2) meteorological factors as concentration (RH), wind speed (WS), atmospheric temperature (T) and atmospheric pressure (P) are directly work on atmospheric visibility, and there is a significant correlation between atmospheric pollutants and atmospheric visibility[6]. In addition, some scholars have pointed out meteorological parameters such as relative humidity (RH) has an indirect relationship on visibility [7-10]. It shows that atmospheric visibility is of great significance as the monitoring index of air environmental quality. However, most studies focused only on 1-3 years of variation, whereas long-term monitoring studies were rarely reported.

Rapid urbanization process and economic development led to a sharp increase in resource consumption, many areas are facing a serious decline in atmospheric visibility challenge. The problem is particularly heavy in the coastal city. Fuzhou is located in the eastern part of Fujian Province, one of the most population density city in China. Fuzhou has experienced a rapid economic development and urbanization. However, the impact of atmospheric pollutants and meteorological factors on atmospheric visibility in this region has never been reported. Fuzhou, as a typical city of Mindong, the study can not only be a data support for improving the atmospheric visibility and air quality, but also provide reference for the other cities of the South China.

2. Materials and methods

2.1. Data

Atmospheric visibility and weather data are available from the National Oceanic and Atmospheric Administration (NOAA, https://www.climate.gov/) and the data monitoring interval is 1 hour.

2.2. Analytical methods

The study used five kinds of statistics and analysis methods: 1) statistical analysis of the annual variation of atmospheric visibility and diurnal variation; 2) analysis of atmospheric visibility monthly "better" (> 19.9 km), "poor" (<4 km); 3) using the SPSS 19.0 software for atmospheric visibility Ridit analysis; 4) calculating the atmospheric visibility Cumulative percentiles; 5) using the method [12], to calculate the atmospheric visibility and atmospheric pollutant regression equation, in order to make the regression equation more typical, eliminate fog, rain and other effects of visibility of the special weather is screen out. Because the higher air humidity has a strong deterrent effect, the relative humidity which > 90% of the monitoring data group is also removed [13].

Extinction coefficient β_{ext} reference Koschmieder's research method to come [1]:

$$\beta_{ext} = 3.912 / V \tag{1}$$

Where β_{ext} is the extinction coefficient (km^{-1}); V is the atmospheric visibility (km).

Data processing

Data analysis and image analysis are performed using SPSS 19.00 and EXCEL 2010.

3. Results and analysis

3.1. Yearly Variation

Table 1 shows the mean value of atmospheric visibility and annual variation trend of Fuzhou during 1957-2016. The average atmospheric visibility in 1957-2016 is 17.98 km and the trend is -2.86 km / decade. Indicating that with the development of economy, atmospheric visibility is gradually reduced. Figure 1 shows the annual average of atmospheric visibility 1957-2016 in Fuzhou, showing a decreasing trend from 1980 to 1990, and the trend of change from 1990 to 2005 is gently. During 2006-2016, due to the increase of government's efforts to control air pollution, the value of atmospheric visibility is increased.

Table 1. Summary statistics of yearly visibility in Fuzhou (1956–2016).

	1957-1964	1973-1976	1977-1986	1987-2006	1997-2006	2007-2016	1957-2016	60-year trend
FuZhou	26.67	22.75	19.42	15.67	15.93	12.03	17.98	-2.86

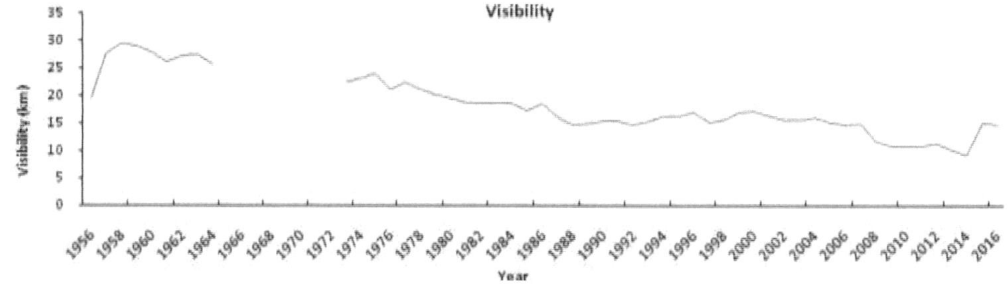

Fig. 1 Variations of yearly value of visibility during 1956-2016 in Fuzhou city

3.2. Seasonal and monthly variation

The highest atmospheric visibility occur in the summer, the lowest value appears in the winter, further observation shows that atmospheric visibility in July showed the maximum value, and in March shows the lowest value. Atmospheric visibility in the summer shows the highest value may be with the heavy rain, high temperature, and frequent typhoon weather.

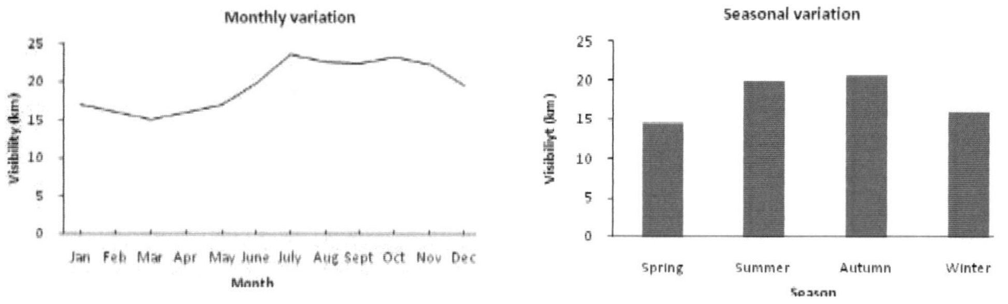

Fig. 2 Variations of the monthly and seasonal value of visibility in Fuzhou city

3.3. Monthly Percentage of Atmospheric Visibility at Different Levels

The ratio of "good (> = 20km)" and "bad (<4km)" levels of atmospheric visibility is quite important for the residents. Figure 3 shows the radio of annual visibility in different levels of visibility during 1956-2016. The "poor" level of atmospheric visibility showed a modest upward trend between 1956 and 2006, and showed a high growth after 2006, indicating that effect of governance to improve the air quality is not obvious, and the trend of "better" level of is decline from nearly 90% in 1956 to nearly 20% in 2006.

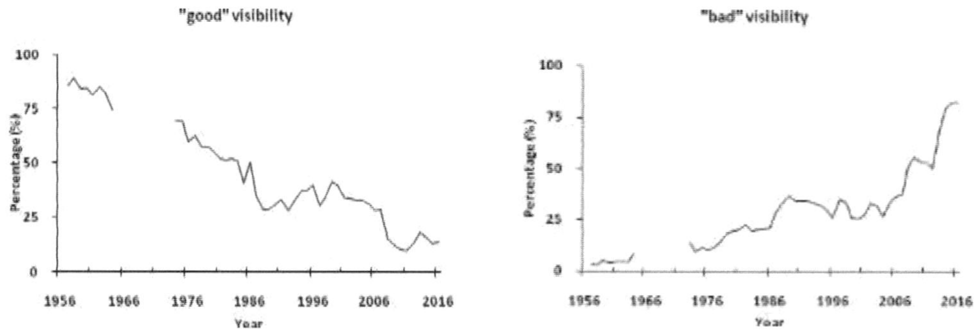

Fig.3. Percentiles of "quit good" and "quit bad" visibility.

3.4. Ridit Analyze

The Ridit value can reflect the trend of atmospheric visibility. When the Ridit value is greater than 0.5, it indicates that the atmospheric visibility in this year is on the rising trend, and if it is less than 0.5, it indicates that it is declining. In 1956-1987 Ridit values are greater than 0.5, the highest value of 0.77, in the upward trend in the fluctuation, in 1987-2005 Ridit value of 0.5 up and down, in a steady period of time the trend is not obvious, and in 2006-2016 Ridit value is less than 0.5, the lowest value of 0.28, indicating that the trend of atmosphere is decline.

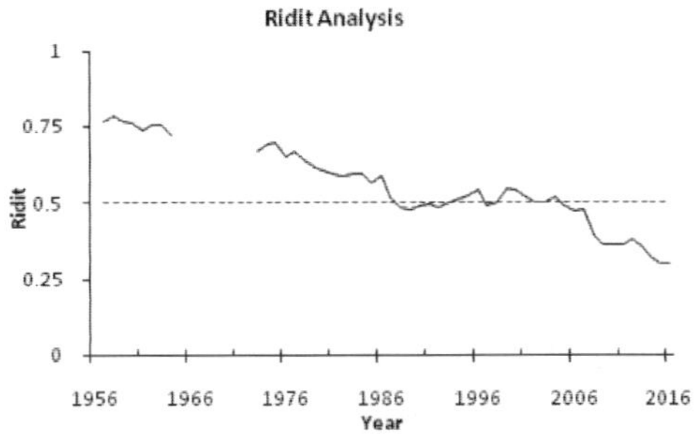

Fig. 4 The Ridit value of Fuzhou city during 1956-2016

3.5. Cumulative percentiles

From Figure 5 we can see that the value of atmospheric visibility about worst 20%, 50% and best 20% is in a downward trend. The best trend of the best 20%, 50% and worst 20% are -0.39, -0.36 and -0.25km/year, respectively. Further analysis of the best 20% in 1987-2006 has a modest upward trend that may be related to the 1980s reform and opening up. In this period, the value of 50% and worst 20% are changing gentle. And in the 1956-1986 and 2006-2016, the value of the worst 20%, 50% and best 20% of the atmospheric visibility are in a sharp downward trend, indicating that the period of pollution control also need to improve.

With the change of extinction coefficient, the concentration of atmospheric particulates is increasing. The worst 20% increase of 0.03 km^{-1}/year is greater than 50% (0.004 km^{-1}/year) and best 20% (0.003 km^{-1}/year). Indicating that the frequency of extreme pollution of the weather is increasing, and better atmospheric visibility frequency of the decline is relatively slow. May be related to the variation of temperature, humidity, and wind speed and so on.

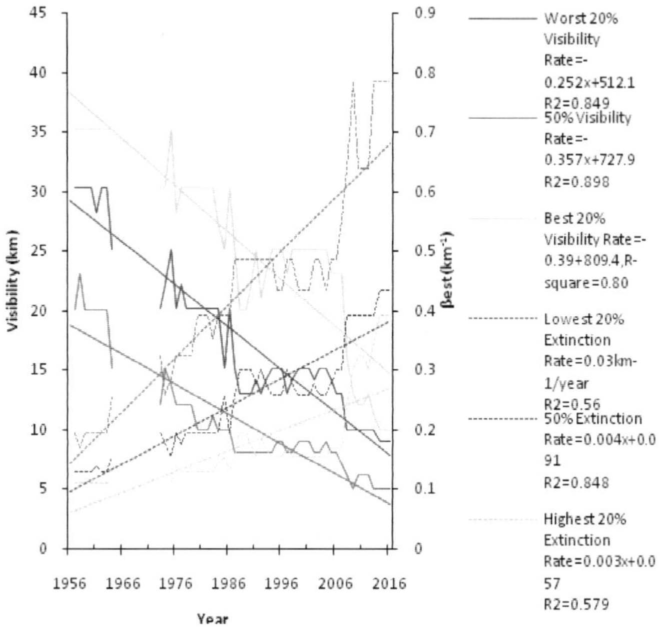

Fig. 5. The variation of the best 20%, 50% and worst 20% visibility in Fuzhou city during 1956-2016.

4. Conclusion

The average atmospheric visibility is 17.98km over Fuzhou city during 1956-2016, the highest value occur in summer and the lowest appears in winter, showing a inverted "V" font. The ratio of "better" and "poor" levels of atmospheric visibility is quite important for the residents. The study find that "poor" atmospheric visibility in the spring and winter appear high frequency, and the in summer and early autumn the ratio of "better" atmospheric visibility is higher than the average ratio.

The results of previous studies show that the atmospheric visibility decreases with the increase of the light scattering effect of the atmospheric particles, and pointed out that the atmospheric pollutant concentration has a certain correlation with the atmospheric visibility [11, 15-18]. Dan Xue et al. [19] found that atmospheric visibility is positively correlated with the concentration of O_3 and negatively correlated with other pollutants. Wei Huang et al. [20] analyzed the relationship between atmospheric visibility, air pollutants and mortality in Shanghai. The correlation between atmospheric visibility and atmospheric pollutant concentration is similar to that of Dan Xue et al [19].

Some researchers have found that atmospheric visibility is closely related to the concentration of atmospheric particulates. The results show that the atmospheric visibility is closely related to the atmospheric particulate matter concentration when the meteorological factors are excluded.

However, due to the limited data acquisition conditions, the research take the data of only one station, further research must set more monitoring station (such as: city center, suburbs, suburbs, etc.) for an in-depth study and related research needs to be further carried out.

References

[1] H. Koschmieder. 1926. Theorie der horizontalen Sichtweite Beit. Phys.Atmos, 12(1926) pp: 33-55.

[2] Watson, J. G. 2002. Critical review discussion-visibility: Science and regulation. Journal of the Air & Waste Management, 52(2002): 973–999.

[3] Doyle M, Dorling S (2002) Visibility trends in the UK 1950–1997. Atmospheric Environment 36: 3161–3172.

[4] Che, H., Zhang, X., Li, Y., Zhou, Z., & Chen, Z. (2006). Relationship between horizontal extinction coefficient and PM10 concentration in Xian, China, during 1980–2002. *China*

Particuology, 4, 327–329.

[5] Brimblecombe, P. (1981). Long term trends in London fog. *Science of the Total Environment, 22*, 19–29.

[6] Wan, J. M., Lin, M., Chan, C. Y., Zhang, Z. S., Engling, G., Wang, X. M., Chan, I. N., & Li, S. Y. (2011). Change of air quality and its impact on atmospheric visibility in central-western Pearl River Delta. *Environmental Monitoring and Assessment, 172*, 339–351.

[7] Du, K., Mu, C., Deng, J., & Yuan, F. (2013). Study on atmospheric visibility variations and the impacts of meteorological parameters using high temporal resolution data: an application of environmental internet of things in China. *International Journal of Sustainable Development & World Ecology, 20*, 238–247.

[8] Tsai, Y. I. (2005). Atmospheric visibility trends in an urban area in Taiwan 1961–2003.*Atmospheric Environment, 39*, 5555–5567.

[9] Deng, J., Wang, T., Jiang, Z., Xie, M., Zhang, R., Huang, X., & Zhu, J. (2011). Characterization of visibility and its affecting factors over Nanjing, China. *Atmospheric Research, 101*, 681–691.

[10] D.O. Lee Trends in summer visibility in London and southern England Atmos. Environ., 17 (1983), pp. 151–159

[11] [12] M. Doyle, S. Dorling 2002Visibility trends in the UK 1950–1997Atmos. Environ., 36 (2002), pp. 3161–3172

[12] D.O. Lee Trends in summer visibility in London and southern England Atmos. Environ., 17 (1983), pp. 151–159

[13] Yuan Chen, Shaodong Xie. 2012Temporal and spatial visibility trends in the Sichuan Basin, China, 1973 to 2010. Atmospheric Research 112(2012):25-34

[14] P. Zhao, X. Zhang, X. Xu, et al.Long-term visibility trends and characteristics in the region of Beijing, Tianjin, and Hebei, China Atmos. Res., 101 (2011), pp. 711–718

[15] D. Chang, Y. Song, B. Liu. 2009. Visibility trends in six megacities in China 1973–2007Atmos. Res., 94 (2009), pp. 161–167

[16] [H. Che, X. Zhang, Y. Li, et al. 2007. Horizontal visibility trends in China 1981–2005 Geophys. Res. Lett., 34 (2007) http://dx.doi.org/10.1029/2007GL031450

[17] A. Molnar, E. Meszaros, K. Imre, et al. 2008. Trends in visibility over Hungary between 1996–2002. Atmos. Environ., 42 (2008), pp. 2621–2629

[18] Dan Xue, Chengfan Li, Qian Liu. 2015 Visibility characteristics and the impacts of air pollutants and meteorological conditions over Shanghai, China. Environ Monit Assess (2015):187:363 http://dx.doi.org/10.1007/s10661-015-4581-8

[19] Wei Huang, Jianguo Tan, Haidong Kan, et al. 2009. Visibility, air quality and daily mortality in Shanghai, China. Science of the Total Environment 407(2009):3295-3330.

IOP Conf. Series: Earth and Environmental Science **64** (2017) 012056 doi:10.1088/1755-1315/64/1/012056

Ecological footprint analysis of environmental impacts by cascaded exploitation of diversion-type small hydropower: a case study in southwest china

Zhenhua Wang[1, 2, 3,*], Qingyun Li[1, 2], Zhuo Huang[1, 2], Xianqiang Tang[1, 2, 3] and Weihua Zhao[1, 2]

[1]Basin Water Environmental Research Department, Changjiang River Scientific Research Institute, Wuhan 430010, China

[2]Key Lab of Basin Water Resource and Eco-environmental Science in Hubei Province, Changjiang River Scientific Research Institute, Wuhan 430010, China

[3]Collaborative Innovation Center for Geo-Hazards and Eco-Environment in Three Gorges Area, Yichang 443002, China

*Corresponding author e-mail: wzh04@163.com

Abstract. Cascaded exploitation of diversion-type small hydropower (SHP) offers a source of new energy as well as socioeconomic benefits; however, it inevitably causes environmental disturbance and damage. Previous studies on the cumulative effect of cascaded diversion SHP rarely discussed using quantitative analysis method. In this paper, the ecological footprint analysis approach is proposed to assess the positive and negative impacts of cascaded diversion SHP on environment of a small-scale river in Southwest China. Positive impact is defined as ecological supply footprint (ESF), which refers to vegetation protection by replacing firewood with SHP. Negative impact is defined as ecological loss footprint (ELF), which includes fish and net primary productivity loss, vegetation destruction and soil erosion. With the raising in the number (n>4) of diversion SHP stations, the difference between ELF and ESF increases remarkably, suggesting that the adverse impacts of cascaded diversion SHP accumulate in the study area. Compared with vegetation destruction and soil erosion, the cumulative loss of fish and net productivity is the most important aspect of the adverse impacts which needs more attentions.

1. Introduction

In China, hydropower with installed capacity not more than 50 MW is officially defined as small hydropower (SHP). The investigation and evaluation result of rural hydropower resources of P. R. China in 2008 indicates that the total technically exploitable SHP potential is 128 GW, located in more than 1,700 counties [1]. By the end of 2015, China has built more than 47,000 SHP stations with total installed capacity of 73 GW and annual electricity generation of 2300 billion kWh, accounting for 25% of the national hydropower [2]. Statistical data indicates that diversion-type SHP stations account for more than 80% of these built SHP [3].

SHP is clean and renewable energy which is internationally recognized [4]. The exploitation of SHP plays an important role in improving energy supply and rural electrification in remote mountainous regions [5]. In China, SHP serves as the main power source in many rural areas [6]. This shifts the pattern of rural energy consumption from biomass-based energy source to a clean energy source. The utilization of SHP is a practical solution to prevent deforestation by reducing the consumption of firewood for cooking and heating [5, 6]. Statistical data shows that the implementation of the *Project of Replacing Firewood with SHP* by the Chinese government reduces the firewood consumption of 6.7 million m^3 and protects forest area of 930 thousand hm^2 [7]. This is obviously beneficial to environmental improvement. However, the development of SHP, especially diversion SHP is also likely to cause environmental disturbance and damage, including changes to the hydrological conditions, aquatic ecosystem structure alteration, vegetation destruction, soil erosion, etc. [8]. These impacts of diversion SHP development could lead to a certain extent loss in the ecosystem [9]. The impact of single diversion SHP on the environment is probably thought to be not severe [10]. But their impacts could possibly increase with cascaded exploitation of river basins.

For a long time, due to lack of single unified term on expressing environmental impacts, previous studies on the environmental effects of diversion SHP mainly discussed using qualitative analysis methods [11-13]. There was little study of the cumulative effect assessment of cascaded diversion SHP using a quantitative analysis method [14]. The *Environmental impact assessment code hydroelectric station project for rural area* (SL315-2005) formulated by the Ministry of Water Resources of China is available only for the environmental effect assessment of single SHP station, and not applicable for the cumulative effect evaluation of cascaded diversion SHP [3]. In order to evaluate and compare the extent of the positive and negative cumulative effects, it is necessary to quantify them using a single unified method [15].

The ecological footprint (EF) analysis method is a simple evaluation tool based on biological physical quantity and closely associated with the sustainable development theory [9]. The EF is expressed in global hectares. A global hectare is one hectare of biologically productive space with world average productivity. The EF can be used to measure any product, activity or impact, at all levels from self to planet. The necessary data can be easily obtained and the operational method can be conveniently operated. It is therefore possible to use the EF in environmental management and as a planning tool [16].

The first aim of the study is to assess the cumulative environmental effects of cascaded diversion SHP stations which in a small-scale river in a mountainous area of Southwest China, using the EF approach. The second objective is to compare the EF components to identify the most important aspects of the environmental impacts which need more special attentions. The EF method could become an effective analysis tool for SHP development policy design and planning.

2. Methodology

2.1. EF Accounts: Accounting Framework

Considering the main positive and negative environmental effects of diversion SHP development discussed above, the EF accounting framework in Figure 1 was proposed. In this present study, positive effect was defined as ecological supply footprint (ESF), which mainly referred to vegetation protection by replacing firewood with diversion SHP. Negative effect was defined as ecological loss footprint (ELF), which included fish and net primary productivity loss in water-reduction section, vegetation destruction and soil erosion by diversion SHP during construction and in the operation period following construction.

2.2. Conversion Factor

The EF account includes a combination of two conversion factors: equivalence factor and yield factor.

2.2.1. Equivalence Factor. Equivalence factor is a scaling factor that converts a specific land type (such as cropland or forest) into a universal unit of bio-productive area, a global hectare. Equivalence factor represents the world average potential productivity of a given bio-productive area relative to the world average potential productivity of all bio-productive areas. For land types (e.g. forest) with productivity higher than the average productivity of all bio-productive area on the earth, the equivalence factor is greater than 1. Different regions in a given year have the same equivalence factors, which only change slightly from year to year. The equivalence factors used in this study are 1.1 for forest and 0.2 for productive waters.

2.2.2. Yield Factor. Because the productive factors (e.g., climate, soil, technology, etc.) vary across regions, the yield capacities of the same land are different in different regions. The yield factor represents the differences between the local yield for one category of productive space (e.g., forests) and the global average yield for the category of productive space (e.g., forests). Each region has its own set of yield factors, one for each type of bio-productive area. In the present study, the yield factors are 0.91 for forest and 1.0 for productive waters.

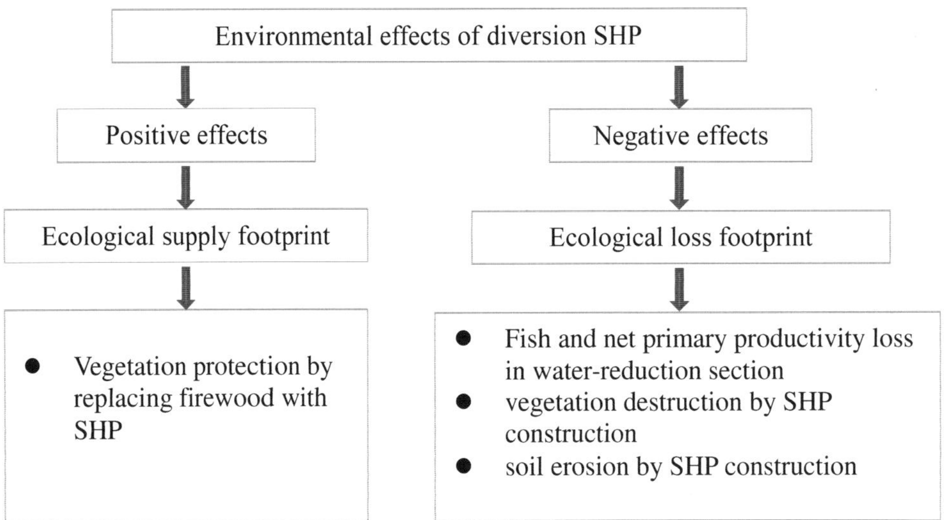

Figure 1. Indicators of ESF and ELF accounts

2.3. Evaluation Models of ESF

2.3.1. Evaluation Model of ESF for Single Diversion SHP. Electricity can be converted into heat. Global average productivity of hydropower is 1000 GJ·ha^{-1}. The conversion relation is as follows: 1 GJ equal to 1×10^9 joule (J) and 1 J equal to 2.778×10^{-7} kW·h. Applying the heat conversion method (converting electricity generation of SHP for replacing firewood into heat) to evaluate the area of protecting vegetation land:

$$EF_f = \varphi\left[\frac{Q_e}{1000\times10^9\times2.778\times10^{-7}}\right]r_3y_3 = 3.6\times10^{-6}\varphi Q_e r_3 y_3 \quad (1)$$

In the above formula, EF_f is the ESF of single diversion SHP (ha); φ is the ratio of electricity generation for replacing firewood to total electricity generation; Q_e is the average annual energy generation; r_3 is the equivalence factor for forest; y_3 is the yield factor for forest.

2.3.2. Evaluation Model of ESF for Cascaded Diversion SHP. The ESF of cascaded diversion SHP is the sum of that for single diversion SHP station:

$$EF_{nf} = \sum 3.6\times 10^{-6}\varphi_n Q_{ne} r_3 y_3 = 3.6\times 10^{-6} r_3 y_3 \sum \varphi_n Q_{ne} \tag{2}$$

Where EF_{nf} is the ESF for n-cascade of diversion SHP (ha); φ_n is the ratio of electricity generation for replacing firewood to total electricity generation in the No. n diversion SHP; Q_{ne} is the average annual energy generation in the No. n diversion SHP.

2.4. Evaluation Models of ELF

2.4.1. Fish and Net Primary Productivity Loss of Single Diversion SHP. The production of the unit weight of fish, consumes at least 25 times as much as the amount of phytoplankton. Thus, there is a quantitative relationship between fish and net primary productivity loss footprint in water-reduction section of diversion SHP:

$$EF_r = EF_a + EF_p = S_r r_5 y_5 + kS_r r_5 y_5 = (1+k) S_r r_5 y_5 \tag{3}$$

Where EF_r is fish and net primary productivity loss footprint of single diversion SHP (ha); k is the ratio of net primary productivity to fish biomass (k =25); S_r is the decrease of water area in water-reduction section (ha); r_5 is the equivalence factor for productive waters; y_5 is the yield factor for productive waters.

2.4.2. Fish and Net Primary Productivity Loss of Cascaded Diversion SHP. Based on theoretical analysis and field research, it was assumed that fish and net primary productivity cumulative loss for n-cascade of diversion SHP increase nonlinearly. With the increase in the number of diversion SHP stations, fish and net primary productivity loss footprint for n-cascade of diversion SHP will increases by λ^{n-1}-fold:

$$EF_{nr} = EF_{(n-1)r} + (1+k) S_{nr} r_5 y_5 \lambda^{n-1} = (1+k) r_5 y_5 \left(S_r + S_{2r}\lambda + S_{3r}\lambda^2 + \cdots\cdots + S_{nr}\lambda^{n-1}\right) \tag{4}$$

Where EF_{nr} is the fish and net primary productivity loss footprint for n-cascade of diversion SHP (ha); $EF_{(n-1)r}$ is the fish and net primary productivity loss footprint for (n-1)-cascade of diversion SHP (ha); k is the ratio of net primary productivity to fish biomass (k =25); S_{nr} is the decrease of water area in water-reduction section of the No. n diversion SHP (ha); λ is river ecological damage degree index (λ =1.3).

2.4.3. Vegetation Destruction of Single Diversion SHP. Vegetation destruction of diversion SHP is mainly caused by the construction of long-distance diversion pipe engineering. Vegetation destruction footprint of single diversion SHP can be calculated as follow model:

$$EF_v = S_v r_3 y_3 \tag{5}$$

Where EF_v is vegetation destruction footprint of single diversion SHP (ha); S_v is the area of vegetation destruction through the construction of long-distance diversion pipe engineering (ha).

2.4.4. Vegetation Destruction of Cascaded Diversion SHP. The vegetation destruction footprint of cascaded diversion SHP is the sum of the vegetation destruction footprint of single diversion SHP station:

$$EF_{nv} = \sum S_{nv} r_3 y_3 = r_3 y_3 \sum S_{nv} \tag{6}$$

Where EF_{nv} is vegetation destruction footprint for *n*-cascade of diversion SHP (ha); S_{nv} is the area of vegetation destruction through the construction of long-distance diversion pipe engineering in the No. *n* diversion SHP (ha).

2.4.5. Soil Erosion of Single Diversion SHP. Soil erosion of diversion SHP is mainly caused by the construction of long-distance diversion pipe engineering. Soil erosion footprint of single diversion SHP can be calculated as follow model:

$$EF_e = S_e r_3 y_3 \tag{7}$$

Where EF_e is soil erosion footprint of single diversion SHP (ha); S_e is the area of soil erosion due to the construction of long-distance diversion pipe engineering (ha).

2.4.6. Soil Erosion of Cascaded Diversion SHP. The soil erosion footprint of cascaded diversion SHP is the sum of the soil erosion footprint of single diversion SHP station:

$$EF_{ne} = \sum S_{ne} r_3 y_3 = r_3 y_3 \sum S_{ne} \tag{8}$$

Where EF_{ne} is soil erosion footprint for *n*-cascade of diversion SHP (ha); S_{ne} is the area of soil erosion due to the construction of long-distance diversion pipe engineering in the No. *n* diversion SHP (ha).

2.5. Judgment of Environment Effects.
The environmental effect of diversion SHP stations can be quantitatively analysed based on the comparison of the ESF and the ELF. A positive result from the numerical value of the ecological supply plus the numerical value of the ecological loss represents a beneficial effect of diversion SHP on the environment, and the greater difference between the two represents the greater positive effect. A negative result represents a negative effect of diversion SHP on the environment, and the greater difference between the two represents the greater adverse effects.

3. Study Area

The study area for the present study was in a small-scale river in Yunnan province, Southwest China. The river runs about 85 km long with a drainage area of 634 km^2 and the altitude of 800-1200 m. Climate in the study area belongs to the subtropical monsoon climate zone, with an annual precipitation of 1200 mm~1600 mm. This river catchment is mainly covered with dense forests and grassland, in addition to a small amount of arable land, houses, roads and water, with vegetation coverage of over 74%. Exposed rock in the basin is relatively fragmented and weathering, coupled with high mountains and steep slopes, geological disasters like landslides, collapse or mudslides occurs in case of heavy rain and flash floods, with an average erosion intensity of 1000 t/km^2.

From the end of last century, seven-cascade of diversion SHP stations have been built on this river, namely one to seven SHP, expressed as H_1, H_2, H_3, H_4, H_5, H_6 and H_7 respectively. For the seven diversion SHP stations, the installed capacity, annual power generation, electricity generation for replacing firewood, and the area of water-reduction section, vegetation destruction area, soil erosion area are listed in the following table (Table 1).

Table 1 The basic situation of the seven-cascade of diversion SHP stations

	H_1	H_2	H_3	H_4	H_5	H_6	H_7
Installed capacity (kW)	2000	630	400	6300	7200	6300	500
Total annual electricity generation (10^4 kW·h)	650	260	180	2100	2300	2000	180
Ratio of electricity generation for replacing firewood to total electricity generation (%)	50	45	50	40	45	45	50
Area of water-reduction section (ha)	0.78	0.56	0.53	2.33	2.74	2.59	0.68
Area of vegetation destruction (ha)	2.68	1.07	0.94	3.14	2.90	2.13	1.08
Area of soil erosion (ha)	2.50	0.90	0.84	2.70	2.50	1.50	0.80

4. Results and Discussion

The ESF and ELF of single diversion SHP are presented in Figure 2, the calculation of which was in accordance with formulae (1), (3), (5) and (7) based on data from Table 1. In this study, the values of ESF range from 3.24 ha to 37.26 ha, and that of the ELF range from 4.88 ha to 19.65 ha. For single diversion SHP of H_1, H_4, H_5 and H_6, the ESF exceedes ELF, and for H_2, H_3 and H_7, the ESF is less than ELF (Fig. 2). From the view of single diversion SHP, most projects have positive impacts on environment; a few have slight negative impacts.

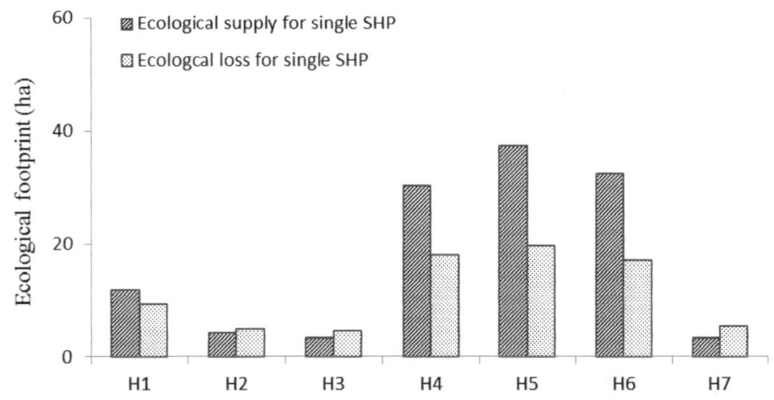

Figure 2. Comparison of ELF and ESF for single diversion SHP

The cumulative ESF and ELF of cascaded diversion SHP are presented in Figure 3, the calculation of which was in accordance with formulae (2), (4), (6) and (8) based on data from Table 1. For cascaded diversion SHP, the cumulative ESF and ELF increases notably with the increase in the number of SHP stations. The values of cumulative ESF range from 15.93 ha to 122.41 ha, and that of the cumulative ELF range from 15.00 ha to 172.47 ha. There is no remarkable difference between the cumulative ecological loss and supply for 2-, 3-, 4-cascade of diversion SHP in the study area. But the imbalance between the cumulative ecological loss and supply is obviously for 5-, 6-, 7-cascade of diversion SHP. With the raising in the number (n>4) of diversion SHP stations, the difference between ecological loss and supply increases clearly, suggesting that the adverse impacts of cascaded diversion SHP would accumulate in the study area.

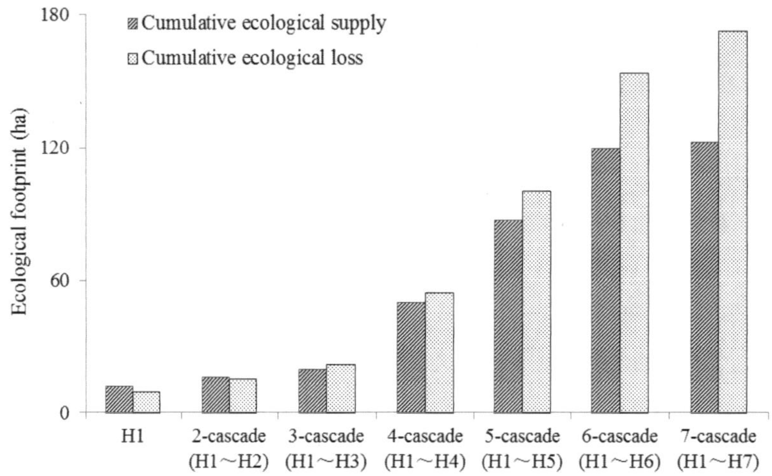

Figure 3. Comparison of ELF and ESF for cascaded diversion SHP

For cascaded diversion SHP, the cumulative loss of fish and net productivity is the largest ecological loss, accounting for 52.3%~85.1% of the total (Figure 4). With the increase in the number of diversion SHP stations, the proportion of fish and net productivity cumulative loss is gradually increasing in the total ecological loss. The increasing cumulative loss of fish and net productivity suggests that more water-reduction sections caused by cascaded diversion SHP could lead to great adverse effects on aquatic ecosystem. Thus, fish and net productivity loss of cascaded diversion SHP development needs more attentions. The cumulative vegetation destruction (8.3%~25.0%) and soil erosion (6.8%~22.7%) are relatively small compared to those of fish and net productivity. This suggests that vegetation destruction and soil erosion are not intensive as fish and net productivity losses during cascaded diversion SHP development in the study area.

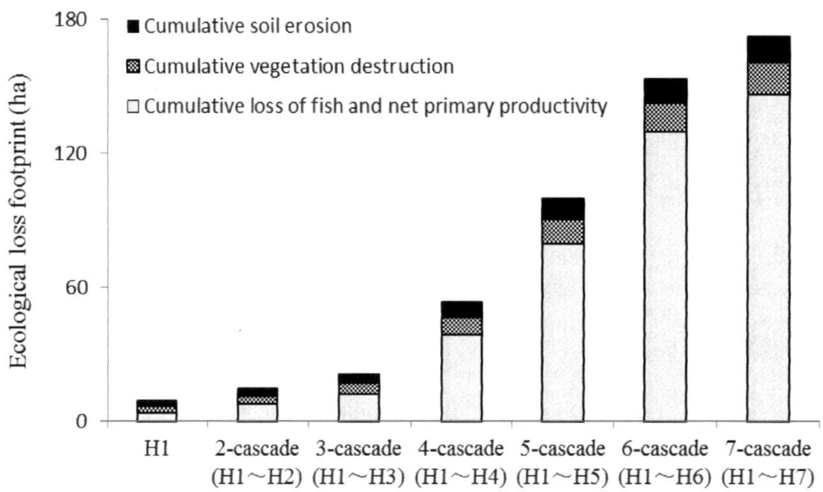

Figure 4. The components of ELF of cascaded diversion SHP

IOP Conf. Series: Earth and Environmental Science **64** (2017) 012056 doi:10.1088/1755-1315/64/1/012056

5. Conclusion

The EF analysis method is available to quantitatively evaluate the extent of cumulative environmental impacts by cascaded exploitation of diversion SHP. Based on the established evaluation models, the ESF and ELF of cascaded diversion SHP stations in Southwest China are calculated. The ESF and ELF show a similar variation tendency for cascaded diversion SHP in this study. For less than 4-cascade ($n \leqslant 4$) diversion SHP, there is no remarkable difference between the cumulative ESF and ELF. With the raising in the number ($n>4$) of diversion SHP stations, the difference between the ELF and ESF increases clearly, suggesting that the adverse impacts of cascaded diversion SHP accumulate in the study area. Compared with vegetation destruction and soil erosion, the cumulative loss of fish and net productivity is the most important aspect of the adverse impacts by cascaded diversion SHP, which needs more attentions.

Acknowledgments

This work was financially supported by Central Public Interest Scientific Institution Research Fund (CKSF2015017/SH and CKSF2017047/SH), the National Natural Science Foundation of China (51209009 and 51409010), the Ministry of Water Resource, Public Interest Scientific Research Fund (201501042) and the National Key Research and Development Program (2016YFC0502201).

References

[1] National Leading Group on rural water resources investigation and evaluation: The investigation and evaluation result of rural hydropower resources of PRC 2009 China Water Power & Press. In Chinese
[2] Ministry of Water Resources in: “12th Five-Year” ten major events of rural hydropower 2016 Small Hydro Power. 2016 p 60. In Chinese
[3] Li Q Y 2014 Ecological environment impact assessment and protection countermeasures of rural hydropower development in the Yangtze River Basin Changjiang Press. In Chinese
[4] Ardizzn G, Cavazzini G, Pavesi G 2014 Renewable & Sustainable Energy Reviews 31 746-761
[5] Darmawi, Sipahutar R, Bernas S M, Imanuddin M S 2013 Renewable & Sustainable Energy Reviews 17 213-215
[6] Liu B X, Chau K, Liao S L 2015 Renewable & Sustainable Energy Reviews 42 43-55
[7] Zhang B T 2016 China Electric Power Enterprise Management 12
[8] Cui G X 2006 China Water Resources 20 23–24. In Chinese with English abstract
[9] Li X J, Zhang J, Xu L Y 2015 Ecological Engineering 76 178-185
[10] Okot D K 2013 Renewable & Sustainable Energy Reviews 26 515-520
[11] Sun H, Gong B W, Hu B 2013 Journal of Yangtze River Scientific Research Institue 30 12-15. In Chinese with English abstract
[12] Huang Z, Li QY 2011 Small Hydro Power 6 47-49. In Chinese
[13] Liang F Q 2010 Small Hydro Power 2 47-49. In Chinese with English abstract
[14] Ding Y F, Tang D S, Wang T 2011 Energy Procedia 5 540-544
[15] Xiao J H, Wang M, Shi G Q, et al. 2008 Journal of Wuhan University of Technology (Transportation Science & Engineering) 32 593-595. In Chinese
[16] Xiao J H, Wang M, Yu Q D, et al. 2015 Acta Ecologica Sinica 35 2726-2740. In Chinese with English abstract

IOP Conf. Series: Earth and Environmental Science **64** (2017) 012057 doi:10.1088/1755-1315/64/1/012057

Carbon association studies of whole industrial chain

Qingmin Yuan, Qingqing Yang *
The Centre of Recycling Economy and Enterprise Sustainable Development, Tianjin University of Technology, Tianjin, China

*Corresponding author e-mail: yangqingqingl@163.com

Abstract. With the coordinated development at "Beijing-Tianjin-Hebei", industrial transfer is bound to bring carbon transfer. Industrial transfer along inter-regional industrial chain has become one of the main forms of industry transfer. So the problems of carbon transfer that happened between industry chain and regional must be pay more attention. This paper based on inter-regional input-output model, counted input-output tables of 2014 on the base of the input-output tables of 2010 to analysis the problems of carbon transfer between upstream and downstream in "Beijing-Tianjin-Hebei" from 2010 to 2014. This paper studied inter-regional, intra-regional, inter-industry, intra-industry carbon linkage in "Beijing-Tianjin-Hebei", in order to provide help for inter-industry cooperative carbon emissions reduction in "Beijing-Tianjin-Hebei"

1. Introduction

Carbon emissions in Beijing-Tianjin-Hebei region accounted for about 20% of the national carbon emissions. Ecological environment problems which were created by high pollutants emission have become an important bottleneck restricting the sustainable development of Beijing-Tianjin-Hebei. Realizing the coordinated development between Beijing, Tianjin and Hebei, carbon transfer problems should be pay more attention.

Economic development in "Beijing-Tianjin-Hebei" is unbalanced, which created inter-regional industrial gradient transfer. Industrial transfer will lead to the spatial transfer of carbon emission (Xu 2010, Chen 2006, Dou 2014, Kheder 2012, Liu 2014), which had done a lot of research by domestic and foreign scholars. Peters believed that international trade made the carbon emissions transfer freely, and producers and consumers of products should bear the responsibility for carbon emissions (Peters 2008). Wang et al estimated commercial carbon emissions. He argued that developing countries as a producer undertook a large number of CO_2 for developed countries (Wang 2011). For the emission problem caused by domestic industrial transfer, Cheng, Wei argued that industry transfer increased carbon emission of transferred area (Cheng 2013).

These views focused on inter-regional carbon transfer generated from the trade and circulation. In addition to, there were also problems of carbon transfer between upstream and downstream. Upstream provides products and services for downstream. This relation made upstream industry undertake a lot of carbon emissions for the downstream industry. That inter-regional industry transfer along the industry chain has become the main form. Therefore, the carbon transfer in this paper refers to the transfer between the upstream and downstream. Definitions are as follows. Industry A invested carbon in industry B. Carbon emission transferred from industry A to industry B. In order to clarify the direction of industrial carbon transfer, this paper studied carbon transfer problem between upstream and downstream based on the multi-regional input-output table. It aimed at adjusting the industrial structure from the perspective of the whole industry chain, to promote synergic development of Beijing-Tianjin-Hebei from industry.

2. Mythology and Date sources

2.1. Estimating carbon emission intensity

This paper used the method which was provided by the United Nations Intergovernmental Panel on climate change (IPCC) to estimate the total CO_2 emissions. We calculated carbon emission intensity as follows:

$$S_j = \frac{\sum_{i=1}^{8} C_{ji}}{E_j} = \frac{\sum_{i=1}^{8} N_{ji} \times c_i}{E_j} \tag{1}$$

Where S is carbon emission intensity; E is total output, in terms of GDP ; j is the jth industry; i is the ith type of energy, which includes coal, coke, crude oil, fuel oil, gasoline, kerosene, diesel oil and natural gas; N is energy consumption; c is emission coefficient of energy, as shown in Table 1.

Table 1 Carbon emission coefficient of different energy

Energy type	coal	coke	crude oil	fuel oil	gasoline	kerosene	diesel oil	natural gas
$c_i(t/t)$	0.7759	0.8550	0.5857	0.6185	0.8150	0.5714	0.5921	0.4483

2.2. Multi-regional input-output model

Multi-regional input-output table can reflect the economic relations between different industries, different regions and different industries. According to the multi-regional input-output table, the increment of intermediate inputs between industries under t_1-t_2 can be expressed as:

$$\Delta x_{ij}^{rh} = x_{ij(t_2)}^{rh} - x_{ij(t_1)}^{rh} \tag{2}$$

Where i and j is the type of industry; r and h is the type of region; x_{ij}^{rh} is industry i of region r invest in industry j of region h ; Δx_{ij}^{rh} is the increment that industry i of region r invest in industry j of region h . Assuming that the carbon emission coefficient is constant, the increment that Industry i of region r invest carbon in industry j of region h is:

$$\Delta C_{ij}^{rh} = \Delta x_{ij}^{rh} S_i^r \tag{3}$$

Where S is carbon emission intensity? Therefore, the total carbon input that region invest in other regions during the period of t_1-t_2 increased:

$$\Delta C^r = \sum_{h=1}^{m} \sum_{i=1}^{n} \sum_{j=1}^{n} \Delta C_{ij}^{rh} \tag{4}$$

The increment that other regions invest carbon in region r during t_1-t_2 is:

$$\triangle C^{r}=\sum_{h=1}^{m}\sum_{i=1}^{n}\sum_{j=1}^{n}\triangle C_{ji}^{hr} \tag{5}$$

2.3. *Date source*

The regional input-output table (2010) was derived from Inter-regional Input-output Table of 30 Provinces, China (2010) (Liu 2011). Assuming that trade relations between industries are invariant, regional input-output table of 2014 was calculated based on regional input-output table of 2010.The date regarding industry energy consumption was derived from the Chain Energy Statistical Yearbook. According to Inter-regional Input-output Table of 30 Provinces, China (2010), industry was divided into six sectors: agriculture, industry, Construction Industry, transportation and storage industry, wholesale and retail industry, other services industry. A, I, C, TS, WR, OS, B, T, H are the abbreviations of agriculture, industry, Construction Industry, transportation and storage industry, wholesale and retail industry, other services industry, Beijing, Tianjin, Hebei.

3. Analysis of industrial carbon correlations

3.1. *Carbon emission intensity for industry*

According to the formula (1), this paper accounted carbon emission coefficient for each industry in "Beijing-Tianjin-Hebei", as shown in figure 1. Carbon emission intensity was largest in Hebei province. And it was least in Beijing. The carbon emission intensity of industrial was biggest among the six sectors. And the industry in Hebei had the largest carbon emission intensity in all sectors.

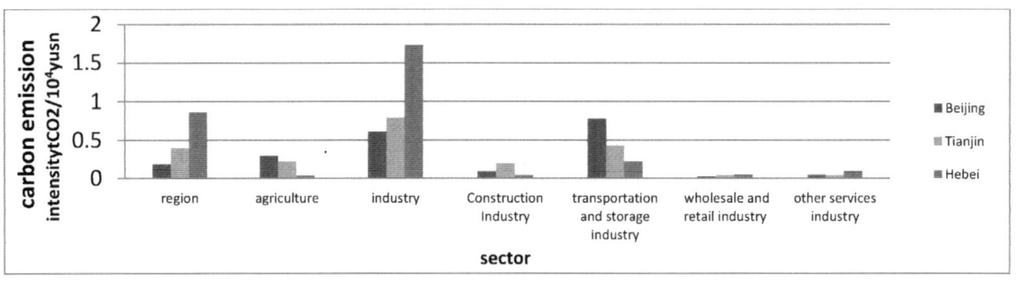

Figure 1 Carbon emission intensity for each industry

3.2. *Inter-industrial carbon linkage in "Beijing-Tianjin-Hebei"*

According to the formula (4) - (5), this paper accounted the changes of carbon input and carbon use in "Beijing-Tianjin-Hebei", as shown in Table 2. Figure 2-4 showed that the change of intrt-industrial carbon input in "Beijing-Tianjin-Hebei", the arrow indicated the direction of carbon input, the width of arrow represented the increased proportion of carbon inputs. For example, increased proportion that industry of Beijing invested carbon in industry of Tianjin is divide the increment of industry of Beijing invested carbon in industry of Tianjin by the increment of Beijing invest carbon in Tianjin.

Table 2 Changes of inter-regional carbon transfer from 2010 to 2014 (million tons).

	increment of carbon use	increment of carbon input
Beijing	50.58	9.94
Tianjin	578.92	582.95
Hebei	584.59	621.20

The increment of carbon use was greater than the increment of input in Beijing from 2010 to 2014. It was the opposite in Tianjin and Hebei. Carbon use of Beijing increased 50.58 million tons from 2010 to 2014. And 48% of these came from increment that Hebei invested in Beijing. Increment of carbon use in Tianjin or Hebei was 10 times more than the increment in Beijing.

3.2.1. Inter-industrial carbon linkage in Beijing. Input that Beijing invested carbon in itself increased by 97.89 million tons. And the increment of carbon input for other services industry was obvious (the proportion is 55%) from 2010 to 2014. For other services industry, 36% of the input increment came from itself, 12% of the input increment came from industry. There was carbon input correlation between industry and other services industry. The carbon input obviously increased during 2010 to 2014. In addition, wholesale and retail industry invested more carbon in industry. For industry, the final result showed that the increase of carbon use was greater than the increase of carbon input.

The increment of input that Tianjin invested carbon in Beijing mainly focused that industry, Construction Industry of Tianjin invested in industry of Beijing , wholesale and retail industry of Tianjin invested in other services industry of Beijing , transportation and storage industry of Tianjin invested in transportation and storage industry of Beijing. The increased input that Hebei invested carbon in Beijing had a lot of transfer path. And increased input that industry, wholesale and retail industry of Hebei invested in industry of Beijing, transportation and storage industry, wholesale and retail industry of Hebei invested in other services industry of Beijing had higher proportion.

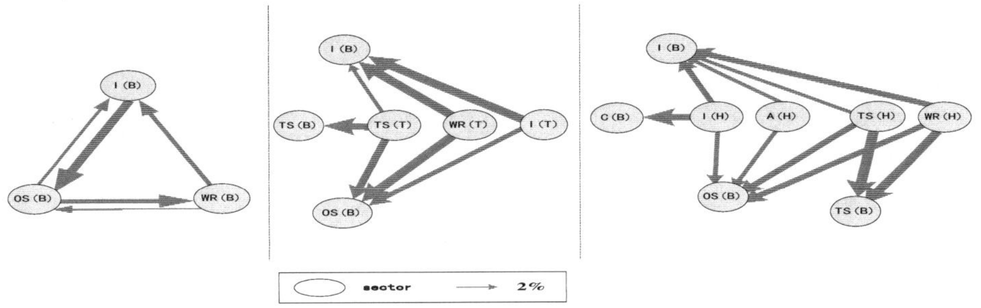

Figure 2 Main transfer paths of carbon transfer in Beijing area from 2010 to 2014.

3.2.2. Inter-industrial carbon linkage in Tianjin. Increment of carbon use in Tianjin region inside mainly because that carbon use of industry increased (industry made up about 61% of the increment). Input that transportation and storage industry, wholesale and retail industry invested carbon in industry obviously increased. The proportion was 19%, 21%. Therefore, Tianjin should strengthen cooperation between wholesale and retail industry and industry, transportation and storage industry and industry to curb the increase of carbon transfer. Carbon use of Construction Industry increased small. The proportion was 8%. While Carbon input of Construction Industry also increased small. Construction Industry had become the sector that the increment of carbon use was more than increment of carbon input.

Input that Beijing invested carbon in Tianjin slightly increased from 2010 to 2014. But the Contribution was negligible. That Hebei invested carbon in Tianjin increased. We can see that the increment mainly embodied that agriculture, industry, transportation and storage industry of Hebei invested in industry of Tianjin, transportation and storage industry of Hebei invested in transportation and storage industry of Tianjin. The proportion was 10%, 11%, 36%, 10%.

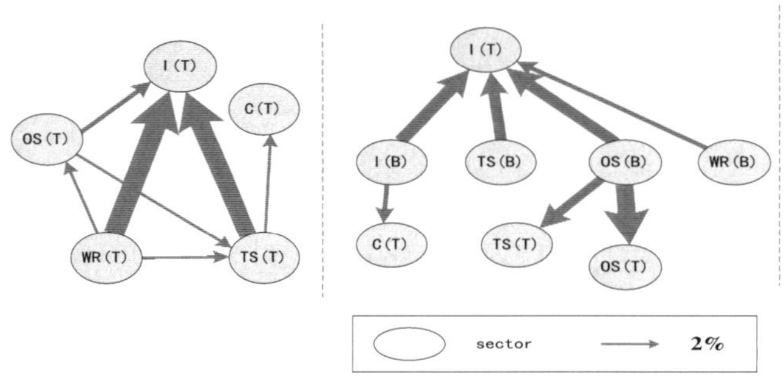

Figure 3 Main transfer paths of carbon transfer in Tianjin area from 2010 to 2014.

3.2.3. Inter-industrial carbon linkage in Hebei. From 2010 to 2014, carbon input which Hebei invested itself region significantly increased. 63% of the increment was that other services industry invested carbon in industry. That carbon use of industry increased mainly due to industry, agriculture, transportation and storage industry, wholesale and retail industry invested more carbon in industry. From 2010 to 2014, carbon input which Beijing invested in Hebei increased by 0.8256 million tons. Increased input that industry and other services industry of Beijing invested in Hebei was in the majority. Carbon input that Tianjin invested in Hebei increased. These mainly were due to the increase that industry, transportation and storage industry of Tianjin invested in industry of Hebei. And the proportion was 30%, 21%.

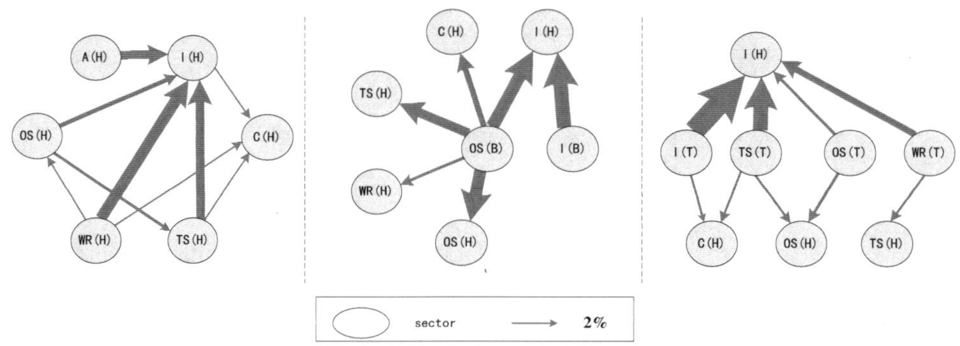

Figure 4 Main transfer paths of carbon transfer in Hebei area from 2010 to 2014.

4. Conclusion

In Beijing, the increment of carbon use was 50.58 million tons, while the increment of carbon input was 9.94 million tons from 2010 to 2014. That indicated that Tianjin and Hebei undertook more carbon emissions for Beijing in 2014. The increase of carbon transfer in Beijing was mainly due to the increase input that Hebei invested in Beijing. Therefore, from the perspective of ecological compensation, Beijing should provide a number of compensation for Tianjin and Hebei. For specific industry, the carbon input increased markedly industry. For example, carbon input among industry increased 86.40 million tons in Tianjin from 2010 to 2014. Therefore, we should continue to adjust the internal structure of the industry, to toward the direction of low energy consumption and low emission. Because of the existence of carbon transfer between industries, different industries need to assume jointly responsibility for emission reduction. We should determine the appropriate ratio of reduction emission between upstream and downstream through the attribution of responsibility, emission reduction capacity and emission reduction responsibility.

Acknowledgments
Supported by the National Social Science Foundation of China (Grant No. 15BGL211), Key Projects of Philosophy and Social Sciences Research, Ministry of Education (Grant No. 15JZD021) and Foundation for University Innovation team by the Ministry of Education of Tianjin Provice(TD12-5013)

Reference

[1] Xu Liu-De. The type of backward area to undertake industrial transfer impact on the environment and prevention[J]. Inquiry into Economic Issues, 2010, (06): 36-39.

[2] Chen Gang, Liu Shan-Shan. The theoretical evaluation of industry transfer: present situation and prediction[J]. Contemporary Finance & Economics, 2006, (10): 91-96.

[3] Dou Jian-Min, Shen Yan-Bin. On the influence of the industrial tran sfer on the environment in the central region of China[J]. China population, Resources and Environment, 2014, (11):96-102.

[4] Kheder S B, Zugravu N. Environmental regulation and French firms location abroad: An economic geography model in an international comparative study[J]. Ecological Economics, 2012, (77):48-61.

[5] Liu Hong-Guang, Wang Yun-Ping, Ji Lu. Characteristics, mechanism and pattern of inter-regional industry transfers in China[J]. Economic Geography, 2014, (01):102-107.

[6] Peters G, Hertwich E. CO2 embodied in international trade with implications for glob al climate policy [J].Environmental Science & Technology,2008,42(05):116-128.

[7] Wang Wen-Ju, Xiang Qi-Feng. Accounting and res ponsibility analysis on carbon emissions embodied in international trade [J]. China Industrial Economics, 2011, (10):56-64.

[8] Cheng Ai-Hua, Wei Hou-Kai. Target design on carbon reduction of promoting regional industrial transfer orderly and coordinated development [J]. China Population, Resources and Environment, 2013, (01): 55-62.

[9] Su B, Ang B. Input-output analysis of CO2 emissions embodied in trade: A multi-region model for China [J]. Energy Policy, 2014, (114):377-384.

[10] Xiao Yan-Fei,Wan Zi-Jie, Liu Hong-Guang. An empirical study of carbon emission transfer and carbon leakage in regional industrial transfer in china: analysis based on inter-regional input-output model in 2001 and 2007[J]. Journal of Finance and Economics, 2014,(2):75-84.

[11] Yang Shun-Shun. Evaluation and prediction on carbon emissions transferring across the industrial sectors in China[J]. China Industrial Economics, 2015, (06):55-67.

[12] Liu Wei-Dong, Tang Zhi-Peng. Inter-regional input-output table of 30 provinces, China(2010)[M].Beijing: China Statistics Press,2014.38-41.

[13] Liu Hong-Guang, Liu Wei-Dong, Liu Zhi-Gao. The quantitative study on inter-regional industry transfer[J]. China Industrial Economics, 2011,(06):79-88.

An analysis of environmental data transmission

Lina Yuan *, Huajun Chen and Jing Gong

Big Data Institute, Tongren University, Tongren, China.

*Corresponding author e-mail: 893422817@qq.com

Abstract. To comprehensively construct environmental automatic monitoring has become the urgent need of environmental management, is a major measure to implement the scientific outlook on development and build a harmonious socialist society, and is an inevitable choice of "building a resource-conserving and environment-friendly society", which is of great importance and profound significance to adjust the economic structure and transform growth pattern. This article first introduces the importance of environmental data transmission, then expounds the characteristics, key technologies, transmitting mode, and design ideas of environmental data transmission process, and finally, summarizes the full text.

1. Introduction

Since the reform and opening-up is for thirty years, with the rapid growth and advance of China's economic, industrialization and urbanization, "made in China" has gone globalization. At the same time, various pollution, such as the construction noise, radioactive source, chemical residues, automobile emissions, the amount of waste gas and effluent outside the industry, is also rapidly increasing, which makes the existing problem of environmental pollution become completely prominent, and severely restricts the further development of our economy and society. Our country has researched and explored the environmental monitoring from the mid-1980s, and after 30 years of development, has made great progress. In recent years, as computer and Internet technology develop swift and violent, adopting the most advanced techniques of communication, electronic and software to design a software of environmental data automatic monitoring, and to realize the automation of environmental data transmission, has laid a good foundation from a higher level to manage environment pollution, and solves the problem of networking inconvenience brought by the scene monitoring stations scattered, which is in favor of realizing the diversification of environmental data transmission modes and resource sharing, greatly improves the automation and informational zed level of environmental monitoring, guarantees for the development of environmental protection work, and provides an important and far-reaching significance of making the laws and regulations of environmental protection [1]-[8].

2. The characteristics and key-technologies of environmental data transmission process

2.1. The characteristics

Environmental protection in the process of data transmission mainly has three characteristics in the following. Firstly, environmental monitoring has the characteristics of complexity and dynamics, referring to multiple departments, multiple regions and multiple fields, to deal with a large amount of data. Secondly, because it needs to realize the networking protocol of environmental data transmission in the various provinces and cities, the software of environmental data transmission tends to be customized, or different communication interfaces, or different networking communication protocols. Thirdly, because the environmental data holds the balance in the process of environmental monitoring, it is the certificate of environmental quality with good or bad, and is also the basis of formulating the relevant laws and regulations [9].

Of course, there are still some problems of environmental data transmission software that need to be solved, for example, too much strong software coupling and poor universality; Long software development cycle; Low reliability of environmental data transmission; difficult to share environmental data.

2.2. The key technologies

Transmission network is the network platform of realizing data interaction between the different roles of different systems in the process of data transmission with environmental protection. The environmental protection bureau (EPA) requires the software of environmental monitoring can interact environmental data with monitoring platform through the way of wired or wireless transmission. This paper mainly expounds two types of technologies about wireless transmission networking.

2.2.1. General Packet Radio Service (GPRS) General Packet Radio Service (GPRS) is a kind of wireless Packet switching technology based on Global System for Mobile Communications (GSM). At present, the transmission speed of GPRS mobile communication network can reach up to 115 kb/s. GPRS is a developed technology based on GSM, and is the technology between the second generation of digital communication and the third generation of grouping type mobile business, so often referred to as 2.5 G. Compared with the traditional transmission way of GSM, GPRS has higher resource utilization, higher transmission rate (the highest up to 171.2 kb/s), and shorter access time, and supports IP protocol and x. 25 protocol, etc [10]-[12].

2.2.2. Code Division Multiple Access (CDMA) Code Division Multiple Access (CDMA) is a technology used on wireless communication, and has all the characteristics of spread spectrum communication. CDMA allows all users to simultaneously use the total bands (1.2288 Mhz), and has the characteristics of the strong ability to be anti-interference and resist decay. Its self-interference system regards signals from the other users as disturbance signals, completely not considering the problem of signal collision, to ensure the stability and reliability of the data in the environmental data transmission process. In addition, in the case of using the same frequency resource, the mobile network capacity of CDMA is 4-5 times larger than that of GSM, and 10 times greater than that of the analog network [9] [13].

2.2.3. The sharing technology Data sharing specifically points all kinds of data that derive from the environmental monitoring scene and can be also obtained by the other related environmental monitoring center. For the implementation of the sharing technology about environmental data, we can adopt two different techniques to achieve, i.e., networking with synchronous forward and "one point conveys more" [9]. Then we explain them as follows.

Firstly, networking with synchronous forward refers to the environmental data transmission software adopting the method of "one point conveys" in the scene of the monitoring, namely, environmental data only directly connects to the networking to transmit with the corresponding master monitoring center by the transmitting network between the monitoring spots and the monitoring center, and accepts its environmental data about the query and instructions to be obtained. The other non-master monitoring center needs to accept, retrieve, and query the site environmental data, must be through the transmission network among the monitoring center, and synchronously forwards data realized by the master control center. To the initiatively uploaded environmental data, the monitoring spot sends the environmental data to the master control monitoring center by the transmission network between the monitoring scene and the monitoring center. After the master monitoring center receives the environmental data, on the one hand, it accesses to the database, on the other hand, through the transmission network among the monitoring center, transmits the environmental data to the other non-master monitoring center. The passively uploaded environmental data, according to the direct transmission network between the master monitoring center and the monitoring spot, sends data obtaining instructions to the monitoring site, after the monitoring scene receives them, it responses the obtaining instructions, and uploads the requested data.

Secondly, the way of "One point conveys more" refers to the software to support the function of "one point conveys more", that is, directly through the transmitting network between the pollution sources in the monitoring site and the monitoring center, sends simultaneously the environmental data to the monitoring centers of multiple EPAs (whether the monitoring center of master control or that of non-master), and accepts the query and the obtaining instructions of its environmental data. For the actively uploading environmental data, transmission software at the same time transmits data to the monitoring center at all levels by the transmission network between the monitoring spot and the monitoring center, and ensures the sent data with consistency and integrity. For the passively uploading environmental data, the equipment operating parameters and the initialization parameter of the monitoring scene, the monitoring centers at all levels can directly send environmental data obtaining and the query instructions to the monitoring site through transmission network. After the software of the monitoring spot receives the obtaining instructions of the environmental data, it responses them, and uploads the request of environmental data.

2.3. The mode of transmission

The networking protocols [14] of environmental transmission are communication agreement between transmission software of environmental data and the environmental monitoring centre, corresponding to the application layer defined 7 layer protocols of ISO/OSI [15] [16]. According to the transmitting networking protocol of the environmental data, the transmitting mode of environmental data can be divided into two types: the mode of active transmission and passive transmission.

Firstly, the mode of active transmission indicates after the environmental data generates, every few seconds or minutes, the environmental monitoring scene automatically sends the environmental data to the environmental monitoring centre through the transmission network.

Secondly, the mode of passive transmission is when the monitoring spot receivers the obtaining instructions of data sent by the environmental monitoring centre, it then responses these obtaining instructions of data, and transmits the specified data frames to the environmental monitoring centre.

In addition, when the spot environmental alarm data and statistics data generate, they need to actively upload to the environmental monitoring centre, meanwhile the monitoring centre also can require the environmental monitoring site to start or stop upload real-time data of environmental monitoring spot, or demand the site to upload the historical data of a certain period of time, mainly according to the content of the transmission communication protocols, thus, it has bigger flexibility.

3. The design ideas of environmental data transmission process

The process of environmental data transmission, from the bottom up step by step, can be divided into three levels: the scene monitoring instrument, the transmission network, environmental monitoring information platform. Industrial personal computer (IPC) through transmission network (such as GPRS, Internet, etc.), according to the communication protocols, uploads the environmental data collected by the underlying analysis instrument to the monitoring center of environmental protection. The network structure of data transmission is as shown in Figure 1 [9].

The design of software framework is a crucial part of the software development, and favorable software architecture is the assurance of succeeding a software development project. Experience shows that the designs of excellent software frameworks have well hierarchical structure, which guides the overall situation of software design. The so-called hierarchical structure

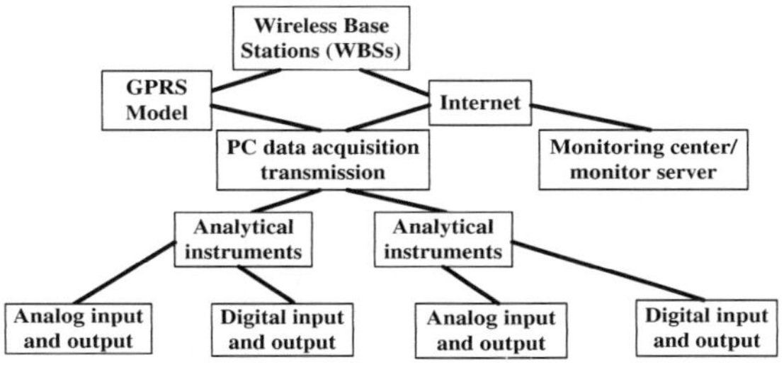

Figure 1. The process of environmental data transmission.

Is a certain structure assembled according to certain level? Hierarchical structure firstly emphasizes the hierarchy, and then stresses the constitutive property among levels. The left of Figure 2 [9] shows the well-informed hierarchical structure. Without doubt, perhaps the actual computer system doesn't conform to the above structure, but it is of universal significance in theory. Firstly, this structure is strictly one-way dependent among the layers from inner to outer, i.e., the application program first depends on the loading links of operating system, and then relies on hardware to execute. In this process, there only exists a kind of dependence, that is, the outer layer relies on the proximate inner layer. Secondly, in this structure, the more the outer layer, and the easier it is for ordinary people to understand. If the hardware details are taken for a kind of concrete and original things, the more the outer layer, the less tangible, and the more independence from the details. In fact, the upper play the role of hidden below details.

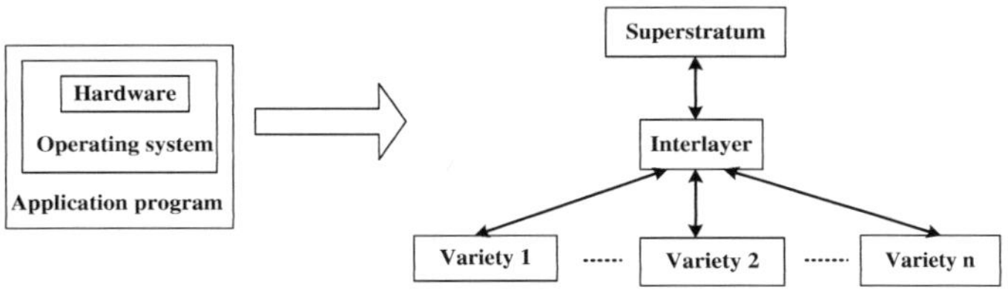

Figure 2. The hierarchical structure model from single to diversity.

Hierarchy is necessary in software design, but need further improvement to be applied in the software design of environmental protection data transmission. Because the way of environmental protection data transmission and networking protocols have the characteristics of "diversification", it needs to have a thorough knowledge of the configurability and scalability among the various modules of transmission software.

Hierarchical structure model for dealing with "diversification" is as shown in the right of Figure 2, the middle layer in the graph plays a key role of unified diversity. It interacts with the multifarious modules, adds a sort of diversity, and then will increase a corresponding processing module. At this point, the work of the middle layer is not easy, but the interface between the middle layer and the upper is unique, which greatly reduces the complexity of the upper. In process of environmental protection software transmission designed by software, provide a unified interface, by modifying the configuration file to compose the connections among various modules, in this way it doesn't lead to customized software due to each module of one-way dependence, also reduce the workload for the later software maintenance and the change of each module, and increase flexibility.

4. Conclusion

In a word, it is necessary for us to research and explore environmental data transmission. Not only the reasons are as shown above, but also environment automatic monitoring of transmission system in China is just getting started so far, there is no unified standard and perfect laws and regulations, and it is temporarily difficult to guarantee the reliability and authenticity of environmental data transmission, which will certainly have great impact on the development of environmental protection work. In the next step we will realize the design idea through the software, and make it applied in the real environment.

Acknowledgments

This work was supported by the Collaborative Fund Project of Science and Technology Agency in Guizhou Province Marked by the word LH on 7487[2014], and the reform project of teaching contents and curriculum system in colleges and universities of Guizhou Province on 2014SJJGXM003, and the project of education and cooperation for talent team word in Guizhou in 2015(NO:[2015]67), and partly supported by the National Natural Science Foundation of China(NO.61562703).

References

[1] Kun Wang, Hui Gao, Xiaoling Xu, Jinfang Jiang, Dong Yue, An Energy-Efficient Reliable Data Transmission Scheme for Complex Environmental Monitoring in Underwater Acoustic Sensor Networks, IEEE Sensors Journal, 16 (2016) 4051-4062.
[2] Evgeni Volpov, Rima Linder, Application of Local Environmental Data to the Lightning Performance Assessment of the IECo Transmission-Line Network, appeared in IEEE Transactions on Power Delivery, 2017, DOI: 10.1109/TPWRD.2017.2661721.
[3] S. Deepa, P. Haritha, S. Haridha, M. Harini, V Kamala Kiran, Energy conservative data transmission using Z-Mac technique in wireless sensor network for environmental monitoring, 2016 IEEE Technological Innovations in ICT for Agriculture and Rural Development (TIAR), (2016) 194-199, DOI: 10.1109/TIAR.2016.7801237.
[4] http://www.zhb.gov.cn, 2006.2.
[5] Modbus Application Protocol Specification 1.1, www.Modbus. org, 2002.12.6.
[6] Carlson, Environmental information organization. Database and expert system. applications,
[7] 1991.
[8] Kefford, Ben J. Effect of saline water disposal: implications for monitoring Programs and management, Environmental Monitoring and Assessment, 63(2002) 313-327.
[9] http://news.bjx.com.cn/html/20160408/723360.shtml.
[10] Dongxu Fu, Study on the key technologies of environmental data transmission, Hangzhou University of electronic science and technology, (2012) 6-13 (In Chinese).
[11] Novato Wireless Inc., GPRS technology overview, Novato Wireless Inc., Nov.2002
[12] Kalden R., Meiriekl., Meyer M., Wireless Internet Access Based on GPRS, IEEE personal Communications, Apr, 2000.
[13] Hou Yaling, The Design of Monitoring System Based on GPRS, 2016 International Conference on Robots & Intelligent System (ICRIS), (2016) 432-435, DOI: 10.1109/ICRIS.2016.91.
[14] Khaled E. Ahmed, Mohamed R. Rizk, Mohammed M. Farag, Overloaded CDMA interconnect for Network-on-Chip (OCNoC), 2016 International Conference on ReConFigurable Computing and FPGAs (ReConFig), (2016) 1-7, DOI: 10.1109/ReConFig.2016.7857179.
[15] Lin Jin-Yong, Dong Wen-Jie, Zhang Shuo, Luo Dan-Yan, Investigation of Network Communication Protocol on Autonomous Aerocraft Flying Formation, 2015 Fifth International Conference on Instrumentation and Measurement, Computer, Communication and Control (IMCCC), (2015) 171-175, DOI: 10.1109/IMCCC.2015.43.
[16] S. Kwong, K. T. Ng, W. N. Chau, Design and implementation of remote procedure call based on ISO/OSI reference model, TENCON '93. Proceedings. Computer, Communication, Control and Power Engineering.1993 IEEE Region 10 Conference on, 1 (1993) 544-548, DOI: 10.1109/TENCON.1993.320047.
[17] Miroslaw Bedzak, Usering. Education self-tuning-recommendations in the 8th level of ISO/OSI model, 2008 International Multiconference on Computer Science and Information Technology, (2008) 79-82, DOI: 10.1109/IMCSIT.2008.4747221.

IOP Conf. Series: Earth and Environmental Science **64** (2017) 012059 doi:10.1088/1755-1315/64/1/012059

Effects of ecological restoration on speciation and release of copper in lake sediments

Weijie Guo*, Liangyuan Zhao, Qingyun Li, Weihua Zhao, Huan Li

Basin Water Environmental Research Department, Yangtze River Scientific Research Institute, Wuhan, 430010, China

*Corresponding author e-mail:guoweijie1986@163.com

Abstract. The conclusion about the effect of macrophytes rhizosphere and mollusks bioturbation on morphology and mobility of heavy metals in the sediments is not entirely consistent. It is necessary to study further on influence mechanism of macrophytes growth and macrobenthos bioturbation on speciation and release of heavy metals in natural water body. This study intends to explore the influence mechanism of *Bellamya aeruginosa* (*Bel*) bioturbation and *Pontederia cordata* (*Pon*) growth on the speciation and release kinetics of Cu in sediment through construction of an aquatic organism-water-sediment system by indoor potted experiment. The results show that the particulate Cu in water of the control group and *Pon* treatment always maintain in lower level and the particulate Cu concentration in water of *Bel* treatment and *Pon+Bel* treatment appeared to a certain level of fluctuations at latter period. There was a significant positive correlation between the content of particulate Cu in water and the water turbidity. *Bellamya aeruginosa* bioturbation has not significantly affected the different speciation of heavy metals in surface sediments and the concentration of reducible Cu and Zn in the rhizosphere sediment was significantly lower than that in the control group.

1. Introduction

The rapid development of industry and agriculture has resulted in more and more heavy metal pollutantshave been discharged into the aquatic environment [1]. Main of them were combined with particulate matter in water by adsorption, complexation, precipitationand other physical/chemical effects, then accumulated and enriched in the sediment, becoming an important accumulation base and potential pollution source [2]. In response to the disturbance of bioturbation, redox conditions change and the growth of macrophytes, the heavy metal in sediments may be released into the water column, causing potential threat to ecosystems and being toxic to the aquatic flora and fauna [3, 4].

At present, different ecological restoration measures including re-establishing the submerged macrophytes and repopulating mollusks were designed and employed in increasing numbers across the world. However, the conclusion about the effect of macrophytes and bioturbation on the morphology and mobility of heavy metals in the sediments is compatible and it is necessary to study further on influence mechanism of macrophytes growth and macrobenthos bioturbation on speciation and release of heavy metals in water body. This study intends to explore the influence mechanism of *Bellamya aeruginosa* bioturbation and *Pontederia cordata* growth on the speciation and release kinetics of Cu in sediment through an indoor pot experiment of constructing an aquatic organism-water-sediment system.

2. Materials and methods

Twelve PVC barrels (d=30cm, h=40cm) were selected to carry out the experiment, and the upper layer of container filled with tap water (h=5 cm) and the sublayer was heavy metal pollution layer (h=30 cm, the design concentration of Cu was 100 mg/kg DW). The physical and chemical properties of sediment were as follows: cation exchange capacity (CEC) was 10.42±1.99meq/100g, organic matter (OM) was 8.25±0.34%, total organic carbon (TOC) was 4.57±0.18%, moisture content was 0.90±0.15%, and pH was 7.22±0.02.

The experiment was designed into three treatments, named *Pontederia cordata* treatment (*Pon*), *Bellamya aeruginosa* treatment (*Bel*), *Bellamya aeruginosa* and *Pontederia cordata* treatment (*Pon+Bel*) and one control treatment (Control), respectively, and there were three replicates for each treatment. The dosage of *Bellamya aeruginosa* was 1.72±0.01 kg/m^2ww (wet weight) and the *Pontederia cordata* treatment was 2.41±0.21 kg/m^2ww. All barrels were placed in the same condition (the average temperature was 22±3℃ and light intensity was 92.5±3.2 $\mu mol/s/m^2$).

Water samples were collected on 0d, 2d, 4d, 7d, 14d, 21d and 28d following start of treatment and the dissolved, particulate and total copper were determined. Sediment samples were collected on 0d and 28d after the experiment commenced and were divided into different layers of 0-2 cm, 2-5 cm and 5-8 cm and different copper forms were determined. Cu concentration was analyzed by ICP-MS. The experiments of sequential extraction procedure were carried out according to Community Bureau of Reference (BCR)'s scheme.

3. Results and discussion

The figure 1 shows that the particulate Cu concentration in water of all treatment showed a decreased tendency after the start of the experiment, due to the effect of natural sedimentation of the water and the effect of the *Bel*. The particulate Cu in the water columnhave shown the biggest reducing trend in *Bel* treatment and *Pon* treatment and the concentration of these two groups had reduced to 0.13±0.02 μg/L and 0.15±0.01μg/L on the second day. With the time goes on, the particulate Cu of the control group and *Pon* treatment always maintain in lower level. But the particulate Cu concentration in water of *Bel* treatment and *Pon+Bel* treatment appeared to a certain level of fluctuations at latter period, which are agree with the results by Ciutat and Boudou [5].

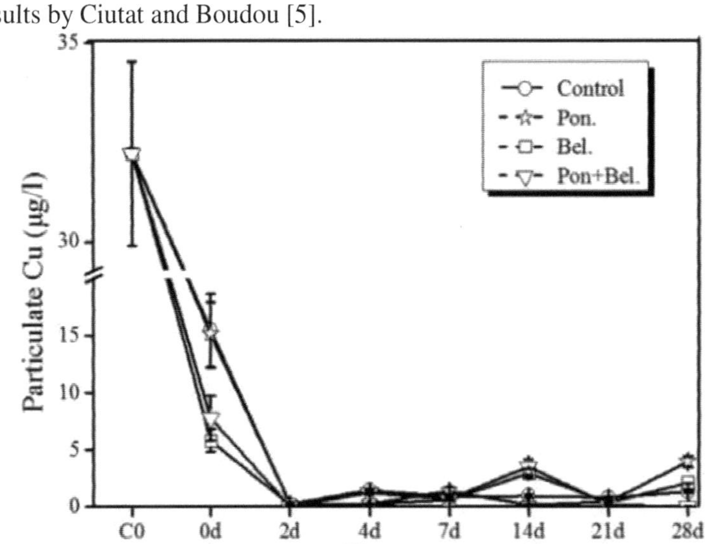

Figure 1. Changes of particulate Cu concentration in water of different treatments

During the study period, there was a significant positive correlation between the content of particulate Cu in water and the water turbidity (Y=0.5449X-0.0876, R^2=0.988), as shown in figure 2. With the sedimentation of the particulate (natural sedimentation and biological action would accelerate sedimentation), the turbidity value of the water decreased, and the content of particulate heavy metal was also decreased. In addition, in the middle and late state of the experiment, the particulate Cu content in the water remained at a very low level under the normal state or the normal activities of the ring snail, and no significant sediment resuspension and release occurred and this was consistent with the previous study [6].

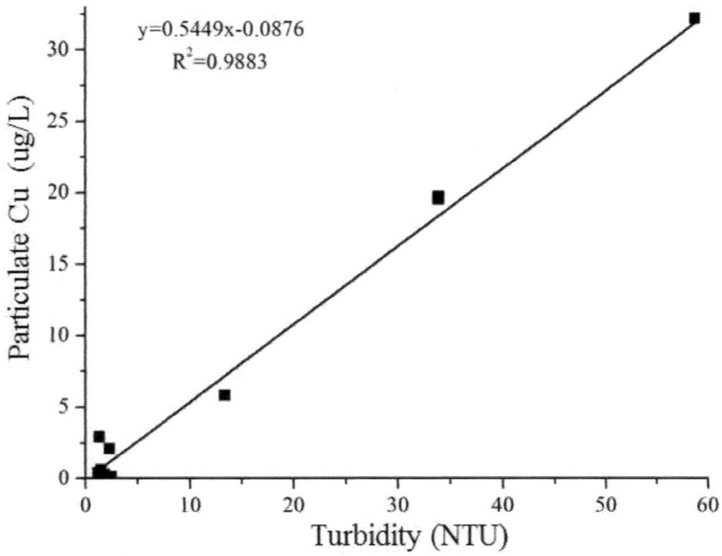

Figure 2. Correlative analysis between particulate Cu concentration and the turbidityin water

In this study, the bioturbation of *Bellamya aeruginosa* has not significantly affected the different speciation of heavy metals in surface sediments (data not shown), which may be related to the mode and intensity of bioturbation and the short experimental period. In the study period, the concentration of reducible Cu in the rhizosphere sediment (5-8 cm) was 73.06±1.87 mg/kg dw (dry weight), significantly lower than that in the control group (P<0.05). The content of labile reduction Cu was 80.09±1.27 mg/kg dw. The content of Zn in sediments was similar to that of Cu (P<0.05), that is, the content of rhizosphere sediment in the treatment group was significantly lower than that in the control group (P<0.05).

Table 1. The effect of *Pontederia cordata* growth on the heavy metal speciation

Fractions (5-8cm)	Cu(mg/kg dw)		Zn(mg/kg dw)	
	Control	*Pon* treatment	Control	*Pon* treatment
Acid extractable	11.22±2.02	13.60±3.28	46.13±1.16	53.08±1.21*
Reducible	80.09±1.27	73.06±1.87*	68.78±2.85	58.66±1.12*
Oxidizable	92.59±2.42	92.96±7.01	33.23±5.08	35.77±5.47
Residual	25.21±1.30	23.74±1.25	29.62±2.69	30.04±3.04

Note: *represents the differences are significant (P<0.05)

4. Conclusions

During the experiment period, the particulate Cu in water of the control group and *Pon* treatment always maintained in lower level and the particulate Cu concentration in water of *Bel* treatment and *Pon+Bel* treatment appeared to a certain level of fluctuations at latter period. There was a significant positive correlation between the content of particulate Cu in water and the water turbidity. The bioturbation of *Bellamya aeruginosa* has not significantly affected the different speciation of heavy metals in surface sediments and the concentration of reducible Cu and Zn in the rhizosphere sediment was significantly lower than that in the control group.

Acknowledgements

The research was financially supported by National Natural Science Foundation of China (Grant No. 51509011), Special fund for basic scientific research business of central public research institutes (Grant No. CKSF2016002/SH and CKSF2016025/SH) and China Three Gorges Corporation Program (0799553).

References

[1] E. Uluturhan, F. Kucuksezgin, Heavy metal contaminants in Red Pandora (Pagellus erythrinus) tissues from the eastern Aegean Sea, Turkey. Water research, 2007, 41(6): 1185-1192.
[2] S. Olivares-Rieumont, D. De la Rosa, L. Lima, et al., Assessment of heavy metal levels in Almendares River sediments-Havana City, Cuba. Water Research, 2005, 39(16): 3945-3953.
[3] T.E. Chow, K.F. Gaines, M.E. Hodgson, M.D. Wilson, Habitat and exposure modeling for ecological risk assessment: a case study for the raccoon on the Savanah River Site. Ecological Modeling, 2005, 189, 151-167.
[4] B. K. Hope, An examination of ecological risk assessment and management practices. Environment International, 2006, 32(8): 983-995.
[5] A. Ciutat, A. Boudou, Bioturbation effects on cadmium and zinc transfers from a contaminated sediment and on metal bioavailability to benthic bivalves. Environmental toxicology and chemistry, 2003, 22(7): 1574-1581.
[6] A. Ciutat, P. Anschutz, M. Gerino, Effects of bioturbation on cadmium transfer and distribution into freshwater sediments. Environmental toxicology and chemistry, 2005, 24(5): 1048-1058.

Comparison research on denitrification efficiency of two types of solid carbon source

Y Cai, J D Zhang *, F Li, Y X Cao, L Y Zhu, Z F Yang
Research Center for Environment and Health, Zhongnan University of Economics and Law, Wuhan 430073, China

*E-mail: jdzhang@zuel.edu.cn

Abstract. C/N rates can greatly influence efficiency of denitrification. It is difficult for current treated effluent to reach GB18918-2002 primary effluent standard because of its low C/N rate. To improve the efficiency of denitrification, the quality of effluent, and realize the waste recycling, this article selected magnolia leaves and degradable meal box as the solid carbon source for periodic denitrification stage to study the change of NO_3^--N, TN, COD, NO_2^--N, NH_4^+, PO_4^{3-} and color. The results showed that in the condition of influent nitrate concentration of 40 mg/L, carbon dosage of 10 g, the reaction temperature of 25℃, the nitrate removal rates of magnolia leaves and degradable meal box reached 89.0% and 56.3% respectively, and the TN removal rates reached 91.7% and 53.9% respectively. But in the aspect of final treated effluent COD concentration level, magnolia leaves experiment (608 mg/L) was much higher than the degradable meal box (78 mg/L). Besides, the accumulation of nitrite, the released concentration of ammonia nitrogen and phosphate of magnolia leaves experiment were also higher than the degradable meal box. Under the integrated analysis, the magnolia leaves are more suitable than the degradable meal box as the denitrification external carbon source.

1. Introduction

In recent years, China's environmental problems are increasing, especially in aspects of air pollution and water pollution [1-4]. In order to reach the primary effluent standard of "the Pollutant Discharge Standard of the Municipal Wastewater Treatment Plant (GB18918-2002)", it is necessary to use the external carbon source to improve the denitrification efficiency because of its low C/N rate. The traditional liquid carbon source such as methanol and acetic acid, has high economic costs or part of biological toxicity [5], and the carbon dosage is usually insufficient or excessive because of the great fluctuation of the quality of effluent, which influences the quality of effluent greatly [6].So the researchers have been working in which would be more suitable to be the external carbon source in recent years. The research of solid carbon source is increasing, especially the organic solid materials with low cost and wide variety of sources, which are generally divided into natural organic solids and synthetic organic solids. Currently some natural organic solid materials are researched as the external carbon source, such as cotton [7, 8], wheat straw [9, 10] and corn [11, 12], and synthetic organic solid materials are mainly biodegradable polymers, such as poly butylene succinate (PBS) [13, 14] polycaprolactone (PCL) [15, 16]. However, most of the existing research is focused on the same type of organic [17-19], and cotton, wheat straw, biodegradable polymer has its own higher economic value [20], which reduces the research value of practical application in a certain extent.

This experiment selected natural organic solid type of magnolia leaves and synthetic organic solid class of degradable meal box as the solid carbon source for periodic denitrification stage to study the change of NO_3^--N, TN, COD, NO_2^--N, NH_4^+, PO_4^{3-} and color, which was to analysis the efficiency of denitrification and influence on the system and to choose the better external carbon source between magnolia leaves and degradable meal box.

2. Methods and materials

2.1. Materials and instruments

This experiment selected natural organic solid type of magnolia leaves and synthetic organic solid class of degradable meal box as the solid carbon source for periodic denitrification stage. The degradable meal box, cut into 1x1 cm, was purchased from Beiyuan supermarket, and magnolia leaves, washed and dried and cut into 1x1 cm, were picked in the woods behind Wenpei building in Zhongnan University of Economics and Law. The activated sludge was taken from the two sedimentation tank of sewage treatment process in Tangxun sewage treatment plant, in Wuhan. The domestic sewage was taken from the sewage outfall of Wenlan building in Zhongnan University of Economics and Law. The quality indicators of domestic sewage are shown in Table 1.

Table 1. Quality indicatorsof domestic sewage

Indicators	COD/ ($mg{\cdot}L^{-1}$)	NO_3^--N/ ($mg{\cdot}L^{-1}$)	NO_2^--N/ ($mg{\cdot}L^{-1}$)	NH_4^+/ ($mg{\cdot}L^{-1}$)	TN / ($mg{\cdot}L^{-1}$)	PO_4^{3-}/ ($mg{\cdot}L^{-1}$)	Color
Concentration	180~300	0.5~0.8	0	7.2~8.4	15.4~17.5	14.3~19.6	21~31
Average	250	0.65	0	7.8	8.3	7.7	25

The reagents used in the experiment are analytical pure. HITACHI U-1900 type ultraviolet spectrophotometer: Xiamen science and Technology Co. Ltd; HACH DR2800 digestion instrument: Shanghai Xinsong Industrial Co. Ltd; HACH DRB200 COD analyzer: Shanghai Xinsong Industrial Co. Ltd.; BOXUN type vertical pressure steam sterilizer: Shanghai Boxun Industrial Company Limited medical equipment factory; HQ30d portable dissolved oxygen meter: Shandong Qingdao environmental protection technology Co., Ltd. Ming Bo; HJ-4A type multi magnetic heating stirrer: Changzhou Guohua Electric Appliance Co. Ltd.

2.2. Experiment methods

The activated sludge was put into the culture container, adding proper amount of domestic sewage for aeration culture. The dissolved oxygen of activated sludge aeration was detected regularly and adjusted appropriately with portable dissolved oxygen meter, to ensure that the dissolved oxygen of the activated sludge was maintained at 3~4 mg/L. Replacing the supernatant daily, the amount of domestic sewage was adjusted according to the clear turbidity of effluent. The quality of effluent was detected regularly to ensure that the concentration of COD is less than 50 mg/L. After cultured for two weeks, the quantity and quality of the activated sludge had grown. Observed by optical microscopy, the number of microbial species changed, and the shaped worms and rotifers appeared.

Taking three parts of activated sludge mixture of 1000 mL (around 4 g/L), the supernatant of the sludge mixture was poured out to ensure the sludge volume in 460 mL. Jar of 1000 mL as the reaction vessel was put on the magnetic heating stirrer, adjusting temperature of 25 ℃, adding sewage of 500 mL (25 ℃), nitrate solution of 40 mL (1000 mg/L, 25 ℃), solid carbon source of 10 g and activated sludge of 460 mL, to ensure that the initial COD/$NO_3^-$was maintained at around 3. After a few seconds, the supernatant of 20 mL was taken as the first sample of 0 h. The sludge in the reaction unitmixed with nutrient solution fully after pushing the rotate button, sampling in 0.5 h, 1 h, 2 h, 3 h (20 mL water supplement after each sampling) to be test.

2.3. *Activated sludge culture*
The activated sludge was put into the culture container, adding proper amount of domestic sewage for aeration culture. The dissolved oxygen of activated sludge aeration was detected regularly and adjusted appropriately with portable dissolved oxygen meter, to ensure that the dissolved oxygen of the activated sludge was maintained at 3~4 mg/L. Replacing the supernatant daily, the amount of domestic sewage was adjusted according to the clear turbidity of effluent. The quality of effluent was detected regularly to ensure that the concentration of COD is less than 50 mg/L. After cultured for two weeks, the quantity and quality of the activated sludge had grown. Observed by optical microscopy, the number of microbial species changed, and the shaped worms and rotifers appeared.

2.4. *Denitrification experiments*
Taking three parts of activated sludge mixture of 1000 mL (around 4 g/L), the supernatant of the sludge mixture was poured out to ensure the sludge volume in 460 mL. Jar of 1000 mL as the reaction vessel was put on the magnetic heating stirrer, adjusting temperature of 25 ℃, adding sewage of 500 mL (25 ℃), nitrate solution of 40 mL (1000 mg/L, 25 ℃), solid carbon source of 10 g and activated sludge of 460 mL, to ensure that the initial COD/$NO_3^-$was maintained at around 3. After a few seconds, the supernatant of 20 mL was taken as the first sample of 0 h. The sludge in the reaction unit mixed with nutrient solution fully after pushing the rotate button, sampling in 0.5 h, 1 h, 2 h, 3 h (20 mL water supplement after each sampling) to be test.

2.5. *Analysis method*
Water quality index analysis method was according to the standard method in "water and wastewater monitoring analysis method (4th edition)" [21]. Nitrate nitrogen (UV spectrophotometry), the total nitrogen (Potassium persulfate oxidation ultraviolet spectrophotometry), nitrite nitrogen (N- (1-naphthyl)-ethylene diamine spectrophotometry), ammonia nitrogen (Nessler reagent photometric method), phosphate (Molybdenum antimony spectrophotometric method), COD (Potassium dichromate colorimetry), Chromaticity (Colorimetric method) and MLSS (Weight method) were detected.

3. Results and discussion

3.1. *Nitrate, total nitrogen concentration changes*
Nitrate removal rate is an important index to measure the performance of denitrifying.Nitrate concentrations in three groups all showed a trend of decline over time in Fig. 1 (a). Specifically, nitrate removal rate of magnolia leaves reached 42.1%within 0.5 h and reached 89.0% in 2 h,then it was basically stable, while the finaleffluent nitrate concentration maintainiedat 3.6 mg/L.Because COD supply was significantly higher than the comparison, which was due tothe fast releasing carbon of early magnolia leaves,the nitrate removal rate ofmagnolia leaves washighly faster than the comparison. Due to the slow releasing carbon and nitrogen ofdegradable meal box within 3 h, itsnitrate removal rate was similar to the comparison, and even lower than the comparison. The nitrate removal rate ofdegradable meal box increased overall but slowly. Its nitrate removal rate (20.0%) was a litter lower than the comparison (26.9%) within 0.5 h, what's even worse was that its nitrate removal rate (56.3%) was also lower than the comparison (64.3%) in the end. The effluent nitrate concentration was maintained at 13.9 mg/L, in the mean time, the nitrate removal rate early was much higher than the late period, which was due to the initialabundant COD supply.

Total nitrogen removal rate is not only an important index to measure the performance of denitrifying, but also an important index to judge the quality of effluent. TN concentrations in three groups all showed a trend of decline over time in Fig. 1 (b). Specifically, TN concentrationof magnolia leaves reduced fromthe original 47.4 mg/L to the final 3.9 mg/L,and its final TNremoval rate reached 91.8%, even 45.6%

within 1 h. However,TN concentrationof degradable meal box reduced fromthe original45.7 mg/L to the final 21.1 mg/L, of which TNremoval rate reached53.9%, even lower than the comparison (65.5%).

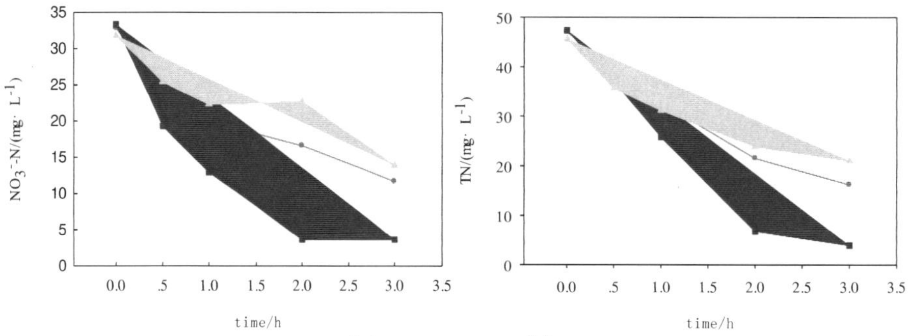

(a) (b)

• Comparison; ■Magnolia leaves; ▲ Degradable meal box

Fig. 1 NO_3^--N and TN changes of different carbon sources

3.2. COD concentration changes

COD is not only a quantitative index of solid material releasing carbon, but also a important index to judge the quality of the effluent.That solid material release carboncan be divided into two stages[22]: firstlythe water-soluble and easy decomposition substances of solid material on the surface are released, then internal refractory substancesof solid material will be releasedslowly, after the small molecules, on the surface and dissolved slowly from the main body, are released in full. Therefore, the COD concentration increase rapidly first and then slowly to stable.

As shown in Fig. 2, the COD concentration ofmagnolia leaves presented a slow increase trendover the reaction, ultimately much higher than GB18918-2002 primary effluent standard, but the COD concentration ofdegradable meal box changed in the trend of slow to decrease, similar to the comparison.Specifically, the COD concentration of magnolia leavesincreased quicklywithin 0.5 h,which period was also the fastest rate of denitrification. Analysised that the small molecules of magnolia leaves on the surface were released rapidly to cause the fast increase of COD concentrationwithin 0.5 h, and the rate of COD increase was much higher than theCOD consumption of denitrification, subsequently, the rate of carbon releasing was more slow, but still higher than the COD consumption. Degradable meal boxwasn't degraded first, after 1 h of reaction, it released carbonslowly, which causedtheslow riseof COD concentration, but thedenitrification rate didn't rise, even lower than that of thecomparison. Analysised that the released substances ofdegradable meal box couldn't be used by denitrifying bacteria, then the COD concentration declined over denitrification reaction.

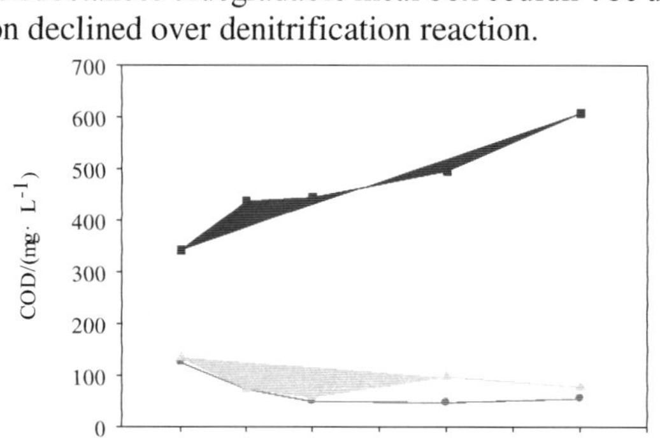

• Comparison; ■Magnolia leaves; ▲ Degradable meal box

Fig. 2 COD changes of different carbon sources

3.3. Nitrite and ammonia nitrogen concentration changes

Denitrification process can be simplified to that NO_3^- is converted to NO_2^- under the reduction of nitrate reductase, and thenNO_2^- is converted to N_2 under the reduction of nitrite reductase. Nitrogen removal process will produce nitrite accumulation, which is due to the competition for electron acceptor between nitrate reductase and nitrite reductase under the situation of lack of organic carbon source, and nitrite reductase at a disadvantage [23]. Therefore, as shown in Figure 3(a), the nitrite accumulation of degradable meal box, which obviously lack oforganic carbon source, reached the maximum amountof 1.0 mg/Lat 0.5 h of reaction, similar to the comparison.However, the nitrite accumulation of magnolia leaves,which full of organic carbon source, was much more than degradable meal box, that's because the activity of nitrite reductase was influenced largely by nitrate [24]. The nitrite accumulation of magnolia leaves rapidly increased to themaximum amount of 7.3 mg/Lat the first 1 h of reaction, which was due to the high concentration of nitrate to seriously inhibit the activity of nitrite reductase. When the concentration of nitrate decreased to about 13 mg/L over the reaction, the inhibition of nitrate to nitrite reductase reduced, which caused that $NO_2^-$was more quickly converted to $N_2$and the nitrite accumulation decreased to 0 mg/L finally.

The ammonia nitrogen concentration of degradable meal box had no significant change, similar to the comparison, asshown in Fig. 3 (b), but the ammonia nitrogen concentration of magnolia leaves showed a trend of rapid growth after 2 h initial stable. Specifically, the ammonia nitrogen concentration of magnolia leaves had been stable at around 7.0 mg/L during the first 2 h, but increased to 18.5 mg/L in the last hour. Analysised thatammonia nitrogen could be released in the leaching process, and ammonia and nitrite might occur in reaction of anaerobic ammonia oxidation under hypoxic conditions [25] because of the presence of nitrite in the reaction system within 2 h, resulting in the consumption of the released ammonia nitrogen. Two hours later, nitrite was exhausted, and the released ammonia nitrogen was accumulated to 18.5 mg/L. Degradable meal box released a bit of ammonia nitrogen in a short time, resulting in the concentration of ammonia nitrogen changing from 3.8 mg/L to 6.9 mg/L in the first 1 h. In the last hour, the released ammonia nitrogen and a small amount accumulation of nitrite occurred in reaction of anaerobic ammonia oxidation, resulting in the concentration of ammonia nitrogen slowly decreasing to 3.9 mg/L.

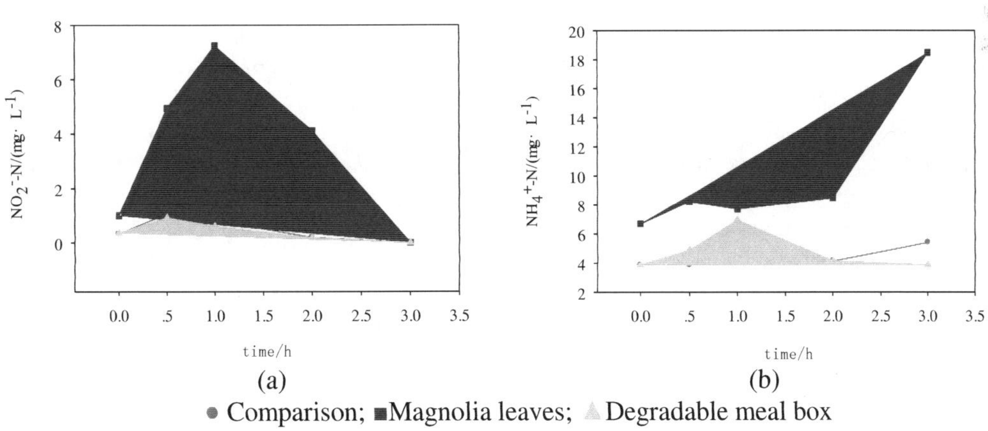

Fig. 3 NO_2^--N and NH_4^+ changes of different carbon sources

3.4. Phosphate concentration and colorchanges

The phosphate concentration of degradable meal box changed in the trend of slow to decrease, similar to the comparison, but the phosphate concentration of magnolia leaves presented a slow increase after a slow decrease trend, as shown in Fig. 4 (a).Specifically, in the denitrification process of degradable meal box and comparison, phosphorus bacteria began to polyphosphate[26], due to a very small amount of nitrite accumulation and nitrate dominating, which caused the concentration of phosphate decreased from 7.1 mg/L to 1.3 mg/L. The same to degradable meal box, the phosphate concentration of magnolia leaves decreased from 8.7 mg/L to 6.5 mg/L in the first hour. Subsequently, nitrite accumulated to dominate, and phosphorus bacteria in the system began to release phosphorus, resulting in phosphate concentration increasing to 10.1 mg/L.

the color of degradable meal box changed in the trend of slow to decrease, similar to the comparison, but thecolor of magnolia leaves presented a rapidincreaseto a stable trend, as shown in Fig. 4 (b). Specifically, with the denitrification, the color of comparisondecreasedfrom initial 22because of the consumption of organic carbon source in sewage. The color of degradable meal boxdecreased and eventually stabilized at about 15,similar to the comparison, due to the low degradation rate of degradable meal box. However, magnolia leaveshad been releasing pigment fast to the maximum of 138 at 0 h, resulting in the color at a high level. Because of the released pigment closing to saturation, the change rang of color decreased in the last hour, finally the color reached 295, 20 times that of the degradable meal box.

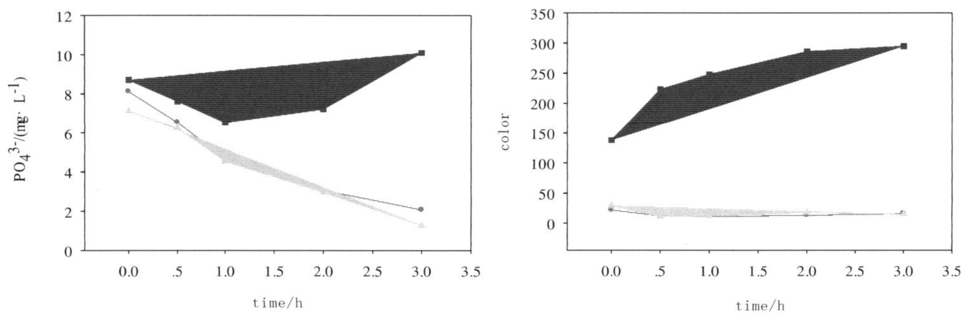

(a) (b)

● Comparison; ■Magnolia leaves; ▲ Degradable meal box

Fig. 4 PO_4^{3-} and colour changes of different carbon sources

4. Conclusion

In the condition of influent nitrate concentration of 40 mg/L, carbon dosage of 10 g, the reaction temperature of 25℃, periodic denitrification time within 3 h, the nitrite maximum accumulation of magnolia leaves and degradable meal box reached 7.3 mg/L and 1.0 mg/L, respectively, but eventually entirely consumed with no nitrite accumulation. At the same time, the final release of ammonia nitrogen and phosphate of magnolia leaves were 18.5 mg/L and 10.1 mg/L, respectively, and the final release of ammonia nitrogen and phosphate of degradable meal box were 3.9 mg/L and 1.3 mg/L, respectively. Magnolia leaves released a large amount of pigment within 3 h, resulting in the color reaching 295, 20 times that of the degradable meal box.

In the condition of influent nitrate concentration of 40 mg/L, carbon dosage of 10 g, the reaction temperature of 25℃, periodic denitrification time within 3 h, the release carbon amount of magnolia leaves was much higher than degradable meal box, which provided sufficient carbon source for denitrification, but caused the high effluent COD concentration. The nitrate and total nitrogen removal rates of magnolia leaves reached 89.0% and 91.7%, respectively, significantly higher than 56.3% and 53.9% of the degradable meal box. Under the integrated analysis, the magnolia leaves are more suitable than the degradable meal box as the denitrification external carbon source.

Acknowledgments

This work was financially supported by This study was supported by the National Natural Science Foundation of China (51308076; 71503268) and the Open Fund from Institute of Wuhan Studies (IWHS2016106).

IOP Conf. Series: Earth and Environmental Science **64** (2017) 012060 doi:10.1088/1755-1315/64/1/012060

References

[1] Li F, Huang J H, Zeng G M, Yuan X Z, Li X D, Liang J, Wang X Y, Tang X J and Bai B 2013 J. Geochem. Explpr. 132 75-83
[2] Li F, Zhang J D, Huang J H, Huang D W, Yang J, Song Y W, Zeng G M 2016 Environ. Sci. Pollut. R. 23 13100-13
[3] Li F, Zhang J D, Yang J, Liu C Y and Zeng G M 2016 Hum. Ecol. Risk Assess. 22 1224-41
[4] Huang J H, Li F, Zeng G M, Huang X, Liu W, Wu H and Li X 2016 Sci. Total Environ. 541 969-76
[5] Zhang Z L 2010. Harbin Institute of Technology
[6] Liu Z X, Sille B L andGert P 2011 J. China water and wastewater. 22 30-35
[7] Volokita M and Abehovich A 1996 J. Water Technol. 34 379-385
[8] Jin Z F, Chen Y X and Okura Kio 2004 J Agr. Environ. Sci. 03 512-515
[9] Shao Y, Xu Z X andJin·W 2009 J. Environ. Sci. 30 1414-1419
[10] Liu J X, Luo Z J and Jin M G 2008 J. Environ. Eng. 02 94-96
[11] Hou W L 2014 D. Hangzhou: Zhejiang University
[12] Zhao W L, Hao R X and Li B 2014 J. Environ. Impact 03 987-994
[13] Wen H, Chen Y F and Gao W2011 J. Chinese J. Environ. Sci. 03 499-504
[14] Zhou H H, Wang J L and Zhao X2006 J. Environ. Sci. 02 290-293
[15] Yu T, Wu2011 J. Chinese J. Ecol. Z1 1127-1132
[16] Ji Q, Zhang Q and Xu X 2014 J. Environ. Sci. Res. 09 1080-1086
[17] Xiong J F 2013 D. Shanghai Normal University.
[18] Li.J M 2015 D. Zhongnan University of Economics and Law.
[19] Ji Q, CHEN S and Yang S B2013 J. China water and wastewater 19 14-17
[20] Xiong J F and Xu H2012 J. Environ. Sci. 11 4057-4061
[21] State Environmental Protection Administration2002 M. China Environ. Sci. Press 216-274
[22] Shao Y, Xu Z X and Wang S2011 J. Environ. Sci. 08 2323-2327
[23] Oh J and Silverstein J1999 J. J. Environ. Eng. 125 234-242
[24] Körner H and Zumft W G1989 J. Appl. Environ. Microbiol. 55 1670-1676
[25] Strous M, Heijnen JJ and Kuenen JG 1998 J Appl. Microbiol. Biot. 50 589-596
[26] Wang Y Y and Wang S Y2005 J. China water and wastewater 07 47-51

IOP Conf. Series: Earth and Environmental Science **64** (2017) 012061 doi:10.1088/1755-1315/64/1/012061

The effect of agricultural non-point Source Pollution of nitrogen and phosphorous on Lake Eutrophication

Hanfeng Xiong[a]

Ezhou University, Ezhou 436000, Hubei, China

[a]xhfeng987@163.com

Abstract. Based on the data from investigation, the evaluation by equal standard pollution loading method was used to study the agricultural non-point source pollution caused by nitrogen and phosphorous from livestock's feces pollution, chemical fertilizer pollution and fish breeding pond pollution in Liangzi Lake wetland. The results revealed that: The lost amount of nitrogen and phosphorous was separately 1276.49T, 103.04T; the equivalent standard pollution loading amount was separately 12.76X10^8 m^3, 5.15X10^8 m^3. The lost amount of nitrogen was highest in chemical fertilizer. Based pollution on the understanding of the cause of agricultural non-point source in Liangzi Lake wetland, some countermeasures were suggested according to different pollution source.

1. Introduction

Eutrophication of water body is one of the most difficult problems in the world. Agricultural non-point source pollution is one of the main causes of water pollution. In order to control the eutrophication of the water body, it is also important to control the agricultural non-point source pollution and control the industrial point source pollution. It is of great theoretical and practical significance to understand the current situation and characteristics of agricultural non-point source pollution in water environment. There are many studies on the effects of agricultural non-point source N and P pollution on the water environment at home and abroad, especially the eutrophication of surface water [1, 2], Study on the environmental capacity of the receiving water, the application of the model to study non-point source N, P pollution load has also been reported [3,4]. This paper used equal standard pollution loading method to evaluate pollution source. This paper will provide the basis for the control of agricultural non-point source pollution in Liangzi Lake wetland.

2. Research method

2.1. Evaluation method

In this study, used equal standard pollution loading method to evaluate pollution source. The equal standard pollution load method is the conversion of various pollutants into a unified comparable amount, the results can reflect the impact of pollutants on quality for the environment [5].

The main content of the method: the pollution source concentration or the total amount, after standardized treatment, it can be transformed to the same scale and can be compared with each other. The main pollutants, the key pollution and the total pollution load can be determined according to the value. The main characteristics of the three identified: Equal standard pollution index, equal standard pollution load and pollution load ratio. Equal standard pollution index refers to the number of pollutants emitted by the pollutant is more than that of the pollution evaluation standard. The equal standard pollution load is the representation of the total pollutant discharge. The pollution load ratio refers to the ratio of the equivalent pollution load of a certain pollution source or a pollutant to the total equivalent standard pollution load, which reflects the pollution contribution of a pollution source [6].

Calculation formula:

The standard emission of pollutants = the amount of pollution loss / threshold concentration of pollutants.

Rate of loss of pollutants = the amount of pollutants entering the water environment / total pollutant emission

Loss of pollutants = Rate of loss of pollutants×emission of pollutants

The diffusion concentration of pollutants = the source of pollution in the region / the total amount of water resources in the region

The equal standard pollution index = standard emission of pollutants/ the total amount of water resources in the region.

2.2. Evaluation factors:

The evaluation factors were TN (total nitrogen) and TP (total phosphorus)

2.3. Evaluation standard

According to the national evaluation standard of "environmental quality standard for surface water (GB3838-2002)". The threshold concentration of TN is 1mg/L, TP is 0.2mg/L.

2.4. Data sources

hemical fertilizer, fishponds area, livestock and poultry breeding stock data from the statistical yearbook of Hubei province 2015.

3. Results Analysis

3.1. Nitrogen and phosphorus pollution of chemical fertilizer

2015 total fertilizer consumption was 20098T. According to the survey the present situation of fertilization in the region, the problem of farmland fertilization are that amount of nitrogen more, phosphorus and potassium less, fertilizer utilization rate is low; the proportion of organic fertilizer and inorganic fertilizer is not reasonable; the unreasonable fertilization method. The crop needs a ratio of N: P205:K20 should be 1:0.3-0.4:0.6-0.7. The actual fertilization ratio $N:P_20_5:K_20$ was 1:0.28:0.24. The part of the fertilizer that is not absorbed by the crops will flow into the water body through the surface runoff, infiltration and so on. According to the nitrogen fertilizer and phosphate fertilizer loss fixed point test report [7, 8] that nitrogen and phosphorus loss accounted were 11% and 1.28%of the year's fertilizer, respectively.

Emissions of pollutants from chemical fertilizer sources= Discharge coefficient×Fertilizer application

The emissions of N and P were 1283.7T and 107.89T (Table 1). The diffusion of N and P were 3.36mg/L and 0.11mg/L, respectively. According to the calculation formula, the equivalent emissions of N and P of chemical fertilizer sources $12.84\times10^8m^3$ and $5.45\times10^8m^3$ were obtained. The standard pollution index of N and P were 5.63 and 2.39, respectively.

Table 1. The pollution of chemical fertilizers

	emissions T	standard pollution index	equivalent emissions 10^8 m^3
TN	1283.7	5.63	12.84
TP	107.89	2.39	5.45

3.2. *Nitrogen and phosphorus pollution of fishpond*

After a period of pond aquaculture activities, the gradual deterioration of water quality, mainly in the water total nitrogen, total phosphorus and BOD, COD were significantly increased, like living in the water with high organic content in algae formed advantage species, leftover feed and fish waste accumulation, water eutrophication [9], easy to outbreak of fish diseases and even cause a lot of dead fish. In this case, the water of fish pond must be changed, the waste water discharged to the nearby streams and lakes will cause some pollution, become an important factor in rural non-point source pollution.

According to the research reports [10], in the normal level of investment, the amount of TN and TP discharged to the environment from the fishponds was 101 kg and 11 kg each hectare in a year

Emissions of pollutants from fishpond= Discharge coefficient×Fishpond area

According to the above data, we can find out the loss of pollution source of the fishpond (Table 2). In 2015, the total amount of pollutants discharged from the fishponds was 958.83 T, N and P were 864.66T and 94.17T, respectively. The diffusion of N and P were 1.70mg/L and 0.18mg/L, respectively. According to the calculation formula of the standard emission, the N, P equivalent standard emissions of $8.65\times10^8 m^3$ and $4.71\times10^8 m^3$ in the pollution source of the fish ponds were calculated. N, P equivalent standard pollution index were 3.79 and 2.07, respectively.

Table 2. The pollution of fish breeding ponds

	emissions T	equivalent standard pollution index	equivalent standard emissions 10^8 m^3
TN	864.66	3.79	8.65
TP	94.17	2.07	4.71

3.3. *Nitrogen and phosphorus pollution of livestock manure pollution sources*

Livestock and poultry are mainly pigs, cattle, sheep, etc. Livestock pollution source is mainly caused by livestock manure. Refer to the domestic research reports [11-13], the annual pollutant discharge coefficient of livestock are shown in table 3.

Table 3. Pollutant discharge coefficients of livestock's feces (kg / year)

	cow dung	Cow urine	sheep manure	pig manure	Pig urine	Poultry manure
TN	31.9	29.2	2.28	2.34	2.17	0.275
TP	8.61	1.64	0.45	1.36	0.34	0.115

According to the results of the survey, the amount of pollutants discharged into the water body is 10%.

There are 244700 live pigs, 13077 cattle, 7743 sheep, and poultry 4740000 in Liangzi lake region.

Emissions of pollutants = Discharge coefficient× feeding livestock.

According to the formula to calculate the pollutants of livestock and poultry manure and the amount of livestock and poultry manure pollutants into the water (Table4). Livestock and poultry pollution sources N, P loss were 121.68T and 35.77T. The concentrations of N and P in livestock and poultry pollution sources were 5.33mg/L and 1.57mg/L, respectively. According to the calculation formula of the equivalent emissions, the N and P emissions of livestock and poultry pollution sources were calculated as $1.21X10^8$ m^3 and $1.79X10^8$ m^3. The N, P equal standard pollution index of livestock and poultry pollution sources were 5.53 and 0.78, respectively.

Table 4. The discharge amount and lost amount of livestock's feces (t/a)

	emissions of pollutants		loss of pollutants	
	TN	TP	TN	TP
cow dung	417.16	112.59	41.72	11.26
Cow urine	381.85	21.45	38.19	2.15
sheep manure	17.65	3.48	1.77	0.35
pig manure	572.60	332.79	57.26	33.28
Pig urine	531.00	83.20	53.10	8.32
Poultry manure	1303.50	545.10	130.35	54.51

3.4. The contribution of pollutants to water pollution

In 2015, the amount of pollutants discharged into water by various agricultural non-point source pollution was 2782.68T, TN and TP were 2470.75T and 311.93T, respectively. Fertilizer emissions of TN, TP amounted to 1391.59T, accounting for 50.01%. The total equivalent load of pollution sources was $40.36X10^8m^3$, TN and TP were $24.71X10^8m^3$ and $15.65X10^8m^3$ (Table 5). The loss of TN in fertilizer is the highest among all kinds of pollution sources (Table 6).

Table 5. Equivalent standard discharged amount of different non-point pollution sources (10^8m^3)

	Fertilizer	livestock manure	fishponds	Total
TN	12.84	3.22	8.65	24.71
TP	5.45	5.49	4.71	15.65

Table 6. Equivalent standard discharged rate of different non-point pollution sources (%)

	fertilizer	livestock manure	fishponds	Equivalent standard discharged rate
TN	31.82	7.98	21.43	61.23
TP	13.50	13.60	11.67	38.77
Equivalent standard discharged rate	445.31	21.58	33.10	

It can be seen from the pollution load ratio (Table 6) that the pollution of chemical fertilizer is the biggest pollution source, the second is the fish pond, and the pollution of livestock and poultry is the least. The main pollution in the chemical fertilizer and the fishponds is TN, and the main pollution is TP in the pollution of livestock and poultry. The equal standard pollution load ratios of TN and TP were 71.25% and 28.75%, respectively. This shows that the water environment is polluted seriously by agricultural nitrogen. The equal standard pollution load ratio of TN was 42.9%, which was the highest among all kinds of pollution sources and pollutants, which indicated that the chemical fertilizer TN had the greatest effect on the water environment.

3.5. Pollution control measures

Because of the complexity of factors affecting agricultural non-point source pollution, it is difficult to control agricultural non-point source pollution. In order to reduce agricultural non-point source pollution, we can start from controlling the source of non-point source pollution and controlling the diffusion path of non-point source pollutants.

3.5.1. Control strategy of fertilizer pollution First of all, according to the requirements of sustainable development, guided by the theory of ecology, through the adjustment of industrial structure, take measures to optimize the allocation of agricultural land resources, the adjustment of crop distribution, radically reduce the input of chemical fertilizer. The terrain around lake is low hilly, the higher to the cultivation of economic forest such as fruit, the lower planting lotus or rice. Secondly, balanced fertilization and nutrient controlled release technology were used to improve the utilization rate of fertilizer. Some measures such as to improved fertilization method, the combination of conventional fertilizer and biological fertilizer, inorganic fertilizer and organic fertilizer.

3.5.2. Control strategy of livestock manure pollution For the pollution prevention and control of livestock and poultry breeding, vigorously promote the livestock and poultry intensive breeding model, this is conducive to centralized treatment of livestock and poultry manure. Using pig- fish - beads and duck - fish - beads and other three-dimensional ecological farming model, the three-dimensional model can also be used for the recycling and full utilization of the nutrition elements. To reduce the content of organic nutrients such as nitrogen and phosphorus in the excreta of livestock and poultry by scientifically formulated feed and change feeding, improve the utilization rate of nutrients.

3.5.3. Control strategy of fish pond pollution The control strategy of fish pond pollution include that vigorously develop the high quality aquatic products, such as mandarin fish, snakehead, crab, reduce the amount of fertilizer and fish medicine. Reasonable bait feeding, attention to the size of the feed particle. Timely change water and silt, reduce internal pollution.

Acknowledgments
This work was financially supported by Hubei science and technology support project (2015BBA151)

Reference

[1] Chen XY, Chen YX, Fang Z., Exotic pollution load analysis of south-west zone in Wuhan East Lake, J. Environmental science and technology, 4(2001)41-43.
[2] Sharpley, A.N. The environmental impact of agricultural nitrogen and phosphorus use, J. Agric. Food Chem. 36(1987)812-817.
[3] Ye M, Yang GS, Zhang WS. Study on Urban Non-point Pollution Characteristics and Pollution Loads Prediction Model, J. Environmental science and technology. 2(2006) 73-75.
[4] Xu XY. A mathematical model of ammonium nitrogen loss from farmland, J. Environmental Pollution and Control .8(1999)34-37.
[5] Lu YS. Assessment of Environmental Quality, M. High education press China, 2001.
[6] Liu lw, Li YB. Thought of the Concept about Equivalent Pollution Load, J. Liaoning urban and rural environmental scientific. 20(2000) 23-26.
[7] Zhuo GD, Zhang JJ, Lu SL. Investigation of Nitrogen Losses and Moving Directions in Shanghai Suburban, J. Shanghai environment science. 15(1996) 37-39.
[8] Zhang HC, Zhang HA, Cao ZH. Research on Phosphorus Runoff Losses from Paddy Soils in the Taihu Lake Region and Its Olsen-P "Change-point", J. Journal of nanjing forestry university. 5(2004):6-10.
[9] Xiong GZ, Dai ZF, Shen B. Pollution in Pounds at Side of Ehai Lake, J. Yuannan environment science.3(2000)32-34.
[10] Zhang DD, Zhang XH, Zhang JJ Integrated research and evaluation on nonpoint source pollution in Shanghai suburbs, J. Acta Agriculturae Shanghai. 13(1997) 31-36.
[11] Liu PF, Chen ZL, Xu SY. The Pollution of Poultry and Animal Feces and the Countermeasures in Yangtze Delta's Suburb, J. Resources and Environment in the Yangze Basin. 11(2002) 456-460.
[12] Lu J, Jin C, Shi HC Pollutant Loading from Agricultural Non-point Sources in Upper Reaches of the Zihe River: Analysis and Control, J. Environmental science and technology. 26(2003)26-28.
[13] Li RG, Xia YL, Wu AZ Pollutants Sources and Their Discharging Amount in Taihu Lake Area of Jiangsu Province, J. Journal of Lake Science. l2(2000)147-153.

IOP Conf. Series: Earth and Environmental Science **64** (2017) 012062 doi:10.1088/1755-1315/64/1/012062

The study of Ni (II) removal from aqueous solution by modified sediment

Shanshan Yu[1], Yang Cheng[2], Yuehong Yang[1] *

[1]Faculty of Environmental Science and Engineering, Kunming University of Science & Technology, Kunming 650500, China
[2]North Water Environment Engineering Co. Ltd., Haerbin, 150000, China

*Corresponding author E-mail: 616805874@qq.com

Abstract. Preparation of high-temperature roasting modified sediment as an adsorbent, to research the effect of adsorption of Ni (II) from aqueous solution through static adsorption test. Through static adsorption experiments, Adsorption kinetics and thermodynamics properties of roasting sediment adsorption of $Ni2^{+}$ were researched. It was observed that the adsorption process was described with the Langmuir and Freundlich theories, and the Freundlich model indicated the best fit to the adsorption process, showing that adsorption is between monolayer and multilayer. The adsorption of the Ni^{2+} onto the roasting sediment was described pseudo-first-order kinetics equation and pseudo-second-order kinetics equation, and the pseudo-second-order kinetics equation showed the best fit to the adsorption process, showing that it is dominated with chemical adsorption. Equilibrium adsorption capacity of roasting sediment for Ni^{2+} was found to be 3.6563mg/g. Thermodynamic parameters indicated (ΔS . ΔH and ΔG)that the adsorption of the Ni^{2+} onto the roasting sediment is endothermic process.

1. Introduction

Nickel is one of the essential elements of the human body [1-2], appropriate amount of intake is beneficial to the health of the human body. However, long-term exposure to drinking water and excessive nickel, will cause serious harm to the human body, and can lead to a variety of lesions. Acute poisoning showed: nausea and vomiting, stomach pain, gastrointestinal mucosa necrosis, severe abdominal pain, bloody stools and other pathological phenomena [3-4].Chronic poisoning symptoms: neurasthenia, neurological dysfunction, pyramidal damage, Parkinson's syndrome and other pathological phenomena. At present, the treatment methods of nickel at home and abroad are mainly contact oxidation method and ion exchange resin. However, these methods are too expensive, complex process, not easy to promote [5-6].Adsorption is one of the most effective methods to remove heavy metal ions in wastewater. It is more mature and stable method and simple operation, low cost, technology is mature.

The traditional methods of sludge treatment are landfill, incineration and dumping, but it will cause serious pollution and damage to the environment, it is the law prohibited. Therefore, it is necessary to find a new way of utilization. Dianchi is rich in sediment resources, it contains a large number of organic

IOP Conf. Series: Earth and Environmental Science **64** (2017) 012062 doi:10.1088/1755-1315/64/1/012062

sediments and putrefaction, also it has large surface area, It has a certain degree of adsorption, but the adsorption effect is more significant, after high temperature modification of the Dianchi sediment.

2. Experimental part

2.1. Main instruments and reagents

722S visible spectrophotometer;PHS-3C type pH meter;SHZ-82 type constant temperature oscillator;LD4-2A low speed centrifuge;AB204-S electronic analytical balance;XL30SEM-TMP scanning electron microscope.

2.2. Experimental materials and methods

The bottom mud is added to the energy saving box type furnace with the initial temperature of 20 ℃ and heated at a heating rate of 5 ℃/min for 85 minutes to about 450 ℃,constant temperature roasting 120min to carbonize, and vibrating sieve machine after cooling, then, select 80 ~ 100 mesh particles for the electric blast oven dry standby.

The surface morphology before and after modification was analyzed by scanning electron microscope (XL30ESEM-TMP).And the results are as follows. As it shown from Fig. a: The crystal structure of the unmodified sediment is smooth and rough, the distance between the pore and larger floc degree is not obvious. As it shown from Fig. b: The crystal structure of the modified sediment was destroyed, higher surface roughness. The pore is well developed and has a wide pore size distribution, a higher degree of flocculation. This makes the sediment has a large specific surface area, which is more conducive to improve the adsorption performance of the sediment.

Fig.1 SEM image of sediment before (a) and after (b) modification

Determination of high-temperature roasting modified sediment by using Bruker TENSOR27 infrared spectrometer in Germany. The results are shown in Figure 2.

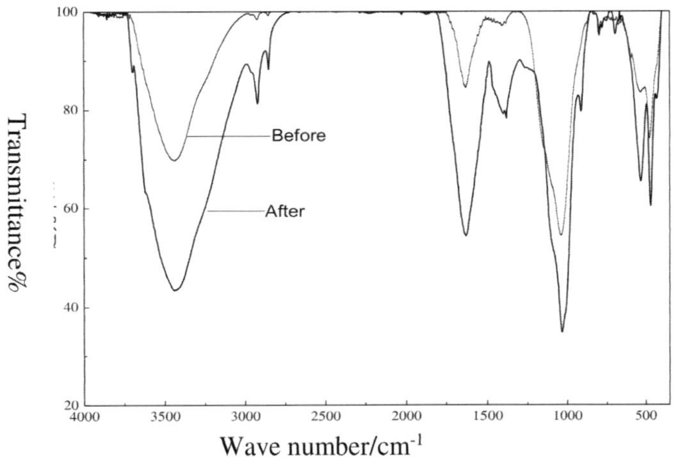

Fig.2 IR spectrum of sediment before and after modification

As can be seen from Fig.2: The original sediment has absorption peak at 3440.95cm^{-1}, this is because of the stretching vibration of O-H. The modified sediment has an absorption peak at 3439.85 cm^{-1}, and the peak shape is sharper than -OH. This is due to high temperature roasting caused by amine N-H and OH bands overlap and association. The absorption peak of original sediment at 2919.79cm^{-}1 is due to the stretching vibration of saturated hydrocarbon C-H and compared with the modified sediment, the dipole moment changes little[7].The modified sediment has an absorption peak at 1630.39 cm^{-1}, which is caused by the stretching vibration of C = O. The presence of a peak at 1039.31 cm^{-1} is due to the appearance of Si-O-C, and the absorption of $V_{Si\text{-}O}$ is due to the presence of Si in the sediment [8] The above analysis shows that the high temperature makes the sediment carbonization and evaporation of large amounts of water and change the internal functional groups, so that the adsorption of heavy metal ions is better.

The effects of Ni^{2+} concentration, adsorption time, pH and the amount of adsorbent on the removal of Ni^{2+} from the modified sediments were investigated by static adsorption method. And then take 100ml of a certain concentration of Ni^{2+} water samples, then, weigh a certain amount of roasted sediment into a 250ml Erlenmeyer bottle to close the bottle. Placed in a thermostat at a temperature of 25℃ for 120 min. The supernatant was transferred to a centrifuge tube. Centrifuge at 3000 r / min for 15 min. The supernatant was took out in turn. The concentration of residual Ni2 + in the water samples after adsorption was determined by the method of spectrophotometric method (GB11910-89).In the course of the experiment, parallel and blank control experiments were carried out.
The formula for calculating the saturation adsorption of Ni^{2+} is [9]:

$$q_e = \frac{(c_o - c_e)V}{m} \tag{1}$$

c_0 and c_e are the mass concentrations of metal ions in the solution before and after adsorption(mg/L);V is the volume of heavy metal solution(L);m is the addition quality of adsorbent(g).
The formula for calculating the removal rate of Ni^{2+} is [9]:

$$\eta = \frac{c_0 - C_e}{c_0} \times 100\% \tag{2}$$

c_0 and c_e are the mass concentrations of metal ions in the solution before and after adsorption(mg/L).

IOP Conf. Series: Earth and Environmental Science **64** (2017) 012062 doi:10.1088/1755-1315/64/1/012062

3. Experimental results and discussion

3.1. Effect of initial concentration on adsorption

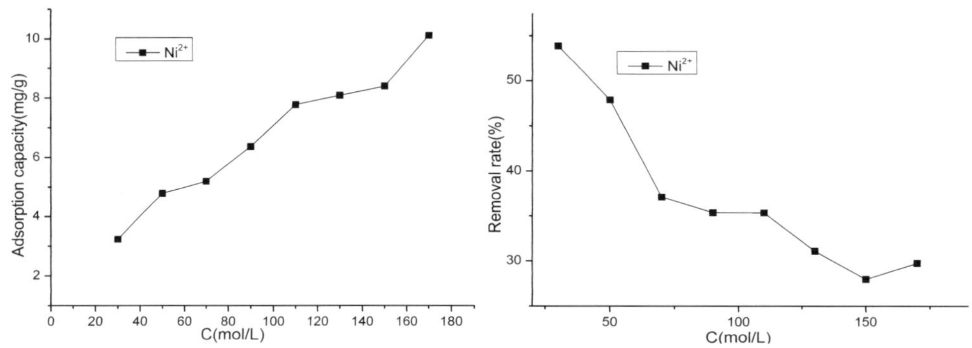

Fig.3 Effect of initial concentration on adsorption

As can be seen from Fig.3: With the increase of ion concentration, the adsorption capacity of modified sediment increased gradually, but the adsorption rate decreased gradually.

This is because the dosage of modified sediment is constant, the surface exchange position is certain, the adsorption capacity is also certain, so the adsorption effect of Ni2+ concentration is much lower than the low concentration of adsorption effect. It can be seen that the modified sediment adsorption of low concentration of wastewater is better, the best range is less than 10mg /L.

3.2. Effect of pH on adsorption

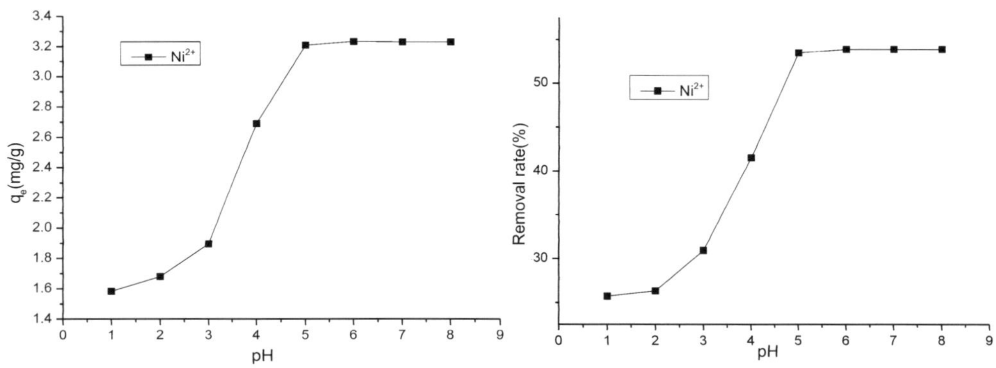

Fig.4 Effect of pH on adsorption

PH is one of the important factors that affect the adsorption and removal rate of heavy metal ions in Wastewater. On the one hand the value of the pH value of the solution directly affects the distribution of the charge on the surface of the adsorbent, thus affecting the combination of ions and adsorbents. As can be seen from the figure, with the continuous increase of the pH value in the system, the removal rate and the adsorption capacity of $Ni2^{+}$ on the whole is gradually increased. When the pH value is low, the adsorption system contains a large number of H+, and attached to the surface of the sediment, and Ni^{2+} formation of intense competition, thus affecting the adsorption of sediment; When the pH value increases gradually, the concentration of H^{+} will gradually decrease, and the competitive relationship between H^{+} and Ni^{2+} will gradually weaken, which is helpful to improve the adsorption capacity of Ni^{2+}.When pH>8, the solution system is mainly OH^{-}present, OH^{-} and heavy metal ions Ni^{2+} precipitation reaction occurs, is not conducive to explore the adsorption effect of adsorbents. Therefore, in weak acid

IOP Conf. Series: Earth and Environmental Science **64** (2017) 012062 doi:10.1088/1755-1315/64/1/012062

and neutral environment is conducive to improve the removal of heavy metals, thus it can be seen that the solution system is the best pH value range of 5 ~ 8, for comprehensive consideration, in the course of the experiment to select pH=7 as the best results.

3.3. Effect of dosage on adsorption

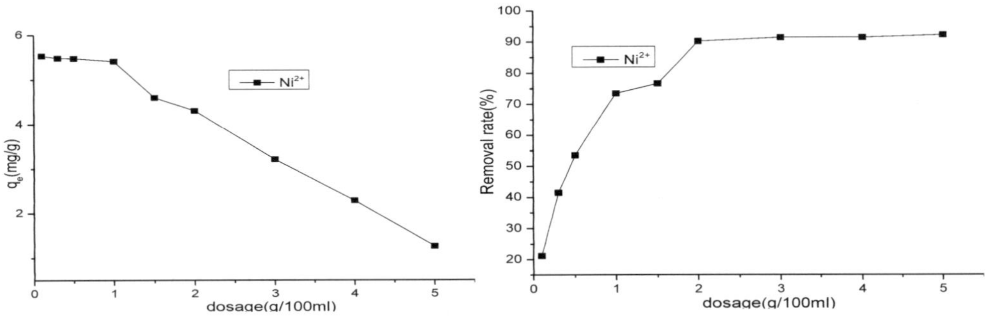

Fig.5 Effect of dosage on adsorption

The dosage of adsorbent is another important factor affecting the adsorption effect. If the dosage of adsorbent is too small, it cannot effectively remove the heavy metal ions in the solution, so it is difficult to meet the emission requirements; if the dosage of adsorbent is too large, it will not only affect the adsorption effect, but also waste the resources and increase the processing cost. It can be seen from the figure that with the increase of the dosage, the removal rate increases gradually, but the adsorption capacity decreases gradually. Comprehensive consideration, the best dosage is 1.0mg/100mL.

3.4. Effect of adsorption time on adsorption

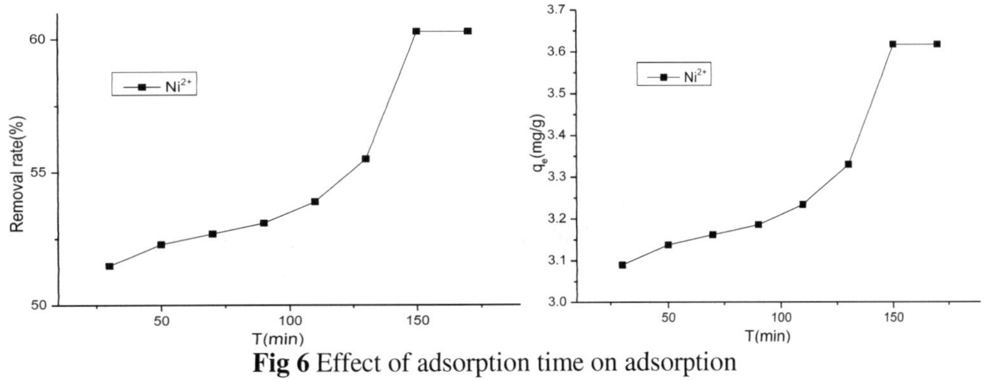

Fig 6 Effect of adsorption time on adsorption

It can be seen from the figure that the sedimentation rate and adsorption capacity of Ni^{2+} modified by high temperature calcination are relatively slow with the increase of time. At 150 min, the adsorption reaches the maximum value and there is almost no analytic phenomenon, which further shows that Ni^{2+} and the adsorption group in the system to maintain a stable adsorption potential, so it is not easy to resolve after adsorption. So the experiment to determine the time is150min.

3.5. Effect of temperature on adsorption

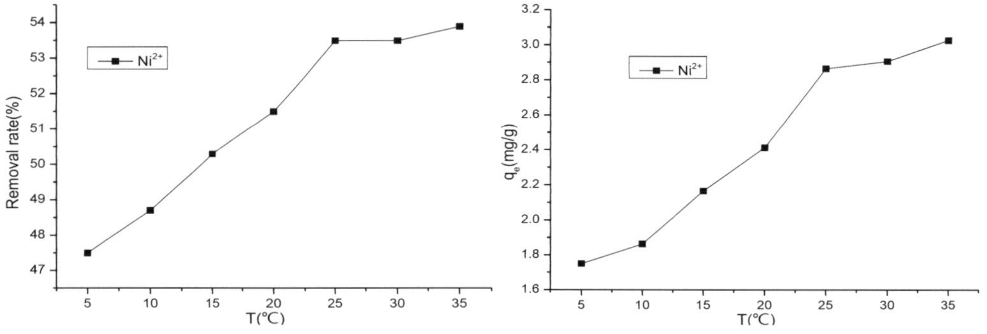

Fig.7 Effect of temperature on adsorption

Since the reaction process has an exothermic and endothermic process, the temperature determines whether the reaction is carried out and the reaction is balanced. Therefore, the temperature is also one of the key factors affecting the adsorption efficiency. It can be seen from the figure that the best temperature is 25℃.

3.6. Adsorption isotherm of modified sediment

Adsorption capacity of high temperature roasting modified sludge adsorption of Ni^{2+} is an important index to measure the sediment adsorption capacity. The adsorption balance between the quantity and the solution in the solid liquid phase system, used to simulate the model of adsorption isotherm and Freundlich isotherm for Langmuir [10].

Langmuir adsorption isotherm: $$\frac{c_e}{q_e} = \frac{c_e}{q_m} + \frac{1}{bq_m} \quad (3)$$

Freundlich adsorption isotherm: $$\lg q_e = \lg k_F + \frac{1}{n} \lg c_e \quad (4)$$

c_e is the mass concentrations of metal ions in the solution after adsorption(mg/L);q_e is adsorption capacity(mg/g);q_m is the maximum adsorption capacity(mg/g);K_F and n are constants.

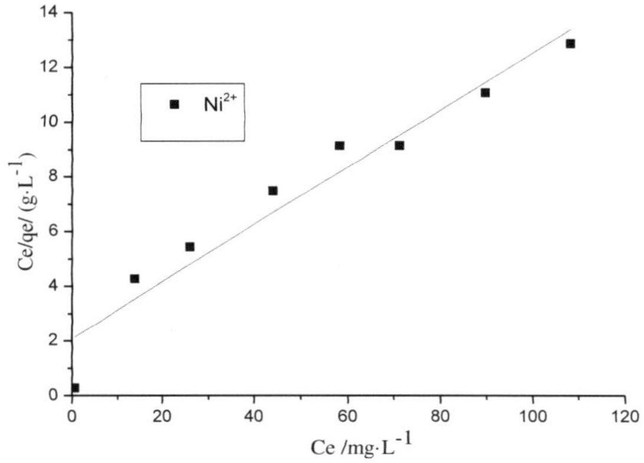

Fig. 8 Langmuir isotherm for adsorption on calcination sediment

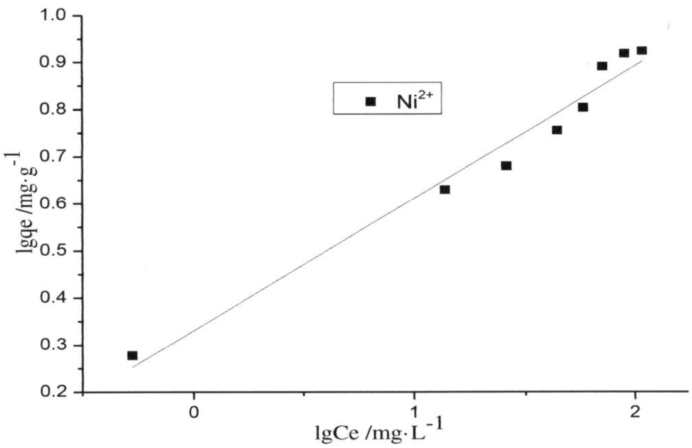

Fig. 9 Freundlich isotherm for adsorption on calcination sediment

Tab.1 Langmuir isotherm and Freundlich isotherm equation parameters for calcination sediment

Heavy metal ion	Langmuir equation			Freundlichequation		
Ni^{2+}	b	q_m	R^2	1/n	K_F	R^2
	0.0462	9.6246	0.9331	0.2880	1.9984	0.9679

The adsorption of Ni^{2+} by high temperature roasting modified sediment is in accordance with Langmuir and Freundlich adsorption isotherm equation. The $R^2$of Freundlich equation is larger than Langmuir equation, we can see that Freundlich equation fitting effect is better than Langmuir equation fitting effect. This indicates that the adsorption of N^{i2+} at high temperature roasting sediment is carried out simultaneously between monolayer and multilayer. The 1/n of the Freundlich equation is related to the adsorption strength and the value is less than 1, which proves that the adsorption process is easier.

3.7. Adsorption kinetics mechanism

Through the study of adsorption kinetics, the adsorption mechanism can be studied more deeply. The relationship between the adsorption capacity and the adsorption time of adsorbent can be expressed by the adsorption kinetic model.At present, many models have been established in the adsorption kinetics, such as, pseudo first order kinetic model, pseudo two order kinetic model, Elvish equation model, intraparticle diffusion equation model, Weber kinetic model, etc.The pseudo first order kinetic equation, pseudo two order kinetic equation, Elovich equation and the intraparticle diffusion equation are widely used in the study of adsorption mechanism[11-12].

Lagergren pseudo first order adsorption kinetics equation:

$$\ln (qe-qt) = \ln qe-k1t \quad (5)$$

Laguerre pseudo two order adsorption kinetics equation:

$$\frac{t}{q_t} = \frac{1}{k_2 q_e^{\ 2}} + \frac{t}{q_e} \quad (6)$$

q_e is the amount of adsorption in equilibrium(mg/g);q_t is the amount of adsorption at time t(mg/g);t is the adsorption time(min);k_1 and k_2 are constants.

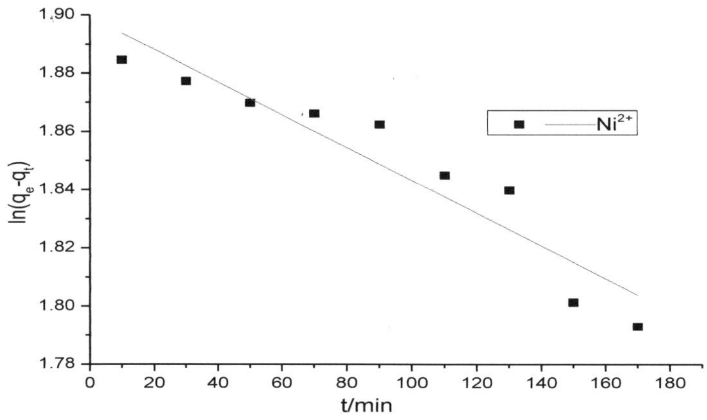

Fig. 10 Pseudo first order kinetic fitting curve

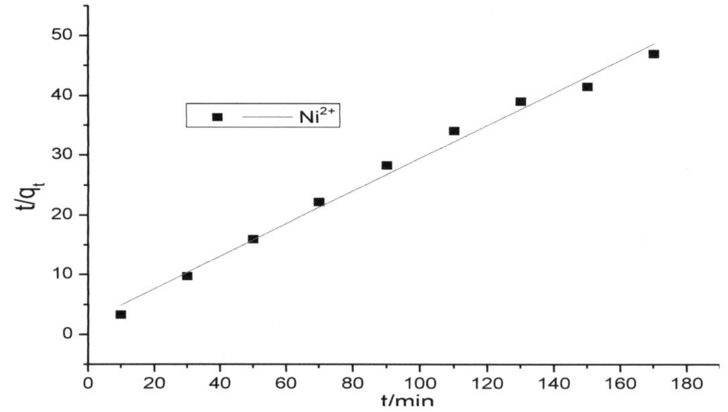

Fig. 11 Pseudo two order kinetics fitting curve

Tab.2 Kinetic parameters

Heavy metal ion	Pseudo first order kinetic equatio			Pseudo two order kinetics equation		
	K_1	$q_e(mg/g)$	R^2	K_2	$q_e(mg/g)$	R^2
Ni^{2+}	0.0006	6.6906	0.8812	0.0349	3.6563	0.9896

From the figure and table we can know that the adsorption curve and the fitting curve have some deviation, which indicates that the pseudo-first-order kinetics is not the main factor controlling the adsorption process during the adsorption process, because the main factors that affect the adsorption rate during the adsorption process are not the diffusion of the liquid film. The pseudo two order kinetic coefficient R^2 is closer to 1.The results show that the adsorption of Ni^{2+} in the high temperature roasting sediment is more in accordance with the pseudo two order kinetics model. The chemical bond is the main factor affecting the adsorption of the adsorbent, so the adsorption process is mainly chemical adsorption. According to table 2, the equilibrium adsorption amounts of Ni^{2+} was 3.6563 mg/g, and the difference between the measured values is less. Adsorption effect is ideal.

4. Conclusion

Compared with the original sediment, high temperature roasting to change the internal functional groups of sediment and produce new groups and chemical bond conversion and generation, and the specific surface area also increased several times, increasing the activity of adsorbents, is conducive to adsorpt heavy metals. The high temperature roasting sediment adsorption of Ni^{2+} data were fitted with Langmuir and Freundlich adsorption isotherm equation, but the fitting effect is better with the Freundlich adsorption isothermal equation, indicating that the adsorption between single and multilayer. The adsorption kinetics of Ni^{2+} adsorption on high temperature roasting sediment is fitted by adsorption kinetics equation. The adsorption process follows Laguerre pseudo-second adsorption kinetics equation, which indicates that chemical adsorption is the main chemical bond, and the equilibrium adsorption capacity of Ni^{2+} was 3.6563mg / g, and the adsorption effect was ideal.

Acknowledgments

Talent research started--Study on the Performance of Typical Heavy Gold Ion in Modified Sediment of Dianchi Lake Sediment (14119031)

References:

[1] Werther J, Ogada T. Sewage sludge combustion[J]. Progress in energy and combustion science, 1999, 25(1): 55-116.
[2] Show K Y, Lee D J, Tay J H, et al. Lightweight Aggregates from Industrial Sludge–Marine Clay Mixes[J]. Journal of Environmental Engineering, 2005, 131(7): 1106-1113.
[3] Jones K W, Feng H, Stern E A, et al. Dredged material decontamination demonstration for the port of New York/New Jersey[J]. Journal of hazardous materials, 2001, 85(1): 127-143.
[4] Dalton J L, Gardner K H, Seager T P, et al. Properties of Portland cement made from contaminated sediments[J]. Resources, conservation and recycling, 2004, 41(3): 227-241.
[5] Tay J H, Show K Y. Resource recovery of sludge as a building and construction material—afuture trend in sludge management[J]. Water Science and Technology, 1997, 36(11): 259-266.
[6] Wang JL, Chen C. Biosorption of heavy metals by Saccharomyces cerevisiae: A review[J]. Biotechnology Advances, 2006,24(5): 427-451.
[7] Droussi Z, D'orazio V, Provenzano M R, et al. Study of the biodegradation and transformation of olive-mill residues during composting using FTIR spectroscopy and differential scanning calorimetry[J]. Journal of hazardous materials, 2009, 164(2): 1281-1285.
[8] Cao X, Ma L, Gao B, et al. Dairy-manure derived biochar effectively sorbs lead and atrazine[J]. Environmental Science & Technology, 2009, 43(9): 3285-3291.
[9] Manzoor Q, Nadeem R, Iqbal M, et al. Organic acids pretreatment effect on Rosa bourbonia phyto-biomass for removal of Pb (II) and Cu (II) from aqueous media[J]. Bioresource technology, 2013, 132: 446-452.
[10] Balsamo M, Di Natale F, Erto A, et al. Arsenate removal from synthetic wastewater by adsorption onto fly ash[J]. Desalination, 2010, 263(1): 58-63.
[11] Shek T H, Ma A, Lee V K C, et al. Kinetics of zinc ions removal from effluents using ion exchange resin[J]. Chemical Engineering Journal, 2009, 146(1): 63-70.
[12] Ho Y S, Ng J C Y, McKay G. Removal of lead (II) from effluents by sorption on peat using second-order kinetics[J]. Separation Science and Technology, 2001, 36(2): 241-261.

Numerical simulation and analysis of characteristics of atmospheric diffusion in coastal area of a site

Zongzhen Yang[1], Xiaofeng Zhang[1], Xinjie Bao[1] and Shuyang Chen[2,*]

[1]Suzhou Nuclear Power Research Institute, Suzhou, China
[2]Suzhou Meteorological Bureau, Suzhou, China

*Corresponding author e-mail: yangcm1229@qq.com

Abstract. A Weather Research and Forecasting Model (WRF) and random walk model have been used to numerical simulate and analysis the characteristics of meso-micro scale wind field and atmospheric diffusion in coastal area of a sit. The results indicated that the atmospheric dispersion in this area is determined both by synoptic scale system and land-sea breeze circulation. Spatial variation of wind field is not obvious in synoptic scale system situation, which leads to the straight dispersion plume. Temporal variation of wind field, particularly the wind direction transition caused by transient of land-sea breezes is obvious. The method of random walk simulation better reflects the characteristics of the air pollutants transportation and diffusion. The atmospheric diffusion parameters obtained with numerical simulation experiment may reflect well the characteristics of the air diffusion in local area.

1. Introduction

The diffusion and transport of air pollutants are affected by local complex terrain and meteorological processes to a large extent. Topography influence includes the disturbance effect of complex terrain to airflow, letdown flow of mountain, sea breeze circulation caused by land-sea difference, and thermal inner boundary layer near coast [1, 2]. Meteorological processes mainly include the synoptic circulation and meso-and-micro scale weather system. Taking into account the complexity of the problems and the local characteristics of atmospheric diffusion, it is very necessary to analyze the meso-and-micro scale wind field and the transport and dispersion of pollutant which is for the actual condition. A lot of work has been done in this field in China, especially power plants and such projects. Various research methods have been used, such as atmospheric boundary layer sounding, turbulence observation, the field tracer test of atmospheric dispersion [3, 4], wind tunnel experiment [5], numerical simulation by [6, 7]. Subject to the weather uncertainty, the experimental conditions and the theoretical limitation, comprehensive conclusion cannot be reached using the former four methods. Numerical simulation has more advantages in studying the characteristics of the planetary boundary layer and atmospheric diffusion, which is good complement for atmospheric experiments [8].

The NCEP/NCAR global reanalysis data and Weather Research and Forecasting Model (WRF) are used here to simulate the meso-micro scale wind field of coastal area of a site. The wind field will be used to drive the meteorological fields, and simulate the transport and diffusion of the atmospheric pollutant in the area of the site, then analyze the atmospheric diffusion characteristics.

2. Numerical Simulation of WRF Program

NCEP FNL (Final) Operational Global Analysis data are on 1-degree by 1-degree grids prepared operationally every six hours, which is used as initial and boundary conditions in WRF. In order to show the meso-micro scale wind field more accurately, four two-way interactive nested domains were selected when WRF running. The spatial grid resolution of 1km(d04). 3km(d03). 9km(d02) and 27km(d01) are used respectively, and outermost domain covers the land and sea. The output result is the hourly wind field in the typical month of January, April, July and October, which represent the four seasons respectively. Terrain data with the resolution of USGS, and land cover type data with the resolution of 1km×1km are used here. The terrain nearby the site is showed in Fig.1.

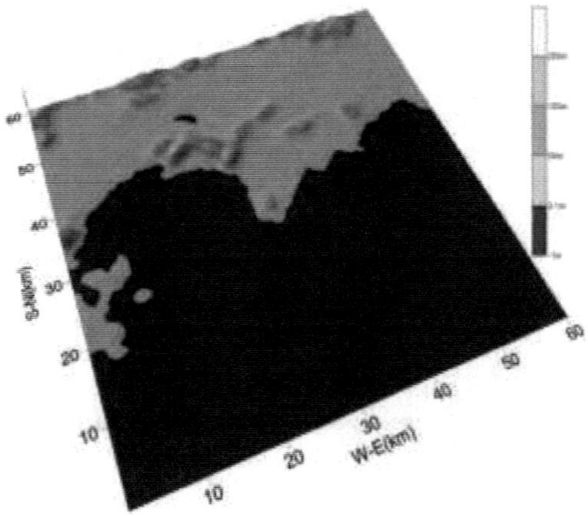

Figure 1. The terrain nearby site

3. Simulated wind field result by WRF

WRF simulated the meso-micro scale three-dimensional wind field at the height of 10m in the area of the site of the typical months. Based on the simulated result, the general characteristics, structures, and the evolution law were gotten as follows.

3.1. Continuous wind field of synoptic system

The system of wind field in spring is the transition field of winter and summer circulation system. The dominant wind direction is mainly east, the wind field distribution is more uniform. The system wind field in autumn as same as spring but the dominant wind direction is north and east. Affected by summer monsoon, southerly wind is dominant in summer. And in winter, Affected by Mongolian high, northeasterly wind is dominant with a higher wind velocity. The typical wind fields of each season are showed in Fig.2.

3.2. Mesoscale local circulation

The local circulation in July 18th is showed in Fig.3. The offshore wind from land to sea was weak at 02am on 18th, and improved until 06am. At around 10am, land breeze transformed to sea breeze gradually. At 02pm, temperature of land rose and the shoreward wind was obvious with increased velocity and land area affected by sea breeze. From 11pm to the next morning, the offshore wind formed again near surface.

Fig.4 shows the North-South vertical cross section of the wind field on July 18th. At 02am, sea breeze circulation started to form. Sea breeze circulation began to form at 02am on 18th July, and low-level offshore winds near the coast began and developed more into sea, and it lasted for about 4 hours. After 10am, affected by land thermal effects and background circulation, there were southerly winds all over the land and obvious upward flows near windward slope. And then during the night, land-sea thermal structure transferred, and northerly winds began to be dominant at low level.

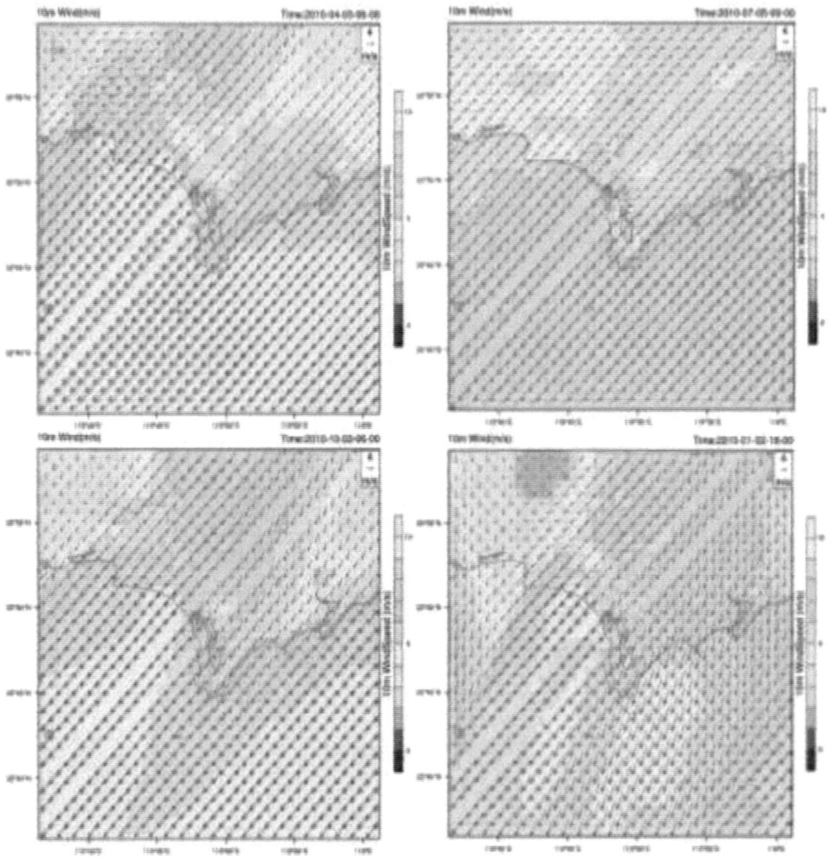

Figure 2. The continuous wind field of the typical synoptic system

IOP Conf. Series: Earth and Environmental Science **64** (2017) 012063 doi:10.1088/1755-1315/64/1/012063

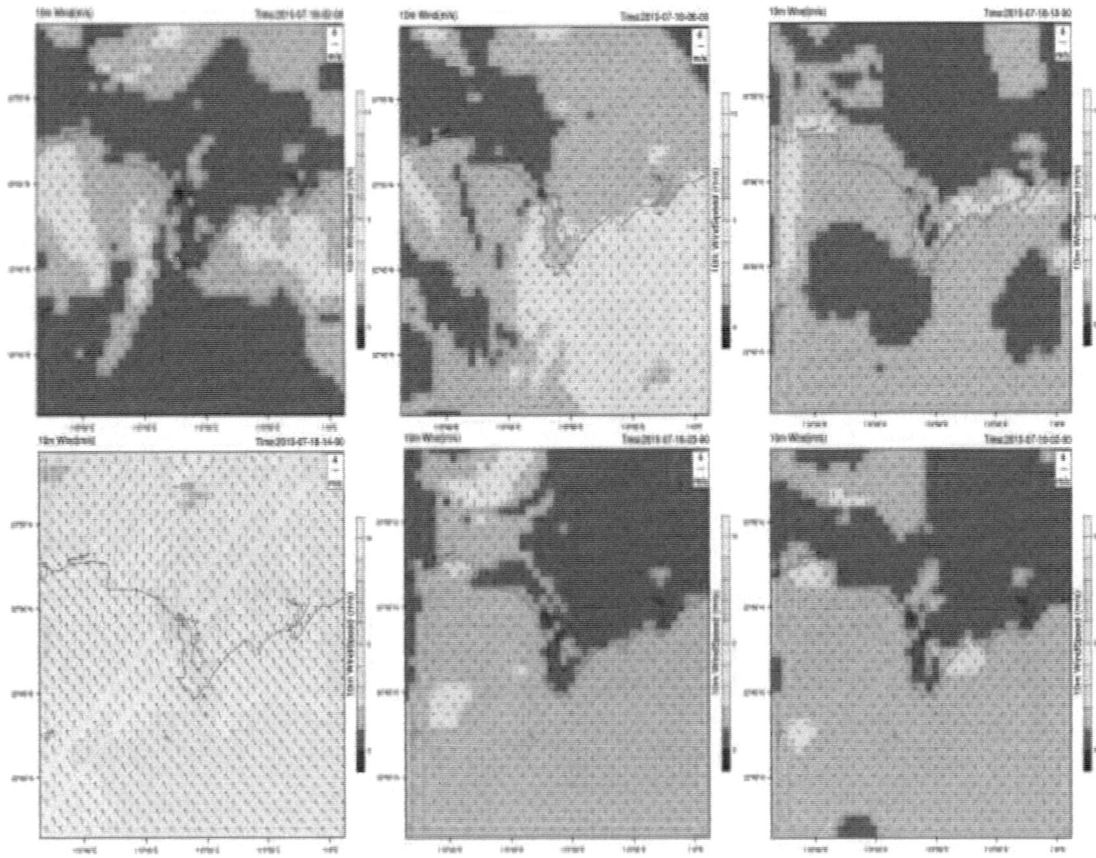

Figure3. The diurnal variation characteristics of the local circulation wind field in Summer

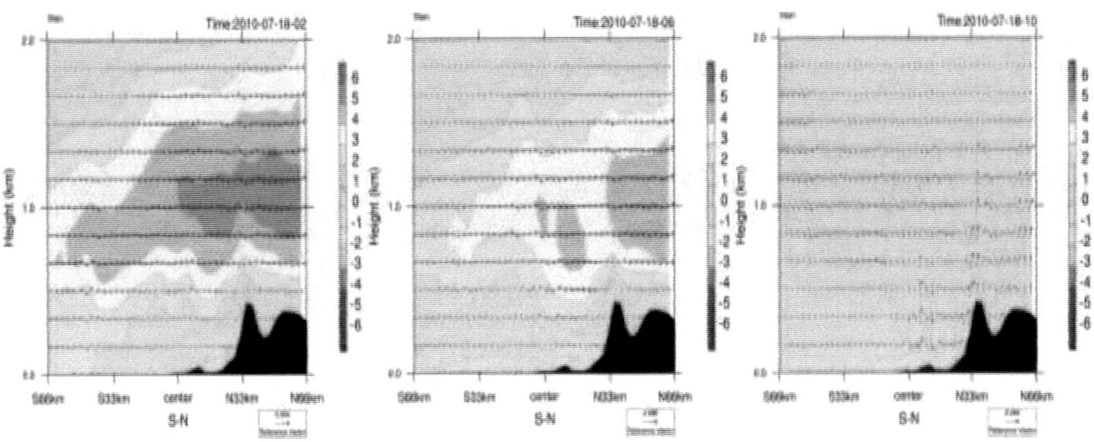

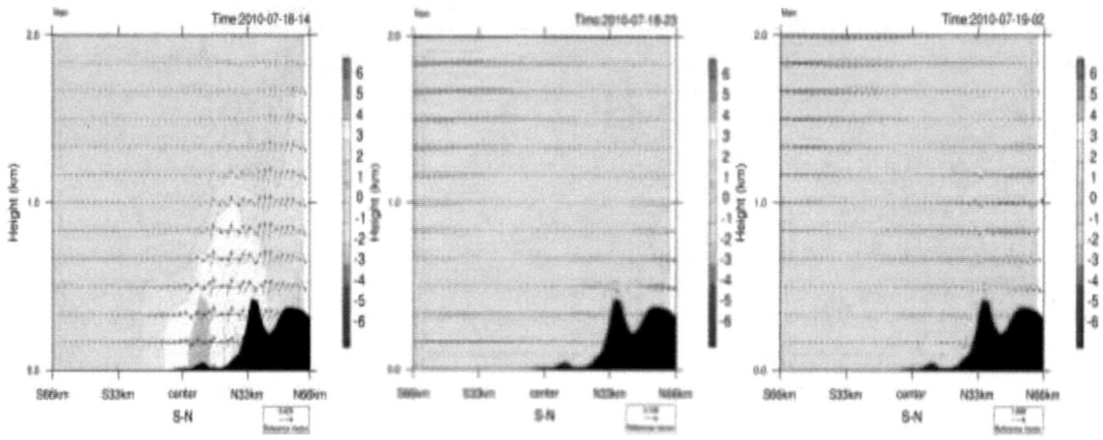

Figure 4. The diurnal variation characteristics of the local circulation vertical wind field in Summer

4. Simulation and analysis of pollutant diffusion

Radioactive pollutants diffusion of the site was simulated by random walk model. The initial wind field of a numerical model was the simulated result of WRF innermost nesting modelling, and the low level wind field had been modified by mass conservation law. The parameters of random walk model were settings as follows. 5km north of the site was set as modelling centre, modelling area was 40km*40km, the horizontal solution was 100m, vertical solution was 10m, and modelling height was up to 800m. Pollution source was set on the height of 10m which was released continuously. The turbulence intensity was used here for model turbulence parameterization, which accorded with the results of the turbulence observation test around the site.

Pollutants diffusion and transportation was mainly affected by background wind field. Northerly winds prevail in winter and autumn, pollutants transport and diffuse from the source to the sea. Easterly winds appear more in spring, pollutants transport to the west. Southwest winds transport the pollutants towards the land. Transportation characteristics of atmospheric pollutants of the seasons were showed in Fig.5. Affected by the mesa-micro scale wind field, transportation of air pollutants appeared obviously diurnal variation which is showed in Fig.6.

Due to the flat terrain of the site, pollutants diffusion appears uniformly distributed. There are east-west hills located 20km north of the site. It caused the higher concentration of pollutants neat the windward slope, and lower concentration near leeward slope. Fig.7 chose two examples whose pollutants were transported to NE and NNW respectively. It was more flat in the direction of NE than NNW. In the former case, the distribution was fairly uniform. And it was showed that there were several hills located on the 12km NNW of the source, the pollutants concentration rose on windward slope much more than leeward slope.

IOP Conf. Series: Earth and Environmental Science **64** (2017) 012063 doi:10.1088/1755-1315/64/1/012063

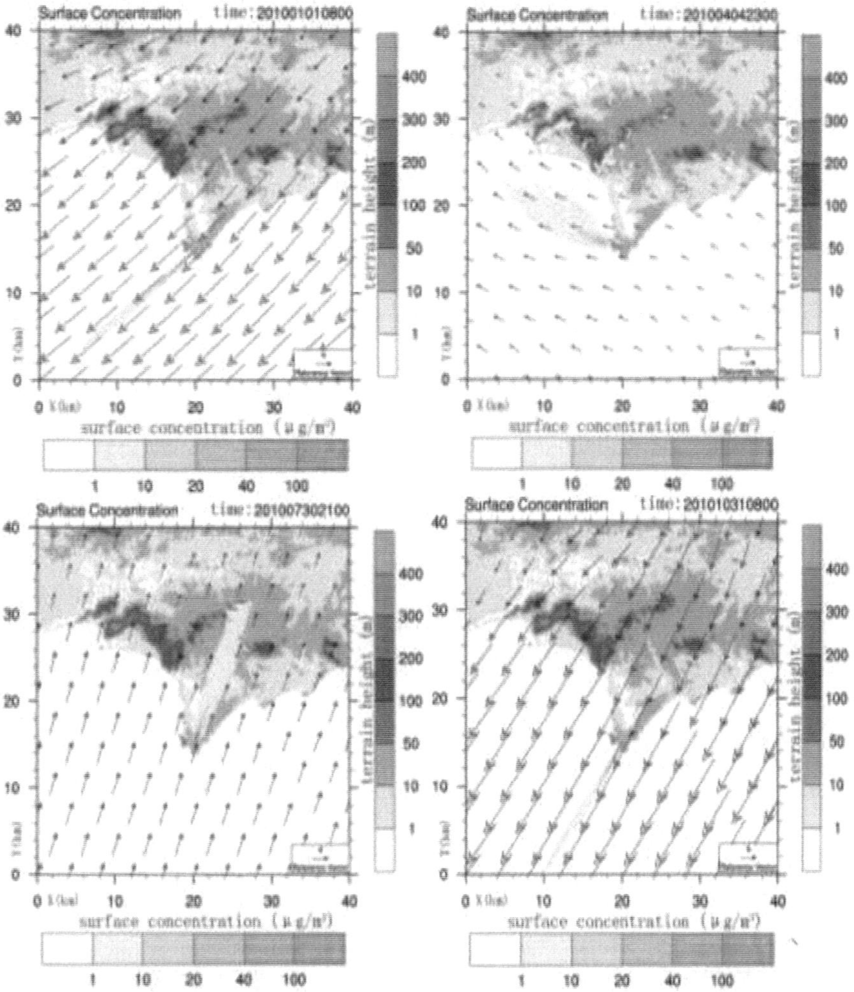

Figure 5. Simulated surface concentration with the system wind circulation in each season

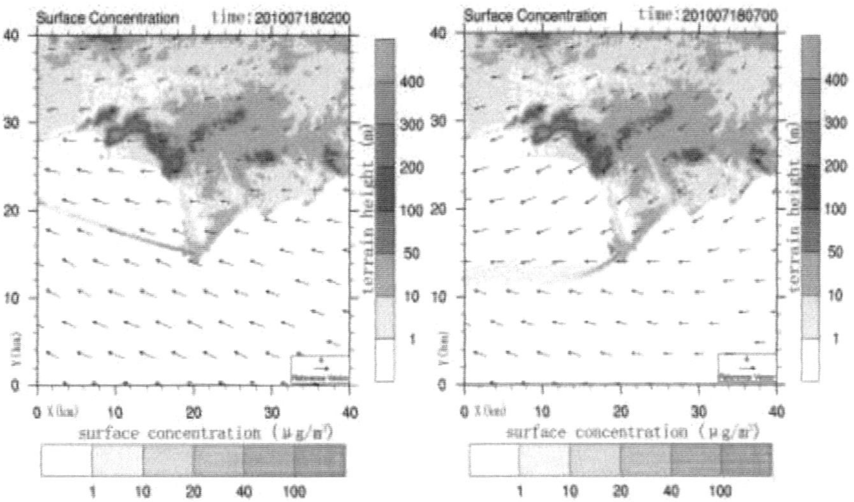

IOP Conf. Series: Earth and Environmental Science **64** (2017) 012063 doi:10.1088/1755-1315/64/1/012063

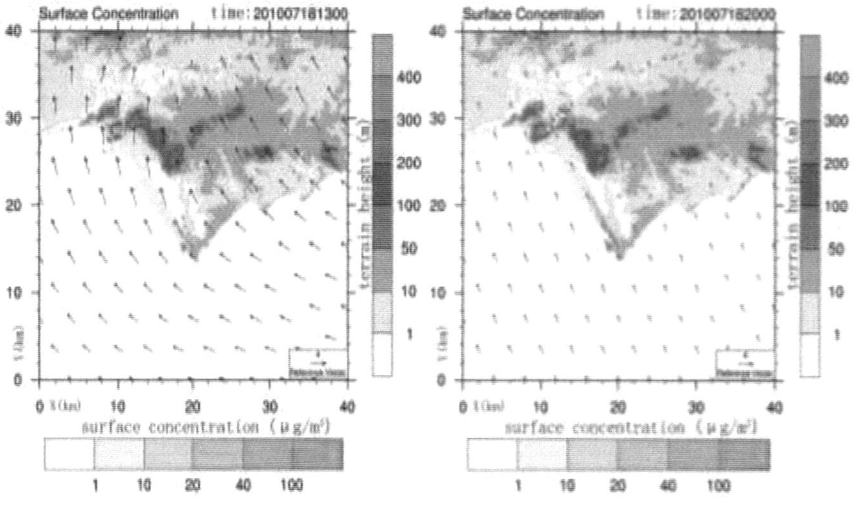

Figure 6. Simulated surface concentration with Summer local wind circulation

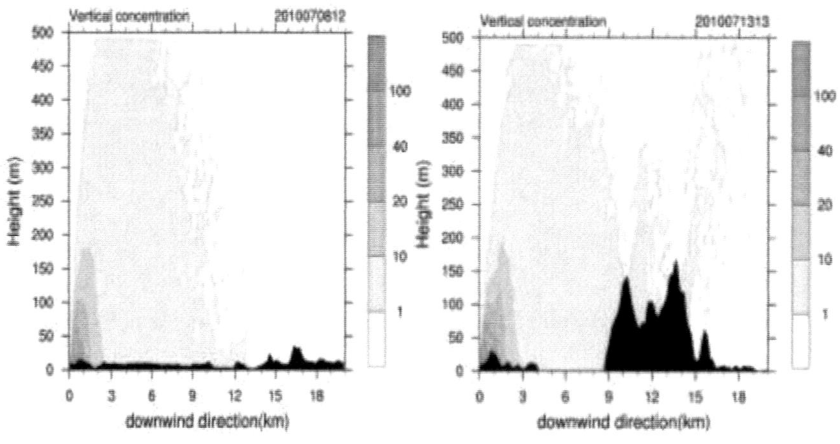

Figure 7. Vertical cross-section concentration fields and corresponding terrain feature

5. Diffusion parameters

The site locates on a peninsula sticking out into the sea. The coastline near the site appears from east to west, so southerly winds could be seen as shoreward winds and northerly winds as offshore winds. We focused on pollutants diffusion on the land and under the southerly winds when wind was weak. Closed several examples of various kinds of stability and statistically analyzed their diffusion parameters. The diffusion parameters were defined here as (1) and (2).

$$\sigma_y = ax^b \quad (1)$$

$$\sigma_Z = cx^d \quad (2)$$

The Table 1 shows the coefficients of the diffusion parameters obtained from numerical model.

Table 1. The coefficients of diffusion parameters

Coefficient \ Stability	A	B	C	D	E	F
a	2.094	2.061	1.710	1.560	1.285	1.238
b	0.706	0.681	0.710	0.676	0.700	0.679
c	1.585	1.215	1.163	1.032	0.896	0.993
d	0.520	0.551	0.529	0.494	0.468	0.402

Fig.8-9 showed the lateral normalized concentration distributions compared with Gaussian distributions of different kinds of stability. The concentration distribution obtained from the experiment was according with Gauss normal distribution under the unstable conditions. And under the stable conditions, the distribution was below the Gauss normal distribution curve, especially in near distance. Since background circulation was relatively weak when atmospheric stratification was usually stable, the distribution of atmospheric pollution was mainly affected by the terrain and underlying surface, so the distribution was more complex. The vertical distribution of the downwind of the source varied greatly, especially the region nearby the site, which was according with the general law of atmospheric pollutants dispersion. Generally speaking, diffusion parameters obtained from numerical model experiments reflected well the characteristic of local atmospheric dispersion.

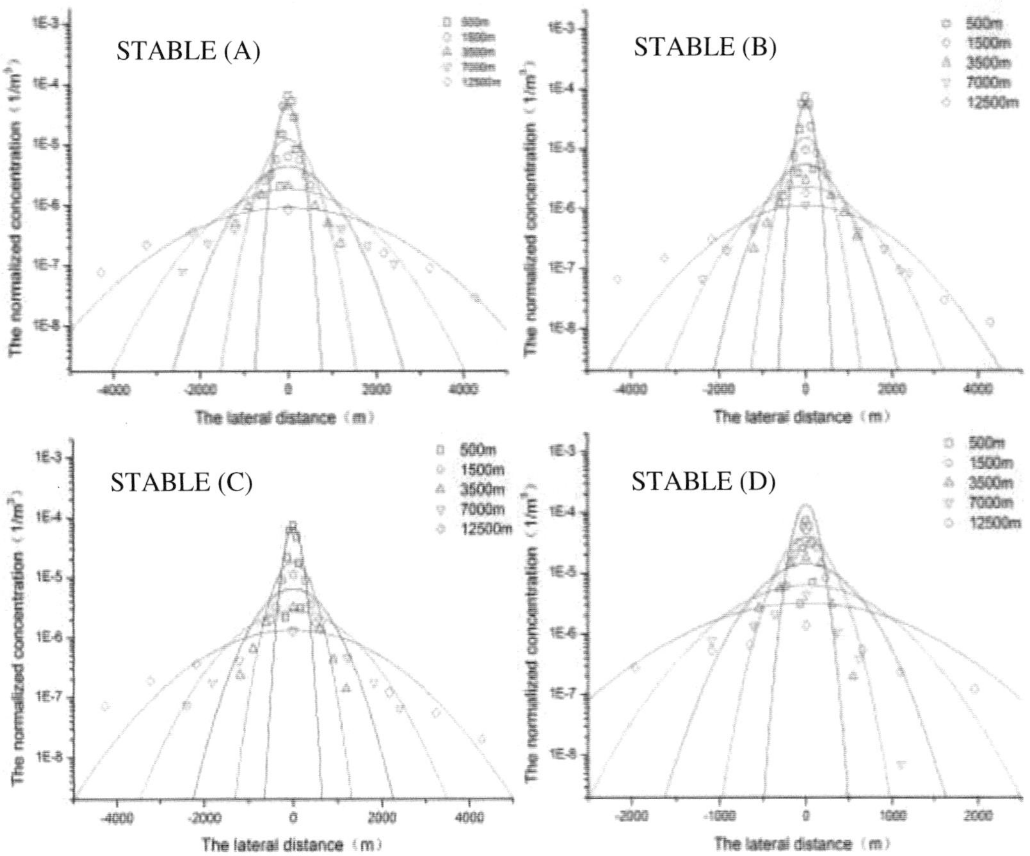

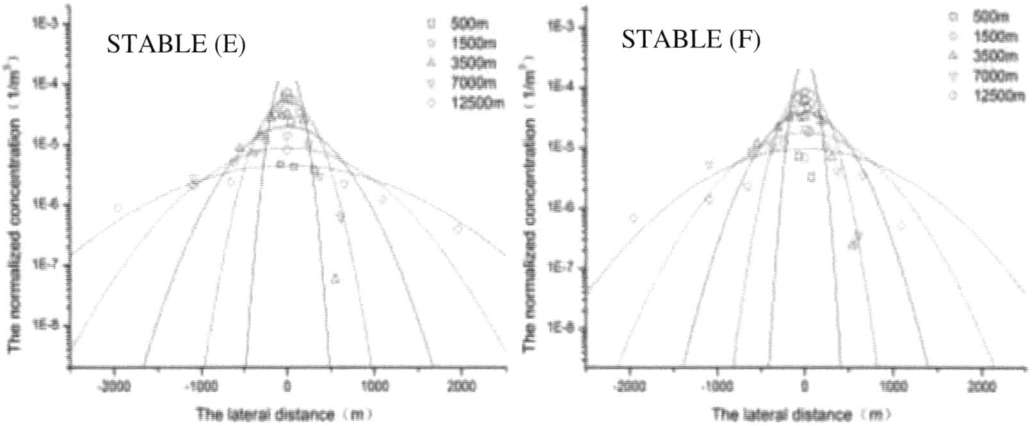

Figure 8. The lateral normalized concentration distribution compared with Gaussian distribution (lines mean Gaussian distribution and scatters mean numerical results in different distances)

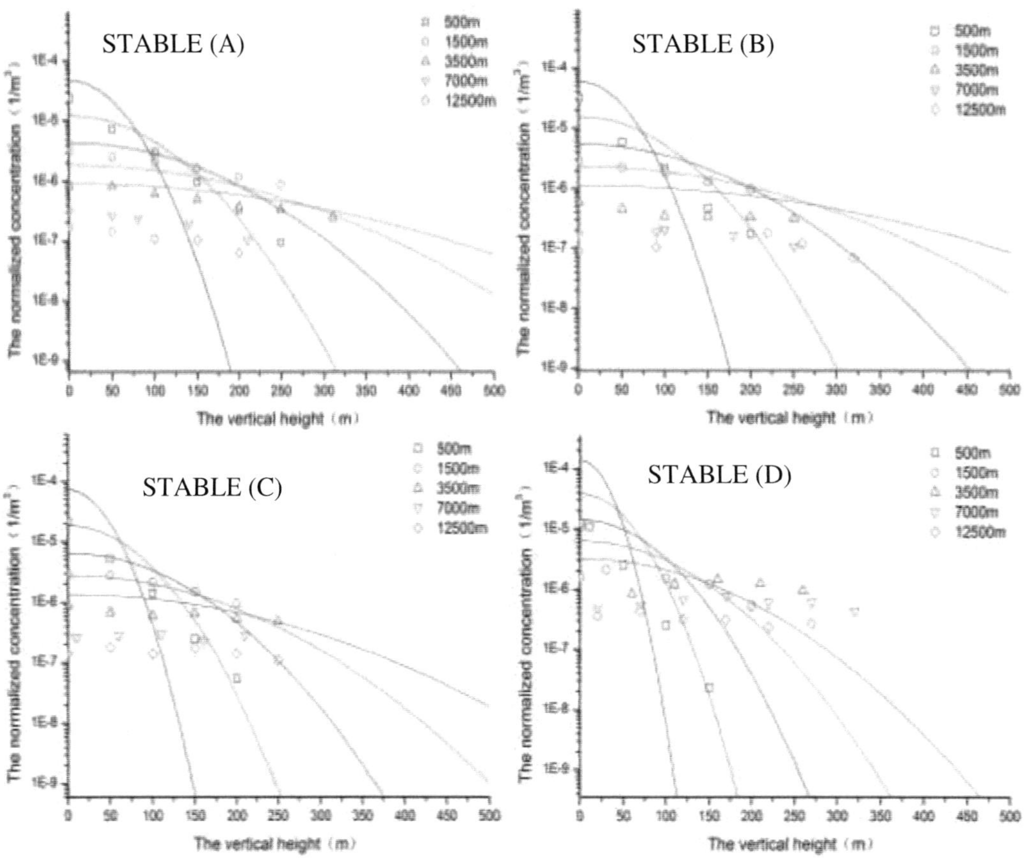

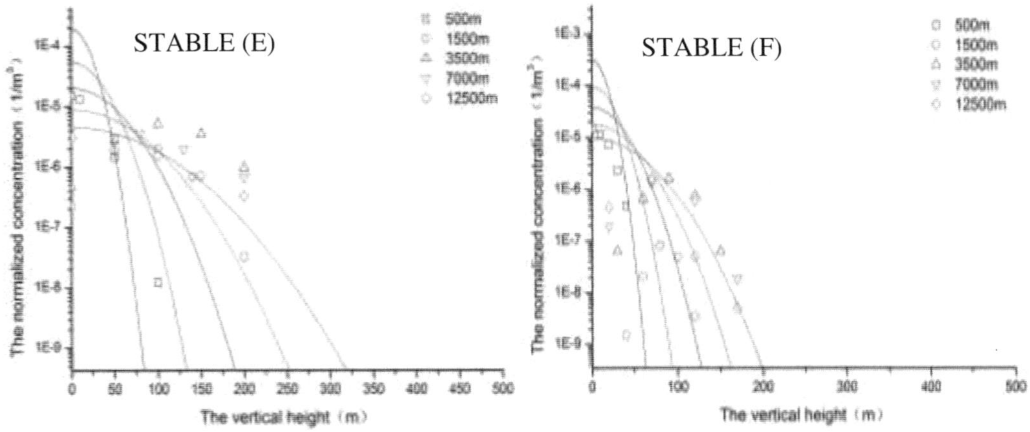

Figure 9. The vertical normalized concentration distribution compared with Gaussian distribution (lines mean Gaussian distribution and scatters mean numerical results in different distances)

6. Conclusion and discussion

In this paper, the wind field of boundary layer and characteristic of atmospheric dispersion in coastal area of a site were simulated and analyzed using WRF and RWM. We find that the site is located at coastal area, meso-micro scale wind field can be broadly classified as systematic wind field and local topographic circulation. The site is affected by synoptic scale system for most of the time, but meso-micro scale circulation still possesses apart of proportion. Under the influence of background atmospheric circulation, pollutant diffusion has significant seasonality characteristics. And under the influence of meso-scale circulation, the transport and diffusion of air pollutants shows diurnal variation, especially in the case of sea breeze circulation. Because the terrain near the site is relatively flat, diffusion of pollutants is relatively evenly. But there are east-west hills located 12km north of the site, which causes higher concentration of pollutants near the windward slope, and lower concentration near leeward slope.

The distribution of air pollutants concentration obtained through numerical simulation accords with Gauss normal distribution, and atmospheric diffusion parameters obtained through numerical model reflect the atmospheric diffusion ability to a certain extent.

The simulation results of wind field reveal the true situation of atmospheric flow field rather well. When simulating surface-layer atmosphere over complex terrain, this numerical simulation optimized the simulated low-level wind field with taking mass conservation law into account. In the research in the future, various physical processes of the WRF model should be focus on, especially optimization and analysis of parameterization scheme for boundary layer simulation, and data assimilation of model initial boundary meteorological field as well, so as to get more accurate results.

References

[1] Cionco R M, Kampe W, Biltoft C, et al. An Overview of MADONA: A multinational field study of high-resolution meteorology and diffusion over complex terrain. Bull. Amer, Meteor, Soc., 1999, 80: 5~20.

[2] Neff W D, King C W, The accumulation and pooling of drainage flows in a large basin. J, Appl. Meteor., 1989, 28: 518~529.

[3] Yongyi Zhang, Jiangyu Yan, Weiqiang Wei. Sulfur Hexafluoride-based Atmosphere Diffusion Trace Experimental Study in Complex Topography. Radiation Protection, 2000,20(03):7-14.

[4] Erbang Hu, Rentai Yao, Zhanrong Gao, et al. Calculation of Atmospheric Dispersion Factor for Accident Release from Coastal Nuclear Power. ACTA SCIENTIAE CIRCUMSTANTIAE, 1999, 19(5): 471~477.

[5] Rentai Yao, Qingdang Qiao, Hongwei Hao. A Wind Tunnel Experimental Study on

Atmospheric Dispersion for Tian Wan Nuclear Power Plant. Radiation Protection, 2001, 21(2): 106~114.

[6] Meigen Zhang, Zhiwei Han, Xiaoen Lei. A Numerical Study of Atmospheric Dispersion in the Area of Nuclear Power Plant Dayawan.Climatic and Environmental Research, 2000, 5(1): 90~96.

[7] Ling Kang, Xuhui Cai, Zhiyuan Wang, et al. Numerical Analysis and Simulation of Characteristics of Atmospheric Diffusion in Coastal Area of Zhangzhou, Fujian.Acta Scientiarum Naturalium Universitatis Pekinensis, 2011, 47(1): 71~78.

[8] Zongzhen Yang, Chunhua You, Xinjie Bao. Characteristics of a Costal Site's Atmospheric Boundary Layer and Atmospheric Diffusion: A Numerical Simulation and Analysis. Radiation Protection, 2015,35(1):9~19.

IOP Conf. Series: Earth and Environmental Science **64** (2017) 012064 doi:10.1088/1755-1315/64/1/012064

Application of BIM technology in green scientific research office building

Xin Ni[a], Jianhua Sun, Bo Wang

China United Northwest Institute for Engineering Design and Research Co., Ltd, Xi'an 710082, China

[a]258503915@qq.com

Abstract. BIM technology as a kind of information technology, has been along with the advancement of building industrialization application in domestic building industry gradually. Based on reasonable construction BIM model, using BIM technology platform, through collaborative design tools can effectively improve the design efficiency and design quality. Vanda northwest engineering design and research institute co., LTD., the scientific research office building project in combination with the practical situation of engineering using BIM technology, formed in the BIM model combined with related information according to the energy energy model (BEM) and the application of BIM technology in construction management stage made exploration, and the direct experience and the achievements gained by the architectural design part made a summary.

1. Features of building project design

The Scientific Research Office Building of China United Northwest Institute for Engineering Design & Research Co., Ltd. is designed with novel shape, complete functions, diverse and varying spaces, to become a green and energy-saving building on the whole. The project plans to occupy a total land area of 39.093mu (1mu=666.67m^2), and a total floor area of 63,644m^2. The design of the project has the following features:

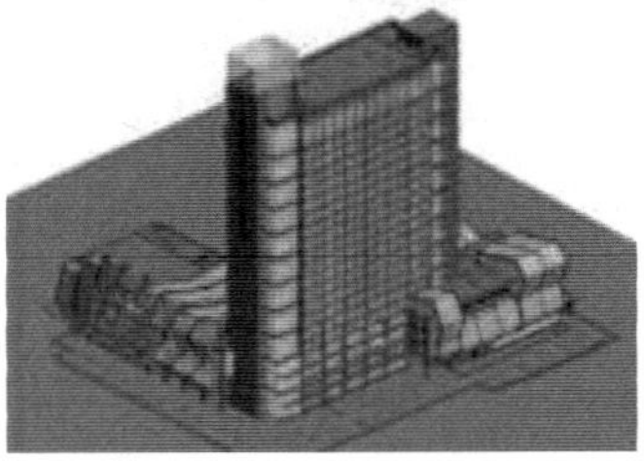

Figure 1. General View of BIM Model

1). Construct and improve the BIM model for all professions gradually from the beginning of design stage to satisfy the needs of applying the BIM technology in different stages.

2). Write the energy information into the BIM model to convert it into an integrated "Energy Model" for analyzing and assessing all kinds of green building indexes.

3). Combine the high-precision BIM model with iTWO software to realize the real-time and interactive management of the project, including progress reporting of key schedule nodes, management of request for payment, and management of application for change at construction site, etc.

2. Establishment of building energy model (BEM) based on BIM model and energy information and data

The BEM model aims to create the electronic energy file of the project with computer technology, and calculate its operations and consumptions related to energy on this basis, so as to simulate the actual operation of the project. The model normally contains the detailed data including geographical and meteorological information, energy properties of components, lighting, temperature control and circulation system, etc. As an ideal carrier, the BIM model classifies all detailed data and implant them category by category into the related materials and components of the project, so as to construct an accurate digital energy model, which provides great convenience and advantages for simulation computing.

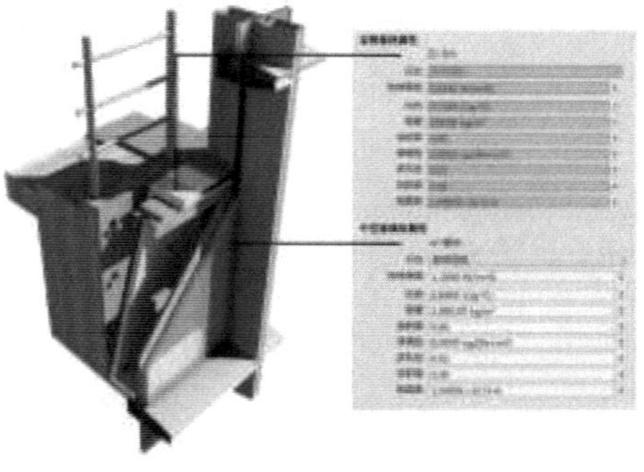

Figure 2. Schematic Diagram of Implanting Thermal Properties of Components

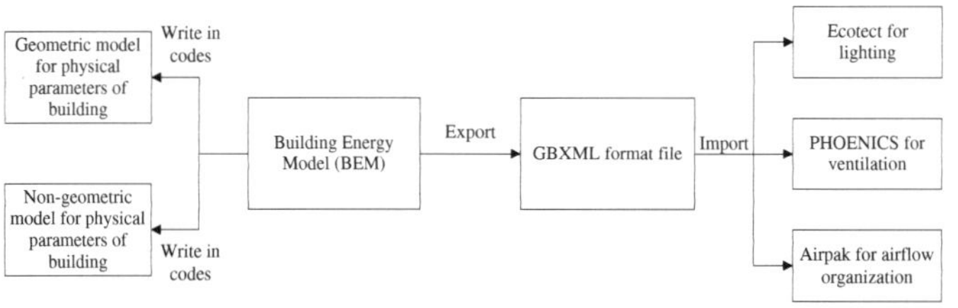

Figure 3. Schematic Diagram of Work Process Flow

A scientific research office building involves complex functions and varying internal spaces, so computer simulation is often utilized to analyze and assess its green building indexes. The common solution is to building models separately with different analysis software (Ecotect for indoor natural lighting analysis simulation, PHOENICS for indoor natural ventilation simulation, PHOENICS for outdoor wind environment simulation and Airpak for indoor airflow organization simulation) for simulation analysis. However, the accuracy of analysis results is often unsatisfying due to the differences between models. In the design, energy information can be written into the BIM model to generate an energy model (BEM), and GBXML format file suitable for all kinds of green analysis software is exported for the purpose of one model for all and accurate information.

The analysis results by importing GBXML format file into different simulation analysis software are presented as follows:

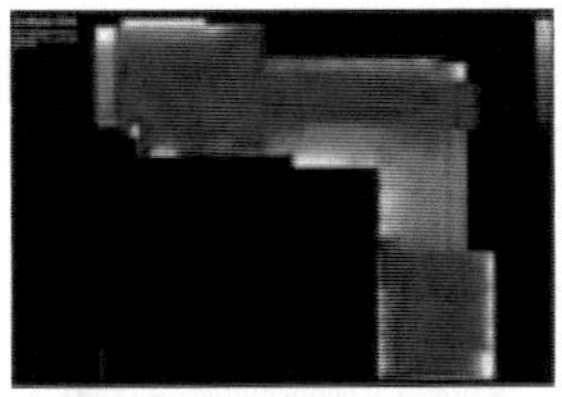

Figure 4. Imported into Ecotect for Lighting Simulation Analysis

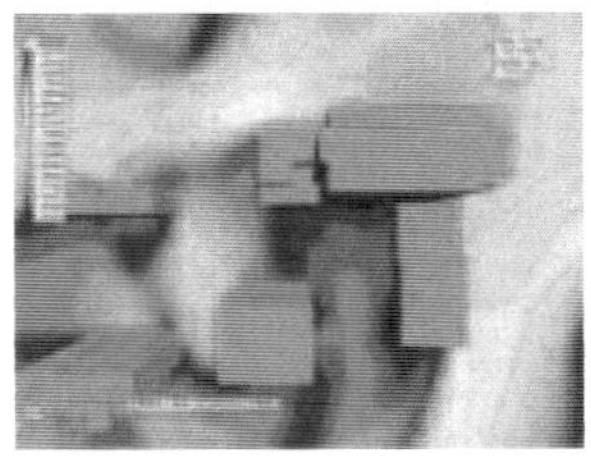

Figure 5. Imported into PHOENICS for Lighting Simulation Analysis

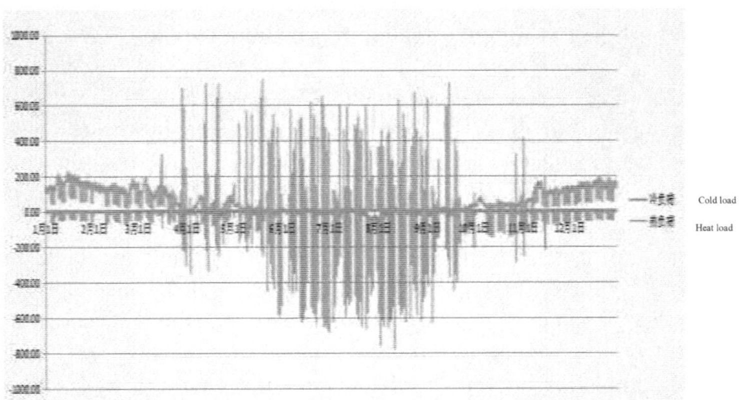

Figure 6. Imported into Hongye HCLoad for Cold and Heat Load Analysis

3. Real-time and interactive project management through combining high-precision BIM model with I TWO platform software

The BIM model is vivid, visible and informative, providing the basic conditions for its application in project management. There are very high requirements for precision of the BIM model in project management. In the modelling stage, we develop the requirements for information granularity and information precision in compliance with iTWO software. Information granularity reflects what components the model consists of, while information precision demonstrates how accurate the information on the components of the model is. Meanwhile, the overlapping relationship between architectural components is defined as well.

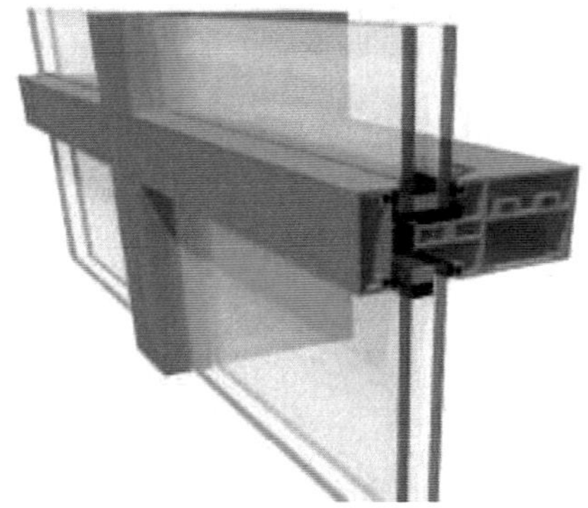

Information Granularity	Information Precision
Frame/panel Filling structure (thermal insulation layer) Installation components	Geographic information: The geographic precision of size and positioning information for modelling shall be 5mm Non-geographic information: information on exterior windows, interior windows, skylights, fireproof windows of all classes, and shutters, etc.

Figure 7. Information Granularity and InformationPrecision of Window Model

High-precision BIM model is combined with iTWO platform software to realize the three-dimensional and dynamic management of transactions generated by the changes of project. In the traditional model of project management, a project change transaction must be completed by multiple participants one by one, including development unit, design unit, costing unit, construction unit and supervision unit, which results in a long cycle and easily causes errors. With this iTWO platform software, multiple participants can work with the same system and the same model to noticeably improve the efficiency. The practice has the following features: 1. Changes are managed by coloring differently; 2. If any change occurs, the corresponding part of project schedule is adjusted to update the revenue and profit of the project automatically; 3. A bill file is generated correspondingly by filling the progress of completing the requested change to the plan; 4. The quantity of changed works and estimated cost can be presented visually to realize the traceability of data; 5. The influence of change can be reflected through 5D simulation; 6. The management of various changes is supported, including hole digging, size change, material change, construction change, demolition change, and list change.

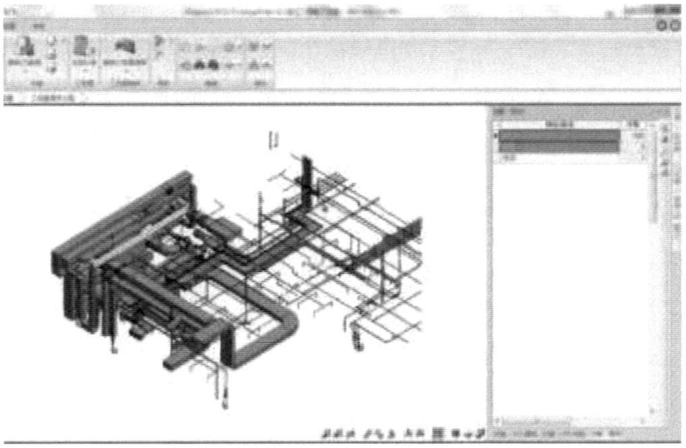

Figure 8. Comparison of Equipment Pipe before and after Change (removal in red and addition in yellow)

Through combination of high-precision BIM model and iTWO platform software, the application for project progress payment can be managed accurately. In the traditional model of project management, the progress payment transaction may be handled after the application is jointly confirmed by multiple participants including development unit, construction unit and supervision unit, etc. and the costing unit determines the specific amount. With this iTWO platform software, all these participants can work with the same system and the same model to improve the efficiency considerably. The practice has the following features: 1. The account is kept based on 3D model and bill of quantities calculated, and the receipts and payments are made visibly; 2. Accounting data and project data can be connected consistently; 3. It can automatically calculate the time needed and the budget of planned activities, and automatically create the file of progress payment application; 4. The statement can be customized as regards requests for payment.

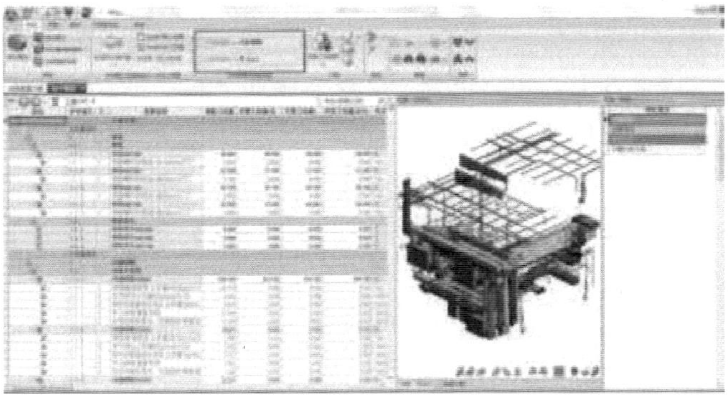

Figure 9. Diagram of Construction Status of Equipment Pipes (constructed pipes in green and not constructed pipes in red)

4. Conclusion on BEM and Real-time and Interactive Project Management

As revealed in the practices, the BIM technology can be greatly expanded into all aspects of building project. To guarantee the transmission and addition of data and information in the whole life cycle of building project, work must be carried out under the unified standard, so the standard followed in the design stage should be based on the unified standard.

At the design stage of the Scientific Research Office Building of China United Northwest Institute for Engineering Design & Research Co., Ltd., the energy information is written into the BIM model for green building simulation analysis, and this application is also expanded into the construction stage. With the 3D model, the design is optimized to improve the quality of design and reduce the unnecessary cost. By combining the high-precision BIM model with iTWO software, the real-time and interactive project management is realized to achieve the intuitive cost control during the process of project implementation.

The application of BIM technology has changed the traditional architectural ideas, and the successful implementation of the project has provided precious experience for the application of the BIM technology in all kinds of comprehensive public buildings.

References

[1] Zhang Hui. (2013). Value, difficulty and development mode of BIM application engineering project management in China. Architecture Technology, 44(10), 870-873.

[2] Ma Xiaoqi. (2015). Application of BIM in cost control. Engineering and Architectural Design, (15).

[3] Lu Wanmei & Wang Wei. (2014). Study on application of BIM technology in the whole life cycle of building—Practice in Tianjin Institute of Architectural Design Research Building Complex Project. Architecture Technique, (2), 99-103.

[4] Tong Jie. (2017). The ideas and application of BIM technology in green energy-saving

buildings. Engineering Technology (Reference Edition), (1), 00074-00074.
[5] Li Xing & Wang Weihong. Study on application of energy-saving design for buildings based on BIM technology. Scientific Research, 00010-00010.
[6] Zhou Fang (2013). China Architecture Design & Research Group: Firm localization of BIM. Architectural Design Management, (8), 28-29.
[7] Wu Cunliang, Song Kang & Zhao Chuan (2016). Perfect combination of BIM technology with scientific & technologic innovation. Architectural Engineering Technology and Design, (1).
[8] Jiang Xi & Wu Jie (2015). Application of BIM in the design of scientific research and test building. Architectural Engineering Technology and Design, (20).

The adjustable intelligent atrium sunshade

Xin Ni[a], Jianhua Sun, Bo Wang

China United Northwest Institute for Engineering Design and Research Co., Ltd, Xi'an 710082, China

[a]258503915@qq.com

Abstract. This article is focused on the specific design techniques of the adjustable atrium sunshade, on the basis of the engineering analyses and practices, it is expected to alter the conventional atrium sunshade design concepts; with its uniqueness and technical excellence, this innovative atrium sunshade system exhibits rich emotions and artistry, creates an inspiring and romantic atmosphere at the atrium area of the building.

1. Introduction

Nowadays, both interior and exterior facilities of a building are controlled by the automatic/intelligent systems in order to save energy and level up building safety. The intellectualization of the A/C and lightning system are being gradually accepted by the society, but the importance of the intelligent sunshade system are being neglected across the country, and yet the intelligent sunshade system is considered as one of the essential elements of the building management system (BMS).

Atrium, it is probably the most common existing form of the public space, many buildings use high light-transmitting glass as its atrium skylight, as the results, more energy could be wasted for cooling due to exposure to direct UV rays; few projects took this issue into consideration and adopted some kinds of sunshade systems in their building designs, but the sunshade design concepts are almost identical in many aspects such as the appearances and the operating modes, and hence weakened the architectural design elements and uniqueness of the building itself. This article is mainly focused on introducing the major factors of the adjustable atrium sunshade system.

2. The concept and features of the intelligent atrium sunshade

Apart from the conventional sunshade panels, the intelligent atrium sunshade system also includes electric motors and control system. The firmware as a part of the control system, embed in the hardware; based on the firmware algorithm, automatically adjust folding/unfolding panel angles according to the atrium temperature; the idea is to limit solar heat gain thereby lowering energy consumptions while improving occupant's comfort by reducing glare.

The intelligent atrium sunshade system is of two different control types; they are manual control and auto. ; the manual control system can be operated by the users, depends on their judgements of the sunlight angle and strength throughout the day, whereas the auto control system is totally related to the atrium air temperatures that detected by the sensors.

3. Natural ventilation and the atrium

There are two different types of micro climate effect that can be utilized at the building atrium: greenhouse effect and stack effect. In order to provide a comfortable environment, different control tactics should be applied for different conditions. A typical public building located in north-west of China, for instance, the skylight was designed as an openable glass roof with the intelligent atrium sunshade system underneath.

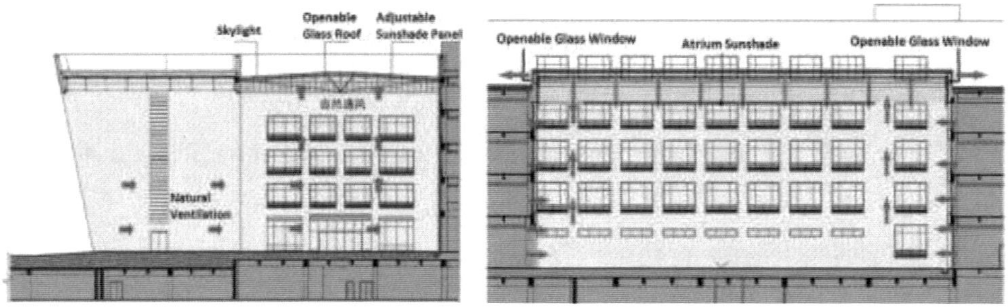

Figure 1. Natural Ventilation at Atrium

During winter daytime, the system was designed to maximize the greenhouse effect by keeping the skylight closed and the sunshade panels folded, and therefore maximize the solar heat gain; unfolded the sunshade panels at night for trapping the heat inside as much as possible. In the summer season, however, it's the opposite way; the solar heat radiations should be kept from entering the building as much as possible while the hot air should be removing continually by natural ventilating

According to the computer simulation results, a proper design of natural ventilation at atrium could lower the energy consumption by around 40%. Also, the air temperature at the atrium may decrease by 5°C if the sunshade panels were unfolded for one hour time. That indicated, the designs of natural ventilations at the atrium are quite effective at lowering the building energy consumptions, and also improve the comfort index.

4. The design concepts of the atrium sunshade

The atrium sunshade design concepts should be considered in the context of the building design styles; also, the sunshade design need to ensure natural ventilations at the atrium, and blocked the unwanted solar heat radiations in order to control the indoor temperatures, and lower the energy consumptions. Control by the intelligent firmware system, the sunshade panels are able to form different pre-defined patterns according to the temperature changes; with the lightning effects, create an inspiring and romantic atmosphere at the atrium area of the building.

5. Intelligent control systems

With intelligent/auto control technologies, the system enables each sunshade panel to rotate individually; the whole system is effective in controlling the solar heat gain throughout the day, and also improve the atrium lightning effects and the artistic expressions while lowered the energy consumptions.

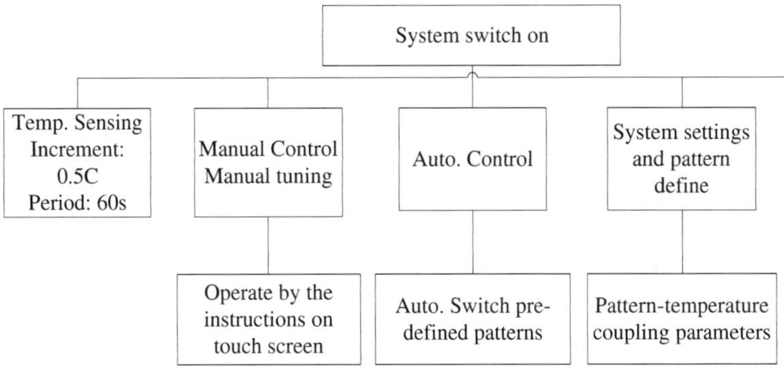

Figure 2. Intelligent sunshade control system process

There were three ways to operate the sunshade panels: first, manually handled by the operator at the control terminal. Second, control by the system based on pre-defined patterns. Third, control by the system based on temperature sensor's feedback signals. The system also kept the indoor temperature data for a calendar month for users' inquiries and control mode optimizations.

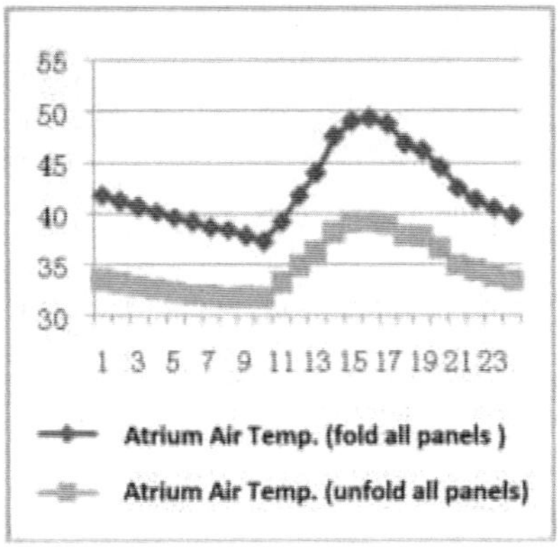

Figure 3. Average **Atrium Air Temperature**

The system has two separate modes for both day time and night time. With the perpetual calendar, it's convenient to pre-define different scheme on the basis of date and hours. For instance, the day time mode starts from 8:00 am in the morning, lasts 10 hours to 6:00 pm, and follow by night time mode till 8:00 am next day. In addition, the system is connected to the skylight control units, enable it to close the skylight when it's raining, and start working as soon as the sun's out; it can also response to emergency events such as fire alarms.

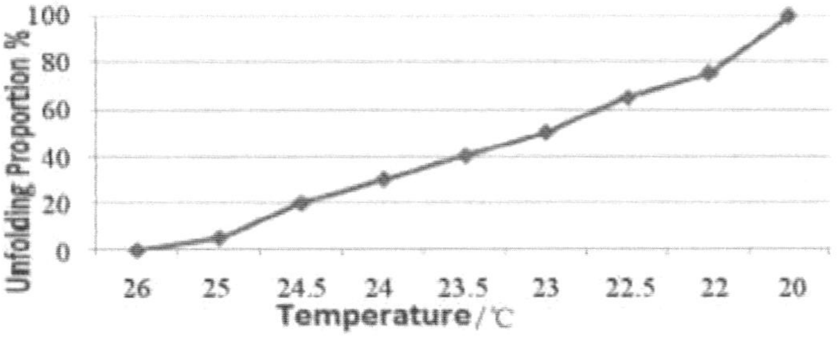

Figure 4. The coupling Curve showed the relationship of the sunshade panel unfolding proportions and the air temperature

6. Temperature curve

The most important algorithm of the intelligent control system is the temperature curve; with it, the system is able to fully control the atrium temperature by precisely describing the unfolding proportion of the sunshade panels.

The temperature curve was generated in the Design Builder software; in a number of real cases, the designers simulated the environment which affected by the sunshade panel folding proportions in different heat conditions. Compared the results to the instant air temperature data, the system controlled by the curve was more reliable in keeping the temp. Fluctuations to minimum; two real case studies proved the control system is more accurate in controlling the temperature while adopting the curve data.

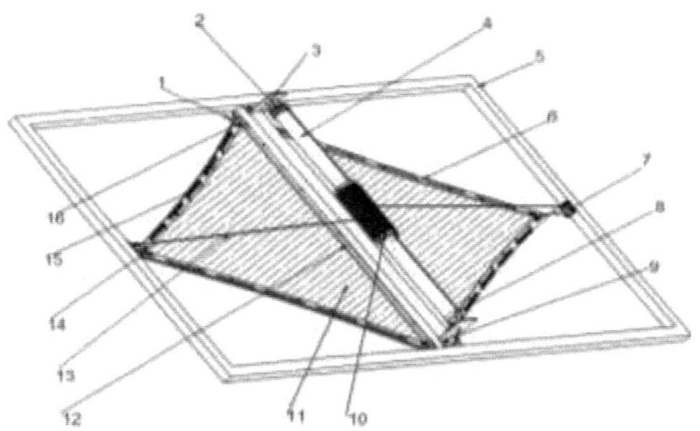

1. Brace; 2. E-motor; 3. 1st Auxiliary brace; 4. Tube; 5. Steel frame; 6. Border edge; 7. Roller; 8. Rear plug; 9. 2nd Auxiliary brace; 10. Cable roller; 11. Sunshade membrane; 12. Mid. Steel frame; 13. Steel-fiber Cable; 14. Outer connection piece; 15. Flexible LED strip; 16. Inner connection piece

Figure 5. Tech. Details of the typical sunshade panel

Another computer simulation mainly analyzed the status of the atrium sunshade panels, in both 100% folding/unfolding situations. First, it simulated the environment with all the panels folded, whenever the temperature reached 26°C, the control system started getting involved; the results showed, the major working period of the system was approx. from 25th April to 22nd September of a year; and the system was most efficient (in energy saving) around 22nd July of a year. Second, the air temperature could be decreased by as much as 12°C in average if the panels were fully unfolded during the period; and there is 6-12°C lower even on the hottest days by doing so. The system limits the solar heat gain effectively.

The further analyses also simulated the conditions when the atrium sunshade system was 20%, 40%, 60% or 80% unfolded; and summarized the calculation results of all the seasons, and finally generated a coupling curve that clearly showed the relationships of the panel unfolding proportions and atrium temperatures. The diagram indicated that, when the atrium temperature is ≥ 26°C, it needs to unfold all panels for keeping the temperature within the comfort zone; and when the temperature is ≤20°C, it needs to fold all panels to increase the solar heat gain. The system will control the panel unfolding proportion automatically if the temperature were in between 20°C~26°C.

7. The technical details of the sunshade panel

A single sunshade panel primarily comprises the steel structure frame, braces and sunshade membranes. The braces are fixed on the steel frame, provide supports to two membranes on each side, these two membranes are connected to the steel frame by a roller on one edge, and fix to the braces on the other edge. There are two sets of auxiliary braces with separated tubes and motors above the steel frame. The membrane is of triangle shape, with LED strips on its hard edge; building up the unique visual effects.

8. The intelligent atrium sunshade system

The temperature curve is combined with modern architectural esthetics, provide the basic control logics to the system firmware, give the sunshade system the ability to form different patterns at any given unfolding proportion; the idea is to design an innovative atrium sunshade system with rich artistry and emotions and finally, create an inspiring and romantic atmosphere at the atrium area of a building.

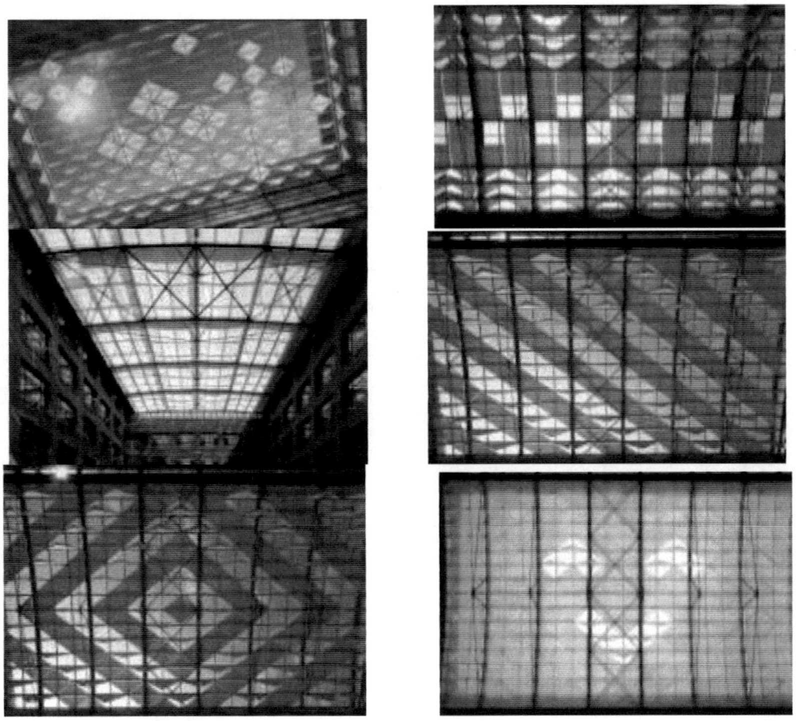

Figure 6. Intelligent Atrium Sunshade System

9. Summary

Although the atrium sunshade system is becoming a popular manifestation of the building public space, more and more atrium sunshade designs tend to be identical in many ways, which may affect the appearance of the whole building. This article emphasized the importance of the atrium sunshade design; introduced a typical intelligent atrium sunshade system, explained the principles, design details, intelligent firmware and the temperature-controlled algorithm of the system. Try to inspire more project owners and architects, to deliver more viable sunshade system designs that rich in era sensations also improve the energy efficiencies of the building.

References

[1] Xin Ni. Key Technology and Application Mode of Green Ecological Building in Northwest China. Xi'An: Xi'an Jiaotong University Press, 2016.12.
[2] Cui Zefeng.Study on the Method of Building Shade. (Doctoral dissertation, Harbin Institute of Technology), 2008.
[3] Cao Yiran. An Experimental Study on Shading Effect of Shade in Different Buildings. Architecture Technology, 2013,44(12), 1099-1102.
[4] Ni Xin, Xing Chao, Wang Fusong,et al. Can transform the pattern of the building atrium shade system. Architecture Technology, 2016,47(9), 794-796.
[5] Ni Xin, Tian Peng, Liu Tao. Application of Electric Shading System in Building Energy Saving

Project.Green Building 2012, (6), 59-61.
[6] Yuan Xiaoping.Development of automatic control system for electric sun visor. (Doctoral dissertation, Tianjin University) 2007.
[7] Zhang Ronglan.Application of Sun-shading Technology in Building Energy Conservation,2006 ,(8), 13-15.
[8] Yang Shichao, Zhou Quan. Application of shading technology in building energy saving.China residential facilities,2012, (8), 19-21.

IOP Conf. Series: Earth and Environmental Science **64** (2017) 012066 doi:10.1088/1755-1315/64/1/012066

Investigation and health risk assessment of heavy metals in soils from partial areas of Daye city, china

M S Xiao, F Li [a], J D Zhang [b], S Y Lin, Z Y Zhuang, Z X Wu

Research Center for Environment and Health, Zhongnan University of Economics and Law, Wuhan 430073, China

[a]lifei@zuel.edu.cn; [b]jdzhang@zuel.edu.cn

Abstract. Heavy metals (Cu and Pb) in four sampling sites from parts areas of Daye city were collected. Concentrations of Cu and Pb in soils in sampling sites were detected, the enrichment degree was measured by geo-accumulation index, and the human health risks were calculated by applying the human health risk assessment model. The results show that the concentrations of Cu and Pb of soils in some areas are much more than Daye City, Hubei Province soil background value. The concentration of Cu and Pb in Xiaganwan soil sample has a higher value and the concentration of Cu (110.17 $mg\cdot kg^{-1}$) exceeds the soil environmental quality standards. The values of I_{geo} of Cu and Pb in the soil in some areas of Daye city are 1 except Xiaganwan sample is 2. For human health risk assessment, the non-cancer risk of Cu in three routes of exposure is less than Pb. The non-cancer risk both adults and children are less than 1 and show a general trend of HQ in oral ingestion exposure pathway > HQ in inhalation exposure pathway > HQ in skin contact exposure pathway. It will not cause significant non-carcinogenic health effects on the human body.

1. Introduction

Heavy metals in soil can cause great harm to the ecological environment for their toxicity, poor-biodegradability and bioaccumulation [1-2]. Heavy metals in soil can be absorbed by plants and accumulated in the animals and human through the food chain [3]. Human will also be harmed when contact with soils containing heavy metals or breathe in dust with heavy metals [4].

With the rapid development of urbanization and industrialization, the pollution of heavy metals in soil has become an urgent problem to be solved [5]. The area of cultivated land polluted by heavy metals including Cd, Pb, Cu, Hg, and Zn reach 20,000,000 square kilometers, which accounts for 20 percent of the total area of China [6]. Heavy metals soil pollution events happened frequently in recent year, such as Cadmium-tainted rice events in 2009, which lead to human health hazards [7].

Daye city is an important mining area with 273 mines in Hubei Province, which is rich in mineral resources, one of the six largest copper ore production bases [8]. Largely mineral extraction in Daye has made a tremendous contribution to local economic development and national construction. But a lot of heavy metals were produced in the process of mineral mining and processing. However, there are quite a few studies on the heavy metals (such as Cu, Pb, Cd, Zn, and as) in soil of Daye, especially on the health risk assessment of human exposure to heavy metals in soil [9].

The objectives of this work were: (1) to measure the concentrations of heavy metals (Cu and Pb) in soils of Daye, (2) to determine the enrichment degree of Cu and Pb in soils, (3) to assess the health risk of human including adults and children exposed to heavy metals applying the health risk assessment model.

2. Methods and materials

2.1. Study area

Daye city is the hinterland of "Metallurgical Corridor" in Hubei, with an area from 29°40′N to 30°15′N, 113°07′E to 114°02′E, located in the hilly land of the north side of Mufu mountain. Daye is the birthplace of Chinese modern industry, one of the earliest mining areas and the cradle of world bronze culture. Tonglv Mountain is the biggest copper mining area in Daye, has a history of over 3000 years, which has been collected into the United Nation's mining heritage list.

2.2. Sampling and analysis

Soils in 4 sampling sites (S1: Yangqiaocun, S2: Xiaganwan, S3: Shangrila community and S4: Yingcai road) were collected from Daye during Aug. 2015, referring to the NY/T1121.1-2006. The soil samples were collected in polytetrafluoroethylene (PTFE) bags and then transferred rapidly to the laboratory in Wuhan. In the laboratory, surface sediments were put evenly on the plastic film to dry naturally in a cool ventilated place. Then, sediments were crushed into small pieces by using pestles and mortals. Next, sediment samples were sifted in 10 mesh nylon sieves to remove stones and plant residue. Finally, all the sediment samples were sifted in 100 mesh sieves and were kept in the plastic bottles prior to analyses.

The soil acidity was measured by pH Meter (Mettler Toledo FE20K FiveEasy, China), using the method in the Industry standard (NYT 1377-2007). For the determination of total heavy metal content, 0.25g treated samples were weighed by an electronic analytical balance (Mettler Toledo-EL204, China). After that, the samples were put into digestion vessels and digested with HCL, HNO_3, HF, and $HClO_4$ by the graphite furnace digestion instrument. Then the solutions were diluted into a final volume of 50 ml with 2% (*v/v*) HNO_3. The heavy metal contents of Cu and Pb were all detected by Atomic Absorption Spectroscopy (AAS, ZEEnit700, and Germany).

To ensure reliability and accuracy of the analysis results, the quality assurance and quality control were assessed strictly by using blank samples, parallel samples and standard reference materials (GB07423). The analysis results were reliable when repeat sample analysis error was below 5%, and the analytical precision for replicate samples was within ± 10%. Accepted recoveries of standard samples ranged from 90% to 108%.

2.3. Geo-accumulation index

To assess the enrichment degree of heavy metals in the soils of Daye, the geo-accumulation index (I_{geo}) was used. I_{geo} can be calculated by the following formula [10]:

$$I_{geo} = \log_2(C_i / B_i) \quad (1)$$

Where, C_i is the actually measured concentration of the heavy metal in the samples. K is corrected coefficient, which take account variation of background value caused by anthropogenic influences or litho logic variations in the sediments (in general k=1.5). B_i is the reference value of heavy metal concentration in sediments.

I_{geo} is classified into seven levels and its corresponding contamination degrees of heavy metal are as follows: (1) $I_{geo} \leq 0$, uncontaminated. (2) $0 < I_{geo} \leq 1$, uncontaminated to moderately contaminated. (3) $1 < I_{geo} \leq 2$, moderately contaminated. (4) $2 < I_{geo} \leq 3$, moderately contaminated to heavily contaminate. (5)

$3 < I_{geo} \leq 4$, heavily contaminated. (6) $4 < I_{geo} \leq 5$, heavily to extremely contaminate. (7) $I_{geo} > 5$, extremely contaminated.

2.4. Health risk assessment

2.4.1. Exposure assessment. In this work, the exposure routes of human exposure to Cu and Pb in soil of Daye were ingestion, skin contact and inhalation. Exposure doses of human exposure to pollutants through these three exposure routes can be calculated as follows [11-12]:

(1) Ingestion

$$ADD_1 = \frac{CS \times IR \times CF \times FI \times EF \times ED}{BW \times AT} \tag{2}$$

Where, ADD_1 is average daily doses in ingestion exposure pathway, CS is chemical concentration in soil ($mg \cdot kg^{-1}$), IR is ingestion rate ($mg \cdot d^{-1}$), CF is conversion factor (10^{-6} $kg \cdot mg^{-1}$), EF is exposure frequency ($d \cdot year^{-1}$), ED is exposure duration (a), BW is body weight (kg), and AT is averaging time (d).

(2) Skin contact

$$ADD_2 = \frac{CS \times AF \times CF \times SA \times ABS \times EF \times ED}{BW \times AT} \tag{3}$$

Where, ADD_2 is average daily doses in skin contact exposure pathway, CS is chemical concentration in soil ($mg \cdot kg^{-1}$), AF is soil to skin adherence factor (mg/cm^2), CF is conversion factor (10^{-6} $kg \cdot mg^{-1}$), SA is the skin area available for contact (cm^2), ABS is absorption factor (unitless), EF is exposure frequency ($d \cdot year^{-1}$), ED is exposure duration (a), BW is body weight (kg), and AT is averaging time (d).

(3) Inhalation

$$ADD_3 = \frac{CS \times PM10 \times DAIR \times PLAF \times FSPO \times CF \times EF \times ED}{BW \times AT} \tag{4}$$

Where, ADD_3 is average daily doses in inhalation exposure pathway, CS is chemical concentration in soil ($mg \cdot kg^{-1}$), PM10 is the concentration of PM10 ($mg \cdot m^{-3}$), DAIR is daily air inhalation rate ($m^3 \cdot d^{-1}$), PIAF is retention fraction of inhaled particulates in body (unitless), FSPO is fraction of soil-borne particulates in air (unitless), CF is conversion factor (10^{-6} $kg \cdot mg^{-1}$), EF is exposure frequency ($d \cdot year^{-1}$), ED is exposure duration (a), BW is body weight (kg), and AT is averaging time (d).

2.4.2. Non-carcinogenic risk assessment

$$HI = HQ_1 + HQ_2 + HQ_3 = (ADD_1 / RfD_1) + (ADD_2 / RfD_2) + (ADD_3 / RfD_3) \tag{5}$$

Where, HI is hazard index of non-carcinogens (unitless). HQ_1, HQ_2 and HQ_3 are hazard quotients of ingestion, skin attach and inhalation. ADD is average daily dose ($mg \cdot kg^{-1} \cdot d^{-1}$), and RfD is reference dose ($mg \cdot kg^{-1} \cdot d^{-1}$). HQ is hazard quotient of non-carcinogens, according to the recommended value by USEPA, $HQ \leq 1$ indicates no risk, $HQ > 1$ indicates that risks do exist [13].

3. Discussion

3.1. Pollution assessment

3.1.1. Concentrations of heavy metals in soils. Table 1 shows concentrations of Cu and Pb in soils from 4 sampling sites in Daye. The concentrations of Cu and Pb from each sampling sites are both decreased in the order of S2 > S4 > S1 > S3. The concentrations of Cu and Pb in all sampling sites are higher than background values of Hubei. Specifically, measured value of Cu in S2 are exceed Grade II standard (GB 15618-1995).

3.1.2. Enrichment degree of heavy metals in soils. The I_{geo} values of Cu in each sampling sites are decreased in the order of S2, S4, S1, S3, which were 1.26, 0.76, 0.49 and 0.42, respectively. Like Cu, the I_{geo} values of Pb in each sampling sites are decreased in the order of S2, S4, S1, S3, which were 1.38, 0.58, 0.27, and 0.07. And for both Cu and Pb, I_{geo} values in S2 are at "uncontaminated to moderately contaminated" level while other sampling sites are at "uncontaminated" level.

Table 1. Concentrations of heavy metals in 4 sampling sites

Sampling sites	pH	Concentration（mg/kg）	
		Cu	Pb
S1	7.45	64.59	48.21
S2	7.24	110.17	104.47
S3	7.99	61.54	41.93
S4	8.18	77.76	60.01
Mean	7.72	78.52	63.65
Background (Hubei)	None	30.70	26.70
Grand II	6.5<pH<7.5	100	300
	>7.5	100	350

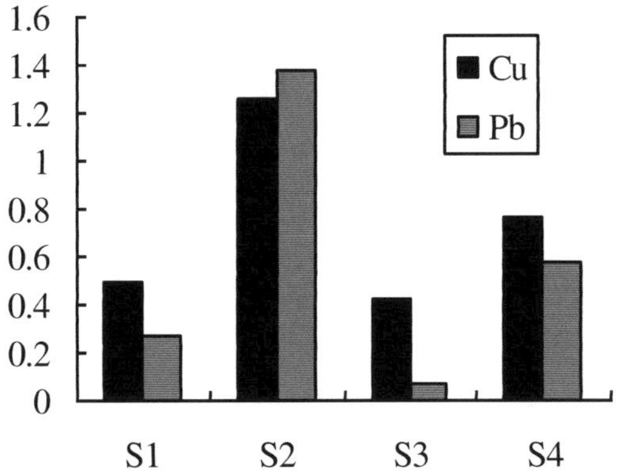

Fig. 1 I_{geo} values of Cu and Pb in 4 sampling sites

3.2. Human health risk assessment

According to exposure factors handbook of USEPA, combining with the actual situation of local residents, the exposure factors mentioned above was determined as in the following Table 2.

According to the arithmetic calculation from (2) to (4) and values of exposure factors listed in Table 2, the calculation results of ADD of Cu and Pb in soils from four sampling sites are in Table 3. As can be seen from Table 3, the values of ADD_1 and ADD_3 of children exposure to Cu and Pb are higher than adults while values of ADD_2 are lower than adults. Overall, children are in the higher level of exposure dose. For both adults and children, the average exposure dose of Cu are higher than Pb, and the average exposure dose of three exposure pathways are decreased in the order of $ADD_1 > ADD_3 > ADD_2$.

The RfD of human exposure to Cu and Pb in different exposure pathways can be seen in Table 4. And the results of human health risk assessment were shown in Table 5. The HI of Cu and Pb in soil of Daye were exceed 1, and the HQ were both decreased in the order of $HQ_1 > HQ_3 > HQ_2$.

For both adults and children, the HQ of Cu under each exposure pathway was lower than HQ of Pb under corresponding pathways. And the HI of Cu and Pb are lower than 1, which means that there were no risks for adults and children. The HI values of children are much higher than adults, and the HI of children exposure to Pb was relatively to 1. It shows that children are at the higher level of health risk and more likely to exposure to heavy metals.

For 4 sampling sites, the HI of both adults and children exposure to Cu and Pb under different routes are decreased in the order of S2, S4, S1,S3.

Table 2. Exposure factors of health risk assessment

Factors	Reference Values	Sources
IR	IR_{adults} =100, $IR_{children}$ =200	[14-15]
CF	10^{-6}	USEPA [16]
FI	1	Maximum limit
EF	365×9/24=200	[14]
ED	ED_{adults} =74, $ED_{children}$ =6	[14]
BW	BW_{adults} =60, $BW_{children}$ =16	[14]
AT	EF×ED	USEPA
AF	AF_{adults} =0.07, $AF_{children}$ =0.2	[14,15]
SA	SA_{adults} =5700, $SA_{children}$ =2800	[14,17]
ABS	0.001	USEPA
PM10	0.3	USEPA
DAIR	$DARI_{adults}$ =20, $DAIR_{children}$ =7.6	USEPA
PIAF	0.75	USEPA
FSPO	0.5	USEPA

Table 3. ADD of Cu and Pb in soils from Daye City

Element	Exposure pathway	Age	S1	S2	S3	S4
Cu	ADD_1	adults	1.08E-04	1.84E-04	1.03E-04	1.30E-04
		children	8.07E-04	1.38E-03	7.69E-04	9.72E-04
	ADD_2	adults	4.30E-07	7.33E-07	4.09E-07	5.17E-07
		children	2.26E-06	3.86E-06	2.15E-06	2.72E-06
	ADD_3	adults	2.99E-05	5.10E-05	2.85E-05	3.60E-05
		children	3.45E-06	5.89E-06	3.29E-06	4.16E-06
Pb	ADD_1	adults	8.03E-05	1.74E-04	6.99E-05	1.00E-04
		children	6.03E-04	1.31E-03	5.24E-04	7.50E-04
	ADD_2	adults	3.21E-07	6.95E-07	2.79E-07	3.99E-07
		children	1.69E-06	3.66E-06	1.47E-06	2.10E-06
	ADD_3	adults	2.23E-05	4.83E-05	1.94E-05	2.78E-05
		children	2.58E-06	5.58E-06	2.24E-06	3.21E-06

Table 4. RfD of human health risk assessment

RfD	Cu	Pb	Sources [16]
RfD_1	4.00E-02	3.50E-03	USEPA
RfD_2	1.20E-02	5.25E-04	USEPA
RfD_3	4.00E-02	3.52E-03	USEPA

Table 5. Human health risks exposure to heavy metals through different routes

Health risk	HQ	Age	S1	S2	S3	S4
HQ_{Cu}	HQ_1	adults	2.69E-03	4.59E-03	2.56E-03	3.24E-03
		children	2.02E-02	3.44E-02	1.92E-02	2.43E-02
	HQ_2	adults	3.58E-05	6.11E-05	3.41E-05	4.31E-05
		children	1.88E-04	3.21E-04	1.79E-04	2.27E-04
	HQ_3	adults	7.47E-04	1.27E-03	7.12E-04	8.99E-04
		children	8.63E-05	1.47E-04	8.22E-05	1.04E-04
HI_{Cu}		adults	3.47E-03	5.93E-03	3.31E-03	4.18E-03
		children	2.05E-02	3.49E-02	1.95E-02	2.46E-02
HQ_{Pb}	HQ_1	adults	2.30E-02	4.97E-02	2.00E-02	2.86E-02
		children	1.72E-01	3.73E-01	1.50E-01	2.14E-01
	HQ_2	adults	6.11E-04	1.32E-03	5.31E-04	7.60E-04
		children	3.21E-03	6.96E-03	2.80E-03	4.00E-03
	HQ_3	adults	6.33E-03	1.37E-02	5.51E-03	7.88E-03
		children	7.32E-04	1.59E-03	6.37E-04	9.11E-04
HI_{Pb}		adults	2.99E-02	6.48E-02	2.60E-02	3.72E-02
		children	1.76E-01	3.82E-01	1.53E-01	2.19E-01

4. Conclusion

The concentration of Cu and Pb in 4 sampling sites were much higher than the background values of Hubei Province, it was shown that soils in parts of Daye was polluted in different degree. The average values of concentration of Cu and Pb not exceeded the national Grade II standard. From the calculated results of geo- accumulation index, Cu and Pb in soils in each sampling sites were enriched in varying degree and S2 was at a higher degree of enrichment. For human health risk assessment, the hazard quotients of human exposure to Cu and Pb under three routes were all lower than 1, it means that there was no risks on human. But the hazard quotients of children were much higher than adults, which were relatively closed to 1. Therefore, children in parts of Daye are probably suffered from heavy metals in soil.

Acknowledgments

This work was financially supported by This study was supported by the National Natural Science Foundation of China (51308076; 71503268) and the Open Fund from Institute of Wuhan Studies (IWHS2016106).

References

[1] Li F, Zhang J D, Yang J, Liu C Y and Zeng G M 2016 Hum. Ecol. Risk Assess. 22 1224-41
[2] Li J L, Jiang X, Wang S H, Wang W W and Chen J Y 2016 China Environmental Science 36 1207-1217
[3] Li F, Zhang J D, Huang J H, Huang D W, Yang J, Song Y W, Zeng G M 2016 Environ. Sci. Pollut. R. 23 13100-13

[4] Li F, Huang J H, Zeng G M, Yuan X Z, Li X D, Liang J, Wang X Y, Tang X J and Bai B 2013 J. Geochem. Explpr. 132 75-83
[5] Li Z Y, Ma Z W, Tsering J V, Yuan Z W and Huang L 2014 Science of the Total Environment 468 843-853
[6] Xiong Y J Mod. 2010 Agri. Sci. Tech. 8 295-297
[7] Wu Y M 2013 Ecol. Environ. 22 901-904
[8] Han X T, Guo Y, Bao Z Y and Zhou S D 2012 Geochimica. 41 585-592
[9] Hu X Y, Sun H F and Chen D L 2007 Ecol. Environ.16 1421-1423
[10] Tang R L, Ma K M, Zhang Y X and Mao Q Z 2013 Appl. Geochem. 35 88–98
[11] Risk Assessment Guidance for Superfund (RAGS): Human health evaluation manual (Part F), Office of Solid Waste Emergency Response, 1989.
[12] Yu Y J, Wang Q and Zhang Y P 2012 Res. Environ. Sci. 25 751-756
[13] Risk Assessment Guidance for Superfund (RAGS): Human health evaluation manual (Part A), Office of Emergency and Remedial Response, 1989.
[14] Duan X L 2012 D. Beijing: Science Press.
[15] Wang Z S, Duan X L, Wang B B, Wang Y N, Qian Y, Nie J and Zhang J L 2012 J. Environment Health. 29 114-117
[16] Exposure Factors Handbook, National Center for Environmental Assessment, 2011.
[17] Zhao X G, Huang N, Duan X L, Wang B B, Cao S Z, Nie J and Jing J L 2012 J. Environment Health. 29 124-126

IOP Conf. Series: Earth and Environmental Science **64** (2017) 012067 doi:10.1088/1755-1315/64/1/012067

Pyrolysis of automotive shredder residue light fraction: maximization of the tar yield using design of experiment

Manuela Anzano[1], Elena Collina[1,*], Elsa Piccinelli[1] and Marina Lasagni[1]

[1]Earth and Environmental Sciences Dept., Milano-Bicocca University, Milano, Italy

*Corresponding author e-mail: elena.collina@unimib.it

Abstract. The general aim of this study is the valorisation of Automotive Shredder Residue (ASR) via pyrolysis. Tar, the condensable gases obtained in the pyrolysis process, is an interesting alternative fuel. Thus, the pyrolysis process was investigated in order to maximize the tar yield. The design of experiment approach was used to plan a series of experiments and to identify which operating variables influence the yield of the process. Temperature and carrier gas flow proved to be significant factors affecting the yield, while the influence of ASR light fraction amount pyrolysed was negligible. In the experimental domain, the maximum response was obtained at 500 °C and 100 mL/min.

1. Introduction

Vehicles at the end of their useful life become end-of-life-vehicles (ELVs), a waste. In 2012, Europe (EU 27) produced 6.2 million tons of ELVs [1]. European directives promote the reuse, recycling and recovery of ELVs. During dismantling, ferrous and non-ferrous metals are removed and recovered in secondary smelter plants. Such large-scale thermal processes can be the source of micropollutants [2, 3], giving rise to public concern about their possible health effects. Automotive shredder residue (ASR) is the remainder which is obtained from the ELVs dismantling process. It accounts approximately for 15 wt. % of the total ELVs and is mainly composed (90%) of a light fraction, also called car fluff. Car fluff usually presents heterogeneous composition: plastics (9-35%), rubber (4-12%), textile (8-40%), wood and paper (1-4%), metals (3-8%), wire (1-5%), foam (4-35%) and fines (up to 75%) [2] With a great amount of polymeric materials. On the basis of this composition, pyrolysis can be regarded as a suitable thermal pre-treatment for car fluff [3].

Several studies reported the pyrolysis as a suitable process for reducing the waste stream, obtaining valuable products: gas, i.e. the non-condensable gases; char, i.e. the solid phase; and tar, i.e. the condensable gases [4, 5, and 6]. Some Authors studied the pyrolysis process of ASR [7, 8], producing tar characterized by a high calorific value, 37-40 MJ/kg [9, 10, 11]. On this basis, the use of car fluff-tar as a fuel seems an interesting alternative [11]. Thus, in the present paper, the pyrolysis process was investigated in order to maximize the tar yield.

The use of design of experiment (DOE) allows to get a clear picture of how the process variables, separately and together, affect the experimental responses and how it is possible to control them in order to make the process more effective [12]. This statistical approach was applied in many different areas. For instance Kazemi et al. [13] adopted DOE to investigate the effects of multiple factors and their interactions on the composting performance of municipal solid waste; Mangili et al. [14]

IOP Conf. Series: Earth and Environmental Science **64** (2017) 012067 doi:10.1088/1755-1315/64/1/012067

employed DOE to investigate the supercritical CO_2 devulcanization process of ground tire rubber; Kumar and Leech [15] reported a design of experiment methodology to investigate and improve the performance of glucose oxidizing enzyme electrodes. In particular, only few papers [16, 17, 18] applied this statistical approach to the pyrolysis process of waste in order to optimize the experimental conditions maximizing the products yield.

In the present paper, the DOE approach was used to investigate the pyrolysis of the ASR light fraction in a bench scale reactor. We considered the process variables temperature, carrier gas flow and ASR light fraction amount, whereas the tar yield was the response. The process variables were optimized with the aim of maximizing the response.

2. Materials and methods

2.1. *ASR sample*

ASR light fraction sample (hereafter named car fluff) was collected at a shredder plant in Italy. The sample underwent a quartering procedure and a thorough characterization. Results for elemental and material composition, proximate analysis, and higher heating value were previously reported [19].

2.2. *Pyrolysis experiments*

Pyrolysis experiments were carried out in a horizontal fixed bed tubular reactor under nitrogen flow (N_2). The tar was collected downstream the oven reactor and weighed. Details about the pyrolysis apparatus and the experimental procedure were described in a previous paper [11].

The tar yield was calculated as the weight percentage with respect to the amount of car fluff pyrolysed.

2.3. *Design of Experiment*

In order to obtain the maximization of tar yield, a two level full factorial experimental design (FF-ED) [12] was applied. Temperature, N_2 flow and car fluff amount were chosen as process variables or factors and the tar yield was chosen as response.

Table 1 shows maximum (coded as +1), minimum (coded as −1) and central (coded as 0) levels for each factor applied to FF-ED.

Table 1. Factors and levels of the FF-ED.

Factor	Minimum level -1	Maximum level +1	Central level 0
Temperature (T), °C	500	800	650
Flow (Fw), mL/min	100	500	300
Car fluff (CF), g	2	6	3

The temperature levels were chosen considering the usual pyrolysis temperature range [20], the N_2 flow levels were chosen considering the operating condition of the flowmeter and the car fluff amount was chosen on the basis of the pyrolysis reactor dimensions.

To perform the FF-ED, the necessary number of experiments is $N = L^k$, where L represents the number of levels (two in our case) and k represents the number of factors (three in our case). So, eight experiments were carried out to investigate the experimental domain; three more experiments were added to investigate the performance in the centre of the experimental domain and to estimate the model validity, reproducibility and experimental error. A total of eleven experiments were carried out. The experiments were randomly performed to minimize the error due to the planning of the experiments.

The dependence of the experimental response (y) on the n variables (x) was modelled applying following equation [12]:

$$y = \beta_0 + \sum_{i=1}^{n} \beta_i x_i + \sum_{i=1}^{n} \sum_{j=i+1}^{n} \beta_{ij} x_i x_j + \varepsilon \quad (1)$$

Where β_0 the constant is term, β_i and β_{ij} are the regression coefficients, and ε is the error. For the determination of the coefficients, multiple linear regression was performed on the scaled and centered variables.

The analysis of variance (ANOVA) was applied to determine the factors and interactions that have a significant influence on the process, on the basis of the *P*-value. Furthermore we calculated the coefficient of determination (R^2), the adjusted R^2 (R^2_{adj}) and the coefficient of determination in prediction (Q^2).

The calculations were carried out with MODDE11 software.

3. Results and discussion

The FF-ED experiments and results are shown in Table 2. The evaluated regression includes the three factors and all two-factor interactions [21]. Higher-order interactions were discarded since their importance decreases with increase of the interaction order [12]. On the basis of this assumption, we excluded three- and four-factor interactions. In this way, we privileged the quality of the model and avoided over-fitting [14].

Table 2. FF-ED experiments and results.

Experiment	T (°C)	Fw (mL/min)	CF (g)	Tar Yield (%)
E1	500	100	2	41.9
E2	800	100	2	14.3
E3	500	500	2	35.0
E4	800	500	2	14.4
E5	500	100	6	42.6
E6	800	100	6	15.5
E7	500	500	6	31.5
E8	800	500	6	10.6
Centre 1	650	300	4	22.0
Centre 2	650	300	4	24.0
Centre 3	650	300	4	21.7

In Table 3 we report the regression coefficients and the *P*-value for the experimental response.

Table 3. Regression coefficients and *P*-values.

	Coefficient	*P*-value
Constant	24.9	<0.001
T	**-12.02**	**<0.001**
Fw	**-2.8**	**0.032**
CF	-0.7	0.488
T*Fw	1.6	0.136
T*CF	0.02	0.979
Fw*CF	-1.2	0.263

The factors and their interactions with *P*-values greater than 0.05 can be considered not significant at 95% confidence level. So, just T and Fw prove to be significant factors [22].

The regression parameters (Table 4) were considered to evaluate the goodness of fit and predictive capacity of the complete model.

Table 4. Regression parameters.

Parameter	
R^2	0.981
R^2_{adj}	0.951
Q^2	0.796

The complete model showed high and comparable R^2 and R^2_{adj}, offering an acceptable explanation of the total variance. Considering the Q^2 value, the predictive ability of the complete model is fair.

The complete model was used to determine the optimal condition range for maximizing the tar yield. To do this, contour plots of tar yield as a function of the temperature, the flow and the car fluff amount are shown in Fig. 1.

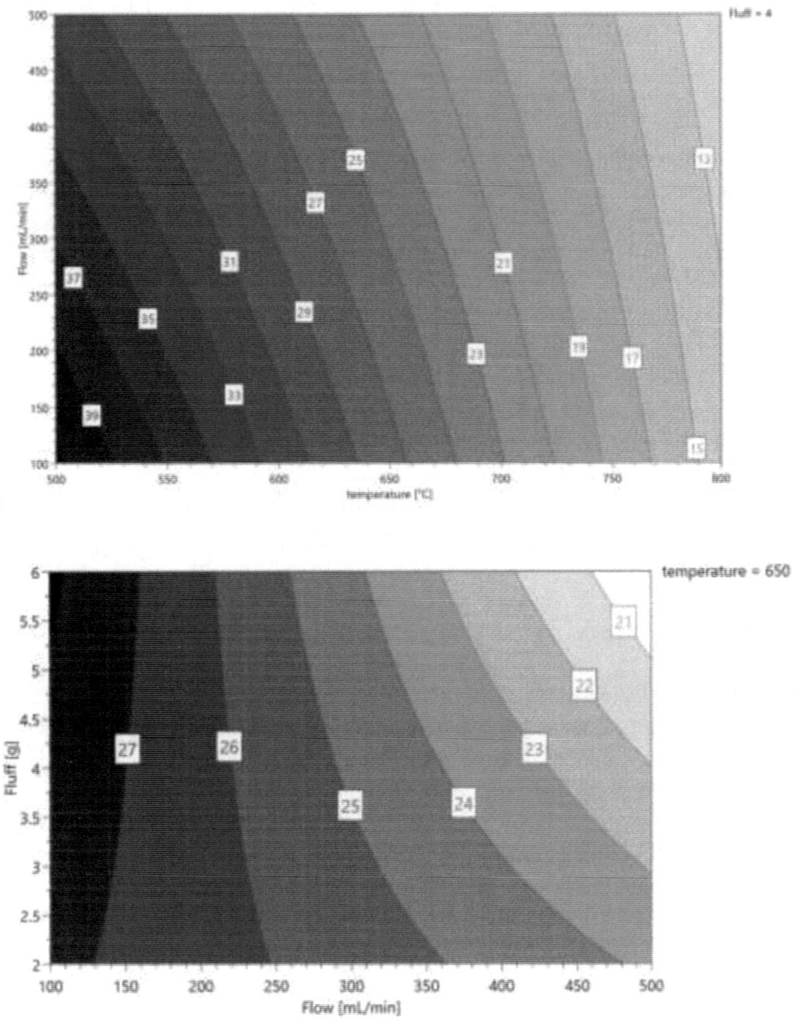

a)

b)

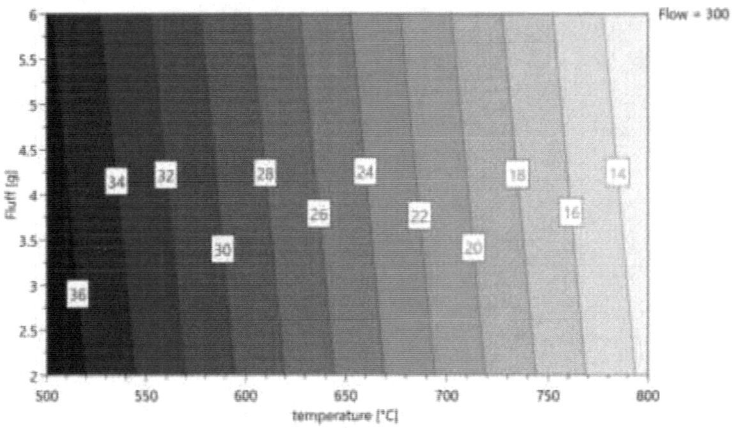

c)

Figure 1. Contour plots of tar yield as function of the three factors.

The black colour represents the optimal condition range for the response and 500°C and 100 mL/min were determined as the optimal conditions (Fig. 1a); whereas the negligible influence of the car fluff amount on the tar yield is confirmed in Fig. 1b and 1c.

4. Conclusions

Tar, the oil fraction produced in ASR pyrolysis, is a valuable alternative fuel. In order to maximize the tar yield, a full factorial design of experiment was used for modelling the process and optimizing the process variables. Eleven experiments were performed to explore the experimental domain. Temperature and carrier gas flow proved to be significant factors affecting the yield, while the influence of ASR light fraction amount pyrolysed was negligible. The maximum response was obtained at 500 °C and 100 mL/min.

As the optimal conditions determined are at the lower limit of the experimental domain, further investigation will be performed in a lower temperature range.

References

[1] Information on http://ec.europa.eu/eurostat/web/waste/key-waste-streams/elvs
[2] K. Everaert, J. Baeyens, The formation and emission of dioxins in large scale thermal processes, Chemosphere 46 (2002) 439-448.
[3] D. Pitea, M. Bortolami, F. Franzoni, E. Collina, G. Cortili, M. Lasagni, E. Piccinelli, Prevention of PCDD/F formation and minimization of their emission at the stack of a secondary aluminium casting plant, Environ. Sci. Technol. 42 (2008) 7476-7481.
[4] I. Vermeulen, J. Van Caneghem, C. Block, J. Baeyens, C. Vandecasteele, Automotive shredder residue (ASR): reviewing its production from end-of-life vehicles (ELVs) and its recycling, energy or chemicals' valorisation, J. Hazard. Mat. 190 (2011) 8-27.
[5] A. López, I. de Marco, B.M. Caballero, M.F. Laresgoiti, A. Adrados, Influence of time and temperature on pyrolysis of plastic wastes in a semi-batch reactor, Chem. Eng. J. 173 (2011) 62-71.
[6] K. Zaafouri, A.B. Hassen Trabelsi, S. Krichah, A. Ouerghi, A. Aydi, C.A. Claumann, Z.A. Wüst, S. Naoui, L. Bergaoui, M. Hamdi, Enhancement of biofuels production by means of co-pyrolysis of Posidonia oceanica (L.) and frying oil wastes: Experimental study and process modeling, Biores. Technol. 207 (2016) 387–398.
[7] I. Hita, M. Arabiourrutia, M. Olazar, J. Bilbao, J.M. Arandes, P. Castaño, Opportunities and barriers for producing high quality fuels from the pyrolysis of scrap tires, Renewable and Sustainable Energy Reviews 56 (2016) 745–759.

[8] I. de Marco, B.M. Caballero, M.J. Chomo, M.F. Laresgoiti, A. Torres, G. Fernandez, S. Arnaiz, Pyrolysis of electrical and electronic wastes, J. Anal. Appl. Pyr. 82 (2008) 179–183.
[9] M. Zolezzi, C. Nicolella, S. Ferrara, C. Iacobucci, M. Rovatti, Conventional and fast pyrolysis of automobile shredder residues (ASR), Waste Manage. 24 (2004) 691–699.
[10] J. Haydary, D. Susa, V. Gelinger, F Cacho, Pyrolysis of automobile shredder residue in a laboratory scale screwtype reactor, J. Environ. Chem. Eng. 4 (2016) 965–972.
[11] H. Joung, Y. Seo, K. Kim, J. Hong, T. Yoo, Distribution and characteristics of pyrolysis products from automobile shredder residue using an experimental semi-batch reactor, Korean J. Chem. Eng. 24 (2007) 996 – 1002.
[12] A. Santini, F. Passarini, I. Vassura, D. Serrano, J. Dufour, L. Morselli, Auto shredder residue recycling: Mechanical separation and pyrolysis, Waste Manage. 32 (2012) 852–858.
[13] M. Anzano, E. Collina, E. Piccinelli, M. Lasagni, Lab-scale pyrolysis of the Automotive Shredder Residue light fraction and characterization of tar and solid products, Waste Manage., accepted for publication (2017) DOI 10.1016/j.wasman.2017.03.013
[14] G.P.E. Box, W.G. Hunter, J.S. Hunter, Statistics for Experimenters, second ed., Wiley, Hoboken, NJ, 1978.
[15] K. Kazemi, B. Zhang, L.M. Lye, Q. Cai, T. Cao, Design of experiment (DOE) based screening of factors affecting municipal solid waste (MSW) composting, Waste Management (2016), DOI 10.1016/j.wasman.2016.08.029
[16] I. Mangili, M. Oliveri, M. Anzano, E. Collina, D. Pitea, M. Lasagni, Full factorial experimental design to study the devulcanization of ground tire rubber in supercritical carbon dioxide, J. Supercrit. Fluids 92 (2014) 249–256.
[17] R. Kumar, D. Leech, A glucose anode for enzymatic fuel cells optimized for current production under physiological conditions using a design of experiment approach, Bioelectrochemistry 106 (2015) 41–46.
[18] S.R. Chandrasekaran, B. Kunwar, B.R. Moser, N. Rajagopalan, B.K. Sharma, Catalytic Thermal Cracking of Postconsumer Waste Plastics to Fuels. 1. Kinetics and Optimization, Energy Fuels 29 (2015) 6068–6077.
[19] F. Pinto, F. Paradela, I. Gulyurtlu, A.M. Ramos, Prediction of liquid yields from the pyrolysis of waste mixtures using response surface methodology, Fuel Processing Technology 116 (2013) 271–283.
[20] M. Sajdak, R. Muzyka, J. Hrabak, K. Słowik, Use of plastic waste as a fuel in the co-pyrolysis of biomass Part III: Optimisation of the co-pyrolysis process, J. Anal. Appl. Pyr. 112 (2015) 298–305.
[21] M. Edo, I. Aracil, R. Font, M. Anzano, A. Fullana, E. Collina, Viability study of automobile shredder residue as fuel, J. Hazard. Mat. 260 (2013) 819-824.
[22] G. Tchobanoglous, H. Theisen, S.A. Vigil, Integrated solid waste management: engineering principles and management issues, McGraw-Hill, USA, 1993.
[23] L. Eriksson, E. Johansson, N. Kettaneh-Wold, C. Wikström, S. Wold, Design of Experiments Principles and Applications, third ed., Umetrics Academy, Umea, 2008.
[24] I.Y. Mohammed, Y.A. Abakr, S. Yusup, F.K. Kazi, Valorization of Napier grass via intermediate pyrolysis: Optimization using response surface methodology and pyrolysis products characterization, J. Cleaner Prod. 142 (2017) 1848-1866.

Effect of COSMOS technologies in detoxifying municipal solid waste incineration fly ash, preliminary results

Elsa Piccinelli[1], Marina Lasagni[1]*, Elena Collina[1], Stefania Bonaiti[1] and Elza Bontempi[2]

[1]Dept. of Earth and Environmental Sciences, University of Milano-Bicocca, Piazza della Scienza 1, 20126 Milano, Italy
[2]INSTM and Chemistry for Technologies Laboratory, University of Brescia, via Branze 38, 25123, Brescia, Italy

*Corresponding author e-mail: marina.lasagni@unimib.it

Abstract. This study investigates the effect of technologies for heavy metal stabilization on the concentration of PolyChlorinatedDibenzo-p-Dioxins (PCDD) and PolyChlorinatedDibenzoFurans (PCDF), abbreviated PCDD/F, in Municipal Solid Waste Incineration (MSWI) fly ash. We determined the variation of the Total Organic Carbon (TOC) and PCDD/F concentration between raw and stabilized material. The technologies, that already proved to be very promising for heavy metal entrapment, showed encouraging results also for PCDD/F detoxification. This result could be very impacting on the management of MSWI fly ash: at the best of our knowledge, there are no methods, in literature, that can provide good results in stabilization of heavy metals, and abatement of chlorinated organic pollutants contained in the same matrix.

1. Introduction

It is well known that organochlorinated micropollutants form in traces in thermal processes, such as waste incineration, iron ore sintering, secondary ferrous and non-ferrous smelters, power plants [1-3]. Municipal solid waste presents an heterogeneous composition, containing also hazardous household waste (as plastic bags, paints, chlorine- and metal-containing substances, discarded electronic waste, etc.), that can be problematic when combined with poor combustion conditions, resulting in significant levels of flue gas contamination [4]. As a consequence Municipal Solid Waste Incinerator (MSWI) fly ash, resulting from the treatment of exhaust fume emissions, is an extremely complex material that includes soluble salts, heavy metals, and organic contaminants such as PolyChlorinated Dibenzo-p-Dioxins (PCDD) and PolyChlorinated DibenzoFurans (PCDF), abbreviated PCDD/F [5]. Consequently, MSWI fly ash is considered a hazardous solid waste and its reuse is limited. To improve its environmental characteristics and enhance the possibility of its recovery, it is necessary to develop an effective and economic treatment technology to stabilize or remove hazardous components.

At the University of Brescia two processes for the stabilization of metals in MSWI fly ash, based on the use of silica, were developed: COSMOS process, which uses colloidal silica (COlloidal Silica Mean to Obtain Safe inert, [6-7]), and COSMOS-RICE process, which uses rice husk ash as silica source (Colloidal Silica Medium, to Obtain Safe inert, from RICE husk ash, [8-9]). Both processes obtained good results in terms of metals stabilization.

In the present work our attention was focused on organic contaminants in MSWI fly ash. In particular, we determined the Total Organic Carbon (TOC) and PCDD/F concentration in raw fly ash as well as in the material stabilized with COSMOS or COMOS-RICE treatment. The PCDD/F data were expressed as concentration, ng/kg, or in terms of equivalent toxicity, ng I-TEQ/kg (NATO/CCMS, 1988). This paper presents the preliminary results to evaluate the effect of these mild and sustainable technologies in terms of PCDD/F reduction.

2. Materials and methods

2.1. Samples
MSWI fly ash was collected at fabric filters hoppers of a MSWI plant located in Northern Italy in two different periods: 2008 (FA1) and 2013 (FA2). Flue gas desulfurization fly ash (FGD_FA) and coal fly ash (C_FA) were collected at a power generation plant located in Northern Italy. Rice Husk Ash (RHA) was collected from the particulate collection equipment at the stacks of rice husk-fired boilers. Colloidal silica precursors (Ludox SM-30, Grace Italy) are aqueous colloidal dispersions of SiO_2 particles [6]. An extensive characterization of FA1 and FA2 was already reported [6,8].

Two fly ash mixtures, FA_mix1 and FA_mix2, were prepared mixing three ashes (FA1 or FA2, FGD_FA and C_FA in the respective weight percent: 65%, 20% and 15%). The two mixtures were stabilized by the COSMOS procedure, the obtained samples were named C_mix1 and C_mix2. FA_mix2 was also stabilized by direct use of RHA (COSMOS Rice procedure), and the resulting sample was named CR_mix2.

2.2. Analytical procedures
The TOC content of the samples was determined via three replicate analyses using a TOC analyser (Dohrmann, Apollo 9000) equipped with the standard module (DC-90), the purgeable organics module (PRG-1), and the sludge sediment sampler accessory (S/SS). Details of the analytical method were previously reported [10]. The variability of the TOC data, expressed as percentage standard deviation, was between 2 and 10.

To determine the PCDD/F concentration, the samples were extracted and purified according to the EN 1948-2/3: 2006 method. The obtained extracts were analysed using a VARIAN CP_3800 high resolution gas chromatograph (Dioxin 2 60 m × 0.25 µm Restek capillary column) coupled to a VARIAN 320_MS triple quadrupole mass spectrometer using the tandem mass (MS-MS) operation mode. The 17 toxic 2, 3, 7, 8-congeners ($PCDD_{2378}$ and $PCDF_{2378}$) and the tetra- to hepta-homologues non-2378 substituted ($PCDD_{non}$ and $PCDF_{non}$) were detected and quantified. To evaluate the repeatability of the complete PCDD/F determination procedure, duplicate analyses were performed for each sample: the variation was less than 5%. The lowest operational detection limit (LOD) was found by injecting the most dilute PCDD/F calibration solution (CS1 purchased at Wellington Laboratories Inc.). The LOD was calculated for each homologue group: TCDD and TCDF, 0.04 ng/kg; PeCDD, PeCDF, HxCDD and HxCDF, 0.08 ng/kg; HpCDD, HpCDF, OCDD and OCDF, 0.16 ng/kg.

3. Results and discussion
The effect of silica on heavy metal stabilization was already reported and discussed [6-9]. To evaluate the influence of the stabilization process on organic compounds concentration, the TOC content was determined in all the matrices (Table 1). FA1 and FA2 have a similar TOC content (33691 ppm and 34416 ppm respectively), whereas TOC content is equal to 9092 ppm for FGD_FA and 150732 ppm for C_FA. These matrices were mixed to obtain the materials afterwards stabilized. On the basis of the percentage contribution of FA1 or FA2, FGD_FA and C_FA, the theoretical TOC concentrations for FA_mix1 and FA_mix2 were equal to 46327 ppm and 46799 ppm, respectively (Table 2). After the stabilization with COSMOS process, the experimental TOC concentration were 47871 and 49029 (Table 2), with percentage variation of 3.3% and 4.8% in C_mix1 and C_mix2 with respect to the theoretical ones. These variations are close to the experimental error. To assess the

percentage variation of TOC after the stabilization using COSMOS-RICE process, it is necessary to consider the contribution of RHA to the organic substance (TOC concentration equal to 81591 ppm). On this basis, it is possible to explain the high TOC percentage variation between FA_mix2 and CR_mix2. Based on these results we concluded that the stabilization processes do not influence the total organic carbon concentration in the matrices. Afterwards, PCDD/F were analysed in all the matrices (Tables 1 and 2). The PCDD/F concentration in FA_mix1 and FA_mix2 was calculated on the basis of PCDD/F concentration in FA1, FA2, FGD_FA and C_FA as explained before.

Table 1. TOC, PCDD, and PCDF concentration in raw matrices.

	FA1	FA2	FGD_FA	C_FA	RHA
TOC (ppm)	33691	34416	9092	150732	81591
$\Sigma PCDD_{2378}$ (ng/kg)	6755	21720	11	7	27
$\Sigma PCDD_{non}$ (ng/kg)	4391	8377	2	< LOD	139
$\Sigma PCDF_{2378}$ (ng/kg)	3868	13844	9	27	4
$\Sigma PCDF_{non}$ (ng/kg)	9639	20034	26	20	22

Table 2. TOC, PCDD, and PCDF concentration in mixtures and treated matrices.

	FA_mix1	FA_mix2	C_mix1	C_mix2	CR_mix2
TOC (ppm)	46327	46799	47871	49029	57714
$\Sigma PCDD_{2378}$ (ng/kg)	4394	14121	1325	12978	9969
$\Sigma PCDD_{non}$ (ng/kg)	2854	5445	1405	5058	4254
$\Sigma PCDF_{2378}$ (ng/kg)	2520	9004	1197	8092	6763
$\Sigma PCDF_{non}$ (ng/kg)	6274	13030	3037	12269	9905

Observing the results obtained, C_FA, FGD_FA and RHA materials were found to contain a very low concentration of total PCDD/F ($PCDD/F_{tot}$, sum of $PCDD/F_{2378}$ and $PCDD/F_{non}$), almost two orders of magnitude lower than the concentrations found in FA1 and FA2. Moreover, the $PCDD/F_{tot}$ concentration in FA2 is two times greater than the concentration found in FA1. The difference in concentration is probably due to the different composition of the waste burned, considering that the two fly ashes were sampled within a temporal interval of five years.

Starting from the data in Table 2, the percentage reductions of $PCDD_{2378}$ in the samples C_mix1, CR_mix2, and C_mix2 were 69%, 29% and, 8% respectively, corresponding to a detoxification (percentage reduction of the toxicological potential of the mixture expressed as ng I-TEQ/kg) of 51%, 25% and 14%, respectively. For $PCDF_{2378}$ the percentage reductions in the samples C_mix1 CR_mix2 and C_mix2 were 52%, 25% and, 10% respectively, corresponding to a detoxification of 51%, 24% and, 10% in the same samples.

To point out the efficacy of the treatment, Fig. 1 reports the experimental concentration of PCDD/F in stabilized materials plotted against the theoretical ones for each congener and homologue group, calculated taking into account the dilution due to the addition of the materials used in the process (FGD_FA, C_FA, colloidal silica and RHA). The ideal line at 45° (not indicated in the Figure) corresponds to coincident theoretical and experimental concentration of PCDD/F. This ideal line is almost coincident with the C_mix2 sample, whose reduction of organic contaminants is very low (few percent).

For the samples C_mix1 and CR_mix2, the percentage reduction of the organochlorinated pollutants is greater than the dilution factor (slope of interpolation line less than 1 in Figure 1). In particular, for sample C_mix1, the reduction is about 60% and for sample CR_mix2 it is about 15%. The stabilization processes do not influence the TOC concentration in the matrix as explained before, whereas PCDD/F concentration variation before and after the stabilization process is greater than the dilution factors. We can suppose, on the basis of our results and previous studies [11], that dehalogenation of organohalogen compounds occurs. Further investigations are needed to verify this hypothesis.

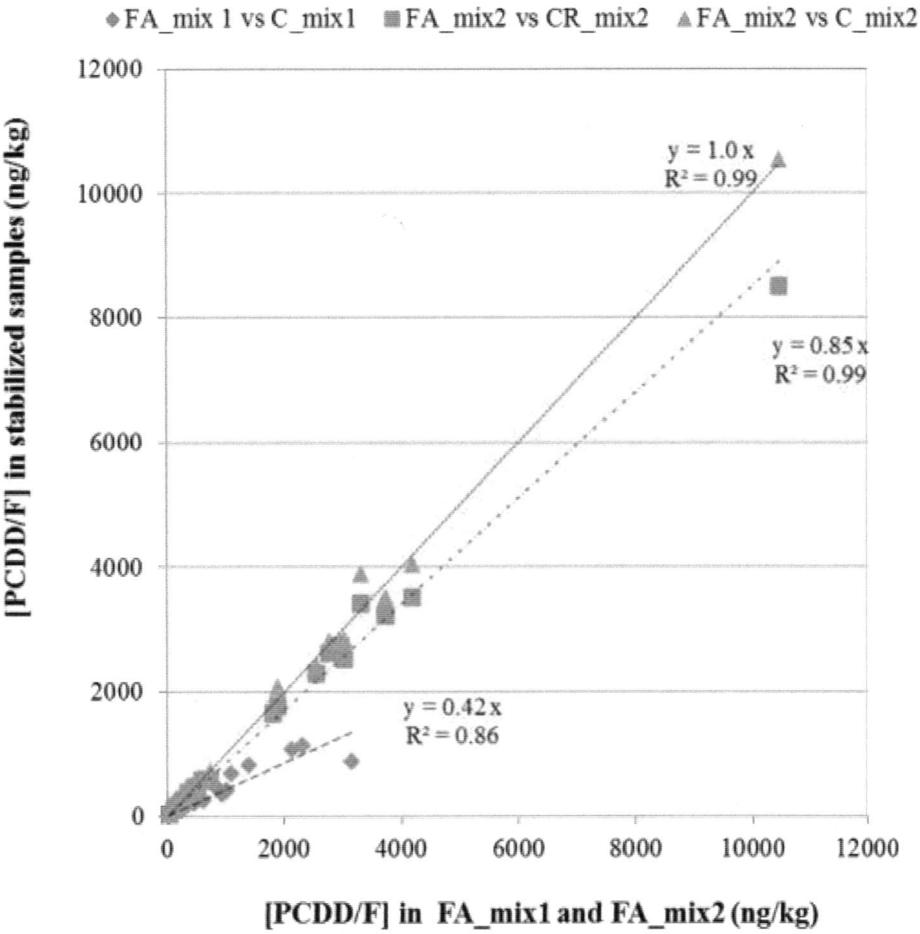

Figure 1. PCDD/F concentration in FA_mix1 and FA_mix2 and the corresponding stabilized material after the treatment with COSMOS and COSMOS-RICE technologies.

4. Conclusion

On the basis of these preliminary results, the COSMOS and COSMOS-RICE technologies, very promising for heavy metal entrapment, proved to be useful also for chlorinated organics reduction. Moreover, these processes are sustainable, as they take place under mild conditions, and all the required reagents are waste materials, except colloidal silica in COSMOS process. This result could be very impacting on the management of the MSWI fly ash: to the best of our knowledge, there are no methods, in literature, that can provide good results not only in salt recovery and stabilization of heavy metals, but also in the abatement of chlorinated organic pollutants contained in the same matrix.

In the next future, the hypothesis of dechlorination taking place during the treatment will be studied in more detail, investigating the role of the reagents, such as colloidal silica, and parameters as the reaction temperature.

References

[1] E. Grandesso, S. Ryan, B. Gullett, D. Touati, E. Collina, M. Lasagni, D. Pitea, Kinetic Modeling of PCDD/F formation based on carbon degradation reactions, Environ. Sci. Technol., 42 (2008) 7218-7224.

[2] D. Pitea, E. Collina, G. Cortili, M. Lasagni, M. Bortolami, F. Franzoni, E. Piccinelli, Prevention of PCDD/F formation and minimization of their emission at the stack of a secondary aluminum casting plant, Environ. Sci. Technol., 42 (2008) 7476-7481.

[3] K. Everaert, J. Baeyens, The formation and emission of dioxins in large scale thermal processes, Chemosphere 46 (2002) 439-448.

[4] M.T. Hu, S.J. Chen, Y.C. Lai, K.L. Huang, G.P. Chang-Chien, J.H. Tsai, Characteristics of Polychlorinated Dibenzo-p-dioxins/Dibenzofuran from joss paper burned in taiwanese temples, Aerosol Air Qual. Res., 9 (2009) 369-377.

[5] A. Zacco, L. Borgese, A. Gianoncelli, R.P.W.J. Struis, L.E. Depero, E. Bontempi, Review of fly ash inertisation treatments and recycling, Environ. Chem. Lett., 12 (2014) 153-175.

[6] E. Bontempi, A.Zacco, L. Borgese, A. Gianoncelli, R. Ardesi, L.E. Depero, A new method for municipal solid waste incinerator (MSWI) fly ash inertization, based on colloidal silica, J. Environ. Monit., 12 (2010a) 2093-2099.

[7] E. Bontempi, A. Zacco, L. Borgese, A. Gianoncelli, R. Ardesi, L.E. Depero, A new powder filler, obtained by applying a new technology for fly ash inertisation procedure, Adv. Sci. Technol., 62 (2010b) 27–33.

[8] A. Bosio, N. Rodella, A. Gianoncelli, A. Zacco, L. Borgese, L.E. Depero, P.A. Bingham, E. Bontempi, A new method to inertize incinerator toxic fly ash with silica from rice husk ash, Environ. Chem. Lett., 12 (2014a) 329−333.

[9] A. Bosio, A. Zacco, L. Borgese, N. Rodella, P. Colombi, L. Benassi, L.E. Depero, E. Bontempi, A sustainable technology for Pb and Zn stabilization based on the use of only waste materials: A green chemistry approach to avoid chemicals and promote CO_2 sequestration, Chem. Eng. J. 253 (2014b) 377−384.

[10] D. Brocca, L. Barilli, E. Collina, M. Lasagni, M. Tettamanti, D. Pitea, A new quantifying TOC and COD methodology for the determination of organic compounds and particulate carbon directly on fly ash of MSWI, Chemosphere, 35 (1997) 2203−2213.

[11] M. Tabata, A. Ghaffar, A. Shono, K. Notomi, Hydrodechlorination/detoxification of PCDDs, PCDFs, and co-PCBs in fly ash by using calcium polysulfide, Waste Manage., 33 (2013) 356−362.

Study on the influence of small hydropower stations on the macroinvertebrates community-Take Nanhe River as a case, China

Weihua Zhao*, Qingyun Li, Weijie Guo, Zhenhua Wang

Basin Water Environmental Research Department, Yangtze River Scientific Research Institute, Wuhan, 430010, China

*Corresponding author e-mail: 524944826@qq.com

Abstract. This study take Nahan River as a case to research the impacts of small hydropower stations on macroinvertebrates community. Results showed that a total of 13 macroinvertebrate samples was collected and contained 56 taxa belonging to 18 families and 35 genera. The influence of runoff regulation was more seriously than hydrological period. There were obvious zoning phenomenon of macroinvertebrates between reservoir, downdam reaches and natural reaches. From reservoir, downdam reaches to natural reaches, species abundance increased in turn. There are the least species in reservoir, the most in natural rivers. The reservoirs had the highest biomass and were quite different from those in downdam and natural reaches. However, there was no significant difference between different periods of hydropower station.

1. Introduction

Macroinvertebrates is one of the most widely distributed taxa of river habitats, and is also an important part of river ecological system (Hall et al., 2003), plays an important role in the process of material circulation and energy flow in the river ecosystem. As of the end of 2015, China has built small hydropower station (installed below 50 thousand KW) more than 4.7 seats. These small hydropower stations had seriously influence on river ecological environment (Sundermann, 2011).If these hydropower stations had a significant influence on macroinvertebrates community structure? And if the influence had differences between runoff regulation and hydrological period? Based on the above questions, this study take Nan River as a case to study the influence of small hydropower stations on macroinvertebrates.

2. Research area and methods

2.1. Research area

Nanhe River, which located in Hubei Province, China. The total length of 255km, the basin area of 6497km^2, the river drop of about 470m, is a subtropical monsoon climate, the average annual rainfall of 917.8mm. There are 4 small hydropower stations located at this rive. In this study, a total of 12 sample sites were set up. The sample sites were showed in figure 1. The investigation were carry out in December, 2013 and July, 2014.

2.2. Methods

1) Macroinvertebrates investigation

The macroinvertebrates were collected by D-frame net (wide 0.30m, 450 m diameter nylon yarn) in river and Peterson grap (1/16 m2) in reservoirs. Samples were sieved in situ and the 420μm fraction was live-picked in field and preserved in 10% formalin. Macroinvertebrates were identified in the laboratory with the aid of a dissecting microscope to the lowest possible tax on using identification keys (Brinkhurst and Jamieson, 1971; Morse et al., 1994).

2) Data analysis

Density (individual's m-2) and biomass (g m-2) of macroinvertebrates within each site were calculated using arithmetic average from all samples and or sites. Nonparametric multivariate analysis of variance (perMANOVA) and gradient ranking (GA) (Canoco 4.5) were used to analysis distribution characteristics of community sites, among them, GA was selected by detrended correspondence analysis (DCA). Excel 2013 and SPSS 17 software were used to data analysis and chart making.One-way ANOVA was used to analyze the significance of different habitat characteristics (p=0.05).

3. Results and analysis

3.1. Species composition and standing crop

A total of 13 macroinvertebrate samples were collected in Nanhe Rivers during the study contained 56 taxa belonging to 18 families and 35 genera. Insects represented the most diverse group and comprised 40 taxa. Oligochaeta comprised 11 taxa, molluscs comprised 3 taxa while other groups (Nematoda, Hirudinea and Crustacea) were represented by 2 taxa. Diptera was the dominant group in aquatic insects. The average densities and biomass were 966±112ind/m^2 and 17.31±1.54g/m^2 respectively.

3.2. Effect of small hydropower station on macroinvertebrate

Runoff regulation and hydrological period (wet season and dry season) are two main flow regulation modes to influence macroinvertebrates. Runoff regulation make out three different habitat type, such as reservoir, downdam reaches and natural reaches. Species abundance difference between different habitat type were analysis using single factor analysis of variance (ANOVA) under two flow regulation modes (Table1).

Table 1. Analysis results of One-way ANOVA of Nanhe river macroinvertebrates abundance between different groups

		Sum of squares	df	Mmean square	F	significance
Runoff regulation	Between groups	81.04	2	40.52	9.46	0.001**
	In groups	89.92	21	4.28		
	Total	170.96	23			
Hydrological period	Between groups	12.04	1	12.04	1.667	0.21
	In groups	158.92	22	7.22		
	Total	170.96	23			

Note: ** Significant at 0.01 level

The results showed that for species parameters, more significant effect of runoff regulation mode, and the influence of hydrologic period is relatively weak, therefore, can be in 2011(dry season) and 2012(wet season), a comprehensive analysis of the data, without considering the influence of different period for hydrological data. For multivariate analysis of nonparametric variance data of the Nanhe River Basin (perMANOVA) also showed that based on the grouping of runoff regulation under the

IOP Conf. Series: Earth and Environmental Science **64** (2017) 012069 doi:10.1088/1755-1315/64/1/012069

different macroinvertebrates community structure significantly (p=0.0036**, 99999 times based on permutation test results), figure1 shows, different sections of macroinvertebrates DCA sequencing results.

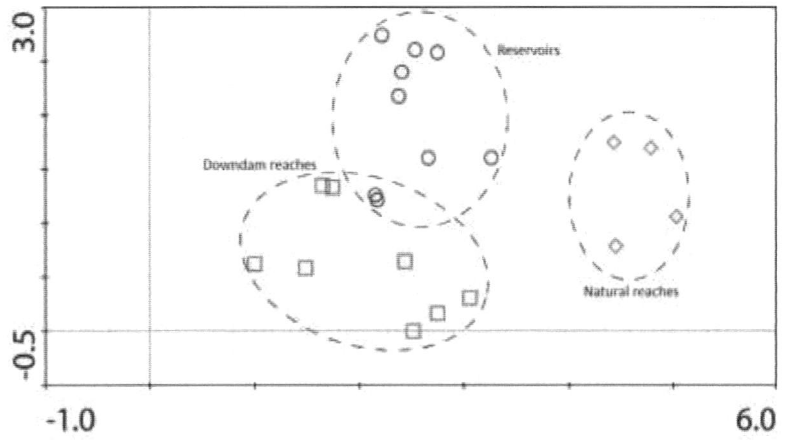

Figure 1. DCA ordination diagram of macroinvertebrates in Nanhe River

The explanation rate of species difference of axis 1 was about 15.5%, and the cumulative interpretation rate of axis 1 and axis 2 was 24.0%, and the maximum gradient length was about 5.033, which indicated that the response of species to the environment was a single peak pattern. Three kinds of samples formed an obvious partition trend in the coordinate system composed of axis 1 and axis 2, and grouping was basically consistent with the runoff regulation model, which mean the distributive characteristics of species community along the environmental gradient in the Nanhe river area, underdam reahces and Natural River have significant difference. The difference between the organisms in the reservoir and the down dam is mainly in the axis 2, and the difference between the natural and non-natural river reaches is mainly in the axis 1.

The structure parameters of macroinvertebrates in the reservoir, downdam reaches and natural reaches were classified and counted (Table2). The results showed that the small deviation fluctuation between species abundance and diversity index (Shannon-Wiener index and Margalef richness index), indicating that the three parameters in the group within each sampling section is relatively stable, on the contrary, the deviation volatility of the biological density and biomass was larger, which showed that the differences of these parameters between the two groups were significant. From the reservoir, downdam to natural river, the species richness increased gradually, the species in the reservoir was the least, and the natural river was the most; macroinvertebrates density in downdam was the highest, followed by natural rivers and reservoir; the reservoirs had the highest biomass and were quite different from those in downdam and natural river; According to the above parameters, we can summarize the structure characteristics of bioroenosis in 3 typical areas: The species of macroinvertebrates in the reservoir was less, the diversity was low, showing a more concentrated single species advantage, and the individual weight of the biological individual was heavier. Under the dam, the species of macroinvertebrates increase slightly, the diversity was improved, the number of species was large, but because of the small weight, the biomass was not large. The species of macroinvertebrates in the natural river was the largest, the diversity was high, and the weight of the individual was smaller.

Table 2. Statistical results of macroinvertebrates structure parameters in Nanhe River

	Reservoir	downdam	natural river
Species richness	4.4±0.7	6.5±0.3	9.5±1.3
Biological density(ind/m^2)	412.0±132.3	1396.0±733.6	825.4±160.3
Biomass (g/m^2)	56.3±20.9	25.24±11.4	5.99±2.6
Shannon-Wiener index	1.06±0.16	1.48±0.07	1.58±0.21
Margalef richness index	6.22±1.05	9.66±0.63	10.46±1.16

4. Conclusion

The influence of runoff regulation was more seriously than hydrological period. There were obvious zoning phenomenon of macroinvertebrates between reservoir, down dam reaches and natural reaches. From reservoir, down dam reaches to natural reaches, species abundance increased in turn. There are the least species in reservoir, the most in natural reaches. The reservoirs had the highest biomass and were quite different from those in down dam and natural reaches. However, there was no significant difference between different periods of hydropower station.

Acknowledgements

The research was financially supported by National Natural Science Foundation of China (No. 51509011), Special fund for basic scientific research business of central public research institutes (No. CKSF2017026/SH and CKSF2016025/SH) and National Key Research and Development Program (2016YFC0502201).

References

[1] Sundermann A, Antons C, Cron N, et al. Hydromorphological Restoration of Running Waters: effects on benthic invertebrate assemblages [J]. Freshwater Biology, 2011, 56(8): 1689-1702.
[2] Bredenhand, E., M.J. Samways. Impact of dam on benthic macroinvertebrates in a small river in a biodiversity hotspot: Cape Floristic Region, South Afric. Insect Consev. 2009, 13(3):297-307.
[3] Barbour M T, Gerritsen J, Snyder B D, et al. Rapid bioassessment protocols for use in wadeable streams and rivers: Periphyton,Benthic Macroinvertebrates and Fish ,2nd edition.Washington ,D. C. : U. S. Environmental Protection Agency ,Office of Water ,1999
[4] Brinkhurst R O, Jamieson, B G M. Aquatic Oligochaeta of the world. Edinburgh: Oliver & Boyd, 1971
[5] Morse J C, Yang L F, Tian L X. Aquatic insects of China useful monitoring water quality. Nanjing: Hohai University Press, 1994, p570
[6] Hall D. L., B. S. Bergthold, R. W. Sites. 2003. The influence of adjacent land uses on macroinvertebrate communities of prairie streams in Missouri. Journal of Freshwater Ecology, 18 (1): 55-67.

IOP Conf. Series: Earth and Environmental Science **64** (2017) 012070 doi:10.1088/1755-1315/64/1/012070

Nitrogen activation of carbon-encapsulated zero-valent iron nanoparticles and influence of the activation temperature on heavy metals removal

Stefania Bonaiti[1]*, Blanca Calderon[2], Elena Collina[1], Marina Lasagni[1], Valeria Mezzanotte[1], Nacho Aracil Saez[2], Andrés Fullana[2]

[1]Dept. of Earth and Environmental Sciences, University of Milano-Bicocca, Milano, Italy.
[2]Dept. of Chemical Engineering, Alicante University, Alicante, Spain.

*Corresponding author e-mail: s.bonaiti3@campus.unimib.it

Abstract. Nanoparticles of zero-valent iron (nZVI) represent a promising agent for environmental remediation. This is due to their core-shell structure which presents the characteristics of both metallic and oxidised iron, leading to sorption and reductive precipitation of metal ions. Nevertheless, nZVI application presents some limitations regarding their rapid oxidation and aggregation in the media which leads to the delivery of the ions after some hours (the "aging effect"). To address these issues, modifications of nZVI structure and synthesis methods have been developed in the last years. The aging problem was solved by using nZVI encapsulated inside carbon spheres (CE-nZVI), synthetized through Hydrothermal Carbonization (HTC). Results showed high heavy metals removal percentage. Furthermore, CE-nZVI were activated with nitrogen in order to increase the metallic iron content. The aim of this study was to test CE-nZVI post-treated with nitrogen at different temperatures in heavy metals removal, demonstrating that the influence of the temperature was negligible in nanoparticles removal efficiency.

1. Introduction

Urban wastewater contains a variety of organic and inorganic substances from domestic and industrial sources. This means that wastewater reuse is associated with certain risks. Even if toxic materials are not present in concentrations likely to affect humans, they might well be at phytotoxic levels and this would limit their agricultural reuse [1, 2]. Agricultural reuse concerns mainly tertiary treated effluents. However, recent research has shown that micropollutants such as pharmaceuticals, heavy metals and other emerging pollutants are variably removed by wastewater treatments and thus, they can be found in treated wastewater even after advanced secondary and tertiary treatments.

In the last two decades nanoparticles of zero-valent iron (nZVI) have proven to be effective at the sequestration of heavy metals from aqueous solutions [3, 4, 5, 6], thanks to both reducing (metallic iron in the core) and adsorbing (iron oxides in the shell) properties [7]. In general, the states of the metals on the nZVI surface exhibit three distinctive responses, depending on their standard oxidation-reduction potential (E^0) [8]:

- Sorption only: for metal cations with E^0 more negative than or close to $E^0(Fe)$; e.g. Zn^{2+}, Cd^{2+}, Ba^{2+};
- Sorption + reduction: for metal cations with E^0 slightly more positive than $E^0(Fe)$; e.g. Ni^{2+}, Pb^{2+};
- Reduction only: for metal cations with E^0 substantially higher than $E^0(Fe)$; e.g. Ag^+, Cu^{2+}, Hg^{2+}.

Nevertheless, nZVI application presents some limitations: they rapidly age in water (oxidation reactions) [9, 10] and, as consequence, a thicker and crystalline shell originates [11]. This structural change leads to a decrease in removal efficiency and metals ions re-dissolve after some hours [9].

To address this issue, a new way of using nZVI is in combination with carbon materials by encapsulation of them (5-10 nm) inside micro- or nano-carbon spheres (6-8 μm) through Hydrothermal Carbonization (HTC) from a moist biomass. HTC is a thermal treatment of an aqueous solution or dispersion of a carbon-containing organic material at moderate temperatures and under pressure, which produces a carbon-rich black solid as an insoluble product [12].

An alternative source to produce carbon substrate is biomass waste, making the synthesis green, sustainable and also inexpensive. A previous research [13] reported the synthesis of carbon-encapsulated nZVI (CE-nZVI) from olive mill waste (OMW) and compared nZVI and CE-nZVI in heavy metals removal. Results showed that using CE-nZVI there was no delivery of heavy metals again to the water. It can be supposed that using CE-nZVI the carbon matrix blocked the delivery and so heavy metals remained adsorbed onto the nanoparticles surface.

The aim of this study was to synthetize CE-nZVI via HTC and activate them with nitrogen at different temperatures in order to increase the Fe(0) content. The synthetized particles were characterized, determining the iron content, the zero-valent iron content, the surface area, the morphology and the crystallinity. The influence of the activation temperature on nanoparticles removal efficiency was investigated.

2. Materials and methods

2.1. Chemicals

The reagents used for CE-nZVI synthesis were Fe(III) as source of iron (22.83 g iron(III) nitrate nonahydrate salt) and clarified olive mill waste (OMW) as source of carbon (300 mL) (CE-nZVI with molar ratio 0.05 nFe/nC).

In order to simulate wastewater conditions, this study was performed with a solution containing a mixture of five heavy metals ions (Zn^{2+}, Cd^{2+}, Ni^{2+}, Cu^{2+} and Cr(VI); 10 ppm each). All the solutions were prepared using purified water.

2.2. nZVI synthesis

During the HTC process, the mixture of iron salts and OMW was heated in a high-pressure reactor at relatively low temperatures (200°C) for 3 hours under inert atmosphere. When the reactor had reached the HTC operating temperature (50 minutes/1 hour), measurement of the residence time (3 hours) began.

The reaction products were gases (mainly carbon dioxide, carbon monoxide, hydrogen, methane, ethane and propene) and a mixture of solid and liquid solution mechanically easy to separate on filter paper (0.2 μm cellulose acetate filter) under vacuum.

Then the filtered part was washed 2 times with a 50/50 ethanol-water solution. After washing, the sample was transferred to an oven (under inert atmosphere to avoid oxidation of Fe(0)) and allowed to dry at 80°C until all water was removed. After being taken out of the oven, the sample was ground into a fine powder.

2.3. Post-treatment with nitrogen

About 1.5 g of sample were put in a quartz tube and blocked with glass wool, then positioned inside a tubular oven and, after having reached the desired temperature (about 1 hour), the nanoparticles were heated under a nitrogen atmosphere (500 mL/min) for 3 hours. The presence of the nitrogen gas created a reducing atmosphere, leading to the reduction of Fe(III) / Fe(II) to Fe(0). Furthermore, the post-treatment also caused the evaporation of tars that may block the pores of the carbon matrix.

After the reaction, zero-valent iron became very reactive and it suddenly burnt in contact with air. Therefore, when the sample was taken out from the oven (after cooling), it was immediately wetted with water in order to protect the zero-valent iron by formation of a thin oxide layer on it.

In this work, temperatures of post-treatment of 600°C, 700°C and 800°C were investigated.

2.4. Batch tests

The post-treated CE-nZVI were added to 700 mL solution containing the mixture of the five metals. About 2 mL of sample from the reaction mixture were taken at different times and filtered through a 0.2 μm cellulose acetate filter (about 20 samples, with a decreasing rate, in about 220 hours, in total). At last nitric acid was added until 2-3% acid concentration to avoid the formation of solid particles.

2.5. Analytical methods

CE-nZVI were characterized using different techniques: zero-valent iron content determination (based on the measurement of hydrogen gas volume, which is evolving during the redox reaction of zero-valent iron and hydrochloric acid), thermogravimetric analysis (to determine the total iron content), X-ray diffraction (to determine the crystalline structure), transmission electron microscopy and scanning electron microscopy (to determine the size, the aggregation and the composition of the nanoparticles and the iron location within the carbonaceous structure) and surface area determination.

The concentration of heavy metals in aqueous samples was determined by inductively coupled plasma-mass spectrometry.

3. Results and discussion

3.1. CE-nZVI characterization

The non-post-treated sample showed a high surface area and a high Fe(0) percentage (4.46%), thanks to the presence of polyphenols in OMW having reducing properties (Table 1). All these features led to an improving of adsorption and reduction mechanisms in heavy metals removal. Furthermore, the total iron percentage was also high (Table 1), index of a good work of encapsulation by carbon from OMW.

The post-treatment with nitrogen proved to increase the Fe(0) content. Results showed that the reduction of the iron increased with the increase of the temperature of the post-treatment (Table 1); however, at 800°C the surface of the nanoparticles became highly crystalline and iron formed agglomerations that led to a decrease in surface area. No great differences were shown in samples post-treated at 600°C and 700°C, both in Fe(0) content and surface area (Table 1). Thus the temperature of 600°C is preferred as it requires less energy and, as a consequence, less costs.

Table 1. CE-nZVI characterization.

T of post-treatment (°C)	Tot Fe content (%)	Fe(0) in tot Fe (%)	Fe(0) in sample(%)	Surface area (m^2/g)
Non-post-treated	42.46	10.51	4.46	168.268
600	52.99	12.72	6.74	194.581
700	56.43	19.49	11.00	219.281
800	73.12	21.05	15.39	161.087

3.2. Influence of CE-nZVI activation temperature in heavy metals removal.

The test aimed to assess how different temperatures of post-treatment with nitrogen could influence the nanoparticles application; the assessed temperatures were 600°C and 800°C. For comparison, the total iron concentration was about the same in both samples (1.018 g/L in 600°C-CE-nZVI and 1.017 g/L in 800°C-CE-nZVI).

Results did not show significant differences, both in heavy metals concentration and in removal rate (Figures 1a, b). To be more precise, there was a slightly higher removal percentage using nanoparticles post-treated at 600°C. This was probably due to the more crystalline surface of 800°C-CE-nZVI, which led to a decrease in adsorbing capacity.

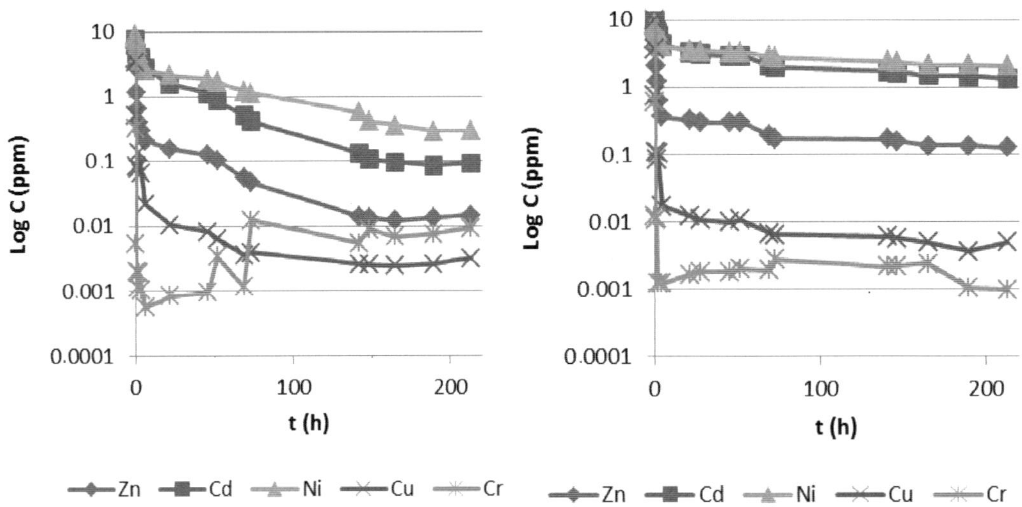

Figure 1. Heavy metals removal using (a) 600°C and (b) 800°C post-treated CE-nZVI.

4. Conclusion

In conclusion, the effectiveness of CE-nZVI in the elimination of heavy metals contaminants has been successfully demonstrated at lab-scale.

The temperature of CE-nZVI post-treatment with nitrogen slightly influenced the removal performance. Even if both satisfactory, the removal percentages using nanoparticles post-treated at 600°C presented slightly higher values, probably due to the more crystalline surface of 800°C CE-nZVI, which led to a decrease in adsorbing capacity. Thus the use of nanoparticles with post-treatment at 600°C is preferred, also because it involves lower synthesis costs, thus economic benefits.

The utilization of Fe-C nanomaterials for the treatment of secondary and tertiary treated effluents is a new technological concept for wastewater remediation, which reflects the originality and the innovative technological character of the project.

Acknowledgments

This work was financially supported by University of Milano-Bicocca fund (2016-ATESP-0597) and University of Alicante (UAFPU2013-5791).

References

[1] I. K. Kalavrouziotis, P. Kokkinos, G. Oron, F. Fatone, D. Bolzonella, M. Vatyliotou, D. Fatta-Kassinos, P. H. Koukoulakis, S. P. Varnavas, Current status in wastewater treatment, reuse and research in some mediterranean countries, Desalin. Water Treat. 53 (2015) 2015-2030.

[2] E. Collina, M. Lasagni, D. Pitea, A. Franzetti, P. Di Gennaro, G. Bestetti, Bioremediation of diesel fuel contaminated soil: effect of non ionic surfactants and selected bacteria addition, Annali di Chimica 97 (2007) 799-805.

[3] B. I. Kharisov, H. V. R. Dias, O. V. Kharissova, V. M. Jiménez-Pérez, B. O. Pérez, B. M. Flores, Iron-containing nanomaterials: synthesis, properties, and environmental applications, RSC Adv. 2, (2012), 9325–9358.

[4] A. N. Bezbaruah, S. S. Shanbhogue, S. Simsek, E. Khan, Encapsulation of iron nanoparticles in alginate biopolymer for trichloroethylene remediation, J. Nanopart. Res. 13, (2011), 6673–6681.

[5] A. B. Cundy, L. Hopkinson, R. L. D. Whitby, Use of iron-based technologies in contaminated land and groundwater remediation: A review, Sci. Total Environ, 400 (2008) 42-51.

[6] F. Fu, D. D. Dionysiou, H. Liu, The use of zero-valent iron for groundwater remediation and wastewater treatment: a review, J. Hazard. Mater, (2014).

[7] R. A. Crane, T. B. J. Scott, Hazard. Mater, (2012) 211, 2931–2942.

[8] X. Q. Li, W. Zhang, Sequestration of Metal Cations with Zerovalent Iron Nanoparticles – A Study with High Resolution X-ray Photoelectron Spectroscopy (HR-XPS), J. Phys. Chem. C 2007, 111, 6939-6946.

[9] B. Calderon, A. Fullana, Heavy metal release due to aging effect during zero valent iron nanoparticles remediation, Water Research 83 (2015) 1-9.

[10] Q. Wang, S. Lee, H. Choi, Aging study on the structure of Fe0-nanoparticles: stabilization, characterization, and reactivity, J. Phys. Chem. C114 (5) (2010) 2027-2033.

[11] S.R. Kanel, B. Manning, L. Charlet, H. Choi, Removal of arsenic(III) from groundwater by nanoscale zero-valent iron, Environ. Sci. Technol., 39 (5) (2005) 1291-1298.

[12] Sevilla M., Fuertes A. B. (2009). Chem. Eur. J. 15, 4195–4203.

[13] B. Calderon, F. Smith, I. Aracil, A. Fullana, "Synthesis of Carbon Encapsulated Iron Nanoparticles from Olive Mill Wastewater for Water Remediation" in proceedings of the 2nd World Congress on New Technologies (NewTech'16), Budapest, Hungary, 2016, Paper ICNFA 144-1, DOI: 10.11159/incfa16.144.

IOP Conf. Series: Earth and Environmental Science **64** (2017) 012071 doi:10.1088/1755-1315/64/1/012071

Green mine research and evaluation based on using solid wastes in urbanization for mine filling

Futong Xu*, **Zaixiang Ren**

Shandong University of Science and Technology, Qingdao, China

*Corresponding author e-mail: 1032522310@qq.com

Abstract. Derived from the disposal problems of construction wastes generated by urbanization and combined with the current situations of mine gob filling, this paper, in accordance with the actual requirements of green mines, probed into the methods of using construction wastes for gob filling, whose technical and economic feasibility was also analyzed. An evaluation system of green-mine filling was established and the goals of green-mine filling were dissected as well. Simultaneously, as a new view and the ultimate evaluation index, environmental protection degree, whose abbreviation is H, was proposed. As for the outcomes of green-mine filling, fuzzy comprehensive evaluation was utilized, whose results can more systematically and factually reflect the environmental protection degree of construction wastes for green-mine filling and thus have high application value.

1. Introduction

As is proposed in *Report on the Work of the Government* delivered in 2017, the emission of the key pollutant should uninterruptedly decrease, the eco-environmental protection and governance should be strengthened and its enhancement should be accelerated as well, which is a goal for the new period. With the continuous advancement of urbanization in China, more and more importance is attached to the disposal of the increasing number of construction waste during urban and rural reforms. As the primary energy of China at present, coal resources play an important role in the development of the national economy. On the other hand, previously, caving mining methods were widely utilized in China, thus making severe destruction on the ground surface and impacting the ecological balance. The popularization of filling mining technology provided underground mining with a new perspective that fully using the construction wastes for filling gob in coal mines is of great significance in putting forward and fulfilling the green-mine construction.

2. Construction wastes in urbanization

2.1. Sources of construction wastes

Construction wastes refer to the wastes produced by constructions (buildings and infrastructures of all types) during constructing, reconstructing, extension and demolishing [1]. City construction wastes in China are mainly originated from the wastes produced by constructions and reconstructions in urbanization and the solid wastes from major disasters. In accordance with function sources, they can be classified into living construction wastes, communal facility construction wastes, industrial construction wastes and other types.

2.2. Disposal methods of construction wastes

The report on the research analysis and the market prospect forecast of China' s construction waste disposal industry from 2016 to 2020, which was released by China Industry Research, opined that the disposal methods of construction waste in China can be generally divided into two categories. The first one is to build some permanent sites for piling up wastes or interring wastes. And the other one is to recycle the wastes by using some recycling equipment which can smash wastes and then regenerate them to be reusable building materials.

Almost all cities adopt the first method which is simple but will take up much urban land during the disposal of the wastes, probably resulting in secondary pollution. In the meanwhile, the disposal in a cursory and coarse style will be detrimental to urban general planning and waste large amounts of resources.

3. Research on filling gob in coal mines

3.1. Current situations of filling coal mine gob

The development of filling techniques in China has progressed along with barren -rock dry filing intended to dispose solid wastes, hydraulic filling using classified tailings and crushed rocks, tailing cemented filling, high-concentration filling and other stages [2]. Nowadays, in China, the traditional filling techniques for gob mainly contain hydraulic filling, coal-ash filling, wind-power filling, waste-rock filling by gravity, coal-wastes strip filling, pasty and pasty-like filling that are developing latterly. At present, many experts are probing into the new-type methods of filing by high-moisture and super high-moisture materials, driving the filling techniques of mine gob into a new development period.

On the other hand, lots of problems, for example, the paucity of filling materials that is a rather serious issue, are still puzzling the mining-gob filling. Waste rocks produced by mining have been far from meeting the requirements of mine gob filling while the filling techniques by new high-moisture materials are still inchoate. Hence, it is necessary to seek more proper materials for mine filling.

3.2. Research on the filling of urban construction wastes

3.2.1 Technical feasibility of the filling of urban construction wastes. With the development of mining science and the improvement of machine manufacturing skills, mine exploitation has proceeded into an era of large-scale mechanized mining, which also laid foundation for mine filling by construction wastes. In the aspect of theoretical research, many scholars have made remarkable accomplishments in ground pressure and strata control and been able to analyze the mechanical deformation after mine gob filling. In the aspect of mechanical haulage, long-distance ribbon conveyors and the application of filling pipelines made possible the gob filling by construction wastes from the ground surface. In the respect of auxiliary equipment, the widespread application of crushers, mixing machines and solid aggregate manufacturing equipment created favorable conditions for gob filling by construction wastes.

As technology is continuously progressing, we have been capable of technically solving all kinds of problems caused by construction waste filling. Meanwhile, there have been engineering practices of underground construction waste filling, which entails the upgrade of the new-generation filling and haulage equipment to achieve high intellectualization and informationization.

3.2.2 Economic feasibility of the filling of urban construction wastes. The underground filling by construction wastes can fulfill the improvement of underground excavation conditions, surrounding rock control and the curtailment of financial loss caused by surface subsidence. In addition, using construction wastes which occupy much urban land for gob filling can save more urban space. What's

IOP Conf. Series: Earth and Environmental Science **64** (2017) 012071 doi:10.1088/1755-1315/64/1/012071

more, construction waste filling is of environmental-friendly significance that the costs for environmental governance can be reduced.

Nowadays, what should be addressed is the cooperation between the places that produce construction wastes and mine companies, where the government can play a coordinating part to help assure best economic benefits at the lowest expense.

4. Establishment of the evaluation system of green-mine filling

With the development of the social economy, people are increasingly concerned about the environmental protection. How to fulfill the green mining of coal sources and the clean utilization of them has been puzzling coal mine researchers. The so-called green mine signifies that in the whole mining process of mineral sources, well-organized mining can be accomplished and great attention is paid to keeping the influence posed by excavation on surroundings in a controllable range. Through using the construction wastes for filling the gob, utilization efficiency can be improved, which is conducive to the green-mine construction. Directing at the filling outcomes of construction wastes in gob, evaluation system of green-mine filling was established.

4.1. Goals of green-mine filling

The usage of construction wastes for filling coal mine gob can realize the goals as follows: (1) acquiring the best filling outcomes at lower economic expenses; (2) recycling sources; (3) controlling the surface subsidence posed by underground mining; (4) harmonious development of both mining and urbanization.

4.2. Evaluation index system of green-mine filling

In accordance with the technical and economic feasibility analysis of the filling using construction wastes and the goals of green-mine filling, the evaluation index system of green-mine filling, which consists of three grades, was established. As is shown in Figure 1, the final evaluation index is the environmental-friendly level, which is a comprehensive evaluation index of green-mine filling [3].

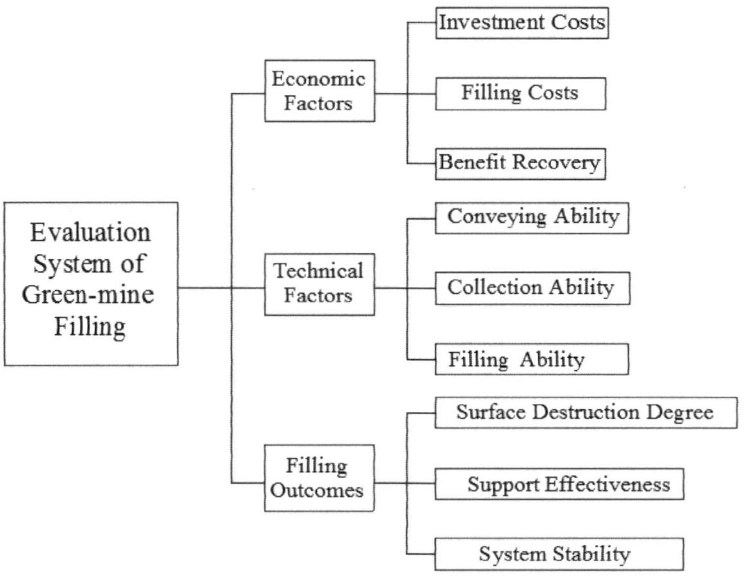

Figure 1. Evaluation system of green-mine filling

H, which stands for the environmental-friendly level in the system, has a range from 1 to 3. And higher value H has, more environmental-friendly the coal mine filling will be, which means more emphasis that has been laid on the environmental protection. We chose three first-grade indexes to measure the main factors impacting environmental-friendly level.

The constitution of the subsystem:

(1) Economic factors: they refer to the cost and benefit factors containing investment costs, filling costs and recycling benefits, which should be taken into consideration when making green-mine filling.

(2) Technical factors: they refer to the necessary technical support containing conveying ability, filling ability and collecting ability when implementing green-mine filling.

(3) Filling outcomes: they refer to three aspects in evaluation of green-mine filling, which are comprised of degrees of surface damages, support effectiveness and system stability.

4.3. Weighted calculation of evaluation index system

Weight is the relative importance of each index in the system. According to the three-grade structure that was established, analytic hierarchy process (AHP) was employed to calculate the weight of each index.

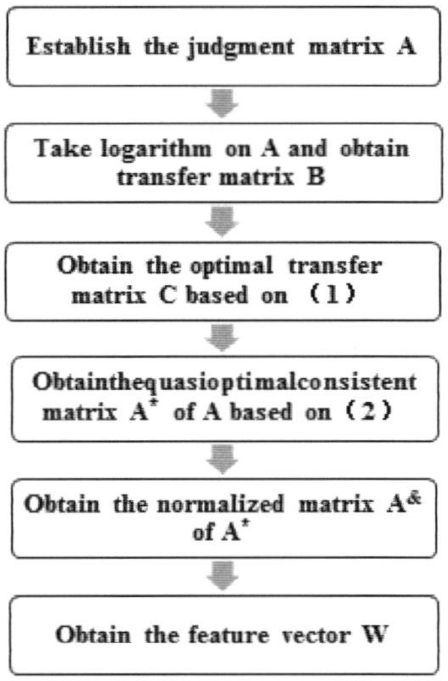

Figure 2. Calculation flow of index weight

$$Cij = \frac{1}{n}\sum_{k=1}^{n}(b_{ik} - b_{jk}) \quad (1)$$

$$a_{ij} = 10^{c_{ij}} \quad (2)$$

Based on the calculation flow shown in Figure 2, the weight of the first-grade indexes was obtained. With the second-grade indexes obeying the laws in the upper grade, AHP method was utilized to construct the judgment matrix. Meanwhile, the weight results were modified according to the actual situations in the green-mine filling, obtaining the weight calculation results of each factor, which are shown in Table 1.

IOP Conf. Series: Earth and Environmental Science **64** (2017) 012071 doi:10.1088/1755-1315/64/1/012071

Table 1. Evaluation index weight of green-mine filling

	Evaluation Indexes	Weight
Economic Factors 0.37	Investment Costs	0.28
	Filling Costs	0.34
	Benefit Recovery	0.38
Technical Factors 0.28	Conveying Ability	0.27
	Collection Ability	0.38
	Filling Ability	0.35
Filling Outcomes 0.35	Surface Destruction Degree	0.37
	Support Effectiveness	0.34
	System Stability	0.29

4.4. *Establishment of membership function of evaluation indexes*

During the green-mine filling, choosing the index parameters is a nonlinear, variable and fuzzy process. In combination with field situation, this paper adopted construction methods of fuzzy set membership functions that are simple and standard [4, 5]. Based on the field experience, on-site statistics, economic and technical indexes and related data, a hierarchical membership function of the second-grade indexes was constructed with reference to the construction methods of standard fuzzy sets. The economic factor sharing the maximum weight was set as an example to construct its reliability indexes.

4.4.1 Investment costs. Investment costs can exert influence on the percentage of the investment from fixed assets to the overall investment in the whole system of green-mine filling. Thus a higher value of this index means more investment from fixed assets. On the basis of the practical experience and related data, its membership degree is shown in Table 2.

Table 2. Membership degree of investment costs

Index k	k≤20%	20%<k≤50%	k>50%
Membership Degree	0	$f = \frac{k - 50\%}{50\% - 20\%}$	1

4.4.2 Filling costs. Filling costs refer to the percentage of the filling costs to the overall costs in the whole green-mine filling process. The greater this index is, the more filling costs will be. With reference to related field experience and data, its membership degree is shown in Table 3.

Table 3. Membership degree of filling costs

Index k	k≤80%	80%<k≤90%	k>90%
Membership Degree	0	$f = \frac{k - 80\%}{90\% - 80\%}$	1

4.4.3 Benefit recovery. Benefit recovery, which refers to the percentage of the economic benefits from green-mine filling to the overall investment costs, is proportional to the generated economic benefits. And its membership degree is shown in Table 4.

Table 4. Membership degree of benefit recovery

Index k	k≤10%	10%<k≤70%	k>70%
Membership Degree	0	$f = \frac{k - 70\%}{70\% - 10\%}$	1

4.5. Disposal of comprehensive evaluation results

The membership degree of all of the second-grade indexes should be calculated when the green-mine filling was being evaluated. Then the comprehensive evaluation was implemented according to the scores of each index. And finally the formula below was applied for calculation:

$$\mathrm{H} = \sum_{i=1}^{4} N_i \times \sum_{i=1}^{j} N_{ij} \times f_{ij} \tag{3}$$

Where, H is environmental-friendly degree which refers to the comprehensive evaluation result of the green-mine filling; N_i is the weight of first-grade (criterion grade) indexes; N_{ij} is the weight of second-grade (factor grade) indexes; f_{ij} is the membership degree of second-grade indexes.

During the real evaluation, the membership degree of each factor belonging to the second grade was obtained in accordance with the specific parameters of the tunneling face. And the comprehensive scores were classified and made statistics, which is shown in Table 5.

Table 5. Evaluation classification of green-mine filling

H	H≥0.80	0.60≤H<0.80	0.40≤H<0.60	H<0.40
Classification	A	B	C	D

Level A stands for the best outcomes of the green-mine filling; level B demonstrates that most of the requirements can be satisfied; level C means the requirements cannot be satisfied with poor outcomes; level D indicates that the whole system should be rectified and reformed in all respects.

5. Conclusion

Stemming from the disposal problem of construction wastes produced by urbanization, this paper took the current situations of coal mine gob filling into consideration, proffered the methods of using construction wastes for gob filling and meanwhile made technical and economic feasibility analysis of this strategy.

In view of the environmental protection concept, this paper established the evaluation system of green-mining filling whose goals were analyzed as well. Simultaneously, a new concept called environmental protection degree, which functioned as an ultimate evaluation index of green-mine filling, was proposed.

Since determining the membership degree of each factor and establishing the judgment matrix can exert great influence on the results, the evaluation of green-mine filling mentioned by this paper is a relatively fuzzy concept which can be improved through adjusting parameters in accordance with the actual situations.

References

[1] Clark C, Jambeck J, Townsend T. A review of construction and demolition debris regulations in United States, Critical Reviews in Environmental Science and Technology. 32(2) (2006) 141-186.

[2] R. Yang, Damages and filling application of urban construction wastes, Shandong Industrial Technology. 15 (2016) 85+73.

[3] Z.L. Song, J.F. Fan, W.H. Qi, X.C. Yang, J.G. Wang, Y.S. Ji, Study on the surface coal mine green mining technology and appraising index system, Journal of China Coal Society, S2 (2016) 350-358.

[4] KUMRAL, MUSTAFA, Reliability Based optimization of a mine production system using genetic algorithms, Journal of Loss Prevention in the Process Industries. 28(3) (2005) 16-18.

[5] WILES T.D., Reliability of numerical modelling pre-dictions, International Journal of Rock Mechanics and Mining Sciences. 21(2) (2006) 31-33.

A new calibration device of particulate matter mass concentration monitor

Junchen Zou[1,*], Wei Fan[1], Yaohua Cui[1], Bing Bao[1], Wenhui Zhou[1], Xingang Zhou[1], Chenglin Lv[1], Jiangtao Liu[1], Zhenqi Ma[1], Yong Wang[2] and Guosheng Gai[3]

[1]Henan Institute of Metrology, Zhengzhou, China
[2]Henan Institute of Product Quality Supervision and Inspection, Zhengzhou, China
[3]College of Material and Mineral Resource, Xi'an University of Architecture and Technology, Xi'an, China

*Corresponding author e-mail: junchen526@163.com

Abstract. All manuscripts must be in English, also the table and figure texts, otherwise we cannot publish your paper. Please keep a second copy of your manuscript in your office. When receiving the paper, we assume that the corresponding authors grant us the copyright to use the paper for the book or journal in question. Should authors use tables or figures from other Publications, they must ask the corresponding publishers to grant them the right to publish this material in their paper.

1. Introduction

Air pollution in developing countries has recently become a serious environmental problem, which needs more active air quality monitoring and analyses [1]. Fine particulate matter (PM) is considered to be the most significant ambient air pollutant in terms of potential health impacts. Since epidemiological studies have shown that exposure to PM with aerodynamic diameters of <10μm (PM_{10}) and especially <2.5 μm ($PM_{2.5}$) induces an increase of lung cancer, morbidity, and cardiopulmonary mortality [2-5]. Therefore, it is important that regulators are able to accurately assess the exposure of PM_{10} and $PM_{2.5}$. The increasing portability of analytical instruments which can measure PM_{10} and $PM_{2.5}$ has appeared, however, there are also a number of technical problems exist, specifically related to calibration.

In this study, a new device for calibrating the PM_{10} and $PM_{2.5}$ mass concentration monitor was designed, however, it is unknown whether the device could be used to calibrate the PM_{10} and $PM_{2.5}$ mass concentration monitor, so the uniformity and repeatability of the device were evaluated.

2. Materials and methods

In this study, some apparatus were used in the process of calibrating the PM_{10} and $PM_{2.5}$ mass concentration monitor, such as dust generator (RBG-1000, PALAS, USA), electrostatic neutralizer (EAN-581, Topas, Germany), PM_{10} cutter (2616, 16.7 L/min, BGI, USA), $PM_{2.5}$ cutter (URG-2000-30ENS, 16.7L/min, URG, USA), vacuum pump (GA-164X, Shanghai guilai industrial co., Ltd), uniform mixing box (AMMC-120, Beijing bundant technology development co., Ltd), integrated work

IOP Conf. Series: Earth and Environmental Science **64** (2017) 012072 doi:10.1088/1755-1315/64/1/012072

station (AMCP-100, Beijing bundant technology development co., Ltd), electronic scales (Dandong Baxter instrument co., Ltd), Microdust pro (880nm, CASELLA) and test dust (ISO 12103-1, A2, USA) .

The test process of the calibration device is showed in Fig.1, and it include air pump (1), the controller (2~4,13~14), dust generator (5), electrostatic neutralizer (6), three-way valve(7), PM_{10} cutter (8), $PM_{2.5}$ cutter (9), uniform mixing box (10) , membrane bracket (11~12), vacuum pump (15) and equipment being calibrated (16~17).

In order to calibrate the PM_{10} and $PM_{2.5}$ mass concentration monitor, the test dust (drying to constant weight) was placed into the dust generator, and was released stablely and uniformly by a certain rise speed and brush speed, then the test dust was neutralized by electrostatic neutralizer, and was diluted by clean air at a certain flow , then went through the PM_{10} cutter and $PM_{2.5}$ cutter in turn, and entered into the uniform mixing box, and the four thief hatch of the uniform mixing box were connected to the membrane bracket and equipment being calibrated, then the mass concentration of the test dust was acquired by comparing the weight difference of the filter paper (drying to constant weight before tested). In this study, equipment being calibrated were instead by the membrane bracket, and the thief hatch which connected to 11, 12, 16 and 17 in Fig.1, were signed by a, b, c, d respectively.

In this study, the calibration device was positioned on the horizontal floor of the selected place. To investigate the uniformity of the device, the feed rate of the test dust was setted at 60-80 mm/h, the speed of the brush was setted at 900 rpm, and sampling time of 30 min was set for each cycle using a constant flow of 16.7 L/min, which both the electrostatic neutralizer branch and the dust diffuser branch were 8.35 L/min, and dilution air flow of 20 L/min, then PM was collected by polycarbonate Track-Etch membrane of 50.0 mm diameter. Most important of all, the device must be placed at specific environment more than 2 hours, which the temperature is (15~30) ℃, the relative humidity is no more than 75%.

To investigate the repeatability of the device, the thief hatch of the calibration device connected to the microdust pro.The device and the microdust pro started at the same time, record data once a minute after the calibration device run steadily. Each thief hatch had three replicates.

3. Results and discussion

Jin et al. [6] measured the concentration of $PM_{2.5}$ in the ambient air by *Determination of atmospheric articles PM_{10} and $PM_{2.5}$ in ambient air by gravimetric method* (HJ 618—2011), analyzed the influence factors, and found that the sampler, filter membrane and the accuracy of the balance had important effect on the measure result, especially, the uncertainty of the sampler mainly contained the deviation of cutter capture efficiency and sampling flow meters. Moreover, some researchers found that the cutting efficiency of the cutter may change the measure result of the instrument [7]. In order to reduce the error, the properties of PM2.5 cutter and PM_{10} cutter were evaluated before starting the test. The result could be seen in Fig.2 and Fig.3, it showed that when the cyclone flow rate was greater than 15 L/min, the cutpoint diameter of PM_{10} cutter and $PM_{2.5}$ cutter were less than 10 μm and 2.5 μm respectively. During the test, the cyclone flow rate was setted at 16.7 L/min, it indicated that PM_{10} cutter and $PM_{2.5}$ cutter could meet the demand of the test.

Figure 1. The test process of the Calibration device.

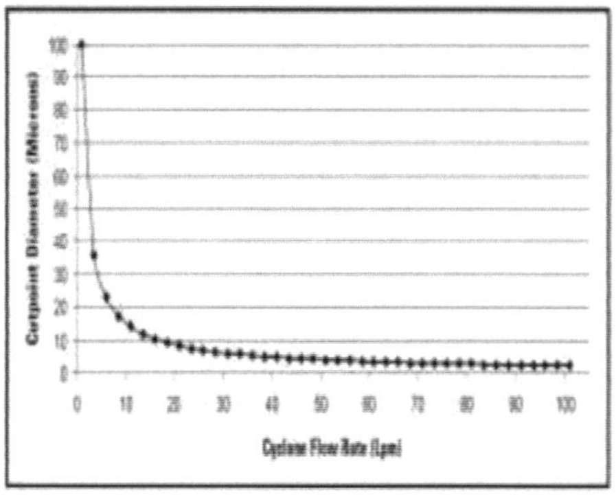

Figure 2. The cutting efficiency Of PM_{10} cutter.

According to the test process of the calibration device, the test was carried out and the uniformity (U) was calculated by

$$U = \frac{1}{1.13} \times \frac{\rho_i - \rho_j}{\bar{\rho}} \times 100\% \tag{1}$$

Where ρ_i and ρ_j showed the measure result of i and j thief hatch respectively, $\bar{\rho}$ showed the arithmetic mean of ρ_i and ρ_j. The data was shown in Fig.4, although the uniformity of the thief hatch a and b, a and c, a and d, b and c, b and d, c and d were different, but all the uniformity was no more than 8%, it indicated indirectly that the uniform mixing box of the calibration device was in good working status.

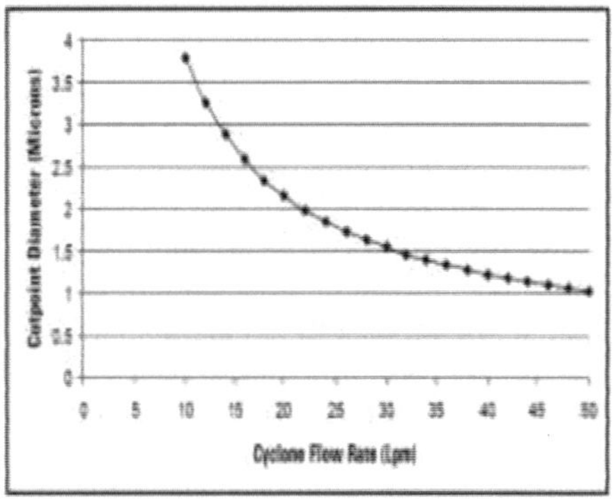

Figure 3. The cutting efficiency Of $PM_{2.5}$ cutter.

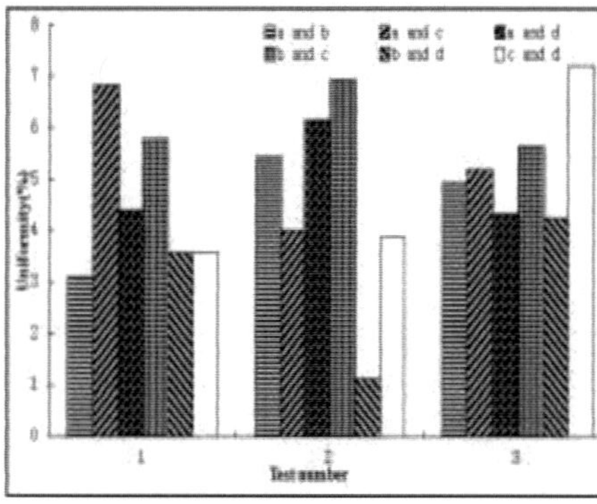

Figure 4. The uniformity of the Calibration device.

Moreover, the obtained repeatability data was shown in Fig.5~Fig.8, and the repeatability (R) was calculated by

$$R = \frac{1}{\bar{\rho}} \sqrt{\frac{\sum_{i=1}^{n} (\rho_i - \bar{\rho})^2}{n-1}} \times 100\% \tag{2}$$

Where ρ_i showed the measure result of i thief hatch, $\bar{\rho}$ showed the arithmetic mean of ρ_i when i started from 1 to n, n showed the total test number. The calculated data of the four thief hatch was 0.42%, 0.49%, 0.51%, 0.59%, 0.55%, 0.97%, 0.75%, 0.80 %, 0.55%, 0.45%, 0.43% and 0.66%, respectively. It was observed that the repeatability of the thief hatch was no more than 5%, and showed that the calibration device had good repeatability.

In this study, the test dust was dried before the experiment, so that it decrease the error which caused by high humidity, it was consistented with previous research [8]. They found that there was a significant discrepancy between the concentration measured for heated and non-heated inlet in high humidity conditions. Without the heated inlet, it was likely that water droplet was contributing a significant proportion of the overall particle count.

In the process of evaluating the repeatability of the standard measurement equipment [9], some researchers found that when the repeatability of the measured objects was as small as possible, the repeatability obtained by experimental standard deviation would be more close to the repeatability of the standard measurement equipment, the data would be more accurate and reliable.

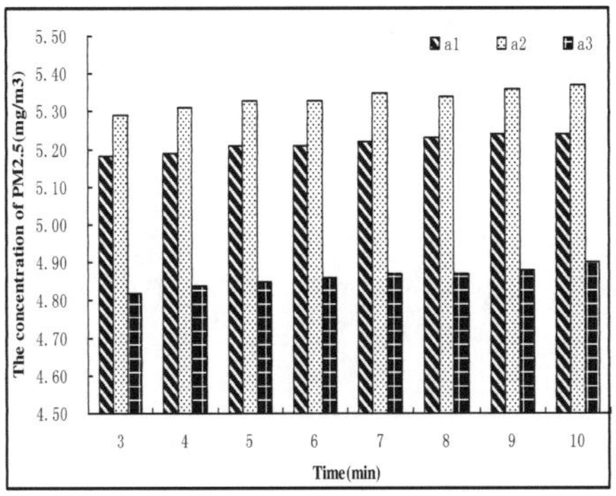

Figure 5. The repeatability of The thief hatch a.

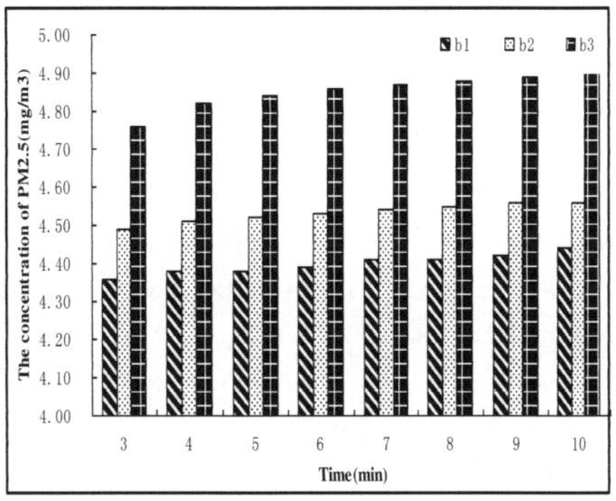

Figure 6. The repeatability of The thief hatch b.

In this study, the uncertainty of PM concentration was from the standard device and the equipment being calibrated. The uncertainty of the standard device was introduced by the sampling flow rate of the standard membrane, time, scales and aerosol blending device. The uncertainty of the equipment being calibrated was mainly depended on its own repeatability.

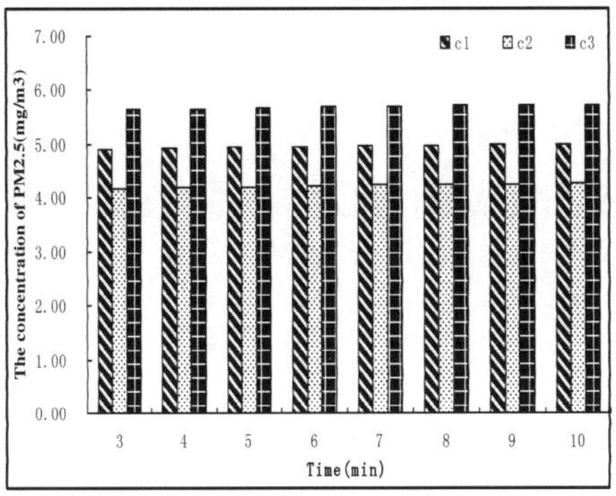

Figure 7. The repeatability of The thief hatch c.

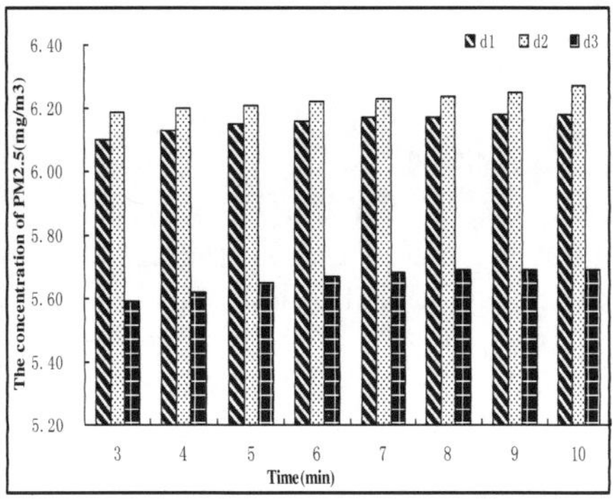

Figure 8. The repeatability of The thief hatch d.

It is known to all, the most common air sampling devices are those that collecting PM which passing through the sampling inlet on the filters. In this filter-based method, error would produce in the filter-handling process because filters must be weighed before and after sampling. When sampling and weighing, it would also lead to the loss of some semi-volatile species including organic compounds and ammonium nitrate. Furthermore, no information was gained about the error of time during sampling [10].

In this study, due to the feed rate was setted at 60-80 mm/h and the brush rate setted at 900 rpm, so the obtained concentration of $PM_{2.5}$ was 4 to 10 mg/m^3, if the feed rate could be decreased furtherly, and the effect of other factors on the experimental result could be controlled better, the obtained concentration of $PM_{2.5}$ would be lower than previous obtained, even could reach the required concentration range of GB 3095-2012 in china. The calibration device could calibrate the PM_{10} mass concentration monitor when $PM_{2.5}$ cutter was removed. Furthermore, the calibration device could real-time measure the concentration of $PM_{2.5}$, PM_{10} and total suspended particulate matter in the air by switching the three-way valve.

4. Conclusion

This work proved that the calibration device had good properties, such as cutting efficiency, uniformity and repeatability, and could be used to calibrate the PM_{10} and $PM_{2.5}$ mass concentration monitor, and would have a good application prospect in real-time measure the PM concentration of the air.

Acknowledgments

This work was financially supported by the Science and Technology Research Project of Henan Province (Nos. 172102210313) fund.

References

[1] R. Alireza, C. Rautenbach, G. Patrik, G. Dimitris, G. Pawan, Temporal changes of particulate concentration in the ambient air over the city of Zahedan, Air. Qual. Atmos. Hlth. 6 (2013) 123–135.

[2] K. Bhaskaran, P. Wilkinson, L. Smeeth, Cardiovascular consequences of air pollution: what are the mechanisms, Heart. 97 (2011) 519–520.

[3] R. Brook, B. Urch, J. Dvonch, R. Bard, M. Speck, G. Keeler, M. Morishita, F. Marsik, A. Kamal, N. Kaciroti, J. Harkema, P. Corey, F. Silverman, D. Gold, G. Wellenius, M. Mittleman, S. Rajagopalan, J. Brooky, Insights into the mechanisms and mediators of the effects of air pollution exposure on blood pressure and vascular function in healthy humans, Hypertension. 54 (2009) 659–667.

[4] G. Sivagangabalan, D. Spears, S. Masse, B. Urch, R. Brook, F. Silverman, D. Gold, K. Lukic, M. Speck, M. Kusha, T. Farid, K. Poku, E. Shi, J. Floras, K. Nanthakumar, Mechanisms of increased arrhythmic risk associated with exposure to urban air pollution, Circulation. 122 (2010) A17901.

[5] M. Stafoggia, E. Samoli, E. Alessandrini, Short-term associations between fine and coarse particulate matter and hospitalizations in Southern Europe: results from the med-particles project, Environ. Health. Persp. 121 (2013) 1026–33.

[6] H. Jin, Q. Fu, X. Wu, Y. Yao, The Uncertainty of Determination of PM2.5 in the Ambient Air by Gravimetric Method, Environ. Monit. in China. 30 (2014) 32-35.

[7] Y. Cai, Y. Zhang, F. Sha, L. Shen, H. Hu, Determination of PM2.5 Concentration by Manual Sampling and Auto Monitoring, Environ. Sci. Mana. 37 (2012) 110-113.

[8] E. Michael, J. Samantha, K. Amy, M. Adam, S. Thomas, Practicalities of mapping PM10 and PM2.5 concentrations on city-wide scales using a portable particulate monitor, Air Qual. Atmos. Hlth. 2016.

[9] Y. Zhang and C. Zhou, The influences on the results of repeatability of measurement standard using different DUT, Metrol. Test. Technol. 40 (2013) 51-52.

[10] Y. Lee, H. Kim, K. Lee, Development of monitoring technology for airborne particulate matter, Environ. Monit. Assess. 70 (2001) 3-20.

IOP Conf. Series: Earth and Environmental Science **64** (2017) 012073 doi:10.1088/1755-1315/64/1/012073

Exploration on the technology for ozone reduction in urban sewage treatment

Min Yang, Yi Sun, Zhicheng Han, Jun Liu

Architects & Engineers Co., Ltd. of Southeast University, Nanjing, 210096, China

Abstract. With the rapid development of China's economy, urban water consumption is increasing. However, sewage treatment plants will produce large amounts of sludge after treatment of sewage. Generally, and the sludge treatment costs are relatively high. Therefore, the problem about how to deal with the sewage sludge becomes the hot issues. Municipal waste water treatment plant produces a lot of sludge. This paper summarized the abroad study of ozonation minimization technology. Introduction and discussion were made on the principle of ozonated efficiency of sludge minimization, the efficiency of sludge minimization and the relationship between efficiency and ozone dosage, as well the effect of return sludge ozonated on waste water treatment running and the sludge setting and the dewatering characteristic. The economic estimation was also made on this technology. It's showed that sludge minimization technology exhibits extensive application foreground.

1. Introduction

With the rapid development of China's economy, urban water consumption is increasing. However, sewage treatment plants will produce large amounts of sludge after treatment of sewage. Generally, and the sludge treatment costs are relatively high. Therefore, the problem about how to deal with the sewage sludge becomes the hot issues. The relevant technical staff should strengthen the research on the sludge deodorization technology of urban sewage treatment plant, reducing the sludge generated by sewage treatment plants. This paper mainly studies the sludge ozone reduction technology in urban sewage treatment plant, which is analyzed and discussed from the ozone sludge reduction principle, the relationship between the sludge reduction and the ozone dose and the the impact of ozone sludge treatment technology on sewage treatment plant. This can enhance the technical level of China's sewage plant [8].

In general, ozone is a potent oxidant that can do the relevant ozone treatment on sludge. First, ozone causes the cell wall of the sludge microbial to crack, causing the organic matter in the cell to be released. When the sludge flows to the biochemical pool, the organic material matrix is metabolized, resulting in H_2O, CO_2, NH_3 and other substances, thus contributing to the reduction of sludge. Second, ozone can produce oxidation on the sludge, making the protein, fat, carbohydrates and other substances oxidizing to carbon dioxide or water, so as to achieve the purpose of reducing sludge.

2. The relationship between sludge reduction and ozone dose

In general, there are a large number of bacteria in the sludge. First of all, the ozone is used as the relevant oxidant entering into the sludge, which can damage and dissolve the cell wall of sludge microbial. After this, the organic matter of microorganisms will be released. These organic matter becomes nutrients of the treatment. Second, in the process of ozone using, ozone will cause another oxidation in the internal sludge. Therefore, the related chemical reaction between the carbohydrates and protein is promoted, which will become nitric oxide and carbon monoxide. In the ozone-related role, the sludge is decomposed into valuable elements, thereby promoting the decrease of the sludge in the sludge treatment plant. Finally, the ozone has a high efficiency on sludge treatment. Generally, the efficiency of the ozone treatment on sludge is related to the ozone dose and the amount of sludge entering the ozone reactor [1]. The relevant studies show that when the ozone-returning sludge is 3.3 times than the total amount of residual sludge and the ozone emission is 0.15kg, it can achieve zero emissions of residual sludge. In addition, the ozone can completely remove the sludge from the sludge pool of sewage treatment plant when it rises to a certain extent. Therefore, the effects of ozonized sludge technology on sludge treatment is very obvious. The sludge reduction efficiency is closely related to the ozone dose. When the ozone dose is as small as the standard dose, the sludge reduction efficiency is positively correlated with the ozone dose. When the ozone dose exceeds the standard dose, the sludge reduction efficiency will not increase significantly after adding the ozone dose [2].

3. Effect of Ozone Sludge Reduction Technology on Sewage Treatment Plant

(1) The impact on water quality

In general, the reduction of the sludge content of the sewage treatment plant should ensure the quality of the effluent from the WWTP. If the quality of the effluent from the sewage treatment plant is deteriorated after the relevant ozone sludge reduction treatment, the meaning of the ozone sludge reduction technology will be lost. According to the relevant reports, the effluent quality after the ozone sludge reduction treatment is basically the same as that of the water after the technical treatment, and the BOD removal efficiency after the ozone treatment is better. Although the SS concentration will increase occasionally with seasonal variation, BOD also increases due to the increase of SS concentration. In addition, BOD and SS basically meet the requirements of local water discharge [3].

(2) The relevant impact on nitrogen and phosphorus removal

Through the biological technology on nitrogen and phosphorus removal, the organic pollutants can be effectively removed, and the plant nutrients such as nitrogen and phosphorus are also removed. Some staff worry that the technology on ozone sludge reduction can affect the effect of nitrogen and phosphorus removal. According to the experiment, it shows that the effluent ammonia nitrogen has no significant change after ozone treatment [4]. The concentration of effluent ammonia nitrogen maintains a relatively low value, which generally is about 4 mg per liter. Therefore, the impact of ozone treatment on nitrogen and phosphorus removal is not worried.

(3) The relevant impact on sludge setting and dehydration performance

After the ozone treatment, whether the sludge sedimentation and dehydration performance caused by a negative impact is also concerned from the relevant issues. The research shows that ozone sludge reduction technology is beneficial for sludge sedimentation and dehydration. First, in general, the sludge will become a small floccule after ozone treatment, and then it will back to the biochemical pool, so as to promote the sludge particles homogenized. In addition, it can promote the sludge settling performance. Second, after the ozone treatment, the volume index of sludge decreased significantly. According to microbiological examination, it showed that the filamentous bacteria significantly reduced after the ozone treatment of sludge [5].

(4) The economic value of ozone sludge reduction technology

After finishing the sewage treatment, it is necessary for us to deal with the sludge in the sewage pool. The sludge treatment costs will be reduced through the ozone sludge reduction technology and the economic benefits of the sewage treatment plant will be enhanced. First, the amount of residual sludge is reduced, and the sludge treatment costs is also reduced. Second, although the ozone and related facilities and electricity is provided in sludge reduction technology to increase the operating costs and investment costs, its high efficiency and processing effect is obvious. After considering the comprehensive effects, ozone technology still has a significant investment value and economic benefits in sewage treatment [6].

4. The methods and results

4.1. Experiment of Sludge ozone treatment

The sludge used in the experiment was taken from the re-fluxing sludge of the secondary sedimentation tank of Beijing Qinghe Sewage Treatment Plant A/A/O. In addition, the MLSS and MLVSS were 8603 and 6185 mg·L-1, respectively. Meanwhile, 1400 mL sludge was taken with aeration for 24 h before each experiment and then diluted to 3500 mL with deionized water. The sludge treatment was carried out using a semi-continuous reaction, which means that sludge was continuously fed into the reactor, and two batches of ozone treatment were completed. The volume of the sludge sample was 500 mL and the flow rate of the ozone gas was 0.2 L·Min-1 by using sand core aeration. In the first batch of experiments, the initial pH is 6.8, and the oxidation time is 10 min. The ozone dosage was 0, 26, 63, 154, 227, and 268 mg·g-1 (in SS dollars). In the second batch, the ozone concentration was 50 mg·L-1, and the oxidation time was 9 min. In addition, the initial pH was adjusted to 3.0, 5.0, 7.0, 9.0 and 11.0 [7].

The ozone concentration was measured by the iodine method, and the residual ozone in the system was purged with nitrogen and collected with 1% KI solution. The ozone actually consumed was calculated by the formula (1).

$$M_{O_{3cons}} = C_{O_{3gasin}} \times Q_{O_{3gasin}} \times \mathrm{T} - \mathrm{M}_{O_{3gasout}} \tag{1}$$

In the formula, $M_{O_{3cons}}$ is the ozone consumption (mg), and C is the ozone concentration in the entrance of the reactor. Q is the flow rate of ozone gas. Measured by gas flow meter, we can obtained that T is ozone oxidation time of activated sludge (min), and M is the amount of ozone that did not participate in the reaction (mg).

The sludge dissolution rate R is used to characterize its lysis degree, and the calculation formula is shown in the follow:

$$\mathrm{R} = \frac{[\mathrm{MISS}]_o - [\mathrm{MISS}]_t}{[\mathrm{MISS}]_o} \times 100\% \tag{2}$$

In the formula, $[\mathrm{MISS}]_o$ and $[\mathrm{MISS}]_t$ represents the sludge MISS (mg L^{-1}) with ozone treatment and without ozone treatment.

4.2. Results and discuss

In the process of sludge ozone treatment, the initial dosage of ozone and the initial pH of activated sludge will affect the dissolution rate of sludge and the dissolution of organic matter, nitrogen and phosphorus, which will affect the effect of sludge reduction. Ozone dosage and sludge initial pH are not only beneficial for sludge reduction, but also to improve the economics of the process. In this experiment, the effect of different ozone dosage on the sludge dissolution rate was investigated. As shown in Fig. 1, the sludge MLSS and MLVSS decreased with the increase of ozone dosage, and the sludge dissolution rate increased gradually.

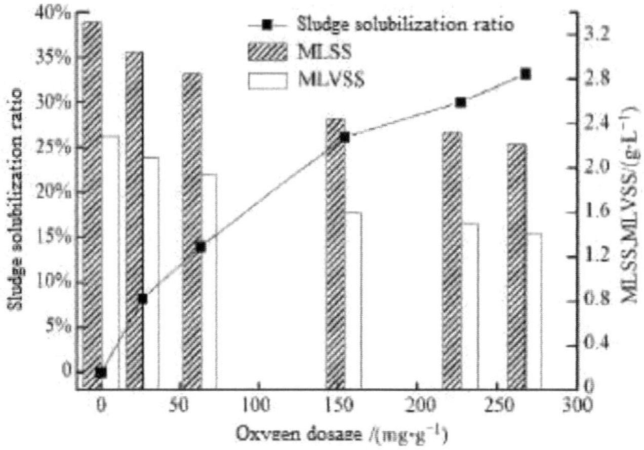

Fig. 1 The effect of ozone dose on sludge solubilization ratio

After the zone treatment, the effect of sludge reduction is more obvious. With the increase of ozone dosage, the MLSS and MLVSS of the sludge gradually decrease, and the sludge dissolving rate gradually increases. Meanwhile, the organic matter, nitrogen, phosphorus and other substances in the microbial cells are released to the sludge liquid. It is suggested that the reasonable ozone dosage is about 150 $mg \cdot g^{-1}$, and the sludge dissolving rate can reach about 26%. The initial pH is alkaline, and the sludge dissolving rate is high. In addition, the organic matter is favorable the dissolution of nitrogen, while the acidic conditions is conducive to the dissolution of phosphorus. Considering the initial pH effect, sludge ozone treatment should be in the condition of neutral or weak alkaline of initial pH.

After the sludge was returned to the biological treatment system, the effect of microbial activity, COD and nitrogen removal was not significantly affected. As the system had no excess sludge discharge, the phosphorus was gradually accumulated in the system, resulting in the decrease of phosphorus removal effect. Therefore, the phosphorus removal process needs to be added.

5. Conclusion

In summary, the technology of ozone sludge reduction can not only effectively deal with the sludge produced by the city's waste water treatment, prompting a significant reduction in the total sludge, but also can promote the oxidation of sludge into organic matter, which is conducive to ecological construction. The relevant technical personnel should strengthen the research on ozone sludge reduction technology. In addition, the technology should be widely used in China's sewage treatment plant, so as to improve the capacity for sewage treatment.

Reference

[1] Yang S, Wang Q. H, Zhang T, et al. Biological nitrogen removal using the supernatant of ozonized sludge as extra carbon [J]. Ozone: Science and Engineering, 2011, 33(5): 410-416.

[2] Saktavwin W, Tsuno H, Nagare H, et al. Advanced sewage treatment process with excess sludge reduction and phosphorus recovery [J]. Water Research, 2011, 39(5): 902-910.

[3] Boteher. M., Siegrist. H, Partial ozonation of activated sludge to reduce excess sludge, improve denitrification and control scumming and bulking [J]. Water Science and Technology. 2004, 49(10): 41-49.

[4] Yan S T, Chu L B, Xing X H, et al. Analysis of mechanism of sludge ozonation by a combination of biological and chemical approaches [J]. Water Research, 2009, 43(1): 195-203.

[5] Ahh K H, Park K Y, Maeng S K, et al. Ozonation of water waste sludge for reduction and recycling [J]. Water Science and Technology, 2001, 46(10): 71-77.

[6] Abraham F J, Md Abul H S. Formation of pyrazines in hydroxyacetaldehyde and glycine nonenzymatic browning Maillard reaction: A computational study [J]. Food Chemistry, 2007, 103(4): 1208-1216.
[7] Zhang G M, Yang J, Liu H Z, et al. Sludge ozonation: Disintegration, supernatant changes and mechanism [J]. Bioresource Technology, 2009, 100(3): 1505-1509.
[8] Yeom I T, Lee K R, Lee Y H, et al. Effects of ozone treatmenton the biodegradability of sludge from municipal wastewater treatmentplants [J]. Water Science and Technology, 2002, 46 (4/5): 421-425.

IOP Conf. Series: Earth and Environmental Science **64** (2017) 012074 doi:10.1088/1755-1315/64/1/012074

Research on information security system of waste terminal disposal process

Chao Zhou, Ziying Wang, Jing Guo, Yajuan Guo, Wei Huang

State Grid Jiangsu Electric Power Research Institute, Nanjing, 211100, China

Abstract. Informatization has penetrated the whole process of production and operation of electric power enterprises. It not only improves the level of lean management and quality service, but also faces severe security risks. The internal network terminal is the outermost layer and the most vulnerable node of the inner network boundary. It has the characteristics of wide distribution, long depth and large quantity. The user and operation and maintenance personnel technical level and security awareness is uneven, which led to the internal network terminal is the weakest link in information security. Through the implementation of security of management, technology and physics, we should establish an internal network terminal security protection system, so as to fully protect the internal network terminal information security.

1. Introduction

Informatization has penetrated the whole process of production and operation of electric power enterprises [3]. It not only improves the level of lean management and quality service, but also faces severe security risks. The internal network terminal is the outermost layer and the most vulnerable node of the inner network boundary [2]. On the one hand, it has the characteristics of wide distribution, long depth and large quantity. The user and operation and maintenance personnel technical level and security awareness is uneven, which led to the internal network terminal is the weakest link in information security. On the other hand, the environment of the internal network terminal is complicated [1]. It is difficult to update various security patches in a timely manner. With the endless stream of new technologies, security technology measures are difficult to fully meet the needs of rapid system changes [4]. In recent years, a variety of network security incidents occurs frequently, especially the large-scale incidents such as distributed denial of service attacks, network worms and botnets, causing a great of economical loss to our society [5]. Electric power enterprises need to implement safety measures in three aspects: management, technology and physics. In addition, enterprises should establish an internal network terminal security protection system in order to fully protect the network information security [6].

2. Internal network terminal security protection system

The internal network terminal security protection system is to complete the security control within the whole life cycle of the network terminal [7]. From the decision-making support, management attention and business layer implementation, enterprises should establish a comprehensive information security system. Specifically, it completes the security work from the three aspects of organization, technology and physics.

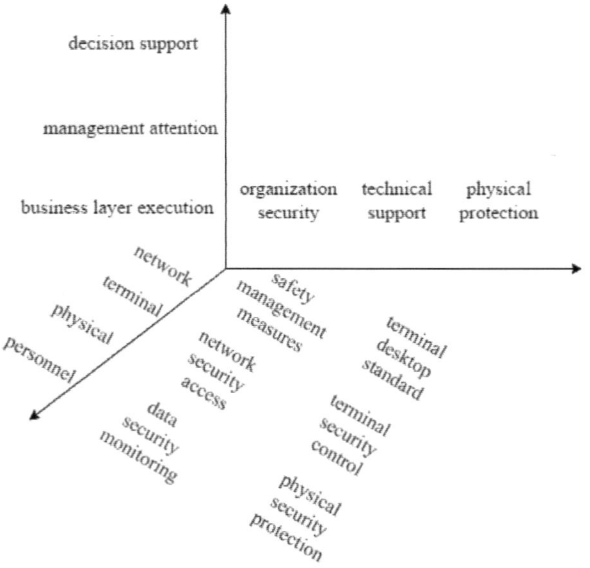

Fig. 1. Internal network terminal information security protection system

The life cycle security of the internal network terminal should complete the standardization and security access control before entering the network, network security supervision and data security monitoring after entering the network, and data security destruction and terminal retirement after get offline [8]. Next, we will describe the security protection system of Intranet terminal from the aspects of safety management, technology and physical security measures.

3. Safety management measures

Safety management measures are the most critical factors in the security system. The organization should pay attention to the safety management in order to provide a strong human and financial guarantee for the safety work. The organization should set up the Information Security Committee, establish the organization security and management measures, improve the safety and technical safeguards measures, complete the operation and maintenance of the internal network terminal, and implement the physical security guarantee of the internal network terminal. In the management, the organization should improve the information security management rules and regulations, strengthen safety training and safety knowledge. In the physical security, through the security label, as well as desktop terminal slogan online push, safety training and assessment, the organization can strengthen the protection. In terms of the system, the organization can establish a post safety responsibility system, and sign the post safety responsibility. For suppliers and external service personnel, the organization should focus on safety training, and sign a security commitment and responsibility for confidentiality. Therefore, the external staff must participate in the company's safety training and through the safety examination before they can engage in related work.

4. Technical and physical safety measures

4.1. Desktop standardization

Desktop standardization is conducive to the establishment of a unified security measures. At the same time, it can prevent and reduce unknown risks. Standardization brings the desktop to a high degree of consistency. Based on a large number of pre-security testing and verification, it forms a fully mature solution to known problems, and it can quickly build an emulation environment and test the solution for unknown problems.

The intranet terminal desktop standardization establishes a standard desktop installation mirror library. The enterprise should unify the operating system, the common application software and each kind of plug-in version and the patch installation, and completes the system and the application security reinforcement. For a typical user, the enterprise should establish a corresponding standard desktop image. When the internal network terminal is connected to the network, the mirror installation can reduce operation and maintenance workload. In addition, for users to install software, the enterprise must carry out the whole process of security management and control. From the application review, the installation of software testing, authorized installation and user feedback improvements, companies must complete the whole process of closed-loop management and update the desktop image source.

In the desktop image security system, companies should strengthen the system, to prevent the system weak password and screen savers are not set or set empty password. In addition, companies have to install a unified anti-virus software, establish enterprise anti-virus server and patch update server, and update the virus database and system patches. At last, enterprises should install the Internet desktop control and confidentiality detection tools to prevent internal network data leakage and terminal illegal outreach.

Intranet terminal desktop standardization can greatly decrease the internal network terminal operation and maintenance workload and problem complexity, and reduce the workload of the operation and maintenance personnel. Standardization can find the existence of internal network terminal security risks in advance. It can improve the quality of desktop operation and maintenance services, and enhance the overall level of corporate information security.

4.2. Network security access

The network security access system should ensure that the terminal of the access network meets the requirements of desktop standardization. Enterprises should clear the responsibility, take great care of the internal network terminal status, to ensure that the network terminal can safely control. Security access is the first step to grasp the security of intranet terminal. It can eliminate the terminal which does not meet the safety requirements, and ensure the access of the internal network terminal to meet the relevant safety requirements, so as to establish a safe and reliable environment for the internal network terminal.

The intranet terminal security must be checked before entering the network. Intranet terminal can be installed with desktop standardization before entering the network. Through the automated detection tools, enterprises can check the internal network terminal operating system, application software, anti-virus software, anti-illegal outreach program, desktop terminal security control client and confidential inspection tools, system patch updates, weak passwords and screensavers, terminal security monitoring client installation, and security reinforcement and so on, automatically generate a system test report. The user submits the application and submits the audit with the test report. After the audit is passed, the internal network access system automatically updates the DHCP server to configure IP / MAC address information.

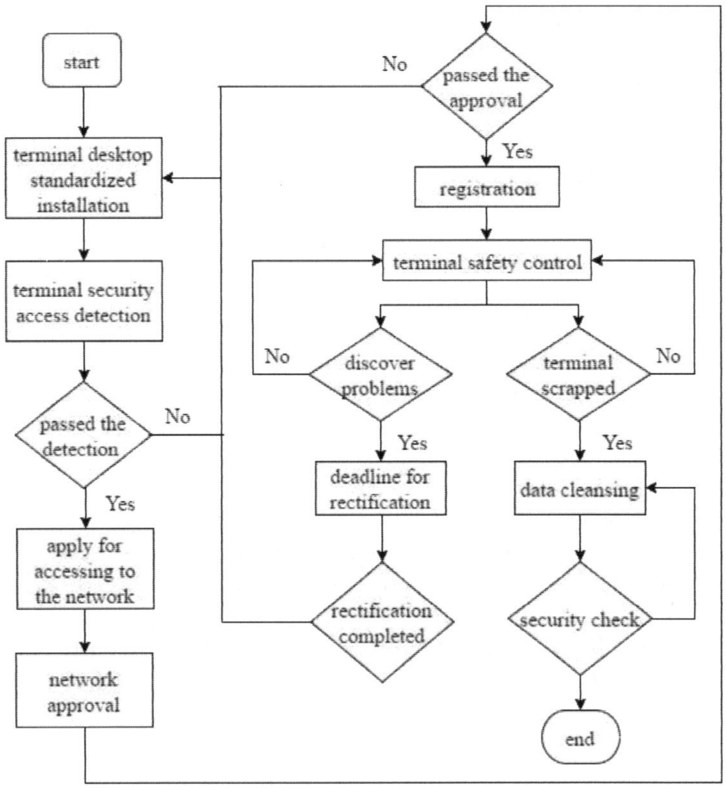

Fig.2. Network terminal access and control process

After the network terminal entering the network, the enterprise can control the security status of the terminal through the network terminal desktop management system, terminal security monitoring tools and security detection tools. After monitoring the system changes within the network terminal that does not meet the safety requirements, the internal security terminal immediately alert rectification. For a limited period of time without timely rectification of the internal network terminal, it will take measures to force off the network. After the completion of the rectification of the internal network terminal, it can re-apply for access according to the network security access process.

4.3. Terminal safety control

The network security control of the internal network terminal prohibits the internal network terminal access to the non-intranet environment, to prevent the internal information leakage or external network Trojan virus placement. Internal network terminal network security control is divided into terminal security control and network security control.

By monitoring the terminal network configuration parameters to change and filter network packets, terminal security management can achieve network security management and control. In general, the IP and MAC addresses of the internal network terminals assigned by the DHCP server through the network security access system will not be changed. It can monitor the change of IP and MAC address of the terminal in real time. At the same time, it can regularly detect the internal and external network targets to monitor the environment within the network terminal. It can set the detection target both inside and outside the network. Once the external network target is detected or the intranet target cannot be detected, it indicates that the terminal is in an insecure network environment and immediately warns the user to take security measures. When it is not connected within the time limit, it immediately disconnects all network connections from the terminal. At the same time, through the driver level network packet filtering technology, all the data packets flowing through the network card are checked, and the data packets that do not meet the company's network address specification are filtered to prevent the internal network information leakage.

By monitoring the network log and behavior audit and anti-virus system log alarm information, network security control system regularly carry out hidden trouble investigation and analysis. It must rectify unregulated terminal, infected with high-risk virus Trojans or other unauthorized access terminal in time.

4.4. Data security monitoring

Data security monitoring of the internal network terminal ensures data storage and transmission security, traces audit data, and monitors sensitive data. It is strictly forbidden to store, process and transmit secret data. The confidential information must use special purpose machine, in order to ensure the security of intranet data information, and to prevent the occurrence of information security incidents.

Enterprises should strictly control the interaction between internal and external network data. All internal network data must be encrypted and copied via secure mobile storage media. External network data must be detected by anti-virus software before entering the network. At the same time, the network terminal must be installed anti-virus software and update the virus database, to prevent the virus Trojan to steal or destroy the network terminal data. Enterprises should unify Internet exports, and deploy sensitive e-mail blocking and network behavior auditing systems. The network behavior audit log should be kept for more than 6 months. The enterprise must conduct a sensitive e-mail review of all outgoing messages, and for the messages that involve sensitive information must be sent with strong password encryption.

4.5. Physical protection measures

Enterprises can take physical measures to ensure the safety of the network terminal. Through the chassis lock, in the case of ensuring physical security while ensuring that the system hardware is not disassembled, enterprise employees can make offline copies of data in order to prevent data leakage. In the cabinet and the display prominently affixed to the security label, as well as clear responsible person information, which can urge the responsible person to improve safety awareness, so as to ensure the physical security of the intranet terminal. In the presence of fingerprint or iris authorization verification conditions, enterprises can take the password and biometric identification of two-factor verification to strengthen the physical protection measures, and strictly guarantee the legitimate use of the network terminal.

5. Conclusion

For the various types of information security issues, the establishment of internal network terminal security protection system needs to organize all levels of linkage and full participation. From the aspects of organization, technology and physics, the enterprise should strengthen the security measures in a comprehensive way, which has an important role in improving the overall level of safety protection, and reducing security incidents and security risks. For the popularization and application of cloud computing, Internet of Things, and large data in the internal network, enterprises should establish a comprehensive security risk early warning mechanism, which is conducive to the establishment of corporate information security top-level protection system.

References

[1] Ahmad, Atif, Sean B. Maynard, and Sangseo Park. "Information security strategies: towards an organizational multi-strategy perspective." Journal of Intelligent Manufacturing 25.2 (2014): 357-370.

[2] Knowles, William, et al. "A survey of cyber security management in industrial control systems." International journal of critical infrastructure protection 9 (2015): 52-80.

[3] Legros, Benjamin, Oualid Jouini, and Ger Koole. "In this paper, an analytic model is built to study the medium access control (MAC) performance and reliability of general two-dimensional (2-D) IEEE 802.11 based wireless networks for one-hop broadcast communications. First, a semi-Markov process (SMP) model is introduced to capture the medium contention and backoff behavior and the impact of hidden terminal problem in IEEE 802.11 broadcast wireless..." Performance Evaluation 99.100 (2016): 37-54.

[4] Lu, Yanrong, et al. "Robust and efficient biometrics based password authentication scheme for telecare medicine information systems using extended chaotic maps." Journal of medical systems 39.6 (2015): 65.

[5] Muttaqi, Kashem M., et al. "Technical challenges for electric power industries with implementation of distribution system automation in smart grids." Renewable and Sustainable Energy Reviews 46 (2015): 129-142.

[6] Sajid, Anam, Haider Abbas, and Kashif Saleem. "Cloud-assisted IoT-based SCADA systems security: A review of the state of the art and future challenges." IEEE Access 4 (2016): 1375-1384.

[7] Silva, Maisa Mendonça, et al. "A multidimensional approach to information security risk management using FMEA and fuzzy theory." International Journal of Information Management 34.6 (2014): 733-740.

[8] Vollmer, Todd, and Milos Manic. "Cyber-physical system security with deceptive virtual hosts for industrial control networks." IEEE Transactions on Industrial Informatics 10.2 (2014): 1337-1347.

IOP Conf. Series: Earth and Environmental Science **64** (2017) 012075 doi:10.1088/1755-1315/64/1/012075

A contrast experiment research on rural wastewater treatment efficiency of BAF with zeolite and zeolite-haydite combined packing

Xuemei Ji[1], Jifu Yang[2], Lirong Xu[1,*], Bin Li[2] and Cui Zhao[2]

[1]School of Resources and Environment, University of Jinan, Jinan 250022, China
[2]Department of Irrigation and Drainage, China Institute of Water Resources and Hydropower Research, Beijing 100038, China

[*]Corresponding author e-mail: stu_xulr@ujn.edu.cn

Abstract. In order to investigate the effect of biological aerated filter (BAF) with zeolite filler and combined packing, a contrastive experimental study was carried out on zeolite BAF and zeolite-haydite combined packing BAF with the operation parameters such as hydraulic retention time (HRT), temperature and DO keeping constantly throughout the experimental process and the removal efficiency of CODcr, NH_4^+-N and SS were analysed during the start-up stage and the steady stage. The results showed that on condition that the average temperature was 27℃ and DO 6mg/L and HRT 8h, BAF with zeolite filler formed film completely in 30 days, with the removal rate of CODcr, NH_4^+-N and SS 59.6%, 94.6% and 87%, while zeolite-haydite BAF formed film completely in 25 days, with the removal rate of CODcr, NH_4^+-N and SS 63.7%, 95.9%, 88.5% respectively, which was better than the former. In summary, BAF with combined zeolite and haydite is superior to BAF with single zeolite for rural wastewater treatment.

1. Introduction

Biological aerated filter (BAF) is a novel biofilm process which has developed in occident countries since 1980s. It has been widely studied and applied in recent two decades because it has advantages of high efficiency, energy-saving, small occupation area and convenience management [1]. The filler is the core part in BAF and previous studies [2, 3, 4] showed that the most popular two kinds of fillers for BAF were zeolite and haydite, which had better application effect than other kinds of fillers. But it was seldom reported that the comparison between BAF filled with single filler and BAF filled with composite filtering material under the same conditions.

In this paper the treatment effects of BAF with single natural zeolite and BAF with combined zeolite and haydite in 1:1 proportion were analyzed, and their removal effects of chemical oxygen demand (CODcr), ammonia nitrogen (NH_4^+-N) and suspended solid (SS) were investigated in both the start-up stage and the steady operation stage. The running laws and treatment effects of two kinds of BAFs will provide a reference standard for its optimal design and operation management.

2. Experimental device and methods

2.1. Experimental device

The BAF design was based on the data correlation of removal performances with BAF dimensions obtained from the previous studies [5]. Two laboratory-scale BAF systems made of transparent polyvinyl chloride (PVC) were designed with the dimensions of 55 cm height (H) ×13cm diameter (D) with an effective working volume of 6 L. The diameter of zeolite particles ranged from 3 mm to 5 mm. Only zeolite particles, in size order smallest on top, were filled into 1# reaction column (zeolite BAF), while zeolite and haydite were respectively layered into 2# reaction column (zeolite-haydite combined packing BAF) in 1:1 proportion from top to bottom. The total filling height of 2# reaction column is about 50cm, with a supporting layer of 8mm-diameter gravels at the bottom. The air inlet is 1cm from the bottom which is for aeration and air-water backwashing. The air flow rate was controlled by an air flow meter. Sampling ports (SPs) were placed along the height of the column at 11 cm intervals to allow biomass and water samplings. A backwash system was periodically operated using a water pump to remove the excess biomass to prevent the reactor from clogging, thereby maintaining biofilm activity [6]. The BAF system was operated in an up flow direction. Rural wastewater was pumped from the influent tank into the BAF column using a peristaltic pump, flowing upward through the medium at a flow rate of 0.3 L/h to occupy the BAF column in about 18 h. The experimental device is shown in Figure 1.

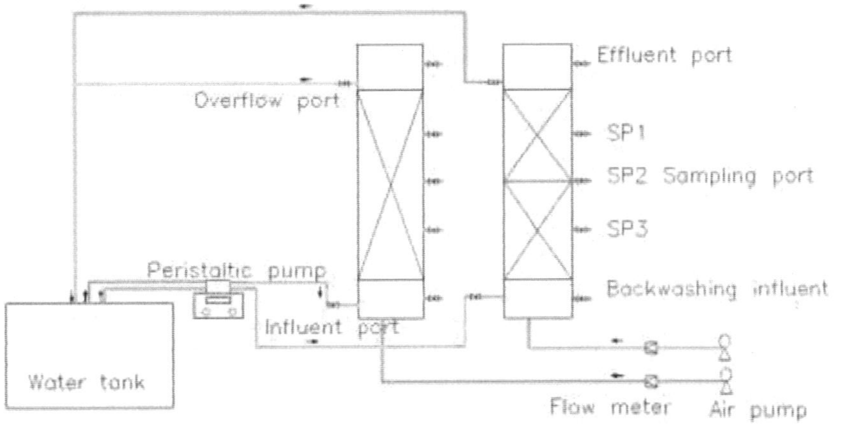

Figure 1. Experimental device.

The experimental device continuously operates in July and August, with the temperature of the laboratory ranging from 23 to 34℃. The influent contains high biodegradable components, so a wide range of microorganisms in raw water would be absorbed onto the surfaces of the filler, and then formed biofilms on it. Flow velocity was as small as 0.3per hour, therefore a large number of microorganisms could be absorbed onto the surfaces of particle easily. The hydraulic loading was gradually increased until biofilm was matured. Previous studies [7] have shown that ammonia nitrogen removal ratio can be regarded as an indicator for biofilm maturation. It takes more time for this kind of biofilm growth manner to finish, but it runs more steadily during the steady operation stage. Water-air combined backwashing was adopted in this experiment, owing to its superior to single water or gas backwashing [8]. The back-washing sequence started with air scour, followed by air scour and water backwash and terminating by water scour, and the air and water flows kept up-flow. Backwashing was carried out when a preset head pressure loss reached. Intense and short backwashing were carried out on the basis of our experience, with 5 L/s of water and 15 L/s of air.

2.2. Filler

Zeolite is a kind of natural porous silicate mineral which has intense adsorb ability and enrichment effect on polar molecule and bacteria. Early experiments suggested that nitrifying bacteria adsorbed on the surface of zeolite could transform ammonia nitrogen into nitrate nitrogen [9].

Ammonia removal by bio Zeolite filter was high because of the dynamic bio-regeneration process. In addition, this system has powerful resistance to impact load. As for the haydite filter, it takes less biofilm maturation time and reaches higher treatment efficiency on treating micro-polluted raw water. Furthermore, it has high resistance to low temperature and erosion [10, 11]. Single filler of BAF has less microbial biomass, resulting in instability treatment effect. This present thesis aims to explore the treatment effects of BAF with single natural zeolite and BAF with combined zeolite and haydite. Physical properties and chemical properties of zeolite and haydite are shown in Table 1.

Table 1. Physical Properties and Chemical Properties of Fillers.

Name	Physical Properties			Chemical Elements /%					
	specific surface / $m^2 \cdot g^{-1}$	total pore volume /$cm^3 \cdot g^{-1}$	loose bulk density /$g \cdot L^{-1}$	Na	Mg	Al	Si	Fe	Others
Zeolite	0.46	0.0269	830	4.25	11.48	18.27	40.28	10.14	15.58
Haydite	3.99	0.103	976	-	1.5	21.5	63.5	6.5	7

2.3. Influent quality

Dynamic experiment was operated under a condition of continuous water input by peristaltic pump from water tank. The wastewater used in this study was collected from wastewater plant of Pangge Town, Beijing. The raw water is characterized by high contents of COD_{Cr} and NH_3-N, thus, the water quality is unstable. Water quality and determination of test sample are summarized in Table 2.

Table 2. Water Quality and Determination of Test Sample.

Parameters	T/℃	PH	$COD_{Cr}/mg \cdot L^{-1}$	NH_3-N/$mg \cdot L^{-1}$	SS/$mg \cdot L^{-1}$
Maximum	36	7.93	386	28.8	240
Minimum	18	6.87	217	18	78
Average	27	7.36	301.5	23.4	159
Determination	Thermometer	pH meter	potassium dichromate method	Nessler's reagent spectrometry	Gravimetric Method

3. Results and discussion

3.1. Removal effect of CODcr

During the start-up stage, influent CODcr concentration was about 217mg/L, while effluent CODcr concentration from both BAFs were not stable, with CODcr removal rate lower than 40%. Effluent quality from 1# reaction column fluctuates greatly, whose CODcr concentration is occasionally higher than influent's. Previous studies [12] showed that biofilm on the zeolite was not mature and stale at this stage, and the biofilm can easily be divorced from the surface of zeolite due to the washing action of influent, so CODcr concentration of effluent increased. On the other hand, 2# reaction column, with a CODcr removal rate above 60%, forms film completely in about 25 days which is shorter than 1# reaction column. It can be seen that 2# reaction column with haydite has higher aerobic biofilm growth rate and more steady bioactivity at the same time.

Varieties and quantities of the organisms increased and basically tended to be stable after 30 days. Removal rate of CODcr of two kinds of BAFs reached about 59%, which showed that heterotrophic bacteria has grown to maturity. However, Filler was gradually clogged by pollutants with increasing operation time, and biofilm was not to be regenerated in time, so the removal rate of

CODcr declined after 30 days. As can be seen from Figure 2, removal rates of 1$^{\#}$ reaction column and 2$^{\#}$ reaction column have declined to 26.3% and 37.1% respectively after 43 days. The removal rate regained its previous level gradually after air-water backwashing.

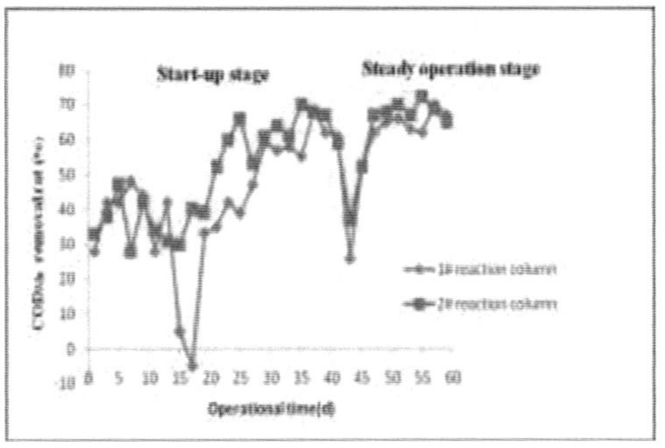

Figure 2. CODcr removal rate.

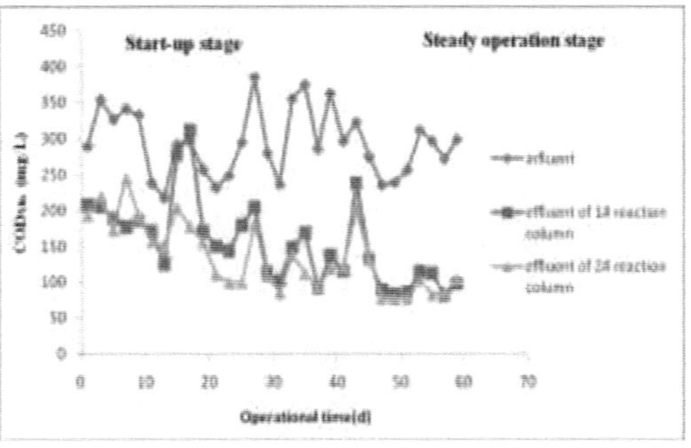

Figure 3. CODcr removal effect.

3.2. Removal effect of NH_4^+-N

NH_4^+-N removal effects of the start-up stage are shown in Figure 4 and Figure 5. In the 9th day of start-up stage, NH_4^+-N removal rates of 1$^{\#}$ reaction column has increased to 96.8%. In 9th ~30th days, the average removal rate of NH_4^+-N was 92.3%. Respectively, in the 13th day of start-up stage, NH_4^+-N removal rates of 2$^{\#}$ reaction has come to 96.7%, in 9th ~30th days NH_4^+-N average removal rate was 95.1%. After the formation of biofilm, NH_4^+-N removal rates of both BAFs reached 90%.

Why could the removal rates of NH_4^+-N keep high at the start-up stage? There was no biofilm formed on the surface of zeolite at the initial stage of starting-up, so high removal rate of ammonia nitrogen on zeolite was mainly due to ion-exchange action [13]. It is estimated that the ability of ion exchange ammonia nitrogen was saturable in the 16th day, which caused the removal rate of influent ammonia nitrogen decreased, however, in 19th day, the removal rate of NH_4^+-N abruptly decreased, because nitrifying bacteria played an important role in removing nitrifying bacteria, which prolonged the regeneration period of biofilm on zeolite. Haydite in 2$^{\#}$ reaction column have no selective adsorption for NH_4^+-N, and zeolite in the upper layer played a main role in removing NH_4^+-N because of the ion-exchange action, whose removal rate was lower than 1$^{\#}$ reaction column. However, haydite filler has advantages of larger interspace and rough surface, and as times goes on, the microorganism grows not only on the surface of the haydite but also in the interspace[14], therefore haydite in the lower layer is more easily to culture biofilm and activated nitrification earlier. So NH_4^+-N removal rate of BAF with combined zeolite and haydite is more stable, whose removal rate of NH_4^+-N generally remains high.

As showed in Figure 4 and Figure 5, NH_4^+-N average removal rate of 1$^{\#}$ reaction column is 94.62±3.42%, NH_4^+-N concentration of effluent ranging from 0.37 mg/ L to 2.25 mg/ L. While NH_4^+-N average removal rate of 2$^{\#}$ reaction column is 95.93±2.15%, NH_4^+-N concentration of effluent ranging from 0.44 to 1.52mg/L. In conclusion, NH_4^+-N removal effect of BAF with combined zeolite and haydite is more stable, whose removal rate of NH_4^+-N is generally higher than the former

At the initial stage of starting-up, adsorption and ion exchange play a part in removing ammonia nitrogen. In contrast, nitrification of microbes played a main role in removing ammonia nitrogen at steady operation stage. Nitrification is a biological oxidation of ammonia-nitrogen which goes through two stages, where NH_4-N in the presence of oxygen, is firstly converted to nitrite-nitrogen (NO_2-N) by the strictly Chemolithotrophic Nitrosomonas, Nitrosococcus and Nitrosospira bacteria, and secondly to nitrate-nitrogen[15]. The reasons for higher NH_4^+-N removal rate of 2$^{\#}$ reaction column is that hydite has larger porosity, which provides larger space for Microorganism's growth and reproduction [16], and ensured better removal effect and stability of effluent quality. Besides, hydite, as a kind of floating media filter, can increase the flow resistance and make distribution of flow even, which is benefit for the growth of nitrifying bacteria. As a result, BAF with combined zeolite and haydite is superior to the BAF with single filler.

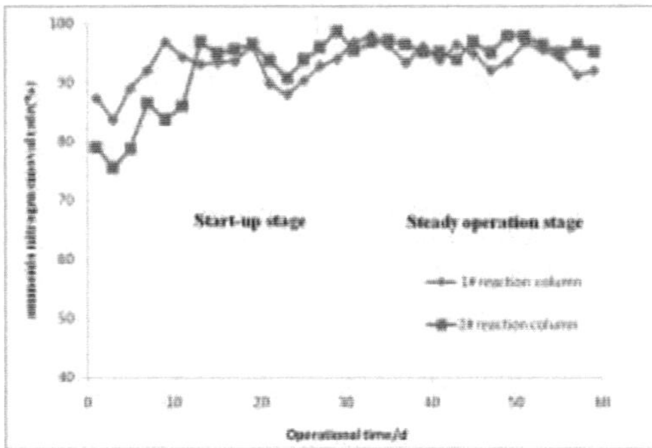

Figure 4. NH_4^+-N removal rate.

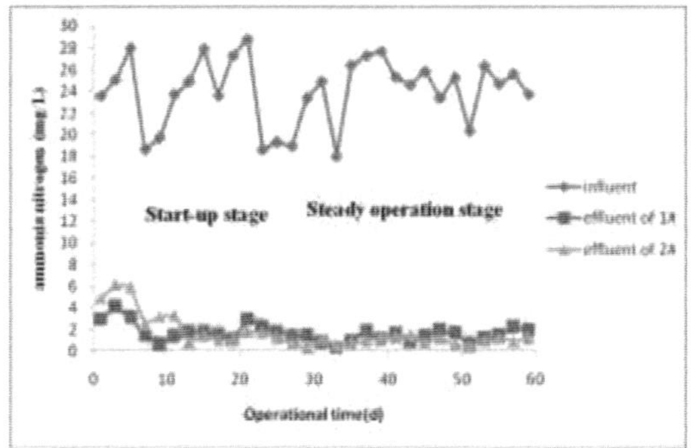

Figure 5. NH_4^+-N removal effect.

3.3. *Removal effect of SS*

The removal effects of SS are shown in Figure 5 and Figure 6. From the 30th day to biofilm culturing ending, effluent SS concentration of 2# reaction column ranged from 3.5 to 18.5mg/L, while effluent SS concentration of 1# reaction column ranged from 8.7 to 49.3mg/L. It can be seen that SS removal rate of 1# reaction column is gradually rising during the start-up stage, however, SS removal rate of 2# reaction column is higher and steadier.

During the steady operation stage, SS removal rates of both BAFs are stable with influent SS concentration ranging from78 to 210mg/L. The effluent SS concentration of 2# reaction column is15.9 mg/L with average removal rate 87%, while the effluent SS concentration of 1# reaction column is19.5mg/L with average removal rate is 88.5%. It can be obviously seen in Figure 4 that SS removal rate of 2# reaction column is higher and more stable, mainly due to the fact that hydite has larger interception capacity and higher efficiency of intercepting sewerage.

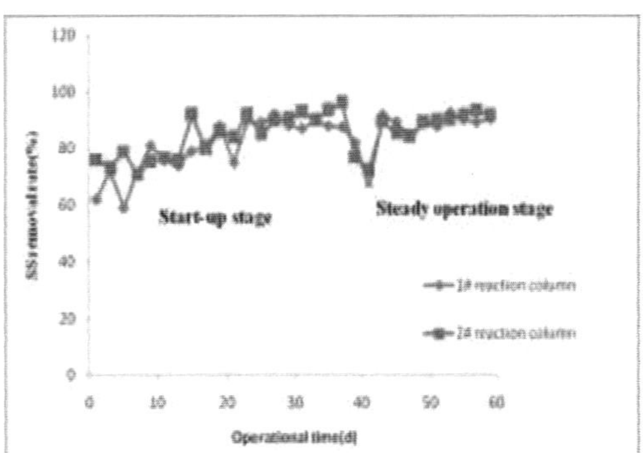

Figure 6. SS removal rate.

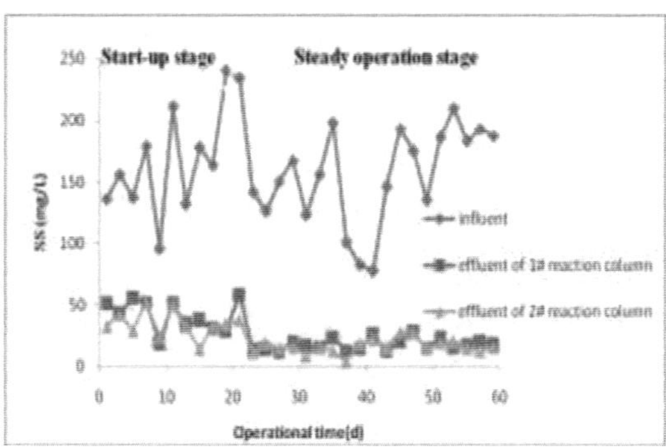

Figure 7. SS removal effect.

4. Conclusion

The results show that the removal rates of CODcr, NH_4^+-N and SS were 63.7%, 95.9% and 88.5% respectively for the BAF with combined packing, which formed film completely in 25 days. while for the BAF with zeolite filler, the removal rates of CODcr ,NH_4^+-N and SS were 59.6%, 94.6% and 87% respectively, which formed film completely in 30 days. When BAF with combined zeolite and haydite was used to treat rural sewage, zeolite and haydite are complementary to get rid of CODcr and NH_4^+-N. It has powerful resistance to impact load of NH_4^+-N. Meanwhile, haydite presenting low density, which helps to reduce energy consumption during backwashing operation; on the other hand, the weight of filler can be reduced, so it will cut down the operational cost. In summary, BAF with combined zeolite and haydite can be widely used to treat rural wastewater, in view of their superior removal rates of CODcr, NH_4^+-N and SS, great reserves and lower cost in China.

Acknowledgments
This work was financially supported by China Institute of Water Resources and Hydropower Research Fundamental Research Funds Program "Study of biological aerated filter and automatic control system for rural sewage treatment" (ID0145B08201500000) and Major Projects for Social Livelihood of Jinan City "Research and demonstration on ecological control technologies of water environment in Jinan" (SHMS2015-301).

References
[1] X.B.Sun, H.Y.Pan, Y.F. Sun, F.Y. Cui and Z.W. Zhao, Comparative experimental research on start-up of biological aerated filter with zeolite filler and combined packing, Journal of Harbin University of Commerce (Natural Sciences Edition). 28 (2012) 171-172.
[2] L.J. Gao, Comparison of biological aerated filtration performance between zeolite media and ceramisite media, Technology of Water Treatment. 35 (2009) 53-54.[In Chinese]
[3] E.D. Du, X. Liu, H. Wang and D. Deng, Ammonia removal behaviour and biological characteristics of zeolite media in a biological aerated filter[J]. Techniques and Equipment for Environmental Pollution Control. 7 (2006) 89-93. [In Chinese]
[4] P. Wang and G.C. Li, Preparation of Biological Aerated Ceramsite Filter Medium and Its Performance, Non-Metallic Mines. 29 (2006) 53-55. [In Chinese]
[5] X.C. Ma and S.T. Liu, Experimental study on Biological Aerated Filter(BAF)Process in Treating Domestic Sewage, Environmental Protection Science. 35 (2009) 34 -36. [In Chinese]
[6] H.J. Wang, W.Y. Dong, T Li. and T.Z. Liu, A modified BAF system configuring synergistic denitrification and chemical phosphorus precipitation: Examination on pollutants removal and clogging development, Bioresource Technology.189 (2015) 44–52.
[7] X.Q. Li, Y.H. Wang and J.D., Zhou The research of the film formation and start-up zeolite biological aerated filter, Environment Science and Management. 33 (2008) 91-93.[In Chinese]
[8] Y. Feng, X. Li, T. Song, L.S. Fan, Y.Z. Yu, J.Y. Qi and X.W. Wang, Effect of backwashing on the microbial community structure and composition of a three dimensional particle electrode coupled with biological aerated filter reactor (TDE-BAF), Ecological Engineering.101 (2017) 21-27.
[9] J.X. Liu, J.S. Lou and C.N. Chen, Test of treatment of slightly polluted source water by zeolite-ceramics biological aerated filter, Industrial Water & Wastewater. 36(2005) 10-12.[In Chinese]
[10] B. B. Baykal, Clinoptilolite and multipurpose filters for up-grading effluent ammonia quality under peak loads, Water Science and Technology. 37 (1998) 235-242.
[11] W.S. Xiao, W.G. Xu and J.C. Yang, Activity of Biomembranes and Biofloccules in Biological Aerated Filters, Transactions of Beijing Institute of Technology. 23 (2003) 656-658. [In Chinese]
[12] Y. Guo, L.J. Chen and D.H. Wen, Study on Nitrite Nitrogen Accumulation in Zeolite Biological Aerated Filter, China Water&Waste Water. 22 (2006) 73-77. [In Chinese]
[13] W.H. Tian, X.H. Wen and Y. Qian, Characteristics of zeolite media biological aerated filter during start-up, Techniques and Equipment for Environmental Pollution Control. 3 (2002) 38-41. [In Chinese]
[14] S.B. Zhang, L.P. Qiu, M.A. Du and J. Ma, Treatment efficiency and nitrification properties in three biological aerated filters (BAF) with different media, Journal of Harbin University of Technology. 41 (2009) 57-60. [In Chinese]
[15] S.I. Abou-Elela, M.E. Fawzya and A.S. El-Gendy, Potential of using biological aerated filter as

a post treatment for rural wastewater, Ecological Engineering. 84 (2015) 53-57
[16] D. Pak, W. Chang and S. Hong, Use of natural zeolite to enhance nitrification in biofilter, Environmental Technology. 23 (2002) 791-798.

Indoor air pollution and preventions in college libraries

Zengzhang Yang*

Library of Shandong Women's University, Jinan 250300, China

*Corresponding author e-mail: wuyegl@163.com

Abstract. The college library is a place where it gets the comparatively high density of students often staying long time with it. Therefore, the indoor air quality will affect directly reading effect and physical health of teachers and students in colleges and universities. The paper analyzes the influenced factors in indoor air pollution of the library from the selection of green-environmental decorating materials and furniture, good ventilation maintaining, electromagnetic radiation reducing, regular disinfection, indoor green building and awareness of health and environmental protection strengthening etc. six aspects to put forward the ideas for preventions of indoor air pollution and construction of the green low-carbon library.

1. Introduction

According to China Standardization Association, 68% diseases are caused by indoor air pollution which has become an "invisible killer" to people's health. America has considered the indoor air pollution as five major environmental factors in damaging public health. Cities residents in China are staying in various indoor environments with around 70% to 90% of time [1].

With the continuous development of the college libraries, the integration services mode of books of collection, borrowing, reading and consulting have been extensively practiced. There is an increasing large scale of libraries with a growing number of in-library readers and rising intensiveness of staff. The opening time has reached to 10 to 13 hours and some have opened round the clock. College students' residence time in the library is 2 to 4 times longer than that in the classroom, averaging to 3 to 6 hours (according to those staying in the library). Indoor air pollution of libraries does a severe harm on physical health of massive readers and staff working in libraries. The main reader groups of college libraries are young students in puberty. Researching and improving the indoor air quality of college libraries is a responsible practice for students and society, which should be given enough attention within it.

2. The testing of indoor air and environmental quality and data analysis of the Shandong Women's University library

Referring to *The Sanitary Standards of the Library, Museum, Gallery, Exhibition Hall* (GB 9669-1996), Shandong Women's University library carries out the testing on the indoor temperature, relative humidity, carbon dioxide, formaldehyde and respirable particulate matter etc, and analyses them. The results of testing and analyzing are as shown in table 2.

Table 1. The Sanitary Standards of the Library, Museum, Gallery, Exhibition Hall (GB 9669-1996)

Subjects	Library
Temperature/ °C (airconditioner turned off)	18～28
Relative humidity/ %	40 ～65
Carbon dioxide/ %	<0.10
Formaldehyde/(mg/m^3)	<0.12
Respirable particulate matter(mg/m^3)	<0.15

Table 2. The analysis results of temperature, relative humidity, carbon dioxide, formaldehyde and respirable particulate matter based on experimental data

Site	Temperature (°C)	Relative humidity (%)	Carbon dioxide (%)	Formaldehyde(mg/m^3)	Respirable particulate matter (mg/m^3)	Remarks
The First Reading &Borrowing Room	22	40	0.05	0.02	0.11	Daily Average
The Second Reading & Borrowing Room	23	41	0.04	0.02	0.05	Daily Average
The Third Reading & Borrowing Room	23	36	0.0	0.01	0.11	Daily Average
The Fourth Reading & Borrowing Room	23	38	0.04	0.01	0.07	Daily Average
The Fifth Reading & Borrowing Room	23	37	0.05	0.02	0.04	Daily Average
The Sixth Reading & Borrowing Room	23	38	0.04	0.02	0.04	Daily Average
The Periodicals Reading Room	24	37	0.11	0.03	0.12	Daily Average
The Self-Studying Room	24	38	0.11	0.02	0.12	Daily Average
The Electronic Reading Room	25	36	0.12	0.03	0.06	Daily Average
The self-studying Zone in the hall	23	41	0.08	0.10	0.10	Daily Average
Ring Gallery surrounding the Hall	23	41	0.08	0.11	0.10	Daily Average
Overdated Periodical Reading Room	23	37	0.05	0.01	0.17	Daily Average
Fundamental Stack Room	23	37	0.05	0.01	0.16	Daily Average
Intensive Stack Room	23	36	0.05	0.01	0.16	Daily Average

According to the table2, the indoor temperature of Shandong Women's University completely accords with the indoor temperature standard without air conditioner working. As for the relative humidity simply the First, Second Reading and Borrowing Room, and the self-studying zone conform to its standard while others do not. It could be attributed to the relatively dry weather of the northern area and consequently understandable the comparatively low relative humidity of the indoor air.

According to the table2, the carbon dioxide in the indoor air reaches the standard except for Periodicals Reading Room, Self-studying Room and Electronic Reading Room of Shandong Women's

University. Carbon dioxide in the air of Periodicals Reading Room and Self-Studying Room reached to 0.11%. The relatively higher concentration belongs to air pollution, as here come a number of students to study around 1000 person-time, and the poor ventilation that cases the disadvantage of carbon dioxide emission. The situation in the Electronic Reading Room is 0.12% containing around 110 persons with 130 computers and 7 windows. But less 3 windows will be opening usually which results in the poor condition to the air flow, belonging to air pollution likewise.

According to the table 2, it tells that the indoor formaldehyde is not beyond the standard in Shandong Women's University Library, except for testing tiny formaldehyde in the zone of hall for self-studying and the ring gallery, others' seldom. The leathered sofas are disposed on the halls from the first to fourth floor. A side wall of ring galleries is being decorated the wood based panels. And the walls of hall and stairs were repainted. Therefore, the formaldehyde is thought as the release from leathered sofas, decorating panels and paintings.

According to the table 2, it is obvious shown only over-dated periodicals reading room, Fundamental Stack Room and Intensive Stack Room failed to accord with the standard on respirable particulate matter while the rest reach it. The Over-dated Periodicals Reading Room, Fundamental Stack Room and Intensive Stack Room are places where are being collecting the "antique-rank" books with quite infrequent reading and flowing. As it is continuing reserved, the volatiles will be increasing with it as well as the floating dust on the books' surfaces which are impossible to clean it totally. In the flowing of the air (such as draught, taking books and walking with wind) will float the dust indoors, causing to indoor air pollution.

3. The influence factors in indoor air pollution of college libraries

3.1. Large intensiveness of students and staff

The library in the college is one of the fondest places for college students. It is opened for a long daily time with around 1000 person-time of readers to a normal reading room per day. Because of the high density of students and staff working in the college library, physical bodies of people themselves are the major source of the indoor air pollution. People indoors will emit heat and moisture and expel a large number polluted substance through respiratory tract, skin and sweat gland such as carbon dioxide, virus and bacteria etc... According to research, an adult will generate around 25 liters carbon dioxide per hour, thereof including over 16 poisonous substances; shed skin to every person by metabolism will reach to total 600 thousand per hour with a great deal of bacteria. In walking into the library, the dust outdoors will be brought into the library together [2]. Undoubtedly, the more people in the library, the severer pollution will be.

3.2. Building decoration materials and interior furniture

Libraries are decorated with decorative materials such as flooring, plywood, carpets, coatings, paints, gypsum, and adhesives etc. to release harmful substances such as formaldehyde, benzene, chloride and so on. Furniture used in the production of a variety of plates, adhesives and paints, it will produce toxic and hazardous substances. According to statistics, 73.6% of indoor pollution is caused by interior decoration and 24.3% by the purchase of furniture. If the library ventilation is not good, variety of materials will release the harmful gas superposition, resulting in excessive harmful gas concentration among the indoor air and the damage to physical health.

3.3. Volatiles and dust of books in stack rooms

Because of the current paper books mainly through the ink, lead printing, ink rich in lead, benzene toxic ingredients, many books are too heavy for oil or improper use of ink, and a lot of ink is filled in the air. As Low flow of books with the continuing preservation, the book gradually increased volatiles. The more infrequent those books are flipped, the more volatile will form while taking books, which results in indoor air pollution. As the books' adsorption to the indoor dust, dust is always on the books, which is hard to clean it out. In air flow the dust will float indoors, leading to the air pollution and effect on physical health [3].

3.4. Bacteria breeding of books in stack rooms
As cellulose, lignin and semi-cellulose basically make up the paper of books, these will offer enough nutrients to bacteria. Mould might have been bred with the warm condition and high humidity. Dust is the transmitter of the mould spores, which will provide the advantage for mould spreading with the high dustiness in the air by the poor sanitary working of the library. There have set up the compact stack rooms in many college libraries where those ancient books lack of circulation are easily parasitic with mites for breeding in a large number. Much of mould will generate poisonous substances with pungent smell by metabolism, causing people's liver, kidney and brain nervous system, but some even will lead to the cancer, damaging on physical health.

3.5. Electromagnetic radiation pollution
With the development of the computer and the network technology, books purchase has been in a relatively small quantity while electronic readings relatively increased. The computer, displayer, typewriter, copier, books monitoring and projector etc are universally applied to library, which of these appliances will produce different wave length and frequencies electromagnetic waves in normal working. According to research by experts, the excessive electromagnetic radiation for a long period time will result in the direct damage on the body's genial system, nervous system and immune system, being the leading precipitating factor of cardiovascular disease, diabetes and cancer mutation. Electromagnetic pollution has been listed in the fourth of the great sources of environmental pollution after water resources, atmosphere and noises by the World health Organization, and become the "invisible killer" of people's physical health. As a result, electromagnetic pollution has become the new threat of indoor environmental pollution of college libraries [4].

4. The prevention methods of indoor air pollution of college libraries

4.1. Selections of environmental protection building materials, decorative materials and furniture
Utilizing the environmental building materials and decorative materials is the crux of retaining the indoor air pollution of libraries. In libraries construction and decoration, it should be in a strict accordance with *Code for Indoor Environmental Pollution Control of Civil Construction Projects* and *Ambient Air Quality Standards* to execute from the sources of materials by selection of unpolluted and harmless decorative materials with environmental protection. In decorative materials and furniture purchasing, it should comply with principles of security, health, comfort and aesthetic appearance. Meantime it is necessary to ensure whether the seller has the IS09001 international quality certificate and ISO14001 international environmental certificate, and assure the "ten rings" mark on the production. In termination of the library decoration, it had better not be urgent to move in but make ventilation for a couple of times to release fully the hazardous gas in materials. After moving into the new library, it should be keeping in ventilation with more windows opened.

4.2. Maintaining of good ventilation
Ventilation is the optimal mean of preventing and controlling indoor air pollution of libraries. Ventilation has been divided into two ways of natural ventilation and mechanical ventilation. Frequent ventilation indoors in libraries can keep in fresh and clean air. According to research, the air qualities of college libraries that have effectively designed the space for draught are better than that of haven't. Under the circumstance of low wind speed in the morning, indoor air can be reached the sanitary standard 2-8 minutes later if there passes the draught; if there is no this convection, it will take as lest 40 minutes.

Many college libraries have adopted the pulling window in order to assure an enough real opening area outside the walls. Sliding windows nearly open flat and the real opening area will be reduced by

40%-50%, so more windows have to be installed on the library's walls. Transoms and skylights are also added available. In addition, shelves should be perpendicular with windows to decrease the wind drag and expedite the air motion.

In the place of people intensiveness, mechanical ventilation system should be used while the natural ventilation cannot meet the requirement. In the author's view, installing intake fans and exhausts fans (or pipes) in an appropriate way will basically meet the requirement for the indoor ventilation.

4.3. The reduction of electromagnetic pollution

Electromagnetic pollution is called the "invisible killer". From current technology levels, it remains difficult to completely eliminate the pollution to the environment from computers. Main approaches to reducing the electromagnetic radiation are as followed: not to use computers for a long time; to keep a certain distance from computers because of the farther from it, the weaker radiation; to purchase the qualified computer and the copier etc productions with the special attention to the graph scope like its security certified marks; not to place the copier in the hall and reading rooms but to the special copying room; to pull the curtains for avoiding the direct exposure of the computer under sunlight; to add a protection screen for the computer etc. The simplest way is to open the door and window or install the exhausts fan to keep ventilation, reducing or eliminating pollution.

4.4. Arrangement for regular disinfection by special staff

Books with bacteria is the main medium that causes the cross infection among readers. To cut the virus and bacteria transmission channel, first of all is to carry out the disinfection to returned books. Currently, disinfection methods by libraries are as followed:

(1) To regularly disinfect with the ultraviolet fluorescent lamps to eliminate bacteria propagules, fungi, mycobacterium tuberculosis etc., and this simple and convenient disinfection method is also economical and practical.

(2) Disinfectant is used for disinfection, such as potassium permanganate solution and 84 disinfectant to be sprayed on where should be disinfected. Wiping the desktop and ground with 500mg / L chlorine disinfectant can help eliminate dust mites, but it should be noticed not to sprinkle directly on the book as chlorine disinfectants cause paper damage.

(3) The air clear bacteria tablets, especially for pneumococcus, influenza virus can eliminate them to the ratio of 100%.

(4) It is available of variety of advanced disinfection equipment, such as electronic disinfection and disinfection cabinet, microwave ovens, paper, microwave insecticide sterilization vehicles.

4.5. The effort to a good job of indoor green building and improvement of indoor environment

Properly placing green plants in the library of not only can properly improve the indoor air quality, but also allows readers to feast for the eyes, feeling the breath of spring. Chlorophyll in green plants absorbs piton of carbon dioxide through photosynthesis and releases oxygen, being carbon dioxide natural consumers. Plants can also absorb the harmful substances in the gas as a nutrient for their own growth within a certain concentration range. As sulfur that is a component of the amino acid of the plant will transform into sulfurous acid and sulfite while being absorbed by the plant, and then into sulphate at a certain rate to form sulfate. Green plants can also purify other harmful gases in the air, such as nitrogen dioxide, carbon monoxide, ozone, lead steam and other heavy metal gases and aldehydes, ketones, ethers, phenol and other carcinogenic substances [5]. Therefore, the growing green plants in the library can reduce benzene in the air, especially from poor print books such as lead, benzene and other harmful substances concentration.

It also need to note too fragrant plant with exaggerating colors, drifting pollen and pests breeding that too strong and hilly, flies flying, insects breeding, smell or smell too thick plants should not be placed in the room. Usually around the reading room door there can be placed upright flowers without blocking sights and ivy and chlorophytum can be placed on top of the shelves, a few pots of aloe vera, bamboo on the windowsill which can get library ambience changed greatly.

4.6. Enhancement of sanitary and environmental awareness of readers and staff

The hygiene and environmental awareness of the staff and the readers and personal sanitary morality can be enhanced through the form of newspaper, column and dissemination of promotional materials by libraries, such as non-smoking, hand washing and changing clothes, etc., which not only enhance the personal physical health, but also improve the indoor air quality of the library. Multimedia reading room staff can stick a health care on the computer table tips to remind the reader to wash their hands, open the window to ventilation, and maintain the line of sight with the computer screen distance.

Besides, staffs working in the library have to often clean dust, keeping hygienic anywhere. In addition to daily hygiene work, it also includes regular general cleaning for fouling; regular cleaning and replacement of the carpet, curtains, regular maintenance, cleaning and disinfection on air conditioning system components and related pipelines, and tuyere. At the same time, developing the relevant environmental health management system improves the indoor health of the library. The cleanliness of the library is conducive to keeping the indoor air fresh.

5. Conclusion

College libraries as a special public places, it is not only a cultural heritage center, the ideological exchange center and knowledge service center, but also one of the three pillars of the college, the second classroom and spiritual civilization construction window for students. The library is a place where college students are more intensive and long residence time. The indoor air quality in the library will directly affect the reading effect and physical health of teachers and students in colleges and universities, and indirectly affect the quality and level of library services. Library indoor air pollution with prevention and control is a comprehensive project, the college library should actively study and take relevant measures to strengthen management, regular disinfection, prevention and control of indoor air pollution, and regularly monitor the air quality, to find out the problem in time and resolve it, constructing the green and low-carbon library.

References

[1] Zhang, J., Wang, L. Indoor air and serving readers in university library. *Journal of Hebei University of Engineering (social science Edition)*, 27(2), 57-58. (2010).

[2] Tang, L. On how to improve a library's interior air quality. *Journal of the Library Science of Sichuan*, 2, 10-12. (2005).

[3] Liu, F. Indoor air quality and readers of university library. *Sci-Tech Information Development & Economy*, 16(22), 98-99. (2006).

[4] Jin, J. The indoor air quality status analysis and optimizing strategy in the university library. *Journal of University Library*, 4, 45-50. (2011).

[5] Huang, R. Air pollution in reading rooms and its control. *Journal of Library and Information Sciences in Agriculture*, 19(12), 62-65. (2007).

IOP Conf. Series: Earth and Environmental Science **64** (2017) 012077 doi:10.1088/1755-1315/64/1/012077

The first flush effect of different urban underlying surfaces through artificial simulated rainfall

Zhao Wang[1, 2, 3, 4, *], Guangyu Lei[1, 2, 3, 4] and Yan Niu[1, 2, 3, 4]

[1]Shaanxi Provincial Land Engineering Construction Group Co., Ltd, Xi'an 710075, China
[2]Institute of Land Engineering and Technology, Shaanxi Provincial Land Engineering Construction Group Co., Ltd, Xi'an 710075, China
[3]Key Laboratory of Degraded and Unused Land Consolidation Engineering, the Ministry of Land and Resources, Xi'an 710075, China
[4]Shaanxi Provincial Land Consolidation Engineering Technology Research Center, Xi'an 710075, China

[*]Corresponding author e-mail: wangzhaotougao@163.com

Abstract. Three kinds of underlying surfaces: asphalt road, impervious tiles and grassland at Chang'an University in south second ring road of Xi'an were chosen as testing sites. Self-made device was used to moderate rainfall with different raining conditions: uniform rainfall intensity of 0.3mm/min, 0.5mm/min, 1.0mm/min; front peak type rainfall and back peak type rainfall. The pollutants in the runoff were measured, that included TN, TP, NO_3^--N, NO_2^--N, NH_4^+-N, PO_4^{3-}-P and PN M(V) curves and MFF_{30} values were analysed to drew these conclusions: under uniform rainfall intensity of 0.3mm/min, 0.5mm/min, 1.0mm/min, the mean values of MFF_{30} from asphalt road runoff were 0.47, 0.42, 0.41, the mean values of MFF_{30} from watertight tiles runoff were 0.40, 0.39, 0.38, the mean values of MFF_{30} from grassland runoff were 0.33, 0.34, 0.32. The first flush magnitude decreased with the increasing of rainfall intensity on asphalt road and watertight tiles, while the first flush intensity first increased and then decreased with the increasing of rainfall intensity on grassland. The mean values of MFF_{30} from the front peak type rainfall were larger than that from the back peak type rainfall on both asphalt road and watertight tile. The mean values of MFF_{30} from the back peak type rainfall were larger than that from the front peak type rainfall on grass. Grassland should be built around asphalt road and watertight tiles to attenuate first flush effect.

1. Introduction

Rainfall runoff through the leaching, erosion, coffin to carry pollutants and it means the main driving force of various pollutants entering the water on the surface sink surface [1]. With the continuous development of urbanization, impervious area of impervious surface gradually increased, making the urban runoff pollution more serious and increasing the city surrounding rivers and lakes sewage discharge pressures [2]. Therefore, the urban surface runoff pollution has become the main form of urban non-point source pollution, and it is the second largest urban water pollution besides domestic sewage and industrial wastewater and gradually deteriorating urban water environmental quality [3-5]. It is found that the concentration of pollutants in the early stage of rainfall runoff is significantly higher than the later stage, and a large amount of pollutants are discharged into the receiving water in a short period of time, causing serious damage to the water environment. This is the FFE (first flush effect), that is, the initial runoff is disproportionately carrying the most of the pollutants in whole runoff [6-9]. It has important economic significance and environmental significance for the management and governance of urban storm runoff [10]. Many domestic and foreign researchers have studied the FFE from various aspects. And the results show that the main factors are rainfall characteristics (rainfall, rainfall duration, maximum rainfall intensity), number of dry days, the type of land use, etc. [11-16]. At present, the research of the FFE is based on natural rainfall, and the monitoring of the FFE under natural rainfall is complex, random and extensive. It has high monitoring time, long cycle and large consumption of human and material resources. So the data obtained is less effective, and the results are also similarities and differences [17]· Such as Li Chunli found that the number of dry days and the degree of FFE has a positive correlation [15], but J.H. Lee found that there is no correlation between the two [18].

The study used self-made artificial rainfall devices to overcome the unfavorable factors of natural rainfall, and to discuss the different types of rainfall on the FFE of common urban surfaces. So as to provide the basis for the evaluation and command of urban rainfall runoff pollution.

2. Materials and Methods

2.1. Study Sites

The study area is Chang'an University campus, which is located in south second ring road of Xi'an. The proportion of land use types in Chang'an University campus is Construction land (66.2%), road (19.9%), Greenland (11.7%), and Watertight tiles (2.2%). Three kinds of typical underlying surfaces, asphalt road, grassland and watertight tiles were selected for artificial rainfall study.

2.2. Research methods

2.2.1. Experimental device. We used the self-made experimental device to study runoff pollution, and the device consists of simulated rainfall and runoff collection. Simulated rainfall device consists of water pump, motor, flow meter, pressure gauge (0-0.6MPa), detachable bracket (3m high, 2m wide), and control box. Nozzle has a large (aperture 4mm), medium (aperture 2.5mm), small (diameter 2mm), and each nozzle is controlled by a separate solenoid valve. The device has a high similarity to natural rainfall in terms of rainfall characteristics. Before the rain, the runoff collection device is placed on the appropriate surface and sealed between the baffle and the underlying surface. So that all the runoff within the baffle from the collection port. The complete set of experimental device diagram shown in Figure 1.

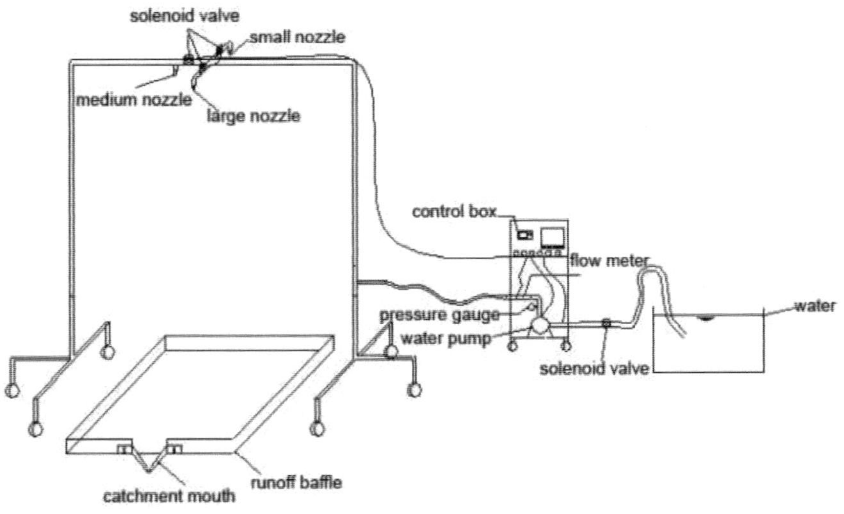

Figure 1. Schematic of the device

2.2.2. Water samples collection and testing. The sampling time is collected at 0 min, 2 min, 5 min, 10 min, 15 min, 20 min, 30 min, 40 min, 50 min, 60 min after runoff, and which collect until runoff ends. Sampling requires recording the volume of water sample and sampling time to calculate the available runoff. Concentrations of TN, TP, NO_3^-N, NO_2-N, PN, NH_4^+-N, PO_4^{3-}-P of water samples were tested when the water samples are collected and immediately sent to the laboratory. The above indicators are measured with reference to national standards [19].

2.2.3. FFE analysis method. At present, there are more analysis methods for the FFE of urban rainfall runoff [20], and the most commonly used is the cumulative pollutant-runoff curve [21, 22], that is according the ratio of cumulative runoff to total runoff (cumulative runoff ratio), the ratio of cumulative pollution load to total rainfall load (cumulative pollution load ratio) and create a dimensionless curve - M(V) curve. According to the curve deviation degree of diagonal above to determine the strength of the FFE. The more the curve deviates from the diagonal, the stronger the FFE.

FFE analysis of M(V) curve is more intuitive, but the lack of quantitative expression, which is not conducive to the application of non-point source pollution assessment and control. Saget conducted a quantitative analysis of the FFE [23], which using a power function to fit the pollutant M(V) curve: $F(X) = X^{b}$. X is the ratio of the cumulative runoff to the total runoff, F(X) is the ratio of the cumulative pollution load to the total rainfall load, b is the FFE coefficient. The b value can be determined by software. When X is 30%, the corresponding pollutant accumulation is called MFF_{30}, and the calculated is:

$$MFF_{30}=0.3^{b} \quad (1)$$

If MFF_{30}> 30%, showing the FFE occurred, and the larger the value that the FFE is more intense. The FFE can be divided into two categories, if 30% <MFF_{30} <60%, showing the FFE is weak; and MFF_{30}> 60%, showing the FFE is strong.

3. Results and discussion

3.1. Analysis of M(V) curve under uniform rainfall

Fig.2 shows the M (V) curves of the pollutants in the runoff of the asphalt road, the watertight tiles and

Greenland underlying surfaces, which under three uniform rain intensity, such as 0.3mm/min, 0.5mm/min, 1.0mm/min.

3.1.1. Asphalt road FFE. It can be seen from Figure 2 that NH_4^+-N, NO_3^--N and PO_4^{3-}-P in the asphalt road occurred in the FFE when rainfall is 0.3mm/min, in which the M(V) curve of NH_4^+-N deviates from the diagonal, and the remaining pollutants has not obvious flush effect. When the rainfall intensity increased to 0.5mm/min, the M(V) curve of NH_4^+-N and NO_3^--N is more obvious deviates from the diagonal, and the M(V) curve of PN is significantly increased in the later stage of rainfall, indicating that the flushing has occurred in the later stage. In addition, PO_4^{3-}-P flush effect was significantly reduced, the remaining pollutants without significant flush amplitude changes. When the rainfall intensity increased to 1.0mm/min, the flush effect of NH_4^+-N decreased, and TP appears more obvious flush effect.

In summary, NH_4^+-N and NO_3^--N on asphalt road are more prone to FFE, and PN is less prone to FFE in the small rainfall intensity. When the rainfall intensity increases, the FFE gradually appears in the late rain, and when the rain continues to increase, the FFE has weakened. Rainfall has washed, leaching, dilution and other multiple effects for pollutants on the underlying surfaces. The stronger the rain, the greater the erosion effect, so it can carry more pollutants, which may be the cause of the PN FFE increased with rain increased. At the same time, when the rainfall intensity increases, the runoff on the pollutant dilution effect also will be strengthened and the pollutant concentration decreases, it will weaken the pollutant flush effect [24]. In addition, PN has two clear flush in the rain which the M(V) curve appears more substantial rise in the late-rain. Asphalt road pavement has strong adhesion to particulate contaminants because of the certain sticky, and particles are not easy to run with the flow of flush. Since then,, the soaking and flush by the early runoff, the particulate gradually released in the late rain, which it may be caused by the above reasons.

3.1.2. Watertight tiles FFE. It can be seen from Figure 2 that NO_2^--N, NH_4^+-N and NO_3^--N in the watertight tiles have a certain degree of flush effect when rainfall is 0.3mm/min, and the remaining pollutants has not obvious flush effect. When the rainfall intensity increased to 0.5mm/min, the flush degree of NO_2^--N also increased, but flush effect of NH_4^+-N and NO_3^--N were weakened. When the rainfall intensity is 1.0mm/min, all of the pollutants showed a weak flush effect. In the study area, the pollution source mainly comes from the sedimentation of pollutants in the atmosphere and the diffusion and transfer of pollutants on the road. Compared with the road, the accumulation of contaminants on the watertight tiles is less, and the dilution effect of the runoff on the pollutants is more obvious. So the bigger the rain is, the harder it is to observe the FFE. In addition, asphalt road water permeability is stronger, which compared with watertight tile. Therefore, it may be the one of the reasons why the FFE of the watertight tiles is weaker than the asphalt road.

3.1.3. Greenland FFE. It can be seen from Figure 2 that NH_4^+-N and NO_2^--N in the grassland have a certain degree of flush effect when rainfall is 0.3mm/min. When the rainfall intensity increased to 0.5mm/min, the flush degree of NH_4^+-N and NO_2^--N also increased slightly, and PN showing late flush effect that the M(V) curve increased at the later stage. When the rainfall intensity is 1.0mm/min, the M(V) curves of the pollutants are close to the diagonal, indicating that no significant flush effect.

Grassland is one of the few in the city and occupies a certain percentage of the underlying surface, so we understand that the FFE on grasslands is important for estimating and controlling urban runoff pollution. Grassland permeability is extremely strong and particle retention is very effective, which is different from poor permeability surface of asphalt road and watertight tiles. The results show that the concentration of runoff pollutants in grassland is always lower than other cities underlying surface, and the flush effect of grassland runoff is small, which has a certain interception effect on pollutants in runoff [25]. Compared with the other two underlying surfaces, the flush degree of the grassland is low, which is the same as the above research results.

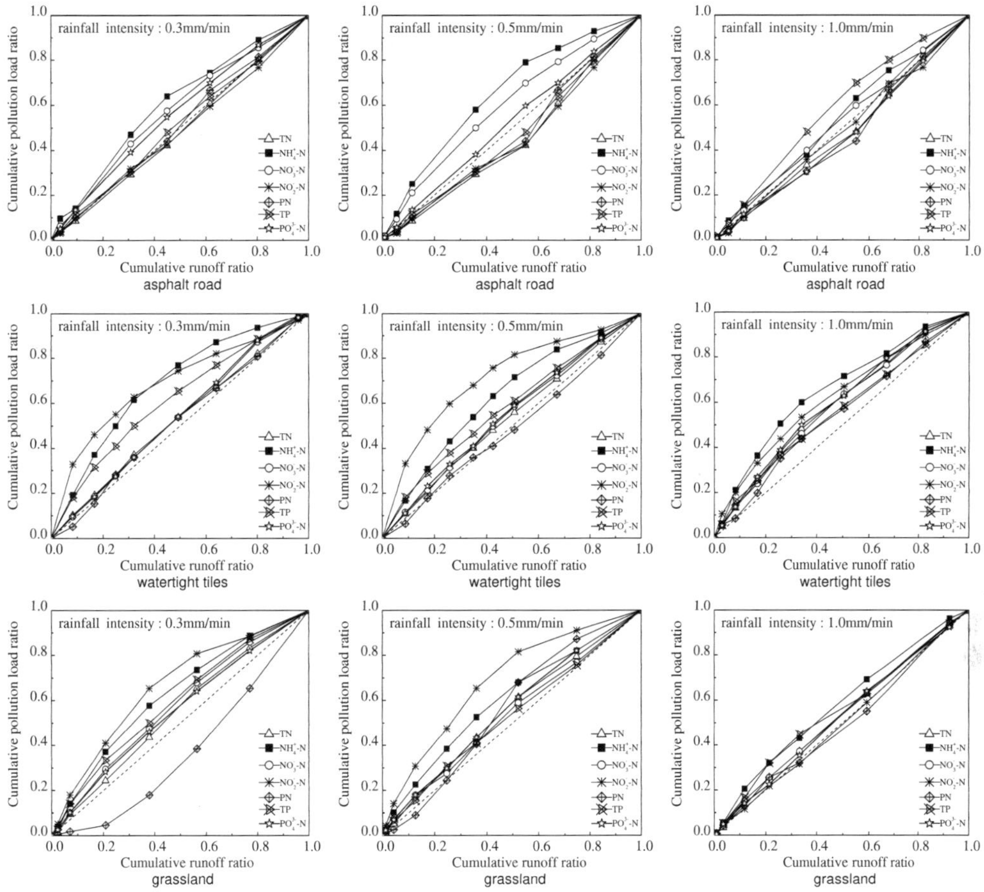

Figure 2. M(V) curves from uniform rainfall.

3.2. M(V) Curve of in unimodal rainfall

In order to be closer to the actual rainfall intensity, we use the Chicago rain-type to set the front and back peaks of two single peak rainfall situation. The Chicago rain-type was a kind of uneven design rain that based on intensity-duration-frequency relationship in 1957. After a lot of simulation and comparison, most domestic and foreign scholars believe that the effect is better and easier to determine the rainfall intensity process, and it is widely used in many research fields at home and abroad [26]. In this study, the uniform rainfall intensity changes every 5mins, and the specific changes is shown in Table 1 and 2.

The artificial rainfall is carried out on each of the underlying surfaces which according to two types in the table, and the resulting of pollutant M(V) curve is shown in Figure 3. Intuitive analysis shows that the contaminants of two peak-type rainfall have occurred in varying degrees of FFE besides NO_2^--N in the asphalt road, and the FFE of the front peak type rainfall is significantly higher back peak type rainfall. The FFE of the front peak type rainfall on watertight tiles is higher than the back peak type, and the output of TP fluctuates obviously in the back peak rainfall. TP on grassland in front type rains showed significant FFE, and TN and NO_3^--N did not occur FFE. While in the back peak rainfall, only the FFE occurred in TP, and TN and NH_4^+-N did not occur FFE, and the output rate of thc remaining pollutants was similar to the runoff rate. In general, front peak rainfall is more likely to cause FFE. Therefore, it can be judged that the rain intensity distribution, especially the maximum rain intensity, is the main factor affecting the FFE.

Table 1. The intensity change process of front peak type rainfall.

t(min)	0-5	5-10	10-15	15-20	20-25	25-30
rainfall intensity(mm/min)	0.20	0.30	0.50	0.65	0.85	0.75
t(min)	30-35	35-40	40-45	45-50	50-55	55-60
rainfall intensity(mm/min)	0.30	0.25	0.20	0.15	0.10	0.05

Table 2. He intensity change process of back peak type rainfall.

t(min)	0-5	5-10	10-15	15-20	20-25	25-30
rainfall intensity(mm/min)	0.10	0.15	0.20	0.25	0.30	0.75
t(min)	30-35	35-40	40-45	45-50	50-55	55-60
rainfall intensity(mm/min)	0.85	0.65	0.50	0.30	0.20	0.15

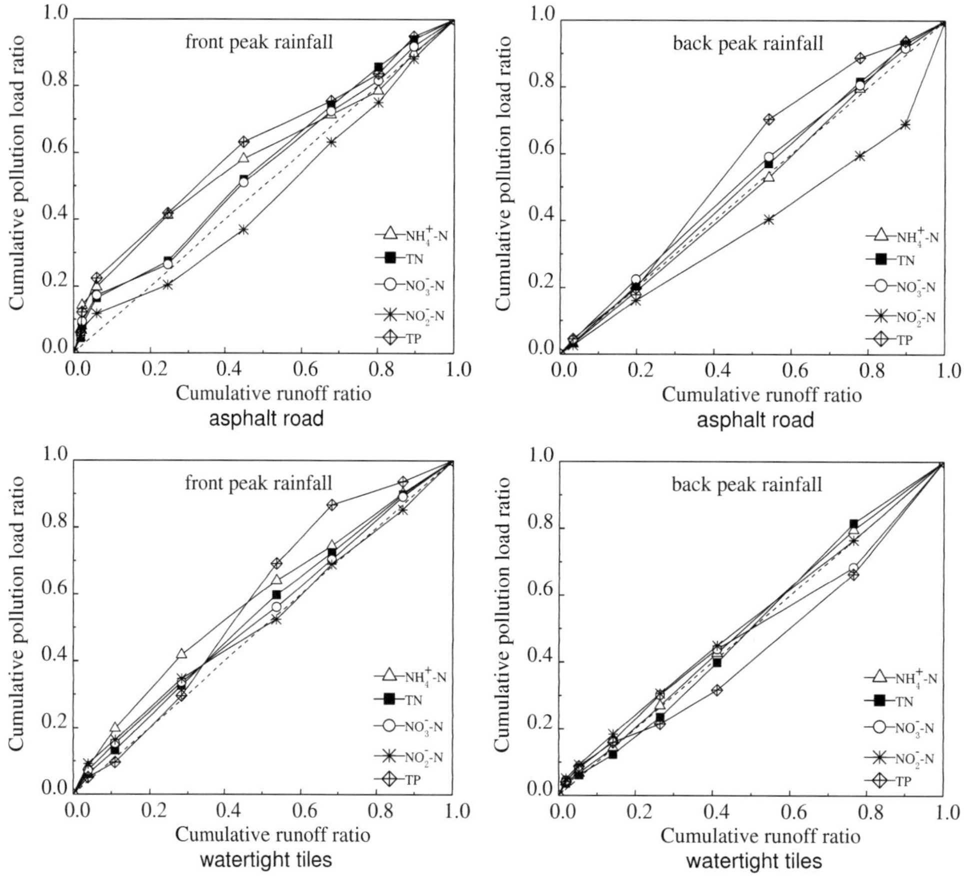

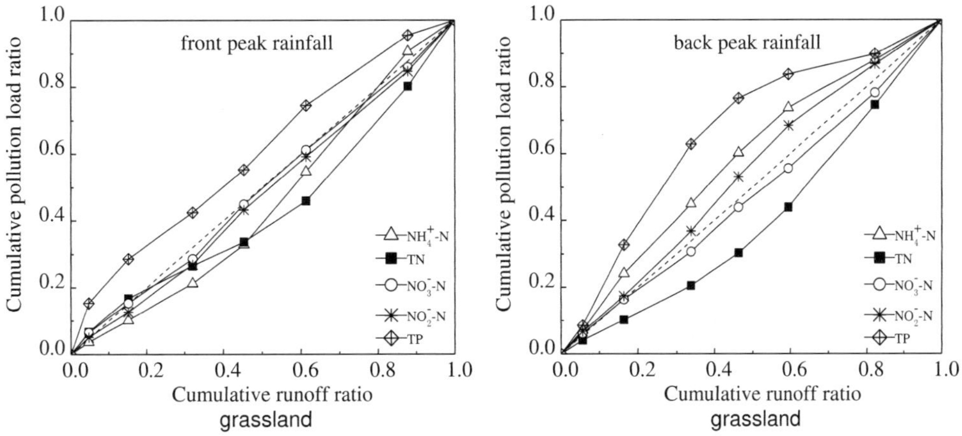

Figure 3. M(V) curves from unimodal rainfall.

3.3. *Quantitative analysis of FFE*

3.3.1. Analysis of uniform rainfall flush effect. After FFE curve fitting by power function, the b value of FFE of different underlying surfaces pollutants in different rainfall events is determined. Then, the MFF_{30} values is shown in Figure 4, which are obtained by the formula $MFF_{30} = 0.3^b$

It can be seen from Figure 4 that most of the rainfall pollutants MFF_{30} are between 0.3 and 0.6. Compared with the underlying surfaces, the FFE of the grassland was weak, and some scenes of rain pollutants MFF_{30} value was less than 0.3. The relation between the strength of FFE on different underlying surfaces is asphalt road> watertight tiles> grassland. However, for the NO_2^--N, the MFF_{30} value on the watertight tiles is significantly larger than the MFF_{30} value on the asphalt road. It can be seen that we should give priority to controlling NO_2^--N pollution of the watertight tiles. It is found that the MFF_{30} values of NH_4^+-N and TP are the highest in each pollutant, indicating that the two pollutants are prone to FFE. In addition, although TP and PO_4^{3-}-P similar in origin, but the latter FFE to be significantly weaker than TP. The MFF_{30} value of TN and NO_3^--N on the asphalt road increased slightly with the increase of rainfall intensity, indicating that the greater the rainfall intensity, the more obvious the flush effect of the two pollutants. And the MFF_{30} of NH_4^+-N, TP and PO_4^{3-}-P decreases with the increase of rainfall intensity, which indicates that these pollutants are not easily washed by runoff, and the runoff dilution effect is stronger. The MFF_{30} of NO_2^--N and TP decreases with the increase of rainfall intensity, but PN increases slightly, and the variation of other pollutants is not obvious. Finally, there was no significant change in the remaining pollutants except that the MFF_{30} of the grassland increased with the increase of the rainfall intensity.

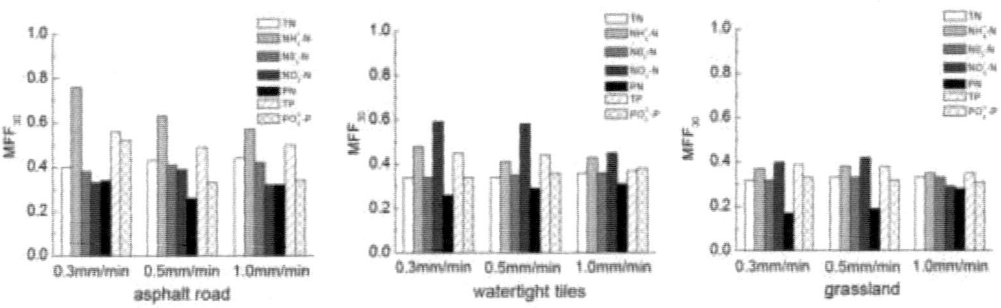

Figure 4. MFF_{30} values from uniform rainfall.

Table 3 shows the MFF_{30} average values of contaminants for different rainfall intensity on each underlying surface. In the table, the FFE of asphalt road and watertight tiles decreases with the increase of rainfall intensity, indicating that the dilution of pollutants on the two underlying surfaces is more pronounced. While the FFE of the grassland increased first and then decreased with the increase of rain intensity, which indicated that the grassland was less prone to dilution.

Table 3. The average values of MFF_{30} from uniform rainfall.

underlying surfaces	0.3mm/min	0.5mm/min	1.0mm/min
asphalt road	0.47	0.42	0.41
watertight tiles	0.40	0.39	0.38
grassland	0.33	0.34	0.32

3.3.2. Analysis of single-peak rainfall flush effect. The pollutants MFF_{30} values of single-peak rainfall on the underlying surface are shown in Figure 5, and the average values of the pollutants MFF_{30} is shown in Table 4. In conjunction with Figure and Table, it can be concluded that the top 30% of the runoff in the pre-peak rainfall is significantly higher than back-type rainfall that on the surface of the asphalt road. Similarly, in addition to NO_2^--N, the contaminants MMF_{30} values of pre-peak rainfall were higher than those the post-peak in the watertight tiles. The reason is that the runoff is often formed at the beginning in the impervious surface, and can carry a large number of pollutants. The results showed that the post-peak rainfall MMF_{30} value of TN, NO_3^--N and TP was higher than the former peak on the grassland, indicating that the FFE of the post-peak rainfall on the grassland was stronger. Because the permeability of the grassland is strong, and the higher descent rate of early rainfall is difficult to form a large number of runoff, and the flush of the pollutants is also less. Therefore, for the impervious surface, we should pay attention to controlling the early pollutants when the pre-peak rainfall. While for the permeable surface, we should focus on pollutant control that late post-peak rainfall.

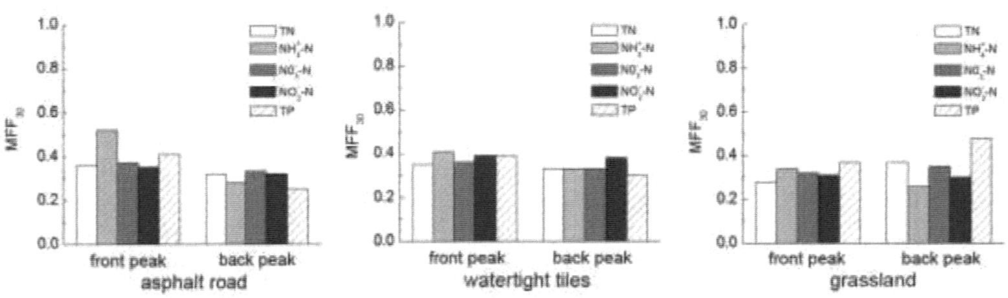

Figure 5. MFF_{30} values from unimodal rainfall.

Table 4. The mean values of MFF_{30} from unimodal rainfall.

underlying surfaces	MFF_{30}	
	front peak	back peak
asphalt road	0.40	0.30
watertight tiles	0.38	0.33
grassland	0.32	0.35

4. Conclusion

4.1 Under the condition of uniform rainfall, the results of the MFF_{30} are similar to those of the M(V) curve. Most of the rainfall pollutants occurred weak FFE, and the relationship between FFE of the different underlying surface is asphalt road> watertight tiles> grassland.

4.2 Intuitive analysis of M(V) curve, all the underlying surface of the FFE is more obvious, when the rainfall is maintained at 0.5mm/min. With the MFF_{30} value analysis, the asphalt road and watertight tiles, all show the rain increased and the FFE intensity decreased.

4.3 In the unimodal rainfall, the MFF_{30} value of each pollutant is generally lower than the MFF_{30} value of uniform rainfall. The FFE of back peak type rainfall of asphalt road and watertight tiles is weaker than that of the former peak type rainfall, and the results on the grassland are the opposite. Asphalt roads and watertight tiles on MFF_{30} of the pollutants is generally higher than the grassland. So grassland should be built around asphalt road and watertight tiles to attenuate FFE.

Acknowledgments
This work was financially supported by Key scientific and technological innovation team planning project of Shaanxi Province (2016KCT-23) fund.

References

[1] Taebi A, Droste R L. First flush pollution load of urban stormwater runoff [J]. Journal of Environmental Engineering & Science, 2011, 3(4): 301-309.
[2] Sønderup M J, Egemose S, Hoffmann C C, et al. Modeling phosphorus removal in wet ponds with filter zones containing sand or crushed concrete [J]. Ecological Engineering, 2014, 66(3): 52–62.
[3] Li J Q, Che W. Rainwater Problems and Countermeasures for Sustainable Development in Cities [J]. Urban Environment and Urban Ecology, 2005, 18(4): 5-8.
[4] Min J P, Park J Y, Shin H J, et al. Projection of future climate change impacts on nonpoint source pollution loads for a forest dominant dam watershed by reflecting future vegetation canopy in a Soil and Water Assessment Tool model [J]. Water Science & Technology, 2010, 61(8): 1975-86.
[5] Kayhanian M, Suverkropp C, Ruby A, et al. Characterization and prediction of highway runoff constituent event mean concentration [J]. Journal of Environmental Management, 2007, 85(2):279-95.
[6] Ren Y F, Wang X K, Han B, et al. Chemical analysis on stormwater-runoffpollution of different underlying urban surfaces [J]. Acta Ecologica Sinica, 2005, 25(12): 3225-3230.
[7] Ma Z B, Ni H G, Zeng H, et al. Function formula for first flush analysis in mixed watersheds: A comparison of power and polynomial methods [J]. Journal of Hydrology, 2011, 402(s 3–4):333-339.
[8] OU Lang-bo, HU Dan, HUANG Ye, et al. First flush analysis of PAHs in runoff in Beijing [J]. Chinese Journal of Environmental Science, 2011, 32(10): 2896-2903.
[9] Hathaway J M, Tucker R S, Spooner J M, et al. A Traditional Analysis of the First Flush Effect for Nutrients in Stormwater Runoff from Two Small Urban Catchments [J]. Water Air & Soil Pollution, 2012, 223(223): 5903-5915.
[10] Li C L, Liu M, Hu Y M, et al. Characteristics of pollutants in urban rainfall runoff in Shenyang city [J]. Chinese Journal of Ecology, 2014, 33(5): 1327-1336
[11] Fam S, Stenstrom M K, Silverman G. Hydrocarbons in Urban Runoff [J]. Journal of Environmental Engineering, 1987, 113(5): 1032-1046.
[12] Artina S, Bolognesi A, Liserra T, et al. Experimental analysis of first foul flush in an industrial area [J]. Water Pollution VIII Modelling Monitoring & Management, 2006, 1: 423-432.
[13] Lee H, Lau S L, Kayhanian M, et al. Seasonal first flush phenomenon of urban stormwater discharges [J]. Water Research, 2004, 38(19): 4153-4163.
[14] Soller J, Stephenson J, Olivieri K, et al. Evaluation of Seasonal Scale First Flush Pollutant Loading and Implications for Urban Runoff Management [J]. Journal of Environmental Management, 2005, 76(4): 309-18.
[15] Li C L, Liu M, Hu Y M, et al. analysis of first flush in rainfall runoff in Shenyang urban city [J]. Acta Ecologica Sinica, 2013, 33(18): 5952-5961.

IOP Conf. Series: Earth and Environmental Science **64** (2017) 012077 doi:10.1088/1755-1315/64/1/012077

[16] Chen Y, Zhao J Q, Hu B, et al. First flush effect of urban trunk road runoff in Xi'an [J]. Chinese Journal of Environmental Engineering, 2012, 06(3): 929-935.
[17] Xu S J, Wei S Q, Xie D T. Analysis of Non-point Source Pollution: Impact Factors [J]. Environmental Science and Technology, 2004, 27(4): 43-45.
[18] Lee J H, Bang K W, Ketchum L H, et al. First flush analysis of urban storm runoff [J]. Science of the Total Environment, 2002, 293(1-3): 163-75.
[19] State Environmental Protection Administration water and wastewater monitoring and analysis methods. Water and wastewater monitoring and analysis methods [M]. BeiJing: China Environmental Science Press, 2002.
[20] Qiang H E, Wang S, Hainan A I, et al. A novel approach in assessing the first flush effect in urban stormwater runoff [J]. Acta Scientiae Circumstantiae, 2011, 31(11): 2432-2439.
[21] Feng W, Wang J L, Che W. Analysis on characteristics of stormwater runoff flush on different land surfaces [J]. Chinese Journal of Environmental Engineering, 2012, 6(3): 817-822.
[22] Wang S M, Guo S G, He Q, et al. Water Quality Characteristics of Stormwater Runoff and the First Flush Effect in Urban Regions [J]. Environmental Science Research, 2015, 28(4): 532-539.
[23] Saget A, Chebbo G, Bertrand-Krajewski J L. The first flush in sewer systems [J]. Water Science & Technology, 1996, 33(33): 101-108.
[24] REN Y F,WANG X K, OUYANG Z Y. Analysis of First Flush Effect of Typical Underlying Surface Runoff in Beijing Urban City [J]. Environmental Science, 2013, 34(1): 373-378.
[25] Hou L Z, Jing Y Y, Feng S Y, et al. Comparison of Water Quality of Rainfall Runoff from Different Underlying Surfaces in Beijing City[J]. China Water and Wastewater, 2006, 22(23): 35-38.
[26] Cen G P, Shen J, Fan R S. Research on Rainfall Pattern of Urban Design Storm [J]. Advances In Water Science, 1998(1): 41-46.

Characteristics of soil salinity in the typical area of Yellow River Delta and its control measures

Mingliang Zhang*, Haixia Wang, Xiaoke Pang, Hui Liu, Qun Wang

School of resources and environment, University of Jinan, Jinan 250022, China

*Corresponding author e-mail: mlzhangsd@126.com

Abstract. The Yellow River Delta is one of important ecological areas in eastern China, however sustainable development of the Yellow River Delta is seriously restricted because of severe soil salinization. The main sources of soil salinity are chloride, sodium and sulfate ions. The distribution of soil salinity in soil profiles showed that surface accumulation of soil salt was significant in the Yellow River Delta. Some control measures including soil improvement and regulation, reasonable combination of salt-fresh water irrigation in farmland, land cover and effective drainage were put forward for soil salinity control.

1. Introduction

The Yellow River Delta has been formed by the sediment materials carried with the Yellow River, which is the youngest large alluvial plain in the world [1]. The Yellow River Delta is one important ecological area in eastern China, and it provides the good condition for the development of ecological, efficient and sustainable industry because of the abundant natural resources, unique ecosystem and good development foundation [2-5]. However, over 90 percent of the land area of the Yellow River Delta has been affected by soil salinization with different degrees of severity [6-10]. Sustainable development of the Yellow River Delta is seriously restricted because of severe soil salinization. The objective of this study was to study spatial distribution of soil salinity in the typical area of the Yellow River Delta. Some control measures including soil improvement and regulation, reasonable combination of salt-fresh water irrigation, land cover and effective drainage were put forward in order to control soil salinity.

2. Materials and methods

2.1. *Studying area*

The studying area is situated on the south bank of the Bohai Bay and the west coast of Laizhou Bay (117°31'-119°18'E; 36°55'-38°16'N). The Yellow River Delta belongs to the warm temperate semi-humid monsoon climate zone with annual average precipitation of 593.6 mm, over 50% of which occurs in July and August. Annual average evaporation is 1944 mm and precipitation-evaporation ratio is 3:1. The annual average temperature is 12.5°C. The capillary uplift strength of the soil is higher than that of the gravity downstream flow, which can create favourable conditions for salt accumulation in the soil surface. The groundwater in the Yellow River Delta is generally shallow (average depth of about 1m). The degree of mineralization is generally high, and the average degree of mineralization greater than 10 g/L, which mostly belongs to salt water and brackish water. As the groundwater is shallow, the salt in the groundwater can easily accumulate to the surface through the rising force of the capillary, forming soil salinization.

2.2. Soil sampling
The soil samples of the typical area of Yellow River Delta were sampled to analyse soil salinity in different depth soil layers (0-100 cm) in May 2015. In sampling sites, surface plants and stone pieces and other shovel were firstly removed, and then a 100 cm depth rectangular pit was dug. The soil profile was divided into five layers (0-20 cm, 20-40 cm, 40-60 cm, 60-80 cm, and 80-100 cm) and soil samples were collected from the bottom layer to top. The soil was air dried in the room, passed through 1mm sieve, and prepared for soil water extract solution (1:5 soil/water ratio) for the measurement of soil salinity. 50 g soil was added into plastics bottle with 250 mL deionized water, and shaked for 3 min, and then filtered with 0.45 um filter paper to get clean leachates. In April 2016, 180 sampling points were sampled from surface soil (0-20cm) in the Yellow River Delta and analysed the distribution characteristics of surface soil salinity. A handheld GPS locator was used to obtain the coordinates and recorded the vegetation type and the surrounding environment of sampling points.

2.3. Soil sampling
The electrical conductivity (Ec) of the soil leachate was determined by conductivity meter and the total amount of salt (soil salinity) was calculated accordingly. The cations of soil leachate were measured by atomic absorption spectroscopy. The concentration of sulfate was measured by ion chromatography and UV/VIS spectrophotometer. The concentration of chloride ion was determined by $AgNO_3$ titration method. Based on the soil salinization classification standard, soil salinity can be divided into five classes (Table 1).

Table 1. Classification standard of soil salinity

Soil salinity (g/kg)	Degree of soil salinization	Response of plants
<1	non salinized	no salt damage to the plant
1-3	mild salinized	salt-sensitive plants may be affected
3-5	moderate salinized	have an impact on salt-sensitive plants, but salt-tolerant plants still grow well
5-10	strong salinized	only salt-resistant plants can grow well
>10	saline soil	only few salt-resistant plants can grow

3. Results and Discussions

3.1. Results
Based on the experimental results of 33 soil profile samples and 181 surface soil samples salinity analysis, the basic situation of soil salinity in the Yellow River Delta was obtained. From the basic composition of soil soluble salt ions, it can be seen that the main sources of soil salinity in the Yellow River Delta were chloride, sodium and sulfate (chlorine and sulfate accounts for more than 50% of total salt). According to the classification standard of soil salinity, there were 53 samples of non-salted soil (total salt content <1 g/kg) in the study area (accounting for 29% of the total samples), and saline soil accounted for about 71% of the total samples. It indicated that most of the soil samples in the Yellow River Delta belonged to saline soil and had a higher degree of salinization.

The soil salinity of soil profiles in the Yellow River Delta was measured by the salt analysis of 0-100 cm depth soil. It showed the soil salinity maximum (48.71g/kg) and the mean (6.21 g/kg) in 0-20 cm soil layer, salinity maximum (31.71 g/kg) and mean (4.06 g/kg) in 20-40 cm soil layer, salinity maximum (28.12 g/kg) and mean (3.80 g/kg) in 40-60 cm soil layer, salinity maximum (21.7 g/kg) and mean (3.09 g/kg) in 60-80 cm soil, salinity maximum (21.80 g/kg) and mean (3.39 g/kg) in 80-100 cm soil. It can be seen that the maximum and mean values of salt content in the surface soil were much higher than those of the salt content of the bottom soil, and the difference between the salt content of the surface soil and the bottom soil can reach more than 40 times, which fully explained the soil salt accumulation in soil profile was significant in the Yellow River Delta. Distribution of total salt and Cl concent and of the soil profiles in four typical sites in Yellow River Delta is shown in Fig.1.

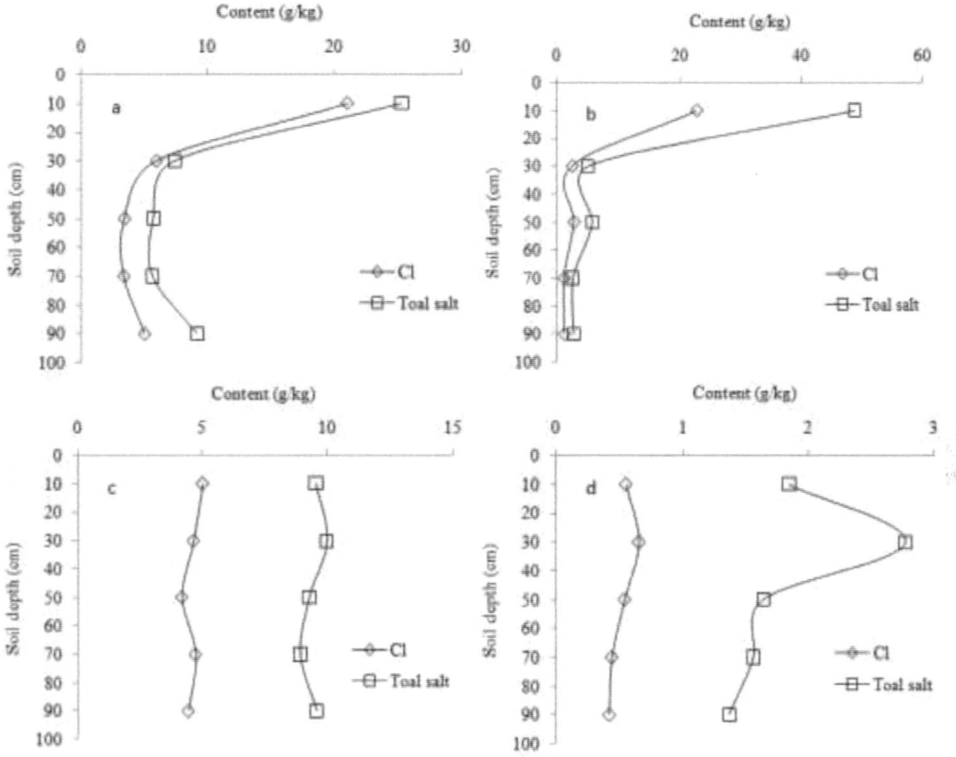

Figure 1. Distribution of total salt and Cl concent of soil profiles in four typical sites (a. Wuhaozhuang, b. Yellow River Delta Geopark, c. Diaokou, D. Yuwa) in Yellow River Delta.

3.2. Discussions

Based on the results of soil salinty in the typical area of Yellow River Delta, some soil salinity control measures were discussed as follows.

(1) Soil improvement and regulation. Chemical improvement measures include the application of amendment substrates such as calcium sulfite, gypsum, phosphogypsum. Biological improvement measures include the cultivation of salt-tolerant plants and pastures, green manure, afforestation and so on. Through fertilization measures (planting green manure, etc.) to supplement and improve soil organic matter and nutrient content, soil structure and physical and chemical properties can be improved.

(2) Measures for improvement of water conservancy. The main fresh water source is Yellow River in the Yellow River Delta, and water supply capacity is very limited. So the use of groundwater as an important measure is very important to alleviate the water crisis. However, groundwater is mainly salt water or brackish water n the Yellow River Delta, the combined irrigation method using brackish water and fresh water during irrigation time is recommended in farmlad of the Yellow River Delta. Many studies have found that the use of various combination irrigation methods (e.g. brackish-fresh water, fresh-brackish-fresh water) was conducive to surface salt leaching or even discharged from soil layer. Although brackish water irrigation will increase the salt content of the soil, fresh irrigation can play the role of salt leaching under the reasonable combination of brackish and fresh water irrigation,

and the salt content of the surface soil can reduce and maintain the crop normal growth. Soil salinity will not cause significant accumulation through the combination irrigation of fresh water and brackish water, plus the rainy season of concentrated leaching salt effect.

(3) Surface covering. Evaporation can affect the redistribution of soil moisture and salinity to a large extent. Ground cover measures can reduce soil surface evaporation. Many studies have found that straw cover or plastic film cover can effectively reduce the evaporation and play a role in water conservation, which can effectively reduce the salt content of surface soil compared with no coverage measures. It was reported that the salinity content of 0-40 cm soil layer using wheat straw cover was reduced by 12.41% than that of control treatment, and it showed that the covering treatment played a good role in the regulation of soil salinity [11].

(4) Drainage. Drainage is one primary method of controlling soil salinity. Effective discharage of excessive salt from soil is one of the important regulatory measures. Pipe drainage technology in the control of salt performance is very good which can quickly remove surplus of water in the soil, reduce the groundwater level to prevent the salt back to soil, and create a good environment for crop growth.

4. Conclusion

Sustainable development of the Yellow River Delta is seriously restricted because of soil salinization. The results showed that saline soil accounted for about 71% of the total samples. The main sources of soil salinity in the Yellow River Delta are chloride, sodium and sulfate ions (chlorine and sulfate accounts for more than 50% of total salt). Surface accumulation of soil salt in soil profile was significant in the Yellow River Delta. Some control measures including reasonable combination of salt-fresh water irrigation, land cover and effective drainage were put forward in order to control soil salinity.

Acknowledgments

This work was financially supported by Shandong provincial natural science fundation (ZR2014EEQ020), Undergraduate research training (SRT) of University of Jinan.

References

[1] Z.R.Wang, G.X. Zhao, M.X. Gao, C.Y. Chang, Spatial variability of soil salinity in coastal saline soil at different scales in the Yellow River Delta, China, Environmental Monitoring and Assessment. 189 (2017): 80.

[2] Y.W. Qin, G.X. Zhao, J. Wang, J. N. Cheng, Y. Meng, C. Dong, T.Lei, Restoration and reutilization evaluation of coastal saline-alkaline degraded landsin Yellow River Delta. Transactions of the Chinese Society of Agricultural Engineering. 25(2009) 306-311.

[3] X. Fan, B. Pedroli, G. Liu, Q. Liu, H. Liu, L. Shu, Soil salinity development in the yellow river delta in relation to groundwater dynamics, Land Degradation & Development. 23(2012), 175-189.

[4] B.H. Li, G.X. Zhao, C.Dong, Y.W. Qin, Dynamic and driving force analysis of salinized land in the Yellow River Delta based on RS and GIS, Journal of Natural Resources. 26(2011), 310-318.

[5] Z.R. Wang, G. X. Zhao, M.X. Gao, C.Y. Chang, S.Q. Jiang, J.C. Jia, J.Li, Spatial variation of soilwater and salt and microscopic variation of soil salinity in summer in typical area of the Yellow River Delta in Kenli County, Acta Ecologica Sinica. 36(2016), 1040-1049.

[6] J.S. Yang, R.J.Yao, Spatial variability of soil water and salt characteristics in the Yellow River Delta, Scientia Geographic Sinca. 27(2007), 348-353.

[7] J.B. Yu, Y.Z. Li, G.X. Han, D. Zhou, Y.Q. Fu, B. Guan, G.M. Wang, K. Ning, H.F. Wu, J.H. Wang, The spatial distribution characteristics of soil salinity in coastal zone of the Yellow River Delta, Environmental Earth Sciences. 72 (2014): 589-599.

[8] Q. Liu, G. Liu, X. Fan, Temporal-spatial distribution of soil salinity profile types in the Yellow River Delta, Shandong Agric Sci. 1 (2010):57-62.

[9] Y. Guan, G. Liu, Q. Liu, Q. Ye, The study of salt affected soils in the Yellow River Delta based on remote sensing. J Remote Sensing. 5(2001): 46-52.
[10] R.J. Yao, J.S. Yang, G.M. Liu, P. Zou, Spatial variability of soil salinity in characteristic field of the Yellow River Delta, Transactions of the Chinese Society of Agricultural Engineering. 22(2006), 61-66.
[11] H.C. Pang, J.S. Yang, H.J. Yan, Effects of irrigation with saline water on soil salinity and crop yield, Plant Nutrition and Fertilizer Science. 10(2004): 599-603.

Research on the removal of nitrogen pollutants in source water by EBO system

Wei Wei, Wenjun Zhang*, **Haoming Yuan, Xiaoyan Zhang, Liwei Geng**

Jilin Normal University, College Environmental Science and Engineering, Siping, 136000, China

*Corresponding author e-mail: 651430136@qq.com

Abstract.This paper describes an innovative process to solve the nitrate build-up problem using enhanced bio-contact oxidation system in situ (EBO system). The novel aspects of the process lay in the EBO system that is enhanced by a multifunctional device for water lifting and aeration (MDWLA) which has widely been applied in improving the source water quality in China. The reservoir water employed to study the nitrogen removal capacity of EBO system. The results showed that the EBO systems filled with Kaldnes suspended packing and sponge balls suspended packing respectively could efficiently remove ammonia, nitrate, nitrite, total nitrogen, COD, and turbidity from polluted surface water under conditions of dissolved oxygen 3.0~4.0 mg L^{-1} and temperature 24~28℃. Particularly, the maximum efficiencies of ammonia, nitrate, total nitrogen and turbidity in EBO system could reach above 71%, 76%, 77% and 90%.

1. Introduction

Until the beginning of 21st century drinking water sources were heavily contaminated by nitrates as a result of excessive fertilization or other pollution—mainly anthropogenic—sources. Elevated nitrate concentration in drinking water was a potential risk to public health since it could directly cause methaemoglobinemia (blue baby syndrome) in infants and its correlation with other diseases, including cancer, was not completely clear [1]. Because of these potential adverse effects, nitrate concentration limits in potable water had been set worldwide. Drinking water regulations were required in order to limit human risks and environmental pollution. While the United States Environmental Protection Agency (USEPA)had set maximum contaminant level goal (MCLG) of 10 mg NO_3-N and 1.0 mg NO_2-N L^{-1}, the World Health Organization and European Economic Community had set standards of 11.3 mg NO_3-N and 0.03 mg NO_2-N L^{-1}. This value was presently not met in a considerable amount of drinking water sources in USA, European Community, Africa, Middle East, Australia, and New Zealand and so on, so it was mandatory to remove the excess nitrate from the contaminated water in order to ensure potable use.

Many researches on the denitrification of source water had been reported, but most experiences were still been stayed on the conversion of different forms of nitrogen, which could not completely remove the nitrogen from water source.

Biological treatment was the most useful and efficient process to remove nitrogen from surface water sources. The full-scale application of this technique were mostly based on off-site bio-contact oxidation (OSBO), whereas very few tests and applications were available with the denitrification of surface water by the enhanced bio-contact oxidation system in situ (EBO system) [1-3]. Due to its overcoming the restriction of biological contact time, an important limiting factor for biological denitrification, the EBO system favored the improvement of nitrogen removal efficiency.

In this research, a multifunctional device for water lifting and aeration (MDWLA) was combined with the EBO system for the denitrification of surface water. The EBO system could be enhanced by MDWLA which had widely been applied in improving the source water quality in China. The main objective of this research was to investigate the feasibility and efficiency of nitrate removal using EBO system for source water denitrification in laboratory-scale. In the experiment, the EBO system with an aeration device acted as the function of MDWLA, was filled with a layer of highly efficient suspended packing materials as biological carrier, and the water samples were taken from a reservoir.

2. Materials and methods

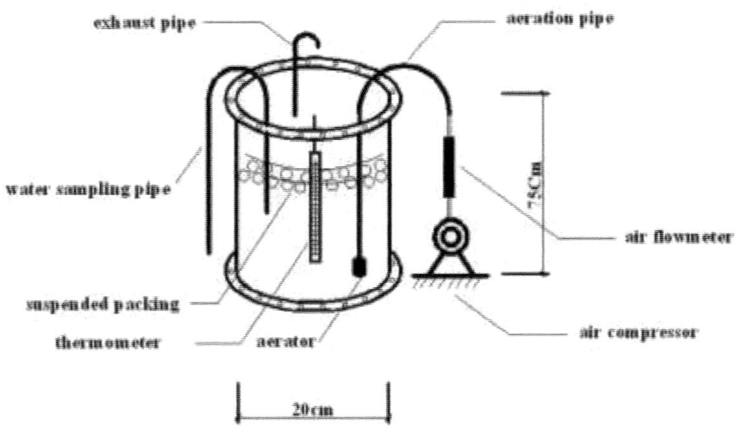

Figure 1. Schematic representation of EBO system

2.1. Experimental system

The experimental set-up included three sets of systems (system 1, system 2 and system 3). Two sets of EBO systems (system 2 and system 3)for the effect experiment were filled with suspended packing and equipped with aeration devices acted as the function of MDWLA. One set of system (system 1) without biological carrier and aeration device was used as the blank control group. The EBO system experimental set-up consisted of a cylindrical organic class reactor, with a 20 cm of inner diameter and 750 cm of height (Fig.1). The EBO system was equipped with a water sampling pipe, an exhaust pipe and an aeration port on its top, and an air flow meter installed at the aeration port. During the entire experimental period, DO concentration and temperature in EBO system was in the range of 3.0~4.0 mg L^{-1} and 24~28℃ respectively.

Two EBO systems for efficiency experiment were filled with the Kaldnes suspended packing (system 2) and the sponge balls suspended packing (system 3) respectively as carrier for microorganism, and the filling rate for both packing materials was 2%.

Prior to operation, the carrier materials were inoculated with microorganisms for 3~4 days by pouring the high efficient bacterial liquid into the systems, then acclimatized to low-nutrient medium prepared in source water taken from a reservoir. The inoculation lasted for about 1 week for microbial growth.

2.2. Source water

In this study, the water sample was taken from a reservoir (water source characteristics: COD 6.3 mg L^{-1}, N-NH_4^+ 0.246mg L^{-1}, N-NO_3^- 1.730mg L^{-1}, N-NO_2^- 1.730mg L^{-1}, TN 1.989 mg L^{-1}, Turbidity 18.5NTU, pH 7.58).

IOP Conf. Series: Earth and Environmental Science **64** (2017) 012079 doi:10.1088/1755-1315/64/1/012079

2.3. Bacterial agent
The bacterial agent used in this study consisted of efficient aerobic denitrifiers isolated and cultured in the low-nutrient medium in order to remove nitrogen desirably under oligotrophic conditions.

A proper amount of the bacterial agent was inoculated in medium solution in 2000 ml bottle with a cotton plug, and cultivated under agitation (120 rpm), at 30℃ and pH 7 for 48h. These pre-cultures were used to inoculate the EBO system.

2.4. Biological packing materials

2.4.1. Kaldnes suspended packing The Kaldnes suspended packing made of the modified high polymer had a diameter of 25 mm and a height of 10 mm. The suspended carrier had a specific gravity of about 1.000 after inoculated, which can rotated freely under the aeration and improve the oxygen utilization by colliding and cutting the air bubbles.

2.4.2. Sponge balls suspended packing The sponge balls suspended packing was prepared by the reticulated polyurethane materials. Due to its bigger porosity the sponge balls suspended packing had a higher specific surface area compared to other packing materials. In this research the sponge suspended packing with a 50 mm of diameter was used in the system.

2.5. Analytical methods
All samples were filtered through a 0.45 μm membrane filter before analysis and were tested within 1 h of collecting. Ammonia was determined by the Nesslerization method, nitrite by the azo dye colorimetric method, nitrate by the Szechrome NAS reagent (diphenylamine sulfonic acid chromogene) method, total nitrogen by the alkaline potassium persiflage digestion-UV spectrophotometric method according to water and exhausted water monitoring analyzed method (National Environmental Protection Committee of China, 4th Edition). NH_4-N, NO_2-N, NO_3-N, TN were analyzed with Hach DR5000 ultraviolet spectrophotometer (Hach Company, Loveland, CO, USA). Turbidity was measured by WGZ-1S turbidometer (Shanghai Leici Instrument Inc., China).

3. Results and discussion
COD concentration in EBO systems during the experiment was presented in Fig. 2. The COD removal efficiency during the entire experiment for both systems was in the range of 8.65% to 25.81%. The consumption of COD mainly resulted from the metabolic activities of microorganism such as oxidation, reduction and synthesis, and the biological flocculation and adsorption process and nitrification also played an important role during the biodegradation of organic substances [4]. At the same time the protozoa and micro metazoa attached in suspended packing also had a certain role in degrading the organic pollutants by phagocytizing algae debris, bacteria and organic debris [5]. Bacterial colony and micro animals constituted together the multifunctional biofilm structure. However, the low utilization percentages of COD implied the scarce availability of organic carbon content in the natural organic substances in source water.

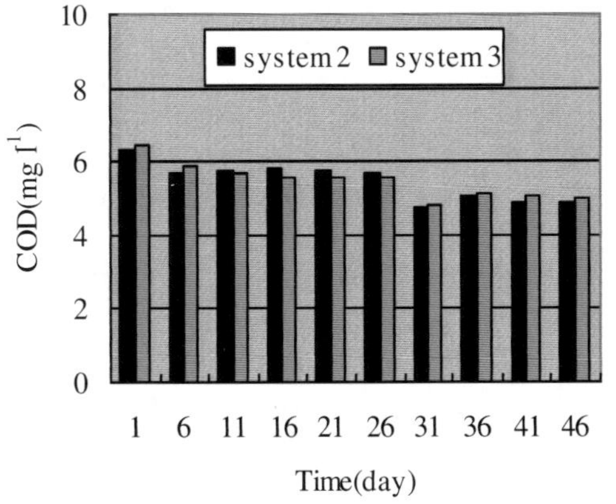

Figure 2. COD concentration variation with operation time for systems 2 and 3

As shown in Fig. 3, the nitrifying percentages of ammonia for both systems 2 and 3 were still desirable compared with system 1 under oligotrophic conditions and low initial ammonia concentration (about 0.25 mg L^{-1}). The highest ammonia removal percentages for systems 2 and 3 were observed at the 15th day, and were 71.32% and 89.59%, respectively. The decline of ammonia concentration in the early operation mainly resulted from the biosorption of the suspended carrier. However, the ammonia values increasing sharply in all system at the 5th day seemed to be caused by the conversion of organic nitrogen to ammonia when the heterotrophs consuming the organic matter. It was evident that the ammonia removal rate at the beginning of experiments was low. It was probably due to the slow growth and adaptability of autotrophic nitrifiers while the nitrification was ineffective. The experimental results showed that the ammonia concentration sharply increased in the late experiment period probably due to the detachment of the ageing biologic membrane. The changes of nitrate concentration in the systems were given in Fig. 4. It was observed that the better removal rates of nitrate were achieved in systems 2 and 3 when compared to the system 1 due to the high denitrification efficiency of bacterial agent including various oligotrophic denitrifiers which had a high tolerance for starvation. Among the three examined systems, the highest denitrification rate with 80.38% was obtained in system 2 at the 31st day. In the experiments of systems 2 and 3, nitrate levels (1.835 mg L^{-1} and 1.910 mg L^{-1}, respectively) were initially higher than the level (1.730 mg L^{-1}) in system 1 as a result of the unimmobilized bacteria falling into water, as explained above. The decrease of ammonia concentration in the earlier operation was mainly caused by the biosorption of biofilm. The conversion of ammonia to nitrate resulted the slightly increase of nitrate at the 15th day as shown in Figs. 3 and 4. However, for approximately 3 weeks after the beginning of operation of the EBO systems, nitrate concentrations had fallen slowly (the concentration changes less than 0.5 mg L^{-1}), which suggested that adaptation of the bacterial population to the sub-nutrition environment occurs in a long time period. The microorganisms acclimatized themselves to the wane of nutriment by reducing their metabolizing levels, extending their specific areas, utilizing more types of substrates. During this period of time, the morphology and physiology of oligotrophic bacteria experienced a significant change, and came to a steady starvation—survival state while the cell size remained almost constant. Also the efficiency of metabolic mechanism of microorganism had been improved due to the energy deficiency [6]. After the adaptation, the nitrate removal was achieved rapidly when the natural substances were utilized as the carbon source, as shown in Figs. 1 and 4. These results indicated the necessity of utilizing higher biological contact time to enhance the denitrification efficiencies, which was the most attractive advantage for the EBO system. After 31 days of operation, the nitrate level rising slightly was probably due to the aging biofilm falling into water. At the end of the experiment, the nitrate concentrations of systems 2 and 3 remained steady and were less than 1.0 mg L^{-1}, probably due to the limitation of usable organic compounds in the

IOP Conf. Series: Earth and Environmental Science **64** (2017) 012079 doi:10.1088/1755-1315/64/1/012079

substrates. This approximation was based on the assumption that denitrification can be modeled as a zero-order reaction with respect to nitrate concentrations down to very low levels [7].

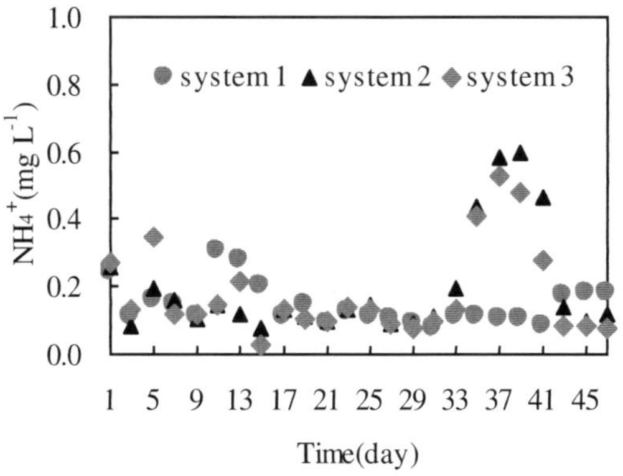

Figure 3. Ammonia concentration variation with operation time for three systems

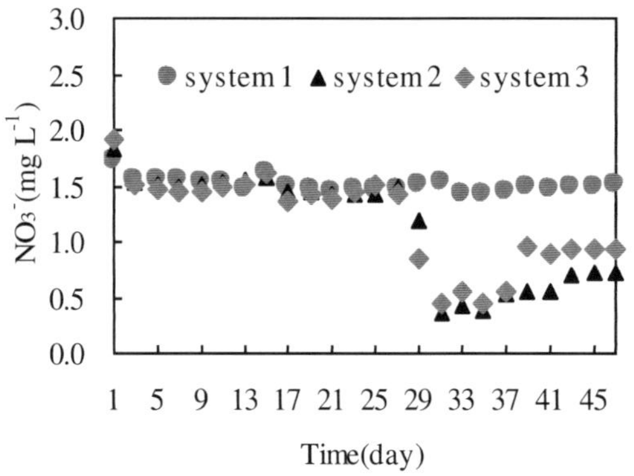

Figure 4. Nitrate concentration variation with operation time for three systems

The experimental results for nitrite variation in the three systems were illustrated in Fig. 5. It was observed that for approximately 4 weeks after the start of the experiment, the nitrite concentrations of systems 2 and 3 did not show significant changes. These findings indicated that the stable and timely conversion of ammonium to nitrate would occur when the ammonia level of source water was low. However, the nitrate concentration rapidly decreased at the 29th day while the nitrite concentration increased as a result of a reduction of nitrate to nitrite, as shown in Figs. 4 and 5. After that, nitrite concentration values started to decrease until they were not detected at the 31st day. It was observed from Figs. 5 and 3 that the nitrite accumulation occurred in systems 2 and 3 in the later period of operation due to the sharply increase of ammonia concentration while the nitrifying bacteria could not accomplish the timely conversion of nitrite to nitrate because of their slower growth rates compared with the ammonia-oxidizing bacteria.

Fig. 6 summarized the total nitrogen variation of the three systems during the entire experimental period. It was indicated that the systems 2 and 3 showed the desirable maximum total nitrogen removal efficiencies (77.62% and 78.17%, respectively) by the action of the oligotrophic denitrifiers which had the obvious advantages of high affinity and strong adsorption capacity in the low-nutrient competition [8]. At the beginning stage ammonia, nitrate, nitrite and biodegradable organic substances were adsorbed quickly by the aerobic and facultative bacteria on the biofilm, and these pollutants were removed by the biodegradation, nitrification and denitrification. It was observed by comparing Figs 6 and 4 that the concentration variation trends for total nitrogen and nitrate were almost coincide. The concentration variation of the total nitrogen was not significant during the first 4 weeks of operation because the bacteria necessitated an adaptation time for the poor nutrient environment, as explained above. At the 31st day, the total concentrations for systems 2 and 3 reached the lowest points (0.470 mg $L^{-1}$and 0.541 mg L^{-1}, respectively) during the entire experiment and did not exceed a maximum limit value of 1.0 mg L^{-1}. However the low denitrification efficiency of total nitrogen seemed to be caused by the lack of organic carbon compounds for use as electron donor at the end of experiments.

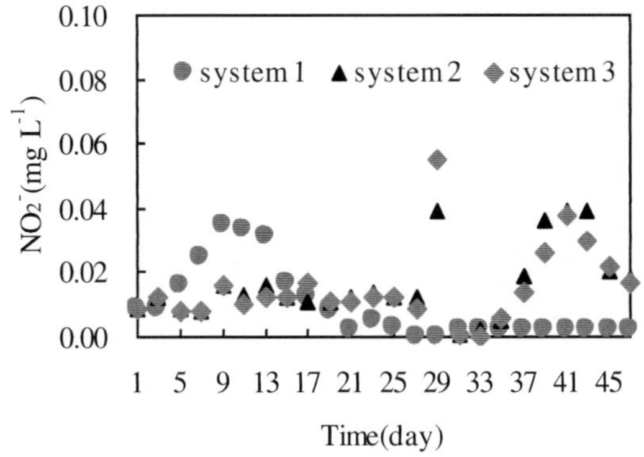

Figure 5. Nitrite concentration variationwith operation time for three systems

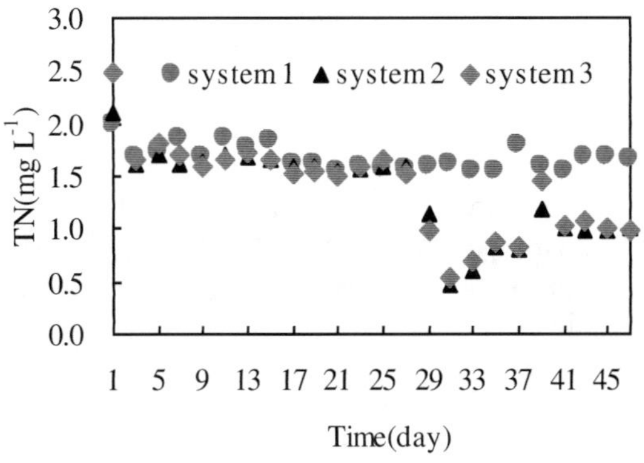

Figure 6. Total nitrogen concentration variation with operation time for three systems

The change of turbidity during the entire experiment was illustrated in Fig. 7. The optimum efficiency of turbidity for systems 2 and 3 could reach above 90% because of the biological adsorption, microbiological flocculation, biodegradation and direct precipitation of the biofilm. The turbidity could be eliminated by the degradation of some organic substances which could produce the turbidity.

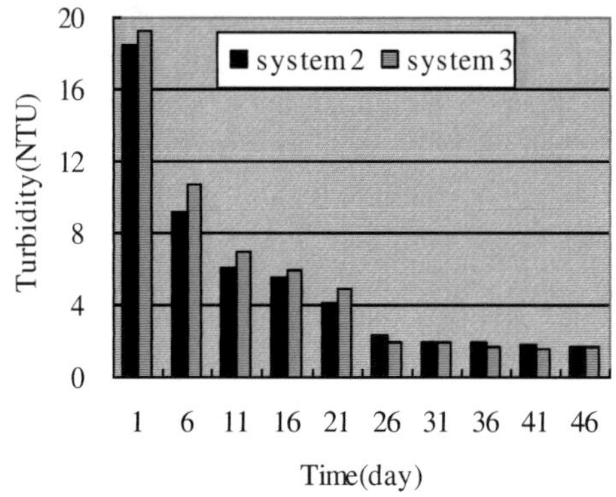

Figure 7. Turbidity variation with operation time for systems 2 and 3

4. Conclusions

EBO system is potentially an interesting alternative for drinking water denitrification, as it reduces the risk of nitrogen contamination and does not require high operational costs. During the entire experiment, the optimum removal rates of COD, ammonia, nitrate, total nitrogen and turbidity reached above 25%, 71%, 76%, 77% and 90%, respectively under the conditions of dissolved oxygen 2.5~3.0 mg L^{-1} and temperature 24~28℃.

In order to achieve desirable denitrificaiton efficiency, the enough biological contact time should be taken into account in EBO system, which was essential for the denitrifiers to adapt the oligotrophic environment.

In future research, the EBO system will be studied for the denitrificaiotn efficiency in pilot scale, which combines the bio-contact oxidation system in situ with the multifunctional device for water lifting and aeration (MDWLA).

Acknowledgments

This research work are financially supported by Doctoral Research Foundation of Jilin Normal University (Grant No. 2016022 and 2016033).

References

[1] Buttiglieri G., Malpei F., Daverio E., et al, Denitrification of drinking water sources by advanced biological treatment using a membrane bioreactor, Desalination, 178(2005), 211-218.
[2] Fabbricino M., Pettab L., Drinking water denitrification in membrane bioreactor/ membrane contactor systems, Desalination, 210(2007), 163-174.
[3] Singer A., Parnes S., Gross A., et al, A novel approach to denitrification processes in a zero-discharge recirculating system for small-scale urban aquaculture. Aquacultural Engineering, 39(2008), 72-77.
[4] Yu-tang Xiao, Jian-hua Xu, Research on the mechanism of micro pollutants removal in the raw water by biochemistry purificaiton process, Environmental Science, 20(1999), 85-88.
[5] Wei-jun Tian, Study on the mechanism and test of biological hardening purification directly for polluted streams, PhD thesis, Environmental Engineering Department, Heihai University, 2005.
[6] Ming Xu, Study on biological contact oxidation process under oligotrophic conditions, PhD thesis, Environmental Engineering Department, Guangdong University of Technology, 2005.
[7] Aslan S., Cakici H., Biological denitrification of drinking water in a slow sand filter, Journal of Hazardous Materials, 148(2007), 253-258.
[8] Kimura Y., Nakano Y., Fujiata K., et al, Isolation and characteristics of yeasts able to grow at low concentration of nutrients, Yeast, 14 (1998) 233-238.

IOP Conf. Series: Earth and Environmental Science **64** (2017) 012080 doi:10.1088/1755-1315/64/1/012080

Effect of a rapid repair mechanism for nitrification capacity in the load impact wastewater treatment

Wei Hong[1, a], Bing Zhang[2], Changdong Sun[3], Xin Tan[1,*], Bo Liu[1] and Xiaofeng Zou[1]

[1]Key laboratory of Shandong Academy of Environmental Science, Jinan, China
[2]Shandong Changle Shikang Water Industry Co., Ltd., Weifang, China
[3]Changle County Bureau of housing and urban rural development, Weifang, China

*Corresponding author email: tanxin_saes@163.com; [a]honw@163.com

Abstract. The nitrification capacity in the wastewater treatment is very important, and is particularly vulnerable to impacts. In this study, a rapid repair mechanism for nitrification was built and the result showing that, with the addition of exogenous nitrifying bacteria and organic nutrient, the nitrification capacity in the pilot scale equipment was restored in 20h, the concentration of NH_3-N in discharge conforms to the one-class A permitted criterion (GB 18918-2002) and remain stable for long time, while the nitrification capacity in control group would not be fixed by itself in 196h. The repaired experimental group has the advantage of strong shock resistance and stable operation, and under the second high impact load, the concentration of NH_3-N in effluent remain stable.

1. Introduction

With the rapid development of world economy, the demand for water in industrial production is increasing, as well as the discharge of wastewater. Excess ammonia nitrogen in wastewater has caused wide attention of researchers and environmentalists advocates. High NH_3-N concentration is known to heavy algal growth thus promoting the eutrophication in water bodies. Elevated NH_3-N concentration also causes certain effect for water treatment equipment and reuse engineering plant [1]. As one of the wastewater treatment processes, microorganism treatment is a method of low cost, high efficiency and pollution free. Nitrification is the predominant removal process of ammonia in the microorganism treatment. While nitrifying bacteria have a low growth rate, and are sensitive to the external stimulus such as temperature, pH, dissolved oxygen and toxic substances, nitrification is also the limiting step of biological denitrogenation, and vulnerable to load shocks[2].

When the nitrification system been impaired and the NH_3-N concentration of discharge water continue to be high, traditional restored approaches, such as stop intake, replace the activated sludge and increase of HRT and aeration, require long stretches of time. Chemistry methods which control the NH_3-N concentration at least temporarily are usually high-cost and generate secondary pollution [3]. It is critical to establish an efficient and effective method to repair the nitrification capacity for the load shocked wastewater treatment. In this study, a rapid repair mechanism is built to restore the nitrification capacity of a load shocked industrial park treatment plant. The park treatment plant was shocked by the upstream factory accident drainage, and the nitrification system of it was nearly destroyed. Pilot scale equipment was built and debug for the reference about the park treatment system, and exogenous nitrifying bacteria and organic nutrient were added to facilitate the reparation of nitrification system. Data and conclusions of this study will provide technical support and theoretical guidance for the treatment plants with the same problem.

2. Materials and methods

2.1. The wastewater treatment plant in the industrial park

The design treatment capacity for the wastewater treatment plant in an industrial park of Suzhou is 8000 m^3/d, and the upstream wastewater was mainly domestic wastewater, printing and dyeing wastewater, tannery wastewater, pesticide wastewater and metal processing wastewater, in which the domestic wastewater accounts for about 40%. The treatment processes in the plant are "Collection-well, Anoxic-tank, Anaerobic-tank, Clarifier, and Effluent". The pollutant concentrations in the inflow and effluent under normal operating conditions are showing in table 1. After been shocked by the accident drainage, the NH_3-N concentration in effluent had increased to 26-31 mg/L and remained high in a long time.

Table 1. Pollutant concentrations in the inflow and effluent under normal operating conditions.

Sample	COD_{Cr} (mg/L)	NH_3-N (mg/L)	TN (mg/L)	pH
Inflow	290-370	23-30	35-43	7.5-8
Effluent	38-46	0.2-0.8	22-32	6.8-7.3

2.2. Equipments and materials for the pilot scale experiment

In this study, two same systems of pilot scale experiment were set up, one was experiment group and other was control group. The treatment processes for the two systems were designed for the reference about the treatment system in the industrial park treatment plant. Experimental studies and debug had been conducted, simulating operating conductions of the park treatment plant. The flow diagram for the pilot scale equipment is showing in Fig. 1. The effective volume of the tank is 0.5 m^3, in which the wastewater from regulation pool in the park treatment plant was pumped as incoming water every 12h. The design treatment capacity for each device is 100L/h. The effective volume of Anoxic-tank is 0.8 m^3 and Anaerobic-tank is 2.4 m^3. The activated sludge of every tank was inoculated from the corresponding treatment tank of plant.

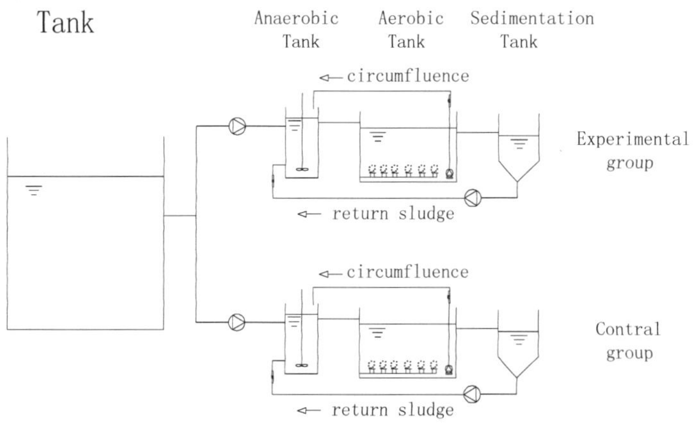

Figure 1. The flow diagram for the pilot scale equipment.

The exogenous nitrifying bacteria added in the experimental group are Novozymes BioRemove^{TM} 5805, and the organic nutrient, which is patented product, was produced by SAES Environmental Science and Technology CO. Ltd. The organic nutrient is to improve the structure of zoogloea granules in activated sludge, and to enhance the attachment and proliferation for the exogenous nitrifying bacteria [4].

2.3. Methods

The rapid repair mechanism was composed of the addition of exogenous nitrifying bacteria and organic nutrient. The dosage of exogenous nitrifying bacteria added into the experimental group's Anaerobic-tank is 100 mg/L, and was provided by twice. At the start of experiment, take time as 0 h and added 60% of nitrifying bacteria. When the experiment time is 6 h provided the other 40% of nitrifying bacteria. The dosage of organic nutrient is 300mg/L and was provided by fifth times. Added 20% of organic nutrient at 0.5h and provided 20% of organic nutrient every 12 h. The control group was added no agents as contrast.

The concentration of NH_3-N in incoming water and effluent from experimental and control group was detected every 4 h according to the Nessler's reagent spectrophotometry [5]. The structure of zoogloea granules in activated sludge from two groups was detected by Olympus BX41 - DP25 microscope.

3. Result and discussion

3.1. Removal of NH_3-N in experiment and control group

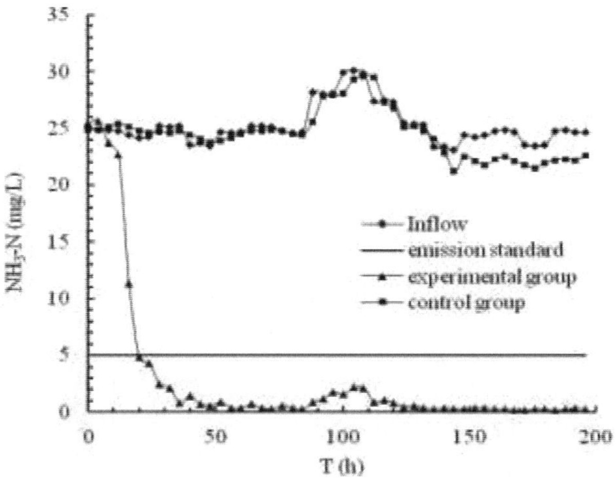

Figure 2. Removal of NH_3-N in experimental and control group.

The NH_3-N concentration in incoming water and effluent from experimental and control group during experiment was shown in Fig 2. As shown in Fig 2, the concentration of NH_3-N in effluent from experimental group increased in 8h after the addition of nitrifying bacteria. Main causes of the fluctuation is that in the medium of nitrifying bacteria, there was a suitable amount of NH_3-N, and the nitrifying bacteria need to take a certain time to attach and grow. After 12h the concentration of NH_3-N in effluent from experimental group decreased with immersing time, and 16h reduced significantly, 24h reduced to standard value (5 mg/L) [6]. As time went on, the concentration of NH_3-N in effluent from experimental group reduced to 1mg/L, and the removal rate was up to 95%. This result suggest that with the assistance of organic nutrients, the nitrifying bacteria were allowed to stay with the zoogloea granules in activated sludge, rapidly multiplying and spreading out to remove NH_3-N, and the repair mechanism for nitrification did work rapidly and effectively in 24h. As a control, the NH_3-N in effluent from control group was almost the same value with that of income water, and the removal rate of control group was still below 5% after 150 h. This results shown that the nitrifying system could not be restored by generation's update during a short time.

At 88 h of experiment time, the concentration of NH_3-N in income water increased to 29 mg/L, which brought second impact load. The concentration of NH_3-N in experiment group had a little volatility and was still below 2 mg/L, conformed to the discharge stand, while that in control group increased rapidly to 28 mg/L during the second impact load. After 24h the concentration of NH_3-N in experiment group revert back to 1 mg/L and 36h to below 0.5mg/L. This result suggest that with the assistance of exogenous nitrifying bacteria and organic nutrient, the nitrifying system in experiment group built stronger stability and shock resistance ability. Similar results have been reported in other studies [7].

3.2. The effect of organic nutrients

With the addition of exogenous nitrifying bacteria and organic nutrient, the nitrification capacity of load shocked treatment system was restored, the structure of zoogloea granules had been changed and activated sludge settleability been strengthened. Fig 3 shows the effect of organic nutrients on the structure of zoogloea granules in activated sludge.

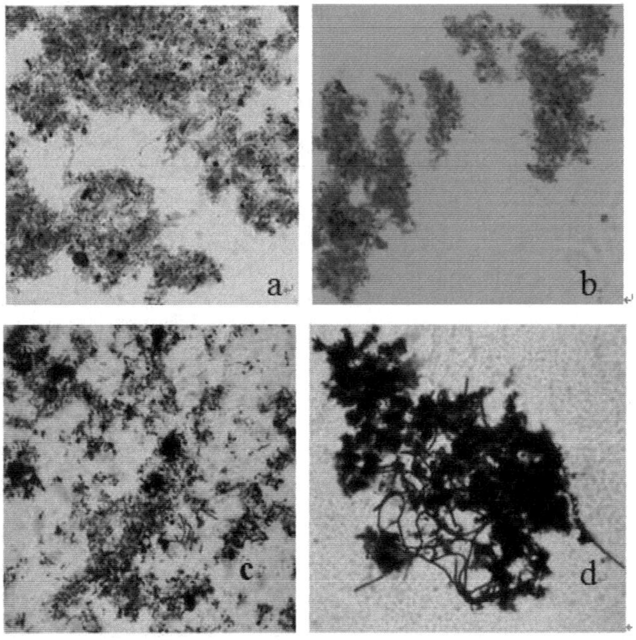

Figure 3. The effect of organic nutrient on the structure of zoogloea granules in activated sludge.

As shown in Fig 3, the microscopy images of 100 times for experimental group is (a) and control group is (b), and 1000 times for experimental group is (c) and control group is (d). Contrast image (a) and (b), with the assistance of organic nutrient, the structure of zoogloea granules in activated sludge from experimental group had been improved and were more compact, which has the advantage of attachment and proliferation for the exogenous nitrifying bacteria. And compare image (c) and (d), the zoogloea granules in experimental group had more filamentous fungi structure, which help to compact the structure of zoogloea granules. These results suggest that the addition of organic nutrient stimulate the proliferation of bacteria in activated sludge. Previous research has proved that the organic nutrient is advantageous to the total number of bacterial species, and the community structure of bacteria [8].

IOP Conf. Series: Earth and Environmental Science **64** (2017) 012080 doi:10.1088/1755-1315/64/1/012080

4. Conclusion

The rapid repair mechanism for nitrification, with the addition of exogenous nitrifying bacteria and organic nutrient, restored the nitrification capacity in the pilot scale equipment in 20h, the concentration of NH_3-N in discharge conforms to the one-class A permitted criterion (GB 18918-2002), and after been repaired for 24h, the nitrification capacity could remain stable for a long time. While only relay on the self-healing capabilities, the nitrification capacity in the control group would not be fixed in 196h.

The structure of zoogloea granules in activated sludge had been improved in the repaired experimental group, which has the advantage of attachment and proliferation for the exogenous nitrifying bacteria. Thus the experimental group had strong shock resistance and stable operation, and under the second high load impact, the nitrification capacity self-restored in 24h and the concentration of NH_3-N in discharge remain stable.

References

[1] Wang Li-ping, Cao Guo-ping, Zhou Xiao-hong.Research Advance in Technology of Treating Ammonia- Nitrogen Wastewater[J] ,Chemical Propellants & Polymeric Materials, 2009, 7(3):26-32.

[2] Wang Jia-ling. Environmental microbiology (second edition)[M], Beijing. Higher Education Press, 2008.

[3] Jin Yuan, Xia Jianxin, Zhang Zijun. Comparative analysis on the treatment of ammonia nitrogen in industrial wastewater[J], Industrial Water Treatment,2013, 33(7):5-10.

[4] SAES Environmental Science and Technology Co. Ltd, A new organic nutrient for wastewater treatment, preparation and application: CN 201510109516.4[P]. 2015-07-01.

[5] HJ535-2009,Water quality-Determination of ammonia nitrogen-Nessler's reagent spectrop hotometry [S].

[6] GB 18918-2002, Discharge standard of pollutants for municipal wastewater treatment plant[S].

[7] Wang Kai,Hong Wei,Liu Bo,et al.The impact of formaldehyde on the coal chemical industry wastewater biochemical treatment system and its faste repair thchnology[J], Environmental Engineering ,2015(09):64-68.

[8] Liu Bo, Hong Wei, Lang Yong-mei,et al.Application of organic nutrition in paper mill effluent treatment[J], China paper industry,2007,(28):80-82.

Munroe effect based on detonation wave collision

Yusong Miao, Xiaojie Li*, Xiaohong Wang, Honghao Yan, Xiang Chen

Faculty of Vehicle Engineering and Mechanics, Dalian University of Technology, Dalian, China

*Corresponding author email: 393291800@qq.com

Abstract. Munroe effect has been more and more used in blasting engineering and most assembling energy technologies use a shaped charge device. In this paper, a new method is used to achieve detonation wave collision by detonating cord initiation system. A numerical simulation using LS-DYNA on detonation wave propagation and collision process caused by different initiation forms is implemented. Numerical results show that peak pressure by this new method can reach 2.42 times than the traditional method, and the growth of specific impulse at the explosive bottom is 49% compared to early results. Based on this numerical simulation, an experiment of explosive-determination of power be implemented, the experiment result can verify the simulation result well.

1. Introduce

Munroe effect is a phenomenon that increases the explosive damage ability by changing the direction of detonation wave propagation. This phenomenon is mainly caused by explosive superposition in the hollow cavity which comes from the use of shaped charge [1-2]. Several studies of Munroe effect are reported in the literature. At present, Munroe effect is mainly caused by metal cover shaped, such as conical, hemispherical and double conical. This effect is widespreadly used in military, exploration and cutting. LI et al [3] applied the Mach reflection principle, to demonstrate that high detonation velocity explosive initiating low detonation velocity explosive can enhance the detonation pressure. Wang et al [4] used meshless MPM methods to simulate the detonation wave. O.Ayisit [5] used the nonlinear dynamics software AUTODYNA to study shaped charge in the process of four different types of asymmetric defects. Baker E L et al [6] had an experiment on the selectable initiation shaped charge to show that jet output is predictable and can be varied significantly for different target. Li et al [7] applied damage mechanics and dynamic caustics transmission to study the crack direction of rock fracture under the shaped charge and describe the physical process of the fracture. J.F. Molinari [8] analysed the formation, fragmentation, and penetration in a plate of a copper jet that develops in a shaped charge by numerical simulation. Qin [9] designed ‘double shaped groove cartridge’ to realize the Munroe effect by a shaped charge device. In this paper, a new method is used to achieve detonation wave collision by high detonation velocity explosive initiating low detonation velocity explosive. Through numerical simulation and standard test, this method can achieve the Munroe effect perfectly. Compared with the former method, this method has advantages of avoiding the shaped device, simplifying the traditional shaped charge, improving the working efficiency, reducing the blasting cost.

2. Detonation pressure increase theory

Detonation pressure increase theory is to use low detonation velocity explosive as the main charge, the high detonation velocity explosive as initiation, causing the detonation wave collision to achieve the Munroe effect. For the detonation pressure P_H of the blasting bore wall, the following formula is often used:

$$P_H = \frac{1}{K+1}\rho_0 D^2 \quad (1)$$

Where K is a constant polytropic exponent, for high detonation velocity explosive K is set to 3, while for low detonation velocity explosive K is set to 1.5~2.8. $P0$ is the initial charge density. D is the explosive detonation velocity.

For the detonating cord, the detonation pressure can reach to 13.74 GPa. However, for ANFO (Ammonium Nitrate / Fuel Oil), the pressure only reaches to 39.87% of the detonating cord. Hence, ANFO is worse than detonation cord if rock fragmentation is the only consideration. But if we use the high-energy explosive (i.e. the detonating cord) to initialize low-energy explosive (i.e. ANFO), may cause to the detonation pressure increase in the hollow cavity where high-energy explosive blast in advance.

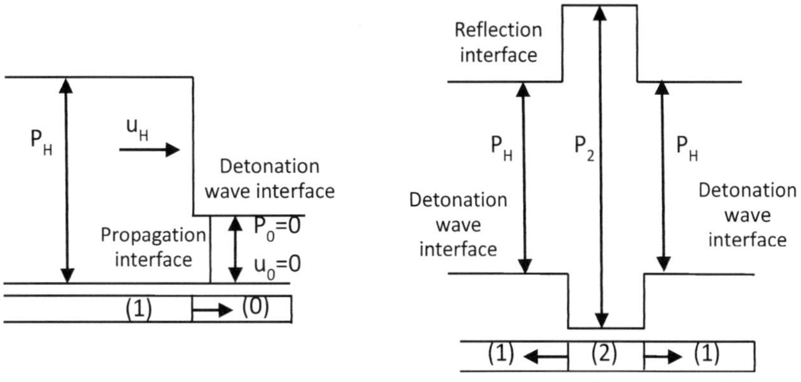

(a) Traditional detonation wave propagation (b) detonation wave collision

Figure 1. Initial parameters of the detonation wave

Figure 1 shows the initial parameters of the detonation wave. As shown in Figure 1(a), the physical state (1) spread to the initial state (0) after the explosive was detonated, the whole is a continuous process, and this is the common detonation wave propagation process. However, the detonation wave reflection will happen when two detonation wave have a collision, it means that if the separate detonating cord initiation is used, the detonation wave will separate from both sides and when the spread to the center, a detonation wave collision will happen, causing the collision state (2) as shown in Figure 1(b). The collision process is satisfied with the conservation equations for mass, momentum and energy. After the collision, part of the detonation wave reflected back in the opposite direction, and the others lead to the Munroe effect.

Firstly, the detonation wave continuously spread from the initiation point when the explosive was detonated. Secondly, the detonation wave gathered in the center of the bore. Finally, the detonation wave spread to the sides along the center line, causing Mach reflection when the angleα reach to a certain value, with the angle's increasing, and the detonation wave can form a trumpet-shaped area where the energy density, pressure and gas density are high. The detonation pressure of this area can be three times than the stability detonation pressure [10]. Therefore, the destructive effect of the hole is stronger.

According to the theoretical calculation, the detonation wave reflection ratio β and polytropic exponent K follow the equation as below:

$$\beta = (5K + 1 + \sqrt{17K^2 + 2K + 1})/4K \tag{2}$$

For a high detonation velocity explosive, K is equal to 3 and the pressure ratio is 2.39 times while for a general explosive, K is equal to 2 and the pressure ratio is 2.44 times.

3. Numerical simulation

3.1. Calculation model

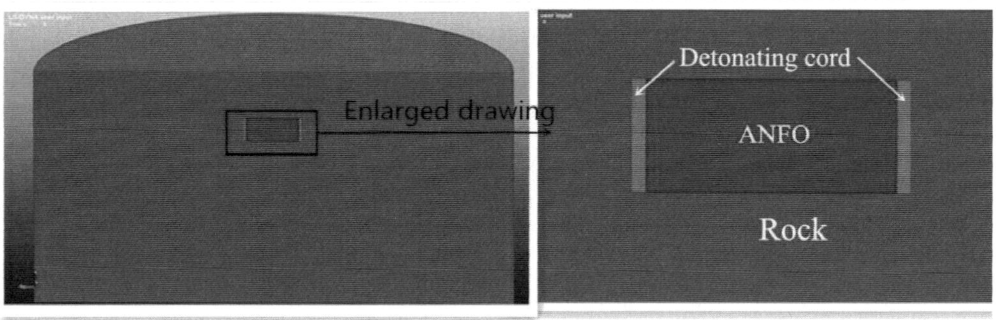

Figure 2. Model for numerical simulation

LS-DYNA is used to simulation the detonation wave propagation and show pressure distribution images under different initiating form. Three initiating forms are chosen such as combine detonating cord, separate detonating cord, and exploder center initiation. Figure 2 is the section view of the calculation model for the present experiment (explosive-determination of power test) [11]. The diameter of lead columnΦ=200mm, height h_1=200mm, the diameter of drillingΦ=24.5mm, loading height h_2=21mm. The FEM formulation for explosive and detonating cord is ALE (Arbitrary Lagrange Eulerian) while lead column is Lagrangian.

3.2. The simulation results and analysis

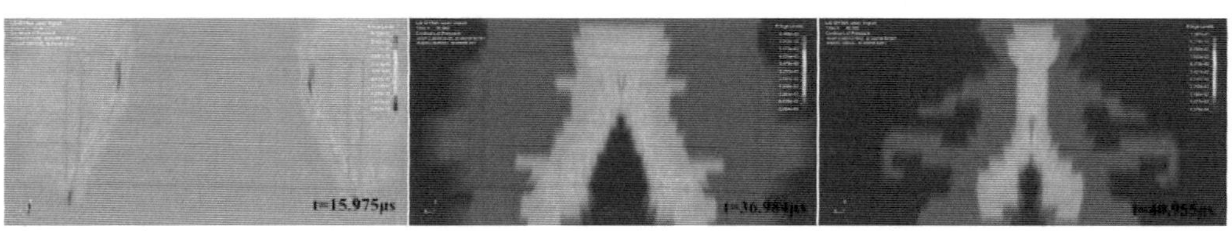

Figure 3. In the process of the separate detonating cord initiation system

Figure 3 shows the process of the separate detonating cord initiation system. At first, the detonation wave propagation along the detonating cord. When t=15.975μs, the detonating cord dies out, and the maximum detonation pressure from the detonating cord transferred to the ANFO. Then, the detonation wave spread to the center line from top to bottom. When t=36.984μs the detonation wave collision for the first time. Whereafter, the detonation wave collision spread to the bottom of the hole. When t=44.987μs, 'Mach reflection' is happened and lead to a high power density, pressure and gas density trumpet-shaped area. Finally, the ANFO die out and assembly energy spread out gradually.

Figure 4 shows the process of the combine detonating cord initiation system. Detonation wave spread to around like the model 'inverted cone'. When the detonation wave spread to bore wall, it will continually spread along the bore wall. When t=37.915μs, the explosive dies out. Subsequently, the detonation wave reflection in the bore wall, and the collision will happen when the spread to the center line. However, the energy of the detonation wave has a great decay at the same time.

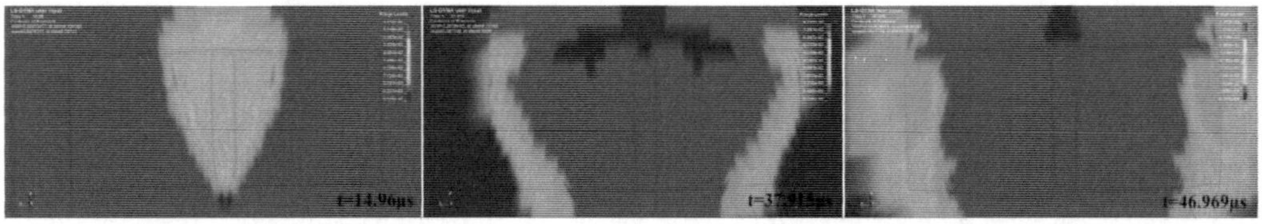

Figure 4. In the process of the combine detonating cord initiation system

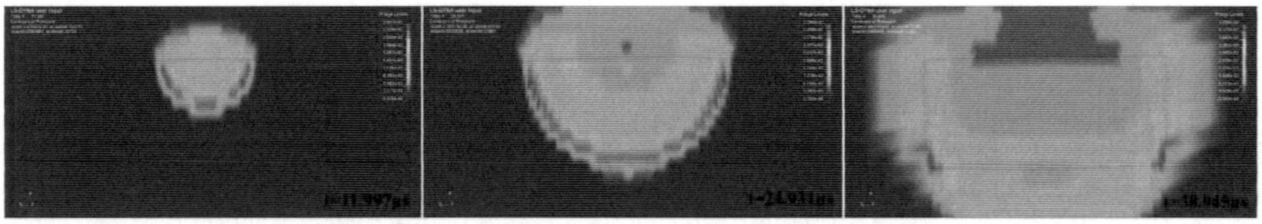

Figure 5. In the process of the exploder center initiation system

Figure 5 shows the process of the exploder center initiation system. The detonation wave spread to the unexploded area as the form of 'spherical'. When t=38.945μs, the explosive dies out. A high pressure region at the base of the explosive will be formed due to collision of the detonation wave. Then, the detonation wave will decay along its spreading direction.

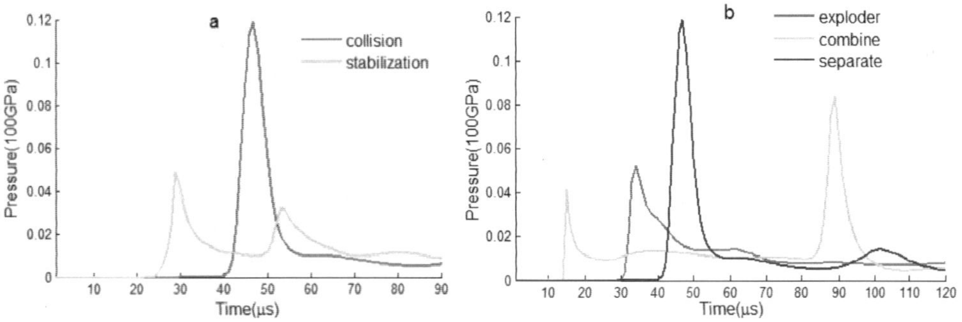

Figure 6. Detonation pressure curve contrast

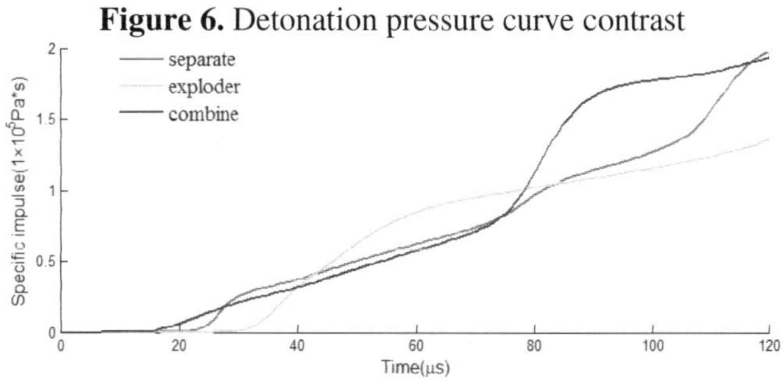

Figure 7. Contrast of impulse at the bottom of explosion centre

IOP Conf. Series: Earth and Environmental Science **64** (2017) 012081 doi:10.1088/1755-1315/64/1/012081

To observe the bore wall detonation wave propagation law under different initiation forms, select the stability point and the collision point, and draw the detonation pressure curve as shown in Figure 6 (a). As can be seen from the curve, the 'Mach reflection point' detonation pressure in the process of the separate detonating cord initiation is up to 13.60 Gpa as 2.42 times as the stability point of the ANFO. The result is consistent with the theoretical calculation. Compare the peak detonation pressure under different initiation forms as shown in Figure 6 (a). The maximum of the separate detonating cord initiation is 19.85% more than combine, and the exploder center initiation only increases 22.7% than the stability pressure, far less than the detonating cord initiation. Therefore, the peak detonation pressure of the separate detonating cord initiation is greater than the combine, and the minimum is the exploder center initiation.

Choose the element at the bottom of the explosive center to draw the specific impulse curve as shown in Figure 7. The specific impulse of the separate detonating cord initiation is value to 1.97×10^5Pa•s increased by 49% than the exploder center initiation. However, the combine only increased by 42% than the exploder. That is to say, the detonating cord initiation system can change the specific impulse distribution, make it increasing dramatically.

4. Lead block test

In order to verify the numerical simulation results. In this paper, 8 groups of experiments had been designed according to the lead block method (The national standard of the People's Republic of China). The sketch of the field test is shown in Figure 8. In order to form contrast test, 2 groups of test were carried out in each initiation forming, taking the mean value making a comparison. The test results are shown in Figure 9.

Figure 8. The photograph of experiment

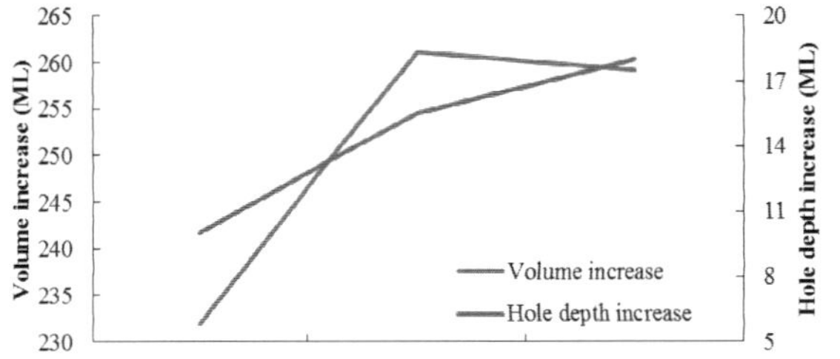

Figure 9. The test result of lead block method

We can learn from the Figure 9. Compared to the exploder center initiation system, the explosive-determination of power has the growth of 12.6% in the combine and 11.8% in the separate detonating initiation system, while the hole depth after blasting increase 55% and 80%. It shows that the detonation wave collision at the bottom of the separate detonating cord initiation system is the most intense. However, the explosive-determination of power in the separate detonating cord initiation system is slightly less than the combine. As a whole, compared to the numerical results, using the detonating cord initiation system can produce the detonation wave collision and form Munroe effect, but different initiation forming can cause different detonation wave propagation forming.

5. Conclusion

In this paper, at first, the detonation pressure increase theory is described in the basis of the Mach reflection. What's more, a numerical analysis on the detonation wave propagation and collision process under different initiation forms using commercial explicit FEM software LS-DYNA is implemented. Finally, a standard test is used to verify the numerical results. On the whole, we come to the conclusions through the present study as follows:

(1) From the comparison of peak pressure, it shows that the peak pressure of the detonating cord initiation system can reach to 2.42 times than the stability pressure.

(2) From the analysis of the element specific impulse at the bottom of the explosive, the separate and the combine initiation system respectively increases 49% and 42% than the tradition. It is preliminarily proved that the detonating cord initiation system can increase the specific impulse at the bottom of the explosive.

(3) From the results of explosive-determination power experiment, the detonating cord initiation system can do 12% more work than traditional initiation system. It can indicate that the detonation wave collision can improve the effect of blasting.

(4) Using the high velocity detonating cord initiation the low velocity explosive ANFO can lead to Munroe effect through the detonation wave collision. Comparing with the relevant literature that using shaped device implementation Munroe effect, this method has a simple process, low cost, low labor intensity and higher work efficiency.

References

[1] C. Yu, T. J. Tong, C. L. Yan. Applied research of shaped charge technology. Int. J. Impact Eng., 23, (1999)981-988.
[2] Masahide Katayama. Analysis of jet formation and penetration by conical shaped charge with the inhibitor. Int. J. Impact Eng., 23, (1999)443-454.
[3] X. J. Li, R. Y. Li, Y. Q, Ma, et al. Development of linear shaped charge loaded with an industrial explosive. Engineering Blasting. 4, (2004)5-8.
[4] Y. X. Wang, X. J. Li, H. H. Yan, et al. Simulation on assembling energy of detonation wave by using MPM. Chinese Journal of Computational Mechanics. 31, (2014)223-227.
[5] O. Ayisit. The influence of asymmetries in shaped charge performance. Int. J. Impact Eng., 35, (2008)1399-1404.
[6] E. L. Baker, A. S. Daniels and J. P. Turci. Selectable initiation shaped charges. Proce. 20th Int. Symp. on Ballistics, 2002.
[7] Q. Li, Y. Liang, K. K. Ren, et al. Experimental study of propagation of directional cracks with shaped charge under blasting load. Chinese Journal of Rock Mechanics and Engineering. 29, (2010)1684-1689.
[8] J.F. Molinari. Finite element simulation of shaped charge. Finite Elem. Anal. Des., 38, (2002) 921-936.
[9] J. F. Qin. Technology of cumulative pre-splitting blasting. Engineering Blasting, 13, (2007)19-24.
[10] X. G. Wei. Technique basis research on the assembling energy blasting of detonation waves. Dalian University of Technology. Dalian, 2012.
[11] GB12436-90, Explosive-determination of power-lead block method.1990.
[12] Stanley P. Marsh. LASL shock hugoniot data. California: University of California Press. 1980.
[13] S. F. Chen, S. W. Xue, W. W. Gao, et al. Experimental study and numerical simulation of rock shaped charge blasting. Blasting, 29, (2012)14-18.

Experimental research on the indoor temperature and humidity fields in radiant ceiling air-conditioning system under natural ventilation

Tao Huang*, Yutong Xiang and Yonghong Wang

China Academy of Building Research, Beijing, China

*Corresponding author e-mail: huangtao_ht@126.com

Abstract. In this paper, the indoor temperature and humidity fields of the air in a metal ceiling radiant panel air conditioning system with fresh air under natural ventilation were researched. The temperature and humidity distributions at different height and different position were compared. Through the computation analysis of partial pressure of water vapor, the self-recovery characteristics of humidity after the natural ventilation was discussed.

1. Introduction

Radiation ceiling air-conditioning system develops rapidly in the residential and office buildings in recent years, gaining more and more attention for its comfortable indoor environment [1, 2]. All the radiation ceiling air-conditioning systems put forward a common requirement that is the doors and windows should be closed to prevent condensation when it was running. So a bottleneck caused by this requirement appears, hindering the further application of the radiation ceiling air conditioning system [3, 4]. So, the influence on the temperature and humidity fields by the opening of windows occasionally becomes the concern of users and researchers. In this paper, the indoor temperature and humidity fields for metal ceiling radiant panel with fresh air conditioning system under natural ventilation was researched [5]. The change law of temperature and humidity fields in the experimental room under natural ventilation was discussed.

2. The experimental system

In the experimental system the construction area of the room is 10.8m2, the surface area of the metal radiation ceiling panel is 8.1m2, the new air outlet dimension is 60 x 12cm, the new outlet is located in the west side ceiling of the experimental room. The south and west sides of the room are the glass curtain walls where the windows able to be opened. The architecture drawing is shown in fig.1. The temperature and humidity recorders were arranged at different position and different height in the room, and the temperature and humidity of the air inside the room were recorded in real time. The positions of the measuring points are shown in fig.2.The position A is near the south outer window and the position B is near the west window. The temperature and humidity recorder were arranged at different height in two locations, with 0.2m, 0.45m, 0.75m and 1.1m distance from the top.

3. Experimental research on the indoor air temperature and humidity field in radiant ceiling air conditioning system under natural ventilation

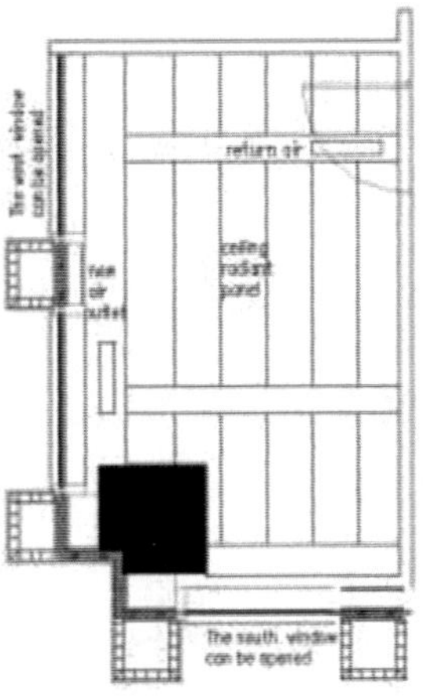

Figure 1. The architecture drawing

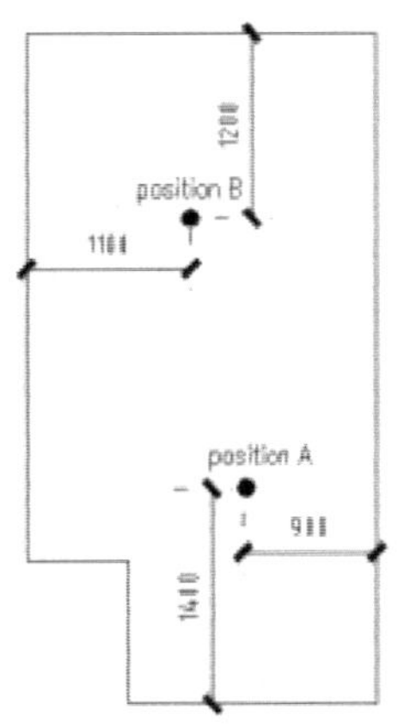

Figure 2. The position of the measuring points

The basic operating condition of the experimental system is described as follows: the surface temperature of the radiant panel is 20℃, the temperature of new air at the outlet is 19℃, the relative humidity of new air at the outlet was 70%, and the wind speed of the new air at the outlet is 0.32m/s. In the experiment, at a certain time the south window or the West window is opened to measure the wind speed of natural ventilation and closed after a period of time. And the system recovered before the windows were opened again.

The variation of the indoor temperature and humidity influenced by the opening of the south window was analyzed as a sample. In this sample, the average wind speed of the window was 0.8m/s.The south window was opened at 15:40pm and closed at 16:35pm.

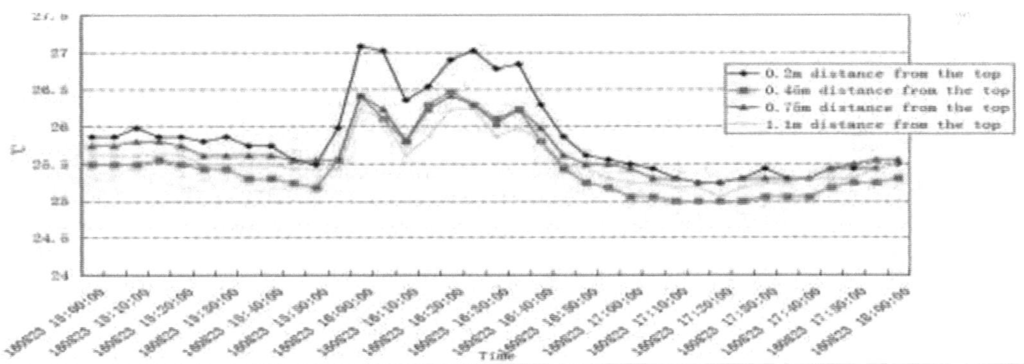

Figure 3. The distribution of temperature at different height at the position A

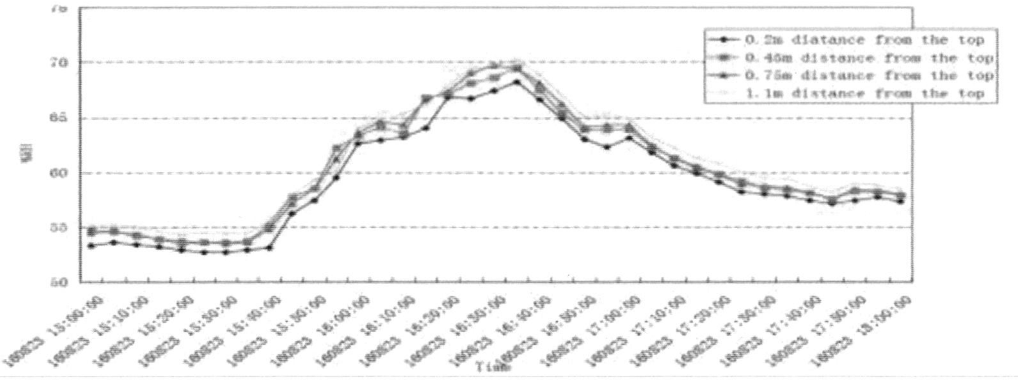

Figure 4. The distribution of humidity at different height at the position A

The changes of temperature and humidity at position A at different height are shown in Fig.3 and Fig.4, it can be seen from the figures that the changing tendencies of temperature and humidity at different height are similar, which when the south window is opened after 15:40, the indoor temperature and humidity increase obviously. The comparison between Fig. 3 and Fig. 4 shows that the increasing of humidity is a gradual process that reaches the maximum value when the window is closed, while the temperature increases rapidly and there is some fluctuation at the high point. The maximum temperature difference at different height at position A is 0.92 ℃ and the maximum difference of humidity is 2.9%.The measuring position A is influenced greatly by the opening of the window, with the higher temperature and lower humidity at the location closer to the ceiling.

The changes of temperature and humidity at position B at different height are shown in Fig.5 and Fig.6, it can be seen that the changing tendencies of temperature and humidity at position B are similar to them at the position A after the opening of window. The maximum temperature difference at different height at position B is 0.8 ℃ and the maximum difference of humidity is 2.6%.The distributions of temperature and humidity at position B is different from those at position A, with the lower temperature and higher humidity at the locations closer to the ceiling. This is because the measuring position B is far from the south window, which is mainly influenced by the top of radiation plate and fresh air. The temperature at the top of the radiation plate and the temperature of new air at the outlet are both lower than the indoor air temperature, leading to the air temperatures in top section of the room are lower than those in the bottom section

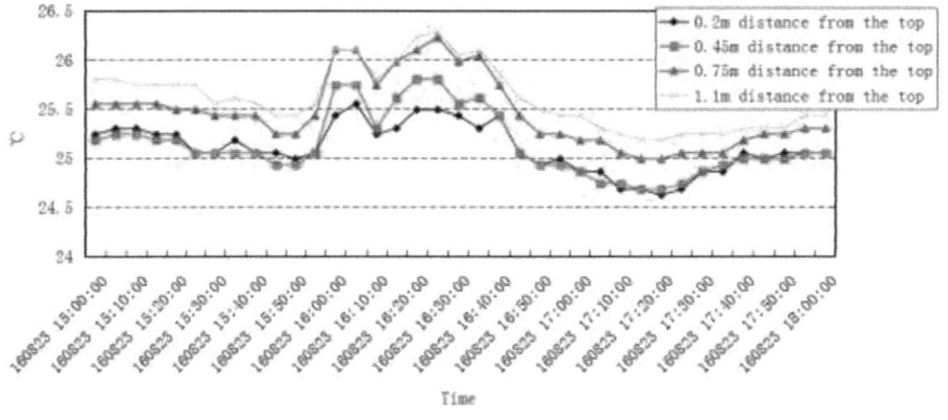

Figure 5. The distribution of temperature at different height at the position B

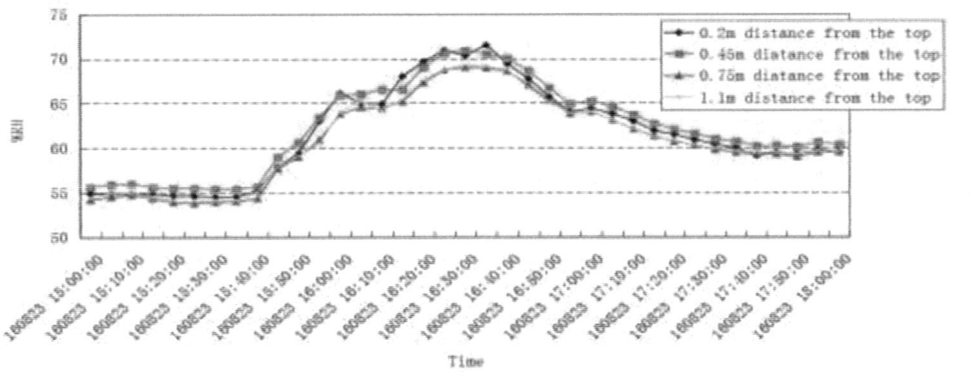

Figure 6. The distribution of humidity at different height at the position B

The data between position A and position B at different height was compared, and it was found that the difference of humidity is small, and the temperature difference is slightly big, with the higher air temperature nearby the opened window. It is indicated that the diffusion of humidity is better than that of the temperature, and the humidity field of the room is more uniform.

4. Study on the self-recovery characteristics of humidity after the natural ventilation

The indoor temperature and humidity field are mainly determined by the outdoor air, indoor fresh air and radiant cooling plate under natural ventilation. Indoor temperature and humidity will change when the window is opened, due to outdoor high temperature and high humidity air flows into the room. After the window is closed, the indoor humidity will be recovered slowly by the fresh air dehumidification system. The diffusion of water molecules in the air is determined by the partial pressure of water vapor, and the partial pressure of water vapor represents the absolute moisture content in the air. So, the self-recovery ability of indoor humidity was studied through analyzing the change rule of average partial pressure of water vapor in this paper.

The partial pressure of water vapor calculation formula is defined as

$$\lg \mathrm{P}=8.10765-{1750.286}/{(t+235.0)} \tag{1}$$

$$\mathrm{Ps}=133.3\times \mathrm{P}\times \phi \tag{2}$$

Where P is water vapor saturation pressure, Ps is partial pressure of water vapor, $\varphi$is relative humidity and t is air temperature.

The changes of partial pressure of water vapor in the experimental room before and after the natural ventilation are shown in Fig.7, It can be seen that change of outdoor partial pressure of water vapor was small, keeping at 2400~2600Pa during the experiment. The indoor partial pressure of water vapor was 1780Pa before the nature ventilation, and half an hour later the indoor partial pressure of water vapor increased to 2130Pa after the opening of the window, for the outdoor pressure is higher than the indoor pressure, resulting in outdoor water vapor flow into the room. The indoor partial pressure of water vapor was 2347Pa and the outdoor partial pressure of water vapor was 2425Pa at the same time after nature ventilation, the partial pressure of water vapor in the indoor air were very close to those in the outdoor air. The ventilation lasted 55 minutes, when the window was closed the indoor water partial pressure of water vapor decreased gradually with self- recovery for the same time. The last indoor partial pressure of water vapor was 1898Pa, which was slightly higher than the pressure before the window opened. It was indicated that under the experiment condition the indoor partial pressure of water vapor cannot completely return to the state before the natural ventilation, and it must take longer time to remove the increasing of indoor humidity caused by the natural ventilation.

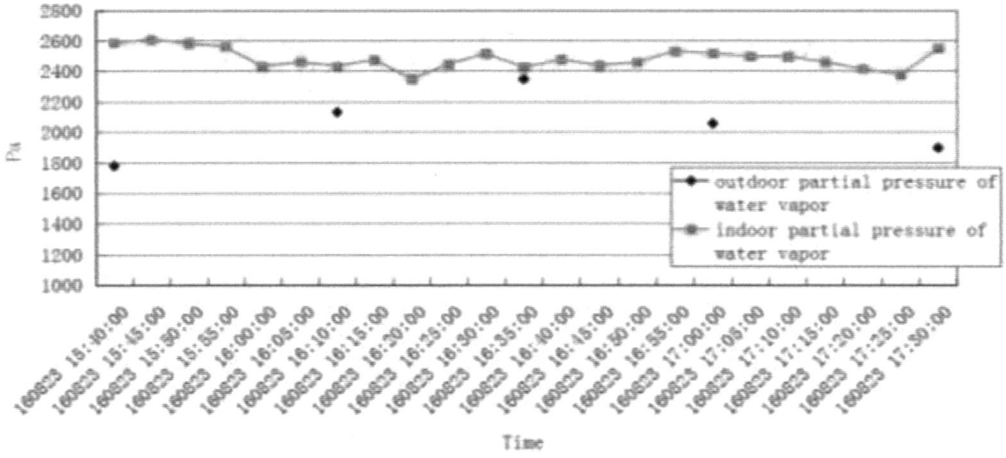

Figure 7. The change of partial pressure of water vapor

5. Conclusion

Indoor temperature and humidity field of radiation ceiling air conditioning system changed obviously under natural ventilation. The temperature near the opened window changed greater than the other. The diffusion of indoor humidity is better than the temperature and humidity changes at different positions in the room evenly. The higher the temperature is, the higher the temperature is, the smaller the humidity is; the other is the contrary. Under natural ventilation, the partial pressure of water vapor in the room increases gradually, until the partial pressure is close to the outside. When the natural ventilation is stopped, the indoor water vapor pressure will gradually restore.

Acknowledgments

This work was financially supported by National Natural Science Foundation (No.51408576)

References

[1] Ma Yuqi,Liu Xuelai,Li yongan,Overview on the Research and Development of Radiant panels,REFRIGERATION,27(2008)29-33.

[2] Huang Tao, Wang Yonghong, LI Na,Experimental Research on Performance of Radiant Ceiling Air-conditioning System in the Office Building in Summer,Building Energy Efficiency, 44(2016)5-7.

[3] Yi Lingli,Sun Tingting,The Design and Control Strategy of Cooling Ceiling Air Conditioning plus an Indenpendent Air System,REFRIGERATION,30(2011) 30-33.

[4] He Jing,Shen Jinmin,Shou qingyun, Chen Xiaoling,Discussion on condensation of radiant cooling system combined with DOAS,HV&AC,38(2008)159-162

[5] LI Na,Yuan Dongli, Establishment of Radiant Air-conditioning System's Laboratory and Sample Room in Office Building,Construction & Design for Project,6(2014)55-57.

The effects of cathodic micro-voltage combined with hydrothermal pretreatment on methane fermentation of lignocellulose substrate

Dianxin Liu, Ping Ning*, Guangfei Qu, Xi Huang, Yuhuan Liu, Jian Zhang

Faculty of Environmental Science and Engineering, Kunming University of Science & Technology, Kunming, 650500, China

*Corresponding author e-mail: ping_ning@163.com

Abstract. The methane fermentation study assisted with cathodic micro-voltage was carried out to investigate the electric field effects on the fermentation of hydrothermally pretreated lignocellulose substrate. It was illustrated that a 0.25V cathode voltage and hydrothermal pretreatment could improve the biogas production, biogas quality and lignocellulose degradation ratio significantly. The cumulative biogas productions in the fermentation of hydrothermally pretreated cow dungs at 50℃, 150℃ and 200℃ with a 0.25V cathode voltage were observed in a total of 6640mL, 9218mL and 9456mL respectively over a detention time of 33 days. In comparison with the fermentation pretreated at 200℃ without any voltage, nearly doubled of cumulative biogas production was obtained in the process of cathode-assisted fermentation. It was also observed that the daily methane content greater than or equal to 70% in the biogas generated with cathode voltage were clearly greater than that without voltages. Furthermore, the fermentation applied with a 0.25V cathode voltage had resulted into significant increases of 12.64% and 9.44% in lignin and cellulose degradation ratio relative to voltage free fermentation. And in the process of fermentation applied with cathode voltage, the final lignocellulose degradation ratio increased with the hydrothermal pretreatment temperature. Thus, the hydrothermal pretreatment and assisting fermentation with low cathode voltage can effectively promote the lignocellulose degradation. All results revealed that cathodic micro-voltage combined with hydrothermal pretreatment can remarkably improve the fermentation of lignocellulosic materials, indicating that a more effective fermentation technology can be developed by applying with cathodic micro-voltage.

1. Introduction

Energy shortage is one of the most serious problems facing mankind today. Making the transition from non-renewable carbon resources to renewable bio resources is absolutely needed. Biomass energy, as a form of solar energy, is the only renewable carbon resource, which attracting more and more attention today.

The fully integrated agro-biomass ways for sustainable technologies are biofuels, heat, electricity and biomaterials [1]. However, methane fermentation is a more efficient way of energy production and promises more extensive application range in comparison to the above technologies. Production of biogas from methane fermentation provides a versatile carrier of renewable energy, as methane can be used for replacement of fossil fuels in both heat and power generation and as a vehicle fuel [2].

Although biomass contains varying amounts of cellulose, hemicellulose, lignin and a small amount of other extractives [3], lignocellulose is the principal part of biomass [4]. The main problems encountered with the biological conversion of lignocellulose arise from its inaccessible structure as cellulose fibre that is partially arranged in a crystalline structure, integrated with hemicellulose and embedded in a matrix of lignin [5]. Therefore, it is particularly significant to improve the lignocellulose degradation rate in the utilization of biomass. The pretreatment methods of lignocellulose [6] include physical (e.g. steam treatment) [7], chemical (e.g. by acid or alkaline hydrolysis) [8], microbial [9], or a combined treatment of these methods [10].

Hydrothermal process is environment-friendly physic-chemical technology that can be conceived as the first step for the chemical utilization of lignocellulose. The effects caused by hydrothermal treatments include hemicellulose deploy-merisation (to yield sugar oligomers, hemicellulosic sugars and degradation products), cellulose alteration (mainly in relation with its physicochemical features) and chemical alteration of lignin (that facilitates further delignification steps with organ solvents and/or alkaline solutions) [11]. In the comparison of liquid hot water and steam pretreatments of sugar cane bagasse for bioconversion to ethanol, liquid hot water pretreatment resulted in much better xylan recovery and ≥80% conversion by simultaneous scarification and fermentation [12]. R. Chandra et al. reported that the hydrothermally pretreated substrate had resulted into an increase of 225.6% in biogas production and 222.0% in methane production relative to untreated rice straw substrate [13]. Mark D. Redwood et al. reported that the extractive fermentation is applied to waste-derived substrates following hydrothermal pretreatment, achieving 83%-99% bio waste destruction [14]. Prasad Kaparaju et al. observed that energy output from the biogas production of hydrothermally pretreated wheat straw was the most energetically efficient and approximately 10% higher than that of untreated wheat straw [15].

After hydrothermal pretreatment, the process of methane fermentation becomes more favorable for the activity of functional microbe. Electricity has long been used to stimulate microbial metabolism [16]. In addition to direct and indirect electron transfer to microbe, other reactions may take place at the electrode surface that can directly impact microorganisms [17]. Thus, the fermentation can be improved by applying with electricity. For example, B. Tartakovsky et al. has enhanced methane production from wastewater in laboratory-scale anaerobic reactors equipped with electrodes for water electrolysis [18].

In the bipolar voltage system, oxygen generated in the anode restrained the methanogen bacteria. Thus, a single cathode may effectively create favorable conditions for methanogens. However, little research had been concentrated on applying with unipolar voltage and combining it with hydrothermal pretreatment. The objective of this work was to investigate the micro-electric field effects on the fermentation of hydrothermally pretreated lignocellulose substrate. Cow dung was considered as the model substrate in the fermentation for its high lignocellulose content. The hydrothermal pretreatment and micro-electric field regulation of fermentation were used to improve biogas production, daily methane content and lignocellulose degradation in the process of fermentation.

2. Experiment

2.1. Materials

The cow dungs and straws used as fermenting substrates were collected from Dengchuan, Yunnan Province, China. The samples were dried in oven and smashed to 60 mesh by grinder. Afterward, they were reserved at dry place for later use. The experiment was conducted under mesospheric digestion condition with the inocula expansively cultivated at 35℃ which was from mesospheric fermentation installation in Dengchuan. The analysis of cow dung composition revealed that the contents of ash and volatile solids were 12.7% and 73.32% respectively. Cow dung contained 62.46% organic matter with 14780J · g-1 of caloric value.

2.2. Experimental apparatus

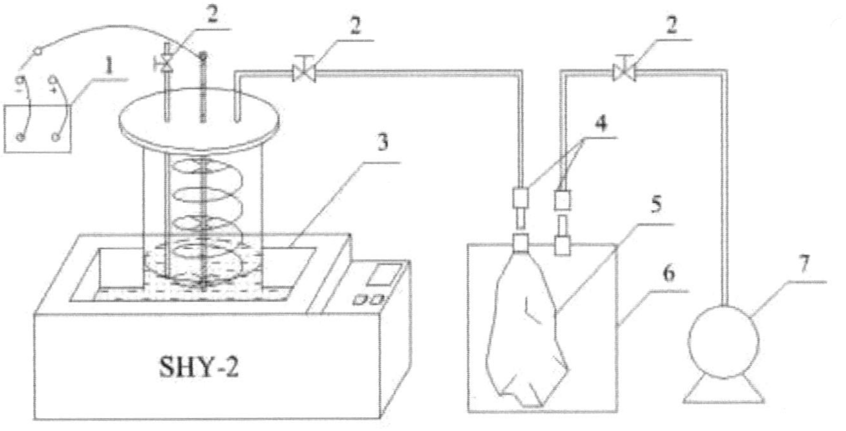

Figure 1. Schematic of experimental setup for methane fermentation.
1. Power supply; 2. Valve; 3. Thermostatic water bath cauldron; 4. Quick joints; 5. Gas collector; 6. Vacuum chamber; 7. Aspirator pump.

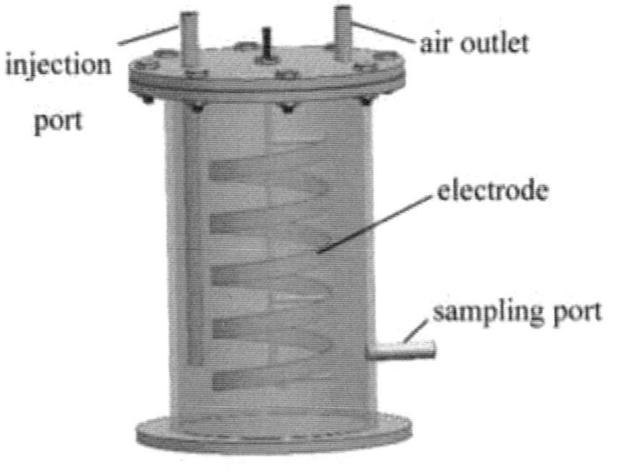

Figure 2. The anaerobic fermentation reactor

Fig.2 showed the schematic of experimental setup for methane fermentation. It was designed as a thermostatic anaerobic fermentation installation, which was composed of a power supply, a fermentation reactor (Fig.2), a thermostatic equipment and a gas collector.

2.3. Experimental setup and analytical methods
The mixture of 30g feedstock which were hydrothermally pretreated in autoclave and 100mL fermentation liquor prepared with 8% total solid mass content was added into fermenter. The process of anaerobic fermentation was maintained at 35℃. The volume of biogas produced on everyday was measured according to the displacement method [19]. The main components of biogas were carbon dioxide and methane, which were analysed by gas chromatograph. VFA was determined by titrimetric and pH was measured by pH meter. Decrement method [20] was used to measure volatile solids mass removal efficiency. The analysis of cellulose and lignin were carried out by spectrophotometry and Klason method [21].

3. Results and discussion

3.1. Cumulative biogas production

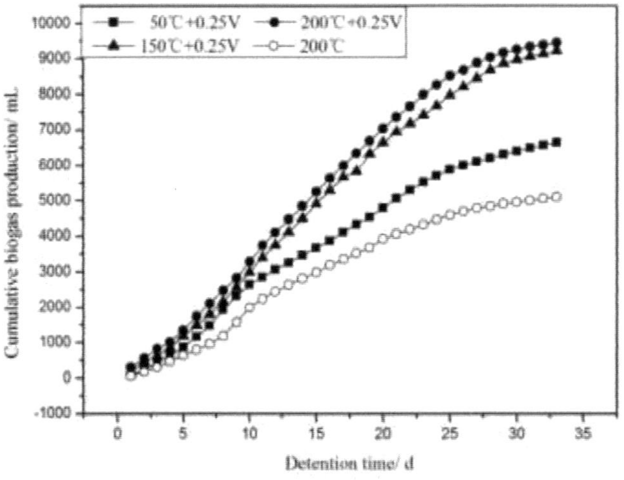

Figure 3. Cumulative biogas production as a function of detention time

As can be seen from Fig.3, in the micro-electric field of a 0.25V cathode voltage, biogas production rates of hydrothermally pretreated cow dungs at 50℃, 150℃ and 200℃ began to decrease after 28 days. The hydrothermally pretreated cow dungs for three temperature levels had resulted in a total biogas production of 6640mL, 9218mL and 9456mL respectively over a detention time of 33 days. With the hydrothermal temperature increased, the cumulative biogas production was raised. These results indicated that hydrothermal pretreatment can promote the cumulative biogas production and the pretreatment temperature is an important factor which can affect the methane fermentation significantly. Similar situations had been observed in other literatures. As R. Chandra described, hydrothermal pretreated substrate had resulted into an increase of 9.2% in biogas production compared to that of untreated wheat straw substrate [22].

The more astonished was that the cumulative biogas production of the hydrothermally pretreated cow dungs at 200℃ was only 5105mL in the absence of micro-electric field. And yet, cow dungs with a cathode voltage of 0.25V had resulted into a total of 9456mL biogas, nearly twice as much as the one without micro-voltage. Thus, a single cathode voltage brought a remarkable improvement for biogas production of fermentation. These promotions may be caused by two reasons. On one hand, the degradation of lignocellulose was actually achieved by the enzymes generated from anaerobe in the process of fermentation. The enzymes may be stimulated by the micro-electric field, especially the hydrolases. Due to the effect of cathode voltage, the change of enzyme activity promoted the combination of reactive sites and substrates. On the other hand, the redox potential of solution which was sensitively affected by several parameters such as pH, temperature and so on, was an essential factor for anaerobe metabolism. The single cathode voltage possibly altered the redox potential to the optimal state of anaerobe, which created favorable conditions for the fermentation.

3.2. Methane content

The methane contents were grouped into three levels: the first level was $C_{CH4}<60\%$; the second was $60\%\leq C_{CH4}<70\%$; and the third was $C_{CH4}\geq 70\%$ respectively. Fig.4 showed the effect of cathode micro-voltage and hydrothermal pretreatment on daily methane content. When cow dungs were hydrothermally pretreated at 50℃ and applied with 0.25V cathode voltage, the daily methane contents were all more than 60% and the maximum days in the third level was found. As the pretreated temperature was raised, a decrease was observed in the number of days that the daily methane contents were greater than or equal to 60%. So it can be concluded that the hydrothermal pretreatment is against the methane content.

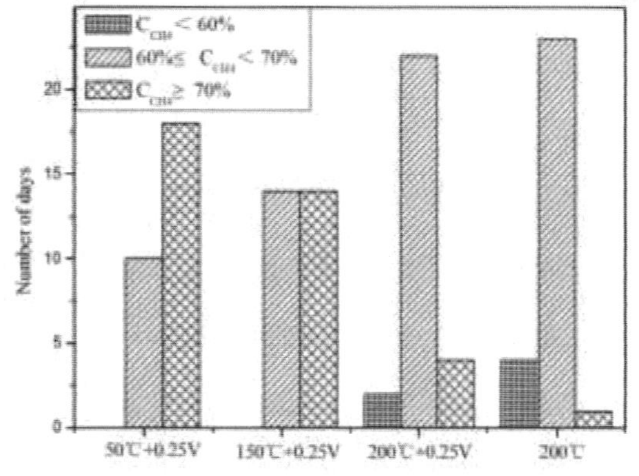

Figure 4. The effect of low cathode voltage and hydrothermal pretreatment on daily methane content

There was a comparison of methane content with or without micro-electric field. When applying a cathode voltage of 0.25V, an increase was observed in the number of days at the third level and the days that the daily methane content was less than 70% were decreased. These results manifested that the methane content can be improved by the applied voltage. There may be two possible reasons. Firstly, the metabolism and reproduction of hydrogenogens were improved by the stimulation of the micro-voltage, as well as the activity of methanogens. Secondly, H_2O, as the hydrogen donor, was more utilized by anaerobe to react with CO_2 in the micro-electric field, leading more generation of CH_4.

3.3. Hydrolysis behaviour of cow dungs

During hydrolysis of hydrothermally pretreated cow dungs over 33 days of detention period, the variations of pH were illustrated in Fig.5.

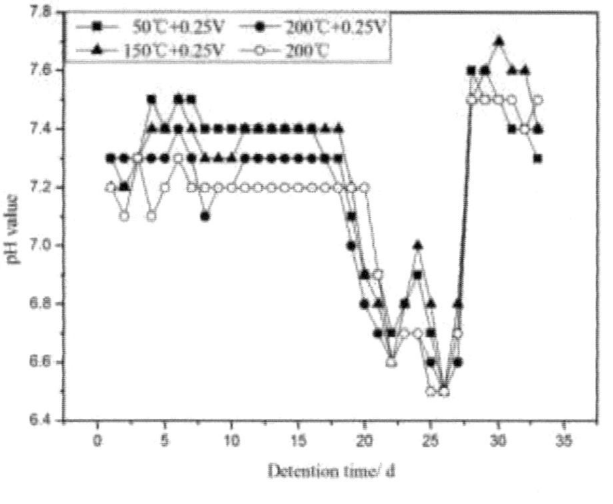

Figure 5. The changes in pH during the fermentation

IOP Conf. Series: Earth and Environmental Science **64** (2017) 012083 doi:10.1088/1755-1315/64/1/012083

As can be seen, the changing trends of pH through the whole hydrolysis process were similar in all experimental conditions. In 15 days, the pH kept in the range from 7.1 to 7.4. This may be due to the equilibrium between the rate of acid production and the rate of acid conversion into methane. Because of the hydrothermal treatment, part of hydrogen bonds in the lignocellulose was broken, leading to the formation of small molecules which are easily digested. Accordingly, inocula well adapted to the fermentation environment and metabolized normally at the initial stage of fermentation. The pH decrease was observed for the first time happening between 17th days to 22th day as the pH dropped to about 6.6. The possible reason was that small molecules had been almost digested and the conversion of carbohydrates and proteins into fatty acids and amino acids had taken place. Afterward, the pH returned to neutral on the 24th day. It was caused by the feedback mechanism of system. Otherwise, the processus of methanogens were inhibited when pH was lower than 6.5. Further, the pH showed a drop again. After 26th day onward, the pH started to increase, which reached to 7.5 and then did not very much. This phenomenon was due to the decomposition of fatty acids and amino acids.

3.4. Volatile solids mass removal efficiency

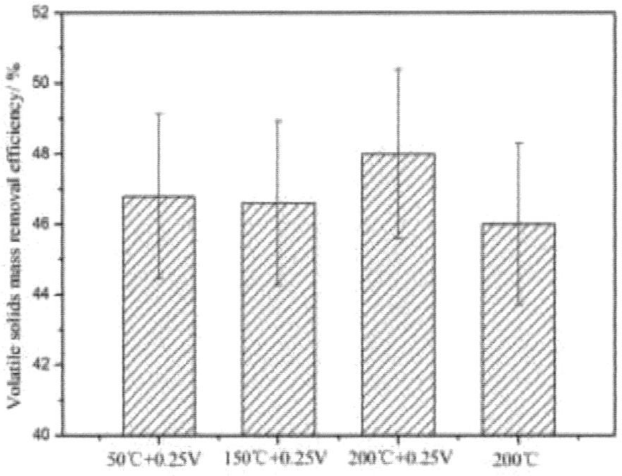

Figure 6. The volatile solids mass removal efficiency of the anaerobic digestion process

Fig.6 showed the effect of hydrothermal pretreatment and the cathode micro-voltage on volatile solids mass removal efficiency. The volatile solids are the effective representation of the content of organic substance in substrate. That is, the more volatile solids were removed, the more volatile solids were converted into methane and carbon dioxide. The hydrothermally pretreated cow dungs at 200℃ with a cathode voltage of 0.25V had resulted into the maximum volatile solids mass removal efficiency as high as 48.0% after 33 days of detention time. Thus the cumulative biogas production should also be the maximum, which was consistent with the result of Fig.3. The volatile solids mass removal efficiencies of hydrothermally pretreated cow dungs at 50℃ and150℃ with a cathode voltage of 0.25V were 46.8% and 46.6% respectively which were much the same. It can be concluded that the hydrothermal temperature had a slightly positive effect on volatile solids mass removal efficiency. While the volatile solids mass removal efficiency of hydrothermally pretreated cow dungs at 200℃ without voltage was only 46% within 33 days, which was 2% less than that with a cathode voltage of 0.25V. Thus the cathode micro-voltage can also improve the volatile solids mass removal efficiency.

3.5. The variation of VFA

Fig.7 illustrated the variation of VFA in the fermentation. The first drop in VFA occurred within 2 days, which indicated that there was no accumulation of fatty acids. The reason may be that the methanogen bacteria were active due to their good adaptability. As a result, the pH kept alkalescent. After that, VFA began to be cumulated from the 14th day, which inhibited the methanogens. Consequently, the pH will decline. As can be seen from Fig.4, the pH was dropped to about 6.6 between 17th day to 22th day of detention time, which was in agreement with the speculation. By the 22th day, a second decrease was observed due to the predominance of methanogens, which was consistent with the variation of pH.

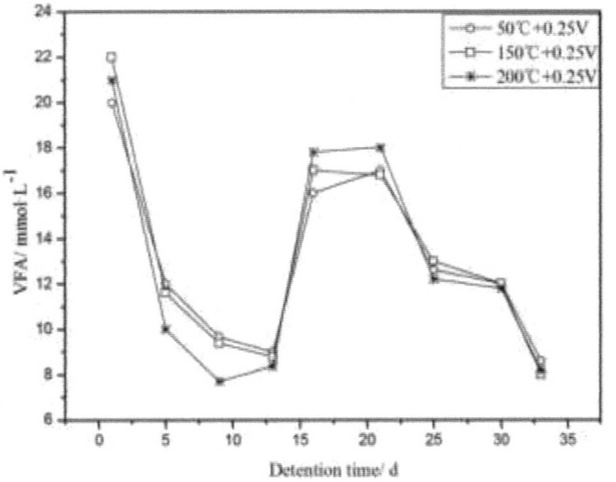

Figure 7. Concentrations of VFA in the fermentation of cow dungs

3.6. The degradation ratio of lignin and cellulose

Fig.8 presented the lignin and cellulose degradation ratio for the substrates after fermentation. The electric-assisted fermentation of hydrothermally pretreated substrates at 50℃ had resulted in 49.18% of lignin degradation and 46.95% of cellulose degradation. The lignin and cellulose degradation ratio of cow dungs hydrothermally pretreated at 100℃ were observed as 52.50% and 54.99% in the fermentation assisted with micro-electric field. When hydrothermally pretreated at 200℃ and applied with 0.25V cathode voltage in the process of fermentation, 57.32% of lignin degradation and 56.60% of cellulose degradation were found. However, for the fermentation of hydrothermally pretreated substrates at 200℃without cathodic voltage, 44.68% lignin and 47.16% cellulose degradation ratio were obtained.

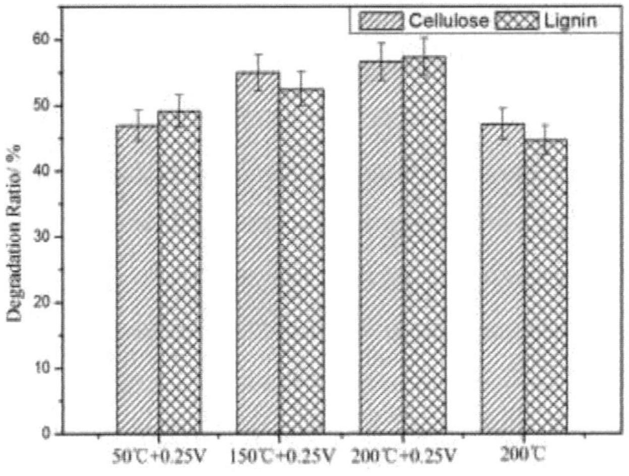

Figure 8. Lignin and cellulose degradation ratio of cow dungs after fermentation

The highest lignin and cellulose degradation ratio was observed for hydrothermally pretreated cow dungs at 200℃ with 0.25V cathode voltage applied in the fermentation. Furthermore, the biogas production and methane content of cow dungs in the same conditions was also the highest. It was possibly caused by the following reasons. Water under high pressure can penetrate into the biomass, hydrate cellulose, and remove hemicellulose and part of lignin in the hydrothermal pretreatment. It enlarges the accessible and susceptible surface area of the cellulose and makes it more accessible to hydrolytic enzymes [23]. In addition, microbes stimulated by cathode voltage in the fermentation become more active, leading to an acceleration of fermentation. Thus, it can be speculated: The higher content of lignocellulose substrate was used for anaerobic digestion, the more bio-methane yield can be obtained, and a cathode micro-voltage and hydrothermal pretreatment can effectively improve the lignocellulose degradation ratio.

4. Conclusion

The experimental results revealed that the hydrothermal pretreatment and assisting fermentation with cathode micro-voltage showed a significant effect on methane fermentation of cow dungs, especially the cathode micro-voltage. The cumulative biogas productions in the fermentation of cow dungs assisted with a 0.25V cathode voltage were increased with the hydrothermally pretreated temperature. In the comparison with the fermentation pretreated at 200℃ without voltage, nearly doubled of cumulative biogas production was obtained in the process of cathode micro-voltage assisted fermentation. The days of daily methane content which is greater than or equal to 0.7 from the fermentation of cow dungs applied with a 0.25V cathode voltage were found more than that of cow dungs applied without voltages. With the hydrothermal temperature raised, the lignocellulose degradation ratios improved. Compared to the fermentation of hydrothermally pretreated cow dungs at 200℃ without voltages, 12.64% and 9.44% improvement in the lignin and cellulose degradation ratio were obtained in the fermentation assisted with micro-electric field.

Acknowledgements

This work was supported by the National major project for water pollution control and scientific management (2014ZX07105-001); and Nan-Qi Ren Yunnan province academician (experts) workstation.

References

[1] Arthur J. Ragauskas, Charlotte K. Williams, et al. The Path Forward for Biofuels and Biomaterials. Science 2006; 311: 484–489.

[2] Peter Weiland. Biogas production: current state and perspectives. Applied Microbiology and Biotechnoloty 2010; 85: 849-860.

[3] Peter McKendry. Energy production from biomass (part 1): overview of biomass. Bioresource Technology 2002; 83: 37–46.

[4] Parveen Kumar, Diane M. Barrett, et al. Methods for Pretreatment of Lignocellulosic Biomass for Efficient Hydrolysis and Biofuel Production. Industrial & engineering chemistry research 2009; 48, 3713–3729.

[5] P. A. M. Claassen, J. B. van Lier, et al. Utilisation of biomass for the supply of energy carriers.Applied Microbiology and Biotechnoloty 1999; 52: 741-755.

[6] Nathan Mosier, Charles Wyman, et al. Features of promising technologies for pretreatment of lignocellulosic biomass.Bioresource Technology 2005; 96: 673–686.

[7] Emma Kreuger, BálintSipos, et al. Bioconversion of industrial hemp to ethanol and methane: The benefits of steam pretreatment and co-production. Bioresource Technology 2011; 102: 3457–3465.

[8] Ender Us, N. Altınay Perendeci. Improvement of methane production from greenhouse residues: Optimization of thermal and H2SO4 pretreatment process by experimental design. Chemical Engineering Journal 2012; 181-182: 120-131.

[9] Shuxia Zhou, Yulin Zhang, et al. Pretreatment for biogas production by anaerobic fermentation of mixed corn stover and cow dung. Energy 2012; 46: 644-648.

[10] C. Sambusiti, F. Monlau, et al. A comparison of different pre-treatments to increase methane production from two agricultural substrates. Applied Energy 2013; 104: 62-70.

[11] G. Garrote, H. Domínguez, J. C. Parajó. Hydrothermal processing of lingo- cellulosic materials.European Journal of Wood and Wood Products 1999; 57: 191-202.

[12] Mark Laser, Deborah Schulman, et al. A comparision of liquid hot water and steam pretreatments of sugar cane bagasse for bioconversion to ethanol.Bioresource Technology 2002; 81: 33–44.

[13] R. Chandra, H. Takeuchi, et al. Hydrothermal pretreatment of rice straw biomass: A potential and promising method for enhanced methane production. Applied Energy 2012; 94: 129-140.

[14] Mark D. Redwood, Rafael L. Orozco, et al. An integrated biohydrogen refinery: Synergy of photofermentation, extractive fermentation and hydrothermal hydrolysis of food wastes. Bioresource Technology 2012; 119: 384–392.

[15] Prasad Kaparaju, María Serrano, et al. Bioethanol, biohydrogen and biogas production from wheat straw in a biorefinery concept.Bioresource Technology 2009; 100: 2562–2568.

[16] Berg H. Electrotransfection and electrofusion of cells and stimulation of their metabolism. Stud. Biophys. 1987; 119(1—3): 17—29.

[17] J. Cameron Thrash, John D. Coates. Review: Direct and Indirect Electrical Stimulation of Microbial Metabolism. Environmental Science and Technology 2008; 42: 3921-3931.

[18] B. Tartakovsky, P. Mehta, et al. Electrolysis-enhanced anaerobic digestion of wastewater. Bioresource Technology 2011; 102: 5685–5691.

[19] Safia Moosvi, Datta Madamwar. An integrated process for the treatment of CETP wastewater using coagulation, anaerobic and aerobic process. Bioresource Technology 2007; 98: 3384–3392.

[20] R. Chandra, V.K. Vijay, et al. Production of methane from anaerobic digestion of jatropha and pongamia oil cakes. Applied Energy 2012; 93: 148-159.

[21] AsriGani, Ichiro Naruse. Effect of cellulose and lignin content on pyrolysis and combustion characteristics for several types of biomass. Renewable Energy 2007; 32: 649–661.

[22] R. Chandra, H. Takeuchi, et al. Improving biodegradability and biogas production of wheat straw substrates using sodium hydroxide and hydrotherm al pretreatments. Energy 2012; 43:

273-282.

[23] Taherzadeh MJ. Karimi K.. Pretreatment of lignocellulosic wastes to improve ethanol and biogas production: a review. International Journal of Molecular Sciences 2008; 9: 1621 - 1651.

IOP Conf. Series: Earth and Environmental Science **64** (2017) 012084 doi:10.1088/1755-1315/64/1/012084

Research on power source structure optimization for East China Power Grid

Lingjun Xu[1], Da Sang[2], Jianping Zhang[2], Chunyi Tang[1] and Da Xu[1,*]

[1]Hydro china Hua dong Engineering Corporation, Hangzhou, China
[2]East China Grid Company Limited, Shanghai, China

*Corresponding author e-mail: xulingjun@163.com

Abstract. The structure of east china power grid is not reasonable for the coal power takes a much higher proportion than hydropower, at present the coal power takes charge of most peak load regulation, and the pressure of peak load regulation cannot be ignored. The nuclear power, wind power, photovoltaic, other clean energy and hydropower, coal power and wind power from outside will be actively developed in future, which increases the pressure of peak load regulation. According to development of economic and social, Load status and load prediction, status quo and planning of power source and the characteristics of power source, the peak load regulation balance is carried out and put forward a reasonable plan of power source allocation. The ultimate aim is to optimize the power source structure and to provide reference for power source allocation in east china.

1. Introduction

The east china power grid is one of the largest provincial power grids in china which is composed of Shang hai, Jiang su, Zhe jiang, and a Hui and Fu jian province. The structure of east china power grid is not reasonable for the coal power [1]. Takes a much higher proportion than hydropower [2], at present the coal power takes charge of most peak load regulation. There are some shortcomings such as the economy is poor and the operation safety and stability of the unit is reduced in peak load regulation by coal power. In order to meet the increasing power demand of East China Power Grid, The coal power, nuclear power, wind power, photovoltaic, other clean energy and hydropower, coal power and wind power from outside will be actively developed in future. For nuclear power is suitable for base load operation, the output of wind power and other new energy [3]. Is randomness, the electric power from outside is aim to send electricity and do not assume the task of peak load regulation, those factors will lead to a phenomenon that the proportion of peak load regulation power source may be seen further declined.

The variation rule [4]. Of electric power in east china power grid is gained by analysis of social economy and energy resources, load and electricity consumption of the power grid. The power demand [5]. In 2025 is predicted combined with load prediction result of east china power grid. The status quo and planning of power source is arranged and optimization model of power source structure is established. Then the electric power market, peak load regulation capacity [6]. And the comparison result of peak load regulation power source are obtained. Based on the optimization for power source structure, the reasonable power source allocation in east china is obtained at last.

2. Load prediction

The load of east china power grid is predicted by analysis of electricity consumption, the relationship between variation rule of electric power structure and economic growth. The result of load prediction is shown as below.

Table 1. The load prediction of east china power grid

Item	2015	2020	2025
The maximum load(MW)	235000	305000	370000
Growth rate (%)		5.35	3.94
Annual power demand(billion kWh)	1400	1800	2150
Growth rate (%)		5.15	3.62

Through analysis of status quo of east china power grid in recent years, It can be found that the peak load in summer usually occurred in July, August and September, mostly in August, and the peak load in winter usually occurred in December; The evening peak load of each month is often higher than early peak load, usually the evening peak load occurred at pm19~20 in summer and pm18~19 in winter. It is predicted that daily load rate of typical days in summer and winter is between 0.798 and 0.834, the daily minimum load rate is between 0.595 and 0.634.

3. The analysis peak load regulation balance

The amplitude of power source in east china power grid are as follows:

(1) Conventional hydropower station: The amplitude of peak load regulation of large and medium-sized hydropower stations can reach to 100%, the small hydropower stations are incapable to regulate peak load.

(2) Pumped storage power station: The amplitude of peak load regulation can reach to 200% of its boot capacity.

(3) Coal power: The amplitude of peak load regulation of large coal units can reach to 40~50%. The small coal units are considered as base load which has no capacity of peak load regulation.

(4) Gas power: The amplitude of peak load regulation can reach to 100% of its boot capacity. Thermal units are considered as base load.

(5) Nuclear power: The nuclear power is considered as base load.

(6) Wind power: The anti-rate of peak load regulation is 45%.

(7) Hydropower from outside: The hydropower from outside bear base load in flood season and take some charge of peak load regulation in dry season.

(8) Coal power from outside: It is considered as base load.

The calculation of peak load regulation balance of east china power grid in 2025 is carried out on the base of above principle, the results are shown in the table below.

The calculated result indicates that the peak valley difference of east china power grid will significantly increase in future With the increase of nuclear power, wind power, other new energy sources and electric power from outside, the amplitude of coal power in future will be high and the pressure of peak load regulation cannot be ignored. It is hard to is difficult to satisfy with the request of Safe, stable and economical operation of power grid. It is an urgent need to build peak load regulation power source.

Table 2. The peak load regulation balance of east china power grid in 2025(Unit: MW)

	item	
1	Maximum load	370000
2	Minimum load rate	0.6340
3	Maximum peak valley difference	136159
4	hot-spare	27900
5	Anti-spare of wind	11025
6	Required capacity of peak load regulation	175084
7	Total boot capacity	399900
	Conventional hydropower	14722
	Pumped storage power station	13220
	Gas power	31169
	Oil power	857
	Nuclear power	39920
	Wind power	1225
	Hydropower from outside	26326
	Coal power from outside	49980
	Small Coal power	24861
	Coal power which can regulate peak load	197620
8	Required capacity of peak load regulation	175084
	Conventional hydropower	6828
	Pumped storage power station	26440
	Gas power	12085
	Coal power which can regulate peak load	129732
9	Amplitude of coal power(%)	65.6
10	Capacity of coal power when amplitude of coal power is 35%	69167
	The profit and loss of capacity of coal power(+ is profit - is loss)	-60565

4. The optimal allocation of power source

On the base of power source which participate in peak load regulation, adjusting the scale of coal power, gas power and pumped storage power stations, maintaining the scale of other power sources to carry out optimal allocation of power source in east china power grid. The result shows that total discounted cost of Pumped storage power station is the least. When the scale of Pumped storage power station is 40000MW, the amplitude of coal power decreased from 65.6% to 44.6%, which means that Pumped storage power station takes a good effect on peak load regulation.

Table 3. Reasonable scale of Pumped storage power station in east china power grid in 2025

scheme	Reasonable scale of Pumped storage power station (MW)	New scale of Pumped storage power station (MW)	Scale of gas power (MW)	New scale of gas power (MW)	scale of coal power (MW)	Total discounted cost (billon RMB)	utilization hours of coal power(h)	Coal consumption (thousand tons)	Gas consumption (billon m3)	Amplitude of coal power (%)	
										summer	winter
M1	13220	0	38961	0	218609	2259.3	4598	313570	27.55	65.6	59.9
C1	22000	8780	38961	0	209411	2238.7	4823	312690	27.55	59.4	53.8
C2	40000	26780	38961	0	191060	2213.7	5339	311020	27.55	44.6	40.0
C3	46000	32780	38961	0	185513	2214.2	5517	310630	27.55	38.9	34.8
R1	13220	0	47741	8780	209411	2287.1	4653	302150	33.75	64.2	58.1
R2	13220	0	65741	26780	190554	2344.6	4783	278920	46.48	60.7	54.1
R4	13220	0	71741	32780	184281	2364.4	4832	271280	50.72	59.5	52.6

The effect of different amplitude of coal power act on the scale of Pumped storage power station is analysed on base of optimization for power source structure. At the same time, the different amplitude

of thermal units, gas power, nuclear power, hydropower from outside, coal power from outside act on the scale of Pumped storage power station is analysed too. When the amplitude of hydropower from outside is 20%, the amplitude of coal power from outside is 30%, the amplitude of thermal units is 20%, the amplitude of nuclear power is 10%, the reasonable scale of Pumped storage power station is 32318MW; When hydropower from outside, coal power from outside, thermal units, nuclear power don't participate in peak load regulation, and take wind power from outside into consideration, the reasonable scale of Pumped storage power station is 60017MW. By analysis of the economic efficiency of power system, capacity and possibility of peak load regulation and uncertainty of electric power from outside, this article presents that the reasonable scale of pumped storage power station is 40000MW, and the proportion of pumped storage power station is 8.1%.
The optimal allocation of power source of east china power grid in 2025 is as follows:

Table 4. The optimal allocation of power source of east china power grid in 2025

Item	2025	
	Scale of power source(MW)	Proportion (%)
Maximum load	370000	
Assemble power	490889	100
Conventional hydropower station	19932	4.1
Pumped storage power station	40000	8.1
Gas powe	38961	7.9
Nuclear power	39920	8.1
Wind power	24500	5.0
Coal power from outside	56300	11.5
Hydropower from outside	40800	8.3
Coal power	230476	46.9

5. Conclusion

The structure of east china power grid is not reasonable for the coal power takes a much higher proportion, The capacity of peak load regulation is obviously different in all kinds of power source, demand of electricity continues to increase, peak valley difference will increase significantly and the pressure of peak load regulation cannot be ignored. By analysis of optimal allocation of power source, it shows that the reasonable scale of pumped storage power station is 40000MW, and the proportion of pumped storage power station is 8.1%. If the power source planning carried out according to the idea of peak load regulation balance, the power source structure of east china power grid in future will be better and better. The method can play an important role in guiding for the planning work in east china power grid.

References

[1] Cong,Y.L.& Niimura,T.Hydro generation scheduling for deregulated and competitive operation [J]. Power System Technology, 2002(1):13-17.
[2] Yeh W.W.G. Reservoir management and operation models:a state of art review.Water Resource Research[J], 1985 ,21(12):1797-1818.
[3] Hernandez, H.M.&DiazJ.A. Operations planning of Colombian hydro-thermal interconnected system [J]. IEEE Transactions on Power Systems, 1991,6(2):778-786.
[4] Georgakakos, A.P. H.Yao,Y.Yu. Control Models For Hydroelectric Energy Optimization [J].Water Resource Research,1997,33(10):2367-2379.
[5] Greg Magsaysay, Thomas Schuette, Russ J Fostiak. Use of a static frequency converter for rapid load response in pumped-storage plants. IEEE Transactions on Energy Conversion, Vol. 10, No.4, December 1995.
[6] Nicolet, C., Avellan, F., Allenbach, P., Sapin, A., Simond, J.-J., "New Tools for the Simulation of Transient Phenomena in Francis Turbine Power Plants". Proceedings of the 21st IAHR Symposium, Lausanne, pp. 519-528, 2002.

IOP Conf. Series: Earth and Environmental Science **64** (2017) 012085 doi:10.1088/1755-1315/64/1/012085

A summary of impacts of wind power integration on power system small-signal stability

Lei Yan*, **Kewen Wang**

School of Electrical Engineering, Zhengzhou University, Henan, China

*Corresponding author e-mail: zzu_dq_yl@163.com

Abstract. Wind power has been increasingly integrated into power systems over the last few decades because of the global energy crisis and the pressure on environmental protection, and the stability of the system connected with wind power is becoming more prominent. This paper summaries the research status, achievements as well as deficiencies of the research on the impact of wind power integration on power system small-signal stability. In the end, the further research needed are discussed.

1. Introduction

The influence of the access of large-scale wind power to the system on small-signal stability is an important subject which can't be ignored. On one hand, randomness is the inherent characteristic of wind power, and it is also the main factor affecting the safe and stable operation of wind power grid. The large capacity power fluctuation caused by the randomness of wind power will lead to a series of problems, such as frequency stability, voltage stability and low frequency oscillation, which are serious threats to the safe operation of the power system [1]. On the other hand, most of the large wind farms in China are hundreds or thousands of kilometers away from load center, so the large scale wind power usually needs to be given across the region, resulting in a loss of regional oscillation damping and intensifying the instability of a large power system.

Under the above background, the research of influence of wind power grid on power system small-signal stability are reviewed. At the beginning, the principle and method of power system small-signal stability analysis are outlined. According to the literature in recent years, then the research of the influence of wind power on power system small-signal stability is reviewed from two aspects of modeling of wind turbine (WT) and the oscillation modes and damping characteristics of power system with wind power integration. At last, the control strategies to improve the system damping characteristics are introduced.

2. Power System Small-signal Stability

Small-signal stability refers to the ability of the power system to maintain synchronism under small disturbances. Such disturbances are related to changes in the grid such as load and generation shifts. This variations occurs constantly in the grid. At the same time this changes are considered small enough for them to be linearized. Small signal stability is largely a problem of insufficient damping of oscillations; therefore instability is normally through oscillations of increasing amplitude.

Small-signal stability analysis methods mainly include the eigenvalue analysis, frequency domain method and the time domain simulation method. In recent years, some scholars put forward some new methods for small signal stability analysis, such as method based on the analysis of the measured signals of power system, the wide-area measurement information analysis method and normal form method based on the nonlinear theory [1].

Eigenvalue analysis method is most widely used in the small-signal stability analysis, which is based on linear system theory and the lyapunov's first law. Because eigenvalue analysis theory is highly reliable and can reveal the nature of stability of the systems with complex dynamic behavior, it is a powerful tool for analyzing system oscillation mode and damping characteristics and has been successfully applied in small signal stability evaluation and the controller's design, configuration, site selection and parameter setting, etc. [2].

3. Impact of Wind Turbine on Power System Small-signal Stability

WT system normally consists of wind turbine, generator and grid interface converters. During the development of the WT techniques, squirrel cage induction generator (SCIG), doubly fed induction generator (DFIG) and direct-drive permanent magnet synchronous generator (DDPMSG) were employed to convert wind power to electrical power, and DFIG is applied most.

3.1. Wind Turbine Modeling

Establishing proper dynamic models of WT and wind farm is the top priority to study the influence of WT on power system small-signal stability, and the accuracy of the models and parameters will be directly related to the precision of the analysis and calculation. Doubly-fed induction generator, as the mainstream of the current wind power equipment, is mostly researched in the wind turbine modeling.

A mathematical model for DFIG was earlier proposed by [3], and this documents simulated the respond of this WT model to two kinds of wind model. [4] established an improved type of DFIG small signal stability model, the authors took into account the characteristics of mechanical system of DFIG, the electromagnetic characteristics of generator of DFIG and the control strategies of the stator and rotor when establishing equation of state of DFIG. Thus, this model can reflect the electromechanical mode for DFIG connected to the power system more accurately. [5] set up a simplified three-order dynamic model of DFIG in the reference frame of stator's circumrotatory magnetic field, which considered rotor field voltage and mechanical torque received by generator as controlled variables, and compared the response of simplified DFIG model's and accurate DFIG model's various state variables to small disturbances by using MATLAB simulation. According to the characteristics of variable speed WT shafting of DFIG, [6] established a dynamic model of DFIG gearbox shafting system, and constructed an eight order simultaneous differential equations based on the dynamic equations of generation. This model can be used to analyze the dynamic behavior of DFIG under various operating conditions. [7] Studied the wind farm equivalent modeling method based on operating data, made comparison with traditional wind farm modeling method, and the analysis shows that the equivalent modeling method based on operating data is more consistent with the physical condition.

Researches on WT have achieved fruitful results, and some commercial software do quite well in WT modeling. However, because the control system of WT is quite complex and each manufacturer applies their own control strategies, among which there are some difference, it has certain difficulty to establish a more realistic model.

3.2. Impact of Grid-connected Wind Farm on Power System Small-signal Stability

Document[8] firstly began to do the research on the impacts of constant speed WT and variable speed WT on power system small-signal stability, and studied the impact of WT on oscillation mode and damping characteristic by using eigenvalue analysis method and time domain simulation; [9] studied the effect of grid-connected wind power in multi-aspects, and pointed out that power system small-signal stability could be effected by the capacity of wind power, access point of wind power and blocking level of the grid, where the influence may be positive or negative depending on the operating conditions. The above conclusions were verified by simulation of an example. [10] Studied the changes of small signal stability and transient stability with the increase of output of DFIG, and analyzed the problem of the selection of wind farm access points by using sensitivity analysis method. [11] Researched power system small-signal stability with high wind power penetration, and draw a conclusion that power system stability could be improved by controlling terminal voltage of DFIG. [12] Calculated the example of WSCC 3-machine-9-bus system to study the impacts of output and terminal voltage control unit of generation on oscillation mode. The author found that the terminal voltage control unit of DFIG had much to do with the characteristics of the regional oscillation mode. Applying inertia control can help DFIG provide the support of inertia to the power system. According to the above theory, document [13] analyzed the mechanism of the effect of the inertia of DFIG on power system small-signal stability, made a quantitative analysis of this effect, and inspected the influence of different control strategies and control parameters on the system small signal-stability by the mode analysis. [14] Researched the influence of wind power fluctuation on the power system small-signal probability stability, which was based on the analysis of probability distribution function of eigenvalues applying Gram-Charlier series expansion, and verified the effectiveness of the above method by using Monte Carlo simulation.

Research on the effect of grid-connected wind power on power system small-signal stability still has many problems. First, while there are many documents researching on wind-power-connected system small-signal stability, most of them are studied based on the simulation of test system and few researches are based on the analysis of the actual system. Second, it's difficult to get consistent conclusion due to varying simplification degree of the examples and the models, the complexity of control of variable speed WT and many other factors which influencing the power system damping. In addition, few papers studied the mechanism of oscillation modal of wind-power-connected system, and so did the analysis of the relation between WT components and power system oscillation mode from physical perspective.

4. Strategies For Improving Power System Small-signal Stability With Wind Power Integration

The strategies for improving power system small-signal stability with wind power integration can be classified into two types, which are the primary system strategy and the secondary system strategy. The primary system strategy include the enhancement of the grid structure and the installation of energy storage devices, etc. The secondary system strategy include the installation of PSS (PSS Power System Stabilizer) and FACTS (flexible AC installed transmission system) devices, and the use of modulation function of HVDC (high-voltage direct current) transmission system, etc[35]. The above strategies are the general measures to improve power system small-signal stability, without considering the influence of wind power access to the system. As for the question of improving power system small-signal stability considering wind power integration, many scholars have made fruitful research [15-43].

To improve the damping of the system, the control of asynchronous WT pitch PID was proposed in [15]. The pitch angle of the blades is adjusted to increase (decrease) mechanical torque and improve (lower) WT power output, thus the system positive damping is increased. This method is similar to the control strategy of power system stabilizer, which can enhance the system damping and improve power system small-signal stability. [16] Proposed the control method of amplitude and angle of rotor field magnetic flux, which can enhance system damping by additional control. Electrical variables related to system oscillation mode were regarded as the input signal and output signal was regarded as the additional control signal of active or reactive power ring, and the low frequency oscillation can be suppressed through adjusting the angle of the rotor field magnetic voltage angle. In paper [17], a new type of power system stabilizer applied to DFIG was designed, and it was different from the traditional power system stabilizer PSS. This type of power system stabilizer was installed inside the WT regarded the local slip signals as additional signals which are transported to the rotor converter of WT to suppress low frequency oscillation. The result showed that it did well in suppressing the active power oscillation of the wind farm. Aiming at the problem of low frequency oscillation between regions, the damping control strategy applying wide-area measurement system (WAMS) has attracted the attention of many scholars. In document [18], the DFIG wide-area damping controller was designed, where the active power reference value of the DFIG rotor-side converter and power angle difference of inter-area synchronous generator were regarded as the input signals. The simulation results showed that the DFIG applied wide-area damping controller could provide more reactive power to the system and provide the system a stronger voltage support, thus improving the system damping and achieving the inhibition of the inter-area oscillation. However, for a large power system, it's difficult to achieve an ideal effect for these reasons: a large amount of system parameters are needed in the PSS design; the structural changes of the system may have a great impact on the damping effect; the PSS of some wind turbines may exist the operating state for some reasons in the actual operation. [19] Designed an adaptive damping controller based on variable frequency transformer, which could automatically calculate and optimize the parameters of the damping controller according to the Prony method.

In general, mechanical control, active power control and reactive power control are applied to improve the damping of power system connected with wind power. As a new control strategy, energy function of power system is also put forward. Wind turbine PSS are researched most in all control strategy, but it's still difficult to achieve an ideal effect due to the nonlinear characteristics of wind power system and the difference of various WT. At the same time, it's difficult to do the commissioning and setting of wind turbine PSS because of the large amount of wind turbines in a wind farm. So,it still needs more effort to study the wind turbine PSS.

5. Conclusion

With the increase of wind power capacity in China, and the intrinsic characteristic of randomness of wind power, it becomes an important subject to research on the impact of grid-connected wind power on power system small-signal stability.

Wind farm modeling is the foundation of the research on wind power, and the accuracy of the model and parameters will be directly related to the accuracy of analysis and calculation. Many freeware and commercial software have obtained certain experience, but it still needs more effort to establish a more realistic model because of the complexity of the wind power system and the diversity of wind turbines applied by the grid.

Squirrel cage induction generator, doubly-fed induction generator and permanent magnet direct-drive generator are employed most in current power system. Constant speed WT itself has no synchronization stability problem, and generally presents positive damping effects on the system. Due to the complexity of control system, variable speed wind power system may present positive or negative damping effect on power system, and there is no certain conclusion. So it still needs further study on the mechanism of the effect of wind turbines on the system damping characteristics.

From the point of WT control strategy, the research mainly focuses on the WT additional damping control strategy. However, because the wind turbine PSS has some inherent shortcomings, the energy storage technology, FACTS and HVDC modulation are widely applied in system damping control, which makes the damping coordination control become a hot issue.

References

[1] Zhou Shuangxi,Su Xiaolin. Advance of Study on Small Disturbance Stability of Power Systems[J]. PROCEEDINGS OF THE CSU-EPSA,2007,19(2): 1-9

[2] Kundur P. Power system stability and control[M]. Mc-Graw-Hill,New York,1994

[3] J G Slootweg,H Polinder,W L Kling. Dynamic modelling of a wind turbine with doubly fed induction generator[C]. Power Engineering Society Summer Meeting,2001,2011,50(6): 1257–1264

[4] Wei Wei,Liu Ying,Ding Lijie,Li Xingyuan. An Improved Small Perturbation Model for Doubly Fed Wind Power System[J]. Power System Technology,2013,37(10): 2904-2911

[5] Li Jing,Wang Weisheng,Song Jiahua. Simplified dynamic model of doubly-fed induction generator and its application in wind power[J]. Electric Power Automation Equipment,

2005,25(1): 58-62
[6] Li Dongdong,Chen Chen. A Study On Dynamic Model Of Wind Turbine Generator Sets[J]. Proceedings Of The CSEE,2005,25(3): 115-119
[7] Yu Yang,Liu Yongguang,Dong Shengyuan. Comparative Study on Wind Farm Equivalent Modeling Based on Operation Data[J]. Advances Of Power System & Hydroelectric Engineering,2009,25(12): 79-83
[8] JG Slootweg,WL Kling. The impact of large scale wind power generation on power system oscillations[J]. Electric Power Systems Research,2003,67(1): 9-20
[9] JL Rueda,I Erlich. Impacts of large scale integration of wind power on power system small-signal stability[C]. International Conference on Electric Utility Deregulation & Restructuring & Power Technologies,2011: 673-681
[10] D Gautam,V Vittal. Impact Of DFIG-based Wind Turbine Generators On Transient And Small Signal Stability Of Power Systems[C]. Power & Energy Society General Meeting .pes.ieee,2009,24(3): 1426-1434
[11] Vittal,Eknath,O'Malley. A small - signal stability analysis of DFIG wind generation[C]. Eighth International Workshop on Large-Scale Integration of Wind Power into Power Systems as well as on Transmission Networks for Offshore Wind Farms,Germany,2009
[12] Yang Lihui,Ma Xikui. Impact of Doubly Fed Induction Generator Wind Turbine on Power System Low-frequency Oscillation Characteristic[J]. Proceedings Of The CSEE,2011,31(10): 19-25
[13] Chen Runze,Wu Wenchuan,Sun Hongbin,et al. Impact of Inertia Control of DFIG Wind Turbines on System Small-signal Stability[J]. Automation of Electric Power Systems, 2014(23): 6-12
[14] SQ Bu,W Du,HF Wang,Z Chen,LY Xiao. Probabilistic analysis of small-signal stability of large-scale power system as affected by penetration of wind generation[J]. IEEE Transactions on Power Systems,2012,27(2): 762-770
[15] C Jauch. Stability and control of wind farms in power system[D]. Denmark: Aalborg University, 2006
[16] Hughes,F.M,Anaya-Lara,O,N Jenkins,G. Strbaca. A power system stabilizer for DFIG-based wind generation[J]. IEEE Transactions on Power Systems,2006,21(2): 763 - 772
[17] O Anaya-Lara,F M Hughes,N Jenkins,G Strbac. Power system stabilizer for a generic DFIG-based wind turbine controller[C]. IEEE Trans on Energy Convesion,2008,23(2): 519 - 528
[18] Liu Qihui,He Yikang,Zhang Jianhua. Operation Control and Modeling-Simulation of AC-excited Variable-speed Constant-frequency(AEVSCF) Wind Power Generator[J]. Proceedings Of The CSEE,2006,26(5): 43-50
[19] Chen Gesong,Zhou Xiaoxin,Song Ruihua. Design of Self-adaptive Damping Controller to Low Frequency Power Oscillation in Interconnected Power Systems Based on Variable Frequency Transformer[J]. Proceedings Of The CSEE,2011,31(16): 1-7

IOP Conf. Series: Earth and Environmental Science **64** (2017) 012086 doi:10.1088/1755-1315/64/1/012086

Development of hybrid electric vehicle powertrain test system based on virtue instrument

Yanmin Xu[1, 2, *], Konghui Guo[1], Liming Chen[2]

[1]School of Mechanical and Automotive Engineering, South China University of Technology, Guangdong, China
[2]School of Automobile, Guangdong Mechanical & Electrical Polytechnic, Guangdong, China

*Corresponding author e-mail: xu.yanmin@mail.scut.edu.cn

Abstract. Hybrid powertrain has become the standard configuration of some automobile models. The test system of hybrid vehicle powertrain was developed based on virtual instrument, using electric dynamometer to simulate the work of engines, to test the motor and control unit of the powertrain. The test conditions include starting, acceleration, and deceleration. The results show that the test system can simulate the working conditions of the hybrid electric vehicle powertrain under various conditions.

1. Introduction

Hybrid electric vehicles, which combine the advantages of traditional fuel vehicles and electric vehicles, can meet the requirements of low emission, low fuel consumption and low cost, and so on [1]. Based on the degree of mixing, hybrid electric vehicles can be divided into three categories such as weak mixing (Micro Hybrid), light mixing (Mild Hybrid) and the whole mixture (Full Hybrid) [2]. In consideration of low investment and relatively high economy and power, hybrid vehicles have obvious advantages comparing to the traditional internal combustion engine vehicles.

The key assemblies of hybrid vehicle powertrain consist of power supply module, motor and motor controller, engine and engine control unit. The verification test of key assemblies' performance and function is conducive to enhance the design goal of hybrid vehicles. The team of author developed a hybrid powertrain test system which can verify the function of integrated starter-generator (ISG), power supply module, motor and motor controller, engine and engine control unit based on virtual instrument.

2. System hardware design

The test system is composed of ISG motor module, dynamometer module, power supply module, auxiliary electric system load simulation module and host computer module, as shown in figure 1. The ISG motor module contains ISG motor and motor controller. The dynamometer module includes power dynamometer, converter / inverter, measuring and controlling instrument, the speed and torque sensor. The power supply module includes a power battery, battery management system, programmable power source. The electric system load simulation module includes DC/DC converter, electric air conditioning compressor and other affiliated electrical system. The host PC module includes data acquisition card, data processing unit, display output unit. BSG motor module, dynamometer module and power supply module communicate respectively with host computer module through CAN bus.

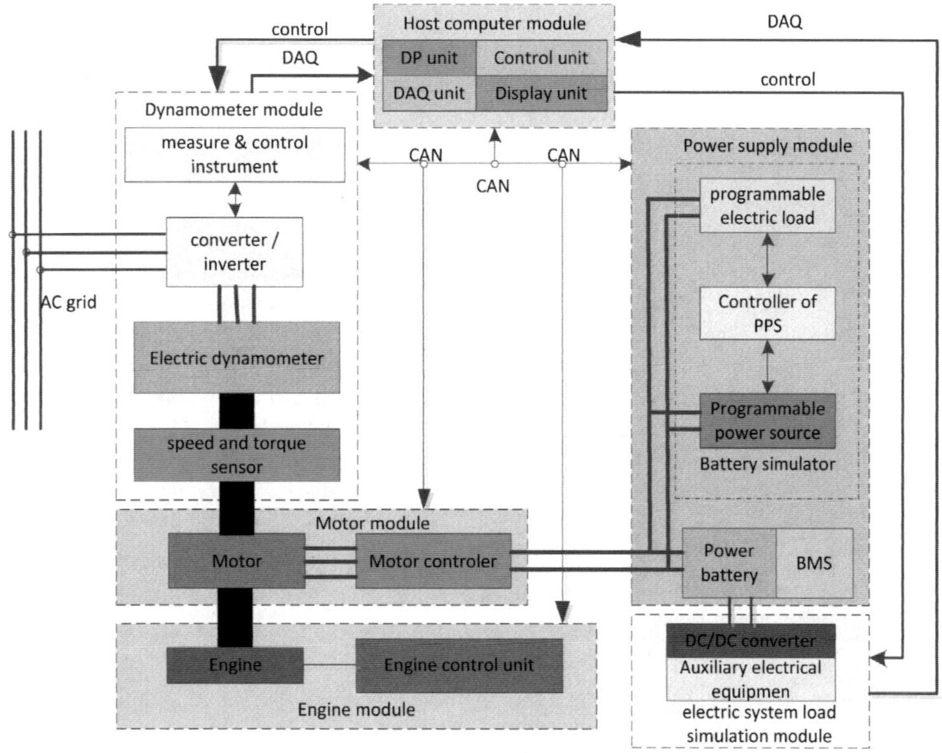

Figure 1. The system hardware block diagram.

The dynamometer module adopts the AC electric dynamometer which has the characteristics of low inertia moment, good regeneration performance, high precision to control and fast response [3]. The dynamometer can be used as the load of the engine to test the performance of the engine, or used as load of the ISG motor to test the performance of the motor. The dynamometer can work in driving mode, taking electric energy from the power grid by the converter, or work in load mode, generating electric energy and feeding to the power grid by the inverter. The control of different working modes is realized by the measuring and controlling instrument. Speed and torque sensor was installed on the output shaft of dynamometer for real-time monitoring. The ISG motor or engine is connected with the dynamometer as the object to be measured.

Power supply module is used to supply power for test system. The power battery is connected with the programmable power source through the common DC bus; the controller of programmable power source is used to control the output voltage and the current of the programmable power source, simulating all kinds of power battery with different specifications. During the test, the programmable power source and the power battery can be used as the power supply of the test system respectively. Programmable power source is integrated with programmable electric load, which consists of discharge modules. The host computer module obtains the working state, voltage and current of the programmable power source. The host computer obtains the working state of the power battery through the battery management system, and sets the terminate condition of charging and discharging, the limit conditions of voltage, current and temperature of the battery.

3. System software design

Virtual instrument technology is a program development environment, it can give full play to the computer's ability to create a stronger function of the instrument, and users can define and develop a variety of instruments according to their own needs [4]. The test system software of hybrid vehicle powertrain was developed based on virtual instrument, using electric dynamometer to simulate the work of engines, to test the motor and control unit of the powertrain. The test conditions include starting, acceleration, and deceleration. Take ISG motor starting engine condition as an example, the test flow chart and human-computer interaction interface design as follows.

3.1. Test flow chart design

The test flow chart of starting condition design is shown in Figure 2.

The initialization parameters can be set in the host computer, and transmit to the controller of programmable power source, measuring and controlling instrument, the battery management system by CAN bus. Simulate torque of the engine using dynamometer; electrify the motor to drive dynamometer. The host computer sends out the control command to measuring and controlling instrument to control the dynamometer according to the feedback of test results and data processing to adjust the power dynamometer through the inverter. If the measured resistance torque is consistent with the set resistance torque, sent the speed feedback to host computer, then the engine is start successfully.

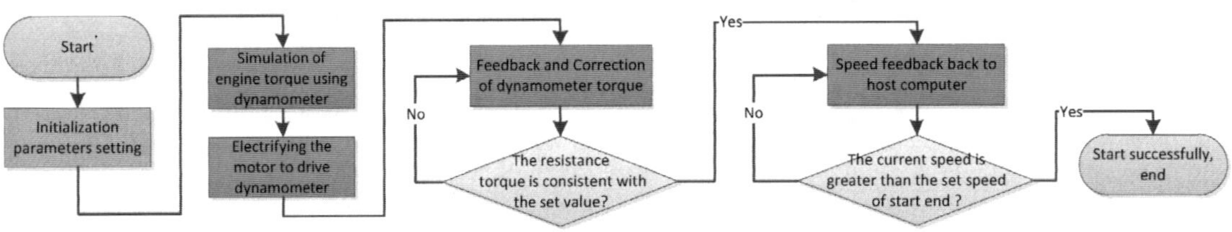

Figure 2. The test flow of starting condition.

Other condition such as acceleration, deceleration, and so on can deduce the rest from the starting condition.

3.2. Human-computer interaction interface design

Human-computer interaction interface of the system was developed based on virtual instrument, which adopted the form of the tab to realize the intuitive and friendly man-machine interactive function. It contains the user login interface, the initial setup interface, the main control program interface, etc..

The initial setup interface, as shown in figure 3, can be used to set and adjust the test parameters of the test system. The set of objects mainly include engine parameters, starting and stopping motor parameters, communication system parameters, cycle test times, power supply parameters, report output path, etc.

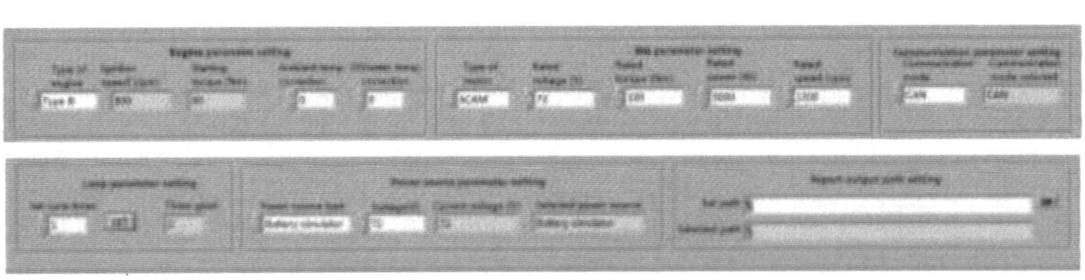

Figure 3. The initial setup interface.

The main control program interface of the test system is mainly composed of the working status indicator, the running parameter output display area and the system control command input area, as shown in figure 4.

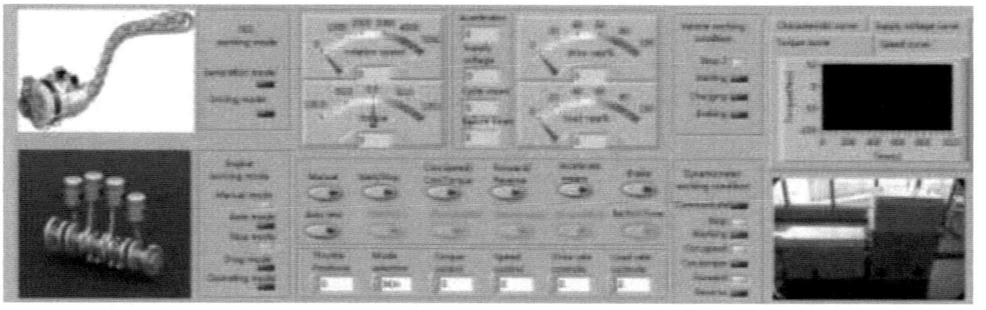

Figure 4. The main control program interface.

4. Conclusion

Hybrid powertrain as the key to improve the automobile's economy, it is necessary to carry out the relevant operation test. After repeated testing and debugging, the developed system has been able to run stably, and achieve the design goal of hybrid powertrain test. Login the test system to verify the system. First set the relevant parameters in the initial settings tab. The result of working condition test of 5 cycles is shown in Figure 5.

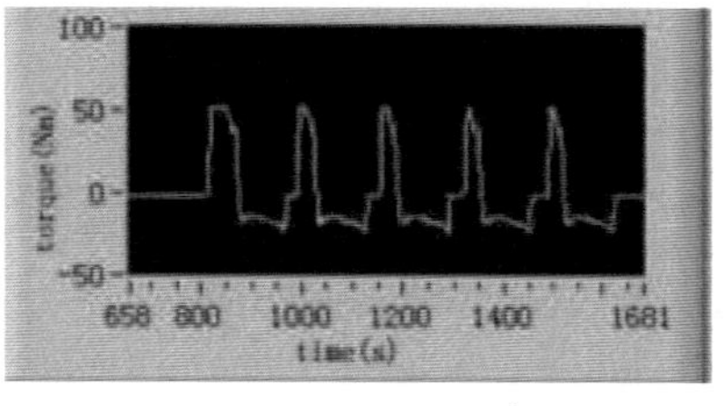

Figure 5. The test result of 5 cycles working condition.

The test system developed in this paper realizes the test function of hybrid powertrain, which can not only test according to the specified conditions, but also be used in cycle testing.

Acknowledgments

This work was financially supported by Talent introduction project in Colleges and Universities [2013] No.246 and YJKJ2016-05 fund.

References

[1] Wang Liping, Li Shoucheng, Wu Haixiao, Yue Hongchao. Study on Start/Stop Control of ISG Hybrid Vehicle, J. Agricultural Equipment & Vehicle Engineering. 9(2011) 38-41. (in Chinese)

[2] Eckhard Karden, Daniel Kok, etc.. Energy Management and Storage Technologies in Micro-Hybrid Vehicles, J. AutoTechnology.5(2005) 56-59.

[3] Tian Ying,Jin Zhenhua,etc..A Study on the Control System of AC Dynamometer, J. Automotive Engineering. 36(2014) 125-128. (in Chinese)

[4] Liang Zhao-hua, Li Zhi.Development of field dynamic balance testing system based on LabVIEW.Modern Electronic Technique, 6 (2014) 110-113. (in Chinese).

IOP Conf. Series: Earth and Environmental Science **64** (2017) 012087 doi:10.1088/1755-1315/64/1/012087

Design and Implementation of Demand Response Information Interactive Service Platform Based on "Internet Plus" Smart Energy

Gaoying Cui[1], Jie Fan[1], Yuchen Qin[2], Dong Wang[2], Guangyan Chen[2,*]

[1]State Grid Jiangsu Electric Power Company Research Institute, Nanjing City, Jiangsu China

[2]Beijing China-Power Grid Information Technology Co., Ltd., State Grid Information & Telecommunication Group, Haidian District, Beijing, China

*Corresponding author e-mail:18500058745@163.com

Abstract. In order to promote the effective use of demand response load side resources, promote the interaction between supply and demand, enhance the level of customer service and achieve the overall utilization of energy, this paper briefly explain the background significance of design demand response information platform and current situation of domestic and foreign development; Analyse the new demand of electricity demand response combined with the application of Internet and big data technology; Design demand response information platform architecture, construct demand responsive system, analyse process of demand response strategy formulate and intelligent execution implement ; study application which combined with the big data, Internet and demand response technology; Finally, from information interaction architecture, control architecture and function design perspective design implementation of demand response information platform, illustrate the feasibility of the proposed platform design scheme implemented in a certain extent.

1. Introduction

Demand response is a means of power demand side management. At present, it is developing rapidly in the standard setting and technical application. The pilot work at home and abroad is fruitful, and gradually realize the development from passive to active. However, with the increasing awareness of the power users, the simple passive demand response cannot meet the needs of power users' electricity cost and comfort. The traditional power supply and demand model is in urgent need of transformation, and the idea of clean alternative and electric energy substitution is urgent to change the traditional energy system.

The rapid development of Internet technology brings an opportunity for the transformation of power supply and demand model. The concept of intelligent energy promotes the transformation of traditional energy system. The development of traditional industrial economy by Internet thinking is the inevitable trend of the development of the times. In 2016, the national development and Reform Commission issued "on the" Internet plus wisdom "energy development guidance", which provides guidance for the integration of Internet information technology and energy industry.

This paper is based on "Internet plus" smart energy technology application on demand response information platform design and implementation. In this paper, we fully consider the application of the Internet, big data and other technical means in the design of demand response information platform, and put forward new ideas for the transformation of the power supply and demand model, with full consideration such as the Internet, big data and other technical means in response to demand information platform design, implementation effect between grid - power users active two-way interaction, in line with the development trend of the reform of the traditional energy system, effectively maintain network security and stable operation and high quality.

2. Research status at home and abroad

2.1. Foreign research status

At present, some research institutions and scholars abroad have played an important role in the design and implementation of demand response information platform.

In 2012, the Laurence Berkeley National Laboratory Research Center released OpenADR demand response 2.0A communication protocol, Auto-DR system installation section in the Seattle area of commercial buildings, as well as the Pacific Gas and electric automation platform to implement demand side bidding project based on Realization of load reduction targets [1]. Matsushita Electric Tokyo headquarters use of new energy-saving technology, using a large data means to analyze the construction of energy saving potential of the building energy managemen system, which can provide users with response to demand for energy-saving programs. Europe relies on the establishment of a wide range of smart energy demand response project features and services platform, they carry out an effective response to the needs of the project [2-4].

Foreign demand response construction application of the information platform is relatively perfect, but cutting-edge technology failed to measure the full integration of applications, a large number of load side data resources have not been fully utilized, the supply and demand side needs to improve the strength of interaction.

2.2. Domestic research status

The construction of domestic demand response information platform combines power service management platform, electricity information collection system and electric power dispatching system, which can realize the real-time online monitoring of user load, using BS mode, APP mobile phone or SMS to interact with users, alleviate the peak, promote the development of energy saving and emission reduction work [5].

At present, the demand response infrastructure includes: power information collection system, electric vehicle charging pile, intelligent park and intelligent community.

iangsu province electric power company intranet deployment of demand response platform, grid information, demand information, solicitation planning through the network service interface onto the APP system (extranet), an invitation sent to the user through the mobile phone APP and information query, and user feedback should be to demand response platform.

In Shanghai and Beijing, the power load management system and the power demand side management integrated service platform are used as the demand response information platform.

Domestic demand in response to the construction of the information platform to meet the needs of the demand response of pilot work to a certain extent, it promotes the establishment and implementation of demand response scheme and strategy, but there are still some problems such as: the load side of large amounts of data are not fully utilized, intelligent demand response is not enough, the lack of interaction between supply and demand side.

3. New demand of electric power demand

With the development of power demand side management and demand response, the traditional demand response mechanism has been unable to meet the development needs of power demand side management and user needs. In the Internet era, the Internet thinking and technology will drive the transformation of traditional power supply and demand model industry.

(1) Development of power demand response industry based on internet.

The Internet represents a kind of advanced productive forces, which promote economy constantly evolve, so as to promote the social economy vitality, it provides a broad platform for the network can reform, development, innovation, which is known as " the new engine" for Chinese upgrade the quality and efficiency of economic.

Currently in the industry, finance, commerce, communications, transportation, people's livelihood, the government is widely used in various industries, and in the field of smart energy industry is only a clue, in the user side needs to respond less.

In addition, with the construction of the global energy Internet system, the demand side management of electricity needs to be integrated into the Internet elements, to promote the demand for the development of the internet.

(2) The demand response visualization applications.

Automation and intelligent demand response field is not a new topic, but the present aggregator fight the enemy separately and the user half believe and half doubt interactive service system incomplete.

(3) The strategy of intelligent demand response

Based on the large data technology, this paper analyzes the implementation effect of the demand response plan, and explores the potential of demand response, which provides support for the intelligent formulation and decision of demand response.

At present, the demand response infrastructure includes: power information collection system, electric vehicle charging pile, intelligent park and intelligent community.

Jiangsu province electric power company intranet deployment of demand response platform, grid information, demand information, solicitation planning through the network service interface onto the APP system (extranet), an invitation sent to the user through the mobile phone APP and information query, and user feedback should be to demand response platform.

In Shanghai and Beijing, the power load management system and the power demand side management integrated service platform are used as the demand response information platform.

Domestic demand in response to the construction of the information platform to meet the needs of the demand response of pilot work to a certain extent, it promotes the establishment and implementation of demand response scheme and strategy, but there are still some problems such as: the load side of large amounts of data are not fully utilized, intelligent demand response is not enough, the lack of interaction between supply and demand side.

References are cited in the text just by square brackets [1]. Two or more references at a time may be put in one set of brackets [3, 4]. The references are to be numbered in the order in which they are cited in the text and are to be listed at the end of the contribution under heading references, see our example below.

4. Intelligent demand response design

4.1. Overall design

Based on the "Internet plus" smart energy demand response information platform is the use of the Internet and mobile Internet technology, we build a wisdom energy information platform to users.

The platform can be used to monitor the energy consumption and energy consumption of each energy consuming equipment and energy consuming unit in the enterprise, and to integrate the demand response technology to realize the reduction and transfer of the power consumption.

IOP Conf. Series: Earth and Environmental Science **64** (2017) 012087 doi:10.1088/1755-1315/64/1/012087

The platform uses the big data technology to construct the multi response model of the demand response to the user side. The model includes the potential demand response resource model, demand response execution model, demand response evaluation model, through large data demand response model, implementation and feedback analysis, explore the potential resources in response to demand response optimization strategy and demand response evaluation of implementation effect. The platform can make use of self-learning ability, pattern recognition technology and so on to realize the automatic optimization and improvement of the model.

The overall architecture design of demand response information platform is shown in figure 1.

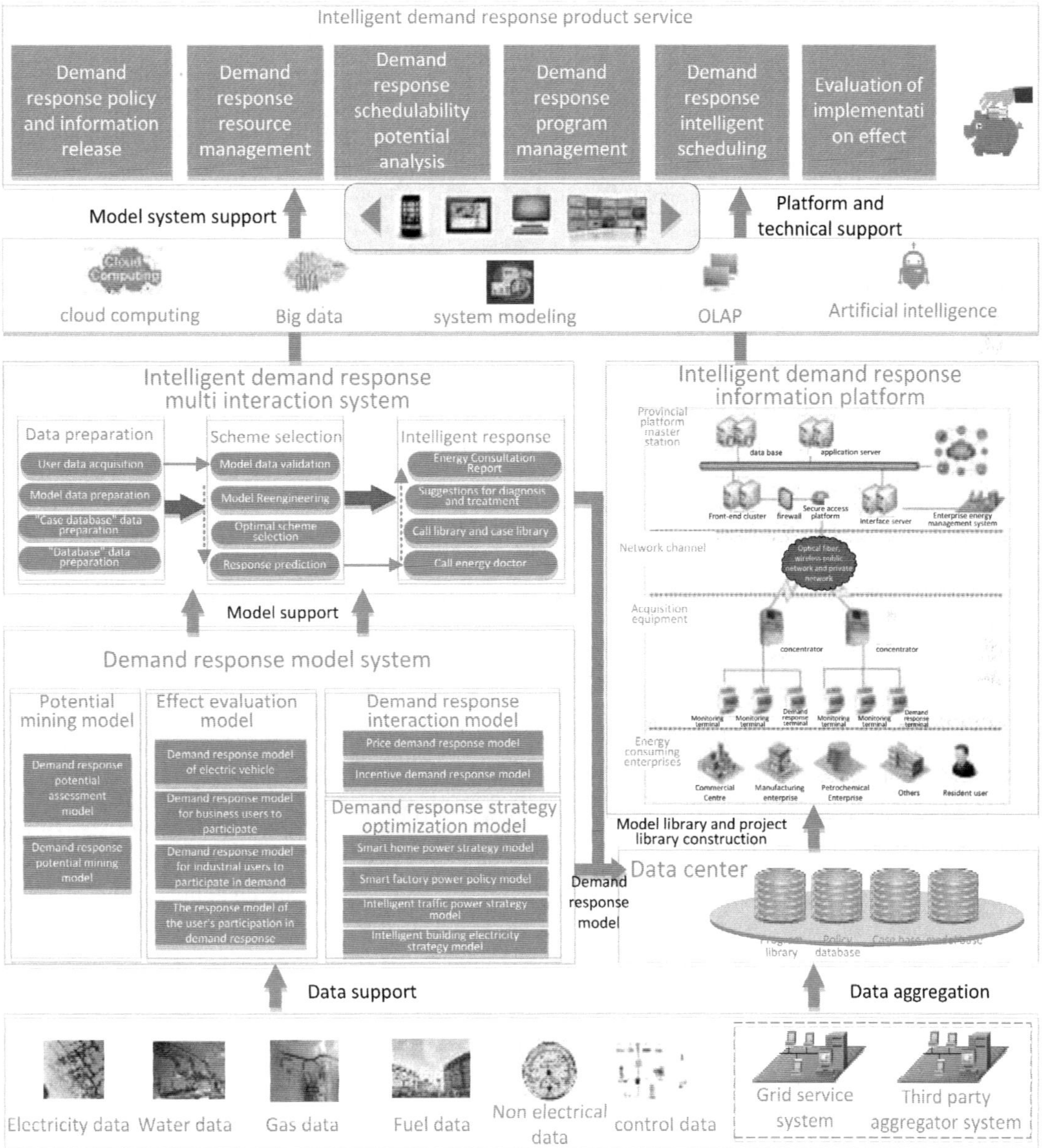

Figure 1. General Architecture of Demand Response Information Platform.

4.2. Demand response strategy

Demand response participation object categories are different in response to load demand in response time [6], response capacity [7] and the scheduling potential of different load characteristics [8], the demand response strategy is the factors that cannot be ignored, can be realized through the optimization of air conditioning [9] and water heater [10] start and stop strategy and electric vehicle [11] charging and discharging strategy to achieve the demand response. Demand of diversity of response, such as in [12] power supply side interests based on the maximization of real-time electricity pricing strategy of demand response, literature [13] considering the minimum delay time and minimum electricity appliances two targets proposed by the power equipment power adjustable scheduling strategy for demand response.

Response to the needs of different users, the demand response strategy is different, its development process is as follows:

(1) In response to user needs clear categories, user participation for industrial users, commercial users and residents to participate in the user load, demand response resource [14] category belongs to a simple adjustable load, energy storage device or distributed power supply;

(2) The load data is obtained by using the electric energy service management platform combined with the electric power dispatching system, and the load characteristics of the demand response are determined by means of data mining and load forecasting;

(3) According to the load characteristics of (2) using the mechanism analysis method to establish the demand response load regulation model;

(4) We need to determine the response objectives and constraints, combined with the demand response load regulation model using intelligent control technology, measurement technology and remote control technology. The demand response strategy is established, and the optimization of the demand response strategy is realized through the multi-objective optimization theory.

Based on the analysis of load characteristics of multi class load in the process of response to different needs, the demand response strategy database is constructed. Relying on the Internet platform to provide customized demand response strategies for specific users, considering the interests of both sides, to recommend the optimal response strategy for users. The demand response strategy formulation process is shown in figure 2.

4.3. Intelligent Implementation of demand response

Response to the needs of the pilot work in the process of promoting the process, the need to respond to the specific implementation of the process can promote a clear response to the need to complete. "Jiangsu province electric power demand response" rules for the implementation of demand response given in the implementation process, to promote the development of experimental work in the construction of demand response plays an important role in promoting. The development of demand response has gone through 3 stages, from manual, semi-automatic to fully automatic [15]. The development of intelligent response to the requirements of the development of automatic data cannot be separated from the use of intelligent methods and applications. Based on the application of demand response to user data and historical load data, by selecting the user performance index, the demand response is automatically selected. In this paper, the requirements of the implementation process of intelligent response shown in figure 3.

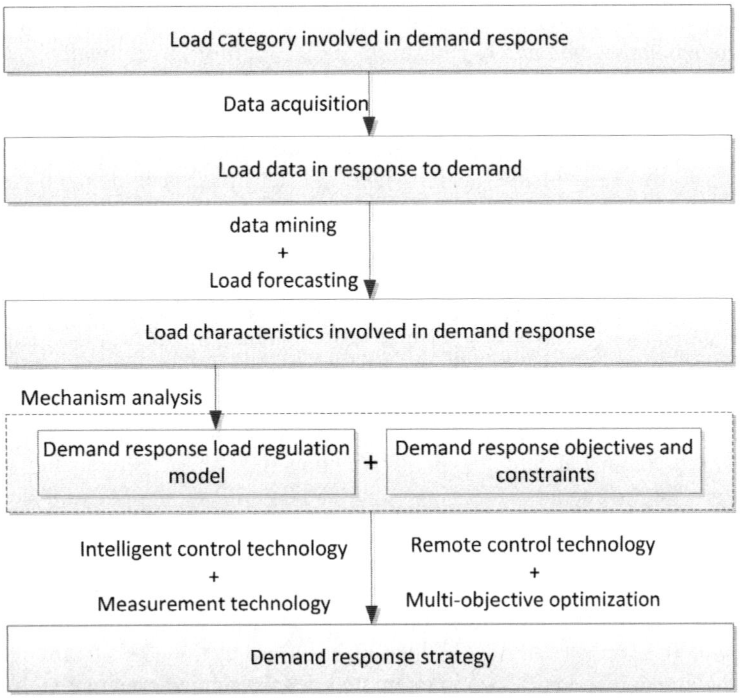

Figure 2. Demand Response Strategy Making Process

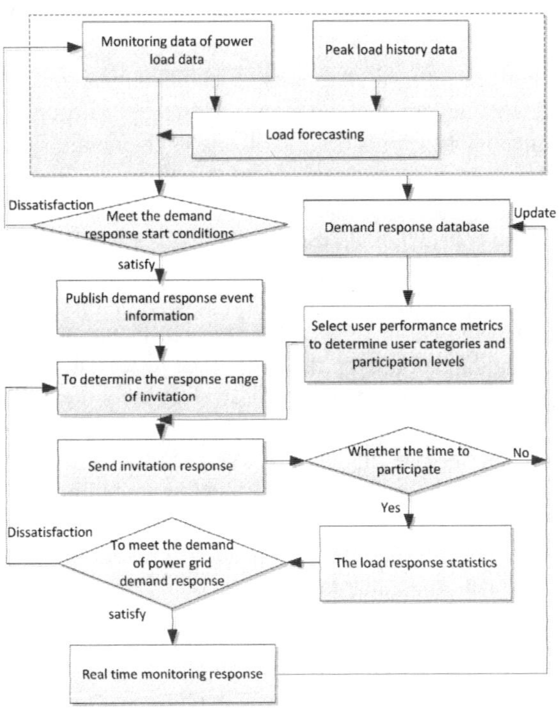

Figure 3. Demand Response Intelligent Implementation Process

(1) Based on the monitoring data of power load data and combining with the historical electricity peak data. By using the time series neural network, regression analysis or load forecasting means to

IOP Conf. Series: Earth and Environmental Science **64** (2017) 012087 doi:10.1088/1755-1315/64/1/012087

determine whether the demand response starting conditions such as power gap, the maximum load and peak valley ratio meet the demand response starting conditions, automatically determine whether should carry out demand in response to the incident.

(2) When the response requirements are met, the response information, such as demand response type, duration and response load. Through the response of user database, the method of index classification is used to deal with the performance index of the user's demand response. In order to determine the level of user participation in demand response, demand response in the solicitation process offer higher level user priority.

(3) In response to a user sends a response to the priority needs an invitation, when the user fails to offer feedback information in a timely manner, to demand response and timely feedback of user database update statistics offer user response to load whether meet the grid load demand response. When does not meet the need to re determine the offer range, according to the updated demand response database to rule out the invitation before the user, recalculate the solicitation of response variables, response level in accordance with the demand for an invitation. To monitor the response time and to update the requirements in response to the user database in order to meet the needs of the next time to prepare for the event.

4.4. Big data demand response

Driven by the Internet and big data era, big data analysis and application of the key factors of the rise of the industry. We consider the demand response load database and the demand response information database, to achieve the application of large data in response to the needs of the response, to promote the development of the demand response in many ways. Specific performance in the following areas:

(1) The application and improvement of relevant policies on demand response.

Demand response in the pilot cities in response to the needs of the relevant policies, such as electricity prices, incentives and other settings are different. However, considering the diversity of user load and power consumption, there are some limitations in the demand response policy. Based on the demand response information platform, the response mechanism model of different requirements is established. Combined with the relevant demand response strategy, through statistical analysis of large data means demand response simulation results, according to the results are retained in the pilot work to promote, promote the improvement of demand response policies.

(2) Refinement of demand response strategies.

At present, there are a variety of demand response strategies based on electricity price policy and incentive policy, which have some adaptability to some extent. From the platform to obtain the specific user load operation data, through large data statistics, mining and analysis of the load characteristics of the load in response to different scenarios, to predict the load response potential. Based on this, the load demand response control model is established, which provides the model support for the formulation of the demand response strategy and the establishment of the demand response strategy.

(3) New demand response service model.

Demand response service model is mainly based on business model. With the liberalization of the electricity market, the business model will not be able to meet market demand, demand response service model needs to break through innovation. Through the analysis and integration of large data understanding of user needs and wishes, to provide users with customized service model, analysis of the role of market diversification. It can provide users with value-added services such as energy management, equipment maintenance and energy saving services.

(4) Build a demand response case database.

Domestic demand response pilot practice work is everything in good order and well-arranged development, plays a promoting role in the peak, etc. In order to promote the needs of the pilot part of the work can be replicated to promote the need to establish a demand response case database. The demand response simulation case is based on the demand response platform, which provides the basis for the demand response scenario setting. At the same time, we need to use large data statistics and integration of demand response pilot example, to provide a basis for the construction of demand response case database.

4.5. Construction of demand response interaction system

Interaction is an essential part of the development of intelligent power, there are related to the relevant literature on smart power interactive system [16], model [17], service [18], etc. And demand response as a means of intelligent power. Study on the interaction of power demand response system to realize high-speed data read, enhance customer service levels, improve the efficiency, reduce the cost of electricity users can promote the same, it is also our country demand response is an indispensable part of pilot project construction work.

In this paper, four levels of demand response interaction architecture are described, the specific interactive architecture as shown in figure 4.

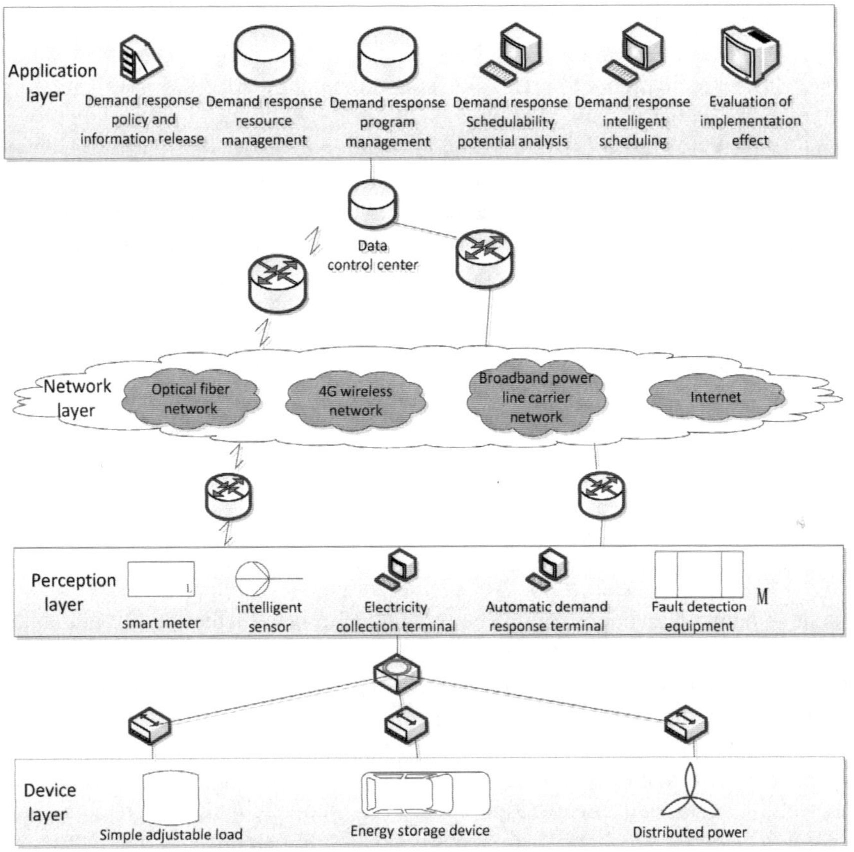

Figure 4. Demand Response Interactive System Architecture

(1) Application layer

Based on large data mining, statistics, analysis and integration of data resources, combined with the example of the pilot work for demand response, demand response situation, explore the new interactive model of demand response, the use of Internet technology, features as a support to provide operational platform for user interaction.

(2) Network layer

Demand response communication network information interactive platform is composed of light network, 3G or 4G wireless network, broadband power line carrier network and Internet, the main link for interactive business and information background.

(3) Perception layer
The sensing layer can be divided into three categories: data measurement, data acquisition and data sensing, which can realize the functions of data collection, aggregation and forwarding.

(4) Equipment layer
The device layer contains elements that are the user side device, the response to the demand for interactive terminals, which are essential elements of the demand response interaction.

4.6. "Internet" + "Demand response"
As a kind of virtual green power plant, the demand response is of great significance to the sustainable development of the intelligent energy. The rapid development of the Internet technology brings the opportunity for the rise of traditional industries. With the support of Internet technology, it can realize the interconnection of demand response resources, the information symmetry of supply and demand, and the energy balance. The combination of Internet and demand response is an inevitable trend, which has the following advantages:

(1) The concept of demand response can be extended to encourage more users to understand the demand response. The application of demand response theory in different industries to promote the effective use of resources, energy conservation and emission reduction;

(2) The demand response resources can be widely understood, so that more users can respond to the needs of the users to participate in the construction of the demand response;

(3) The combination of the Internet and demand response can promote the development of high quality demand response service, by means of large data related to the integration of data resources, the user can be provided to satisfy the different requirements of the demand response service, while improving the user awareness of science of electricity and electricity comfort;

(4) The combination of the Internet and the demand response can promote the development of automatic demand response. The use of the Internet means, within a certain range of independent access to data information and to determine the response to the needs of the event information, to provide a response to the needs of human strategy, while the realization of intelligent response to demand.

5. Design and implementation of demand response information platform

5.1. Design of data acquisition and interactive architecture
The development of demand response cannot be separated from the effective use of data resources. At present, the intelligent electric meters and other measuring devices collect a large number of load side data, and the utilization ratio of the data information resources is low. However, the study of information interaction can promote the sharing of information resources and the interaction between supply and demand. The information interaction between the grid side and the user side consists of three parts. Interactive information is the content of the demand response interaction, independent of the implementation of interaction. Interaction depends on whether the user is involved and the degree of user participation, interactive channel for the smooth implementation of interactive support path. Specific information interaction architecture shown in figure 5.

(1) Interaction
Interaction is divided into manual and automatic mode. Manual mode is the user to view the demand response event information, selectively participate in the demand response; Automatic mode is to set the user needs to participate in response to the automatic mode, set a certain demand response capacity, according to the response to the needs of the event information received, to determine whether to participate in the demand response; Semi-automatic mode is similar to automatic mode, but it is manually operated when deciding whether to participate in the demand response.

(2) Interaction channel

Interactive channel for demand response event information sending and receiving the network form, divided into the power line carrier network, light network, 3G or 4G wireless public network and internet.

(3) Interactive information

Demand response interactive information includes information resource base, load resource pool and response information. The information resource library includes user information, environmental information and policy information; the load resource library includes simple adjustable load, energy storage device and distributed power supply information; the response information includes the user's participation in demand response history information and real-time information. Through the realization of information interaction, we can realize the efficient optimization of demand response.

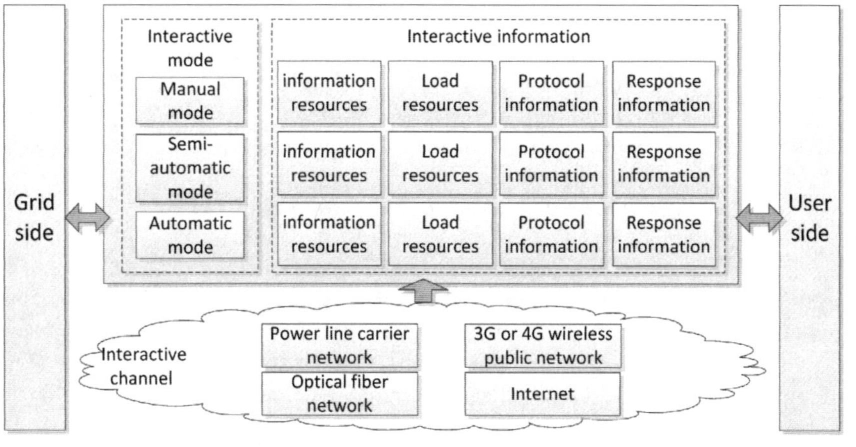

Figure 5. Information Interaction Architecture

5.2. Demand response control architecture design

With the popularization of the Internet, the real-time monitoring of electricity load increase. The application of the Internet of things sensing technology to the real-time monitoring of the load equipment operation state information, the determination of the demand response load resources participation, and the establishment of the demand response business control plan has the support function. At the same time, the design of the control structure can help to sort out the demand response event information as well as the Internet, the Internet of things and the demand response.

Through the management of the control log, we can understand the operation of each module, and put forward the demand response control framework. The architecture design is shown in figure 6.

(1) Communication layer

The communication layer includes platform communication and sensor network communication. Platform communication is responsible for docking with the various systems, data receiving and transmitting information, sensor network communication is responsible for communication with the demand response perception layer devices.

(2) Business logic layer

The business logic layer implements the control business processing function, which is composed of platform interaction, information interaction, real-time acquisition and control, online monitoring, data archiving and external interface module. The main function of this system is to get the load data in real time, and to transfer the data through the interface.

(3) Platform support layer

The platform support layer is provided with three groups of modules, which are a protocol library module, a device management module and a data interface module. Protocol library module provides protocol parsing function. The device management module provides the sensing layer communication equipment status query and the related operation interface function. The database interface module provides the function of database access interface.

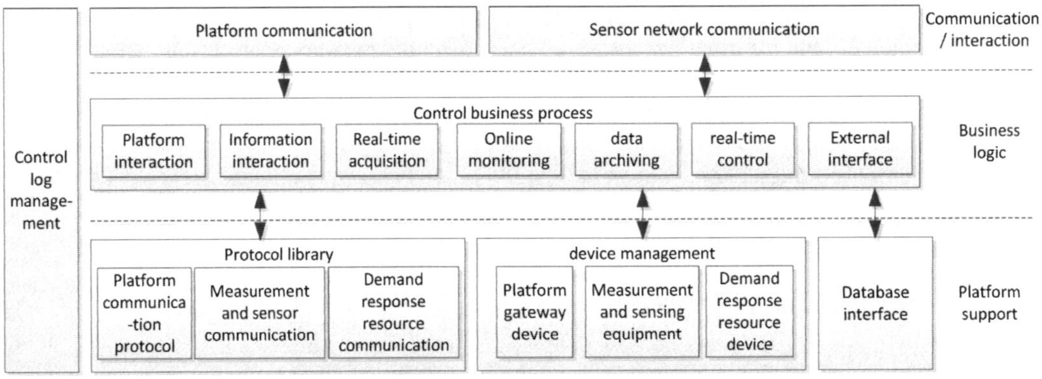

Figure 6. Demand response Control Architecture

5.3. Functional design

In order to promote the interaction between supply and demand, and realize the visualization of demand response information. The platform has the functions of the demand response policy and information release, resource management, schedulability analysis, program management, intelligent scheduling and implementation effect evaluation. Specific function structure shown in figure 7.

(1) Demand response policies and information dissemination

Demand response information platform to provide information and policies to respond to the needs of the release of information. Demand response policy information includes: legal regulations and policies, implementation policy, price policy and incentive policy; demand response event information including notification time, the implementation of demand response type, duration, response capacity etc. According to the platform to provide policy and information, the user needs to understand the dynamic response of timely information and analysis of demand in response to the incident response demand environment, selective response policies and event types, this can promote the active participation of users. At the same time, according to the feedback result of the demand response, the response information of the specific environment and the user group is adjusted, and the information is published in the demand response information platform.

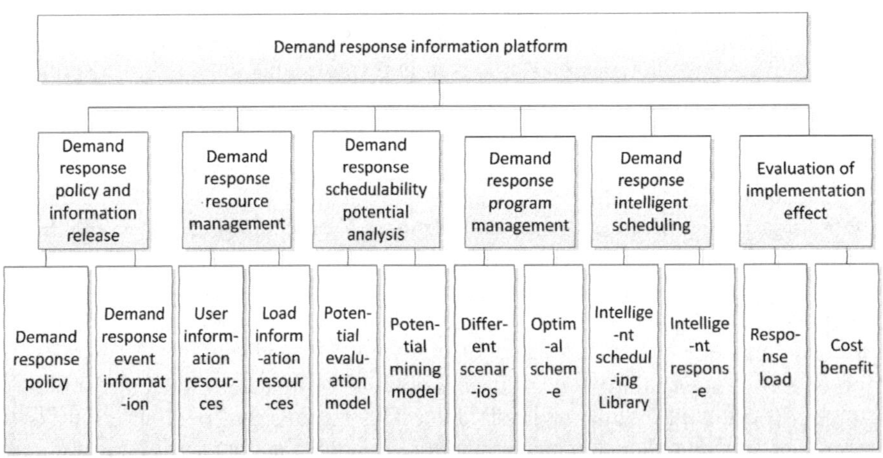

Figure 7. Functional Structure of Demand Response Information Platform

(2) Demand response resource management

The demand response information platform provides the resource archive management function, and the demand response resources are divided into two forms: demand response user information and load resources. Load resources mainly include simple adjustable load, energy storage device and distributed power supply. Control center data classification storage of user information and load information resources, to understand the needs of users in response to implementation of the use of the Internet online real-time dynamic, the use of networking means to monitor the equipment running information, and save data information for demand response scheduling potential demand analysis and response strategy use.

(3) Demand response potential analysis

Based on the model of load potential mining and the evaluation model of the load potential, the demand response information platform provides the potential function of load dispatch. Electrical operation characteristics, adjustable load has the characteristics of diversity, by means of large data mining analysis of user load data characteristics, selected measure of load characteristics index, using the relevance assessment of technical analysis tools evaluate construction load, load model and load model to assess the potential of mining potential, realize the analysis of user load potential, reasonable load scheduling.

(4) Demand response program management

Demand response information platform provides a demand response program management function, according to the traditional data, historical data and other means, relying on big data and Internet technology, built a variety of demand response operation scene and simulation test, the establishment of suitable for different application scenarios of demand response plan. And storage management program can be run by a small range of pilot, the implementation of the program and correction. Through the Internet real-time understanding of meteorological, environmental information. According to the different demand response events, we can develop a user-friendly response plan which can be selected by users.

(5) Demand response intelligent scheduling

The demand response information platform provides the demand response intelligent scheduling function, which is triggered by the demand response. And the application of user information to participate in the use of intelligent means for screening. The demand response intelligent scheduling library is established, and the response scheduling is carried out. At the same time, the data in the scheduling process is aggregated to the demand response database.

(6) Performance evaluation

The performance evaluation system is shown in figure 8. The integration of different areas of performance evaluation scheme, through screening the different evaluation index of different demand response events, select the corresponding performance evaluation scheme, combined with the use of large data analysis and evaluation results of evaluation schemes are constantly updated, to plan specific evaluation form specific environment, realize the scientific and reasonable evaluation.

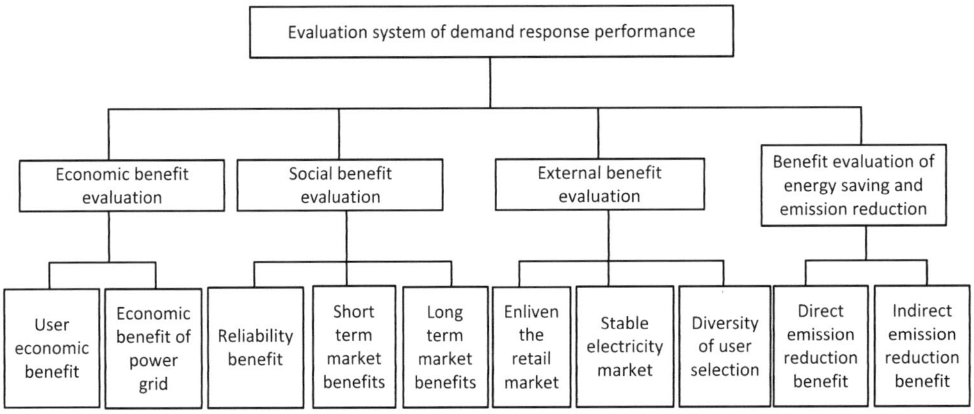

Figure 8. Implementation Effect Evaluation System of Demand Response

6. Conclusion

In order to adapt to the development trend of the Internet, to promote the development of demand response information visualization and good interaction between supply and demand side, this paper is based on "Internet plus" smart energy construction demand response information platform, the Internet, networking and big data and other methods combined, realize the interaction of supply and demand side, to provide quality services and electricity to provide platform to support fine demand response strategies for the user, can be copied, promotion.

Acknowledgments

This work was financially supported by State Grid 2016 technology guide project “Research on demand response mechanism, model and application system and multi-time scale resource scheduling potential analysis” fund.

References

[1] ZHANG Jing, WANG Ting, LI Bin. Research on Technical Standardization for Electric Demand Response[J]. Proceedings of the CSEE, 2014,22:3623-3629(in Chinese).
[2] WANG Like, YANG Shengchun, CAO Yang, YAN Mifang, YAN Chunhua, FAN Chen. A Study on Demand Response International Standard Architecture[J]. Proceedings of the CSEE, 2014,22:3601-3607(in Chinese).
[3] LIU Dongqi, WANG Liancheng, PANG Jinhai, WU Yugaung. Study on the Automated Demand Response Technology System for Smart Grid[J]. Process Automation Instrumentation, 2013,11:76-79(in Chinese).
[4] PAN Xiaohui, WANG Beibei, LI Yang. Demand response technology abroad and its practice[J]. POWER DSM, 2013,01:58-62(in Chinese).
[5] TIAN Shiming, WANG Beibei, ZHANG Jing. Key Technologies for Demand Response in Smart Grid[J]. Proceedings of the CSEE, 2014,22:3576-3589(in Chinese).
[6] CHEN Xinhui,OUYANG Sen,BAI Shan,LI Qi. Demand side response in smart grid environment and the interactive information platform construction in Guangzhou[J]. POWER DSM, 2014, 03:36-40+44(in Chinese).
[7] YAN Huaguang, CHEN Songsong, ZHONG Ming, JIANG Limin. Research and Design of Demand Side Energy Efficiency Management and Demand Response System[J]. Power System Technology, 2015,01:42-47(in Chinese).
[8] Shen Minxuan, Sun Yi, Li Bin. Study on The Architecture and Application of The Automatic Demand Response System Based on the Technology[J]. Electrical Measurement & Instrumentation, 2014,24:96-100(in Chinese).
[9] Yan Chunhua, Cao Yang,, Tian Wei. Design of demand response management system [J]. [9],

2013, S1:239-242.

[10] FU Wei, JING Zhanghao, LUO Zhiyong, HUANG Xiumou, WU Youyi. A Time-of-Use Pricing Based Control Scheme for Intelligent Household Appliances[J]. Power System Technology, 2015,03:717-723(in Chinese).

[11] CAO Jia, MA Hongyan, LIU Yang, YAN Zheng, LIU Fubin, YANG Libing. Research on Demand Response Strategy Based on Nodal Price[J]. Power System Technology, 2016, 05:1536-1542(in Chinese).

[12] DING Yi, JIA Heping, SONG Yonghua, LI Yaping, JIN Wende. Review of Reliability Evaluation Methods for the Smart Grid Considering the Interaction Between Wind Power and Flexible Demand Resources[J]. Proceedings of the CSEE, 2016,06:1517-1526(in Chinese).

[13] ZHAO Huiying, LIU Guangyi, JIA Hongjie, et al. Analysis of Demand Response Program Based on Refined Models[J]. Power System Protection and Control, 2014,42(1):62-68(in Chinese).

[14] PAN Zhanghui,GAO Ciwei. Economic Dispatch of Electric Vehicles Mohammad Shahriar Rahman, Anirban Basu, Shinsaku Kiyomoto. Privacy-friendly Secure Bidding Scheme for Demand Response in Smart Grid[C]//IEEE First International Smart Cities Conference.

[15] CHEN C,KISHORE S,SNYDER L.An innovative RTP-based residential power scheduling scheme for smart grids[C]. Acoustics, Speech and Signal pow-Processing (ICASSP), 2011 IEEE International Conference 2011:5956-5959.

[16] MOHSENIAN-RAD A,LEON-GARCIAA. Optimal residential load control with price prediction in real-time electricity pricing environments[J].Smart Grid,IEEE Transactions on,2010,1(2):120-133.

[17] GAO Ciwei, LI Qianyu, LI Huixing, ZHAI Haibao, ZHANG Liang. Methodology and Operation Mechanism of Demand Response Resources Integration Based on Load Aggregator[J]. Automation of Electric Power Systems, 2013, 17:78-86(in Chinese).

[18] GAO Ciwei, LIANG Tiantian, LI Yang. A Survey on Theory and Practice of Automated Demand Response[J]. Power System Technology, 2014,38(2):352-359(in Chinese).

[19] HUANG Li, WEI Zhinong, WEI Yanfang, SUN Guoqiang, SUN Yonghui, LIU Jinsong. A Survey on Interactive System and Operation Patterns of Intelligent Power Utilization[J]. Power System Technology, 2013,37(8):42-47(in Chinese).)

[20] XIA Ye, KANG Chongqing, NING Bo, DING Maosheng. Generation and Load Integrated Sastic Security Checking on Interaction Mode on Customer Side[J]. Automation of Electric Power Systems, 2012,36(9):35-41(in Chinese).

[21] LIN Hongyu, ZHANG Jing, XU Kunpeng, PI Xuejun. Design of Interactive Service Platform for Smart Power Consumption[J]. Power System Technology, 2012,36(7):255-259(in Chinese).)

IOP Conf. Series: Earth and Environmental Science **64** (2017) 012088 doi:10.1088/1755-1315/64/1/012088

A quick method to estimate low voltage problem

Yuqing He[1,*], Hongbin He[2], Cong Liu[1], Zhuohan Jiang[1] and Bo Liu[1]

[1]State Grid Hunan Electric Power Corporation Economical & Technical Research Institute, Changsha, China
[2]State Grid Hunan Electric Power Corporation, Changsha, China

*Corresponding author e-mail: iheyuqing@163.com

Abstract. In order to solve the problem in the prediction of low voltage problem in distribution network, a method of estimating low voltage problem is proposed from two aspects: network simplification and load simplification. In the basis of the difference construction of the backbone and branch line, a backbone-branch network simplified model is proposed, and also the large input parameters problem is solved through the parameter estimation. In the basis of the division of the trunk that a branch load model is structured to realize a rapid distribution of the load. And finally, by using the voltage droptheoretical, a simple and practical low voltage loss quick check is formed to make it easy for grassroots staffs to use.

1. Introduction

Under the dual action of the continually improvement of national economy and the strong support for agriculture, the rural economy has developed rapidly and the living standard of the peasants has been significantly improved. However, the contradiction of incompatible between the development of power and the economic development especially for the problem of low voltage becomes more and more prominent, which has seriously impact on the normal production and consumption of rural areas and restricted the further development of rural power. Therefore, a scientific, reasonable and practical low voltage problem quantification method needs urgently to be studied and promoted to solve the problem of low voltage effectively.

The load moment theory is widely used in low voltage problem prediction. In literature [1], it proposes method of the rapid determination of low-voltage lines in zone area based on the load moment theory in accordance with the actual situation of production and related design rules and regulations. In literature [2], the load moment is used to analyse the intrinsic relationship between the line load moment and the cross section of the wire, and the different diameters are normalized to determine the optimal section for the wire diameter replacement inzone area. Literature [3] studies about the selection problem of calculating coefficient related to voltage of power supply system and wire material happening at the time of load moment calculation. Literature [4] puts forward the concept of equivalent load moment and tests the rationality of the concept according to the actual project.

The judging method of low voltage problem based on the load moment to a certain degree solves the problem of large demand for data based on the voltage loss data completely calculated but still can't solve the following problems: 1) The load moment method dose not fundamentally reduce its requirement for the line account, load and other data which is actually difficult to obtain. 2) The load moment method has a complete theoretical logic that requires high about the user's theoretical quality in practical use, which brings only wake effect in rural areas with a salient problem of low voltage. 3) Professional software and computers and other equipment are needed in the use of the load moment method, and this brings an inconvenience for grass-roots staffs to use in the field survey and review.

In order to solve the problem in the prediction of low voltage problem in distribution network, a method of estimating low voltage problem is proposed from two aspects: network simplification and load simplification. In the basis of the difference construction of the backbone and branch line, a backbone-branch network simplified model is proposed, and also the large input parameters problem is solved through the parameter estimation. In the basis of the division of the trunk that a branch load model is constructed to realize a rapid distribution of the load. And finally, by using the voltage drop theoretical, a simple and practical low voltage loss quick check is formed to make it easy for grassroots staffs to use.

2. Backbone-Branch Network Model

2.1. Model construction

The load (line current) of any branch of the medium and low voltage network in the non-distributed power is determined by its own branch load and has no relationship with any other branch, and the voltage drop of one branch is only decided by its head voltage and electrical characteristics. While, the voltage drop of the trunk line in the network is affected by load of all branches. In the basis of characters stated above, the backbone-branch network model shall be constructed at the moment of conducting medium-low voltage analysis in the medium and low voltage networks.

The method for constructing backbone-branch network model as below: simplify the original distribution network into a network that is constructed by the branch element which composed by each branch line and the trunk component composed by trunk line.

According to above analysis, any medium and low voltage network can be simplified.

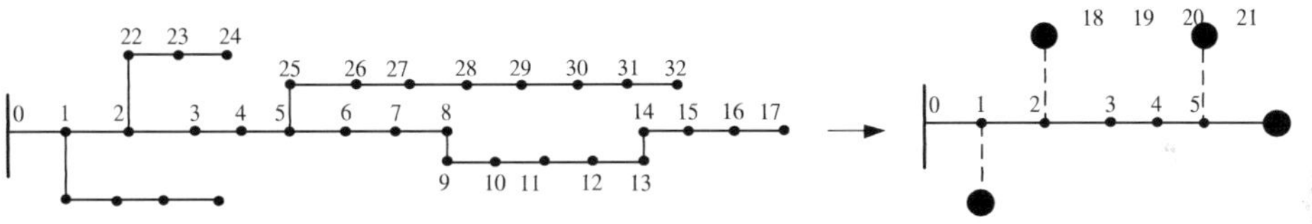

Figure.1 Simplified network of the trunk-branch model

2.2. Parameter estimation

Main parameters of trunk-branch network model's parameter estimation includes: trunk line length, trunk line diameter, branch line length and branch line diameter.

Estimation of the length of trunk line means to estimate the length of the trunk line, which can be completed more accurately by checking the accounting data or by the rough estimation from the grassroots operating personnel experience.

Estimation of the trunk line diameter means to estimate its average diameter, which can be obtained by checking the accounting data and calculating the average diameter or be gotten from grass-roots staff's estimation of the length of kinds of line diameter and calculations on them.

Estimation of the length of branch line means to estimate the length of the branch line, which can be obtained more accurately by checking the accounting data or by the rough estimation from the grass-roots operating personnel experience.

Estimation of the branch line diameter means to estimate the average diameter of it, which can be obtained by checking the accounting data and calculating the average diameter or be gotten from grass-roots staff's estimation of the length of kinds of line diameter and calculations on them.

IOP Conf. Series: Earth and Environmental Science **64** (2017) 012088 doi:10.1088/1755-1315/64/1/012088

3. Branch Load Model

3.1. Category of the medium branch load

According to the different types of power supply of the medium-voltage lines, medium voltage lines are divided into two categories as urban and rural.

According to the characteristics of the power load, the urban areas subdivided into three categories as industrial type, residential type and commercial type, among which the industrial type means urban area power supply mainly with industrial electricity load, the residential type means urban area power supply mainly with residential electricity load and the commercial type means urban area power supply mainly with commercial electricity load.

The rural areas according to economic development and production and living standards are divided into three categories as class I village group, class II village group, and class III village group, among which class village group means generally town, township government seat, urban and rural areas or relying type economic development zone, the class II means generally small processing industry or more developed areas of agricultural production and the class III means all apart areas of rural basic production and living.

3.2. Category of the low branch load

According to the different types of power supply, the low voltage branch load is divided into three categories as life class, production category and commercial category, among which the life class electricity supply load means mainly about residents living electricity load, the production one means mainly about electricity load of small processing industry, aquaculture, irrigation and small mining (digging) and the commercial one means mainly about commercial electricity load.

3.3. Load distribution coefficient

Load district but ioncoefficientrefers to the proportion of each branch loading the line or zone area, the branch load distribution coefficient has the following characteristics:

$$\begin{cases} 0 \le \sigma_B \le 1 \\ \sum \sigma_B = 1 \end{cases} \tag{1}$$

Accurate load distribution coefficient must be collected and calculated through load measurement on the same time, but actually the measurement is not necessary because it is difficult to be realized under the situation of many branches and low degree automation or to reflect the line load distribution

Characteristics over a period of time. This paper is estimating load distribution coefficient σ_{S_i} of each branch through methods stated below:

$$\sigma_{S_i} = \bar{\mu}_{S_i} \bullet t_{S_i} \bullet \frac{P}{T_{S_i}} \tag{2}$$

In this formula, t_{S_i} is the estimated capacity of the branch S_i, T_{S_i} is the total capacity of the line (or zone), P is the active load for the line (or zone), $\bar{\mu}_{S_i}$ equals to the normalized load category factor which is calculated by the load factor of all branches:

$$\bar{\mu}_{S_i} = \frac{\mu_{S_i}}{\sum \mu_{S_i}} \tag{3}$$

IOP Conf. Series: Earth and Environmental Science **64** (2017) 012088 doi:10.1088/1755-1315/64/1/012088

In this formula, the load classification factor is the load characteristics of different load categories. The larger the load factor, the greater the corresponding load, otherwise vice versa. The load factor is not static but changed along with the actual situation of each branch and can be estimated through the experience of grass-roots operators.

3.4. Load distribution strategy

The load $P_B$at the maximum load time point of each branch can be obtained by calculating the maximum load $P_{max}$of the line (or zone area) and the load distribution coefficient of the branch, which is:

$$P_B = \sigma_B \bullet P_{\max} \tag{4}$$

4. Analysis process

In the basis of the abovementioned trunk-branch simplification network model and branch load model, the practical method of rapid estimation of low voltage problem is constructed as below:

1) Construct the trunk-branch network simplified model of a line (or zone) according to the situation of the backbone and branch line;
2) Estimate the main line length, main line diameter, branch line length, branch line diameter and other parameters;
3) Determine the load category of each branch and the load category factors;
4) Calculate the load distribution coefficient of each branch;
5) Estimate the numbers of the users of the low voltage according to the quick check manual of voltage drop;
6) Calculate the quantitative impact of the project on the low voltage problem.

5. Typical examples

For the sake of verifying the validity of the method proposed in this paper, the zone area of Shizishan village will be analysed. By parallel comparing and analysing the voltage loss of Power System Calculation and Drawing Software Package developed by Zhengzhou Defang Company, the model is calculated.

(1) Network simplification

According to the method proposed in this paper, backbone-branch network simplified model is firstly constructed, comparing to backbone-branch network simplified model of the zone area as shown in Figure2.

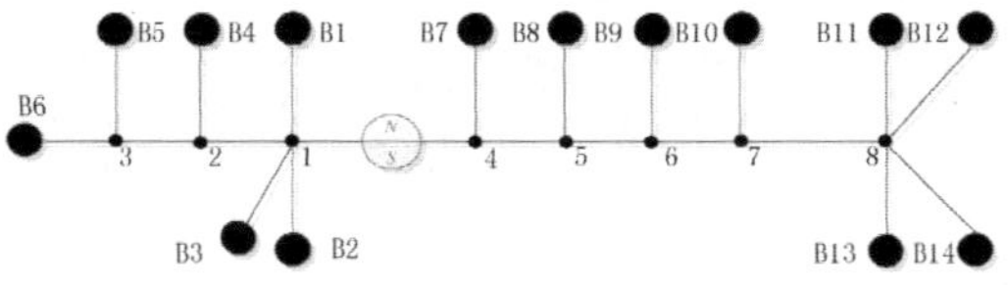

Figure.2 Backbone-branch network

(2) Parameter estimation

It is to use the experience of the grassroots operating personnel to estimate the length of the trunk line, the main line diameter, branch line length, branch line diameter and other parameters. As can be seen from the data in Table1, there is no deviation between the estimated data of the branch line diameter and the actual data. As for branch line length, the maximum and minimum deviation is 23 meters and 0 meter and the maximum error rate is 8.3%. The total error is 57 meters and the total error rate is only 1.6%, which means a small range of error between the actual data and estimated data.

According to the model proposed in this paper, the load category and load size of each branch is determined. Table1shows the load category and load size of each branch.

Table.1 Load category and load size of each branch

Branch	Power supply type	Maximum load	Branch	Power supply type	Maximum load
B1	Commercial type	5.5	B8	Commercial type	24.6
B2	Commercial type	5.5	B9	Commercial type	10.9
B3	Commercial type	16.4	B10	Residential type	3.3
B4	Residential type	14.7	B11	Industrial type	7.1
B5	Residential type	14.7	B12	Industrial type	46.1
B6	Residential type	3.3	B13	Industrial type	28.4
B7	Commercial type	5.5	B14	Industrial type	14.2

(3) Deviation analyze

In Table 2, it states the comparison between the proposed method and the voltage loss calculation method (VLC method). According to the data in the table, there are two branches have deviation after separately calculating by the proposed method and by VLC method, besides, other branches' results have consistent results. Among which, the maximum error of branch 12 is two users who were found to truly have a low voltage situation at the peak load period after communicating with these two users. The main reason for miscalculation of voltage loss calculation method is its analysis only on single load state while the method proposed in this paper has overcome this limitation.

Table.2 Comparison between the proposed method and VLC methods

Branch	Proposed method	VLC method	Branch	Proposed method	VLC method
B1	0	0	B8	0	0
B2	0	0	B9	0	0
B3	0	0	B10	0	0
B4	0	0	B11	0	0
B5	5	4	B12	8	6
B6	1	1	B13	0	0
B7	0	0	B14	4	4
Total	18	15			

6. Conclusion

In order to solve the problems in the prediction of low voltage problem in distribution network, a method of estimating low voltage problem is proposed from two aspects: network simplification and load simplification. In the basis of the difference construction of the backbone and branch line, a backbone-branch network simplified model is proposed, and also the large input parameters problem is solved through the parameter estimation. In the basis of the division of the trunk that a branch load model is structured to realize a rapid distribution of the load. And finally, by using the theoretical analysis of voltage drop, a simple and practical low voltage loss quick check is formed to make it easy for grassroots staffs to use.

Acknowledgments

This work was financially supported by State grid science and technology projects 5216A016000K and 5216A2160007.

References

[1] Fengchao, Qianjing, Huang Hong sheng. Distribution line planning and distribution transformer location optimization based on load moment [J]. Zhejiang electric power, 2015, 02:5-8.

[2] Zhengpeixiang, Xuanjuqin, Wang kai. Government of low voltage area based on normalized load moment for distribution networks [J]. Journal of electric power science and technology, 2015, 02:80-86.

[3] Baolifang, Li wen. Discussion on calculation coefficient in calculation of voltage loss via load moment [J]. Building electricity, 2012, 04:31-33.
[4] Li binghua, Yueyuntao, Wang likun. Research and application of equivalent load moment [J]. Building electricity, 2012, 12:3-9.

Overview of HVRT technical measures for DFIG

Haiyang Yu

Heilongjiang Electric Power Research Institute, Harbin 150030, China

Abstract. A With the rapid increase of the proportion of double-fed induction wind turbines generator (DFIG), higher requirements are proposed for the continuous operation ability under external power grid failure. The paper first analysed the transient process of DFIG in cases of power grid voltage surge, proposed the related standard requirements for high voltage ride-through ability of several countries, including China, and then summarized and evaluated the various high voltage ride-through technical measures suitable for DFIG, and finally forecasted the high voltage ride-through technology development trend.

1. Introduction

Northwest China Grid suffered large-scale trip accident of 598 turbines in 2011 which caused a loss of 837.34MW wind power, and made the whole grid frequency of Northwest China Grid cut down to 49.854Hz [1]. As for the causes of the accident, firstly, the turbines did not have the ability of low voltage ride-through; secondly, after some turbines snapped off, the excess reactive power made system voltage rise, but the turbines did not have the ability of high voltage ride-through. As a result, 274 turbines snapped off because of low voltage network, and 324 turbines snapped off because they failed to tolerate high voltage when the voltage in the machine rose to 110% of the rated voltage in the accident. Through the analysis of the large-scale turbine accident, great importance must be attached to the high voltage ride-through ability of wind turbine generators.

2. Transient process analysis of DFIG

In order to study the high voltage ride-through characteristics of DFIG, it was very necessary to analyze the transient response process of DFIG when power grid failure caused voltage surge.

It was assumed that power grid voltage had symmetry surge at t_0, DFIG stator voltage was [2]; when $t < t_0$, $u_s = V_{se} e^{j\omega_s t}$; When $t > t_0$, $u_s = V_{se} e^{j\omega_s t} + pV_{se} e^{j\omega_s t}$.

Where $p = (u_s - V_{se})/V_{se}$ is the size of power grid voltage surge.

The stator voltage and flux linkage equations are as follows:

$$\begin{cases} u_s = R_s i_s + \dfrac{d\psi_s}{dt} \\ u_s = L_s i_s + L_m i_r \end{cases} \tag{1}$$

The following stator flux state equation could be obtained through equation (1):

$$\frac{d\psi_s}{dt} = u_s - \frac{R_s}{L_s}\psi_s \tag{2}$$

Then, in turn, stator flux linkage equation when the grid voltage soared could be obtained:

$$\psi_s = (1+p)\frac{V_{se}}{j\omega_s}e^{j\omega_s t} - p\frac{V_{se}}{j\omega_s}e^{j\omega_s t_0}e^{-(t-t_0)/\tau} \tag{3}$$

When grid voltage soared, the open circuit voltage corresponding to the rotor (ignoring $1/\tau$ item) was as follows:

$$u_{r0} = \frac{L_m}{L_s}[s(1+p)V_{se}e^{j\omega_s t} + (1-s)pV_{se}e^{j\omega_s t_0}e^{-(t-t_0)/\tau}] \tag{4}$$

Analysis results were as follows: when the power grid failure cause voltage soar, DC component of the stator flux linkage and the negative sequence components produced by asymmetric fault would induce the larger voltage and impulse current in the rotor circuit; if no reasonable protective measure was taken, it might damage the rotor side converter; it might also make the converter DC side bus voltage rise, and burn up DC side capacitor.

3. High voltage ride through grid specifications

There are relatively rare studies of high voltage ride-through and the running characteristics of DFIG in Chin. Because European and American countries such as the United States, Canada, Australia, Denmark and the Scottish developed wind power earlier, Europe and the United States and grid operators, including WECC of the United States, AEMC of Australia, AESO of Canada, Energinet.dk of Denmark and Scottish Power of Scotland have formulated the relevant standards of wind power high voltage ride-through ability.

WECC of the US. Requires for the ability of running for 2s each time when the power grid voltage soars to 118% of the rated voltage, the ability of operating for 1s each time when the power grid voltage soars to 120% of the rated voltage, and continuous operation when the power grid voltage soars to 110% of the rated voltage [3]. AEMC of Australia regulates the ability of running for 0.06s each time when the power grid voltage soars to 130% of the rated voltage. Besides, AESO of Canada requires all of the above 5MW wind turbines should be able to run continuously when grid-connected site voltage is 90%~110% of the rated voltage [4]. Energinet.dk of Denmark requires for the ability of operating for 0.2s each time when the power grid voltage soars to 120% of the rated voltage [5]. Scottish Power of Scotland requires for the ability of operating for 0.8s each time when the voltage soars to 120% of the rated voltage, and continuous operation when the voltage soars to 110% of the rated voltage [6].

State Grid Corporation in China also developed the corresponding standards of high voltage ride-through for wind power plants in December 2016. It requires wind turbines to keep continuous operation when the per unit value of grid-connected site power frequency voltage $U_T \leq 1.1$, the ability of operating for 10s each time when $1.1 < U_T \leq 1.15$, the ability of operating for 200ms each time when $1.15 < U_T \leq 1.2$, and full-load operation when $U_T > 1.2$ [7]. High Voltage ride through grid specifications is show in fig.1.

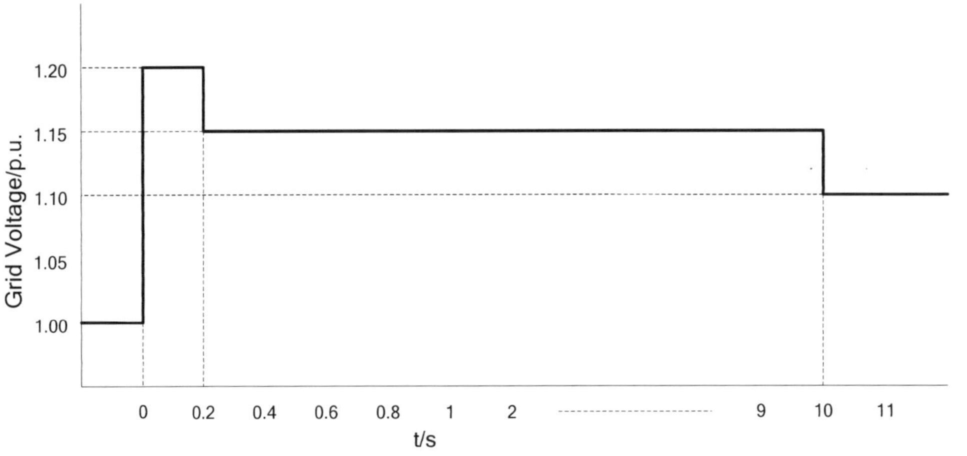

Fig.1 High voltage ride through grid specifications

4. The existing high voltage ride through technologies

Based on the analysis from the perspective of energy, when power grid failure caused voltage surge, DFIG could not conduct normal transmission to power grid. Because DFIG had large inertia, variable propeller system required a certain time to adjust, so short-term adjustment range was limited, and caused excess wind energy, while it was obvious that the wind energy could not be effectively released only through regulating ability of the DFIG itself.

Currently, there are two kinds of common methods. One is to realize high voltage ride-through ability of DFIG through the control method technique of improving rotor side converter; Another one is to increase hardware auxiliary circuit to release excess energy of the system. The section first introduced the improved control method technology without increasing hardware auxiliary circuit, and then introduced stator side and rotor side and other methods of voltage ride-through by adding hardware auxiliary circuit respectively according to different installation positions of hardware auxiliary circuit.

4.1. The control method of improving rotor side converter

Many experts and scholars at home and abroad are studying and exploring the methods of high voltage ride-through of DFIG by only improving the rotor converter and grid side converter control strategy, without increasing hardware circuit. Literature [8] proposed the compensation dosage control method of adding stator exciting current change on DFIG controller, which could make necessary corrections for decoupling circuits. Literature [9] proposed the direct dynamic current control technology that used stator terminal voltage disturbance to compensate the dynamic impact of rotor current, which could limited the dynamic oscillation of rotor current to a certain extent. Literature [10] proposed a kind of degaussing control method suitable for the oriented vector control of DFIG power grid voltage, which could reduce the oscillation of stator and rotor current. Besides, such method could effectively reduce the impact of transient DC component in stator flux linkage on DFIG when the power grid voltage surged slightly. Although the control strategies proposed in the above literatures improved rotor current control ability during power grid fault, they also increased the input voltage of rotor, and no extra energy of DFIG was released. Thus, these methods could play a role only when power grid voltage surged within small range.

Literature [11] proposed the compensation technology based on transient state flux linkage of DFIG that introduced same frequency voltage in rotor terminal voltage. The technology could eliminate the induction electric potential of DC component and negative sequence components of stator flux linkage in the rotor circuit when power grid voltage surged, and then control the rotor loop current, but the method had weak compensating control ability. Literature [12] improved the method.

Crowbar was input when voltage surged instantaneously; rotor side converter began operating when rotor side current attenuated to the allowable scope; the degaussing current made flux linkage DC component attenuate rapidly, and sent out a certain amount of reactive power to the grid at the same time. In such method, Crowbar action time was only 11.16ms, which could guarantee rotor side converter restore control timely, thereby helping power grid voltage recover, and achieve high voltage ride-through of DFIG.

Literature [13] proposed the rotor side current phase angle compensation control technology. The technology could solve the problem of possible damage of the converter caused by rotor over-current during the grid voltage recovery period after Crowbar exited. Such technology aimed at symmetrical fault of power grid, but it ignored the influence of the transient process of transition of stator flux linkage when voltage soared. In fact, there are more asymmetric faults in practice, so the method application condition is limited.

Literature [14] proposed the method in which the asymmetric system was decomposed into positive and negative sequence component, and then positive and negative sequence current double d-q control was conducted on their respective d-q axis decoupling control. There was bigger error in thee negative sequence current extracted by the method; the application of wave trapper in positive and negative sequence decomposition caused the breakdown time delay, so the dynamic performance of the converter in the process of recovery was affected. Literature [15] designed a kind of unbalanced control method of DFIG, and put forward new harmonic current control technology. The technology simplified the positive and negative sequence current instruction value algorithm of the rotor under the various enhanced control ability control target through positive sequence d+ shaft power grid voltage orientation. Besides, it didn't need unified regulation on the current, negative sequence decomposition of rotor side, and strengthened the continuous operation ability of DFIG under the asymmetric fault.

Under larger range of grid voltage surge, DFIG machine terminal voltage fluctuations will cause sharp oscillation in flux linkage, so the reactive power control technology is also very necessary for realizing the high voltage ride-through of DFIG. Literature [16] proposed the new reactive power optimization control technology that could make the some excitation control of rotor side converter shift to grid side converter. Such technology increased the capacity of the grid side converter, could improve the power factor and voltage stability, effectively reduced DFIG loss, and enhanced the high voltage ride-through ability of DFIG.

4.2. Stator side method

1. Series dynamic braking resistor (SDBR) between stator and power grid

During normal operation, R_{SDBR} doesn't work when bypass switch is closed. When power grid failure makes voltage soar, R_{SDBR} can accelerate the stator flux linkage attenuation, and then accelerate the attenuation of voltage and current induced by rotor side, and protects DFIG. However, series R_{SDBR} operation at stator side will make the machine side voltage of DFIG increase and resistance loss increase. Therefore, the method is only applicable to the small grid voltage gain. Its topology structure is show in fig.2.

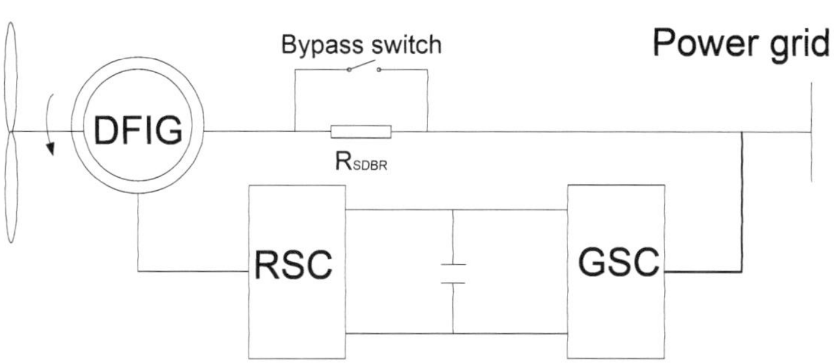

Fig.2 Series dynamic braking resistor between stator and power grid

IOP Conf. Series: Earth and Environmental Science **64** (2017) 012089 doi:10.1088/1755-1315/64/1/012089

2. Series dynamic voltage restorer (DVR) [17]

During normal operation, DVR doesn't work. When power grid failure makes voltage soar, DVR guarantees the voltage stability of the DFIG machine side by making series transformer compensate corresponding fault phase voltage. The series DVR can ride through any fault type. Besides, in the whole process of fault, rotor side and grid side converter can provide DFIG with excitation, but it needs additional voltage type inverter and series transformer, which increase the hardware cost of DFIG, and is more difficult to implement. Its topology structure is show in fig.3.

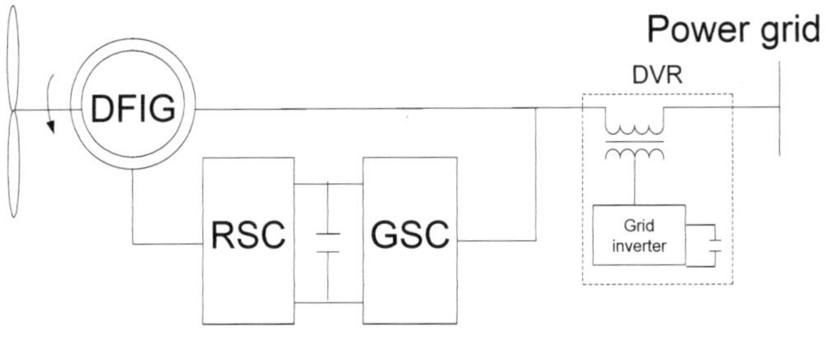

Fig.3 Series dynamic voltage restorer

4.3. Rotor side method

1. Install crowbar circuit

For DFIG, the most commonly-used method is to increase protection circuit (such as a Crowbar circuit) in rotor side. The choice of reasonable crowbar resistance, the control strategies and switch can improve the high voltage ride-through ability of DFIG, and reduce the bad influence of wind power plant on system voltage stability after inputting Crowbar circuit in the wide range input. The method is simple, effective, easy to implement, and economical. However, when Crowbar circuit inputs, DFIG is at the uncontrolled state of induction motor. During grid voltage recovery period, DFIG will absorb large amounts of reactive power from the power grid, so it is bad for power grid voltage recovery. Besides, the actual effect after the installation is quite very dependent on operating conditions and fault characteristics, and had very limited role of asymmetric fault. Its topology structure is show in fig.4.

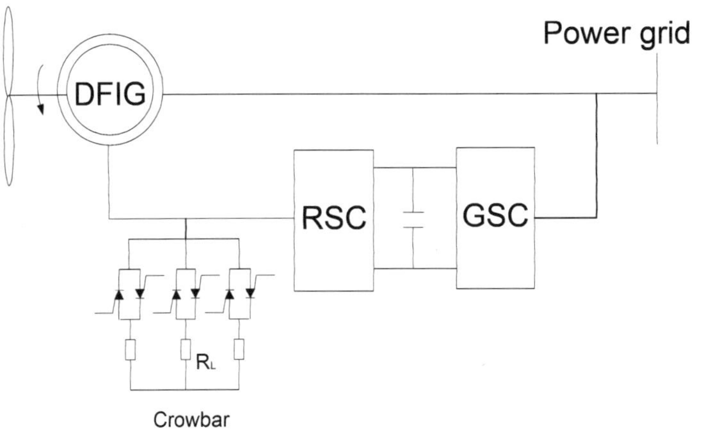

Fig.4 Install crowbar circuit

2. Install improved Crowbar circuit (SDBR + Crowbar) [18]

During normal operation, RSDBR doesn't work. When grid voltage surge causes rotor side over-flow, R_{SDBR} is input, when R_{SDBR} limits surge current and accelerate rotor current attenuation. When voltage surge makes rotor current be beyond the adjusting range of R_{SDBR}, Crowbar bypass the rotor side converter was input. The combined action of R_{SDBR} and R_L can make the rotor over current decay faster; when current attenuates to the allowable scope of rotor side converter and rating value Crowbar and R_{SDBR} exited respectively. The improved control technology reduces the input frequency and time of Crowbar, but it increases the hardware input and complexity of control. Its topology structure is show in fig.5.

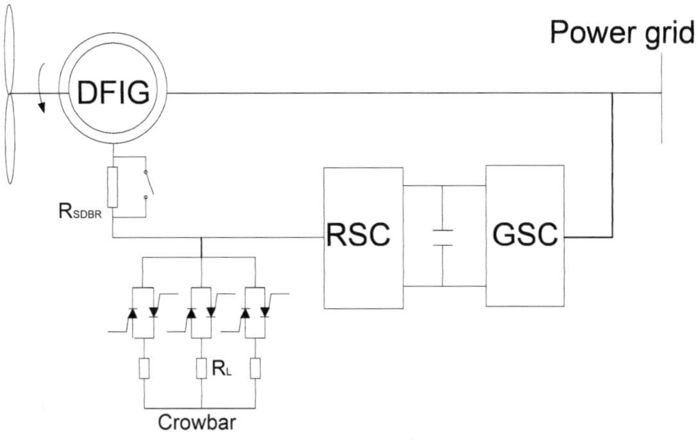

Fig.5 Install improved Crowbar circuit (SDBR + Crowbar)

3. DC side series chopped wave energy dissipation device (Chopper circuit) [19]

When grid voltage surges, rotor side converter over-current limits grid side converter power output, which will make DC side voltage rise, and possibly cause damage of DC side capacitor and power device. In order to solve the problem, a chopper circuit can be added in DC side; when grid voltage surges, unloading load is input to consume excess energy, and maintain a constant DC voltage. Its topology structure is show in fig.6.

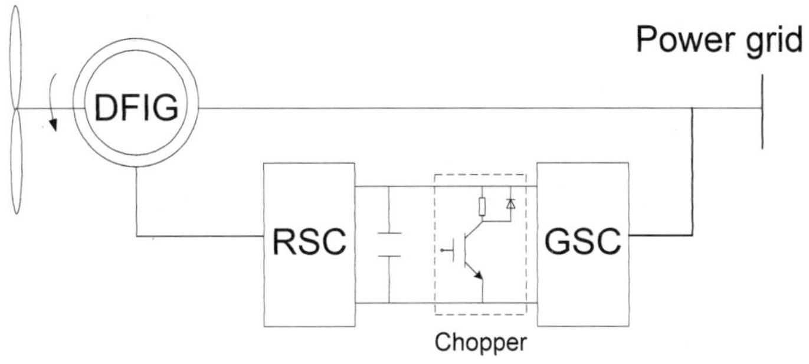

Fig.6 DC side series chopped wave energy dissipation device

4. DC side series storage protection circuit (ESS) [20]

Series storage protection circuit can store the excess energy during voltage surge, and send the stored energy to power grid at the end of the fault. The advantages of the method is as follows: without switching under different running conditions, it can avoid transient process caused by switch, and conduct continuous control on the system. The disadvantage is as follows: ESS can not effectively control rotor side current; only by increasing the capacity of rotor side converter, can the rotor be ensured no damage due to over current of converter. Increasing hardware and improving control strategy increase the total cost, but it also creates better conditions for the realization of high voltage ride-through. Its topology structure is show in fig.7.

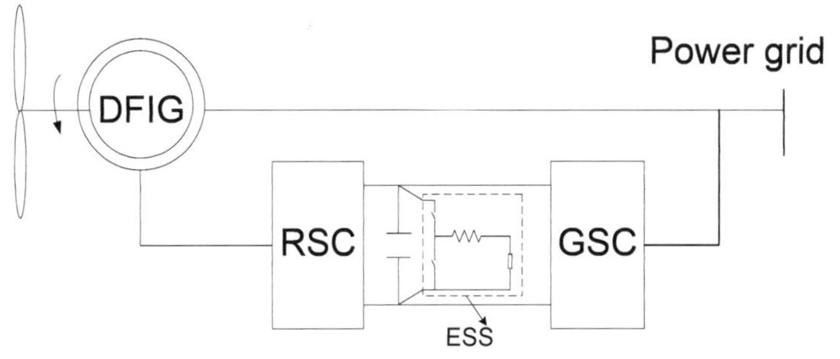

Fig.7 DC side series storage protection circuit

5. Development direction of high voltage ride-through technology

After analyzing the technologies of improving DFIG high voltage ride-through, and by combining with the present development situation of wind power technology in China, it is expected the technology will revolve around the following several aspects of research in the future:

(1) State Grid Corporation has made enterprise standard in view of the high voltage ride-through ability of wind power plants, while national standard has not yet introduced. Therefore, at present, it is necessary to implement the relevant requirements of the standards in terms of manufacturing and wind power operating as soon as possible; especially the regions with weak power grid in China should perform high voltage through standard as soon as possible to ensure the safe operation of power grid.

(2) To achieve the accuracy of the control strategy and the effectiveness of the protection device, it is necessary to improve the accuracy of transient mathematical model of DFIG and excitation frequency converter. How to build the transient mathematical model containing protective device (such as Crowbar, SDBR + Crowbar, Chopper and ESS) will be the research focus of DFIG high voltage ride-through technology.

(3) Accurate judgment of power grid failure and rapid detection technology are the foundation of achieving DFIG high voltage ride-through operation control, so it is one of research directions of realizing DFIG fault crossing ability more effectively.

(4) The above-mentioned protection devices have their own advantages and disadvantages. How to make full use of them and make them complement each other to achieve very good effect of fault ride-through will be the focus of future research. There are fewer literature researches currently.

6. Conclusion

At present, most of the wind power plant machine model in China use DFIG as the main body, so it is particularly important to improve the high voltage ride-through ability of DFIG under power grid failure. The paper systematically summed up and evaluated domestic and overseas related standards, the latest technology and research results in terms of DFIG high voltage ride-through ability, and forecasted the development direction of high voltage ride-through technology, and provided reference for the subsequent research work and engineering applications.

References

[1] SUN Huadong, ZHANG Zhenyu, LIN Weifang, TANG Yong, LUO Xuzhi, WANG Ansi. Analysis on Serious Wind Turbine Generators Tripping Accident in Northwest China Power Grid in 2011 and Its Lessons[J], Power System Technology,2012,36(10).

[2] QU Tingyu. Research on High Voltage Ride Through of DFIG for wind power generation [D].Hefei, Hefei University of Technology,2012.

[3] Aisikaer, ZHU Yongli, TANG Binwei. Summarizing for fault ride through characteristics of wind turbines[J]. Power System Protection and Control,2013,41(19): 147-152.
[4] XIE Zhen, ZHANG Xing, YANG Shuying, SONG Haihua, QU Tingyu, SHI Quanli. An Improved Control Strategy for Doubly Fed Induction Wind Generators Under Unbalanced Grid Voltage Swell [D]. Proceedings of the CSEE,2012,36(3): 39-45.
[5] LIU Xuejing, ZHU Dan, SONG Fei, WANG Tong, MIAO Ya. Feasibility analysis on high voltage ride-through of wind turbines [J]. Renewable Energy Resources,2013,31(11): 34-38.
[6] Xie Zhen, Zhang Xing, Song Haihua, Qu Tinyu, Yang Shuying. High Voltage Ride-through Control Strategy of a Doubly Fed Induction Generator Wind Turbine Based On Variable Damping. CSEE power system automation professional committee of three session of a conference and academic exchange in 2011,2011(6).
[7] State grid corporation of enterprise standard Q/GDW 1392-2015《Technical rule for connecting wind farm into power grid》.
[8] Jiaqi Liang, Wei Qiao, Ronald G Harley. "Direct Transient Control of Wind Turbine Driven DFIG for Low Voltage Ride-Through", Power electronics and machines in wind application,2009,pp.1-7.
[9] Ling Peng, Bruno Francols, Yongdong Li. "Improved crowbar control strategy of DFIG based wind turbines for grid fault ride-through" ,Applied Power Electronics Conference and Exposition,2009.APEC 2009.Twenty-Fourth Annual IEEE.2009,2.pp.1932-1938.
[10] Nian Heng, Cheng Peng, Ziqiang Zhu. Optimized Control Strategy of Rotor Current for Doubly Fed Induction Generators During Symmetrical Voltage Fault[J]. Transactions of China Electrotechnical Society,2014,29(7): 200-208.
[11] Graham Pannell, David J.Atkinson, Bashar Zahawi. "Minimum-Threshold Crowbar for a Fault-Ride-Through Grid-Code-Compliant DFIG Wind Turbine",IEEE Transactions on Energy Conversion,Volume 25,Issue 3,2010Page(s):750-759.
[12] Francisco K.A.Lima,Alvaro Luna,Pedro Rodriguez. "Rotor voltage dynamics in the doubly fed induction generator during grid faults",IEEE trans on power electronics.vol.25.NO.1.2010.1.
[13] HU Jiabing, HE Yikang, GUO Xiaoming, NIAN Heng. Modeling and Control of the DFIG Based Wind-power Generation System Under Unbalanced Grid Voltage Conditions. AUTOMATION OF ELECTRIC POWER SYSTEMS,2007,31(14):47-56.
[14] Hu Jiabing, He Yikang, Nian Heng. "Enhanced control of DFIG used back-to-back PWM voltage-source converter under unbalanced grid voltage conditions",Journal of Zhejiang University:Science A,2007,8(8):1330-1339.
[15] WANG Junrui, ZHONG Yanru, SONG Weizhang, YANG Bo. Proportional Resonant Control for Two Stage Matrix Converter Excited Doubly-fed Wind Power Generation System. High Voltage Engineering,2013,39(5):1210-1217.
[16] JIA Junchuan, LIU Jin, ZHANG Yigong. Novel Reactive Power Optimization Control Strategy for Doubly Fed Induction Wind Power Generation System [J], PROCEEDINGS OF THE CHINESE SOCIETY FOR ELECTRICAL ENGINEERING,2010,30(30):87-93.
[17] XU Jianbing, JIANG Quanyuan, SHI Qingjun. Coordinated Control of Voltage Ride Through for DFIG Wind Turbine Systems Using Energy-based DVR [J], Automation of Electric Power Systems,2013,37(4):14-20.
[18] Chang Jing, Huang Jiadong, Bai Kai, Song Peng, Wang Jiageng. Discussion of Doubly-fed Wind Turbines High Voltage Ride-through Solutions [J]. North China Electric Power,2014(6): 33-37.
[19] Omar, Abdel-Baqi, Adel Nasiri. "A dynamic LVRT solution for doubly fed induction generators",IEEE Transactions on Power Electronics. 2010,25(1): 193—196.
[20] Xiangwu Yan, Giri Venkataramanan, Yang wang, Qing Dong, Bo zhang. "Grid—Fault Tolerant Operation of DFIG wind Turbine Generator Using a Passive Resistance Network", IEEE Transactions on Power Electronics. Volume PP,Issue 99,2010 Page(s)1.

Research on improvement of power quality of Micro - grid based on SVG pulse load

Chuang Lv, Pu Xie

Department of Vehicle and Electrical Engineering, Ordnance Engineering College, Shijiazhuang 050003, China

Abstract. Pulse load will make the micro-grid public bus power to produce a high peak pulse due to its cyclical pulsation characteristics, , and make the micro-grid voltage fluctuations, frequency fluctuations, voltage and current distortion, power factor reduction and other adverse effects. In order to suppress the adverse effects of the pulse load on the microgrid and improve the power quality of the microgrid, this paper established the SVG simulation model in Matlab / Simulink environment, the superiority of SVG is verified by comparing the improvement of power quality before and after adding the SVG to microgrid system. The results show that the SVG model can suppress the adverse effects effectively of the pulse load on the microgrid, which is of great value and significance to the reactive power compensation and harmonic suppression of the microgrid.

1. Introduction

Military micro-grid is based on photovoltaic, diesel generator, energy storage systems in one of the power supply system. As a weapon and equipment of the military micro-grid, energy protection, it is essential to our military field conditions and protect the power quality power of the supply equipment. With the development of power electronics technology, some special loads, especially pulsed loads, such as phased array radar, electromagnetic track launch weapons, pulse generators, have been widely used in various fields. The Pulse load's pulse cycle is usually tens or hundreds of milliseconds, power consumption is not stable and present typically pulse characteristics, with the average power is low and the peak power of the characteristics of large. The pulse load is divided into AC pulse load and DC pulse load. This paper focuses on the DC pulse load. In literature [1] three kinds of AC pulse load models are established and their influence on diesel generator sets is studied. Impulsive load can cause military micro-grid voltage fluctuations, frequency fluctuations and harmonic distortion and cause serious power quality problems. The existence of reactive power and harmonics poses a serious threat to the stability and safe operation of the power supply system. Therefore, it is really important to realize the real-time fast and accurate compensation of reactive power.

As a new type of reactive power compensation device, the static reactive power generator (SVG) has the advantages of fast response speed, good compensation ability and good reliability, and has become the development trend of reactive power compensation.

At home and abroad in the micro-grid system compensation control research, In the literature [2] proposed the use of active filter as a limited capacity system filtering means to suppress the pulse-type harmonics on the power supply system and the literature [3] proposed a power factor correction technique based on single cycle control to improve the power factor of the system and to suppress the pollution of the harmonics;[4] designed an improved power allocation method based on the traditional hybrid capacitor system composed of supercapacitor and battery to improve the effectiveness of the fluctuation power compensation; In [5], the droop controller in the microgrid is improved to ensure that the amplitude and frequency of the microgrid voltage fluctuate within the allowable range; and [6] through the introduction of virtual synchronous generator technology to suppress the frequency of fluctuations. However, the above documents are for a micro-grid power quality problems in a certain aspect of the solution, not through a compensation device or control strategy to solve the power quality of multiple aspects of the problem; and the load for the load is mostly stable power of the conventional load but not on the pulse load of this power repeatedly mutated load to do in-depth study.

This paper in the microgrid system with pulse load as the research background. According to the problem of voltage quality, frequency fluctuation and voltage and current distortion caused by pulsed load, by adding the SVG compensation device to microgrid system, the reactive power current detection method based on dq transform and the hysteresis comparison control strategy are used to suppress the power quality problem caused by pulse load on microgrid. Finally, the model is validated sexuality and superiority by simulation.

2. Microgrid system structure and working principle

2.1. Microgrid system architecture

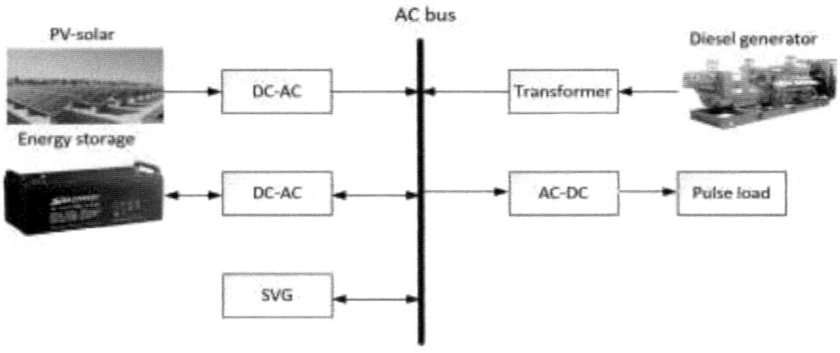

Figure 1. Joined the SVG military micro-grid system structure

Figure 1 is the military micro-grid system structure which join the SVG. In the power supply system, the solar photovoltaic power generation as a new energy, lithium iron phosphate battery as a system of energy storage units, diesel generators as emergency power supply equipment, the three common in parallel with an exchange bus to form a military micro-grid power supply system. The pulse load is connected to the DC bus via AC-DC. The duty cycle can be adjusted by adjusting the opening and closing time and the ratio of the entire switching period. The switching period is adjusted by simulating the periodic load on the pulse load. By adjusting the pulse load resistance to adjust the peak power. SVG connected to the AC bus by adjusting the bridge circuit AC side of the output voltage phase and amplitude, so that SVG absorption or issued to meet the requirements of the reactive current to achieve the purpose of dynamic reactive power compensation.

2.2. Pulse load simulation structure

The pulse load simulation structure is mainly composed of a controllable rectifier unit, a filter unit, a DC switch unit and a load cell. The pulse load simulation model shown in Figure 2, Where in the DC switch is provided with an action command from the pulse control unit, and the control part of the controllable rectifier unit utilizes the synchronous 6 pulse generator to generate a pulse trigger signal to drive the thyristors of the three-phase bridge rectifier circuit.

IOP Conf. Series: Earth and Environmental Science **64** (2017) 012090 doi:10.1088/1755-1315/64/1/012090

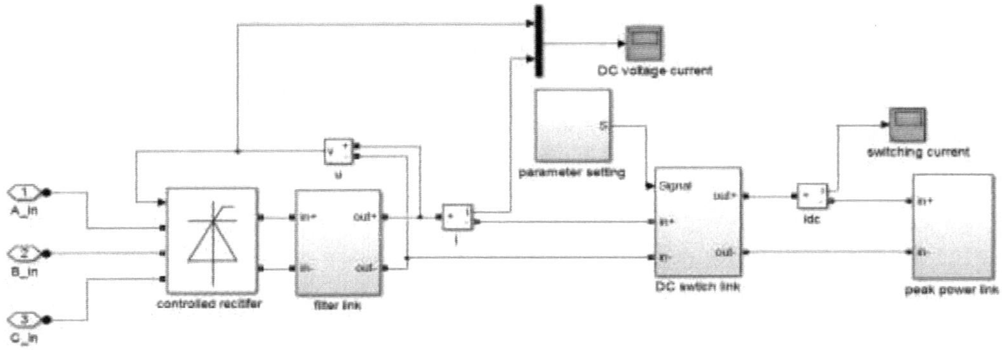

Figure 2. Pulse load simulation structure

In order to obtain an intuitive output DC voltage, the control circuit with output voltage feedback is added to the model. The input signal is the ideal DC voltage, and then compared with the circuit output value, the PID control is adopted to obtain the phase shift control angle input signal, and then directly output DC bus voltage. This model can set the filter capacitor, duty cycle, duty cycle, peak power, different parameters under the pulse load to show the different operating characteristics.

2.3. SVG simulation structure

SVG simulation model mainly includes reactive current detection part and control part. The key to establish SVG model lies in the selection of reactive current detection method and its control strategy. P-q detection method and ip-iq detection method in the synchronous rotation coordinate transformation need to know the phase of the three-phase power grid, and pulse load will produce a large number of micro-grid harmonic, when the power grid distortion, positive and negative sequence voltage and current Harmonic voltage and current interaction will make the reactive power fluctuations caused by the test results are not accurate. The dq detection method is a further optimization design of the first two detection methods, which reduces the computational complexity and can be combined with the control strategy mentioned later. The pulse load can achieve accurate and fast reactive power compensation. Dq detection method simulation structure shown in Figure 3. In this paper, we use the tracking hysteresis comparison method to control the SVG, the simulation structure shown in Figure 4. Hysteresis comparison method is real-time control mode, the current response is really fast; does not need the carrier and the output voltage does not contain the specific frequency harmonic component; if the hysteresis width is fixed, the current tracking error range is fixed. So we choose this method as the SVG control strategy.

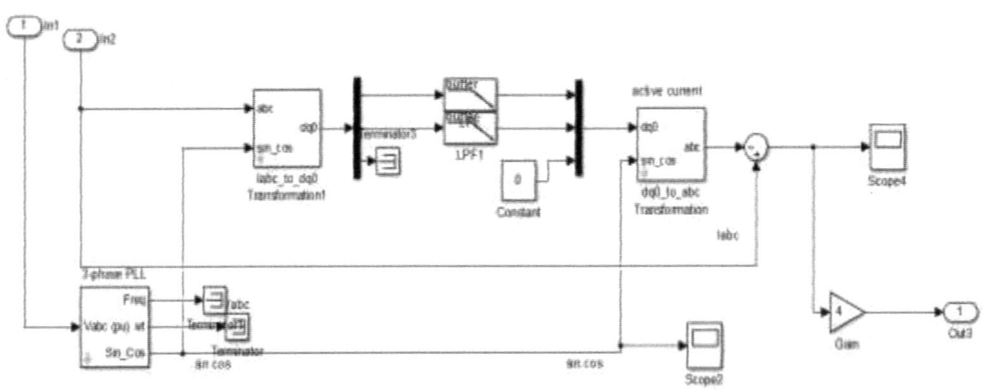

Figure 3. dq detection method simulation structure

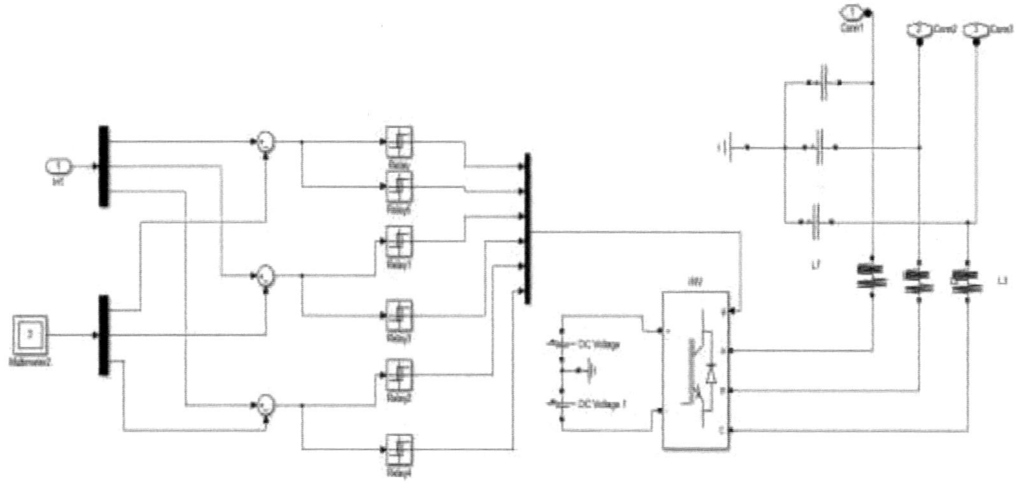

Figure 4. Hysteresis comparison method simulation structure

3. Simulation results and analysis

In order to verify the correctness and validity of the proposed algorithm, this paper built the SVG simulation model in Matlab / Simulink environment and carried on the simulation experiment. The microgrid mainly consists of 5kW photovoltaic power generation system, 30kWh energy storage system and 50kW diesel generator system. The pulse load is connected to the DC bus via AC / DC, and set the pulse load switch cycle is 56ms, duty cycle is 0.4, the peak power is 50kw, the filter inductance is 0.125mH, the filter capacitance is 4000μF. The simulation results are shown in the following figure:

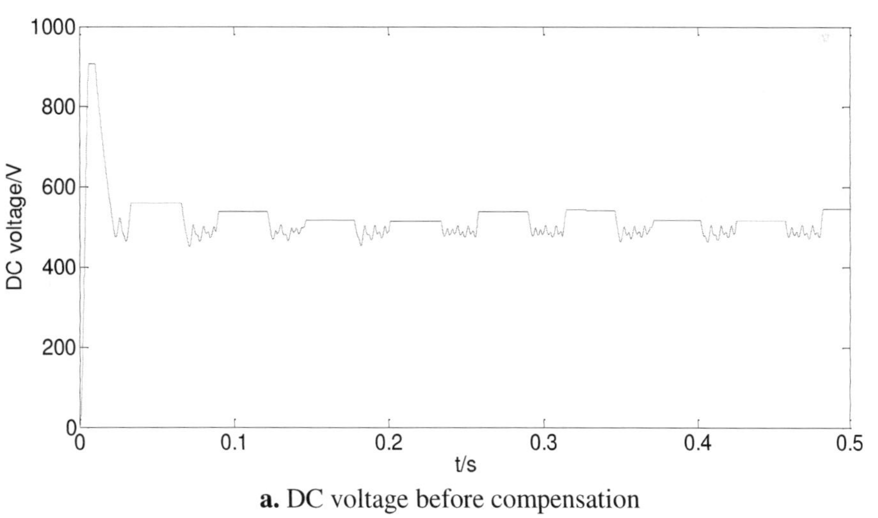

a. DC voltage before compensation

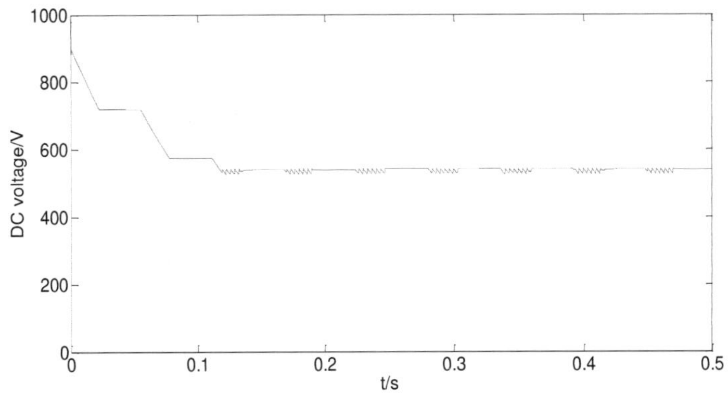

b. Compensated DC voltage

Figure 5. DC voltage before and after Join SVG

Figure 5 is the compensation before and after the DC bus voltage waveform, it can be seen from the figure that the DC bus voltage fluctuations strongly before the compensation, and the DC bus voltage fluctuation significantly reduced after adding SVG, the DC side voltage reach the set value after about 0.1s, and after a small overshoot, the stability of the set value of 500 V. We can see the SVG with a fast compensation speed and the system can work safely, improve the quality of power supply.

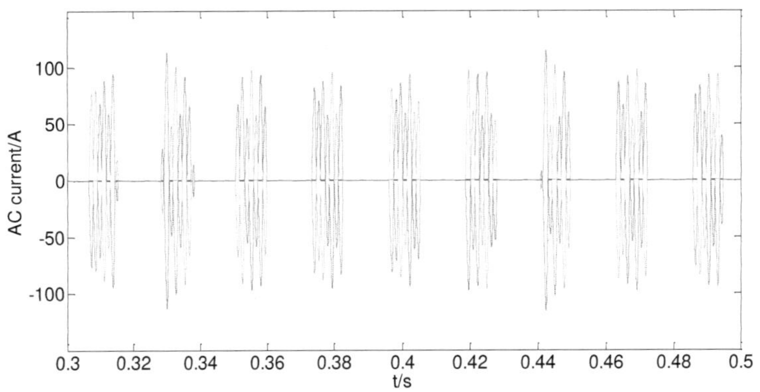

a. AC current before compensation

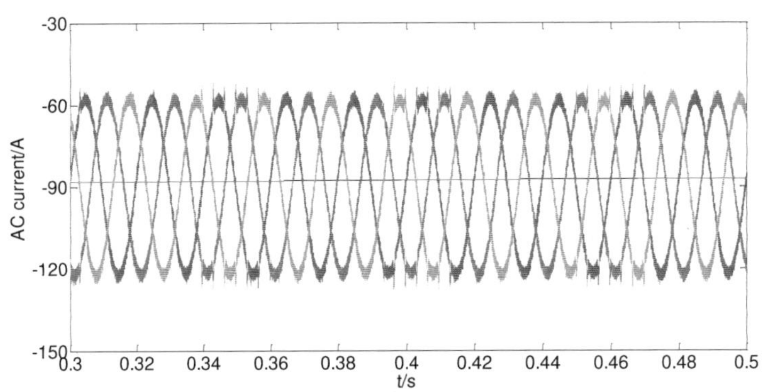

b. Compensated DC voltage

Figure 6. AC current before and after Join SVG

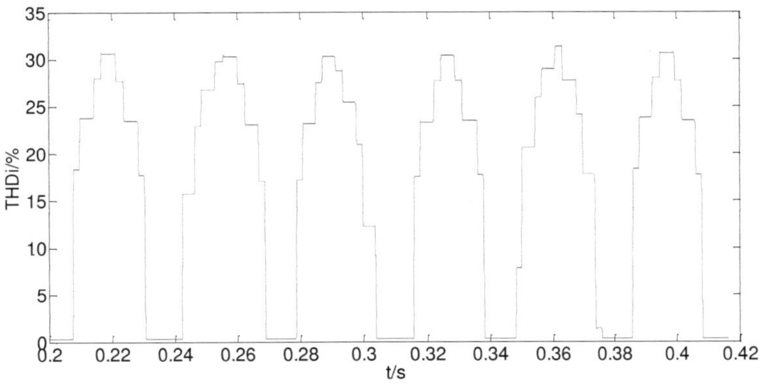

a. THD_i before compensation

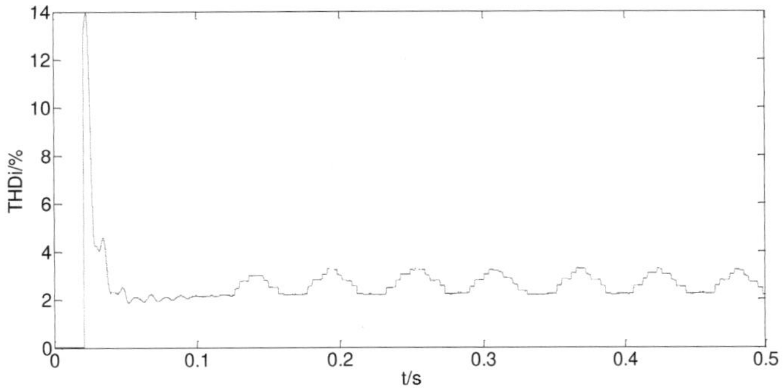

b. Compensated THD_i

Figure 7. The THD_i before and after Join SVG

Figure6 and Figure7 respectively before and after compensation for AC current and THD_i waveform, it can be clearly seen before the compensation AC current distortion is serious, when the pulse load peak power comes, the current change is most obvious, before adding the SVG, the compensation current is provided by the filter capacitor, the compensation effect is not significant; and after adding SVG, the net current is quickly compensated, basically reached with the AC voltage in phase. After 0.05 s, the network side phase current quickly stabilized, making the system to be fully compensated reactive power. Before the compensation of the total harmonic current distortion rate of about 30%, seriously affected the system power supply quality. After compensation, the total harmonic distortion of the current is reduced to about 2%, the current distortion is significantly reduced and reached the national standard. Greatly improving the quality of power supply.

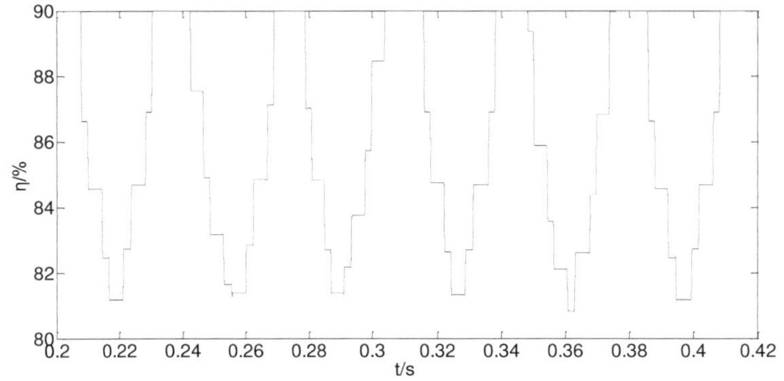

a. the power factor before compensation

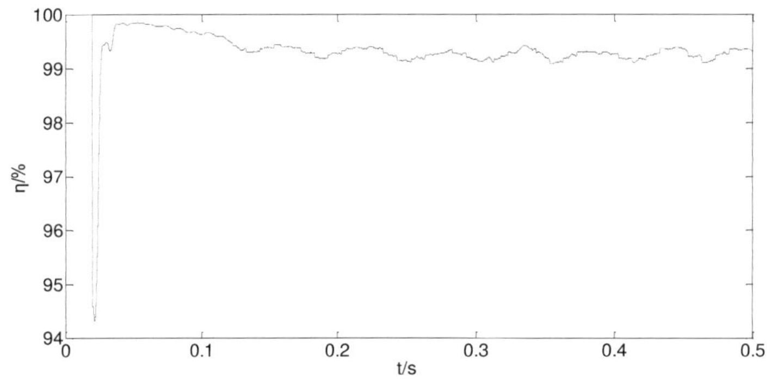

b. Compensated the power factor

Figure 8.the power factor before and after Join SVG

Figure 8 is to compensate before and after the system power factor waveform, from the figure can be clearly seen the system power factor is about 83% before the compensation pulse load peak power comes, and after adding SVG compensation system power factor significantly increased, after a slight fluctuation power factor reached 99%, it achieved a complete compensation for reactive power and basically reached the ideal expected value.

4. Conclusion

In this paper, SVG is introduced into the military microgrid with pulse load. By using the reactive power current detection method based on dq transform and the control strategy of hysteresis comparison, it can meet the instantaneous pulsation performance of the load and improve the power supply of the microgrid. The feasibility and superiority of the SVG model are verified by constructing the MATLAB simulation model. The simulation results show that the reactive current of the system can be compensated rapidly after the SVG compensation is added. The DC voltage fluctuation of the DC bus is slowed down, the current distortion is suppressed, the system power factor is improved, and the power quality of the military microgrid is improved, the compensation effect is expected.

Acknowledgments

This work was financially supported by National Natural Science Foundation of China (No. 51307184) and Science and technology project of Hebei Province (No.15210701D).

References

[1] Liu zheng chun, Wang yong, Yin zhi yong. Modeling and simulation of pulsed power load limited capacity system, J. Journal of north China electric power university, 2014, 41(1): 33-

37.
[2] Guo Xiao Liang, Ma Li Shuang, The Influence of High-capacity Impact Loads on the Generator Set,J. Electrical engineering and automation, 2010, 276(30): 121-122.
[3] Zhao C, Chen J H, Wang J, et a1. Research on Technology of Lightning Disaster Risk Assessment for Power System [J]. High Voltage Engineering, 2011, 37 (12): 3012-3021.
[4] Hou Yang Jun, Chen Tian Ying, Power Quality Signal Detection Analysis Based on HHT, J. Shaanxi power, 2014, 42(2): 51-55.
[5] Pan Cong Mao. The Detection and identification of Transient Power Quality Based on Wavelet Transform, J. The electric measurement and instrument, 2013, 50(11): 69-72.
[6] Li Yue Hua, Chen Gui Hui. Study of Detecting Method of Harmonics and Reactive Current of Power System, J. Electrical applications, 2006, 25(11): 87-89.

IOP Conf. Series: Earth and Environmental Science **64** (2017) 012091 doi:10.1088/1755-1315/64/1/012091

Experimental study of the distillation process in a tilted cavity

Xihui Wang[1,*], Houtao Chen[1], Zhijie Wang[1], and Gang Xiao[2]

[1]State Grid Hu Nan Electric Power Company Research Institute, Changsha 410007, China

[2]State Key Laboratory of Clean Energy Utilization, Institute for Thermal Power Engineering, Zhejiang University, Hangzhou 310027, China

*Corresponding author e-mail: wangxihui0601@163.com

Abstract. Heat and mass transfer correlations in a basin solar still are not applicable for brine-flowing type solar still. Thermal efficiency is hard to determine explicitly due to the varying temperature of brine along the flow direction. A simple method is proposed in the present work to evaluate the performance of the brine-flowing type still through the energy balance equation of brine. Experiments in a tilted cavity which is used to simulate a brine-flowing type solar still are conducted, and it is found that thermal efficiency of the simulated still is always less than 76.3%. An empirical equation is obtained, which may provide a clue for the development of an explicit theoretical model to predict fresh water production of the brine-flowing solar still.

1. Introduction

Solar stills are widely used in remote areas to produce fresh water by brine, e.g. sea water, brackish water, alkaline water, because of simple structures and easy operating skills. Main factors on fresh water production in basin solar still were studied, including input amount of solar radiation, depth of brine layer, thickness of insulation, temperature difference between brine and condensation surface, wind speed and etc. [1-5]. Besides fresh water production, thermal efficiency is another important index to evaluate a solar still. Dunkle [6], Adhikari and Kumar [7] and Zheng et al. [8] proposed theoretical models to describe heat and mass transfer process in basin type solar stills, respectively. Kumar et al. [9] proposed an empirical correlation to predict fresh water production by hot brine temperature. However, all the correlations mentioned above are applicable under conditions that the temperatures of brine in the still are uniform.

Another kind of solar still, called brine-flowing solar still in the present work, allows a thin brine layer flowing on a heating basin, including weir-type inclined solar still [10], weir-type cascade solar still [11], floating cum tilted-wick type solar still [12], inclined solar water distillation system [13] and etc. Brine flows over a basin and its temperature increases along the flow direction. Heat and mass transfer models of a basin type solar still are not applicable to the brine-flowing solar still due to the varying temperature. This paper attempts to propose a simple method to evaluate the performance of brine-flowing solar still through energy balance equations and find out quantitative relationship between fresh water production and influencing factors. An empirical relation is obtained to provide a clue for the development of explicit theoretical model to predict fresh water. To ensure the repeatability of experiments, electrical heating plate is used to simulate solar radiation.

2. Experimental setup and methods

Figure 1 shows a schematic of experimental equipment. It mainly consists of a two-layer box without a cover, the up-layer surface of which acts as an evaporation basin, a condenser in which cooling water passes through and an electrically heating plate. The condenser is connected to the box by screws with a distance of 0.08 m to the evaporation basin which has a length of 0.5 m and width of 0.3 m. Wicks are uniformly paved on the evaporation basin. A water-feeding tube of 8mm diameter with 15 holes of 0.2 mm diameter upside is at the upstream and 10 mm over the evaporation basin. 18 K-type thermocouples are arranged along the centric line of the evaporation basin and the centric line of the condensation surface. The heating plate with the same size of the evaporation basin is closely fixed beneath the up-layer of the box and is controlled by a transformer. The input voltage and electric current of the heating plate are measured by a voltmeter with accuracy of 0.1 V and an ammeter with accuracy of 0.01 A, respectively. The space between the heating plate and the box is tightly filled with insulation 1 which has a thermal conductivity of 0.036 W/m·K and thickness of 0.3 m. To reduce heat loss, an insulation 2 which has a thermal conductivity of 0.046 W/m·K and thickness of 0.5 m is wrapped outside the box. Tap water with salt is used to simulate brine in this study. Brine from the feeding tube flows along the evaporation basin as a uniform thin film due to the capillary force of the paved wicks and is heated by the heating plate. Vapor goes up to the condensing surface and condenses into water drops that are collected in the fresh water tank. The remained hot brine is drained out. Fresh water production is measured with a weighing machine with sensitivity of 10^{-4} kg and a stopwatch. All the thermocouples are connected to an Agilent 34970A data acquisition and then to a computer.

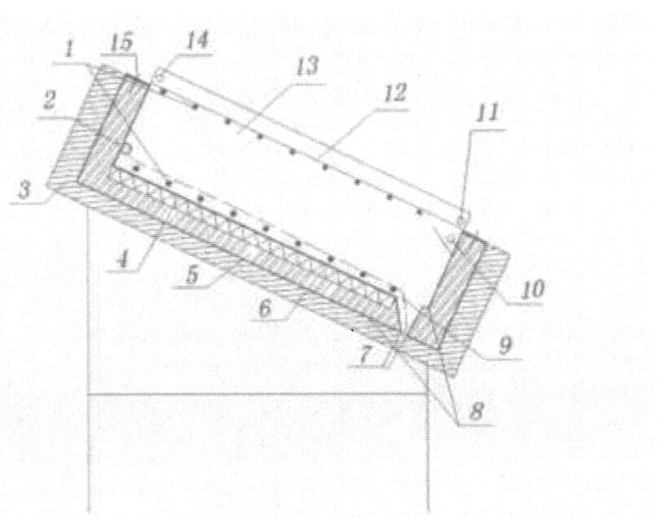

1-thermocouples 2-feeding water tube 3-insulation 2 4-wicks 5- insulation 1 6-heating plate 7-feeding water outlet 8-two-layer stainless box 9-water film 10- fresh water outlet 11-cooling water inlet 12-condenser 13-space between evaporation basin and condenser 14- cooling water outlet 15-sealing pad

Fig.1 Schematic of experimental device

As shown in Figure 2, flowing brine is heated by the heating plate, and at the same time heat is transferred through evaporation, convection and radiation from hot brine to the condensation surface. It is assumed: evaporation, convection and radiation heat fluxes are uniform and temperature of brine is equivalent along the *x* direction; vapor leakage and heat loss is neglected. A steady-state energy balance equation is given by,

$$q_f dy = \frac{1}{3.6} M_{in} c_p dT + (q_e + q_c + q_r) dy \tag{1}$$

$$\overline{q_e + q_c + q_r} = \frac{\int_0^l (q_e + q_c + q_r) dy}{l} \tag{2}$$

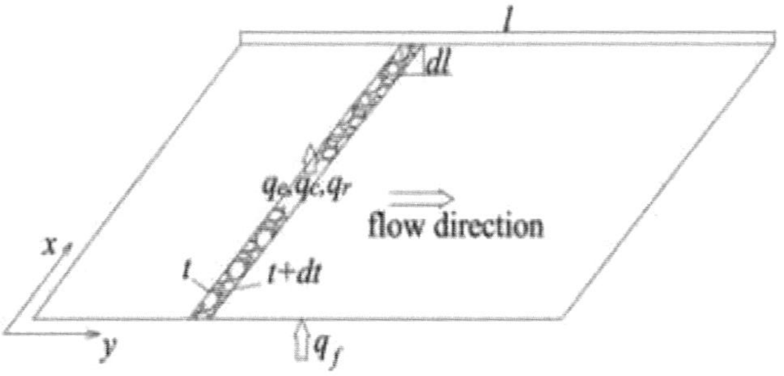

Fig. 2 Schematic illustration for energy balance analysis

α_1 is defined as a ratio of the mean total heat flux including evaporation, convection and radiation to the input heat flux, thus

$$\alpha_1 = \frac{\overline{q_e + q_c + q_r}}{q_f} \quad q_f > 0 \tag{3}$$

α_2 is defined as a temperature increase by unit length along *y* axis, thus

$$\alpha_2 = \frac{3.6 q_f}{M_{in} c_p} \ (M_{in} > 0) \tag{4}$$

δ is defined as:

$$\delta = \frac{\overline{q_c + q_r}}{q_f} \tag{5}$$

In which:

$$\overline{q_c + q_r} = q_f - m_d \cdot \bar{\gamma} / 3.6 \tag{6}$$

Combining equations (1), (3) and (4), the following equation is obtained:

$$\alpha_2 l-(T_o-T_{in})=\alpha_1 \cdot \alpha_2 l \tag{7}$$

Values of T_{in} and T_o are measured by thermocouples.

3. Results and Discussion

3.1. Analysis of thermal efficiency

Assuming $\alpha_2 l$ is an abscissa and $\alpha_2 l-(T_o-T_{in})$ is an ordinate, experimental data are shown in Figure 3 and a linear equation is fitted,

$$\alpha_1 \cdot \alpha_2 l = 0.763\alpha_2 l - 10.155 \tag{8}$$

Based on equations (3) and (4), equation (8) is rewritten by,

$$\frac{\overline{q_e+q_c+q_r}}{q_f} = 0.763 - \frac{2.82M_{in}c_p}{q_f l} \tag{9}$$

Thermal efficiency is defined as:

$$\eta = \frac{q_e}{q_f} \tag{10}$$

which is apparently less than the left side of equation (9). This indicates that the thermal efficiency of a brine-flowing type solar still could hardly exceed 76.3%.

Figure 4 shows temperatures of feeding brine along y axis. The maximum temperature difference between the inlet and the outlet in the figure is ~40 °C and the minimum difference is ~15°C. Since vapour latent heat changes less than 5.3% when the temperature of brine ranges from 30 °C to 80 °C, which is the interesting range for solar still desalination. As a result, it is acceptable to choose a mean vapour latent value (e. g. 2346.25 kJ/kg which is a value of saturation vapour latent heat corresponding to a temperature of 55 °C) to represent the evaporation heat flux.

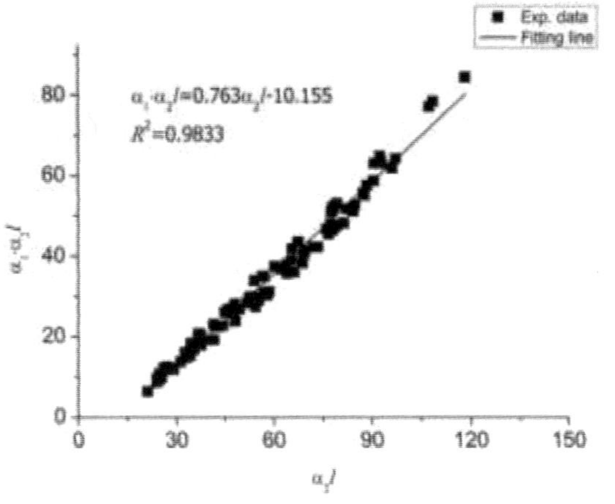

Fig. 3 Function of $\alpha_2 l-(T_o-T_{in})$ vs. $\alpha_2 l$. Experimental condition: feed brine flow rate 5.47, 6.67, 7.67, 8.80 kg/h m respectively; inlet temperature of feed brine: 13-18 °C; heating flux: 400-1500 W/m^2

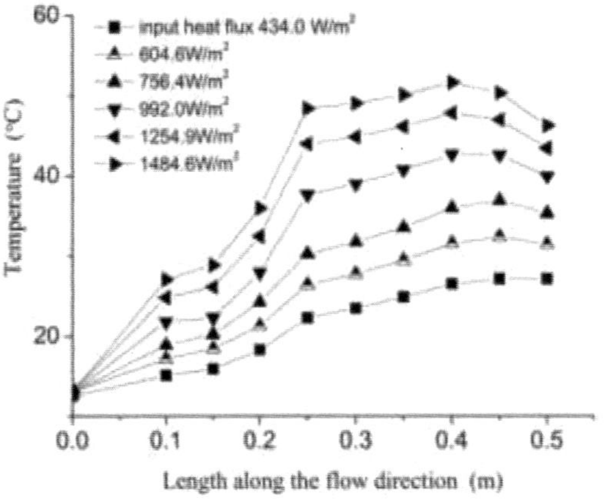

Fig. 4 Temperatures along y axis when feed brine flow rate is 7.67 kg/h m

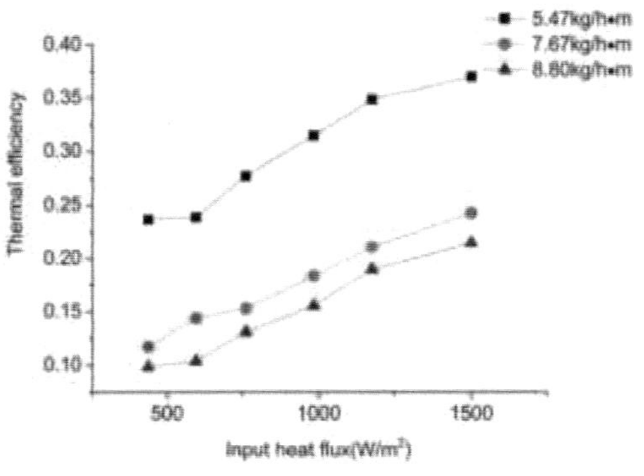

Fig. 5 η vs. input heat flux and feeding brine flow rate.

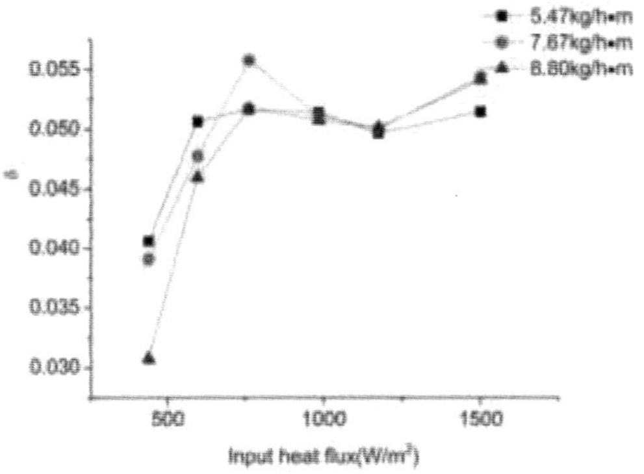

Fig. 6 δ vs. input heat flux and feeding brine flow rate.

Figure 5 shows that η increases with the increase of input heat flux and the decrease of feed brine flow rate. This indicates that a solar concentrator is practical for brine-flowing solar still. Figure 6 shows that δ increases with the increase of input heat flux when the input heat flux is less than 750 W/m^2. When the input heat flux is greater than 750 W/m^2, δ maintains unchanged at a value of ~5%. It is also seen that feed brine flow rate has little influence on the value of δ . This indicates that heat transferred from hot brine to condensation surface by convection and radiation is no more than 6% of the input heat flux for a brine-flowing type solar still. The sum of δ and η is no more than 45%. It suggests that most of input heat wastes with hot brine drained out, which should be avoided in practical application. An external pump which re-imports hot brine at the outlet of the device into still will increase the thermal efficiency.

3.2. Empirical relations of fresh water production

$Nu=C\ (GrPr)^n$ is not applicable due to the varying temperature (as shown in figure 4). In order to find out the relationship between the fresh water production and factors include input heating flux, feed brine flow rate, condensation condition and the inlet temperature of feeding brine, lots of experiments at conditions that feed brine flow rate are 5.47, 6.67, 7.67, 8.80 kg/h m respectively, flow rates of cooling water changes from 10 to140 kg/h m inlet brine temperature changes from 13 to 18 °C and heating flux changes from 400 to 1500 W/m^2 are conducted. The results show that varying cooling water flow rate brings little effect on fresh water production. This may because that the condensation surface is capable enough of condensing the vapour produced in the still. In these cases, input heat flux, inlet temperature and feed brine flow rate are the three most important factors to fresh water production. Assuming α_2 and m_d are x and y axis, respectively, experimental data and fitting lines are plotted in Figure 7 (a) and (b). The fresh water production can be expressed as: $m_d = 2\times10^{-4}\alpha_2^{1.553}$ for inlet temperature of 13-15 °C, and for inlet temperature of 15-18 °C : $m_d = 2\times10^{-4}\alpha_2^{1.577}$. According to the fitted expression and equation (4), the fresh water production is expressed as: $m_d=2\times10^{-4}\times(3.6q_f/M_{in}c_p)^b$. The value of b is related to the inlet temperature of the brine. Much more experiments are under going to find out the quantitative relationship between the inlet brine temperature and value of b.

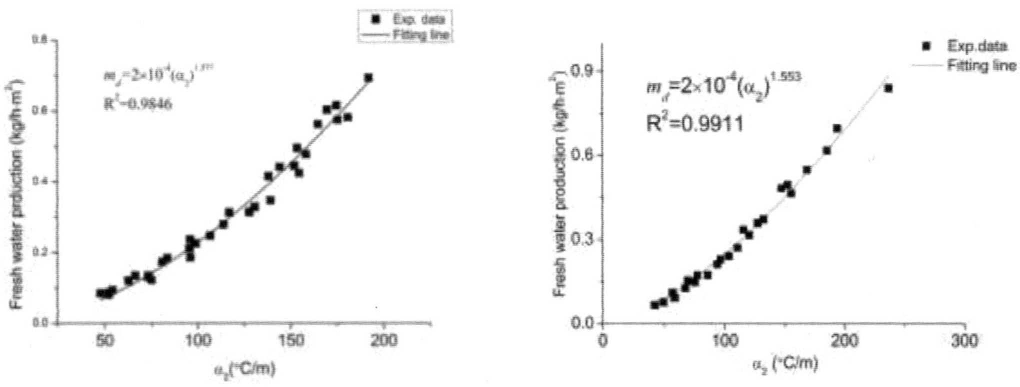

(a) inlet brine temperature: 15-18 °C (b) inlet brine temperature: 13-15 °C

Fig. 7 Fresh water production vs. α_2.

4. Conclusion

Based on the experimental data in a simulated still, it is found that the thermal efficiency of a bring-flowing type solar still, being less than 76.3%, increases with the increase of input heat flux and decrease of feed brine flow rate. An external assisting system which re-imports hot brine into still will increase the thermal efficiency. Empirical relations are fitted to predict fresh water production by input heat flux, inlet temperature and feed brine flow rate, which are regardless of comprehensive heat and mass transfer analysis.

Acknowledgments
This work was financially supported by National Natural Science Foundation of China (51476140) and science research fund of Hunan Xiangdian test and research institute Co. Ltd.

References

[1] A. A. Al-Karaghouli, W. E. Alnaser, Performances of single and double basin solar-stills, Applied Energy 78 (2004) 347-54..

[2] S. Kumar, G. N. Tiwari, Life cycle cost analysis of single slope hybrid (PV/T) active solar still, Applied Energy 86 (2009) 1995–2004..

[3] K. Abdenacer, S. Nafila, Impact of temperature difference (water-solar collector) on solar-still global efficiency, Desalination 209 (2007) 298–305.

[4] R. Dev, S. A. Abdul-Wahab, G. N. Tiwari, Performance study of the inverted absorber solar still with water depth and total dissolved solid, Applied Energy 88 (2011) 252-64..

[5] K. K. Murugavel, S. Sivakumar, J. R. Ahamed et al., Single basin double slope solar still with minimum basin depth and energy storing materials, Applied Energy 87 (2010) 514-23..

[6] R. V. Dunkle, Solar water distillation: the roof type solar still and a multi-effect diffusion still, International developments in heat transfer. A.S.M.E, Proceedings of International Heat Transfer, part V, University of Colorado. (1961).

[7] R. S. Adhikari, A. Kumar, Estimation of mass-transfer rates in solar stills, International Journal of Energy Research 14 (1990) 737-744.

[8] H. F. Zheng, X. Y. Zhang, J. Zhang, Y. Y. Wu, A group of improved heat and mass transfer correlations in solar stills, Energy Conversion and Management 43 (2002) 2469-2478.

[9] S. Kumar, G. N. Tiwari, M. K. Gaur, Development of empirical relation to evaluate the heat transfer coefficient and fractional energy in basin type hybrid (PV/T) active solar still, Desalination 250 (2010) 214–221.

[10] S. B. Sadineni, R. Hurt, C. K. Halford, R. F. Boehm, Theory and experimental investigation of a weir-type inclined solar still, Energy 33 (2008) 71-80.

[11] F. F. Tabrizi, M. Dashtban, H. Moghaddam, K. Razzaghi, Effect of water flow rate on internal heat and mass transfer and daily productivity of a weir-type cascade solar still, Desalination 260 (2010) 239–247.

[12] B Janarthanan, J. Chandrasekaran, S. Kumar, Performance of floating cum tilted-wick type solar still with the effect of water flowing over the glass cover, Desalination 190 (2006) 51-62.

[13] H. S. Aybar, F. Egeliofglu, U. Atikol, An experimental study on an inclined solar water distillation system, Desalination 180 (2005) 285-289.

Experimental study on temperature distribution of membrane water wall in an ultra-supercritical pressure once-through boiler burning zhundong coal

Honghao He, Wenjun Li, Jun Zeng, Guohong Xie, Min Peng, Xuenong Duan

State Grid Hunan Electric Power Corporation Research Institute, Changsha, China

Abstract. Taking an ultra-supercritical pressure once-through boiler as an example, the temperature distribution of the lower membrane water wall is investigated experimentally, the conclusion reveals that increasing the proportion of Zhundong coal can effectively reduce the district heat load, which benefits the temperature uniformity in the lower membrane water wall. When the boiler being operated at middle load, the temperature deviation in lower membrane water wall increase simultaneously, one of the reasons is that the restriction orifice could not adjust the flow rate of working fluid as expected. By adjusting boiler performance, the temperature uniformity of lower membrane water wall can be improved to a certain degree.

1. Introduction

Coal, which is the main primary energy in China, is found a major reserve in the west of China, Zhundong, located in Xinjiang province, with the forecast reserves of 390 billion tons is the largest intact coalfield in the world, it is reckoned that the Zhundong coal can meet the needs for coal consumption in China for the next 100 years [1, 2]. The Zhundong coal is characterized by high volatile content, which leads to high ignition quality, and low exploitation cost. However, due to the process of coal formation is under a special natural environment, the content of sodium in Zhundong coal ash is exceptional high, generally in the range of 5% to 10% [3]. The high content of sodium usually results in severe slagging and fouling on heat transfer surfaces, in order to relieve the deposition, the boiler is specially designed with low district heat capacity, and Table.1 gives district heat capacity comparison between boilers burning different coal.

The membrane water wall safety is an essential aspect in plant boiler operation, especially for the ultra-supercritical pressure boiler, when the steam is being in the supercritical condition, the property is unstable, additionally, the variation of unit load or operation conditions may lead to saltation of steam property, and then induce problem of overheating of membrane water wall. As the Zhundong coal is exploited continuously, more number of boilers designed for burning high-sodium coal are put into operation. However, there is no research investigates temperature distribution of membrane water wall in boilers burning Zhundong coal. In this paper, experiment study is conducted to obtain characteristics of temperature distribution of membrane water wall in a commercial 660MW boiler with low district heat capacity.

IOP Conf. Series: Earth and Environmental Science **64** (2017) 012092 doi:10.1088/1755-1315/64/1/012092

Table.1 The comparison of district heat capcity of boilers burning different coal

Items	Unit	Common boiler A	Common boiler B	Boiler burning high-sodium coal
Volume heat capacity	kW/m^3	62.51	63.9	68~80
Section heat capacity	MW/m^2	3.83	3.915	4.0~5.0
Burner district heat capacity	MW/m^2	1.25	1.185	1.0~1.6

2. Experimental

2.1. Experimental object

The experimental object is a commercial 660MW boiler with Π type arrangement, single furnace, corner tangential firing, low NOx combustor, and the main design parameters is given in Table 2. The analyses of coal and ash is given in Table 3, the design coal and the testing coal A (coal A), with high content of sodium, are typical Zhundong coal, while the testing coal B (coal B), with relative low content of sodium, is a kind of common Chinese coal. Since there is no boiler that could completely be operated on Zhundong coal, the coal A was co-fired with coal B in this study.

Table.2 The main design parameters of the boiler

Items	Unit	Values
Main steam temperature	℃	605
Main steam pressure	MPa	28.25
Main stream flow	t/h	1948.38
Reheat steam temperature	℃	613
Reheat steam pressure	MPa	2.50
Reheat steam flow	t/h	1960.93
Section heat capacity	MW/m^2	3.83
Volume heat capacity	MW/m^3	62.51
Burner district heat capacity	MW/m^2	1.25

The furnace, consisted of membrane water wall, has a 19230×20336mm cross-section area, the high membrane water wall and the low membrane water wall is connected by mixture headers at height of 55600mm. Restriction orifices are designed and installed at entrance of the lower membrane water

Table. 3 Analyses of coal and ash

Items		Unit	Design coal	coal A	coal B
Ultimate analysis	C_{ar}	%	42.51	46.74	64.37
	H_{ar}	%	2.13	2.96	3.68
	O_{ar}	%	11.36	13.09	10.22
	N_{ar}	%	0.48	0.47	0.82
	$S_{t.ar}$	%	0.52	0.31	0.36
Proximate analysis	M_t	%	24.0	26.6	11.0
	M_{ad}	%	14.51	15.61	4.84
	A_{ad}	%	21.37	20.82	15.71
	V_{daf}	%	39.60	42.85	35.27
$Q_{net. ar}$		MJ/kg	14.81	13.51	22.57
Content of sodium in ash		%	3.70	3~4	1.28

Wall to regulate mass flux, the location of the restriction orifice is schematically shown in Fig 1. The low NOx combustor consists of six layers combustor, whose angle can be adjusted in range of -20°~20° in vertical direction, each layer combustor refer to a set of coal pulverizer.

2.2. Temperature measurement

In this study attention are mainly paid on characteristics of temperature distribution of low membrane water wall, and the outside temperature of tubes, near the inlet of the mixture header, is

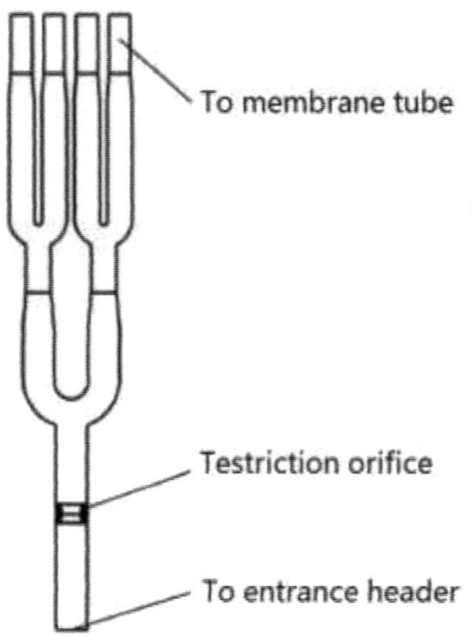

Fig.1 The schematic of restriction orifice location

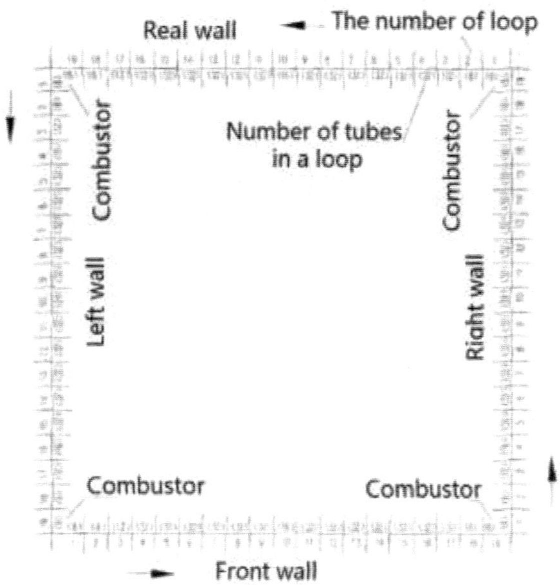

Fig.2 The schematic of loops in lower membrane water wall

Measured, temperature measurements are installed every 8 tubes. The low membrane water wall consists of numbers of loops, and the loop schematic is shown in Fig 2. The average temperature of tubes in a loop is calculated from temperature measurements in the same loop.

3. Results and discussion

3.1. Effect of coal A ratio

As illustrated in Fig 3, all the temperature distributions of the four walls display a bell-shaped appearance with a peak at the middle, which can be explained by heat load distribution in horizontal direction [4]. The coal powder need a certain time to be ignited after ejected into the furnace, then release heat at location near the number 13, 14 loop, which leads to a relative high heat load in this region. Consequently, large size of orifice is chosen in these loops to ensure sufficient mass flux of working fluid, which is illustrated in Fig 3. However, the temperature profile curves uncover a bad uniformity of temperature distribution in practice, especially for the front wall and the left wall.

In this study the coal A was co-fired with coal B, and the temperature distribution of low membrane water wall is affected by the coal A ratio, which can be indicated in Fig 3. When the boiler was operated with coal A ratio being 0%, the maximum temperature was 435℃ with deviation of 25℃ in the front wall, while the maximum temperature was 425℃ with deviation of 21℃ in the left wall. When the ratio of coal A increased to 50%, the maximum temperature descend to 414℃ with deviation of 14℃ in the front wall, while the maximum temperature descend to 419℃ with less change of deviation. In addition, the temperature level in the rear and right wall decline in a certain degree. When the boiler was operated at supercritical pressure, the thermophysical property of the working fluid is unstable, which may leads to hydraulic instability or deteriorates the profile of mass flux, these may even results in overheating or departure from nucleate boiling [5]. The district heat capacity near burner is higher than that of design when burning coal B exclusively, these would further aggravate the condition of membrane water wall.

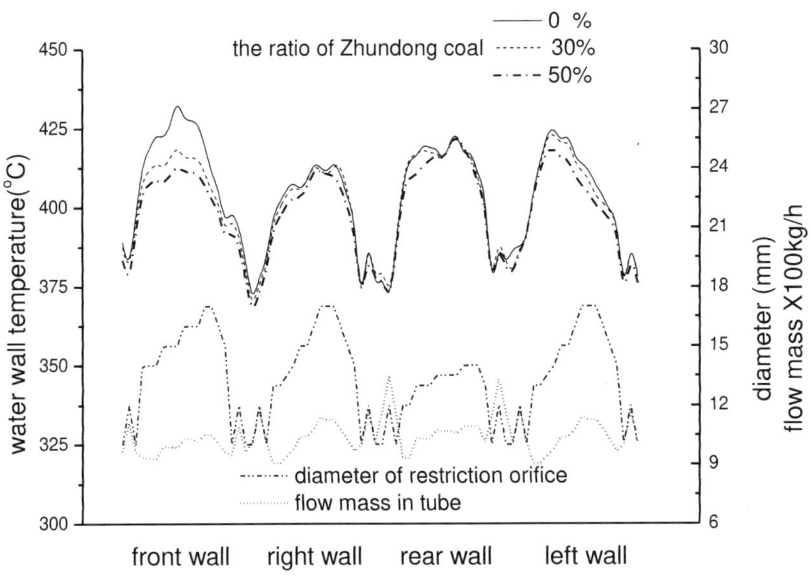

Fig.3 The temperature distribution of lower membrane water wall under tests with different ratio of Zhundong coal

3.2. Effect of unit load

When the boiler was operated at different load, the thermophysical property of the working fluid changes remarkably, as well as the flow field distribution in the furnace, those may leads to variation of temperature distribution of membrane water wall. Fig 4 shows the effect of unit load on the temperature of membrane water wall, with the boiler operating at 340 MW load, the working fluid is being at subcritical condition, leading to a homogeneous temperature distribution on the membrane water wall, especially for the right and left wall. As the unit load increase, the working fluid approaches supercritical condition, and the temperature distribution curve becomes similar to that of design.

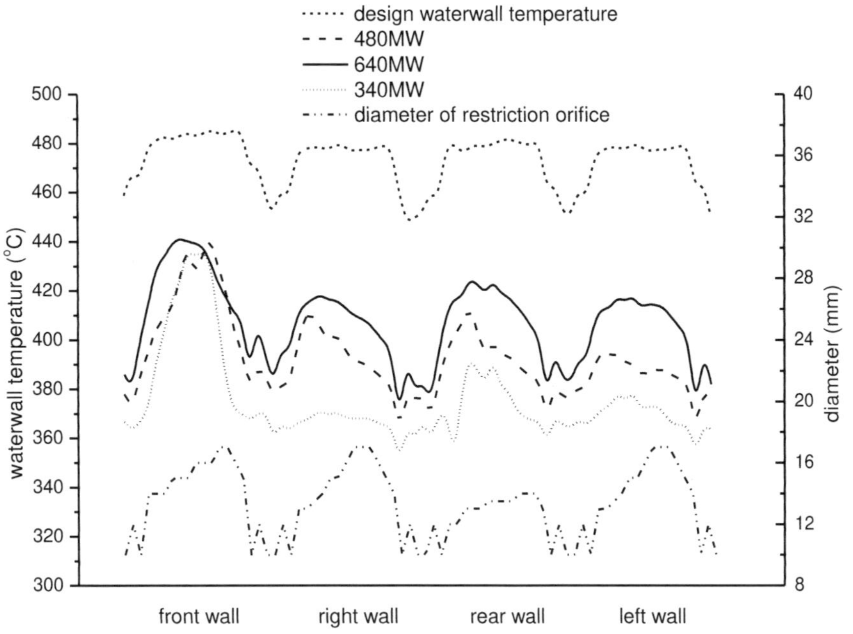

Fig.4 The temperature distribution of lower membrane water wall under different unit load

It worth mentioning that the maximum value of wall temperature increases with unit load, however, the uniformity improves notably. When the unit load increases from 480MW to 640MW, the maximum deviation of temperature reduce from 37℃ to 23℃ for front wall, 21℃ to 16℃ for right wall, 26℃ to 19℃ for rear wall.

The temperature decrease seems to benefit the safety of the membrane water wall at low unit load, nevertheless, regarding the temperature deviation, which is higher at low unit load, the threaten of membrane water wall restriction crack increase due to nonhomogeneous expansion. At present, more plant boilers of large capacity and high parameter have to undertake the mission of peak regulation as a result of oversupply in Chinese power market, this may lead to an increase of running time for the boilers operating at middle load, simultaneously, the unit load varies more frequently, so the augment of temperature deviation of membrane wall tubes should be paid more attentions.

Indicated from the temperature distribution of membrane water wall under 640MW unit load, the maximum temperature appears in number 7, 8 and 9 loop, while the restriction orifices of large size were installed in number 13, 14 and 15 loops, apparently, the restriction orifice do not achieve the mass flux regulation efficiently.

3.3. Effect of operation condition

Fig 5 reveals the temperature distribution of membrane water wall under different secondary air pressure. As can be seen in Fig 5, the overall temperature level decreases with increasing secondary air pressure, for the front wall, the maximum temperature present a decrease by 5℃, and the maximum deviation declines from 36℃ to 28℃, for the rest wall, the maximum value of both temperature and deviation present a certain decline. These can be explained as follow, the increase of secondary air pressure enhances the air momentum, then improves the fullness of flue gas in the furnace, which optimizes the temperature distribution.

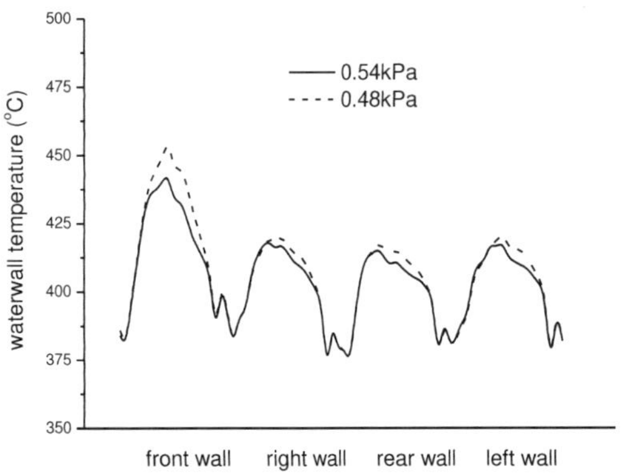

Fig.5 Influence of the secondary air on the temperature distribution of lower membrane water wall

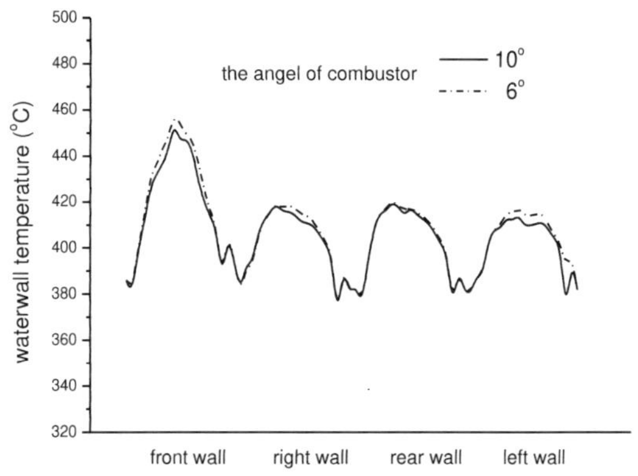

Fig.6 Influence of the combustor angle on the temperature distribution of lower membrane water wall

Fig 6 shows the effect of combustor angel on the temperature distribution of membrane water wall. In Fig 6, the overall temperature of the four wall rises with the combustor angel varies from 10° to 6°, this is because that the flame center of furnace ascends with increasing the combustor angel, which diminishes the district heat load of low membrane water wall. Similarly, the feed-coal quantity profile along vertical direction and the operation pattern of coal pulverizers may influence he temperature distribution of membrane water wall in the same way.

4. Conclusion

(1) When co-fired with common coal, the uniformity of temperature distribution of membrane wall tubes can be improved by increasing the ratio of Zhundong coal. The increase of district heat capacity usually lead to hydraulic instability or deteriorates the profile of mass flux, which may enlarge the membrane wall temperature deviation.

(2) When the boiler is operated at high unit load, the temperature distribution of membrane water wall is more similar to that of designed and the uniformity is better, by contrast, the temperature deviation increase at low unit load. The restriction orifices fail to regulate the flow mass as expect, especially for the restriction orifices in the front wall.

(3) The increase of secondary air pressure is beneficial to the temperature uniformity of membrane water wall. The increase of flame center of furnace, due to the adjustment of boiler operation conditions, can decline the district heat load of low membrane wall and temperature uniformity is improved.

References

[1] J. Xu, D. Yu, B. Fan, X. Zeng, W. Lv, J. Chen, Characterization of ash particles from co-combustion with Zhundong coal for understanding ash deposition behavior, Energy Fuels, 28 (2013) 678-684.

[2] L. Yan, X, Xia, S, Lv, et al. Great promotion of development of large scale integrative energy

base in Xinjiang [J]. Advanced Technology of Electrical Engineering and Energy,2011, 30(1): 1-7.
[3] S. Zhang, C. Chen, D. Shi, et al. Situation of combustion utilization of high sodium coal[J]. Proceedings of the CSEE, 2013, 33(5): 1-12.
[4] C. Jiang, M. Zhang, J. Xu, et al. Distribution characteristics of outlet temperature of working medium in lower waterwall of 600 MW ultra-supercritical boilers[J]. Journal of Power Engineering, 2009, 29(4): 335-341.
[5] G. Quan, W. Yan. Boiler Principle [M], Beijing: China Electric Power Press, 2004.

Application of a method for calculating coal quantity feed into boiler for double inlet and outlet ball mill in the coordinated control system

Houtao Chen*, Xihui Wang, Zhijie Wang, Liang Peng

State Grid Hunan Electric Power Corporation Research Institute, Changsha, China

*Corresponding author e-mail: 2004xiao_chen@163.com

Abstract. Accuracy of the measurement of coal quantity feed into the boiler furnace has great influence on the combustion and coordinated control of power plant. According to the characteristics of double inlet and outlet mill pulverizing system, the method of calculating the amount of coal into furnace using the correlated variable was presented. On this basis, the coordinated control system was designed. Running results show that the method of calculating the fuel quantity accurately and effectively guarantee the stability of coordinated control system. It has a high adjustment quality.

1. Introduction

Double inlet and outlet mill pulverizing system (DIOMPS) has been widely used because of its characteristics of high reliability, good adaptability to various coals, and rapid response to the varying load of the boiler [1, 2]. In contrast to the medium speed mill pulverizing system, the coal quantity feed into the boiler furnace of the DIOMPS is calculated from capacity air volume rather than the coal quantity of the feeder [3, 4]. However, because of the influences of poor installation and calibration, bad operating environment of the measuring equipment, and block of the sampling hole, the measuring result of the capacity air volume of the mill is not reliable, and there is a great error between the calculating and practical quantity of the coal supply. Therefore, this error easily leads to great fluctuation of the main operating parameters and even failures of the coordination system, which greatly affects the stability and safety of the thermal power unit [5, 6]. It is helpful to improve the combustion and coordination control and enhance the stability and economic efficiency of a thermal power unit by solving the problem of coal measurement of the DIOMPS.

2. Characteristics analysis of the DIOMPS

In the DIOMPS, cold and hot primary air is mixed at the inlet of the mill. The temperature and velocity of the coal and air mixture at the outlet of the mill is controlled by adjusting the ratio of the cold to hot primary air. The capacity air flows through the mill and brings the coal to the separator. Big coal cakes go back into the mill to be grinded and fine coal particles are carried into the boiler furnace by the primary air. The quantity of the coal feed into the boiler furnace is related to the primary air volume (capacity air volume) in the mill. The coal quantity feed into the boiler furnace is nearly proportional to the capacity air volume when the material level in the mill is steady. The capacity air volume depends mainly on the capacity valve opening and pressure of the primary air. Consequently, the coal quantity feed into the boiler furnace from the DIOMPS is mainly relative to the capacity valve opening, pressure of the primary air, and material level in the mill. So far, coal quantity of a DIOMPS is usually monotonously characterized by capacity air volume of the mill, and the calculating result is used for coordination control [7, 8]. The quality of the coordination control is greatly affected by the accuracy of the measurement of the capacity air volume, which should be improved.

3. Method for calculating the coal quantity feed into boiler of the DIOMPS

According to the analysis of the operating parameters of the unit and features of the equipment, the coal quantity feed into the boiler furnace from the DIOMPS is strongly related to the capacity valve opening, pressure of the primary air, and material level in the mill [9]. As a result, the coal quantity, defined as simulating coal quantity, is calculated by coupling the capacity valve opening, pressure of the primary air, and material level in the ball mill.

The simulating coal quantity is calculated by multiply the function of the capacity valve opening, function of the pressure of the primary air and function of the material level. The functions of the three variables are used to correct the simulating coal quantity by an inertial element. The calculating formula is as follows:

$$F=f(C)\times f(P)\times f(L)\times LAG(C)\times LAG(P)\times LAG(L) \quad (1)$$

Where, F is simulating coal quantity; C is capacity valve opening; P is pressure of the primary air; L is material level in the mill; LAG(C), LAG(P) and LAG(L) is the inertial element of capacity valve opening, pressure of the primary air and material level in the mill, respectively. The quality of the measuring pressure of the primary and feedback of the capacity valve opening is justified before simulating coal quantity calculation. The hot primary air pressure replaces the primary air pressure at the mill inlet when the quality of the latter is bad. The command of the capacity valve opening replaces the feedback of the capacity valve opening when the quality of the latter is bad. When the quality of the material level becomes bad, the measurement value remains as same as the one of the front moment. The coal quantity is regarded as zero when the corresponding mill does not run. The total coal quantity is the sum of simulating coal quantity of the all operating mills.

By using the simulating coal quantity as the actual coal quantity feed into the boiler furnace, the coordinated control system of power plant unit was designed. The strategy of boiler master control, turbine control, and fuel control is similar to that of the conventional coordinated control system, which will not be explained in detail.

4. Application of the coordinated control system based on simulating coal quantity

The coordinated control system based on simulating coal quantity has been commissioned and applied in a 630MW supercritical unit. Running results show that the coordinated control system based on simulating coal quantity has excellent regulation quality and can meet the requirements of unit operation.

4.1. Results of steady state test

The unit was maintained at the stable operation in coordinate mode. The main automatic control system of coordination, fuel, and boiler feed water and steam temperature kept operating. The unit load was about 480MW. The setting value of main stream pressure was 19.63MPa and main steam temperature was 565°C. The trend of main parameters of the unit was shown in Figure 1. The total simulating coal quantity is basically same as the total feed coal quantity during steady state operation (Fig.1), indicating that the model of simulating coal quantity was accurate. During steady state test, the maximum deviation between main steam pressure and its setting value was -0.23MPa, the maximum deviation between main steam temperature and its setting value was -1.8℃, and the maximum deviation between reheated steam temperature and its setting value was 3.2℃. The above indexes are superior to the relevant technique regulations requirements

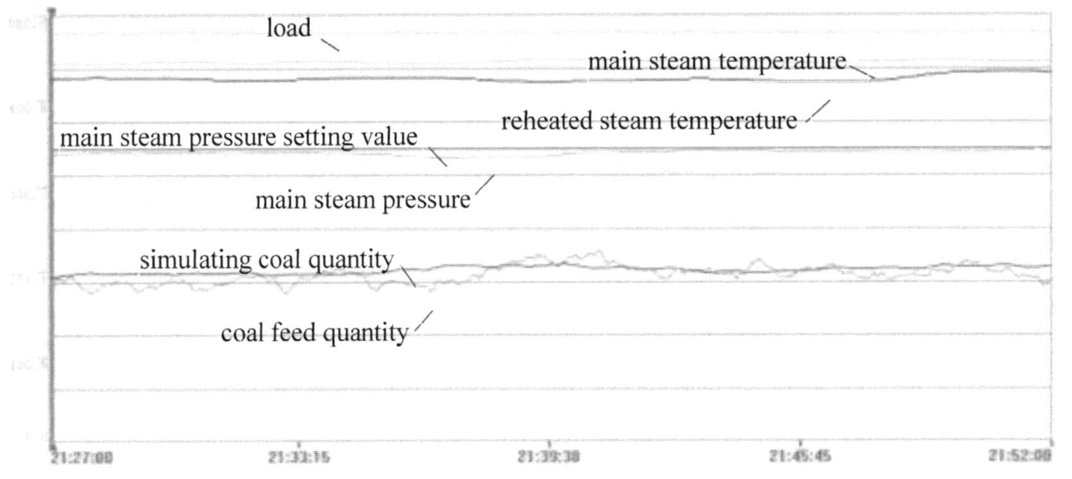

Figure 1. Trend of main parameters of steady state test

4.2. *Results of down load test*

The load was decreased from 630MW to 530MW in coordinate mode, of which the variation amplitude was about 100MW, and varying rate was 10MW/min. The trend of main parameters of the unit during the down load test is shown in Figure 2. Figure 2 shows that the total simulating coal quantity is close to the total coal feed quantity when load is stable. As the load changed, there is a large deviation between the total simulating coal quantity and the total coal feed quantity, and the maximum deviation is 88t/h. The total coal feed quantity is no longer able to characterize the coal quantity feed into the boiler furnace. During the load down test, the maximum deviation between load and load command is -3.2MW, the maximum deviation between main steam pressure and its setting value is -0.50MPa, the maximum deviation between main steam temperature and its setting value is 4.8℃, and the maximum deviation between reheated steam temperature and its setting value is -6.0℃. The results are better than the standards of the relevant technique regulations. The test shows that the coordinated control system based on simulating coal quantity accurately calculates the fuel and meets the requirements of thermal power unit operation.

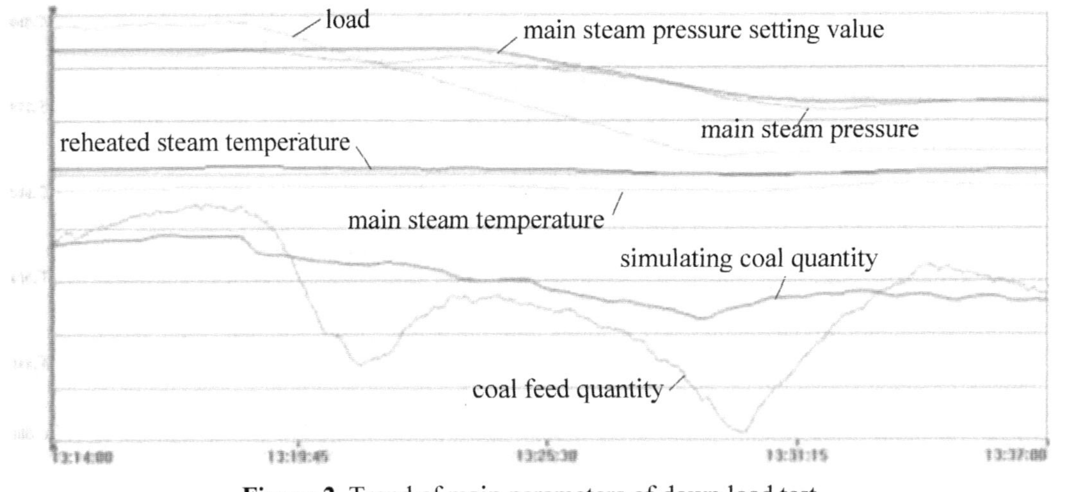

Figure 2. Trend of main parameters of down load test

4.3. Results of up load test

The load was increased from 450MW to 550MW in coordinate mode, of which the variation amplitude was about 100MW and varying rate was 10MW/min. The trend of main parameters of the unit during the up load test was shown in Figure 3. The total simulating coal quantity was very close to the total coal feed quantity when the unit is running at stable state without load change. The deviation between the total simulating coal quantity and total coal feed quantity becomes large when the load is changed. During the test, the maximum deviation between load and load command is -3.2MW. The maximum deviation between main steam pressure and its setting value is -0.45MPa. The maximum deviation between main steam temperature and its setting value is -4.5℃. The maximum deviation between reheated steam temperature and its setting value is -5.5℃. It meets the requirements of unit operation.

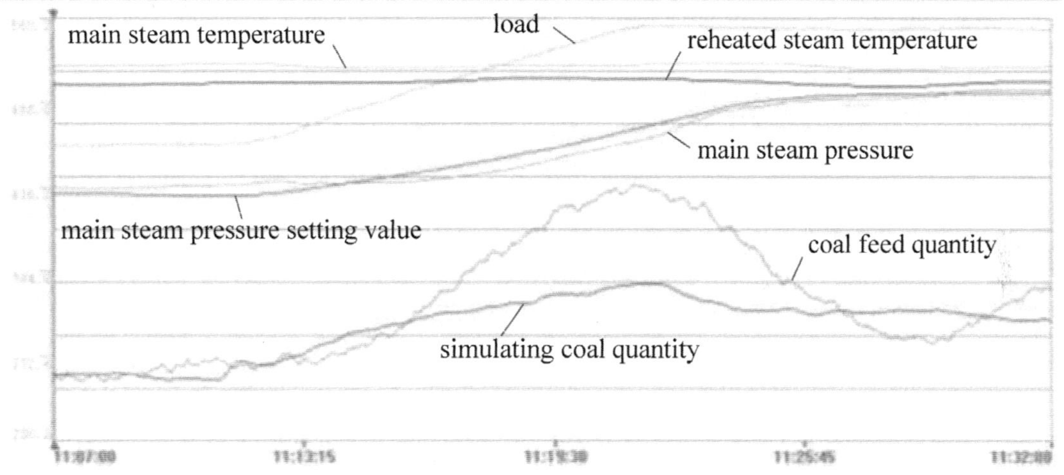

Figure 3. Trend of main parameters of up load test

5. Conclusion

A method of calculating the coal quantity feed into the boiler furnace by multiply the function of the capacity valve opening, function of the pressure of the primary air and function of the material level is presented, and the method is applied to the coordinated control system in a 630MW supercritical unit.

The application results show that the total simulating coal quantity is almost the same as the total coal feed quantity and the deviation between them is slight under the stable state. However, the coal feed quantity is unable to characterize the coal quantity feed into the boiler furnace when the load is changing. During the test of down load and up load, the maximum deviation between main steam pressure and its setting value are -0.50MPa and -0.45MPa, the maximum deviation between the main steam temperature and its setting value are 4.8℃ and -4.5℃, respectively. The test results show that the proposed coordinated control system based on the simulating coal quantity has high quality and can meet the requirements of normal operation of the unit, which can provide a reference for the similar type of thermal power unit.

Acknowledgments

This work was financially supported by science research fund of Hunan Xiangdian Test and Research Institute Co. Ltd.

Reference

[1] H. D. Chen, H. Y. Huang, Characteristics and schemes of control system of 600MW supercritical strait boiler, Boiler Technology. 2 (2006) 26-30.

[2] Y. S. Hao, Z. Z. Lu, Intelligent control and optimizing decision on ball mill pulverizing system based on soft sensing technique, Energy Research and Utilization. 4 (2003) 21-24.

IOP Conf. Series: Earth and Environmental Science **64** (2017) 012093 doi:10.1088/1755-1315/64/1/012093

[3] X. Lu, C. L. Hu, Optimization of automatic fuel control system of 660MW supercritical unit, Water Conservancy & Electric Power Machinery. 11 (2011) 6-8.
[4] Y. H. Tang, H. B. Liu, F. Zhu, N. M. Yan, Calculating as-fired coal quantity for double-inlet and double-outlet mill based on soft sensing, Instrument Technique and Sensor. 5 (2012) 27-28.
[5] J. J. Gu, Y. Zhang, Z. Z. Bai, W. S. Yu, Real-time prediction model of feed coal quantity for boilers based on power to coal coefficient, Thermal Power Generation. 7 (2016) 61-66.
[6] J. Sun, The research of multivariable PID neural network in BBD ball mill control system, Master Degree, Shenyang University of Technology, Shenyang, China, 2011, PP. 5-6.
[7] W. J. Wan, H. Liu, W. J. Huang, S. H. Chen, Dynamic characteristics and mathematic models of process parameters of ball mills with double inlets and outlets, Thermal Power Generation. 11 (2015) 58-61.
[8] Z. D. Zhao, J. L. Han, L. X. Lian, J. X. Wang, etc., Control strategy optimization for supercritical unit equipped with double-inlet and double-outlet ball mill, Thermal Power Generation. 11 (2013) 16-21.
[9] L. H. Feng, W. H. Gui, F. Yang, Soft sensor modeling for mill output of direct fired system based on improved least squares support vector machines, Electric Machines and Control. 11 (2011) 79-82.

IOP Conf. Series: Earth and Environmental Science **64** (2017) 012094 doi:10.1088/1755-1315/64/1/012094

Optimization based inversion method for the inverse heat conduction problems

Huaiping Mu[1,*], Jingtao Li[1], Xueyao Wang[2] and Shi Liu[1]

[1]School of Energy, Power and Mechanical Engineering, North China Electric Power University, Beijing, China
[2]Institute of Engineering Thermophysics, Chinese Academy of Sciences, Beijing, China

*Corresponding author e-mail: muhuaipingdr@126.com

Abstract. Precise estimation of the thermal physical properties of materials, boundary conditions, heat flux distributions, heat sources and initial conditions is highly desired for real-world applications. The inverse heat conduction problem (IHCP) analysis method provides an alternative approach for acquiring such parameters. The effectiveness of the inversion algorithm plays an important role in practical applications of the IHCP method. Different from traditional inversion models, in this paper a new inversion model that simultaneously highlights the measurement errors and the inaccurate properties of the forward problem is proposed to improve the inversion accuracy and robustness. A generalized cost function is constructed to convert the original IHCP into an optimization problem. An iterative scheme that splits a complicated optimization problem into several simpler sub-problems and integrates the superiorities of the alternative optimization method and the Broyden-Fletcher-Goldfarb-Shanno (BFGS) algorithm is developed for solving the proposed cost function. Numerical experiment results validate the effectiveness of the proposed inversion method.

1. Introduction

The IHCP analysis method is often employed to estimate the thermal physical parameters of materials, boundary conditions, initial conditions, heat flux, heat sources, etc., from the given temperature measurement data in the field of the thermal engineering. In light of providing an effective way for acquiring such parameters, the IHCP method has attracted growing attentions.

The efficiency of the inversion algorithms has a great influence on applications of the IHCP method. A variety of algorithms have been developed for solving the IHCP, e.g., the Levenberg-Marquardt method [1, 2],the conjugate gradient technique [3, 4], the standard Tikhonov regularization (STR) algorithm [5], the particle swarm optimization technique [6, 7], etc. The interesting readers can refer to [8-15] for more details. In light of the complexity of the problem, generally, figuring out an inversion algorithm with low computational cost and high inversion accuracy remains a crucial issue.

The ill-posed nature of the IHCP will give rise to a formidable quandary, i.e., small perturbations on the input data may lead to lager fluctuations of the inversion solutions, which may make the final results meaningless. Conventional inversion algorithms emphasize the measurement errors and fail to take into account the inaccurate properties on the forward problem derived from the facts such as (1) the assumptions and simplifications of a real physical problem; (2) the imprecise initial conditions, boundary conditions, geometric conditions and thermal physical properties of materials; and (3) the discretization of the original problem and the approximation of the numerical computation. As a result, developing an algorithm that simultaneously highlights the inaccurate properties of the measurement data and the forward problem may be essential for improving the inversion precision. In this work an inversion model is put forward to simultaneously underline the inaccurate properties of the forward problem and the measurement data is proposed to improve the inversion accuracy and robustness. A cost function is constructed to convert the original IHCP into an optimization problem. An efficient algorithm is developed for searching for the optimal solution of the proposed cost function. Numerical experiment results validate effectiveness of the proposed method.

2. Mathematical model

There are two key steps in seeking for the solution of the IHCP: the forward problem and the inverse problem. In this section, we first revisit the mathematical models for the IHCP.

2.1. Forward problem

By means of employing an effective numerical method, e.g., the finite element method, the finite volume method, etc., the forward problem estimates the temperature distribution from the given conditions of the determined solutions, which can be specified as follows [16]:

$$\rho c_p \frac{\partial T}{\partial t} = \nabla \cdot (\lambda \nabla T) \tag{1}$$

where ρ , c_p and λ mean the density, heat capacity and heat conductivity coefficient, respectively; T represents the temperature distribution and t implies the time index.

2.2. Traditional inversion model

The inverse problem estimates initial conditions, boundary conditions or thermal physical parameters of materials from the given temperature measurement data. For compact notation, the inverse problem can be written as:

$$\boldsymbol{T}(\boldsymbol{\vartheta}) = \boldsymbol{y} + \boldsymbol{r} \tag{2}$$

where $\boldsymbol{T}(\boldsymbol{\vartheta})$ implies the predicted temperature distribution data from equation (1); $\boldsymbol{\vartheta}$ stands for unknown variables; $\boldsymbol{y}$ means the temperature measurement data; $\boldsymbol{r}$ represents the measurement noises. In light of the ill-posed attribute, the major challenge in solving the IHCP stems from the solution of the inverse problem. In accordance with the optimization theory and the Tikhonov regularization method, the solution of equation (2) is often cast into an optimization problem.

2.3. Generalized inversion model

In equation (2), the inaccurate property of the measurement data is highlighted. In practical applications, however, the computation of the forward problems may be inaccurate derived from the facts such as (1) the assumptions and simplifications of a real physical problem; (2) the imprecise initial conditions, boundary conditions, geometric conditions and thermal physical properties of materials; and (3) the discretization of the original problems and the approximation of the numerical computation. In this study, an inversion model is put forward to emphasize the above inaccurate attributes, which can be formulated as:

$$\boldsymbol{T}(\boldsymbol{\vartheta}) + \boldsymbol{B} = \boldsymbol{y} + \boldsymbol{r} \tag{3}$$

where $\boldsymbol{B}$ represents the inaccurate attributes of the forward problem. It is necessary to mention that in the field of the econometrics equation (3) is also called as the semi-parametric regression model.

3. Design of the cost function

In accordance with the Tikhonov regularization method, the solution of equation (6) can be formulated into the following optimization problem:

$$\min_{\boldsymbol{\vartheta},\boldsymbol{B}} \left\{ E(\boldsymbol{\vartheta},\boldsymbol{B}) + \alpha_1 \Omega_1(\boldsymbol{\vartheta}) + \alpha_2 \Omega_2(\boldsymbol{B}) \right\} \tag{4}$$

where $E(\boldsymbol{\vartheta},\boldsymbol{B})$ measures the data fidelity; Ω_1 and Ω_2 mean the regularizers; α_1 and α_2 are the regularization parameters.

In this study, the combination estimation that integrates the superiorities of the least squares estimation and the M-estimation is used to measure the data fidelity [17], which is specified as:

$$\min_{\boldsymbol{\vartheta},\boldsymbol{B}} \left\{ (1-\delta) \| T(\boldsymbol{\vartheta}) + \boldsymbol{B} - \boldsymbol{y} \|^2 + \delta \sum_{j=1}^{m} \rho(T(\boldsymbol{\vartheta})_j + \boldsymbol{B}_j - \boldsymbol{y}_j) + \alpha_1 \Omega_1(\boldsymbol{\vartheta}) + \alpha_2 \Omega_2(\boldsymbol{B}) \right\} \tag{5}$$

where $0 \le \delta \le 1$; $\| \cdot \|$ defines the 2-norm and $\rho(\cdot)$ means an M-estimation function [17-19].
In this study, the regularizers are defined as:

$$\Omega_1(\boldsymbol{\vartheta}) = \sum_{j=1}^{m} | \boldsymbol{\vartheta}_j |^p \tag{6}$$

$$\Omega_2(\boldsymbol{B}) = \| \boldsymbol{B} \|^2 \tag{7}$$

where $p > 0$, and $| \cdot |$ is an absolute value operator.
For convenient computations, the absolute value function is approximated by [20]:

$$| x | \approx (x^2 + \varepsilon)^{1/2} \tag{8}$$

where $\varepsilon > 0$ is a predetermined parameter. As a result, equation (9) can be approximated by:

$$\Omega_1(\boldsymbol{\vartheta}) \approx \sum_{j=1}^{m} (\boldsymbol{\vartheta}_j^2 + \varepsilon)^{p/2} \tag{9}$$

In this work, the G-M function is employed, which can be formulated by:

$$f(x) = 0.5x^2 / (\beta + x^2) \tag{10}$$

where $\beta > 0$ is a scaling parameter.
Finally, we can obtain a generalized cost function for solving the IHCP, i.e.

$$\min_{\vartheta,\boldsymbol{B}}\left\{(1-\delta)\|\boldsymbol{r}\|^2+\delta\sum_{j=1}^{m}\frac{1}{2}\frac{\boldsymbol{r}_j^2}{\beta+\boldsymbol{r}_j^2}+\alpha_1\sum_{j=1}^{m}(\vartheta_j^2+\varepsilon)^{p/2}+\alpha_2\|\boldsymbol{B}\|^2\right\} \tag{11}$$

where $\boldsymbol{r}_j=T(\vartheta)_j+\boldsymbol{B}_j-\boldsymbol{y}_j$.

4. Solving of the cost function

The direct solution of equation (11) is challenging. In accordance with the computational strategy reported in [21-23], equation (11) can be decoupled as the following sub-problems:

$$\vartheta^{k+1}=\min_{\vartheta}\left\{(1-\delta)\|\boldsymbol{r}\|^2+\delta\sum_{j=1}^{m}\frac{1}{2}\frac{\boldsymbol{r}_j^2}{\beta+\boldsymbol{r}_j^2}+\alpha_1\sum_{j=1}^{m}(\vartheta_j^2+\varepsilon)^{p/2}\right\} \tag{12}$$

$$\boldsymbol{B}^{k+1}=\min_{\boldsymbol{B}}\left\{(1-\delta)\|\boldsymbol{r}\|^2+\delta\sum_{j=1}^{m}\frac{1}{2}\frac{\boldsymbol{r}_j^2}{\beta+\boldsymbol{r}_j^2}+\alpha_2\|\boldsymbol{B}\|^2\right\} \tag{13}$$

where superscript k represents the index of iterations.

Due to the excellent numerical performances, the BFGS algorithm is used to minimize equations (12) and (13), and the computation flowchart can be outlined as follows [24, 25]:

Step 1. Determine the cost function $\psi(\boldsymbol{x})$, and specify the initial value $\boldsymbol{x}^{(0)}$ and other algorithmic parameters.

Step 2. Set $\boldsymbol{H}_0=\boldsymbol{I}_n$, estimate the gradient $\boldsymbol{g}_0=\nabla\psi(\boldsymbol{x}^{(0)})$, and let $k\leftarrow 0$.

Step 3. Update the search direction:

$$\boldsymbol{d}^{(k)}=-\boldsymbol{H}_k\boldsymbol{g}_k \tag{14}$$

Step 4. Set

$$\boldsymbol{x}^{(k+1)}=\boldsymbol{x}^{(k)}+\alpha_k\boldsymbol{d}^{(k)} \tag{15}$$

$$\boldsymbol{g}_{k+1}=\nabla\psi(\boldsymbol{x}^{(k+1)}) \tag{16}$$

where α_k can be computed by solving the following minimization problem:

$$\psi(\boldsymbol{x}^{(k)}+\alpha_k\boldsymbol{d}^{(k)})=\min_{\gamma\geq 0}\psi(\boldsymbol{x}^{(k)}+\alpha\boldsymbol{d}^{(k)}) \tag{17}$$

Step 5. If $\|\boldsymbol{g}_{k+1}\|\leq\varepsilon$, terminate the iteration and output the solution; otherwise return to Step 6.

Step 6. Set $\boldsymbol{u}^{(k)}=\boldsymbol{x}^{(k+1)}-\boldsymbol{x}^{(k)}$, $\boldsymbol{\ell}^{(k)}=\boldsymbol{g}_{k+1}-\boldsymbol{g}_k$, and compute $\boldsymbol{H}_{k+1}$ using equation (18); set $k\leftarrow k+1$, and loop to Step 3.

$$\boldsymbol{H}_{k+1}=\boldsymbol{H}_k+\left(1+\frac{\boldsymbol{\ell}^{(k)T}\boldsymbol{H}_k\boldsymbol{\ell}^{(k)}}{\boldsymbol{u}^{(k)T}\boldsymbol{\ell}^{(k)}}\right)\frac{\boldsymbol{u}^{(k)}\boldsymbol{u}^{(k)T}}{\boldsymbol{u}^{(k)T}\boldsymbol{\ell}^{(k)}}-\frac{\boldsymbol{u}^{(k)}\boldsymbol{\ell}^{(k)T}\boldsymbol{H}_k+\boldsymbol{H}_k\boldsymbol{\ell}^{(k)}\boldsymbol{u}^{(k)T}}{\boldsymbol{u}^{(k)T}\boldsymbol{\ell}^{(k)}} \tag{18}$$

Eventually, we obtain the following iterative technique to solve equation (11):

Step 1. Specify the algorithmic parameters and initial solutions of the unknown variables $\boldsymbol{\vartheta}$ and $\boldsymbol{B}$.

Step 2. Update variable $\boldsymbol{\vartheta}$ via solving equation (12) using the BFGS algorithm.

Step 3. Update variable $\boldsymbol{B}$ via solving equation (13) using the BFGS algorithm.

Step 4. Return to Step 2 until a predetermined stopping condition is met.

5. Numerical experiments and discussions

The proposed inversion procedure is called as the generalized semiparametric inversion (GSI) algorithm. In this section, numerical experiments are implemented to test the performances of the GSI algorithm, and the inversion quality is compared with the STR method.

All algorithms are implemented using the MATLAB software. The forward problems are calculated by the finite element method. We use the relative errors (RE) to assess the inversion quality, which can be specified as follows:

$$\mathrm{RE} = \frac{\| \boldsymbol{\vartheta}_{\text{True}} - \boldsymbol{\vartheta}_{\text{Estimated}} \|}{\| \boldsymbol{\vartheta}_{\text{True}} \|} \times 100\% \tag{19}$$

where $\boldsymbol{\vartheta}_{\text{True}}$ and $\boldsymbol{\vartheta}_{\text{Estimated}}$ mean the true values and the estimated values, respectively.

5.1. Case 1

An inversion task of the three boundary conditions is implemented to test the efficiency of the GSI algorithm, and the forward problem is described by:

$$\rho c_p \frac{\partial T}{\partial t} = \lambda \left(\frac{\partial^2 T}{\partial x^2} + \frac{\partial^2 T}{\partial y^2} \right) \tag{20}$$

$$\begin{cases} T(x,y,t) = 520, & x = 0,\ 0 \le y \le 0.6 \\ T(x,y,t) = 620, & y = 0.6,\ 0 \le x \le 0.8 \\ T(x,y,t) = 400, & x = 0.8,\ 0 \le y \le 0.6 \\ -\lambda \partial T / \partial y = \varphi(x), & y = 0,\ 0 \le x \le 0.8 \\ T(x,y,0) = 300 & 0 \le x \le 0.8,\ 0 \le y \le 0.6 \\ T(x,y,t) = 400, & (x-0.4)^2 + (y-0.3)^2 = 0.1^2 \end{cases} \tag{21}$$

where $\varphi(x) = 8000 + 3000\sin(2.5x\pi/0.8)$, $\rho = 8700\ \mathrm{kg/m^3}$, $t \in [0,300]$, $c_p = 385\ \mathrm{J/(kg \cdot K)}$ and $\lambda = 400\ \mathrm{W/(m \cdot K)}$.

The inverse problem estimates the boundary conditions, a, b and c, at $x=0$, $y=0.6$ and $x=0.8$ from the given temperature data, and the true values for a, b and c are 520, 620 and 400. The simulated data is used to serve as the temperature measurement data. To simulate the inaccuracy of the forward problem, boundary condition at $(x-0.4)^2 + (y-0.3)^2 = 0.1^2$ is perturbed into 780. The algorithmic parameters of the GSI algorithm are defined as $\delta = 0$, $\alpha_1 = 0.0001$, $\alpha_2 = 0.04$ and $p = 1$. The regularization parameter for the STR method is 3×10^{-6}. The initial values of variables a, b and c for the both methods are 710, 720 and 700, respectively. Figures 1-4 are the estimation results and the REs for the competitors.

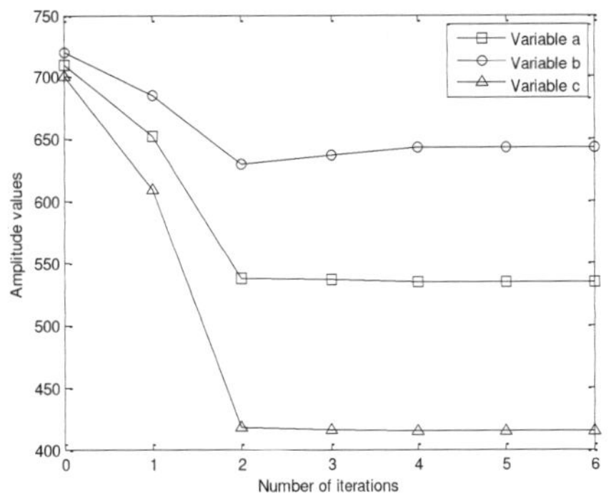

Figure 1. Estimated results for the STR method.

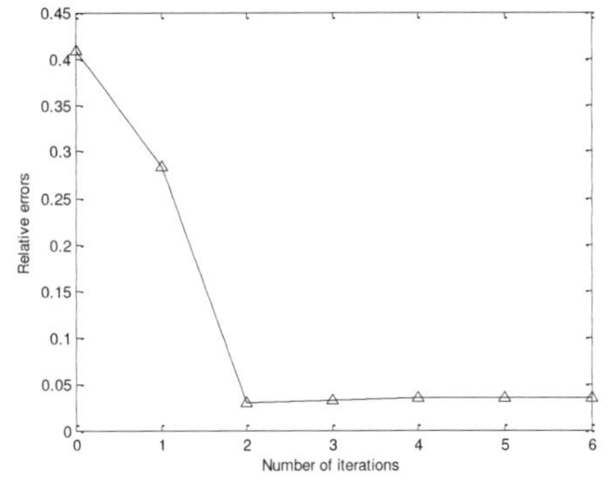

Figure 2. Relative errors for the STR method.

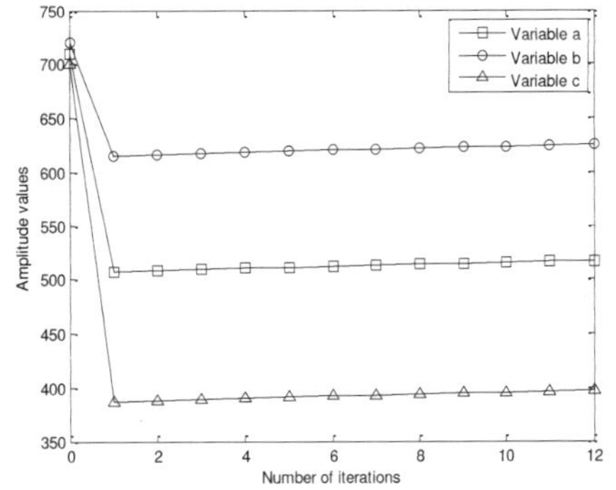

Figure 3. Estimated results for the GSI algorithm.

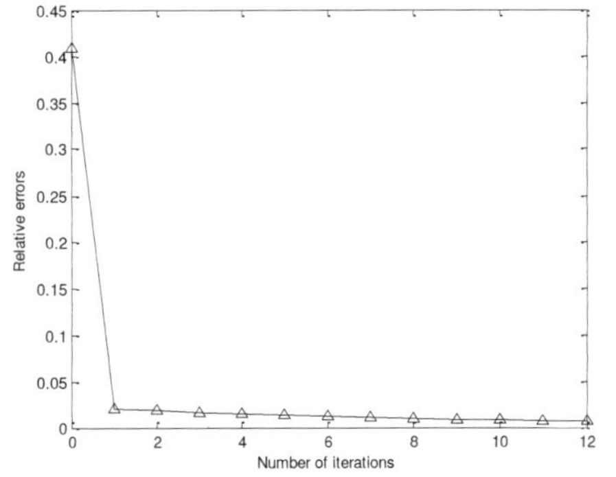

Figure 4. Relative errors for the GSI algorithm.

The estimation results and the relative errors from the STR technique are illustrated in Figures 1 and 2. Numerical simulation results indicate that the STR method can ensure a stable inversion solution. Nevertheless, the final inversion results are not satisfactory. Especially, the final estimation results for a, b and c are 537.44, 629.41 and 417.94, and the RE, 2.96%, is higher than the GSI technique. Besides, in Figure 2, it can be found that the inaccuracy of the forward problem results in the oscillation of the solutions in the process of the numerical computation.

The result estimated by the GSI algorithm is shown in Figure 3. Numerical simulation results confirm that the GSI algorithm can alleviate the numerical instability of the IHCP in light of the fact that the Tikhonov regularization method is introduced to the cost function. With the consideration of the inaccuracy of the forward problem, the GSI algorithm shows satisfactory numerical performances, and the estimation results of variables a, b and c gradually approximate the true values with the increment of the number of iterations. In Figure 3, the final inversion results for variables a, b and c are 517.01, 625.14 and 397.70, which closely approximates the true values, 520, 620 and 400. In Figure 4, the REs monotonously decrease with the increase of the number of iterations, and the final RE of the GSI method is 0.74%, which is far lower than that of the STR technique. The encouraging results indicate that the GSI algorithm is competent in solving the IHCP.

5.2. Case 2

In this case, a complex inversion problem of the time-varying nonlinear boundary condition is simulated. The boundary condition at $y=0$ is $\varphi(t)=a+b\sin(\pi t/150)$, in which $a=500$ and $b=400$. The forward problem is described by:

$$\rho c_p \frac{\partial T}{\partial t} = \lambda \left(\frac{\partial^2 T}{\partial x^2} + \frac{\partial^2 T}{\partial y^2} \right) \tag{22}$$

$$\begin{cases} T(x,y,t) = 400, & x = 0, 0 \le y \le 0.6 \\ T(x,y,t) = 450, & y = 0.6, 0 \le x \le 0.8 \\ T(x,y,t) = 700, & x = 0.8, 0 \le y \le 0.6 \\ T(x,y,t) = \varphi(t), & y = 0, 0 \le x \le 0.8 \\ T(x,y,0) = 300 & 0 \le x \le 0.8, 0 \le y \le 0.6 \\ T(x,y,t) = 800, & (x-0.4)^2 + (y-0.3)^2 = 0.1^2 \end{cases} \tag{23}$$

where $\lambda = 400\ \mathrm{W/(m \cdot K)}$, $\rho = 8700\ \mathrm{kg/m^3}$, $t \in [0,150]$ and $c_p = 385\ \mathrm{J/(kg \cdot K)}$.

The inverse problem estimates the boundary condition at $y = 0$. For easy calculation, the variables a and b in the formula of the boundary condition are estimated. Boundary conditions at $x = 0$, $y = 0.6$, $x = 0.8$ and $(x-0.4)^2 + (y-0.3)^2 = 0.1^2$ are perturbed into 380, 480, 690 and 780 to simulate the inaccuracies of the forward problem. The regularization parameter of the STR method is defined as 1×10^{-5}. The algorithmic parameters of the GSI algorithm are specified as $\delta = 0$, $\alpha_1 = 0.01$ and $\alpha_2 = 1 \times 10^{-4}$. The initial values for the both algorithms are 660 and 300. The estimation results and the REs for the STR technique and the GSI algorithm are shown in Figures 5-8. The REs for the both methods are presented in Figure 9.

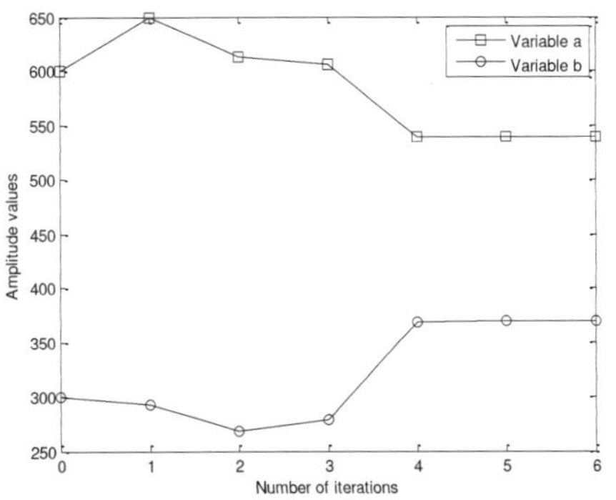

Figure 5. Estimated results by the STR method.

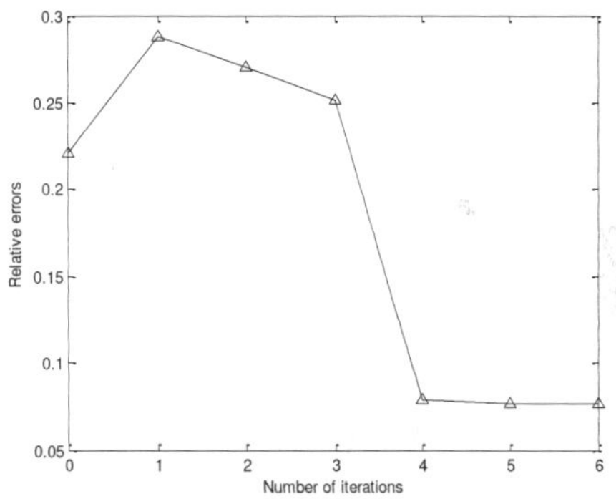

Figure 6. Relative errors for the STR method.

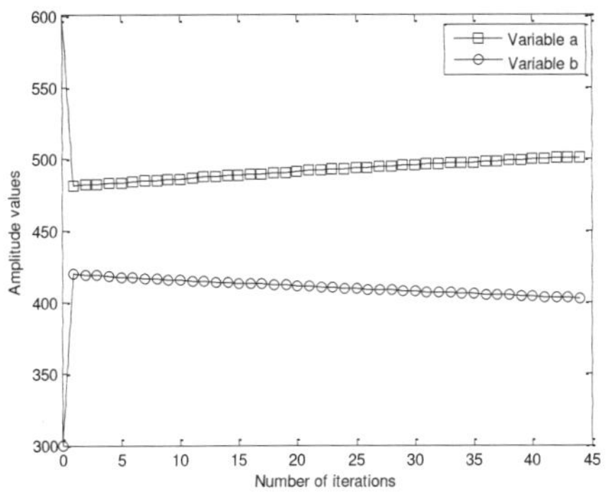

Figure 7. Estimated results by the GSI algorithm.

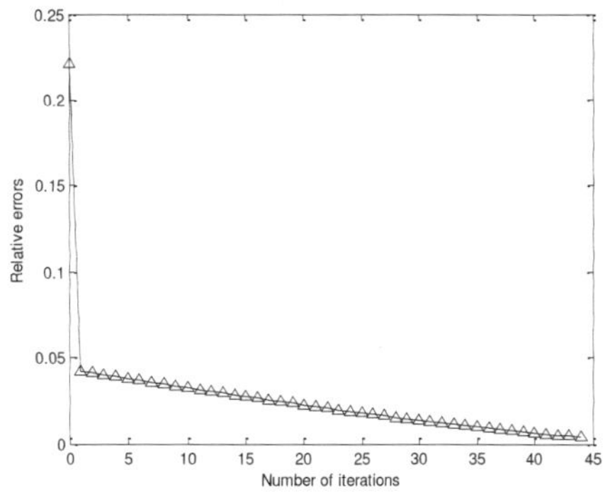

Figure 8. Relative errors for the GSI algorithm.

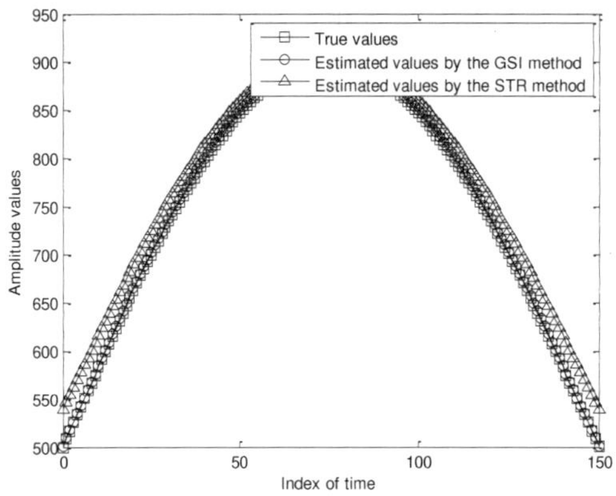

Figure 9. Final estimation results of the GSI method and the STR algorithm.

The estimation results and the REs for the STR method are presented in Figures 5 and 6, respectively. For the STR method, the final estimation results for unknown variables a and b are 539.24 and 370.42, and the RE is 7.67%. Obviously, the estimation accuracy of the STR method is lower than that of the GSI method.

The estimation results of the GSI technique for variables a and b are presented in Figure 7. As can be expected, for the time-varying inversion problem, the GSI algorithm shows satisfactory numerical performances. The final estimation results for variables a and b are 500.83 and 402.50, which approximates the true values, 500 and 400. Meanwhile, it can be seen from Figure 8 that the REs gradually decreases with the increase of the number of iterations, and the final RE is 0.41%, which is far lower than the STR technique.

Figure 9 presents the final boundary condition estimation results for the both methods. Obviously, when the inaccuracy of the forward problem is emphasized, the final estimation results from the GSI algorithm are in a better agreement with true boundary conditions as compared with the STR method, which indicate that the GSI method is highly suitable for solving the IHCP.

6. Conclusion

Different from traditional inversion methods, in this study a new inversion model that simultaneously highlights the measurement errors and the inaccurate properties of the forward problem is proposed to improve the inversion accuracy and robustness. With the assistance of the Tikhonov regularization method, a cost function is constructed to convert the original IHCP into an optimization problem. An iterative scheme that splits a complicated optimization problem into several simpler sub-problems and integrates the superiorities of the alternative optimization technique and the BFGS algorithm is developed to search for the optimal solution of the proposed cost function. Numerical experiment results indicate that the proposed method can ensure a stable numerical solution and improve the inversion accuracy and robustness, and the numerical implementation is easy. For the cases simulated in this study, excellent numerical performances and satisfactory results are observed, which indicates that the proposed algorithm is competent in solving the IHCP. As a consequence, a promising inversion algorithm is introduced for the solution of the IHCP.

References

[1] K.W. Kim, S.W. Baek, Inverse radiation-conduction design problem in a participating concentric cylindrical medium, International Journal of Heat and Mass Transfer. 50 (2007) 2828-2837.

[2] S.K. Singh, M.K. Yadav, R. Sonawane, S. Khandekar, K. Muralidhar, Estimation of time-dependent wall heat flux from single thermocouple data, International Journal of Thermal

Sciences. 115 (2017) 1-15.
[3] T. Liu, B. Liu, P.X. Jiang, Y.W. Zhang, H. Li, A two-dimensional inverse heat conduction problem in estimating the fluid temperature in a pipeline, Applied Thermal Engineering. 30 (2010) 1574-1579.
[4] C.H. Huang, S.P. Wang, A three-dimensional inverse heat conduction problem in estimating surface heat flux by conjugate gradient method, International Journal of Heat and Mass Transfer. 42 (1999) 3387-3403.
[5] A.N. Tikhonov, V.Y. Arsenin, Solution of Ill-posed Problems, V.H. Winston & Sons, Washington, 1977.
[6] F.B. Liu, Particle swarm optimization-based algorithms for solving inverse heat conduction problems of estimating surface heat flux, International Journal of Heat and Mass Transfer. 55 (2012) 2062-2068.
[7] K.H. Lee, S.W. Baek, K.W. Kim, Inverse radiation analysis using repulsive particle swarm optimization, International Journal of Heat and Mass Transfer. 51 (2008) 2772-2783.
[8] B. Czél, G. Gróf, Inverse identification of temperature-dependent thermal conductivity via genetic algorithm with cost function-based rearrangement of genes, International Journal of Heat and Mass Transfer. 55 (2012) 4254-4263.
[9] C.S. Liu, A self-adaptive LGSM to recover initial condition or heat source of one-dimensional heat conduction equation by using only minimal boundary thermal data, International Journal of Heat and Mass Transfer. 54 (2011) 1305-1312.
[10] J.A.M. García, J.M.G. Cabeza, A.C. Rodríguez, Two-dimensional non-linear inverse problem heat conduction problem based on the singular value decomposition, International Journal of Thermal Sciences. 48 (2009) 1081-1093.
[11] F.B. Liu, A modified genetic algorithm for solving the inverse heat transfer problem of estimating plan heat source, International Journal of Heat and Mass Transfer. 51 (2008) 3745-3752.
[12] S. Deng, Y. Hwang, Applying neural networks to the solution of forward and the inverse heat conduction problems, International Journal of Heat and Mass Transfer. 49 (2006) 4732-4750.
[13] M.N. Ozisik, H.R.B. Orlande, Inverse Heat Transfer, Taylor & Francis, New York, 2000.
[14] O.M. Alifanov, Inverse Heat Transfer Problems, Springer-Verlag, New York, 1994.
[15] J.V. Beck, B. Blackwell, C.R. Clair, Inverse Heat Conduction: Ill-posed Problems, Wiley, New York, 1985.
[16] R.T. Lin, Theory and Methods on the Heat Conduction, Tianjin University Press, Tianjin, 1992.
[17] Y. Dodge, J. Jureckova, Adaptive Regression, Spring-Verlag, New York, 2000.
[18] P.J. Huber, Robust Statistics, John Wiley & Sons, New York, 1981.
[19] T. Wang, Methods on robust estimation and detection of multiple outliers in linear regression, Ph.D. Dissertation, The Fourth Military Medical University, 2000.
[20] R. Acar, C.R. Vogel, Analysis of bounded variation penalty methods for ill posed problem, Inverse Problems. 10 (1994) 1217-1229.
[21] J. Lei, W.Y. Liu, Q.B. Liu, X.Y. Wang, S. Liu, Shearlet regularization and dimensionality reduction for the temperature distribution sensing, Measurement, 82 (2016) 176-187.
[22] Y. Xiao, T. Zeng, J. Yu, M.K. Ng, Restoration of images corrupted by mixed Gaussian-impulse noise via L1–L0 minimization, Pattern Recognition. 44 (2011) 1708-1720.
[23] X.G. Lv, Y.Z. Song, S.X. Wang, J. Le, Image restoration with a high-order total variation minimization method, Applied Mathematical Modelling. 37 (2013) 8210-8224.
[24] J. Nocedal, S.J. Wright, Numerical Optimization, Science Press, Beijing, 2006.
[25] Z.H. Ma, Handbook of Modern Applied Mathematics: Operation Research and Optimization Theory, Tsinghua University Press, Beijing, 1998.

IOP Conf. Series: Earth and Environmental Science **64** (2017) 012095 doi:10.1088/1755-1315/64/1/012095

Dynamic reconstruction algorithm of temperature field based on Kalman filter

Yanqiu Li[1,*], Shi Liu[1] and Ren Han[2]

[1]School of Energy Power and Mechanical Engineering, North China Electric Power University, Beijing, China
[2]Tonghua Electric Power Survey Design CO.Ltd, Tonghua, China

*Corresponding author e-mail: lyq8@ncepu.edu.cn

Abstract. Development of temperature reconstruction algorithm plays an important role in the application of temperature field measurement by acoustic tomography. A dynamic model of temperature field reconstruction by acoustic tomography is established. A dynamic reconstruction algorithm based on Kalman Filter (KF) is proposed considering both acoustic measurement and the dynamic evolution information. An objective function fusing space constrain with dynamic evolution information is designed. Simulation results of three temperature field distribution models show that the reconstruction quality of dynamic reconstruction method based on KF is better than those of static reconstruction methods.

1. Introduction

Information of the temperature field distribution plays an important role in heat balance calculation, combustion efficiency evaluation, saving energy and reducing the pollution. Acoustic pyrometry technology is very important in many real industrial needs and applications because of the advantages as non-contact measurement, wide measuring range, easy for real time continuous measurement, convenient maintenance. The technology has been used in many areas, such as deep-sea hydrothermal temperature field measurement, industrial furnace combustion temperature field distribution for reconstruction and detection, atmospheric temperature field distribution measurement of the storage of grain temperature [1-3].

The classical reconstruction algorithms can be classified into two categories, i.e. two dimensional and three dimensional reconstruction algorithms. The main algorithms in two dimensional temperature field reconstruction are least squares[4-5], Fourier regularization method[6], gaussian function and regularization algorithm[7], algebraic reconstruction algorithm[8], Tikhonov regularization algorithm[9] etc.. The main algorithms of three dimensional temperature field reconstruction include the least square and SVD algorithm [10], CGLS and LSQR algorithms [11], methods based on RBF neural network [12], ART and SIRT algorithms [13], Tikhonov and TSVD regularization algorithm [14] and so on. The algorithms mentioned above only consider the measurement information and not focus on the temperature field dynamic information in dynamic process, it is more reasonable to adopt dynamic reconstruction algorithm to obtain the distribution of temperature field. A dynamic reconstruction algorithm is present to reconstruct the three-dimensional temperature field based on

Kalman filter (KF). The numerical simulation results show that this method can reconstruct temperature field more accurately than the static algorithms.

2. The principle and the static model of acoustic pyrometry

The basic principle of acoustic pyrometry is based on that the propagation speed of sound wave in gas medium is the function of gas temperature [15]

$$v = \sqrt{\frac{\gamma RT}{M}} = Z\sqrt{T} \tag{1}$$

Where v is the propagation speed of acoustic wave in gas medium, m/s; R is molar gas constant, J/mol.k; γ is gas specific heat ratio, T is the absolute temperature of gas, K; M is molar mass of gas, g/mol; Z = $\sqrt{\gamma R/M}$, it is a constant to specified gas.

Time of flight (TOF) that acoustic wave is propagated along a sound ray is represented

$$y_i = \int_{L_i} \frac{1}{v_j(x,y,z)} dl + n_i = \int_{L_i} f_j(x,y,z) dl + n_i \tag{2}$$

in which y_i represents the time of flight of the wave along the ith sound path; L_i is the transmission path of the ith sound wave ray; (x,y,z) is the location of units; $v_j(x,y,z)$ is the sound speed of the jth imaging unit; $f_j(x,y,z)$ is the slowness of the jth pixel units (i.e., the reciprocal of velocity), n_i is the measurement noise. Then the formula (2) can be simplified as static reconstruction model in the form of a matrix equation:

$$Y = AF + n \tag{3}$$

In which, A $\in R^{M \times N}$ represents the line segment length, $Y \in R^M$ is the TOF vector measured practically, $F \in R^N$ represents the space state factor, i.e. the reciprocal of velocity. M is the total of independent TOF measured across the temperature field, N is the number of pixels, $n \in R^M$ represents the noise vector in TOF measurement data. F is calculated by proper reconstruction algorithm, and then the temperature T(x,y,z) is obtained according to (4).

$$T(x,y,z) = \frac{1}{F(x,y,z)^2 Z^2} \tag{4}$$

3. Reconstruction method of three-dimensional temperature field based on KF

The inversion problem above is expressed as state estimation problem in time-varying estimation method [16]. It is assumed the measured value Y_k at moment t_k is obtained, where k is the discrete time step. The information provided is used to update the sate estimation at the moment t_k. It is also assumed the problem is discrete with time variable. The state equation describing the time evolution information of the temperature field and the measurement equation expressing the relationship between temperature field distribution and measured value are needed in state estimation problem. The time evolution information equation of the temperature field distribution T_k (or slowness distribution F_k) is shown in (5)

$$F_{k+1} = g(F_k, w_k) \tag{5}$$

Where F_k is the slowness variable at moment k; g (.) describes the dynamic evolution information expressed by a series of partial derivative equations in temperature field measurement; w_k represents

IOP Conf. Series: Earth and Environmental Science **64** (2017) 012095 doi:10.1088/1755-1315/64/1/012095

the uncertainty of the dynamic evolution equation; and subscript k is the discrete time index. Equation (5) can be approximated to a linear equation

$$F_{k+1} = B_k F_k + w_k \tag{6}$$

Where B_k is the evolution operator (state transformation matrix), equation (6) can be considered as pure random evolution model because there is not process dynamic information and B_k is assumed as a unit matrix. It is assumed w_k is Gaussian white noise and the covariance matrix determines the time evolution rate.

The measurement equation is

$$y_k = A_k F_k + n_k \tag{7}$$

A_k is measurement operator, n_k represents the uncertainty of dynamic evolution equation.

The estimation value can be calculated by minimizing following objective function at the assumption of Gaussian assumed condition

$$J(F_k) = \left\| F_k - F_{k|k-1} \right\|^2 + \left\| y_k - A_k F_k \right\|^2 \tag{8}$$

The recursion KF method is obtained by minimizing the objective function of (8) which consists of time updated equation (9) and (10) and measurement updated equation (11), (12) and (13).

$$F_{k+1|k} = B_k \cdot F_{k|k} \tag{9}$$

$$D_{k+1|k} = B_k \cdot D_{k|k} \cdot B_k^T + S_k^w \tag{10}$$

$$K_k = D_{k|k-1} \cdot A_k^T \cdot [A_k \cdot D_{k|k-1} \cdot A_k^T + S_k]^{-1} \tag{11}$$

$$F_{k|k=} F_{k|k-1} + K_k \cdot [y_k - A_k \cdot F_{k|k-1}] \tag{12}$$

$$D_{k|k} = [I_N - K_k \cdot A_k] \cdot D_{k|k-1} \tag{13}$$

The estimation value $F_{k|k}$ of real state vector is obtained according to the recursive minimum mean square error by using initial value $F_{1|0}$ and $D_{1|0}$. The gain K_K can be calculated off-line in advance so that the calculation burden on-line is decreased as only (9) and (11) are needed to calculate on-line. The previous measurement data saved to estimate current state is decreased. Simple diagonal matrix is used to be the state and observation variance.

4. Numerical simulation of the reconstruction method of temperature field

Numerical simulation is adopted to verify the feasibility of the dynamic reconstruction algorithm based on KF, the reconstruction quality is compared with least squares, algebra reconstruction technique (ART) method. A 12 m×12 m×12 m cube space area is chosen and 20 acoustic sensors are arranged in three different planes respectively to obtain 58 the acoustic sound rays independently. The whole image reconstruction area is divided into 3×3×3=27 sub areas. The bicubic interpolation of 31×31×31 are obtained after getting the temperature in each subdomain.

Three classical temperature fields are chosen to simulate for verifying the feasibility of the proposed algorithm, which are single-peaked symmetric, double peak symmetric and four peak symmetric, the model temperature field distributions are shown in Fig.1.

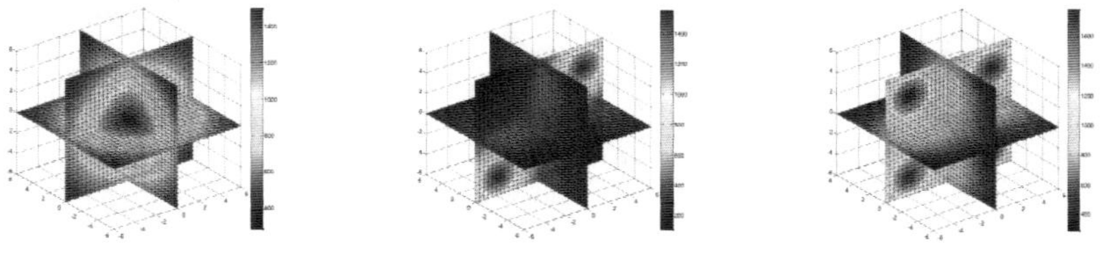

(a) single-peaked symmetric (b) double peak symmetric (c) four peak symmetric

Figure 1. Diagrams of model temperature fields

The temperature field reconstruction results of least square method, ART and the dynamic algorithm based on KF are shown in Figure 2, Figure 3 and Figure 4. The reconstruction images show that three algorithms can reconstruct the distribution images according with model temperature field. Three errors are adopted to evaluate the reconstruction quality of three algorithms that are the maximum relative error E_1, average relative error E_2 and root mean square error E_3. The maximum relative error, relative error of the mean and the root mean square error are defined as (14), (15) and (16).

$$E_1 = \frac{|T_{M\max} - T_{R\max}|}{T_{M\max}} \times 100\% \tag{14}$$

$$E_2 = \frac{|T_{Ma} - T_{Ra}|}{T_{Ma}} \times 100\% \tag{15}$$

$$E_3 = \frac{\sqrt{\frac{1}{N}\sum_{j=1}^{N}[T_M(j) - T_R(j)]^2}}{T_{M\max}} \times 100\% \tag{16}$$

where N is the number of pixels in reconstructed temperature field; T_M (j) is the average temperature of model temperature field; $T_{M\max}$ is the maximum temperature of pixels in model temperature field; T_R (j) is the temperature of pixels in reconstructed temperature field, T_{Ra} is the average temperature in reconstructed temperature field; $T_{R\max}$ is the maximum temperature of pixels in reconstructed temperature field.

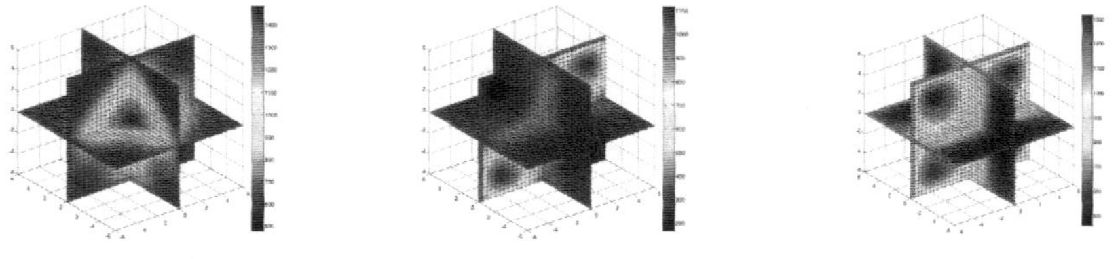

Figure 2. Diagram of reconstructed temperature field by least squares

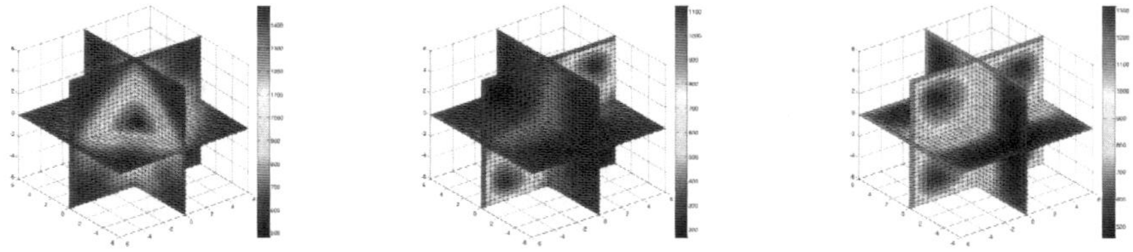

Figure 3. Diagram of reconstructed temperature field by ART

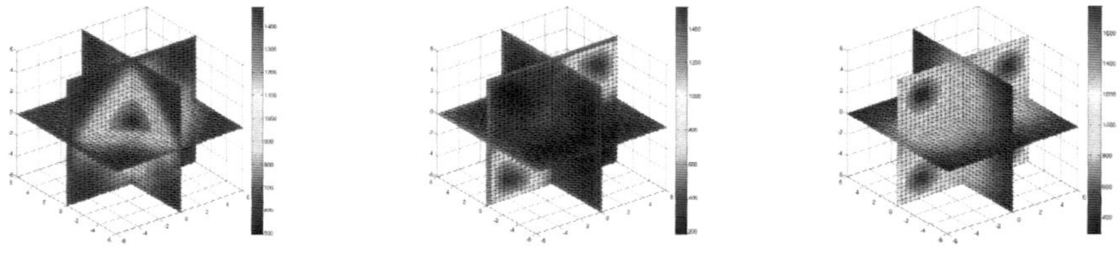

Figure 4. Diagram of reconstructed temperature field by dynamic reconstruction algorithm

The root mean square error of reconstruction by least square reconstruction for three model temperature fields that are single-peaked symmetric, double peak symmetric and four peak symmetric temperature field are 0.0389, ,0.0845 and 0.1256 respectively. The root mean square error of ART method are 0.0335, 0.0812 and 0.1232 respectively. The root mean square error of dynamic reconstruction method bases on KF are 0.0126, 0.0438, and 0.0755 respectively. The image quality is improved by adopting the dynamic reconstruction method.

5. Conclusion

The measurement technology of temperature field by acoustic tomography is attached important to widely. The reconstruction algorithm plays an important role in temperature field reconstruction by acoustic technology which affects the reconstruction accuracy and practicability. The past algorithms focus on static reconstruction, but the temperature field is the object varying with time. The dynamic reconstruction algorithm of temperature field based on KF is proposes and simulation has been carried out with three typical temperature fields. The results show that the precision of algorithm based on KF is increased significantly compared with those of the least squares and ART algorithm. It provides a highly effective method for acoustic temperature field reconstruction.

Acknowledgments

This work was financially supported by the "111 Plan " (B13009)

References

[1] Yan Hua, Cui Kexin, Xu Ying, Temperature field reconstruction based on a few sound travel-time data, J. Chinese Journal of Scientific Instrument. 2 (2001) 470-475.

[2] An Liansuo, Li Gengsheng, Shen Guoqing, et al. Application of Acoustic Pyrometer in a 200 MW Power Plant Boiler, J. Journal of Chinese Society of Power Engineering. 12 (2011) 928-932.

[3] Kan Zhe, Meng Guoying, Wang Xiaolei, et al. Research of boiler temperature field reconstruction algorithm based on genetic algorithm, J. Journal of Electronic Measurement and Instrumentation, 10 (2014) 1149-1154.

[4] Fumio I, Masayasu S. Fundamental studies of acoustic measurement and reconstruction combustion temperature in large boilers, J. Trans. Japan Soc. Mech. Eng., B53(1985) 1610-1614.

[5] Lu J; Wakai K;Takahashi S. Acoustic computer tomographic pyrometry for two-dimensional

measurement of gases taking into account the effect of refraction of sound wave paths, J. Sci. Technol. 11 (2001) 692-697.

[6] Mauro Bramanti, Emanuele A, Salerno, Anna, et al. An Acoustic Pyrometer System for Tomographic Thermal Imaging in Power Plant Boilers, J. IEEE Transactions on Instrumentation and Measurement, 45 (1996) 159-167.

[7] TIAN Feng, SUN Xiaoping, SHAO Fuqun, et al. A Study of Complex Temperature Field Reconstruction Algorithm based on Combination of Gaussian Functions with Regularization Method, J. Proceedings of the CSEE, 5 (2004) 212-220.

[8] SHEN Guo-qing, AN Lian-suo, JIANG Gen-shan, et al. Simulation of Two-dimensional Temperature Field in Furnace Based on Acoustic Computer Tomography, J. Proceedings of the CSEE, 2 (2007) 11-14.

[9] Yan Hua, Li Kun, Wang Shanhui. A 3D acoustic temperature field reconstruction algorithm based on tikhonov regularization[C]. Advances in Intelligent and Soft Computing, 1 (2012) 365-370.

[10] WANG Ran, AN Liansuo, SHEN Guoqing, et al. Research on simulation of furnace three-dimensional temperature field reconstruction by acoustics based on singular value decomposition, J. Proceedings of the CSEE, z1 (2014) 147-152.

[11] AN Liansuo, WANG Ran, SHEN Guoqing, et al. Research on reconstruction of three-dimensional temperature field in furnace by acoustics based on iterative algorithm, J. Journal of North China Electric Power University, 1 (2015) 69-73.

[12] Zhou Xian, Wang Qiang, Miao Zhinong, et al. Three-dimensional temperature field reconstruction algorithm based on RBF neural network, J. Instrument Technique and Sensor, 5 (2013) 99-102.

[13] AN Liansuo, WANG Ran, SHEN Guoqing, et al. Simulation study on reconstruction of 3D temperature field in boiler furnace by acoustic CT algorithm, J. Journal of Chinese Society of Power Engineering, 1 (2015) 13-18.

[14] WANG Ran, AN Liansuo, SHEN Guoqing, et al. Three-dimensional temperature field reconstruction with acoustics based on regularized SVD algorithm, J. Chinese Journal of Computational Physics, 2(2015) 195-201.

[15] Du Gonghuan,Zhu Zemin,Gong Xiuqin, Fundamental of Acoustic. Nanjing University Press, Nanjing, 2001, pp. 186-188.

[16] Fu Mengyin,Deng Zhihong and Yan Liping. Kalman filter theory and its application in navigation system, Science press, Beijing, 2010, pp. 12-16.

IOP Conf. Series: Earth and Environmental Science **64** (2017) 012096 doi:10.1088/1755-1315/64/1/012096

In-situ synthesis of sulfur-TiO_2 hollow shell materials for high-performance lithium-sulfur batteries

Bo Hai[1], Litong Ma[1], Hui Yan[1] and Hang Wei[2,*]

[1]School of Chemistry and Chemical Engineering, Inner Mongolia University of Science & Technology, Inner Mongolia, China
[2]College of Chemistry and Chemical Engineering, Inner Mongolia University, Inner Mongolia, China

*Corresponding author e-mail: weihang@imu.edu.cn

Abstract. Lithium-sulfur batteries with higher energy density are highly attractive, but the practical applications have been greatly affected by their poor cycle performance. Despite much effort has been devoted to design the structure of sulfur cathode to suppress polysulfide dissolution, relatively little emphasis has been placed on in-situ immobilizing the sulfur atoms. Herein, we demonstrate a new approach of in-situ immobilizing the sulfur atoms into the TiO_2 host, in which, the polysulphides can localized in the cathode side and efficiently reused during cycling due to the novel S-TiO_2 hollow shell structure. The battery based on the well-designed S-TiO_2 cathode can deliver a discharge capacity of 601 mA h g^{-1} at 0.5 C after 100 cycles. The good electrochemical performance could be attributed to the homogeneous dispersing of sulfur in the TiO_2 host in the in-situ formation process, and the hollow structure of the S-TiO2 materials. The economical and simple strategy to overcome the polysulfide dissolution issues provides a commercially feasible way for the construction of lithium-sulfur batteries.

1. Introduction

Rechargeable batteries with superior performance are very attractive and important in advanced energy storage systems [1]. In a wide variety of cathode materials, sulfur as an abundant and nontoxic element can deliver a theoretical specific capacity of 1675 mA h g^{-1}. Therefore, Li–S batteries have attracted increasing interest in recent years and been considered as one of the most prospective candidates for the next generation Li-ion battery [2].

Despite of the considerable advantages, there are still many issues in blocking the Li-S battery's commercialization process, such as, the insulating nature of the sulfur and the final discharge product lithium sulfides, and the sulfur cathode proceeds through multiple steps and involves a series of intermediate products (Li_2S_x, x=8, 6, 4, and 2) during discharging. The dissolved high order species can diffuse and react with metal lithium, then the low order products can diffuse back to the cathode side, which called "shuttle effect" [3]. To tackle the drawbacks mentioned above of Li-S batteries, many approaches have been developed on structure design of the cathode electrode, and different approaches always have their own advantages and limitations, also rapid capacity fading during cycling has not completely avoided, especially the high cost and the complexity of the synthesis process prevent their large-scale applications.

As is well-known, $Na_2S_2O_3$ which always used as sulfur precursor in many researches can be reduced to form sulfur particles in the acid solution according to reaction formula (1) [4]. However, little work focus on the in-situ modification in this reaction process. Moreover, when the $Ti(SO_4)_2$ is dissolved in water, hydrogen ions will be released and anatase TiO_2 which can bind favourably with polysulphide anions due to hydrophilic Ti–O groups and surface hydroxyl groups can be formed after hydrolysis reaction (reaction formula (2, 3)) [5]. As the result, the two reaction can occur at the same time, and sulfur particles can in-situ dispersed in the TiO_2 structure. In view of its advantages, herein, we report a facile and effective method for in-situ modifying the sulfur cathode and suppressing the shuttle effect in lithium-sulfur batteries by synthesizing S-TiO_2 hollow core-sell composites. The sulfur can in-situ dispersed in the TiO_2 networks, which may shorten the migration pathway of lithium ions as well as a trap for dissolved polysulfides.

$$S_2O_3^{2-} + 2H^+ = SO_2 + S + H_2O \quad (1)$$

$$Ti^{4+} + 4H_2O = Ti\,(OH)_4 + 4H^+ \quad (2)$$

$$Ti\,(OH)_4 = TiO_2 + 2H_2O \quad (3)$$

2. Experimental section

2.1. Sample preparation

S-TiO_2 hollow core-sell composites were synthesized by slowly adding concentrated Ti $(SO_4)_2$ (0.133 g, 100 mL) to an aqueous solution of $Na_2S_2O_3$ (1.033 g, 100 mL). After reaction in ice-water bath for 2 h, the S-TiO_2 hollow core-sell composites were washed by centrifugation and then the as-synthesized S-TiO2composites dried under vacuum overnight.

2.2. Characterization and electrochemical measurements

Powder XRD patterns were collected on a Bruker D8 Advance diffractometer using Cu K α irradiation (λ = 1.5406 Å) and the morphology were obtained by a Hitachi s-4800 high resolution scanning electron microscope coupled with a BRUKER QUANTAX EDS spectrometer.

The cathode electrodes used in the experiments was prepared by mixing S-TiO_2 hollow core-sell particles, super P and polytetrafluoroethylene (PTFE) binder at a 7:2:1 mass ratio. Then the as-prepared cathode was dried in a vacuum oven at 60 ° C for 24 h. After that, CR2320 coin cells were assembled in a glove box filled with argon using lithium metal as the counter electrode. The electrolyte was 1, 3-dioxolane and 1, 2-dimethoxyethane (1:1 in volume) with 1 M lithium LiTFSI and 0.1M $LiNO_3$ and the separators was Celgard 2320. Cycling was carried out between 1.7 and 2.8 V on a Neware multi-channel battery tester at room temperature.

3. Results and discussion

Figure 1 shows the X-ray diffraction patterns of the as-prepared S-TiO_2 hollow shell composite, sulfur and TiO_2. The peaks of crystalline sulfur and TiO_2 (JCPDS No. 21-1272) clearly indicated that the composition of S-TiO_2 composite. The broadening peaks of the TiO_2 anatase demonstrates that the resulting TiO_2 host are very small indicating good dispersion of sulfur within the TiO_2 host.

The SEM images of S-TiO_2 hollow shell composite are shown in Figure 2. It can be clearly identified that the S-TiO_2 composite is globular, and the typical diameter of it is in range of 500 nm – 1μm. Moreover, from Figure 2c, the morphology of the S-TiO_2 composites is truly hollow shell structure and the rough shell of hollow shell can be clearly observed from the Figure 2d, which is conducive to the contact between cathode materials and electrolyte. To further determine the distribution of different element in the S-TiO_2 composites, the results of element mappings were obtained for the composites. Figure 3b-3d show the element mappings for oxygen, sulfur and titanium based on the SEM image which shown in Figure 3a. The element mappings result of oxygen, sulfur and titanium from Figure 3 demonstrate a very similar intensity distribution, which testifying that sulfur and TiO_2 are uniformly distributed, and since the in-situ formation of the sulfur in the metatitanic acid, the sulfur is tightly bonded with TiO_2 in the S-TiO_2 composites. Consider to the TiO_2 possess hydrophilic Ti–O groups and surface hydroxyl groups [4, 6], which can bind with polysulphide anions, hence when the S-TiO_2 hollow shell composite used as cathode materials in Li-S battery, the specific capacity of the Li-S battery would be raised.

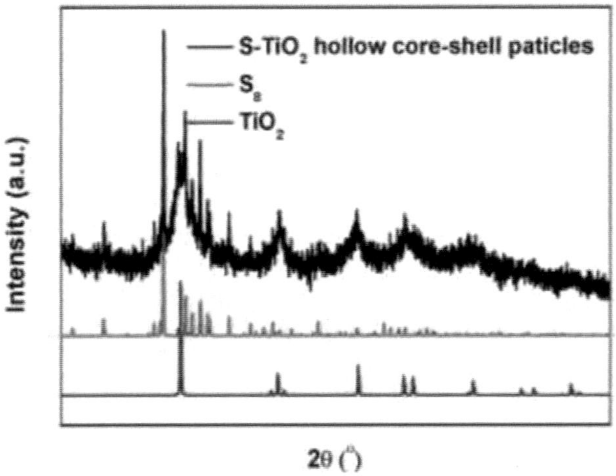

Figure 1. X-ray diffraction pattern of elemental sulfur, TiO_2, and S-TiO_2 hollow core-shell composites.

Figure 2 SEM images of as prepared S-TiO_2 hollow shell composites at different magnification scale.

Cycling performance of the S-TiO_2 hollow shell composites cathode are presented in Figure 4, and the capacity result in this experiment was calculated based on sulfur mass. The sulfur content was obtained to be ~52 wt% in the hollow shell nanostructures according to the TGA measurement. Figure

4 shows the cycling performance of S-TiO_2 hollow shell composites cathode at a current rate of 0.5 C (1C = 1675 mA g^{-1}). An initial discharge capacity of 848 mA h g^{-1} was observed, even after 100 cycles, 601 mA h g^{-1} discharge capacity was also achieved. Meanwhile, the average coulombic efficiency in the 100 cycles reaches about 96%. The good charge and discharge stability can be assigned to the in-situ formation of the sulfur which have great dispersion in the TiO_2 hosts. In addition, TiO_2 which possess hydrophilic Ti–O groups and surface hydroxyl groups can tightly interact with sulfur, effectively preventing the diffusion of polysulfide anions and minimizing the "shuttle effect" [4, 6].

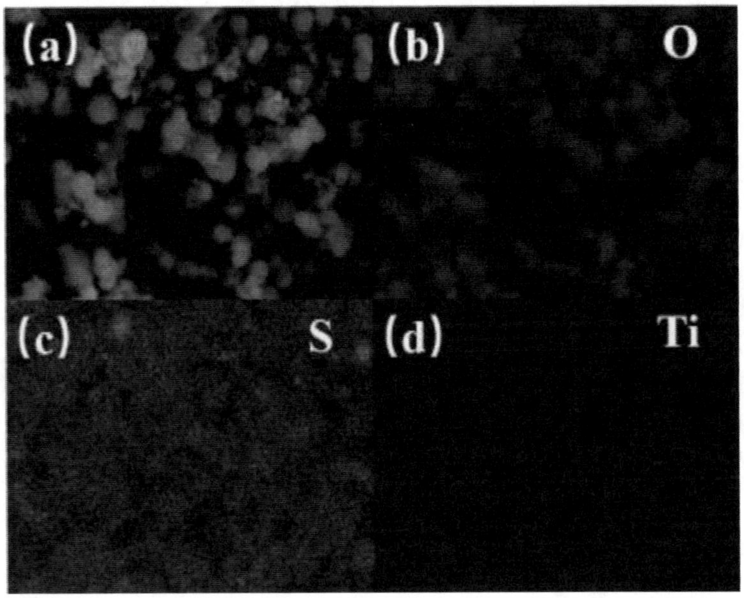

Figure 3 X-ray diffraction patterns of elemental sulfur, TiO_2, and S-TiO_2 hollow shell composites.

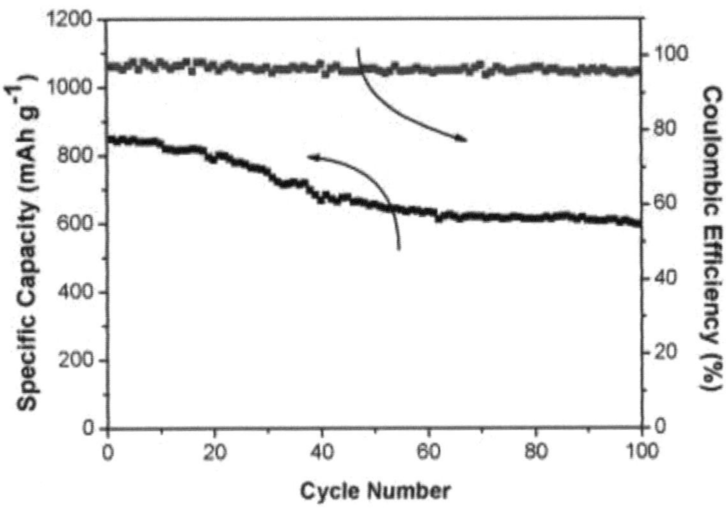

Figure 4 Cycling performances of S-TiO_2 hollow shell composites cathode.

4. Conclusion

In summary, a novel S-TiO_2 hollow shell composites cathode which consisted of small dimension TiO_2 and homogeneously distributed sulfur, has been successfully fabricated by an in-situ co-hydrolysis strategy. The S-TiO_2 hollow shell composite shows good cycling stabilities duo to the in-situ sulfur formation in the TiO_2 hosts, which displays the 601 mA h g^{-1} at 0.5 C after 100 cycles. This result demonstrates that the TiO_2 hosts play an important role in preventing the dissolution of the polysulphide anions to the electrolyte and minimizing the "shuttle effect". Furthermore, this novel synthetic method provides an excellent choice to the future cathode design of Li-S batteries.

Acknowledgments

This work was financially supported by the Inner Mongolia University of Science and Technology Innovation fund (2016QDL-B05).

References

[1] A. Manthiram, S.H. Chung, C. Zu, Lithium-sulfur batteries: Progress and Prospects, Adv. Mater. 27 (2015) 1980-2006.

[2] M.A. Pope, I.A. Aksay, Structural design of cathodes for Li-S batteries, Adv. Energy Mater. 5 (2015) 1500124.

[3] H. Wei, J. Ma, B. Li, Y. Zuo, D. Xia, Enhanced cycle performance of Lithium-sulfur batteries using a separator modified with a PVDF-C layer, ACS Appl. Mater. Interfaces 6 (2014) 20276-20281.

[4] Z.W. She, W. Li, J.J. Cha, G. Zheng, Y. Yang, M.T. McDowell, P.C. Hsu, Y. Cui, Sulphur-TiO_2 yolk-shell nanarchitecture with internal void space for long-cycle lithium-suphur batteries, Nat. Commun. 4 (2013) 1331.

[5] X. Fu, B. Wang, Z. Ren, C. Fan, G. Qian, Z. Wang, Controllable synthesis of TiO_2 hierarchical nanostructures and their applications in lithium ion batteries, RSC Adv. 4 (2014) 42772-42778.

[6] K. Xie, Y. Han, W. Wei, H. Yu, C. Zhang, J.G. Wang, W. Lu, B. Wei, Fabrication of a novel TiO2/S composite cathode for high performance lithium-sulfur batteries, RSC Adv. 5 (2015) 77348-77353.

Evaluation on the implementation effect of wind power policies in china

Zhuhan Long[1,*], Yuqing Wang[1], Jiawen Ye[1], Ming Zeng[1], Xiaorong Wang[2] and Xinying Wang[2]

[1]School of Economics and Management, North China Electric Power University, Beijing, China
[2]Technology Strategy Research Center, China Electric Power Research Institute, Beijing, China

*Corresponding author e-mail: 532300376@qq.com

Abstract. To promote the development of wind power generation, China has introduced a series of supporting wind power policies. However, it is difficult to quantitatively evaluate the implementation effect of these policies. In order to evaluate the implementation effect of these policies, this paper puts forward a comprehensive evaluation method of policy implementation effects; it firstly quantitatively evaluates the weak link of effect through the effect indexes, and then evaluates the policy implementation and the existing problems through the action weight of policies. The conclusion can provide a reference to improve the wind power policy-making.

1. Introduction

Wind is a rich, clean natural and renewable energy, with no environmental pollution problems. It is of great development potential, which has aroused widespread concern at home and abroad, so it is developing rapidly these years. In order to promote the development of wind power, our government plays an obvious role. In recent years, a large number of relevant policies have been issued to accelerate the innovation and development of wind power industry with the government's attention and promotion, and to access the industry backward advantages. Therefore, a scientific and rational policy evaluation can provide scientific basis for the further formulation of wind power generation policies, which is benefit to the rapid and ordered development of wind power.

In the development of wind power industry, problems and solutions are inseparable from the relevant government and policy. Government planning and management and policy have important impacts in the Chinese wind power industry. At present, most of the researches on the wind power industry policy are based on the analysis of the wind energy resources, the current situation and the difficulties of wind power industry, so as to put forward policy suggestions which is corresponding and can promote the development of the wind power industry [1-2], while studies on the quantitative evaluation of the effect of wind power generation policy are scarce. Literature [3] analyzes the characteristics of some national wind power industry policy, combined with the status of China's tax policy; it puts forward proposals to improve China's wind power industry tax policy. Literature [4] analyzes the wind power price policy at home and abroad, and evaluates the different policy types respectively. In this paper, the evaluation index system of wind power generation is established, based on which the evaluation method of policy implementation effect is put forward. Then, the effect of policy implementation in recent three years is evaluated quantitatively, and the existing problems are found out, then the suggestions are put forward.

2. Wind power generation policy evaluation method

2.1. Evaluation index system

The principles of the evaluation index system should be comprehensive, hierarchical, relatively independent and operational. In this paper, the method of qualitative and quantitative analysis is used to construct the evaluation index system which is suitable for the evaluation of implementation effect of wind power policy from the four aspects of development speed, development order, cost efficiency and grid-connected operation. There are 3 levels, 4 aspects and 15 indexes in this system, as shown in Table 1.

Table 1. Evaluation index system of the implementation effect of wind power generation policies.

Target layer	Criteria layer	Index layer	Index explanation
Implementation effect of wind power generation policy A	Development speed B1	Wind power installed capacity C1	Wind power total installed capacity / National power installed capacity
		Production speed C2	Growth of Wind power installed capacity / Growth of power installed capacity
		Development level of wind power industry C3	Domestic / Global production growth
	Development order B2	Standard specification C4	Experts score according to the standard number and improvement degree
		Technique level C5	Fan conversion efficiency
		Layout coordination C6	Deviation between actual layout and theoretical layout / Total development scale
		The deviation rate between planning scale and completion C7	Deviation between actual installation and planning installed capacity / Planning installed capacity
	Cost efficiency B3	Wind power project yield C8	Provincial weight ×internal income of the province
		Tax preferential margin C9	Experts score according to tax preferential margin
		Financing level C10	Experts score according to the financing preferential measures and financing difficulties
		Subsidy releaseC11	Full payment ratio× proportion of the arrears
	Grid-connected operation B4	Operation level C12	Actual power generation / Theoretical power generation (Key provinces)
		Adequacy of power degree C13	Total development scale / Acceptance ability
		Grid-connected service level C14	Experts score according to grid service speed and quality
		Business model C15	Experts score according to the maturity of business model

2.2. Comprehensive evaluation method of policy implementation effect

The evaluation index system above can fully reflect the effect of the wind power related policies, but it cannot reflect the effect of the specific policy implementation. The ultimate goal of the comprehensive evaluation of the wind power policy implementation is not only to analyze the overall implementation of the policy, but also to analyze and comb those specific policy which results in the ineffective aspects, so as to provide a direct reference for further developing of the policy.

Based on the reasons above, this paper presents a policy implementation effect evaluation method based on the AHP, called Assessment for Policy Implementation Effect (APIE).

(1) Evaluation weight of policy. As is shown in Figure 1, the judgment matrix is used to calculate the weight of the policy importance degree, that is, according to the order of importance level, we get the importance of one aspect relative to the target layer (development speed, development order, cost-effectiveness, grid-connected operation), and the importance of one policy relative to the overall implementation effect. The policy classification covers 17 areas that promote development of wind power policies.

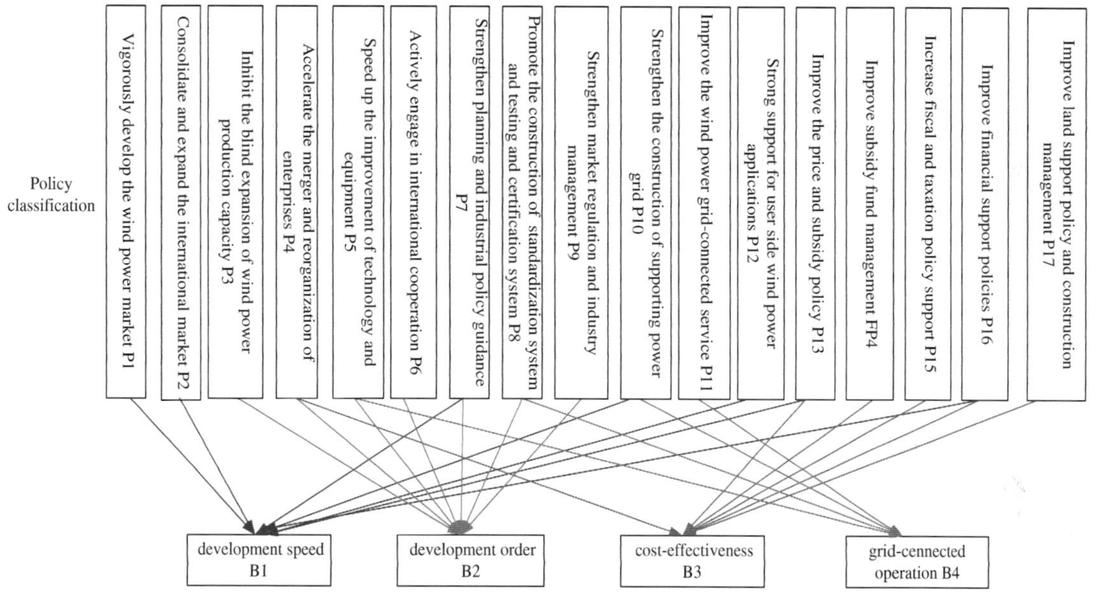

Figure 1. The relationship between wind power generation policy and implementation effect

(2) Index evaluation. According to the index system in Table 1, we select data in recent years to quantitatively calculate the index weight and index value, and then evaluate the changes and trends of the wind power policy in all aspects of wind power development.

(3) Comprehensive evaluation. According to the implementation effect, we evaluate the calculation results, identify weaknesses, and compare with the importance order of policy evaluation, then sort out the existing problems, analyse the implementation situation and the difficulties, finally put forward suggestions.

3. Evaluation on the implementation effect of wind power policy

3.1. Evaluation of policy weight

Since the APIE method evaluates the importance of policy impact, it is necessary to firstly calculate the weight of the policy, including the weight of criteria layer and the weight of policy layer. The calculation of the criteria layer weight needs to list the judgment matrix of the development speed B1, the development order B2, the cost effectiveness B3, and the grid-connected operation B4 relative to

IOP Conf. Series: Earth and Environmental Science **64** (2017) 012097 doi:10.1088/1755-1315/64/1/012097

the objective layer A, and then calculate the maximum feature root and weight vector of the judgment matrix, with conducting a consistency test. The policy weight calculation is based on four judgment matrixes consist of 17 policy classifications relative to the 4 criteria of criteria layer in Figure 1, with calculation of the weight vector and consistency test. According to the calculated monolayer weight of the policy, we calculate the weight of each policy to the total target and sorts. In this paper, all the judgment matrixes of policy weight evaluation are formed of the tone method synthesized with the expert scoring results, as shown in Table 2.

Table 2. Weight of wind power policy.

Policy layer	B1	B2	B3	B4	Total
	0.351	0.351	0.189	0.109	sorts $\sum_{j=1}^{4} b_j f_{ij} (i=1,2,\ldots,17)$
P1	0.205	0	0	0	0.072
P2	0.067	0	0	0	0.024
P3	0	0.20	0	0	0.070
P4	0	0.071	0.051	0	0.035
P5	0	0.157	0	0.077	0.064
P6	0	0.048	0	0	0.115
P7	0.114	0.214	0	0	0.099
P8	0	0.143	0	0.231	0.075
P9	0	0.167	0	0	0.059
P10	0.259	0	0	0.346	0.129
P11	0	0	0	0.346	0.038
P12	0.091	0	0	0	0.032
P13	0.132	0	0.231	0	0.090
P14	0	0	0.154	0	0.029
P15	0	0	0.205	0	0.039
P16	0.132	0	0.231	0	0.090
P17	0	0	0.128	0	0.024

The size of the policy weight can reflect the relative importance of the 17 policies relative to the 4 criteria for evaluating the effectiveness of wind power policies and the overall target of implementation effect.

3.2. Evaluation of Index

Based on the index system in Table 1, we use the judgment matrix to calculate the weight of the index layer relative to the criterion layer.

IOP Conf. Series: Earth and Environmental Science **64** (2017) 012097 doi:10.1088/1755-1315/64/1/012097

Table 3. Weight of wind power index.

Criteria layer and its weight	Index layer	Monolayer weight	Total sorts
B1 0.351	C1	0.429	0.151
	C2	0.429	0.151
	C3	0.142	0.050
B2 0.351	C4	0.126	0.044
	C5	0.234	0.082
	C6	0.435	0.153
	C7	0.205	0.072
B3 0.189	C8	0.508	0.096
	C9	0.079	0.015
	C10	0.218	0.041
	C11	0.195	0.037
B4 0.109	C12	0.227	0.025
	C13	0.434	0.047
	C14	0.227	0.025
	C15	0.112	0.012

We select three years of wind power development-related evaluation indexes from 2013 to 2015 to calculate and score, and then carry out uniformization and dimensionlessness on the calculated value, getting three-year evaluation value of each index. The evaluation value is multiplied by each value of the index and monolayer weight. The comprehensive evaluation value of the criterion is added by the value of each index in the criterion layer, eventually the evaluation value of the four criteria multiply its weight, which is the comprehensive evaluation value of year's wind power industry policy implementation effect, as shown in Table 4.

Table 4. Comprehensive evaluation results of the implementation effect of wind power policy.

Criteria layer	Criteria layer weight	Index name	Index layer weight	Evaluation result 2013		2014		2015	
Development speed B1	0.351	Wind power installed capacity C1	0.429	0.335	0.299	0.450	0.370	0.513	0.423
		Production speed C2	0.429	0.277		0.314		0.365	
		Development level of wind power industry C3	0.142	0.258		0.297		0.323	
Development order B2	0.351	Standard specification C4	0.126	0.018	0.119	0.023	0.116	0.034	0.115
		Technique level C5	0.234	0.025		0.027		0.028	
		Layout coordination C6	0.435	0.178		0.166		0.158	
		Planning scale completion the deviation rate C7	0.205	0.165		0.170		0.173	
Cost efficiency B3	0.189	Wind power project yield C8	0.508	0.188	0.129	0.184	0.134	0.185	0.140
		Tax preferential margin C9	0.079	0.033		0.058		0.072	
		Financing level C10	0.218	0.055		0.074		0.089	
		Subsidy releaseC11	0.195	0.099		0.103		0.106	
Grid-connected operation B4	0.109	Operation level C12	0.227	0.086	0.079	0.072	0.065	0.060	0.052
		Adequacy of power degree C13	0.434	0.057		0.043		0.028	
		Grid-connected service level C14	0.227	0.141		0.123		0.102	
		Business model C15	0.112	0.020		0.023		0.028	
Comprehensive evaluation				0.180		0.203		0.221	

The comprehensive evaluation results are affected by varieties of factors, the specific analysis is as follows:

(1) During 3 years, evaluation values of the development speed index are the highest, and it shows an increasing trend year by year, indicating that in the development process of wind power industry, the development trend is relatively ideal, the implementation effect of relevant policies is well, which promote the rapid development of wind power industry and provides a basic guarantee for it.

(2) The evaluation value of development order index is relatively stable, it mainly has declined in the layout coordination degree, while the growth of remaining several indexes is not obvious, indicating that the scale and speed of wind power development has a significant increase relative to the entire power system and has improvement in the quantity level, but the quality level did not reach the expected one, adjustment of relevant policies is needed.

(3) The valuation value of the cost-effectiveness index is slightly higher, indicating that the investment attractiveness level in wind power projects has become stronger. Under the positive policy, the tax concessions and financing levels of wind power projects have increased significantly, which has promoted the development of wind power projects and have a positive impact on the investment in wind power industry.

(4) The evaluation value of grid-connected operation index decreases obviously, indicating that there is a big problem in grid-connected operation. The remaining indexes have all declined except the business model, the main factor may be the contradiction between the rapid development of wind power industry and the planning and construction of the power grid, which results in a large number of nest abandoned wind. Therefore, supporting power grid construction and other relevant policies need to speed up.

3.3. Policy issues and related recommendations

According to the policy weight, it can be seen that in the policies development speed, "strengthening the supporting power grid construction" (P10) is the main influencing factor; in the policies of development order, "strengthening the planning and industrial policy guidance" (P7) is the main influencing factor; In the cost-effective policies, "perfecting price and subsidy policies" (P13) and "perfecting financial support policies" (P16) are the main influencing factors; in the grid-running policies, "strengthening the supporting power grid construction" (P10) and "perfecting the wind power grid-connected operation services "(P11) have significant impacts. Therefore, it is recommended that the relevant policy making departments focus on the following aspects:

1) Strengthening the guidance of development planning. Strengthening the national wind power in a coordinated way between overall planning and special planning, power planning and grid planning, national planning and local planning, enhancing the consistency, coordination and seriousness of the planning, and releasing information in advance to guide the wind power to a balanced and orderly development.

2) Innovating and perfecting the supporting policies. Establishing a dynamic adjustment mechanism of subsidy policies, which adjusts the standard of subsidies dynamically and utilizes the limited subsidies efficiently according to changes in power generation costs; focusing on the matching of the total amount of annual subsidies and the scale of development to prevent the delinquent subsidies, untimely payment and other issues; strengthening financial innovation and exploring new commercial development model.

3) Strengthening the construction of supporting power grids. Enhancing the overall capacity of wind power and establishing effective mechanisms for synchronous planning and simultaneous production of wind power projects and grid engineering; establishing and improving the technical standards and strengthening the operation and monitoring system to gradually realize the on-line monitoring of the wind power grid running and to control the running at a certain degree.

4. Conclusion

Aiming at the problem that implementation effect of wind power generation policy lack comprehensive quantitative evaluation, this paper puts forward a comprehensive evaluation method called APIE. The evaluation results show that in order to promote the development of wind power in China, we need to focus on the development planning, standard specification, grid operation and so on, to further improve or introduce relevant policies.

The APIE method proposed in this paper not only applies to wind power, but also can establish similar evaluation index systems as required to use on other forms of energy, comprehensively evaluating their implementation effect. What should be noted is that since the implementation of policy is cyclical, it is necessary to dynamically adjust the policy evaluation and index evaluation in this method according to the actual policy function.

Acknowledgments
This work was financially supported by State Grid Science and Technology Project (JS71-16-005).

References

[1] Zhou Haili, Current situation and development trend of wind power generation, J. Gansu Science and Technology. 24 (2008) 9-11.
[2] Hong Jun, Current situation and development trend of wind power generation, Modern Business. Wide Angle, 251-254.
[3] Wang Xi, Li Guijun, Policy Innovation Promoting the Development of New Energy Industry — A Study about Tax Policies towards China's Wind Power Industry, China Soft Science. 12(2014) 170-178.
[4] Pei Xiaoni, The Analysis and Evaluation of The Wind Power Tariff Policy, B.S. Changsha University of Science & Technology, 2006.

The study on working fluids of airborne power generation system based on Rankine cycle by heat energy

Yuan Guo*

Beijing Power Machinery Research Institute, Beijing 100074

*Corresponding author e-mail:tomoranger@163.com

Abstract. This paper proposed a new concept named airborne power generation system based on Rankine cycle by heat energy, namely, the presented system combined the Rankine cycle with environmental control system in aircraft to recycle the waste heat of engine bleed air with high temperature and generate power. This paper mainly discussed the choosing of optimum working fluid which could apply in the combined power generation system mentioned above when the temperature of the coming bleed air was about 400 degree centigrade.

1. Introduction

The core part of the airborne environmental control system is the refrigeration system, which is used to secure the safety and comfort, guarantee the normal operation of electronic equipment. At present, air turbine refrigeration technology is widely applied to both the advanced civil aircraft and military aircraft [1]. Furthermore, this technology derives from the high-temperature bleed air of the airborne engine. It's generally accepted that there is much thermal energy contained in the bleed air, and the temperature is roughly 400 degree centigrade under cruising condition. According to the above situation, the part of high-temperature gas have to be cooled preliminary by ram air. This traditional method not only causes the waste of the thermal energy of bleed air, but also threatens the stealth performance seriously due to the use of ram air. Therefore, with the high-speed development of modern aircraft, how to improve traditional aircraft environmental control system reasonably becomes the most urgent 'thermal problem'.

Under this circumstances, this paper proposed the power generation system by waste heat energy in bleed air based on Ranking cycle. This combined system can refrigerate the engine bleed air when recycling and utilizing the thermal energy of bleed air, which can not only improve the level of thermal energy, but also guarantee the modern aircraft's excellent performance owning to decrease the use of ram air. As mentioned above, the power generation system by waste heat energy in bleed air is an effective way to solve the 'thermal problem'. As working fluid is the essential factor to power generation system based on Ranking cycle, this paper discussed the choosing of optimum working fluid applied to this specific application technology in aircraft.

2. Calculation principles of Rankine cycle system.

As shown in Fig.1 (a) and (b), the working fluid at point 1 comes into evaporator to absorb the heat energy and begins to evaporate. Afterwards, it becomes superheated vapor with both high temperature and high pressure at point 2, and then the vapor comes into the turbine to expand to point 3. Next, it enters condenser to release heat and begins to condense until it gets the state of point 4. In the end of a cycle, it comes into the pump to increase fluid's pressure to the evaporation pressure and arrives back to point 1.

The expanding power in turbine:

$$w_g = h_2 - h_3 \tag{1}$$

The compressing power in pump:

$$w_h = h_1 - h_4 \tag{2}$$

The net power of cycle:

$$w_{net} = w_g - w_h = (h_2 - h_3) - (h_1 - h_4) \tag{3}$$

The heat absorption of system:

$$q = h_2 - h_1 \tag{4}$$

The thermal efficiency of cycle:

$$\eta = \frac{w_{net}}{q} = \frac{(h_2 - h_3) - (h_1 - h_4)}{(h_2 - h_1)} \tag{5}$$

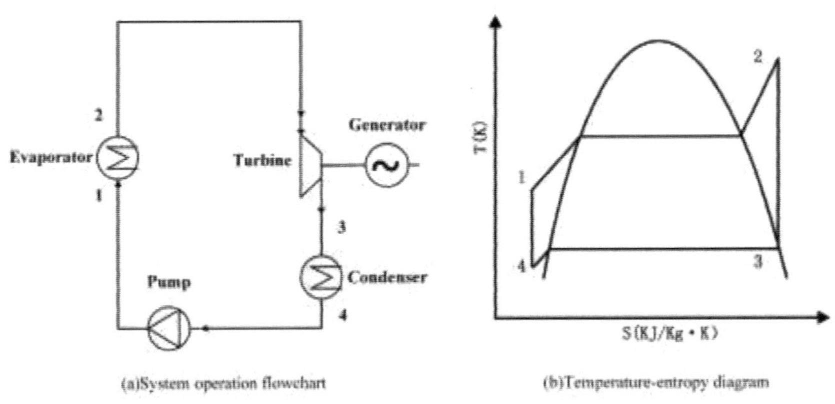

Fig.1 System operation flowchart and the temperature-entropy diagram

3. Selection method of the optimum working fluid.

3.1. The preliminary selection.

As mentioned, the temperature of waste heat source is about 400 degree centigrade. We may just as well select the working fluids whose critical temperature is larger than 150 degree centigrade from the common fluids[2] attached to the Rankine cycle according to the rule that fluids with high critical temperature usually have the better performance during Rankine cycle showed by several references [3,4,5,6].

This paper selected 19 working fluids such as the following, toluene, n-butane, n-pentane, n-hexane, n-heptane, n-octane, n-nonane, n-dodecane, isopentane, cyclohexane, methanol, ethanol, R11, R113, R114, R123, R141b, and R365mfc, water. Among these fluids, water has the largest critical temperature which could reach 374 degree centigrade, and it also can satisfy the temperature requirement of the combined power generation system.

3.2. The method of calculation.

As mentioned, the temperature of the aircraft engine bleed air is nearly 400 degree centigrade under cruising condition. So we can pick the evaporation temperature of working fluids as 20 degree centigrade lower than their respective critical temperature. Considering that the heat sink of aircraft are cabin air and low temperature fuel oil, we may make the condensation temperature of working fluids be 50 degree centigrade. Furthermore, we could control the superheating temperature and the undercooling temperature to be 10 and 5 degree centigrade respectively. We must point out that it is necessary to observe the dryness of steam at point 3 and make it higher than 0.86[7] all the time to ensure the safe operation of turbine, even all the system.

On this foundation, we can acquire the optimum working fluid on the basis of the calculation principles of the Rankine cycle system mentioned above with the help of MATLAB for the proposed system attached to different calculation aims such as the most net power of cycle, the most heat absorption of system, and the highest thermal efficiency of cycle.

3.3. Calculation results based on the most net power.

As shown in Fig.2, we can see the trend on net power of different working fluids.

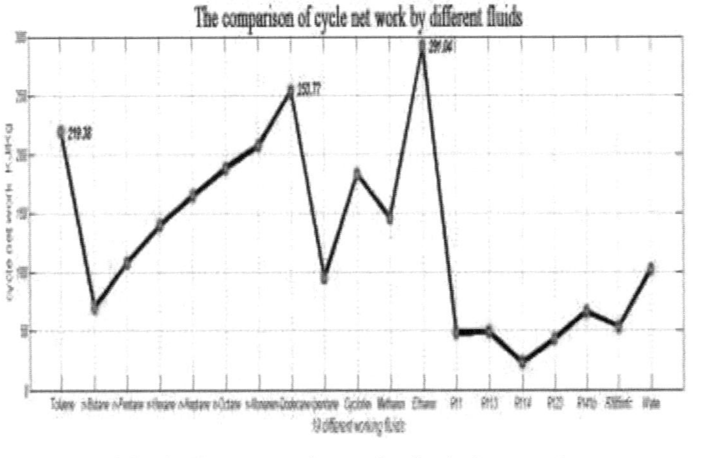

Fig.2 The comparison of calculation results

In cycle net work

According to Table.1, among 19 fluids, ethanol has the most net power to 291.04 KJ/Kg. And n-dodecane and toluene are following with 253.77 KJ/Kg and 219.38 KJ/Kg respectively. According to the engineering experience, when we pick ethanol as the working fluid of this combined power generation system and make the mass flow be 1000Kg/h, then the electricity output could approach up to 80.8KW if the loss of generator is ignored.

Table.1. The results of cycle network and output power

Fluids	cycle net work /(KJ/Kg)	output power by 1000kg/h mass flow /KW
toluene	219.38	60.94
n-butane	69.98	19.44
n-pentane	108.35	30.10
n-hexane	139.63	38.79
n-heptane	165.08	45.86
n-octane	188.20	52.28
n-nonane	207.74	57.71
n-dodecane	253.77	70.49
isopentane	95.43	26.51
cyclohexane	183.33	50.93
methanol	146.03	40.56
ethanol	291.04	80.84
R11	48.38	13.44
R113	49.00	13.61
R114	23.05	6.40
R123	43.33	12.04
R141b	65.85	18.29
R365mfc	53.52	14.87
Water	102.00	28.33

3.4. Calculation results based on the most heat absorption.

Under some certain conditions, it is significant to recycle most heat energy to the extreme extent to obtain the most temperature drop of bleed air. Then what we need is not the most power output but the most heat absorption in the evaporation process.

As shown in Fig.3, we can see the trend on heat absorption of different working fluids.

Fig.3. The comparison of calculation results

In cycle heat absorption

According to Table.2, among 19 fluids, water has the most net power up to 1339.94 KJ/Kg. And ethanol and n-dodecane are following with 1102.11 KJ/Kg and 1085.96 KJ/Kg respectively.

Table.2. The results of cycle heat absorption

Fluids	cycle heat absorption /(KJ/Kg)
toluene	757.18
n-butane	469.71
n-pentane	573.47
n-hexane	664.52
n-heptane	747.48
n-octane	827.12
n-nonane	899.22
n-dodecane	1085.96
isopentane	534.80
cyclohexane	726.26
methanol	974.39
ethanol	1102.11
R11	228.60
R113	234.39
R114	167.62
R123	229.97
R141b	306.26
R365mfc	303.95
Water	1339.94

3.5. Calculation results based on the highest thermal efficiency.

As shown in Fig.4, we can see the trend on thermal efficiency of different working fluids.

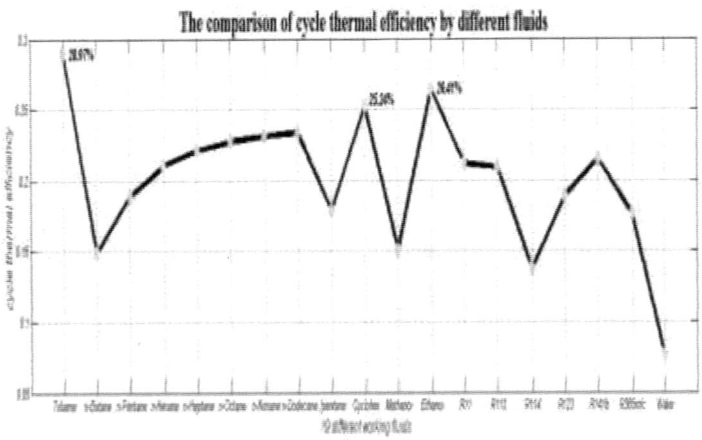

Fig.4. The comparison of calculation results

In cycle thermal efficiency

According to Table.3, among 19 fluids, toluene has the highest thermal efficiency up to 28.97%. And ethanol and cyclohexane are following with 26.41% and 25.24% respectively.

Table.3. The results of cycle thermal efficiency

Fluids	cycle thermal efficiency /%
toluene	28.97
n-butane	14.90
n-pentane	18.89
n-hexane	21.01
n-heptane	22.08
n-octane	22.75
n-nonane	23.10
n-dodecane	23.36
isopentane	17.84
cyclohexane	25.24
methanol	14.99
ethanol	26.41
R11	21.16
R113	20.90
R114	13.75
R123	18.84
R141b	21.50
R365mfc	17.61
Water	7.61

3.6. Thermodynamic performance analysis of moist working fluids.

We can distinguish easily from Fig.4 that as the typical moist fluid, water has the lowest thermal efficiency as low as 7.61%.However, we can see that the net power of water is not the minimum value from Fig.2. The main cause is that when we use water as the system working fluid, we must control the condensation pressure to ensure that the dryness of steam at point 3 is larger than 0.86 and thus to ensure the safe operation of turbine. Compared with dry working fluids, the higher condensation pressure will decrease the net power and the thermal efficiency. Furthermore, methanol is also a moist fluid under this temperature range, according to the calculation, we can see that the condensation temperature must be 130 degree centigrade at least to guarantee the value of the dryness of steam at point 3 as mentioned above.

4. Summary

1) Combining Ranking cycle with aircraft environment control system can not only refrigerate the engine bleed air to make it meet the system operation criteria, but also decrease the utilization of ram-air and ensure the performance of modern aircraft. Furthermore, this method can generate power by using waste heat energy, transfer thermal energy to electronic energy and upgrade the level of energy.

2) Combining Ranking cycle with aircraft environment control system can not only refrigerate the Targeting at the largest cycle net power, calculation result showed that ethanol performed best upon 19 working fluids, and the acquired recycling net power of unit mass ethanol was 291.04KJ/Kg. n-dodecanal and toluene took the following position, were 253.77KJ/Kg and 219.38KJ/Kg respectively. The cycle net power of R114 was the least, and the acquired net power of R114 is just 23.05KJ/Kg.

3) Combining Ranking cycle with aircraft environment control system can not only refrigerate the Targeting at the largest cycle heat absorption, calculation result showed that water performed best upon 19 working fluids, and the acquired value of unit mass water was 1339.94KJ/Kg. And ethanol and n-dodecanal took the following position, were 1102.11KJ/Kg and 1085.96KJ/Kg respectively.

4) Combining Ranking cycle with aircraft environment control system can not only refrigerate the Targeting at the highest cycle thermal efficiency, calculation result showed that toluene performed best upon nineteen working fluids, and this maximum value was 28.97%. Ethanol and cyclohexane

were following with 26.41% and 25.24% respectively. The value of cycle thermal efficiency by water was the least with just 7.61%.

5) Combining Ranking cycle with aircraft environment control system can not only refrigerate the As far as moist working fluids, controlling condensation pressure to ensure that the dryness of steam at point 3 in Fig.1 is larger than 86% is needed to secure the system's safety. But it is necessary to point out that this method will weaken the cycle net power and thermal efficiency.

6) Combining Ranking cycle with aircraft environment control system can not only refrigerate the what we must point out is that the selection methods must consider the realistic application situation. And even if ethanol, toluene, n-dodecane and cyclohexane have the satisfied thermodynamic properties, there are also so many restrictive conditions due to their unstable chemistry properties.

References

[1] SHOU Rong-zhong, HE Hui-shan. Environmental Control of Aircraft. Beihang University Press, 2003, pp. 143-145.(In Chinese)

[2] FENG Xun: Theoretical research of the mid-low-temperature waste heat generation using organic rankine cycle. (M.E.Degree, Shanghai Jiao Tong University, 2011). pp.22-23. (In Chinese)

[3] ZHANG Hong. Power generation technology based on organic rankine cycle by waste heat used working fluids with low boiling point. Cement. (2006) No.8, pp.11-15. (In Chinese)

[4] GU Wei: Theoretical and experimental study of organic rankine cycle for low and medium grade heat source utilization. (Ph.D. Degree, Shanghai Jiao Tong University, 2010).pp.30-39. (In Chinese)

[5] ZHANG Xin-xin, HE Mao-gang. Selection of working fluid used in vapor power cycle for waste heat recovery of vehicle energy. Journal of Engineering Thermophysics. Vol.31(2010)No.1,pp.15-18. (In Chinese)

[6] KANG Le-ming, TAN Yu-fei, WU Jia-zheng, ZHU Tong. Engineering Thermodynamics. China Architecture

IOP Conf. Series: Earth and Environmental Science **64** (2017) 012099 doi:10.1088/1755-1315/64/1/012099

Fault diagnosis of power transformer based on fault-tree analysis (FTA)

Yongliang Wang[1], Xiaoqiang Li[2], Jianwei Ma[3] and SuoYu Li[2]

1State Grid NingXia Electric Power Company, China
2State Grid Electric Power Research Institute, China
3State Grid Corporation of China, China

Abstract. Power transformers is an important equipment in power plants and substations, power distribution transmission link is made an important hub of power systems. Its performance directly affects the quality and health of the power system reliability and stability. This paper summarizes the five parts according to the fault type power transformers, then from the time dimension divided into three stages of power transformer fault, use DGA routine analysis and infrared diagnostics criterion set power transformer running state, finally, according to the needs of power transformer fault diagnosis, by the general to the section by stepwise refinement of dendritic tree constructed power transformer fault

1. Introduction

Power transformer is an important power transmission equipment in power grid. Its operation is directly related to the safety and power supply reliability of hair and power supply system [1-2]. With the increasing of power system scale and single transformer capacity, the losses to the national economy caused by the fault of power transformer is also growing. Thus making reliability analysis and risk assessment of transformer is an increasingly important issue [3-5].

Fault diagnosis of power transformer is under the situation of device hanging hood and not disintegration, based on experience and data, determine the state of transformer through certain technical means and track diagnosis and estimate the existing fault and evolution of transformer [1, 6]. Fault detection and diagnosis of the running transformer is a key channel to improve the safety and power supply reliability of power system [1, 6-7].

References [8-10] summed up the failure modes of power transformer, and fuzzy control carried out hazard assessments. References [11-12] discussed the way to build transformer fault diagnosis system based on artificial neural networks and fuzzy mathematics. References [13] discussed the application of oil gas analysis in fault diagnosis of transformer. References [14-15] researched fault diagnosis and condition-based maintenance of power transformer, preliminary exploration the online monitoring and repair methods. Above-mentioned researches were restricted to one or several troubleshooting methods, but it's lack of research on comprehensive and three-dimensional hierarchy application, unrealized high degree of unity echelon management and integrated scheduling, and it's lack of energy efficiency data comprehensive assessment system and dynamic optimization integration strategies.

The paper summarizes the five power transformer fault types according to the site firstly, then divided power transformer fault into three stages from the time dimension. Then set a criterion of

IOP Conf. Series: Earth and Environmental Science **64** (2017) 012099 doi:10.1088/1755-1315/64/1/012099

power transformer running state using DGA conventional analysis and infrared diagnostics. Finally, according to the demand of power transformer fault diagnosis, built power transformer fault tree by pressing dendritic stepwise refinement from overall to section.

2. Fault types and stages

2.1. The structure of power transformers
Power transformer can be divided into the body and attachments from the structure.

(1) Body

The body of power transformer can be divided into three parts: 1) transformer body, includes a core, a coil and a clamping device; 2) tank, to place the body and transformer oil; 3) transformer oil, for the insulation, cooling, corrosion of body.

(2) Attachments

The attachments of power transformer mainly includes:1) transformer oil pillow; 2) cooling system; 3) casing; 4) on-load tap-changer; 5) testing instrument; 6) body guard.

2.2. Fault types of power transformers
The main fault types of power transformer can be divided into ten categories according to different parts, such as core fault, casing fault, cooling system fault, lead fault, winding fault, insulation fault, radiator fault, tap changer fault, seal fault, and other fault.

There are five main fault as follows:

(1) Lead fault

Lead is the wire to connect the packaging body and external of power transformer. Lead fault is affected by welding quality, and it can be subdivided into poor contact, open circuit and short circuit. Lead phase short circuit may further lead to a winding short circuit. The harm of poor contact and ground short circuit is relatively small.

(2) Winding fault

Winding is the main component of power transformer. Winding fault is mainly in the terminal, longitudinal insulation and coils. It includes winding loose, open circuit, short circuit, and deformation.

(3) Insulation fault

Insulation system mainly includes insulating paper, board, transformer oil and insulating member. Insulation fault is the fault of insulation system caused by insulating paper thin and brittle.

(4) Tap changer fault

Tap changer can be divided into on-load tap-changer and off-circuit tap-changer. It can change the number of turns of the high voltage coil. On-load tap-changer is the high incidence site of power transformer fault because of its run regulating device and the switching operation is very frequent.

(5) Seal fault

The consequence of seal fault is more serious. The main reasons of seal fault are bushing rupture, pump seal bad, welding bad and seal aging.

2.3. Fault stages of power transformers
With the increasing of use time, the fault stages of power transformer can be divided into early fault period, accidental fault period, loss fault period.

(1) Early fault period

Early fault period is the 1-3 years after power transformer was put into use. The fault rate is relatively high at this stage. The fault rate is lower with the operation time is longer.

(2) Accidental fault period

Accidental fault period is the remaining useful life after early fault period. The fault rate is steady at this stage. Fault doesn't occur under normal circumstances unless the accident occurred.

(3) Loss fault period

Loss fault period is on the eve of retirement. The fault rate has a substantial increase with the operation time is longer.

3. Running state criterion

3.1. DGA routine analytical methods

Currently, most power transformers are still oil-immersed transformers. The main method of operation condition monitoring of oil-immersed transformers is dissolved gas analysis (DGA). The data sampling accuracy is high, the efficiency is high, and the timeliness is good.

Three ratio method and characteristic gas method are the most common methods. Artificial intelligence analysis methods such as fuzzy diagnosis, artificial neural network and expert diagnostic system have been applied gradually.

There are two methods of fault diagnosis of power transformer.

(1) The normal value of the dissolved gas in oil is standard, determinate dissolved gas content in power transformer oil. The normal value of the dissolved gas in oil is as shown in table I. The limit values of hydrogen and hydrocarbon gas content are listed in table I. The dissolved gas in oil should be less than this value under normal operating conditions.

Table I The normal value of the dissolved gas in oi**l**

Gas Composition	H_2	C_2H_2	Total Hydrocarbons (C_1+C_2)
Normal Limit value (μL/L)	150	1-5	150

(2) Determinate gas production rate of power transformer. The gas production rate of power transformer is affected by factors such as fault location, fault energy consumption and temperature. Specific determination methods are shown as table II.

Table II determination methods of gas production rate

Serial Number	Phenomenon	Conclusion
1	$a<b, c<d$	Normal
2	$b<a<3b, c<d$	Slight Fault
3	$b<a<3b, d<c<2d$	Moderate Fault
4	$a>3b, c>3d$	Serious Fault

a: the absolute value of total hydrocarbons actual content;
b: the normal value of total hydrocarbons content;
c: the actual gas production rate of total hydrocarbons;
d: the normal value of the gas production rate of total hydrocarbons

3.2. Infrared diagnosis methods

Fault infrared diagnosis technology is a emerging technology area. It used modern infrared technology and artificial intelligence technology to solve the problem of equipment diagnosis in modern equipment management system engineering.

Infrared diagnosis technology can depth analysis of the heat distribution of power transformer fault. It's easy to clear the surface temperature of the heater case. The analysis of temperature changes very accurate. In addition, its data source is wide, the automation degree is high, and it's stable and reliable. Thus, infrared diagnosis technology is an important technology of fault diagnosis of temperature rise type of power transformer.

IOP Conf. Series: Earth and Environmental Science **64** (2017) 012099 doi:10.1088/1755-1315/64/1/012099

4. Fault tree diagnosis

4.1. Fault tree of power transformer

Fault tree analysis (FTA) is an important method of fault diagnosis of power transformer. It is a special logical causal diagram, and it analysis from the whole to the local level like a inverted tree.

The main purpose to construct fault tree of power transformer is analysis the proportion of each part step by step by expression such as event code and logic gate symbols. And then the technology or management tools can be put forward to management fault hidden troubles.

Fault tree of power transformer includes 8 subtrees such as "current carrying system", "core", "insulating oil", etc. The symbols of fault tree of power transformer are shown as table III.

Table III The symbols of fault tree of power transformer

Event Symbols			
Serial Number	Name	Symbols	Explanation
1	Bottom Event		Unable to continue decomposition of the underlying event
2	Top Event	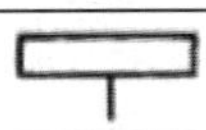	Event at the top of the fault tree
3	Interme-diate Event	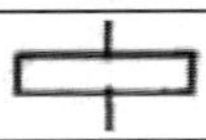	Event between bottom event and top event
Logic Gate Symbols			
Serial Number	Name	Symbols	Explanation
1	Or Gate	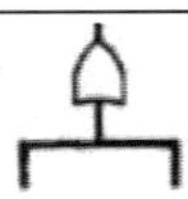	Multiple input events, if one input event or multiple input events occur, output event occur
2	And Gate	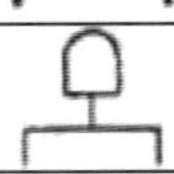	Multiple input events, all input events occur at the same time, output event occur

The subtrees such as "cooling system", "current carrying system" and "insulation damage" are constructed as follow figures:

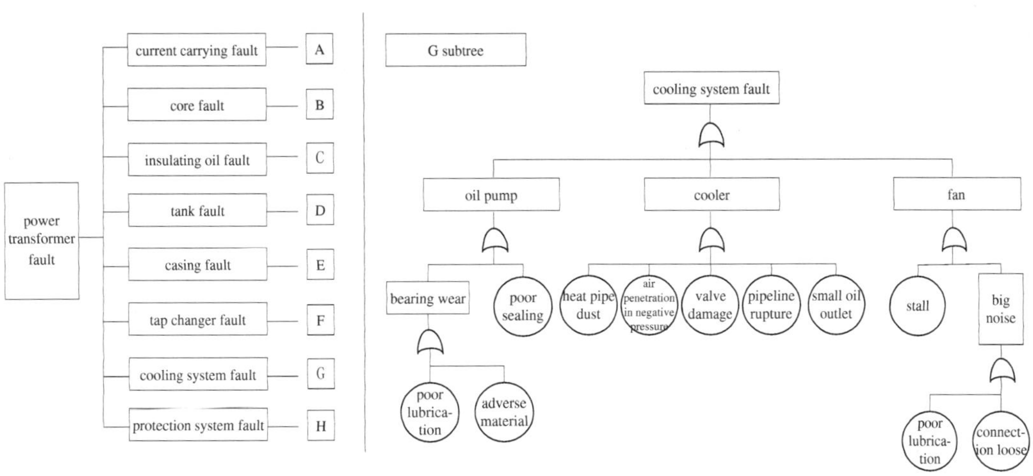

Fig. 1. Cooling system fault tree

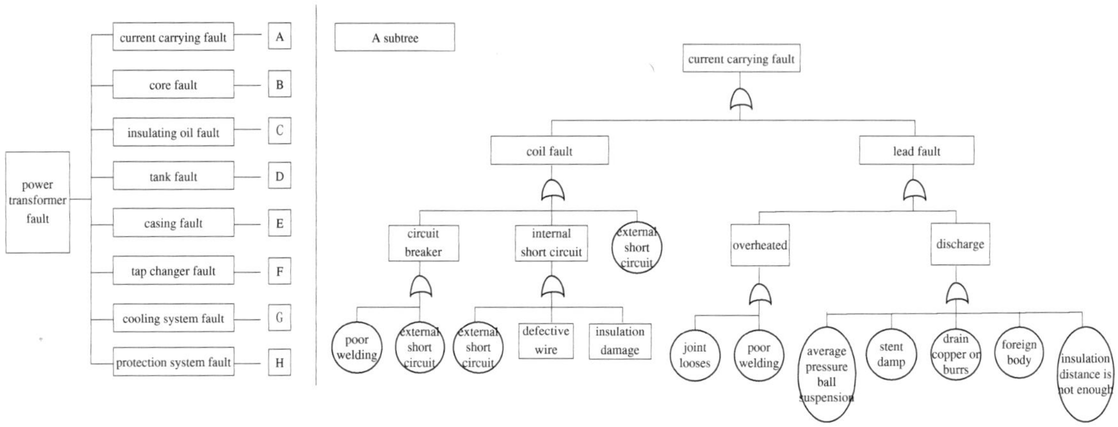

Fig. 2. Current carrying system fault tree

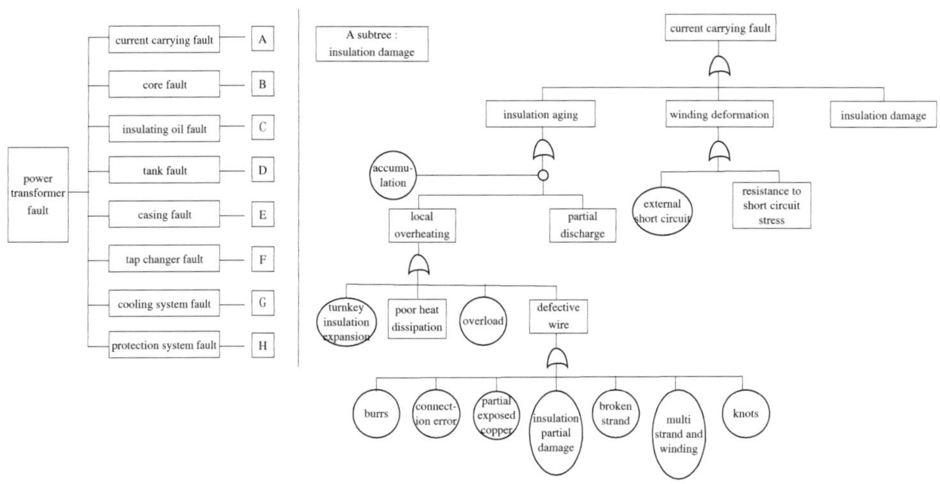

Fig. 3. Insulation damage fault tree

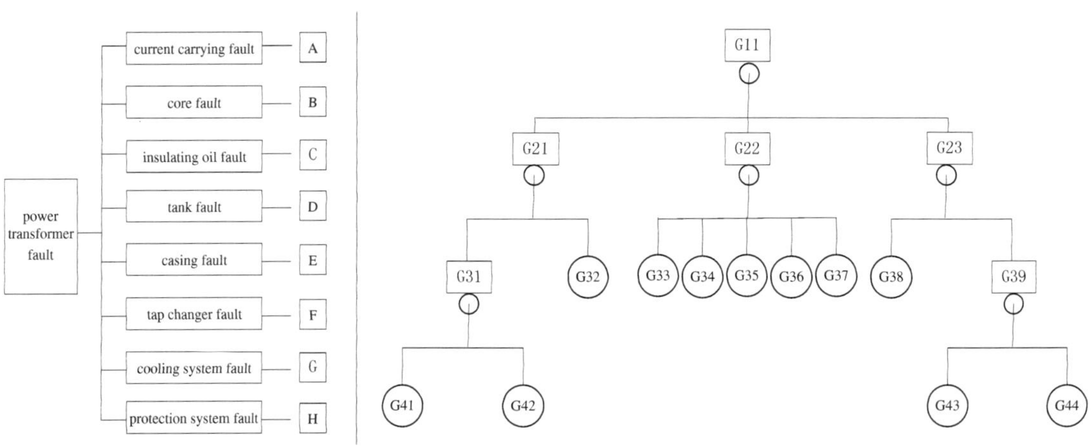

Fig. 4. Cooling system fault tree draft according to fault tree files

4.2. Analysis modules of fault tree

Analysis modules of fault tree of power transformer include probability file module, correction module, qualitative and quantitative analysis module, reliability analysis module and display module.

(1)Probability file module

Probability file module includes some information such as fault name, fault probability and fault code. It calculated the probability of bottom events through annual fault rate, probability value, operation time, etc.

(2) Correction module

Correction module mainly used to modify annual fault rate and operation time.

(3) Qualitative and quantitative analysis module

Qualitative and quantitative analysis module mainly carry out qualitative and quantitative analysis of subtrees.

(4) Reliability analysis module

Reliability analysis module combined analysis results of subtrees before analysis the entire fault tree.

(5) Display module

Display module is used to intuitive display the results of fault diagnosis of power transformer. It can display the logical relationship between events of fault tree comprehensively and clearly.

5. Conclusion

With the development of economy and society, people are getting more and more high demand for power supply. Power transformer is an important equipment of power plants and substations. Simultaneously, it's an important hub of send transmission substation link of power system. The performance quality and operating condition of power transformer direct impact the reliability and stability of power system. Therefore, the importance of fault diagnosis of power transformer is becoming increasingly prominent.

The fault types of power transformer are relatively complicated. Summary the fault types of power transformer is an important reference to design, manufacture and inspection power transformers. This paper summed up five fault types of power transformer firstly. Then divided fault stages of power transformer into three stages from time dimension. Afterwards, set running state criterion of power transformer by DGA routine analytical methods and infrared diagnosis methods. Finally, constructed fault tree of power transformer by the general to the section by stepwise refinement of dendritic tree according to the needs of power transformer fault diagnosis. The fault tree of power transformer provides an important basis for state monitoring and fault diagnosis of power transformer.

6. References

[1] G.Dliang, "The malfunction analysis and research on power transformer," master dissertation, Hefei Polytechnic University, 2004.

[2] W.Njiang, "Fault diagnosis and analysis of power transformer," master dissertation, Wuhan University, 2004.

[3] Y. Hu, L. Cheng, "Statistical analysis of large power transformer fault," Electric Power Safety Technology, vol. 1, pp. 20-22, 2003.

[4] W. Guan, "Research on fault mode and consequence analysis of power transformer," master dissertation, North China Electric Power University, 2008.

[5] Van Gestel T, Suykens A K, Baesens B, et al, "Benchmarking least squares support vector machine classifiers," Machine Learning, vol. 54, pp. 25-32, 2004.

[6] H. Qzhang, "Academic discussion on fault and fault diagnosis technology of power transformer," Business Herald, vol. 6, pp. 290-292, 2012.

[7] Watkins A, Timms J, Boggess L, "Artificial immune reconition systems(AIRS): an immune-inspired supervised learning algorithm," Genetic Programming and Evolvable Machines, vol. 5, pp. 291-317, 2004.

[8] Mathur A, Foody G M, “Multiclass and binary SVM classification: implications for training and classification users,” IEEE Geoscience and Remote Sensing Letters, vol. 5, pp. 241-245, 2008.
[9] J. Flu, “Diagnosis of power transformer fault,” Jiangxi Electricity Vocational and Technical College Journal, vol. 1, pp. 10-13, 2004.
[10] N. Liu, G.Dliang, “Failure mode effect and criticality analysis for condition maintenance of power transformer,” High Voltage Engineering, vol. 29, pp. 3-8, 2003.
[11] Q. Pyang, W. Dxue, “Fault diagnosis of power transformer based on artificial neural network,” Transformer, vol. 3, pp. 84-87, 2000.
[12] J. Wzhang, D. Gzhao, L. Wdong, “Transformer fault diagnosis system based on Fuzzy Mathematics,” High Voltage Engineering, vol. 12, pp. 9-12, 1998.
[13] Q. Pyang, W. Dxue, “Application of gas analysis in transformer fault diagnosis,” Shanghai Electric Power University Journal, vol. 3, pp. 24-27, 1996.
[14] D. Wan, J. Mwang, Y. Mwu, “Faults diagnosis and maintenance strategy of power transformer,” Jiangsu Power Engineering, vol. 6, pp. 37-42, 2003.
[15] G. Zguan, J. Lhe, “On-line insulation monitoring techniques and condition based maintenance,” Electric Power, vol. 33, pp. 46-50, 2000.

IOP Conf. Series: Earth and Environmental Science **64** (2017) 012100 doi:10.1088/1755-1315/64/1/012100

Study on dynamic performance of SOFC

Haiyang Zhan*, Qianchao Liang, Qiang Wen and Runkai Zhu

Naval University of Engineering, Wuhan 430033, China

*Corresponding author e-mail:13349923382@163.com

Abstract. In order to solve the problem of real-time matching of load and fuel cell power, it is urgent to study the dynamic response process of SOFC in the case of load mutation. The mathematical model of SOFC is constructed, and its performance is simulated. The model consider the influence factors such as polarization effect, ohmic loss. It also takes the diffusion effect, thermal effect, energy exchange, mass conservation, momentum conservation. One dimensional dynamic mathematical model of SOFC is constructed by using distributed lumped parameter method. The simulation results show that the I-V characteristic curves are in good agreement with the experimental data, and the accuracy of the model is verified. The voltage response curve, power response curve and the efficiency curve are obtained by this way. It lays a solid foundation for the research of dynamic performance and optimal control in power generation system of high power fuel cell stack.

1. Introduction

Solid oxide fuel cells can convert chemical energy directly into electrical energy, which breaks the limitation of the cycle thermal efficiency of Kano. It has the incomparable efficiency of traditional heat engine, and the combined cycle efficiency of thermal power is as high as 50% [1]. The application of fuel cell in the next generation distributed generation device and portable mobile power supply has the advantages of high efficiency, environmental protection, safety and so on. SIEMENS in 2000 to achieve the 100kW level of the fuel cell stack power generation system running [2], but the majority of domestic fuel cell stack is still in the theoretical stage.

At present, the technology of SOFC is not mature, and the price is very high. When the load changes, it will cause the irreversible loss of the fuel cell plate. In order to solve the load and fuel cell power real-time matching problem, the urgent need to explore the high temperature solid oxide fuel cell dynamic response process of mutation in the load; urgent need construct a simple, accurate and reliable mathematical model, saving the cost, shorting the test cycle. In this paper, a one-dimensional mathematical model is established, which lays a solid foundation for the design of the stack structure and the construction of the combined power generation system [3].

2. Constructing mathematical model

The fuel cell plate is divided into three layers: anode, electrolyte and cathode. The energy of any node in the stack is:

$$E_{\text{total}} = \Delta H \cdot I/nF \tag{1}$$

The output power in the form of electricity, can be expressed as:

$$W = -IV \tag{2}$$

The heat loss of the node is:

$$Q = \Delta H \cdot I/nF - W \tag{3}$$

It is concluded that the loss of the heat dissipation is the formation of polarization heat generation and entropy production. Using the polarization, diffusion, forced convection heat transfer, mass conservation and other factors as the constraint conditions, the iterative algorithm is used to reduce the error [4]. The logical structure of the mathematical model is shown in Figure 1:

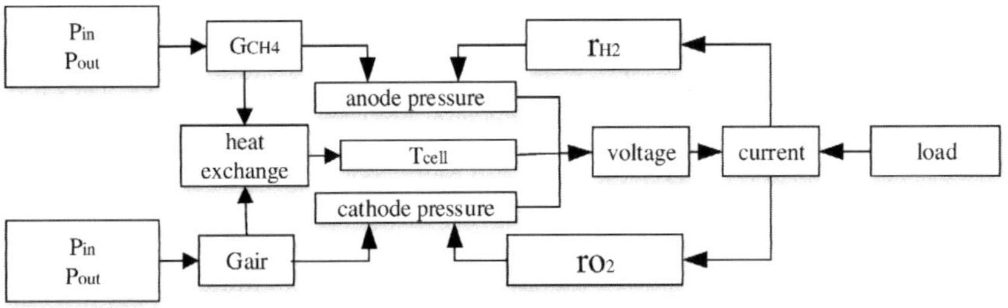

Figure 1. Schematic diagram of fuel cell mathematical model

2.1. Reformer model

The fuel gas of the fuel cell stack is methane and steam reforming reaction mixture, and the steam -to-carbon ratio should be greater than 3 in order to ensure that no carbon deposition phenomenon, There are two reversible reactions in the reformer under the action of catalyst [5]:

$$CH_4 + H_2O \leftrightarrow CO + 3H_2,\ \Delta H = 206.2\ KJ/mol \tag{4}$$

$$CO + H_2O \leftrightarrow CO_2 + H_2, \Delta H = -41.2\ KJ/mol \tag{5}$$

The rate of methane consumption was determined by reforming reaction rate:

$$r_r = K_r P^{\alpha}_{CH_4,tpb} P^{\beta}_{H_2O,tpb} \exp(-\frac{E_r}{RT_{fuel}}) \tag{6}$$

Reforming coefficient α=0.85, β= -0.35, activation energy of reforming reaction Er=95 kJ/mol, Reforming reaction rate coefficient Kr=8542. The residual amount of methane in the exhaust gas can be neglected, and the reforming reaction of methane reaches equilibrium state [6]:

$$K_r = \exp(A_r T^4_{fuel} + B_r T^3_{fuel} + C_r T^2_{fuel} + D_r T_{fuel} + E_r) \tag{7}$$

The reaction speed is faster and the equilibrium state is more rapid than reforming reaction [7]:

$$K_s = \exp(A_s T^4_{fuel} + B_s T^3_{fuel} + C_s T^2_{fuel} + D_s T_{fuel} + E_s) \tag{8}$$

2.2. Electrochemical model

Electrochemical reactions are as follows in the anode and cathode [8]:

$$H_2 + O^{2-} \rightarrow H_2O + 2e^- \tag{9}$$

$$\frac{1}{2}O_2 + 2e^- \rightarrow O^{2-} \tag{10}$$

The experimental results show that the electrochemical reaction is extremely fast enough to consume all reactants reaching the reaction interface [9]. The voltage of SOFC can be expressed by the Nernst equation:

$$E^0 = 1.2723 - 2.7645 \times 10^{-4} T_{\text{cell}} \tag{11}$$

Actual stack voltage:

$$v_{\text{fc}} = E - \eta_{\text{ohmic}} - \eta_{\text{conc}} - \eta_{\text{act,a}} - \eta_{\text{act,c}} \tag{12}$$

Ohmic polarization indicates the resistance of electrons to the electrolyte; Concentration difference polarization indicates that the polarization resistance formed by the transfer process of the gas phase material through the electrode can be observed only under the high current density or the low concentration of the segregation; Anode activation polarization and cathode activation polarization indicate that the electrochemical reaction must overcome an energy barrier for the activation energy of the reaction[10].

2.3. Diffusion Model

The diffusion process plays an important role in the process of dynamic response. This is because the actual reaction takes place in the three-phase interface layer, and the reactants must diffuse through the porous electrode to the three-phase interface layer. In the first phase the material flow must pass through the interface layer to the surface of the electrode; in the second stage, the material flows through the porous electrode to the reaction interface [11]. Diffusion are major means of transmission, and the driving force of diffusion is a concentration gradient, Fick's law shows that [12]:

$$\frac{\partial C}{\partial t} = -D\frac{\partial^2 C}{\partial x^2} \tag{13}$$

The diffusion coefficient can be used to express the diffusion velocity, and the pressure of the gas components in the three-phase interface can be obtained:

$$P_{\text{H}_2,\text{tpb}} = P_{\text{H}_2} - \frac{RTl_{\text{a}}I}{2FD_{\text{eff(a)}}} \tag{14}$$

$$P_{\text{O}_2,\text{tpb}} = P_{\text{air,in}} - (P_{\text{air,in}} - P_{\text{O}_2}) \times \exp\frac{RTl_{\text{c}}I}{4FD_{\text{eff(c)}}P_{\text{air,in}}} \tag{15}$$

Anode plate channel gas contains CH_4, CO, CO_2, H_2, and H_2O. The material flow constraints are obtained by mass conservation:

$$\frac{\mathrm{d}G_i}{\mathrm{d}t} = A_\mathrm{a}\frac{P_\mathrm{fuel,in}}{RT_\mathrm{fuel,in}}u_\mathrm{fuel,in}x_{i,\mathrm{in}} - A_\mathrm{a}\frac{G_i}{dV_\mathrm{a}}u_\mathrm{fuel} + R_i \tag{16}$$

$$R_i = [-r_\mathrm{r}; r_\mathrm{r} - r_\mathrm{s}; r_\mathrm{s}; 3r_\mathrm{r} + r_\mathrm{s} - r_3; -r_\mathrm{r} - r_\mathrm{s} + r_3] \tag{17}$$

$$\frac{\mathrm{d}G_{O_2}}{dt} = A_\mathrm{c}\frac{P_\mathrm{air,in}}{RT_\mathrm{air,in}}u_\mathrm{air,in}x_{O_2,\mathrm{in}} - A_\mathrm{c}\frac{G_{O_2}}{dV_\mathrm{c}}u_\mathrm{air} - r_3 \tag{18}$$

Here r is electrochemical reaction rate.

2.4. Temperature Model

The fuel cell plate is very thin along the radial direction, and the temperature difference is very small. There is a temperature gradient along the direction of the flow, but the temperature of the reactor cannot be measured directly. The average temperature of the outlet is used as the temperature of the cell [13]. The main heat transfer in the fuel cell is the forced convection in the gas flowing through the channel:

$$q_\mathrm{conv} = hA\left(T_\mathrm{cell} - T_\mathrm{fuel}\right) \tag{19}$$

Here h is convective heat transfer coefficient; A is heat exchange area; Tcell is plate temperature; Tfuel is gas temperature. The experimental results show that the gas flow in the channel is laminar, and the Reynolds number of the anode and cathode channel is 66 and 287. The relationship between Reynolds and Planck is used to determine the heat transfer coefficient:

$$m_\mathrm{cell}C_\mathrm{cell}\left(T\right)\cdot\frac{\mathrm{d}T_\mathrm{cell}}{\mathrm{d}t} = -P_\mathrm{fc} - r_3\Delta H_3\left(T\right) - q_\mathrm{conv} \tag{20}$$

$$\begin{aligned} C_\mathrm{re}\left(T\right)\cdot\frac{\mathrm{d}T_\mathrm{out}}{\mathrm{d}t} = {} & \sum\left(G_{\mathrm{in},i}^\mathrm{air}\cdot h_{\mathrm{in},i}^\mathrm{air}\right) + \sum\left(G_{\mathrm{in},i}^\mathrm{fuel}\cdot h_{\mathrm{in},i}^\mathrm{air}\right) - \sum\left(G_{\mathrm{out},i}^\mathrm{air}\cdot h_{\mathrm{out},i}^\mathrm{air}\right) - \\ & \sum\left(G_{\mathrm{out},i}^\mathrm{fuel}\cdot h_{\mathrm{out},i}^\mathrm{fuel}\right) - r_\mathrm{r}\Delta H_\mathrm{r}\left(T\right) - r_\mathrm{s}\Delta H_\mathrm{s}\left(T\right) + q_\mathrm{conv} \end{aligned} \tag{21}$$

Here P_{fc} Output power.

Based on the above analysis, the dynamic simulation model of fuel cell is shown in Figure 2:

IOP Conf. Series: Earth and Environmental Science **64** (2017) 012100 doi:10.1088/1755-1315/64/1/012100

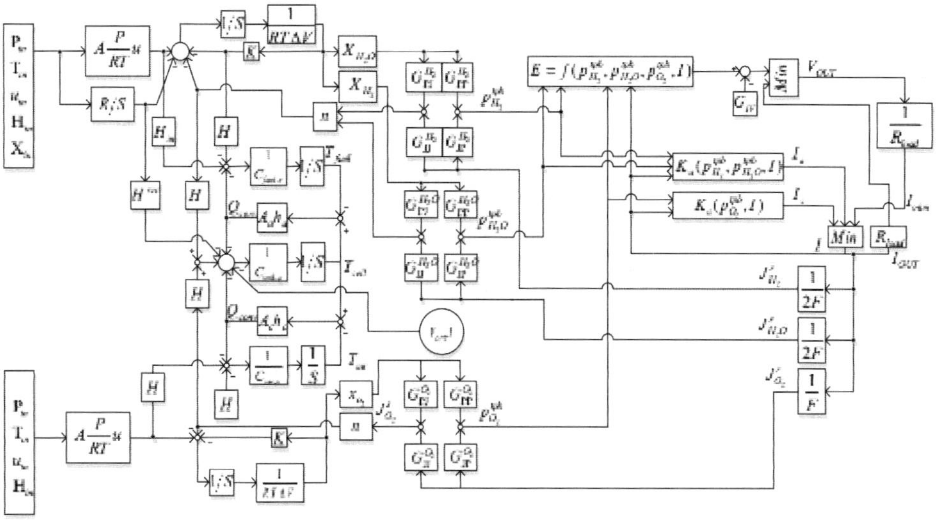

Figure 2. Schematic diagram of dynamic model

3. Dynamic performance analysis

3.1. Test data

In this paper, 30 identical fuel cell plates are used to form a stack. The anode of cell is negative electrode, and the cathode is positive electrode. The collector is a double-layer silver mesh to increase the electrical conductivity between the junction and the stack electrode. Sealing ring is used for sealing anode and bottom connection, same as cathode and top connector. Specific structure is shown in Figure 3, 4:

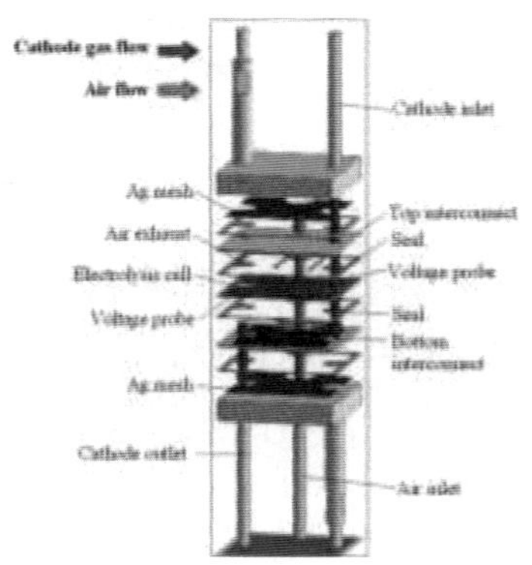

Figure 3. The structure of reactor connection

Figure 4. Fuel cell structure

The test system is composed of three modules: monitoring system, reforming control system and stack system:

Figure 5. Experimental test system

The experiment was carried out by using the methane flow rate 3.5SLM, the excess air 58SLM, the steam-to-carbon 3, and the constant current discharge mode of electronic load. Specific experimental parameters are shown in Table 1:

Table 1. Experimental parameters

Parameter	Numerical value	Note
$P_{air,in}$	11.7 kPa	Cathode side air inlet gauge pressure
$P_{fuel,in}$	6 kPa	Anode side CH_4 inlet gauge pressure
G_{CH4}	3.5 SLM	Reformer inlet CH_4 flow rate
G_{air}	58 SLM	Cathode side inlet air flow
G_{H2O}	8.435 ml/min	Reformer inlet water flow
T_{cell}	732.2℃	Stack mean temperature
P_{sofc}	388kg/cm^2	Sealed fuel cell stack pressure
N	30	The stack consists of 30 identical plates

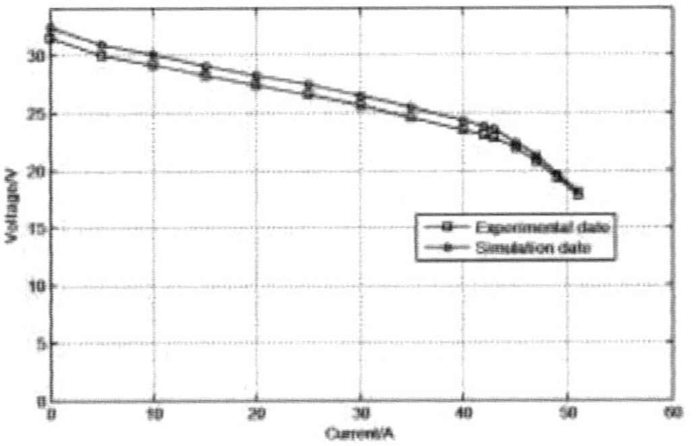

Figure 6. V-A characteristics

It is concluded from Fig. 6 that the polarization loss leads to the nonlinearity of the volt ampere characteristic in the initial stage, and then the ohmic loss is dominant. When the discharge current is large, the reactant is too late to diffuse into the three-phase interface layer, which makes the polarization of the concentration difference cause great loss to the voltage. In the process of increasing current, the simulation results are basically consistent with the experimental data. It is good to reflect the change law, and the maximum error is only 2.3%.

The experimental test was carried out on 30 pieces of plate type fuel cells produced by the SOFC company. The experimental results show that the characteristic curves of cell are obtained under different CH_4 flow rate, as shown in figure 7-9:

Table 2. Experimental parameters

Number	1	2	3	4
G_{CH4}	3 SLM	3.5 SLM	3.9 SLM	4.4 SLM
G_{air}	40 SLM	46.6 SLM	52 SLM	58.6 SLM
G_{H2O}	7.23ml/min	8.4 ml/min	9.4 ml/min	10.6 ml/min
T_{cell}	748℃	748℃	748℃	748℃

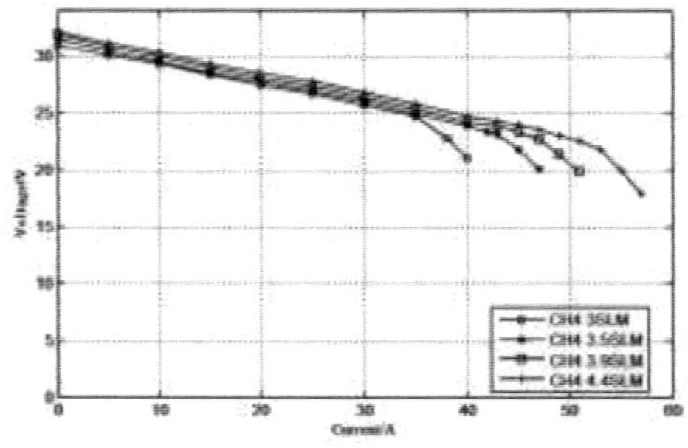

Figure 7. V-A characteristics curve

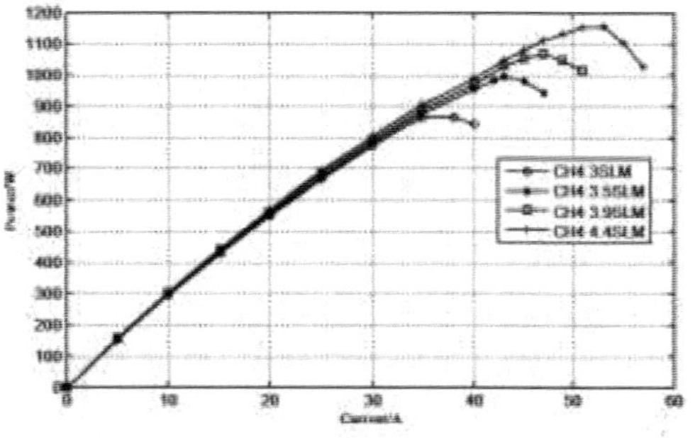

Figure 8. Power characteristic curve

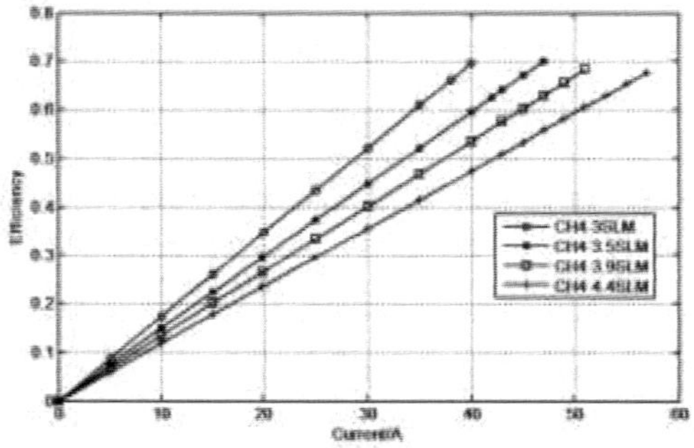

Figure 9. Characteristic curve of fuel utilization

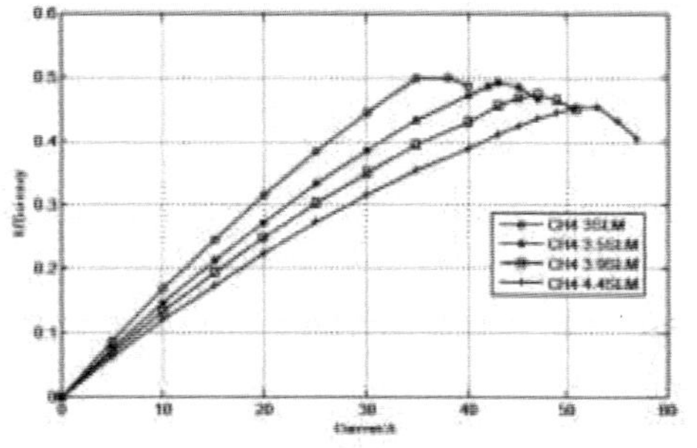

Figure 10. Efficiency characteristic curve

Figure 7-9 shows that the voltage attenuation is alleviated with the increase of methane flow rate. The power increases first and then decreases with the increase of current. The fuel utilization rate increases with the increase of the discharge current, but the excessive fuel utilization will lead to the irreversible decay of the cell. The experimental fuel utilization rate is controlled within 65%. When the power reaches the maximum, the electrical efficiency reaches the maximum value. When the fuel utilization ratio is 60%~65%, the power reaches the maximum, and the electrical efficiency reaches 48%~50%.

3.2. Dynamic performance simulation

In the simulation of dynamic performance, the discharge current is simulated by 5A, 10A, 15A, 20A, 25A, 30A, 35A, 40A, 43A for 60 seconds. Simulation of dynamic response process as shown in figure 11-13:

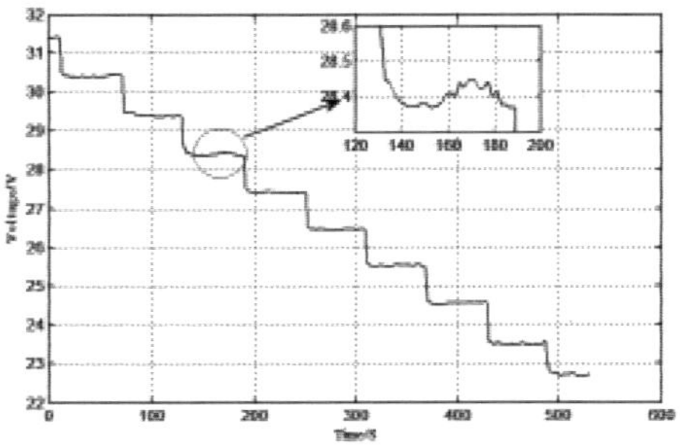

Figure 11. Voltage dynamic response process

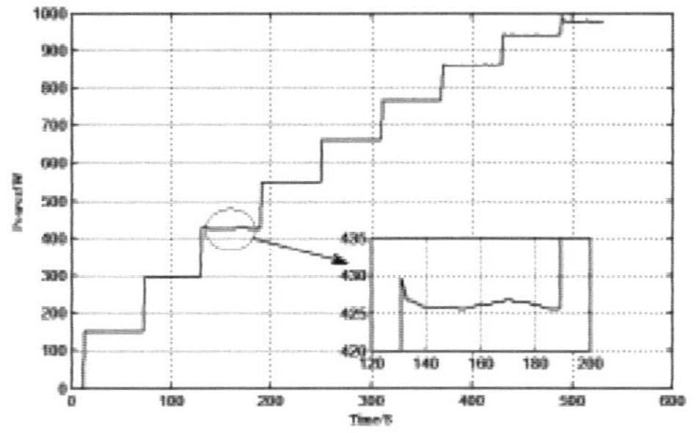

Figure 12. Power response process

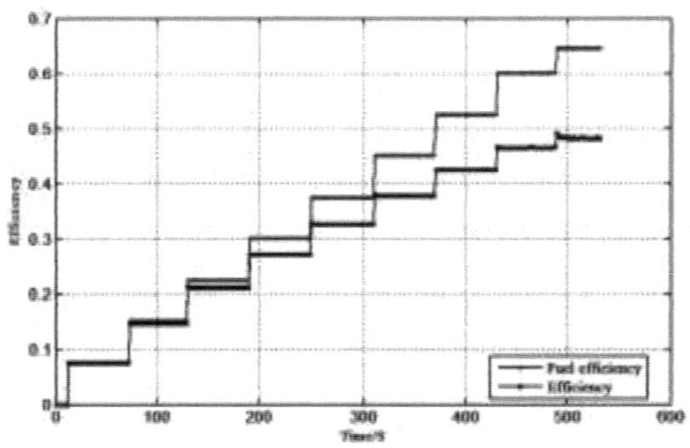

Figure13. Efficiency response process

Figure 11-13 shows that the existence of hysteresis in the voltage response. Because the electric double layer capacitance on the interface between the ion conductor and the electron conductor leads to the voltage cannot change suddenly. With the increase of the discharge current, the loss of cell increases and the voltage decreases. Due to the decrease of the fuel concentration in the three-phase interface layer, the diffusion effect is enhanced. At the same time, increasing the discharge current will cause the cell temperature to rise, and the voltage will increase slightly. With the increase of the discharge current, the power increases at the same time, but the power will have a certain amount of overshoot. The fuel utilization efficiency and the electric efficiency increase with the increase of the discharge current.

In the simulation of dynamic performance, the discharge current is simulated by 43A, 40A, 35A, 30A, 25A, 20A, 15A, 10A, 5A for 60 seconds. Simulation of dynamic response process as shown in figure 14-15:

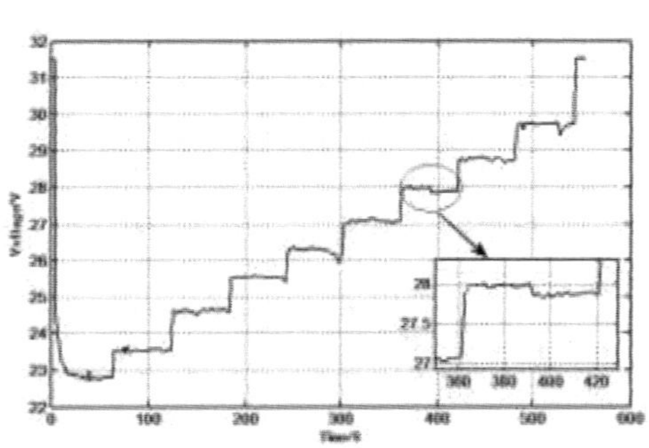

Figure 14. Voltage dynamic response process

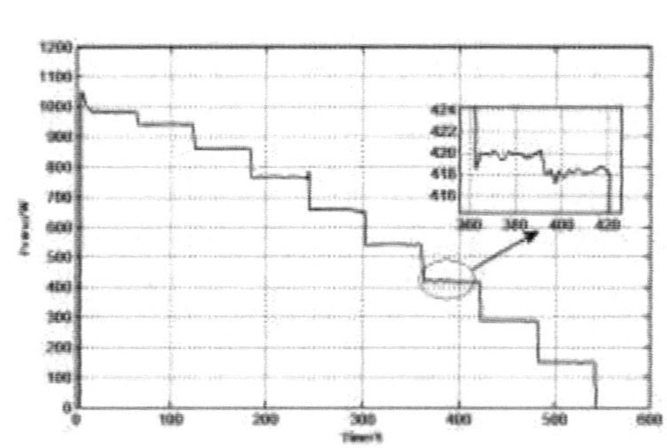

Figure 15. Power response process

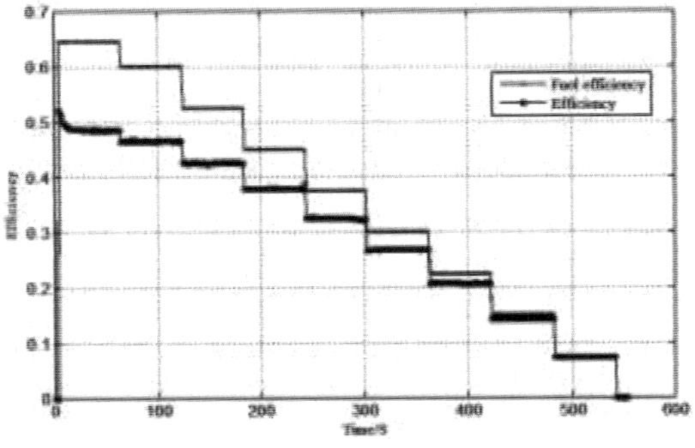

Figure 16. Efficiency response process

Figure 14-16 shows that the voltage response also has hysteresis, when the discharge current is reduced. The reduction of the discharge current will reduce the cell loss, so that the voltage can be improved. At the same time, the small current discharge will cause the decrease of the cell temperature, which makes the voltage drop slightly. Power, fuel utilization and electrical efficiency decrease with the decrease of discharge current.

4. Conclusion

Based on the mathematical model of fuel cell established in this paper, the simulation results are in good agreement with the experimental data, and the reliability of the model is verified. In this paper, when the fuel utilization ratio is 60%~65%, the power and the efficiency reach the maximum; When the discharge current increases, the voltage response has hysteresis, which will cause the transient overshoot of power; When the discharge current is reduced, the voltage response is delayed, which results in the decrease of temperature. In order to realize the real-time matching of power and load, it is the most feasible way to adjust the methane flow rate. By studying the dynamic response process of the cell, it provides a way to solve the problem of load and power real-time matching. At the same time, it can lay a good foundation for the popularization and application of fuel cell distributed generation equipment.

Acknowledgments

Thanks to the support of the Hubei Innovation Fund Project (HX2015B1003) and the 12th Five-Year Defense Advanced Research Foundation (No.40103030403).

References

[1] Bove R, Lunghi P, Sammes NM. SOFC mathematic model for systems simulation –Part2: Definition of an analytical model [J]. International Journal of Hydrogen Energy, 2005, (30): 189-200.

[2] Wu XJ, Zhu XJ, Cao GY, Tu HY. Predictive control of SOFC based on a ga-rbf neural network model [J]. Journal of power Energy, 2008, 179(1): 232-239.

[3] Yang J, Li X, Mou HG, Jian L. Predictive control of solid oxide fuel cell based on an improved takegi-sugeno fuzzy model [J]. Journal of power Energy, 2009, 193(2): 699-705.

[4] L Jin, WB Guan, X Ma, C Xu, WG Wang, A chieving hydrogen production through solid oxide electrolyzer sack by high temperature electrolysis [J]. ECS, 2012(41): 103-111.

[5] Florian Leucht, Wolfgang G. Bessler. Fuel cell system modeling for solid oxide fuel cell/gas turbine hybrid power plants, Part I: Modeling and simulation framework [J]. Journal of Power Sources, 2011 (196): 1205–1215.

[6] Yifeng Zheng, Qingshan Li, Wanbin Guan, Cheng Xun, WeiWu, WeiGuo Wang. Investigation of 30-cell solid oxide electrolyzer stack modules for hydrogen production [J]. Journal of

Power Energy, 2014(40): 5801-5809.

[7] Qingshan Li, Yifeng Zheng, Wanbing Guan, Le Jin, Cheng Xu, Wei Guo Wang. Achieving high-efficiency hydrogen production using planar solid-oxide electrolysis stacks [J]. Journal of Power Energy, 2014(39): 10833-10842.

[8] Le Jin, Wanbing Guan, Xiao Ma, Huijuan Zhai, Wei Guo Wang. Quantitative contribution of resistance sources of components to stack performance for planar solid oxide fuel cells [J]. Journal of Power Energy, 2014(253): 305-314.

[9] Guan WB, Zhai HJ, Jin L, Xu C, Wang WG. Temperature measurement and distribution inside planar SOFC stacks [J]. Fuel cells, 2012, 12(1): 24-31.

[10] Tan Xun-qiong, Wu Zheng-qiu, Zhou Ye, Zhong Hao, Li Jun-jun. Solid Oxide Fuel Cell Lumped Modeling and Simulation [J]. Proceedings of the CSEE, 2010, 6(17): 104-110(in Chinese).

[11] Tang Gen-tu, Luo Zhong-yang, Ni Ming-jiang, Yu Chun-jiang, Cen Ke-fa. Numerical simulation and performance analysis of planar anode-supported solid oxide fuel cell [J]. Proceedings of the CSEE, 2005, 5(10): 116-121 (in Chinese).

[12] Sun Xiu-yu, Wang Hao. Dynamic modeling and predictive control of solid oxide fuel cell [M]. Beijing: China Machine Press, 2015:153-173 (in Chinese).

[13] Lu Xiao-Jing, Lu Chao-Hao, Geng Xiao-Ru, Zhu Xin-Jian, Weng Yi-Wu. Effect of Steam on the Performance of an IT-SOFC/GT Hybrid System [J]. Journal of engineering thermal physics, 2016, 4(4): 705-710 (in Chinese

Security attack detection algorithm for electric power gis system based on mobile application

Chao Zhou[1], Renjun Feng[2], Liming Wang[1], Wei Huang[1], Yajuan Guo[1]

[1]State Grid Jiangsu Electric Power Research Institute, Nanjing, 211100, China
[2]State Grid Suzhou Power Supply Company, Suzhou, 215004, Chian

Abstract. Electric power GIS is one of the key information technologies to satisfy the power grid construction in China, and widely used in power grid construction planning, weather, and power distribution management. The introduction of electric power GIS based on mobile applications is an effective extension of the geographic information system that has been widely used in the electric power industry. It provides reliable, cheap and sustainable power service for the country. The accurate state estimation is the important conditions to maintain the normal operation of the electric power GIS. Recent research has shown that attackers can inject the complex false data into the power system. The injection attack of this new type of false data (load integrity attack LIA) can successfully bypass the routine detection to achieve the purpose of attack, so that the control center will make a series of wrong decision. Eventually, leading to uneven distribution of power in the grid. In order to ensure the safety of the electric power GIS system based on mobile application, it is very important to analyze the attack mechanism and propose a new type of attack, and to study the corresponding detection method and prevention strategy in the environment of electric power GIS system based on mobile application.

1. Introduction

Geographic Information System (GIS) [2] is widely used as a means of resource management and visual analysis of resources in power generation, transmission, substation, distribution, electricity, communication and other professional. It can effectively improve the data collection, storage efficiency, improve fault location, analysis and processing capacity, and provide support for decision support for the operation of the power grid [3]. At the same time, it can effectively reduce the risk of production and operation. Electric power GIS is a virtual representation of all aspects of electricity, which encompasses all systems including human society, and is expressed in multi-dimensional, multi-scale, multi-temporal, multi-faceted information facilities. So, people can browse, interact, and use a large number of concentrated, power-related elements, such as tower, insulator, substation, etc. With the deepening of the power grid engineering strategy and the deepening understanding of the framework of the electric power GIS system and key technologies, data reception and rapid processing, multi-source heterogeneous spatial data integration, spatial information database, map services and virtual reality and other fields have made corresponding progress

With the development of portable devices, mobile devices are used more and more widely in the daily life and work. In the process of continuous advancement of information technology, the use of mobile terminals to carry out business operations on site has become one of the important work of the power industry. As a convenient data acquisition and field operations tool, the electric power GIS system based on the mobile application [4] can significantly improve the efficiency of power grid production and management. The introduction of the electric power GIS system based on mobile application is an effective extension of the geographic information system which has been widely used in the power industry.

For now, the power GIS system is not very mature. It carries out the combination and development basically based on the digital earth platform, and then to achieve power production, power management, power engineering and other aspects of the application. Because the openness and sensitivity of the system to the user, it is vulnerable to outside attacks. A new type of false data injection attack [5], which is collecting and analyzing information from a large number of intelligent instruments, can inject the false data into the smart meter to change the information collected by the power SCADA system, and then change the power system state estimation results, to change the load integrity of the power grid. This new type of false data injection attack is also known as load integrity attack (LIA). This type of attack can lead to an erroneous state estimation of the power system, which makes the load distribution unbalanced and further misleads the operation and control functions of the energy management system (EMS) in the electric power GIS system, resulting in catastrophic consequences for the grid. In the face of this kind of load integrity false data injection attack, the current false data detection method is not enough to detect this new type of attack. Therefore, according to the characteristics of the attack, it is imperative to analyze the mechanism of attack and put forward the appropriate countermeasures to improve the operational safety of the electric power GIS.

2. Methods

For the security of social domain strategy: including the staff of the power grid security knowledge and safety skills training, the user safety education for the use of electricity to improve people's awareness of security; the development of safety regulations and policies and regulations to enhance the safety of the power grid Of the system to ensure that; social engineering methods to combat social engineering attacks.

Security protection strategy for information domain: The traditional IT security technology and means can be used to secure the security of intelligent grid information domain, such as encryption and decryption mechanism, secret key management technology can be used to ensure the confidentiality of data, integrity and availability; Access control, authentication and auditing can be used to ensure the normal operation of various services in the smart grid and its security and controllability; IPSec, SSL, TLS, etc. can be in the protocol layer to strengthen the smart grid security; and firewall, intrusion detection, virus check Kill and other technical means can be achieved in a timely manner to detect security threats and its dissemination to suppress, to minimize the destructive.

False data injection attacks belong to a kind of network attack. With the continuous development of attacks, security threats are emerging and evolving continuously. This is essentially an asymmetric technical game between attackers and defenders. Combined with the existing information security defense technology, we use the honeypot technology to break the situation of the false data injection attack and defense asymmetry in order to achieve the purpose of active defense. Honeypot can be defined as a security resource that can be used to capture and analyze attack behavior by deploying some bait hosts or some network service information in the control center to deceive attackers and induce attackers to attack honeypots, and to speculate the attack intentions and motivations. So, the defenders can clearly understand the security threats they face, and then develop technical and management tools to enhance the smart grid system security capabilities.

The basic process of honeypot technology in preventing false data injection attacks is as follows:

1) False data injection attack must first invade the specific host of the control center. The current intrusion detection system can detect the illegal intrusion function. After the detection of the invasion [6], we should start using the redirection function, and let the attack stream redirect to the honeypot host;

2) Because the honeypot host configures the software to imitate the attack, so it is relatively easy to attack, so that it can attract the attacker, and then to dilute the attack and protect the normal host of the control center;

3) Then the attacker will be attracted to attack the honeypot, and the control center host will be replaced by the honeypot system to cut off the control information purposes;

4) The security log function will be deployed in the honeypot system to record different intrusion attacks for learning;

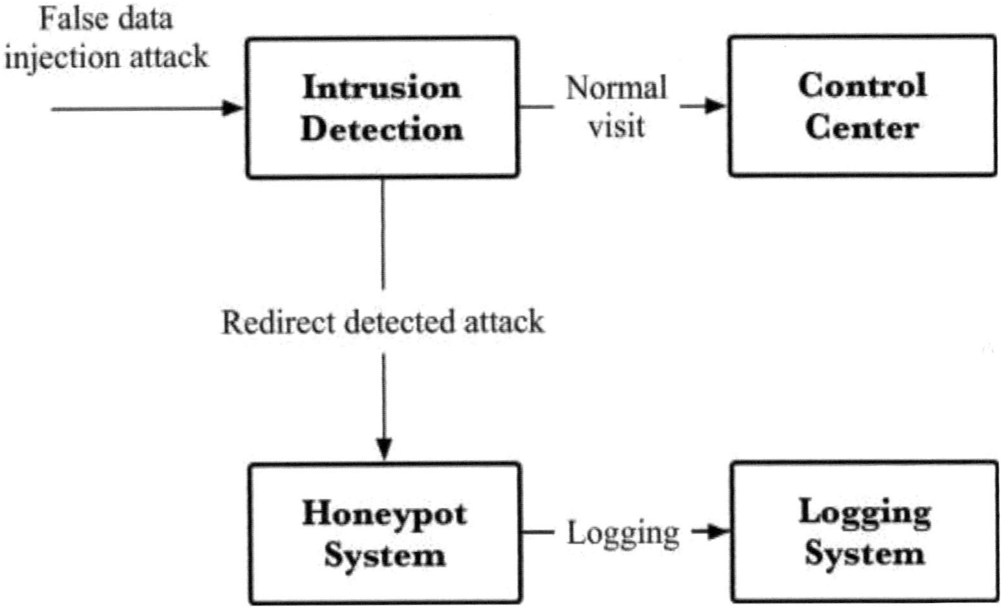

Figure 1. Honeypot mechanism system diagram

The principal of the use of honeypot technology to prevent false data injection attack is shown in the following figure. The intrusion detection system directs the attack flow to the honeypot. The honeypot system will respond and record the log. Once the intrusion detection system finds that the attacker is in the intrusion control center, the intrusion will be fed back to the corresponding honeypot by address redirection. The honeypot will send an attacker's desired response to the attacker based on the initial setup, making the honeypot think the honeypot is the controlled host, and send control information to the honeypot. Assuming that the probability that the attacker is found by the intrusion detection system and redirects to the honeypot is P_1, the number of hosts in a network is k, the probability of each host being attacked is P_2, and the number of honeypots is h, then the probability of successful intrusion to host and the probability of intrusion honeypot are:

$$P_2 (1 - P_1)k / (h + k) \tag{1}$$

$$P_1 + (1 - P_1)h / (k + h) \tag{2}$$

Analysis of log information can continuously improve the probability P_1 that the intrusion detection system finds the attacks, and reduce the probability P_2 that the control center host is attacked illegally, to take the initiative to prevent false data injection attack.

3. Results

The simulation results is shown in figure 1 and figure 2.

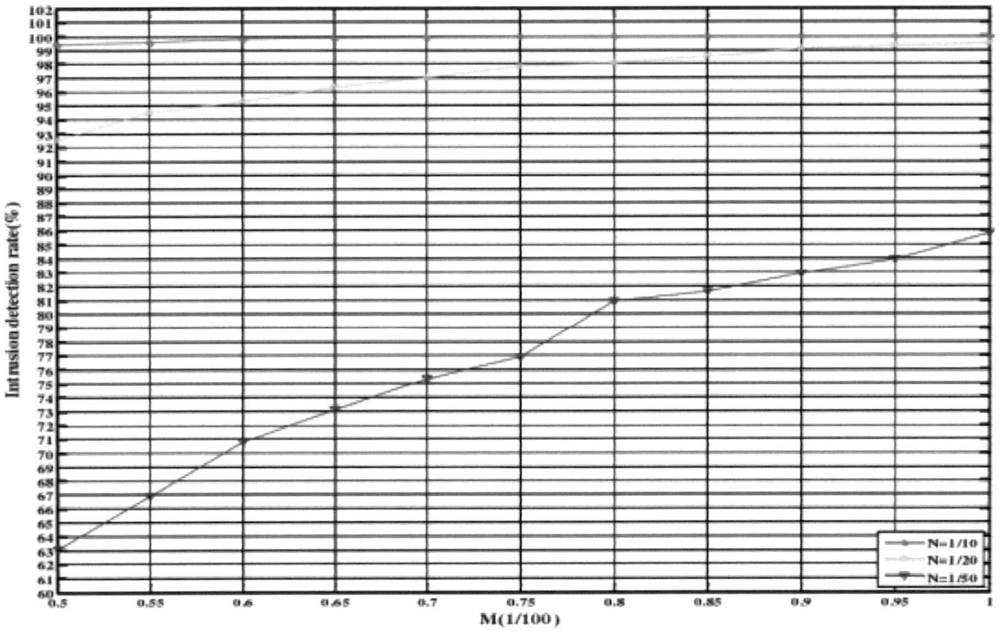

Fig. 2. The change of intrusion detection rate with M under different N values

Figure 2 shows the change in intrusion detection rate with M under the first set of experimental conditions. The results show that with the increase of M, the intrusion detection rate is increasing. In particular, when the detection rate of M03 / 100 and N = 1/10 is 100%, the intrusion detection rate under N = 1/20 is more than 97%, and the intrusion detection rate under N = 1/50 More than 75%. That is, when M = 0.75 / 100, the experimental results meet all of the test criteria, so to a certain extent, we can say that spying domain is effective and low cost. Figure 28 also shows that the M value of spying domain should be set to 0.75 / 100 because 0.75 / 100 is the smallest of all M that meets all evaluation criteria.

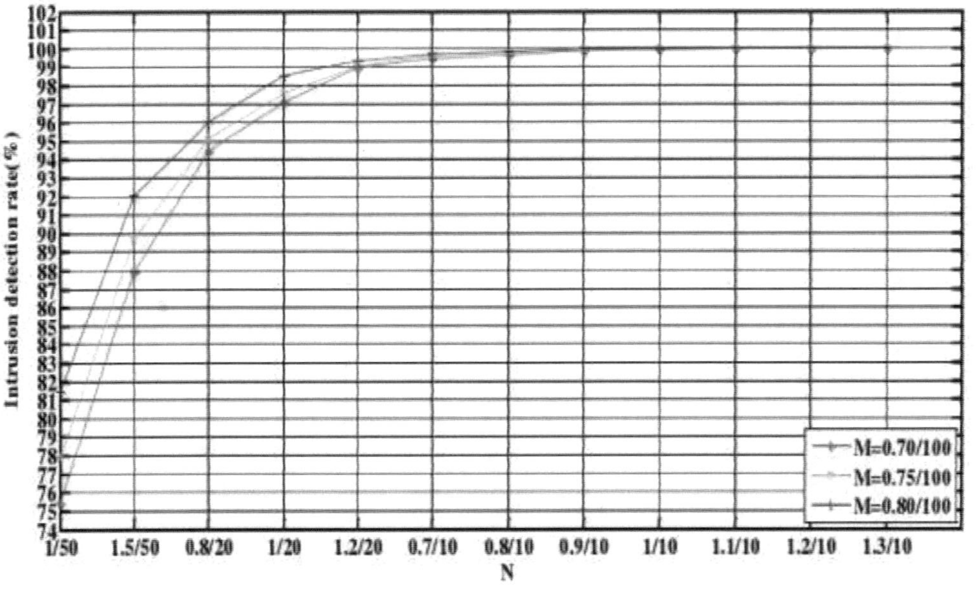

Fig. 3. The change of intrusion detection rate with N under different M values

Figure 3 shows the change in intrusion detection rate with N under the second set of experimental conditions. When N = 1/10, when M = 0.70 / 100 and M = 0.75 / 100, the intrusion detection rate does not all reach 100%. For example, when N = 1.1 / 10, the detection rates for M = 0.70 / 100 and M = 0.75 / 100 are 99.96% and 99.99%, respectively. In contrast, when M = 0.80 / 100, N = 1/10 under the intrusion detection rate of 100%, to meet the standard 2. In addition, the attack rate of spyingdomain under M = 0.80 / 100 satisfies all criteria.

4. Conclusion

In this paper, from the perspective of prevention, we analyzed the measures of the principle of false data injection attack in electric power GIS based on mobile application and the necessary conditions for attack initiation, and put forward the honeypot technology method to protect against attack. Analysis of log information can continuously improve the probability P1 that the intrusion detection system finds the attacks, and reduce the probability P2 that the control center host is attacked illegally, to take the initiative to prevent false data injection attack and to ensure the security of the electric power GIS based on mobile application.

This paper opens a new era of deep learning to detect false data injection attacks. In the future, we will carry out relevant work in depth. The intelligent grid load integrity detection and prevention technology still needs to be improved.

Reference

[1] Li H., Lai L., Zhang W. Communication requirement for reliable and secure state estimation and control in smart grid[J]. Smart Grid, IEEE Transactions on. 2011,V2(3):476-486
[2] Qiao Shiming. Design and Implementation of Visualization System for Power Network Planning and Design Based on 3D GIS Technology [J]. Science and Technology Innovation Guide. 2015 (02): 118-223
[3] Dan G., Sandberg H. Stealth attacks and protection schemes for state estimators in
[4] power systems[A]. Smart Grid Communications (SmartGridComm), 2010 First IEEE International Conference on[C]. 2010:214-219
[5] Kosut O., Jia L., Thomas R. J., et al. Limiting False Data Attacks on Power System State Estimation[A]. the 44th Information Sciences and Systems (CISS) Conference[C].2010: 1-6
[6] Kosut O., Jia L., Thomas R., et al. Malicious data attacks on the smart grid[J]. Smart Grid, IEEE Transactions on. 2011,V2(4):645-658
[7] Yuan Y , Li Z., Ren K. Quantitative Analysis of Load Redistribution Attacks in Power Systems[J]. Parallel and Distributed Systems, IEEE Transactions on. 2012,V23(9):1731-1738

Numerical simulation transient electron distribution of direct current negative corona discharge in air

Xinghua Liu[1,*], Richang Xian[2], Peng Yu[3], Ying Pei[4], Xuebin Lv[5], Xuefeng Sun[1], Tao Wang[1], Shangyuan Ning[1], Shikun Wang[1]

[1]State Grid Zibo Power Supply Company, Shandong 255000, China
[2]Shandong University of Science and Technology, Shandong 255000, China
[3]State Grid Shandong Electric Power Research institute, Shandong 255000, China
[4]State Grid of China Technology College, Shandong 255000, China
[5]State Grid Shandong Power Company, Shandong 255000, China

[*]Corresponding author e-mail: 30280241@qq.com

Abstract. In order to explore the characteristics of electron in DC negative corona discharge, microcosmic process of negative corona discharge in air is simulated in this paper. The numerical computation is established with a bar-plate electrode configuration with an inter-electrode gap of 3.3 mm, the negative DC voltage applied to the bar is 5.0 kV, the pressure in air discharge is fixed at 1.0 atm, and the gas temperature is assumed to be a constant (300 K). By solution the system of electron conservation equation, the electron mean energy conservation, the heavy species multi-component diffusion transport equation, and the Poisson's equation, characteristics of electrons (electron mean energy, electron density, and generation and dissipation performances of electrons) at 6 representative time points during a pulse are obtained and then discussed emphatically.

1. Introduction

Corona discharge is a non-equilibrium low-temperature plasma discharge process and often occurs close to the point electrode with large radius of curvature. With the development of electric power industry in China, high-pressure and ultra-high voltage power transmission lines become the main wire frame and corona loss is assignable. Meanwhile, the audible noise, electromagnetic interference, circuit corrosion, insulation degradation and public in harmony which are brought by corona discharge have attracted high attentions. Various excited particles, ions and free radicals are formed in the microscopic process of corona discharge, which can make complicated physical and chemical reasons. The diversity of plasma species determines the diversity of particle interaction and the law of dynamics is extremely complicated [1]. The impact ionization of electrons and neural electrons is the dominant ionization process in the air discharge process. Additionally, electrons are main carrier of energies between extra electric field and heay ions [2]. Therefore, studying behavioral characteristics of electrons in corona discharge process is of important significance to explain mechanism of corona discharge.

In early numerical studies, scholars often view gas discharge equal to a circuit and simulate arious gas discharge equations by analyzing circuit parameters [3]. With the development of plasma simulation technology, people's understanding on gas discharge are not limited in discharge current and discharge voltage of circuit simulation, but goes deep into th distribution law of charged particles produced in the discharge process. Many deeper theoretical discoveries have been achieved. The atmospheric air discharge that considers chemical reaction between particles and non-equilibrium state has become one very important direction of scientific research in future [4]. Nahorny et al. [5] considered 430 chemical reactions when studying the gas discharge process. It is the study that considers the most chemical reactions at present and lays good foundations for follow-up scientific researchers. Pancheshnyi [6] chosen 10 main particles based on Nahorny and proposed a two-dimensional physical model of streamer discharge when the pressure of N2: O2 (9:1) mixed gas is 760Torr. He analyzed discharge parameters (e.g. electric field distribution, charged particle density and reaction coefficient) thoroughly by using this model, which are in high accordance with test data. Liu Xinghua et al. [7] established the fluid-chemical kinetics model of positive corona discharge by Comsol, which was verified by experimental data of Antao [8], achieving high consistency. Key characteristic parameters like current density, electric field/potential distribution and positive/negative ion density distribution in early positive corona discharge were analyzed based on this model. According to existing research fruits, there' no model that can reflect electron characteristics during the negative corona discharge. Discussions on microscopic characteristics (e.g. electron mean energy, electron density, electro generation and dissipation characteristics) in the corona discharge have important theoretical and actual significance.

In this paper, the electron mean energy transport equation was considered based on the two-dimensional fluid kinetics model of corona discharge and some important processes like photo ionization, secondary emission and chemical reactions between particles were discussed. An improved numerical model of negative corona discharge was constructed. Corona discharge under 3.3mm bar-plane distance and -5.0kV applied voltage was calculated based on this model, getting the single-discharge pulse current waveform of negative corona discharge. The electron mean energy distribution, electron density distribution, electron generation and dissipation characteristics in the negative corona pulse process were analyzed by choosing 6 typical time points in one discharge pulse period.

2. Model

The solving of this model is actually to convert the electron continuity equation, heavy particle transport equation, electron energy equation and Poisson's equation to appropriate system of partial differential equations. Then, the model was solved by discrete numerical difference form after normalization of these equations.

2.1. Control equations

The electron continuity control equations [9] is:

$$\frac{\partial n_e}{\partial t} + \vec{\nabla} \cdot \vec{\Gamma} e = R_e \tag{1}$$

where ne is the electron number density, Re is the electron generation and dissipation rate, and $\vec{\Gamma}_e$ is electron flux.

$$\vec{\Gamma}_e = -\vec{\nabla}(D_e n_e) + \mu_e n_e \vec{\nabla}\varphi \tag{2}$$

where μ_e is the electron mobility, D_e is the electron diffusion coefficient and φ is potentials.
The electron mean energy can be described as [10]

$$\frac{\partial}{\partial t}(n_e\vec{\varepsilon})+\vec{\nabla}\cdot\vec{\Gamma}_\varepsilon=-e\vec{\Gamma}_e\cdot\vec{\nabla}\varphi+e\sum_{i=1}^{I_g}\Delta\varepsilon_i^e R_i \tag{3}$$

where $\vec{\varepsilon}$ is the electron mean energy, e is the elementary charge (1.602x10-19 C), $\Delta\varepsilon^e{}_i$ is energy loss of single electron in one impact during the impact reaction process, R_i is the reaction rate of the chemical reaction i and $\vec{\Gamma}_\varepsilon$ is the electron mean energy flux[10].

$$R_i=\int_0^\infty(\vec{\varepsilon})^{1/2}f(\vec{\varepsilon})\sigma(\varepsilon)d\varepsilon \tag{4}$$

$$\vec{\Gamma}_\varepsilon=n_e\varepsilon\mu_\varepsilon\vec{\nabla}\varphi-D_\varepsilon\vec{\nabla}n_e\vec{\varepsilon} \tag{5}$$

where $\vec{\varepsilon}$ is the electron mean energy. $f\left(\vec{\varepsilon}\right)$ is the electron energy distribution function (EEDF). In this paper, EEDF was gained from solving the Boltzmann equation by constructing a one-dimensional model based on Comsol Multiphysics[11]. σ(ε) is the impact section area of the reaction and is determined according to Reference [12] in this paper. μ_ε and D_ε are electron mean energy migration rate and the electron mean energy diffusion coefficient[10]:

$$\mu_\varepsilon=\frac{5\mu_e}{3},\ D_\varepsilon=\frac{5D_e}{3} \tag{6}$$

The single discharge is very short (generally within hundreds of seconds), during which the influences of particle interaction on particle transport almost could be neglected. Therefore, existing researches on corona discharge simplify the heavy particle transport process. The evolution equation of heavy particles in the discharge process after simplification is:

$$\begin{cases}\dfrac{\partial n_k^+}{\partial t}=R_{ion}+R_{ph}-R_{rec}^{ei}-R_{rec}^{ii}\\ \dfrac{\partial n_k^-}{\partial t}=R_{att}-R_{rec}^{ii}\\ \dfrac{\partial n}{\partial t}=R_{rec}^{ei}+R_{rec}^{ii}-R_{ion}-R_{ph}-R_{att}\end{cases} \tag{7}$$

where nk+, nk- and nn are positive ions, negative ions and neural particles. *Rion*, *Rph*, $R^{ei}rec$, $R^{ii}rec$ and *Ratt* are reaction rates related with impact ionization, photo ionization, electron-ion, ion-ion and electron adsorptions.
Finally, the Poisson's equation is:

$$\varepsilon\nabla^2\varphi=-e(\sum_k n_k^+-\sum_k n_k^--n_e) \tag{8}$$

where ε is the vacuum dielectric constant.

2.2. Chemical reactions of particles

Tremendous types of particles will be formed in the corona discharge process and the reaction equation is relatively complicated. It is difficult and unnecessary to analyze the reaction between electrons and all particles. Based on our previous researches [7, 13], the micro mechanism of corona discharge is described by particle reactions in Table 1.

Table 1 Kinetic processes considered in the model

No.	Reaction equation	Reaction rate [11,13]
R_1	$N_2 + e \rightarrow 2e + N_2^+$	$f(\varepsilon)$
R_2	$O_2 + e \rightarrow 2e + O_2^+$	$f(\varepsilon)$
R_3	$N_2^+ + 2N_2 \rightarrow N_4^+ + N_2$	5.0×10^{-41}
R_4	$N_2^+ + N_2 + O_2 \rightarrow N_4^+ + O_2$	5.0×10^{-41}
R_5	$N_4^+ + O_2 \rightarrow O_2^+ + 2N_2$	2.5×10^{-16}
R_6	$N_2^+ + O_2 \rightarrow O_2^+ + N_2$	$1.04\times10^{-15}T^{-0.5}$
R_7	$2N_2 + O_2^+ \rightarrow N_2O_2^+ + N_2$	$8.1\times10^{-38}T^{-2}$
R_8	$N_2O_2^+ + N_2 \rightarrow O_2^+ + 2N_2$	$14.6\ T^{-5.3}\exp(-2357/T)$
R_9	$N_2O_2^+ + O_2 \rightarrow O_4^+ + N_2$	1.0×10^{-15}
R_{10}	$O_2^+ + O_2 + N_2 \rightarrow O_4^+ + N_2$	$2.04\times10^{-34}T^{-3.2}$
R_{11}	$O_2^+ + 2O_2 \rightarrow O_4^+ + O_2$	$2.04\times10^{-34}T^{-3.2}$
R_{12}	$O_4^+ + e \rightarrow 2O_2$	$1.4\times10^{-12}(300/T_e)^{0.5}$
R_{13}	$O_2^+ + e \rightarrow 2O$	$2.42\times10^{-13}(300/T_e)$
R_{14}	$2O_2 + e \rightarrow O_2 + O_2^-$	$2.0\times10^{-41}(300/T_e)$
R_{15}	$O_4^+ + O_2^- \rightarrow 3O_2$	1.0×10^{-13}
R_{16}	$O_4^+ + O_2^- + N_2 \rightarrow 3O_2 + N_2$	2.0×10^{-37}
R_{17}	$O_4^+ + O_2^- + O_2 \rightarrow 3O_2 + O_2$	2.0×10^{-37}
R_{18}	$O_2^+ + O_2^- + N_2 \rightarrow 2O_2 + N_2$	2.0×10^{-37}
R_{19}	$O_2^+ + O_2^- + O_2 \rightarrow 3O_2$	2.0×10^{-37}
R_{20}	$O + O_2 + N_2 \rightarrow O_3 + N_2$	2.5×10^{-46}
R_{21}	$O + 2O_2 \rightarrow O_3 + O_2$	2.5×10^{-46}
R_{22}	$e + N_2^+ + N_2 \rightarrow 2N_2$	$6.07\times10^{-34}T_e^{-2.5}$
R_{23}	$2e + N_2^+ \rightarrow N_2 + e$	$5.651\times10^{-27}T_e^{-0.8}$
R_{24}	$O_2^+ + O^- \rightarrow O + O_2$	$3.46\times10^{-12}T_e^{-0.5}$
R_{25}	$N_2 + e \rightarrow e + N_2$	$f(\varepsilon)$
R_{26}	$O_2 + e \rightarrow e + O_2$	$f(\varepsilon)$
R_{27}	$O_2 + e \rightarrow O + O^-$	$f(\varepsilon)$

Due to the neutralization reaction on electrode surface, positive and negative ions are changed into neutral particles on electrode surface through following reactions

$$\begin{gathered} N2+\rightarrow N2,\ N4+\rightarrow 2N2, \\ O2+\rightarrow O2, O4+\rightarrow 2O2, \\ N2O2+\ \rightarrow O2+N2,\ O2-\rightarrow O2,\ O-\rightarrow 0.5O2 \end{gathered} \quad (9)$$

Generally, the electrons generated by photo ionization are far less than those generated by impact ionization. However, the photo ionization can trigger new electron avalanche at the electron avalanche head and facilitate the discharge develop forward. Photo ionization under atmosphere pressure plays an important role in air discharge. Zheleznyak et al. believed that photo ionization in air discharge is the process that photos released when the excited nitrogen molecules ($b^1\Pi$, $b'^1\Sigma_u^+$ and $c_4'^1\Sigma_u^+$) attenuate into steady $N_2(x^1\Sigma_g^+)$ are absorbed by oxygen molecules and form oxygen ions.

IOP Conf. Series: Earth and Environmental Science **64** (2017) 012102 doi:10.1088/1755-1315/64/1/012102

$$N_2(b^1\Pi、b'^1\Sigma_u^+、c_4'^1\Sigma_u^+) \rightarrow N_2(x^1\Sigma_g^+) + hv \tag{10}$$

Equation (9) represents the process that excited nitrogen molecules release photons and hv means releasing photons. Equation (10) represents the photo ionization of oxygen molecules. In this paper, the photo ionization coefficient Sph was calculated by Eddington method [15] in this paper:

$$S_{ph} = \frac{1}{4\pi} \cdot \frac{p_q}{p + p_q} \int_V d^3 r_1 \frac{S_{ion}(\vec{r_1})}{\left|\vec{r} - \vec{r_1}\right|^2} h(\left|\vec{r} - \vec{r_1}\right| \cdot p) \tag{11}$$

where $\vec{r}$ is the distance between photo radiation and absorption; $\vec{r_1}$ is the photo ionization source point, p_q is the attenuation pressure of N2 molecules, p is gas pressure, and $h(\left|\vec{r}-\vec{r_1}\right| \cdot p)$ is the photo ionization absorption coefficient.

2.3. *Boundary conditions*

The external circuit and computational domain of negative corona discharge are shown in Fig.1, where radius of curvature of bar electrode is 0.4mm, the radius of plane electrode is 5.0cm, and the bar-plane distance is 3.3mm. The external circuit is composed of DC power supply, capacitance C and protection resistance R. In this model, DC voltage, C and R are -5.0kV, 1.0pF and 5.0kΩ, respectively. The environmental conditions are set T=300K and p=1.0atm. The model is solved by the plasma module based on Comsol Multiphysics software. The physical computational domain of corona discharge is simplified into a two-dimensional bar-plane structure through axial symmetric rotation (the dotted frame in Fig.1). The computational domain is divided into 244,884 solving units. The boundary conditions of this model mainly include:

Voltage at the bar end is 5.0kV and the plane electrode has zero potential because of the ground connection. The voltage boundary condition of open boundary is [16]:

$$\vec{n} \cdot (\varepsilon \vec{\nabla} \varphi) = 0 \tag{12}$$

Boundaries of electron flux on electrodes are: [16]

$$\vec{\Gamma} \cdot \vec{n} = \frac{1-\gamma_e}{1+\gamma_e} \left[-(2\alpha_e - 1)\mu_e \vec{E} \cdot \vec{n} n_e + \frac{1}{2} v_{e,th} n_e \right] \tag{13}$$

The boundary condition of electron mean energy flux that the electron mean energy equation meets at electrode surface is:

$$\vec{\Gamma}_\varepsilon \cdot \vec{n} = \frac{1}{3} v_{e,th} \vec{\varepsilon n_e} \tag{14}$$

where $\vec{n}$ is normal vector on electrode surface, γ_e is the secondary electron yield coefficient, and $v_{e,th} = \sqrt{8k_BT / \pi m_e}$ is the electron thermodynamics speed. When the electron flux orients to electrode, a_e =1; otherwise, a_e =0.

The boundary condition of ion is [16]:

$$\vec{\Gamma}_i \cdot \vec{n} = \frac{\gamma_i}{4}\sqrt{\frac{8k_B T}{\pi m_i}} n_i \tag{15}$$

where $\vec{\Gamma}_i$ and n_i are density flux and number density of the particle i, γ_i is the surface neural reaction coefficient of the particle i, and mi is the weight of the ith type of particles.

At bar tip, suppose the particles before the air discharge are in Gaussian distribution with maximum value of $10^{16}m^{-3}$[17].

$$N_{e,i} = N_{max} \times \exp(-\frac{(r-r_0)^2}{2s_0^2} - \frac{(z-z_0)^2}{2s_0^2}) \tag{16}$$

where $N_{max} = 10^{16}m^{-3}$, $r_0 = 0\mu m$, z_0 is position of the bar tip, and $s_0 = 25\mu m$.

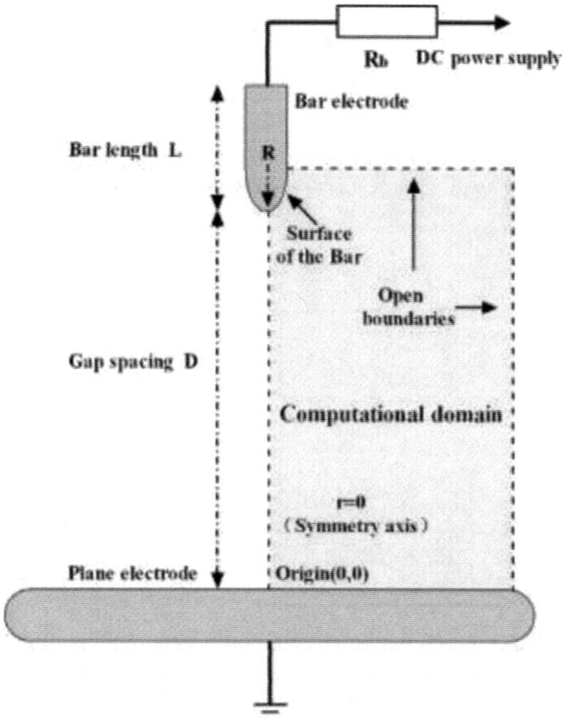

Figure. 1 Schematic diagram of the negative corona discharge

3. Results and discussion

3.1. Current waveform of single discharge

The negative corona pulse waveform is shown in Fig.2. It is dark current (uA) at the initial discharge period, which is caused by few space charges in the electrode gap and very high equivalent resistance in the migration area. Current increases to 0.195A quickly during t_1-t_2. This is mainly because electric field concentrated region is mainly close to the bar electrode during this period. The field-induced ionization generates abundant electrons and positive ions close to the bar electrode, and positive ions make neural reaction (Equation (9)) quickly after gained electrons at the cathode. Subsequently, current begins to decrease during t_2-t_4. Since abundant charged ions will be produced by adsorption when the electron avalanche moves toward the anode, the internal field formed by ion cloud and electrons causes distortion of electric field in the electrode gap, thus weakening the field-induced ionization. Only few electrons have to enter into the anode surface in order to keep equilibrium of electric field. Therefore, the current during t_4-t_6 is maintained at about 1mA. In this paper, electronic properties in the negative corona discharge were analyzed by 6 time points (t_1-t_6) in single pulse process.

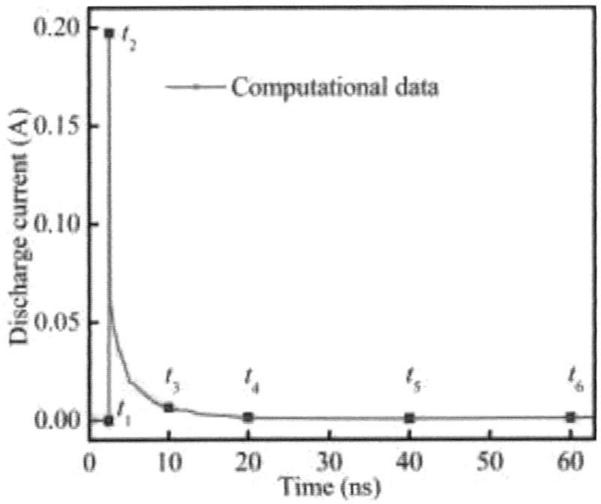

Figure. 2 The discharge current waveform in the electrode gap

3.2. Electron mean energy distribution

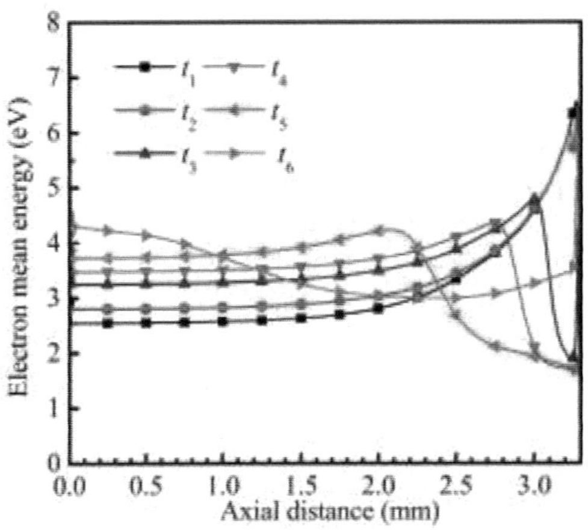

Figure. 3 Electron mean energy distribution along the axis

The electron mean energy determines the reaction rate and energy transfer rate at impact diffusion of electrons and other particles. Studying electron mean energy in corona discharge is conducive to understanding the impact diffusion (ionization and excitation) between electrons and other particles during the corona discharge process and has important significance to reveal the microscopic physical mechanism of streamer discharge. Electron mean energy distribution curves along axis at different discharge moments are shown in Fig.3. "0mm" in the x-axis is the position of ground electrode (anode) and "3.3mm" is the position of bar tip (cathode). The electron mean energy distribution trend in Fig.4 is similar with the electric field distribution trend in negative corona pulse process in the Reference [18]. Obviously, the energy acquisition of electron is accelerated through the electric field. As the discharge process continuous, the distortion of space electric field in the electrode gap caused by space charges is the main cause of changes of electron mean energy. This can explain following phenomena in Fig.3: in the initial pulse stage (t_1), the electron mean energy mainly concentrated close to the bar electrode, which is determined by the electrode gap structure. As the discharge continues (t_2-t_6), the maximum of electron mean energy occurs in the field-induced ionization region and the energy decreases, accompanied with the field-induced ionization zone moving toward the anode. This because impact and adsorption reactions will generate plenty of charged ions when the electron avalanche moves toward the anode and the internal field formed by ion cloud and electrons causes electric field distortion in the electrode gap. The electron mean energy close to the anode (0- 1mm) increases gradually during t_1-t_6. This is because electrons arriving at the anode increase continuously and the internal field between the anode and electron avalanche enhances the electric field close to the anode.

3.3. Electron density distribution

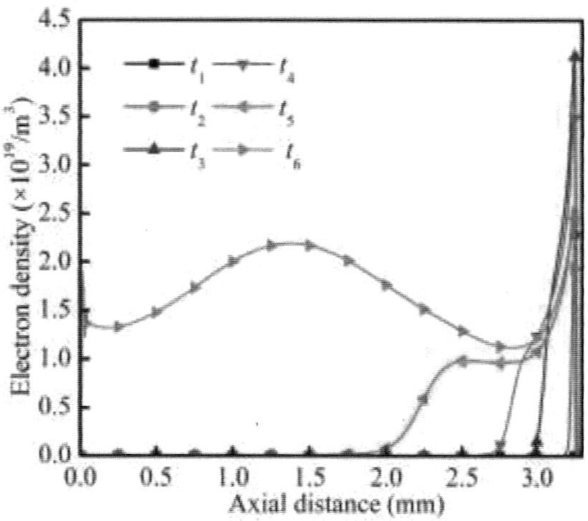

Figure. 4 Electron density distribution along the axis

The electron density distribution along the axis at 6 time points in one corona pulse process is shown in Fig.4. The electron density close to the cathode sheath is approximately 0, because electrons are kept off from the cathode. The maximum electron density at outer layer of the cathode sheath, which is related with the strongest field emission effect close to the bar electrode where has the most concentrated electric fields. In one discharge pulse period, electron density distribution develops toward the anode and the electron density in the electrode gap increases gradually during t_1-t_6. This is caused by continuous occurrences of impact and ionization reactions as electrons move toward the anode, construction of basic electron avalanche towards the anode, and continuous excitation reactions and ionization reactions.

3.4. Generation and dissipation characteristics of electrons

Table 2 The key reactions at different instants

Time	The key reactions at different instants
t_1	R_1, R_2, R_3, R_7 R_{10}, R_{22}, R_8...
t_2	R_1, R_{22}, R_2, R_7, R_{10}, R_3, R_8...
t_3	R_1, R_2, R_{22}, R_7, R_3, R_{10}, R_8...
t_4	R_2, R_1, R_{22}, R_7, R_8, R_{10}, R_3...
t_5	R_2, R_1, R_{22}, R_7, R_8, R_{10}, R_3...
t_6	R_1, R_{22}, R_2, R_7, R_8, R_{10}, R_3...

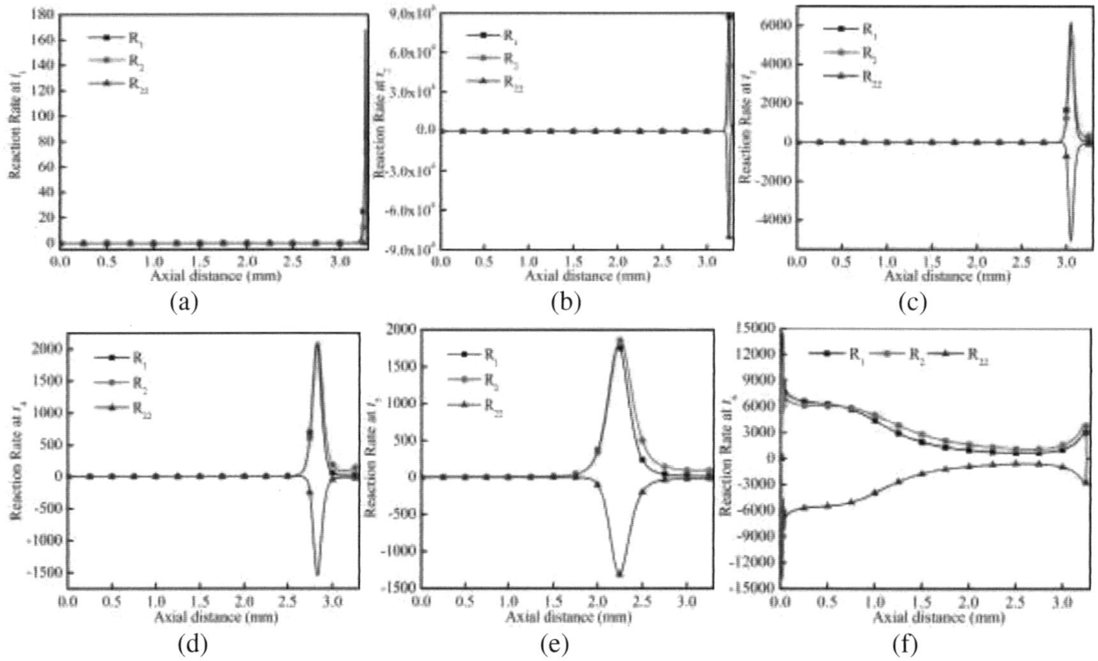

Figure. 5 Top 3 reaction rates at 6 typical time

Fig.5 Chemical reaction rates of main particles at different time points: (a) t_1, (b) t_2, (c) t_3, (d) t_4, (e) t_5, (f) t_6("+" is production process of electrons and "-" is the diffusion process of electrons. The unit of monomer reaction rate constant is s^{-1}, the unit of disome impact rate constant is m^3s^{-1}, and the unit of trisomy impact rate constant is m^6s^{-1}.)

It can be known from Table 1 that R_1, R_2 and R_{17} are the three key reactions in the whole pulse discharge process. The first two represents impact ionization reactions related with N_2 and O_2 and are the main process of electron reproduction. The reaction rates of R1 and R_2 are almost constant throughout the discharge process (Fig.5). R_{17} is the composite reaction involving N_2, N_2+ and electrons and occupies absolute advantages in electron diffusions. This is because the ionization threshold energy (12.06 eV) of O_2 is lower than that (15.6 eV) of N_2, thus inhibiting the impact ionization of N_2. In addition, R_6 and R_7 are charge transfer reaction containing position N ions and have relatively higher reaction rate. This is why positive ions containing O have far higher content than position ions containing N in the electrode gap throughout the negative corona discharge process[17]. Electron production and diffusion rate at different time points are shown in Fig.5. Development trend of electric field intensity under same bar-plane electrodes with time was computed in Reference [16], finding that electron production and diffusion mainly concentrated in field-induced ionization region, because electric field attenuation also concentrated in this region. In the early pulse stage, particle reaction increases to the maximum during t_1-t_2. In this period, abundant electrons and ions will be produced, bringing a sharp rising edge of discharge current. Subsequently, the electric field amplitude of field-induced ionization region reduces during t_2-t_5 and the particle reaction rate decreases gradually. At the end of single pulse period (t_6), aerial drainage will occur in the bar-plane gap, thus making the electrode gap recover to neutral quickly. This causes serious distortion of field intensity in the electrode gap, manifested by an abnormal growth trend of particle reactions in the whole gap.

4. Conclusion

1) In the early pulse period, electron mean energy mainly concentrate at places close to the bar electrode. The maximum electron mean energy occurs at the field-induced ionization region and

moves toward the anode as the discharge continues, accompanied with reduction of energy. In addition, the electron mean energy close to the anode increases gradually.

2) The electron density in cathode sheath is approximately 0 and achieves the maximum at outer layer of the cathode sheath. In one discharge pulse period, electron density distribution develops toward the anode as the discharge continues and the electron density in the electrode gap increases gradually.

3) R_1, R_2 and R_{17} are the three key reactions in the whole pulse discharge process. The first two represents impact ionization reactions related with N_2 and O_2 and are the main process of electron reproduction. The reaction rates of R_1 and R_2 are almost constant throughout the discharge process. R_{17} is the composite reaction involving N_2, N_2^+ and electrons and occupies the absolute advantages in electron diffusions.

References

[1] Du Hongliang, He Liming, Ding Wei, at el. Calculation of non-equilibrium plasma under air discharges [J]. High Power Laser and Particle Beams, 2011, 23(4): 1077-1081(in Chinese).

[2] Fofana I, Beroual A. A model for long air gap discharge using an equivalent electrical network [J] IEEE Transactions on Dielectrics Electrical and Insulation, 1996, 3(2): 273-282.

[3] d'Agostino R, Favia P, Oehr C, at el. Low-temperature plasma processing of materials: Past, present, and future [J]. Plasma Processes and Polymers, 2005, 2(1): 7-15.

[4] Nahorny J, Ferreira C M, Gordiets B, at el. Experimental and theoretical investigation of a N2-O2 DC flowing glow discharge [J]. Journal of Physics D: Applied Physics, 1995, 28(4): 738-747.

[5] Pancheshnyi S V, Starikovskii A Y. Two-dimensional numerical modelling of the cathode-directed streamer development in a long gap at high voltage [J] Journal of Physics D: Applied Physics, 2003, 36(21): 2683-2691.

[6] Liu X H, He W, Yang F, Wang H Y, et al. Numerical simulation and experimental validation of direct current air corona discharge under atmospheric pressure [J]. Chinese Physics B. 2012, 21(7): 075201.

[7] Antao D S, Staack D A, Fridman A, et al. Atmospheric pressure dc corona discharges: operating regimes and potential applications [J]. Plasma Sources Science and Technology, 2009, 18: 035016.

[8] Georghiou G E, Papadakis A P, Morrow R, et al. Numerical modelling of atmospheric pressure gas discharges leading to plasma production[J] Journal of Physics D: Applied Physics, 2005, 38(20): R303-R328.

[9] Li C, Ebert U, Hundsdorfer W. 3D hybrid computations for streamer discharges and production of runaway elctrons [J]. Journal of Physics D: Applied Physics, 2009, 42(20): 202003.

[10] Liu Xinghua, He Wei, Yang Fan, et al. Simulation and analysis of electron transport paramenters in air discharge [J]. High Voltage Engineering, 2011, 37(7): 1614-1619.

[11] Y. Itikawa. Cross sections for electron collisions with oxygen molescules [J]. J. Phys. Chem. Ref. Data, 2009, 38: 3025886.

[12] Zheleznyak M B, Filimonova E A, Simulation of a gas-Phase chemical reactor for removal of toxic impurities, based on the use of a pulsed streamer discharge [J]. High Temperature, 1998, 20 (4): 533-540.

[13] Segur P, Bourdon A, Marode E, et al. The use of an improved Eddington approximation to facilitate the calculation of photoionization in streamer discharges [J]. Plasma Sources Science and Technology, 2006, 15(4): 648-660.

[14] Simulation of dc atmospheric pressure argon micro glow-discharge [J]. Plasma Sources Science and Technology, 2006, 15(4): 676-688.

[15] Tran T, Golosnoy I O, Lewin P L, et al. Numerical Modelling of Negative Discharges in Air with Experimental Validation [J]. Journal of Physics D: Applied Physics, 2011, 44(1): 015203.

[16] Liao R J, Wu F F, Yang L J, et al. Investigation on Microcosmic Characteristics of Trichel Pulse in Bar-plate DC Negative Corona Discharge based on a Novel Simulation Model [J]. International Review of Electrical Engineering, 2013, 8(1): 504-413.
[17] Sima W X, Peng Q J, Yang Q, et al. Study of the characteristics of a streamer discharge in air based on a plasma chemical model [J]. IEEE Transactions on Dielectrics Electrical and Insulation, 2012, 19(2): 660-670.

The pressure cold wind system on the impact of industrial boiler economy and security

Henan Li[1, *], Hongli Qian[1], Lei Jiang[2], Dekai Yu[1], Guannan Li[3] and Hong Yuan[1]

[1]Dalian Boiler and Pressure Vessel Inspection Institute, Dalian 116000, China
[2]Ningxia Boiler and Pressure Vessel Inspection Institute, Yinchuan 750001, China
[3]Shanghai University of Engineering Science, Shanghai, 200000, China

[*]Corresponding author e-mail: 172579206@qq.com

Abstract. Industrial boiler is one of the most energy-consuming equipment in china, the annual consumption of energy accounts for about one-third of the national energy consumption.Industrial boiler in service at present have several severe problems such as small capacity, low efficiency, high energy consumption and causing severe pollution on environment, the average industrial boiler operation efficiency is only 65%.If the efficiency increased by 15% ~ 20%, which meet the international advanced level, each year 70 million tons of coal saving and reduce environmental pollution[1]. As energy conservation and emissions reduction becomes the basic national policy of our country, improving the efficiency of industrial boiler energy is facing opportunities and challenges, optimizing the operation mode of the existing units, it is necessary to increase the flexibility of the boiler control.

1. Introduction

This paper, taking the BG - 220/9.81 - M in one power plant boiler as an example, to adjust the pressure of air cooling caused by a series of problems were analysed, and used to find the reasonable operation mode of the cold wind pressure, in order to improve the safety and economy of boiler.

2. The introduction of a thermoelectric boiler and operation mode

2.1. Introduction of the boiler unit

This boiler is Beijing boiler plant design and manufacture of BG - 220/9.81 - M boiler. This boiler is a type of π, natural circulation, according to the open layout 7 degree earthquake resistant design. Huaibei burning lean coal boiler design, the use of ball mill feeding system, intermediate storage, hot air is four dc burner, one, two, three times the wind 400 / phi 750 hypothetical double tangential combustion organization.

Boilers are all steel structure architecture. Furnace adopts single drum, concentrated drop tube, natural circulation system. Near the front of the boiler furnace, be full of membrane type water wall, decorate the furnace exit platen super heater, the level of high and low temperature of flue furnish two levels of convection superheated and the tail shaft staggered two levels of economizer and two levels of air preheater. Horizontal flue and turned to the chamber are arranged around the diaphragm wall bag wall tube to ensure the boiler seal [2].

Boiler two cross heat exchange by radiation, convection, two-stage spray desuperheating system of superheater, the superheater including ceiling pipe, wall, superheater, low temperature superheater and high temperature superheater, and high-temperature superheater and points of cold and heat. Boiler using water seal machine in dregs at the bottom of the slag discharge to slag bucket, slag underground wells installed slag door, adopts hydraulic control.

2.2. Operation situation

A thermal power plant has three BG - 220/9.81 - M boiler, a 50 mw condensing steam turbine generator set and a 25 mw back pressure turbine generator set. These make up the factory production process, to customers for centralized heating and power cogeneration process. According to heating the off-season and busy season, and the heating load is different, different operation mode, as shown in the following table 1.

Table 1. Operation mode under different heating load.

Season	Heat power load Q (t/h)	The number of boiler	The number of turbine
Summer	Q≤150	1	1
Spring and autumn	150≤Q≤250	2	1
Winter	Q≥250	3	2

3. Boiler operation before modification are faced with the problem

3.1. The primary air temperature control difficult

Boiler load change on the size of the primary air temperature influence is not big. Under a certain load, if the coal properties change, primary air temperature is not easy to adjust. If because of a high wind temperature, and make the boiler burner fire or tempering is serious, can only load down or shut down the machine to protect the burner. On the one hand, increase the frequency of the change of working condition of boiler, make the heating surface fatigue damage, cause compression part leak or tube; on the other hand, the change of the load will change once the wind temperature, frequent changes will lead to the steam turbine generator electrical load and corresponding thermal efficiency index decline.

3.2. The boiler combustion efficiency decline

In the process of operation, in order to prevent the burner with fire, guarantee the normal work of the burner, to reduce the fineness of pulverized coal, the pulverized coal ignition delay. Not only worsen the operation condition of the pulverized coal ignition phase, and to reduce the retention period of pulverized coal in the furnace, the chemical and mechanical incomplete combustion incomplete combustion loss increases, the boiler efficiency is reduced [3].

3.3. Heating surface coking seriously

When the cold wind of the boiler is not injection, heat load of burn center is higher, fly ash is easy to soften, make to coking heating surface more opportunities.

3.4. Boiler coal adaptability is poor

In the process of operation, Because of the air preheater exports of primary air temperature is difficult to adjust, to cooperate the efficient operation of the burner, prevent burner fire cause part of the burner interrupt operation, affect the boiler load. Before using pressure air cooling systems, boiler using coal is unitary, the high requirement of carbon and volatile content in raw coal, coal source is relatively small, increasing the cost of fuel.

4. The influence of pressure cold wind on boiler security and economy

4.1. The pressure cold wind effects on a wind temperature control properties

Under high loads, the temperature of primary air is a bit too high. Under low loads, the speed of primary air is a bit too low. This can lead to fire in advance, will affect the safe operation of the boiler coal powder, concentration of pulverized coal, the temperature of primary air, the speed of primary air, the burner geometry size and combustion condition play an important role in the process of the steady combustion of boiler. Under a certain load, the pressure cold wind is main measures to prevent the burner ignition [4].

Through experiments it is concluded that: Under the same load, the pressure cold wind adjusting characteristics of primary wind, the results are shown in table 2 below.

Table 2. The pressure cold wind effects on primary air temperature.

The opening of the pressure cold wind (%)	The primary air temperature
0	305
50	284
100	276

From table 2 it can be seen that the primary air temperature is inversely proportional to the opening of the cold wind. As the change of the opening, the primary air temperature gradient change greatly. The pressure cold wind has an important effect on ignition and combustion stability.

4.2. The primary air temperature, pulverized coal temperature and Pulverized coal concentration

The creditable measurement of pulverized coal concentration in primary air is the basis of combustion optimization and safe operation of utility boilers. The specific heat of coal has an enormous impact on gas-solid mixture temperature [5].

Table 3. When the primary air temperature is 300 degrees, pulverized coal concentration and gas-solid mixture temperature.

The Pulverized coal temperature (℃)	The concentration of pulverized coal (Kg/Kg)		
	0.15	0.40	0.70
70	270	234.36	205
75	270.65	235.71	207.36

Table 4. When the Pulverized coal temperature is 70 degrees, pulverized coal concentration and gas-solid mixture temperature.

The primary air temperature (℃)	The concentration of pulverized coal (Kg/Kg)		
	0.15	0.40	0.70
300	270	234.37	205
305	274.34	237.88	208.28

As you can see in table 3 and table 4, pulverized coal concentration is one of important factors which have influence on gas-solid mixture temperature. With the increase of pulverized coal concentration, gas-solid mixture temperature decrease. Gas-solid mixture temperature changes significantly as the load of a supercharged boiler changes. Pulverized coal temperature has relatively little impact on the primary air temperature. Therefore, boiler under certain load, pressure air cooling to the safe operation of the boiler has great benefits.

4.3. The Influence of Pressure Cold Wind on the Heating Surface Coking Properties

The pressure cold wind quantity increase under a certain load, wind powder mixture temperature is reduced, the burning zone temperature is reduced, the burner surface heat load is reduced, the ash softening of smaller share, make the burner coked the possibility of a smaller.in addition, the pressure cold wind increases, the increase of wind speed, a wind to enhance rigidity, increasing distance of ignition and burner; Cold wind pressure increase, a volume increase, increase oxygen around the burner, reduced the reducing atmosphere, ash softening temperature increase, so the pressure cold wind quantity increase, greatly reduce the possibility of a burner coking.

4.4. The Influence of Pressure Cold Wind on Operation of Coal Pulverizing System

By table 4 to see, the pulverized coal temperature, the wind powder mixture of temperature change is very small, cold wind baffle opening pressure, appropriate can stable wind powder mixture temperature, working condition of stable combustion chamber of a stove or furnace [6]. So, in the coal pulverizing system security under the premise of coal mill can maintain relatively high grinding temperature after operation, in this way, improve the dry coal mill output, and reduce powder consumption [7].

4.5. The Pressure Cold Wind Effect on Combustion Chamber of a Stove or Furnace Area

The pressure cold wind increase, on the one hand, reduces the temperature of the wind powder mixture, the pulverized coal and oxygen should be temperature drop, the family of the activation and collision opportunities to reduce the number of molecules, oxidation reaction time delay; On the other hand, a relative increase air flow and a wind speed increases, power combustion of pulverized coal area, make pulverized coal can't timely transition to higher oxygen diffusion combustion zone, some large particles can not make full combustion, coal powder boiler efficiency decline[8].

4.6. The Influence of Pressure Cold Wind on the Boiler Thermodynamic Characteristic

Under a certain boiler load, different pressure of the opening of the cold wind, the boiler's main thermodynamic parameters, data calculation and drawing curves, such as table 3-4.

Table 5. The pressure cold wind on the influence of main parameters

The open degree of cold wind (%)	0	50	100
Furnace temperature (℃)	714	718	720
Superheater outlet temperature (℃)	477	482	485
Flue gas emission temperature (℃)	130	132.5	134
The main steam pressure (MPa)	8.70	8.65	8.63
The flow of lukewarm water (t/h)	1.10	1.41	1.60
The oxygen content (%)	5.17	5.24	5.27
The primary air temperature (℃)	306	283	275
Boiler efficiency (%)	86.9	86.4	86.2

IOP Conf. Series: Earth and Environmental Science 64 (2017) 012103 doi:10.1088/1755-1315/64/1/012103

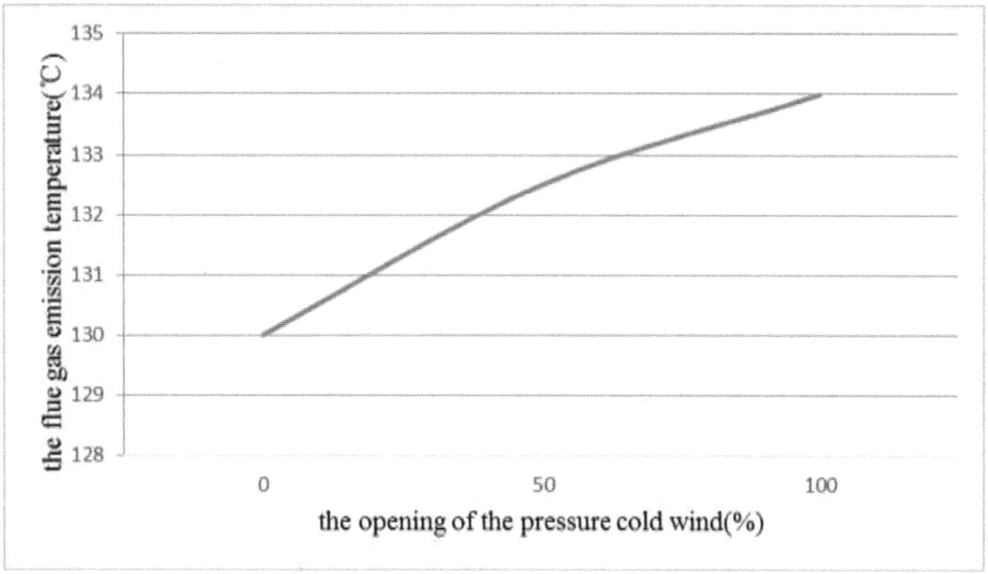

Figure 1. The flue gas emission temperature changes with the opening of pressure cold wind.

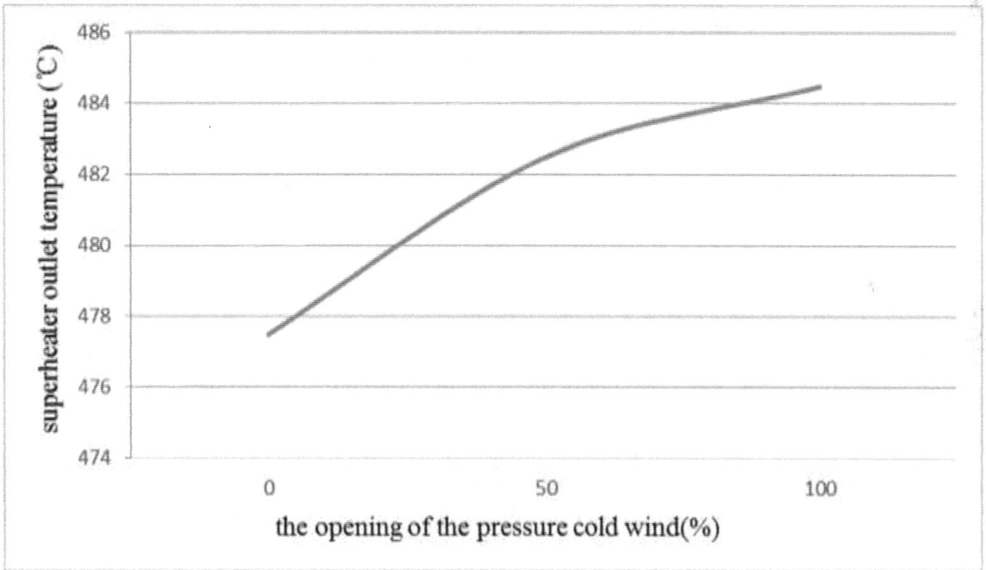

Figure 2. The superheater outlet temperature changes with opening of the pressure cold wind.

By figure 1, 2 can be seen: the superheater outlet temperature and the flue gas emission temperature with the increase of the pressure cold wind opening increased.

IOP Conf. Series: Earth and Environmental Science **64** (2017) 012103 doi:10.1088/1755-1315/64/1/012103

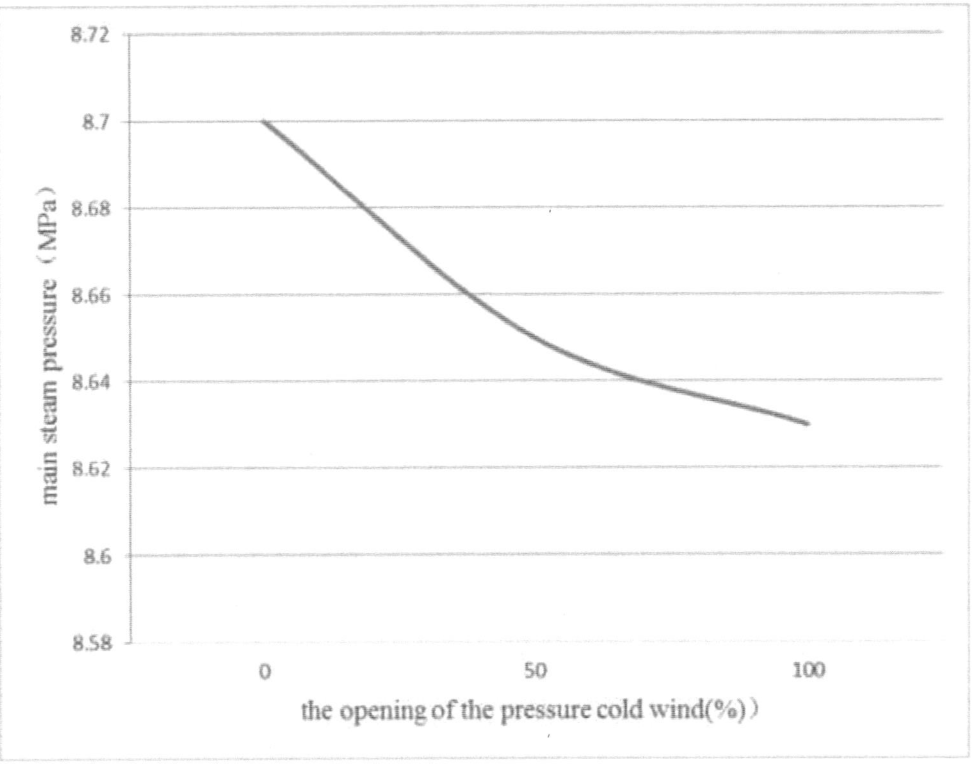

Figure 3. The main steam pressure changes with the opening of the pressure cold wind.

By figure 3 can be seen: the main steam pressure is lower with the amount of cold wind.

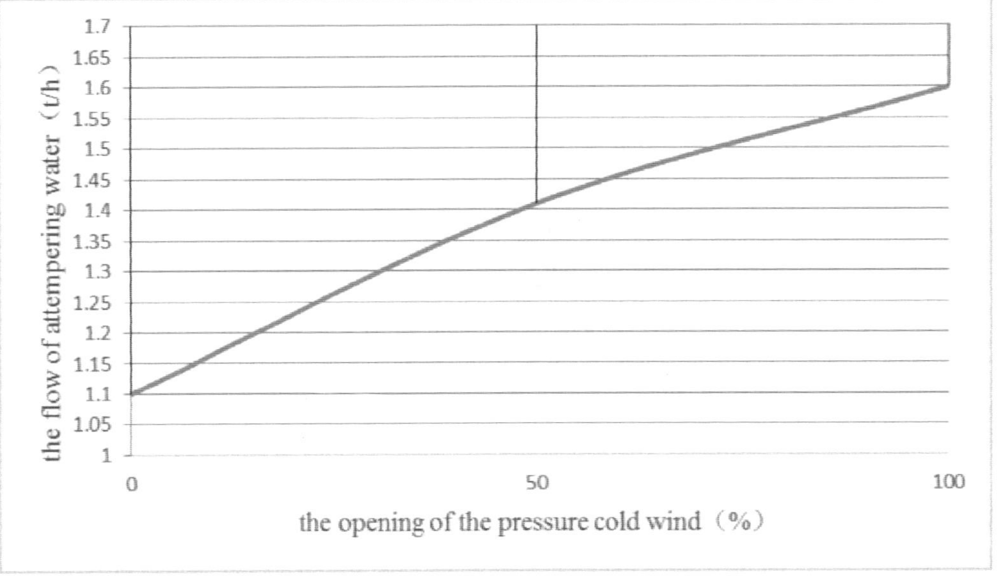

Figure 4. The flow of attempering water changes with the opening of the pressure cold wind.

By figure 4 can be seen: the flow of attempering water increased with increased with the amount of pressure cold wind.

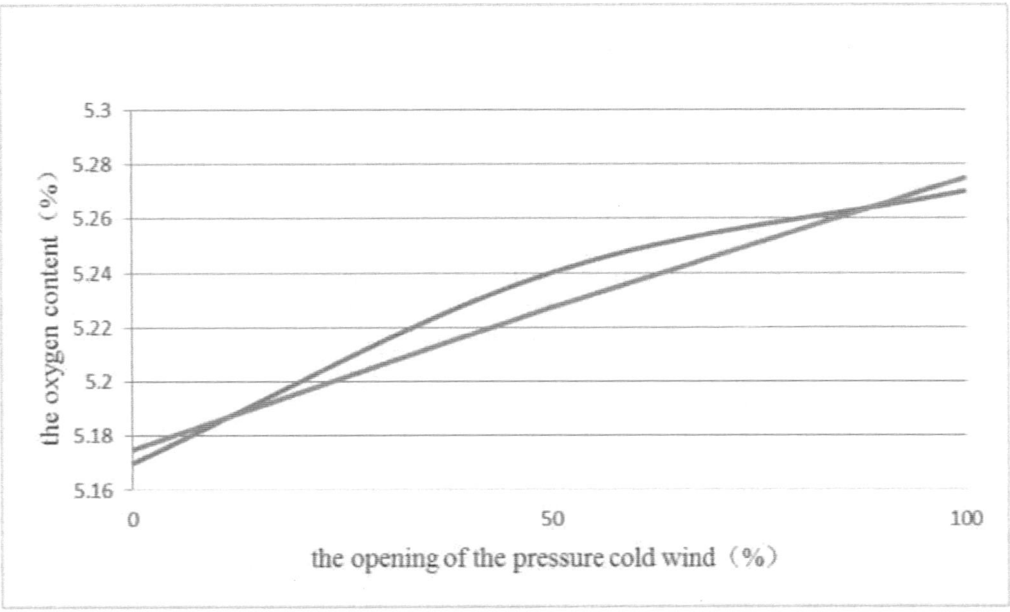

Figure 5. The oxygen content changes with the opening of the pressure cold wind.

By figure 5can be seen: the oxygen content increased with increased with the amount of pressure cold wind.

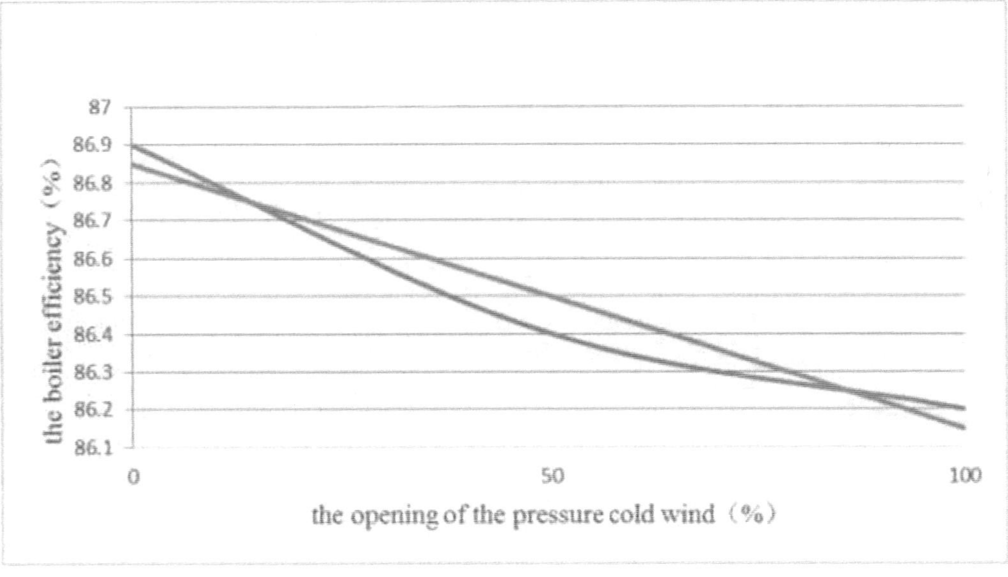

Figure 6. The boiler efficiency changes with the opening of the pressure cold wind.

By figure 6 can be seen: the boiler efficiency is lower with the amount of cold wind.

4.7. Interpretation of Result

a. Increase with the cold wind, a wind speed increases, the reduced wind temperature, ignition delay, flame center up, relative to reduce time spent in pulverized coal furnace, furnace outlet temperature increasing [9].

b. From the analysis on the main steam pressure, on the basis of the external load, the cold wind pressure increase, the flame center move, water wall and the contact time of high temperature flue gas decreased; Chamber of a stove or furnace section heat load is reduced, the average temperature of the high-temperature flue gas; Unit of heat radiation is reduced, part of the hot flue gas heat cannot fully with water wall, lowered the evaporation of water wall unit time, the boiler output drop, characterized by steam pressure drop.

Look from the desuperheating water flow and outlet temperature of the superheater, cold wind pressure increases, is the furnace smoke gas exports increase, coupled with the furnace exit the superheater import smoking temperature on the high side, superheater of convective heat transfer enhancement; External load constant, unit time through the superheater steam content is constant, increase the heat absorption capacity of unit mass steam, overheating temperature increase. If you want to maintain the temperature constant, so the desuperheating water flow increases, the steam quality to drop, lower steam turbine efficiency.

c. From the exhaust temperature, pressure from the cold wind barrier to fully open, exhaust smoke temperature 3 ~ 4 ℃.On the one hand, economizer inlet temperature increases, Increase pressure, on the other hand, the cold wind, that cut through the air preheater air volume is relatively, so the exhaust temperature, exhaust temperature generally rise every 12 ~ 15 ℃, smoke damage increased by 1%.

d. From the perspective of the oxygen pressure cold wind increases, primary air and secondary air volume and reduce, but by the flow characteristics of parallel line, total air volume is increased, so the excess air coefficient was increased, higher performance for oxygen[10].

e. The table 3.1 parameters are calculated and can be seen from the figure 3.6: pressure of the opening of the cold air from close to open, the comprehensive efficiency boiler by 0.5% or thereabouts.

5. Conclusion

Through the analysis, we can see that the cold air has a major effect on slagging of boiler and temperature of burner. For coals that contained more volatiles and high load of the unit, the cold air had more obvious effect on the security of the boiler. In consideration of the Boiler economic operation, the pressure cold wind can reduce energy consumption of milling system and the probability of boiler slagging. Boiler is one of key equipment in power plant; the cold air channel system will directly affect the work of boiler, By means of reconstruction, the run mode of the installation was optimized, the performance became steady, the quality of product was improved, and more economic achievements were acquired.

References

[1] V. N. Shemyakin, A. E. Karapetov, and S. V. Krylov,"Experience Gained with Practical Use of the Fluidized Bed Technology at Industrial and Municipal Power Facilities," Trudy TsKTI, No. 298, 57–58 (2009).J

[2] NING Ping, REN Bing-nan. Comprehensive Utilization and the Ways to Purify of Yellow Phosphorus Tail Gas [J]. Environmental Science, 2003, 22(3): 149-151.

[3] Joris Roels Willy Verstraete. Occurrence and Origin of Phosphine in Landfill Gas [J]. Science of the Total Environment, 2004, 327: 185-196.

[4] Siyuan C Morton; Marc Edwards. Reduced Phosphorus Compounds in the Environment [J]. Critical Reviews in Environmental Science and Technology, 2005, 35(4): 333-364.

[5] Tan W, Marquez HJ, Chen T, Liu J .Analysis and control of a nonlinear boiler–turbine unit. J Process Control.2005, 15:883–891.

[6] Lo KL, Zeng PL, Marchand E, Pinkerton A, Modelling and state estimation of power plant steam turbines.2006, IEEE Proc Part C 137(2): 80–9.

[7] Liu XJ, Kong XB, Hou GL, Wang JH .Modeling of a 1000 MW power plant ultra super-critical boiler system using fuzzy-neural network methods. Energy Convers Manag, 2013, 65: 518–527.

[8] Fang F, Wei L. Backstepping-based nonlinear adaptive control for coal-fired utility boiler–

turbine units. J Appl Energy, 2011, 88: 814–824.
[9] Orikasa H, Tomita A. NO and N 2 formation behavior during the high-temperature O2 gasification of coal char. Energy Fuels, 2003, 17(2): 405–411.
[10] Spinti JP, Pershing DW. The fate of char-N at pulverized coal conditions. Combust Flame, 2003, 135(3): 299–313.

International Symposium on Resource Exploration and Environmental Science IOP Publishing
IOP Conf. Series: Earth and Environmental Science **64** (2017) 012104 doi:10.1088/1755-1315/64/1/012104

Analysis on energy consumption index system of thermal power plant

J B Qian[1, 2, 3], N Zhang[1] and H F Li[1]
[1]Department of Power Engineering, North China Electric Power University, No.689 Huadian Road, Baoding City, 071003 China
[2]Reserch Institute Of Bit in ZhongShan, No.23 Boai Road, Zhongshan City, 528400 China

Abstract. Currently, the increasingly tense situation in the context of resources, energy conservation is a realistic choice to ease the energy constraint contradictions, reduce energy consumption thermal power plants has become an inevitable development direction. And combined with computer network technology to build thermal power "small index" to monitor and optimize the management system, the power plant is the application of information technology and to meet the power requirements of the product market competition. This paper, first described the research status of thermal power saving theory, then attempted to establish the small index system and build "small index" monitoring and optimization management system in thermal power plant. Finally elaborated key issues in the field of small thermal power plant technical and economic indicators to be further studied and resolved.

1. Introduction

With the implement of the policy of "power plants separated from electric network, competing electric price" in China power industry, the power generation enterprises make efforts to reduce cost and improve efficiency of power generation to succeed in the competition. Meanwhile , along with the increasing emphasis on environmental protection, as the primary energy guzzler and pollution emitters, thermal power plants has to consider how to achieve high efficiency and low emissions on the basis of providing reliable power for the sustained and healthy development of national economy[1]. Therefore, under the premise of safe and reliable operation, it is of great significance to reduce the coal consumption rate of thermal power units, and to improve the thermal economy of the thermal power units [2]. The operation and management of thermal power plants is mainly to control the main indicators of thermal power system, if these indicators are controlled, the unit can run safely and economically.

In this article, we reviewed and discussed the monitoring and optimizing management of small index of thermal power unit from following aspects: First, development status of performance monitoring of power plant thermodynamic system at home and abroad, and construction method of small index system are introduced. Then, sorting out the meaning and algorithm of all the small indexes involved and pointing out the problems existing in the research work and the key problems to be studied in the next step, in order to provide a scientific basis for the establishment of small index monitoring and optimization management system for thermal power units.

2. Research Status

2.1. *Abroad Research Status*

In early 1970, The United States, Britain, France and other countries have been pumping money for the research into performance monitoring of power plant thermodynamic system, and lay a foundation for the operation optimization of thermal power unit. In the beginning, the performance monitoring of the unit is mainly used as an additional part of the data acquisition and data processing system [3]. From the beginning of 80s, the performance monitoring of power plants is gradually developing in the direction of guiding the on-site operation and completing the performance and overall performance analysis and diagnosis[4].After years of development, a large number of software of thermal power plant operating optimization management are constantly emerging at abroad. Among them, the function of some software package, which has been applied in a number of foreign plants, is outstanding and complete, such as the Synergy system of German Siemens company and the Optima system of Swiss ABB company, which can be supplied as a whole and divided into different needs as well[5].

2.2. *Domestic research status*

Because of the implementation of the reform and opening policy in 1980s, the automatic information technology of thermal power plant in China has entered a period of rapid development on the basis of large-scale electric power construction. At present, many researches have been done on the performance monitoring of thermal power plant, and some research institutes and universities have done a lot of research in the field of operation management software. In order to make the performance analysis and optimization of the operation of the unit reach the international advanced level, such as Tsinghua University, Wuhan university, North china electric power university and Southeast university. Some thermal power plants in China have put the real time optimization management software in actual production activities, for example, the on line energy loss analysis and optimization system for thermal power plant, which was developed by the Institute of thermal power and automation in Zhejiang university, has put into use in the 300MW unit of Guangdong Shajiao Power Plant [6].

3. Construction of small index system

At present, DCS, MIS, SIS and other intelligent optimization software in China's thermal power enterprises, which can supervise and manage the operation index value of the unit in real-time or post-processing, have entered the stage of mature application. Thus, the control level of the unit has been more precise and reduced the cost of power generation. For the monitoring software, whether the evaluation index is scientific or not has great influence on its accuracy and practicability[7].Therefore, the study of the technical and economic indexes of thermal power plant is quite necessary.In this paper, the small index system of thermal power plant is constructed based on the coal-saving index and electricity-saving index . As shown in Table 1.

3.1. *Coal-saving index*

3.1.1. The main boiler performance index (1) Boiler thermal efficiency (anti-balance method):

$$\eta_{b1}=100-\left(q_2+q_3+q_4+q_5+q_6\right) \tag{1}$$

Where, q_2 is boiler exhaust heat loss, q_3 is heat loss of chemical incomplete combustion, q_4 is heat loss of mechanical incomplete combustion, q_5 is heat loss of dispelling heat of boiler, q_6 is physical heat loss of ash.

Table 1. The small index system of thermal power plant

Ttarget layer	Ffirst-class index	Ssecond-class index	Tthird-class index
The small index system of thermal power plant	Coal-saving index	The boiler performance index	Boiler thermal efficiency, Boiler exhaust heat loss, Heat losss of chemistry incomplete combustion, Heat loss of mechanical incomplete combustion, Heat radiation loss, Heat loss due to sensible heat in slag, Excessive air coefficient, Superheated steam pressure, Superheated steam temperature, Reheat steam pressure, Reheat steam temperature, Leakage rate of air preheater
		The steam turbine performance index	Steam turbine heat rate, Steam turbine thermal efficiency, Turbine steam rate, Turbine back pressure, Heater' s upper terminal temperature difference, Heater' s bottom terminal temperature difference, Feed water temperature of high pressure heater outlet , The vacuum of the steam condenser, Inlet temperature of circulating water, Temperature rise of circulating water, Terminal temperature difference of steam condenser, Exhaust steam temperature
	Electricity-saving index	The unit consumption and electricity consumption rate of auxiliary machine	The unit consumption and electricity consumption rate of coal mill, The unit consumption and electricity consumption rate of powder evacuation machine, The unit consumption and electricity consumption rate of air feeder, The unit consumption and electricity consumption rate of induced draft fan, The unit consumption and electricity consumption rate of water feed pump, Electricity consumption rate of circulating water pump, Electricity consumption rate of condensate pump, Electricity consumption rate of coal handling system, Electricity consumption rate of dashing device, Eelectricity consumption rate of desulphurization equipment system,

(2)Superheated steam pressure and temperature

The superheated steam pressure and temperature refer to superheated steam pressure and temperature of boiler final superheated outlet, if the final superheated outlet of the boiler is provided with a plurality of main stream pipes, we should take its arithmetic mean value.

(3)Excessive air coefficient

The ratio of the actual air supply to the theoretical air volume is called the excess air coefficient. In order to make the coal can complete combustion in the furnace, reduce loss of incomplete combustion, the actual amount of air into the furnace is higher than the theoretical air volume, so that the combustion reaction can be carried out in the presence of excess oxygen.

(4)Leakage rate of air preheater

$$A_L=\frac{RO_2^{'}-RO_2^{''}}{RO_2^{''}}\times 90\% \tag{2}$$

Where, $RO_2^{'}$ is the volume percentage of three atomic gases in the inlet of air preheater, the unit is %; $RO_2^{''}$ is the volume percentage of three atomic gases in the outlet of air preheater, the unit is %.

3.1.2. The main steam turbine performance index (1) Steam turbine heat rate

For reheat steam turbine, the calculating formula as follow:

$$q=d\left[(h_{ma}-h_{fw})+\frac{G_{rh}}{G_{ms}}(h_{rhr}-h_{rhl})\right] \tag{3}$$

Where, q is heat rate, the unit is kJ/kW·h; h_{rhr} is the enthalpy of reheater outlet steam, h_{rhl} is the enthalpy of reheater inlet steam, the unit is kJ/kg; h_{ms} is the enthalpy of live steam, h_{fw} is the enthalpy of boiler feedwater, the unit is kJ/kg; G_{ms} is main stream flow, G_{rh} is reheated steam flow, the unit is kg/h; d is steam rate, the unit is kJ/kW·h.

(2) Steam turbine thermal efficiency

$$\eta=\eta_t\eta_i\eta_m\eta_g=\frac{3600}{q}\times 100\% \tag{4}$$

Where, η_t is cycle thermal efficiency of steam turbine, the general value is 36%-54%; η_i is relative internal efficiency of steam turbine; η_m is mechanical efficiency of steam turbine; η_g is generator efficiency, high-power generator efficiency is generally 98%-99.5% .

(3) Turbine steam rate

$$d=\frac{D_{zq}}{P_{fd}} \tag{5}$$

Where, d is turbine steam rate, the unit is kg/kWh; D_{zq} is the main stream flow entering the turbine, the unit is kg/h or t/h; P_{fd} is the generation load of steam turboset, the unit is kW.

(4) Heater's terminal temperature difference

Heater's upper terminal temperature difference is the difference between the saturation temperature under heater inlet steam pressure and the outlet temperature of the heater heating, the unit is ℃. Heater's bottom terminal temperature difference is the difference between the drainage temperature of the working fluid into the drain cooler and out the drain cooler, the unit is ℃.

(5) The vacuum of the steam condenser

$$V_{zkd}=1-\frac{P_{dq}-P_{zk}}{98.1}\times 100 \tag{6}$$

Where, V_{zkd} is the vacuum of the steam condenser, the unit is %; P_{dq} is the atmospheric pressure, the unit is mmHg; P_{zk} is the absolute value of condenser vacuum, the unit is mmHg.

(6) Exhaust steam temperature

Exhaust steam temperature is the temperature when steam worked in the steam turbine and discharged into the condenser, which also named saturation temperature, is an important indirect index which is often ignored but important in steam turbine operation. There is a one-to-one relationship between exhaust temperature and exhaust pressure, when the exhaust temperature changes 1℃, the coal consumption changes about 1.1g/(kW▪h).

$$t_{pq}=t_{xh}^{rk}+t_{xh}^{ws}+t_{nq}^{dc} \quad (7)$$

Where, $\mathbf{t_{pq}}$ is the exhaust steam temperature, the unit is℃; t_{xh}^{rk} is the inlet temperature of circulating water, the unit is℃; t_{xh}^{ws} is the exhaust steam temperature of condenser, the unit is℃; t_{nq}^{dc} is the terminal temperature difference of condenser, the unit is℃.

3.2. The main electricity-saving index

(1) The unit consumption and electricity consumption rate of coal mill

$$L_{mm}=\frac{W_{mm}}{W_{fd}}\times 100\% \quad (8)$$

Where, L_{mm} is the electricity consumption rate of coal mill, the unit is %; W_{mm} is the amount of electricity consumed during the grinding of pulverized coal, the unit is kWh; W_{fd} is the generating capacity during given period, the unit is kWh.

$$Dh_{mm}=\frac{W_{mm}^{yd}}{M_{mf}} \quad (9)$$

Where, Dh_{mm} is the unit consumption of coal mill, the unit is kWh/t; W_{mm}^{yd} is the amount of electricity consumed during the grinding of pulverized coal, the unit is kWh; M_{mf} is the amount of pulverized coal produced by coal mill during given period, the unit is t.

(2) The unit consumption and electricity consumption rate of powder evacuation machine

The electricity consumption rate of powder evacuation machine refers to the percentage of electricity consumption of powder evacuation machine out of the generating capacity during given period. The unit consumption of powder evacuation machine is the electricity consumption while powder evacuation machine exhausting the mixture of air and dust produced by per ton pulverized coal, characterizing the operation economy of powder evacuation machine.

(3) The unit consumption and electricity consumption rate of air feeder

The electricity consumption rate of air feeder is the percentage of the amount of electricity that the air feeder consumed by producing steam in the boiler out of the generating capacity during given period.

The unit consumption of air feeder refers to the amount of electricity consumed by the air feeder producing per ton steam in the boiler.

(4) Electricity consumption rate of coal handling system

$$L_{sm}=\frac{W_{sm}}{W_{fd}}\times 100\% \quad (10)$$

IOP Conf. Series: Earth and Environmental Science **64** (2017) 012104 doi:10.1088/1755-1315/64/1/012104

Where, L_{sm} is the electricity consumption rate of coal handling system, the unit is %; W_{sm} is the amount of electricity consumed by coal handling system, the unit is k·Wh.

(5) Electricity consumption rate of desulphurization equipment system

$$L_{tl} = \frac{W_{tl}}{W_{fa}} \times 100\% \tag{11}$$

Where, L_{tl} is the electricity consumption rate of desulphurization equipment system, the unit is %; W_{tl} is the amount of electricity consumed by desulphurization equipment system, the unit is k·Wh.

4. Conclusions

The main work and important conclusions of this paper are as follows:

With the gradual implementation of energy saving and emission reduction policies and the construction of energy-saving society, the development of China's thermal power will tend to be more efficient, more energy saving, and more environmentally friendly.

In this paper, the small index system of thermal power plant is constructed based on the boiler performance index, the steam turbine performance index, and the unit consumption and electricity consumption rate of auxiliary machine. It has a certain reference value for the establishment of "small index" monitoring and optimization management system to further regulate the energy conservation management and tap the energy saving potential of the thermal power plant.

The future direction of this paper: The following research will be based on the small index system established in this paper, and combined with usual operation experience, establishing a comprehensive small finger library, covering all the systems and equipment involved in the operation of the thermal power plant, including turbine, boiler and others. Then according to the requirements of the user and the actual operation of the unit, select all or some of the small indicators, to complete the small index monitoring and optimization management system implementation case based on the DCS and SIS system of operation unit. To realize the real-time monitoring, operation optimization and statistical management of unit operation small index, provide guidance for the operator to implement the "small index" energy-saving operation, help the technicians and managerial personnel control the "small index" running state of the unit, provide guidance and suggestions for the development of energy saving technical innovation and operation management system for power plant.

Acknowledgements

This research is supported by Fundamental Research Funds for the Central Universities (Grant No.2016MS103, 2015MS107) and Science and Technology Planning Project of Guangdong Province, China (Grant No.2015B010101002).

First name	Surname	Institution	Email address	Title (Prof/ Dr./Student)
Qian	Jiangbo	Department of Power Engineering, North China Electric Power University	qianjiangboo@gmail.com	Pro
Zhang	Nan	Department of Power Engineering, North China Electric Power University	hddlzhangnan@126.com	Student
Li	Hengfan	Department of Power Engineering, North China Electric Power University	Lhfdt100@163.com	Dr

References

[1] Yuan G L, Zhang J H, Wang T H, et al. Comprehensive energy-saving evaluation of thermal power plants based on TOPSIS gray relational Projection and the Weight Sensitivity Analysis[J]. Journal of Chinese Society of Power Engineering, 2015, 35(5): 404-11

[2] Hu Y, Liu J ZH, Zeng D L, et al. Thermo-economic analysis for water make-up system of power plant condensers[J].Journal of Chinese Society of Power Engineering, 2013, 33(5): 370-74

[3] Xu X B. On-Line Calculation on Thermo-economic indicators for a large fire power generating unit[D]. Taiyuan: Taiyuan University of Technology, 2008:1-2

[4] Sun L W. Investigation of energy-saving emission reduction and optimization real-time system for power plants[D]. Taiyuan: Shanxi University, 2010:1-3

[5] Shi B. The research on operation performance appraisal system of large-scale thermal power plant[D]. Harbin: Harbin Engineering University, 2007: 5-7

[6] Li B, Ren R, Ren H R, et al. Online energy-loss analysis & optimization system for fossil-fuel power plants[J]. Energy Engineering, 2001, 28(6):34-36

[7] Zhou Y B, Hu X L, He T, et al. Index management system and its implementation for thermal power station[J]. Electronic Power, 2002, 35(10): 77-80

[8] Jiang M C.Handbook of energy consumption index analysis of thermal power plant [M]. Beijing: China power press, 2011

[9] Li Q, Gong W P. Energy saving and index management technology of thermal power plant M]. Beijing: China power press, 2006

Potentiality Prediction of Electric Power Replacement Based on Power Market Development Strategy

Bo Miao[1, a], Shuo Yang[2, b], Qiang Liu[3, c], Jingyi Lin[1, d], Le Zhao[4, e], Chang Liu[1, f], Bin Li[1, g]

[1]China Electric Power Research Institute, Beijing, China
[2]State Grid Hebei Electric Power Company, Hebei, China
[3]State Grid Zhejiang Electric Power Company, Zhejiang, China
[4]State Grid Beijing Electric Power Company, Beijing, China

[a] miaobo@epri.sgcc.com.cn, [b] yangs@he.sgcc.com.cn, [c] liu_qiang_yx@zj.sgcc.com.cn, [d] linjingyi@epri.sgcc.com.cn, [e] zhaole@bj.sgcc.com.cn, [f] liuchang3@epri.sgcc.com.cn, [g] libin5@epri.sgcc.com.cn

Abstract. The application of electric power replacement plays an important role in promoting the development of energy conservation and emission reduction in our country. To exploit the potentiality of regional electric power replacement, the regional GDP (gross domestic product) and energy consumption are taken as potentiality evaluation indicators. The principal component factors are extracted with PCA (principal component analysis), and the integral potentiality analysis is made to the potentiality of electric power replacement in the national various regions; a region is taken as a research object, and the potentiality of electric power replacement is defined and quantified. The analytical model for the potentiality of multi-scenario electric power replacement is developed, and prediction is made to the energy consumption with the grey prediction model. The relevant theoretical research is utilized to realize prediction analysis on the potentiality amount of multi-scenario electric power replacement.

1. Introduction

In recent years, energy crisis and environmental pollution have aroused high attention from all social members. To further promote the rapid expansion of energy conservation and emission reduction in our country, the State Grid Corporation of China timely puts forward the development strategy of electric power replacement, which is not only an important measure to increase the consumption proportion of renewable clean energy, but also a significant means to quicken energy structure adjustment of our country. Therefore, relevant theoretical research on the aspect is urgently required to be carried out on the basis of the current situation of the electric power replacement in our country. References develop a potentiality model of electric power replacement in rural areas according to their consumption characteristics, and make relevant case analysis; References develop an analytical model of energy environment emission reduction benefits of electric power replacement in Beijing according to the correlativity between energy consumption and environment emission reduction, and discuss the energy conservation, emission reduction and economic benefits of electric energy replacing fuel coal. References make analytical prediction on the future demand of main final energy in Beijing by building grey energy demand prediction model, and conduct a research on the potentialities of electric energy replacing coal and petroleum based on the predicted result. The consumption of the energies such as GDP, coal and petroleum etc is taken as the evaluation indicator of electric power replacement potentiality in every region in the thesis. The principal component factors are extracted with PCA, and the variance contribution rate of all factors is the weight. The comprehensive evaluation indicator function is obtained through linear combination of all factors, and analysis ranking is made to the potentiality of electric power replacement in various regions. A region is taken as a research object, and quantification and definition are made to the potentiality amount of electric power replacement. On the basis of the different regional promoting factors such as economic development, policy support and technical level of electric power replacement etc, diverse development scenarios of electric power replacement are constructed and prediction analysis is made to the potentiality of electric power replacement.

2. Analysis on the Potentiality of Electric Power Replacement in the National Various Regions

In general, the higher the regional energy consumption is, the larger the potential value of electric power replacement is. Moreover, the regional energy consumption is in direct proportion to its economic factors. In other words, the potential value of electric power replacement in a certain region increases with the high regional energy consumption and is in direct proportion to the regional economic factors. Therefore, the consumption of GDP, coal, petroleum and gas energies is selected as the potentiality evaluation indicator of regional electric power replacement, and integral analysis is made to the potentiality of all regions in the national various regions.

2.1. Build Potential Model

(1) Build data matrix

Assuming that there are n samples, and each sample has p variables, thereby constituting a $n \times p$ data matrix, namely:

$$X = \begin{pmatrix} x_{11}, & x_{12}, \cdots, x_{1p} \\ x_{21}, & x_{22}, \cdots, x_{2p} \\ \cdots & \cdots \quad \cdots \quad \cdots \\ x_{n1}, & x_{n2}, \cdots, x_{np} \end{pmatrix} \tag{1}$$

(2) Calculate correlation coefficient

The correlation coefficient r_{ij} of x_i, x_j in the original data is calculated. The specific calculation formula is as shown below:

$$r_{ij} = \frac{\sum_{k=1}^{n}(x_{ki}-\bar{x}_i)(x_{kj}-\bar{x}_j)}{\sqrt{\sum_{k=1}^{n}(x_{ki}-\bar{x}_i)^2 \sum_{k=1}^{n}(x_{kj}-\bar{x}_j)^2}} \tag{2}$$

The correlation coefficient matrix R is obtained according to the above formula:

$$R = \begin{pmatrix} r_{11}, & r_{12}, \cdots, r_{1p} \\ r_{21}, & r_{22}, \cdots, r_{2p} \\ \cdots & \cdots \quad \cdots \quad \cdots \\ r_{n1}, & r_{n2}, \cdots, r_{np} \end{pmatrix} \tag{3}$$

The solution to the above matrix can determine the eigen value λ_i of correlation coefficient matrix and corresponding eigenvectore$_i$.

(3) Determine the number of principal components

S_i is defined as the sum of variance contribution rate of the first $m(m \leq p)$ principal components.

$$S_i = \frac{\sum_{k=1}^{m} \lambda_k}{\sum_{k=1}^{p} \lambda_k} \tag{4}$$

The number of new variables in principal component is generally determined by calculating the contribution rate of principal component and accumulative contribution rate, and the number is determined by variance contribution rate $S_i \geq 85\%$.

(4) Calculate principal component load

$l_{ij} = \sqrt{\lambda_i} e_{ij}, z_1, \ z_2, \ldots, z_m (m \leq p)$ are new variable indicators. The variation form is as shown below:

$$\begin{cases} z_1 = l_{11}x_1 + l_{12}x_2 +, \ldots l_{1p}x_p \\ z_2 = l_{21}x_1 + l_{22}x_2 +, \ldots l_{2p}x_p \\ \ldots \\ z_m = l_{m1}x_1 + l_{m2}x_2 +, \ldots l_{mp}x_p \end{cases} \tag{5}$$

The above model is represented in matrix as follows:

$$Z = A * X \tag{6}$$

Where $Z = (z_1, z_2, \ldots, z_m)^T$

$$X = (x, x_2, \ldots, x_p)^T$$

$$A = \begin{pmatrix} l_{11}, & l_{12}, \ldots, l_{1p} \\ l_{21}, & l_{22}, \ldots, l_{2p} \\ \ldots & \ldots \quad \ldots \quad \ldots \\ l_{n1}, l_{n2}, & \ldots, l_{np} \end{pmatrix}$$

A becomes coefficient matrix of principal component.

(5) Solve rotation matrix

The coordinate system turns to an angle θ through orthogonal rotation, so that its major axis of an ellipse is in the direction of coordinatey_1, and the minor axis of an ellipse is in the direction of coordinatey_2. The rotation formula is as follows:

$$\begin{cases} y_{1j} = l_{1j}cos\theta + l_{1j}sin\theta \\ y_{2j} = -l_{2j}sin\theta + l_{2j}cos\theta \end{cases} \tag{7}$$

U is rotation matrix of coordinates.

$$U = \begin{pmatrix} cos\theta, sin\theta \\ -sin\theta, cos\theta \end{pmatrix} \tag{8}$$

(6) Calculate factor score

Function F with synthesis score is determined through the above formula:

$$F = \frac{\omega_1 f_1 +, \ldots, + \omega_m f_m}{\omega_1 +, \ldots, + \omega_m} \tag{9}$$

Where:

f_m—factor function;

ω_m—factor proportion weight

2.2. Analysis on the Potentiality of Electric Power Replacement in the National Various Regions

The PCA is used to process consumption data of energies in every region such as GDP, coal and petroleum etc, obtaining the total variance of all variables.

Table 1. Circumstance of Factors Explaining the Total Variance of Original Variables

Components	Initial Eigen Value			Load of Rotation Quadratic Sum		
	Total	% of Variance	% of Accumulation	Total	% of Variance	% of Accumulation
1	2.45	61.2	61.2	1.7	42.6	42.6
2	0.82	20.5	81.7	1.1	26.1	68.7
3	0.52	13.1	94.8	1.1	26.1	94.8
4	0.21	5.3	100.0			

It's seen from the above table that the characteristic root value of the 1st factor is 2.45, and the contribution rate of its variance is 61.2%; the characteristic root value of the 2nd factor is 0.82, and the contribution rate of its variance is 20.5%; the characteristic root value of the 3rd factor is 0.52, and the contribution rate of its variance is 13.1%. The accumulated variance contribution rate of the first three factors is 94.8%≥85%. Therefore, the number of the principal components is determined 3.

Table 2. Rotation Load Matrix of Factors

Potential Index Factors	1	2	3
GDP A1	0.851	0.282	0.277
Coal consumption A2	0.244	0.967	0.059
Petroleum consumption A3	0.922	0.163	0.191
Gas consumption A4	0.266	0.060	0.961

It's seen from the above table that GDP and petroleum consumption have higher load in the 1st factor. Two variables are mainly explained in the 1st factor, thereby obtaining factor score function:

f_1=0.553A1-0.267A2+0.728A3-0.288A4

f_2=-0.052A1+1.090A2-0.240A3-0.004A4

f_3=-0.085A1-0.003A2-0.242A3+1.113A4

Synthesis score and ranking are eventually made to the potentialities of electric power replacement in all regions.

The following conclusion can be drawn according to the analysis:

(1) The potentialities of electric power replacement in Shandong, Guangdong and Jiangsu are the largest, respectively ranked in the 1st, 2nd and 3rd places.

(2) The potentialities of electric power replacement in Hainan, Ningxia and Qinghai are the lowest and ranked in the last three places.

3. Analysis on Multi-Scenario Potentialities of Regional Electric Power Replacement

3.1. Quantum Chemical Calculation for the Potentialities of Electric Power Replacement

There is certain correlativity between the regional electric power replacement amount and power consumption. Therefore, the electric power replacement amount cannot serve as its evaluation indicators according to the increase in power consumption. Meanwhile, regional economic development, policy support of electric power replacement and the promotion role of the technical level of electric power replacement need to be considered in the potentialities of electric power replacement.

IOP Conf. Series: Earth and Environmental Science **64** (2017) 012105 doi:10.1088/1755-1315/64/1/012105

To achieve the quantum chemical calculation for the potentialities of electric power replacement, a certain year can be selected as a reference year JZ. Assuming that the consumption proportion of electric energy in final energy is the same as the reference year, and the influencing parameters of policy support, technical level and economic development etc are represented in A, B and C, respectively. Therefore, the calculation formula of the potentiality amount of electric power replacement in the tth year is as shown below:

$$F_t = \left(Y_{Dt} - \frac{Y_{DJZ}}{Y_{NJZ}} Y_{Nt}\right)[(1 + A)(1 + B)(1 + C)]^t \tag{10}$$

Where:

F_t—the substitution amount of potentialities of electric power replacement;

Y_{Dt}—the total power consumption amount of the tth year;

Y_{Nt}—the total energy consumption amount of the tth year;

3.2. Analysis on Multi-Scenario Potentialities of Electric Power Replacement

In the process of potentiality analysis, 3 different replacement scenarios are constructed as background support of potentiality analysis of electric power replacement, based on diverse driving factors such as economic development, policy support and technical level etc. See Table 4 for the setting of the specific scenarios.

Table 3. Analysis on Scenarios of Electric Power Replacement

Scenario Analysis	Specific Scenario
Scenario I	Relevant promotion measures are not taken for the development of regional electric power replacement.
Scenario II	The development of electric power replacement is impelled by issuing relevant policies of electric power replacement.
Scenario III	The foreign advanced electric power replacement technology is introduced to improve regional electrification level.

Electric power industry is a pillar one in our country. When electric power supply is insufficient, it will seriously restrict economic development. When economy develops rapidly, power demand will also increase. Therefore, there is a certain correlation between economic growth and the development of electric power replacement. The basic basis for scenario analysis on the potentialities of electric power replacement can be obtained by reference to relevant theoretical research in Reference:

(1) $(1 + C) > (1 + A)(1 + B)$

It indicates that the execution of electric power replacement need depend on economic development, and that the relevant policies and technical level of electric power replacement have a less important effect on promoting electric power replacement.

(2) $(1 + C) = (1 + A)(1 + B)$

It indicates that the speed of economic development is basically consistent with the process of electric power replacement. At the moment policy support is a major factor to promote electric power replacement.

(3) $(1 + C) < (1 + A)(1 + B)$

It indicates that the implementation and popularization of electric power replacement doesn't need to rely on economic development. At the moment the policy support and technical level are the key factors to affect electric power replacement.

IOP Conf. Series: Earth and Environmental Science **64** (2017) 012105 doi:10.1088/1755-1315/64/1/012105

3.3. Energy Consumption Prediction Model

Grey system theory is a theoretical method to research and solve grey system analysis, modeling and prediction, characterized by limited modeling information, convenient calculation and high accuracy etc. Therefore, it is widely applied in various prediction fields. During the analysis of regional potentialities, a grey prediction model is used for predicting the total regional energy consumption and total power consumption. The specific steps to build a model are as follows:

(1) Establish differential model equation

In consideration of $x_1, x_2, x_3, \ldots, x_n$ n variables, i.e.

$$x_i^{(0)} = \left\{x_i^{(0)}(1), x_i^{(0)}(2), \ldots x_i^{(0)}(\mathrm{p}) \right\} \tag{11}$$

Accumulation is made to $x_i^{(0)}$ generating accumulative sequence, i.e.

$$x_i^{(1)}(\mathrm{k}) = \sum_{m=1}^{k} x_i^{(0)}(m) \tag{12}$$

A differential model is built according to the treated sequence. Its matrix form is:

$$\frac{dX^1}{dt} = aX^1 + b \tag{13}$$

(2) Determine equation coefficient:

The undetermined coefficients a and b in the formula can be solved with least square method. The specific steps are as follows:

$$C = (a, \mathrm{b})$$

It is analyzed from the above formula that:

$$\mathrm{C} = (L^T\mathrm{L})^{-1}\mathrm{L}^T\mathrm{Y}_i$$

Where:

$$\mathrm{L} = \begin{pmatrix} \frac{1}{2}(x_1^{(1)}(2) + x_1^1(1), & \frac{1}{2}(x_2^{(1)}(2) + x_2^1(1)), & 1 \\ \frac{1}{2}(x_1^{(1)}(3) + x_1^1(2), & \frac{1}{2}(x_2^{(1)}(3) + x_2^1(2)), & 1 \\ \cdots & & \cdots \\ \frac{1}{2}(x_1^{(1)}(m) + x_1^1(m-1), & \frac{1}{2}(x_2^{(1)}(m) + x_2^1(m-1)), & 1 \end{pmatrix}$$

$$Y_i = (x_i^0(2), x_i^0(3), \ldots x_i^0(m))^T \tag{14}$$

(3) Solve response equation

The numerical values a and b are substituted into the above formula to obtain time response equation as follows:

$$X^{(1)}(\mathrm{k}+1) = \left[X^{(1)}(1) - \frac{b}{a}\right] e^{-ak} + \frac{b}{a} \tag{15}$$

3.4. *Case Analysis*

Hebei Province is taken as an example for potential analysis of electric power replacement, and Year 2014 is selected as a reference year. It's seen from data analysis that the power consumption in final energy in the reference year is about 2%.

(1)Energy consumption prediction

A grey prediction method is adopted to predict the power consumption and final energy consumption of Hebei Province, thereby obtaining the following data.

Table 4. Hebei Predicted Value of Final Energy Consumption during 2016-2020

Unit: standard coal with millions of tons

Year	2016	2017	2018	2019	2020
Predicted value of final energy consumption	3.06	3.11	3.16	3.21	3.38

Table 5. Hebei Predicted Value of Power Consumption during 2016-2020

Unit: standard coal with ten thousand tons

Year	2016	2017	2018	2019	2020
Predicted value of power consumption	884	1058	1266	1515	1872

(2)Setting of Analysis Parameters of Multi-Scenario Potentialities

The rapid economic development plays an active propelling role in electric power replacement. It is seen from the analysis on *Hebei 2015 Economic Yearbook* that the economic growth of Hebei Province was 6.5% in 2014, and it kept 6.8% during 2015-2016, so the economic operation is generally steady. It is predicted that the economic growth rate of Hebei Province will be about 8% during 2016-2020; in the aspect of policy support, in Scenario I, assuming that no promoting measures are taken for the development of electric power replacement in Hebei, so its implementation depends on economic development. In Scenarios II and III, on the basis of the relevant policies of electric power replacement issued by Hebei; in technical level and Scenarios I and II, the impact of technical level of electric power replacement is not considered. In Scenario III, based on the promotion and electrification level of electric power replacement in Hebei, the influence of electric power replacement on implementing electric power replacement is increased.

According to the above theoretical basis, setting is made to the parameters of economic development, relevant policy support and technical level etc in Scenarios I, II and III. The parameters of the three aspects are represented with A, B and C, respectively. See Table 7 for more information.

Table 6. Basic Parameters of Multi-Scenario Analysis of Electric Power Replacement

Setting of Scenarios	A/ (%)	B/ (%)	C/ (%)
Scenario I	0	0	8
Scenario II	3.5	0	8
Scenario III	3.5	4.3	8

(3)Potentiality calculation and analysis

The above predicted values of energy consumption, along with the parameters of relevant policy support, technical level and economic development etc, are substitutes into the formula, thereby obtaining the following result:

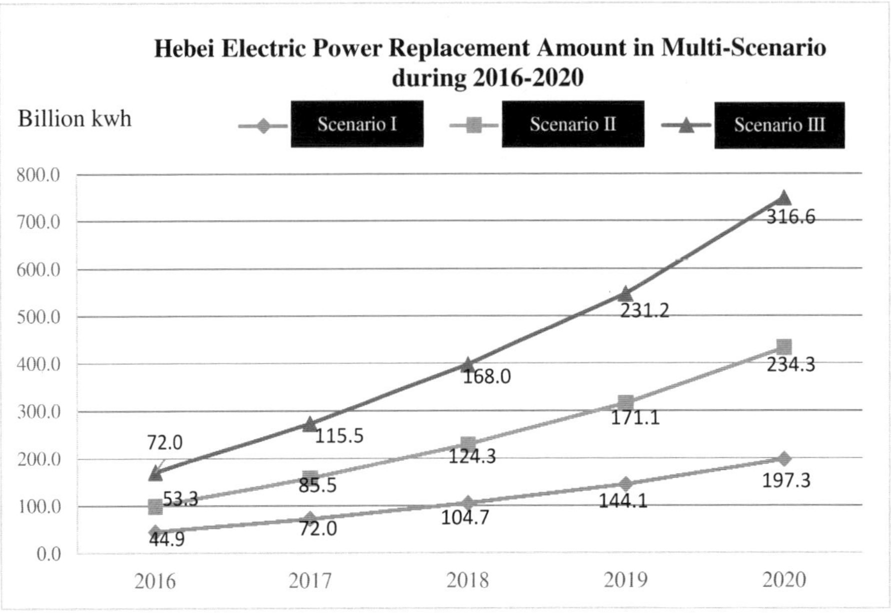

Fig. 1 Hebei Electric Power Replacement Amount in Multi-Scenario during 2016-2020

The following conclusion can be drawn according to the analysis on the potentiality amount of electric power replacement in different scenarios:

(1) In Scenario I, the electric power replacement amounts of Hebei are 4.49 billion kWh, 720 million kWh, 10.47 billion kWh, 14.41 billion kWh and 19.73 billion kWh, respectively. The growth rate of the electric power replacement amount is slower;

(2) In Scenario II, the electric power replacement amounts are 5.33 billion kWh, 8.55 billion kWh, 12.43 billion kWh, 17.11 billion kWh and 23.43 billion kWh, respectively. Compared with Scenario I, its growth rate is about 18.7%, indicating that electric power replacement promotes its continuous development;

(3) In Scenario III, the electric power replacement amounts are 720 million kWh, 11.55 billion kWh, 16.8 billion kWh and 31.66 billion kWh, respectively. Compared with Scenario II, its growth rate is about 35.2%, indicating that the electric power replacement amount is largely raised driven by policy support and technical level of electric power replacement.

4. Conclusion and Prospect

Compared with the energies such as coal, petroleum and gas etc, electric energy is cleaner, more convenient and safer. The governments in various regions shall, according to their own geographical locations, environmental conditions, economy and energy consumption, formulate policies to support electric power replacement and guide the social members to select on their own the energy using mode of high pollution and low efficiency of electric power replacement. Meanwhile, the governments shall also constantly improve the relevant technology of electric power replacement, widely apply foreign relevant electrification technology and improve regional electrification level.

Acknowledgments
This research was supported by the Science and Technology Project of State Grid Corporation Headquarters (Grant No. YD71-16-014).

References

[1] Zheng Jincheng. Analysis on the Potentiality and Environmental Benefits of Electric Power Replacement in Rural Areas [D]. Beijing: Master's thesis of North China Electric Power

IOP Conf. Series: Earth and Environmental Science **64** (2017) 012105 doi:10.1088/1755-1315/64/1/012105

University. 2015.3.
[2] Zhao Yinhui. Analysis on Energy Replacement and Study on Optimized Model under the Objective of Environmental Emission Reduction in Beijing [D]. Beijing: Doctorate's thesis of North China Electric Power University. 2014.6.
[3] Yin Hang. Research on Evaluation Methods of Electric Energy Replacing Other Energies under the Environment of Energy Conservation and Emission Reduction [D]. Beijing: Master's thesis of North China Electric Power University. 2013.3
[4] He Youshi, Xu Wenqin. Application of Factor Analysis in Comprehensive Economic Benefit Evaluation of Industrial Enterprises [J]. Mathematical Statistics and Management. 2003, 22(1):19-22.
[5] Sun Yi, Zhou Shuang, Shan Baoguo et. al. Analysis on the Potentiality of Electric Power Replacement in Multi-scenarios [J]. Power Grid Technology, 2017(1):118-124.
[6] Lou Wei, Li Meng. Application of Scenario Analysis in energy Planning Research [J]. China Power, 2012(10): 17-22.

Effects of different mechanized soil fertilization methods on corn nutrient accumulation and yield

Qingwen Shi[1,a], Chunming Bai[2,a], Huixin Wang[2,a], Di Wu[1,a], Qiaobo Song[1,a], Zengqi Dong[1,a], Depeng Gao[1,a], Qiping Dong[1,a], Xin Cheng[1,a], Yahao Zhang[1], Jiahui Mu[1], Qinghong Chen[1], Wenqing Liao[1], Tianru Qu[1], Chunling Zhang[1], Xinyu Zhang[1], Yifei Liu[1,*] and Xiaori Han[1,*]

[1]Shenyang Agricultural University, Shenyang, China
[2]Liaoning Academy of Agricultural Sciences, Shenyang, China

[a]Joint first authors
*Co-corresponding author e-mail: liuyifeiscience@163.com

Abstract. Aim: Experiments for mechanized corn soil fertilization were conducted in Faku demonstration zone. On this basis, we studied effects on corn nutrient accumulation and yield traits at brown soil regions due to different mechanized soil fertilization measures. We also evaluated and optimized the regulation effects of mechanized soil fertilization for the purpose of crop yield increase and production efficiency improvement. Method: Based on the survey of soil background value in the demonstration zone, we collected plant samples during different corn growth periods to determine and make statistical analysis. *Conclusions:* Decomposed cow dung, when under mechanical broadcasting, was able to remarkably increase nitrogen and potassium accumulation content of corns at their ripe stage. Crushed stalk returning combined with deep tillage would remarkably increase phosphorus accumulation content of corn plants. When compared with top application, crushed stalk returning combined with deep tillage would remarkably increase corn thousand kernel weight (TKW). Mechanized broadcasting of granular organic fertilizer and crushed stalk returning combined with deep tillage, when compared with surface application, were able to boost corn yield in the in the demonstration zone.

1. Introduction

Nowadays, poor mechanization applicability and unreasonable use of organic fertilizer prevail in agricultural production. Extensive cultivation and abuse of chemical fertilizer lead to soil nutrient imbalance and soil fertility decline, so crop nutrient accumulation and yield are impacted[1,2]. Therefore, it is very pressing to study and implement effective measure to improve soil fertility. Alfred Gathorne-Hardy, based on his study from a prospective of sustainable development, found out that mechanization was featured with higher efficiency than cattle ploughing as labor-saving, and both were not so significantly different in terms of greenhouse gas emission [3, 4]. Mechanized deep placement could facilitate rice tillering, favorable for rice root system elongation to absorb nutrients, so rice yield increased consequently [5]. Mechanized deep placement as adopted in Japan proved that it was effective for yield increase. It was reported that as for deep placement 35 days prior to rice

IOP Conf. Series: Earth and Environmental Science **64** (2017) 012106 doi:10.1088/1755-1315/64/1/012106

heading at the agricultural experimental station of Aomori, Japan, nitrogen use efficiency even amounted to 86%. Koyama and others (1973) found that the applicability of nitrogen for whole layer placement was twice as much as that for surface application. Vachhani (1952) stated out that as for granular fertilizer, deep placement around paddy plants was more effective than surface application. Prasad and others (1970) also confirmed that deep placement of nitrogenous fertilizer was better than surface application both in terms of paddy yield and nitrogen uptake [6]. As some studies indicated, when organic granular fertilizer (Aike brand) was combined with urea and potassium sulfate as a practice of mixed application, it was able to increase paddy yield; while there was not so significant difference in paddy yield between organic granular fertilizer of exclusive application (Aike brand) and conventional fertilization [7]. Liu Shujie during his study confirmed that mechanical deep placement of chemical fertilizer was able to increase paddy yield remarkably, and this practice would save fertilizer as well [8]. Research showed that mechanize deep placement of chemical fertilizer was able to increase paddy yield and save fertilizer, too [9, 10].

2. Material and method

2.1. Experimental material

2.1.1. Corn variety for experiment: Shenghe 18

2.1.2. Soil type: It is brown soil, located in Lujiafangshen Village, Dagujiazi Town, Faku County. The physical and chemical properties of the soil for test are listed in the following tables:

Table1 The physical-chemical properties of soil before seeding in 2014

Treat.	Organic matter	Total N	Total P	Total K	Alkali-N	Olsen-P	NH_4OAc-K	Bulk density	PH
	(C $g{\cdot}kg^{-1}$)	(N $g{\cdot}kg^{-1}$)	(P $g{\cdot}kg^{-1}$)	(K $g{\cdot}kg^{-1}$)	(N $mg{\cdot}kg^{-1}$)	(P_2O_5 $mg{\cdot}kg^{-1}$)	(K_2O $mg{\cdot}kg^{-1}$)	(g/cm^3)	
Content	11.89	2.27	0.12	15.93	100.78	19.28	86.83	1.39	5.06

2.1.3. Fertilizer for test: (1)Organic fertilizer: decomposed cow dung at the local cattle farm (nutrient content: organic carbon 33.81%, N content 2.41%, P2O5 content 1.919%, K2O content 1.445%, moisture 58.29%).

(2)Corn stalk: local corn stalk (nutrient content: organic carbon 43.54%, N content 0.863%, P content 0.260%, K content 0.834%, moisture 9.07%).

(3)Granular fertilizer as biochar-based: granular fertilizer as biochar-based from the Plant Nutrition and Fertilizer Research Institute of Shenyang Agricultural University (moisture 25-30%, organic matter≥45 % , NPK≥5%, nitrogen 2.06%, phosphorus 1.82%, kalium 1.78%, secondary and microelements≥26%).

(4)Conventional chemical fertilizer: slow release fertilizer with the brand of Qingyuan, which was used as the base fertilizer for deep placement, 50kg per mu (≈667m2, Chinese traditional unit area and the same below) (NPK content of 27-11-13, zinc 0.02% for the product made by Changjiang Controlled Release Fertilizer Company Ltd., Jilin). Seed fertilizer for surface application: Compound fertilizer of 15-15-15 with the brand of Yanyangtian.

(5)Microbial agent: stalk decomposing agent applied for stalk returning to field.

2.1.4. Machinery for test: (1) HP 55/60 tractors and plowing, harrowing, ridging as well as sowing equipment.

(2)Stalk returning equipment: mechanical stalk cutting (5-10cm long), returning to field by manual work.

IOP Conf. Series: Earth and Environmental Science **64** (2017) 012106 doi:10.1088/1755-1315/64/1/012106

(3)Broadcaster for granular chemical fertilizer: single disc broadcaster as developed by Nanjing Agricultural Mechanization Research institute of the National Agriculture Ministry.

(4)The spraying machines and equipment for field fertilizer, pesticide and microbial agent shall be operated conventionally as farmers usually do.

2.2. Test Method

On the premise of the same plot with the same soil fertility under the same tillage conditions, five experimental zones were established for our experiments: Zone I (with mechanical broadcasting of decomposed cow dung, 500kg each mu as air-dried basis, moisture 20%, rotary tillage for 20cm on ridges); Zone II (mechanical cutting and crushing for stalk returning and surface application, 400kg per mu, with stalk decomposing agent and chemical nitrogen fertilizer to be used together and sprayed at the same time according to the single factor, then rotary tillage on ridges for 20cm); Zone III (mechanical cutting and crushing for stalk returning first and deep tillage for 30cm further, 400kg per mu, with stalk decomposing agent and chemical nitrogen fertilizer to be used together and sprayed at the same time according to the single factor); Zone IV (mechanical broadcasting of granular fertilizer as biochar-based) and Zone V (conventional management of tillage and fertilizer application, for which 18kg nitrogen, 5kg phosphorus pentoxide, 6kg potassium oxide, urea and potassium chloride 20kg+10kg as the basic fertilizer, 10kg ammonium phosphate as the seed fertilizer, extra fertilization for 225kg/hm at bell stage, or special fertilizer for corns would be selected, and due to stalk returning practice, potash fertilizer amount could be properly reduced).

These experiments started from the year of 2013, and then soil samples were collected in autumn to determine their volume weight, organic matter content and nutrients. Then fertilizer and stalk composition were under analysis. Tasks after crop harvesting or before sowing included stalk returning, broadcasting of organic fertilizer, chemical fertilizer and microbial agent, and soil tillage as well. Soil samples and plant samples were collected according to growth periods (before fertilization and tillage, at elongation stage, at bell stage, at filling stage and at ripe stage).

3. Results and analysis

3.1. Mechanization effects on nutrient accumulation for corns during their different growth periods

Table2 Effects of different treatments on N accumulation in different growth stages (kg/mu)

Treat.	Total N(kg/mu)			
	Jointing stage	large bell stage	Filling stage	Mature stage
I	0.429±0.019a	8.509±0.378c	7.939±0.886b	13.574±0.347a
II	0.360±0.011c	9.265±0.394bc	7.327±0.733b	10.983±0.092bc
III	0.296±0.007d	10.375±0.907ab	6.625±0.403b	12.246±0.800b
IV	0.388±0.033bc	9.867±0.605ab	9.678±0.694a	10.908±1.206c
V	0.409±0.015ab	10.722±0.496a	10.552±0.341a	11.515±0.160bc

Note: Different letters in the same column mean significant difference at 5% level.

Table3 Effects of different treatments on P accumulation in different growth stages (kg/mu)

Treat.	Total P(kg/mu)			
	Jointing stage	large bell stage	Filling stage	Mature stage
Ⅰ	0.075±0.010a	2.516±0.129ab	3.070±0.815a	3.521±0.563ab
Ⅱ	0.068±0.006a	1.981±0.182c	2.313±0.362b	3.001±0.039bc
Ⅲ	0.040±0.003b	2.420±0.321b	1.874±0.126b	4.212±0.423a
Ⅳ	0.064±0.005a	2.785±0.103a	1.924±0.187b	2.650±0.252c
Ⅴ	0.065±0.007a	2.553±0.176ab	2.295±0.313b	3.179±0.337bc

Note Different letters in the same column mean significant difference at 5% level.

Table4 Effects of different treatments on K accumulation in different growth stages (kg/mu)

Treat.	Total K(kg/mu)			
	Jointing stage	large bell stage	Filling stage	Mature stage
Ⅰ	0.750±0.099a	19.297±0.274a	2.548±0.676a	15.277±0.445a
Ⅱ	0.602±0.008b	14.591±0.152c	1.919±0.301b	9.479±0.437c
Ⅲ	0.453±0.016c	14.017±1.617c	1.555±0.105b	12.564±0.215b
Ⅳ	0.775±0.062a	16.707±0.712b	1.597±0.156b	8.537±0.246d
Ⅴ	0.627±0.026b	9.759±0.272d	1.904±0.259b	8.296±0.301d

Note Different letters in the same column mean significant difference at 5% level.

As indicated from table 2, 3 and 4, we could perceive that nitrogen accumulation and potassium accumulation per unit area in treatment I was 17.88% and 84.15%, respectively higher than N and K accumulation per unit area in treatment V. Treatment III, when compared with treatment II, remarkably increased phosphorus and potassium accumulation per unit area at corn ripe stage. Both treatment I and treatment III remarkably increased nitrogen, phosphorus and potassium accumulation per unit area inside plants at corn ripe stage, when compared with conventional cultivation and fertilization.

3.2. Mechanization effects on yield traits for corns at their ripe stage

Table5 Effects of different treatments on yield characters of maize in mature stage

Treat	Bald tip length/cm	Ear length/cm	Ear diameter/cm	Corn cob diameter/cm	Rows per ear	Kernel numbers per ear	1000-kernels weight/g	Yield/kg/mu
Ⅰ	0.716±0.095a	18.49±0.32ab	5.387±0.065a	2.911±0.199a	18.0±0.40a	671.7±6.0b	390.33±2.75c	843.80±13.23b
Ⅱ	0.443±0.124b	18.86±0.47ab	5.381±0.044a	2.797±0.028a	18.1±0.46a	653.3±12.5c	397.33±1.53b	903.07±9.98a
Ⅲ	0.307±0.073b	18.71±0.38ab	5.422±0.073a	2.912±0.037a	18.1±0.46a	685.3±3.2b	413.00±1.80a	807.72±2.95c
Ⅳ	0.257±0.055b	18.12±0.87b	5.415±0.154a	2.898±0.112a	18.7±0.83a	671.3±7.1b	399.17±3.21b	836.93±3.87b
Ⅴ	0.270±0.068b	19.30±0.06a	5.405±0.045a	2.875±0.045a	18.0±0.40a	704.0±5.0a	385.77±0.75d	845.73±6.85b

Note: Different letters in the same column mean significant difference at 5% level.

Indication from Table 5: bare top length in treatment I was remarkably different from that of the other four treatments. Ear length of treatment I, treatment II, treatment III and treatment V were roughly identical. Ear diameter, cob diameter and ear rows for each treatment were basically identical. Grain number per panicle of treatment V was remarkably larger than that of the other four treatments. Thousand kernel weight (TKW) of treatment I, treatment II, treatment III and treatment IV were remarkably higher than that of treatment V. This means that these four methods of mechanized soil fertilization were all able to remarkably increase corn TKW, compared with conventional cultivation and fertilization. Among them, treatment II enjoyed the highest yield, remarkably different from that of the other treatments. This is also to indicate that crushed stalk returning with surface application, when compared with deep tillage or conventional cultivation and fertilization, could better increase corn yield in the demonstration zone.

4. Conclusion

1. Decomposed cow dung, when under mechanical broadcasting, was able to increase nitrogen and potassium accumulation per unit area in plants for corns at their ripe stage. Crushed stalk returning combined with deep tillage would remarkably increase phosphorus and potassium accumulation content in plants for corns at their ripe stage. For decomposed cow dung under mechanical broadcasting and stalk returning combined with deep tillage, both were able to remarkably increase nitrogen, potassium and phosphorus accumulation per unit area in plants for corns at their ripe stage.

2. Compared with conventional cultivation and fertilization, decomposed cow dung and organic granular fertilizer under mechanical broadcasting, as well as crushed stalk returning with deep tillage and surface application, were able to remarkably increase corn TKW. Compared with deep tillage and conventional cultivation and fertilization, crushed stalk returning with surface application could remarkably increase corn yield in the demonstration zone.

Acknowledgments

This work was financially supported by National science and technology support project (2013BAD08B04), Natural Science Foundation of China (31301842; 31601627), Cultivation Plan for Youth Agricultural Science and Technology Innovative Talents of Liaoning Province (2014043), and Program of Liaoning Excellent Talents in University (LJQ2015100), China Scholarship Council (CSC NO. 201708210143) and National Peanut Research System (CARS-14).

References

[1] Matthews R B, Pilbeam C. Modelling the long-term productivity and soil fertility of maize/millet cropping systems in the mid-hills of Nepal[J]. Agriculture, Ecosystems & Environment, 111(2005)119-139.
[2] Cheng Hu, Zhiping Cao, Yanrui Luo,et al. Effects of long term application of bio organic fertilizer on soil fertility and microbial biomass carbon. Chinese Journal of Eco-Agriculture, 03(2007)48-51.
[3] Alfred Gathorne-Hardy. The sustainability of changes in agricultural technology: The carbon, economic and labour implications of mechanisation and synthetic fertiliser use. Ambio, 45(2016)885-894.
[4] Shuxia Liu, Haiyan Wu, Lanpo Zhao, et al. Effect of potassium application on potassium uptake and utilization of Maize. Maize Science, 04(2008)172-175.
[5] Manhong Zhang.The analysis and comparison of testing results of the deep application of chemical fertilizer applicator. Fujian Agricultural Machinery. 02(2004)25-26.
[6] Dianping Li. Study on the effect of deep fertilization on yield and quality of rice. Northeast Agricultural University, 2004.
[7] Jianhua Zhang, Xiangqian Fang, Yujie Liu, et al. Effect of Aike particulate organic fertilizer on the biological traits and yield of rice. Anhui Agricultural Science Bulletin, 03(2009)108-109.
[8] Shujie Liu. The use of new technologies of wheat planter fertilizer applicator. Hebei Agriculture, 05(2014)32-33.
[9] Renchao Liu. Comparative experimental study on the application of deep mechanized rice fertilizer technology. Agricultural extension, 06(2001)18-19.
[10] Deshui Tan, Jiyun Jin, Shaowen Huang, et al. Effects of long-term application of K fertilizer and straw returning on crop yield and soil potassium under different cropping systems. Chinese Agricultural Science, 01(2007)133-139.

Responses of heterotrophic bacteria abundance and activity to Asian dust enrichment in the low nutrients and low chlorophyll (LNLC) region of the Northwestern Pacific Ocean

Dongwan Shi[1], Kuiran Li[1], Yanzhao Tian[2], Xiaohao Zhang[2], Jie Bai[2,*]

[1]College of Marine Life Science, Ocean University of China, Qingdao266000, China
[2]Key Laboratory of Marine Environment and Ecology, Ministry of Education of China, Ocean University of China, Qingdao266100, China

*Corresponding author e-mail: baijie@ ouc.edu.cn

Abstract. Bacteria, as an essential part of microbial food web, play a significant role in the marine ecosystem. Dust deposits into the surface ocean carrying with vital nutrient such as Inorganic nitrogen and phosphorus etc., which has an important influence on the life activities of heterotrophic bacteria. The microcosm experiments with Asian dust deposition was carried out on board in the station K3 (26.18°N, 136.73°E) in April 2015, aiming to estimate the impact of dust deposition on the oligotrophic Northwestern pacific Sea, the main goal of the present paper was to assess how dust deposition events affect the abundance and activity of heterotrophic bacteria in low nutrient and low chlorophyll (LNLC) sea area. Station K3 located in the central northwestern Pacific Ocean, which has the characteristic of low nutrient and low chlorophyll. The study shows that there was an N–P co-limitation in station K3, and the deposition of Asian dust can increase the abundance, and promote the activity of heterotrophic bacteria in the station K3.

1. Introduction

Heterotrophic bacteria play a central role in the microbial loop, and increasing evidence indicates that not only phytoplankton but also heterotrophic bacteria are limited by inorganic nutrients, mainly P and N, in oligotrophic oceanic systems [1, 2]. There are many factors regulating bacterial distribution and growth, such as grazing by micro zooplankton, viral infection, and the supply of these resources like P, N, and effects of temperature.

Dust deposition is recognized as a significant source of nutrients to the surface ocean [3], with a great impact on autotrophic production by changing the phytoplankton community structure, inducing algal blooms and enhancing primary production [4]. And with recent study, dust deposition also impact heterotrophic bacterial dynamics in the surface ocean. Considering the central role of bacteria in the microbial loop, it is likely that dust deposition also affects the structure and the functioning of the whole microbial food web [5].

The environmental condition of the Northwestern Pacific Ocean is changeable and complex, which significantly influences the abundance and activity of bacteria. And the central region of Northwestern Pacific Ocean is a typical LNLC region, particularly well adapted to study the question of the impact of Asian dust deposition in microbial food web[5, 6], especially the bacterial dynamic. To better understand the response of bacteria in to the change of environmental conditions, it is very important to explore the responses of heterotrophic bacteria abundance and activity in this area.

In this study, onboard experiments were performed in LNLC region in Northwestern Pacific in April 2015 to explore the responses of heterotrophic bacteria abundance and dynamics of added Asian dust.

2. Materials and methods

2.1. Microcosm set-up.

The microcosm experiments were carried out in the station K3 (26.18°N, 136.73°E) of the Northwestern Pacific onboard the R/V *Dongfanghong II* during April 2015(Table 1). Four 2L aseptic incubation bottles were used in this microcosm, filled with the 0.8-μm pre-screened surface seawater to exclude zooplankton and phytoplankton, with Controls (no addition), Dust, P, N+P (Table 2). The incubation was conducted for 7 days at station K3.

Table 1. Initial environmental factor of the station K3

station	Temperature (℃)	Salinity	Conductivity	DOC (mg/L)	NO^{3+} (mg/L)	NO^{2+} (mg/L)	PO_3^{4+} (mg/L)	Bacteria abundance(cell/mL)
K3	19.75	34.859	4.75	1.583	--	--	0.007	4.08×10^5 cell/mL

2.2. Heterotrophic Bacteria Abundance.

Seawater (10mL) was fixed with buffered paraformaldehyde (0.5% final concentration) and stored at -80℃ until flow cytometric analysis. Heterotrophic bacteria abundance was measured by a Becton-Dickinson FACSCalibur cytometer.

2.3. Heterotrophic bacteria activity.

Flow cytometry enabled the discrimination of two Bacterial cell groups: bacterial cells with high nucleic acid (HNA) and low nucleic acid (LNA) contents. The proportion of high nucleic acid bacteria (HNA %) can indicates the activity of Heterotrophic bacteria.

Table 2 Ambient nutrient concentration

Incubation Bottle No.	Treatment	Added Dust and Nutrients Concentration
C	Control	------
D	Dust	2 mg/L
P	NaH_2PO_4	0.2 uM/L
N+P	$NaNO_3$ +NaH_2PO_4	**2** uM/L+0.2 Um/l

2.4. Statistical analysis.

All samples had three replicates at least, with results presented as mean SD (standard deviation). Statistical analysis was performed using one-way analysis of variance (ANOVA) with the Fisher LSD post hoctest. Differences were considered statistically significant when $p<0.05$.

3. Result and Discussion

As a typical station in oligotrophic sea area, the seawater at K3 stations was oligotrophic as indicated by low macronutrient and Chl a concentrations. From table 1, the temperature of station K3 was relatively high, about 19.75℃.And the salinity of the seawater was 34.86, while the concentration of DOC was 1.58mg/L; as well nitrate concentration was below the limit of detection of the used technique. In addition, phosphate concentration averaged 0.007mg/L, which showed that the stationK3 was in a typical oligotrophic region. The shortage of dissolved inorganic phosphate (DIP) and dissolved inorganic nitrogen (DIN) suggested that the growth of heterotrophic bacteria in this area might be limited by phosphorus and nitrogen.

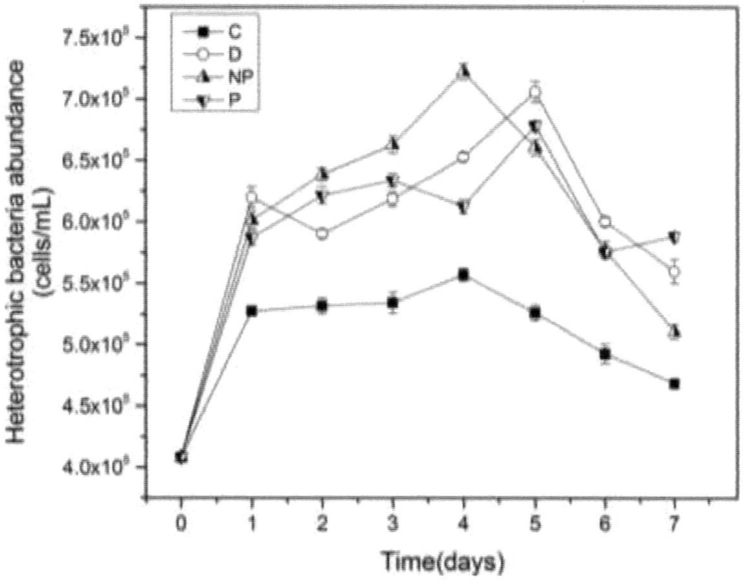

Figure 1. Variation of the heterotrophic bacteria abundance among different treatments

Heterotrophic bacteria abundance increased obviously during the incubation period in Dust deposition enrichment culture experiment at Station K3. On the fifth day of incubation, the abundance of heterotrophic bacteria increased to 7.06×10^5 cell/mL, the maximum over the incubation, which was 1.34 times higher than the abundance of the Controls. Then the bacteria abundance of Dust group began to decline, and decreased to 5.60×10^5 cell/mL, which was 1.20 times higher than abundance of the control group.

There was significant difference in bacteria abundance between the Dust group and the Control ($P<0.05$). It demonstrated that the dust deposition can impact heterotrophic bacteria abundance. We can see from the graph that the dust deposition played an important role in promoting growth of heterotrophic bacteria in the early stage of the incubation. The growth of bacteria abundance in the Dust deposition-microcosm was similar to it in the P-microcosm and NP-microcosm, suggesting that both the dissolved inorganic nitrogen and phosphorus from the dust can promote the increase of heterotrophic bacteria abundance. But compared to the NP-microcosm, the maximum of bacteria abundance in Dust deposition-microcosm appeared one day later, simultaneously with the maximum of P-microcosm. It may prove that the dissolved inorganic phosphorus from the dust facilitate the growth of heterotrophic bacteria abundance more, and the growth of heterotrophic bacteria abundance in this area is limited by the shortage of dissolved inorganic nitrogen and phosphorus.

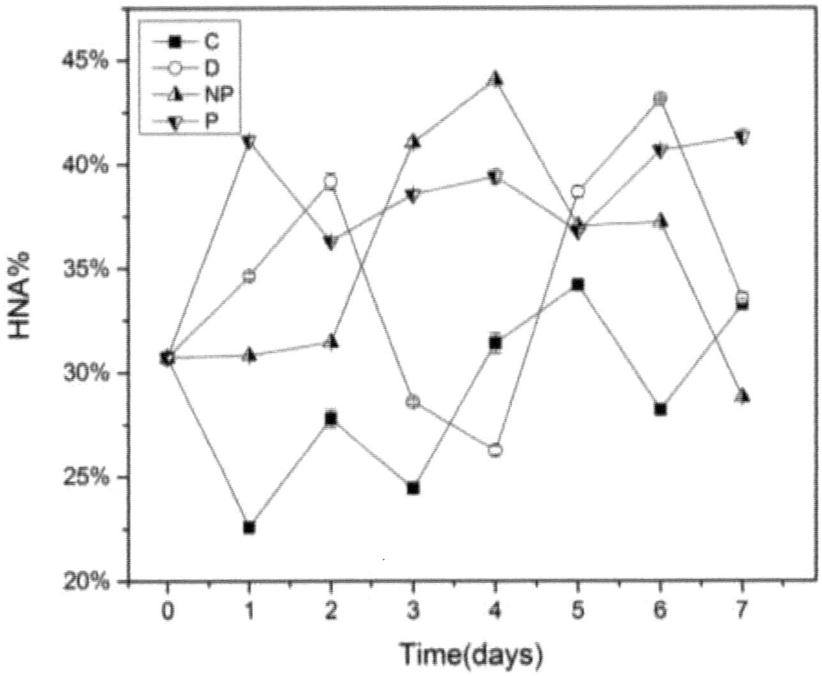

Figure 2. Variation of the HNA% of heterotrophic bacteria among different treatments

The HNA% of heterotrophic bacteria varied during incubation time in dust addition experiment at Station K3 (Fig.2). Compared to the controls, the HNA% of bacteria showed slightly increase in Dust deposition-microcosm at the first 2 days, and then started to decrease. At the 4th day, the HNA% of heterotrophic bacteria in Dust deposition-microcosm decreased to 52.53 %(0.84-fold), this is the minimum value during the incubation. And at the 6th day, the HNA% increased to the maximum 86.24 %(1.53-fold). When the area is in the condition of N–P co-limitation, the Dust has a visible impact on the HNA% of heterotrophic bacteria.

4. Summary

Our study shows that dust deposition to the Northwestern pacific surface oligotrophic region can promote the growth of abundance and activity of heterotrophic bacteria. And for the bacteria in Station K3, there growth was N–P co-limited. During the incubation, the addition of dust promoted the abundance and activity of bacteria by increase the concentration of inorganic nutrients, especially for P.

Acknowledgments
This work was supported by the National Nature Science Foundation of China (No. 41210008) and the Major Science and Technology Program for Water Pollution Control and Treatment (No. 2013ZX07202-007).

References

[1] Pulido-Villena, E., et al., Microbial food web dynamics in response to a Saharan dust event: results from a mesocosm study in the oligotrophic Mediterranean Sea. Biogeosciences Discussions, 2014. 11(1): p. 337-371.
[2] Welschmeyer, D.K.R.K., Biomass and production of heterotrophic bacterioplankton in the oceanic subarctic Pacific. 1993.
[3] Wang, F., et al., Characterization of carbonaceous aerosols over the East China Sea: The impact of the East Asian continental outflow. Atmospheric Environment, 2015. 110: p. 163-173.

[4] Guo, C., et al., Effect of East Asian aerosol enrichment on microbial community composition in the South China Sea. Journal of Plankton Research, 2013. 35(3): p. 485-503.

[5] Pulido-Villena, E., V. Rérolle and C. Guieu, Transient fertilizing effect of dust in P-deficient LNLC surface ocean. 2010. 37(1).

[6] Schwier, A.N., et al., Primary marine aerosol emissions from the Mediterranean Sea during pre-bloom and oligotrophic conditions: correlations to seawater chlorophyll a from a mesocosm study. Atmospheric Chemistry and Physics Discussions, 2014. 14(19): p. 26187-26230.

Study on the syhthesis process of tetracaine hydrochloride

Wenli Li*, Jie Zhao, Yujie Cui

Chong Qing Energy College, Jinan Depton Chemical Technology Co., Ltd, China

*Corresponding author e-mail: 332458254@qq.com

Abstract. Tetrachloride hydrochloride is a local anesthetic with long-acting ester, and it is usually present in the form of a hydrochloride salt. Firsleb first synthesized the tetracaine by experiment in 1928, which is one of the recognized clinical potent anesthetics. This medicine has the advantages of stable physical and chemical properties, the rapid role and long maintenance. Tetracaine is also used for ophthalmic surface anesthesia as one of the main local anesthetic just like conduction block anesthesia, mucosal surface anesthesia and epidural anesthesia. So far, the research mainly engaged in its clinical application research, and the research strength is relatively small in the field of synthetic technology. The general cost of the existing production process is high, and the yield is low. In addition, the reaction time is long and the reaction conditions are harsh. In this paper, a new synthetic method was proposed for the synthesis of tetracaine hydrochloride. The reaction route has the advantages of few steps, high yield, short reaction time and mild reaction conditions. The cheap p-nitrobenzoic acid was selected as raw material. By esterification with ethanol and reaction with n-butyraldehyde (the reaction process includes nitro reduction, aldol condensation and hydrogenation reduction), the intermediate was transesterified with dimethylaminoethanol under basic conditions. Finally, the PH value was adjusted in the ethanol solvent. After experiencing 4 steps reaction, the crude tetracaine hydrochloride was obtained.

1. Introduction

Tetracaine hydrochloride is also known as ropivacaine hydrochloride, pantocaine, pantocaine and four ropivacaine hydrochloride. The molecular formula is $C_{15}H_{25}ClN_2O_2$, and the molecular weight is 300.82. The melting point is 147-150 ℃. Its English name is tetracaine hydrochloride CAS NO: 136-47-0. It is easily soluble in water and ethanol, but insoluble in ether, benzene. Tetracaine hydrochloride is a local anesthetic with long-acting ester. In addition, procaine is widely used in clinical, but because of its poor permeability and poor narcotic performance, a small tetracaine was found in 1930 found. Its fat solubility is 100 times than that of procaine, and its effectiveness and maintenance time also correspondingly increased. Compared to procaine, it has high performance on fat-soluble and anesthetic performance, and the anesthesia time is longer. Commonly, the tetracaine with 2% concentration is used for surface anesthesia, and 4%-5% ointment is formulated with an anhydrous machine to apply for skin surface anesthesia. In asdition, it is also used for the tracheal catheter surface to reduce intraoperative airway irritation. Tetracaine hydrochloride can be used for topical anesthesia, and 0.05%-0.1% tetracaine hydrochloride can provide long-term and good effect of surface anesthesia. Meanwhile, tetracaine hydrochloride is still widely used in ophthalmic surface anesthesia, which has a long time for surface anesthesia and sensory resistance.

2. Experiment part

2.1. Experimental reagents and instruments

The required reagents for the synthesis process of tetracaine hydrochloride are shown in Table 1 and the required equipment for the synthesis process of tetracaine hydrochloride is shown in Table 2:

Table 1. Reagents of experiment

Reagent name	Purity	Manufacturer
Vitriol (98%)	AR	Yantai Sanhe Chemical Reagent Co., Ltd
Muriatic acid (36%)	AR	Yantai Sanhe Chemical Reagent Co., Ltd
zinc powder	Industrial grade	Tianjin Huirui Chemical Technology Co., Ltd
glacial acetic acid	AR	Tianjin Bodi Chemical Co., Ltd
N-butyl aldehyde	AR	Tianjin Guangcheng Chemical Reagent Co., Ltd
p-nitrobenzoic acid	AR	Shanghai Puzhen Biological Technology Co., Ltd

Table 2. Experiment equipment

Equipment	Type	Manufacturer
Rotary evaporator	RE-52AA	Shanghai Yarong Biochemical Equipment Factory
Constant temperature heating magnetic stirring bar	GS-2	Gongyi Instrument Co., Ltd.
Thermostatic water bath	ZXYY-0.5L	Zhengzhou Yingyu Instrument Co., Ltd.
Magnetic stirring bar	84-3	Gongyi Yingyu Instrument Factory
Circulating water vacuum pump	SHZ-D	Gongyi Yingyu Instrument Co., Ltd.
Electronic balance	JA2003N	Shanghai Precision Science Instrument Co., Ltd
Vacuum drying oven	DZF-6020	Shanghai Kesheng Instrument Co., Ltd
Micro melting point apparatus	SGW-4	Shanghai Electric Optical Instrument Co., Ltd
Nuclear magnetic resonance	BRUKER AVANCE500MHz	Germany - Switzerland Brooke Spectrum Instrument Company

2.2. Experimental steps

2.2.1. Synthesis of ethyl ethyl 4-nitrobenzoate In a 500 mL four-necked flask equipped with a stirrer, thermometer, spherical condenser and constant pressure dropping funnel, 83.5g (0.5 mol) p-nitrobenzoic acid and 115g (2.5 mol) anhydrous ethanol were added. The solution was stirred and mixed evenly, and then 6.8 g (0.05 mol) potassium bicarbonate was dropped. After finishing the adding process, the temperature was raised to about 80℃ to re-flux. After 3h of the reflux reaction, the mixture solution was cooled to room temperature. 8% sodium carbonate solution was added the reaction solution to adjust the pH value to the range of 7-8, while the water is added to wash. The solution was placed at 0 to 5℃ for low temperature precipitation. Meanwhile, the filtration and washing carried out three times. 93.9g white crystalline solid was obtained after drying. The yield was 93.6%. The nuclear magnetic resonance spectroscopy of the target product was shown in Fig. 1.

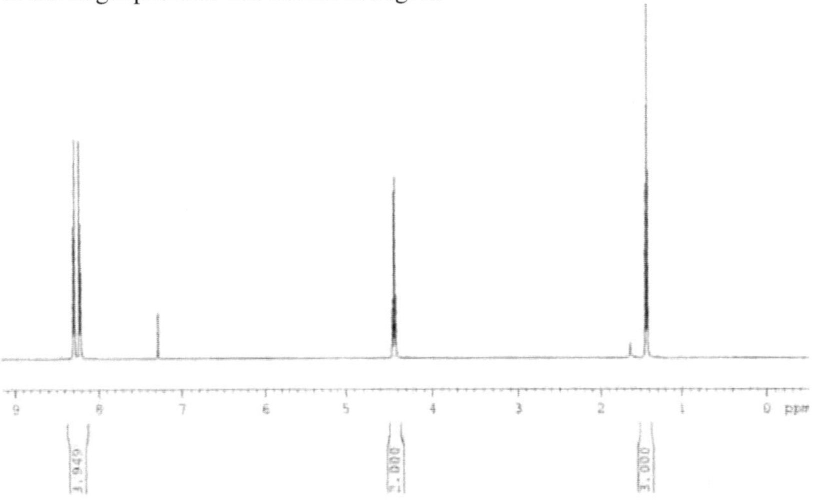

Fig. 1 HNMR spectrum of ethyl 4-nitrobenzoate

2.2.2. Synthesis of ethyl 4-(butylamino)benzoate 93.9g (0.48 mol) p-nitrobenzoate and 138.2 g (4.32 moles) methanol were added to a 1L four-necked flask equipped with a stirrer, a spherical condenser, a thermometer and a constant pressure dropping funnel. And then 38.1g (0.53 mol) n-butyraldehyde and 124.8g (1.92 mol) zinc powder were added there too. After cooling to 0-5℃, 230.4g (3.84 mol) acetic acid was added to the system with 0.5h. And then room temperature reaction and TLC tracking reaction stared, and this process spent about 2h. After the reaction, light brown yellow transparent liquid can be obtained by direct pumping, and the filter cake was washed with water (50 mL × 3). In addition, the generated salt was removed and the zinc powder was recovered.meanwhile, the mother liquor obtained by suction filtration was evaporated under reduced pressure and the methanol was used for recovery. The remaining part was dissolved using 200mL ethyl acetate, and the salt and acetic acid were removed by washing. And then the organic phase was dried. 100.8g ethyl p-butylaminobenzoate crude can be obtained with 94.7% yield. The target product of nuclear magnetic resonance spectrum was shown in Fig. 2.

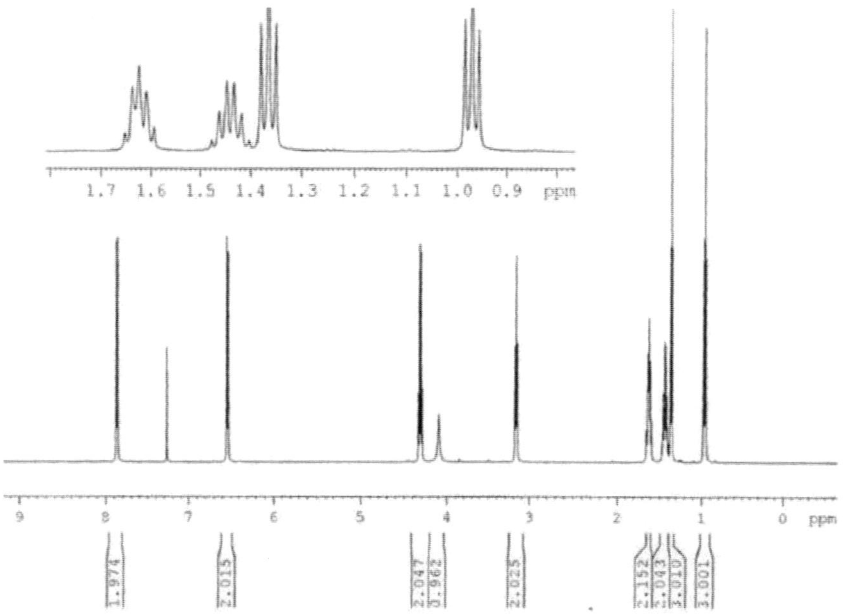

Fig. 2 HNMR spectrum of ethyl 4-(butyl amino) benzoate

3. Results

(1) The synthesis of p-nitrobenzoic acid ethyl acetate: The p-nitrobenzoic acid was selected as raw material and ethanol vinegar in this reaction under the temperature of 80℃. In addition, potassium hydrogen sulfate was taken as a catalyst, p-nitrobenzoic acid, ethanol, and the optimum reaction molar ratio of p-nitrobenzoic acid, anhydrous ethanol and potassium hydrogen sulfate was 1: 5: 0.1. The reaction yield was 3%, and the final yield was 96.3%. It was a simple acetic acid reaction with high yield. After cooling, the low concentration of sodium carbonate was added to remove the p-nitrobenzoic acid, filter, drying and the drying product p-nitrobenzoic acid.

(2) Synthesis of p-butylbenzoic acid and acetylene: The optimum reaction molar ratio of p-nitrobenzoic acid ethyl acetate, n-butyraldehyde, zinc powder and acetic acid was 1:1.1:4:8 at the temperature of 25℃. And the reaction yield was 94.7%. The reaction time is short, and the condition is mild. The product is simple after the reaction. The filter work was completed after the reaction. Meanwhile, the excess zinc powder can be washed, and the oxide layer can be removed through acidification. The zinc power can be recycled after the drying, while the application effect was better. However, the mother liquor is filtered under reduced pressure, which can recover methanol with 80% recovery rate.

4. Conclusion

In this paper, tetracaine hydrochloride is studies as a local anesthetic. It is widely used in clinic, which is often used for surface, infiltration and other local anesthesia. There are few manufacturers of tetracaine hydrochloride in the domestic market, and most of tetracaine hydrochloride products in the market come from abroad. The domestic production is far from meeting the market demand, and there are many problems existing in the production process, such as low yield, high cost, and harsh conditions and so on. By verifying the feasibility of the original route, a new synthetic route is proposed according to the experimental conditions. The synthetic route was optimized, and the effects of reaction factors on the reaction were determined by repeated experiments. Combined with the actual production conditions of industrialization, the industrial production line is determined. Compared with the prior art, the method has the advantages of low cost, easy and safe operation, low requirement of reaction equipment, low cost, short production period, high yield and high purity. Reagents that are less harmful to the environment are selected as far as possible in the reaction process. These measures reduce the harm to the human body. The raw materials are cheap and the production cost is reduced. According to the experimental reagent, the reaction device is simplified to make it easier for industrial production. From the whole synthetic route, this kind of synthetic process is suitable for industrial production.

Acknowledgments
Supported by the Scientific and Technological Research Program of Chongqing Municipal Education Commission (No. KJ1505501)

Reference

[1] Ethington J, Goldmeier D, Gaynes B I. Exponential Decay Metrics of Topical Tetracaine Hydrochloride Administration Describe Corneal Anesthesia Properties Mechanistically[J]. Cornea, 2017, 36(3): 363-366.
[2] Lin, H., Zhang, Y., Hou, M., & Wang, Y. L. Transdermal Local Anaesthetic Films of Tetracaine in Comparison to the Conventional Gel in Effective Management of Pain[J]. International Journal of Pharmacology, 2016, 12(6): 669-674.
[3] Wu W, Wang S. A Novel Sensitive Electrochemiluminescence Method for Determination of Tetracaine Hydrochloride Based on ZnO/RGO Modified Electrode [J]. Sensor Letters, 2016, 14(11): 1067-1073.
[4] Pang X, Fan T J. Cytotoxic effect and possible mechanisms of Tetracaine on human corneal epithelial cells in vitro[J]. International journal of ophthalmology, 2016, 9(4): 497.
[5] Yuan M L, Xie J H, Zhou Q L. Boron Lewis Acid Promoted Ruthenium - Catalyzed Hydrogenation of Amides: An Efficient Approach to Secondary Amines[J]. ChemCatChem, 2016, 8(19): 3036-3040.
[6] Pek Y S, Wu H, Mohamed S T, et al. Long - Term Subconjunctival Delivery of Brimonidine Tartrate for Glaucoma Treatment Using a Microspheres/Carrier System[J]. Advanced healthcare materials, 2016, 5(21): 2823-2831.
[7] Yuan M L, Xie J H, Zhou Q L. Boron Lewis Acid Promoted Ruthenium - Catalyzed Hydrogenation of Amides: An Efficient Approach to Secondary Amines[J]. ChemCatChem, 2016, 8(19): 3036-3040.
[8] Rawat P, Verma S M. Design and synthesis of chroman derivatives with dual anti-breast cancer and antiepileptic activities [J]. Drug Design, Development and Therapy, 2016, 10: 2779.

IOP Conf. Series: Earth and Environmental Science **64** (2017) 012109 doi:10.1088/1755-1315/64/1/012109

Effects of different mechanized soil fertilization methods on corn soil fertility under continuous cropping

Qingwen Shi[1,a], Huixin Wang[2,a], Chunming Bai[2,a], Di Wu[1,a], Qiaobo Song[1,a], Depeng Gao[1,a], Zengqi Dong[1,a], Xin Cheng[1,a], Qiping Dong[1,a], Yahao Zhang[1], Jiahui Mu[1], Qinghong Chen[1], Wenqing Liao[1], Tianru Qu[1], Chunling Zhang[1], Xinyu Zhang[1], Yifei Liu[1,*] and Xiaori Han[1,*]

[1]Shenyang Agricultural University, Shenyang, China
[2]Liaoning Academy of Agricultural Sciences, Shenyang, China

[a]Joint first authors
*Co-corresponding author e-mail: liuyifeiscience@163.com

Abstract. Experiments for mechanized soil fertilization for corns were conducted in Faku demonstration zone. On this basis, we studied effects on corn soil fertility under continuous cropping due to different mechanized soil fertilization methods. Our study would serve as a theoretical basis further for mechanized soil fertilization improvement and soil quality improvement in brown soil area. Based on the survey of soil physical characteristics during different corn growth periods, we collected soil samples from different corn growth periods to determine and make statistical analysis accordingly. Stalk returning to field with deep tillage proved to be the most effective on available nutrient improvement for arable soil in the demonstration zone. Different mechanized soil fertilization methods were remarkably effective on total phosphorus improvement for arable soil in the demonstration zone, while less effective on total nitrogen or total potassium, and not so effective on C/N ratio in soil. Stalk returning with deep tillage was more favorable to improve content of organic matter in soil, when compared with surface application, and organic granular fertilizer more favorable when compared with decomposed cow dung for such a purpose, too.

1. Introduction

Since soil is the material basis of crop production, soil fertility magnitude will determine crop growth and yield. Organic fertilizer applied in agricultural production plays its roles mainly in soil improvement, fertilization betterment, and yield gain and quality improvement. Different tillage methods and soil fertilization methods both would affect soil fertility greatly. Mechanized fertilization worked nine times faster than traditional fertilization, so it was able to save labor resource, and both were not so significantly different in terms of greenhouse gas emission [1]. Therefore, mechanized fertilization is of higher promotional value in a way. Stalk returning to field was able to increase organic matter in soil and improve nitrogen use efficiency for the purpose of high yield eventually [2,3]. Compared with the use of stalk, farmyard manure was more favorable to the accumulation of potassium and SOM in limestone soil [4, 5]. As research indicated, culture substrate with decomposed cow dung as the main material was able to improve substrate volume weight, total porosity, along with

IOP Conf. Series: Earth and Environmental Science **64** (2017) 012109 doi:10.1088/1755-1315/64/1/012109

the growth and physiological indicators of pepper seedlings [6, 7]. Chemical fertilizer when applied could improve soil fertility to some degree, but it was not so remarkably effective. As some studies indicated, normal farmyard manure when applied was featured with more greenhouse gas emission than chemical fertilizer. However, organic granular fertilizer was effective to increase yield with persistent fertilizer efficiency for soil fertilization, and this was not only favorable to increase after crop production proportionately, but to optimize ecology as well [1,8-9]. Broadcast application will not give full play to the organic granular fertilizer, and it will only reduce application effects. Then if furrow application or whole application is assumed, better effects may be expected.

2. Material and method

2.1. Experimental material

2.1.1. Corn variety for experiment: Shenghe 18

2.1.2. Soil type: It is brown soil, located in Lujiafangshen Village, Dagujiazi Town, Faku County. The physical and chemical properties of the soil for test are listed in the following tables:

Table1 The physical-chemical properties of soil before seeding in 2014

Treat.	Organic matter (C g·kg^{-1})	Total N (N g·kg^{-1})	Total P (P g·kg^{-1})	Total K (K g·kg^{-1})	Alkali-N (N mg·kg^{-1})	Olsen-P (P_2O_5 mg·kg^{-1})	NH_4OAc-K (K_2O mg·kg^{-1})	Bulk density (g/cm^3)	PH
Content	11.89	2.27	0.12	15.93	100.78	19.28	86.83	1.39	5.06

2.1.3.Fertilizer for test: (1)Organic fertilizer: decomposed cow dung at the local cattle farm (nutrient content: organic carbon 33.81%, N content 2.41%, P2O5 content 1.919%, K2O content 1.445%,moisture 58.29%).
(2)Corn stalk: local corn stalk (nutrient content: organic carbon 43.54%,N content 0.863%, P content 0.260%, K content 0.834%, moisture 9.07%).
(3)Granular fertilizer as biochar-based: granular fertilizer as biochar-based from the Plant Nutrition and Fertilizer Research Institute of Shenyang Agricultural University (moisture 25-30%,organic matter≥45%,NPK≥5%,nitrogen 2.06%,phosphorus 1.82%,kalium 1.78%, secondary and microelements≥26%).
(4)Conventional chemical fertilizer: slow release fertilizer with the brand of Qingyuan, which was used as the base fertilizer for deep placement, 50kg per mu (≈667m2, Chinese traditional unit area and the same below) (NPK content of 27-11-13, zinc 0.02% for the product made by Changjiang Controlled Release Fertilizer Company Ltd., Jilin). Seed fertilizer for surface application: Compound fertilizer of 15-15-15 with the brand of Yanyangtian.
(5)Microbial agent: stalk decomposing agent applied for stalk returning to field.

2.1.4. Machinery for test: (1) HP 55/60 tractors and plowing, harrowing, ridging as well as sowing equipment.
(2)Stalk returning equipment: mechanical stalk cutting (5-10cm long), returning to field by manual work.
(3)Broadcaster for granular chemical fertilizer: single disc broadcaster as developed by Nanjing Agricultural Mechanization Research institute of the National Agriculture Ministry.
(4)The spraying machines and equipment for field fertilizer, pesticide and microbial agent shall be operated conventionally as farmers usually do.

2.2. Test Method

On the premise of the same plot with the same soil fertility under the same tillage conditions, five experimental zones were established for our experiments: Zone I (with mechanical broadcasting of decomposed cow dung, 500kg each mu as air-dried basis, moisture 20%, rotary tillage for 20cm on ridges); Zone II (mechanical cutting and crushing for stalk returning and surface application, 400kg per mu, with stalk decomposing agent and chemical nitrogen fertilizer to be used together and sprayed at the same time according to the single factor, then rotary tillage on ridges for 20cm); Zone III (mechanical cutting and crushing for stalk returning first and deep tillage for 30cm further, 400kg per mu, with stalk decomposing agent and chemical nitrogen fertilizer to be used together and sprayed at the same time according to the single factor); Zone IV (mechanical broadcasting of granular fertilizer as biochar-based) and Zone V (conventional management of tillage and fertilizer application, for which 18kg nitrogen, 5kg phosphorus pentoxide, 6kg potassium oxide, urea and potassium chloride 20kg+10kg as the basic fertilizer, 10kg diammonium phosphate as the seed fertilizer, extra fertilization for 225kg/hm at bell stage, or special fertilizer for corns would be selected, and due to stalk returning practice, potash fertilizer amount could be properly reduced).

These experiments started from the year of 2013, and then soil samples were collected in autumn to determine their volume weight, organic matter content and nutrients. Then fertilizer and stalk composition were under analysis.Tasks after crop harvesting or before sowing included stalk returning, broadcasting of organic fertilizer, chemical fertilizer and microbial agent, and soil tillage as well. Soil samples and plant samples were collected according to growth periods (before fertilization and tillage, at elongation stage, at bell stage, at filling stage and at ripe stage).

3. Results and analysis

3.1. Effects on available nutrients of arable soil for corns during their different growth periods due to different mechanized soil fertilization methods

At ripe stage, contents of alkali-hydrolyzable nitrogen, available phosphorus and available potassium in topsoil in zone III were all remarkably higher than those with other treatments. This was because stalk returning with deep tillage was able to improve soil structure and accelerate microorganism reproduction and stalk decomposition, then facilitating release of available nutrients. In case of stalk returning and surface application, release of alkali-hydrolyzable nitrogen was quicker, though not so persistent. Available potassium in the decomposed cow dung enjoyed quicker release than the available potassium in the organic granular fertilizer did. Therefore, when compared with surface application, deep tillage helped to improve available nutrients in soil after stalks were crushed and returned to field.

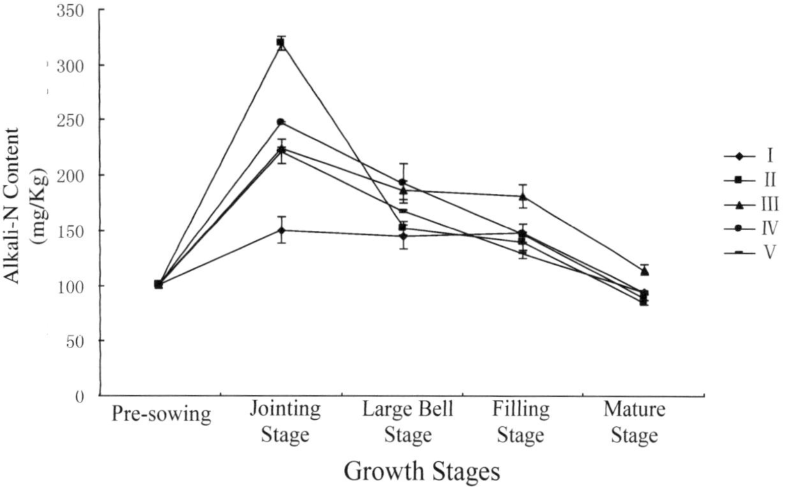

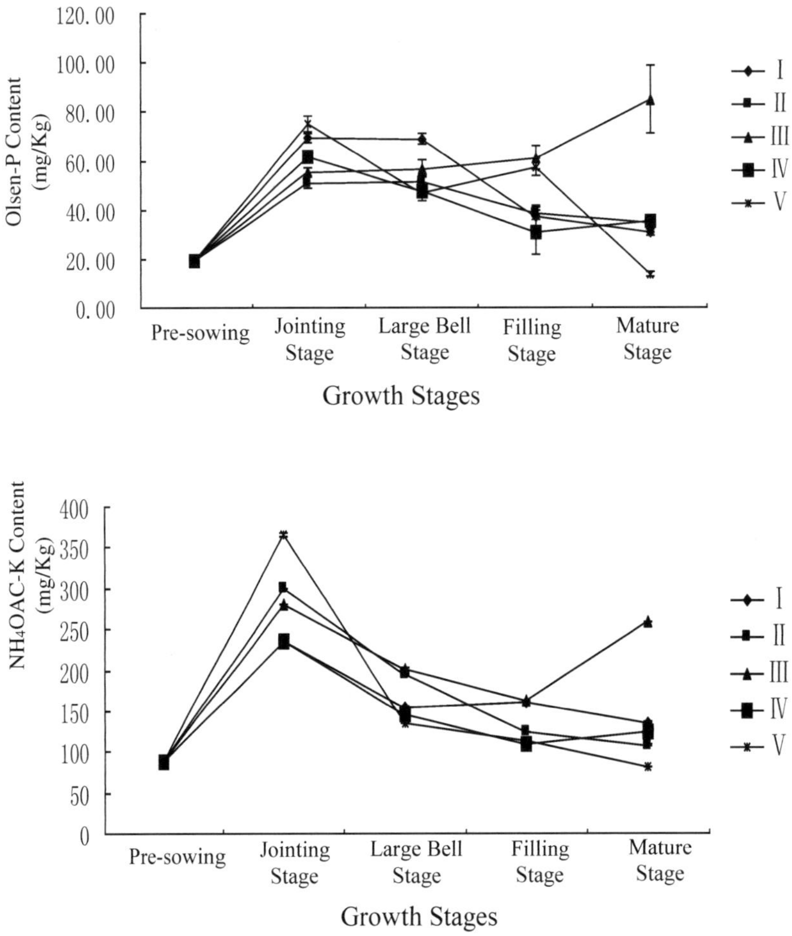

Figure 1. Changes of the contents of soil available nutrients in each growth stage of maize

3.2. Effects on soil total nutrients, organic matter and C/N for corns at their ripe stage due to different mechanized soil fertilization methods

As compare with situations before fertilization and tillage, total nitrogen and total phosphorus at the arable soil in zone I, zone II, zone III and zone IV increased for 3.19% and 127.71% respectively, while total potassium reduced for 0.91% on the average. Mechanical deep tillage, if combined with crushed stalk returning, was beneficial to improve total nitrogen content at topsoil. Additionally, stalk returning when combined with surface application was able to remarkably improve total phosphorus content at topsoil.

Mechanical deep tillage, when compared with surface application, was able to remarkably improve organic matter content in soil. So were organic granular fertilizer when compared with decomposed cow dung, and organic granular fertilizer when compared with conventional fertilizer application. organic matter content in arable soil in zone I, zone II, zone III and zone IV increased for 21.32% on the average, when compared with the situations before fertilization and tillage. Mechanical deep tillage and fertilization did not remarkably influence soil C/N ratio, when compared with conventional fertilization.

Table2 Effects of different treatments on Total nutrient of soil in maize mature stage

Duncan[a,b]

Treat.	Total N(g/kg)		Total P(g/kg)		Total K(g/kg)	
	0-20cm	20-40cm	0-20cm	20-40cm	0-20cm	20-40cm
I	1.89±0.11b	2.19±0.28a	0.321±0.012b	0.157±0.009a	15.794±1.055a	15.238±1.100a
II	2.32±0.51ab	2.13±0.13a	0.372±0.003a	0.170±0.005a	15.239±0.858a	15.017±0.144a
III	2.65±0.20a	2.41±0.13a	0.221±0.002c	0.155±0.016a	16.732±0.435a	15.160±1.335a
IV	2.51±0.58ab	2.08±0.12a	0.179±0.002d	0.161±0.007a	15.377±0.491a	15.137±0.225a
V	2.27±0.32ab	2.08±0.09a	0.176±0.007d	0.154±0.005a	16.474±0.767a	15.839±0.435a

Note: Different letters in the same column mean significant difference at 5% level.

Table3 Effects of different treatments on topsoil Organic matter and C/N in maize mature stage

Duncan[a,b]

Treat.	Organic matter(g/kg)		C/N	
	0-20cm	20-40cm	0-20cm	20-40cm
I	13.41±0.42cd	10.90±0.76c	4.12±0.27a	2.93±0.58b
II	12.07±0.67d	9.80±0.27d	3.13±0.78a	2.67±0.23b
III	17.01±0.59a	11.14±0.43bc	3.74±0.42a	2.69±0.25b
IV	15.21±1.07b	13.09±0.13a	3.67±1.01a	3.67±0.25a
V	13.52±0.66c	11.96±0.13b	3.51±0.61a	3.33±0.18ab

Note: Different letters in the same column mean significant difference at 5% level.

4. Conclusion

4.1. Mechanization remarkably was effective on available nutrient increase in soil in the demonstration zone.

Alkalihydrolyzable nitrogen enjoyed its quicker release in case of stalk returning and surface application, though not so persistent. Available potassium in the decomposed cow dung enjoyed quicker release than the available potassium did. Stalk returning with deep tillage, if compared with surface application, was more beneficial to improve available nutrient content in soil.

4.2. Mechanization remarkably affected total phosphorus improvement for the arable soil at the demonstration zone.

However, it was less effective on and total nitrogen and total potassium. Crushed stalk returning combined with deep tillage, in addition to decomposed cow dung application, was able to remarkably improve total phosphorus content at topsoil.

4.3. .Mechanization did not remarkably influence soil C/N ratio for the arable soil in the demonstration zone.

As compared with conventional fertilization, crushed stalk returning with deep tillage, when organic granular fertilizer broadcasting was also adopted, was able to remarkably improve organic matter content at topsoil. Mechanical deep tillage, when compared with surface application, was more beneficial to improve organic matter content in soil, and so was organic granular fertilizer when compared with decomposed cow dung application.

Based on the analysis above, it is concluded that crushed stalk returning with deep tillage for 30cm is able to remarkably increase the available nutrient content in brown soil at the corn continuous cropping area. Organic granular fertilizer, when compared with decomposed cow dung, is more beneficial to improve organic matter content at topsoil. Thus, for brown soil at the corn continuous cropping area, it is advised that crushed stalk returning combined with deep tillage, in addition to organic granular fertilizer broadcasting by agricultural machinery shall be adopted for management. This practice shall not only remarkably increase soil available nutrient in arable soil, but also improve organic matter content to the effect of loose soil structure and local soil fertility betterment.

Acknowledgments
This work was financially supported by Natural Science Foundation of China (31301842; 31601627), National science and technology support project (2013BAD08B04), Cultivation Plan for Youth Agricultural Science and Technology Innovative Talents of Liaoning Province (2014043), Program of Liaoning Excellent Talents in University (LJQ2015100), China Scholarship Council (CSC NO. 201708210143) and National Peanut Research System (CARS-14).

References

[1] Alfred Gathorne-Hardy. The sustainability of changes in agricultural technology: The carbon, economic and labour implications of mechanisation and synthetic fertiliser use. Ambio, 45(2016)885-894.

[2] Jianling Pan, Wanan Dai, Zhanhuan Shang, et al. Effects of straw returning on soil organic matter and nitrogen availability and its mechanism.Chinese Journal of agricultural ecology. 05(2013)526-535.

[3] Ru Zhang. Design of multifunctional straw returning machine and experimental study on its performance. Northeast Agricultural University, 2014.

[4] Wei Zhang, Hongsong Chen, Yirong Su, et al. Effects of different fertilization on crop and soil fertility in newly reclaimed lime. Chinese Journal of Soil Science, 04(2013)925-930.

[5] Jie Li, Mingliang Wu, Yuanju Tang,et al. Research status and development trend of organic fertilizer application machinery. Journal of Hunan Agricultural University(Natural Science Edition), S1(2013)97-100.

[6] Jing Li, Jihua Yu, Jianming Zhe,et al. Effect of seedling amount of cow manure compost on pepper seedling quality. Journal of Gansu Agricultural University, 04(2012)38-42.

[7] Chikowo R, Mapfumo P, Nyamugafata P, et al. Maize productivity and mineral N dynamics following different soil fertility management practices on a depleted sandy soil in Zimbabwe[J]. Agriculture, Ecosystems & Environment, 102(2004)119-131.

[8] Yongling Sun, Fengxiang Chen. Preliminary report on high yield cultivation of hybrid rape with particulate organic fertilizer.Anhui Agricultural Science Bulletin, 17(2009)74-75.

[9] Lyimo H J F, Pratt R C, Mnyuku R S O W. Composted cattle and poultry manures provide excellent fertility and improved management of gray leaf spot in maize[J]. Field Crops Research, 126(2012):97-103.

IOP Conf. Series: Earth and Environmental Science **64** (2017) 012110 doi:10.1088/1755-1315/64/1/012110

The influence of soybean new blade's sugar content variation to its apparent growth

Yiqiong Wu[1], and Yun Gong[2,*]

[1]School of Landscape Architecture, Zhejiang A&F University, Lin'an, China
[2]School of Engineering, Zhejiang A&F University, Lin'an, China

*Corresponding author e-mail: gongyun204@163.com

Abstract. To a certain extent, leaf sugar content shows the plant growth status. At the beginning of the rapid growth of the seedlings of soybean, internal physiological indexes, such as the correlation between sugar content and external apparent growth is not very clear. This experiment measured the soybean seedling stage in different gradient of the soluble sugar content and the correlation of the apparent growth indicators, including leaf weight, leaf width, leaf length, root length, number of radicle, etc. The correlation between soluble sugar content and the apparent growth studies to preliminary discussion on this issue is analyzed. The data shows that in the seedling stage, leaf soluble sugar content increase quickly, and the apparent growth has different correlation with indicators. The weight of leaf has a intimate relativity with soluble sugar content, but leaf width and length have weaker correlation with it. And the root length and number of young root show little correlation with soluble sugar content. Experiment results indicate that in soybean seedling stage, internal physiological indexes do have a specific relationship with external growth index in some degree.

1. Introduction

Glycine max (L.) Merr is generally called soybean, is an annual leguminous herb attributes to bean. What's more, it is one of the important food crops in China which has a five thousand years of cultivation history and it was called SHU in ancient times [1]. Soybean is one kind of comprehensive nutrient food, and compared with the same amount of pork, the protein content, and calcium content, iron content of soybean is 1 times, 33 times, 26 times more respectively, while the price is lower than pork. Soybean protein contains the all kinds of amino acids needed in the body, especially the essential amino acids, such as lysine, leucine, and threonine. Only the content of methionine is relatively few [2]. This is the opposite of general cereal. Therefore, soybean mixed with other grain to eat can complement each other and greatly improve the nutritional value of them. Soybean contains a lot of fat which are unsaturated fatty acids, especially the content of linolenic acid which plays a great role to prevent atherosclerosis is found most abundantly. Soybean also contains about 1.5% of phospholipids which are the basal components of cells and play an important role in maintaining the health of nerves, liver, bones and skin.

All aspects of agricultural production of soybean are paid great attention to because of high nutritional value of it. At present, the research of soybean at home and abroad mainly focuses on the cultivation and transgenosis [2, 3]. There are some reports about the relationship between the internal physiological indexes and the external growth index. For instance, studies on the correlation between the sugar content of the beetroot and the growth amount of the plant have been carried out [4]. In addition, the changing trend of sugar contents during the growth of the soybean seedlings is preliminarily discussed [5]. These research results help us to understand the changes of some important nutrients during the growth of plants[6, 7]. However, these studies were mainly focused on a single index, and there was no report on the correlation between the apparent growth and the sugar content in the early stage of the development of soybean seedling leaves. The sugar content and growth of soybean leaves were measured at different time in this study, including leaf length, leaf weight, leaf width, plant height, and root length and so on. The main research methods of this thesis are documentation method and experimental method [8]. The discussion on the relationship between the sugar content and the apparent growth during the growth of soybean was based on a mathematical model which was established for the dynamic change of sugar content and its apparent growth in different time periods.

2. Expriment

2.1. Experiment Reagents and Instruments

723 Visible Spectrophotometers (Shanghai optical instrument factory), Electronic Analytical Balance (Beijing Jingke RitaTechnologies co., ltd), Thermostatic Water Bath (M344133, Beijing ZhongXi Yuanda Technology Co., Ltd.). Phenol Solution, Trichloroacetic acid (TCA), concentrated sulphuric acid, Hoagland nutrient solution, ether, acetone, anhydrous ethanol and other reagents are all analytical pure, purchased from Sinopharm Chemical Reagent Co., ltd. "Kenfeng series" soybean seeds purchased from Heilongjiang Province Academy of Agricultural Sciences and Crop Research Institute.

2.2. Experimental Procedure

2.2.1. Cultivation of soybean seedlings The cultivation methods of soybean can refer to the previous literature reports [9]. Buy "Kenfeng series" soybean seeds from seed company, disinfect with mercury solution (10 mL 0.1%) for 10 minutes, and then filter the mercury solution, wash soybean seeds with running water for half an hour, rinse with distilled water for three times. The seeds should be evenly placed in a Petri dish covered with two layers of wet filter paper, and then put into constant temperature incubator after covering. Accelerating germination for 48 h under the dark condition of 25 ℃, then select seeds of same germination to sow in plastic basin nylon net, use hoagland's nutrient solution to culture, the culture temperature is 25-28 ℃. Shine a light on them for 12 hours a day, transfer to the plastic basin (12 cm in diameter) when seedlings grow to the stage of seedling with one leaf. Single basin area (about 200 plants per pot). The culture solution should be changed every two days, then sample and measure each index after a week.

2.2.2. Determination of total soluble sugar content and its standard curve soluble sugars mainly includes soluble monosaccharides and oligosaccharides which are soluble in water and ethanol. The sugar is dehydrated to produce furfural or hydroxymethyl furfural under the action of concentrated sulfuric acid, which can react with phenol to form an orange red compound. In the range of 10 to 100 mg, the color depth is directly proportional to the sugar content, and has a maximum absorption peak at the wavelength of 490 nm, so the total sugar content can be determined by colorimetry at the wavelength of 490 nm. Sulfuric acid - phenol method can be used for the determination of methylated sugar, pentose and polysaccharide, this method is simple and sensitive, and the color stability time is above 160 min. The measured data is absorbance, and then use the standard curve method, the sugar content can be obtained through the standard curve. Prepare a series of standard solutions of different concentrations, measure the absorbance of the standard solution respectively. Then, the standard curve can be drawn, with the concentration as abscissa and the corresponding absorbance as ordinate. Measure the absorbance of the solution under exactly the same conditions, and obtain the solution concentration from the standard curve. Take six test tubes (20ml), numbered from 0-6, add solution and water according to table 1, then add phenol solution (1 mL 6%) to the test tube in sequence and shake well. Add 5 mL of concentrated sulfuric acid from test tube walls in 5-20 s time and shake well. The total volume of the color standard solution is 8 mL, place 30 min at room temperature, cooling and the color development reaction will happen. Then take the blank as the reference, colorimetric determination at 490 nm wavelength, take the sugar content as the abscissa, the light density as the ordinate, draws the standard curve, the standard line equation can be obtained.

Table 1 sugar content of glucose standard solution (ug)

tube number	1	2	3	4	5	6
glucose standard solu-tion (mL)	0.0	0.8	1.0	1.2	1.4	1.6
distilled water (mL)	2.0	1.2	1.0	0.8	0.6	0.4
sugar content (g)	0	40	50	60	70	80

2.2.3. Determination of apparent growth index The experiment will use vernier caliper to measure the leaf length, width and root length; and will use analytical balance to record the weight of young leaves and the number of radicles (Recording time from the stage of seedling with one leaf, take 5, 7, 9, 11, 13 and 15 days for the time gradient).

2.2.4. Determination of total soluble sugar content The method for determination of total soluble sugar is based on the experimental methods reported previously [13, 14], but in the specific operation process, we improved some experimental parameters, and achieved the desired results.

1) By taking new leaves of soybean, grinding, centrifugation, speed 3000 r/min, a total of three times. The first time was 15 minutes, take the supernate. The latter two was about 5 mins respectively, the supernatant was taken into the cone type colorimetric tube (25 mL). The final filtrate kept around 18 mL.

2) Constant volume sampling after 2 hours of water bath. After the bath, added 2 mL 6 mol/L sodium hydroxide and shaked well after cooling with running water. Constant-volumed with a 25 mL volumetric flask. Draw 0.2 mL of sample solution and add distilled water to make a total volume of two mL. Then add 1 mL 6% of phenol and 5 mL of concentrated sulfuric acid, shake well and cool to room temperature for 20 minutes. Use the spectrophotometer to measure the absorbance at 490 nm wavelength. Each measurement should take dual control, the content of polysaccharide was calculated by standard curve.

3. Results and discussion

3.1. Standard linear equation

Soluble sugar content $X=(C\times D)$

X - total sugar content in each sample

C - the sugar content that shows on the standard curve (g)

D -dilution ratio

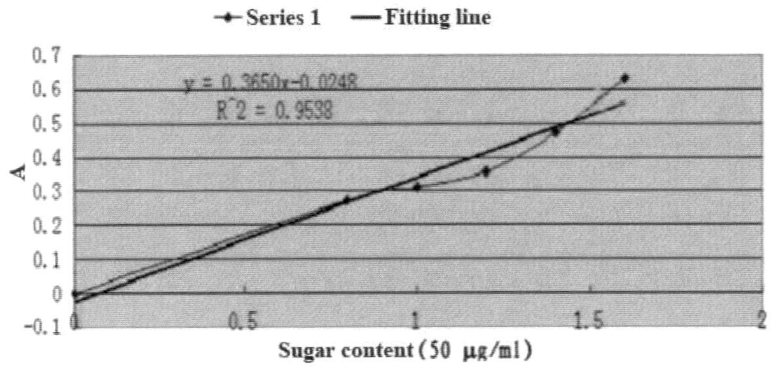

Figure 1 standard linear equation

3.2. The changing characteristics of the growth amount of the newly-growing soybean leaves

Table 2 Data summary table of growth amount

Days (d)	5	7	9	11	13	15
Leaf length (CM)	0.99	1.46	2.31	3.01	3.48	3.50
Leaf width (CM)	0.62	0.92	1.79	2.34	2.86	3.52
Leaf weight (g)	0.0183	0.0425	0.074	0.121	0.173	0.21
Root length (CM)	4.84	5.49	5.51	10.10	7.12	8.20
The number of radicles (root)	15	16	29	45	34	36

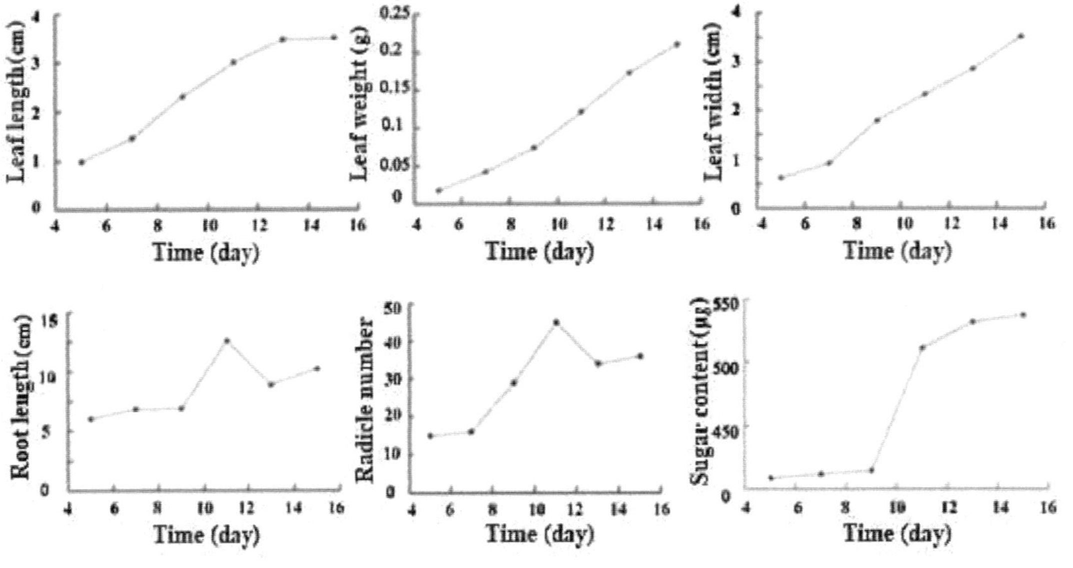

Figure 2 The dynamic changes of leaf width, leaf weight, leaf length, root length, the number of radicles and sugar content of leaves over time

3.3. Correlation between growth amount and sugar content

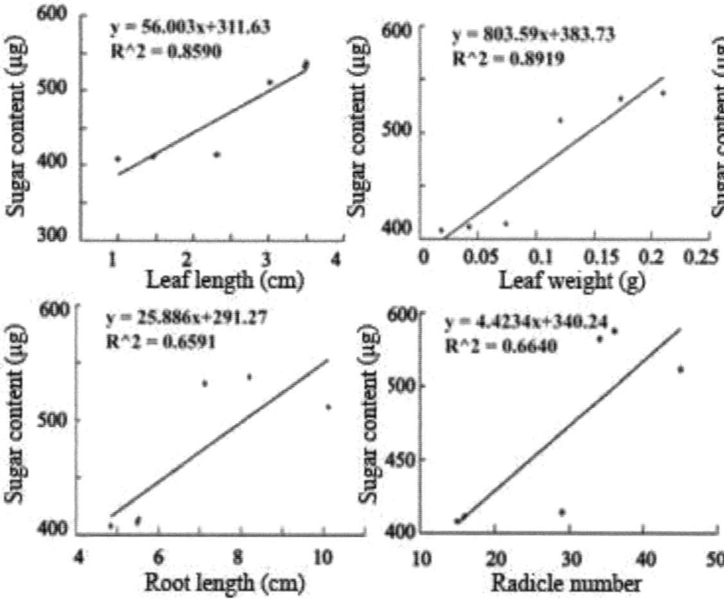

Figure 3 Correlation between the sugar content of the newly-growing soybean leaves and its leaf length, leaf weight, leaf width, root length and the number of radicles

3.4. Standard curve of total soluble sugar

According to the preparation scheme in the table 1, the standard curve (Figure 1) can be drawn by measuring the absorbance value at the wavelength of 490 nm, taking the sugar content as the abscissa and the absorbance as ordinate, and the equation is y = 0.3650x-0.0248. The standard equation provides an important basis for the determination of sugar content in different organs of soybean leaves.

3.5. Change rules of the growth amount of new soybean leaves

"Kenfeng series" of soybean seed began to grow to the phase of seedling with one leaf after five days, then we measured the soybean seedling leaf length, leaf weight, leaf width, root length, the number of radicles and the sugar content of leaves from the fifth day. It can be seen from Figure 2 that the leaf length, leaf weight, leaf width were steadily increasing in the 5-15 days, only the increment speed of leaf length noticeably slowed down, basically reached the platform, grow to 3.5 cm in 15 days. The leaf weight and leaf width had been going up all the time, which increased from 0.62 cm to 3.52 cm, and increased from 0.0183 g to 0.21 g respectively. The root length growed rapidly in 9-11 days, reached 10.10 cm, the number of radicles increased rapidly from the seventh day, and had been at a high level of quantity. The total sugar content did not change much from 5-9 days, however, the sugar content increased sharply from 9-11 days, and then flatten, the total sugar content increased from 408.3 g to 537.6 g within 15 days.

3.6. Correlation between growth amount and sugar content

We can see from Figure 3: The content of soluble sugar in the leaves of soybean seedlings increased within 15 days under normal conditions, increased from 408.3 g to 537.6 g while leaf length increased from cm to 3.5 cm, the leaf weight increased from 0.0183 g to 0.21 g, leaf width increased from 0.62 cm to 3.52 cm. The linear equation of soluble sugar content and leaf length, leaf width and leaf weight were obtained by linear fitting according to the experimental results: y=56.003x+331.61, R^2=0.8590; y=803.59x+383.73, R^2=0.8918; y=52.598x+363.65, R^2=0.8485. Analysed the equations and the growth trend of total sugar and leaf length, leaf weight , leaf width can obtain: during the period of seedling, when the leaf length, leaf weight and leaf width increased, the sugar content in the leaves also increased over time, showed a positive correlation, and a strong correlation. During the 15 day test, the root length increased from 4.84 cm to 8.2 cm, the number of radicles also increased from 15 to 36. The linear equation of soluble sugar content and root length or the number of radicles can be obtained by linear fitting according to the experimental data: $y=25.886x+291.27$, $R^2=0.6591$; $y=4.4234x+340.27$, $R^2=0.6640$. Although with the increase of soluble sugar content,the root length and the number of radicles increased too, but it can be seen from the equation that the correlation between the sugar content and the root length or the number of radicles is very week.

4. Conclusion

The changes of apparent growth and total sugar content of soybean seedlings were measured in the early development process (5-15 days), and the relationship between the different apparent growth indexes and the total sugar content were linearly fitted. The results showed that with the increase of the leaf length, leaf weight and leaf weight, the sugar content in leaves also increased over time, and there is a linear relationship because of the large R square value. And for the root length or the number of radicles, the results of linear fitting with the total sugar content show that the R square value is small, illustrated that the root length and the number of radicles can account for the small change of total sugar content. Further experiments are needed, and other factors should be taken into account in future.

References

[1] Zuo J, Dong H, Hou H. Porduction and application situaiton of soybean protein [J]. Cereals & Oils. 2007, 5: 12-15.

[2] Liu X, Tu C, Zhang L, Lu J. Research on nutrition and health benefits of soy protein [J]. Journal of Beijing Technology and Business University (Natural Science Edition). 2012, 30 (02): 1-6.

[3] Dornbos Jr D, Mullen R. Influence of stress during soybean seed fill on seed weight, germination, and seedling growth rate[J]. Canadian Journal of Plant Science. 1991, 71 (2): 373-83.

[4] Xie J. Study on the growth and development of sugar beet [J]. Sugar Crops of China. 1980 (01): 7-16.

[5] Huber SC. Biochemical basis for effects of K-deficiency on assimilate export rate and accumulation of soluble sugars in soybean leaves[J]. Plant Physiology. 1984, 76 (2): 424-30.

[6] Le J, Song S, Lehne P, Huang D, Wu C. Fruit development and sugar accumulation of muskmelon in organic farming system [J]. Acta Agriculturae Universitatis Jiangxiensis. 2007, 29 (04): 526-532.

[7] Fredeen AL, Terry N. Influence of vesicular-arbuscular mycorrhizal infection and soil phosphorus level on growth and carbon metabolism of soybean[J]. Canadian Journal of Botany. 1988, 66 (11): 2311-6.

[8] Gao C, Yuan H, Wang R, Zhang J, Dai X. Improvement of extraction and determination method of reducing sugar content in flue-cured tobacco leaves [J]. Journal of Southwest China Normal University (Natural Science Edition). 2010, 35 (05): 87-90.

[9] Ma C, Yao Y, Huo H, Gong Z, Wei D, Chi F. Dynamic changes of sucrose synthase in leaves among different soybean genotypes [J]. Crops. 2011, 04: 32-37.

[10] Zhou C, Zhong X, Fan H, Lv L, Su H, Li Y. Evaluation of different determination of water-soluble total sugar and reducing sugar in fruit and vegetable [J]. The Food Industry. 2012, 33 (05): 89-92.

[11] Li B, Zhang L, Guo S, Liu M, Hao X. Studies on the method of joint determinetion of soluble sugar and amylum in Chinese chestnut seeds [J]. Hebei Journal of Forestry and Orchard Research. 1997, 12 (02): 27-30.

[12] Thomas J, Boote K, Allen L, Gallo-Meagher M, Davis J. Elevated temperature and carbon

dioxide effects on soybean seed composition and transcript abundance[J]. Crop Science. 2003, 43 (4): 1548-57.

[13] Thomas J, Boote K, Allen L, Gallo-Meagher M, Davis J. Elevated temperature and carbon dioxide effects on soybean seed composition and transcript abundance[J]. Crop Science. 2003, 43 (4): 1548-1557.

[14] Ma C, Guo H, Gong Z, Wei D, Chi F. Sugar accumulation among different soybean genotype (I)

Fabrication of solid alginate particles containing functionalized MWCNTs using a microfluidic technique

Rulu Yang, Yanan Li, Weiwei Cai, Hua Zhang and Jiao Li*

School of Materials Science and Engineering, Shandong University of Technology, Zibo, PR China

*Corresponding author e-mail: haiyan9943@163.com

Abstract. A series of solid alginate particles containing different amounts of functionalized multi-walled carbon nanotubes (MWCNTs) have been successfully prepared by microfluidic technique. The influence of MWCNTs doping on the surface morphology, structure and properties of calcium alginate (CA) particles has been investigated using SEM, FTIR, XRD, TGA and UV-Vis spectra. SEM indicates that the obtained hybrid particles are of good morphological characteristics, and its color changes from yellow to black which proves MWCNTs is successfully dispersed in CA particles. Moreover, the particles surface changes rough from smooth after MWCNTs doping. FTIR results verify the interactions between functional groups of CA and MWCNTs. Besides, the incorporation of MWCNTs to CA particles can efficiently enhance the thermal stability of CA particles at high temperatures, which is ascribed to the better thermal stability of MWCNTs as well as the interactions between MWCNTs and CA particles. The removal efficiency of CA particles doped with a lower percentage of MWCNTs (0.06 wt. %) for methylene blue is 96.38%, which is much higher than that of pure CA particles (61.21 %).

1. Introduction

Sodium Alginate (SA), a linear polysaccharide, is naturally occurring polymer composed of α-guluronate and β-mannuronate residues. Its solutions can be transformed into a hydrogel on addition of metallic divalent cations such as Ca^{2+}. Moreover calcium alginate (CA) hydrogel particles can be used as good adsorbents for removal of different metal cation or organic pollutes due to the presence of carboxyl and hydroxyl groups along its chains [1-2]. However, some prepared CA particles still have some drawbacks, such as low adsorption capacity or poor thermal stability [2-3], which make them difficult to be further used in environmental pollution management.

To solve problems above mentioned, alginate particles have been blended with some especial filler to improve its adsorption capacity. Sun et al [4] reported that the GO/CA compostite particles had higher adsorption capacities for acridine orange than those without GO. Hassan et al [5] proposed the use of activated carbon/CA hybrid particles for the removal of arsenic from aqueous solutions. The results indicated that the adsorption capacity of composite particles has higher adsorption capacity for arsenic comparing with pure CA particles or activated carbon. Besides these, other studies [6-7] also reported the enhancement of the adsorption capacity of CA particles for Pb (II) or methylene blue could be achieved by adding some magnetic materials such as glycine-Fe_2O_3 or Cyanex 272.

Carbon nanotubes (CNTs) is a more promising nano-filler for polymers compared to above-mentioned filler. It has been reported that incorporating of nanotubes into some adsorbents can effectively improve their ability to capture dyes or metal ions from aqueous phase [8-9], which can be attributed to nanotubes' large specific surface area, exceptional hollow and layered structures. Our previously studies also proved that CNT impregnation is a good method to enhance adsorption capacity of chitosan hydrogel particles [10]. However, to the best of our knowledge, the detailed study on the effect of CNTs doping on the morphology and structure of CA particles has hardly been reported. Moreover, no study about the thermal stability of CNTs/CA has been reported. On the other hand, functionalized microspheres consisting of various organic or inorganic materials can be prepared using a microfluidics technology [11-12], but the fabrication of solid alginate particles containing functionalized MWCNTs by the microfluidic technique has not received much attention.

In this study, a series of solid alginate particles containing different amounts of functionalized multi-walled carbon nanotubes (MWCNTs) are fabricated by microfluidic technique. The hybrid particles are characterized by means of SEM, FTIR, XRD, TGA and UV-Vis spectra. The effect of MWCNTs doping on the structure and properties of CA particles is discussed in detail. The possible influence mechanism is also investigated

2. Experimental

2.1. Materials and device

Sodium alginate (SA) and methylene blue (MB) are obtained from Sigma Aldrich (USA). MWCNTs (diameter: 10~30 nm, length: 5~15 μm) are purchased from Beijing Nachen Nanotech Co.Ltd and MWCNTs are further acid-treated by the previously established methods [10]. The acid-MWCNTs can be dispersed into sodium alginate solution and no precipitation is observed in the solution after several weeks. All other agents are of analytical grade. Deionized (DI) water is used in all experiments. The microfluidic device used in this study is supplied by Beijing CapitalBio Corporation.

2.2. Preparation of particles

Fig. 1 shows the schematic for the production of particles. 1.5 g SA are added into 100mL DI water in a beaker to obtain SA solution and then stirred for 12 h at 60 °C. A certain amount of acid treated-MWCNTs are added to the SA solution in an ultrasonic bath and loaded into a single out-put syringe as the disperse phase, and peanut oil is loaded into a double output syringe as the continuous phase. Then the disperse phase and continuous phase are simultaneously injected into the microfluidic chip through Teflon tubes by using Syringe pumps (WHP01 Model, Baoding Longer, and CHI). A water-in-oil (w/o) emulsion is formed at the cross-junction by using two continuous peanut oil streams to shear the SA and MWCNTs-SA solution. The droplets are dripped into a stirring calcium chloride solution through the other Teflon tube. The water-soluble droplets are transformed into water-insoluble MWCNTs-calcium alginate particles (Marked M-CA particles) in the condition of calcium chloride. Finally, the obtained MWCNTs-CA particles are washed with deionized water several times to remove residual calcium chloride from particle surface and then air-dried for 24 h, followed by drying at 55 °C for 12 h. In comparison, the pure CA particles are also prepared according to the above procedure. (Marked CA particles).

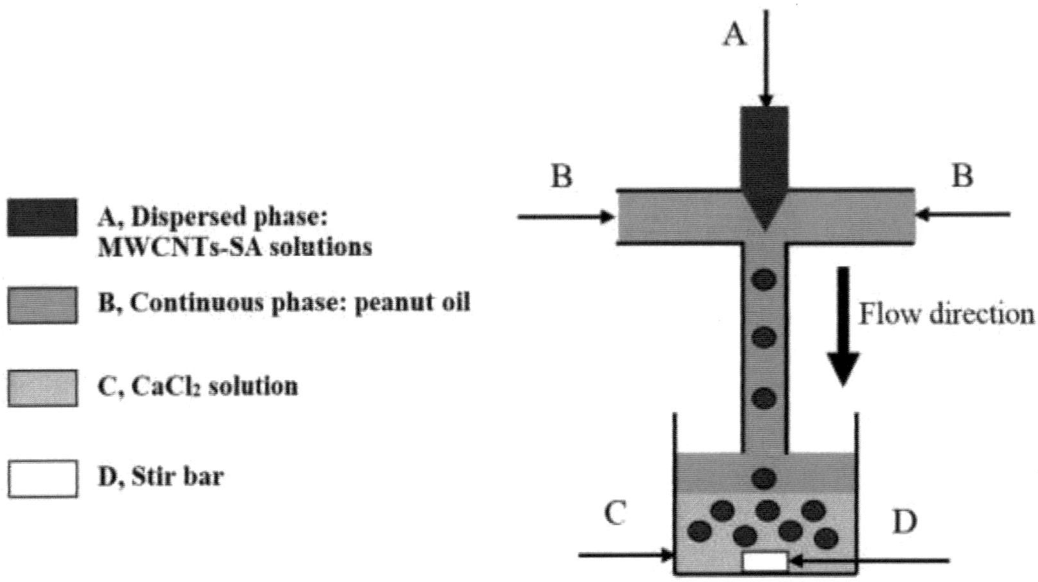

Figure 1. Schematic drawing of MWCNTs-SA composite droplet formation in a microchannel cross-junction and solidification with the crosslinking reaction process between the MWCNTs-SA and $CaCl_2$ solution

2.3. Measurements

The molecular structure analyses of all the samples are carried out with FTIR (Nicolet 5700, USA) with the wavenumbers recorded from 400 to 4000cm^{-1} at a 1 cm^{-1} resolution. Sample structures are analyzed by XRD (D8 Advance, GER) with Cu $K_{\alpha 1}$ radiation. The surface morphology of the dried particles is measured by SEM (FEI Sirion 200, NLD). TGA is performed on a TG/DTA thermal system (Model HCT-2, CHI) with a heating rate of 5 °C min^{-1} over the temperature range of 0~950°C under air atmosphere.

Batch adsorption experiments are conducted according to the following procedure. 100 mg of hydrogel particles are added into erlenmeyer flask containing 25 mL of MB solution (initial concentration 200 mg/L, initial pH 8.0). After shaken at 30 ℃ overnight to establishment the adsorption-desorption equilibrium, and the dye solution was separated from the adsorbent by centrifugation at 4000 rpm for 30 min. The concentrations of MB in filtrate are determined by a double beam UV-Visible spectrophotometer (Model TU-1901, CHI) at 666 nm with a linear regression equation ($y = 0.1043x - 0.0133$, $R^2 = 0.9998$), which is obtained by plotting a calibration curve for dye over a range of concentrations. All experiments are conducted in triplicate.

The removal percentage (R %) of MB by hydrogel particles at the equilibrium can be calculated using Eq (1).

$$R(\%) = \frac{C_0 - C_e}{C_0} \times 100\% \qquad (1)$$

Where C_0 (mg/L) and C_e (mg/L) are the liquid phase concentration of MB at initial and equilibrium, respectively.

3. Results and discussion

3.1. Morphology

The surface morphology of CA and M-CA particles are shown in Figure 1. Macroscopically, the particles obtained are all spherical in shape, and the addition of MWCNTs does not alter the macroscopic surface morphology of the particles. However, one noticeable characteristic of the particles is its yellow to black color (Inset in Fig 2), which confirms MWCNTs is successfully dispersed in CA particles.

The microscopic surface morphology of particles was observed by SEM. The CA particles have a tight and smooth surface (Fig. 2a) with some concave structures spotted on the surface of particles, which might be caused by partly collapsing the polymeric gel network during drying [13]. The surface of M-CA particles is, however, highly rough although the spherical shape is maintained (Fig. 2b). Another noticeable characteristic of M-CA particle surface is very homogeneous and has not any concave structures. These obvious surface morphological changes might be due to the homogeneous dispersion and supportive role of MWCNTs in CA particles.

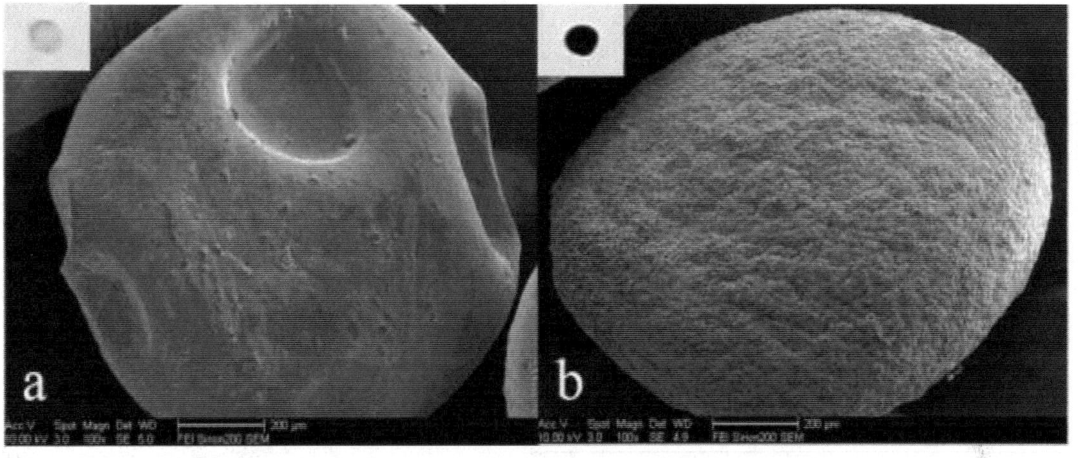

Figure 2. The surface micrographs of CA and M-CA particles
(a) CA particles, (b) M (0.06*wt* %) -CA particles

3.2. FT-IR analysis

Fig. 3 shows the FTIR spectrum of CA, MWCNTs and MWCNTs/CA in the region of 4000 to 400 cm^{-1}. The FTIR spectra of CA particles shows the characteristic peaks at 3415, 1612, 1430, 1087 and 1033 cm^{-1}, which are due to the stretching of O-H, -COO (asymmetric), -COO (symmetric) and C-O-C, respectively (Fig. 3a). In the FTIR spectrum of acid-MWCNTs, the principal absorption peaks appearing at 3434 cm^{-1} for -OH stretching, 1630cm^{-1} for C=C skeleton vibrations, 1710 and 1590cm^{-1} for -COO stretching in the -COOH groups reveal the -COOH functionalization of MWCNTs (Fig. 3b). Incorporating functionalized MWCNTs into CA particles causes an obvious shift to lower wavenumber and a decrease in intensity of -OH stretching peaks of CA particles (Fig. 3c). This indicates that an intermolecular hydrogen bonding is formed between CA and functionalized MWCNTs. And the formation of hydrogen bonding can possibly occur between the carboxyl groups on the MWCNTs and hydroxyl groups of CA. Moreover, it can also be seen that the intensity of the band representing -COO (1597 and 1417 cm^{-1}) stretching of carboxyl groups becomes weaker and has a noticeably red shifting when CA particles doped with MWCNTs. It is possible to explain that the negative charge of the carboxyl group in the alginate has some electrostatic interaction with the positive charge of MWCNTs.

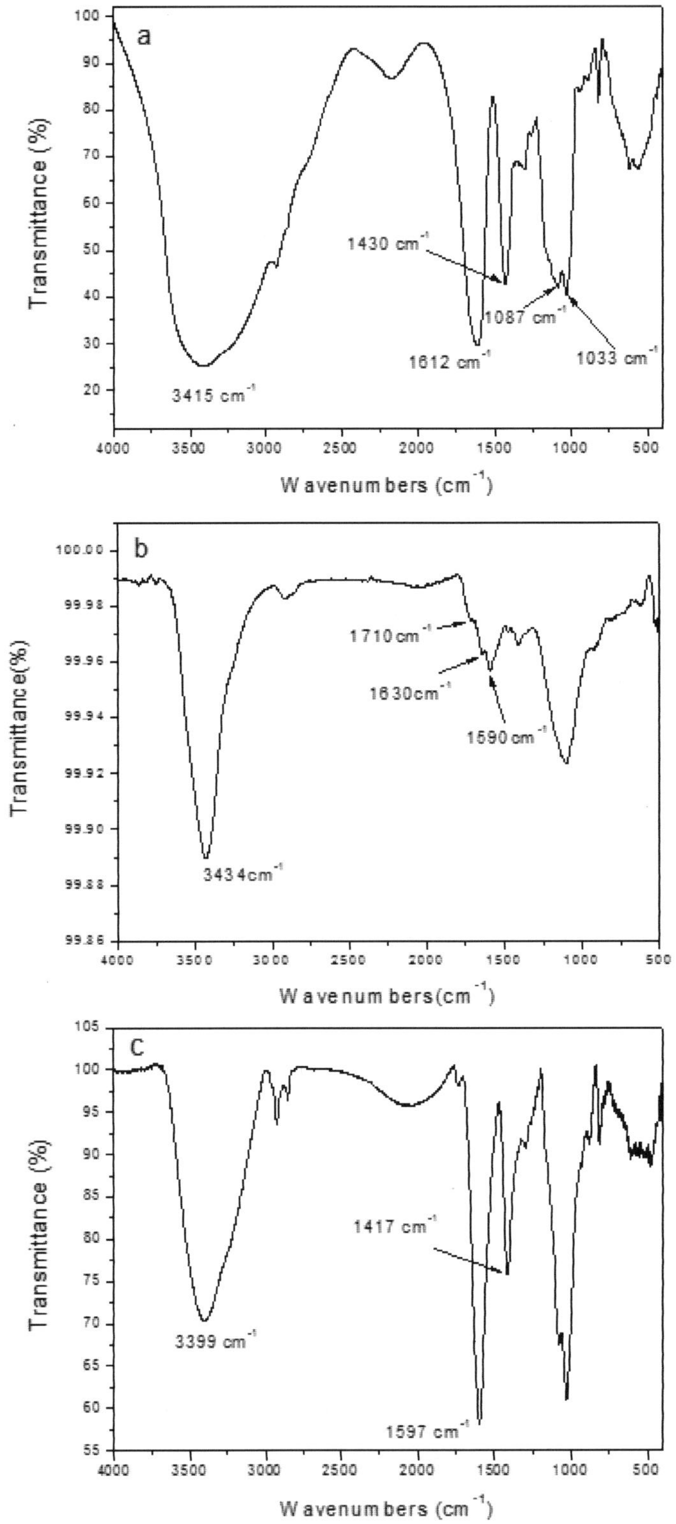

Figure3. The FTIR spectra of (a) the CS particles, (b) acid -MWCNTs and (c) the M (0.1 *wt* %) -CS composite particles

IOP Conf. Series: Earth and Environmental Science **64** (2017) 012111 doi:10.1088/1755-1315/64/1/012111

3.3. XRD analysis

Molecular interaction of CA and MWCNTs in hybrid particles is also investigated using XRD. CA gel particles only has a broad diffraction (0 0 1) peak around 10.4 (Fig 4a), indicating the defective crystals and a low crystallinity of CA particles [14]. The pure MWCNTs shows a sharp peak centered on 2θ value of 25.6 corresponding to the (0 0 2) planes of MWCNTs (Fig 4b). The composite particles (Fig 4c) present the characteristic peaks of both MWCNTs and CA without any additional new bands, which represent the absence of covalent interactions between the phases. Moreover, the crystallization behavior of the CA particles is not changed.

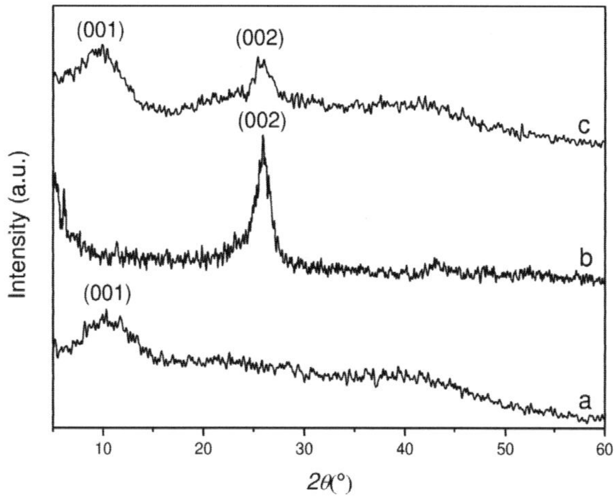

Figure 4. X-ray diffraction patterns of (a) the CA, (a) acid -MWCNTs and (c) M (0.06 *wt* %) -CA particles

3.4. Thermal stability

Thermal stability of different hydrogel particles (Fig. 5) are analyzed by the TGA test. The TGA thermogram of pure CA particles shows four distinct stages at 80-125 °C, 220-480 °C, 510-750 °C and 750-790 °C, corresponding to the loss of residual water, the destruction of glycosidic bonds, the oxidation of the intermediate carbonaceous char material, and the partial decomposition of calcium carbonate, respectively. Similar TGA data were previously reported [15]. The TGA curve of M-CA particles is almost the same as that of pure CA particles between 50-250°C. However, when the temperature is higher than 250°C, the TGA curve of M-CA particles becomes slightly different from that of pure CA particles. The TGA curve of M-CA particles shows that the residue weight is increased by adding MWCNTs. The residue weight of M-CA particles at 840°C is about 20.38%, higher than that of pure CA particles (12.45%), which indicate that the existence of MWCNTs in the composite particles can enhance the thermal stability of CA particle at high temperatures. This is ascribed to the better thermal stability of MWCNTs [16] as well as the special interactions, such as inter-molecular hydrogen-bonding interaction and the electrostatic interaction between MWCNTs and CA lattices, which limit the motivation of CA.

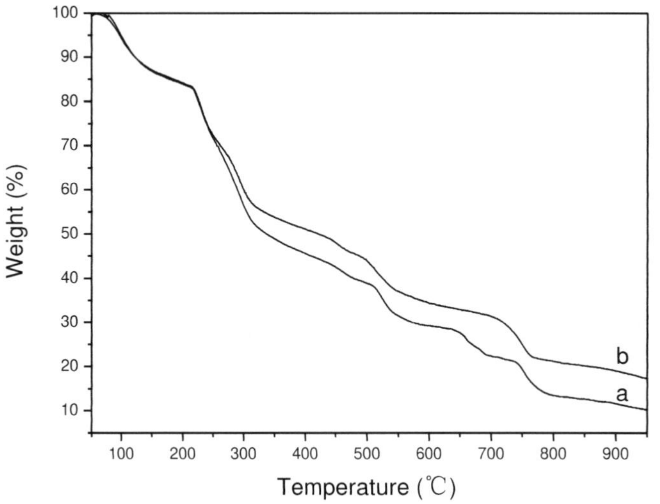

Figure 5. TGA thermograms of (a) pure CA and (b) M (0.06 *wt%*) -CA particles

3.5. MB removal efficiency

As shown in Fig. 6, the MB removal efficiency of hydrogel particles increased with an increase in the concentration of MWCNTs. The CA particles doped with 0.06*wt%* MWNCTs have higher removal efficiency (96.38%) than that of pure CS particles (61.21%). The good MB removal efficiency of composite particles suggested the strong affinity of M-CA particles towards dye molecules, which is likely to be due to the following reasons. On one hand, as a nano-filler, MWCNTs has good adsorption capacity for dyes because their large specific surface area, hollow and layered nanosized structures [17-18]. Moreover, the abundant oxygen-containing functional groups on surfaces of functionalized MWCNTs also can provide numerous chemical adsorption sites to increase the adsorption capacity of particles. On the other hand, as shown in SEM, the addition of MWCNTs into CA particles made the surface of M-CA particles much larger wrinkles. The presence of these wrinkles indicates that there is a large exposed surface area of hybrid particles, which may be convenient for the dye molecules to penetrate into the polymeric network of the composites, leading to the increase in the absorption capacity of MB to a certain extent.

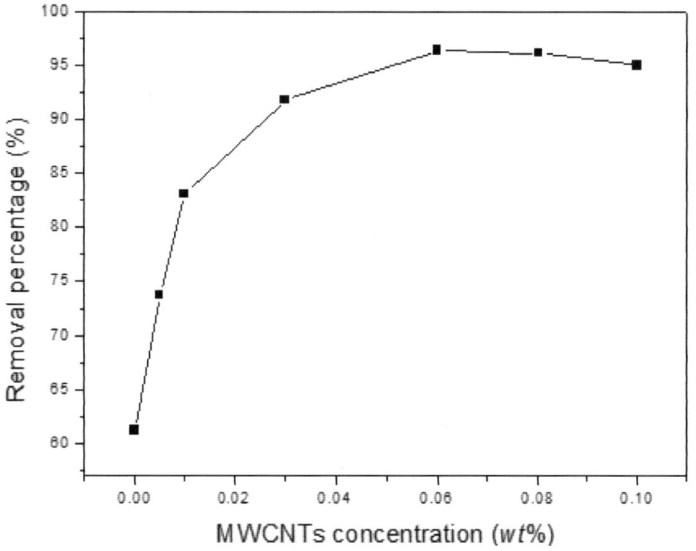

Figure 6. Effect of the MWCNTs concentration on adsorption capacity of particles for MB Adsorption experiments: dye concentration: 200mg/L; sample dose: 0.1g/25.00mL; natural pH 8.0; temperature: 30℃; equilibrium time: 3 hour.

4. Conclusions

In summary, functionalized MWCNTs/CA hybrid particles have been successfully prepared using a microfluidic technique. The results suggest that MWCNTs doping has a significant influence on the morphologies, thermal stability and MB removal efficiency of CA particles. The obtained hybrid particles are of good morphological characteristics, and the color change from yellow to black proves MWCNTs is successfully dispersed in CA particles. The incorporation of MWCNTs to CA particles can efficiently enhance the thermal stability of CA particles at 840°C, which is ascribed to the better thermal stability of MWCNTs as well as some special interactions between MWCNTs and CA particles. The removal efficiency of CA particles doped with a lower percentage of MWCNTs (0.06 *wt %*) for methylene blue is 96.38 %, higher than that of pure CA particles (61.21 %). Therefore, we consider that the MWCNTs/CA hybrid particles prepared by microfluidic technique might be used as a promising, effective adsorbent for the removal organic dye from wastewater.

Acknowledgments

We gratefully acknowledge the support of the College Technology Development Project of Shandong Province (J15LA08) and the Support Development Project for Young Teachers of Shandong University of Technology (4072-114019) on this work.

References

[1] Z.Q. Wang, Y.G. Huang, M. Wang, G.H. Wu, T.M. Geng, Y.G. Zhao, A.G. Wu, Macroporous calcium alginate aerogel as sorbent for Pb2+ removal from water media, J. Environ. Chem. Eng. 4 (2016) 3185-3192.

[2] R. Aravindhan, N.N. Fathima, J.R. Rao, B.U. Nair, Equilibrium and thermodynamic studies on the removal of basic black dye using calcium alginate beads, Colloids Surf., A. 299 (2007) 232-238.

[3] W. Block, Water status and thermal analysis of alginate beads used in cryopreservation of plant germplasmq, Cryobiology . 47 (2003) 59-72.

[4] L. Sun, B. Fugetsu, Graphene oxide captured for green use: Influence on the structures of calcium alginate and macroporous alginic beads and their application to aqueous removal of

acridine orange, Chem .Eng. J. 240 (2014) 565-573.

[5] A.F. Hassan, A.M. Abdel-Mohsen, H. Elhadidy, Adsorption of arsenic by activated carbon, calcium alginate and theircomposite beads. Adsorption of arsenic by activated carbon, calcium alginate and theircomposite beads, Int. J. Biol. Macromol. 68 (2014) 125-130.

[6] R. Verma, A. Asthana, A. K. Singh , S. Prasad, M.A.B.H. Susan, Novel glycine-functionalized magnetic nanoparticles entrapped calcium alginate beads for effective removal of lead, Microchem. J. 130 (2017) 168-178.

[7] A. F. Ngomsika, A. Beea, J. M. Siaugue, D. Talbot, V. Cabuil, G. Cote, Co(II) removal by magnetic alginate beads containing Cyanex 272®, J. Hazard. Mater. 166 (2009) 1043-1049.

[8] S. Chatterjee, M.W. Lee, S.H. Woo, Adsorption of congo red by chitosan hydrogel beads impregnated with carbon nanotubes, Bioresour. Technol. 101 (2010) 1800-1806.

[9] M. Inyang, B. Gao, A. Zimmerman, M. Zhang, H. Chen, Synthesis, characterization, and dye sorption ability of carbon nanotube–biochar nanocomposites, Chem. Eng. J. 236 (2014) 39-46.

[10] J. Li, Q.C. Gao, H. Zhang, Y. Zhou, H. D. Wang, Effect of MWCNTs Doping on the Morphology, Structure and Properties of Chitosan Beads, J. Macromol. Sci. A. 49 (2012) 372-379.

[11] C. H. Yang, C. Y. Wang, K. S. Huang, C. P. Kung, Y. C. Chang, J. F. Shaw. Microfluidic one-step synthesis of Fe3O4-chitosan composite particles and their applications. Int. J. Pharmaceuti. 463 (2014) 155-160.

[12] X. T. Sun, M. Liu, Z. R. Xu, Microfluidic fabrication of multifunctional particles and their analytical applications, Talanta 121 (2014) 163-177.

[13] G. Pasparakis, N. Bouropoulos, Swelling studies and in vitro release of verapamil from calcium alginate and calcium alginate–chitosan beads, Int. J. Pharmaceut 323 (2006) 34-42.

[14] L. Li, Y. Fang, R. Vreeker, I. Appelqvist, Reexamining the Egg-Box Model in Calcium-Alginate Gels with X-ray Diffraction, Biomacromolecules 26 (2007) 464-468.

[15] J. Wang, X. Ying, X. Li, W. Zhang, Preparation, characterization and swelling behaviors of polyurethane-grafted calcium alginate hydrogels, Mater. Lett. 126 (2014) 263-266.

[16] S.M. Sohel Murshed, C.A. Nieto de Castro, Superior thermal features of carbon nanotubes-based nanofluids-A review, Renewable Sustainable Energy Rev. 37 (2014) 155-167.

[17] Y. Yao, F. Xu, M. Chen, Z. Xu, Z. Zhu, Adsorption behavior of methylene blue on carbon nanotubes, Bioresource. Technol. 101 (2010) 3040-3046.

[18] C.Y. Kuo, C.H. Wu, J.Y. Wu, Adsorption of direct dyes from aqueous solutions by carbon nanotubes: determination of equilibrium, kinetics and thermodynamics parameters, J. Colloid Interface Sci. 327 (2008) 308-315.

IOP Conf. Series: Earth and Environmental Science **64** (2017) 012112 doi:10.1088/1755-1315/64/1/012112

Study on rapid bio-drying technology of cow dung with CaO_2

Xiaotian Chen, Guangfei Qu *, Shugen Liu, Ruosong Xie, Yanhua He

Kunming University of Science and Technology, Kunming 650000, China

*Corresponding author e-mail: qgflab@sina.com

Abstract. Effect of CaO_2 on cow dung rapid bio-drying technology was researched. A static aerobic composting system was applied to this experiment which combining natural ventilation with Turing in the process of composting. The physical characteristics of cow dung was observed and the compost temperature, moisture content, organic matter, total nitrogen, total phosphorus, potassium content was determined which in order to study the effect of CaO_2 on rapid drying of cattle in the compost. In the initial stage of compost, adding CaO_2 groups compared with the control group, the temperature rise faster, 4-6 days in advance to the thermophilic phase; at the end of composting, the CaO_2 composition and moisture content decreased significantly to below 30%. The addition of CaO_2 in fertilizer was shorten the composting time, extend the thermophilic phase, to provide sufficient oxygen meeting the growth needs of aerobic microorganisms. It convinced that the rapid bio-drying of dairy manure has a good effect and provided a new idea for the effective treatment of cow dung.

1. Introduction

The intensity and concentrated activity of the livestock industry generate vast amounts of biodegradable wastes [1]. Cow breed in accounts for a large proportion of livestock and poultry breeding, dairy manure can turn into an important pollution source without treated properly. In recent years, cow dung can convert into organic fertilizers by composting technology, which realized harmlessness, reduction and resource utilization of the cow dung [2]. Composting of organic wastes is a biooxidative process involving the mineralisation and partial humification of the organic matter, leading to a stabilised final product, destruction of pathogens and with certain humic properties [3]. The aerobic composting of cow dung was dicussed in this paper.

According to the determination, the water content of the bedding and non-flush faeces is generally 65% ~ 85%, while the water content of flush livestock has come up to 90%. The higher water content has more influences to the storage, processing and transportation of faeces, it's an important step in faeces treatment to reduce the water content of livestock faeces. Nowadays, there are a lot kinds of rapid desiccation of livestock faeces, including physical methods, chemical methods and biological methods, among them, the application of biological drying technology is more extensive, mainly because of the low cost, the easy to operate and the high fertilizer efficiency of biological drying technology. In 1984, American Cornell University scientist Jewell W J[4] proposed the word of feces of livestock and poultry bio drying (Bio-drying) firstly, and put forward the principle of biological drying is high temperature aerobic microbial composting process and it can quickly decompose organic matter and produce a lot of heat. The temperature of the stacking material, in the case of ventilation or turning, to promote the distribution of water in the pile, dry dung to reduce the purpose of water. Choi [5] et al. (2013) and Richard [6] et al. (1997) proposed continuous batch or semi-continuous composting techniques on the basis of previous batch composting work and designed and manufactured Oxygen bioreactor. Many trials have shown that composting products have a moisture content of more than 30% at the end of composting. Singh [7] (2009)

the use of vegetable waste, sawdust and cow dung as raw material for aerobic composting, composting for 20 days, the water content of each treatment group decreased from 63.5% ~ 76.0% to 47.5% ~ 72.0%, The effect is not very obvious. Roca-Perez [8] et al. (2009) in the sludge and straw mixed aerobic composting test, after 90 days of composting, composting material moisture content from the initial 60.3% to 39.9% ~ 42%, moisture content decreased obviously, but the composting cycle is longer.

There's little research of utilize of composting technology in the disposal and drying of livestock and poultry manure in China, compare to it's utilize in the application of sludge. Similarly, the research of livestock and poultry manure is less in bio drying technology aspect, Jiangsu Academy of Agricultural Sciences soil and fertilizer Chang zhiz hou [9] and so on (2000) using pig manure for biological drying test, the results shows that the use of batch aerobic composting, adding additives and other methods, after 20 d of pig manure moisture content decreased from 700 g/Kg down to 550 g/Kg below, reduced by 150 g/Kg. Chen Tian rong developed a chicken manure fermentation drying equipment, using the equipment of biological fermentation processing of chicken manure, 25 ~ 35d after, the water content from 650g/Kg down to 200g/Kg, the following effect is obvious. This experiment uses cow dung and tobacco smoke as raw materials to study the effects of temperature, water content, pH, nitrogen, phosphorus and potassium on the composting of cow dung under the action of CaO_2 by adding CaO_2, in order to explicit the influence of CaO2 to the dry cow dung compost rapid biological technology, and to provide guidance of theory and technology used in harmless treatment of cow dung compost and manure production.

2. Materials and methods

2.1. Compost raw materials

Fresh cow dung (Dali backyard cow dung), tobacco waste (factory smoke rod), CaO2 (industrial grade). The properties of composting raw materials are shown in Table 1.

Table 1 Physical and chemical properties of the composting materials

materials	moisture content /%	pH	organic matter %	N/%	P/%	K/%
cow dung	84.60	7.42	93	1.8	1.7	1.5
tobacco smoke	22.76	5.66	83	1.3	0.6	6.8

2.2. Composting design

According to the mass ratio of 5:2 (the fresh cow dung: tobacco waste) well mixed (the proportion of integrated organic fertilizer factory production and pre-experiment), which also added to bacteria, CaO_2, microbial inoculum and CaO_2, microbial inoculum and CaO_2 (continuous addition), the treatment group was divided in Table 2, All the windrow fermentation piles were 1.5 meters tall, 3 meters long and 3 meters wide. Which turned and mixed once every three days until composting was over, the fermentation period is 35 days.

Table 2 Formula for different treatments

Treatments	
A:	3t a + 1.2t b
B:	3t a + 1.2t b + 15kg c
C:	3t a + 1.2t b + 100kg d
D:	3t a + 1.2t b + 100kg d + 15kg c
E:	3t a + 1.2t b + 100kg d + 15kg c(10kg CaO2 once per heap)

a=: cow dung; b= tobacco smoke; c=bacteria; d= CaO_2

2.3. Determination index and methods

2.3.1. Temperature measurement. During the composting process, the temperature was measured at 8:30 am every day. The measured temperature was the temperature of 5 The positions of 20 cm, 35 cm and 50 cm below the surface of the each piles was measured by thermometer, the average of was regarded the compost temperature of each piles in one day.

2.3.2. Determination of pH, moisture content, organic matter, TN, TP, TK. Sampling was carried out at 9:00 am every day from five sample points of each piles by quartation, which was equilibrated at a distance of 30 cm at 50 cm far from the surface of the each piles, 50 cm and in different orientations, separate the two diagonals by four equal portions until the required sample size is 500 g each. Moisture content, pH, organic matter, TN, TP, TK are determinated by organic fertilizer agricultural standard NY525-2012 [10].

3. Results and Analysis

3.1. The changes and analysis of physical properties of compost materials

At the beginning of the composting, the contrast compost group A, B ware brown, and have a great smell of cow dung; With the addition of CaO_2, the color of the treatment group C, D, E compost mixture shown dark brown, the smell of cow dung was reduced significantly due to the formation of alkali. After 6 days of composting, the surface color of the treatment group C, D, E were deepen, which appeared white hyphae, the contrast group B group also appeared white mycelium, but its surface color change was not obvious, due to the agent added, while at ninth days the contrast group A appeared white mycelium on the surface, it delayed 4 days compared to the treatment group. After about fifteenth days of composting the piles body of the treatment group C, D, E tended to be uniform and loose without caking phenomenon. At the end of composting, the color of the compost material of the contrast group A, B presented light brown, have caking phenomenon but not obviously, the surface color of the treatment group C, D, E shown black brown, and piles are loose, uniform and without caking. The effect of group E was obvious. The color of the decayed fertilizer products became dark brown, odorless, with obvious smell of humus (soil flavor), its structure is loose granular. The color of the under decayed fertilizer finished is brown, can smell the odor, with the phenomenon of caking, and the particles is larger. Thus, the contrast group A, B have not reached maturity and the treatment group C, D, E have reached maturity at the end of composting. Then fater lasted about 7 days of composting the time contrast group A, B reached maturity.

3.2. The changes and analysis of composting temperature.

The change of temperature is an important indicator of microbial activity during composting, which can reflect the degree of maturity of compost [11]. On the one hand, the temperature chang have an effect on the rate of biological drying directly. Temperatures are essential for composting, range of optimum temperature is 50°C~ 60°C during aerobic composting, which can result in the death of pathogens and weed seeds on this condition. The composting temperature below optimum temperature have an effect on microbial activity, which can reduce the rate of organic matter decomposition. The composting temperature above optimum temperature can inhibit aerobic microbial activity in the aerobic thermophilic composting and lead to composting piles carbonization [12]. The composting experiment was carried out in winter and the environment temperature was low, the change of temperature as shown in Fig.1 (a) and Fig. (b), group A and group B need 12 days and 11days up to the thermophilic phase respectively and the peak temperature above 55°C. On the one hand, group C, group D and group needs 6 days, 7 days and 7 days respectively. In the thermophilic phase, group a, group B, group C, group D, E lasted for 22 days, 22 days, 28 days, 28 days and 27 days respectively, and then entering cooling stage. The results indicated that the addition of CaO_2 can shorten the heating period in the compost, which enter the thermophilic phase 5~ 7 days in advance and prolong the time of thermophilic phase, make the pile composting more thoroughly. According to group B and group C, it is concluded that the pile before 6 days into the thermophilic phase due to the addition of CaO_2 instead of the effect of microbial inoculum. The mechanisms can explained that CaO_2 reacts with the water in the piles to produce oxygen with release of energy. The combined effect leads to the increase of the temperature of the composting piles and avoid the production of anaerobic environment. The temperature fluctuation is caused by the turning and ventilation.

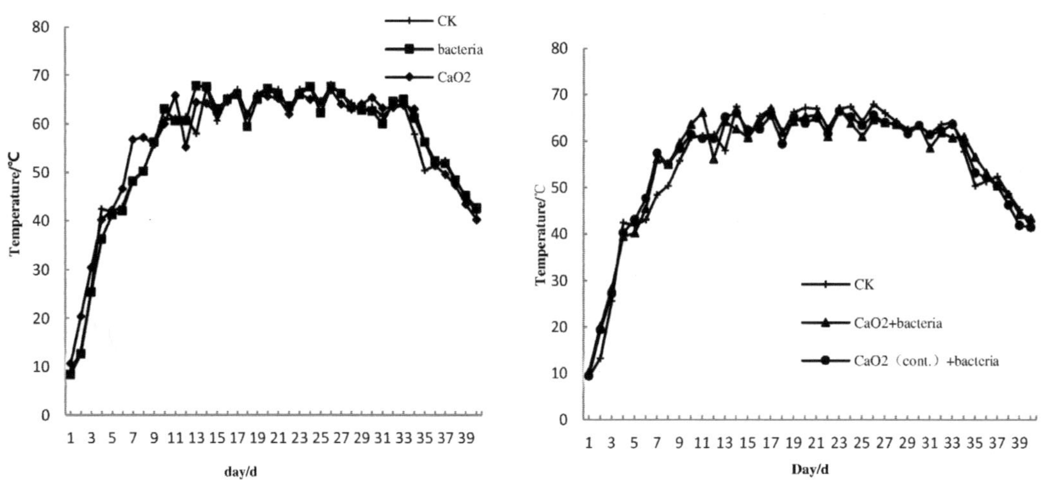

(a) (b)

Fig.1 Temperature changes of different treatments.

3.3. The changes and analysis of pH value during composting process

During composting, pH is an important parameter for aerobic microbial activity. Under normal circumstances, the most suitable microbial pH value is between 6 ~ 9, when bellows 4.5, it will seriously affect the activities of microorganisms, when bigger than 10.5, the most of bacterial activity decreased. PH value is too high or too low, it will lead to protein denaturation, and is harmful to the survival of microbes [13]. The change in pH during composting was result of the organic acids from microbial decomposition and nitrogen-containing organic compounds, ammonia and protein from piles worked together [14].

Fig.2 reflected that the trend of pH about each pile during whole composting process. The initial pH value of each treatment was 6.57, 6.52, 6.7, 6.46 and 6.61 respectively, which all satisfied the growth conditions of aerobic microorganisms. With the change of composting time, the pH began to rise, the treatment with addtion of carbon dioxide rised obviously, it may due to the reaction of CaO_2 with water to produce $Ca(OH)_2$ and alkaline substances. This alkaline environment provided by CaO_2 contributes to the biodegradation of refractory organic matter. With the composting reaction continues, the pH value of group C, D, E both decreased, CaO_2 addition went into the thermophilic phase in advance, which is a result of a large number of microbial metabolic accumulation. Since group E with the continuous addition of CaO_2, the pH value presented fluctuant change.

3.4. Changes and analysis of moisture content during composting process

Compost moisture content is an important parameter which affect composting, moisture content directly influences the activity of the cow dung aerobic microorganisms and the degree of maturity at the end of the composting process, too high or too low of moisture content have a negative impact on aerobic microbial decomposition and metabolic activities, which is not conducive to aerobic composting process [15, 16]. Generally, initial moisture content is controlled at 50% ~ 60% on initial composting [17].

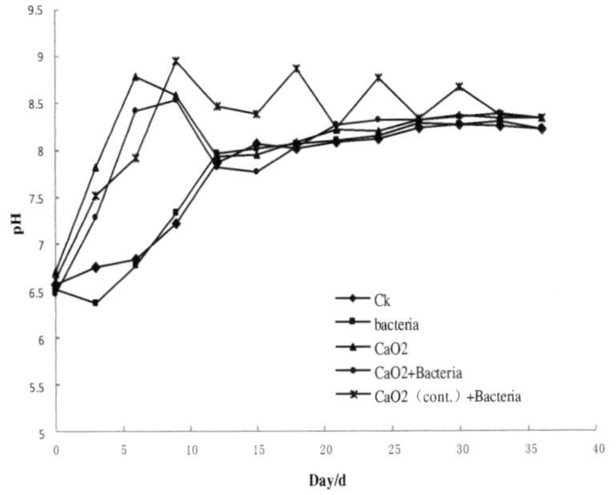

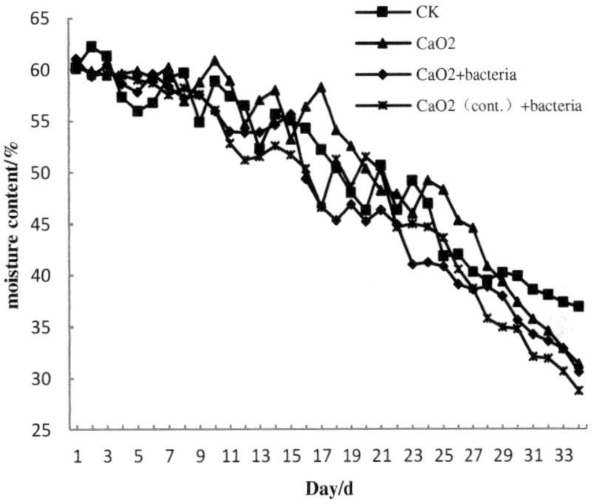

Fig.2 pH value changes of different treatments. **Fig.3** Moisture content changes of different treatments.

The change of moisture content with different treatments during composting process as showen in Fig.3, the initial moisture content of group A, C, D, E. is 60.2%, 60.1%, 61.05% and 60.2% respectively by processing. The moisture content of each treatments components is declining with the composting process. Moisture content of group E dramatically decreased, which declined to 28.76% at the end of the composting and group A, C, D fell to 36.87%, 36.87% and 30.35% respectively, although the control group did not reach the standard organic agriculture (≤30%) [10]. Moisture content of the groups have fallen by 23.33%, 29.75%, 30.49% and 23.33% respectively. It may due to CaO_2 with water by reaction, which released the oxygen and energy, furtherly promoted the moisture evaporation. Consequently, moisture content of three groups (C, D, E) are lower than the control group after the end of composting cycle. Fluctuation of descent curve is affected by the rainfall.

3.5. Changes and analysis of organic matter during composting

Figure 4 shows the variation curves of organic matter in the compost. Organic matter changes mainly due to aerobic microbial decomposition, it is transformed into the necessary nutrients for its own growth, On the one hand, organic matter was decomposed into water, organic acids, carbon dioxide, minerals, etc. then microorganisms transformed anti-product into humic substances[18]. At the end of composting, the moisture content of organic matter decreased to 60.92%, 61.16%, 61.11%, 60.28% and 56.87% respectively. From the present national standards in agricultural organic fertilizer, the organic matter content of each treatment accorded with the organic matter content standard (≥45%). The results show that there is an increase in the content of organic matter in each treatment group during composting. The reason may be that thermophilic aerobic microorganisms have a high degree of activity, which consumes water and organic matter in the middle stage of composting. Although the composting piles moisture and organic matter decreased gradually, but compare to the loss of piles volume and weight, the content of organic matter increased. When the volume and weight of composting piles keep relatively stable, the content of organic matter is decreasing and maintain to the end of composting. The organic matter value declined 74.99% to 56.87% in group E, reduce the amount of 18.12%. The amount of reduction more than other treatments, it may due to the continuous addition of carbon dioxide replenishes continuous oxygen for composting piles, improved the rate of decomposition and metabolism of aerobic microorganisms, which can decompose more organic matter.

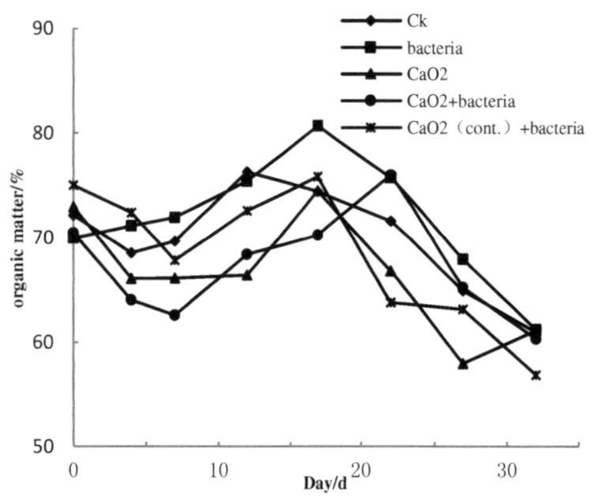

Fig.4 The organic matter changes during composting

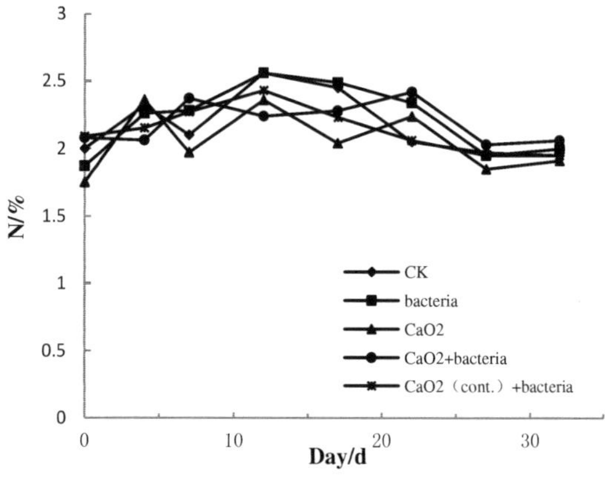

Fig.5 The total nitrogen changes during composting

3.6. Changes and Analysis of TN, TP, TK during composting

3.6.1. Changes of total nitrogen content. As shown in Fig. 5, the total nitrogen content of each treatment showed a relatively stable trend with the composting time, and the total nitrogen content did not decrease significantly due to nitrogen volatilization. Although the addition of CaO_2, the pH value increased accordingly, the oxygen content of the composting piles is supplemented, which promote the degradation of protein in the piles to produce ammonia and ammonia oxidation into nitrate furtherly; On the other hand, the reduction of the volume and weight of the piles lead to the increase of total nitrogen content during composting. The combination of these two aspects and nitrogen volatilization showed that the trend of total nitrogen content is relatively stable, there is no significant downtrends. The change of total nitrogen between the different treatments was consistent, and there was no obvious discrepancies. Therefore, CaO_2 have a weak nitrogen retention effect during composting, but the effect is not obvious, which needs further study.

3.6.2. Changes of total phosphorus content. The volume and weight of the composting piles decreased due to the organic matter decomposition by microorganisms during composting, however, generally the total phosphorus and total potassium concentration increases during composting due to the concentration effect [19]. As show in Fig.6, the content of total phosphorus presented slowly raise in the compost. At the end of composting, the final value of the total phosphorus in every treatment is higher than the initial value. Furthermore, it is also observed that the total phosphorus content of the group with addition of CaO_2 was higher than control group, it may due to the addition of CaO_2 prolonged the thermophilic period, increased the consumption of organic matter and promoted the "concentration effect", which resulting in a higher final value than the control group.

3.6.3. Changes of total potassium content. The change in total potassium for different treatments are shown in Fig.7. It can observed that the total potassium of each treatment are higher than initial value, which increased by 0.73, 0.54%, 0.89%, 1.25% and 1.23 % respectively. The increase of total potassium in treatment group B was smaller than that in control group, The increase of total potassium in treatment group D was much higher than that in control group, it may due to the addition of CaO_2, which supplemented the content of oxygen, prolonged thermophilic phase, be conducive to the evaporation of water and the consumption of organic matter, and promoted the "concentration effect". It is observed that microbial inoculum have no effect on the increase the total potsium compared the treament only with CaO_2, the fore, it can be concluded that the addition of CaO2 was related to the increase of total potassium content instead of microbial inoculum.

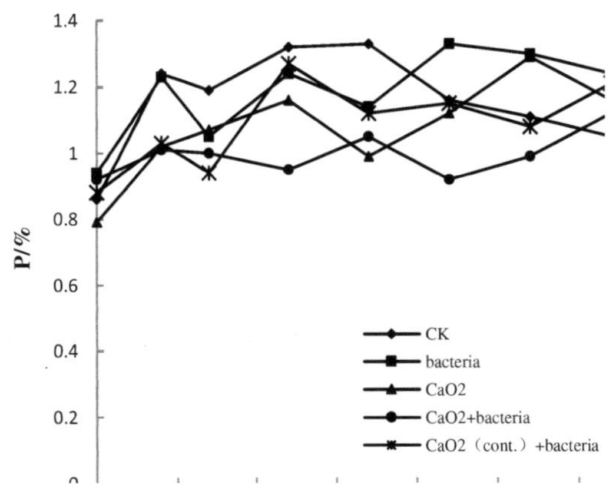

Fig.6 The total phosphorus changes during composting

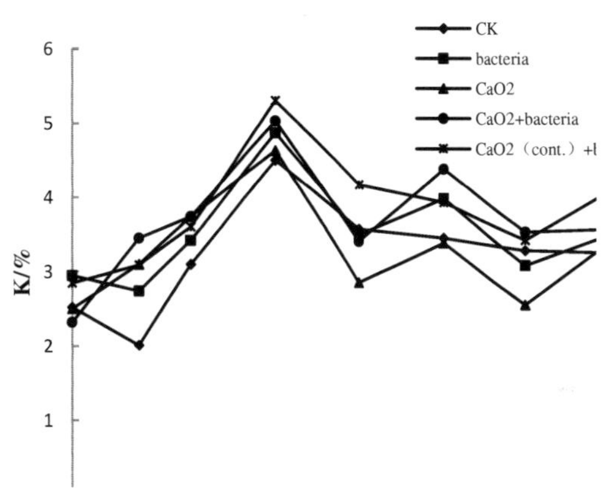

Fig.7 The total potassium changes during composting

3.6.4. Changes of total nutrient content. The total nutrient content affected of total potassium content, which reached the peak value on the 12th day of composting, and then the decline eventually stabilized. At the end of composting, the final values of total nutrient contents were higher than the initial values, increasing by 0.87%, 0.97%, 1.42%, 1.42% and 1.42% respectively. It can be observed that the total nutrient content higer than control group and CaO2 contributes to the increase in total nutrient content. The total nutrient content of E was 7.24%, which was much higher than the total nutrient content (≥5%) stipulated by the national organic fertilizer standard.

4. Conclusion

The addition of CaO_2 could promote the rapid bio-drying of cow manure and play an emphasis on the Whole composting process. This study determined common indices of compost maturity from cow manure. The results showed that the final pH value of compost dropped around 8, which within the optimum PH range. The moisture content decreased to below 30%, which decreased by 6% ~ 8% compared with the control group, which was conducive to aerobic microbial decomposition and metabolism. Total nitrogen, total phosphorus, total potassium and total nutrient content were increased compared to the initial stage of composting, which can enhance the fertilizer efficiency. However, The organic matter decreased but met organic fertilizer requirements. Through the research, it can be explained that the function of CaO_2 include there aspects: CaO_2 released oxygen slowly by chemical reaction, which continuously replenished the oxygen content in the piles, so that promoted the microbial activity and avoid the formation of anaerobic environment, it can make the composting process become more thoroughly, furthermore, continuous adding CaO_2 is best; CaO_2 released energy by chemical reaction, it accelerate into the thermophilic phase and shorten the composting cycle; the generation calcium hydroxide by chemical reaction with CaO_2 and water, which can cause the micro-alkaline environment and be beneficial to the biodegradation.

Acknowledgments
The authors are grateful for the National Major Projects of Science and Technology for Water Pollution Control and Management (No. 2014ZX07105-001).

References
[1] Burton, H., Turner, C., 2003. Manure management, second ed.. Treatment Strategies for Sustainable Agriculture Silsoe Research Institute, Lister and Durling Printers,Flitwick, Bedford, UK.
[2] Shi C, Zhang X, Jiamali S, et al. Dynamic changes of total N, P, K and organic matter in different treatments on cattle manure composting.[J]. China Cattle Science, 2010.
[3] Zucconi, F., de Bertoldi, M., 1987. Compost specifications for the production and characterization of compost from municipal solid waste. In: de Bertoldi, M.,Ferranti, M.P., L'Hermite, P., Zucconi, F. (Eds.), Compost: Production, Quality and Use. Elsevier, Barking, pp. 30-50.
[4] Jewell W J, N C Dondero P J. etal. High temperature stabilization and moisture removal from animal wastes for by-product recovery. final report prepared for the Cooperative State Research service, USDA, Washington, DC. Project number SEA/CR 616-15-168.1984.169.
[5] Choi H L, Richard T L, Ahn H K. Composting High Moisture Materials: Biodrying Poultry Manure in a Sequentially Fed Reactor[J]. Compost Science & Utilization, 2013, 9(4):303-311.
[6] Richard T L, Choi H L. Optimizing the composting process for moisture removal[J]. Korean Journal of Animal Science, 1997.
[7] Singh Y K, Kalamdhad A S, Ali M, et al. Maturation of primary stabilized compost from rotary drum composter. J]. Resources Conservation & Recycling, 2009, 53(7):386-392.
[8] Roca-Pérez L, Martínez C, Marcilla P, et al. Composting rice straw with sewage sludge and compost effects on the soil-plant system.[J]. Chemosphere, 2009, 75(6):781-787.
[9] Chang Z Z, Zhu W B, Xiao mei Y E, et al. A Biological Technique for Drying Animal Excreta[J]. AGRO-ENVIRONMENTAL PROTECTION, 2000.
[10] The Ministry of Agriculture of the People's Republic of China. NY 525-2012,Organic manure[S]. Beijing: China Agriculture Press,2012
[11] Brito L M,Coutinho J,Smith S R. Methods to improve the composting process of the solid fraction of dairycattle slurry[J]. Bioresource Technlogy, 2008,99,8955-8960
[12] Tiquia S M,Tarn N E Y,Hodiss I J,Microbial activities During composting of pig manure sawdust litter at different moisture contents [J].Biores Trchnol, 1996, 55(2):201-206
[13] Gareia C, Hernandez T, eta1.Study on water extract of biosolids composts[J].Soil Science Plant Nutrition, 1991, 37: 399-408
[14] ZHANG Shuo."Four modernizations"treatment of livestock waste[M].Beijing: China's agricultural science and technology press,2007
[15] LI Guoxue, ZHANG Fusuo. Solid waste composting and organic compound fertilizer production[M]. Beijing: Chemical Industry Press, 2000:190-191(in chinese)
[16] RYNK R.Monitoring moisure in composting system[J].Biocycle, 2000, 41(10): 53-58
[17] Bernal M P,A, Moral R.Composting of animal manures and chemical criteria for compost maturity:A review[J].Bioresource Technology,2009,100(22): 5444-5453
[18] Shi C, Zhang X, Jiamali S, et al. Dynamic changes of total N, P, K and organic matter in different treatments on cattle manure composting.[J]. China Cattle Science, 2010.
[19] NIU Junling, LIANG Lizhen,LAN YanPing. The substance changes duing the composting of cow dung and envelope splits of chestnut[J]. J.Agro-Environ.Sci. , 2009,28(4): 824-827(in Chinese)

IOP Conf. Series: Earth and Environmental Science **64** (2017) 012113 doi:10.1088/1755-1315/64/1/012113

Characterization of sulfate reducing bacteria isolated from urban soil

Mingliang Zhang*, Haixia Wang

School of resources and environment, University of Jinan, Jinan, China

*Corresponding author e-mail: mlzhangsd@126.com

Abstract. Sulfate reducing bacteria (SRB) was isolated from urban soil and applied for the remediation of heavy metals pollution from acid mine drainage. The morphology and physiological characteristics (e.g. pH and heavy metals tolerance) of SRB was investigated. The SRB was gram-negative bacteria, long rod with slight curve, cell size 0.5× (1.5-2.0) μm. The pH of medium had significant effect on SRB growth and the efficiency of sulfate reduction, and it showed that the suitable pH range was 5-9 and SRB could not survive at pH less than 4. The maximum tolerance of Fe (II), Zn (II), Cd (II), and Cu (II) under acidic condition (pH 5.0) was about 600 mg/L, 150 mg/L, 25 mg/L and 25 mg/L, respectively. The result indicated that SRB isolated in this study could be used for the bioremediation of acid mine drainage (pH>4) within the heavy metals concentrations tolerance.

1. Introduction

Acid mine drainage (AMD) can be formed by the oxidation of sulfide minerals when they are exposed to water and air. AMD is one of the most serious environmental issue to the mining area because of its strong acidity and high concentrations of heavy metals. Microbial remdiation by sulfate reducing bacteria (SRB) has been considered as one interesting biological process. SRB can reduce sulfate to hydrogen sulfide when supplied with reliable organic carbon, and metal sulfide precipitation can be formed during the remediation process. However, one reported problem associated with sulfate reducing process is the negative effect of metal toxicity which can inhibit SRB growth and affect AMD remediation. SRB growth will be completely inhibited if heavy metal ions concentrations reach a certain extent. In this study, one strain of SRB was isolated from urban soil and physiological characteristics of SRB were studied. The effect of pH and heavy metal to SRB was also investigated. The study will be helpful for SRB used for the remediation of heavy metals pollution from acid mine drainage.

2. Materials and Methods

2.1. Isolation of SRB

In this experiment, one strain of sulfate-reducing bacteria was isolated from Jinan urban soil. A modified Postgate growth medium was used in all experiments. The medium was composed of: KH_2PO_4 0.5 g/L; NH_4Cl 1.0 g/L; Na_2SO_4 4.5 g/L; $CaCl_2 \cdot 6H_2O$ 0.06 g/L; $MgSO_4 \cdot 7H_2O$ 0.06 g/L; sodium lactate 6 g/L; yeast extract 1 g/L; $FeSO_4 \cdot 7H_2O$ 0.5 g/L; sodium citrate$\cdot 6H_2O$ 0.3 g/L. The pH of

IOP Conf. Series: Earth and Environmental Science **64** (2017) 012113 doi:10.1088/1755-1315/64/1/012113

the medium was not adjusted at about 5.5. The experiment was conducted in anaerobic conditions at 30°C in a biochemical incubator for 7 days. The soil soaking solution was added into a stoppered flask filled with Postgate liquid culture, sealed until the culture medium became dark black, and lead acetate test paper turned black, and strong smell could be smelled. It indicated that SRB growth was good. Enrichment and isolation described above were repeated 3-5 times.

Spread plate method was used for isolation of SRB, and the plate was placed in the biochemical incubator for about 5-7 days for black colonies growth, and then pick well-growing single colony, inoculated in liquid medium for enrichment. The isolation described above was repeated 3-5 times. Finally the well-grown sulfate-reducing bacteria were isolated.

2.2. Physiological Characteristics of SRB

The morphology of the SRB strain was observed by scanning electron microscopy (SEM, FEI Quanta FEG250). Test of SRB growth curve: Postgate medium with 5% inoculation amount with the initial pH of 5.0 and 7.0, was placed in a biochemical incubator at 30°C for 7 days. The concentration of sulfate was collected and the growth of bacteria was measured periodically. Test of different carbon sources for SRB: sodium lactate, yeast extract, sodium acetate, ethanol or sodium citrate as a single carbon source was used as the potential carbon source. Considering acid mine drainage remediation during sulfate reducing process, the acidic environment and heavy metals in acid mine drainage have inhibitory effect on SRB activity, so the effect of pH and heavy metal ions on SRB growth will be analyzed.

2.3. Effect of pH on SRB growth

SRB is very sensitive to the pH of the growing environment, which directly affects SRB growth and the reduction efficiency of sulfate. In the Postgate medium with 5% bacterial solution was inoculated with the initial pH of 2, 4, 5, 6, 7, 8, 9, and 10 in the biochemical incubator at 30 °C for 4 days. The concentration of sulfate and SRB growth were measured periodically.

2.4. Effect of heavy metals on SRB activity

The effect of heavy metals on SRB activity was studied by the following batch test. Modified Postgate medium described above was used as control medium for SRB's tolerance toward different concentrations of heavy metal ions. The experiments were conducted in anaerobic conditions at 30°C in a biochemical incubator for 14 days, using 150 mL Erlenmeyer flask with stopper containing 100 mL of growth medium with 5 mL of SRB inoculum of log phase cells (pH around 5.5, not adjusted). The pH, Eh and sulfate concentration of solution were measured to study the tolerance of SRB toward heavy metals. The effects of pH and temperature on SRB activity was also studied by similar batch tests.

2.5. Analytical methods

The pH and Eh of samples were measured immediately after collection. Sulfate concentration was determined using the turbidimetric method. OD_{600} was measured by UV-VIS spectrophotometer (Shimadzu). The samples were filtered and acidified with one drop of nitric acid and atomic absorption spectroscopy (AAS) (AA-7000 model spectrometer, Shimadzu) was used for the measurement of the concentrations of Fe, Zn, Cu and Cd.

3. Results and Discussions

3.1. The morphology of the SRB

The soil soaking solution was added to the liquid medium and sealed for 7 days for SRB enrichment. The SRB were grown in large quantities (Fig.1). Gram staining indicated that the strain was gram-negative. It was found that the isolated SRB strain had a slightly curved rod shape with a size of about

0.5 × (1.5-2) μm by scanning electron microscopy. The identified SRB strain belongs to *Desulfovibrio*, and it can be used sodium lactate, ethanol, yeast extract and sodium citrate as the electron donor.

3.2. Growth curve

The growth curve of isolated SRB strain is shown in Fig.2. The strain growth was in lag phase in16h without H_2S production and lead acetate test paper did not turn black. Sulfate concentration did not significantly decrease. In 16-48 h the strain shifted from lag phase to the logarithmic growth phase, and the concentration of sulfate decreased sharply from 3372mg/L to 964mg/L (at pH 7.0), and a large amount of H_2S was generated. After 48-72h, the OD_{600} reached the maximum (0.66) and entered the stable period, and the Eh remained -380 ~ -400 mV. After 72 hours, OD_{600} showed a downward trend (0.46-0.53), indicating the strain entered the decay period. At pH 5.0, the lag phase of SRB strain was significantly increased to 24 h, and the reduction rate of sulfate were decreased compared with the condition at pH 7.0.

Figure 1. The isolated SRB in liquid medium

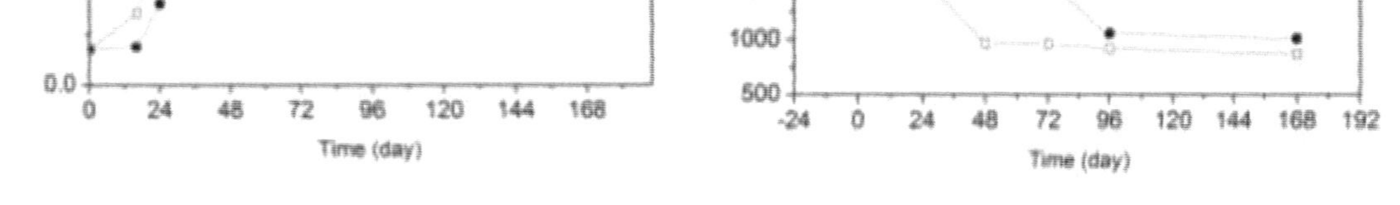

Figure 2. Growth curves of SRB

3.3. Effect of pH

Solution pH has an important effect on the growth of SRB. Although eosinophilic SRB strains in extreme acidic wastewater were isolated by some scholars, most of SRBs have a suitable pH range of 5-8. The effect of pH on the growth of SRB strains isolated from this experiment is shown in Fig.3.

The results showed that the optimum growth pH of the isolated SRB was 7, and the SRB could grow well at pH 5-9, and the concentration of sulfate decreased obviously. When the pH was less than 4 and more than 10, the growth of the strain was inhibited, and no H_2S was produced, and the sulfate concentration did not decrease obviously. It was found that although the isolated SRB was able to grow under the conditions of pH 5-9, lag phase was significantly different. The lag phase was only 16 hours at pH 6-8, and it was 24 hours pH 5 or 9. Therefore, the provision of appropriate pH is an important prerequisitesto improve the SRB growth activity and effective remediation of acid mine drainage.

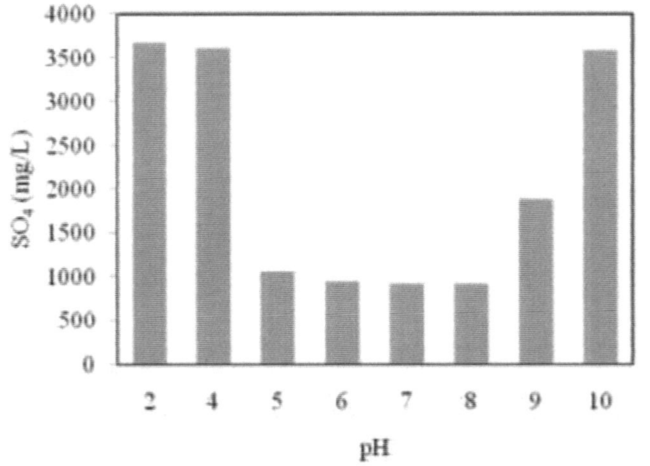

Figure 3. Effect of pH on sulphate reduction by SRB

3.4. Heavy metals tolerance

The effects of different concentrations of Fe (II), Zn (II), Cd (II) and Cu (II) on SRB growth are shown in Fig.4. It showed that the removal of sulfate was generally decreasing with the increase of heavy metal ion concentration. The maximum tolerable concentrations of SRB to Fe (II), Zn (II), Cd (II) and Cu (II) were 600 mg/L, 150 mg/L, 25 mg/L and 25 mg/L. In tolerable concentrations, SRB growth was good which was evidenced that sulfate concentration decreased significantly, the solution pH increased to 6 and the Eh decreased to below -100 mV.

When the concentration of Fe(II) was 800 mg/L, the solution Eh was postive (91 mV) and pH was 4.98 (slightly lower than the initial pH 5.0), indicating that the SRB strain was completely inhibited and could not tolerate Fe (II) at 800mg/L. It was found that the lag phase of SRB growth was prolonged with the increase of heavy metal concentration. The lag phase of SRB growth was 1 day at Fe 100 mg/L, lag phase was 4 days with Fe 600 mg/L, and SRB will stop growing at Fe 800 mg/L. The lag phase of SRB growth was 3 days at Zn 25 and 50 mg/L, and SRB displayed a prolonged lag phase of 4 days at Zn 100 mg/L. SRB will stop growth at Zn 150 mg/L. It indicated that heavy metal ions had a significant impact on the SRB growth.

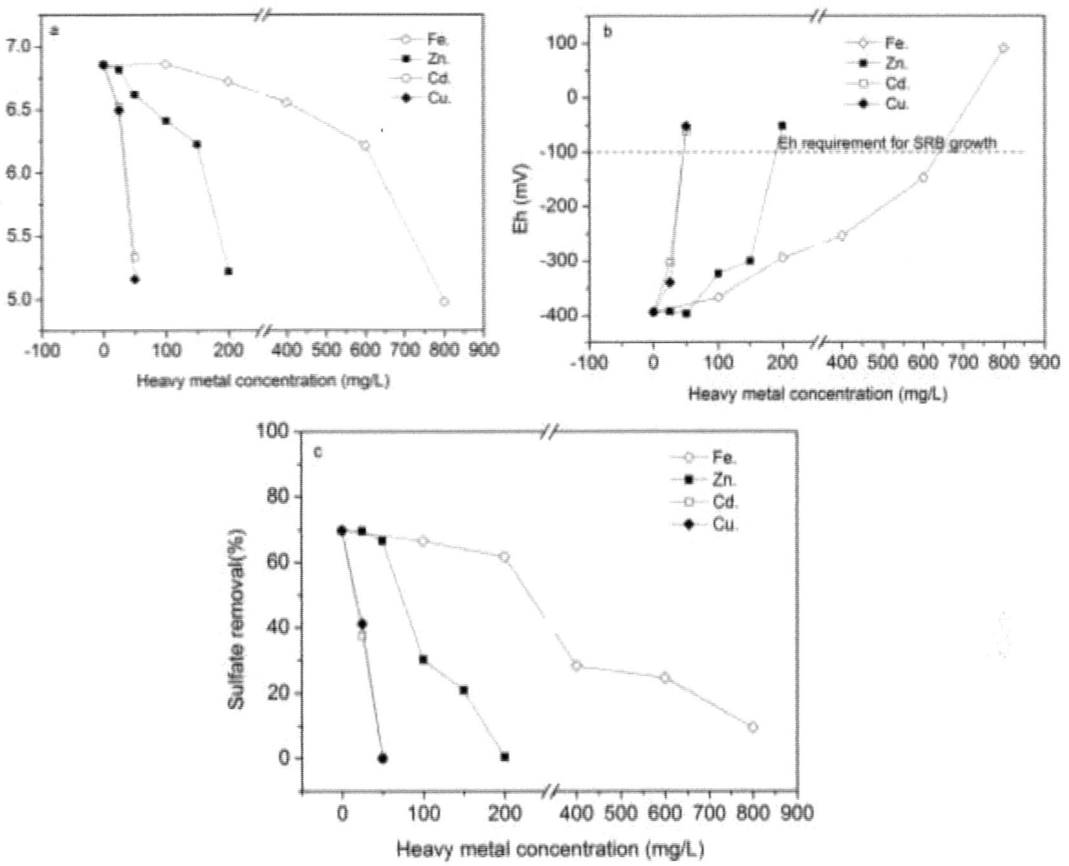

Fig. 4 The tolerance of heavy metals, (a) pH variation, (b) Eh variation, (c)Sulphate reduction varitation

4. Conclusion

The SRB isolated from soil could not survive at pH less than 4, and the suitable pH range was 5-9. The maximum tolerance of Fe (II), Zn (II), Cd (II), and Cu (II) under acidic condition (pH 5.0) was about 600 mg/L, 150 mg/L, 25 mg/L and 25 mg/L, respectively. The result indicated that SRB isolated in this study may be used for the bioremediation of acid mine drainage (pH >4) within the heavy metals concentrations tolerance.

Acknowledgments

This work was financially supported by Shandong Provincial Key Research and Development Program (2015GSF117014).

References

[1] R.P.Choudhary, A.S.Sheoran, Performance of single substrate in sulphate reducing bioreactor for the treatment of acid mine drainage, Miner. Eng. 39(2012) 29-35.

[2] C.D.McCullough, M.A.Lund,. Bioremediation of Acidic and Metalliferous Drainage (AMD) through organic carbon amendment by municipal sewage and green waste. J. Environ. Manage. 92(2011) 2419-2426.

[3] T.Hao, P.Xiang, H.R.Mackey, K.Chi, H.Lu, H.Chui, M.C.M.van Loosdrecht, G.H.Chen, A review of biological sulfate conversions in wastewater treatment, Water Res. 65(2014), 1-21.

[4] D.B.Johnson, K.B.Hallberg, Acid mine drainage remediation options: a review, Sci. Total

Environ. 338(2005), 3-14.
[5] A.S.Sheoran, V.Sheoran, R.P.Choudhary, Bioremediation of acid-rock drainage by sulphate-reducing prokaryotes: a review, Miner. Eng. 23(2010)1073-1100.
[6] M.C.Costa, E.S.Santos, R.J.Barros, C.Pires, M.Martins, Wine wastes as carbon source for biological treatment of acid mine drainage. Chemosphere 75(2009), 831-836.

IOP Conf. Series: Earth and Environmental Science **64** (2017) 012114 doi:10.1088/1755-1315/64/1/012114

Impacts of ammonia on zinc oxide nanoparticle toxicity to *Nitrosomonas europaea*

Junkang Wu[1, 2], Yan Chang[1, 2], Huan Gao[1, 2] and Ran Yu[1, 2,*]

[1]School of Energy and Environment, Southeast University, Nanjing, China
[2]Key Laboratory of Environmental Medicine Engineering (Southeast University), Ministry of Education, Nanjing, China

*Corresponding author e-mail: yuran@seu.edu.cn

Abstract. A Although the toxicity effects of engineering nanoparticles (NPs) in biological wastewater nitrogen removal (BNR) system have been extensively attracted, the impacts of co-existing contaminants from wastewater on NP toxicity have been less addressed. In this study, the effects of ammonia on ZnO NP toxicity to typical ammonia oxidizing bacteria-*Nitrosomonas europaea* were investigated, as indicated by the cell density, membrane integrity, ammonia oxidation rate, and AMO activity. After 6-h's exposure to 10 mg/L ZnO NPs, the cell density, membrane integrity, ammonia oxidation rate, and AMO activity was dramatically suppressed despite of the increasing ammonia loading. Ammonia at varying concentrations did not obviously affect ZnO NPs toxicity to cell density. The presence of ammonia at 100 or 200 mg/L significantly alleviated the antibacterial effects of ZnO NPs on cells. The reduction of the concentration of released Zn^{2+} might be responsible for the compromised ZnO NPs toxicity. However, the presence of extremely dosed ammonia at 200 mg/L imposed restrictions on further alleviation of ZnO NPs toxicities probably due to the production of free ammonia and acclamation of nitrite. All these findings would provide new insights for risk assessment of the combined effects of NPs with other co-existing contaminants in the BNR system.

1. Introduction

With the rapid development of nanotechnology, diverse nanomaterials featuring at least one dimension with less than 100 nm in size have been increasingly employed in various fields due to their unique physicochemical properties, such as surface effect, interface effect, quantum size effect, and antibacterial performance [1, 2]. Among them, Zinc oxide nanoparticles (ZnO NPs) with some special properties (e.g. catalytic activity, electrical conductivity, dissolved ability and antibacterial effects) have been increasingly utilized in chemical industry and commercial products [3]. As listed as one of the most used engineering NPs in the world, the number of ZnO NPs products have increased to more than 38 according to Consumer Products Inventory (CPI) [4]. The introduction of massive ZnO NPs have raised remarkable concerns on their ecotoxicity and biosafety in the surroundings [5, 6].

Wastewater treatment plants (WWTPs) have been considered as an essential intermediary for NPs transport into ecosystems [7]. By modeling prediction, 1.7 ~ 2.1 μg/L ZnO NPs passed through WWTPs, and 17 ~110 mg/Kg ZnO NPs were retained in the activated sludge [8]. The presence of these emerging contaminants in the biological treatment process would pose a great threat to metabolic activities of the critical functional bacteria. It has been extensively reported that ZnO NPs exerted negative effects on the ammonia removal and total nitrogen removal efficiency, the related ammonia oxidizing or nitrifying activity, and the abundance of related ammonia oxidizing bacteria (AOB) or nitrifying bacteria in biological nitrogen removal (BNR) system [9, 10, 11]. In particular, ammoxidation, which is acknowledged as the primary and the rate limiting step for nitrification, are highly sensitive to ZnO NPs stress [10, 12].

Currently, the toxicity effects of NPs on ammoxidation have been widely addressed, which is considered to be induced by physical interruption, reactive oxygen species (ROS) generation and dissolved zinc release [13, 14]. However, most of these studies excluded the combined effects of diverse existing contaminants in wastewater. Ammonia, as one of the main contaminants in wastewater, could theoretically interact with released zinc ions, affect the physicochemical properties of ZnO NPs, and therefore their antibacterial effects. Kostigen et al found that the presence of ammonia affected the silver NPs dissolution and toxicity to *Nitrosomonas europaea* [15]. Furthermore, the concentration of ammonia in the influent of WWTPs varies from circumstance to circumstance. During the ammoxidation, the variation of concentration of oxidation product-nitrite will in turn affect the metabolic activity of AOB [16, 17].

The object of this study is to investigate the effects of ammonia on the physiological responses of the typical AOB-*N. europaea* to ZnO NPs and the potential mechanisms. *N. europaea* is a model chemolithotrophic ammonia oxidizer, widely present in biological nitrogen removal process and highly sensitive to toxicants stress. In this study, the cell density, ammonia oxidizing rate, cell membrane integrity and ammonia oxidase (AMO) activity was determined to indicate the bacterial physiological signal variations.

2. Materials and methods

2.1. Bacterial cultivation

N. europaea (ATCC 19718) was constantly cultured in a chemostat bioreactor (working volume = 3 L, hydraulic retention time = 2.3 d) in the dark at 28 ℃. The dissolved oxygen (DO) concentration was kept at (0.50 ± 0.02) mg/L by constant aeration. Compressed air was pumped into the culture after filtered through a 0.22 μm filter (Merck Millipore, Billerica, MA, USA). The feeding culture medium containing 280 mg-N/L was designed according to our previous publications [9]. The pH in the bioreactor was maintained (7.50 ± 0.10) by a pH controller (Meller, Taiwang, China). The buffer of 80 g/L $NaHCO_3$ was automatically added into the culture while pH decreased.

2.2. Nanoparticle characterization

ZnO NPs were bought from Sigma-Aldrich (St. Louis, MO, USA). The size distribution and morphology of ZnO NPs were characterized using a scanning electron microscope (SEM, Japan Electronics Co., Ltd, Japan). The average size of ZnO NPs was (96 ± 25) nm in length. The particle size distribution (PSD) and ζ potential were determined by dynamic light scattering (DLS) using a Malvern nano ZS90 zetasizer. The ZnO NPs suspensions at 10 g/L was ultrasonicated for dispersion (17 W, 40 kHz, 20 min) before exposure experiment.

2.3. Nanoparticle exposure experiment setup

300 mL of cultures were withdrew from the bioreactor into 500 mL flask for bath assays. A series of flasks containing 300 mL cultures were added with 15 g/L NH_4^+-N to reach the final concentrations at 25, 50, 100, and 200 mg/L, respectively, and acted as controls. Another two series of 4 flasks containing both 300 mL cultures and 25, 50, 100, or 200 mg/L NH_4^+-N was loaded 10 mg/L ZnO NPs and 3.4 mg/L Zn^{2+}, respectively. All the three series were incubated in the dark at 25 ℃ for 6-h. The designed concentration of added Zn^{2+} at 3.4 mg/L was obtained according to our previous study [18]. After 6-h exposure, cultures were sampled to determine the physiological signals of cells.

2.4. Analytical methods

The concentrations of NH_3-N and NO_2^--N of all samples were determined referring to the Standard Methods [19]. The cell density was measured with bacterial count methods under a OLYMPUS-BX41 microscope (OLYMPUS-BX41, Olympus, Japan). The cell membrane integrity was obtained using a LIVE/DEAD® BacLightTW Bacterial Viability Kit (Life Technologies, Waltham, MA, USA) according to the manufacturer's instructions. The specific activity of AMO enzyme in charge of ammonia oxidation, was examined according to the procedure described previously, and indicated by the nitrite production rate per unit of protein. The dissolved Zn (DZn) concentrations were measured by a AAnalyst 400 Atomic Absorption Spectrophotometer (PerkinElmer, Norwalk, CT, USA).

2.5. Statistics analysis

All the results were expressed as mean ± standard deviation. A Student's t-Test method was applied for statistical analysis by Microsoft Excel software (Microsoft Corp., Redmond, WA, USA). The tests were considered to be significant when P-value ≤ 0.05.

3. Results and discussions

3.1. Impacts of ammonia on zinc ion release

In our previous study, the concentration of released Zn^{2+} from 10 mg/L ZnO was determined to be (3.4 ± 0.9) mg/L [18], therefore, we set up a series of cell culture exposed to 3.4 mg/L Zn^{2+} to determine the toxicity effects exerted by ZnO NPs dissolution. As presented in Figure 1, the concentration of Zn^{2+} displayed no obvious variations between these two groups. In addition, the presence of ammonia at 25 and 50 mg/L did not significantly affect the concentration of Zn^{2+} released from ZnO NPs, while ammonia dosed at 100 and 200 m/L remarkably lower the Zn^{2+} concentration by (36.7 ± 6.0) % and (56.3 ± 1.9) % after exposure to 10 mg/L ZnO NPs when compared with the control (Figure 1). NH_3 is considered to be able to react with Zn^{2+} to form (Zn $(NH_3)_4)^{2+}$. This reaction would probably lower the Zn^{2+} concentrations and potentially affect ZnO NPs toxicity.

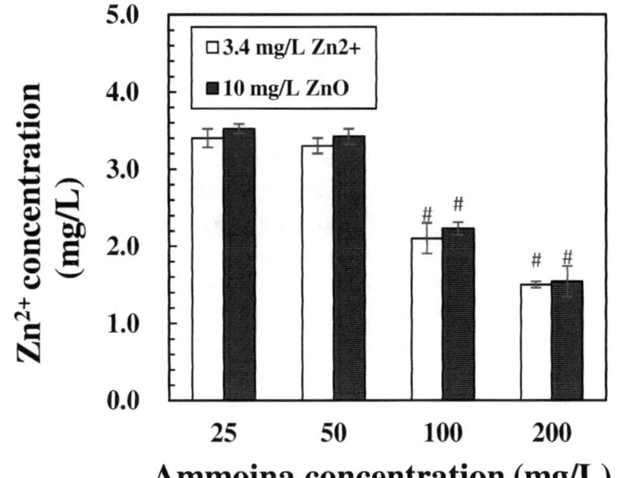

Figure 1. Impacts of ammonia on zinc ion release. # indicates the statistically significant difference for the samples in the presence of ammonia at different concentrations when compared with those loaded by 25 mg/L ammonia.

3.2. Impacts of ammonia on NP toxicity to cell density and membrane integrity

It can be seen from Figure 2A that cell density significantly decreased under both Zn^{2+} and ZnO NPs stress despite of the increasing ammonia loading. The ammonia did not obviously impact ZnO NPs inhibition on cell density. However, 10 mg/L ZnO NPs exerted more negative effects on cell density than 3.4 mg/L Zn^{2+}, which confirmed our previous findings that the size-effect of NP it-self could still be mainly responsible for Zn NPs toxicity although the released Zn^{2+} contributed part of the NP toxicity effects [18].

The interruption of membrane structure and function have been considered as main toxicity mechanisms of NPs [9]. In this study, the cell membrane integrity was inhibited by (7.8 ± 0.1) %, (7.4 ± 0.1) %, (5.4 ± 0.1) %, and (5.2 ± 0.1) % respectively after 6-h's exposure to ZnO NPs with ammonia concentration increasing from 25 to 200 mg/L (Figure 2B), which suggested the cell membrane damage caused by ZnO NPs or generated ROS [14]. However, the damage of membrane integrity was significantly alleviated when ammonia was dosed up to 100 or 200 mg/L. This might be explained by the reduced Zn^{2+} concentration [20]. It is noteworthy that there is no significant differences on the membrane integrity in the presence of ammonia at 100 and 200 mg/L, which implied that the presence of extremely high concentration of ammonia did not always decrease the NP toxicity.

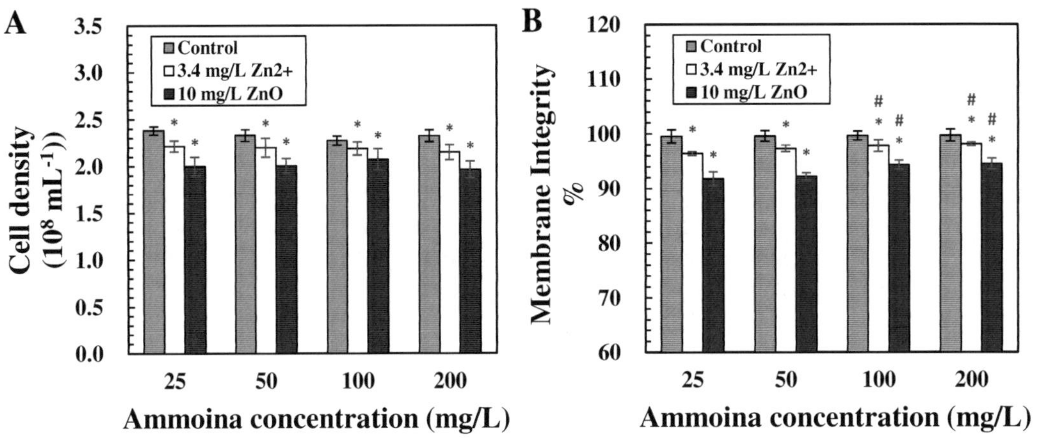

Figure 2. Impacts of ammonia loading on ZnO NPs toxicity to cell density (A) and membrane integrity (B). * represents the statistically significant difference for samples when compared with control. # indicates the statistically significant difference for the samples in the presence of ammonia at different concentrations when compared with those loaded by 25 mg/L ammonia.

3.3. Impacts of ammonia on NP toxicity to ammonia oxidation rate

N. europaea derives energy only from ammonia oxidation to assimilate inorganic carbon for substance synthesis [21]. Therefore, the concentration of ammonia plays a role in ammonia oxidation rate. In is study, the ammonia oxidation rate of *N. europaea* dramatically decreased by (42.9 ± 0.4) %, (35.0 ± 0.3) %, (13.4 ± 0.2) %, and (21.2 ± 0.2) % respectively under ZnO NPs stress when compared with the control (Figure 3A). As ammonia loading concertation rose from 25 to 100 mg/L, the inhibition of ZnO NPs on ammonia oxidation rate was increasingly compromised (Figure 3A). Since the cell was withdrew from the chemostat with high and stable activity, the "starving" cells in the presence of high dose of ammonia would accelerate the ammonia oxidation [22], which resulted in the obvious lightened NP suppression. However, when ammonia dosed at 200 mg/L, the inhibition of ammonia oxidation rate increased to (21.2 ± 0.2) % although the ammonia oxidation rate was enhanced without adding NPs or Zn^{2+}. This might be related to more production of $(NH_4)^+$, as well as the accumulation of nitrite. It has been found that the presence of higher concentration of free ammonia and nitrite would suppress the ammonia oxidation and metabolic activity of AOB [10, 12].

3.4. Impacts of ammonia on NP toxicity to AMO activity

As a membrane protein in charge of ammonia oxidation [23], AMO activity was seriously inhibited by (56.5 ± 2.1) %, (51.6 ± 1.2) %, (47.0 ± 0.6) %, and (50.7 ± 1.0) %, respectively compared to the control with the concentration of ammonia arranging from 25 to 200 mg/L (Figure 3B). Ammonia loaded at 25 or 50 mg/L did not induce significant variations of cellular AMO activity, while the addition of ammonia at 100 or 200 mg/L remarkably increased the AMO activities. In accordance with the impacts of ammonia on NP toxicity to ammonia oxidation rate, the inhibition of NP on AMO activity was lightened as ammonia concentration increased from 25 to 100 mg/L. Nevertheless, the addition of ammonia at 200 mg/L aggravated the NP toxicity to the AMO activity, which further demonstrated that the extremely high concentration of ammonia would make AOB more sensitive to ZnO NPs stress.

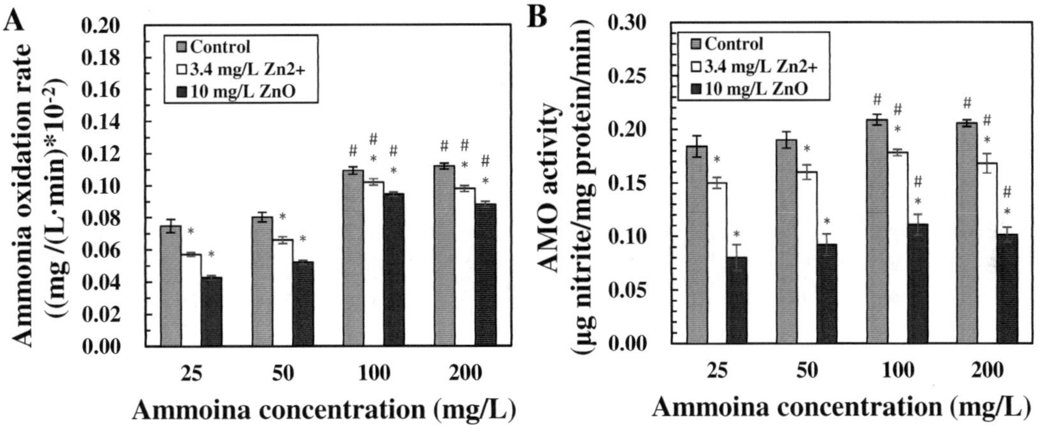

Figure 3. Impacts of ammonia loading on ZnO NPs toxicity to ammonia oxidation rate (A) and AMO activity (B).s * represents the statistically significant difference for samples when compared with control. # indicates the statistically significant difference for the samples in the presence of ammonia at different concentrations when compared with those loaded by 25 mg/L ammonia.

4. Conclusion

In this study, the impacts of ammonia on the NP toxicity to typical AOB *N. europaea* were investigated, as indicated by the cell density, the membrane integrity, the ammonia oxidation rate, and the AMO activity. After 6-h's exposure to 10 mg/L ZnO NPs, the cell density, the membrane integrity, the ammonia oxidation rate, and the AMO activity was significantly inhibited despite of the increasing ammonia loading. Ammonia at varying concentrations did not obviously influence ZnO NPs toxicity to cell density. The presence of ammonia at 100 or 200 mg/L significantly alleviated the antibacterial effects of ZnO NPs on cells. The reduction of the concentration of released Zn^{2+} might contribute to the compromised ZnO NPs toxicity. However, the presence of extremely dosed ammonia at 200 mg/L restricted further alleviation of ZnO NPs toxicities probably owing to the production of free ammonia and acclamation of nitrite. Our findings would provide deep understandings of the influences of wastewater constituents on the NP toxicity effects in BNR system.

Acknowledgments
This work was financially supported by National Natural Science Foundation of China (No. 51678134 and No. 51208092), the Fundamental Research Funds for the Central Universities, Innovative Graduate Student Project of Jiangsu Province (KYLX16_0282), and Scientific Research Foundation of Graduate School of Southeast University.

References

[1] F. Piccinno, F. Gottschalk, S. Seeger, B. Nowack, Industrial production quantities and uses of ten engineered nanomaterials in Europe and the world, J. nanopart. Res. 14 (2012) 01-11.

[2] A. Kumar, X. Zhang, X.-J. Liang, Gold nanoparticles: Emerging paradigm for targeted drug delivery system, Biotechnol. Adv. 31 (2013) 593-606.

[3] S. Ji, C. Ye, Synthesis, growth mechanism, and applications of zinc oxide nanomaterials, J. Mater. Sci. Technol. 24 (2008) 457-472.

[4] M.E. Vance, T. Kuiken, E.P. Vejerano, S.P. Mcginnis, H.M. Jr, D. Rejeski, et al., Nanotechnology in the real world: Redeveloping the nanomaterial consumer products inventory, Beilstein J. Nanotechnol. 6 (2015) 1769-1780.

[5] A. Nel, T. Xia, L. Mädler, N. Li, Toxic potential of materials at the nanolevel, Science, 311 (2006) 622-627.

[6] G. E. Schaumann, A. Philippe, M. Bundschuh, G. Metreveli, S. Klitzke, D. Rakcheev, et al., Understanding the fate and biological effects of Ag- and TiO2-nanoparticles in the environment: The quest for advanced analytics and interdisciplinary concepts, Sci. Total Environ. 535 (2015) 03-19.

[7] D. Wang, Y. Chen, Critical review of the influences of nanoparticles on biological wastewater treatment and sludge digestion. Crit. Rev. Biotechnol. 36 (2016) 816-828.

[8] T.Y. Sun, F. Gottschalk, K. Hungerbuhler, B. Nowack, Comprehensive probabilistic modelling of environmental emissions of engineered nanomaterials, Environ. Pollut. 185 (2014) 69-76.

[9] R. Yu, J. Wu, M. Liu, G. Zhu, L. Chen, Y. Chang, H. Lu, Toxicity of binary mixtures of metal oxide nanoparticles to Nitrosomonas europaea, Chemosphere 153 (2016) 187-197.

[10] S. Wang, M. Gao, Z. She, D. Zheng, C. Jin, L. Guo, et al., Long-term effects of ZnO nanoparticles on nitrogen and phosphorus removal, microbial activity and microbial community of a sequencing batch reactor, Bioresour. Technol. 216 (2016) 428-436.

[11] S.T. Wang, S.P. Li, W.Q. Wang, H. You, The impact of zinc oxide nanoparticles on nitrification and the bacterial community in activated sludge in an SBR, RSC Adv. 5 (2015) 67335-67342.

[12] G. Liu, D. Wang, J. Wang, C. Mendoza, Effect of ZnO particles on activated sludge: Role of particle dissolution, Sci. Total Environ. 409 (2011) 2852-2857.

[13] N.Q. Puay, G. Qiu, Y.P. Ting, Effect of ZnO nanoparticles on biological wastewater treatment in a sequencing batch reactor (SBR), J. Clean. Prod. 88 (2015) 139-145.

[14] Zheng, X.O., R. Wu, and Y.G. Chen, Effects of ZnO nanoparticles on wastewater biological nitrogen and phosphorus removal, Environ. Sci. Technol. 45 (2011) 2826-2832.

[15] Kostigen Mumper, C., et al., Influence of ammonia on silver nanoparticle dissolution and toxicity to Nitrosomonas europaea, Chemosphere 93 (2013) 2493-2498.

[16] L. Zhang, S. Zhang, Y. Peng, Review of study on the effects of free ammonia inhibitions on nitrifying bacteria in wastewater treatment, J. Harbin Inst. Technol. 44 (2012) 75-79.

[17] C. Wei, G. Luo, The influence of free ammonium on stable short-cut nitritying system, Chongqing Environ. Sci., 25 (2003) 50-52.

[18] M. Liu, R. Yu, L. Chen, J. Wu, Biological stress effects of typical metal oxide nanoparticles on Nitrosomonas europaea, China Environ. Sci. 35 (2015) 190-195.

[19] SEPAC, Methods for monitor and analysis of water and wastewater, 4th ed., China Environment Science Press (in Chinese), Beijing, 2002.

[20] M.H. Li, S. Pokhrel, X. Jin, L. Madler, R. Damoiseaux, E.M.V. Hoek, Stability, bioavailability, and bacterial toxicity of ZnO and iron-doped ZnO nanoparticles in aquatic media, Environ. Sci. Technol. 45 (2011) 755-761.

[21] M. Whittaker, D. Bergmann, D. Arciero, A.B. Hooper, Electron transfer during the oxidation of ammonia by the chemolithotrophic bacterium Nitrosomonas europaea, Biochim. Biophys. Acta, 1459 (2000) 346-355.

[22] J.A. Torà, J. Lafuente, J.A. Baeza, J. Carrera, Combined effect of inorganic carbon limitation

and inhibition by free ammonia and free nitrous acid on ammonia oxidizing bacteria, Bioresour. Technol. 101 (2010) 6051-6058.

[23] A.B. Hooper, T. Vannelli, D.J. Bergmann, D.M. Arciero, Enzymology of the oxidation of ammonia to nitrite by bacteria, Antonie Van Leeuwenhoek 71 (1997) 59–67.

IOP Conf. Series: Earth and Environmental Science **64** (2017) 012115 doi:10.1088/1755-1315/64/1/012115

Effect of three kinds of surfactants and β-cyclodextrin on the phytoremediation of BDE-209 contaminated sediment

Liangyuan Zhao*, Weijie Guo, Weihua Zhao, Meng Long, Huan Li

Basin Water Environmental Research Department, Yangtze River Scientific Research Institute, Wuhan, China

*Corresponding author e-mail: zhaoliangyuannew@163.com

Abstract. Plant-accelerated removal of BDE-209 from sediment by aquatic macrophyte *Scirpus validus* Vahl in the presence of a cationic-surfactant (CTAB), an anionic-surfactant (SDS), a nonionic-surfactant (Tween 80) and β-cyclodextrin (β-CD) at the concentrations ranged from 300 mg/kg to 1000 mg/kg were investigated. Significantly negative effect were not observed for the growth of *S. validus* in terms of plant height and stem diatemeter, which indicated that it is preferable for CTAB, SDS, Tween 80 and β-cyclodextrin to be utilized as the BDE-209 phytoremediation amendment. Furthermore, CTAB, SDS and Tween 80 in the certain concentrations significantly enhanced the phytoremediation efficiencies and 11.78-19.33% of increase in BDE-209 removal rates was obtained. Significantly enhance of BDE-209 phytoremediation efficiency was not observed in the added β-CD concentration ranges. Results obtained from this study provided some insight with rcgard to the feasibility of phytoremediation for BDE-209 contaminated sediments with addition of suitable solubilizers, especially Tween 80.

1. Introduction

Soil or sediment contamination with organic pollutants(OPs) has increasingly become a serious global environmental issue in recent years[1]. A large number of studies have shown that the slow release of OPs such as polycyclic aromatic hydrocarbons(PAHs), polychlorinated biphenyl s(PCBs) and polybrominated diphenyl ethers (PBDEs) in the aqueous phase resulted in their limited bioavailability, thereby reducing the degradation rate of these pollutants in the soil[2]. The use of solubilizers capable of increasing the solubility of OPs for contaminated soil or sediment is a very promising remediation technique that increases the apparent solubility of OPs in soil or sedimentary facies[3].

The technology of solubilizer enhanced phytoremediation is widely used in phytoremediation of OPs contaminated soil or sediment, and the two kinds of solubilizers, surfactants and cyclodextrins, are widely used in the OPs remediation application [4]. The main principle of this technique is that surfactant can desorb the OPs adsorbed on the soil, solubilize it in the soil solution, improve the plant and rhizosphere microbial availability of OPs and promote the OPs degradation through the rhizosphere microorganisms, thereby improving the bioremediation efficiency[5]. Cyclodextrins are cyclic, no reducing malt oligosaccharides produced from the enzymatic degradation of starch and related compounds by certain bacteria that contain the cyclodextrin glycosyltransferases [6].

IOP Conf. Series: Earth and Environmental Science **64** (2017) 012115 doi:10.1088/1755-1315/64/1/012115

Cyclodextrins have been used in environmental applications to improve the remediation of contaminated soil and groundwater[7].

Decabromodiphenyl ether (BDE-209) is a member of polybrominated diphenyl ethers (PBDEs), which are becoming increasingly prevalent in the environment[8]. The ubiquitous distribution of PBDEs in the environment as well as their potential risk to animals and humans has become an increasing concern [9]. Therefore, it is necessary to remove this pollutant from the environment.

In the present research, the potential and effectiveness of solubilizers-enhanced phytoremediation of BDE-209 contaminated sediment in the presence of three typical kinds of surfactants, cationic-surfactant (cetyltrimethyl ammonium bromide, CTAB), anionic-surfactant (sodium dodecyl sulfate,SDS), nonionic-surfactant (Tween80) and β-cyclodextrin(β-CD), was investigated. Results obtained from this study are expected to provide some insight with regard to the feasibility of phytoremediation for PBDEs contaminated sediment with solubilizers and to provide a theoretical basis for the development of practical techniques for efficient phytoremediation of PBDEs contaminated sediment.

2. Materials and methods

2.1. Chemicals

Decabromodiphenyl ether (BDE-209, $C_{12}OBr_{10}$) was purchased from Alfa Aesar (Johnson Matthey, USA) with 99% chemical purity (GC). Dichloromethane and n-hexane of HPLC grade were used in the extraction and cleanup. CTAB, SDS, Tween80 and β-cyclodextrin used were of analytical grade.

2.2. Plant materials

Scirpus validus Vahl (*S. validus*) used in the experiment were collected from the Tangxun Lake in Wuhan, Hubei Province, China (30°29′04.61″ N, 114°21′30.33″ E). Plants with initial height of about 60 cm were cultured in the artificial pond for 10 days before experiments.

2.3. Sediments

Sediments without detectable BDE-209 were collected from the 0-20 cm surface sediments at a local lake in Wuhan, Hubei province, China. Physical and chemical properties of sediment were: organic matter, 10.23%; pH, 5.88; total nitrogen, 0.48mg/g; total phosphorus, 0.27 mg/g.

2.4. Phytoremediation treatments setup

15 kg of contaminated dry sediments (BDE-209 at approximately 2 mg kg^{-1}) were loosely packed into cultivation box and then equilibrated for 7 days before the addition of water. The top of each pot was covered to minimize the evaporation and photolysis of BDE-209 in sediments. Each treatment had triplicates and *S. validus* with initial plant height of about 60 cm were transplanted into each box. Tap water was then added and kept at the level 4 cm above the sediment surface. The exposure experiment was carried out in natural conditions for 180 days. Five treatments were established and displayed in Fig.1.

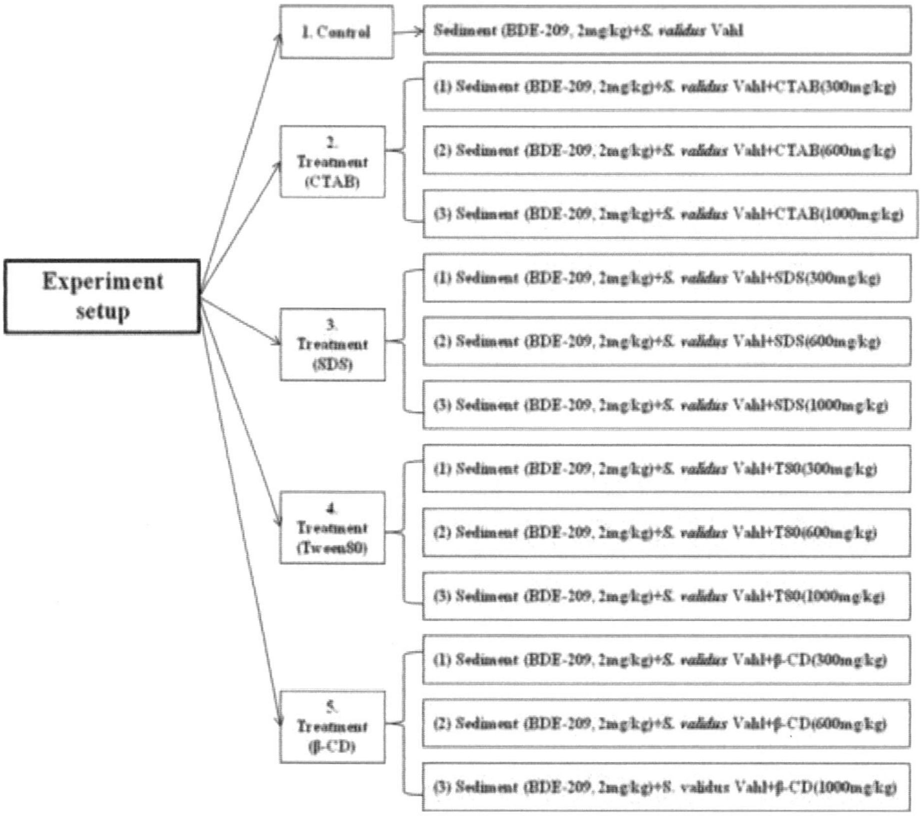

Fig.1 Phytoremediation experiment setup

2.5. Sample preparation

Sediments were collected using a sampling spade. Rhizosphere sediment samples were collected by vigorously shaking and rolling the roots to remove residual sediments. After collection, the sediment samples were oven-dried, ground, homogenized, passed through a 80-mm standard sieve, and then stored in glass containers at -20 °C until analysis.

2.6. BDE-209 analysis

2.6.1. BDE-209 extraction and cleanup. Three replicate sediment samples (dw, 3 g) were extracted with 30 mL of n-hexane/dichloromethane (1:1, v: v) for 30 min by ultrasonic extraction. This extraction was performed for three times. The combined 90 mL of extract was deeply filtrated by glass microfiber Filters (GF/C, whatman) and then concentrated to 3 mL by rotary evaporation. The extracts were then partitioned with concentrated H_2SO_4 to remove proteins, lipids, and atrace water. After centrifugation (3000 rpm, 5 min), the organic phase was collected and transferred. The acidic phase was re-extracted with 1 mL of n-hexane and the extracts were also collected and combined with the former collection.

2.6.2. BDE-209 analysis. Quantitative analysis of BDE-209 was performed by an Agilent 6890 gas chromatograph (GC) equipped with a ^{63}Ni-electron capture detector (GC-μECD). DB-5 HT column (15 m Length × 0.25 mm I.D. × 0.10 μm Film, J & W Scientific, Folsom, CA) was selected as the analytical column. The GC-ECD was operated as follows: injection port, 280°C with splitless mode; Ultrahigh nitrogen (99.99%) at a constant flow rate of 1.5 mL min^{-1} was used as carrier gas. Detector temperature was 320 °C. The flow rate of make-up gas was 60 mL min^{-1}. The oven program was set for: 2 min at 110 °C, ramp at 15 °C/min to 315°C (10-min hold). The extracted sample (1 μL) was injected into the DB-5 HT column.

2.7. Statistical analysis.

One-way analysis of variance (ANOVA) was used to examine the significance in plant growth (plant height and stem diameter) and BDE-209 dissipation rates among different treatment. Differences between two parameters were compared by independent-samples t-test (two-tailed). All data was presented as mean± SD (n≥3) and statistical analyses were performed by using GraphPad Prism 5.0.

3. Results and discussions

3.1. Effect of Surfactants and β-cyclodextrin on the growth of S. validus

As for the growth, *S. validus* grew well in the process of entire experiment. Compared to the control, *S. validus* in the treatments of CTAB, SDS, Tween80 and β-cyclodextrin in the range of 300-1000 mg/kg showed no significant poisoning symptom in terms of plant height (p>0.05) and stem diameter (p>0.05) (Table 1). The result of growth indicated that the concentration range of CATB, SDS, Tween 80 and β-CD is suitable for use as a synergist for phytoremediation.

Table 1. The growth of *S. validus* Vahl in the presence of CATB, SDS, Tween80 and β-cyclodextrin

Treatment	Concentration	Plant height (cm)	Stem diameter (mm)
Control	-	205.1±11.2	13.3±0.1
CTAB	300 mg/kg	194.2±12.2	13.0±0.8
	600 mg/kg	191.5±12.5	13.2±0.9
	1000 mg/kg	187.1±10.9	12.3±0.6
SDS	300 mg/kg	196.1±12.9	13.5±0.5
	600 mg/kg	196.5±7.9	12.8±0.8
	1000 mg/kg	198.7±6.2	12.9±0.7
Tween 80	300 mg/kg	204.2±7.9	12.9±1.2
	600 mg/kg	208.5±5.5	13.5±0.8
	1000 mg/kg	202.0±8.0	12.8±0.6
β-CD	300 mg/kg	201.8±7.7	13.4±0.6
	600 mg/kg	205.4±8.1	13.0±0.8
	1000 mg/kg	208.2±7.0	13.8±0.4

3.2. Effect of surfactants and β-cyclodextrin on the phytoremediation

Surfactants have the ability to increase aqueous concentrations of low solubility compounds enhancing their availability to microorganisms[8]. Effect of different concentrations of CTAB on the phytoremediation rate of BDE-209 was displayed in Fig.2a. With the increase of CTAB concentration in sediments, the phytoremediation effiency increased. After 180 days, the concentration of BDE-209 in the control sediments was reduced by 34.93% and CTAB of 300mg/kg slightly increased the degradation rate of BDE-209(2.85%), but the enhanced efficiency was not significant(p>0.05). CTAB with concentration of 600-1000mg/kg significantly enhanced the phytoremediation rate of BDE-209 in sediment, which were respectively 6.93% and 11.78% higher than the control sediments (p<0.05). The above results indicated that cationic-surfactant (CTAB) in the suitable concentration of 600 to 1000 mg/kg can significantly promoted the removal of BDE-209 in the phytoremediation.

The effect of different concentrations of SDS on the removal rate of BDE-209 was displayed in Fig.2b. The phytoremediation effiency increased as the increase of SDS concentration added. The BDE-209 degradation rate was increased by 4.52% in the presence of SDS with concentration of 300mg/kg, but the enhanced efficiency was not significant (p>0.05). CTAB with concentration of 600-1000mg/kg significantly enhanced the phytoremediation rate of BDE-209 in sediment, which were respectively

11.03% and 14.25% higher than the control sediments. This indicated that anionic-surfactant (SDS) in the suitable concentration of 600 to 1000 mg/kg can significantly promoted the removal of BDE-209 in the phytoremediation.

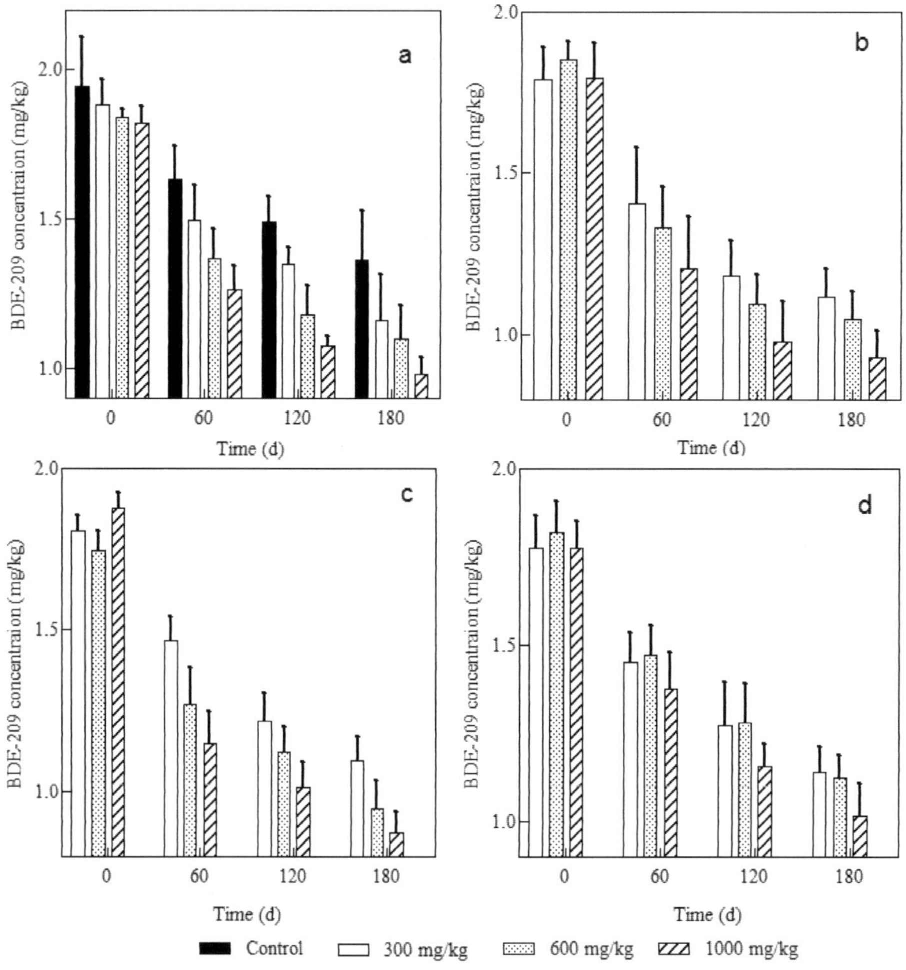

Fig.2 Changs of residual BDE-209 concentrations in sediment as a function of the added CTAB(a), SDS(b), Tween80(c) and β-cyclodextrin (d) concentrations(Mean±SD)

In the three typical kinds of surfactants, nonionic-surfactant (Tween 80) had the greatest enhanced effect on the removal of BDE-209 in the phytoremediation and was displayed in Fig.2c. With the increase of Tween 80 concentration in sediments, the phytoremediation effiency increased. When the concentration of Tween 80 was 300mg/kg, the degradation rate of BDE-209 increased by 7.37%, the enhanced efficiency was significant ($p>0.05$). Tween 80 with concentration of 600-1000mg/kg significantly enhanced the phytoremediation rate of BDE-209 in sediment, which were respectively 14.81% and 19.33% higher than the control sediments. Nonionic surfactant(Tween 80) can increase the solubility of hydrophobic organic compounds such as BDE-209 may due to the formation of micelles in the solution[8].The above results indicated that nonionic-surfactant (Tween 80) in the suitable concentration of 600 to 1000 mg/kg can significantly promoted the removal of BDE-209 in the phytoremediation.

Effects of different concentrations of β-CD on the removal rate of BDE-209 were presented in Fig.2d. When the concentration of β-CD was 300mg/kg, the degradation rate of BDE-209 was increased by 1.03%-7.78% in the presence of β-CD with concentration of 300-1000mg/kg, although the removal rate was further improved, but there was no significant difference ($p > 0.05$). The above results indicated that β-cyclodextrin 300-1000 mg/kg did not improve the efficiency of BDE-209 in sediments.

4. Conclusion

The present study demonstrated the use of surfactants and β-cyclodextrin for enhanced phytoremediation of BDE-209 contaminated sediment. Growth results indicated that CTAB, SDS, Tween 80 and β-cyclodextrin can be utilized as the BDE-209 phytoremediation amendment. Among all, CTAB, SDS, Tween 80 significantly enhanced the phytoremediation efficiencies, especially Tween 80, 11.78-19.33% of increase in BDE-209 removal rates was obtained. β-cyclodextrin at the concentration range of 300-1000mg/kg did not significantly enhance the BDE-209 phytoremediation efficiency in the added concentration ranges and was not the suitable solubilizers for phytoremediation.

Acknowledgments

This work was financially supported by Special fund for basic scientific research business of central public research institutes (Grant No.CKSF2017045+SH, CKSF2016270/SH), Non-profit Industry Financial Program of MWR (Grant No.201501042, 201501019) and China Three Gorges Corporation Program (0799553).

References

[1] Zhang X , Wang H , He L , Lu K , Sarmah A , Li J , Bolan NS ,Pei J , Huang H,Using biochar for remediation of soils contaminated with heavy metals and organic pollutants. Environ Sci Pollut Res,20(2013) 8472-8483.

[2] Mulligan C.N., R.N. Yong, B.F. Gibbs, Surfactant-enhanced remediation of contaminated soil: a review. Eng Geol,60(2001) 371-380.

[3] Bagga, D.K. and S. Peterson, Phytoremediation of arsenic-contaminated soil as affected by the chelating agent CDTA and different levels of soil pH. Remediation Journal, 2001. 12(1): p. 77-85.

[4] Pilon-Smits E., Phytoremediation. Annu. Rev. Plant Biol.,56(2005) 15-39.

[5] Gao Y, Ling W, Zhu L, Zhao B, Zheng Q,Surfactant-Enhanced Phytoremediation of Soils Contaminated with Hydrophobic Organic Contaminants:Potential and Assessment. PEDOSPHERE,17(2007) 409-418.

[6] Velusamy P, Pitchaimuthu S, Rajalakshmi S, Kannan N, Modification of the photocatalytic activity of TiO2 by β-Cyclodextrin in decoloration of ethyl violet dye. J Adv Res, 5(2014) 19–25.

[7] Tick G R ,Lourenso F,Wood A L , Mark L, Brusseau M L, Pilot-scale demonstration of cyclodextrin as a solubility-enhancement agent for remediation of a tetrachloroethene-contaminated aquifer. Environ. Sci. Technol.,37(2003) 5829-5834.

[8] Zhou J , Jiang W , Ding J, Effect of Tween 80 and beta-cyclodextrin on degradation of decabromodiphenyl ether (BDE-209) by White rot fungi. Chemosphere, 70(2007) 172-177.

[9] Hites R A, Polybrominated Diphenyl Ethers in the Environment and in People: A Meta-Analysis of Concentrations - Environmental Science & Technology (ACS Publications). Environ. Sci. Technol., 38(2004) 945-956.

IOP Conf. Series: Earth and Environmental Science **64** (2017) 012116 doi:10.1088/1755-1315/64/1/012116

Pyrosequencing reveals benthic bacteria changes responsing to heavy deposition of *Microcystis* scum in lab — searching bacteria for bloom control

Yali Tang[1*], Dongmei Cheng[1], Baohua Guan[2], Xiufeng Zhang[1], Zhengwen Liu[1, 2] and Zejun Liu[1]

[1] Department of Ecology and Institute of Hydrobiology, Jinan University
[2] State Key Laboratory of Lake and Environment, Nanjing Institute of Geography and Limnology

*Corresponding author e-mail: litangyali@163.com

Abstract. Bacteria capable of degrading cyanobacteria Microcystis are crucial for determining the ecological consequences of Microcystis blooms in freshwater lakes. Scum derived from Microcystis blooms tends to accumulate in bays of large lakes and then sink to the sediments where it is finally consumed by benthic bacteria. Understanding the response of benthic bacterial communities to massive Microcystis deposition events may help identify the bacteria best suited to Microcystis hydrolyzation and even bloom control. For that purpose, an experimental system was set up in which intact sediment cores were incubated in the laboratory with normal and heavy deposits of Microcystis detritus. Pyrosequencing was performed in order to describe a phylogenetic inventory of bacterial communities in samples taken at 0–1, 1–2 and 2–3 cm depths in incubated sediments and in original untreated sediment. A hierarchical cluster tree was constructed expose differences between sediments. Similarity percentage calculations were also performed to identify the bacterial species contributing to variation. The results of this study suggest that: (1) deposition of Microcystis scums exerts a strong effect on the bacterial community composition in the surface (0–1 cm) and sub-surface (1–2 cm) sediment layers; (2) bacterial community responses to Microcystis detritus deposition vary across vertical gradients. A list of bacteria with potential roles in Microcystis degradation was compiled. These findings may inform the development of future measures for Microcystis bloom control in lakes.

1. Introduction

Microcystis blooms are very commonplace in eutrophic freshwater lakes, ponds and rivers, where their effects include increased turbidity, smothering of macrophyte growth and negative impacts on important habitats for invertebrates and fish [1]. Some *Microcystis* species produce toxins like microcystin, which is potentially lethal to diverse organisms including humans [2]. In some large lakes, *Microcystis* blooms drift and accumulate in nearshore areas as dense scum layers up to 30cm thick [3-4]. In such dense accumulations, only a small proportion of *Microcystis* cells are decomposed or consumed in the water. The majority sink into the sediments as detritus [5-6], where they are ultimately consumed by benthic bacteria causing oxygen depletion, endotoxin release and an unpleasant smell [7-8]. Elucidating the structure of bacterial communities in these sediments and observing their response to massive depositions of detritus may improve understanding of the effects of *Microcystis* blooms. Furthermore, identifying bacteria capable of hydrolyzing *Microcystis* may inform future efforts to manage and control bloom events.

Scientists have been actively seeking bacteria capable of killing phytoplankton or degrading phytoplankton-derived detritus efficiently. These bacteria sometimes increase in abundance as algal blooms decline, and may play a significant role in bloom dynamics [9]. Previous studies have identified a number of pelagic bacteria important in *Microcystis* degradation [4, 10-13], including several species with proven algicidal properties and applicability in the management of blooms to avoid bloom causing nuisance issues [14-16].

In sediment, as in most ecosystems, in situ organic resources are generally exploited by relatively few species of abundant microbes; the remaining species lie dormant until roused by favourable environmental conditions [17-19]. It was hypothesized that massive inputs of fresh organic matter in the form of *Microcystis* detritus may trigger an explosion of activity and growth in otherwise cryptic sediment microbes, some of which may be *Microcystis* hydrolyzation specialists.

An experimental system was devised in which intact sediment cores were incubated the laboratory with differing loads of *Microcystis* detritus (normal and heavy). Pyrosequencing was applied to bacterial communities sampled from these incubated sediments and also from untreated sediment in order to reveal the benthic bacteria able to take advantage of Microcystis deposition. Pyrosequencing is a DNA sequencing technique is based on the detection of pyrophosphates released during DNA synthesis [20] and can offer a hundred times more phylogenetic information than previous PCR-DGGE (Polymerase Chain Reaction and Denatured Gradient Gel Electrophoresis) methods [21]. The twin objectives of this bacterial inventory were to determine how sediment bacterial community structure changes in response to the deposition of cyanobacterial scum across vertical gradients and to identify bacteria potentially specialised in *Microcystis* degradation.

2. Materials and Methods

2.1. Study area and experimental design

When Lake Taihu (30°55′–31°32′N, 119°52′–126°36′E) is the third largest lake in China, with a surface area of ~2338 km^2 and an average depth of 1.9 m. The sampling site lies in Meiliang Bay in the northern part of the lake (31°28′39″, 120°12′35″) in Wuxi, an area characterised by high densities of *Microcystis* species [3]. The water temperature at the time of study was 28.5 °C, and the illumination intensity at the bottom of the lake was 0.

On August 22, 2012, three intact sediment cores (inner diameter 15 cm) were sampled to a depth of 20 cm depth with 20cm overlying water and transported to the laboratory. After a 24 h acclimation and sediment stabilisation period, one core was sampled to represent the initial, untreated condition of all sediment communities (group O). The next core (group C) was supplied with 24.3 μg *Microcystis* detritus, representing the natural conditions of detritus deposition in Meiliang Bay. The third core (group H) was treated with 242.5 μg of *Microcystis* detritus to simulate heavy deposition comparable to that previously recorded after a severe bloom event in Lake Taihu with thick accumulations of scum [11]. All three chambers were incubated in the dark at 28.5 °C for 5 days. The cores were subsampled at depths of 0–1, 1–2 and 2–3 cm, pH tested and then freeze-dried and stored at -80 °C for later analysis of DNA.

2.2. DNA extraction, 16Sr RNA gene construction and pyrosequencing

Total microbial community DNA was extracted from approximately 1 g of material per subsample using the MoBio UltraClean Soil DNA isolation kit (MoBio Laboratories, USA) and following the manufacturer's protocol. DNA was used as templates for library construction. DNA concentrations were quantified using a NanoDrop spectrophotometer (NanoDrop Technologies Inc., DE). V3–V5 regions of the 16S rRNA gene (*Escherichia coli* positions 357 to 926) are targeted using barcoded primers.

Sequences of the partial 16S rRNA genes were determined using a GS-FLX 454 sequencer (Roche). Raw sequence reads were retrieved from all sediment samples and processed with the PyroNoise algorithm to remove 454 sequencing errors. Sequences were analyzed for the presence of chimeras using the UChime algorithm on the Mothur platform. Pairwise distances between aligned sequences were calculated at a 97% similarity (or 0.03 distance) cutoff and were then clustered into unique operational taxonomic units (OTUs) in Mothur. Taxonomic assignation was performed by the RDP classifier, and Fastaq files containing sequences were submitted to the NCBI Sequence Read Archive (SRX 1462880). A Bray–Curtis dissimilarity matrix was calculated using PRIMER v5.0 software, and a hierarchical cluster tree was constructed at phylum level based on the resulting matrix. PRIMER v5.0 was also used to perform similarity percentage calculations (SIMPER).

3. Results and Discussion

The pyrosequencing analysis of 16 S rRNA gene amplicons from nine sediment samples produced 204,034 sequence reads, with an effective read number of 124,701 after noise removal. The average sequence length was 435 nucleotides, excluding the adaptor and barcode primer sequences. A total of 8606 operational OTUs were obtained at the 97% sequence identity threshold. The number of OTUs is more than an order of magnitude greater than previously reported for benthic bacterial communities in Lake Taihu [22-23] and for other sediment bacterial groups associated with phytodetritus deposition [24], providing a much more detailed bacterial profile.

OTUs from all samples were assigned to 23 categories at phylum level, 71 at class level, 117 at order level, 217 at family level and 437 at the level of genus. At phylum level, *Proteobacteria* made the largest contribution to all samples, ranging from 25.96% to 45.43% of identified OTUs, with the smallest relative abundances observed in the 1–2 cm sediment layer of the high biomass group (H2). Other phyla contributing more than 1% of recorded OTUs were *Bacteroidetes* (average proportion of 5.58%), *Actinobacteria* (5.22%), *Chloroflexi* (2.98%), *Firmicutes* (1.08%) and *Acidobacteria* (1.04%). *Proteobacteria* dominated in all sediment samples and have previously been reported as the dominant group in freshwater, intertidal wetland and marine sediments [25]. *Bacteroidetes*, *Actinobacteria*, *Chloroflexi*, *Firmicutes* and *Acidobacteria* are also common phyla in eutrophic lake sediments [22, 26].

A hierarchical cluster tree was constructed from the phylum profiles of all samples (Figure 1). Groups O and C were clustered together by similarities in bacterial community composition at phylum level for each depth (O1 and C1, O2 and C2 and O3 and C3). The 2–3 cm subsample of the heavy deposition group (H3) also exhibited similarities with groups O and C, with H3, O3 and C3 forming a distinct cluster. The upper layers of group H (H1 and H2), however, occupy two different branches, indicating changes in the structure of the bacterial community consistent with the differences in community composition observed at phylum level.

IOP Conf. Series: Earth and Environmental Science **64** (2017) 012116 doi:10.1088/1755-1315/64/1/012116

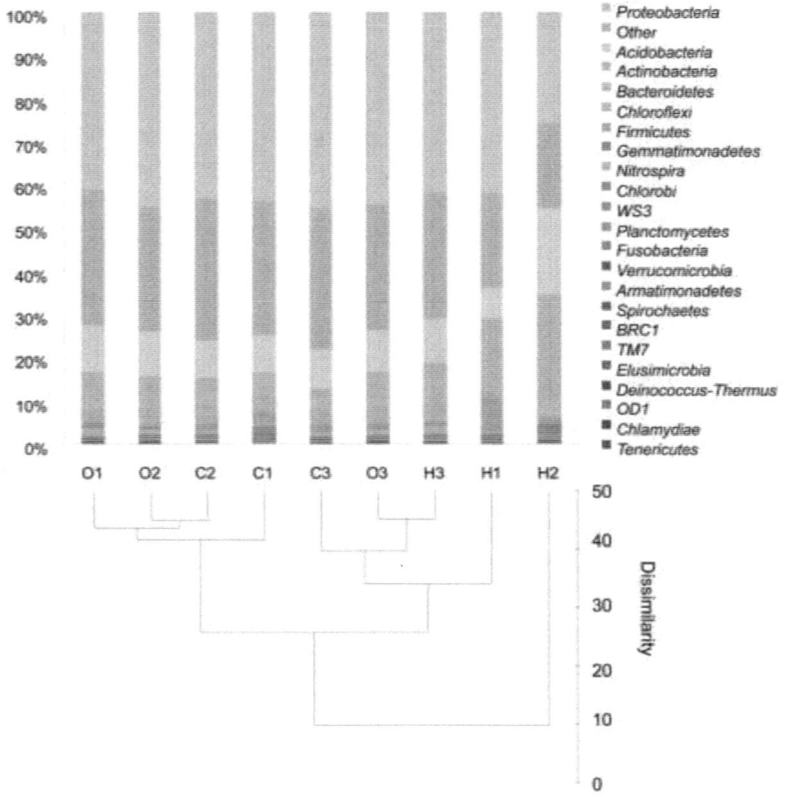

Figure 1. Visual representation of the similarity between samples based on bacterial community composition at phylum level, including a cluster tree calculated using Bray–Curtis dissimilarity and a bar chart displaying the community composition of each sample, giving the 22 recognised taxa in all samples, with unidentified taxa included as 'Other'. Note: O1, O2 and O3 represent samples taken at 0–1, 1–2 and 2–3 cm from the untreated sediment core; C1, C2 and C3 represent samples from 0–1, 1–2 and 2–3 cm in the control treatment; H1, H2 and H3 represent samples at 0–1, 1–2 and 2–3 cm in the treatment subject to heavy *Microcystis* deposition.

Fewer than 1% of OTUs recorded in this study contributed more than 50% of bacterial abundance in each sample. Data from H3, O3 and C3 are not shown because of their lack of variability.

The ten most dominant OTUs in group O and their variations in the treated samples are shown in Table 1. SIMPER analysis was conducted to compare species variation in sediments with heavy *Microcystis* deposition with the untreated samples and the control treatment. The OTUs that contributed most significantly to differences between bacterial communities are shown in Table 2.

Table 1. Ten most dominant OTUs of the untreated sediment (group O), experimental control (group C) and heavy deposition groups (group H) at 0–1 and 1–2 cm. Taxa are given at the level to which they could be identified.

Depth	Dominant OTU	Taxa	Group O		Group C		Group H	
			No	Percentag	No	Percentag	No	Percentag
0-1 cm	OTU0000	Unidentified	13	2.27%	12	2.08%	11	0.73%
	OTU0000	*Steroidobacter*	12	2.18%	61	1.04%	11	0.72%
	OTU0000	*Desulfobacteracea*	10	1.82%	82	1.39%	10	0.61%
	OTU0002	*Betaproteobacteria*	88	1.50%	40	0.68%	68	0.42%
	OTU0002	*Betaproteobacteria*	78	1.33%	53	0.90%	76	0.46%
	OTU0000	*Acidobacteria*Gp17	73	1.24%	45	0.76%	78	0.48%
	OTU0001	*Betaproteobacteria*	72	1.23%	16	2.71%	11	0.68%
	OTU0000	*Betaproteobacteria*	64	1.09%	40	0.68%	52	0.32%
	OTU0002	*Betaproteobacteria*	64	1.09%	40	0.68%	82	0.51%
	OTU0001	Unidentified	64	1.06%	46	0.97%	57	0.28%
1-2 cm	OTU0001	*Betaproteobacteria*	19	3.32%	60	1.05%	16	0.04%
	OTU0000	*Desulfobacteracea*	13	2.24%	13	2.40%	94	0.24%
	OTU0000	*Burkholderiales*	11	1.87%	10	1.84%	30	0.07%
	OTU0000	*Steroidobacter*	10	1.80%	91	1.61%	97	0.23%
	OTU0002	*Betaproteobacteria*	10	1.71%	42	0.74%	66	0.16%
	OTU0000	*Betaproteobacteria*	95	1.61%	50	0.88%	89	0.21%
	OTU0000	Unidentified	87	1.47%	89	1.57%	76	0.18%
	OTU0001	*Betaproteobacteria*	80	1.35%	80	1.41%	15	0.36%
	OTU0000	*Acidobacteria*	73	1.24%	63	1.11%	83	0.20%
	OTU0001	*Thiobacillus*	71	1.21%	99	1.75%	23	0.05%

Table 2. SIMPER analysis results displaying the top OTUs contributing to differences in bacterial community composition between untreated and control sediments (groups O and C respectively) and the treatment with heavy *Microcystis* deposition (group H) at 0–1 and 1–2 cm and their sequence abundance in all sediments. Taxa are given at the level to which they could be identified.

Depth	OTU	Taxa	Initial Abundanc	Control Abundanc	Bloom Abundanc	Contributio n (%)
0–1 cm	OTU0004	*Bacteroidetes*	0	0	853	6.01
	OTU0000	*Hydrogenophaga*	8	84	863	5.75
	OTU0007	Unidentified	7	6	559	3.89
	OTU0000	Unidentified	24	116	586	3.63
	OTU0007	*Acidovorax*	1	8	388	2.70
	OTU0014	*Holophaga*	3	4	325	2.26
	OUT0004	Unidentified	14	8	277	1.87
	OTU0006	*Comamonadaceae*	0	6	258	1.80
	OTU0014	*Dechloromonas*	1	6	252	1.75
	OTU0015	*Clostridium* III	3	4	213	1.48
1–2 cm	OTU0002	*Acidobacteria* Gp6	7	4	1284	3.14
	OTU0004	*Bradyrhizobium*	3	0	961	2.36
	OTU0005	Unidentified	1	0	957	2.35
	OTU0004	*Acidobacteria* Gp6	5	2	926	2.27
	OTU0005	*Micromonosporacea*	0	0	919	2.26
	OTU0001	Unidentified	31	31	730	1.72
	OTU0007	*Rhizobiales*	1	0	674	1.65
	OTU0008	*Pseudonocardia*	0	0	663	1.63
	OTU0001	*Acidobacteria* Gp6	47	29	622	1.43
	OTU0010	*Microlunatus*	0	0	548	1.35

The dominant OTUs from the untreated lake sediment remained active in incubated samples regardless of the level of *Microcystis* detritus added. This group is accustomed to daily deposition of phytodetritus, which constitutes an effectively inexhaustible supply of organic material allowing the bacteria to remain continuously active. These components of the bacterial communities share similar traits to soil local bacteria as depicted by Fontaine *et al.* [27], in that that they grow slowly and maintain continuous activity regardless of level of fresh organic matter input.

The bacteria dominating in group H samples (heavy deposition) contributed most to community structure variation at the sediment surface (0–1 cm layer) (Table 1, 2), with most presenting in only very small numbers (close to the detection limit) in the untreated (O) and control (C) sediments. This pattern is consistent with a microbial dormancy strategy as described by Jones and Lennon [17, 18]. The sudden superabundance of *Microcystis* detritus biomass allows enhanced activity and growth of these previously starving microbial populations, which expand rapidly to exploit the new substrate and result in remarkable changes in the structure of the microbial community [28-29]. In the present context, group H conditions might be expected to promote the growth of bacteria specialising in the decomposition of *Microcystis*, and indeed, most of the bacteria dominating this group are known to be associated with the pelagic *Microcystis* degradation. For example, members of the phylum *Bacteroidetes* can degrade biopolymers, in particular all kinds of polysaccharides [30]. The increased abundance and concentration of the unidentified bacteroidete OTU00043 may be indicative of a role in the degradation of *Microcystis* sp. mucilage, which consists mainly of polysaccharides [31]. The bacteria *Hydrogenophaga*, Acidovorax, *Dechloromonas*, *Comamonadaceae* and *Clostridium* have previously been reported with algicidal properties [4, 12-13, 16]. The current study is the first to indicate a role in phytodetritus degradation for the genus *Hophaga*.

The shifts in dominant bacteria in the subsurface (1–2 cm) sediment layer with heavy deposition were totally different to those at the surface (Table 2). Remarkable increases in abundance were observed among members of the *Acidobacteria* GP6 subgroup. *Acidobacteria* GP6 are considered to be ubiquitous and abundant in soil but are rarely cultured and thus remain poorly studied [32]. Naether *et al.* [33] reported that members of the *Acidobacteria* GP6 subgroup exhibit their highest relative abundances in soils with high nutrient levels. This being the case, the increases observed in heavily deposited sub-surface sediments may be a response to nutrient released from *Microcystis* detritus during degradation by surface sediment bacteria. Members of the order *Rhizobiales* including *Bradyrhizobium*, can serve as denitrifiers [34-35], and are likely to flourish in response to the release of nitrites from deeply degraded cynobacteria detritus. The genus *Microlunatus* comprises a number of highly competitive polyphosphate-accumulating bacteria [36] and is likely to serve in phosphate removal. Members of the Actinomycetales such as the *Micromonosporaceae* and *Pseudonocardia* sp are potentially involved in the mineralisation of refractory organic compounds [37]. Thus members of the *Micromonosporaceae* and *Pseudonocardia* sp. may benefit from the availability of some recalcitrant forms of *Microcystis* carbon that persist after surface sediment bacteria have assimilated the more accessible forms.

In general, the bacterial communities recorded in our sediment samples can be divided into three functional groups: local dominant bacteria, bacteria activated by massive deposition of *Microcystis* detritus, and dormant bacteria. Totally different groups of bacteria were activated by phytodetritus deposition at the sediment surface (0–1 cm) and sub-surface (1–2 cm) layers, resulting in increased differentiation. We suggest that the bacteria dominating in different sediment layers may reflect different processes of *Microcystis* degradation. Furthermore, we suspect that some of the bacteria thriving in the 1–2 cm sediment layer were not be involved directly with *Microcystis* detritus degradation but responded to the nutrients released by other bacteria. The low levels of species variation observed in the 2–3 cm layer between treatments may indicate the exhaustion of the organic resources.

4. Conclusion

In conclusion, by setting up an experimental system with intact sediment cores incubated with different levels of *Microcystis* detritus and then pyrosequencing the resulting bacterial communities, we found that the deposition of *Microcystis* scums exerted a strong influence on community composition in the surface (0–1 cm) and sub-surface (1–2 cm) sediments. These microbial community compositions reflect the activation of bacteria adapted to exploit the organic resource of *Microcystis* detritus. The bacterial community response to heavy deposition of detritus varied across vertical gradients, possibly reflecting different degradation processes. We conclude that the unidentified bacteroidete bacteria OTU00043, the *Hophaga* sp OTU00146, *Hydrogenophaga* sp OTU00009, *Acidovorax* sp OTU00075, *Dechloromonas* sp OTU00145, *Clostridium* sp OTU00158 and the comamonadacean OTU00069 are potentially involved in the early stage of *Microcystis* degradation, and may be crucial in *Microcystis* bloom termination. Pure cultures of these bacterial strains will be isolated in order to testify their ability to hydrolyze *Microcystis* cells.

Acknowledgments
This work was financially supported by National Natural Science Foundation of China (31000219, 31570456).

References

[1] H.W. Paerl, J. Huisman, Bloom likes it hot, Science 320 (2008) 57-58.
[2] D.R. de Figueiredo, U.M. Azeiteiro, S.M. Esteves, F. Goncalves, M.J. Pereira, Microcystin-producing blooms--a serious global public health issue, Ecotox. Environ. Safe. 59(2004) 151-163.
[3] B.Q. Qin, P. Xu, Q. Wu, L. Luo, Y. Zhang, Environmental issues of Lake Taihu, China Hydrobiologia 581 (2007) 3-14.
[4] P. Xing, L. Guo, W. Tian, Q.L Wu, Novel Clostridium populations involved in the anaerobic degradation of Microcystis blooms, ISME. J. 5 (2011) 792-800.
[5] R.D. Fallo, T.D. Brock, Planktonic blure-green algae: production, sedimentation, and decomposition in Lake Mendota, Wisconsin, Limnol. Oceanogr. 25 (1980) 72-88.
[6] N. Takamura, M, Yasuno, Sedimentation of phytoplankton populations dominated by Microcystis in a shallow lake, J. Plankton. Res. 10 (1988) 283-299.
[7] L.A. Meyer-Reil. Benthic response to sedimentation events during autumn to spring at a shallow water station in the Western Kiel Bight, Mar. Biol. 77 (1983) 247-256.
[8] B.C. Sander, J. Klaff 1993 Factors controlling bacterial production in marine and freshwater sediments, Microb. Ecol. 26 (1993) 79-99.
[9] X. Mayali, F. Azam, Algicidal bacteria in the sea and their impact on algal blooms, J. Eukaryot. Microbiol. 51 (2004) 139-144.
[10] K.Christoffersen, S. Lyck, A. Winding Microbial activity and bacterial community structure during degradation of microcystins, Aquat. Microb. Ecol. 27 (2002) 125-136.
[11] H. Li, P. Xing, M. Chen, Y. Bian, Q.L. Wu, Short-term bacterial community composition dynamics in response to accumulation and breakdown of Microcystis blooms, Water. Res. 45 (2011) 1702-1710.
[12] H. Li, P. Xing, Q.L. Wu Characterization of the bacterial community composition in a hypoxic zone induced by Microcystis blooms in Lake Taihu, China, FEMS. Microbiol. Ecol. 79 (2012) 773-778.
[13] H. Cai, H. Jiang, L.R. Krumholz, Z. Yang, Bacterial community composition of size-fractioned aggregates within the phycosphere of cyanobacterial blooms in a eutrophic freshwater lake Plos One 9 (2014) e102879.
[14] P.M. Manage, Z. Kawabata, S.I. Nakano, Algicidal effect of the bacterium Alcaligenes denitrificans on Microcystis spp, Aquat. Microb. Ecol. 22 (2000) 111-117.
[15] K.K. Rashidan, D.F. Bird. Role of predatory bacteria in the termination of a cyanobacterial bloom, Microbiol. Ecol. 41(2001) 97–105.
[16] Y.-K. Kang, S.-Y. Cho, Y.-H. Kang, T. Katano, E.-S. Jin, D.-S. Kong, Isolation, identification and characterization of algicidal bacteria against Stephanodiscus hantzschii and Peridinium bipes fot the control of freshwater winter algal blooms, J. Appl. Phycol. 20 (2008) 375-386.
[17] S.E. Jones, J.T. Lennon. Dormancy contributes to the maintenance of microbial diversity, Proc. Natl. Acad. Sci. USA. 107 (2010) 5881-5886.

[18] J.T. Lennon, S.E. Jones, Microbial seed banks: the ecological and evolutionary implications of dormancy, Nat. Rev.Microbiol. 9 (2013) 119-130.
[19] M. de Nobili, M. Contin, C. Mondini, P.C. Brookes, Soil microbial biomass is triggered into activity by trace amounts of substrate, Soi. Biol. Biochem. 33 (2001) 1163-1170.
[20] M. Ronaghi, Pyrosequencing sheds light on DNA sequencing, Genome Res. 11 (2001) 3-11.
[21] P.N. Polymenakou, C.A. Christakis, M. Mandalakis, A. Oulas, Pyrosequencing analysis of microbial communities reveals dominant cosmopolitan phylotypes in deep-sea sediments of the eastern mediterranean sea, Res. Microbiol. 166 (2015) 448-457.
[22] K. Shao, G. Gao, B. Qin, X. Tang, Y. Wang, K. Chi, J. Dai. Comparing sediment bacterial communities in the macrophyte-dominated and algae-dominated areas of eutrophic Lake Taihu, China, Can. J. Microbiol. 57 (2011) 263-272.
[23] K. Shao, G. Gao, Y. Wang, X. Tang, B. Qin, Vertical diversity of sediment bacterial communities in two different trophic states of the eutrophic Lake Taihu, China, J. Environ. Sci. 25 (2013) 1186-1194.
[24] M.A. Franco, I. de Mesel, M.D. Demba, K. van der Gucht, D. van Gansbeke, P. van Rijswijk, M.J. Costa, M. Vincx, J. Vanaverbeke, Effect of phytoplankton bloom deposition on benthic bacterial communities in two contrasting sediments in the southern North Sea, Aquat. Microb. Ecol. 48 (2007) 241-254.
[25] Y. Wang, H.F. Sheng, Y. He, J.Y. Wu, Y.X. Jiang, N.F. Tam, H.W. Zhou. Comparison of the levels of bacterial diversity in freshwater, intertidal wetland, and marine sediments by using millions of illumina tags, Appl. Environ. Microbiol. 78 (2012) 8264-8271.
[26] H. Tamaki, Y. Sekiguchi, S. Hanada, K. Nakamura, N. Nomura, M. Matsumura, Y. Kamagata, Comparative analysis of bacterial diversity in freshwater sediment of a shallow eutrophic lake by molecular and improved cultivation-based techniques, Appl. Environ. Microbiol. 71 (2005) 2162-2169.
[27] S. Fontaine, A. Mariotti, L. Abbadie, The priming effect of organic matter: a question of microbial competition? Soil. Biol. Biochem. 35 (2003) 837-843.
[28] B. Behera, G.H. Wagner, Microbial growth rate in glucose-amended soil, Soil. Sci. Soc. Am. J. 38 (1974) 591-594.
[29] B.S. Griffiths, K. Ritz, N. Ebblewhite, G. Dobson Soil microbialcommunity structure: effects of substrate loading rates, Soil. Biol. Biochem. 31(1998) 145-153.
[30] J.I.Schwarz, W. Eckert, R. Conrad. Response of the methanogenic microbial community of a profundal lake sediment (Lake Kinneret, Israel) to algal deposition, Limnol. Oceanogr. 53(2008) 113-121.
[31] Y. Amemiya, K. Kato, O. Nakayama, Changes in the chemical composition of carbohydrates and proteins in surface water during a bloom of Microcystis in Lake Suwa. Ecol. Res. 5 (1990) 153-162.
[32] W.J. Sul, S. Asuming-Brempong, Q. Wang, D.M. Tourlousse, C.R. Penton, Y. Deng, J.L.M. Rodrigues, S.G.K. Adiku. Tropical agricultural land management influences on soil microbial communities through its effect on soil organic carbon, Soil. Biol. Biochem. 65 (2013) 33-38.
[33] A. Naether, B.U. Foesel, V. Naegele, P.R. Wüst, J. Weinert, M. Bonkowski, F. Alt, Y. Oelmann, A. Polle, Environmental factors affect acidobacterial communities below the subgroup level in grassland and forest soils, Appl. Environ. Microbiol. 78 (2012) 7398-7406.
[34] A.H. Treusch, S. Leininger, A. Kletzin, S.C. Schuster, H.P. Klenk, C. Schleper, Novel genes for nitrite reductase and Amo-related proteins indicate a role of uncultivated mesophilic crenarchaeota in nitrogen cycling, Environ. Microbiol. 7 (2005) 1985-1995.
[35] C.M. Jones, B. Stres, M. Rosenquist, S, Hallin. Phylogenetic analysis of nitrite, nitric oxide, and nitrous oxide respiratory enzymes reveal a complex evolutionary history for denitrification, Mol. Biol. Evol. 625 (2008) 1955-1966.
[36] K. Nakamura, A. Hiraishi, Y. Yoshimi, M. Kawaharasaki, Y. Kamagata, Microlunatus phosphovorus gen. nov. sp. nov. a new gram-positive polyphosphate-accumulating bacterium

isolated from activated sludge, Int. J. Syst. Evol. Microbiol. 45 (1995) 17-22.
[37] M.D. Chengalroyen, E.R. Dabbs, Characterization of rubber degrading isolates, J. Microbiol. 2 (2012) 872-885.

Toxic effects of heavy metal Cu^{2+} on the pacific oyster *Crassostrea gigas*

Ceng Gao[a], Xinxin Zhang[b], Xiumei Li[c] and Xuexi Tang[*]

College of Marine Life Science, Ocean University of China, Qingdao 266003, China

[a]gokeystar@163.com, [b]Zhangxinxin0423@163.com, [c]Nxm210@yahoo.com,
[*]Corresponding author e-mail: Tangxx@ouc.edu.cn

Abstract. The effects of different concentrations of heavy metal ions on the survival of the Pacific oyster *Crassostrea gigas* were studied by using experimental ecology method in 96 h. The results showed that the LC50 of copper ion was 21.748mg/L and the safe concentration was 2.1748mg/L mg/L. Under the condition of laboratory, under laboratory conditions, the research of Cu^{2+} Stress on the *C. gigas* gill and digestive gland and adductor muscle tissue SOD, GPx and the induction of CAT activity. The results showed that the activities of SOD, GPx and CAT in the *C. gigas* were significantly changed by copper ion + stress. The results showed that in the low concentration Cu^{2+} treatment could induce the three kinds of enzymes, in the high concentration Cu^{2+} treatment group, SOD and CAT and GPx on the inhibition of the effect. The sensitivity of the three antioxidant enzymes to copper ion showed a certain difference. The sensitivity of the three kinds of tissue enzymes to Cu^{2+} treatment was digestive gland> fascia> gill. The experimental results show that the single factor for copper in water pollutants, the *C. gigas* digestive gland tissue SOD, GPX and CAT activity has certain significance to it, but will it as index applied to the actual water need further study.

1. Introduction

Crassostrea gigas, commonly known as oyster, is the world's largest aquaculture shellfish, is also one of China's well-known aquaculture octopus one of China's *C. gigas* farming in the world's first production [1]. *C. gigas* have a very high accumulation of heavy metals, which are 1 to 2 orders of magnitude higher than other aquatic organisms [2-3]. Under normal circumstances, living in heavy metal pollution in the sea area of *C. gigas*, the body will enrich the relatively high concentrations of heavy metals. As a result, many countries have adopted *C. gigas* as indicators of organisms, which are used to monitor marine heavy metal pollution [4]. Studies have shown that domestic *C. gigas* in the presence of heavy metals in varying degrees, reflecting the existence of heavy environmental pollution in the breeding environment [5-6]. Although Cu is a necessary nutrient in the body, but when the body absorbs excess Cu, the organism will cause poisoning [7]. In this study, gill and digestive gland tissues and adductor tissue of *C. gigas* were selected as the object of study. The gills were directly exposed to the tissues of the water bodies. The digestive glands were the main detoxification organs in the body. The effects of copper stress on the physiological status of the two tissues were and the possible mechanism was discussed. In order to screen out the biomarkers this could reflect the representative and high sensitivity of lead in the sea area of Jiaozhou Bay, and to explore the possible differences in the antioxidant responses of different tissues.

2. Materials and methods

2.1. Experimental materials

The experimental use of *C. gigas* was purchased from Yantai near the sea oyster farms freshly caught just for less than 12 hours. Before the experiment, *C. gigas* were washed thoroughly before the experiment, and the attachments were removed and placed in a large culture box of 0.8 × 0.5 × 0.8 (m) for about 3 days, domesticated during the feeding without feeding, day and night continuous oxygenation.

C. gigas will be starved for 3 days, the larvae were treated with semi-static water for 7 days, and the culture density was 50-60 individuals/m^3, continuous inflatable, the culture temperature was (20±1)℃, the ratio of light to dark was 12h: 12h, and control the light intensity is below 10 $\mu mol \cdot m^{-2} \cdot s^{-1}$, the salinity of the aquaculture water is 31, the pH is 8.0, and the water is changed at a fixed time every day. Daily feeding the right amount of bait, bait for the algae and algae Qinghai mixed. After holding for 7 days, the healthy individuals with the length of 7-8 cm and the body weight of 16-19 g were selected for the experiment.

Experimental contaminants were aqueous lead nitrate solution with a mother liquor concentration of 10 mg/ml. Dose into the set concentration.

2.2. Experimental methods

2.2.1. Acute toxicity test. After the pre-test to determine the experimental drug mass concentration range (96 h after the maximum concentration of live and the total lethal concentration limit), According to the logarithmic interval, five concentration groups were determined to be 15.00 mg/L, 20.25 mg/L, 27.34 mg/L, 36.91 mg/L, 50.00 mg/L, one control groups were set up, each group has three parallel samples.

The experimental containers were 3 L glass cylinders and each group was randomly placed in 50 domesticated *C. gigas* and cultured in a semi-static manner. During the experiment, the water was changed every 24h. After adding water, the concentration of Cu^{2+} solution was kept constant, and the dead individuals were removed at any time. For the long-time *C. gigas*, the individuals who did not respond within 5 min And the number of deaths in each group of *C. gigas* at 24 h, 48 h, 72 h and 96 h were recorded.

2.2.2. Determination of antioxidant enzyme activity. According to the results of acute toxicity test, two experimental gradients, namely 0.04 mg/L (low concentration treatment group) and 0.1 mg/L (high concentration treatment group) were designed. The natural seawater was used as blank control group. And the concentration of water in each concentration group was 3L. Thirty domesticated C. gigas were randomly distributed with the same method. The experimental time was 25 days. The samples were sampled at 0 d, 3 d, 7 d, 12 d, 18 d, and 25 d, respectively, and the samples were tested in parallel for the antioxidant enzyme activity of C. gigas. No oyster died during the experiment.

2.3. Determination of enzyme activity

2.3.1. Sampling and sample pretreatment. Two oysters were randomly selected from each of the two experimental groups. The capsules, gills and digestive gland tissues were preserved in liquid nitrogen. Before the measurement, approximately 0.5g of the samples were taken from the caudal muscle, the gill and the digestive gland, and 1: 3 (w/v) of the pre-cooled 0.86% saline was added to the pre-cooled glass homogenizer. Ice bath homogenate 6~8 min; centrifuge at 10 000rpm for 20 min at 4°C. Take the supernatant and store it in a refrigerator at -80°C for activity determination of SOD, GPx and CAT.

2.3.2. Determination of the activity of peroxidase GPX. Use the Nanjing Institute of Bioengineering Institute of the determination of the kit.

According to GPx can promote the reaction of H_2O_2 and GSH to produce H_2O and GSSG, so the activity of GPx can be expressed by the speed of its digestion, and the activity of GSH can be obtained by measuring the consumption of GSH in this enzymatic reaction. The activity unit of GPx in the tissue is defined as the tissue protein per milligram, and the non-enzymatic reaction is deducted per minute. The concentration of GSH in the reaction system is reduced by 1 μmo/L to one enzyme activity unit.

2.3.3. Catalytic activity of catalase CAT. Use the Nanjing Institute of Bioengineering Institute of the determination of the kit.

The reaction of catalase decomposition and H_2O_2 can be quickly stopped by the addition of ammonium moly date, and the remaining H_2O_2 reacts with ammonium moly date to produce a complex that defines the amount of H_2O_2 decomposed by 1umol per gram of tissue protein for 1 hour as a viable unit (U).

2.4. Data processing

The experimental data were analysed by spss 13.0 and plotted with sigma plot 10.0. Using spss13.0 statistical software, the probability of the concentration of Cu^{2+} as the abscissa and the probability of death of *C. gigas* was plotted as the ordinate, and the semi-lethal concentration and 95% confidence interval of 96h were calculated and the safety concentration and the calculation Half of the lethal concentration[8]. The results were expressed as mean ± standard error (mean ± SD); the data between groups were compared by double tailed test, $p < 0.05$, was considered significant difference; $p < 0.01$, was considered significant difference.

3. Experimental results

3.1. Acute Toxicity of Cu^{2+} to C. gigas

Table 1. Acute toxicity of $Cu^{2+}$to *C. gigas*

Cu^{2+} concentration (mg/L)	Stress time			
	24h	48h	72h	96h
	mortality rate %	mortality rate %	mortality rate %	mortality rate %
0	0.00	0.00	0.00	0.00
15.00	0.00	0.00	5.00	10.00
20.25	10.00	10.00	25.00	45.00
27.34	10.00	20.00	55.00	75.00
36.91	17.50	30.00	65.00	95.00
50.00	40.00	75.00	100.00	100.00

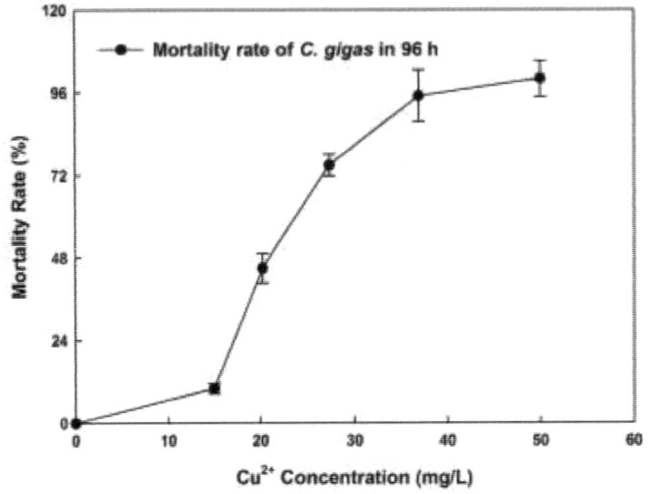

Fig. 1 Mortality rate of *C. gigas* under Cu^{2+} in 96 h

The mortality of *C. gigas* increased with the increase of Cu^{2+} concentration, the lowest concentration group (15.00 mg/L) within 72 h of the death of individuals, and the highest concentration group (50.00 mg/L) 24 h that is dead. Over time, the mortality rate increased by 50% at a dose of 50.00 mg/L at 96 h. The control group was normal and no death. Fig.1 shows the acute experimental toxicity curve of *C. gigas* 96 h. Linear regression analysis showed that the regression equation was Y = 7.434X - 9.942, where p <0.01 and half of the lethal concentration LC_{50} = 21.748 mg/L.

3.2. Effects of Copper on Antioxidant Enzyme Activities in Different Tissues of C. gigas

3.2.1. Effects on SOD activity

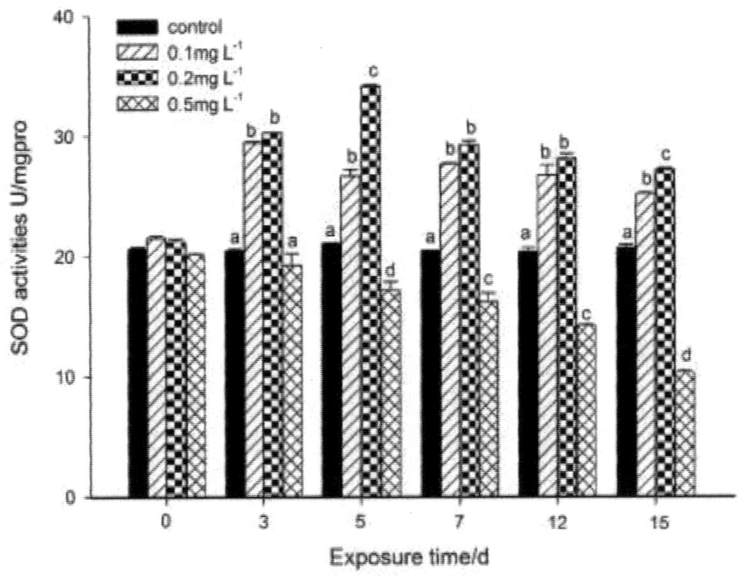

Fig. 2 Cu^{2+}effects on the SOD activity in the muscle tissue

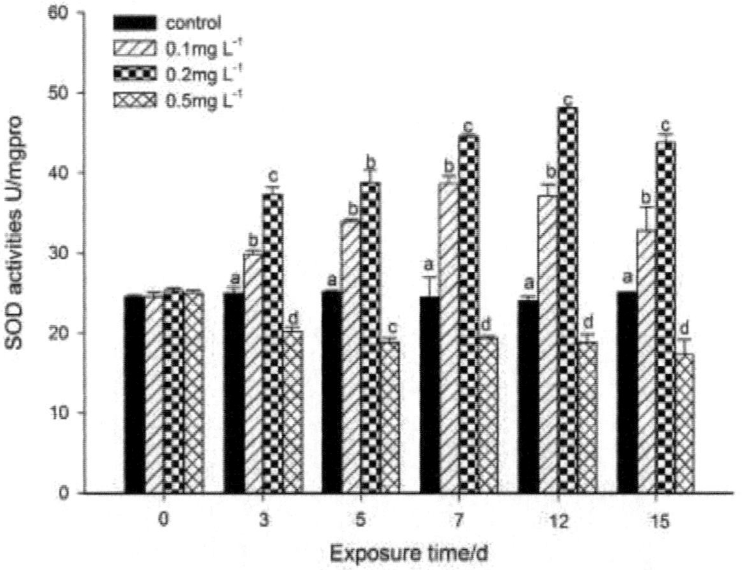

Fig. 3 Cu^{2+} effects on the SOD activity in the visceral mass

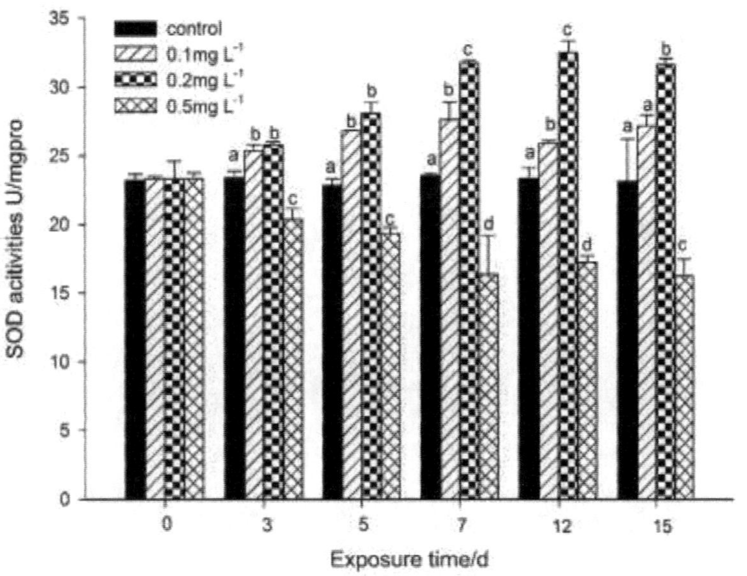

Fig. 4 Cu^{2+} effects on the SOD activity in the gill

It can be seen from Fig. 2-4 that, as time goes on, it can be seen from the above graph that the changes of SOD activity in the three tissues during the stress of Cu^{2+} against *C. gigas* are very different, Where the gill tissue is least sensitive to Cu^{2+}, digestive gland tissue sensitivity of the strongest, closed muscle tissue between the two. The activity of SOD in the digestive gland of *C. gigas* was significantly increased ($p < 0.05$) under the low concentration (0.1 mg/L and 0.2 mg/L) Cu^{2+} stress, and the uptake was increasing in the first 25 days ($p < 0.05$), but there was a trend of ascending and descending. The concentration of SOD in gills was changed, but the change was not as obvious as the above two tissues.

In the high concentration of Cu^{2+} (0.5 mg/L), the SOD activity of the three tissues decreased in different degrees.

3.2.2. Effects on GPX activity

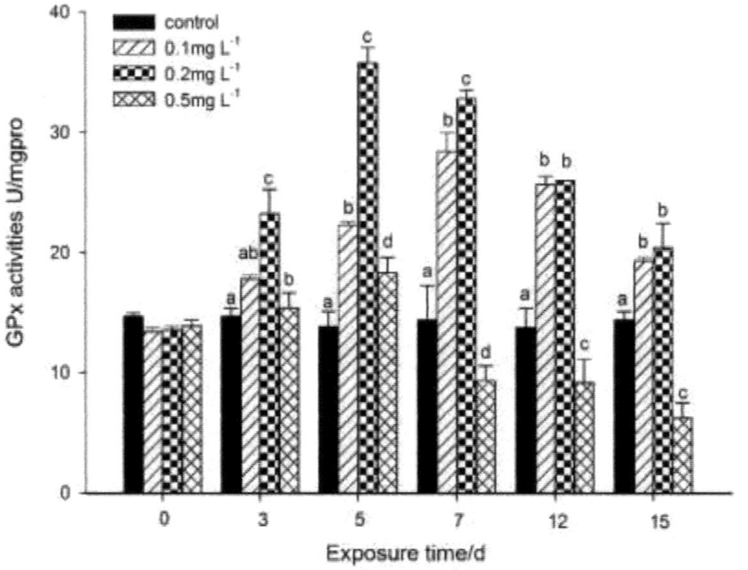

Fig. 5 Cu^{2+} effects on the GPx activity in the muscle tissue

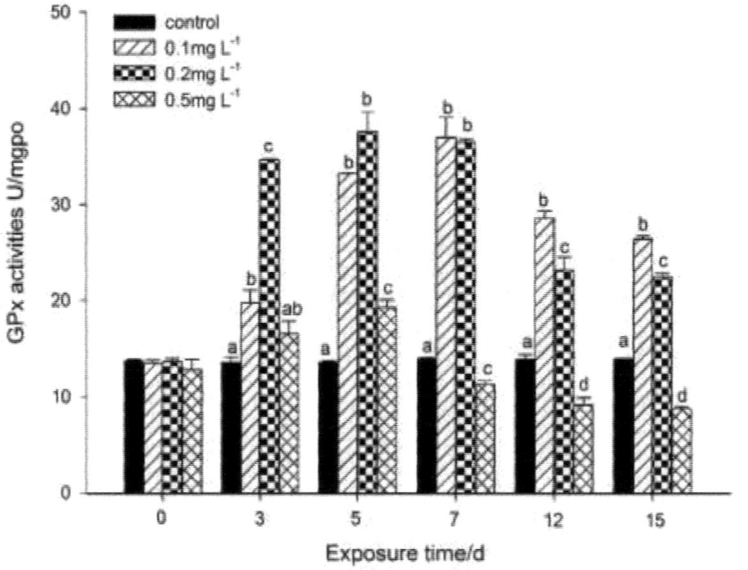

Fig. 6 Cu^{2+} effects on the GPx activity in the visceral mass

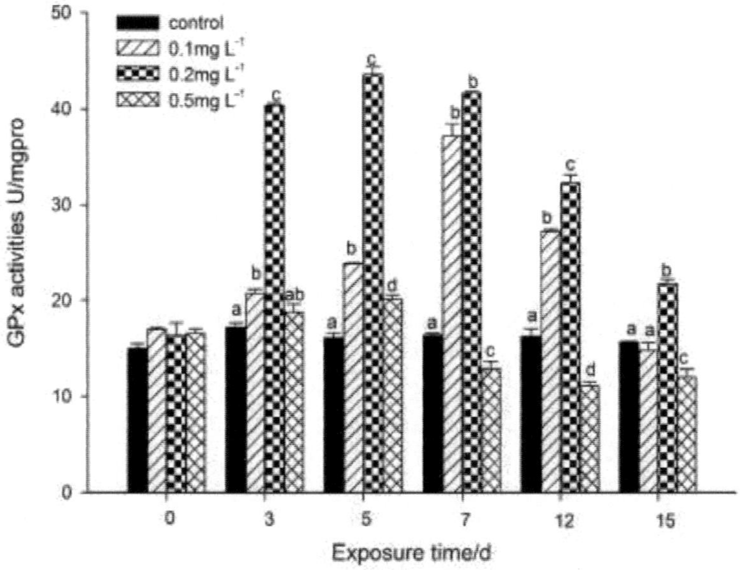

Fig. 7 Cu^{2+} effects on the GPx activity in the gill

The fluctuation of GPx activity in control organism of Fig. 5-7 was much higher than that of SOD and Cu^{2+} stress on the activity of GPx. The activity of GPx in digestive gland and cranial muscle and gill of *C. gigas* was significantly increased (p <0.05) under the stress of two low concentrations of Cu^{2+} (0.1 mg/L and 0.2 mg/L), and the initial stress was increased When the stress reaches a certain degree, it begins to fall again. In the high concentration (0.5 mg/L) Cu^{2+}, the GPx activity of the three tissues began to increase slightly. But soon fell below normal.

3.2.3. Effects on CAT activity

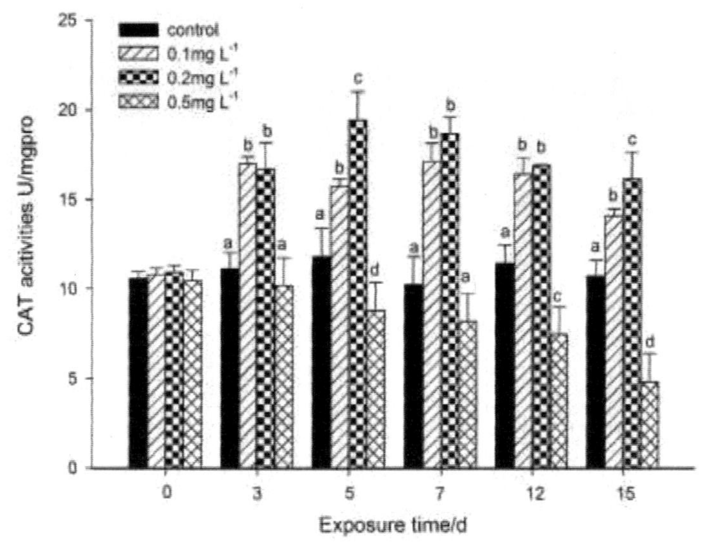

Fig. 8 Cu^{2+} effects on the CAT activity in the muscle tissue

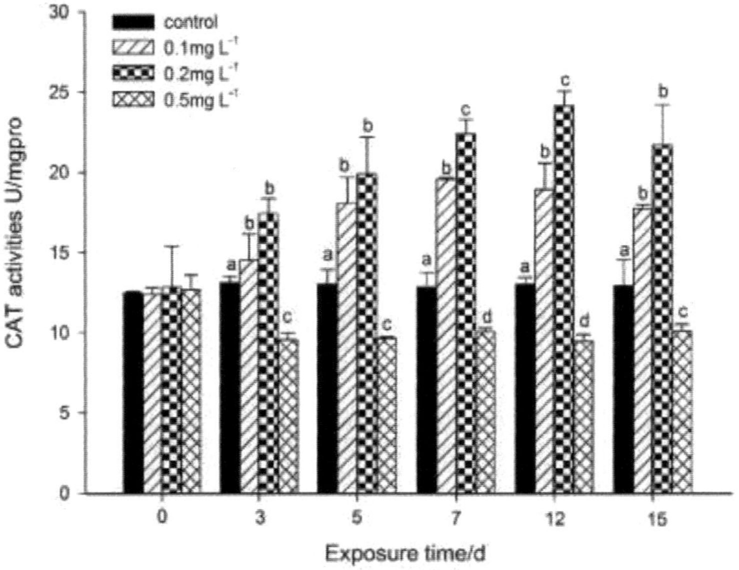

Fig. 9 Cu^{2+} effects on the CAT activity in the visceral mass

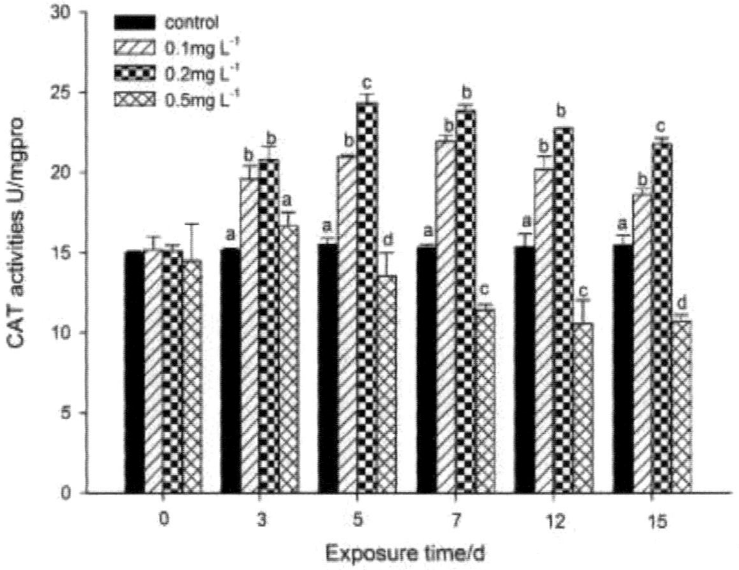

Fig. 10 Cu^{2+} effects on the CAT activity in the gill

It can be seen from Figure 8-10 that under Cu^{2+} stress, in the process of Cu^{2+} stress on *C. gigas*, the changes of CAT activity in the three tissues were very different, and the gill tissue is the least sensitive to Cu^{2+}, and the digestive gland the most sensitive, closed muscle tissue between the two. The activity of CAT in the digestive gland was significantly increased (p <0.05) under the stimulation of two low concentrations of Cu^{2+} (0.1 mg/L and 0.2 mg/L), in the first 25 days of metal accumulation stage has been on the rising trend, but there was a trend of ascending and descending (p <0.05). The concentration of CAT in the gills was changed, but the change was not as obvious as the above two tissues. In the high concentration of Cu^{2+} (0.5 mg / L), the CAT activity of the three tissues decreased in different degrees.

4. Discussion

Studies have shown that dissolved oxygen, pH, temperature and other changes and heavy metal pollution will make the body of superoxide dismutase (SOD) activity significantly changed[9-10]. Therefore, for aquatic organisms, SOD is a class of sensitive molecular Eco toxicological indicators. In this experiment, Cu^{2+} can induce SOD activity to enhance the body's antioxidant capacity.

CAT catalyzes the formation of H2O, which has the same function as SOD, and cleanses the oxygen free radicals produced in the living body. Therefore, the activity of CAT is similar to that of SOD, which is related to the activity of SOD in the liver tissue CAT activity changes in the trend is the same [11].

Glutathione peroxidase (GPx) is one of the important active oxygen free radical scavengers in the body. When oysters are under Cu^{2+} stress, they produce a lot of peroxy radicals, and the peroxyl radicals are disrupted by superoxide (SOD), which is decomposed into H_2O_2, and the increase of H_2O_2 induces the increase of glutathione peroxidase (GPx) activity and also affects the activity of catalase (CAT) [12].

Oysters produce excess reactive oxygen species (ROS) in the body at the beginning of Cu^{2+} stress, and the amount of reactive oxygen species (ROS) produced at high concentrations is the largest and the lowest concentration is produced. Therefore, the levels of superoxide dismutase (SOD), catalase (CAT) and glutathione peroxidase (GPX) activity in the gills, Glutathione peroxidase (GPX) activity levels are induced to increase due to increased activity of reactive oxygen species (ROS), which is consistent with the "toxic effects" between toxic contaminants and organisms. But the stress continues to proceed, with the passage of time, in the high copper concentration group, due to oyster body produced a large number of reactive oxygen species (ROS), when the production rate of these reactive oxygen species (ROS) exceeds the rate of clearance of antioxidant enzymes in oysters, the "toxic inhibitory effect" between toxic and stressed organisms is produced, reactive oxygen species (ROS) on the cells caused by the oxidation reaction, the cells of unsaturated fatty acids into saturated fatty acids, causing changes in cell structure and the destruction of some physiological life activities, causing cell aging and death, and then All three tissues and organs caused damage, so that the three antioxidants were so reduced and dropped below the control group level.

The antioxidant enzyme activities of different tissues showed a certain difference in the sensitivity of copper ions, and the sensitivity of the three tissues was digestive gland> hamster muscle> gills. Although the three kinds of tissues and organs of *C. gigas* have a certain "toxic excitatory effect", but "toxic excitatory effect" on the digestive gland tissue far more than the gill tissue. This has nothing to do with the gill tissue detoxification function.

The activity of SOD, GPX and CAT in the digestive gland of Pacific oyster has some indication for the single factor of copper pollutants in water, but it is necessary to apply it as an indicator to the actual aquaculture water.

References

[1] Z.Y. Qi. China's economic mollusks. 1998. Beijing: China Agricultural Publishing House, 194-199.

[2] W.X. Wang, J.F. Pan. Transfer of Heavy Metals in Marine Food Chains. Acta Ecologica Sinica, 2004, 24 (3): 599-604.

[3] S.Y. Gao, D.L. Zou. Research on the content of heavy metals in muddy bay biology and its evaluation [J]. Marine Environment Science, 1994, 13 (l): 39-45.

[4] C.H. Lu. Near Oyster as a heavy metal contaminant species preliminary study. Taiwan Strait, 1994, 13 (l), 14-20

[5] Fang Z Q, Cheung R Y H, Wong M H. Heavy metal concentrations in edible bivalves and gastropods available in major markets of the Pearl River Delta. Journal of Environmental

IOP Conf. Series: Earth and Environmental Science **64** (2017) 012117 doi:10.1088/1755-1315/64/1/012117

Sciences, 2001, 13(2): 210-217.

[6] Fang Z Q, Cheung R Y H, Wong M H. Heavy metals in oysters, mussels and clams collected from coastal sites along the Pearl River Delta, South China. Journal of Environmental Sciences, 2003, 15(1): 9-24.

[7] J.Y. Dai, W.Y. Zheng, S.H. Wang. Study on the combined toxicity of heavy metals and organophosphorus pesticides to true gill and gill larvae. Environmental Science, 1997, 9: 44-94.

[8] J.H. Chen, B.H. Yang, Z.Y. Deng. Environmental Statistics Application. Beijing: Chemical Industry Press, 2005, 141-145.

[9] C.L. Yao, W.N. Wang, A.L. Wang. Advances in studies on superoxide dismutase in aquatic animals. Marine Sciences, 2003, 27 (10): 18-21.

[10] L.H. Xu, Y.Y. Zhang, Y.Y. Chen. Advances in molecular ecotoxicology and its significance in water environment protection [J]. Journal of Hydrobiology, 1995, 19 (3): 171-185.

[11] Chinese Journal of Applied Ecology, 2001, 12 (3): 422-424.

[12] Q.X. Zhou, F.X. Kong, L. Zhu. Ecological Toxicology. Beijing: Science Press, 2004

The changes of proteins and polysaccharides in extracellular polymeric substance for Spirogyra fluviatilis under different salinity

Yichao Lee, Shuiping Chang*

Department of Environmental Engineering Kun Shan University, Tainan City, Taiwan (R.O.C.)

*Corresponding author e-mail: spchang@mail.ksu.edu.tw

Abstract. Spirogyra is a genus of widely distributed, large green fresh water algae. This study discovered that changes in salinity can induce Spirogyra fluviatilis to produce amounts of extracellular polymeric substance (EPS) when controlling other environmental conditions. If culturing S. fluviatilis with salinity greater than a 3.0‰ medium for 4 hours, the secretion EPS will be changed. And the level of polysaccharides and proteins, the primary components of EPS, is slightly increased in accordance with the increase in the salinity. But the proteins to polysaccharides ratio changes are not significantly

1. Introduction

As recent research has shown, the EPS of the algae is highly sensitive to salinity. Once the salinity of the growing environment is increased, the species of *S. fluviatilis* starts to change the composition of polysaccharides and protein in EPS. In recent years, interest in the exploitation of valuable EPSs has been increasing for various industrial applications and renewable natural resources representing an important class of biotechnological importance [1, 2]. Therefore, making the in-depth study of EPS is also important in term of development of value-added EPS applications [3, 4, 5]. Algal is the photosynthetic organism can grow in the different environment and EPS production which makes algae a perfect candidate for biotechnological exploration. [6]

This study investigated EPS production under different salinity by exploring the features of *S. fluviatilis*. This study also investigated the changes of polysaccharides and proteins, the primary components in EPS, and hopefully the study results can be used as a reference for valuable EPS.

2. Methods

2.1. Algae

Spirogyra (Chloropyta, Zygncmatale, Zygnemaceae) is a filamentous macro algal algae. It is one of the commonest fresh water algae which is worldwide in distribution. It occurs in extensive masses, free floating on the surface of water, forming familiar pond scum, water silk or pond silk. The bright green slippery shining tangled masses of filaments are found in the stagnant water of ponds, puddles, ditches, tanks, and also slow flowing rivers. The alga appears in great abundance during and after rains [7].

IOP Conf. Series: Earth and Environmental Science **64** (2017) 012118 doi:10.1088/1755-1315/64/1/012118

2.2. Influence of salinity on S. fluviatile EPS
The *S. fluviatile* was harvested from the campus of Kun-Shan University (N 23°25′, E 120°21′) in Taiwan. The pond water was also brought back to the laboratory to prepare a culture medium with different concentrations of salinity, including 8 different salinity levels such as unadjusted (assuming it was 0.0 ‰), 0.5‰, 1.0‰, 3.0‰, 5.0‰, 10.0‰, 20.0‰, and 30.0‰, so as to culture *S. fluviatile*. Individual aquariums contained 10 liters of river water with different salinity, and the water was circulated using a small motor to create an adequate environment in which to culture *S. fluviatile* Samples were then taken at 4h, 8h, 12h, 24h, 48h and 72h to measure the changes of polysaccharides and proteins in EPS.

2.3. EPS extraction using ultra sonication
This study used ultra sonication to extract EPS. The algae was harvested from the aquarium, washed twice with tap water, and drained in a strainer at room temperature for 2 h until it dried naturally. Afterward extraction procedures were started (as indicated in Fig. 1).

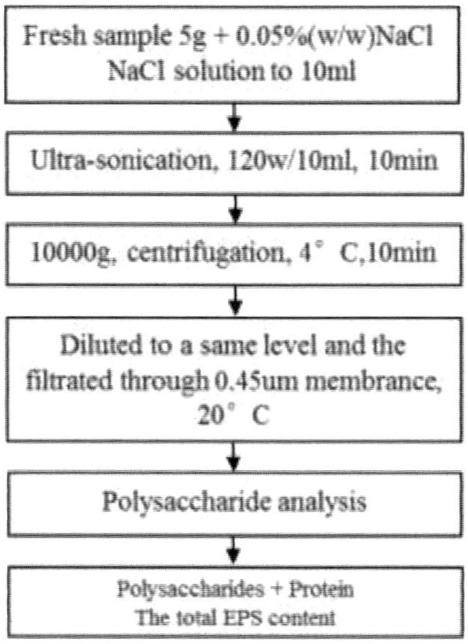

Figure 1. Procedure for the extraction process of *S. fluviatilis*.

2.4. Analysis method
All chemicals used in this work were of analytical grade. Polysaccharide content was determined by the phenol-sulfuric acid [8] using glucose as a standard. Protein content was determined according to Bradford (1976) [9] with bovine serum albumin BSA, Sigma A2153 100mg/ml) [10]. The total EPS content was measured as the sum of these two substances.

3. Results and discussion

3.1. The influence of different salinity on the level of polysaccharides in EPS
The changes in polysaccharides levels in EPS under different salinity and culture time are shown in Fig. 2. From the results, salinity impacts of *S. fluviatilis* EPS production. The polysaccharides is slightly increased in accordance with the increase (about 15% increase) in salinity.

IOP Conf. Series: Earth and Environmental Science **64** (2017) 012118 doi:10.1088/1755-1315/64/1/012118

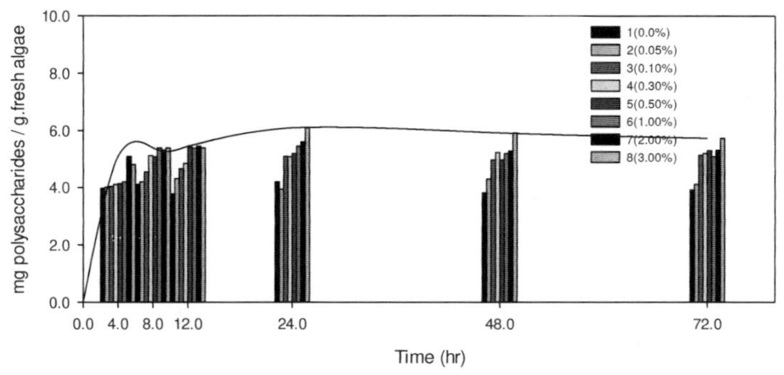

Figure 2. Change of polysaccharide level under different salinity and culture time.

3.2. The influence of different salinity on the level of proteins in EPS

In this study, we isolated EPS from *S. fluviatilis* cultured under different salinity. The extracellular salt stress affects the protein content in EPS, and significant advancement about increast 9% in the protein production was observed at higher salinity (Fig 3), is contrarily slightly impact by culture time.

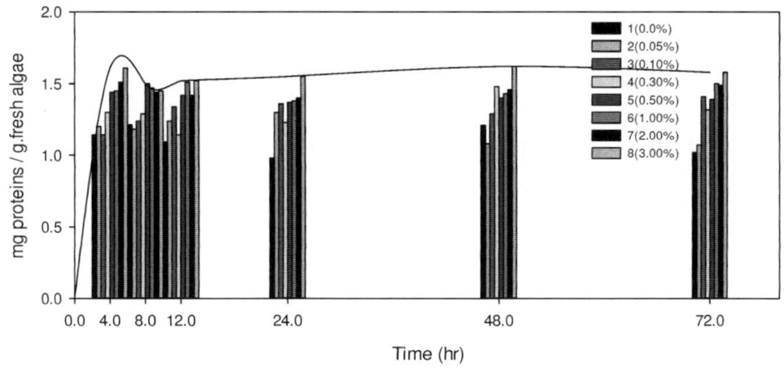

Figure 3. Change of proteins level under different salinity and culture time.

3.3. The influence of salinity on the S. fluviatilis EPS

Table 1 and Fig. 4. Demonstrate the changes of *S. fluviatilis* EPS amounts under culture conditions with different salinity. The results indicated that although EPS secretion was not obvious change if the *S. fluviatilis* was cultured lower 3.0‰ condition, once the culture salinity increased (≧3.0‰), EPS secretion also greatly increased accordingly. However, the increment of EPS secretion was not in a portion to the increase in culture time. In this case, *S. fluviatilis*, start to change EPS in accordance with the increase in salinity. To protect itself, the algae could change its physiological conditions to adapt to the environmental changes. But the study results implied that critical salinity for creating significant changes in EPS production is 3.0‰.

Table 1. Amounts of EPS and polysaccharide proteins ratio in different salinity.

Salinity	Culture time					
	4hr	8hr	12hr	24hr	48hr	72hr
1(0‰)	5.12/0.29*	5.32/0.29	4.87/0.29	5.18/0.23	5.03/0.32	4.93/0.26
2(0.5‰)	5.21/0.30	5.38/0.28	5.550.29	5.24/0.33	5.38/0.25	5.18/0.26
3(1.0‰)	5.18/0.28	5.78/0.27	5.99/0.29	6.46/0.27	6.27/0.26	6.55/0.27
4(3.0‰)	5.41/0.32	6.41/0.25	5.98/0.24	6.32/0.24	6.72/0.28	6.52/0.25
5(5.0‰)	5.58/0.35	6.68/0.31	6.84/0.26	6.57/0.26	6.38/0.28	6.69/0.26
6(10.0‰)	5.65/0.35	6.87/0.27	6.89/0.28	6.83/0.25	6.63/0.28	6.59/0.29
7(20.0‰)	6.61/0.30	6.76/0.27	6.88/0.26	7.01/0.25	6.75/0.28	6.8/0.28
8(30.0‰)	6.41/0.34	6.85/0.27	6.91/0.28	7.65/0.25	7.54/0.27	7.32/0.28

*EPS/ratio：EPS= polysaccharide + proteins, ratio= proteins/ polysaccharide

In this study, we discovered that ratio of proteins to polysaccharides between 0.23-0.35(mean=0.29), the changes are not significant, an increase in salinity can accelerate the secretion of EPS, which thickens the extracellular mucilage layer.

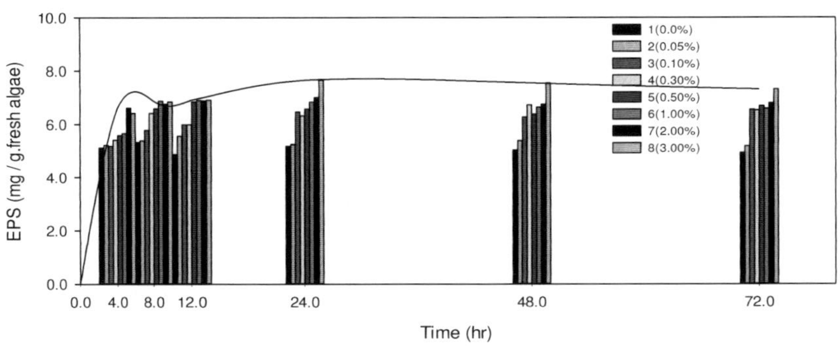

Figure 4. Changes of the EPS level under different salinity and culture time.

4. Conclusion

Spirogyra is a genus widely spread all over the world and is also a dominant species, large green algae in fresh water. The species of *S. fluviatilis* used in this study originally grew in freshwater areas, EPS could be secreted due to the changes in salinity in the habitat. Consequently the mucilage layer outside of the cell walls of the algae becomes thicker. This study revealed that EPS secretion was increased when the *S. fluviatilis* was cultured in 3.0‰ of salinity for 4 hours, but the amount of EPS merely slightly increased in accordance with culture time.

The total amount of EPS is slightly increased in accordance with the increase in salinity; however, the effects for ratio of proteins to polysaccharides not significant. This study may allow further exploration of *S. fluviatilis*. Studies on the biotechnological importance and ecological significance of the extracellular polymeric substances of *S. fluviatilis* deserve further attention.

References

[1] R. Jain, S. Raghukumar, R. Tharanathan, and N.B. Bhosle, Extracellular polysaccharide production by Thraustochytrid Protists, Marine Biotechnol. 7 (2005) 184-192.

[2] A. Mishra, and B. Jha, Isolation and characterization of extracellular polymeric substances from micro-algae Dunaliella salina under salt stress, Bioresource Technol. 100 (2005) 3382-3386.

[3] Mishra, A., Jha., Isolation and characterization of extracellular polymeric substances from micro-algae Dunaliella salina under salt stress, Bioresour. Technol. 100:13(2009) 3382-3386.

[4] Raposo, M.F.d.J., de Morais, R.M.S.C., Bernardo de Morais, A.M.M., Bioactivity and

Applications of Sulphated Polysaccharides from Marine Microalgae, Mar. Drugs 11:1 (2013) 233-252.

[5] Yalcin, I., Hicsasmaz, Z., Boz, B., Bozoglu, F., Characterization of the Extracellular Polysaccharide from Freshwater Microalgae Chlorella sp., LWT - Food Science and Technology 27:2 (1994) 158-165.

[6] Zheng, Y., Xiao, R., Roberts, M., Polymer-enhanced enzymatic microalgal cell disruption for lipid and sugar recovery, Algal Res. 14 (2016) 100-108.

[7] J. Simons, A. P. VanBeen, *Spirogyra* species and accompanging algae from pools and ditched in the Netherlands, Aquat. Bot. 37 (1990) 247-269.

[8] H. Xiaomeng, W. Zhiwei, Z. Chaowei, and Z. Wu, Effect of ultrasonic power density on extracting loosely bound and tightly bound extracellular polymeric substances, Desalinat. 329:15 (2013) 35-40.

[9] M. M. Bradford, A rapid and sensitive method for the quantitation of microgram quantities of protein utilizing the principle of protein-dye binding, Anal. Biochem. 72 (1976) 248-254.

[10] M. Dubois, K.A. Gilles, J.K. Hamilton, P.A. Rebers, and F. Smith, Colorimetric method for determination of sugars and related substances, Anal. Chem. 28:3 (1956) 350-356.

IOP Conf. Series: Earth and Environmental Science **64** (2017) 012119 doi:10.1088/1755-1315/64/1/012119

Effect of ultrasound pretreatment on hot air drying rate of scallop muscles

Yaxiang Bai*, Xuejie Fan and Yujing Bi

School of Science, Dalian Ocean University, Dalian, China

*Corresponding author e-mail: Byx0671@163.com

Abstract. To improve the drying rate and to reduce the energy consumption of aquatic product, scallop muscles were treated by ultrasonic with different frequency, time and different temperature. The hot air drying rate of treated scallop muscles was determined, and, meanwhile, several quality parameters such as the shrinkage and rehydration rate of samples with pretreatment were determined to compare with those of untreated samples. The results indicated that the drying rate of scallop muscles can be improved by the ultrasonic pretreatment without affecting the quality. Under a 70 kHz for 30min at 25°C, the drying rate of pretreatments is 1.29 times higher compared with that of the control group.

1. Introduction

Scallop is a popular aquatic product available in China. Since 1980s, cultivation of scallops has developed rapidly. In recent years, the production of scallop has been in excess of 100, 0000 tons [1]. Scallop muscle contains many desirable flavor components. As an important marine product, dried scallop is also regarded as healthy food that is gaining popularity. Hot air drying is one of the most commonly used procedures in dried scallop muscle, with main disadvantages being prolonged duration and high energy consumption [2]. Therefore it is necessary to explore low power consumption processing for hot air drying.

Dry pretreatment technology is utilized to treat food through some physical methods before drying it. Pretreatment can not only improve the quality of dried food but also speed up its drying and thus reduce energy consumption. For example, during the process of drying and dehydration for fruits and vegetables, high pulse electric field pretreatment at certain intensity will enhance their drying rate and thus reduce the drying energy consumption [3-7].

Ultrasound has been used in the food industry for many purposes. Low-frequency, high-energy ultrasound has been applied to determine food properties. However, this technology in food industry is relatively new and it has not been explored in-depth until recently [8].

Studies in recent years have demonstrated that as a pre-processing step for fruits and vegetables in the drying process, a ultrasound is able to cause reversible membrane perforation, and it is hoped that it will help to increase dryer efficiency, save energy, and enhance product quality [9-11]. For example, Nowacka et al. used ultrasound to pretreated apples, and found the ultrasound treatment caused reduction of the drying time by 31% in comparison to untreated tissue. The ultrasound treated apples exhibited between 9% and 11% higher shrinkage, 6-20% lower density, and porosity of 9-14% higher than untreated samples [12].

IOP Conf. Series: Earth and Environmental Science **64** (2017) 012119 doi:10.1088/1755-1315/64/1/012119

At present, there are only few reports on drying seafood products using ultrasound pretreatment. In this paper, scallop muscle, as a representative of seafood products, were treated in an ultrasound system, in different frequency, time and different temperature. The hot air drying rate of treated scallop muscle was determined and, meanwhile, several quality parameters such as the shrinkage and rehydration rate of samples with pretreatment were determined to compare with those of untreated samples.

2. Experimental procedure

2.1. Experimental equipment

The ultrasound pretreatment system shown in figure1 was developed by the authors. Its main components are an ultrasonic generator and a constant temperature water bath. The range of frequency of which can be adjusted from 50 to 100 kHz by a controller.

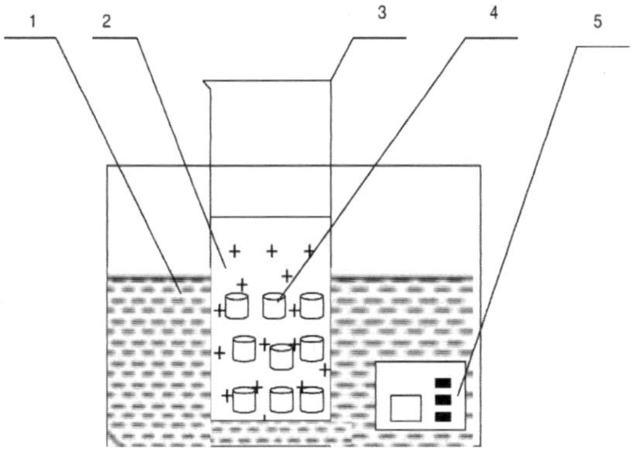

1. Constant temperature water bath 2. Water
3. Beaker 4. Samples 5. Ultrasonic generator

Figure 1. Schematic of the ultrasound pretreatment system

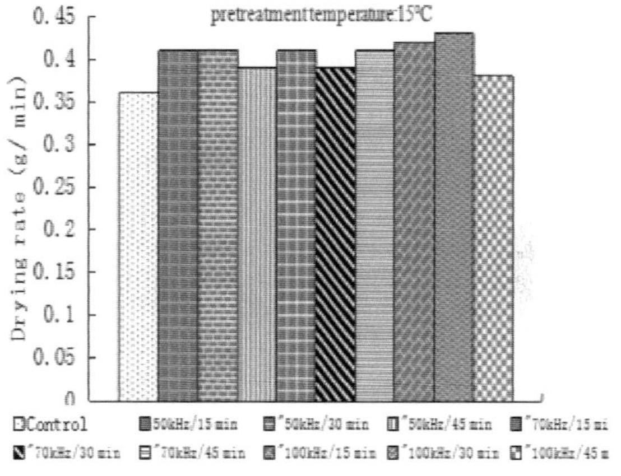

Figure 2. Effect of ultrasonic frequency and pretreatment time on average drying rate of scallop muscle at 15°C.

2.2. Preparation and drying of scallop muscles

Fresh scallops were procured from the local market near Dalian Ocean University, China. Before drying, scallops were washed with tap water and blanched in boiling water. The scallops were quickly taken out as soon as their shells opened; they were then cooled in tap water. The scallop muscles were cut and rinsed. After rinsing, the scallop muscles were dipped in NaCl solution at 90~95 °C for 10 min. They were then taken out and drained. The initial moisture content of scallop muscle was 71±1%. The samples were divided into 28 groups with similar weights. One was the control group and the others were experimental groups. The experimental groups were subjected to ultrasound pretreatment with different time (15, 30, 45 min), different ultrasonic frequency (50, 70, 100 Hz) and different temperature (15, 25, 35 °C), respectively. Then, all samples were dried by hot air (60℃) until they reached the final moisture content (15±1%).

2.3. Shrinkage measurement

Shrinkage determination was made by manually measuring the sizes (average diameter and average thickness) of the fresh scallop muscle and the dried scallop samples obtained from different drying processes using a vernier caliper. The equivalent column volume of each sample was then determined and compared. Ten samples were measured under each experimental condition. Percentage of shrinkage was calculated using the following equation:

$$\text{Percentage shrinkage} = (Z_0 - Z_f)Z_0^{-1} \times 100 \quad (1)$$

Where Z_0 (cm^3) and Z_f (cm^3) are the volume of each group of scallop muscle at the beginning and at the end of each drying experiment, respectively.

2.4. Rehydration ratio measurement

Rehydration ratio was determined by soaking dried scallop muscle in hot water at 100°C. The soaked scallop muscles were blotted with a paper towel of good quality to remove excess water. They were then weighed and placed back into the soaking water. Rehydration ratio value was expressed in percentage and calculated as mass of water absorbed per 100 g of dried scallop muscle using the following formula.

$$\text{Percentage rehydration ratio} = (m_g - m_0)m_0^{-1} \times 100 \quad (2)$$

Where m_0 (g) and m_g (g) are weights of the samples before and after rehydration, respectively.

3. Results and discussion

3.1. Influence of different ultrasonic frequency and pretreatment time for different temperature on drying rate

The influence of ultrasonic frequency and pretreatment time for different temperature on drying rate was shown in Figure 2-4. It can be seen from the graphs that the drying rate of ultrasonic pretreatment scallop muscle was higher than that of the control.Under a 70 kHz for 30min at 25°C, the drying rate of pretreatments is 1.29 times higher compared with that of the control group.

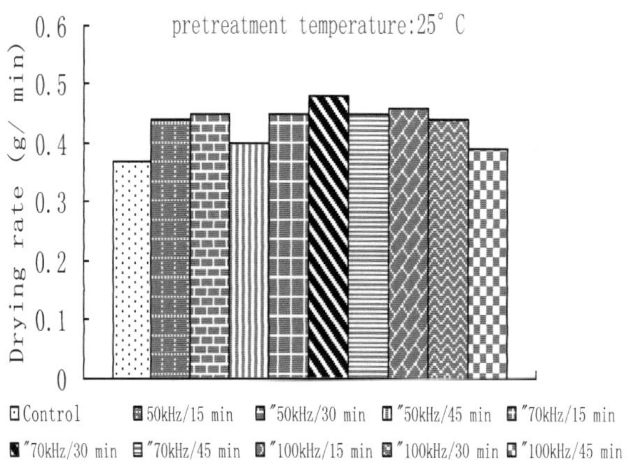

Figure 3. Effect of ultrasonic frequency and pretreatment time on average drying rate of scallop muscle at 25°C.

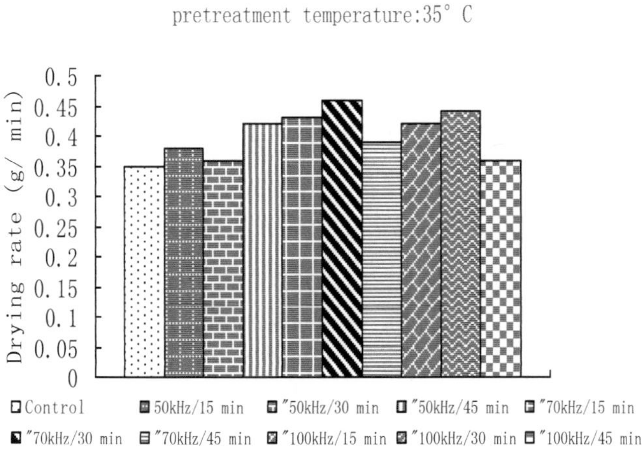

Figure 4 Effect of ultrasonic frequency and pretreatment time on average drying rate of scallop muscle at 35°C.

3.2. Influence of ultrasonic pretreatments on shrinkage and rehydration ratio of scallop muscle

The influence of ultrasonic frequency and pretreatment time for different temperature on shrinkage and rehydration ratio of scallop muscle was shown in Table 1. It can be seen from the table that the shrinkage of ultrasonic pretreatment scallop muscle was lower than that of the control. But, rehydration ratio results showed that the pretreatment improves rehydration capability of scallop muscle.

Table 1. Three Scheme comparing.

Shrinkage		Rehydration ratio	
Pretreatment group (70 kHz for 30min at 25°C)	control group	Pretreatment group (70 kHz for 30min at 25°C)	control group
71.1	71.3	125.7	122.2

4. Conclusion

The drying rate and quality of scallop muscle can be improved by ultrasonic pretreatment, and while ultrasonic pretreatment was applied by 70 kHz for 30min at 25°C, the drying rate could be increased approximately 29%.

Acknowledgments

This work was supported by the Scientific Research Fund of Liaoning Provincial Education Department under contract No (L2013273)

References

[1] G.C. Zhang, Z.H. Mao, C.X. Mu, Combined vacuum-microwave and hot-air drying of scallop, Trans. CSAE. 21 (2005) 144-147.
[2] Y.X. Bai, G.J.Yang, Y.C. Hu, M. Qu, Physical and sensory properties of electrohydrodynamic (EHD) dried scallop muscle, J. Aquat. Food Prod. Technol. 21 (2012) 238-247.
[3] N.I. Lebovka, I. Praporscic, E. Vorobiev, Enhanced expression of juice from soft vegetable tissues by pulsed electric fields, Consolidation stages analysis, J. Food Eng. 59 (2005) 309-317.
[4] M. Fincan, F. DeVito, P. Dejmek, Pulsed electric field treatment for solid-liquid extraction of red beetroot pigment, J. Food Eng. 64 (2004) 381-388.
[5] E. Amami, L. Khezami, E. Vorobiev, N. Kechaou, Effect of pulsed electric field and osmotic

dehydration pretreatment on the convective drying of carrot tissue, Drying Technol. 26 (2008) 231-238.
[6] Y.L. Wu, Y.M. Guo, D.G. Zhang, Study of the Effect of High-Pulsed Electric Field Treatment on Vacuum Freeze-Drying of Apples. Drying Technol. 29 (2011) 1714-1720.
[7] R. Soliva-Fortuny, A. Balasa, D. Knorr, Effects of pulsed electric fields on bioactive compounds in foods: A review, Trends Food Sci Tech. 20 (2009) 544-556.
[8] D. Gennaro, S. Cavella, D. Silva, The use of ultrasound in food technology I: Inactivation of peroxidase by thermosonication, J. Food Eng. 39 (1999) 401-407.
[9] D. Yun, Y.Y. Zhao, Effects of pulsed-vacuum and ultrasound on the osmodehydration kinetics and microstructure of apples (Fuji), J. Food Eng. 85 (2008) 84-93.
[10] A.N. Fabiano, L.F. Fernandes, S. Rodrigues, Ultrasound as pre-treatment for drying of pineapple, Ultrasonic Sonochemistry, 15 (2008) 1049-1054.
[11] K. Schossler, H. Jaeger, D. Knorr, Novel contact ultrasound system for the accelerated freeze-drying of vegetables, Innov Food Sci Emerg. 16 (2012) 113-120.
[12] M. Nowacka, A. Wiktor, M. Sledz, D. Witrowa-Rajchert, Drying of ultrasound pretreated apple and its selected physical properties, J. Food Eng. 113 (2012) 427-433.

IOP Conf. Series: Earth and Environmental Science **64** (2017) 012120 doi:10.1088/1755-1315/64/1/012120

Preparation and Physiological activities of sulphated derivative extracted from corn bran

Qing Mo[1], Linghao Dai[2], Jianjun Ma[1], Xiaojing Zhao[1] and Linghui Zhu[2,*]

[1]Hangzhou Medical College, Hangzhou, China
[2]Zhejiang Chinese Medical University, Hangzhou, China

*Corresponding author e-mail: jhzlh1125@163.com

Abstract. In the present study, the sulphated derivative (S-CBP) with the degree of substitution (0.46) was successfully prepared from the polysaccharide extracted from corn bran. Compared with native polysaccharide, the structures of the sulphated derivative were confirmed by FT-IR and SEC-LLS and the molecular weight were changed by chemical modification. Sulfation enhanced the antioxidant activities in a dose-dependent way, which seemed to be dependent on the character of the substituted group. The results suggest that the sulphated derivative, extracted from corn bran, are potential natural antioxidant and blood fat reduce agent.

1. Introduction

Now days, an increasing number of studies have been published concerning the biological activities of polysaccharides from low-valued agricultural industry [1-3]. Researchers of corn have increased because of its uses as adhesive, thicker and stabilizer [4] and a film former and emulsifier [5-7].

To promote comprehensive utilization of corn bran, it is critical to understand the effects of chemical modification on physiological activities in detail. However, no efforts have yet been made to improve the physiological activities. In the present work, the polysaccharide, extracted from corn bran, were modified by sulfation. Instantly, their physiological activities such as antioxidant and bile acid binding capacity were investigated as a function of the effect of modified group

2. Materials and methods

2.1. Material and reagents

Corn bran was collected in Hangzhou, Zhejiang Province, China in 2015, and de-oil, de-starched according to Yadav *et al.* [7]. The powder of the materials was screened through a 40 mesh sieve and stored in a desiccator at room temperature. All the reagents were purchased from Aladdin Industrial Co. and used without further purification.

2.2. Extraction of polysaccharide

On the basis of previous method [7], the de-oil and de-starched material was extracted and centrifuged. The supernatant was concentrated, deproteinized using the Sevag method and then precipitated upon addition of absolute ethanol. The residue was collected by centrifuged at 5000rpm for 10min, dialyzed, lyophilized and then reserved in a dryer.

2.3. Preparation of sulphated derivative

Sulfation of the polysaccharide was carried out using the chlorosulfonic acid-pyridine method [7]. The sample obtained was coded as S-CBP.

2.4. Physicochemical characterization

2.4.1 Degree of substitution of sulphated derivative. The sulfur content of the sulphated polysaccharide was estimated using the benzidine method [1].The DS, which is the average number of sulfate group on each sugar residue, was calculated from the sulfur content using the following equation.

$$DS = (1.62 \times S\ \%)/(32 - 1.02 \times S\ \%) \quad (1)$$

Where, S% was the mass fraction of S atom.

2.4.2 FT-IR analyses. Fourier transform infrared spectra (FT-IR) of the derivatives were obtained by Nicolet FT-IR spectrometer (Magna-IR 760 E.S.P, Nicolet Instrument Corp., Madison, WI). Samples were ground with potassium bromide (KBr) at a ratio of 1:20 and pressed into a thin pellet for FT-IR analyses.

2.4.3 SEC-LLS measurement. According to our previous research, the Homogeneity and molecular mass were measured Size-exclusion chromatography combined with laser light scattering measurements were performed on the eight-angle laser photometer [9].

2.5 In vitro antioxidant activity assay

2.5.1 Hydroxyl radical scavenging assay. The hydroxyl radical scavenging activity was analyzed according to our previous researcher [9, 10].

$$\text{Scavenging effect } (\%) = (1 - (A_1 - A_2)/A_0) \times 100\% \quad (2)$$

2.5.2 Superoxide anion scavenging assay. The assay was based on the capacity of the sample to inhibit the photochemical reduction of nitroblue tetrazolium in NADH-NBT-PMS according to our previous research [9, 10].

$$\text{Scavenging effect } (\%) = (1 - A_1/A_0) \times 100\% \quad (3)$$

Where A_0 was the absorbance of the control (without sample) and A_1 was absorbance in the presence of the sample.

2.6 Bile acid binding capacity

Based on the method of our previous research [9], the effect of the samples on the bile acid capacity was investigated *in vitro*.

Statistical Analysis: All experiments were carried out in triplicate. Statistical analysis was performed using Analysis of Variance (ANOVA) and the significances of the differences between samples were determined using Duncan`s multiple range test. Statistical significance was set at a level of $P < 0.05$.

3. Results and discussion

3.1 FT-IR spectroscopy

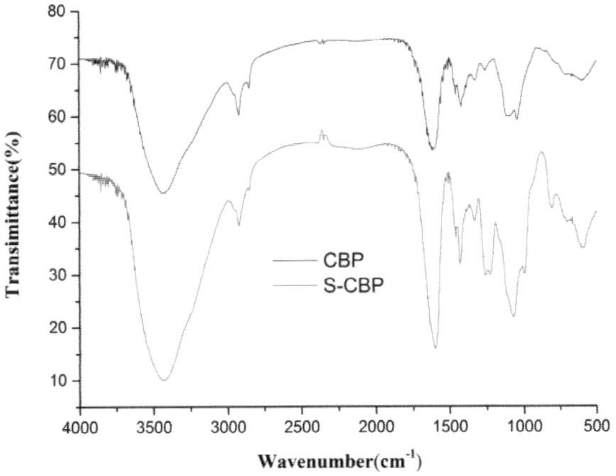

Fig.1 FT-IR spectrum of CBP and S-CBP.

The characteristic absorption peaks of polysaccharide were recorded, such as the broad intense peak at around 3400cm^{-1} related to stretching vibration of –OH and the weak peak at around 2900cm^{-1} attributed to the stretching vibration and deformation vibration of C-H. The peak at 1039cm^{-1} was due to C-O-C unsymmetrical stretching and also confirmed the presence of monopyranose component [12,13].After sulfation, two new characteristic absorption bands at 1200cm^{-1} described an asymmetrical S=O and another at 810cm^{-1} represented a symmetrical C-O-S, confirming the existence of sulfate group. The results were similar to the observations reported in many researches [12].

3.2 Homogeneity and molecular mass

Table 1. The molecular mass and degree of substitution of the samples

	M*w*/kDa	M*w*/Mn	R(avg)	DS
CBP	335	1.78	28.9	/
S-CBP	178	1.82	27.7	0.44

The profiles of SEC-LLS chromatograms all showed two symmetrical peaks as showed, which indicated each sample was a homogenous polysaccharide [11]. Table 1 revealed that the weight-average molecular weight (M*w*) of CBP and S-CBP were 335kDa and 178kDa, respectively. Compared with the native polysaccharide, the molecular mass of the sulphated derivative degraded apparently [14].

3.3 Antioxidant activities

3.3.1 Hydroxyl radical scavenging activity. The hydroxyl radical scavenging effect is one of the most crucial indexes for the antioxidant activity [9, 10]. Fig.2 described that the inhibitory effects on the hydroxyl radical of the polysaccharides with ascorbic acid used as the reference. The hydroxyl radical scavenging effect was in the order of S-CBP>CBP, indicating chemical modification could improve the hydroxyl radical scavenging activity. Since the native polysaccharide and the sulphated derivatives exhibited little ferrous ions chelating activity, the mechanism of hydroxyl radical scavenging were mainly related to scavenge the hydroxyl radicals directly [9]. As a consequence, the sulfation would be an efficient way to improve the hydroxyl radical scavenging activity.

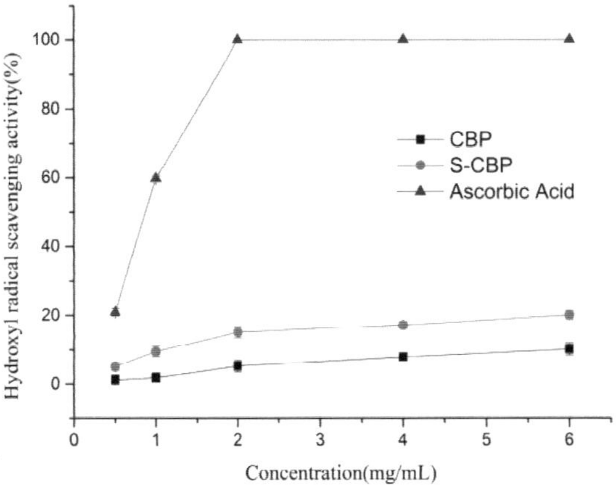

Fig.2 Hydroxyl radical-scavenging activity of CBP and S-CBP. Each point is the mean ± SD of triplicates.

3.3.2 Superoxide anion-scavenging activity. Fig.3 showed the percentage inhibition of superoxide anion radical generation for different concentrations of the polysaccharides, compared with ascorbic acid. The S-CBP showed the higher scavenging ability (85.09 ± 1.26%) at the concentration of 6mg/mL, while the ascorbic acid scavenging ability was 65.73%, which were in accordance with Wang *et al.* [9]. The enhanced superoxide radical scavenging ability was related to the characters of the substituted groups. Consequently, the sulphated derivative would be one of the most efficient ways to improve the superoxide radical scavenging ability.

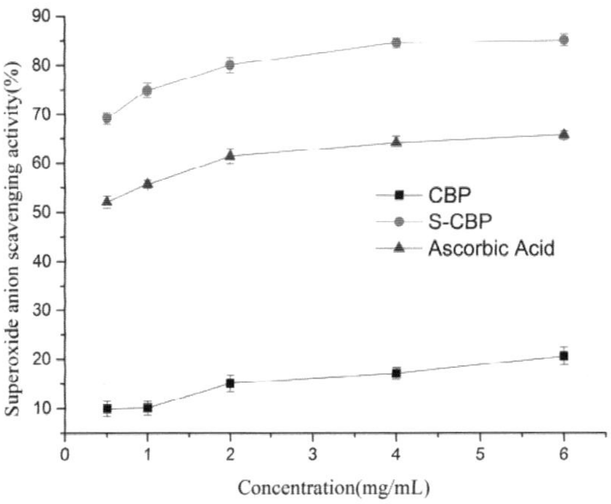

Fig.3 Superoxide anion-scavenging activity of CBP and S-CBP. Each point is the mean ± SD of triplicates.

3.4 Bile acid binding capacity

The Fig.4 described the bile acid binding capacities of the samples at the concentration of 6mg/mL. And, S-CBP showed the better bile acid binding capacity (2.98μM). Compared with native polysaccharide, the bile acid binding capacity S-CBP strengthened moderately. Except for water solubility, numerous factors such as ionic interaction, hydroxyl group interaction and trapping in the overall bile acid-binding capacity of polymers [9, 15], might explain the enhanced bile acid-binding capacity of the derivatives *in vitro* in the present study.

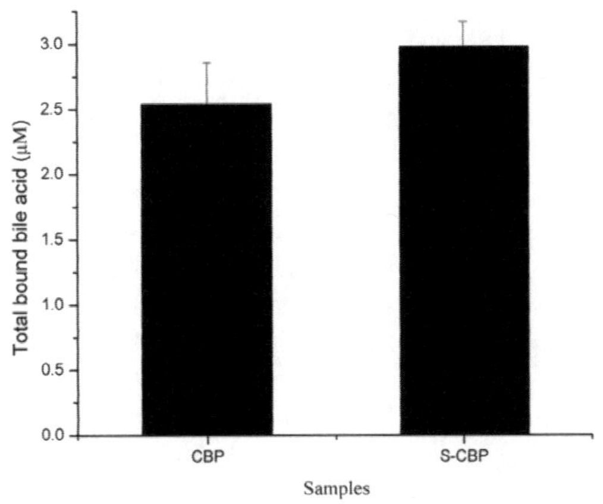

Fig. 4 Bile acid binding capacity of CBP and S-CBP.

4. Conclusion

In the present study, the sulphated derivative (S-CBP) with the degree of substitution (0.46), were successfully prepared from the polysaccharide extracted from corn bran (CBP). Compared with the native polysaccharide, the molecular weights were changed by chemical modification and the structures were confirmed by FT-IR and SEC-LLS. The sulfation enhanced the antioxidant activities in a dose-dependent way, which seemed to be dependent on the character of substituted group. And the sulfation of CBP strengthened the bile acid binding capacity moderately. To promote comprehensive utilization of the corn bran, more researches on modification and the toxicological experiments need to be further focused on.

Acknowledgments

This work were financially supported by Zhejiang Chinese Medical University fund (No. 2015ZY30) and Zhejiang Medical College fund (No. 2015XZA03).

References

[1] J.T. Wang, Q. Wang, J.R. Han, Yield, polysaccharides content and antioxidant properties of the mushroom Agaricussubrufescens produced on different substrates based on selected agricultural wastes. Scientia Horticulturae, 157(2013):84-89.
[2] A. Slavov, H. Kiyohara, H. Yamada, Immunomodulatingpectic polysaccharides from waste rose petals of Rosa damascene Mill. International Journal of Biological Macromolecules, 59(2012):192-200.
[3] S.B. Liang, A.G. McDonald, E.R. Coats, Lactic acid production with underfined mixed culture fermentation of potato peel waste. Waste Management, 11(2014):2022-2027.
[4] F. Zhang, L.N. Tu, D. Kang, Q.W. Jin, H.B. Zhang, M.P. Yadav, Viscosifying properties of

corn fiber gum with various polysaccharides. Food Hydrocolloids, 43(2015):218-227.

[5] M.P. Yadav, D.B. Johnston, K.B. Hicks, Corn fiber gum: new structure/function relationships for this potential beverage flavor stabilizer. Food Hydrocolloids, 23(2009):1488-1493.

[6] M.P. Yadav, D.B. Johnston, K.B. Hick, E.A. Nothnagel, The role of lipids and protein components in the emulsification properties of gum Arabic and corn fiber gum. Foods and Food Ingredients Journal, 3(2006):245-252.

[7] M.P. Yadav, R.A. Moreau, A.T. Hotchkiss, K.B. Hicks, A new corn fiber gum polysaccharide isolation process that preserves functional components, Carbohydrate Polymers, 87(2012):1169-1175.

[8] Y. Chen, H. Zhang, Y.X. Wang, S.P. Nie, C. Li, M.Y. Xie, Sulfated modification of the polysaccharides from Ganodermaatrum and their antioxidant and immunomodulating activities. Food Chemistry, 186(2015):231-238.

[9] Y.J. Wang, Q. Mo, Z.N. Li, H.W. Lai, J. Lou, S.W. Liu, J.W. Mao, Effects of degree of carboxymethylation on physicochemical and biological properties of pachyman. International Journal of Biological Macromolecules, 51(2012):1052-1056.

[10] C.Y. Huang, S.J. Wu, W.N. Yang, A.W. Kuan, C.Y. Chen, Antioxidant activities of crude extracts of fucoidan extracted from Sargassumglaucescens by a compressional-puffing-hydrothermal extraction process. Food Chemistry, 197(2016):1121-1129.

[11] Y.B. Shen, H. Zhang, L.L. Cheng, L. Wang,. H.F. Qian, X.G. Qi, In vitro and in vivo antioxidant activity of polyphenols extracted from black highland barley, Food Chemistry, 1(2016):1003-1012.

[12] S. Bagchi, K.J. Kumar, Studies on water soluble polysaccharides from Pithecellobiumdulce (Roxb.)Benth. Seeds, Carbohydrate Polymers, 138(2016):215-221.

[13] X.Y. Li, L. Wang, Effect of extraction method on structure and antioxidant activity of Hohenbueheliaserotina polysaccharides. International Journal of Biological Macromlecules, 83(2016):270-276.

[14] Y. Liu, C. Liu, H. Tan, T. Zhao, J. Cao, F. Wang, Sulfation of a polysaccharide obtained from Phellinus ribis and potential biological activities of the sulfated derivatives. Carbohydrate Polymers, 77(2009):370-375.

[15] J.L. Wang, W. Yang, Y.Y. Tang, Q. Xu, S.L. Huang, J. Yao, J. Zhang, Z.Q. Lei, Regiselectivesulfation of Artemisia sphaerocephala polysaccharide: Solution conformation and antioxidant activities in vitro. Carbohydrate Polymers, 136(2016):527-536.

IOP Conf. Series: Earth and Environmental Science **64** (2017) 012121 doi:10.1088/1755-1315/64/1/012121

Toxic effects of combined effects of anthracene and UV radiation on *Brachionus plicatilis*

Ceng Gao[a], Xinxin Zhang[b], Ningning Xu[c] and Xuexi Tang[*]

College of Marine Life Science, Ocean University of China, Qingdao 266003, China

[a]gokeystar@163.com, [b]Zhangxinxin0423@163.com, [c]xuning0901@163.com,
[*]Corresponding author e-mail: tangxx@ouc.edu.cn

Abstract. Anthracene is a typical polycyclic aromatic hydrocarbon, with photo activity, can absorb ultraviolet light a series of chemical reactions, aquatic organisms in the ecosystem has a potential light induced toxicity. In this paper, the effects of anthracene and UV radiation on the light-induced toxicity of *Brachionus plicatilis* were studied. The main methods and experimental results were as follows: (1) The semi-lethal concentration of anthracene in UV light was much lower than that in normal light, The rotifers have significant light-induced acute toxicity. (2) Under UV irradiation, anthracene could induce the increase of ROS and MDA content in *B. plicatilis*, and the activity of antioxidant enzymes in *B. plicatilis* significantly changed, Where SOD, GPx activity was induced within 24 hours of the beginning of the experiment. And the content of GPX and CAT was inhibited after 48 hours. Therefore, the anthracite stress induced by UV radiation could more strongly interfere with the ant oxidative metabolism of *B. plicatilis*, and more seriously cause oxidative damage, significant light-induced toxicity.

1. Introduction

Polycyclic aromatic hydrocarbons are difficult to degrade in nature, can exist for a long time in the environment, and with the atmosphere, such as migration around the way. At present, PAHs are not only present in the environment of human activities, but also in the Arctic, deep sea and other remote ecosystems can also find its traces [1]. Because of its strong fat-soluble and high n-octanol /water partitioning coefficients, PAHs are easily absorbed and accumulated, and are enriched by the food chain, resulting in a threat to human health [2]. PAHs have photo activity, can absorb ultraviolet light to produce a series of chemical reactions, the aquatic organisms in the ecosystem there is a potential light induced toxicity [3]. At present, there have been some researches on the role of PAHs in bio-light-induced toxicity at home and abroad [4-6]. The results show that the photo-sensitization is one of the important mechanisms of PAHs light-induced toxicity, and the photo-sensitization reaction causes a large number of active oxygen clusters (ROS) [7]. When the ROS production rate exceeds the rate of removal of antioxidant system in the body, it will cause anti-oxidation reaction, it will cause damage to the biofilm, but also and the body of protein, nucleotides and other biological macromolecules interaction, Damage and death of living cells [8-10].

It has an important role in the circulation and energy flow of marine ecosystems because of its advantages such as extensive distribution, short life cycle, fast breeding speed, easy cultivation and sensitive reaction to exogenous toxicants [11, 12]. However, there is almost no study of PAHs on light-induced toxicity of rotifers. In this experiment, the most active anthracene was used as the stress factor under UV irradiation, and the effect of anthracene on the light induced toxicity of *B. plicatilis* was studied under the condition of ultraviolet radiation. In this study, the blanks of anthracene on the light-induced toxicity of C. mongolicus were studied, which provided a basic theoretical basis for comprehensively evaluating the effect of anthracene on the photoinduced toxic effects of marine zooplankton.

This study fills up the blank of the photoinduced toxicity effect of anthracene on the rotifer, and provides a theoretical basis for the comprehensive evaluation of the photoinduced toxicity of anthracene to marine zooplankton.

2. Experimental materials and methods

2.1. Experimental materials

In this experiment, we selected the lively and robust *B. plicatilis* from the Ocean University of China. The container was cultured with 1.5 L beaker and placed in a GXZ-3008 intelligent light incubator. The incubation temperature was maintained at 20 ± 1 ° C and the light intensity was 1313.8 ± 54.8 μW/(cm^2•s^{-1}), the illumination period was 12 h: 12 h. Before the start of the formal experiment, *B. plicatilis* had been acclimated for three months under experimental ecological conditions.

2.2. Light source settings and experimental devices

The acute toxicity test was carried out in a constant temperature light incubator. The UV radiation system mainly consisted of two UV-A fluorescent tubes and one UV-B fluorescent tube with visible light irradiance of 1313.8 ± 54.8 μW/cm^2; UV-A irradiation UV-B and UV-B can effectively simulate the sunlight in the natural environment. The UV-B and UV-B can be used to simulate the sunlight in the natural environment.

2.3. Experimental design

2.3.1. Determination of the lethal concentration of anthracene in B. plicatilis under different light conditions. A total of 480 healthy larvae of B. plicatilis were selected from the pre-cultured B. plicatilis and were exposed to anthracene solutions at different concentrations of different light sources (Table 2.3.1) The Each concentration gradient takes 20 larvae, each with three parallel samples. 24 hours to record the number of dead insects in each group of rotifers.

2.3.2. Effects of anthracene on the microstructure of B. plicatilis induced by UV radiation. The morphological structure of rotifers in 10 μg/L anthracene-treated group and 10 μg/L anthracene-treated group induced by UV radiation was observed in the control group after 24 hours of stress.

2.3.3. Determination of antioxidant enzyme activity. In this part, the concentration of two anthracene solutions was 0.4 μg/L and 4 μg/L, respectively. After pre accumulation of anthracene, transfer to the new medium, with / without UV radiation treatment. The larvae of Brachionus plicatilis were exposed to UV radiation and UV-free radiation respectively, and quantified to each experimental concentration. Each concentration was three parallel, the rotifer density was 100/mL, and no bait was given during the application. Each parallel had a total of 10,000 fold Brachionus (about 0.2 g). Exposure experiments were carried out at 24 h and 48 h, respectively, and the Brachionus plicatilus was collected. Add the concentration of 0.86% saline, in the ice bath under the conditions of ultrasonic cell breaker broken, every 5 s, interval 10 s, 5 times after the break completely. After crushing, centrifuge at 0 ° C and 2500 r / min for 10 minutes. The supernatant was then used to determine the total protein TSP and MDA contents and the activity of SOD, CAT and GPX in the crude enzyme solution.

(1) Determination of protein content

The protein concentration of *B. plicatil* was determined by BCA Protein Concentration Kit. The operation is carried out according to the product brochure. The procedure is as follows: ① 50 volume BCA reagent A plus 1 volume BCA reagent B (50: 1) preparation of BCA working solution 5mL, is now with the use; ② Take 10μL protein standard diluted to 100μL, the final concentration of 0.5 mg/mL; ③ the standard by 0,1,2,4,8,12,16,20 μL added to the 96-well plate of the standard hole, plus the standard for the dilution of the standard set to 20μL; ④ add appropriate volume of sample to the sample hole of 96-well plate, dilute the standard to 20mL; ⑤ each hole to join 200μL BCA working fluid, room temperature placed 10min; ⑥ Determination of absorbance at wavelength 562nm, according to the standard curve to calculate the protein concentration. The protein concentration of each group of samples was diluted with PBS buffer to determine the protein concentration in each sample.

(a) MDA content detection

Use the Nanjing Institute of Bioengineering Institute of the determination of the kit. Principle: MDA in lipid peroxides can be condensed with thiobarbituric acid (TAB) to form a red product with a maximum absorption peak at 532 nm and then quantitatively determined by colorimetric method. Defined as per mg of tissue protein MDA in the nanomolar number.

(b) Determination of SOD activity

Use the Nanjing Institute of Bioengineering Institute of SOD kit determination. When the sample contains SOD, xanthine and xanthine oxidase reaction system produced by the superoxide anion radicals have a specific inhibitory effect, so that the oxidation of hydroxylamine formed nitrite reduced, can be colorimetric determination, thus SOD activity was determined. The amount of SOD corresponding to the amount of SOD per mg of tissue protein in a 1 ml reaction solution was 50%, and the activity was defined as a viable unit (U).

(c) Determination of POD activity

Refer to Nanjing Institute of Bioengineering POD kit. POD activity determination principle: in the presence of H2O2, POD can catalyze the oxidation of guaiacol to produce brown material, and then by measuring its absorbance at 420 nm to obtain POD activity. The POD activity in the tissue is defined as the activity of 1μg of guaiacol substrate per mg of protein per minute at 37°C for a POD activity unit.

(d) Determination of CAT activity

CAT determination using catalase test box, purchased from Nanjing built bioengineering research institute. The main principle is that the addition of ammonium molybdate in the process of H_2O_2 formation of H_2O to H_2O_2, the rapid termination of the reaction, the remaining H_2O_2 and ammonium molybdate action to produce a light yellow complex, at 405 nm to detect the absorption of light Value, can calculate the vitality of CAT. A viable unit is defined as the amount of H_2O_2 in the tissue homogenate that decomposes 1μmol per second of tissue protein per second.

2.4. Data processing

The data were analyzed by software SPSS19.0, and the data were analyzed by software Sigmaplot12.5. The data of the experimental group and the blank group were analyzed by one-way ANOVA. Use multiple comparisons (LSD) to analyze the variance of the data. Use Dunnett's T3 to analyze variance data. When $P < 0.05$, the difference was significant, $P < 0.01$, the difference was significant. The change of each index is given by mean ± standard error (mean ± SE) (n = 3).

3. Experimental results

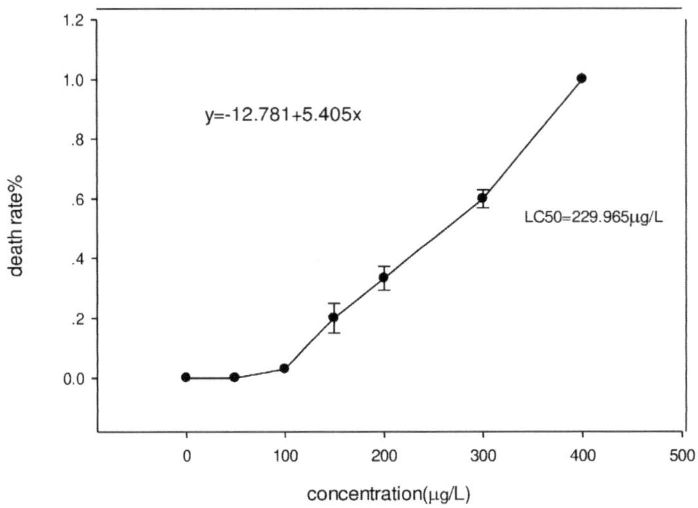

Fig 1 Effect of different concentrations of ANT on the dead rate of B. plicatilis individuals at 24 h

As shown in Figure 1, with the increase in anthracene concentration, the number of dead individuals of the *B. plicatilis* increased and the mortality rate increased. Linear regression analysis of anthracene concentration and mortality at 24 hours of rotifers under normal culture conditions (Chen Jianhong et al., 2005). The regression equation is: Y = 5.450X-12.871, R^2 = 0.976 where P <0.01, half of the lethal concentration LC50 = 229.97 μg/L.

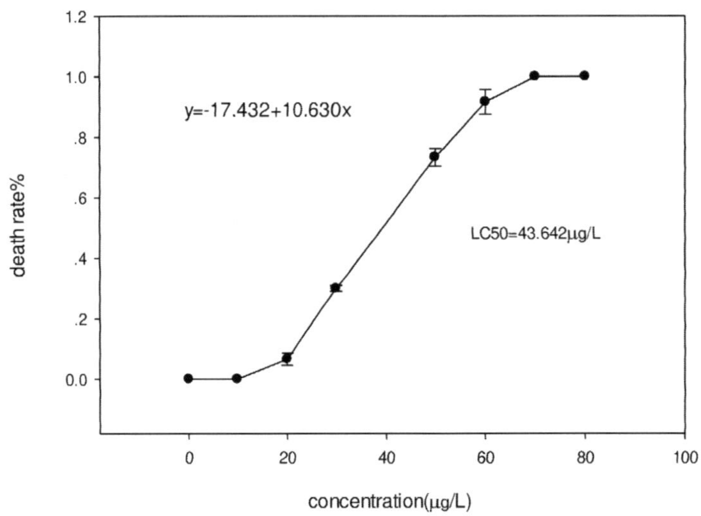

Fig 2 Effect of different concentrations of ANT on the death rate of *B. plicatilis* individuals under UV-A and UV-B radiation at 24 h

It can be seen from Figure 2 that the acute toxicity of anthracene to *B. plicatilis* has a significant concentration-effect relationship under UV-A and UV-B radiation conditions. The regression equation was obtained as follows: Y = 10.63X - 17.43, R^2 = 0.944 where P <0.01, half of the regression equation for the death rate and mortality of rotifers induced by UV radiation (Chen Jianhong et al., 2005) Lethal concentration LC50 = 43.64 μg/L.

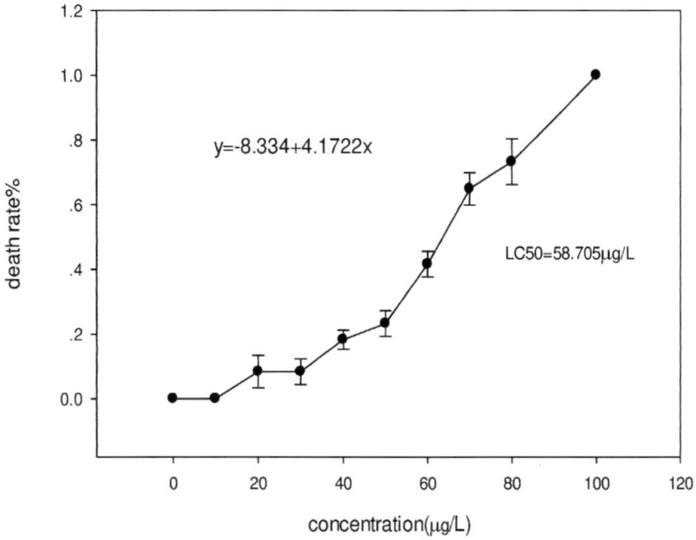

Fig 3 Effect of different concentrations of ANT on the death rate of *B. plicatilis* individuals under the UV-A radiation at 24 h

Figure 3 shows the effect of different concentrations of anthracene on the survival rate of *B. plicatilis* under UV-A induction. In combination with UV-A and UV-B, the lethal rate of anthracene to rotifers was dose-dependent only under UV-A radiation. According to the linear regression analysis of anthracene concentration and mortality of rotifers for 24 hours under UV-A irradiation conditions (Chen Jianhong et al., 2005), the regression equation was: Y = 4.17X-8.33, $R^2 = 0.921$ where $P < 0.01$, the half lethal concentration LC50 = 58.71 μg/L.

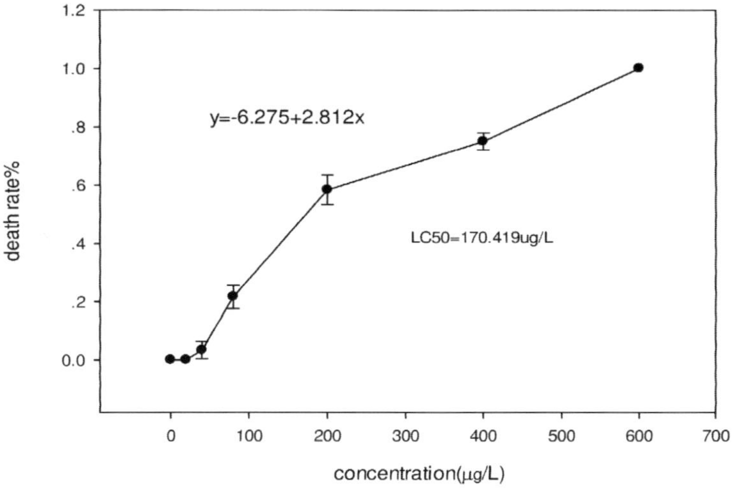

Fig 4 Effect of different concentrations of ANT on the death rate of *B. plicatilis* individuals under the UV-B radiation at 24 h

Under the condition of UV-B radiation, the mortality of *Brachionus plicatilis* increased with the increase of anthracene concentration (Fig. 4). Linear regression analysis was performed on 24 hours of death and mortality in rotifers under UV-B exposure (Chen et al., 2005). The regression equation was: Y = 2.81X -6.27, $R^2 = 0.98$ where $P < 0.01$, the median lethal concentration LC50 = 170.42 μg/L.

3.1. Effects of anthracene on the morphological structure of Brachionus plicatilis induced by UV radiation

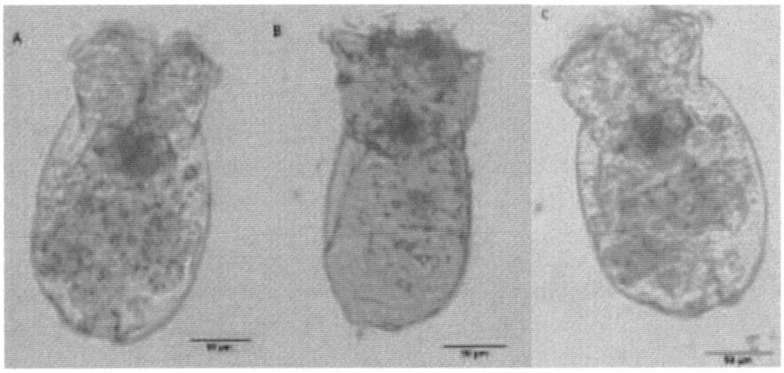

Fig 5 Effect of ANT with different concentrations on B.plicatilis morphology under inverted microscopy at 24h. A: Blank control; B: treantment with 10 μg /L ANT under UV irradiance C: treantment with10μg /L ANT.

As shown in Figure 5, the morphological structure of the rotifers in the 10μg/L anthracene-treated group and the 10μg/L anthracene-treated group, which was induced by UV radiation in the control group, was observed under the inverted microscope for 24 hours. The results showed that in the control group, the rotifers were full and spherical, and the inside of the body cavity was enriched, and the internal tissues were clearly discernible. Compared with the normal individuals, anthracene stress had little effect on the microstructure of rotifers. Under the induction of UV radiation, the individual morphology of rotifers changed significantly under anthracnogenic stress, which showed that the rotifers were thin and trendy. And the internal blur, the color becomes darker; the organ tissue shrinks to the middle of its body, and the opacity increases.

3.2. Effects of UV radiation and anthracene on the total protein content (TSP) of B. plicatilis

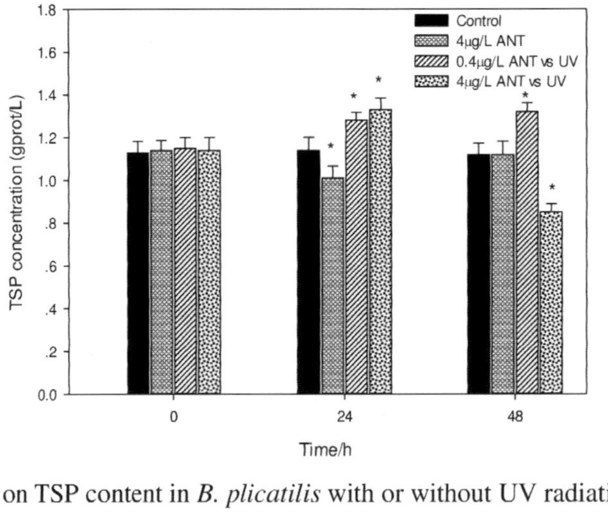

Fig 6 Effect of ANT on TSP content in *B. plicatilis* with or without UV radiation exposure. Note: * represents statistically significant differences between control and treatments at the $P < 0.05$ level.

IOP Conf. Series: Earth and Environmental Science **64** (2017) 012121 doi:10.1088/1755-1315/64/1/012121

The effect of anthracene stress on TSP content in *B. plicatilis* was shown in Fig. 6. Compared with the control group, the total protein content of *B. plicatilis* decreased significantly at 24 h in the 4 μg/L anthracene treatment group, which was only 88.59% of the control group, and the difference was significant ($P < 0.05$) and then rose to the control group level at day 48. When the UV radiation and 0.4μg/L anthracene stress, the content of TSP in *B. plicatilis* was induced at the 24th hour, which was about 1.11 times ($P < 0.05$), and the TSP ($P < 0.05$); In the UV radiation and high concentration anthracene compound treatment group, the content of TSP in *B. plicatilis* increased to the highest level in the initial stage, and reached the initial value of the control group ($P < 0.05$) ($P < 0.05$), and then decreased rapidly at 48 hours, only 74.56% of the control group ($P < 0.05$). The results showed that the content of TSP in Cynodon dactylon was increased first and then decreased under the condition of UV radiation. The change of TSP content was related to anthracene concentration.

3.2.1. Lipid Peroxidation of B. plicatilis by UV radiation and anthracene

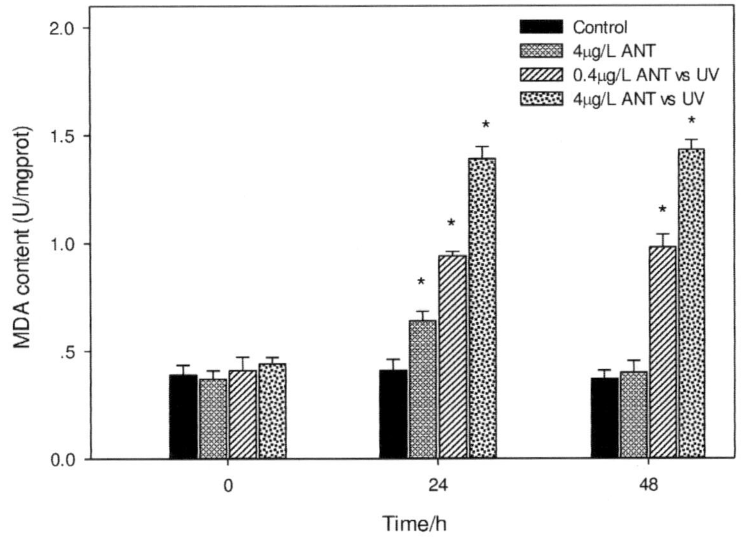

Fig 7 Effect of different concentrations on MDA levels affected in *B. plicatilis* with or without UV radiation exposure. Note: *represent statistically significant differences between control and treatments at the $P < 0.05$ level.

According to Figure 7, it is not difficult to find that the changes of MDA level and the change of ROS content in *B. plicatilis* have similar trends. In the 4 μg/L anthracene treatment group, the MDA in the *B. plicatilis* increased significantly ($P < 0.05$) at 24 h and then decreased to the control group at the 48th hour. When the UV radiation was applied, the content of MDA in the *B. plicatilis* increased gradually, which was about 2.29 times and 2.39 times higher than that of the control group at the initial stage of stress, 0.4 μg/L and 4 μg/L anthracene treatment. The content of MDA in the *B. plicatilis* increased ($P < 0.05$). The results showed that UV radiation and anthracene complex induced more MDA in the *B. plicatilis*. Therefore, the anthrax stress under UV radiation increased the lipid peroxidation of *B. plicatilis* and caused more severe oxidative damage.

3.2.2. Effects of UV radiation and anthracene on SOD activity in B. plicatilis

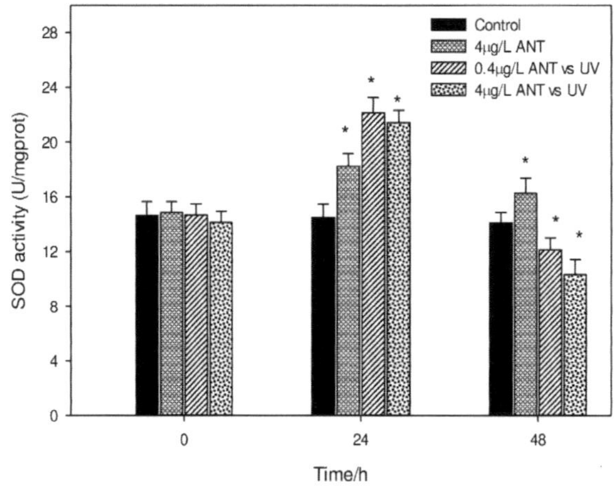

Fig 8 Effect of different concentrations of ANT on SOD activities in *B. plicatilis* with or without UV radiation exposure. Note: * represent statistically significant differences between control and treatments at the $P<0.05$ level.

The effect of anthracene stress on SOD activity in *Brachionus plicatil* was shown in Figure 8. After 24 hours, the activity of SOD in *B. plicatilis* was significantly increased by 4 μg/L anthracnose stress, which was 1.23 times of that of the control group, which was significantly different from that of the control group (P <0.05), and then inhibited at 48th hour Higher than the control group (P <0.05). When the UV radiation was applied, the activities of SOD in the B. plicatil were significantly increased with the increase of the stress time, and then the inhibitory trend was significantly inhibited. The SOD activity of *B. plicatilis* in the anthracene-treated group was higher than that in the non-UV-irradiated anthracene-treated group, and the activity of SOD was inhibited at 48 hours, which was much higher than that in the non-UV-irradiated 4 μg/L anthracene Group. The results showed that the SOD activity induced by UV radiation decreased the SOD activity of Brachionus plicatili significantly more strongly.

3.2.3. Effects of UV radiation and anthracene on POD activity of B. plicatilis

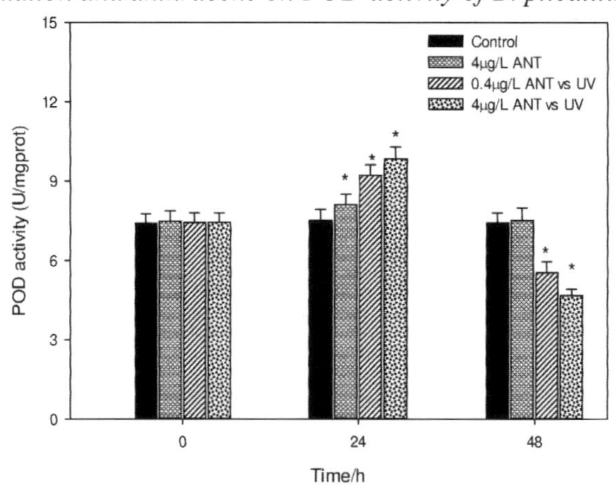

Fig 9 Effect of different concentrations of ANT on POD activities in *B. plicatilis* with or without UV radiation exposure. Note: * represent statistically significant differences between control and treatments at the $P<0.05$ level.

The effect of anthracene stress on POD activity in *B. plicatil* was shown in Figure 9. Compared with the control group, the POD of *C. dactylon* in 4 μg/L anthracnon stress increased first and then decreased to the control group (P>0.05). The activity of POD in *B. plicatilis* was strongly induced at 24 hours in the two anthracene-treated groups induced by UV radiation, which were 1.24,1.33 for the initial values of the control group (P <0.05), and it was strongly inhibited at 48th day, only 74.56% and 62.69% of the control group (P <0.05). At 24 o, the POD activity of *B. plicatilis* in the 4 μg/L anthracene group induced by UV radiation was higher than that in the 4 μg/L anthracene group without UV, while the 48% inhibition of POD activity was much greater No UV radiation in the 4 μg/L anthracene group. The results showed that anthracene stress had stronger effect on POD activity under UV radiation.

3.2.4. Effects of UV radiation and anthracene on CAT activity of B. plicatilis

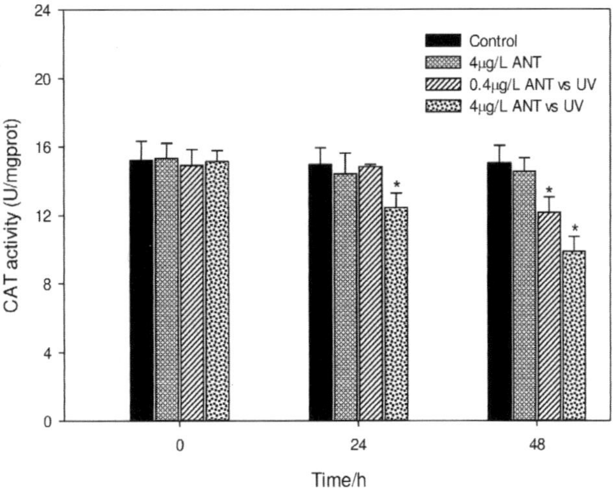

Fig 10 Effect of different concentrations of ANT on CAT activities in *B. plicatilis* with or without UV radiation exposure. Note: * represent statistically significant differences between control and treatments at the $P<0.05$ level.

Effects of UV and anthracene on CAT activity in *B. plicatil* were shown in Figure 10. The CAT activity of *B. plicatilis* did not change significantly with time in the 4 μg/L anthracene treatment group (P> 0.05). The activity of CAT in the *B. plicatilis* did not change significantly at 24 hours compared with the control group (P> 0.05), and the difference between the control group and the control group was not significant (P> 0.05). At the 48th hour, the activity was strongly inhibited, only 81.89% of the control group (P <0.05). In the UV radiation and 4 μg/L anthracene complex stress group, the CAT content of *B. plicatilis* was always strongly inhibited (P <0.05). The results showed that the activity of CAT was lower than that of anthracene in the anthracene-treated group, and the effect of anthracene on CAT activity was related to anthracene concentration.

4. Discussion

Pterodactylus is a zooplankton that lives in water for life. Exogenous poisons are introduced into the *B. plicatilis* by direct absorption. Therefore, exposure to acute toxicity test is an important aspect of the acute toxicity test of *B. plicatilis*. Through preliminary understanding of the characteristics and intensity of the action of exogenous pollutants on *B. plicatilis*, the target organ of the action can be judged, and its toxicity and potential risk are evaluated. In this study, it was found that anthracene had a significant effect on the survival of *B. plicatilis* in the case of UV radiation, and its effective concentration range was far lower than the effective concentration range without UV radiation. It was concluded that anthracene had acute light-induced toxicity to *B. plicatilis* under UV conditions. The intensity of UV-A and UV-B radiation in this experiment is basically set according to the ratio of sunlight. Compared with UV-B radiation, UV-A radiation triggers anthracene to produce more intense phototoxicity due to UV-A is more likely to cause the photo-sensitized material to produce reactive oxygen species, which in turn causes damage to the organism [13, 14].

Under normal physiological conditions, the production and clearance of ROS in the living body maintain a dynamic equilibrium state. But in the external environment, may lead to metabolic disorders and suddenly produce a large number of ROS. When the ROS exceeds the antioxidant defense system clean-up capability, it will attack the polyunsaturated fatty acids of the *B. rotundus* biofilm, the formation of lipid peroxides, causing cell damage, which is an important aspect of the negative impact of ROS. Malondialdehyde (MDA) is one of the most common products, is an important indicator of the level of biological peroxidation. The degree of lipid peroxidation can be detected by measuring the content of MDA in the organism, and the degree of oxidative damage of the organism cells is reflected from the side. In this study, the accumulation of anthracene in rotifers can increase the content of MDA and absorb UV light after the anthracene can induce more ROS production, which led to a more significant increase in MDA levels, which means that more serious oxidative damage occurred. Therefore, the increase of ROS-induced membrane lipid peroxidation may be the main mechanism of the anthracene on the phototoxicity of *B. plicatilis* [15].

Antioxidant defense system is a line of defense against ROS peroxidation in vivo, mainly including antioxidant enzymes and antioxidant small molecules. Therefore, it is helpful to study the process and mechanism of the activity of anthracene on the photoinduced toxicity of the anthracene to the *B. plicatilis* by the change of the enzyme activity in the anti-oxidative defense system after the UV radiation and no UV radiation. The The elimination of superoxide anion radicals ($O^{2-}\cdot$) produced by the main catalyzed biodegradation of SOD is a kind of oxidoreductase formed by the combination of protein and metal ions. CAT is erythropoietin, which exists in biological cells that can breathe. It can be used with POD to remove SOD catalyzed $O^{2-}\cdot$product H_2O_2, and H_2O_2 decomposition of water and oxygen, thus eliminating its oxidative toxicity to organisms. In this experiment, the activity of SOD in *B. plicatilis* in the single anthracene treatment group was inhibited before induction and the degree of induction and inhibition was much higher than that of UV radiation, which indicated that during the process, SOD In the anthracene induced by light induced ROS play an important role. CAT activity was not sensitive to the response of single anthracene stress compared with SOD, but the activity of CAT was significantly inhibited by UV radiation and anthracene complex. It is generally believed that low-toxic pollutants have an induced effect on SOD and CAT, which is the adaptive response of organisms to pollutants, mainly to enhance the ability of the body to eliminate reactive oxygen free radicals, to avoid damage to the body; high toxic pollutants against oxidation Inhibition of enzyme activity is the role of pollutants on the body has been more than the body's ability to adapt, the resulting reduction in enzyme activity is a precursor to the poisoning reaction.

References

[1] Koike T, Koike H, Kurumisawa R, et al. Distribution, source identification, and historical trends of organic micropollutants in coastal sediment in Jakarta Bay, Indonesia. Journal of hazardous materials. 217(2012) 208-216.

[2] Valavanidis A, Iliopoulos N, Gotsis G, et al. Persistent free radicals, heavy metals and PAHs generated in particulate soot emissions and residue ash from controlled combustion of common types of plastic. Journal of hazardous materials. 156(2008) 277-284.

[3] Thomas R E, Lindeberg M, Harris P M, et al. Induction of DNA strand breaks in the mussel (Mytilus trossulus) and clam (Protothaca staminea) following chronic field exposure to polycyclic aromatic hydrocarbons from the Exxon Valdez spill. Marine Pollution Bulletin. 54(2007) 726-732.

[4] Petersen M R, Hecht M W, Wingate B A. Efficient form of the LANS-α turbulence model in a

primitive-equation ocean model. Journal of Computational Physics. 227(2008) 5717-5735.
[5] El-Alawi Y S, McConkey B J, Dixon D G, et al. Measurement of short-and long-term toxicity of polycyclic aromatic hydrocarbons using luminescent bacteria. Ecotoxicology and environmental safety. 51(2002) 12-21.
[6] Lesser M P. Elevated temperatures and ultraviolet radiation cause oxidative stress and inhibit photosynthesis in ymbiotic dinoflagellates. Limnology and oceanography. 41(1996) 271-283.
[7] Parikh S J, Chorover J, Burgos W D. Interaction of phenanthrene and its primary metabolite (1-hydroxy-2-naphthoic acid) with estuarine sediments and humic fractions. Journal of contaminant hydrology. 72(2004) 1-22.
[8] Calabrese E J, Baldwin L A. Peptides and hormesis. Critical reviews in toxicology. 33(2003) 355-405.
[9] Christa L S, Stephen M B, Sean R A. The effect of ultraviolet radiation on the vertical distribution and mortality of estuarine zooplankton. Journal of Plankton Research. 22(2000) 2325-2350.
[10] Ankley G T, Tietge J E, Holcombe G W, et al. Effects of laboratory ultraviolet radiation and natural sunlight on survival and development of Rana pipiens. Canadian Journal of Zoology. 78(2000) 1092-1100.
[11] Nikkilä A, Penttinen S, Kukkonen J V K. UV-B-induced acute toxicity of pyrene to the waterflea Daphnia magna in natural freshwaters. Ecotoxicology and environmental safety. 44(1999) 271-279.
[12] Chen Y, WAng J, Wang Y, et al. Toxicity and population growth effects of microcystin on the rotifer Brachionus plicatilis. Zhongguo Huanjing Kexue. 22(2001) 198-201.
[13] Wondrak GT, Jacobson MK, Jacobson EL. Endogenous UVA-photosensitizers: mediators of skin photodamage and novel targets for skin photoprotection. Photochem Photobiol Sci. 5(2006) 215-237.
[14] Knak A, Regensburger J, Maisch T, Bäumler W. Exposure of vitamins to UVB and UVA radiation generates singlet oxygen. Photochem Photobiol Sci. 13(2014) 820-829.
[15] Wang Y Q, Tao S, Jiao X C, et al. Polycyclic aromatic hydrocarbons in leaf cuticles and inner tissues of six species of trees in urban Beijing. Environmental Pollution. 151(2008) 158-164.

Comparison of two chemical extraction methods for proteins and polysaccharides of Spirogyra fluviatilis in extracellular polymeric substances

Shuiping Chang, Yichao Lee*

Department of Environmental Engineering Kun Shan University, Tainan City, Taiwan (R.O.C.)

*Corresponding author e-mail: lyc007@mail.ksu.edu.tw

Abstract. Under natural circumstances, *Spirogyra fluviatilis* excretes large amounts of extracellular polymeric substances (EPS) and forms a mucilage layer. This study used two algal EPS extraction methods such as NH_4OH and formaldehyde-NaOH to compare the extraction yields of EPS and the levels of its primary components of polysaccharides and proteins. The results indicated that the extraction yields of EPS using formaldehyde-NaOH and NH_4OH are 8.05 mg EPS/g-fresh algae and 5.26 mg EPS/g-fresh algae, respectively. The ratio of proteins to polysaccharides in EPS is around 0.24 to 0.34, which is not a significant difference between the two extraction methods. In other words, the level of polysaccharides is always higher than the level of proteins in EPS extraction with both two methods

1. Introduction

EPS are metabolites accumulated on the surface of microcells. They not only stabilize the structure of plasma membranes but also protect cells against external harsh environments. EPS are high molecular isomeric polymers, which primarily consist of polysaccharides, proteins, nucleic acids and fats [1]. Therefore, they can be transformed and become the source of carbon and energy during starvation [2].

EPS can be found in marine microorganisms and fungi [3, 4]. *Spirogyra* is a genus of large filamentous green algae commonly found in freshwater all over the world. The cell surface of *S. fluviatilis* can excrete large amounts of EPS and then form a unique mucilage layer. However, the extraction methods of *S. fluviatilis* and the features of EPS were not explored.

The application of microbial EPS in the industry has been valued highly [5]. The demands for natural polymers in various industrial applications have increased lately, and therefore, novel algae have been used to synthesize high-value products [6].

This study compared the EPS extraction yield of the large freshwater algae *S. fluviatilis* and the levels of its primary major components of polysaccharides and proteins with two different chemical extraction methods of NH_4OH and formaldehyde-NaOH. We expect to establish an EPS extraction-based biotechnology from *S. fluviatilis* and explore its potential in industrial development.

2. Materials and methods

2.1. Algae

The algae that were used in this study the species *S. fluviatilis* were collected from the campus of Kun-Shan University (N23◦25, E120° 21') in Taiwan. The algae were washed twice with tap water, placed in a strainer at room temperature for 2 hrs to naturally drain the water, and then used for EPS extraction.

2.2. EPS extraction methods

In extraction methods, cationic exchange resin (CER) is not suitable for EPS extraction from large Filamentous algae. Therefore we used two chemical methods, formaldehyde-NaOH, and NH_4OH, to extract EPS.

(1) NH_4OH: 50g fresh algae+0.05%(w/w) NaCl solution to 100ml+50ml 25% NH_4OH at 4℃ for 24h.

(2) Formaldehyde-NaOH: 50g fresh algae+0.05%(w/w) NaCl solution to 100ml+50ml 36.5% 1N Formaldehyde-NaOH at 4℃ for 1h.

The treated suspension was centrifuged (10,000 × g) for 10 min with the suspension being collected. All supernatants were diluted by Mili-Q water to at the same level and then filtrate using 0.45μm cellulose nitrate membrane filter with filtrate collected.

The flowchart of extraction procedures are shown in Fig. 1.

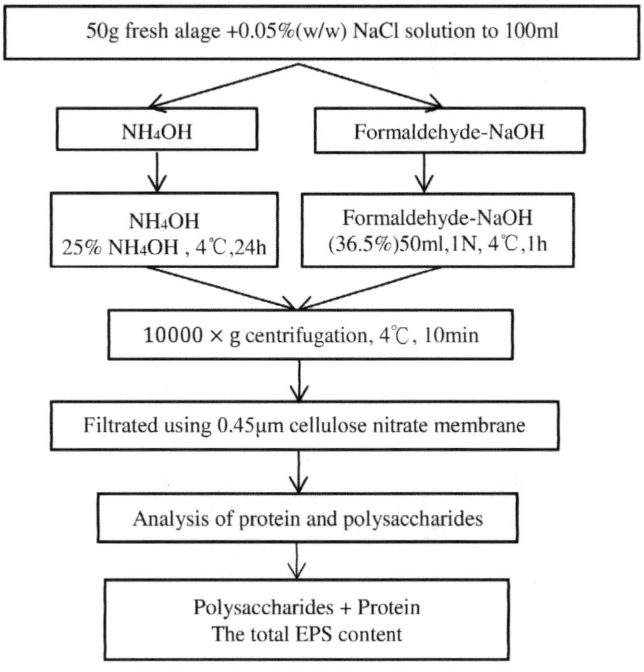

Figure 1. Procedures for the 2 extraction methods of *S. fluviatilis* EPS.

2.3. Analysis method

All chemicals used in this work were of analytical grade. Polysaccharide content was determined by the phenol-sulfuric acid [7] using glucose as a standard. Protein content was determined according to Bradford (1976) [8] with bovine serum albumin BSA, Sigma A2153 100mg/ml) [9]. The total EPS content was measured as the sum of these two substances.

3. Results and discussion

3.1. Comparison of different chemical extraction methods for assessing the levels of proteins and polysaccharides of EPS

The protein levels from different chemical extraction methods were measured using common testing kits. The results were shown in Fig. 2. The results revealed that the protein level with the formaldehyde-NaOH method was significantly higher than the protein level with the NH_4OH method. Similarly, the polysaccharides levels from different chemical extraction methods and are shown Fig. 3. The data reflect the average results from 6 repeated measurements, which showed that the average polysaccharides weight of EPS extraction with the formaldehyde-NaOH method was 6.07 mg-polysaccharides/g-fresh algae, w/w, but the average weight of polysaccharides with NH_4OH was only 4.08 mg-polysaccharides/g-fresh algae, w/w. Based on the aforementioned data, we can conclude that formaldehyde-NaOH also has a greater extraction yield of polysaccharides than NH_4OH.

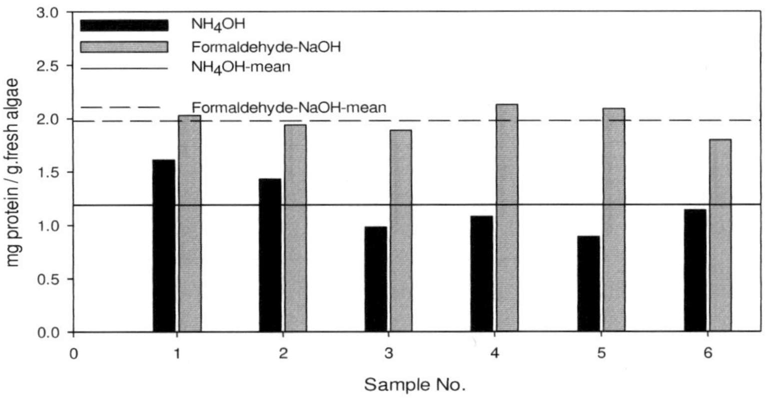

Figure 2. Protein level for formaldehyde-NaOH and NH_4OH EPS extraction method.

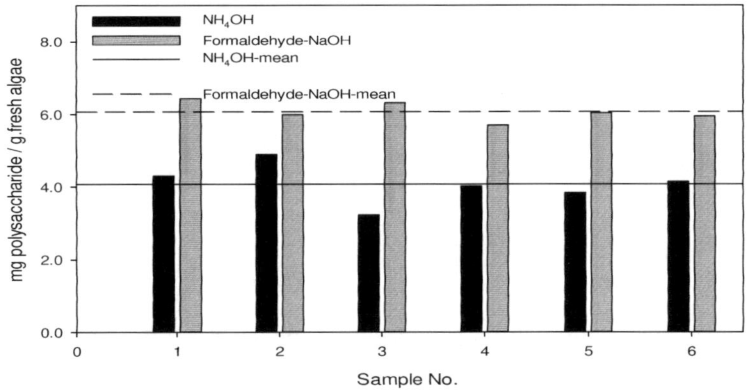

Figure 3. Polysaccharides level for formaldehyde-NaOH and NH_4OH EPS extraction method.

3.2. Comparison of formaldehyde-NaOH and NH4OH extraction methods for the extraction yield of EPS

The results are shown in Table 1 and Fig. 4. The results demonstrated that formaldehyde-NaOH has a greater extraction yield of EPS than NH_4OH.

Table 1. Amounts and compositions of EPS extracted by different methods

Extracted method	Organic matter contents (mg g-1 fresh algae)			
	proteins	polysaccharide	EPS(proteins+ polysaccharides	poteins/polysaccharides
Formaldehyde-NaOH	1.98±0.13	6.07±0.26	8.05±0.27	0.327± 0.028
NH_4OH	1.19±0.28	4.08±0.55	5.26±0.77	0.291± 0.047

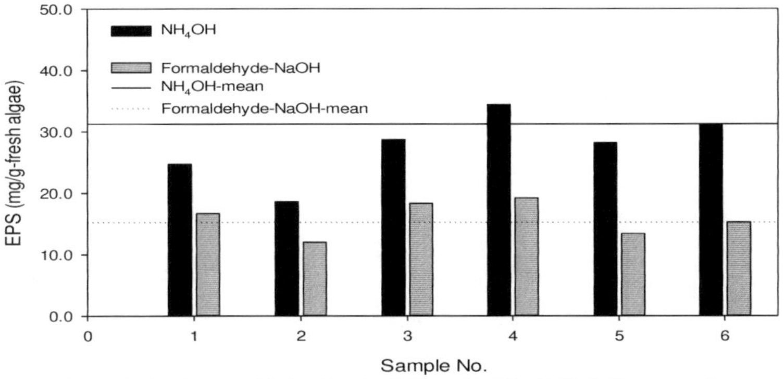

Figure 4. EPS for Formaldehyde-NaOH and NH_4OH EPS extraction method.

This study further compared the primary components of EPS (polysaccharides and proteins) with different extraction methods. The results showed that although different extraction methods can lead to different extraction yields of EPS, the ratio of proteins to polysaccharides in EPS also significant differences.

4. Conclusion

To compare the extraction yields of EPS from large *S. fluviatilis* and the changes in its primary major components of polysaccharides and proteins with different chemical extraction methods, the study results showed that: (1) Formaldehyde-NaOH has a greater extraction yield of EPS than NH_4OH method. (2) The ratio of proteins to polysaccharides in EPS is around 1.05 to 1.38, which is not a significant difference with the two extraction methods. In other words, the level of polysaccharides is always higher than the level of proteins in EPS extraction with the two methods.

Genus of *Spirogyra* is a biomaterial with great potentials. In its natural habitat, *S. fluviatilis* can produce large amounts of EPS, therefore it has great potential in EPS extraction. These preliminary study results helped us understand the extraction methods of EPS applicable to large algae, and the influences on its primary components. We conclude that NH_4OH and Formaldehyde-NaOH are both suitable for EPS extraction methods for large algae.

References

[1] B.S. McSwain, R.L. Irvine, M. Hausner, and P.A. Wilderer, Composition and distribution of extracellular polymeric substances in aerobic flocs and granular sludge, Appl. Environ. Microbiol. 71 (2004) 1051-1057.

[2] L. Gao, X.L. Pan, D.Y. Zhang, S.Y. Mu, D.J. Lee, and U. Halik, Extracellular polymeric substances buffer against the biocidal effect of H2O2 on the bloom-forming cyanobacterium Microcystis aeruginosa, Water Res. 69 (2015) 51-58.

[3] Z. Chi, C.D. Su, and W.D. Lu, A new exopolysaccharide produced by marine Cyanothece sp. 113., Bioresource Technol. 98 (2007) 1329-1332.

[4] M. Lundkvist, U. Gangelhof, J. Lunding, and M.R. Flindt, Production and fate of extracellular polymeric substances produced by benthic diatoms and bacteria: A laboratory study, Estuar. Coast. Shelf Sci. 75:3 (2007) 337-346.
[5] M. Avinash, K. Kumari, and J. Bhavanath, Characterization of extracellular polymeric substances produced by micro-algae Dunaliella salina, Carbohyd. Polym. 83 (2011) 852-857.
[6] Y. Zheng, R. Xiao, M. Roberts, Polymer-enhanced enzymatic microalgal cell disruption for lipid and sugar recovery, Algal Res. 14 (2016) 100-108.
[7] H. Xiaomeng, W. Zhiwei, Z. Chaowei, and Z. Wu, Effect of ultrasonic power density on extracting loosely bound and tightly bound extracellular polymeric substances, Desalinat. 329:15 (2013) 35-40.
[8] M. M. Bradford, A rapid and sensitive method for the quantitation of microgram quantities of protein utilizing the principle of protein-dye binding, Anal. Biochem. 72 (1976) 248-254.
[9] M. Dubois, K.A. Gilles, J.K. Hamilton, P.A. Rebers, and F. Smith, Colorimetric method for determination of sugars and related substances, Anal. Chem. 28:3 (1956) 350-356.